# MOSCOW BOUND

## POLICY, POLITICS AND THE POW/MIA DILEMMA

### JOHN M.G. BROWN

(With research assistance of Thomas V. Ashworth)

AN ARCHIVAL, LITERARY AND HUMAN SOURCE INVESTIGATION INTO THE FATE OF AMERICAN PRISONERS OF WAR HELD BY SOVIET RUSSIA AND OTHER COMMUNIST NATIONS, FROM 1918–1993.

Veteran Press
Eureka, California
1993
2nd edition, 2024

Veteran Press
Eureka, California
Copyright John M. G. Brown, 1991, 1992, 1993
All Rights reserved.
Inquiries should be addressed to P.O. Box 30, Petrolia, CA. 95558
First published as Moscow Bound by the author in 1992
for the U.S. Senate Select Committee on POW/MIA Affairs
Library of Congress Registration Number: TXU 545 2326,
Library of Congress Catalog Card number: 93-61625
ISBN: 979-8-218-34778-9 (print edition)
Also available in eBook edition


Library of Congress Cataloging in Publications Data
Brown, John M.G., 1947-
Moscow Bound, Policy Politics and the POW/MIA Dilemma
Includes Index
1. United States-Relations Soviet Union. 2. Prisoners of War-Soviet Union.
3. Prisoners of War-United States. 4. World War I, 1914-1918, Prisoners,
Soviet. 5. World War II, 1939-1945, Prisoners, Soviet. 6. Korean War, 1950-
1953, Prisoners, Soviet, Chinese, North Korean. 7. Vietnam War, 1961-1975,
Prisoners, Soviet, Chinese, North Vietnamese, Laotian, Cambodian, Cuban.
8. Intelligence service-Soviet Union. 9. Soviet Union-Politics and Gov't.

Printed in the United States of America

Publication managed by AuthorImprints.com

# TABLE OF CONTENTS

*To all the Americans left behind in captivity and
to Josie*

# PREFACE

A condensed version of MOSCOW BOUND (Copyright 1991) was supplied by the author as a 133-page briefing-book edition under the title: POLICY, POLITICS AND THE POW/MIA DILEMMA, to Senators John Kerry and Robert Smith of the Senate Select Committee on POW/MIA Affairs in March 1992, and as an interim report on the POW/MIA matter to President George Bush on April 9, 1992. Subsequent July 4, 1992, and October 1992 full-length editions of MOSCOW BOUND contained corrections, additions and a number of substantiating notes, and were printed and distributed by the author since no U.S. publisher contacted by the author expressed interest in publishing the book, in any form. All documentary and human-source evidence included within this work was delivered by the author to the Senate Foreign Relations Committee Republican staff in 1990 and 1991, and to the Senate Select Committee on POW/MIA Affairs in 1992, before which the author testified in July and again in November of that year, with research colleague Thomas V. Ashworth. Thousands of additional and supporting archival documents, which have recently been declassified, could not be utilized due to space limitation imposed by a single volume.

## ACKNOWLEDGEMENTS

Most of all, I wish to thank my patient and loving wife of 24 years, Josephine Duke Brown, better known from Alaska to Amazonia as "Josie," my editor and the mother of my five understanding children, Moses, Jack, Earl, Cordelia and Ben; Also: My father, John M. G. Brown Sr., who died in 1978 but left me with a solid grounding in history, my mother, Joan Scull Brown and my father-in-law Anthony Drexel Duke, both of whom taught me something about compassion and humanity; my sister Joan and my brother Bobby; Ambassador Angier Biddle Duke, John Ordway Duke, Margaret Biddle Robbins, Ambassador George Kennan, Carl Heinmiller—an old line infantry officer of Haines, Alaska, whom I had the honor to assist in Boy Scouts of America; General William C. Westmoreland, USA, for consistent encouragement; General James Van Fleet, USA and to Lt. General Eugene Tighe, USAF, for upholding the honor of the armed forces and the nation; U.S. Representatives John LeBoutillier, Bill Hendon, Frank McCloskey, Bob Dornan and others who tried to prove that the U.S. Congress can do its job as the founding fathers intended; Senators Charles Grassley and Robert Smith, who did their utmost for the POWs and MIAs—with the help of dedicated Senate staffers and investigators Kris Kolesnik, Daniel Perrin, Tracy Usry (CID) and John McCreary (DIA); and others; POW/MIA family members Marcia Welch, Marian Shelton, Ann Holland, Dolores Alfond, Dianne Van Renselaar, the Standerwick sisters, Colonel Earl Hopper and many others; former U.S. Marine captain, researcher, author and friend Thomas V. Ashworth, his wife Jane and their whole family; Alexander Grube (in memorium), John Noble, Martin Siegel, Captain Sidney Miller, Lt. Col. Delk Simpson, Lt. Col. Phillip Corso, Serban Oprika, Jerry and Barbara Mooney and others who knew the truth and assisted me; Ross Perot of Texas-for refusing to abandon American prisoners of war; Assistant Archivist of the United States Trudy Huskamp Peterson, John and Margaret Nevin, Michael Caron; for special research, the many old grunts, sailors, airmen and marines who encouraged me along the endless way, including: Francis "Franny" Homsher, Tommy Donaghue, Reuben "Pancho" Carrera and Roy Inman; Mike Booth and Colonel Bill LeGro—for years of special service, Ed "Gino" Casanova, John Molloy, Carl Rice, Bill Anderson, "Chuck" Gage, Richard Keeton, Anthony J. Drexel Biddle III, Ted Sampley, Carl Rice, Jerry Kiley, Lt. Col. "Bo" Gritz, John Heyer, Terry Pardee, Maj. Mark Smith, Col. Bob Howard, SFC Mel McIntire, SGM John Holland, Bobby Garwood, J.P. Mackley, SGM Norman A. Doney, John Biddle Brock, Tom Flaherty, Terry Minarcin, Al Santoli, "Ned" Tuthill, Major General George S. Patton

III, Lt. Commander Dan Weaver, Willie Ziolkowski, Ed Smith, Mike Manzi, Larry Smith and others; Editors: George Schwab, F.R. Duplantier, Richard Kolb, William Hilliard, Peter Thompson, Donald E. Graham, Stephen Rosenfeld, Tony Diamond and Mike Milne, other researchers and journalists—Bill and Monika Jensen-Stevenson, Larry J. O'Daniel, Rod Colvin, Bill Paul, Mike Van Atta, Nigel Cawthorne, Jim Sanders, Mark Sauter, Ted Grevers and Mike Blair; Thomas Sgovio—who told me about surviving in the Gulag; pioneer film producer Ted Landreth, insiders John Whitehead of State, Andy Gembara, Doug Dearth, Jim Wink and others who must remain nameless; Mark Waple, an honorable lawyer; Dr. Stephen Johnsson, Marion Lelong for special research, Carlos Benemann the bookman, Kris Nelson for special research, Lee McSherry—a scholarly Florida farmer-and his wife December, Marilu Duke Cluett for liaison, Rex Sinclair, Melissa Ledesma, Wade Ladue-VFW, Joseph Andry and David Givans-DAV, Dick Christian-American Legion; the sometime-saviour of my computer files-John Wentworth; "Arney's" family of Fortuna, Dr. John Stewart, the Thomas Malmberg family of Port Bailey, Kodiak Island, Alaska; the Abner Nelsons' and Tim Ward family of Port Lions, Alaska; the Jim Moore family of Haines, Alaska; the Perry Bolsters', Mike and Sue Schroeder, Chris Pinnell, Tom and Chris Frank and other POW/MIA activists of Sequim, Washington; and Ken Roscoe, (in memorium) an old time American rancher, historian and author of the northern California Coast Range-with his wife Marie, who helped to keep me centered as I wrote this book; also the people of the land—Greg and Margie Smith, "Herky" Lawrence (in memorium)-who inspired me to work harder; the Shorts and Chambers, "Curley" Wright, David and Jane Simpson, Dan Austin, Ellen Taylor and Ed Gilda; all from the Valley, and to many others, too numerous to mention, who assisted or encouraged me in various ways.

HAINES, ALASKA AND MATTOLE VALLEY, CALIFORNIA, 1987-1993

# INTRODUCTION

### *"YE SHALL KNOW THE TRUTH AND THE TRUTH WILL MAKE YOU FREE."*

(Motto cut in stone at CIA Headquarters, Langley, Virginia)

It has long remained official US policy to refuse public payment of ransom to Communist nations holding American prisoners of war (POWs) or civilians as hostages. This policy evolved from Soviet conduct with U.S. POWs and missing in action (MIAs) of the 1918-1920 American Intervention in Russia, the Allied response to Lenin's withdrawal of Russia from WW I, after the 1917 Bolshevik Revolution. Subsequent Soviet actions in retaining thousands of missing U.S. POWs of WW II, Korea and Vietnam for intelligence purposes, and as forced-labor were dictated by Russian national interests in what became a death-struggle between Soviet Communism and western democratic Capitalism, led by the United States. This example was followed by subsequent Soviet-surrogate regimes in eastern Europe, Communist China, North Korea, Cuba, Vietnam and Laos. The announced ending of the Cold War may ultimately reveal the fate of many of these lost American POWs and could result in the return of survivors to the United States.

During the Revolutionary War of 1775 to 1783, American prisoners of war had been held in appalling conditions on British prison ships or in dungeons and many American POWs, denied the most basic necessities and care, died in British captivity. This contributed to the great bitterness felt in the newly-free nation towards British-American loyalists, who were subsequently mistreated and expelled from their communities in the 13 former colonies. During the American Civil War, from 1861-1865, both Union, and Confederate prisoners of war were mistreated, starved and even murdered by their guards, but at the end of that war the survivors were released. As the victors, U.S. authorities subsequently conducted investigations of Confederate war crimes against Union prisoners of war, and carried out reprisal executions.

Official American policy toward military and civilian hostages seized by a foreign state, for use in diplomatic or monetary blackmail, may be said to descend from U.S. reaction to the Barbary pirates of North Africa illegally seizing American prisoners and boasting of it. At first, from 1795 to 1801, large amounts of money were paid by the United States for protection against the pirates and as ransom, but under President Thomas Jefferson, the U.S. went to war against Tripoli from 1801 to 1805, and subsequently against Algeria. After this, the seizing of American sailors on the high seas by Great Britain led to the War of 1812. These attitudes reflected more than two centuries of American experience at frontier Indian wars, conducted by descendants of European immigrants on the margins a vast continent, in which the often rude and unlettered settlers were actually outnumbered by the indigenous inhabitants and sometimes, as at the time of the 1675-1678 Narragansett, or King Phillip's War, were in danger of being driven into the sea by the natives. In one Indian war after another that followed, from the 1600s to the late 1800s, known American captives had been ransomed whenever possible, or tracked down and liberated by the regular army or volunteer citizen-scouts, if they could be found.

　　Subsequent experiences in America's minor foreign wars of the late 19th century, in Cuba and the Philippines, did not call for implementation of a different policy. The natural American reaction to public knowledge of U.S. prisoners being held hostage was expressed by President Theodore Roosevelt during the turn-of-the-century era when he quoted a West African proverb: "Speak softly and carry a big stick." Roosevelt believed in using the threat of American military force to carry out U.S. foreign policies. In its youth and vigor as a new nation that had achieved world power status by the early 20th century, America had bypassed some hard-learned lessons which had resulted in the subtleties of European and Asian diplomacy regarding prisoners of war and hostages. These experiences extended back over two millenniums, from the time of the Persians, Greeks, Romans and Muslims, and Europe had since gone through other evolutions in the treatment of war prisoners from the Dark and Middle ages, through the Renaissance.

　　The Russian Revolution and subsequent Bolshevik triumph in the civil war resulted in a return to a bygone age in which all war prisoners became hostages, to be secretly held for future use. Since the time of Czar Ivan the Terrible, who created the Oprichnina political police in 1565, state-imposed terror had been a fact of life in Russia. Carrying a dog's head and broom, representing their authority to sweep away traitors, thousands of these agents, dressed in black and riding black horses, roamed Russia in the 1570s, administering death sentences under authority of the Czar. Secret confinement, torture and execution of suspects became commonplace in Russia. This traditional oppression of the landed peasants and city dwellers was continued by Czar Peter the Great and his Romanov successors up until the time of Nicholas 11. Under the Communists after 1918, the state terror apparatus was enormously expanded to levels of persecution and mass-death never equaled, before or since, in human history. The American prisoners of war, from the fabled and far-away continent of emigrant dreams, added a mere rivulet to the vast flood of millions of state prisoners who perished in the Soviet slave camps over the next seven decades.

America had not faced a foreign power governed by an ideology dedicated to overthrowing both its representative form of government and its capitalist economic system until the end of the First World War, with the emergence of Soviet Communism in Russia, under Vladimir I. Lenin. American and Allied war prisoners of 1918-1920 were treated as potential political assets to be exploited for use in a world-wide propaganda war. For the first time, the United States was confronted with an ideological enemy that held certain American prisoners of war secretly, to gain the use of their minds in understanding the enemies of Bolshevism, and which also held other U.S. prisoners of war and civilians openly, for use as hostages to gain diplomatic concessions and ransom. The official American response to this situation in 1919 was to "classify" all information about the existence of such prisoners of war, or missing men suspected of being held prisoner, as a state secret. The public was not allowed to know this fact, which resulted in many misconceptions about Soviet methods and intentions. Admission of their existence would have revealed the impossibility of recovering the prisoners without resorting to another major war, less than a year after the 1918 Armistice. An ill and exhausted President, Woodrow Wilson, sought to avoid prolonged warfare in Russia in 1919 by urging international cooperation in settling disputes, but an isolationist Congress rejected American participation in the League of Nations. The American prisoners of war and the missing in action captured by the Bolsheviks in North Russia and Siberia, who remained behind in captivity, became a secret source of contention between the two countries until the United States extended diplomatic recognition to Stalin's Soviet government in 1933.

At the end of World War II, as the Red Army overran eastern Europe, hundreds of thousands of western Allied and American prisoners of war and interned civilians were liberated by the Soviets, but many thousands disappeared while under Russian control. Faced once again with impossible Soviet demands for diplomatic concessions, financial aid and other conditions for the release of prisoners they publicly denied holding, President Harry Truman chose once again to avoid a military confrontation so soon after the Second World War. The entire body of information concerning U.S. POWs and MIAs, who had been in eastern European German POW camps when the Soviets took control, was again 'classified' in the name of national security, and remained secret for more than four decades to come. The emboldened Soviets, under dictator Josef Stalin and his successors, thenceforth kidnapped American military and civilian personnel, including aircrews from

U.S. Cold War reconnaissance missions, almost with impunity, secure in the knowledge that an unspoken "wink and nod diplomacy" existed between the superpowers: that prisoners thus captured would remain secret, and unless publicly acknowledged, were "gone forever." The results of the Korean War further indicated that the United States was willing to abandon U.S. prisoners of war known to have been alive in Soviet, Chinese and North Korean control, rather than submit to blackmail demands that would establish a precedent that the United States was willing to 'trade in human lives.' President Dwight D. Eisenhower, the former U.S. commander in Europe during World War II, this time made the decision that the families of POWs and MIAs held in China and Russia should not be informed about the fate of their loved ones, since it was then believed that the KGB would never release them.

Five years after the end of the Korean War, Colonel Archibald Roosevelt, a son of former President Theodore Roosevelt, addressed a meeting of Republican leaders and others at Sagamore Hill, New York, on Independence Day, July 4, 1958, saying: "This is a sad Fourth of July for me, but most recent Fourths have been sad ones..." He observed that when Theodore Roosevelt was President, an American citizen was safe anywhere in the world, but today, he said, "Communist tyrants are allowed to hold Americans prisoner with impunity."[i] Although the speaker was the son of a distinguished American President, with an honorable military record in his own right, the United States Government officially denied that what he had said about Communist-held American POWs from Korea was true. This book attempts to explain how such a situation came about.

When America took over the defense of Indochina from the French, against the Vietnamese surrogates of Communist Russia and China, U.S. experience with secretly-held war prisoners consisted of the ideologically-motivated confrontations with Lenin and Stalin and their successors which had followed 1919, 1945 and 1953. As has been demonstrated throughout this book, these experiences were 'classified,' and thus kept secret from the American people by a small core of diplomatic, military and intelligence officials, who believed that by so doing, they were avoiding more prolonged warfare with potentially much greater losses, in each case. Thus, the American people were led into yet another military confrontation with a Communist power that had already secretly withheld French prisoners of war, still holding the simple and ignorant beliefs of a frontier Indian-fighting society, in regard to prisoners of war or hostages held illegally by an enemy power.

The American foreign policy, military and intelligence elite, however, who collectively knew that U.S. war prisoners had been abandoned alive in Communist control three times within living memory, and believed that this had been a tragic but necessary evil, chose to enter yet another anti-Communist foreign war fully prepared, institutionally, to do the same thing again. The mechanisms for such action had long been in place and had been perfected since 1919. The methods of disinformation practised by America's totalitarian adversaries in three foreign wars of the 20th century had, out of perceived necessity, been adopted by the diplomatic, military and intelligence establishment of the United States, because a loosely-run democratic method of government was inherently ill-prepared to confront the decisive, amoral actions of such dictatorial regimes. Each succeeding President, being only a temporary Commander-in-Chief, with executive powers limited by a representative and elective form of government, and guided by this small elite who actually formulated U.S. foreign policy, was expected and mandated to place "national security" ahead of any heartfelt personal concerns for American prisoners of war reported by eyewitnesses to be in Communist hands. This necessitated adoption of a policy that held the testimony of eyewitnesses who had seen Communist-held U.S. prisoners of war to be insufficient evidence of the existence of such prisoners, resulting in an automatic debunking of virtually all human sources of intelligence, and rigid classification of virtually all electronic or "special intelligence." This policy was to be upheld after WW II, Korea and Vietnam, as revealed within this book.

The age of nuclear warfare ushered in at the end of WW II had lent much weight to this official but 'classified' method of dealing with such a volatile issue as the deliberate abandonment of those Americans who had put their lives on the line for democracy, and had been captured in combat. As in the 1950s Cold War reconnaissance shootdown incidents, when prisoners known or believed to be alive in Soviet control were written off as "presumed dead," such decisions in Indochina were to be made in secret, by American officials who believed they were acting for the 'greater good,' in order to avoid any sudden escalation of tensions with the main adversary, the hostile, nuclear-armed Communist superstate of the Union of Soviet Socialist Republics. This was the only world power which could legitimately be said to threaten the very existence of the United States and its more than 200,000,000 citizens, whose lives above all else, the American leadership was pledged to defend, while simultaneously confronting the main adversary's Vietnamese Communist surrogates in another 'limited war.'

---

i    Human Events, July 14, 1958

# PART I

# THE BOLSHEVIK REVOLUTION
# AND
# THE RISE OF SOVIET POWER

## PROLOGUE

America and Russia, the two most powerful nation-states of the 20th century, both arose in frontier borderland regions, which, during their formative years, were often exposed to hostile attack. Russia's neighbors in Europe and Asia remained formidable forces however, while America was allowed to develop from a much later start on a newly-discovered continent, relatively unhindered by devastating foreign invasions. A history of the fate of American war prisoners held by Soviet Russia and it's later allies must therefore include a brief summary of the development of Russia to the time of the First World War, when American and Russian forces first collided in an ideological confrontation that was to continue for over seventy years, with a brief, enforced interlude of cooperation during the Second World War, from mid-1941 to late 1945.

The Russian nation developed from Slavic tribal settlements along the broad rivers of the European-Asiatic border region, which flowed south into the Black Sea and the Caspian Sea, and upon the rivers draining from this region to the north, into the Baltic Sea and the Arctic Ocean. The Slavs had supplanted or taken over earlier peoples in this region, the Scythians, Sarmatians, Goths, Huns and Avars, who had for centuries assaulted the frontiers of Mediterranean civilization, and who had ultimately overrun the ancient Western Roman Empire in the 5th century AD. Exploring Scandinavian Vikings, called Varangians, pushed their longships up these north-flowing rivers under their leader Rurik in AD 862, when, according to the earliest written Russian history in the Primary Chronicle, they contacted the Slavs at Novgorod in what came to be called the "land of the Rus." The Chronicle says that the principal Russian settlement at Kiev was captured by the Varangian Prince Oleg in 882. Vikings and Slavs intermingled in Kiev, which was linked by the Dnepr River to the Black Sea and thus to the ancient culture of the east Roman imperial capital at Constantinople, which was also known by its Greek name, Byzantium. Graeco-Roman city-states had existed on the northern coast of the Black Sea for centuries, protected from the northern barbarians first by Rome and then by Constantinople. It was from here that Orthodox Christianity was brought by Byzantine missionaries to the wilderness inhabited by the Slavs, whose Grand Prince, Vladimir I, converted Russia to Christianity in 988. Under centuries of Byzantine influence, the Russian state led by Kiev developed into a trading nation of substantial power, until it was suddenly crushed by a massive Mongol invasion from the east, of the "Golden horde," from 1237-1240, under Batu, who was a grandson of Genghis Khan. The ruthless Mongols exacted tribute and soldiers from the shattered Russian state and about the year 1318 they promoted Grand Prince Yuri of Moscow to be their surrogate ruler of the Slavs, to enforce their control. This led to the rise of Moscow as a military power, but It was not until the reign of Ivan III, in 1480, that the Mongol rule over Russia was broken. The Mongol conquest and centuries of virtual enslavement that had followed contributed to shaping the Russian national character. Yet with the fall of Constantinople to the Ottoman Turks in 1453, Russia had become the primary defender of the Orthodox Christian faith.

Under Czar Ivan IV, (the Terrible) who ruled from 1547 to 1584, the Russians fought the Tartars at Astrakhan and Kazan in the southeast, and conquered much territory there and beyond the Ural Mountains in western Siberia, but czarist became more oppressive. Ivan the Terrible was a brutal dictator who created a special internal police force to punish suspected traitors among the landed aristocrats, churchmen and the peasants, and murdered his own son along with thousands of others, in a reign of terror. Ivan the Terrible also passed laws that bound the Russian peasants to the land as serfs, and this system was to become even stricter under his successors, at a time when Europe was undergoing the Renaissance and serfdom was dying out in the West. In his only major defeat, Ivan's military expansion toward the northwest and the Baltic Sea was halted by the armies of Lithuania, Poland and Sweden; enemies who would repeatedly block Russian ambitions in that area.

In 1610, during a period of weakness known as the Time of Troubles, the then-powerful Kingdom of Poland invaded Russia and captured Moscow, which was occupied by Polish forces until 1612. Frontier stock raisers and warriors known as Cossacks, who lived freely on the steppes of the south and east, led peasant revolts against the powerful boyars and aristocrats. But the Polish attack helped unite the Russian forces, which drove the Poles from Moscow. After this, the Zemskii Sobor, a primitive parliament with little real power, elected a new czar named Michael Romanov, who founded a dynasty that was to last until the czarist system was overthrown more than three centuries later. Under Peter the Great, who ruled from 1682-1725, Russia fought Sweden for access to the Baltic Sea and that czar established a new capital on the Neva River, which was named St. Petersburg. Peter traveled to Europe in search of new technology and ideas for developing Russia and constructed Western-style factories and schools, but he also consolidated more power in the hands of the czar. By the end of his reign the Russian empire had been expanded east as far as the Pacific Ocean, opposite Japan and Alaska, and the Russians began to explore that sea while the eastern part of North America was being colonized and fought over by the British, French and Spanish.

The autocratic rule of Catherine the Great resulted in a great peasant revolt in 1773-1774, led by a Cossack named Emelian Pugachev, that engulfed the region between the Volga River and the Ural Mountains and almost reached Moscow before being crushed by Russian Army troops. Catherine was a German princess who married into the Romanov family; she subsequently brought thousands of German settlers into Russia and established them along the Volga River, where they multiplied and became known as the "Volga Germans." German military methods and modern European arms were also adopted by the Russian Army under Catherine. In this period Russia conquered the Crimea during a war with Turkey, and much of present day Byelorussia was seized from Poland. In 1784 Catherine authorised the founding of a colony at Kodiak, in Alaska, to exploit the local resources and cement Russian claims to the northwestern coast of North America. The American Revolution of 1775-1783, and the subsequent French Revolution, sent tremors as far as Russia. The radical idea that individual liberty is an inalienable human right was considered a threat to czarist rule, but a famous American naval hero of the War for Independence, John Paul Jones, came to Russia in 1788 and gained victories for the Empress in a naval war with the Turks on the Black Sea.

In 1809, not long after American overland explorers had finally reached the Pacific coast of North America, the first United States Ambassador to Russia, John Quincy Adams, who had been appointed by President James Madison, arrived in Saint Petersburg on an American sailing ship. He presented his credentials as the Minister Plenipotentiary of the United States of America, to the government of Czar Alexander I, Catherine's grandson. Three years later, Russia faced an invasion by 600,000 French troops under the French Emperor Napoleon I, who wished to halt Russian trade with France's major enemy, England, and stop the Russian advance into the Balkan region, in their wars against Turkey. Although Napoleon's "Grand Armée" captured Moscow in September 1812, the city was largely destroyed by a fire, Napoleon ordered a withdrawal from Russia that fall, to avoid becoming trapped by the approaching Russian winter. Under constant Russian attack and abandoned by Napoleon who hastened to Paris with a small escort, the retreating French Army lost an estimated 500,000 men of several nationalities who were killed and captured, or who deserted during the campaign, most of whom disappeared forever inside Russia. Only 1,000 men reached Paris still in military formation in early 1813.[1]

After Nicholas I succeeded to the throne of the czars, in December 1825, a military revolt of some 3,000 soldiers, led by 30 army officers, called the "Decembrists" attempted to seize control of the government and were immediately put down, but the czar then instituted even more repressive methods of control, including the creation of a separate department for a secret police force. Revolts in Poland and Hungary were also put down with great brutality by Russian troops. America had meanwhile survived a second war with Great Britain, from 1812 to 1814, and was expanding it's frontier settlement westward across the continent of North America to the Mississippi River. In 1835 Alexis De Tocqueville, a French author and traveler, wrote with great clairvoyance about these two emerging superstates:

"There are at the present time two great nations in the world which seem to tend toward the same end, although they started from different points: I allude to the Russians and the Americans. Both of them have grown up unnoticed; and while the attention of mankind was directed elsewhere, they have suddenly assumed a most prominent place among the nations...All other nations seem to have nearly reached their natural limits, and only to be charged with the maintenance of their power; but these are still in the act of growth... The Anglo-American relies upon personal interest to accomplish his ends, and gives free scope to the unguided exertions and common sense of the citizens; the Russian centers all authority of society in a single arm; the principal instrument of the former is freedom; of the latter servitude. Their starting point is different, and their courses are not the same, yet each of them seems to be marked out by the will of heaven to sway the destinies of half the globe.[2]

A German-born philosopher and revolutionary named Karl Marx, who lived from 1818-1883, was to have an enormous influence on the future of Russian history, and of the world, in the century to come. Marx believed that the individual was the highest being, not God, and that through a collective effort of production people made the necessities of life, which came from natural resources, factories and labor-also called the "proletariat." Marx taught that the private ownership of the means of production by the "bourgeoisie" was the basis of the class system, and that for humans to be free, the means of production must be owned by the community as a whole, that is, the government. Social equality for all mankind would follow, in a world in which all persons would be able to pursue their desires and be creative with their personal lives. Marx said that all of human history is a conflict between the ruling class, or bourgeoisie, and the working class, or proletariat. Marx became political writer in the early 19th century, who greatly influenced peoples thinking in Europe and around the world. In America, Marx wrote for the New York Tribune as a political reporter. Just before the outbreak of the German revolution of 1848, Marx published the "Communist Manifesto," which was a statement of the theories that later came to be called Marxism. In that year serious revolts occurred in France, Germany and Austria. With the collapse of the 1848 German revolution, Marx fled to London where he remained an exile the rest of his life. Karl Marx founded the International Workingmen's Association in 1864, which later became known as the "First International."[3] Thus, the philosophy and political movement called Marxism took root in Russia.

England and France had gone to war with Russia from 1853-1856 in the Crimea, to halt Czarist expansion in the Black Sea region and thereby limit the potential of Russia becoming a world-class sea power. In 1861 Czar Alexander II had freed the serfs of Russia by a decree, as the United States split into two warring parts, largely over the issue of slavery. In 1863 at the height of the American War between the States, The Imperial Russian fleet visited New York, San Francisco and Washington. This was widely perceived at the time to be a Russian signal to Britain and France, that their meddling in North America was not unnoticed, although later research was to indicate that other Russian national interests were at stake. Despite Alexander II's reforms in education, press control and the courts, many radical Russians felt the czar had not gone far enough, and an attempt was made to kill him in 1866. A terrorist group called the "Will of the People" tried unsuccessfully to assassinate this czar several times. The eventual assassination of Alexander II, by a terrorist bomb in St. Petersburg in 1881, led to even more repressive measures under his successor Alexander III, who further limited freedom of the press and local self government of villages and towns. The czarist "Ochrana" secret police infiltrated university movements and workers groups, and political dissidents and revolutionaries were executed, imprisoned or exiled to Siberia. An American traveler and author of the 1880s, George Kennan (the elder), wrote for "Century Magazine," and published a book called "Siberia and the Exile System," which documented the czarist oppression, and according to the later Bolshevik leader Mikhail Kalinin, it later became something of a "Bible" for Russian revolutionaries.[4]

Nicholas II became czar in 1894, and he was to be the last of the Romanov line to rule all Russia. Revolutionary sentiment grew rapidly in Russia during the 1890s with increased industrialization, spreading discontent among city workers and the middle class, and bad harvests causing starvation among the peasants. A depression after 1899 added to the misery of the common people. Anti-Jewish pogroms, abetted by czarist intelligence, military and police, made scapegoats out of an oppressed minority, to deviate popular discontent. Karl Marx's thinking influenced many disaffected Russians of the 1880s and 1890s, including future Soviet leader Vladimir I. Ulyanov, who later took the name Lenin. The Marxists established the Russian Social Democratic Labor Party in Russia in 1898, as America was engaging in it's first imperialist war, against Spain, in Cuba and the Philippines. This Marxist party split into two factions in 1903; the majority, called "Bolsheviks," wished to limit party membership to a small number of revolutionaries who would guide the proletariat, and a minority, called "Mensheviks," who desired a more democratic leadership and much wider party membership.

Reacting to Russian imperial expansion in the far east, Japan had attacked Russian forces at Port Arthur, in Manchuria, which had been leased from China by the czarist government. Russia was severely defeated in the ensuing Russo-Japanese War of 1904-1905. The American President, Theodore Roosevelt, helped to end the Russo-Japanese War by bringing the warring sides together in a peace conference, and in 1906 he became the first American to win the Nobel Prize. The defeat by Japan increased public disaffection in Russia. Liberal constitutionalists desired to replace the czarist autocracy with a parliamentary government, while the social revolutionaries worked towards an uprising by the peasants, and the Marxists organized for a proletarian revolution among city industrial workers. On January 22, 1905, which came to be known as "Bloody Sunday," thousands of unarmed workers marched on Czar Nicholas II's Winter Palace in St. Petersburg and Russian troops fired into the crowd, killing and wounding hundreds of demonstrators. After this the revolutionary movement grew rapidly as strikes broke out and peasant and military groups revolted. In October 1905 there was a general strike and the revolutionaries formed the first revolutionary "Soviet" (Russian for a council) in St. Petersburg. Although Czar Nicholas II eventually agreed to form a "Duma," or parliament, many of the revolutionaries were dissatisfied, and the uprisings continued. In December 1905, the Russian Army crushed a serious revolt in Moscow, which broke the back of resistance to czarist authority. Russia returned to a restless and uneasy peace, but the revolutionaries used this time to reorganize for another try when the time was right.

It was at the beginning of the 1905 revolution that two future Soviet leaders, Vladimir I. Lenin and Josef Stalin first met, albeit in a foreign country. Stalin had been born Iosif Ivanovich Djugashvili, in 1879, the son of a shoemaker at Gori, near Tbilisi, Georgia, in the Caucasus region of southern Russia. Stalin had joined a revolutionary Marxist group in 1898, while studying at a seminary and was subsequently expelled. Stalin was arrested in 1902 for continued revolutionary activity and in 1903 transferred from prison to exile in Siberia. He escaped in 1904, and joined the Bolsheviks; he first met Lenin in Finland in 1905. He had already taken the name "Koba" as a revolutionary, to confuse the Ochrana, and about 1913, he began to use the name Stalin. At this time he served briefly as editor of the Bolshevik newspaper Pravda (Truth), prior to being arrested and exiled once again on the eve of World War I.

In June 1914, following the assassination of the Austro-Hungarian Archduke Francis Ferdinand and his wife in Sarajevo, Bosnia, by a Serbian radical, Russia sided with the Serbs against Austria-Hungary during the ensuing crisis. This led to the Germans under Kaiser Wilhelm II siding with Vienna, and British and French support for Russia. Austria-Hungary declared war on Serbia on July 28th, and Russia mobilized for war in support of Serbia. Germany declared war on Russia on August 1, 1914. France had signed a mutual defense treaty with Russia in 1894 and 1904 and Britain had done the same in 1907, which inevitably led to their involvement in the war against Germany. German troops crushed a huge Russian army at Tannenberg, in East Prussia, later that year, but the Russians defeated an Austrian army in several battles at Lemberg, in the Galicia region of Austria-Hungary. In 1915 Austrian and German forces pushed the Russians back, but in 1916 the Russians attacked again and advanced in Galicia. In 1917 the Russian army advanced into the Carpathian Mountains but were driven back by the Germans and Austrians. The Great War was largely stalemated on the eastern and western fronts by 1917, when America reluctantly entered the European war on the side of Britain, France, Russia and the other Allies.

# CHAPTER ONE

## WORLD WAR I AND THE ALLIED-AMERICAN
## INTERVENTION IN RUSSIA 1918-1921

By early 1917, a vastly destructive World War and the oppressive social conditions in Russia had created an explosive climate in the world's largest nation. While the Russian Army continued to suffer heavy losses in the war against Germany, the aristocracy and wealthy bourgeoisie were almost oblivious to the widespread discontent, although the majority of literate and thinking people had turned against the czarist government. The Russian Revolution of March 15th, 1917 (called the February Revolution because it took place during February on the old Russian calendar), was actually a series of riots and strikes which grew into an uprising against the czarist government. Compared to the brutally suppressed 1905 Revolution and post-1917 events in Russia, it was a relatively mild affair resulting in the proclamation of a new government of Liberals, Social Revolutionaries and Constitutional Monarchists under Prince Lvoff.[5]

Czar Nicholas II abdicated on March 16, 1917, but the interminable war against Germany continued. Over 2,000,000 Russian prisoners of war languished in Germany and Russian attacks on the eastern front made little headway. Communist literature and newspapers were spread in the army and throughout Russia by the Bolsheviks, and in the coalition government the Social Revolutionaries gradually gained control under Alexander Kerensky. A "Soviet" of workers and soldiers was also formed in Petrograd in March 1917, acting on its own authority as an unofficial part of the provisional government. Other soviets were then formed in cities across Russia and soon many of them came under control of Bolsheviks and other radical socialists who were, at first allied with them.[6]

Through the summer of 1917 food shortages and famine spread across the country, while at the same time the Russian Army facing the Germans began to break up with mass-desertions and mutinies in which soldiers killed their officers and enlisted men took control. The collapse of the army was aided and abetted by the Military Revolutionary Committee of the Soviets, which was controlled by Bolshevik and Menshevik revolutionaries. Vladimir I. Lenin returned from a long exile in Western Europe in September 1917, aided by the Germans in the hope that if he succeeded in gaining power he would take Russia out of the First World War. In an attempt to maintain power, Kerensky outlawed the Military Revolutionary Committee, and Lenin chose this time to attack. He was assisted in the planning and execution of a coup d'état by Leon Trotsky, who was a Russian Jew actually named Lev Davidovitch Bronstein. Trotsky became a master strategist while serving as Lenin's highly effective Commissar for War.[7] Many of the early Bolshevik leaders, as well as Communist Party members and secret police operatives were Jews, who had been among the most persecuted groups in Czarist Russia. For centuries, Russian and Ukrainian Jews had alternated between brief periods of relative prosperity as merchants and absentee landlords, and longer intervals of confinement in ghettos, "beyond the pale," often subjected to fierce pogroms in which many lost their lives. Such anti-Semitic attacks by the Russian peasantry and city dwellers were sometimes encouraged by czarist Ochrana agents. This ultimately resulted in a large number of Jewish-born recruits to the radical Socialist and Bolshevik parties in pre-Revolutionary Russia, and an exodus of hundreds of thousands of Russian Jews to America, many of whom would spread their radical political beliefs in the teeming slums of the new world. These would-be revolutionaries also formed a substantial ethnic group within the American Socialist and Communist movement during the WW I period and thereafter, a fact that was to arouse the ire of the Justice Department, John Edgar Hoover and other U.S. security officials which would affect American political and social life for decades.

The impact of the Bolshevik "October Revolution," of 1917 was to rock the world for seven decades to come. It was thus named for its date (on the 24th of that month) on the old Russian calendar, which was not to be changed until 1918. A force comprised of Bolshevik-led soldiers and sailors, called the "Red Guard," attacked the Petrograd headquarters of the Provisional Government in the Winter Palace, which was lightly defended by young students from a military school. From this somewhat inglorious beginning stemmed the Bolshevik seizure of power and the imposition of Leninist Soviet rule. Most members of the provisional government were arrested, except Kerensky, who commandeered an automobile of the American Embassy and fled. The October Revolution was chronicled by an American graduate of Harvard named John Reed, who had been born in Portland, Oregon, and had become a leading spokesman for radical causes in the United States. As a war correspondent who was originally sympathetic to Germany because of his anti-British colonialist convictions, Reed had gained some notoriety during WW I by firing on the French Army with a Mauser rifle from the German trenches, whereupon he was barred from ever entering France. Reed had married a radical journalist named Louise Bryant in 1916 and together they sailed for Russia in August 1917. Reed was on hand for many events of the Russian Revolution and the Bolshevik Coup d'état, and soon wrote and published a book called: "Ten Days That Shook The World," a highly sympathetic account of the Bolshevik seizure of power. By November 8, 1917, Lenin and Trotsky controlled Petrograd, and the Kerensky government was either in prison or in flight. (Petrograd was later to be renamed Leningrad in honor of the first Bolshevik dictator, but 70 years later, by a vote of its citizens, the city assumed its original name.)

A nearly simultaneous and bloodier attack occurred in Moscow and by November 15, 1917, Lenin's Communists were in control of the political heart of Russia. Within a short time, Lenin's new government was spreading Bolshevik control to the rest of Russia through the many local soviets, which by and large followed the central Soviet's lead.[8]

The new Communist government took over the industries of Russia, placing them under control of central management bureaus, but for the time being Lenin allowed the Russian peasants to seize much of the farmland. This last action was allowed by Lenin to enable the Bolshevik government to consolidate its power. Lenin had no intention of abrogating his Marxist principles by allowing peasant ownership of land for long. To insure that the Bolsheviks would permanently retain their power, Lenin set up a secret police and security force to succeed the czarist Ochrana which included both Bolsheviks and former Ochrana agents who switched sides. This agency soon came to be known and feared as the Cheka, an acronym for the "Extraordinary Commission to Fight Counterrevolution and Sabotage." Although this acronym was to be changed to GPU, NKVD, MVD, KGB and others in the decades to come, Soviet secret police would always call themselves "Chekists."[9]

The strongest "White" force confronting the Bolsheviks arose in the south, where General Kornilov's "Volunteer Army" encouraged establishment of an opposition democratic-type government. The Don and Kuban Cossacks, who had once fought against and then served the Russian czars, later allied themselves with this White Russian Army, as powerful cavalry arms. They thereby earned everlasting hatred from the then-tottering Bolsheviks, for which the surviving Cossacks were to pay a devastating price after a second world war, in 1945. Other strong opposition to the Bolsheviks arose in the Ukraine with creation of a nationalist Green Army under Nestor Makhno. Kornilov was killed-in-action at Ekaterinodr and General Anton Denikin took command of the White army of the south. At about the same time the Don Cossacks had rebelled, elected General Petr Krasnov as Ataman, and thereafter served the White armies.[10]

As challenges to the Bolsheviks arose on every side in early 1918, Lenin was forced to re-mobilize hundreds of thousands of former-czarist officers and NCO's to reconstitute a Russian-Red army under control of Trotsky. Instilling the discipline in this army sufficient to ensure a communist victory was to require mass numbers of executions. Lenin had, officially, abolished capital punishment at the insistence of more liberal members of the coalition government on October 28, 1917. But the Bolsheviks began breaking this new law soon after their successful November 1917 revolution. At the beginning of 1918, Admiral Aleksai Schastny was publicly sentenced to be shot for his refusal to scuttle the Black Sea Fleet. The sentence was carried out despite protests from the Left Social Revolutionaries and others. Many of these protesters were soon to be shot by the Cheka themselves. In Moscow and other cities throughout Russia, "Revolutionary Tribunals" were set up to deal with "counterrevolutionaries" and "spies." The Bolshevik Council of People's Commissars had created the tribunals and the Cheka secret-police.[11]

The Soviet representatives of Lenin negotiated a separate peace with Germany which was signed at Brest-Litovsk on March 3, 1918, by Soviet Foreign Minister Georgi Chicherin, Leo Karakhan and G.I. Sokolnikov. Lenin had held out against German demands for months in hope that the Allies would recognize and support his government, in the interest of keeping Russia in the war. The Allied leaders, however, had no trust in the Bolsheviks, who continually broadcast their utter contempt for all "bourgeois capitalist regimes," including the German enemy and Russia's Western Allies.[12]

The Bolsheviks had supporters within the Allied governments; among them a young American diplomat named William C. Bullitt. The 27-year-old Bullitt was a Yale graduate of 1910, of mixed Philadelphia blueblood and Jewish descent, who had become known as a radical voice within the U.S. State Department, calling for diplomatic recognition of the Bolshevik regime. As a war correspondent in Europe during World War I, he had shamelessly vended intelligence information and privileged conversations to Frank L. Polk (#2 in the State Department) and to "Colonel" Edward House, the former Texas political operator who was President Wilson's top personal adviser and a founder of the influential Council on Foreign Relations. As a reward, through House's considerable influence on the President, Bullitt had been appointed as an Assistant Secretary of State in December 1917, nominally under Joseph C. Grew of the Division of Western European Affairs, but actually reporting directly to Edward House and the President.

As early as February 3, 1918, William Bullitt had been proposing in memos for House and the President that the United States extend diplomatic recognition to Lenin's new Communist regime. In proposing an address by the President to Congress which would place American war aims, "on the most liberal and lofty plane," Bullitt added in a letter to Edward House:

"...It is obvious that no words could so effectively stamp the President's address with uncompromising liberalism as would the act of recognizing the Bolsheviks."[13]

A German attack between February 18th and 24th had brought German forces to the eastern border of what would become Latvia and Estonia and to Narva which was within a hundred miles of Petrograd. It was clear the Germans could occupy the capital of Russia if they desired, and people began to flee that city. The Soviet treaty delegation started on it's journey to meet the German negotiators. On March 6th the seat of the Soviet government was moved to Moscow, with most Bolshevik offices reopened there by the middle of that month.

Both the American agent Raymond Robins and the British agent Bruce Lockhart were queried by Lenin just prior to his agreement to ratifying the treaty with Germany, on their governments' willingness to come to terms with him. The United States Government refused to recognize the authority of the Bolshevik regime as the de-jure government of Russia, but at the suggestion of William C. Bullitt and the president's adviser, Colonel Edward House, Woodrow Wilson finally sent a very late but conciliatory communication to Russia, that was addressed to the Congress of Soviets (which the Bolsheviks by now, however, viewed as a potential threat to their revolution). The message was sent by cable to Moscow on March 11th, where Raymond Robins personally handed it to Lenin on March 12th. A copy of it was also sent to U.S. Ambassador Francis, then in Vologda. It read:

"May I not take advantage of the meeting of the Congress of the Soviets to express the sincere sympathy which the people of the United States feel for the Russian people at this moment when the German power has been thrust in to interrupt and turn back the whole struggle for freedom and substitute the wishes of Germany for the purposes of the people of Russia. Although the Government of the United States is unhappily not now in a position to render the direct and effective aid it would wish to render, I beg to assure the people of Russia through the Congress that it will avail itself of every opportunity to secure for Russia once more complete sovereignty and independence in her own affairs and full restoration to her great role in the life of Europe and the modern world. The whole heart of the people of the United States is with the people of Russia in the attempt to free themselves forever from autocratic government and become the masters of their own life. Woodrow Wilson"[14]

Bullitt actually wrote some substantial portion of Wilson's message to the Congress of Soviets, but Acting Secretary of State Frank Polk had substituted the words: "struggle for freedom" in place of the original term used: "Revolution."[15]

The Soviet Congress opened on March 14th, and was addressed by Lenin, who argued for ratification of the Brest-Litovsk Treaty. According to Raymond Robins, who on the second night of the Congress (on March 15th) was seated on a step of the platform where Lenin himself was sitting, Lenin asked him: "What have you heard from your government?" Robins had to answer, "Nothing." Lenin asked Robins what Lockhart had heard from Lloyd George's government and Robins again said, "nothing." Robins recalled later that Lenin then said: "I am now going to the platform and the peace will be ratified." After a one hour and twenty minute speech by Lenin, the peace was ratified by a two and one half to one vote.[16]

An insulting March 15th response of the Congress of Soviets, under Chairman I. Sverdlov, to President Wilson's message, which obviously reflected Bolshevik pressure, left little room for further diplomacy:

"The Congress expresses its appreciation to the American people and in the first instance to the toiling and exploited classes of the United States of North America, for the expression by President Wilson, through the Congress of Soviets, of sympathy for the Russian people in these days when the Soviet Socialist Republic of Russia is undergoing heavy trials. The Russian Socialist Soviet Federated Republic avails itself of this communication from President Wilson to express to all those peoples perishing and suffering under the horrors of the imperialist war its warm sympathy and its firm confidence that the happy time is not far distant when the toiling masses of all bourgeois countries will throw off the yoke of capitalism and will establish a socialist order of society, which alone is capable of assuring a just peace as well as the cultural and material well being of all the toilers."[17]

Despite the urgings of Bullitt and a few other radical voices, Wilson and Lloyd George had remained silent on the subject of officially recognizing the Bolshevik regime, and the Brest-Litovsk treaty was ratified by the Congress of Soviets, then the highest structure of government in Russia, on the night of March 15th-16th, 1918. The Germans were allowed to occupy the Baltic states, the Ukraine and part of Byelorussia. Last minute Turkish demands to recover lands in the south seized by Russia in the 1870s, were also acceded to. Lenin called it "a shameful peace," although he had planned all along to take Russia out of the war. Although reportedly more reluctant, Leon Trotsky agreed with this move. (After the later surrender of Germany to the Western Allies, much of the this land was returned to Russia.) By early 1918 the Russian people had, in any case, lost the will to wage war against Germany. The regular Russian Army was continuing to collapse and melt away in mass-desertions as the Red Army was formed around a nucleus of dependable foreign mercenaries such as the Latvian Rifles, and a core of Bolshevik soldiers, sailors and revolutionary civilians.

Hundreds of thousands of German and Austro-Hungarian prisoners of war were released over the months following the Brest-Litovsk treaty. While many of the ex-POWs trekked slowly toward their homelands, others became outlaw bands which terrorized local regions. Some German and Austrian prisoners also reformed themselves into military units and began taking sides in the struggle between the Red Communists and White Russian anti-communists. The Bolsheviks had made a concerted effort to propagandize the German and Austro-Hungarian prisoners while they were in the camps, and it was they who made the most use of former prisoners of war in the coming struggle. The Bolsheviks quickly armed large groups of the POWs and formed them into "International" units of the newly created Red Army. Many of these men, disaffected by years of war and privation, were encouraged by the Communists among them to believe that serving the Bolshevik government would lead to the establishment of revolutionary regimes in their own homelands. Meanwhile, a Czechoslovak Corps which had been fighting for the Allies on the Russian Front was trapped inside a now-neutral Russia, and the Germans demanded they be disarmed and interned by the Bolshevik government. The Czechs refused to comply and open hostilities broke out. The Czechs were a well-disciplined, hard fighting force, and the Bolsheviks were unable to take them under control. The western allies viewed the Lenin's capitulation to German peace terms and the unfolding anarchy in Russia with alarm, as potentially detrimental to the American, British and French war effort.[18]

THE SOVIET CIVIL WAR<br>AND THE 1918 ALLIED-AMERICAN INTERVENTION IN RUSSIA

The separate Russian peace with the Germans by signed by Lenin's Bolshevik government in early March 1918 greatly angered Allied leaders, who faced a major offensive by the now united German Army on the Western Front later that same month. American, British and French soldiers suffered severe losses and were driven back in many areas. The Brest-Litovsk treaty was viewed as a treacherous act and its signing had greatly increased demands in London and Paris for intervention in Russia against the Bolsheviks, whom it was felt had betrayed the Allied cause for Germany. Some British and French leaders were convinced that Lenin was actually a German agent, as he had benefitted from their assistance at a crucial time in his struggle to gain power. On the American and British home fronts, early public sympathy for the revolutionary ideals of the Bolshevik revolutionaries was giving way to a widespread feeling of betrayal by Russia.[19]

In France, as the bloody trench-warfare involving millions of men dragged on, the Allies faced a political problem with the symbolic 20,000-man Russian Expeditionary Force which had been fighting for the Allies on the Western Front. Even prior to the Brest-Litovsk Treaty, the signing of a December 15, 1917 Armistice between Russia and the Central Powers of Germany, Austria-Hungary and Turkey, had caused these Russians to correctly conclude they were no longer at war, and they refused to continue fighting. The French thereupon chose to consider them traitors and cowards and interned them in early 1918.[20]

The British Prime Minister, Lloyd George, was in a quandary over events in Russia. In January 1918, he had sent the former British Consul General in Moscow, R.H. Bruce Lockhart, to establish communication with Lenin's Bolsheviks. Lockhart and other British agents and also French intelligence in Russia warned their home governments against Allied intervention without the approval of the Bolsheviks. They felt that unilateral intervention would strengthen the position of the Bolsheviks as defenders of Russia's sovereignty. Nevertheless, with the growing German strength in the West the demand in Allied capitals for intervention solidified into action.

After the Brest-Litovsk treaty was signed, the Murmansk Soviet, with the authority of the Central Soviet, began negotiations with Allied representatives in north Russia, asking for Allied military intervention to protect the Murmansk railroad from possible capture by the Finnish White Guard. Lenin approved of this action. British and Allied troops were sent to north Russia in the spring of 1918 ostensibly to prevent large quantities of supplies from falling into German or Finnish White Guard hands, and to Siberia to protect vital railways and the evacuation of Czechoslovak POWs in Russia. (This military aid had been sent to encourage Russia to stay in the war so that the Allies would not have to face the entire German Army on the Western Front. In the turmoil following the revolution much of this material reportedly disappeared.) The initial March 1918 commitment of British troops to Murmansk and Archangel in North Russia was followed by increments of French and Canadian soldiers (together with an armed shore party of U.S. Navy sailors from the American fleet), with the Allies taking the side of the White armies against the Bolsheviks in active combat with elements of War Commissar Leon Trotsky's 6th Red Army.[21]

Two thousand British troops, including Royal Marines and engineers, were first landed at Murmansk in March 1918. The Soviet Commissar for war, Leon Trotsky, had ordered the Murmansk Soviet to cooperate with the Allies because the Bolsheviks had felt menaced by a German-Finnish offensive which threatened Murmansk.

The Russian withdrawal from the war in March of 1918 had allowed the German Army to mass more troops for a major attack on the Western Allied front and still maintain forces on the Russian-Finnish front. With the hostility of Western Allied governments to the Bolshevik regime for taking Russia out of the war with Germany, the Allied forces, particularly in the Murmansk (and later Archangel) area, quickly became identified with the anti-communist White Russian armies in the sectors they occupied.

Also in March 1918, the Allies received word, on the 23rd, that a force of 25,000 stubborn Czechoslovak soldiers were proceeding across Russia to Cheliabinsk from the Ukraine, hoping to reach the North Russian ports for shipment to France and service in the Allied armies. Now technically neutral, the Soviets attempted to prevent this, attacking the Czechs at Ekaterinburg in June and blocking their escape toward Murmansk and Archangel. The Czechs then turned toward Siberia where an estimated 200,000 Czech POWs remained on Russian territory.

More British troops arrived in Murmansk on June 23 and fighting broke out between the British and the Bolsheviks in North Russia on June 28, 1918. As the British Army began taking casualties, their commander, Major General F. C. Poole, pressed for more troops and the United States was asked to assist by the British and French.

In the American government, one of the influential voices against U.S. intervention in Russia was that of William C. Bullitt. On June 24th, just before the deployment of U.S. troops to Russia, he wrote President Wilson's advisor Edward House:

"...I feel that we are about to make one of the most tragic blunders in the history of mankind...The Russian upper classes are about to appeal to Germany to overthrow the Soviets. Let them appeal...Let us join the Soviets."[22]

President Woodrow Wilson, who had long vacillated on the subject of Russia, somewhat reluctantly agreed to furnish a small contingent of American troops on July 17, 1918. His desire appears to have been that US soldiers should not take sides against the Bolsheviks but merely protect American interests and citizens. On the same date Wilson wrote an aide-memoir which illustrates the confusion between his personal thoughts and official policy about to be executed, and which may indicate uncertainty generated by reading Bullitt's memos and by their influence upon the President's principal adviser, Edward House:

"Military intervention there would add to the present sad confusion in Russia rather than cure it...Whether from Vladivostok or from Murmansk and Archangel, the only legitimate object for which American or Allied troops can be employed...is to guard military stores which may subsequently be needed by Russian forces and to render such aid as may be acceptable to the Russians in the organization of their own defense."

This high-minded goal for the intervention was to prove very different from the actual military conduct in Russia, which was under overall command of British officers. American troops were not to reach Archangel in force until September 1918, although a contingent of 50 American sailors from the USS Olympia had accompanied General Poole's force which started for Archangel from Murmansk on July 31.

The United States Ambassador to Russia from 1916-1918 was David Francis, an anti-Bolshevik former businessman, who had assured President Wilson that Kerensky would be able to maintain control and had been taken by surprise when the Bolsheviks seized power. Ambassador Francis had left the old capital of Petrograd in case of a German occupation of Moscow at the time of the Brest-Litovsk treaty and taken up diplomatic residence 300 miles to the east in Vologda. Vologda was on the main railroad line between Petrograd and the port of Archangel on the White Sea of north Russia, a point of escape for foreigners attempting to leave Russia. After the murder in Moscow of the German Ambassador on July 4th, 1918, the Soviet leaders began clamoring for all western diplomats to move to Moscow, the new Soviet capital, ostensibly for safety. The Allied ambassadors suspected the Bolshevik leaders might take them hostage in the event of a full-fledged intervention by the British and Americans in the civil war.

In July 1918 the Western diplomats in Vologda received a secret message from General Poole, the British military commander in north Russia (Murmansk), that Britain was going to formally intervene at the end of July, with the arrival of a fleet at Archangel. The American and several other Ambassadors succeeded in reaching Archangel just before the British fleet, and an anticipated outbreak of fury by the Bolshevik leaders. Ambassador Francis was reported evacuated by ship on the 29th of July. Lenin, Trotsky and Dzerzhinsky were well aware that Western intelligence agents had been plotting a coup to overthrow the Bolshevik regime. Also, in mid-July 1918, Czar Nicholas II and his family were killed by the Bolsheviks at Yekaterinburg and their bodies reported variously as burned and thrown into a coal mine. The Bolsheviks were afraid of their power as living symbols to the anti-communist White opposition.

Just prior to arrival of the British force under Poole, a coup d'état against the Soviet authority in Archangel was carried out, on August 1, by a Russian naval officer who had been in communication with the British command. Poole's fleet landed in Archangel on August 3, 1918. The landing was preceded by a British naval bombardment of the Russian positions defending the entrance to Archangel. After the landing of Allied forces Poole ordered British and French troops to follow the retreating Bolshevik forces south along the railroad toward Vologda. The 50 American sailors from the USS Olympia served in the vanguard of this pursuit on a machine gun-armed train and engaged the Red Army 75 miles south of Archangel, supported by French infantry. This appears to have been the first instance of American troops in combat on Russian territory. The mass of Allied war supplies which was supposed to be protected by Poole's intervention force had been largely removed to the south by the Bolsheviks on the railroad.[23]

Most embassy, military and consular staff people of the Allied missions in Petrograd and Moscow were unfortunately not forewarned of the military intervention. They were left behind and many were arrested, along with hundreds of other Westerners. Captain Francis Cromie, the British naval attaché in Petrograd, was murdered by a Bolshevik-led mob at the British Embassy. Cromie had been deeply involved in schemes with other British intelligence officers, Bruce Lockhart, Captain George Hill, and Russian-born British secret agent Sidney Reilly, but his murder infuriated the British government.[24]The assassination of the German Ambassador, Count Mirbach was followed by an uprising by the Left Social Revolutionaries that captured secret police chief Felix Dzerzhinsky and the Cheka's Lubyanka headquarters. In a subsequent purge, the Left SR's were done away with and Dzerzhinsky was restored to a thoroughly Communist Cheka.[25]

On August 30, 1918, an attempt was made to murder Lenin by a disenchanted Social Revolutionary woman named Dora (Fanny or Fanya) Kaplan. Lenin was shot by her several times at close range and was severely wounded. His life was said to be in danger, and in addition, the head of the Petrograd Cheka was murdered. The next day the "Red Terror" was unleashed by the Cheka under the Polish Communist Felix Dzerzshinsky and his deputy, Peters. Thousands of Russians under any suspicion of anti-state activity, no matter how trivial, were arrested and eventually paid with their lives for "oppositionism." The killing squads of Felix Dzerzhinsky's Cheka executed some 1,200 enemies of the Bolsheviks soon after the shooting of Lenin. Some 700 of these people were shot in Petrograd and about 500 more in Moscow. (The Cheka admitted to only 600 executions in the retaliation.)

Hostage-taking became a fundamental part of the Red Terror, and Dzerzshinski issued instructions that hostages were always to be seized from the most influential part of the population, especially officials, landowners, academics and others of high social position. As has been noted by Heller and Neckrich, "This hostage system, unknown in pre-revolutionary Russia, was supplemented by another instrument of repression new to the country, the concentration camp. The notoriety stemming from its use by Hitler should not obscure the fact that the Soviet state was the initiator of this institution."[26]Half a century after these events, the author met a former officer in the Chevalier Guards Regiment of the Czar's Army, Serge Obolensky, who had fought against the Bolsheviks, and who also confirmed to the author the Communist methods described earlier, which were used in winning the Russian Revolution and prosecuting the civil war against the White forces. (Obolensky was to subsequently serve in the American OSS in Europe during WW II.)[27]

After the shooting of Lenin, the Bolshevik press kept up a tirade of demands for vengeance:

"...Without mercy, without sparing we will kill our enemies in scores of hundreds; let them drown themselves in their own blood. Let their be floods of blood of the bourgeois..."[28]

The murder by a Cheka-led mob of the British Naval Attaché, Captain Cromie, who had in reality been working as a secret agent with Bruce Lockhart, Sidney Reilly, Captain George Hill, the American secret agent Xenophon Kalamatiano and others, against the Bolshevik regime, was followed by other Cheka arrests and executions of foreigners. The British agent for Prime Minister Lloyd George, R.H. Bruce Lockhart, was arrested and confined in the "kennels" of the Cheka's Lubyanka Prison, with Lenin's attempted assassin, Dora Kaplan, just prior to her execution. Although released after an initial period of interrogation, Lockhart was soon rearrested and this time held by the Bolsheviks as a hostage. Dora Kaplan was executed. The head of the American secret service in Russia, Kalamatiano, was also arrested and although reported to have been executed by the Cheka, he was later officially listed as released in August 1921.[29]

It has been estimated by Russian historians that some 8,000 persons were executed throughout Russia soon after this time. Even Latsis, the Cheka historian, admits to 6,000 shot in the last half of 1918. It is not known how many foreigners lost their lives as such records remain hidden in the secret archives of the KGB. According to the later US State Department Russian expert, George Kennan, "most" of those imprisoned from the Western diplomatic missions got out of Russia by late 1918. Kennan does not record how many of them disappeared forever.[30]

At this time more Allied troops were landing in Russia, in what would come to be known as the Intervention. To the Bolsheviks, everything seemed to point toward a coordinated Allied plan to overthrow their regime, and indeed there were leaders in the West and secret allied agents in Russia who worked hard to accomplish this end. Winston Churchill, soon to be appointed the British Secretary of State for War, was a major force behind actual military intervention on the side of the Whites who opposed Bolshevism.

Although the Intervention was to have bitter and long-lasting results for the United States and its allies, it was in reality a very minor military operation, and in not a seriously threatening Allied assault on Bolshevik power. The total American force in Siberia was to average about 7,000 men for a year and a half, and about 5,000 on the Archangel front in North Russia, for less than one year. The Red Army in this period contained over a million men.[31]

During the revolution and the early stages of the civil war the most loyal military units to the Bolsheviks were foreign troops such as the Latvian Rifles, and also Polish, Czech, Finnish and Chinese soldiers who were known as "internationalists." The Bolsheviks were infuriated by the success of Lockhart, Cromie and Reilly at subverting some of the Latvians, who remained crucial to the survival of Lenin and Trotsky. There were many more foreigners fighting for the Bolsheviks than those eventually fighting against them as "interventionists," but the foreigners on the Bolshevik side were looked upon as heroes of the coming world revolution. The implication was that the foreign "internationalists" were fighting for a progressive, new society, while the other foreigners in Russia, the "interventionists" who opposed the Bolsheviks, were nothing more than hired foreign mercenaries fighting for an evil, reactionary cause. Many of these foreign troops fought bravely for the Bolsheviks out of an idealistic commitment to the promised "new world" that they would inherit upon victory. According to Soviet sources, the foreign "internationalists" numbered 50,000 in the fall of 1918 and some 250,000 by the summer of .1920.[32]The ultimate thanks received by the surviving Latvian riflemen who had saved the revolution early-on, was to be arrested en masse in Stalin's purge of 1937, and mostly disappear into forced-labor camps of the Gulag.[33]

## THE SIBERIAN INTERVENTION

By July 25th, 1918, before the Allied intervention in Archangel, some 1,270 American troops were ashore in Siberia, and by September 13, some 3,315 Americans were ashore. They were to serve under overall command of British Major General Alfred Knox, who was assisted by the British High Commissioner, Sir Charles Eliot.

Shortly after the initial landing of part of the U.S. 27th Infantry (later named "Wolfhounds"), in August 1918, an American detachment saw brief combat with a Bolshevik force north of Vladivostok. The Americans were added to a British force, including a partial battalion of the Middlesex Regiment and Japanese troops and attacked a Bolshevik force later in August 1918. This engagement was to clear the railroad line to Irkutsk, and was the initial commitment of US troops in combat in the Siberian "Intervention." American casualties for this action are unclear.

The American troops in Siberia do not appear to have been intentionally committed to battle with Red Army or Bolshevik troops while under the command of American Major General William S. Graves, who arrived later. They were primarily deployed to guard a portion of the Trans-Siberian railroad, and the Suchon mines, which provided coal for the engines, against what were reported as "anti-Kolchak" rather than "Bolshevik" forces. According to General Graves, Admiral Kolchak's White Russian army engaged in a brutal partisan warfare campaign in Siberia, which included many massacres of hostages and attacks on civilians by both sides. General Graves operated under extreme pressure from the British and the Japanese army in Siberia to directly support Kolchak's and other White Russian forces, but he refused to exceed the mandate given to him by President Wilson to take neither side in the struggle.

American forces in Siberia averaged 7,000 to 8,000 men from late 1918 until early 1920, and most of these men served in the 27th and 31st U.S. Infantry regiments. The total American losses of the Siberian Intervention are somewhat obscure, according to War Department records in the National Archives. A few casualties were lost in the initial fight before Graves' arrival and in minor skirmishes and at least one sharp battle with raiding anti-Kolchak bands thereafter. Both Kolchak's White Russians and his Bolshevik-partisan enemies demanded unrestricted use of the railroad, placing the Americans guarding it, and trying to adhere to Wilson's mandate, in an impossible situation. In addition, large numbers of former POWs attempting to reach Vladivostok for repatriation after the Armistice also demanded use of the railroad, sometimes commandeering trains by force. In military records of the U.S. Army Chief of Staff the total American casualties are recorded as 36 combat deaths (28 killed in action and 8 died of wounds) up until October 1, 1919.41 more Americans are listed as died of disease, 5 as suicides and 27 more dead of accidents or "other causes" (which are not explained). The total number of American dead in Siberia up to October 1919 was officially reported to be 109. The number of wounded in action is not given but was no doubt (as in Archangel) on a ratio of at least 2:1 in relation to killed in action/died of wounds, indicating perhaps 70 or more Americans wounded.[34]

Thousands of American troops remained in Siberia from October 1st, 1919 into March of 1920, BUT NO CASUALTIES ARE GIVEN FOR THESE LAST SIX MONTHS OF SIBERIAN SERVICE IN U.S. ARMY RECORDS, although records exist of combat action resulting in American killed and wounded at least into January 1920. Apparently for this reason, several Army sources, including records of the U.S. Army Medical Department, state that casualties of the Siberian

Intervention are "incomplete" or "confused."[35]In a separate source found in the National Archives (1930 classified War Department documents cited herein) Seven Americans were reported as MIAs or taken prisoner, but the circumstances of their loss and subsequent recovery within a few days is clearly recorded by the colonel commanding their regiment the 31st U.S. Infantry.

A decade later, a November 8, 1930, US War Department document stated:

"I understand that one officer and six enlisted men were taken prisoners in Siberia, BUT I UNDERSTAND THAT THEY WERE REPATRIATED FROM CAPTURE BY THE FORCES OPERATING AGAINST THE AMERICANS IN SIBERIA." A later US War Department document of November 12, 1930 reported: "1 US officer and 6 US enlisted men carried on the records as missing in action from the American Expeditionary Forces in Siberia, all of whom were subsequently repatriated."[36]On the original document, now in the National Archives, the phrase "missing in action" has a line through it and an addition of "captured by enemy." This refers to the 1 officer and 4 U.S. enlisted men captured and released within 5 days at Novitskaya in June 1919, and the two other enlisted men captured and released in a separate incident, also in June 1919, at Kraevski, Siberia.

In Grave's memoirs he noted a March 16, 1919 cable, from the U.S. Military Attaché in Tokyo stating: "Japanese General Staff reports ten American Army deserters have joined Red Guards northeast of Vladivostok, and one has been captured..." Graves replied that he knew of only, "SIX AMERICAN DESERTERS AT LARGE...The one reported as captured has not been delivered to me. False statements have been so common that I will not believe that they have captured one unless I see him..."[37]

When the 27th and 31st U.S. Infantry regiments returned from their arduous Siberia service in early 1920, it is interesting to note that the former director of U.S. Military Intelligence, Major General Ralph Van Deman, known later as the "father of U.S. Military intelligence," was sent to take command of the Post of Manila and the 31st U.S. Infantry after their return from Siberia, where he remained until 1923.[38]

It is not as yet known whether some or all of those Americans reported as deserters in Siberia were actually Missing in Action (MIAs), or men reported killed, body not recovered (KBNR), who were secretly taken prisoner by the Bolsheviks, or others who remained behind voluntarily in Siberia after their units sailed for home. Graves' memoirs are incomplete on the subject of MIAs, deserters, or Americans left behind in Siberia, as U.S. documents not declassified until 70 years later in the National Archives, illustrate. After U.S. forces had evacuated Vladivostok in 1920, one U.S. officer, Lieutenant Frank B. Coughlin, who, according to a memorandum of

General Graves, was discharged from the U.S. Army in Vladivostok, and three American enlisted men, remained behind in Siberia voluntarily in 1920, claiming they were on a secret mission to locate other Americans, either missing or called "deserters." Two of the enlisted men were reported to be Privates of the 27th U.S. Infantry, which had already sailed for home. The U.S. Consul in Harbin, Manchuria reported later that he believed that a pass signed by General Graves which Coughlin was using was a forgery, and that the ex-Lieutenant was a "thoroughly bad character." The Consul also remembered: "It was...reported that he was an officer on special work of some sort-that he was in search of deserters..."[39]

The four Americans and an interpreter traveling with them were captured by anti-Bolshevik Semenov partisans and three were shot by a firing squad next to the tracks of the trans-Siberian railroad, near Chita. Lieutenant Coughlin was reported to have escaped into Manchuria-only to be labeled as a renegade and a liar by U.S. government officials there, for claiming to be on a mission. A Russian porter on the Trans-Siberian Railway was a witness to the fate of the executed Americans and Colonel B.O. Johnson of the War Department later confirmed the accuracy of this account after interviewing another American named Robbins, who also escaped, but who thought Lieutenant Coughlin had been killed in the escape attempt. A copy of this information was sent to General Graves, but Colonel Johnson reported that no reply had been received. Yet, on September 16, 1920, the Chief of Staff at the War Department received a letter from General Graves at Fort William McKinley in the Philippines dated August 2nd, in which Graves said nothing about Coughlin being a "thoroughly bad character" or that Coughlin had forged an order with the General's signature ordering him to Zabaikal Province for the purpose of arresting deserters of the American Army." Instead Graves wrote:

"These four Americans have lost their lives because of resentment against the United States. This resentment is due to the failure of the U.S. military representatives to take part in the murdering and robbing of Russians by Semionoff and his gang. IN ADDITION TO ENLISTED MEN ON ENCLOSED LIST, LEFT IN TRANSBAIKAL SECTION, LIEUTENANT FRANK B. COUGHLIN WAS DISCHARGED AND REMAINED IN SIBERIA." It may never be known whether Lieutenant Coughlin's mission to locate deserters inside Bolshevik-occupied territory was actually a cover for locating American Army prisoners of war or missing in action in Siberia or from among the 200-300 U.S. citizens serving in the U.S. Railway Corps in Russia.[40]

# THE NORTH RUSSIA-ARCHANGEL ALLIED INTERVENTION

The decisive theater of the Russian civil war in 1918-1919 was far to the west of Siberia, and it was in northwestern Russia, where the Allied military presence threatened Bolshevik control of Petrograd (Leningrad) and Moscow, that most of the casualties of the American Intervention occurred. The 1919 Annual Report of the Secretary of War briefly outlined the official U.S. version of American participation in the Allied intervention in North Russia:

"During the summer of 1918 the situation in Russia presented a problem which demanded serious consideration. The chaotic state of the Government, the distress among the Russian people, the intrigues and activities of German prisoners of war, had produced a situation which had to be met and solved. After careful consideration the Supreme Inter-allied War Council decided to send an expeditionary force to Archangel to prevent the Germans from using that port as a submarine base, and to guard allied supplies which might be there, or which would be landed there in the future. The Government of the United States approved of this project and decided to cooperate in it. Accordingly it mobilized at Aldershot, England, a force consisting of the 339th Infantry, the first Battalion of the 310th Engineers, the 337th Field Hospital, the 337th Ambulance Company, and the 310th Sanitary Train. Colonel George E. Stewart (339th Inf.) was placed in command of this force and reported for duty to General Poole, commanding officer of the British forces at Archangel on September 4, 1918."[41]

Five thousand American troops arrived as reinforcements at Archangel, in north Russia, in September 1918, two months before the Armistice which ended World War I. They were soon fighting against the new Soviet Army, alongside British and French troops in an offensive aimed towards Vologda on the trans-Siberian railroad, and Moscow, with the idea of eventually linking up with the Czechs and anti-Bolshevik White Russians. American soldiers began to disappear on these battlefields in small numbers in the autumn of 1918.

Among the first U.S. soldiers listed as missing in action in Russia prior to the Armistice which ended World War I, was Corporal Herbert A. Schroeder of Company B, 339th U.S. Infantry., a red-haired, fair complexioned soldier who in civilian life had been a linotype operator from Detroit, Michigan, and who disappeared in combat action on September 20, 1918 near Seltzos, south of Archangel, Russia. Although the Soviets later denied it and the U.S. Army eventually changed his status from MIA to "presumed Killed in Action," Corporal Schroeder had in fact been captured alive and secretly transported to the new Soviet capital of Moscow, where he was secretly confined for future use, unknown to his own

Government until 1921. Schroeder, among the first of thousands of "Moscow-bound" American POWs of four wars, was to be reported by the U.S. State Department three years later as alive and in Bolshevik captivity in Moscow. He was never repatriated, and a decade later, in 1931, he was still listed in War Department documents as Missing in Action—remains not recovered, presumed killed, by the Adjutant General of the U.S. Army on the date he disappeared during World War I.[42]

In the fall of 1918 Lenin expressed his personal view of the American government and ruling class:

"The American millionaires, those modern slave owners, have opened a particularly tragic page in the bloody history of bloody imperialism by giving their consent...to the armed campaign of the Anglo-Japanese beasts for the purpose of crushing the first Socialist republic...We are in the position of a beleaguered fortress until other detachments of the international socialist revolution come to our aid. Such detachments exist, they are more numerous than ours...We are unconquerable because world proletarian revolution is unconquerable."[43] Two days before the Armistice ending World War I was announced, a republic was proclaimed in Germany and workers' and soldiers' councils on the Soviet model were formed. Although short-lived this development caused a stir around the world and inspired Lenin.

On the North Russia war front, some of the American soldiers who had willingly signed up to fight the Kaiser Wilhelm's German Army in France were unhappy with their role in opposing the first "workers state." Many of the draftees and volunteers in the 339th U.S. Infantry had been exposed to the radical Socialist politics of early-20th century America, represented by militant labor groups like the IWW (Industrial Workers of the World). Founded in 1905, the IWW combined the tough independence of western miners and loggers with a 19th century, midwest populist tradition, and radical Socialist ideas imported from Europe. Their object was the overthrowing of the "Capitalist system" in the United States and institution of a Socialist state run by American workers. As a militant vanguard in sometimes-violent strikes, the IWW gained a measure of respect among many American workers, and the organization reached it's greatest size and influence just before WW I. The IWW opposed America's participation in the "capitalist" war, and Woodrow Wilson's Justice Department administration had indicted 166 IWW leaders, of whom 93 were convicted and given heavy sentences. These immediately became martyrs to many militant American workers, who were torn between patriotism felt towards their country and the aspirations of a long-neglected underclass.

Such contemporary issues made some of the soldiers of the

339th susceptible to Marxist-Leninist propaganda, if they were captured by the Bolsheviks. In the taiga forest south of Archangel, where the blockhouses and trenches of the Americans, British, Canadians and French faced the "Bolo" lines, leaflets, booklets and tracts appeared or were dropped from Soviet airplanes on the U.S. and Allied troops below. The content of some propaganda leaflets left by Cheka operatives in the Red Army for Allied soldiers struck a responsive chord among some embittered Americans who wanted to go home after the Armistice that had ended World War I, on November 11th, 1918. For example, one aimed at Americans urged:

"You will be fighting not against enemies but against working people like yourselves. We ask you, are you going to crush us?...Be loyal to your class and refuse to do the dirty work of your masters... Go home and establish industrial republics in your own countries, and together we shall form a world-wide cooperative commonwealth."[44]

On the war front south of Archangel, American troops in company and platoon-size formations were holding a dangerously lengthy front, with Allied companies of the British Liverpools and Royal Scots, but in widely separated locations. Under War Commissar Trotsky's leadership, the Bolsheviks grew steadily in numbers and their organization and weapons were improved. It also became clear that they were making every effort to capture American and British soldiers, who sometimes disappeared when outnumbered U.S. and Allied patrols were ambushed and forced to retreat. Two weeks after the Armistice ending WW I had been signed one such action occurred, which was recorded by the 339th Infantry's medical officer, Major John Hall in his war diary:

"Arrived at Shenkursk November 11th. Over one hundred patients...contagious ward was full of 'flu' and influenza cases... People were dying by hundreds in the neighboring villages...Tried to segregate cases in Shenkursk and immediate vicinity as much as possible...I proceeded to villages not yet reached by others. REPORT FROM UST PADENGA THAT LIEUTENANT CUFF AND FOURTEEN ENLISTED MEN KILLED OR MISSING ON PATROL NOVEMBER 29th, SOME OF THE BODIES RECOVERED. Weather growing colder. Twenty degrees below zero, with snow four inches deep. Evacuated sick and wounded from Ust Padenga eighteen versts beyond Shenkursk in sleds filled with hay and blankets necessary for warmth...Action reported on Dvina and hospital captured; later retaken. Slight action every day or so at Ust Padenga..."

Many of these 15 American MIAs of November 29th, 1918, were never seen again, and few bodies were found. Eleven years later, a combined U.S. Army Graves Registration and Veterans of Foreign Wars expedition to north Russia allegedly recovered the skeletal remains of 1st Lieutenant Francis W. Cuff and returned them to the U.S. in 1929, together with several other skeletal remains purported to be members of this patrol. (This expedition is described in more detail later.) But some of these soldiers remained missing in action forever, and their "remains" were not reported as "recovered," when the United States finally recognized the Soviet government of Russia, in 1933. One of these MIAs was Private Elmer W. Hodge, a farmer in North Dakota when the war broke out, who had been born in Michigan and ended up in the 339th Infantry after his induction in North Dakota. Hodge was one of those Americans who vanished in the fight on November 29th at Ust Padenga, as was Private Nicolas Jonker, a ruddy-complexioned bookkeeper from Grand Rapids, Michigan; Private Thurman L. Kissick, a 30-year-old farm laborer from Fleming County, Kentucky, Henry R. Weitzal, also from Michigan, an American citizen who had been born in Warsaw, Poland, when it was part of the Russian Empire. It is likely that given the circumstances of their disappearance, these men became secret prisoners of the Soviets for the rest of their lives.[45]In the capture of the hospital on the Dvina River other Americans disappeared.

During a December 1918 meeting of the Petrograd Soviet, which led to establishment of the Comintern (Communist International), published Soviet sources record that American, English and Scottish POWs captured on the Archangel Front were used for propaganda purposes. Boris Reinstein, the Russian-born representative of the American Socialist Labor Party (an ancestor of the American Communist Party), had earlier supported a general resolution advocating "a revolutionary struggle" and support for, "the October revolution and the Soviet government." But it was viewed as a much greater propaganda victory by Lenin's Bolsheviks to win over converts among native-born American prisoners of war who had been captured in battle with the 6th Red Army, south of Archangel. An "international meeting," presided over by the well-known Socialist writer Maxim Gorky, was convened by the Petrograd Soviet on December 19, 1918, and was attended by top Bolshevik leaders, including Grigory Zinoviev, who announced: "We have among us today guests who are neither Marxists nor Communists, but all of us here are agreed on one point, in our hatred of the bourgeoisie, in our hatred of a class guilty of the death of millions of men in the interests of a small group."

The Socialist leader Boris Reinstein, a trusted member of Lenin's inner circle, "represented" the United States, while others pretended to speak for British, French, Serbian, Bulgarian, Turkish, Chinese, Hindu, Persian and Korean workers. In addition, "SPEECHES WERE ALSO DELIVERED BY SCOTTISH, ENGLISH AND AMERICAN PRISONERS OF WAR CAPTURED ON THE ARCHANGEL FRONT (the first was introduced as the 'delegate for Scotland')..." The ultimate fate of these American, British and Scottish POWs in Petrograd remains unknown.[46]

As many Americans at home wondered why U.S. forces were fighting the Bolsheviks in Russia, after the Armistice that had supposedly ended World War I, the morale of the American infantry in the lines south of Archangel began to sag. In a coded cable to the Military Staff Washington, the U.S. Military Attaché in Archangel, Colonel James A. Ruggles, loyally asked for Christmas greetings to be sent to the American soldiers fighting in Russia, from President Woodrow Wilson, Secretary of War Newton Baker and General Pershing, Commander of the American Expeditionary Force (AEF) in Europe:

"...In behalf of American troops, Christmas greetings be cabled addressed to them from President of the United States if possible and from Secretary of War and General Pershing. They have been obliged to endure great hardships under most trying conditions and such message would have most beneficial effect. Pershing informed. Ruggles-Archangel."

At this time, according to now-declassified telegrams of Colonel James Ruggles, the mustachioed and observant young U.S. Military attaché in Archangel, who had served at the American Embassy in St. Petersburg under Brigadier General William V. Judson (recalled to the U.S. in January 1918) and Ambassador Francis, during the stirring events of late 1917 and early 1918, U.S. POWs were used by Bolshevik agents in the trench lines, in front-line propaganda attempts at encouraging U.S. soldiers to defect to the Communists. Coded Milstaff cable No. 861 reported:

"An American Armenian appeared on the night of the 28th (of December) with a White flag and stated that he had a Scotch and an American prisoner to exchange. Agreements were made by the Americans to parley and the two men in question actually appeared, SHOUTING ACROSS AT THE AMERICANS TO THROW UP THE JOB AND JOIN THE BOLSHEVIKS; THAT THEY WERE WELL FED AND TREATED. The Commanding officer Major Donaghue, told his men to disregard such nonsense and the two prisoners were greeted with jeers, called renegades, and the party eventually driven back. The Americans the following morning attacked Kadish gallantly, capturing the village and inflicting many casualties on the Bolsheviks. Out of the 300 attacking, the Americans lost 7 killed and 32 wounded. You may obtain from this some idea of to what an extent the command here has been affected by Bolshevik contamination." (The tone of this cable suggests that the American and Scottish POWs and their Bolshevik captors may have been fired upon. under a flag of truce.)[47]

It is noteworthy that the American combat soldiers then actually serving at the front in north Russia gave a somewhat different version of this incident than Major Donaghue's, which was in reality an attempted prisoner-of-war exchange initiated by the Bolsheviks:

"It had been in the early weeks of winter, during the time that Captain Heil with "E" Company (339th U.S. Infantry) and the first platoon machine-gunners were holding the Emsta bridge line, that the Bolsheviks almost daily tried out their post-Armistice propaganda. The Bolo commander sent his pamphlets in great profusion; he raised a great bulletin board where the American troops and Canadian artillery forward observers cold read from their side of the river his messages in good old I.W.W. style and content; he sent an orator to stand on the bridge at midnight and harangue the Americans by the light of the Aurora Borealis. He even went so far as to bring out to the bridge two prisoners whom the Bolos had for many weeks. One was a Royal Scot lad, the other was Private George Albers of "I" Company (339th U.S. Infantry) who had been taken prisoner one day on the railroad front. These two prisoners were permitted to stand near enough their comrades to tell them they were well treated. Captain Heil was just about to complete negotiations for the exchange of prisoners one day WHEN A PATROL FROM ANOTHER ALLIED FORCE RAIDED THE BOLOS IN THE REAR AND INTERRUPTED THE CLOSE OF THE DEAL. THE BOLOS WERE OCCUPIED WITH THEIR ARMS. AND SHORTLY AFTERWARD DONAGHUE HEARD OF THE NEGOTIATIONS AND THE WILEY PROPAGANDA OF THE REDS AND PUT A STOP TO IT..."[48]

The State Department expressed great concern about this propaganda war, requiring General Marlborough Churchill, the successor to Major General Ralph Van Deman as Director of U.S. Military Intelligence, and the American Charge D'Affairs in Archangel, Dewitt Clinton Poole, to supply further information and copies of Soviet propaganda to Washington for analysis. Declassified U.S. documents reveal that important Russian intelligence information during the winter of 1918-1919 was cabled directly to "Colonel" Edward House, President Wilson's personal political adviser of the time.[49]Meanwhile, an uprising in Berlin supported by the German Communist Party was brutally suppressed and German Communist leader Rosa Luxemburg, who had criticized Lenin for creating a dictatorship, was killed. Fighting continued in north Russia, where waves of Red Army infantry were sent against the trenches and blockhouses of the Americans, British and French soldiers, who shot them down with Vickers machine-guns, rifle fire and Canadian artillery shells. In action after action Americans disappeared, usually alone, or in small groups of a few at a time, and their fate remained unknown.

According to the now-declassified, coded Milstaff cables from Colonel Ruggles, and subsequent War Department documents (not declassified until the late 1980')s, 71 American soldiers were actually reported as "Missing in Action" (MIA) on the Archangel front, while 56 more were listed as killed, body not recovered (KBNR). A single February 4, 1919 Milstaff cable alone lists 64 U.S. soldiers in the MIA category, and other Americans disappeared in action during February, March and April of 1919. On the February 4th cable (No. 159), 28 of the American enlisted men (and two officers) missing in action from January 19th to 31st are carried separately in the missing category, opposite a question mark, indicating uncertainty about their fate.[50] However, NO U.S. MIAS WERE LISTED IN THE OFFICIAL 1919 REPORT OF THE SECRETARY OF WAR, which lists 144 U.S. killed in action or dead of wounds, including 28 "presumed killed," out of a total of 244 U.S. dead of all causes on the Archangel front. These 28 soldiers appear from declassified records to be the U.S. enlisted men who disappeared on the 19th of January 1919, during vicious defensive fighting in below-zero weather, when American positions at Ust Padenga and Nijni Gora, south of Archangel, were overrun by the Red Army. Unfortunately, the contents of U.S. coded cables from Archangel listing missing Americans as MIAs rather than as known POWs were also available to the Bolsheviks at this time, as Felix Dzerzhinsky's Cheka had succeeded in breaking the American codes; therefore the Soviets were able to choose how many American MIAs they would hold secretly for future use, while a handful of others, some of them considered "progressives" by the Communists, would eventually be released as part of a propagandized sign of conciliation.

Many American prisoners from among the 71 U.S. soldiers originally listed in cables from Archangel as MIA, or the 56 carried as KBNR, were in fact captured alive (some of them wounded and reported dead by their comrades) and secretly confined by the Red Army between January and April 1919 on the northern front, at Ust Padenga and Nijni Gora, and later Vistavka, Seletskoe and Bolshierzerka. Much of this fighting was defensive in character, with the Bolsheviks losing an aggregate of thousands of soldiers killed while attacking outnumbered American and Allied troops, armed with machineguns and automatic rifles and supported by Canadian artillery, who were dug into strong defensive positions. But as successive American positions were overrun, U.S. soldiers disappeared in the confusion of retreats. These men were initially reported as MIA, but were later to be declared killed in action by the U.S. Army. U.S. officers reported on one of the worst actions in this fighting that lasted from January 19th to the end of that month:

"To withdraw we were compelled to march straight down the side of this hill, across an open valley some 800 yards or more in the terrible snow under the direct fire of the enemy. There was no such thing as cover, for this valley of death was a perfectly open plain, waist deep with snow. To run was impossible, to halt was worse yet…One by one, man after man fell wounded or dead in the snow...OF THIS ENTIRE PLATOON OF FORTY-SEVEN MEN, SEVEN FINALLY SUCCEEDED IN GAINING THE SHELTER OF THE MAIN POSITION UNINJURED. During the day a voluntary rescue party... went out into the snow under continuous fire and brought in SOME OF THE WOUNDED AND DEAD, BUT THERE WERE TWELVE OR MORE BRAVE MEN LEFT BEHIND IN THAT FATAL VILLAGE WHOSE FATE WAS NEVER KNOWN AND STILL REMAINS UNKNOWN TO THE PRESENT DAY, THOUGH LONG SINCE REPORTED BY THE UNITED STATES WAR DEPARTMENT AS KILLED IN ACTION."[51]

Seventeen Americans from this platoon remained missing in action while at least six more who did not return were carried as "presumed killed," whose bodies were not recovered. Other Americans also disappeared in this several days retreat, bringing the total U.S. missing to at least 28 enlisted men and two officers.

The desperate fighting continued and at Vistavka, on February 4th, Private Thomas Keefe of Company C, 339th Infantry disappeared and was listed as "missing in action." Keefe was a very short man, only 5 feet, 1 1/2 inches tall, according to Army records, who had been a steamfitter in civilian life in Chicago, Illinois. A decade later his "remains" had still not been found, and although he had been presumed dead by the War Department, he still remained an MIA. In yet another of the nearly continual Bolshevik attacks, on February 7th, at Seletskoe, a respected and popular officer, Lieutenant Clifford Bateman Ballard, of the 339th's Machine Gun Company, disappeared. An Army Reserve officer who was commissioned in August 1917, Ballard had been a teacher and a social worker in civilian life, and his caring for other people obviously extended to his soldiers, who idolized him. He had gone forward towards the Bolshevik positions during an attack, with a squad of allied Russian Lewis machinegunners, to locate a better position for his own American machinegunners to protect the flank of the King's Own Liverpools:

"THE RUSSIAN LEWIS GUNNER WAS THE ONLY ONE TO GET OUT. HE RETURNED WITH HIS GUN AND DROPPED AMONG THE AMERICANSKI MACHINE GUNNERS TELLING OF THE DEATH OF BALLARD AND THE RUSSIAN SOLDIERS, AT THE POINT OF THE BOLSHEVIK BAYONETS...HIS BODY WAS NEVER SEEN OR RECOVERED. HOPE THAT HE MIGHT HAVE BEEN TAKEN AS A WOUNDED PRISONER BY THE REDS STILL LIVED IN THE HEARTS OF HIS COMRADES..."

What actually happened to Lieutenant Ballard was never revealed by the Soviets and his remains were not recovered by the subsequent 1929 VFW-U.S. Graves Registration expedition. An American officer of Ballard's character would have been a prize for the Cheka secret police operatives who accompanied the Red Army everywhere. On the Archangel front during the early months of 1919, it was clear to American and British officers, that some number of their missing men had actually been captured by the Bolsheviks, who held many of them secretly, without revealing to the Allies their names or locations. This had become the standard Soviet practise with foreign war prisoners, and remained Soviet policy hereafter.

In February 1919, at least partially in retaliation, and also, according to now-declassified U.S. documents, to prevent many released Russian POWs from joining the Red Army, the Allies took control of many thousands of former Russian prisoners of the Germans, then still in western Europe. The Soviet government in turn, demanded their immediate release, as communist-recruited European POWs then formed a major portion of the Red Army in "Internationalist" units, a mainstay in keeping Lenin's Bolsheviks in power. Some Russian prisoners did not wish to return to communist rule, which would have necessitated their forcible repatriation.[52]

As the Allied Intervention forces were still battling the Red Army in early 1919, there were several attempts by Allied leaders involved with the ongoing Paris Peace Conference to deal with Russia. Some, like U.S. Ambassador to Russia David R. Francis, urged the dispatch of a far more powerful Allied Intervention force of 150,000 men from the U.S., Britain and France, and occupation of Petrograd. Winston Churchill, who had been appointed Secretary of State for War in Lloyd George's cabinet on January 18th, also supported the use of many more British, French and American soldiers in Russia to overthrow Lenin's Bolsheviks. However, Lloyd George, Wilson and Clemenceau realized how tired of war their peoples were and how unpopular a full scale attack on the first "workers state" would be.

The more liberal view, represented in the extreme by William Bullitt, of allowing Russia to evolve on it's own, was to prevail in the end. In the interim a series of peace proposals would be advanced to Lenin's regime in hope of securing some concessions for ultimately recognizing the Bolsheviks as the lawfully constituted government of Russia. All involved the planned withdrawal of the relatively small Allied Intervention forces. In all of these efforts, freedom for American and Allied prisoners in Russia was an important consideration. The Allies were at the point when their relatively minor intervention forces had begun to suffer painful casualties for what seemed to many, a meaningless cause. There was a powerful domestic cry after the Armistice, to "bring the boys home." There were also strong Socialist and workers movements in the United States, Britain and France, which by and large strongly supported the existence of Lenin's experimental communist state.

In Russia, aside from the relatively small Allied forces, there was a White Russian Army of 12,500 fighting the Bolsheviks in North Russia. 85,000 White soldiers in the south, including 20,000 under Wrangel in the Don River basin, also opposed the Red Army.

40,000 under Nestor Makhno fought the Reds in the Ukraine. White forces totaled 104,000 on the western front and 140,000 on the eastern front, including 10,000 under Kolchak in Siberia.[53] From the Paris Peace Conference at Versailles, an invitation was extended to send representatives of all Russian factions to meet with Allied representatives at Prinkipo, Turkey, in the Black Sea region; an initiative which eventually failed utterly. British Prime Minister Lloyd George had hoped to invite Lenin and other Bolshevik leaders to Paris to discuss a settlement, but the French government of Clemenceau and British conservatives led by Winston Churchill argued against this.

President Wilson had attempted the compromise, suggesting that all hostile Russian factions meet in a remote location to negotiate with Allied representatives. Wilson personally typed the proposal that the meeting occur at Prinkipos Island, on the Turkish coast of the Sea of Marmara. The proposal was sent to Moscow, and although Lenin accepted it on February 4th, French, British and American governmental and public pressure against the plan caused it's abandonment.[54] A coded U.S. cable (No.66) dated January 31, 1919 (from Robbins in Peking), stated concerns about the effect of peace negotiations on soldiers fighting the Bolsheviks:

"Churchill and Graves have been informed of the following: A ruinous effect on the morale of the Russian Army has resulted from the invitation to Bolshevik leaders to attend the Princess (Prinkipo) Island Conference. Soldiers refuse to fight believing an armistice has been declared. Full advantage is being taken of the situation by the Bolshevik commissars. Omsk government is in a desperate condition. To prevent catastrophe immediate action by the Allies is urgently needed. Urgent request from General Knox for British and Canadian troops has been refused by the British government."[55]

At this time Winston Churchill opposed what he saw as vacillation and weakness on the part of British Prime Minister Lloyd George. Churchill was all for a ten-day ultimatum to Lenin, to cease fighting or face an immediate and powerful increase in the Allied intervention forces, with the object of causing the speedy overthrow of the Bolshevik regime. Churchill, however was rebuffed by both Lloyd George and Woodrow Wilson.

In north Russia, morale of American British and French troops was sinking as their position became more difficult and the fighting dragged on interminably, without any clear-cut goal that the soldiers could recognize. Bolshevik propaganda that had been scorned or ignored before the Armistice was becoming a topic of serious conversation among the combat troops by February 1919. Anonymous petitions and printed criticism by soldiers of Allied policy in Russia had begun to appear in the dugouts and blockhouses south of the gay and unreal city of Archangel, where hangers-on and profiteers held high carnival with rear-echelon British, French, American and White Russian officers. One leaflet circulated in the front line areas, allegedly written by a U.S. combat officer, raised issues that many understood:

"The manner in which this expedition has been mishandled is a disgrace to the civilized world...The majority of the people here seem to prefer Bolshevism to British intervention...WHERE IS OUR MONROE DOCTRINE?...We have no heart for the fight...We have earnestly endeavored to find some justification for our being here, but have been unable to reconcile this expedition with American ideals and principles instilled within us..."[56]

The high command in Archangel, London and Washington became alarmed as symptoms of trouble appeared. One fear was that Allied prisoners of war would be infected with communist dogma at the Bolshevik propaganda school in Moscow and then be released among British and American troops to sow seeds of revolt. This fear had a solid base in fact as an intelligence summary by General Headquarters Intelligence-Archangel dated February 4, 1919, signed by Captain L. Hodson, reveals:

"Three or four British soldiers are said to have been sent from Niandoma (Naundoma) to Moscow to be prepared as propaganda. On completion of course of training they are to be sent back through the lines to work amongst British units."[57]

A coded cable to the War Department (No. 856) from Slocum, the U.S. Military Attaché in London, dated February 18, 1919, indicated the high level of concern with Bolshevik attempts to indoctrinate Allied POWs for communist propaganda purposes:

"The following report dated Feb. 14, has been received by the British War Office from the Commanding General Archangel (Ironsides):

'A number of returned prisoners of war have come in during the last week on the S.S. Bolshieozerka. There have been discovered among these, and arrested, several returned prisoners and Bolsheviki educated at the propaganda school in Moscow.'"

A message from Colonel Ruggles in Archangel dated February 24th, answered official questions about past Soviet propaganda attempts involving U.S. and Scottish POWs, but quelled concerns in the General Staff that American soldiers were being contaminated:

"The following is the gist of General Ironsides report on a rumor from a Russian source that American troops had been fraternizing with Bolshevik troops.

"A Bolshevik speaking English came down to the bridge over the Emsta on Christmas Day and stood out well in the open singing an English carol. Having listened to him for a short time, the American sentry, having ordered him to go away without effect, finally stoned him to get rid of him..."

Such intelligence as this continued to reach Paris where William Bullitt was still Chief of the Division of Current Intelligence Summaries, with the rank of Assistant Secretary of State. (Among those serving under Bullitt in the State Department's Division of Current Intelligence Summaries in 1919 were Allen Dulles, a future OSS officer in WW II and later CIA Director, and Christian Herter, a future Secretary of State under President Eisenhower.) By February 16th President Wilson had agreed to the plan advanced by Bullitt and promoted by Presidential advisor Edward House to send a diplomatic mission to Russia to explore the possibility of a peace settlement. Although the youthful Bullitt was considered by many to be a radical pro-Bolshevik, House persuaded Wilson to send him as the head of the mission. His position as a junior statesman gave the mission deniability in case of embarrassment. It was also hoped that the mission would suffice, in the words recorded on February 16, 1919 by Secretary of State Lansing in his diary, "To cure him of his Bolshevism."

Winston Churchill, now the British War Secretary in Lloyd George's cabinet, had confronted President Wilson at a meeting of the Supreme Allied Council on February 14th, urging that a decision be reached on policy toward Russia and the intervention. Wilson said he felt "guilty" about sending "insufficient forces" to Russia, but that no matter how badly they were needed to bolster the White cause they couldn't remain "forever." Wilson left the conference that night to return to the United States, leaving his personal advisor Edward House (Bullitt's ally) in his stead, to represent the United States with Secretary of State Lansing. Lloyd George also left the Conference and returned to Britain.

House and Lansing stalled Winston Churchill in his move toward increased Allied intervention until February 17th, when they gained further support from President Wilson in a cable message: "...Greatly surprised by Churchill's Russian suggestion. I distinctly understood Lloyd George to say there could be no thought of military action there..."[58]

The President's stand forestalled Churchill's plan to increase the intervention forces. Wilson followed up by approving Edward House's plan to send William Bullitt on a then-secret peace mission to the Bolshevik government, and by publicly announcing his decision for "the withdrawal of American troops in North Russia at the earliest possible moment." This announcement was heard in Russia, causing a great sense of relief and impatience among American troops there. An ecstatic William C. Bullitt and two aides started for Russia in late February, with the pro-Soviet journalist Lincoln Steffens, a few days after Wilson's decision.

On the Archangel front, the morale problem among the Allied soldiers was evidenced by the refusal of a battalion of the Yorkshire

Regiment of British infantry to go into the battle line on February 26th, at Kadish. The unit had just been dispatched as reinforcements to Archangel from Murmansk (outfitted and accompanied by the veteran Antarctic explorer, Sir Ernest Shackleton). Two sergeants who acted as spokesmen for the mutinous soldiers were arrested, court-martialed and sentenced to be shot. (The sentence was altered to life imprisonment because King George V had halted executions of British soldiers following the Armistice.) The British incident was followed by a minor revolt of French troops who had been ordered to the front from Archangel on March 1st, in which the ringleaders were arrested.

Meanwhile, American losses continued as the Soviets renewed their attacks in the Vistavka area on the Vaga River, south of Archangel. White Russian forces under Kolchak in the east had begun a drive toward Viatka on the Trans-Siberian Railway, where the line branched north to Archangel and south to Moscow, and the Bolsheviks were determined that no link up with the Allied forces would ever occur.

On March 1, 1919 a combat patrol of Company B 339th U.S. Infantry was ambushed near Toulgas by a Bolshevik force on skis. The patrol leader, Corporal Arthur Prince, ordered the other six men to crawl back under the heavy rifle fire and escape. In the attempt, two of the Americans were initially reported killed and three more seriously wounded while Prince never reappeared. A rescue party was pinned down in trying to reach the wounded and dead until the Bolsheviks were driven off by supporting Canadian artillery fire. Although blood in the snow where Prince had been indicated he had been wounded, he could not be found, and was believed to have been captured by the Bolsheviks. (Eventually he was presumed dead, but he was to mysteriously turn up alive and free in Germany, nearly a year and a half later, in August of 1920, at a time when the Bolsheviks were in dire need of American and Allied aid.)[59]

The military attaché in London reported to Washington on casualties in early March: "British War Office reports following cable from Commanding General dated March 6: 'Fighting ceased on Vaga yesterday and the enemy has withdrawn some versts leaving many dead and wounded. I do not think he will try again for a week or two and I am sure he received a heavy check. Our casualties in the four days from March 1 to March 4 were about 50. All is quiet on other fronts.' Churchill informed." Colonel Ruggles also called attention to the fact that Bolshevik General Samoilo had recently taken command of the 6th Red Army on the Archangel front under overall authority of War Commissar Leon Trotsky, and his deputies were Barzakovsky and Philipovsky.

Evidence that American soldiers reported killed in this early March fighting were actually missing in action is contained in a cable (No.189) from Colonel Ruggles in Archangel: "...To the list in the last sentence of my 178, March 3, 9 p.m., add 'Missing, all enlisted men, number of wounded unknown." This would indicate that the two other American soldiers thought to have been killed with Corporal Arthur Prince had also vanished and were now being carried as MIAs. In another fight in the Viskavka area on March 7, 1919 Corporal Albert E. Moore of Company A (339th) disappeared and was reported killed.[60]

On March 9th the Soviets commenced a heavy artillery attack at Vistavka, followed by a Red infantry assault by several hundred troops who almost reached the blockhouses and dugouts of the Americans. The attack was broken by rifle and machine-gun fire, and Canadian artillery shells which killed and wounded hundreds of Soviet soldiers urged on by their armed political commissars. Corporal Bernard F. Kenny, Private Earl D. Sweet and Private Walter J. Wolstead, all of Co. A (339th Infantry), were reported as killed, body not recovered, and a dozen more Americans were wounded. On the same day another soldier of Co. A, Private Dennis Trummell, of Clio, Kentucky, disappeared at Maximoskaya and was reported killed. The outnumbered Americans and Canadians, with their White Russian allies, soon abandoned the shattered ruins of Vistavka and pulled back to Kitsa. The Red Army was urged on in its attacks by Cheka "OO's" (Osobye Otdel or "OO"), who had been authorized by Dzerzhinsky on February 3, 1919, and adopted by the All-Russian Executive Committee on February 6th. These political commissars were assigned to every unit down to company and platoon level, and aside from monitoring the political thinking of all Red Army officers, they were empowered to shoot soldiers on the spot who lagged during attacks, and did so.[61]American troops reported that they could hear defeated Red Army troops being met with machine-gun fire when they reached their camps. Such ruthless discipline enforced by Dzerzhinsky and Trotsky, explains the enormous losses sustained by Soviet troops attacking fortified American and Allied positions in north Russia.

During this same period, President Wilson had authorized the peace negotiating mission to Moscow by the State Department's William C. Bullitt, whose credentials were signed by the Secretary of State on February 18th, and who had left Paris for Russia on February 22nd.[62]Bullitt had entered Russia on March 6, 1919, after sailing from Britain to Finland. He was accompanied by the radical journalist Lincoln Steffens, whose pro-Bolshevik stance was well known to Lenin and his colleagues, a U.S. military intelligence officer named Captain W. W. Petit traveling as an aide, and his private secretary, R.E. Lynch. After meetings with Lenin (on March 14), Chicherin and Litvinov, the Bolsheviks presented Bullitt with a draft agreement they had drawn up for ending the Intervention and the civil war in Russia, which would have also led to a release of all American, Allied and Russian prisoners. Included was a proposal leaving large areas under White control in return for Allied recognition of the Bolshevik regime. These areas included the Archangel-Murmansk area, the Baltic States and Finland, much of the Ukraine, the Caucasus, the Urals and all of Siberia. White governments under Deniken and Kolchak would have been permitted.[63]

Lenin's final written condition was to impose a deadline on the Allies: "The Soviet Government of Russia undertakes to accept the forgoing proposal provided it is made not later than April 10, 1919."

Bullitt returned to Paris from Russia on March 25, 1919, not only with this document but greatly excited by what he had seen there. On his way home he had sent a cable to Edward House from Helsinki on March 18th urging that a peace settlement for Russia be reached:

"You must do your utmost for it, for if you had seen the things I have seen during the past week and talked with the men I have talked with, I know that you would not rest until you had put through this peace."

In another message to Paris he portrayed a state of order and normalcy, rather at odds with some other contemporary accounts of conditions in 1919 Russia:

"The Soviet government is firmly established, and the Communist Party is strong politically and morally...There is order in Petrograd and Moscow. There have been no riots and no uprisings for many weeks. Prostitution has disappeared. Robberies have almost ceased. One feels as safe as in Paris. The opera, theaters and ballet are performing as in peace."[64]

Bullitt blamed the Allied blockade, "true cause of the miseries of Russia," for any problems existing under the Bolsheviks. He and Lincoln Steffens openly proselytized influential Americans for Lenin's Bolshevik regime upon their return to Paris. Steffens later remembered one incident (in which he used a slogan also used by Bullitt):

"'So you've been over into Russia?' said Bernard Baruch, and I answered very literally, 'I have been over into the future, and it works.'"[66]

Steffens also recalled that Captain Petit, the U.S. Army officer who had accompanied Bullitt (and no doubt ultimately reported to U.S. Military Intelligence Division (MID) of the War Department), was similarly impressed with the revolutionary ideals of Bolshevik Russia:

"Our return from Moscow was less playful than the coming. Bullitt was serious. Captain Petit was interesting on the hunger and the other sufferings of Petrograd, but not depressed As he would have been in New York or London. 'London's is an old race misery,' he said. 'Petrograd is a temporary condition of evil, which is made tolerable by hope and a plan.'"[65]

Bullitt reported immediately to Wilson's personal advisor, Edward House, pressing for U.S. acceptance of Lenin's terms and for American diplomatic recognition of the Soviet regime.

President Wilson refused to receive Bullitt following his return, probably because acknowledging the Soviet-drafted document would have placed the Allies in an uncomfortable diplomatic position. Wilson's adviser, Colonel Edward House, received Bullitt but could not prevail upon Wilson to do so. Although Lloyd George spoke to Bullitt privately, he too refused to seriously consider the Bolshevik terms, for fear of criticism from the right wing at home.

The spread of Bolshevik-inspired revolutionary socialist uprisings to Europe by communist-propagandized war prisoners returning from Russia and professional Soviet agitators, had alarmed many western statesmen. Although the communist-supported Sparticist revolt in Berlin had been crushed in January 1919, a communist-led radical regime came to power in Hungary under Bela Kun (a repatriated POW from Siberia who had been present during the Bolshevik revolution), on March 21st of that year, even as Bullitt was returning from Moscow. The Hungarian Soviet Republic was toppled by a Romanian invasion. In April 1919, an uprising led by two Russian Socialists in Bavaria resulted briefly in a Bavarian Soviet Republic, only to be crushed in a few weeks by the regular army and irregular forces.

The New York Tribune reported in front-page headlines on March 27th: "Red Revolt in Hungary Provokes Near Panic In Peace Conference." The newspaper also reported that Bolshevik rebels were undermining governmental authority in Romania and Poland. In Moscow, meanwhile, the Third Workingmen's International, called the "Communist International" by Lenin, was meeting during March to plan the further spread of world revolution. On March 31st, the same New York newspaper headlined: "Wilson Blocks War on Lenine; Bolshevism Divides Council, France Begins to Act Alone." President Wilson and Prime Minister Lloyd George were reported to be clinging to a policy of conciliation with the Bolsheviks, while Clemenceau and Italy's Orlando were in favor of using heavy Allied force to crush the Bolsheviks. Although France had been told again that America would not assist an increased intervention with U.S. soldiers and money, and the French were being urged by the American delegation to lift the blockade of Russia, Clemenceau had already decided to act alone and had sent a strong military force to Odessa on the Black Sea under General d'Espry. But two regiments of French troops refused to proceed against Bolshevik forces in the Ukraine, leaving the French government in a quandary.[66]

Bullitt's peace mission thus ultimately failed, although Lenin had offered terms for the cessation of hostilities. Woodrow Wilson, Lloyd George and Clemenceau had rejected the Bolshevik conditions, but the Allied leadership could not agree about what to do next. (The later publicizing of confidential negotiating positions by the embittered Bullitt caused the resignation of Secretary of State Robert Lansing.) Lincoln Steffens later remembered:

"It was a disappointing return diplomatically. Bullitt had set his heart on the acceptance of his report; House was enthusiastic, and Lloyd George received him immediately at breakfast the second day and listened and was interested. Of course. Bullitt had brought back all the prime minister had asked. And the same morning I was received and questioned, very intelligently, by 'British information, Russian section.' I had learned to despise the secret service; they were so un- and mis-informed; but these British officers knew and understood the facts...."

"No action was taken on the proposal Bullitt had brought back from Moscow, and after a few weeks of futile discussion the Bullitt mission was repudiated. I heard that the French, having got wind of it, challenged Lloyd George; he and Wilson had gone back of the French to negotiate with the Russians, they charged. And Lloyd George took the easiest way out. He denied Bullitt in Paris, and when there were inquiries in London, he crossed the Channel to appear before the House of Commons to declare explicitly and at length that he knew nothing of the 'journey some boys were reported to have made to Russia.' ...It was a political custom in British parliamentary practise to use young men for sounding or experimental purposes, and it was understood that if such a mission became embarrassing to the ministry, it was repudiated; the missionaries lay down and took the disgrace till later, when it was forgotten, they would get their reward."

"I had been pretty outspoken in my hopefulness of the Bolshevik government, and our secret service was suspicious and active. Once I invited to our table the young officers who followed me and Jo Davidson into a restaurant and set about converting them to Bolshevik communism. I could not quite win them over, but they reported, I heard afterward, that there was no use shadowing me; I would tell them more to their faces than they could possibly overhear.

"...Secretary of War Baker, who was in Paris, invited me to his room and asked me to tell him just what I had seen, heard and thought about in Russia. I did. I had noticed that there was somebody in his bathroom stripped to the waist and shaving, but I let Baker draw me out, and when I had ended my story a colonel in full uniform appeared, all shaved and fine, to be introduced as the chief of one branch of the secret service. I think this was Baker's characteristic way of dealing with his officer's suspicions; he disarmed by confirming them."[67]

In a memoir published in 1920 Lenin was to write: "...When we proposed a treaty to Bullitt a year ago, a treaty...which left tremendous amounts of territory to Denikin and Kolchak, we proposed this treaty with the knowledge that if peace was signed, those governments could never hold out."[68]

Lenin must have been supremely confident, since at the time of the Bullitt mission the Bolshevik position had been anything but secure. Deniken's White army in the Azov region, Yudenich's Northwestern White Army in the Baltic region, Kolchak's White Army in the Omsk area of western Siberia and the White forces in the Archangel area under Miller all threatened the Red heartland around Moscow and Petrograd. From late March to early April 1919, as Kolchak's White Russian forces pushed west of the Ural Mountains, the Soviets mounted a series of attacks on American and Allied positions in the Bolshiezerka area south of Archangel. In two weeks of fighting against the heavily dug-in Americans, supported by Canadian artillery and White Russians, they lost, according to the Communists' own newspapers, 2,000 men in killed, wounded and missing.[69]

A March 17th, 1919 cable #CO 263) to the American Military Mission in London from the 339th U.S. Infantry's Commander, the Australian-born Colonel George Stewart, was concerned with the fate of some of the reported U.S. MIAs and POWs in Bolshevik hands:

"In reply to HQ.116434, 34 Americans reported missing and unaccounted for. HOW MANY OF THESE ARE ALIVE IN HANDS OF BOLSHEVIKS UNKNOWN AND WE HAVE NO MEANS OF FINDING OUT. INFORMATION FROM A BOLSHEVIK PRISONER RECENTLY IN HOSPITAL ON VAGA FRONT TO THE EFFECT THAT THERE ARE TEN AMERICANS OF COMPANY A 339th INFANTRY MISSING FROM UST PADENGA FIGHT OF JANUARY 22, WOUNDED AND IN BOLSHEVIKI HOSPITAL AT VALSK ON THE VAGA. THIS REPORT...IS HIGHLY PROBABLE AS 17 MEN WERE MISSING AFTER THIS ACTION. THE SAME PRISONER STATED SOME AMERICANS WERE PRISONERS AT MOSCOW..." The distribution of this message included the Commanding General in Europe, General Pershing.[70]

Meanwhile, fierce fighting had continued through late March at Bolshieozerki. On March 16-17, a French position was overrun and annihilated, and a convoy was captured, accounting for more of the French POWs and MIAs of the Archangel front.[71] American troops in this area were also engaged and lost casualties from March 21st to the 23rd, and in this fighting Private John Frucce, of Company H, 339th Infantry, a machinist in civilian life, disappeared and was listed as "missing in action," as was Corporal Earl W. Collins, a former bank clerk from Detroit, and Private August B. Peterson a peacetime farmer from Michigan. While the Bolsheviks later claimed in POW negotiations that Peterson had died, they never supplied any proof of this, while Frucce and Collins vanished and were never heard of again; nor were their "remains" reported recovered prior to U.S. diplomatic recognition of the Soviet Union in 1933.

Company H of the 339th U.S. Infantry and some of the British Yorkshires attacked the Bolsheviks during the first two days of April 1919, losing more three Americans killed and three additional U.S. MIAs. Among those Americans who had disappeared and very possibly been captured by the Bolsheviks on April 2nd, at Bolshieozerka, was Private Floyd R. Auslander of Company H, 339th infantry, a farmer-thresher from Decker, Michigan, who had a birth-mark below the small of his back on the right and scars on his forehead and back of the neck. Allied losses were relatively light, but in the context of ongoing efforts for a diplomatic solution, painful, as the Bolsheviks no doubt intended.[72]

At Odessa on the Black Sea, the French intervention forces had meanwhile become so mutinous that Clemenceau was left with no alternative but to evacuate that strategic port. On April 6th, Bolshevik troops entered Odessa in triumph. A few days later, the American media luridly reported that U.S. Army troops had mutinied in Archangel. The New York Tribune of April 11th: "American Draft Troops Mutiny On Archangel Front...'NOT AT WAR; WON'T FIGHT,' SAY SOLDIERS." The newspaper cited an official dispatch received by the War Department on April 10th in it's reporting:

"A company of infantry (339th) refused to obey the orders of their officers to prepare for movement to the front lines and went forward finally only under the urging of their officers and after one enlisted man, placed in confinement for disobedience of orders, had been released, according to advice received by the War Department... The mutineers, while moving forward, persisted that they would not go to the front lines...and predicted general mutiny in the American forces on the Russian front if a statement was not forthcoming from Washington regarding the withdrawal of American troops from Russia."[73] Although veterans of the 339th U.S. Infantry later disputed the American media's sensationalized reporting of what was in reality a trivial incident by exhausted and discontented soldiers, who did not understand why they were being required to fight in Russia in an undeclared war, the widespread reporting of this and other incidents of the time left many Americans uncertain about the actual policy of the United States towards Bolshevik Russia.

For weeks after Bullitt's secret mission, Allied-Bolshevik POW/MIA negotiations continued on the Archangel front. Repeated Soviet demands for diplomatic recognition were linked with promises for information on several hundred American, British and French MIAs who had disappeared on the Archangel front.

Army 1st. Lieutenant Dwight Fistler later remembered an April 5, 1919, POW/MIA negotiation with the Soviets on the front lines, south of Archangel: "We had 500 Russian prisoners. They had seven of ours. WE WERE WORRIED ABOUT HUNDREDS OF MISSING FROM OUR RANKS..."Behind us our men trained Vickers guns on them, and behind them, their outposts trained machine guns on us. If there was to be treachery both of us were prepared. "Negotiation was difficult. Interpreters were not very efficient. But the Reds learned what we were up for, and haggled. The end was, they traded us two of the seven Americans for the 500 Russian soldiers., and we had to toss in a round of cigarettes to seal the bargain. WE NEVER DID LEARN WHAT BECAME OF THE MISSING. The five Americans later were exchanged through Moscow in Finland."[74]

Official documents declassified in the 1980s now in the National Archives indicate that prisoner negotiations during April 1919 on the Archangel front were halted by April 14th, on direct orders from the U.S. commander in Europe, General John J. Pershing, due to Bolshevik linking of promises for information on missing American and Allied soldiers to simultaneous demands for diplomatic recognition by the U.S. of Lenin's Moscow regime:

"...Negotiations for the exchange of prisoners have been terminated by orders from General Pershing, after having been delayed, although under discussion by both sides, through failure of the Bolshevik commander to obtain authority from Moscow. As far as the negotiations went they showed the presence of wounded Allied prisoners, apparently receiving good treatment (in order to strengthen the effect of Bolshevik propaganda), in hospital at Vologda and in prison at Moscow. Have informed Warburton."[75]Nothing was said of the hundreds of American, British and French MIAs, of whom many were thought by their fellow soldiers to be alive in Soviet control.

Lenin and Trotsky held out for diplomatic recognition of their regime and it soon became clear that the Bolshevik commanders on the 6th Red Army front were not empowered to give up any more than a token number of Allied prisoners of war or secretly-held Allied MIAs until this main political object was attained.

On April 23rd, the U.S. attaché in Archangel reported to the War Department that, "Negotiations with the Bolshevik government for the exchange of Allied prisoners are to be resumed. One American Captain...one French captain and one British Captain will form the delegation. Harzfeld has been selected."

The U.S. State Department reacted to Bolshevik moves by reminding the Secretary of War, in writing, that the U.S. did not recognize Lenin's Communist regime. Acting Secretary of State Frank Polk wrote Secretary of War Newton Baker on April 28, 1919:

"IN CONNECTION WITH THE NEGOTIATION FOR THE EXCHANGE OF

PRISONERS...I HOPE THE AMERICAN MILITARY AUTHORITIES BEAR IN MIND THAT THE UNITED STATES HAS NOT RECOGNIZED THE BOLSHEVIK REGIME AS A GOVERNMENT."[76]

This warning was transmitted to Archangel by General Churchill on May 1, 1919. Meanwhile, under local authority of the formidable Allied Commander in North Russia, British General Edmund Ironside, the Allied Commission had met with the Bolsheviks between April 26 and 28, who had promised to deliver four POWs, but in the exchange freed only two Americans for 500 Red Army soldiers. One of the Bolshevik POW negotiators was a young Red Army Colonel, S.K. Timoshenko (who was eventually promoted to Marshal of the Red Army by Stalin, during WW II.)

In the now-declassified U.S. report of these negotiations, the American, British and French POWs of the Bolsheviks are identified as "hostages," and the presence of many other WOUNDED Allied prisoners in Vologda hospitals AND IN MOSCOW PRISONS was confirmed. On May 3rd the Bolshevik negotiators announced that they were only empowered to negotiate for 22 wounded Allied prisoners then in Vologda, ignoring the many more then in Moscow prisons, including a number whose identities were kept secret by the Communists. While the negotiators gained the release of one Scottish and one American POW, two Americans were supposed to be released at this time. The Bolsheviks however, announced at the last minute that one of the Americans, August Peterson of the 339th U.S. Infantry, had "died," and provided a piece of paper stating so. This was accepted without further evidence or protest by the American government. The only American POW released, Earl Fulcher, was a wounded soldier of Company H, 339th U.S. Infantry.

At the time of these negotiations the Soviets also released five other American POWs (NONE OF WHOM WERE WOUNDED) through Finland, on April 25, 1919. These were Private George Albers, captured on November 3rd; Private Mike Hurilik, captured on November 28th; Private Walter Huston, captured on November 28th; Private William Schuelke, captured on March 17th; and Private Anton Vanis, captured on January 23rd.[77] Nearly a month after the negotiations were broken off, in an apparent propaganda ploy, the Bolsheviks suddenly released four more UNWOUNDED U.S. POWs who had been held in Moscow, along with two captured YMCA workers. These POWs are reported to have arrived in Stockholm via Finland on May 25, 1919. They were: Sergeant Glenn Leitzell, captured March 31st; Corporal Jens Laursen, captured March 31st; Private Freeman Hogan, captured March 31st; and Private John Triplett, captured November 29th All 9 of these returned POWs were of the 339th Infantry, and since none had been wounded they were not from the other groups of wounded American prisoners reported under Soviet control in Moscow and north Russia.[78]Most of these men had disappeared in combat action and been carried as MIAs or presumed dead, and were not definitely known to have been POWs. It appears that some of these released American prisoners had participated in Bolshevik propaganda efforts, (possibly in order to survive) and were relatively well treated prior to their release. One of the released prisoners, Sergeant Glenn Leitzell of the 339th Infantry, was given, "wireless proposals to our government sealed by the Bolsheviki foreign office." Leitzell later recorded how they got out of Russia:

"...THE SOVIET GOVERNMENT WAS SATISFIED THAT WE WERE GOOD BELIEVERS IN BOLSHEVISM. AFTER A NUMBER OF LECTURES, AND AS THEY WISHED TO BE RECOGNIZED BY THE U.S., THEY GAVE US PASSPORTS TO FINLAND. THE PROPAGANDA DID NOT DECEIVE US."[79]

At least 40-50 other American POW/MIAs (and possibly as many as 75-100), no doubt including the most recalcitrant, were kept separate from those to be released and secretly imprisoned in Moscow and elsewhere, for future use. A casualty memorandum of American Headquarters in Archangel, dated May 21, 1919, still lists 43 U.S. MIAs, not including presumed killed. The 339th Infantry Regiment and attached service companies had lost 244 U.S. dead of all causes, of which 144 were killed in action, permanently missing in action and presumed killed, or died of wounds. More than 300 other Americans, mostly of the 339th Infantry, had been wounded in action from September 1918 to May 1919.[80]

Following the battle of Bolshiezerka, in early April, General Ironside began gradually replacing the American infantry in the lines with White Russian and British troops who continued fighting the Bolsheviks in a holding action that eventually led to an embarrassed withdrawal of all Allied troops. The arrival of 4,000 fresh British volunteer soldiers (including some Australians), on May 27th, only postponed the inevitable, although White Russian victories by Denikin and Wrangel in the south during late May and June 1919 again raised hopes of an eventual Red collapse. The retreat of Kolchak's Army in the east during June of 1919, and a mutiny by soldiers of a White Russian battalion near Archangel, who murdered 5 of their British officers on July 7 and went over to the Bolsheviks, presaged the future.

Most of the weary Americans had sailed on transports from Archangel during the month of June 1919. A few U.S. Engineers and railroad transportation troops remained through July, and in August 1919, as British troops mounted a last major attack upon the Bolsheviks, US Graves Registration units were disinterring American dead from their graves for shipment in coffins back to the United States. Despite Winston Churchill's best efforts, the decision had already been made in London to withdraw from north Russia before the winter of 1919-1920 set in. The last British forces withdrew from Archangel by September 27,1919. The White Russian commander there, General Miller, refused to be evacuated and remained to continue the fight against the Bolsheviks until February 21, 1920 when with the collapse of resistance the 154th Infantry of the Red Army entered Archangel. (Long afterward General Miller was to be murdered by Soviet GPU in Europe.) The retribution of Lenin and Trotsky by the Cheka death squads began. Two senior Chekist agents, the American specialist Aleksandr Eiduk, and Mikhail S. Kedrov were in charge of the mass-executions of thousands of Russians who were charged with collaborating with the Allies in Archangel and Murmansk. The senior US Naval Intelligence officer in Russia at the time, Admiral Newton McCully, had tried to save these people, who begged to be rescued when the Allied forces were withdrawing from Archangel by sea. He had requested urgently that 96,000 "seriously compromised" White Russians be evacuated by the Allies. His request was refused and the frantic and terrified Whites remained behind to face the Cheka death squads as the Allies sailed away. More than half a century later, on the other side of the world in Indochina, a similar ending to an American foreign intervention was to occur, and as in 1919, American prisoners were to be secretly left behind in captivity.[81]

This conduct with U.S. and other war prisoners was to be repeated by successive Communist regimes for over half a century to come. When U.S. forces evacuated North Russia in the summer of 1919, under pressure from the Red Army, the bodies or persons of 127 missing or presumed killed U.S. soldiers, over 50% of the total of 244 reported American dead of all causes in North Russia, remained behind in areas under Communist control.[82]Many may have been wounded in the actions where they disappeared, and were considered by their fellow soldiers to be dead. A party of U.S. Army Graves Registration troops left behind with the British Army (which held Archangel for a few more months) removed some remains from the Archangel area, but was unable to locate most of the missing Americans. When the British withdrew from Archangel in the fall of 1919, they took Bolshevik prisoners as hostages, in the hope that their own estimated 150 POW/MIAs would later be exchanged. Published Soviet sources revealed that Lenin's intelligence commissar, Felix Dzerzhinsky, had specifically targeted the American military personnel for capture, and had assigned Aleksandr Eiduk as the chief Chekist (later GPU-NKVD-KGB) agent for operations against Americans in the Archangel area during 1918-1919.[83]The kidnapping of thousands of Russian and foreign hostages had been a Cheka-Bolshevik trademark since the institution of the "Red Terror" following a nearly-successful attempt on Lenin's life in August 1918, in which the Soviet leader had been shot and seriously wounded by a Social Revolutionary woman named Dora "Fanya" Kaplan. Foreign-born or politically-active Americans were viewed as particularly valuable for possible intelligence use. All U.S. military prisoners were valuable to Soviet intelligence in order to obtain a thorough analysis of the American military and intelligence establishment, which was already perceived as a major threat to Bolshevism.

## Notes on Chapter One

1    *With Napoleon in Russia, Memoirs of General de Caulincourt, Duke of Vicenza from the original Diaries*, edited by Jean Hanoteau, William Morrow and Company, NY, 1935.; Also: *War and Peace*, by Leo Tolstoy.

2    Alexis De Tocqueville, *Democracy in America*, 1835

3    The Second International met from 1889-1914, again in 1920, and then became the Socialist and Labor International in 1923. The Third International was founded by the Communists in Russia in March 1919, and was always controlled by them from Moscow. This was the "Comintern," which Moscow announced was officially disbanded in 1943, as a result of the World War II alliance. It was replaced by the Cominform in 1947, during the Cold War between Russia and America.

4    *America and Russia*, op cit., pg. 119

5    Solzhenitsyn, *Gulag Archipelago*; and original War Department intelligence reports in the National Archives.

6    *Service With Fighting Men*

7    *Utopia in Power, Service With Fighting Men; Gulag Archgipelago*

8    *Utopia in Power*

9    *Forced Labor in Soviet Russia*, Dallin and Nikolaevsky; Gulag Archipelago, Aleksandr Solzhenitsyn; *Utopia in Power*, Heller and Nekrich; *KGB, The Inside Story*, by Andrew and Gordievsky, Harper Collins, 1990; *The New KGB*, Corson and Crowley, William Morrow, NY, 1985

10   *Utopia In Power*, op. cit.

11   *Gulag Archipelago*, by Aleksandr Solzhenitsyn; *The New KGB; KGB, The Inside Story*

12   *Ace Of Spies*, and others

13   *So Close to Greatness*, pp. 65-67

14   Foreign Relations of the United States, 1918, Russia, Volume I

15   *Russia Leaves The War*, by George Kennan, Princeton University Press, 1956

16   Kennan, *Russia Leaves the War*, Volume I, p. 515-516, who cites Bolshevik Propaganda hearings before the Subcommittee of the Judiciary Committee, third session of the 65th Congress, 1919

17   Kennan, *Russia Leaves the War*, op cit., Kennan's translation; Sumner's translation is in Foreign Relations of the United States, 1918

18   *Service With Fighting Men, Gulag Archipelago, Utopia in Power*; and U.S. War Department intelligence reports in RG 165, National Archives

19   *Utopia in Power, The History of the Soviet Union From 1917 to the Present*, by Mikhail Heller and Aleksandr M. Nekrich, Summit Books, 1986,; also, *Russia and the West Under Lenin and Stalin*, by George Kennan, Little, Brown, 1960

20   *Service With Fighting Men, Volume II*, Editorial Board Chairman, William Howard Taft, Managing Editor, Frederick Harris, Association Press, New York, 1922

21     *Utopia in Power*, p. 89 and original War Department documents in the National Archives

22     *So Close to Greatness*, p.69

23     *Ignorant Armies*, E.M. Halliday; also: original War Department reports in the National Archives

24     Kennan, *Russia and the West*;, also *Ace of Spies* by Robin Bruce Lockhart

25     *KGB: The Inside Story*, op. cit.

26     *Utopia in Power*, p. 66, op. cit.

27     These conversations occurred in the New York home of Obolensky's long- time friend, Cordelia Biddle Robertson, in 1969, while the author was a member of the U.S. Army.

28     *Ace of Spies*, op. cit.

29     *Ace of Spies*, Stein and Day, 1967 and Aug., 1921 A.P. Reports

30     *Gulag Archipelago* and *Utopia in Power*, op. cit.

31     *Utopia in Power and* War Department records in the National Archives

32     *Utopia in Power*, p. 92, Heller and Nekrich, Summit, NY, 1986; also, L. I. Zharov and V.M. Ustinov, The International Units of the Red Army in the Battles for Soviet Power, Moscow, 1960.

33     Aleksandr Solzhenitsyn, *Gulag Archipelago*

34     War Department records, Reel 113, Chief of Military History, Title: Historical material collected by the Historical Division Army War College, from folder #3362 through Folder 3384, Siberia- Folder 3374

35     The Medical Department of the U.S. Army in the World War, Volume XV-Statistics, part II: Medical Casualty Statistics, Table 45, Strength, Russia and Siberia 1918- 1919, For Russia, Office of the Adjutant General, for Siberia, Medical Reports of the Office of the Surgeon General, U.S. Government Printing Office, 1925, a footnote in this text says the figures for casualties are incomplete or confused

36     POW/MIA Policy and Process, p. 667

37     Graves, *Siberian Adventure*, p. 168.

38     *The Armies Of Ignorance*, op. cit.

39     Memorandum dated June 29, 1922, Russian Division, Department of State to the Military Intelligence Division, War Department, in the National Archives

40     Memorandum of personal interview with Peter A. Peluggaystchaff, U.S. Consulate, Harbin, Manchuria, July 6, 1920; Letter from Major General William S. Graves, August 2, 1920 in microfilmed War Department records, 1920, National Archives

41     Official records of the War Department, RG 165, National Archives

42     National Archives RG 165 10110- 2623/4, Adjutant General 383.6, Russia, Nov. 1930, also RG 57, RG 59, and "American Expedition Fighting the Bolsheviki, by Captain Joel Moore, Lieutenant Harry H. Mead and Lieutenent Lewis E. Jahns, 339th U.S. Infantry, The Polar Bear Publishing Co., Detroit, Michigan, 1920

43     *A History of Soviet Russia*, Volume Three, Edward Hallett Carr quotes Lenin, Sochineniya, xxiii, 176-189

44     *The Bolshevik Revolution*, Volume 3; Edward Hallett Carr cites M. Fainsod, International Socialism and the World War, Harvard, 1927

45     *American Expedition Fighting the Bolsheviki;*", p. 99, also: National Archives, RG 165 10110-2623/4, The Adjutant General's, " List of Officers and Enlisted Men Reported Killed and Missing in Action in the North Russia Expedition, Archangel, For Whom No Grave Location Has Been Found;" Also: Quartered in Hell: Lists of the Fallen, including 1929 remains recovery

46    *A History of Soviet Russia*, Volume Three, The Bolshevik Revolution, by Edward Hallett Carr, MacMillan, New York, 1953: "The records of the meeting were published in German (Sowjet-Russland und die Volker der Welt, Petrograd, 1920 and in French and presumably also in Russian and English; an earlier 'International meeting' held in Moscow and presided over by Kamenev was reported in Izvestiya, December 7, 1918

47    Code Milstaff cable No. 861, February 24, 1919, in microfilm records of the Military Intelligence Division, War Department, RG 165, National Archives

48    *Fighting the Bolsheviki.* op. cit., p. 127

49    Cable message from the American Minister at Stockholm, No. 3267, December 3, 1918, RG 165, in the National Archives: "The above for the information of Colonel House. Copies have been communicated to the American Ambassadors at London and Paris.

50    Microfilmed coded telegrams of the U.S. Military Attache-Archangel, Records of the War Department General and Special Staffs, Record Group 165, Military Intelligence Division- MID; a copy of the original February 4, 1919 cable in the National Archives is reproduced in the Appendix to this book.

51    *The American Expedition Fighting the Bolsheviki*, Captain Moore, Lt. Mead and Lt. Jahns, Polar Bear Publishing, Detroit, 1920, p. 137

52    *Service With Fighting Men* (YMCA in WW I, Volume II, Association Press, 1922.

53    *Utopia In Power*, p. 103

54    *So Close to Greatness*, p. 79; also: Russia and the West Under Lenin and Stalin.

55    MID cables, RG 165, NA

56    RG 165 NA

57    Ibid., See also: Archangel 1918-1919,pp. 58-59, 125, etc., by Edmund Ironside, Constable, London, 1953. The location of many British prisoners of war behind the Archangel front or in Vologda was not known for certain, but the U.S. Ambassador to England (Davis) learned from the British government that, "there is but one British civilian prisoner at Petrograd and 50 or 60 military officers and men at Moscow." See: Foreign Relations of the United States-1919, pp. 169-170, #789, to Acting Secretary of State Frank Polk.

58    *Ignorant Armies*, p. 173

59    Milstaff cables of the Military Attache, Archangel, RG 165, NA; *American Expedition Fighting the Bolsheviki; The Ignorant Armies; Quartered in Hell*, op. cit.

60    Ibid.

61    *The New KGB*, op. cit.

62    *America and Russia*, Edited by Oliver Jensen for American Heritage Magazine, Simon and Schuster, NY, 1962

63    Foreign Relations of the United States, 1919; records a cable from William C. Bullitt to the Peace Commission in Paris dated March 16, 1919, conveying the Bolshevik proposals, which included a demand that the "Allied and Associated Governments...give a general amnesty to all political opponents, offenders and prisoners," and repatriate all "Russian prisoners of war in whatever foreign country they may be, likewise all Russian soldiers and officers abroad and those serving in all foreign armies." In return, the Bolsheviks promised "facilities for repatriation...(for)PRISONERS OF WAR OF NON-RUSSIAN POWERS DETAINED IN RUSSIA (and) ALL NATIONALS OF THOSE POWERS NOW IN RUSSIA." The Bolsheviks also demanded the withdrawal all Allied troops from Russia, cease aiding the White Russian regimes and establishment of diplomatic relations with Lenin's Communist government.

64    *So Close to Greatness*, p. 88

65   *The Autobiography of Lincoln Steffens*, Harcourt, Brace and Company, p.799, According to Steffens, this occurred in Jo Davidson's studio, where Baruch was sitting for a portrait.

66   Ibid., p. 798

67   New York Tribune, March 27 and March 31, 1919

68   *The Autobiography of Lincoln Steffens*, op. cit., pp. 799-801

69   *So Close to Greatness*, op. cit.

70   *Fighting the Bolsheviki*, p. 192, op.cit.

71   From microfilmed records of the Military Intelligence Division of the War Department in RG 165, National Archives. The French had already negotiated independently with the Bolsheviks for an exchange of prisoners of war by mid-February, but despite reaching an agreement more French POWs and civilians continued to vanish. See; Foreign Relations of the United States- 1919, p. 172, #7224 Ambassador Sharp to the Acting Secretary of State, February 14, 1919.

72   *Fighting the Bolsheviki*, p. 189.

73   Adjutant Generals list of missing Americans in North Russia, op cit.

74   New York Tribune, April 11, 1919

75   *"Quartered in Hell," The Story of American North Russian Expeditionary Force*, 1918-1919, by Dennis Gordon, Chief of Research, Hayes Otoupalik, The Doughboy Historical Society and G.O.S., Inc. Missoula, MT, ISBN: 0:942258-00-2, p. 196

76   Microfilmed Coded cables from Archangel in RG 165, Military Intelligence Division of the War Department, in the National Archives

77   Microfilmed records of the Military Intelligence Division, War Department, RG 165, NA

78   NA, RG 165, Report # 1031, April 28, 1919, "American Prisoners Out Of Russia." Copies distributed to Director of Military Intelligence, American Expeditionary Forces, Military Attache, Archangel, M.I. Branch-OCS, Military Intelligence Division, War Department

79   NA RG 165, Report # 1055, May 26, 1919, "American Prisoners From Russia Passing Through Sweden," Marked as distributed to: Office of the Chief of Staff-Executive Division, Military Intelligence Branch; M.I.2; M.I. 5; 3 copies to Director of Military Intelligence; 2 copies to the Peace Commission (at Versailles) A total of 22 American prisoners were released by the Soviets on April 25, 1919, many of them civilians, including the U.S. Consul at Tashkent, Roger C. Tredwell, who had been arrested in October 1918 and whose release had been demanded by the U.S. ever since. See: Foreign Relations of the United States, 1919, #254 Consul at Helsingfors to the Acting Secretary of State (Frank Polk), April 25, 1919. The Soviets also falsely claimed that the United States forces in Siberia under General Graves were holding Russian prisoners of war. A cable of May 27, 1919 (#1637) from Polk to the U.S. Embassy in Copenhagen, Denmark, stated: "Department has positive report from American Consul Vladivostok, confirmed by General Graves, that Americans have not taken any Bolshevik prisoners in Siberia..." (FRUS-1919, p. 185)

80   *Quartered in Hell*, op. cit.

81   Annual Report of the Secretary of War, RG 165 NA

82   *Secret Intelligence*, by Ernest Volkman and Blaine Bagget, Doubleday, NY, 1989; also: original reports of Admiral McCully in RG 165, National Archives; *The Ignorant Armies; Quartered in Hell*; Foreign Service (later called Veterans of Foreign Wars Magazine).

83   List of the dead and dates of remains recovered in *Quartered in Hell*, op cit.

# CHAPTER TWO

## THE U.S. HOSTAGES AND AMERICAN RECOGNITION
## OF THE SOVIET UNION 1921-1933

The combination of the Cheka secret police and Bolshevik Revolutionary Tribunals ensured swift and merciless execution for tens of thousands of early Russian opponents to Communism. According to Russian historian Aleksandr Solzhenitsyn, some 16,000 persons were shot by execution squads in the twenty central provinces of Russia during the period of June 1918 to October 1919. These were largely reprisals against "kulaks" and other peasants resisting Communist confiscation of grain and other food supplies, part of Lenin's policy of feeding his favored city proletariat at all costs. After this the number of executions increased enormously as the Bolsheviks retaliated on those Russians who had aided the Western Intervention. In the Archangel area alone, in early 1920, as many as 30,000 people may have been executed by Cheka firing squads following the withdrawal of American and British troops (in late 1919). According to careful researchers, the total number of persons executed by the Cheka between 1919 and 1921 exceeded 250,000.[1] This of course does not include hundreds of thousands of others killed by the Red Army in offensives of the civil war.[2]

News of the ruthless methods of the Bolsheviks spread through their own announcements, escapees, neutral observers and Allied intelligence agents in Russia, causing deep apprehension in Western capitals. The foreign prisoners being held as hostages were valuable to the Bolsheviks however, and most were not executed but secretly imprisoned to be used for intelligence purposes and in future diplomatic negotiations. Since the formal founding of the Comintern in Moscow in March 1919, despite the fact that only five delegates attended from abroad, the Soviets had begun to act on their belief that the world was ready for a proletarian revolution. Bolshevik agents such as Mikhail Borodin and Salme Pekkala entered the United States and Britain carrying Czarist jewels to finance local Communist parties, and security officials in the west feared serious trouble in the postwar atmosphere of widespread political and labor discontent and high expectations. Influential pro-Bolshevik Western journalists like Arthur Ransome of Britain acted as agents of influence in shaping public perceptions about Soviet Russia. Ransome had been a correspondent for the (London) Daily News in Russia during the revolution and the civil war, who moved to, Riga, Latvia in 1919, but continued to travel in and report on Russia for the Manchester Guardian. He knew Lenin and Trotsky (whose secretary he married) and many other Bolshevik leaders, who grew to trust his reliable sympathy for the revolution and valued his intelligence and opinions. Ransome was constantly urging British diplomatic recognition of Lenin's regime and regularly met with senior Cheka officials under Dzerzhinsky, such as Peters and Unshlikht, and also N.K. Klyshko, who was a Cheka agent in the Soviet delegation that negotiated a resumption of trade with Britain.[3] He developed a close friendship with a British diplomat then negotiating with the Soviets, Robert Hodgson, MP, who was apparently ignorant of Ransome's connections to Soviet intelligence. The Soviet government denied holding any American MIAs as prisoners of war through Lenin's designated representative in the United States, Ludwig C.A.K. Martens, on November 15, 1919:

"American soldiers taken prisoners in the Archangel district, which was invaded by American troops without a declaration of war, have been treated in Soviet Russia with especial consideration, and were unconditionally released as soon as it was practicable to send them home, SO THAT THERE REMAIN TODAY NO AMERICAN PRISONERS OF WAR IN RUSSIA...The lot of thousands of Russians in the United States today is exceedingly unhappy...They are indiscriminately accused in the most sweeping terms by government officials of criminal and subversive acts...They have been arrested without warrant and subjected to oppressive treatment...Within the past few days great numbers of Russian citizens in the city of New York and elsewhere have been arrested...! THEREFORE RESPECTFULLY SUGGEST THAT THE UNITED STATES GOVERNMENT COULD EASILY BE RELIEVED OF THE PRESENCE OF UNWELCOME RUSSIAN CITIZENS, IF ALL THOSE CITIZENS OF RUSSIA WHOSE LIVES ARE BECOMING UNBEARABLE IN THE UNITED STATES WERE PERMITTED TO DEPART..."[4]

Thus, it was Lenin's own appointed ambassador to the United States who denied that any American military POWs were being held in Russia on the one hand, but requested the immediate repatriation of a large number of Russian-American radicals at the same time. The United States Government was only too happy to comply with this Soviet demand, even if it meant the Wilson Administration would be castigated by liberals and human rights advocates for the act. Ludwig Martens himself was to be deported from the United States, along with hundreds of other pro-Bolshevik Russian-Americans whose release had been demanded by the Soviet government. This was done under authority of Attorney General Palmer and the U.S. Department of Justice, and was ordered by the Secretary of State, Charles Hughes. Among those sent back to Russia was the famous radical union activist Emma Goldman, also known as "Mother Jones," together with her friend Alexander Berkman, who had recently been imprisoned. Within a short time Goldman became thoroughly disillusioned by the harsh reality of Communist Russia, where, according to her own account, she nearly perished, and she begged to be allowed to return to America. Lincoln Steffens, the radical American journalist later wrote:

"It was harder on the real reds than it was on us liberals. Emma Goldman, the anarchist who was deported to the socialist heaven, came out and said it was hell."[5] Unfortunately, Secretary of State Hughes had not insisted that the Soviets reciprocate after the first 150 Russian-American Bolsheviks had been deported.[6]

In early 1920 the French government decided to repatriate all remaining Russian POWs, and many were transported to the Black Sea region, site of an abortive 1919 French intervention, and released in South Russia. The French government was eventually forced to resort to threats of naval attacks on Soviet positions in the Black Sea region to obtain release of some of their war prisoners and civilian hostages in Russia. How many French military and civilian prisoners were exchanged for the repatriated Russians was not publicized, although at least 240 were reported to have reached Paris. Other Russian prisoners of war were returned in Allied ships to Baltic Republic ports, as the Russo-Polish War still continued to block most overland shipment. Still other POWs were taken by ship all the way to Vladivostok and exchanged for Czechs, Germans, Austro-Hungarians and Turks, who returned in the same ships.[7]

In the United States the head of the anti-radical Division of the Justice Department's Bureau of Investigation, John Edgar Hoover, had been working closely with Van Deman's successor as chief of the War Department's Military Intelligence Division (MID), Brigadier General Marlborough Churchill, who had received all important communications of the military attaché in Archangel during the fighting in north Russia. Churchill and Hoover together had been collecting files on hundreds of thousands of Americans considered to be subversives or Communists. A new Attorney General, Alexander Palmer, who had been a moderate until his home was bombed by radicals in June 1919, had authorized a mass arrest of alien radicals in the United States, based on intelligence collected by Hoover and the MID. The Military Intelligence Division of the War Department had been working closely with Hoover's agents to build up files on hundreds of thousands of Americans suspected of disloyalty. The MID and Bureau of Investigation had been greatly assisted in their anti-radical activities by two nationwide civilian organizations which functioned as a volunteer secret police force during World War I. These were the American Protective League (APL) and the anti-German American Defense Society (ADS), which had been founded in 1914 with support from Theodore Roosevelt and Charles J. Bonaparte (a descendent of the French Emperor who had served as Roosevelt's Attorney General). APL supplied intelligence on radical activities to U.S. Government agents from a nationwide network of informants, by infiltrating Socialist, Union and particularly IWW groups, and also provoked violence with police on occasion. Later, with the blessing of MID and the Bureau of Investigation APL began a nationwide campaign to track down arrest draft evaders, in which the civil rights of many Americans were violated. Some APL chapters became, in effect, vigilante organizations, operating with official sanction. By 1921, using such sources of intelligence, much of it possibly based on inaccurate information, J. Edgar Hoover had assembled a card file on more than 500,000 individuals. A professional military Intelligence officer of the time, Colonel Nicholas Biddle, had warned the previous director of U.S. Military Intelligence, Major General Ralph Van Deman:

"...The power of these volunteer organizations is tremendous, and it would seem to me advisable that some steps be taken to curb or control their activities."[8] Eventually steps were taken to curb APL, and some ADS members were prosecuted by the Justice Department and convicted of supplying false intelligence.[9]

The so-called Palmer raids officially began on January 2, 1920, in 33 cities across the U.S., and over 4,000 alien radicals and Communists were subsequently arrested and jailed, but this was in addition to many hundreds of Russian-American Communists and Socialists who had already been arrested in November and December of 1919. When American civil liberties advocates began to protest the tactics used by federal authorities, Hoover and the MID agents began researching them too, looking for connections to the revolutionaries.[10] Many of these Socialists and Communists were of Russian and Polish birth, both Jewish and non-Jewish, and those without U.S. citizenship were ultimately deported from America, in complying with Soviet demands.

In relation to the November 15th letter from Lenin's ambassador in America, Ludwig Martens, about American prisoners of war in Russia and Russians imprisoned in America, the Secretary of State had notified the U.S. Ambassador in London on December 23rd, of the impending deportations, and the message was given the widest distribution to other embassies:

"There are being deported from the United States to Soviet Russia about 250 citizens of Russia who are undesirable here...The deportation is in accordance with the law. Precaution has been taken to request for them safe conduct and humane treatment at the hands of the authorities under whose jurisdiction they will pass enroute to Soviet Russia. Repeat to Stockholm, Christiana, Copenhagen, The Hague and Brussels, sent direct to Paris, to Sofia, American Commissioner Riga, Warsaw, Berne, Prague, Rome, Madrid, Lisbon with instruction to present to foreign office and to release for publication...repeat to Berlin, Vienna, and Budapest to release for publication. Lansing"

This cable from Lansing indicates that the United States was returning the undesirable Russian Communist immigrants as a gesture to Lenin, in hopes of receiving the American prisoners held in Russia in return. Subsequent statements of the Soviet Foreign Minister, Georgi Chicherin, were to bear this out. Lansing also cabled the U.S. Commissioner in Finland on January 13th about a Finnish request to retain 10 of the Russian-American Bolsheviks being deported, as hostages for retrieving Finns held in Russia: "The Department regrets to be unable to comply with the request of the Finnish Government to retain 10 Bolsheviki as hostages. The terms of the law under which these aliens are deported make it necessary to send them to their country of origin..." The Allied Supreme Council voted on January 16th to cease the economic blockade of Soviet Russia, and on January 26th the American Ambassador in Britain cabled the Secretary of State that, "The British government has no objection... in the initiation of negotiations...for the release from Soviet Russia of...American citizens provided that such negotiations should not be prejudicial to those already initiated on the behalf of British prisoners of war and civilians in Soviet Russia..."

During March of 1920 the United States Government began deportation proceedings against the officially-unrecognized Soviet ambassador in America, Ludwig Martens, who had claimed in November 1919 that all American prisoners of war in Russia had been released. The Bolshevik government had authorised Martens to act for it in the continued absence of diplomatic relations between Washington and Moscow. The case against Martens was handled by John Edgar Hoover, the Justice Department lawyer and head of the Bureau of Investigation's anti-radical Division.

Ludwig Martens was an important Soviet connection to American Socialists and Communists who supported the ideals of the Bolshevik Revolution, such as Dr. Julius Hammer. Hammer was a Russian-born medical doctor whose political mentor was Boris Reinstein, the American Socialist who had taken part in the 1918 and 1919 founding-meetings of the Comintern.

Julius Hammer had known Lenin personally since attending a Socialist conference in Germany in 1907, and after that he had recruited many members for the Socialist Workers Party. In early 1919 Hammer was involved in founding a radical Leninist wing of the Left Socialists in America, and he was under surveillance by federal agents of the Justice Department's Bureau of Investigation. Julius Hammer had worked closely with Martens at his "Russian-Soviet Bureau" in the U.S., in supplying medical supplies and pharmaceuticals for the Red Army during the civil war and the Allied intervention, through a company he controlled with his son, Armand Hammer: Allied Drug and Chemical Company. Armand Hammer had been studying medicine at Columbia University in New York during World War I, but also was involved in the family business ventures, from which he obtained an income of over $1,000,000 during 1919. In the summer of 1919, his father Julius Hammer had performed an illegal abortion on the wife of a Czarist diplomat in New York, and the patient subsequently died. Julius Hammer was later convicted of first-degree manslaughter and imprisoned at Sing Sing Prison in 1920. Soviet dictator V. I. Lenin was later to write that he believed Julius Hammer had been imprisoned, "in revenge for his Communism." His son, Armand Hammer (who had been recruited into the Socialist party in 1916), was soon to visit Russia and meet with Lenin personally, and for seven decades to come he was to act as a sympathetic American emissary to Soviet leaders.[11]

Meanwhile, on May 26, 1920, the Secretary of State had cabled the U.S. Commissioner in Riga, Evan E. Young: "Department of Labor desires to ship more deportees to Russia and inquires possibility of landing vessel at Reval (Estonia)..." The number of this proposed shipment was given a few days later as "about 500," and Young was asked to find out if the Estonians would allow the deportees to cross into Soviet territory. This resulted in a reply from Estonia on June 21st that identified a problem with such repatriations that was to reoccur on an enormous scale two decades later after another world war:

"Reply just received from Estonian authorities states that as the Government has been repeatedly reproached for forcible deportations into Soviet Russia, it requests assurances that deportation 'is not effected against the will of all persons concerned.'"[12]

On June 25th, Acting Secretary of State cabled the American Minister in Switzerland (Gary), about the missing U.S. war prisoners in Russia, whom the United States Government officially denied existed, while trying to negotiate their release:

"Press reports Nansen leaving for Russia...as head of Red Cross delegation to repatriate prisoners 'including some Americans. MAKE EVERY EFFORT TO ARRANGE THROUGH INTERNATIONAL RED CROSS FOR NANSEN TO NEGOTIATE FOR REPATRIATION AMERICAN PRISONERS AND CITIZENS HELD BY BOLSHEVIKS."[13]

This message obviously meant to differentiate between U.S. military prisoners and civilian citizens held as hostages. On July 19th, the American Charge d'Affairs in Stockholm relayed a message from Nansen after his return from Moscow:

"The International Red Cross in Geneva sent me a telegram on behalf of the American government asking that I take up with the Russian Soviet government the question of relieving Americans who are prisoners and are being detained in Russia. I ALSO LOOKED UP EIDUK WHO IS THE HEAD OF THE CENTRAL ORGANIZATION IN CHARGE

OF PRISONERS. THEY SAID THAT THEY WERE WILLING TO FREE MOST AMERICANS AND PERMIT THEM TO LEAVE THE COUNTRY. THEY WISHED VERY MUCH, HOWEVER, TO NEGOTIATE ABOUT IT DIRECTLY WITH THE AMERICAN GOVERNMENT..."[14]

An agreement for freeing all the British prisoners in Soviet control was also under negotiation in July 1920, with the Bolsheviks substantially agreeing to British terms, which included an ending of hostilities and propaganda warfare by the communists. During the negotiations conducted with Maxim Litvinov by the British Member of Parliament James O'Grady, respecting the exchange of British war prisoners, Secretary of State Lansing had cabled the U.S. Ambassador in England (on January 10th):

"In case the negotiations by O'Grady may involve...the repatriation from Russia of others than British subjects, the Department desires that you present to the British Government the question of including Americans in any arrangement that may be made."[15] As the U.S. continued to negotiate for the return of war prisoners and missing soldiers held in Russia, the Bolshevik state grew stronger and fears that it would spread Communism to Western Europe, prompted the United States to encourage peace negotiations between Russia and it's neighbors, particularly Poland.

The first American Minister to newly independent Poland, Hugh Gibson, a Californian who had assisted Herbert Hoover in WW I relief efforts in western Europe and had become a close friend of the future President, was appointed to Poland by Woodrow Wilson through the influence of Hoover on the President's confidante, "Colonel" Edward House. Gibson had arrived in Warsaw in the spring of 1919, assisted by his second secretary, Arthur Bliss Lane. (Lane would one day also represent the U.S. Government in Warsaw, after another world war.) In the summer of 1919 a secret truce between the Poles and Lenin's government had freed the Red Army in Ukraine to fight the White Armies in the south and force the final retreat of the Allies from Archangel, in north Russia. The British military was then clandestinely assisting the Polish Army in confronting the Soviets, in a bid to restore Poland's borders, which had been pushed westward by repeated "partitions" between Germany and Russia from the late 1700s to the pre-WW I era. The United States and British governments urged the Poles to make peace on terms offered by Lenin, but Polish national ire had been aroused after centuries of czarist threats to their nation's existence. In late April of 1920 a renewed Polish offensive against the Red Army, ordered by Marshal Joseph Pilsudski, captured Kiev in the Ukraine. A few American military officers from World War I were serving as volunteer advisers against the Bolsheviks in the Polish forces at this time, including a veteran U.S. World War I pilot, Captain Marion Cooper, who was among those captured by the Red Army and transported to a

Moscow prison. (A young British officer also serving in Poland as an advisor at this time, Major Harold Alexander, would one day confront the Red Army as a British field marshal, on the Danube River in Austria.) Gibson already knew the truth about Bolshevik methods, and would be an opponent of diplomatic recognition of the Soviet government in the years to come.

Gibson traveled to Washington for consultation in April 1920, and to urge American aid for the Marshal Joseph Pilsudski's army, on the grounds that Poland's defeat would expose Germany and the rest of western Europe to direct Bolshevik attack. This was at a time when the Communist International was calling for a world revolution, which had alarmed Western diplomats and security officials. The fact that many of the Russian Bolshevik leaders, such as Trotsky, Kamenev, Zinoviev, Radek, Chicherin and Litvinov were Jews, as were many early party members and Cheka operatives, had contributed to a glaring anti-Semitic attitude within the American mission in Warsaw, and elsewhere within the U.S. foreign service in this period. Severe oppression and murders of Jews were belittled or went unreported in some cases, despite requests from the Department of State in Washington to supply such information. In Bolshevik Russia, anti-Semitism was now officially nonexistent, but widely practised, and Lenin's up-and-coming successor Josef Stalin, a Georgian with centuries of prejudice ingrained in his soul, was to one day vent his hatred on countless Soviet Jews, whom he referred to as "cosmopolitans." The prejudice in the American Foreign Service resulted, to some degree, from the exclusive, upper-class background and anglophilia of many foreign service officers at the time, and was aggravated by the fact that at home in the United States, the same situation existed with regard to the ethnic origin of many of the foreign-born Socialists and Communists in the teeming slums of New York and other cities. Research by a careful Harvard scholar, Martin Weil, has documented how this attitude affected Gibson, Moffatt, Grew, and William Castle Jr. In Poland, the disaffected and sometimes-persecuted Jewish minority also formed a large percentage of the Communist activists, as Moffatt, the new second secretary in the U.S. Embassy, wrote about in a description of a 1920 May Day parade in Warsaw:

"May Day passed off on the whole rather quietly. There were great processions all morning with red banners, about 75% of which were covered with Yiddish inscriptions. You would have thought from seeing the streets that it was a Jewish holiday. The main focus of infection was in the main square just in front of the Legation."[16]

President Wilson had authorized a special commission to examine the plight of Jews in newly-independent Poland, which included Brigadier General Edgar Jadwin, Homer Johnson and Henry Morgenthau. Morgenthau represented a different, well-established

Jewish constituency in America, which extended back long before the late 19th and early 20th century migrations; in the cases of some families, to the Revolutionary War period. While sympathetic to the plight of the recently-arrived Jews and those still subjected to pogroms in Central Europe and Russia, they were more at ease with the non-Jewish liberal elite who frequented the halls of power in Washington, than in the crowded tenements of New York, where the immigrant Socialists and Communists from Russia and Poland organized and schemed about the coming revolution. While Morgenthau called attention to beatings and summary executions of Polish Jews, the other two members of the commission were more critical and cited provocations by Jews. Such religious and ethnic hatreds in Poland and the rest of Europe were only to increase in the years to come, and would ultimately be exploited by both the Germans and the Soviets, while the Americans and their allies continued to look the other way. The anti-Semitic attitude of Foreign Service and State Department officers and other U.S. officials of this era was to have a profound impact on a group of liberal leaders such as Morgenthau, who would come to be called "New Dealers" and a decade later, under Franklin D. Roosevelt, would succeed in capturing the White House. The resulting exclusion of many senior State Department officials from important policy decisions about American-Soviet relations was to affect a coming post-Second World War confrontation with Soviet Russia over American prisoners of war.

While in Washington D.C. for consultations about Poland, Hugh Gibson wrote a memorandum about the issue of U.S. prisoners in Russia:

"The French Ambassador called yesterday afternoon and stated that the Austrian government is anxious to get rid of Bela Kun and desires to send him back to Soviet Russia with some of his companions...The French government feels that it is desirable to get all the Bolshevists possible into the territory occupied by the Soviet government but has informed Dr. Renner that Bela Kun can be allowed to return to Russia ONLY IF ALL EUROPEAN AND AMERICAN PRISONERS HELD BY THE SOVIETS ARE SAFELY RETURNED. After consultation with Mr. Polk today I notified the Ambassador that we were in agreement with the French attitude."[17]

This was followed up by a message to Paris from Bainbridge Colby, acting secretary: "United States Government hopes that transfer to Russia of Hungarian Communists now held by Austrian authorities will not take place except in connection with release of all Americans now detained in or out of prison in Soviet Russia. United States Government understands that other governments are assuming similar position..."[18] Thus the United States was now willing to condone the use of Communist hostages held by friendly powers, to obtain the return of American prisoners in Russia. On July 8, 1920, Gibson wrote to Arthur Bliss Lane's successor as second secretary, Jay Pierrepoint Moffat, about a major American policy change towards Soviet Russia, that reflected the ongoing progress in negotiations over trade and a prisoner exchange by Britain:

"We yesterday announced the removal of trade restrictions with Russia, a step that was taken over my dead body...I have been pushing like hell for all I am worth to get the President to make a statement in regard to Poland and Bolshevism in general..."[19]

Minister Hugh Gibson pressed for a clear expression of American policy towards Polish independence and Lenin's government, and was rewarded for his persistence by a subsequent policy statement on Poland and Soviet Bolshevism from the acting Secretary of State, Bainbridge Colby, in a message to the Italian Ambassador on August 10, 1920:

"This Government believes in a united, free and autonomous Polish state...While deeply regretting the withdrawal of Russia from the war at a critical time, and the disastrous surrender at Brest-Litovsk, the United States has fully understood that the people of Russia were in no wise responsible...That the present rulers of Russia do not rule by the will or the consent of any considerable portion of the Russian people is an incontestable fact...At the moment when the work of creating a popular representative form of government based upon universal suffrage was nearing completion, the Bolsheviki...by force and cunning seized the machinery and powers of government and have continued to use them with savage oppression to maintain themselves in power...It is not possible for the Government of the United States to recognize the present rulers of Russia as a government with which the relations common to friendly governments can be maintained... The existing regime in Russia is based upon the negation of every principal of honor and good faith, and every usage and convention, underlying the whole structure of international law; the negation, in short, of every principal upon which it is possible to base harmonious and trustful relations, whether of nations or of individuals..."[20]

Hugh Gibson and the new government of Poland became further involved in the issue of American prisoners in Russia a few months later, as noted in a cable from Warsaw signed by Gibson on December 6th:

"Polish Government has expressed willingness to discuss with Soviet delegation, Riga, exchange Americans in Soviet Russia. Young has informally furnished list of names and suggests I ask Department as to how far matter should be pressed."

Gibson's second secretary at the Warsaw Embassy, Arthur Bliss Lane, went on to become a special assistant to Undersecretary of State Joseph Grew, an old line diplomat who had entered the

Department of State in 1904 and had served in Japan, St. Petersburg, and at other important posts. Joseph Grew was one of the "favored few" who had come into the government service through Groton and Harvard, but he was to be a strong opponent of Soviet Communism all his life, and he would make an honorable mark in history in 1945, at the end of the Second World War. Grew was to influence many of the young foreign service officers like Arthur Bliss Lane, who would have to deal first-hand with the Soviets at that time.

On August 10, 1920, meanwhile, the U.S. Ambassador to Great Britain (Davis) had sent the substance of an intercepted Soviet radio message for Dr. Nansen from the Soviet Foreign Minister, Georgi V. Chicherin, to the Secretary of State:

"We note that the American Government ask for the repatriation of Kalamatiano who has been convicted as a military spy and a criminal...WE DO NOT FIND IN YOUR COMMUNICATION ANY PROMISE FROM THE AMERICAN GOVERNMENT TO PERMIT THE DEPARTURE OF RUSSIANS WHO HAVE BEEN SENTENCED IN MASSES IN AMERICA TO LONG TERMS OF PRISON FOR NO OTHER OFFENSE THAN THEY REMAINED LOYAL TO SOVIET RUSSIA...THE SOVIET GOVERNMENT WILL GLADLY SEEK A SATISFACTORY SOLUTION OF THE QUESTION RAISED BY YOU IF REAL RECIPROCITY... CAN BE REACHED...IF THE AMERICAN GOVERNMENT REMOVE ALL OBSTACLES NOW PREVENTING RUSSIANS FROM LEAVING AMERICA, INCLUDING ALL THOSE WHO HAVE BEEN IMPRISONED FOR POLITICAL OFFENSE, SUCH PERSONS TO BE NAMED BY OUR REPRESENTATIVE IN THE UNITED STATES, LUDVIG MARTENS..."[21] Thus the Soviets were actually demanding the return to Russia of many of the radicals that the Justice Department and J. Edgar Hoover wanted to deport from America.

In what next appears to have been a significant Soviet diplomatic signal, a U.S. Army prisoner of war from Archangel (since March 1919), Corporal Arthur Prince, of the 339th Infantry, who had long been missing in action and presumed dead, was suddenly released by Lenin's government and appeared on August 20, 1920 in Germany, very much alive. No public explanation by the Soviet government is recorded, or was asked for by the United States Government. The significance of the incident was hardly noted in the American press.[22]

The U.S. Commissioner in Riga, Latvia (Young) cabled the Secretary of State on September 7th:

"Estonian Minister for Foreign Affairs informs me Soviet Russia agrees to admit 500 deportees provided they are visaed by (Ludwig) Martens. It is also desired that permission be granted for departure to Russia of 50 non-Russian citizens now in United States, names to be furnished later. The Soviet Government willing to release Mrs. Kennedy and permit her departure from Russia if United States will release Larkin..."[23]

The Soviets were clearly interested in trading hostages as James J. Larkin was an Irish radical terrorist, imprisoned in the U.S. and wanted by Great Britain. A telegram from the U.S. Charge d'affairs in Oslo, Norway, of September 25th, conveying more information from Nansen, left little doubt of the reciprocal action Lenin desired from America:

"On his last trip to Russia he went to Kovno only, where he met Eiduk who has charge of questions as to the departure of foreigners. THE LATTER SAID THAT ALL THE AMERICANS WOULD BE PERMITTED TO LEAVE RUSSIA IF ALL RUSSIANS WERE PERMITTED TO LEAVE THE UNITED STATES...EIDUCK COMPLAINED OF THE RETENTION OF RUSSIANS IN THE UNITED STATES SAYING 'SOME OF OUR VERY GOOD MEN ARE IN PRISON THERE.' HE ADDED THAT THERE WERE ABOUT A THOUSAND RUSSIANS WHOM IT WAS DESIRED TO BRING; NANSEN SUGGESTED THAT SOVIETS CHARTER A STEAMER TO REPATRIATE RUSSIANS NOW IN THE UNITED STATES..."[24]

This was the same Aleksandr Eiduk of the Cheka who had been an American specialist on the Archangel front in 1919. Following the gesture with Corporal Prince, on the first of October 1920, another "messenger," an American civilian mining engineer named Alfred Wood Stickney, was allowed to leave a Moscow prison, and he brought word of other Americans who were still confined in Russia. One of those reported as still held alive was the 1917-1918 American intelligence agent Xenophon B. Kalamatiano, whom the Unites States Government wished to recover for debriefing.[25]

Ambassador Wallace in Paris cabled the Secretary of State on October 5th, that press reports of French threats to take action with their Black Sea fleet against the Soviets unless all French prisoners were repatriated by October 1st, were correct, and he reported further, "...The arrival in Paris yesterday of 240 repatriated French citizens." Four days earlier he had reported Chicherin's message of Soviet capitulation to the French threat of naval force: 'We bow to brute force.'[26]

A detailed chronicle of the north Russia fighting endured by the 339th U.S. Infantry was written and published in the fall of 1920 by three U.S. combat officers who had served there, who did not forget their fellow soldiers still missing in Soviet Russia. Although the list of U.S. missing in action was incomplete in their book, it nevertheless contained twenty-nine American names, while the book also reported that the bodies were never found of many more of those, "reported killed." In a chapter entitled: "Captive Doughboys in Bolshevikdom," they wrote:

"The Bolsheviki took a British Chaplain...Father Roach...Thinking they had a convert, the Soviet Commissar gave Father Roach his freedom and sent him through the lines at the railroad front in April. NEWS WAS BROUGHT BACK BY FATHER ROACH THAT MANY AMERICAN

AND BRITISH AND FRENCH PRISONERS WERE AT MOSCOW OR ON THEIR WAY TO MOSCOW...THE BOLSHEVIK MILITARY AUTHORITIES WERE UNABLE TO TRACE ALL THEIR PRISONERS...AN AMERICAN SERGEANT...SAYS THAT WHILE HE WAS IN MOSCOW SIX BRITISH SOLDIERS WERE LUCKILY DISCOVERED BY THE RED AUTHORITIES IN A FOUL PRISON WHERE THEY HAD BEEN LOST TRACK OF...Corporal Arthur Prince...was, finally in August 1920, released from hospital and prison in Russia, and crippled and sick joined American troops in Germany. His...stamina must have been one hundred percent to stand it all those long seventeen months. Herbert Schroeder, "B" Company, who was captured on the 21st of September (1918), has never been found. HIS COMRADES STILL HOPE THAT HE WAS THE AMERICAN PRINTER WHOM THE REDS DECLARED WAS PRINTING THEIR PROPAGANDA IN ENGLISH FOR THEM AT VIATKA...EVEN AS THIS BOOK GOES TO PRESS WE ARE STILL HOPING THAT OTHERS OF OUR OWN AMERICAN COMRADES AND OF OUR ALLIES WILL YET COME TO LIFE OUT OF RUSSIA..."[27]

British prisoners of war and civilians were also held hostage by the Soviets, among them the British agent and former embassy official, R.H. Bruce Lockhart. Some, including Lockhart, had been exchanged for Soviet officials held in retaliation in Britain, including the later Soviet Foreign Minister, Maxim Litvinov, but British MIAs on the Archangel front had also disappeared, and were secretly held as prisoners of war. Soviet documentary confirmation of this fact occurred more than seven decades later, when "Task Force Russia," a special U.S. Army intelligence team formed to analyze Russian intelligence on prisoners of war, received from the Russian State Military Archives, official notice of the locating of: "LISTS OF BRITISH OFFICERS IMPRISONED AFTER THE ANGLO-AMERICAN INTERVENTION OF 1918-1920.)[28]

From 1920-24 British political leaders occasionally flaunted their penetration of Soviet codes and ciphers at times when the value of revealing such a priceless secret outweighed Soviet countermeasures. One such incident took place while British-Soviet trade negotiations were occurring and over the final draft of a prisoner repatriation agreement. On September 10, 1920, Prime Minister Lloyd George accused Lev Kamenev, head of the Soviet delegation in Britain of "gross breach of faith" and subversion, and claimed to have "irrefutable evidence of this," a deliberate signal that Soviet ciphers were broken. Kamenev, planning to return to Moscow the next day, was effectively expelled from Britain by being told that he would not be allowed back.[29]

Such a demonstration of superior intelligence and firmness may have assisted British negotiators over the next few weeks, as the British and Soviet governments announced on October 14 that they had signed a prisoner-repatriation agreement and that all

British prisoners would be released beginning October 20, 1920.[30] How many ultimately remained behind in secret captivity is still a Soviet, and possibly a British state secret. Some were declared dead, though they were in fact being used by Soviet intelligence. Writing from Britain in 1987, KGB defector Ilya Dzhirkvelov revealed his work in the secret KGB Archives in the early 1950s. He discovered documentary evidence that the famous British secret agent of this period, Sidney Reilly, long remained alive in Soviet secret police control after news of his death had been published and accepted in Britain. These reports were shown to him to gain his ultimate capitulation. Artuzov, his secret police interrogator, had then told Reilly: "You no longer exist in this world. Your hopes of being freed will come to nothing..."[31]

This tactic for gaining the cooperation of prisoners was repeatedly used by the Cheka and it's successor organizations, the GPU, NKVD and KGB, in succeeding years; and there is every reason to believe that it was used on the secretly-held U.S. Army prisoners of the 1918-1919 Intervention and on those held after each succeeding war involving Americans.

Despite the now obvious Allied intention of repatriating all Russians, the Bolshevik authorities continued their policy of seizing POWs as hostages for political purposes. A January 16th, 1921, news report illustrates one such action: "When a train carrying German prisoners of war back from Russia was about to reach the city of Narva it was stopped and sixty Bavarians were taken from it by Soviet authorities, said a report reaching Munich on December 18. (1920). THE BAVARIANS WERE TO BE HELD AS HOSTAGES BECAUSE THE FOOD SERVED IN THE CAMP OF RUSSIAN PRISONERS IN ERLANGEN, BAVARIA, WAS REPORTED TO BE POOR AND A SOVIET COMMISSAR HAD BEEN INSULTED BY A GERMAN OFFICER."

It was only a day after this latest hostage-taking, on December 19th, that the Soviets finally realized how extensively their code and cipher systems had been penetrated by the British signals intelligence team led by former czarist cryptographer Ernst "Petty" Fetterlein. Mikhail Frunze, the Soviet commander of the Southern Red Army Group who defeated the White forces in the Crimea under Baron Wrangel, sent an alarming message to Moscow:

"It emerges from a report furnished to me today...that absolutely all our ciphers are being deciphered by the enemy in consequence of their simplicity...The overall conclusion is that all our enemies, particularly England, have all this time been entirely in the know about our internal, military-operational and diplomatic work."[32] The British Government Code and Cipher School was able to decrypt most of the Soviets high-grade diplomatic communications during the Anglo-Soviet negotiations. In the United States at this time the Soviet codes had been cracked by two experts, Major Herbert Yardley and Captain William Freidman, whose contribution is covered in more detail later in this book. Despite evidence of Bolshevik perfidy gained by British (and American) decryptions and the continued Soviet hostage-taking, by the spring of 1921 practically all Western-held Russian prisoners had been repatriated.[33]

By late 1920 the regular White Russian armies had been defeated everywhere and the Civil War had been practically won by the Bolsheviks. The Polish forces on the west had also been driven back and much of the old western Russian border restored. Yet the nationwide repression and the mass executions and confiscations of food by the Communist authorities, had produced a whole new series of armed peasant revolts. Severe and widespread food shortages and mass starvation in several regions again threatened the central Soviet authority. Peasant armies, formed into partisan military units, arose from Siberia to Tambov, Byelorussia and the Ukraine. Aleksandr Antonov led a 50,000-man peasant partisan army in the Tambov area by the beginning of 1921, and some 40,000 fighting men were serving under the Ukrainian anti-Bolshevik Green leader Nestor Makhno, at this time.

Lenin and Trotsky unleashed the most successful Red Army commanders on the armed peasant bands. They were accompanied into each subdued district by the Revolutionary Military Tribunals and the omnipotent Cheka killing squads. Hundreds of thousands of people were killed in the fighting or executed, and hundreds of thousands of other men, women and children were deported from their homelands to the desolate arctic north or to Siberia. Hostage-taking of families became a standard Cheka method to obtain the surrender of armed anti-communist partisans. Lenin was determined to destroy the independent landholding Russian farmers, known as the "kulaks," as a class (thus previewing Stalin's later purges of the 1920s and 30s. They and remnant bands of White military forces who joined together fought in many desperate stands, but were inexorably destroyed.

The years of war, repression and destruction had produced a major crisis of mass starvation and revolt in several regions of Russia by the winter of 1920-21. On January 22, 1921 Soviet authorities decreed that the bread ration for city workers had been cut again, this time by one third. Demonstrations and strikes immediately spread among the Bolsheviks cherished proletarian workers throughout Petrograd, causing real fear in Lenin's regime.

The rebellion spread to the naval base of Kronstadt, where some 4,000 to 5,000 sailors of the battleships Petropavlovsk and Sevastopol, and other units of the garrison, endorsed a resolution calling for freedom of speech, release of all political prisoners, voting by secret ballot, review of all cases of concentration-camp inmates and other democratic reforms. The Kronstadt sailors and garrison, who had been a mainstay of support to the Bolshevik revolution since the beginning, were joined by civilians of the town in the protest. The delegation sent from Kronstadt to Petrograd was arrested by the Cheka and on March 2, 1921, Lenin and Trotsky signed an order for the Red Army to assault the base. The Bolshevik leaders reacted with overwhelming force against this threat to their control. An army of 50,000 Red Army troops was massed against the 4,000-5,000-man Kronstadt garrison and attacked the defenses on March 17-18, 1921. Some of the Kronstadt sailors escaped to Finland, including their chosen leader, Petrichenko. Hundreds of others were eventually shot by Cheka death squads and the remaining thousands of survivors were deported to the White Sea concentration camps. Kronstadt sailors were to be reported in the Gulag labor camps for many years to come.[34] In the West, pro-Bolshevik journalists like Arthur Ransome, of the (British) Manchester Guardian, defended the brutal suppression of the Kronstadt revolt.

The 1920-21 peasant uprisings and the Kronstadt revolt had a major impact on Lenin, however, as he now finally realized that Soviet repression had gone too far. The result was to be "a step back" from rapid communization, called by Lenin the New Economic Policy (NEP), which was to continue until after his death and his successor Josef Stalin's initial consolidation of power. (This temporary relaxation of total centralized control over the economic marketplace was later to be viciously reversed by Stalin with the mass-arrests of hundreds of thousands of "NEP-men."[35]

## THE 1921 U.S.-SOVIET AID FOR HOSTAGES EXCHANGE

When the Bolsheviks finally triumphed in the Civil War, Russia faced the greatest famine to that time in its history, from which over 5,000,000 Russians and Ukrainians eventually starved to death. Lenin's regime had run out of any other domestic options to avert a possible general uprising against their rule. It was at this point that the Allied offer to supply aid to Russia in return for Bolshevik fulfillment of certain conditions, including freeing selected American prisoners, began to seem attractive to Lenin and his cohorts. The Soviet leaders knew that the Americans were concerned about their prisoners still held in Cheka prisons, and were also anxious to demonstrate U.S. good will toward the Russian people. Thus, the American civilians and US military prisoners of the intervention were used by Lenin as hostages in 1921 to obtain desperately needed food and medical supplies under authority of war relief director (later President) Herbert Hoover, during the administration of President Warren Harding. Herbert Hoover later described how the subject of American aid to Russia was raised:

"In July 1921, just as we were preparing to wind up our activities in the rest of Europe, Maxim Gorky, the Russian author, addressed an appeal to me and the American people for aid in the stupendous famine among Russian people in the Ukraine and the valley of the Volga. This had been due partly to freaks in the weather, but mostly to a halt in agricultural production while the Soviets were communizing the Russian peasants..."

Gorky appealed for U.S. aid, particularly on behalf of Russia's children, and as if on cue, this theme was immediately embraced by American Communists and many liberals of the time. The pro-Bolshevik New York "Nation" ran full-page ads of the New York Committee for Russian relief, pleading: "Do you wish to relieve the suffering children of Russia...?" Influential liberals, including Mrs. August Belmont and Felix M. Warburg lent their names to an organization headed by Charles Sabin and William Chadbourne, to raise private funds and to push the United States Government into aiding the starving Russians.[36] Many of these funds were later demonstrated to have been used to spread Bolshevik propaganda. American journalists who sided with the Communists, such as the Hearst reporter Louise Bryant, reported to their newspapers in the United States that the American prisoners in Russian prisons were extremely well treated, and that tales of torture and executions were exaggerations or fabrications. Louise Bryant was the widow of the American Communist John Reed, a radical Harvard graduate who had written a sympathetic eyewitness account of the Russian Revolution, entitled: "Ten Days That Shook The World," but who had died in Russia at the age of 32 in October 1920, and was eventually buried in the Kremlin wall. Granville Hicks, a Communist contemporary of Reed's who later wrote a biography about him, stated that Reed was about to disavow Lenin just prior to his untimely death.[37] Louise Bryant subsequently married the pro-Bolshevik former Assistant Secretary of State, William C. Bullitt, who was to become the first U.S. Ambassador to Soviet Russia in 1933. (Bullitt later was to become an outspoken anti-Communist) In an August 12, 1921 letter, published in the New York Times under the heading "Prisoners of the Soviet," Louise Bryant wrote about her meeting in a Soviet prison with an American Red Cross official and former U.S. Army officer, Captain Emmitt Kilpatrick (who later reported he was starved and brutalized by his Communist captors at the time):

"The camp where Kilpatrick was imprisoned was once an old monastery. It was a beautiful place and had a lovely old garden. The prisoners were allowed to go about quite freely from room to room... Personally, I regarded Kilpatrick a 'mamma's boy,' who ought never to have ventured so far from home..."

Lewis S. Gannett contributed to the Soviet disinformation campaign in "The Nation," a publication that was to sympathize with Soviet aims for decades to come:

"There were, so far as I could discover, eight Americans in prison or partly confined in Moscow...most of them were accused of some kind of espionage....In fact the Americans held in Moscow were receiving better nourishment than those who had returned to Moscow to serve the government that was to them a promise of a brighter future...There has been nothing in Russia to compare with Deer Island or Detroit, or even with the raid on the Russian Peoples House...in New York. Mistakes have occurred in Moscow, but as yet nothing to compare with the Palmer raids...Industrial depression sent thousands of Russian-Americans back to Russia. In four months, December 1920 to March 1921, 15,000 poured in through Libau alone...Some were, despite American prohibition, habitual drunks. Russia is, by the way, a country of real prohibition. I did not see an even slightly intoxicated person in an entire month in Moscow..."[38]

On August 24, 1921, Associated Press reported that some 50 Americans were being held in a "sort of concentration camp in Moscow..." Yet, according to an October 21, 1920, Associated Press report, the total number of American citizens in Russia was estimated at 3,000 (including some families of naturalized Americans). The US State Department had denied this and said that it knew of only 35 Americans in Russia. The Soviets had concealed the identity of many of their prisoners and it appears that hundreds of native-born American civilians and 60 to 100 U.S. military prisoners were being held at this time. Some were held in Moscow prisons with other foreigners and word of their presence had gradually leaked out, ensuring their eventual escape. But others, particularly the U.S. military prisoners from Archangel and Siberia were mostly held in other, secret locations. One of the quasi-military civilian hostages, the Red Cross official (and former U.S. Army intelligence officer in WW I) named Emmitt Kilpatrick, who had been captured with men of Wrangel's White Army, had smuggled a March 11th letter out of a Cheka prison, which eventually reached the U.S. Department of State: "...I am now held in prison as a hostage for one Jim Larkin (a communist agitator imprisoned in the U.S.) now serving a sentence of twenty years, and the same has been awarded me."[39]

Kilpatrick also succeeded in getting the names of other imprisoned Americans out of Russia, as did a number of other prisoners, and these names eventually reached the State department and the Military Intelligence Division of the War Department. Scores and probably hundreds of British and other Western European war prisoners and civilians were also being held hostage with the Americans by the Soviets at this time, in an effort to gain concessions from their home governments.

Hoover made it clear to Lenin's government that the highest priority reciprocal action expected from the Bolsheviks in return for massive American aid was: FREEDOM OF ALL AMERICAN PRISONERS IN RUSSIA..."[40] This demand, which was considered nonnegotiable, was passed by Hoover's European director of the ARA, Walter L. Brown, to Maxim Litvinov.

The Washington Post had published the official United States Government response to the Bolshevik request for aid in an Associated Press report, under the heading: "Insists Americans in Russia be Freed...State Department, in curt note, demands prisoners be released at once...U.S. unable to countenance relief while citizens are held, Moscow is told...Formal demand for release of American prisoners in Russia has been made on the Soviet authorities by Secretary Hughes. The State Department was advised that the communication had been handed to the Soviet representative at Reval yesterday by Consul Albrecht...The action was taken in the name of humanity, and because all efforts to secure the release of the Americans, made through Dr. Nansen of the Red Cross, have failed... The communication was sent to Consul Albrecht July 25. THE CURT DEMAND MADE UPON RUSSIAN AUTHORITIES FOR THE RELEASE OF AMERICAN PRISONERS IS THE FIRST OFFICIAL REPRESENTATION MADE UPON THE SUBJECT, although Lenine and Trotzky and their associates have been personally advised informally of the determination of the United States not to consider closer relations with Russia until this was done. The dispatch of the communication was timed so it would reach the Russians...simultaneously with the message sent by Secretary Hoover in response to the appeal made by Maxim Gorky..."[41] Other reports spoke of a possible exchange of Americans held in Moscow, for Russian-born Communists held in American jails.

Approximately 100 Americans, nearly all them civilian hostages, including Emmitt Kilpatrick and a former U.S. Army intelligence officer, Captain W.H. Estes, were finally released between August and September 1921, upon promise of massive aid shipments from the United States. On August 20th, before the release of some of the American prisoners had been completed, the first American ARA representatives entered Russia. The U.S. aid from 1921-1923 saved the lives of an estimated 10,000,000-15,000,000 Russians.

Two U.S. soldiers, labeled by the Army as deserters from the Siberian expeditionary force, were also reported freed, although one, Thomas Hazlewood (who claimed his real name was Russell Pattinger) of San Francisco, stated to another prisoner while in a Moscow prison that he was actually captured in January 1920, and subjected to severe tortures. A few other American military POWs may have been released, but if so, the U.S. Government remained silent about them. The freed prisoners revealed horrifying details of mass-executions in Soviet prisons, which were largely down-played or ignored by pro-Soviet journalists of the period, such as Walter Duranty of the New York Times and Louise Bryant, the young widow of American communist John Reed, and later wife of William C. Bullitt.

The Soviet government swore eternal friendship to the American people for their generosity at the time, but simultaneously the Bolshevik leaders Lenin and Trotsky had secretly withheld many American prisoners, including kidnapped civilians and most of the US Army POW/MIAs of the Archangel Intervention. Captain Emmitt Kilpatrick, the American Red Cross official who had served as a U.S. intelligence officer during WW I, was among those released through Reval in the Baltic region. The Bolsheviks had suspected Kilpatrick of espionage; although even if this was true, they had certainly given the United States a legitimate reason for spying inside Russia, by illegally retaining American prisoners of war and civilian hostages in secret locations. Upon reaching freedom, Kilpatrick revealed that the Soviets had refused to admit the grain-for-hostages exchange negotiated by Hoover which gained the release of some Americans, and he told of actual conditions in Soviet prisons:

"When we were released, officials told us that the Soviet had granted us an amnesty. We did not discover the real reason for our freedom until we reached Petrograd. There we saw newspapers, which Kalamatiano translated, containing Hoover's note. I talked with men from Kuban, Crimea, Omsk, Tomsk, Siberia and other provinces. These men reported that the famine...was largely brought about because the Communists confiscated last year's grain crop, including the seed....The most horrible experience I was forced to go through was when I was confined to a basement cell next to the execution room. I was under death sentence once and was taken into the hall room. The water runs over the floor constantly to wash away the blood. I could not sleep nights because of the constant shooting between 2 and 4 in the morning. There is more terror in Russia than ever before.

"Practically the entire staff of the Soviet Foreign Office arrived at the prison one day because of information which had leaked out. Mrs. Louise Bryant was brought to my cell. She questioned me about Captain Marion Cooper, saying he had done wrong in aiding the Poles and would be punished. Later I read her signed article in the Siviata praising the prison treatment accorded the Americans. She told of the good food we were receiving when we were starving. She told of how all we had to do was sleep, whereas we were too weak to stand. If America starts relieving the Russians, I hope it will watch the food go down the throats of the hungry, otherwise the Communists will be the only ones to be provided for. Already the Soviet is sending out propaganda urging that only the fit should get food. All others must die."

Captain Emmitt Kilpatrick made his most significant statement following his release when he alluded to the missing Allied and other European POWs in Soviet captivity:

"A HORRIBLE TRAGEDY IS BEING ENACTED IN RUSSIA TODAY. PRISONERS CAPTURED ON THE BATTLEFIELDS OF THE LATE WAR ARE ROTTING IN PRISONS, ALL RECORDS OF THEM BEING DESTROYED AND THEIR EXISTENCE FORGOTTEN. THEY ARE NOW ACTUALLY STARVING BECAUSE THE QUARTER OF A POUND OF BREAD GIVEN THEM DAILY IS NOT ENOUGH TO KEEP LIFE IN THEIR BODIES."[42]

The U.S. Government agent, Kalamatiano, who knew more than probably any other American prisoner, said very little to reporters when he returned from captivity in Russia, but immediately began writing a full (but classified) report of his experiences. He was credited by other Americans with having helped keep them alive in the Lubyanka and Butyrka prisons in Moscow, but was subsequently ignored by the United States Government and died in relative obscurity as a tutor at the University of Chicago, in 1923, of blood poisoning resulting from a hunting accident.[43]

By 1921 all the missing U.S. military POWs had been officially "presumed dead" by the War Department, although one more U.S. soldier, Private Sidney Vikoren of the 31st Infantry, who claimed he was captured in Siberia in March 1920, was returned through the Baltic coast region in November 1921, only to be labeled by the War Department as a "deserter." His appearance proved that the transfer of American military prisoners from Vladivostok, Siberia to Moscow had occurred. Private Vikoren's return was noted in a telegram for Alexander R. Magruder, Charge d'Affairs at the American Legation at Helsingfors (Helsinki), from Secretary of State Charles Hughes, of 28 November 1921: "You may obtain funds for Vikoren from Commissioner, Riga for REPATRIATION or TRANSPORTATION Coblenz, if upon examination you believe Vikoren is bona fide in his claims and is not a deserter. Consult military attaché." Corporal Herbert Schroeder, the WW I soldier of the 339th U.S. Infantry who had been missing in action in north Russia since September 1918, was reported held alive in a Moscow prison in 1921, but was not released with the other American hostages in September of that year.[44] In the subsequent aid to Russia program of Herbert Hoovers American Relief Administration (ARA) at least one American ARA worker vanished in the Ukraine and was never heard of again, while others were constantly hounded by Dzerzhinsky's Cheka operatives, who arrested Russian ARA workers at will and stole entire trainloads of relief supplies. Hoover later wrote in his memoirs that American ARA veterans were never able to locate any of their Russian ARA colleagues in subsequent years.

The massive American relief effort in Russia from 1921-1923, together with Vladimir Ilyich Lenin's "New Economic Policy," encouraged American businessmen, of different political persuasions, but all eager to trade with the Bolsheviks, in traveling to Russia. One American who arrived in 1921 was Armand Hammer, son of the radical Socialist physician who had known Lenin since before the First World War. Armand Hammer shipped with him on his trip to Russia a U.S. surplus field hospital, a new ambulance, medical supplies and surgical instruments, as a gift to the Soviet government. His father's old colleagues, Reinstein and Martens, arranged a personal interview with Lenin for the 23-year-old American, who was rewarded afterward with an asbestos mining concession in the Ural Mountains. Lenin not only knew and trusted Hammer's father, but his desire for U.S. diplomatic recognition and trade led him to encourage joint American-Soviet business ventures. Hammer raised the issue of possible labor problems with Lenin, but the Bolshevik leader said this problem could be taken care of rapidly. In the agreement Lenin also authorised Armand Hammer to "...Take Urals valuables on Commission for sale in America..."[45]

Armand Hammer was to turn this authorization into a multimillion dollar business of exporting Russian art treasures to America, which had been confiscated from the Russian bourgeoisie and aristocracy, many of whom ended their lives in Soviet forced-labor camps. The asbestos concession actually got into production, employing over 1,000 workers for a time, but Hammer was to spend years living in a sumptuous mansion in Moscow which had been assigned for his use, and he developed a major fur trading business in Soviet Russia for export to the West, with contacts across Russia and Siberia. The American upper class developed a nostalgic fondness for expensive pre-Soviet Russian art, and Armand Hammer thus expanded his circle of influence in the United States. From the time of Lenin and Stalin until the rise of Mikhail Gorbachev, Armand Hammer would act as a messenger for American leaders who needed a friendly contact with Soviet rulers.[46]

Another young American, of vastly different background, who conducted business with the Bolsheviks as early as 1922, was Averell Harriman. Harriman was born in 1891, the multi-millionaire son of railroad magnate, E. H. Harriman, whose Union Pacific and Southern Pacific railroads tied the continent together. Edward Harriman had once been described by President Theodore Roosevelt as a "malefactor of great wealth," and an "enemy of the Republic," during Roosevelt's "trust-busting" campaign, because of his high-handed ways against competitors. Averell Harriman had boarded at Endicott Peabody's exclusive English-model prep school, Groton, which produced other illustrious graduates, including President Franklin D. Roosevelt and future State Department officials Sumner Welles, Joseph Grew and Alexander Kirk. From Groton, both Harriman and his boyhood friend, Robert Lovett, went on to Yale, where both were inducted into the secretive "Skull and Bones" society, which produced such leaders as William Howard Taft, Henry Stimson, Henry Luce (of Time Magazine), William and McGeorge Bundy, and eventually, President George H.W. Bush. A large and powerful man known for his somewhat plodding but imperial ways, Harriman helped coach younger students on the Yale rowing crew, including one named Dean Acheson, a future U.S. Secretary of State. During World War I, while many of his classmates served in France, the younger Harriman secured a large and profitable government contract to produce ships in Philadelphia, but they were not completed in time for use in the war.[47] In contrast, his friend Robert Lovett, son of an executive in the senior Harriman's financial empire and a future secretary of defense, served as a combat bomber pilot on the Western Front in World War I, and was awarded the Navy Cross.

After the First World War, Harriman began to build upon his father's fortune by expanding into shipping lines and founding an investment banking firm in New York in November 1919, to finance maritime securities, which he called W.A. Harriman & Company. Although the U.S. Government had instituted a trade embargo against Soviet Russia and was opposed to American citizens doing business with the Bolsheviks, Harriman formed a partnership with the German Hamburg-American Steamship Company to supply it ships, and to jointly form a shipping line to Russia, called Deutsch-Russiche Transport Company. It was a joint venture with the Soviet regime, which retained half the capital of the company that began operations in November 1922. Harriman's investment bank also teamed with a German bank in financing critical Soviet imports by purchasing discounted Russian notes from companies doing business with the Bolsheviks. As a result of Lenin's New Economic Policy and the Soviet need for hard currency to buy critically-needed Western equipment and industrial technology, the Soviet government was offering other concessions similar to Armand Hammer's asbestos mine.

Lenin had suffered a series of paralyzing strokes beginning in May 1922, that may have resulted from the wounds he received in August 1918, and this led to his gradual withdrawal from direct running of the Soviet government. His lieutenants, including War Commissar Leon Trotsky, and a scheming, power-hungry former political commissar of the southern war front, Josef Stalin, began struggling for control of Soviet Russia. Although Lenin finally died in 1924, his policy of offering concessions continued for a time, because of the Soviets desperate need of convertible currency. Thus, Averell Harriman began secret negotiations with Soviet officials in 1924, to obtain a manganese concession in the Caucasus Mountains of Georgia, including one of the largest mines in the world of this critical alloy in steelmaking. In 1926, Harriman made an agreement with German exporters to market $42 million in bonds to finance long-term credit for the recently-formed Union of Soviet Socialist Republics. The State Department opposed him, but Harriman argued against isolating the USSR. Meanwhile, the British journalist and agent-of-influence, Arthur Ransome's single-minded pursuit of diplomatic recognition of the Soviet Union by England, which had continued until 1924, was finally rewarded when the new Labor government of Ramsay MacDonald normalized relations in January, more than 9 years before the United States did so.[48]

As Josef Stalin struggled to gain total control over the Soviet state, in December 1926, Averell Harriman went to Moscow, to renegotiate the terms of his manganese mining concession in the Caucasus Mountains. By this time Stalin had maneuvered himself into the position of Secretary of the Communist Party, and was engaged in fomenting distrust and fear among Lenin's old Bolshevik comrades, Radek, Kamenev, Trotsky and others. Stalin had maneuvered the appointment of Leon Trotsky, the former Commissar for War who, unlike the feral little Georgian, had contributed substantially to the Bolshevik triumph in Russia, as chairman of the concessions committee. This had exposed Trotsky to criticism from fanatical Communist supporters of Stalin within the Party, and led to his banishment in the east by Stalin, and ultimately to his exile and 1940 murder by a Chekist agent in Mexico. Harriman met with Trotsky in Moscow but found him formal and strict in adhering to established Party doctrine on the Western concessions. Harriman traveled to south Russia and Georgia by railway in a czarist palace car, to view his manganese mines and to deal with local Communist officials. He subsequently expressed the belief that the Russian peasants had more freedom than under the czars, and were as well off, financially, and that Josef Stalin was, "...not a dictator in any sense of the word... but he is a political boss in the sense of Charles Murphy of Tammany Hall."[49]

Perhaps because he had always had access to enormous wealth, Harriman exhibited all his life a trait of believing that any person with whom he dealt, whether a friend or an enemy, would strike a bargain and could be held to it. His early experiences at trying to extend credit to the Bolsheviks in Russia seemed to bear this out, as the Soviets eventually bought out his investment in the manganese mining concession for $3.5 million in Soviet bonds, which were later honored by Stalin's government. Thus the Soviets convinced Averell Harriman that they would honor their commitments, and his belief in their sincerity would prove critical when he later served as Roosevelt's and Truman's Ambassador to Russia.

One of Averell Harriman's executives in this period was a

young graduate of Saint George's and Yale and future U.S. senator named Prescott Bush, who had served as an Army captain in the 1918 Meuse-Argonne Offensive, and was the son-in-law of George H. Walker, President of W. A. Harriman & Company. His son, George Herbert Walker Bush, born in 1924, would one day become President.[50] W.A. Harriman & Company merged with an old established firm, Brown Brothers, and remained a powerful and influential concern, while Harriman retained his interest in the Soviet Union and his philosophy that the USSR should not be isolated economically from the world and that trade and credit with the Soviets would gain concessions from them. Through his activist sister, Mary, Harriman became politically involved in the Democratic Party, in which an older friend of the family's, Franklin D. Roosevelt, Governor of New York and former Assistant Secretary of the Navy under Woodrow Wilson, was a rising star.[51]

Armand Hammer and Averell Harriman were representative of a group of influential American businessmen who argued that increased trade with the new Bolshevik regime would bring better relations between the United States and the world's first Communist state. This view was to ultimately prevail over those in the Department of State and other areas of the government, or those in the media, who felt that previous illegal Bolshevik actions demanded their isolation from the world community of nations. The outcome of this struggle was influenced by the fact that from 1919 onward, the harsh realities of Soviet conduct with American and Allied war prisoners and other sensitive but substantive matters, were "classified," and thus kept secret from the American people. Thus, Americans of all political persuasions were denied knowledge of important facts about Bolshevik Russia which may have altered the perception of many.

Publicly at least, the books were now closed on the missing U.S. military POWs and MIAs from Archangel, and those other missing Americans who had been labeled as "deserters" in Siberia. Yet a classified "confidential" inter-department memo from the War Department to the Department of State, dated November 9, 1923 (not declassified until 1990), HAD STILL LISTED 43 U.S. SOLDIERS AS MISSING IN ACTION IN NORTH RUSSIA, and said nothing about U.S. missing in Siberia. This official (but classified) 1923 U.S. Government figure represented the total of 71 U.S. missing in action soldiers listed in the 1919 cables from the Archangel front, minus the 28 U.S. enlisted men missing in action at Nijni Gora in January 1919, who were later moved to the category of "presumed killed" in the 1919 Annual Report of the Secretary of War, which listed no U.S. MIAs in north Russia. It also appears that some of the 230 or more civilian American railway employees in Russia and Siberia, some of whom were ex-U.S. military personnel, also vanished.[52]

Following the ransoming of the last American and western European prisoners who were allowed out by the Soviets, a silence had ensued on the fate of the remainder. Any livesightings of American prisoners in Russia were secretly classified, as were deciphered Soviet messages which may have revealed their ultimate fate. Given British expertise in this area at the time, it is likely that today, somewhere in hidden archives protected by Britain's severe Official Secrets Act, there still exist decoded, secret Russian messages of 1918-1929 which would shed light on the fate of missing British, American and French prisoners in Russia.

The fate of some of them may have been execution The American crypt-analyst (Major) Herbert Yardley achieved fame in the intelligence world for his uncanny ability to break any foreign code. Aside from mastering German, Japanese and other codes he broke the Russian codes after WW I and the Bolshevik Revolution. Yardley's WW I team at the MI-8 section of US Military Intelligence, and after the war as a covert section of the US State Department, was to produce outstanding intelligence for the United States until 1929, by which time he was reported to have cracked over 45,000 cryptograms, including many of Soviet Russia's.[53] When Henry Stimson was appointed Secretary of State by Herbert Hoover he was provided with a file of fresh intelligence from decryptions of cables by Yardley and his team. Stimson inexplicably ordered the immediate separation of Yardley's team from the State Department. The funding was cut off and Yardley and his assistants were fired. As Yardley remembered it the Department of State defined policies:

"If the Department considered the code messages of foreign governments inviolate, then inviolate they must remain. It would be a usurpation of power on the part of the War Department if it engaged in activities against the policies of the State Department."[54]

Actually, the U.S. government had two other code-breaking intelligence organizations at this time, although constant interservice rivalry reduced the effectiveness of all three. ONI (Office of Naval Intelligence) which gathered intelligence and conducted some covert operations, had maintained a secret code-breaking section since 1924. This section had achieved success in decoding secret Japanese messages. Also, in 1929, as Yardley's operation was about to be shut down, the US Army appointed William Friedman as head of a new Army Intelligence organization: the Signals Intelligence Service. Friedman, who had worked for the Army's Code and Signal Section in WW I and as a civilian employee afterward was fully equal to Yardley in proficiency. He had known Yardley during the war, but the two didn't get on well. Friedman preferred to work anonymously, and specialized also in training young replacements. He continued working for the Army and later the National Security

Agency (NSA) until the 1970s.[55]

The United States was partially blinded, however, by Yardley's departure, and the only explanation Stimson (later Secretary of War during WW II under Roosevelt) ever gave was: "Gentlemen do not read each other's mail." This answer was no doubt a source of sardonic amusement to the Soviet and other hostile intelligence services. As one example, the efforts of the Navy Code and Signal Section, which attempted to decode 3,000 subpoenaed, coded Amtorg telegrams from 1930 onward, in an effort to learn more about the first major Soviet intelligence effort against the United States, between 1924 and 1930. Amtorg had deployed industrial and military spies across the United States. According to the U.S. Navy cryptanalysts, "The cipher used by the Amtorg is the most complicated and possesses the greatest secrecy within their (the Navy cryptanalysts) knowledge." Hamilton Fish later passed these same encrypted communications of Amtorg to the War Department, also without success. He later stated on the floor of the Congress: Not one expert-and they had from six months to a year-succeeded in decoding a single word of those cablegrams, although they assured me they could decode them."[56] Yardley, penniless and out of work in the great depression, wrote a memoir of his intelligence career which the US State Department subsequently attempted to suppress. With its publication in 1931, Yardley revealed that he had broken the secret Soviet codes during the 1919-1921 period, and had been forwarding this intelligence to the State and War Departments, and through an associate to the Justice Department, housing the Bureau of Investigation (FBI). He revealed a document which outlined Soviet intelligence methods in use in the 1920s for penetrating foreign embassies and intelligence networks and how in some cases spies were recruited by Cheka officers posing as agents of other nations, and those recruited did not know they were actually working for Russia. Following this Yardley revealed a significant clue to the possible fate of some of the American POWs and other foreign hostages remaining in the USSR in the 1919-21 period:

"The Soviet authorities in Moscow will be loud in their denial of the authenticity of this document, but they will recognize it; and having recognized it, will know that I must also possess other Soviet documents of a more sensational character. For instance, INSTRUCTIONS FOR THE MASSACRE OF FOREIGN NATIONALS, etc."

In a footnote to this revelation Yardley added:

"Soviet agents, please note. Yes, I once had copies of these documents, but I don't care to have my throat cut and do not plan to publish them. In fact they have been destroyed..."[57]

In the period after 1921 mass-executions by Soviet secret police and military remained commonplace, and foreign prisoners and hostages had disappeared. Their fate was indeed uncertain, without eyewitness confirmation, which if offered, would in turn would become classified information. It was also in 1921 when the Soviet security police began concentrating enemies of the Bolshevik regime into newly formed, permanent forced labor camps on Solovetsky Island, northeast of Archangel in the White Sea. Here, at the site of another old Christian monastery, the Communists confined captured Czarist and White Army officers and soldiers, "counterrevolutionaries," socialist and rightist enemies and foreign prisoners sentenced as spies. This early Bolshevik "corrective labor" camp established a system envisioned by Lenin and enormously expanded under Stalin after 1924, to turn the regime's enemies into useful state-slaves, whose labor production was turned into exports to obtain Western products requiring hard, convertible currency.

Although it is evident that some number of foreigners were executed by the Soviets for various reasons, it is also clear from existing livesighting reports of American prisoners in the National Archives, that some number were retained alive in captivity at least until the 1930 period. It is also very likely that in some hidden file in Washington, DC, or perhaps at National Security Agency (NSA) Headquarters in Fort Meade, Maryland, there are many more decoded messages concerning the ultimate fate of the American, British, French and other foreign nationals imprisoned in the USSR after 1921.

Despite such U.S. Government duplicity and secrecy, the memories about U.S. MIAs in Russia lingered among the American soldiers who had fought on the Archangel front. Under pressure from veterans groups during the Hoover Administration, in 1929 a combined VFW/U.S. Graves Registration expedition to north Russia attempted to identify and recover American "remains" of the 127 MIAs and presumed dead. The members of the Polar Bears, an association of the Veterans of the 1918-1919 Archangel Intervention force, based in Michigan, began to clamor for the return of their comrades remains from Russia. For years there had remained a nagging doubt in the minds of some of these men as to the fate of 127 of their fellow American soldiers who had been left behind in the Soviet Union. While some were known to have been killed and others were believed to be dead, 70 or more had been listed as missing in action or had simply disappeared. The uncertainty of their fate rankled many of the Polar Bears, who began to lobby for an expedition to find and return to America as many remains of these missing soldiers as possible. The State Legislature of Michigan appropriated $16,000. for this purpose and U.S. Senator Arthur H. Vandenberg succeeded in gaining a further Congressional appropriation of $80,000. on March 4, 1929 (later raised to $200,000).

The Soviet government of Josef Stalin, however, refused to

allow the veterans to enter Russia, on the grounds that the United States had not extended formal diplomatic recognition to his communist regime. At this point the Polar Bears began looking for further support in their quest.

The Veterans of Foreign Wars assumed a leadership role in this affair, after receiving an appeal by the Polar Bear veterans. VFW National Commander Eugene Carver and VFW National Legislative Chairman and National Service Bureau Director, Edwin S. Bettleheim, went through private channels to gain Stalin's permission for entry into Russia, in what was a humanitarian concern. The expedition was under the ostensible command of Bettleheim who went to Paris and there met 4 men of the US Army's Graves Registration Service, who were also VFW members. (This constituted an early form of "sheepdipping" military personnel to accomplish a sensitive mission.) In Moscow, Bettleheim and the four U.S. Army Graves Registration men met with a group of 5 members of the Polar Bears under the Association's president, Walter Dundon. Their expenses had been paid out of the Michigan appropriation. Soviet officials then directed the group to proceed to Archangel by the railroad.

The group split into different parties in Archangel Province, some searching battlefields of the railroad front while others searched the Dvina River-Vaga River areas. Bettleheim and the 4 U.S. Army Graves Registration men used a river steamer to cover the Dvina and Vaga river front, where most of the American MIAs had disappeared.[58] In addition the American searchers used horses and carts and also traveled on foot to reach the often remote sites where U.S. soldiers had disappeared in 1918 and 1919. After 10 years many of the remains were scattered and difficult to identify. While Bettleheim claimed to have made impressive discoveries of some of the remains of the 28 MIAs of Nijni Gora, the bodies of Americans, English, French and Russians were often mingled and identification by "dog tags" was sometimes impossible as the metal had rusted away or disappeared. Although a great effort was made for two months, the majority of remains collected consisted of a few bones and some rotten cloth. Members of the expedition recorded the American searchers were split by dissension and that the Russian peasants were sometimes uncooperative, or led the searchers to graves of other nationalities.[59]

The bones that were collected were disinfected, wrapped in cloth and deposited in coffins for shipment back to the United States. A total of 86 sets of remains were reported to have been recovered, of which 16 could not be identified and were recorded as "unknown," and 14 more reported as "accounted for" but not returned. After the 72 "known" and "unknown" sets of remains were brought to France, 11 were buried in Europe at the requests of family members. Upon reaching the United States, amidst great ceremony, three of the remains were sent to Arlington National Cemetery and some were returned to their families, while most of the remainder were shipped by train to Detroit. Here they were placed in a mausoleum, with further ceremony, at White Chapel Cemetery. The following Memorial Day, the 56 remaining coffins were buried and a great Polar Bear Monument was dedicated on the site, in their honor. Among those remains "identified" were those of William Martin and Lindsay Retherford, missing since January 1919.[60]

Officially only 41 Americans now remained unaccounted for in Russia (although the 16 unknowns and the 14 "accounted for" but not returned, raise additional questions as to this figure). After publicity over the Michigan event subsided, American prisoners of war in Russia who were thought to be dead, and now buried, toiled on in Stalin's labor camps, unknown and all but forgotten.[61]

## THE 1927-1929 LIVESIGHTINGS OF US POW'S IN RUSSIA

It wasn't until 1930 that some escaped prisoners from the Soviet gulags, bearing tales of Stalin's terror, reached Britain and Scandinavia as stowaways on ships and succeeded in gaining attention in the Western media. As more escaped and corroborated the revelations of the first arrivals, the stories caused a stir which led to a protest of Soviet slave-labor usage in England and America. As Dallin and Nikolaevsky noted:

"In the United States the first reports on forced-labor in Russia coincided with the arrival of Russian lumber and manganese and matches. The deep impression created by the news led not only to a vigorous debate in the press but also to government action. Section 1307 of the new Tariff Act of 1930...prescribed that all goods...mined, produced or manufactured...by convict of forced labor...shall not be entitled to entry at any of the Ports of United States" Later, at the insistence of Senator Burton K. Wheeler and others, the embargo of Soviet goods was dropped.[62]

The American Secretary of the Treasury, Andrew Mellon, who had recently negotiated an art purchase totaling millions of dollars with Soviet officials, assisted in blocking U.S. embargoes against Soviet products.[63] Other influential Americans favoring trade with Russia such as Averell Harriman and Armand Hammer, also used their influence to soften the American stance.

Among those prisoners who escaped from Stalin's Gulag and reached Britain to tell the Western World the truth about Soviet reality in 1930 was Alexander Grube. Born in Riga, Latvia in 1901, he had immigrated to the United States in 1920, and had first taken out American citizenship papers in New York in 1925. A sailor by occupation, like many Baltic immigrants, Grube had signed up at the port of New York on the American Export Line's S/S "Winona," and had sailed for Black Sea ports on that ship in February 1926. By his own later admission in a sworn Bureau of Investigation (FBI) affidavit, Alexander Grube had heard propaganda "that living conditions and employment were very desirable in Russia," and that, he did, on March 5, 1926, desert the S/S "Winona" while said ship was harbored in the port of Constantinople, Turkey, and did thereafter, on April 2, 1926, sign on as able seaman aboard the S/S "dekabrist," a Russian freighter bound for Russian ports."

Upon the arrival of his ship in the Siberian port of Vladivostok, Alexander Grube was immediately arrested by the GPU (later NKVD-KGB), and thrown into prison charged with being an American spy. For nine months of his imprisonment in the GPU prison in Vladivostok, as was usual for all prisoners of the secret police, he was tortured to extract a confession, often being beaten unconscious. The only evidence against him was the fact that he had deserted an American ship on which he was being paid $62.50 a month to join a Russian ship for $35.00 a month, and also that: "...upon being searched, certain papers were found on his person indicating prior service on his part on ships of the American Transport Line, which Russian officials believed to be part of United States forces." Grube does not say if he ever "confessed," but on December 22, 1926,"...he received notice from the Collegia of the GPU at Moscow that he had been sentenced to serve ten years at hard labor on Solovetz Island and thereafter five years in Siberia."

Thereafter, the immigrant American sailor Alexander Grube joined the endless convoys of prisoners being shipped from prison-island to prison-island of Stalin's gulags. He arrived at the notorious Lubyanka Prison in Moscow, on March 1, 1927. There, Grube later related he first heard of, and later: "...OCCASIONALLY SAW FOUR AMERICAN ARMY OFFICERS AND FIFTEEN UNITED STATES ARMY SOLDIERS WHO HAD, SINCE 1919, BEEN THERE IN PRISON."

While Alexander Grube was being confronted in the Lubyanka Prison with the stark reality of Soviet perfidy, the now-famous leftist American "journalist" Walter Duranty, who had free run in Russia and was a friend of Josef Stalin, was telling the American people a different story. 1927 New York Times articles by Duranty included such headings as: "Soviets Growing Spirit of Compromise" and "Dawning of Friendlier Days" (March 27, 1927). With a reporter like this in Moscow, it is no wonder then, that the truth of Stalin's concentration camps, and the U.S. soldiers within them, were hidden from the American people.

Soviet prisons were notorious for the rigid enforcement of silence between all prisoners. At this time, and for many decades after this prisoners were rarely, if ever allowed by guards to communicate with each other, and in fact were usually required to even avoid looking at other prisoners when met with in the

Lubianka's corridors. The commonly heard command according to Solzhenitsyn and other former prisoners, was a shrill: "EYES ON THE FLOOR!" According to surviving witnesses, infractions were usually punished by immediate, vicious beatings. Even so, Grube was able to learn the name of, or a corruption of the name of one of these American Army prisoners: Alfred LINDSAY. He was able to communicate with this one American because in 1927 LINDSAY was working as a cook in the prison kitchen of the Lubianka. (LINDSAY'S job as a cook would have enabled him to stay alive longer and possibly even assist some of the other American POWs in the Lubyanka.)

From the Moscow's Lubyanka prison in 1927, Alexander Grube was sent on another prisoner transport to Solovetz Island, just below the Arctic Circle in the White Sea between Archangel and Murmansk, where he was to serve his "tenner" as a slave for Stalin; a term which few survived.

At this time the Solovetsky Islands, also known as "Solovki," were the site of one of the first OGPU "Corrective Labor" camp systems. These camps were an outgrowth of the concentration camps that were located throughout Russia during and after the civil war, wherein prisoners from the White armies and other "anti-Soviet" elements had been confined. By this time Stalin was greatly expanding the "corrective labor" camps which Lenin had created. A former GPU official of the northern labor camps, Kiseliov-Gromov, stated that some 30,000 men were held in the Solovetski slave camps in 1928. The numbers of prisoners grew rapidly after this time with Stalin's forced-collectivizations and purges. By 1930 at least, other camps were located on Kond and Myag Islands in Onega Bay of the White Sea, in Karelia south of Murmansk, in the Archangel region and in Turkestan (Kazakh SSR). These camps were in addition to the GPU prisons existing in every major city in Russia. According to Kiseliov-Gromov the population of forced-laborers had grown to 662,257 by 1930. According to a sworn statement of another GPU official who escaped to Finland in 1930, "734,000 prisoners were employed under the OGPU in the autumn of 1929." This is to be compared with the Czarist record of 142,399 prisoners in 1916.[64] After 1930 the number of Gulag prisoners climbed rapidly into the millions. Future students who probe the secret history of the Gulag and question how the Soviets could have later hidden hundreds of thousands of foreign prisoners, would do well to observe the minuscule size of the Solovetskiy Archipelago within the vast geography of the USSR.

The Solovetsky Islands Cheka-GPU forced labor camps were on the site of earlier Czarist prisons dating back to the time of Ivan the Terrible. After centuries of use they were closed by the Czar in 1905, only to be reopened by Lenin in 1921 and 1922 for commanders, officers and other military prisoners from the White Russian armies. Others imprisoned there by the Bolsheviks included sailors of the Kronstadt Revolt of 1921 (who had not been executed), Orthodox Christian priests and bishops from all over Russia, czarist, White and other right-wing enemies of the communist regime and later Socialists and Anarchists of subsequent purges. At least one prisoner was known to be held only because he had worked for Herbert Hoover's American Relief Administration.[65]

The transformation of the Soviet penal system from an originally high-minded "corrective labor" experiment, as envisioned by Marxist revolutionaries out to remake their political enemies into useful bodies for the new workers state, into a mass forced-labor empire serving the Soviet elite, was first accomplished in the Solovetski Islands GPU slave camps. The forced-labor prisoners became permanently integrated into the state economy and soon had to meet ever-increasing quotas of production. On the forested Solovetz Islands and adjacent camps of the White Sea mainland, logging and sawmill lumber production, much of it for export to the West for hard currency, was the primary slave-labor industry. The labor of logging was performed with crude hand axes and saws, often in winter temperatures far below zero, by starving prisoners wrapped in rags. Gulag survivors report those who were thus intentionally worked to death in winter were stacked like cordwood in neat, frozen piles until the taiga's spring thaw permitted burial in hidden, mass graves. Other important slave labor included peat production, brickmaking, rock quarrying and the loading of ships, all with a further, planned attrition rate.

Food meant life, and was rarer in the Solovsky camps then in most of malnourished Russia. The main ration was a portion of thin "soup" each day, a pound or a little more of bread and a small amount of porridge. In the earliest period of forced-labor some British Army canned meat captured following the retreat of the Intervention forces from Archangel was occasionally added to the gruel. In later years no meat or vegetables were issued, resulting in widespread scurvy among the prisoners. The prisoners slept on plank beds so closely spaced on the floors that sometimes there was not room for people to lay on their backs. The below-zero cold in winter and filthy sanitary conditions in summer contributed to disease and a constantly high death rate. Mass graves of thousands of prisoners were scattered throughout the archipelago.

The camp administration of USLON, and the GPU guards, exercised the right to execute prisoners on their own authority and outside any judicial system. Others were sent to the notorious Sekirny mountain penal institution of the Solovetz camps, under a commandant named Antipov. Offending prisoners were beaten, hanged from trees or shot. Over a period of many years the Solovetsky

Archipelago became a vast burial ground for untold thousands of secretly murdered enemies of the Soviet state.

It was in the main Solovetsky Island prison, during the spring of 1927 and until his escape in August 1929, according to Alexander Grube's 1930 sworn affidavit, "...WHERE, AS FELLOW PRISONERS, HE MET MANY AMERICAN SOLDIERS AND CIVILIANS WHO WERE THERE IN PRISON, AMONG WHOM HE PARTICULARLY MENTIONS THE FOLLOWING: A MR. MARTIN OR MARTEN AND A MR. G. NEINAINKRUK (THIS LAST NAME BEING SPELLED AS NEARLY LIKE THE NAME AS IT SOUNDED TO THE AFFIANT), BOTH OF WHOM HE THINKS WERE AMERICAN ARMY OFFICERS WHO HAD BEEN SENT TO SOLOVETZ ISLAND FROM VLADIVOSTOK AND WHO WERE HERE WORKING AS LABORERS. THAT IN ADDITION, HE MET ONE ROY MOLNER WHOM, HE STATES, HAD BEEN A SERGEANT IN THE UNITED STATES ARMY, FORMERLY LOCATED IN ARCHANGEL, FROM WHICH CITY HE HAD BEEN SENT AS A PRISONER TO SOLOVETZ ISLAND. AFFIANT FURTHER STATES THAT THE TERMS OF IMPRISONMENT OF ALL THESE AMERICAN SOLDIERS ARE INDEFINITE. MOLNER IS 35 YEARS OF AGE."

What number of US prisoners would Mr. Grube mean by "MANY AMERICAN SOLDIERS AND CIVILIANS," after he had previously remembered the specific number of 19 U.S. military prisoners he'd seen in the Lubyanka? It would clearly seem to have been at least 25 or more, or he would certainly have used a more specific number when being interviewed by the FBI agent Francis X. O'Donnell in New York a eighteen months later. Were there also other American POWs of the Intervention alive in camps which Grube didn't see? If 40-50 American military POWs of 1918-19 were still alive in two Soviet prisons in the 1927-1930 period, how many had died in captivity since 1919? According to Gulag survivors, the death rates in the labor camps were often 50%-100% per year. Even if special care ensured that the Americans survived, this leads to obvious questions: Were the actual American casualties of the U.S. Intervention in Russia higher than the official figure? Or, were many more of the some 200 Americans officially reported as killed in action, died of wounds, missing in action or dead of "other causes" not related to disease, in fact captured alive?

Alexander Grube does not reveal how he escaped from Solovetz Island in August 1929. Other prisoners had stated escape was impossible and he may have wished to protect his route for others. Perhaps he infiltrated a work detail sent to load ships (a common practise) at the USLON Administration transit center at Kem. In any case it was at Kem on the White Sea that he was "SMUGGLED AS A STOWAWAY BY THE GERMAN CREW OF SAID SHIP (S/S "ERIC LARSON") AT THE PORT OF KEM ON THE WHITE SEA."

Alexander Grube, the Baltic-American immigrant who had lived through the worst of Stalin's early death camps, now had certain knowledge of a great secret. Upon reaching Liverpool, England, on board the German ship, he stowed away on another ship, bound for the U.S., so that he could reveal the awful fate of the missing American prisoners of the 1918-1919 Intervention. He returned to a United States of 1930 that was little prepared to believe what he had seen or perhaps care much, with the Great Depression underway. In Charleston, South Carolina he was discovered as a stowaway and informed he was to be deported. Before this could happen he fled overland to Baltimore where he succeeded in getting a job on another ship. After months of enforced travel and sickness in a hospital he finally made his way to New York.

While residing at the Seaman's Church Institute in New York, Grube sought out an anti-Communist publisher named Gregory G. Bernadsky who, with another man named Trevor, was to help him reveal his discovery in relative safety. Berna(r)dsky, as president of a firm called Research Publishing Corporation, was aware of the magnitude of Grube's revelations and wished to insure the information was immediately available to the US Government. The Bureau of Investigation (FBI) of the Justice Department was contacted and a Special Agent named F. X. O'Donnell was assigned to the case. J. Edgar Hoover, formerly the head of the Bureau's Intelligence Division, had taken over as Director of what would soon be renamed the FBI, in 1924. Hoover was a dedicated enemy of the Soviet communists and American "subversives" who supported them, and had imbued his agents with his distrust of any communist actions or motives.

Special Agent O'Donnell had Alexander Grube swear out an official affidavit, witnessed by himself for the FBI, Gregory G. Bernadsky and notary public Anette E. Fleser on the 25th of October 1930. Grube remembered several of the Americans names, two of which appear to have been deliberately masked as partial names, probably for protection from GPU retaliation. In addition, he identified an important Soviet GPU (KGB) agent named Karklin, operating in the U.S. under cover of the Russian Amtorg trading organization and reporting to a New York-based senior GPU officer, Gregori Grafpen. (Gregory Bernadsky, an active anti-communist, was later to be reported in the New York Times of April 4, 1931 as urging investigation of the Soviet espionage front, Amtorg, the Russian "trading firm" in New York which had been founded in 1924. He suspected that Soviet agents connected to Amtorg had murdered one A.I. Pogojeff.)[66]

Special Agent E.J. Connelly, in charge of the (FBI) New York office, considered the information on the American POWs of such importance that he immediately telephoned Director J. Edgar Hoover, personally, in Washington, D.C. He followed this up with a 25 October 1930, memo for Hoover marked "personal and confidential," and enclosed the sworn affidavits of Alexander Grube, witnessed by (FBI) Special Agent F.X. O'Donnell.

On October 31, 1930, Director J. Edgar Hoover sent the sworn affidavit of this eyewitness concerning the American POWs in Russia to the G-2 Intelligence Division of the War Department for action. The lag of several days may indicate that the Director brought the matter to the attention of President Herbert Hoover (who would not have been surprised, given his bitter personal experiences with the Communists).[67]

The Secretary of War under President Herbert Hoover in October 1930 was Patrick J. Hurley of Oklahoma, a prominent lawyer and former U.S. Army officer in World War I. Hurley was later to become an influential diplomat who, from 1942-45, served variously as General Marshall's personal representative in the Far East, as President Roosevelt's personal representative to Russia and other nations, and finally as U.S. Ambassador to China during the Second World War.

The State Department, then under Henry L. Stimson, also received a copy of the affidavit. While Stimson may have believed that, as he said during this period, "gentlemen do not read each others mail," the contents should have hardly surprised veteran diplomats such as William R. Castle or Robert Kelley, then at State, who knew something of actual Soviet methods. Some idea of the quantity of information supplied to the Department of State by the chief of the Justice Department's Bureau of Investigation, was revealed by Arthur Bliss Lane, later a U.S. Ambassador to Poland. Lane had been a Special Assistant (U-2) to Undersecretary of State Joseph Grew, in 1925, and remembered, "The volume of U-2's correspondence with (J. Edgar) Hoover of Justice is perhaps not realized. During the month of May I received 27 letters from him and sent 65 personal letters, excluding those letters transmitting Scotland Yard reports on revolutionary activities in the United Kingdom."[68]

Among the first to receive the information on U.S. prisoners of war in Moscow's Lubyanka prison and at Solovetsky, was the Intelligence Branch of the Military Intelligence Division (G-2) of the U.S. Army General Staff. One November 4th note from this branch stated: "...It is not at all a remarkable possibility...that some of the missing...may have landed in Bolshevik prisons." Colonel Robert Cherry Foy, the acting chief of G-2 passed the memoranda, affidavits and notes to the Adjutant General of the Army on the 4th, with the note: "...THIS PAPER WAS INFORMALLY REFERRED TO THE WORLD WAR RECORD DIVISION FOR A QUICK CHECK AMONG THE REPORTED CASUALTIES. NOTHING WAS FOUND."

Yet a crucial (now-declassified) November 8, 1930 War Department memorandum by R.J. Drown, the Chief Clerk charged with investigating some of the specific names reported in Alexander Grube's sworn affidavit, indicates something quite different:

"I HAVE LOOKED INTO THIS QUESTION AND FIND THAT AT LEAST ONE CASE HAS AN IMPORTANT BEARING ON IT, NAMELY THE CASE OF WILLIAM J. MARTIN, COMPANY A, 339TH INFANTRY... UNDER DATE OF FEBRUARY 3, 1919, A REPORT FROM ARCHANGEL SHOWED MARTIN MISSING IN ACTION. THIS WAS THE LAST INFORMATION RECEIVED FROM EUROPE. UNDER DATE OF MARCH 14, 1921, WE MADE A DETERMINATION SHOWING: 'WAS KILLED IN ACTION JANUARY 19, 1919.' THIS DETERMINATION WAS NO DOUBT PREDICATED ON THE UNEXPLAINED ABSENCE OF THE SOLDIER FOR ABOUT TWO YEARS..."

"I also find another case which may possibly be involved., it is that of Lindsay Retherford. LINDSAY Retherford was reported missing and a similar determination was made in his case. I showed to Colonel Parrott the Martin case and he is very much impressed by the thought that Martin may possibly be alive..."[69]

The subsequent War Department investigation merely affirmed the Army's 1921 "presumed finding of death" for the American soldiers who were very likely still alive in Josef Stalin's gulags. On November 12, 1930 the Adjutant General of the Army ruled on the case, affirming a precedent which would be followed for U.S. POW/MIAs through 3 succeeding wars:

"In the case of those reported as missing in action, an administrative determination had been placed on each of their records that they were killed in action on the date they were reported as missing. This action was taken in accordance with office precedent based on the opinion of the Judge Advocate General of the Army under date of April 12, 1923, (Dan Mathews, #2,649,999) that when the latest authentic report indicates soldier was in action and when sufficient time has elapsed so that report of him would have been received through various organizations and there is no information found as to his status, the only reasonable presumption is that he is dead."[70]

Since 86 of the missing Americans had later been "identified" by matching their names with fragmentary "remains" recovered in 1929, all of these 86 Americans were automatically removed from consideration as possibilities for being alive in Soviet prisons. Only the remaining 41 "unidentified" missing in action and presumed dead U.S. soldiers were permitted to be candidates for the names reported by the eyewitness.

Recovered human skeletal remains had been assigned to the names of the missing William Martin and Lindsay (Retherford) as their "remains." Accordingly, by War Department bureaucratic reasoning, they could not be alive in a Russian forced-labor camp, even though both had actually been declared dead in 1921 by a purely "administrative" process.

Congressman Hamilton Fish of New York, a heroic commander of black U.S. combat troops in WW I and a founder of the American Legion, was then chairing a Congressional Committee investigating Communist subversion in the United States. "Ham" Fish was fully informed of the eyewitness livesighting of American POWs in Russia, having received Alexander Grube's affidavit, and he had requested an explanation from the War Department on November 10th.

On November 14, 1930, Colonel Foy of U.S. Military Intelligence wrote to Congressman Hamilton Fish and established a pattern of official lying to Congress on the subject of U.S. MIAs that was to be repeated for decades to come, by degrading Grube's affidavit: "...In which he alleged that certain persons named therein are probably American officers and soldiers, YOU ARE INFORMED THAT A CAREFUL SEARCH OF THE RECORDS IN THE WAR DEPARTMENT FAILS TO IDENTIFY ANY OF THE PERSONS NAMED THEREIN AS HAVING SERVED IN THE ARMY."

Since the War Department controlled the lists of casualties (which in fact did show that several of the names matched missing Americans of 1919), Congressman Fish had no choice but to accept Colonel Foy's word. This appears to have ended Congressional involvement in the case. Declassified U.S. documents indicate that U.S. Army Intelligence took great interest in the report of Fish's Committee, which ultimately did not contain Grube's affidavit on the U.S. POWs of 1919, still in Russian prisons. (It is doubtful that Congressman Fish ever saw these classified documents prior to his death in January 1991.)

The day following Colonel Foy's inaccurate response to Congressman Fish, on November 15th, Director J. Edgar Hoover appears to have tried to keep the case alive by sending, under his own signature, further reports (#136872) of Special Agents Connelley and O'Donnell which were "RELATIVE TO ONE ALEXANDER GRUBE, WHO HAS FURNISHED INFORMATION CONCERNING RUSSIAN PRISON." These concerned an unsuccessful November 7th attempt by an Estonian agent (who may have been working for the GPU) to bribe the impoverished Alexander Grube into repudiating his statements about the American POWs in Russia. The War Department had already made it's decision however, and the case was officially closed. Although attempts were initiated by the U.S. Immigration Commissioner Uhl, to deport Alexander Grube from the United States, he was able to prove his prior residence in America and service in the Merchant Marine for three years. (Declassified U.S. documentary evidence indicates that an FBI agent and Mr. Gregory Bernardsky assisted Grube in gaining employment in the Sikorsky Aircraft Company plant, in Connecticut so that he could remain in the U.S.)[71]

First-hand livesightings from human intelligence sources of American officers and enlisted military POWs in Soviet prisons thus became known to the American government by, at latest, 1930, but appear to have been officially ignored, while Congress was deliberately misled by the military. One reason for this, and also a possible explanation for ignoring livesighting reports of later American POWs in Communist captivity, can be found in the original affidavit of the eyewitness. The escapee, Alexander Grube, who had learned the actual identities of some of the American POWs, warned in the strongest terms that the United States government asking for the soldiers return, by name, "WOULD RESULT IN THEIR IMMEDIATE EXECUTION." This episode proved to be a severe (if secret) lesson to the United States on the value of Soviet promises, and no doubt infuriated President Herbert Hoover.

Although author George Kennan was later to write that the U.S. soldiers who fought at Archangel "were mostly young Polish-American boys," the names on the list do not reflect this. Instead they appear to represent a wide cross-section of the population from the States of Michigan, Ohio, Illinois, Indiana, Kentucky and North Dakota. The recorded names, many of which are the names of America's first POW/MIAs to remain secretly held in Soviet captivity, are: John Agnew, Myron J. Asire, Arthur Dargan, Floyd Auslander, William Babinger, Clifford Bateman Ballard, Helmer Bloom, James Carter, Charles F. Chappel, Joshua Clark, Earl W. Collins, Arthur A. Frank, John Frucce, Marlie H. Hester, Elmer W. Hodge, Nicholas Jonker, Thomas H. Keefe, Bernard F. Kenny, Simon P. Kieffer, Thurman L. Kissick, Edward Kreizinger, John Kroll Jr., Alfred E. Lyttle, Joseph Marchlewski, Edward L. Martens, Albert E. Moore, August B. Peterson, Josef Remtowski, Herbert A. Schroeder (reported alive in Soviet captivity, 1921-22), Perry C. Scott, Wilbur R. Smith, Earl D. Sweet, Otto V. Taylor, Dausie W. Trammel, George E. Van Devanter, Charles J. Votja, Henry R. Weitzel, Walter J. Wolstead, John J. Westerhof. Others who must also be included are: William J. Martin, Lindsay Retherford, Sgt. Roy Molner, or a correlation thereof (aged about 35 in 1929), whose place of initial capture was confused by Grube with the accounts of other POWs from Siberia) and a US soldier possibly named Crook(e) or Clark; or some sailor or soldier missing from the Army Transport Ship CROOK" in Siberia ("Nei-nam-kruk," "My_name-Crook (?)") during the spring of 1920. In addition there were the 16 unidentified remains returned in 1929, who may have been remains of other nationalities and represent that many more potentially living prisoners.[72]

The places of internment for these American soldiers were reported as Seletskoe, Vistavka, Ust Padenga, Shenkurst, Archangel, Yakolevskaya, Kitsa, Kadish, Maximovskaya and Bolhieczerska. Nearly all of these places were actually the battlefields where these men were last seen alive. The actual place of final rest for many of these men was the ever-smoking crematorium of the Lubyanka Prison or the frozen taiga of the Solovetsky Archipelago.[73]

The hideous conditions of the "Solovki" Islands forced-labor logging camps gradually became known through the crews of timber ships, released prisoners and a few who later escaped the White Sea area to the outside world. Mounting criticism of Soviet methods from "abroad" stung the Communists, who responded typically with a form of controlled madness which was later to become commonplace in the gulags. To counter what was becoming public knowledge a propaganda campaign was orchestrated around a visit to Solovki by the celebrated Soviet writer, Maxim Gorky (for whom the ancient Russian city of Nizhnii Novgorod was later renamed), from the 20th to the 22nd of June 1929. Gorky had been in an out of favor with the Soviet regime since his use as a Russian spokesman calling for American aid in 1921. He had been permitted to live abroad and return to the Soviet Union when he desired. In 1928 he was living on the Island of Capri but was urged to return to the USSR by many of his friends and by official Soviet organizations. This was part of Stalin's campaign to woo back expatriate Russians who were viewed as pliable and essentially supportive of the Communists. Gorky returned to Moscow in 1928 and was given a luxurious house in Moscow and two country dachas, including one in the sunny Crimea. Stalin allowed development of a cult around Gorky, who not only had a city renamed in his honor but also schools, factories and even "corrective labor" camps were named after him.[74] As part of his willing campaign of supporting Stalin, Gorky had visited the Solovetsky Archipelago forced labor camps. Alexander Grube and the abandoned American Army prisoners of the 1919 U.S. intervention force were still there.

The USLON and GPU administrators ordered the prisoner-slaves to thoroughly clean the barracks, remove all punishment signs and regulations and furnish false reading-rooms called Krasnyi Ugolok, meaning "red corners." The Solovki prisoners were paraded before the novelist Gorky and ordered to sing a song, the words of which have fortunately been recorded for posterity:
"Deported as we are for our deeds,
We still enjoy many a right.
We publish newspapers which everybody reads,
We stage performances-a lovely sight!
We write, and our songs we sing.
Abroad they've never dreamed of such a thing!"[75]

Shortly after Maxim Gorky left, conditions at Solovki returned to their former horrifying state. And before long, Gorky was himself to fall under Stalin's suspicion and was to die under questionable circumstances. Reports that he was actually murdered by Stalin have persisted to the present time.[76] Gorky went on to write propaganda for Stalin's brutal regime until his death in 1936, under suspicious circumstances. Gorky's later-published letter, "To the Women Shockworkers at the building site of the Moscow-Volga Canal," illustrates his complicity in the deaths of thousands of forced-laborers whom he knew were being worked and starved to death in "corrective labor" camps on the grandiose projects envisioned by Stalin and his lieutenants:

"Your labor once again demonstrates to the world what a healthy effect work can have upon people, work that has been given meaning by the great truth of Bolshevism, and it demonstrates how splendidly the Lenin-Stalin cause has done in organizing women."[77]

By deceiving the literate part of the populace of the USSR and much of the gullible Western media, Gorky and others of his ilk helped anaesthetize the Soviet masses and much of the world's population to the monstrous crimes against humanity that were being carried out by Stalin and his GPU Commissar, Henrich Yagoda. But the bloody 1930-31 collectivization campaign in the Ukraine and elsewhere shocked Gorky. As Alexander Orlov, a former Soviet diplomat and counterintelligence official, wrote later: "...An announcement appeared in the Soviet press that forty-eight men, who were accused of causing a famine by their wrecking activities, had been executed. When Gorky read about it he became hysterical and in his conversation with Yagoda reproached the government for shooting innocent people with the intention of shifting the responsibility for the famine from itself to those unfortunate men."[78] Yet after this time Gorky continued to produce utterly false propaganda in support of Stalin's regime.

Alexander Grube, an extraordinary man, escaped from Solovki in August 1929, only weeks after Gorky's visit, and at the very time that the U.S. VFW/Graves Registration Expedition under Bettleheim was in the White Sea area of north Russia to search for the missing American soldiers. It appears likely that given the extended time the VFW expedition spent in the Archangel-White Sea area, less than 150 miles from the Solovetski Archipelago, the American POWs and other foreign prisoners confined there would have heard of the expeditions presence through the prisoner and GPU convoy guard grapevine. Not long after Grube's escape, in the fall of 1929, a plotted revolt of prisoners on Solovetsky Island was uncovered, either unwittingly or by a traitor. The plan was to disarm the guards and take all the weapons and the ships belonging to the USLON Administration and escape on them to the West. According to Nikoov-Smorodin, in the first reprisals for the plot, 53 prisoners were executed by the GPU, and later 140 more were shot.[79]

It is a melancholy possibility that some of the U.S. Army

prisoners of war whom Alexander Grube had met, imbued with a lifelong love of the American liberty they had known in their youth, took part in this failed plot and lost their lives. In any case, after the escape of Alexander Grube and of other prisoners who may have seen American POWs in Russia in 1929 or 1930, the Americans lives would have already been in jeopardy. Alexander Grube points this out himself in his later sworn statement warning of possible executions of Americans, and added:

"...CONDITIONS AT SOLOVETZ ISLAND ARE SUCH AS TO SOONER OR LATER GUARANTEE THE DEATH OF ANYONE THERE INCARCERATED, WHICH ACCOMPLISHES THE SAME RESULT AS THAT OBTAINED BY A FIRING SQUAD WITHOUT THE DISADVANTAGE WHICH WOULD ACCRUE TO PUBLICITY CONCERNING THE LATTER METHOD."

According to survivors, the winter following these events, of 1929-1930, was the worst in the history of the Solovetsky forced-labor camps. 20,000 more deportees, mostly peasants who were victims of Stalin's forced-collectivizations in Russia and Ukraine, were sent to the Solovetsky Archipelago, and a great typhus epidemic occurred from which many thousands of prisoners died.

Stalin's ruthless and bloody collectivization of the peasants through the winter and spring of 1929-30 had taken the form of mass terror against the common people. By the spring of 1930 some fourteen million Russian and Ukrainian peasants had been herded onto heavily policed collective farms, or shipped to forced labor camps by the GPU and the Red Army. Villages which resisted were sometimes surrounded by Red Army units and Chekist secret police and massacred as an example to others.

Communist officials made their political strongholds in the much-heralded "Machine Tractor Stations" (MTS's) serving the new collective and state farms. Here orders were received from Moscow and issued as directives to the terrorized peasants. Here also, the records were kept, and dossiers built up on those peasants and villagers perceived as "Kulaks," "class enemies" or "counterrevolutionaries."

Those constantly being sent to the forced labor camps of the White Sea or in Siberia replaced the hundreds of thousands deported before them, who were continually dying. George Kitchin, a citizen of Finland whose mother was British, and who had been falsely convicted of spying for the British and the Finns in 1929, was being shipped in a railway car to the White Sea area at this time:

"Our car held sixty-four prisoners. Some of them were from our Lefortovsky prison and some from other Moscow prisons...'Wait till we arrive, then you will see,' came a voice from an upper shelf. It belonged to our taciturn Lefortovsky bandit, whose cold grey eyes were staring at us significantly out of his pockmarked face. The meaning intonation of his speech disturbed us, but even he would not tell us anything definite except that he had already finished a three-year term in the Solovetski camp, had escaped from his transport into exile, had returned to Moscow, and had there again been caught in a 'messy' affair. He spoke of the Solovetski camp with hatred and loathing:

"'They beat you there like dogs, the amount of work exacted is beyond all endurance, and the prisoners die like flies,' he said. 'Prisoners live in tents on the swampy ground, the food is nothing but slops, they mock at you and beat you at every occasion. Just wait and see, it will be the same here.'"

"...Finally the train slowed down and stopped. The door was instantly opened and a command resounded: 'Get Out!' All sixty-odd passengers scrambled and hurried to the exit, prompted by the cries and commands outside. Both inside the car and out there was a loud hubbub of cries, urgings, protests and cursings. Old Timofeyich was ahead of me in the passage-way. At the door he involuntarily stopped. Right under him in the snow sprawled Pevny, who had just jumped out of the car...'Jump, you old fogey,' shouted the convoy-soldier.

"'But excuse me, can't you see there is a man down there, how can I jump right on top of him...'

"His tirade was cut short by a curse and a strong blow in the neck which threw him out of the car and sent him sprawling right on top of Pevny...Not waiting for a further invitation I jumped out...

People continued to pour out of the cars. They joined the crowd and were irresistibly drawn on with it. Here and there men fell under their heavy loads on the uneven path. Butts of rifles urged them up and in despair they left half their belongings in the snow and trudged on without them. The crowd was milling about much as a herd of sheep does...'Go ahead, go ahead, don't stop or we'll shoot,' constantly filled the air...'Two in a row, two in a row,' we were ordered. This command was difficult to fulfill as so many of the men stopped and fell from exhaustion. Close to me I saw Granovsky's face, convulsed with rage. He was pushed forward by a soldier who poked a gun in his side. 'Move on, you son of a bitch,' he bawled at him..."

After arrival at the guarded camp the prisoners were forced to wait interminably in freezing temperatures:

"...We continued 'cooling off' in the yard. After several minutes Lyskin came out of the office with two fellows in army greatcoats and felt boots. A command rang out:

'Divide! Criminals right, others left!'

"The crowd separated into two groups. There were about two hundred criminals. Our group numbered some three hundred.

'Criminals, follow me," commanded the first greatcoat. The criminals filed by, casting unfriendly glances at us. The unexpected prospect of finding themselves in a tent in midwinter did not please them, but no protest was heard. Even a tent was preferable to the open air. Our group was moved in the opposite direction. Turning the corner, we entered a large yard bounded on one side by a long wooden barracks with several entrances. On the other side of the yard were several smaller buildings. We...were led to a small hut. A sentry stood at the door, gun in hand. 'Hurry up, enter quickly'..."

"...People were still coming in, crowding, pushing, arranging themselves...Swearing and abuse was heard...The rooms were jammed chock full. With people sitting and standing everywhere, there was not a square foot of empty space...

"Suddenly the door opened with a bang. Several men, their greatcoats covered with snow and with lanterns in their hands, roughly pushed their way through the crowd to the doorway of the second room. 'Silence! Everybody listen! bawled a stentorian voice. The head man jumped on top of somebody's bundle, placed his lantern on a hook and looked us over. Small green eyes looked at us from under his low-fitting helmet. A strong, coarse, square jaw protruded above the upturned collar of his greatcoat.

"'Silence! Stop that noise!...You forget where you are...You are addressed by the warden. Note this and remember once and for all. The penal camp is quite different from what you know as prison. Here you have military discipline and martial law. The smallest disobedience is punished severely...

"'You have arrived in the Northern Penal Camps of the OGPU. There is no district attorney here, you cannot complain to anybody. Therefore I advise you to work conscientiously and not to make any row. There can be no counterrevolution here. For attempting counterrevolution we line people up against the wall and shoot them. For rows, thieving, insubordination—also to the wall. I advise you to realize this and remember it. Not auntie's house party, but a penal camp. Forget all your intelligentsia's grievances and other tricks, otherwise we shall bend and break you, you 'intelligentsia'. More than one of you has already departed for better worlds. For refusal to work-the dungeon, and for a second offense-shooting. Understand?'

"'Beginning tomorrow,' continued the warden, 'you will start work. Up at five, drink your tea, receive your porridge and out with you into the yard. Your squad commander will be this comrade-Grigoriantz...'

George Kitchin later recalled the beginning of his forced labor routine:

"...The thirty rows from the front and the rear, surrounded by the convoy, went out of the camp gates...A large group was sent to the river bank to free a large float of timber from the ice...My lot fell with men shoveling snow in the yard. I soon learned from those who had arrived before us that we were not to remain long at the transfer point. From here we were to be dispatched to our permanent work in one of the camp divisions. Almost daily fresh bands of prisoners arrived in Kotlas from all parts of Russia. The supply was not diminishing in spite of the fact that prisoners in the Northern Penal Camps already numbered over forty thousand. The Kotlas Transfer Station of the Northern Penal Camps of the OGPU was located five kilometers from the town of Kotlas, on the steep bank of the northern Dvina River at its junction with the Sukhona River...The Transfer Station itself occupied some 15 acres and was surrounded by barbed wire. The long barracks were located near the center and served to house transient prisoners...Besides the barracks the buildings comprised storehouses, kitchen, hospital, dungeon, office and the bathhouse..."[80]

The United States Government knew the true conditions in Soviet Russia, although no formal diplomatic relations existed between the two countries, but the most sensitive information involving Americans remained classified. At the Department of State in Washington, a Harvard graduate who had specialized in the study of Russian foreign relations, named Robert Kelley, became the chief of the Russian Division. Kelley was totally opposed to U.S. diplomatic recognition of the Soviet government, and he brought into the Department an investigator named Ray Murphy, who was then trained to become an expert on Soviet subversion throughout the world. A constant liaison was maintained with J. Edgar Hoover's Bureau of Investigation at the Justice Department. Kelley and Murphy built up important files on the Soviets from diplomatic and military cables, police reports, eyewitness accounts and published matter relating to Russia. The conditions in the Soviet mass forced-labor camps were well known within the Department of State. Another Foreign Service Russian expert at this time was Loy Henderson, who had served as a member of a Red Cross relief mission to the Baltic States and western Russia in 1919 and 1920, before entering the Foreign Service in 1922. Assigned to the Eastern European (EE) Division, Henderson was given the job of tracing the connections between the American Communist Party and the Communist International. Murphy concentrated on the domestic connections while Henderson specialized in international Communist ties.[81]

Unfortunately much of this information was not made known to the public at the time by the major news media in America, which often pretended an uncertainty about the mass use of slave labor by Stalin and his henchmen. A Time Magazine article of December 22, 1930, reported on the rumors of slave labor in the USSR, by reviewing the 10,000-mile journey of American journalist Hubert Knickerbocker: "...Yes, some Soviet timber is cut by 'forced labor' but of a peculiar kind. Diarist Knickerbocker reported that these cutters appear to receive the same wages as other Soviet workmen. They are forced not to chop wood but to live in certain forest regions where such labor is the only sort in demand. In a word, the 'forced laborers' are former KULAKS (rich farmers) dispossessed of their land by the Soviet program of 'collectivizing farms.'" Knickerbocker visited many of Stalin's major projects, including Azbest in the Urals, scene of Armand Hammer's early venture in mining, where 13,000 workers labored under a Princeton mining engineer from the U.S.; New Nizhnii Novgorod, where Soviet Ford vehicles were built under American guidance; the Dneiprostroy hydroelectric dam construction, guided by another American, Hugh L. Cooper; Magnitogorsk, a new steelmaking center called the "largest construction camp in the world," and many other areas. But the vast human cost in lives lost attaining Stalin's goals was largely downplayed or ignored.

Two months later, in a cover story entitled "Gay-pay-oo,"[82] which drew from sources such as Walter Duranty in Moscow, Time reported on the Cheka's successor organization, the GPU, then under control of Viacheslav Rudolphovitch Menzhinsky, who, like his "late, great predecessor," Dzerzhinsky, Time reported, was a Pole. In a deceptively vague and light-hearted report, the most influential news magazine in America painted a false picture of the human characteristics of those GPU operatives then engaged in the torture and mass-murder of their fellow countrymen, and any foreigners unfortunate enough to fall into their control:

"...The second ranking official of the G.P.U. is Comrade Yagoda. He is supposed to be 'Stalin's man'...and also to direct a sort of private espionage service in the sole interest of the Dictator...Unlike the celibate and sour Menshinsky, Yagoda is married, happily it is said. About one quarter of the G.P.U. staff in Moscow are women. They are, on the whole, more cheerful then the men...Horrors laid at the door of the G.P.U. run the whole gamut from individual rape and extortion to general massacre of rebellious villagers. ONE CAN BELIEVE LITTLE OR MUCH. AS IN THE CASE OF 'GERMAN ATROCITIES' AND THE 'THIRD DEGREE' ADMINISTERED BY U.S. POLICEMEN...AT THE RECENT 'RADIO TRIAL' IN MOSCOW, PROFESSOR LEONID RAMZIN AND THE OTHER 'COUNTER REVOLUTIONARIES' BORE NO MARKS OF TORTURE WHATEVER AND WERE CERTAINLY IN POSSESSION OF BOTH HANDS. THE POWER OF THE G.P.U. LIES LESS IN HORROR THAN IN THE INFINITE RAMIFICATIONS OF ITS NET OF SPIES...If President Hoover knew as much about every U.S. citizen as Dictator Stalin knows...about any Russian, the political future would hold for him few mysteries—for Knowledge is Power."[83] Thus, the true nature of Stalin's terror regime, which was then in the process of killing millions of Soviet people by either execution or starvation-slavery, was hidden from the American people by those whose journalistic duty it was to fully expose it.

Lenin's brutal successor, the ruthless dictator Josef Stalin, had published his "Dizziness From Success" article in Pravda, on

March 2, 1930, ludicrously cautioning party operatives against the use of excessive force in collectivizing millions of Russian and Ukrainian peasants, which he himself had ordered the Party and the GPU to carry out.

One witness to the fate of these formerly-free small farmers was Anatoli Granovsky, then a schoolboy who eventually was to become an NKVD agent. In April of 1930 he joined his father, a Communist official who had been appointed chief of construction and development of the great Berezniakovsky chemical-manufacturing combine. Forced labor was used extensively in the construction, as it also was at Magnitogorsk and all other major Soviet projects which were then being praised in influential American publications such as Time Magazine. With his brother Valentin, "Tolya" Granovsky got his first look at the GPU's slaves shortly after he arrived:

"...Across a stretch of open ground and about eighty yards from where we stood was the stone face of a quarry, at the foot of which lay great jagged chunks of stone that had been blasted out. All around these boulders were gangs of people working, breaking and hauling the stone. They were not ordinary laborers. None was fully clothed and many were almost entirely naked. There were almost as many women as men. Most of them were bent and thin and moved with a grotesque sluggishness like people on the brink of death. It was gruesome and oddly fascinating. These were Ulonovsti. Standing among them and around them in strategic positions were uniformed men of the OGPU armed with rifles and pistols.

"What we were witnessing was a detachment of slave laborers at work. Little thought was given to the nourishment of the Ulonovtsi, and their shelter was worse than miserable. It did not matter if they died off like flies, for there were millions of them to draw from. The were petty landowners and their families (kulaks), erstwhile employers, political prisoners of all sorts, convicted enemies of the people. Many thousands worked on the Berezniakovsky project under control of the ULON (Section of Special Purpose Camps).

"While we watched we heard a high-pitched wailing noise and saw that it issued from a ragged bundle at the foot of a dead tree. At the sound a young woman dropped her wheelbarrow and started running toward the tree. I realized it was her child that was crying. There was something in the air. Suddenly fascinated, I watched her hesitate as a guard called loudly for her to take up her wheelbarrow. But she ignored the guard and ran on. The distance was not above twenty yards and everything happened in the time it took her to run that distance.

'Come back, you bitch!' the guard called out angrily.

"But the woman, with a deaf, desperate stubbornness, ran on and bared her breast to suckle the child. She reached the tree and dropped, kneeling, one white breast round and ready for the hungry child. The guard raised his rifle and shot her, twice. She fell without a sound. No one went to her, no one looked. Even the guard had already turned his back. And the child continued its nagging wailing. There was a sound beside me and I turned to see Valentin very white in the face and about to vomit. I took him by the arm and we started to walk home in silence. Finally he spoke.

'How awful, Tolya,' he said. 'It made me feel sick.'

'You are a baby Valentin,' I answered.

Anatoli Granovsky, destined for arrest and imprisonment and then many years of employment by the NKVD, was learning quickly. Granovsky also remembered a small prisoner revolt at Berezniakovsky:

"...The Ulonovtsi were confined in a cluster of desperately overcrowded shacks surrounded by barbed wire in the taiga. The camp was some four miles from where we lived and the slaves were marched out to their work early in the morning and back at night in long, slow columns, the guards marching alongside them with guns at the ready and trained dogs on leashes. In the camp itself, at intervals along the high barbed-wire fence, there were observation towers with searchlights and machine guns mounted on them. Every reasonable precaution was taken to ensure that no break-outs should occur. Nevertheless, one night something did happen.

"A small detachment of prisoners was being marched back to camp exceptionally late one night in the charge of three guards with one dog. The guard with the dog...tripped over the leash and fell to the ground, dropping his rifle. In a flash, almost as if he had acted with animal instinct, the prisoner marching closest to the guard broke ranks, kicked the sprawling guard in the face and snatched his rifle. The dog attacked him but the prisoner clubbed it to death with furious blows of the rifle butt. By the time the second guard had realized something was wrong and run round the squad to see, the prisoner had the rifle ready and shot him. But the third and last guard, approaching from a different direction, was able to shoot the prisoner in the back.

"Control over the prisoners was lost, but most of them were too stunned to do anything but stand still and stare. Not all, however. One man flung himself from behind on the one remaining guard while another picked up one of the abandoned rifles and removed the guard's weapons at gun point. They then clubbed the guard to death, and they were free-at least for a while...

"Of the forty odd men and women remaining in the detachment, only seven men and two women decided to avail themselves of the situation and escape. The rest were so beaten already, so degraded and so hungry that they marched on alone to the camp.

"Of the nine who ran off two, a man and a woman, were caught the next day in Zirianka village, across the river from the camp. The thirty-one who returned to camp were punished for having allowed their comrades to escape without hindrance. The two who were caught were executed. And the other seven? I don't know. I was told many times, and believed, that they would assuredly be caught. But I never actually heard that they had been. It is possible, just very remotely possible, that they were not."[84]

Despite the secretly-known existence of the American military and civilian prisoners in Russian labor camps such as these, the newly-elected President, Franklin D. Roosevelt, established diplomatic relations with the Soviet Union in 1933.

Hyperinflation in Germany in the 1920s, which resulted from Allied insistence on war reparations, was followed by a world-wide depression, that began in America with the stock market crash of October 1929. Herbert Hoover's reelection was doomed by the resulting impoverishment and upheaval of many millions of Americans, while the President insisted that soon all would be well. As the depression deepened, shanty towns called "Hoovervilles" arose in many parts of the United States. In 1932 a mostly peaceful demonstration by veterans of World War I, who were demanding promised war service benefits all at once because of their financial destitution, were attacked by regular troops under General Douglas MacArthur and by Washington, D.C. police, and driven from the Capitol. This unnecessarily brutal action was due to false intelligence supplied by FBI Director J. Edgar Hoover, that there were many dangerous Communists among the veterans in the "Bonus Army," when in fact there were very few, and the war veterans were well-policed by their own internal organization. The Secretary of War, Patrick Hurley, Army Chief of Staff General MacArthur and J. Edgar Hoover of the FBI all maintained it was necessary to crush the Red menace represented by the peaceful marchers, most of whom had in reality defended the nation under dangerous wartime conditions.[85] The resulting violence by bayonet-wielding regular troops and the shooting deaths of several Bonus marchers at the hands of the Washington D.C. Metropolitan Police, further eroded what popular support remained for Herbert Hoover's reelection that year.

Franklin Delano Roosevelt, a liberal Democrat who was a younger cousin of the former President, was elected in the fall of 1932, and he brought into office with him a coterie of left-tending Democrats and Socialist sympathizers who came to be known as the "New Dealers," after the name of the Roosevelt economic recovery program. These included men like Harry Hopkins, a left-leaning, former social worker in the slums of New York who was a protege of First Lady Eleanor Roosevelt, and who would ultimately become the President's closest advisor and confidante. Others like Henry Wallace, Secretary of Labor and later Vice President, Henry

Morgenthau Jr., a neighbor of Roosevelt's who became Secretary of the Treasury from 1934-45, and a Harvard professor and future Supreme Court Justice, Felix Franfurter, both of whom represented a powerful labor and liberal Jewish constituency that had assisted Roosevelt in his climb to the White House, had great influence on Roosevelt's thinking. Morgenthau initially served as head of the Farm Credit Administration and pushed the idea of seeking new markets for American grain and farm equipment in Soviet Russia. Other Roosevelt friends included Patrick Hurley, former Secretary of War under Hoover and William Donovan, a New York lawyer who had become a hero as commander of the 69th New York Regiment in France, during World War I. Averell Harriman was one of the first American multi-millionaires to support Roosevelt's candidacy this was to ultimately result in his appointment to high posts in the Administration and his key role as U.S. Ambassador in Moscow during World War II. Other wealthy supporters of Roosevelt in the beginning included Joseph Davies and Edward Settinnius, who would also play important roles in American-Soviet relations during the approaching Second World War.[86] All these men were to have a powerful impact on American-Soviet relations at a critical juncture in 1945. Their foreign policy ideas were instinctively liberal and reactive to the Hoover Administration's, and they did not yet have access to the great secrets, such as the American war prisoners of 1919 secretly held by Soviet Russia.

A young Philadelphian named Anthony J. Drexel Biddle Jr., of a family line descended from the first defenders of American independence, also contributed heavily to the Democratic Party, was an early supporter of Roosevelt, and campaigned for him. A St. Paul's boy who did not go on to college, Biddle had been affected by the patriotism of his legendary father, Anthony Drexel Biddle, who had mobilized thousands of young men for service in WW I through his "Athletic Christian" movement. The younger Anthony Biddle had fought in the Army under Pershing in France, and reached the rank of captain at the age of 22, in 1918. After the war he had entered the business world and married Mary L. Duke, heiress to the enormous Duke tobacco fortune (whom he later divorced). After being introduced to the new Secretary of State, Cordell Hull, Biddle was to receive a diplomatic appointment as his reward that would ultimately lead to his occupying a unique posting at the outbreak of the Second World War, as U.S. Ambassador to Poland.[87]

For his Secretary of State Roosevelt had chosen Cordell Hull, a Tennessean whose appointment appeased the more conservative southern wing of the Democratic Party. Hull adopted as his protege in the Department of State, a young working class foreign service officer, James Clement Dunn. Roosevelt acted as his own Secretary of State in foreign affairs, and Hull was satisfied to concentrate on his own interest, which was trade treaties. Roosevelt was a liberal Democrat who had surrounded himself with advisers on Russia who had, in some cases, been sympathetic to the Soviet regime since it's inception. One of those who spent the most time closeted with FDR was Walter Duranty of the New York Times, whose biased reporting on Stalin's rule in Russia had earned him an ill-deserved Pulitzer Prize. Another influence on Roosevelt about Russia at the time was William C. Bullitt, who was appointed by that President to be the first U.S. ambassador to Stalin's government later that year.

More importantly, however, Roosevelt wanted to send an unmistakable signal to a resurgent Germany in Europe and to an expansionist Japan in the Pacific.[88] By recognizing the Soviet government despite bonafide and longstanding U.S. grievances, some of which, like the issue of American prisoners in Russia, remained secret, Roosevelt intended to keep these potential aggressors uncertain about American intentions. As a former Assistant Secretary of the Navy, Roosevelt viewed America as a maritime power facing two oceans, while he saw Russia as a land-based power facing both Germany and Japan, and he intended to implement his new policy whether or not the State Department cooperated. Roosevelt initiated discussions with Soviet officials about American diplomatic recognition through his friends Henry Morgenthau and William Bullitt, thus bypassing the Department of State.[89] Despite frantic efforts by the experienced Russian experts, led by Robert Kelley and supported by Undersecretary William Phillips, to forestall Roosevelt's extending American diplomatic recognition to Stalin's government in Russia, the new President decided to do so on October 23, 1933, and asked the Soviets to send a representative to the United States. Stalin sent Maxim Litvinov, and subsequent negotiations over Russian debts and claims were largely staged to give the public an impression that such matters were being settled, while American negotiators suspected, correctly as it turned out, that Soviet compliance would be more difficult to obtain.[90] A young foreign service Russian expert of the time, George Kennan, was to write later:

"F.D.R. did insist, however, that Litvinov sign a number of other 'assurances" dealing with subjects on which, it was assumed, Americans might have particular anxiety in connection with the resumption of relations: propaganda, the legal protection of Americans in Russia, the right of such Americans to have religious liberty on Russian soil, and the right to seek information on economic conditions in Russia. These assurances were very curious documents. No one who knew anything about the Soviet Union could have imagined for a moment that they would restrain or modify the established behavior of the Soviet government in any way. Two of them—those dealing with economic espionage and with the legal protection of Americans in Russia—were ones the Soviet government had already signed in dealing with the Germans. It was, as it happened, I myself who had drawn the attention of the United States Government to those clauses—as examples of valueless undertakings on the part of the Soviet government which had failed to inhibit in any way that government's behavior in practise."[91]

William Bullitt presented his credentials to the Soviet government in December 1933 and then returned to collect his staff. The foreign service officers chosen to serve in the new embassy to be established in Moscow included Loy Henderson, Elbridge Durbrow, George Kennan and Charles "Chip" Bohlen. Henderson, with his Soviet experience that dated back to the intervention period, would serve as Third Secretary of the Embassy. George Kennan, a shy and introspective Princeton graduate, whose namesake uncle had been a 19th century American expert on czarist Russia and Siberia, had joined the foreign service in 1927, and became the first graduate of a training program initiated by the "Russian," Eastern European (E.E.) Division chief, Robert Kelley. During the late 1920s, while serving with the U.S. missions at Tallin and Riga in the Baltic republics, Kennan had been immersed in the culture and intrigue of the exiled White Russian community, which was a constant target of Soviet GPU agents. Bohlen had entered the foreign service in 1929 after graduating from St. Paul's and Harvard, where he developed a strong interest in Soviet Russia. Bohlen was another of Robert Kelley's newly trained Russian specialists, who served first in Prague and then studied the Russian language in Paris with a friend from Harvard named Edward Page who also became a translator. Durbrow had joined the foreign service in 1930 and served in central European embassies at Warsaw and Bucharest before going to Moscow. A young graduate of West Point named Charles Thayer, who, like his friend and future brother-in-law, Bohlen, was from Philadelphia and had attended St. Paul's, decided to resign his military commission to study Russian. In his eagerness to join the foreign service, Thayer traveled to Moscow ahead of the new mission, to be there when they arrived and thus secure employment. Bullitt would be well-served when the American Embassy in Moscow was opened the next year.[92]

Meanwhile, Corporal Herbert Schroeder, the red-haired printer and World War I American soldier of the 339th U.S. Infantry, who had been missing in action in north Russia since September 20, 1918, and who had been reported alive in a Moscow prison in 1921, remained on a "classified" 1931 War Department list of 41 missing and presumed dead U.S. soldiers whose remains had never been found, officially or otherwise.

Among the most important provisions desired by President Roosevelt, and agreed to by the Soviets, was the pro-forma recognition of the rights of any Americans inside the USSR. In an exchange of letters of November 16, 1933, between President Franklin D. Roosevelt and Maxim N. Litvinov, the Soviet Commissar for Foreign Affairs, provisions were included in which each government agreed to adopt measures to inform representatives of the other government as soon as possible, and in any case within seven days, whenever a national of the other country is arrested. The Soviet government violated this agreement from the beginning by continuing to hold American prisoners of war from the 1918 intervention, and repeatedly violated it after three more wars to come.[93] However, in response to a citizen's query to the Department of State about U.S. military POWs of 1919 remaining in Soviet control, Robert Kelley, recognized as the foremost Soviet expert in the Division of Eastern European Affairs, revealed the post-recognition U.S. position on American POW/MIAs of the World War I intervention in Russia in the following letter of December 29, 1933:

"IN REPLY TO YOUR LETTER OF NOVEMBER 17, 1933, IT MAY BE STATED THAT THIS DEPARTMENT HAS NO INFORMATION TO THE EFFECT THAT ANY AMERICAN OFFICERS OR SOLDIERS OF THE MILITARY FORCES OF THIS GOVERNMENT SENT TO NORTH RUSSIA AND SIBERIA ARE HELD AS PRISONERS BY THE SOVIET GOVERNMENT."[94]

This statement can be viewed as a pattern for similar responses to Americans requesting information about communist-held U.S. prisoners for three more wars, over half a century to come. A second U.S. Graves Registration expedition sent to Russia to recover American remains in 1934 resulted in questionable identifications of 19 more sets of human bones as specific, missing American soldiers. 20 U.S. soldiers officially remained missing in Russia, although the Soviet GPU knew their bitter fate.

---

## Notes on Chapter Two

1    *KGB: The Inside Story*, op. cit.

2    *Gulag Archipelago; Russia and the West; Utopia in Power*, op. cit.

3    *KGB: The Inside Story*, op. cit.

4    Foreign Relations of the United States, 1920, Volume III, pp. 687-690

5    Contemporary news accounts and The Autobiography of Lincoln Steffens, p. 799, Harcourt, Brace and Company, NY, 1931

6    *The New KGB*, op. cit.

7    *Service With Fighting Men*, Volume II, Association Press, 1922

8      *The Armies of Ignorance*, William Corson, Dial Press, NY, 1977

9      Ibid.

10     *Armies of Ignorance*, op. cit.

11     *Armand Hammer, The Untold Story*, by Steve Weinberg, Little, Brown, 1989; also: V.I. Lenin, Collected Works, Volume 45, Progress Publishers, 1970

12     FRUS, Vol.III, 1920, p. 698

13     Foreign Relations of the United States, Volume III, 1920

14     Ibid.

15     FRUS, 1920

16     *A Pretty Good Club*, p. 39

17     FRUS, 1920, Vol III

18     Ibid.

19     *A Pretty Good Club*

20     Foreign Relations of the United States, 1920, Volume II, pp. 463-468

21     FRUS, vol. 3, 1920

22     *Fighting the Bolsheviki* and other sources

23     FRUS, Vol. III, 1920, p. 700

24     FRUS, Vol.III, 1920

25     NYT, Chicago Tribune and others,; also declassified intelligence reports in RG 165, National Archives

26     FRUS, 1920, Vol.III

27     *Fighting the Bolsheviki*, P.P. 273-74, Detroit, 1920.

28     5-6 April 1993 Report of TFR-M member Johnson, of a meeting at Russian State Military Archives, Moscow, on file in the author's records

29     *KGB*, Andrew and Gordievsky, p. 78

30     New York Times, October 21, 1920

31     *Secret Servant, My Life With the KGB & The Soviet Elite*, by Ilya Dzhirkvelov, Harper and Row, NY, 1987, pp. 92-93

32     KGB, Andrew and Gordievsky, pp. 78-79

33     *Service With Fighting Men*, Vol.11, p. 314-327, YMCA Association Press, NY, 1922

34     *Gulag Archipelago; Utopia in Power*

35     Solzhenitsyn, *Gulag Archipelago; Utopia in Power*, and others

36     Volume 113, No. 2929, The New York "Nation", August 1921

37     *The New KGB*, Op Cit.

38     The Nation, August 17, 1921, p. 167

39     NYT, August 23, 1921

40     *The Memoirs of Herbert Hoover*, Volume II, p. 23, MacMillan, NY, 1952

41     Washington Post, July 28, 1921

42   Published interview in the Chicago Tribune, also published on the front page of the New York Times, August 12, 1921

43   *The New KGB*, op. cit.

44   Memorandum, American Mission and Legation Riga, #310, 1922, National Archives

45   V.I. Lenin, *Collected Works*

46   *Armand Hammer*, by Weinberg, op. cit.

47   *The Wise Men*, Isaacson and Thomas, Simon and Schuster, NY, 1986

48   Wise Men; *KGB: The Inside Story*, op. cit.

49   *The Wise Men*, p. 103, op. cit.

50   *George Bush, An Intimate Portrait*, by Fitzhugh Green, Hippocrene, NY, 1989

51   Ibid., and Wise *Men*, op. cit.

52   The Decision to Intervene, Vol. II., pp. 64&281, by George Kennan, Princeton, 1958

53   *Diplomat*, by Charles Thayer, Harper's, NY, 1959

54   *American Black Chamber*

55   *Armies of Ignorance*, by William Corson, op. cit.

56   *The Codebreakers*, by David Kahn, MacMillan's, NY, 1967, p. 635

57   Yardley, *The American Black Chamber*, p.162.

58   *Quartered in Hell*, p. 293

59   See original reports by Bettelheim, Dundon and others; also: "Foreign Service," the 1929-1930 VFW Magazine, 1929-1930 and *Ignorant Armies*, p. 216-217

60   Foreign Service, the VFW's Magazine, 1929-1930; also: Ignorant Armies; and declassified War Department records in the National Archives

61   War Department records, RG 165, NA

62   *Forced Labor in Soviet Russia*, Dallin and Nikolaevsky, Yale, New Haven,1947.

63   *Russian Art and American Money*, Robert C. Williams, Cambridge, Mass., 1980.

64   *Utopia*, p.219, op. cit.

65   *Forced Labor in Soviet Russia*, p.52-54, 169-85, Yale University, 1947, op. cit.

66   NA RG 165 10110-2623. Some of the original documents pertaining to this matter were declassified in the mid-1980s and placed in the National Archives, where the author and Ashworth located them in 1988, and published an analysis of them in the May 1989 U.S. Veteran Special Report: "A Chain of Prisoners, From Yalta to Vietnam; A Secret That Shames Humanity. Reference to these documents had been made in the footnotes of William R. Corson's and Robert T. Crowley's excellent study of Soviet intelligence, *The New KGB*, William Morrow and Co., 1985. The authors, however, neglected to investigate the fate of the American MIAs and POWs of 1919, who had been secretly held by the Soviet regime.

67   Ibid.

68   *A Pretty Good Club*, op. cit

69   A copy of the original document in the National Archives supplied by the author

is reproduced on page 668, Volume II, POW/MIA Policy and Process, Hearings of the Select Committee on POW/MIA Affairs, United States Senate, November 1991, U.S. Government Printing Office, 1992, ISBN 0-16-038479-6

70    A copy of the original document in the National Archives supplied by the author to the U.S. Senate is reproduced on p. 667, Volume II, POW/MIA Policy and Process, U.S. Government Printing Office, 1992

71    Original documents in the National Archives, RG 165 10110-2623

72    *Quartered In Hell* for complete list of missing men's remains reported returned from north Russia in 1929

73    NA RG 165 10110-2623/4

74    *Utopia in Power*, p. 272

75    *Forced Labor in Soviet Russia*, Dallin and Nikolaevsky, op. cit.

76    *The Secret History of Stalin's Crimes*, by Alexander Orlov, Random House, NY, 1953

77    *Utopia in Power*, p. 276, op. cit.

78    *The Secret History of Stalin's Crimes*, op. cit.

79    *Forced Labor in Soviet Russia*, Dallin and Nikolaevsky, op. cit.

80    *Prisoner of the OGPU*, Longmans, Green & Co., London, 1935

81    *Pretty Good Club*, by Martin Weil, op. cit.

82    Time Magazine, Foreign News, December 22, 1930

83    Time Magazine, February 23, 1931

84    Granovsky, *I Was An NKVD Agent*, pp. 14-17, Devin-Adair, New York, 1962.

85    *The Armies of Ignorance*, p. 70,op. cit.

86    *Roosevelt and Hopkins; The Wise Men; A Pretty Good Club*, op. cit.

87    *Poland and the Coming of the Second World War*, The Diplomatic Papers of A.J. Drexel Biddle Jr., Edited by Philip V. Cannistraro, Edward Wynot and Theodore Kovaleff, Ohio State University,1976

88    Kennan, *Memoirs and Russia and the* West, op. cit.

89    *A Pretty Good Club*, op. cit., p. 69

90    *The Wise Men; A Pretty Good Club*, op. cit.

91    *Russia and the West Under Lenin and Stalin*, pp.298-99,op. cit.

92    Bohlen, *Witness to History; Weil, A Pretty Good Club*, op. cit.

93    This reference was used more than two decades later in a U.S. State Department note No. 30 to the Soviet government, about the detention of American prisoners in the USSR, dated July 16, 1955

94    Original letter in the Department of State records in the National Archives

# PART II

# THE DICTATORSHIP OF JOSEF STALIN AND THE ORIGINS OF THE COLD WAR

## CHAPTER THREE

### THE SECOND WORLD WAR FROM STALIN'S PURGES TO YALTA

Throughout the 1930s Lenin's successor, dictator Josef Stalin, imprisoned and killed millions of Russian and Ukrainian peasants and other minority peoples of the USSR, who resisted forced collectivization and harsh Communist rule. Later researchers, using eyewitness accounts, Soviet documents and reports from Western officials in Russia at the time have estimated that the cost in lives of Stalin's collectivization of the peasant farms and villages and the great industrial projects accomplished by mass forced labor amounted to some 11 million peasant dead and 3.5 million more dead among those who had been arrested by the OGPU and sent to labor camps between 1930 and 1937. Of these 14.5 million dead, 6.5 million died as a result of dekulakization, 1 million died in Kazakhstan, 5 million died in 1932-33 in the famine in Ukraine, 1 million died in 1932-33 in the famine in the Caucasus region and 1 million more died during the same famine years elsewhere in Russia.[1] Also shot or worked to death were hundreds of thousands of "counterrevolutionaries," purged Soviet officials and Red Army officers, and thousands of visiting or resident foreign socialists and communists suspected, ironically, of espionage. Stalin's exploitation of millions of state-slaves in a continent-wide empire of thousands of forced-labor camps provided a costless source of exportable and domestic wealth for the Communist Party elite. The network of Gulag forced-labor camps became a major Soviet state industry, requiring ever more slaves acquired by force, but requiring practically no capital investment.[2]

Stalin felt threatened by events in the Far East area of the Soviet Union during this period, which ultimately encouraged him to ally with Hitler in 1939, out of distrust for the British, French and other Western Capitalist nations, including the United States. The Soviets had retained rights to the Chinese Eastern Railway through Manchuria and had fought nationalist Chinese warlords in 1929 to keep them, but with a Japanese invasion, Moscow agreed to negotiate for the sale of those rights to Japan. The Japanese attack on Manchuria began in September 1931 and they set up a puppet regime called "The Government of Manchukuo," while the League of Nations debated the crisis but substantively did nothing.

In China at this time, Chiang Kai-shek launched what he called "extermination campaigns" against the Chinese Communists, beginning in November 1930 and continuing to October 1933, even as his Nationalist forces were fighting the Japanese. Among those arrested and interrogated, who provided information leading to the smashing of a Communist spy ring was one Nguyen Ai Quac, posing as a Chinese, but in reality a Vietnamese activist for the Communist International who would become known to the world as "Ho Chi Minh," leader of two Indochinese wars against France and the United States.[3] In October 1934, Chinese Communist leaders Mao Tse Tung and Chu Teh led some 90,000 men west and north from Kiangsi on a 6,000 mile trek to Yenan in Shansi Province, that later became known as "the long march." For the 20,000 of these men who still survived a year later, Yenan became a new base, backed by the Communist infiltrated province of Sinkiang and the sanctuary of Soviet-occupied Mongolia. Mao Tse Tung, a former librarian who had been born to a Hunan peasant family, was destined to found a new Chinese dynasty and to expand his power until he was confronted on China's southern frontier by the Americans and their European Allies, 16 years later, in Korea.[4]

## THE GREAT PURGE

Stalin's wife, Nadezhda Allilueva, depressed about the terror and repression in Russia, had committed suicide in November 1932. That same year the old Bolshevik leaders Zinoviev and Kamenev were exiled to Siberia, but after recanting their anti-Stalin heresies they were allowed to return, although their rehabilitation was to be brief. During the XVII Party Congress in 1934, Stalin announced that the victory had been won and "there was no one to fight." Although all external enemies of the party had been officially liquidated or terrorized into submission, there still remained those opposed to Stalin from within the Party hierarchy, and it was to these that the dictator now turned his baleful attention. Meanwhile, in July 1934, the name of the OGPU secret police was changed to: Narodnyi Kommissariat Vnutrennykh Del, known colloquially as the NKVD (meaning People's Commissariat of Internal Affairs), but the remaining under control of Stalin's henchman, Genriikh Yagoda.[5]

Upon arrival at the new U.S. Embassy in Moscow in March 1934, in company with Charles Bohlen, who was fluent in Russian, Ambassador William C. Bullitt was determined to make every effort to get along with Stalin's government on behalf of President Franklin D. Roosevelt, who wanted better relations with Soviet Russia. The rise of Hitler in Germany and the aggressive moves of Japan in the Pacific had caused Roosevelt to desire improved relations with Russia as a counterpoint, in order to keep the totalitarians guessing about American intentions.[6] The embassy Counselor was a career diplomat, John C. Wiley, who was assisted by Loy Henderson, Kennan, Bohlen, Thayer, Durbrow and others. A military aide to Bullitt was U.S. Navy Lieutenant Roscoe Hillenkoetter, who acted as a "spy" for the Ambassador, and 13 years later would be named as a Director of the CIA. Carmel Offie, from an upstate Pennsylvania Italian immigrant family, was subsequently sent to Moscow to act as Bullitt's personal secretary. He had previously worked in the U.S. Embassy in Honduras, and would end up working for Bullitt for many years to come, as well as serving in the OSS during WW II and after assisted another U.S. Ambassador, Anthony J. D. Biddle Jr. American tourists were few in the first year, so that embassy personnel had little to concern themselves with other than supporting the negotiations and gathering information about conditions in Russia.[7]

Relations deteriorated rapidly over debt negotiations involving czarist bonds from WW I held by Americans, as well as unfulfilled obligations of the Kerensky government. It has not been revealed whether the secretly-held American POWs of the intervention were part of these negotiations. Maxim Litvinov demanded a straight cash loan from the United States, but the U.S. Senate had passed a bill in April authored by Hiram Johnson of California, which prohibited loans to any foreign government in default of it's debts to the U.S. Bullitt returned to Washington for consultations in October.[8] The Soviet Ambassador in the United States, Aleksandr Troyanovsky, had met with Roosevelt on April 30th, and threatened to initiate counter claims by the Russian government for damages caused by the U.S. intervention in Archangel and Siberia from 1918-1920. Roosevelt had countered by saying that the American troop presence had "saved Siberia from being taken over by the Japanese."[9] Cordell Hull, cut out of the loop on Soviet affairs, believed that Stalin's government would never come to any reasonable agreement with the U.S. Bullitt's stalled negotiations with Litvinov bore this out. Bullitt's and Litvinov's relations had evolved into a feeling of mutual dislike. The American ambassador felt that the Litvinov had broken his word on the agreed terms for U.S. diplomatic recognition, and that relations should be severed. Bullitt was finally being "cured of his Bolshevism."

Returning to the Soviet Union in April 1935, Bullitt found the situation worsening. On the 23rd he cabled Roosevelt: "...There can be no possibility of private conversations. The terror...has risen to such a pitch that the least of the muscovites as well as the greatest, is in fear. Almost no one dares to have any contact with foreigners..." Bullitt left again for the U.S. in November, stopping in Paris where his ex-wife, the pro-Soviet journalist Louise Bryant, died in January 1936. Bullitt returned to Moscow in February 1936 but left in May, with the understanding that Roosevelt wanted him to resign. He was subsequently appointed U.S. Ambassador to France, and would play a role in the Second World War, but as he grew older he became more anti-Soviet, and embittered about the way Stalin had forsaken a chance to fundamentally improve Soviet-American relations.

In Germany, meanwhile, in what became known as the "Night of the Long Knives," Nazi leader Adolph Hitler had ordered the murder of hundreds of his old SA Nazi Party colleagues, including Ernst Roehm. Stalin is now believed to have been inspired by Hitler's ruthlessness toward his Nazi colleagues in ordering a perceived rival, Sergei Kirov, to be assassinated in Leningrad on December 1, 1934. This act had signaled the start of a great new purge. From 1935 to the middle of 1936, untold thousands of "supporters" of Zinoviev, Kamenev, Trotsky and other leaders within the party were purged and shot, or deported to the GPU's slave camps. A public trial of sixteen "Old Bolsheviks," including those who had led the Revolution such as Zinoviev and Kamenev, was used to create the illusion of a great conspiracy against the Soviet state and Josef Stalin. After the initial purge and executions, Yagoda, the murderous Chief of the NKVD, was himself purged and shot, along with many of his closest secret police henchmen, apparently to remove key witnesses to Stalin's diabolical methods. He was succeeded as Chief of the GPU by N.I. Ezhov (Yezov), who was to become known among the Zek prisoners of the state security organs as "the bloody dwarf." Ezhov announced that the victims of the 1936 purge had implicated Bukharin and other old party leaders in conspiracies. More public trials were held in 1937 and many of the defendants, including Bukharin, Piatakov and Radek, were shot. In June 1937 the Red Army underwent a severe purge in which Marshal Tukhachevsky, many senior generals, and thousands of colonels and other officers were ruthlessly executed. Stalin thus nipped in the bud any chance of a "Bonapartist" coup, a military takeover, one of the great fears of all Soviet leaders. Subsequent evidence indicated that German intelligence abetted Stalin's suspicions of his own officers by covert actions which seemed to cast doubts on the loyalty of the Red Army leadership.

After an interval, Roosevelt had appointed a wealthy political supporter, Joseph Davies, as Bullitt's successor in Moscow. Davies had married Marjorie Post, an heiress to the Post Cereal fortune, and chose to remain totally ignorant of the brutal methods Soviet

Communism; he was determined to follow Roosevelt's mandate to improve U.S. relations with Russia at any cost. The senior U.S. Military Attaché in the Moscow embassy under Davies was Colonel Philip R. Faymonville, who had been a pro-Bolshevik since he had served as an intelligence officer on the staff of General Graves during the American intervention in Siberia in 1919-1920, and was fluent in Russian. Faymonville reported directly to Harry Hopkins in Washington which gave him direct access to President Roosevelt. Bohlen, Kennan and others at the embassy detested Davies and deplored Faymonville's unswerving support of Stalin's regime, and his reports indicating that victims of Stalin's purges were guilty as charged. Charles Bohlen later wrote:"

Davies, who had succeeded Bullitt, made matters even worse. He had gone to the Soviet Union sublimely ignorant of even the most elementary realities of the Soviet system and of its ideology. He was determined, possibly with Bullitt's failure in mind, to maintain a Pollyanna attitude. He took the Soviet line on everything except issues between the two governments. He never even faintly understood the purges, going far toward accepting the official Soviet version of a conspiracy against the state...Colonel Faymonville...was not very useful because he was inclined to favor the Soviet regime in almost all its actions. AT STAFF MEETINGS...HE WOULD STOUTLY DEFEND THE PURGES, INSISTING THAT THEY WERE UPROOTING TRAITORS AND ENEMIES OF THE PEOPLE FROM THE RED ARMY, AND THEREFORE DID NOT WEAKEN BUT, ON THE CONTRARY, STRENGTHENED THE MILITARY..."[10]

Stalin's purge prosecutor-General was Andrei Vyshinsky, a late joiner of the Communist party who was anxious to please his Georgian master and who conducted the public trials with a viciousness that endeared him to the dictator. Nearly all the defendants repeatedly confessed their "crimes," of conspiring with foreign secret agents to overthrow Stalin and the Communist Party, to restore capitalism and to give up Soviet territory to fascist Germany and Japan. In his memoirs, "Chip" Bohlen, who acted as Ambassador Davies translator during the purge trials, later recalled in his memoirs, the nature of this man who would become a senior Soviet diplomat during the Second World War:

"Vyshinsky was one of the most unsavory products of Bolshevism. Originally one of the opposition Mensheviks, he had written an article about Bolshevik atrocities in the early days of the Revolution. Much of his bloodthirsty career can be traced to his desire to compensate for his anti-Bolshevik past...Anyone who saw him, as I did, mercilessly pursuing, mocking and prodding defendants will never forget the ferret-like quality of Vyshinsky."[11]

George Kennan, the career State Department Russian expert who served in Moscow at this time, later recorded Stalin's actions in the 1930s toward his own colleagues in the higher echelons of the Soviet communist party:

"By way of response, apparently, to what seems to have been some opposition to his purposes on the part of the Seventeenth Party Congress in 1934, Stalin killed, in the ensuing purges of 1936 to 1938, 1,108 out of a total of 1,966 of the members of the Congress. Of the Central Committee elected at that Congress and still officially in office, he killed 98 out of 139...these deaths were only a fraction, numerically, of those which resulted from the purges of those years. Most of the victims were high officials of the Party, the Army, or the Soviet government apparatus. All this is aside from the stupendous brutalities which Stalin perpetrated against the common people...The number of victims here-the number, that is, who actually lost their lives-runs into the millions...and the millions who were half-killed."[12]

Actually it was to be decades before the vast number of people done to death in the purges and gulag slave camps was known. Careful researchers later compiled facts which indicated that in January 1937 the Soviet prisons and labor camps held about five million people, and that seven million more were arrested throughout the Soviet Union from January 1937 to December 1938. This did not include common criminals, who were in a class by themselves in the gulags. In this same period an estimated 1 million were shot and 2 million more died in the gulags and prisons, or during transports.[13] Russian author Aleksandr Solzhenitsyn, however, estimated that 1,700,000 people had been shot by January 1939.[14] British author Robert Conquest pointed out the Western complicity in this mass death by noting that all Soviet ships carrying prisoners to the far eastern Gulags of Kolyma were insured by Lloyd's of London.[15] And the deaths continued after this date; in 1939, 1940, 1941, all through the Second World War and into the 1950s. The total cost of the Soviet Gulag prison system has been variously estimated to be between 20,000,000 and 50,000,000 dead.

While wholesale terror reigned in the USSR, a civil war had broken out in Spain, and leftists and Communists in Europe and America rallied in support of the Republican cause, while anti-Communists and Fascists supported Francisco Franco. Hitler sent the latest German war equipment and aircraft to Spain for testing in combat, along with Nazi military and intelligence advisers to support Franco. The Soviets did the same thing for the Republican side, and many idealistic Americans, British and others served the Communists in international volunteer units, including the Abraham Lincoln Brigade. Those who failed to perform satisfactorily were shot by firing squads under authority of NKVD agents, giving many of their compatriots a harsh lesson in Soviet methods which were then being practised in Russia. The wives of many Spanish Republican soldiers were sent to the USSR, for "education," but remained as hostages, and their children, and other orphans from Spain brought to Russia, became known as "Los Niños," some of whom later became agents for the NKVD/KGB. Others were reported years later to be still alive in Gulag slave camps of Siberia or the far north. The GPU/NKVD chief in Spain was L.L. Felbin, better known as "General Alexander Orlov," while Soviet generals Malinovsky, Rokossovsky, Konev and others served as advisers to the Republican forces, and gained additional combat experience with new war equipment that would prove useful during WW II.

In western Europe NKVD agents engaged in a massive espionage operation and a series of assassinations of anti-Communist Russian emigres. One of those enemies of Soviet power who was done away with was the former White Army commander in Archangel, General Yevgeni Karlovich Miller. General Miller vanished from a Paris street in September 1937. He had left a note saying that he was to meet with an NKVD agent, Skoblin, who also disappeared and was probably killed by the NKVD himself, in Spain. The French Sûreté investigation revealed that General Miller had been kidnapped and his body shipped to Russia in a large steamer trunk. However, it was later revealed that General Miller had been alive and drugged inside the trunk. Upon arrival in the Soviet Union Miller was tortured to obtain all possible information about the White Russian emigree organizational structure, and after a secret trial he was shot.[16]

In the United States meanwhile, following Franklin D. Roosevelt's reelection by a landslide in 1936, that President had determined to remake the Department of State into an institution that would serve his views rather than long-established policy goals as interpreted by professional diplomats and foreign service officers. Sumner Welles, a Groton and Harvard man like the President, who had been a page boy at Franklin and Eleanor's wedding, was appointed Undersecretary of State, with a mandate to take control of the Department from the group of anti-Soviet "Europeanists." He had the backing of the "New Dealers" in the Administration like Harry Hopkins, Henry Wallace, Felix Frankfurter, Henry Morgenthau, and of course, Eleanor Roosevelt. In 1937, at the behest of FDR, Welles abolished the old Division of Eastern European Affairs and transferred the anti-Soviet Robert Kelley to Turkey. The Division was merged with the Western European Division and became the Division of European Affairs. The "Russian Division" thus became a single desk in the Department, with another desk for the Baltic Republics. It was at this point that the incomparable archives known as the "Russian collection" was broken up and dispersed, thus partially blinding future American policymakers. The White House group responsible for this action wanted the collection destroyed, but reportedly were unsuccessful. Bohlen later succeeded in getting much of the collection transferred to the Library of Congress, where it remained. Extensive files and dossiers on foreign and domestic Communists were ordered to be destroyed, also.[17]

Moscow Embassy staffer Elbridge Durbrow, a close friend of Bohlen's, later remembered how the Soviet government urged foreign Communists to move to Russia: "Bring your family, we'll give you good housing and we'll pay you $500. a month."[18] Many of the idealistic immigrants threw away their passports upon arrival in Soviet Russia, or voluntarily turned them over to Soviet authorities-who in some cases, used them subsequently to insert espionage agents into the United States. Many of these Americans and other foreigners who had come to the USSR to "build socialism," quickly became disenchanted with the repression and appalling conditions there and tried to leave, which the Soviet government did not want them to do for fear they would spread anti-Soviet propaganda. Some had witnessed the mass deaths during the collectivization of the Ukraine, with bodies stacked like cordwood and empty villages where all the inhabitants had been massacred or deported to Siberia by the NKVD. During the first year at the American Embassy in Moscow, Durbrow himself interviewed 300 Americans who wanted to go back to the U.S.[19]

During the 1930s, due to extremely depressed economic conditions, there was substantial growth of the communist parties in western Europe, the United States and elsewhere. Many persons believed the capitalist system was failing and that communism would bring a fairer way of life for the disadvantaged of the world. Surrounded by millions of suffering, unemployed workers, it was easier for many to dismiss the tales of escapees from Stalin's camps as "capitalist propaganda." Domestic communists and sympathizers strictly adhered to the party line laid down by Moscow. When an American communist, Fred Beal, returned from the USSR in 1935 and told of his experiences as a prisoner to American communists, they refused to believe him. Beal had become a hero to American communists when he was arrested during the Gastonia, North Carolina textile strike in the late 1920s. (A bomb thrown into a crowd during a strikers demonstration had killed several police and citizens. Beal had subsequently jumped bail and fled to the Soviet Union on orders from the Communist Party.) After his voluntary return from Russia he began to write and speak the truth about the oppressive conditions under Stalin, and was judged a traitor by American communists.

Among those who read in the Hearst newspapers of Beal's experiences in Russia and rejected them as lies were the Sgovios', an Italian-American immigrant family, imbued with the father's Marxism. Despite this prescient warning, in August of 1935 they started their journey to work in the Soviet Union. 25 years later,

Thomas Sgovio, a 19 year old youth in 1935, would at last escape a living death in Stalin's Gulags, where millions had been consumed all around him. He would survive to record his experiences in a memoir that few bothered to study, but which revealed the true horrors of Stalin's death camps.

After three years living in dismal, oppressive Moscow, young Thomas Sgovio remembered that in 1938, "doubts about the deification of Stalin had begun to creep into my twenty-one year old mind...The Soviet propaganda machine was waging a furious war against the enemies of the people. Almost everyone was a spy, wrecker or saboteur. The Moscow trials had recently taken place. Overnight, Genriikh Yagoda, Chief of the NKVD, became a traitor. Karl Radek, the patriarch of communist philosophy, and Piatakov, Deputy Chief of the Heavy Metals Commissariat, confessed to their crimes and were executed. Many other Old Bolsheviks vanished. Only yesterday they were our beloved leaders..."

Sgovio was still not yet cured of his years of communist indoctrination, though. He was later to remember:

"I regarded myself as an American revolutionary living in the Soviet Union where the capitalist system had already been overthrown. I had come to the USSR on the assumption that I would study art for four years, receive a higher education, and return to the United States. My dream had turned out to be an illusion...I was homesick...the purges had begun. I could not understand what was happening...It was the time of the communist inspired Spanish Civil War. I followed the events from the communist press and propaganda. Foreign communists were fighting against the Franco forces under the banner of the International Brigade. Among them were several of my American and Italian comrades, recently recruited in Moscow by the International Red Aid. As a true communist, I too, wanted to do my bit.

"In December 1937, I applied to the Central Committee of the International Red Aid, asking to be sent to Spain. Refused by Elena Stassova, the General Secretary, I decided to go on my own. The next step—that of renouncing my Soviet citizenship, took much thinking and hesitation. I went for the first time to the United States Embassy to file an application for an exit visa and renunciation of my Soviet citizenship in January 1938. I was interviewed by Mckee, the First Secretary, who assured me the Embassy officials would write to the Soviet Commissariat of Foreign Affairs to find out how I had become a Soviet citizen."[20]

After witnessing the brutal arrest and disappearance of a close, innocent friend in a neighboring apartment, by the NKVD, Sgovio was more resolved than ever to leave the Soviet Union. On March 21, 1938 he went again to the U.S. Embassy in Moscow to push for assistance. While there he caught the eye of a well-dressed

Russian employee of the Embassy, who soon afterward proved himself
to be an agent of the Soviet secret police. Sgovio was arrested outside
the U.S. Embassy by this same man and another plainclothes NKVD
man after leaving the building. In due course he was incarcerated in
the Lubyanka prison on Dzerzhinsky Square, for interrogation.

Months later, as a convicted prisoner who had been sentenced
in absentia by the OSO, Thomas Sgovio was carried with other
prisoners in a disguised "black maria" van, to a deserted railway siding
on the outskirts of Moscow. There a gauntlet of blue-capped NKVD
soldiers armed with bayonetted rifles, holding leashed, snarling
Russian wolfhounds, forced them into freight-cattle cars in the
charge of convoy guards. More vans unloaded a stream of new fodder
for the far eastern gulags, mostly they were Russians but there were
also Poles, Latvians and other foreigners. Sgovio remembered:

"Being young, agile and unencumbered by belongings, I missed
the blows which the elderly received. They could not run fast enough
or climb by themselves into the red cattle cars already half full with
prisoners from Buktirki Prison...I was surrounded by several English-
speaking prisoners from Buktirki. There was a Kaplan who had lived
in Canada, Schwartz from New Jersey, and finally Banchik, who had
resided in Philadelphia for about two years, but spoke English fluently.
Upon hearing my name, Banchik asked, 'Is your father also in Russia?'
"'Yes,' I answered.

"'In Buktirki there was an Italian political immigrant in the same
cell with me. His last name was Sgovio...first name and surname were
Giuseppe Tomasovitch. He was...arrested in broad daylight on the
Moscow streets about a year ago. Could it be your father?'

"It was then I knew my father was also a prisoner. I did not see
the others. Unable to hold back the tears, a kaleidoscope of childhood
reminiscences flashed through my mind..."[21]

Both innocent and serious communist sympathy and cases
of active subversion also existed in the West among the educated
elite. Socialist or communist central planning offered the possibility
of experimenting with new social concepts, for those envisioning
higher goals for humanity.

The Soviet government and Comintern recognized the
advantageous trend and moved quickly to exploit it for intelligence
purposes. The result was the "Red Orchestra," a massive foreign
intelligence network rigidly controlled by Moscow. This network
inserted agents and recruited spies and moles such as H.R. "Kim"
Philby Guy Burgess Donald Maclean and Anthony Blunt for Moscow
Center.; gathered information, stifled or deflected criticism of the
Soviet regime and affected Western policy toward Russia up to and

during the Second World War.

George Kennan later wrote of this problem in the United States: "The penetration of the American governmental services by members or agents (conscious or otherwise) of the American communist party in the late 1930s was not a figment of the imagination of the hysterical right-wingers of a later decade. Stimulated and facilitated by the effects of the Depression, particularly on the younger intelligentsia, it really existed; and it assumed proportions which, while never overwhelming, were also not trivial...Our efforts to promote American interests vis-a-vis the Soviet government came into conflict at many points with the influences to which this penetration led; and our own situations were sometimes affected by it...We were inclined to suspect, for example, that the sudden abolition of the old Russian Division of the Department of State, in 1937, was the result, if not of direct communist penetration, then at least of an unhealthy degree of communist influence in higher counsels of the Roosevelt administration."

This action resulted in the breaking up of the Russian Collection in the State Department, containing a great quantity of priceless knowledge of the Soviet Union, gained over many years, as noted in the previous chapter. The effect of this purge by the "New Dealers" in the Roosevelt Administration was to cause a further blinding of the policymakers of the United States, in the coming critical stage of Soviet-American relations. Author William Corson shed more light on the phenomena of high-level Soviet influence on American policy by interviewing former State Department official Loy W. Henderson (in 1982). Henderson had been a force within the department of the 1930s for acting with the utmost caution toward Soviet proposals and promises, stated that he was eventually removed from the State Department through the influence of Maxim Litvinov on Harry Hopkins and Eleanor Roosevelt, and ultimately on President Franklin Roosevelt.[22] The influential columnist Drew Pearson had led the attack on the anti-Soviet elements in the Department of State, urging that the New Deal be expanded to include American foreign policy, particularly after the latter part of 1937, and the theme was taken up by influential editors at the New York Times and other publications.

The Italian Fascist dictator Benito Mussolini's mechanized army, backed by air power, had invaded Ethiopia during 1935, in a bid to create an Italian colonial empire in Africa. Yet when Britain and Italy signed a treaty, signaling British approval, and Roosevelt approved of the action, Drew Pearson had blamed the conservatives in the State Department for the President's action. Fears of another world war had increased when German dictator Adolph Hitler ordered

German troops into the demilitarized zone of the Rhineland in March 1936, and when later that year Germany and Japan signed an anti-Comintern pact to oppose Communism. A skirmish occurred between Soviet and Japanese forces on the Amur River border between Manchuria and the USSR in June 1937, and more fighting broke out near the Soviet-Korean-Manchurian border during 1938, and still later on the Mongolian-Manchurian border, in May 1939. Although war between the two Pacific powers was not to occur until August 1945, the border skirmishes kept tensions high and required both the Soviets and Japanese to divert large forces to this remote frontier area.[23]

In his later book "Mission to Moscow," published in 1941 and made into a film during WW II, Ambassador Davies included a 1938 dispatch for Roosevelt about a personal meeting he had with Stalin, Molotov and Kalinin in the Kremlin, after the purge trials, which illustrates his apparent total ignorance of the Soviet system of mass state-slavery and reveals a little about the Soviet dictator's thinking at the time:

"Scarcely had we been seated, when I was startled to see... Mr. Stalin come into the room alone...I noticed he was shorter than I conceived and he was quite 'slight' in appearance. His demeanor is kindly, his manner almost deprecatingly simple, his personality and expression of reserve strength and poise very marked as we arose he came forward and greeted me cordially...Meeting Mr. Stalin, I then said, was a great surprise, and that I was very much gratified to have this opportunity...I then went on to say that I had personally inspected typical plants of practically all the heavy industries of the Soviet Union, as well as the great hydraulic developments of the country; that these extraordinary achievements, which had been conceived and projected in the short period of ten years, had commanded my great admiration; that I had heard it said that history would record Stalin as the man who was responsible for this achievement and that he would be recorded as a greater builder than Peter the Great or Catherine; that I was honored by meeting the man who had built for the practical benefit of common men.

"To this Stalin demurred and stated that the credit was not his; that the plan had been conceived and projected by Lenin...and above all it was the 'Russian people' who were responsible...He gave me the impression of being sincerely modest...He then went on to say that the reactionary elements in England, represented by the Chamberlin government, were determined upon a policy of making Germany strong, and thus place France in a position of continually increasing dependence upon England; also with the purpose of ultimately making Germany strong as against Russia..."[24]

Roosevelt recalled Joseph Davies as U.S. Ambassador to Russia in early 1938, after realizing that, although the Soviets were understandably very fond of him, he still could make no progress on the outstanding issues between the two governments. Davies went to Belgium and was replaced in Moscow by a career diplomat from the Foreign Service, Laurence Steinhardt. In August of 1939 the Soviet Union and Nazi Germany signed a mutual non-aggression treaty with attached secret protocols dividing up Poland and parts of eastern Europe once again in a "partition." This was the Hitler-Stalin pact negotiated by Soviet Foreign Minister Vyacheslav Molotov and German Foreign Minister Von Ribbentrop, which was not to become fully known in the west until the Nuremberg trials. (The Soviets were to deny existence of the secret protocols dividing up Poland and the Baltic republics until the 1990s.

Ambassador Steinhardt appropriately represented official U.S. displeasure at this turn of events, and filed numerous protests with Stalin's government about subsequent Soviet actions such as the detention of the American merchant ship City of Flint, and its crew. The American left largely continued its blind support for Stalin's regime, which had resulted in published excuses for the great purges of the 1930s, which had killed millions in Russia, and its attacks on those liberals who dared to expose the truth of Stalin's terror. In "The Great Deception," author Moshe Dector pointed out later: "In August 1939, just nine days before the signing of the Hitler-Stalin pact, some four hundred leading American intellectuals of the arts, sciences, and professions published a long 'Open Letter' branding as fascists and reactionaries all those who expressed the 'fantastic falsehood that the U.S.S.R. and totalitarian states are fundamentally alike' in their suppression of cultural freedom, civil liberties, and free trade union activity."[25] When it became known, the Hitler-Stalin pact caused some disarray within the American left, but Soviet Communism continued to have many influential defenders.

The Hitler-Stalin pact helped precipitate Germany's attack on Poland and the Second World War. Following subsequent Soviet military invasions from the east, Stalin was to order the deportation of over 2,000,000 Polish, Baltic and Finnish people to forced labor camps in Siberia and throughout Russia. While some survived the war, many disappeared forever, leaving a legacy of bitterness still strong today in Poland, Latvia, Lithuania, Estonia and Finland.[26]

THE OUTBREAK OF THE SECOND WORLD WAR

The Germans attacked Poland on September 3, 1939, bombing cities from the air and invading the country in a mechanized assault which soon became known to the world as "blitzkrieg," or "lightning war." The Polish Army fought stubbornly and courageously but was no match for the Nazi war machine that Hitler had created in spite of limitations imposed by the Allies after WW I.

One of the first Americans to actually bear the brunt of WW II was the U.S. Ambassador to Poland, Anthony J. Drexel Biddle Jr. A vigorous and athletic man, Biddle exuded vitality and took the most minute interest in his work, although he had originally been a political appointee. The son of a famous Philadelphian who had distinguished himself in WW I combat, Anthony Biddle Jr. himself had also served in France as an infantry captain in 1918. A friend of William C. Bullitt, Biddle had been an early supporter of Franklin D. Roosevelt's candidacy for President and was eventually rewarded with the Ambassadorship to Norway in 1935.

With the steadily increasing tensions arising from Hitler's expansion of German power, Roosevelt had Biddle, who was a trusted friend, transferred to Poland, which was already perceived as the potential flash-point for another world war. Anthony Biddle was a brilliant choice because of both his devotion to duty and his characteristic charm which made him a welcome addition to meetings with the highest officials, thereby producing valuable intelligence for the U.S. As the New York Times was to report much later, "...His reports of the oncoming Nazi assault on Poland and France remain in the State Department files as models of prescience and accurate information."[27]

Later Polish Ambassador to the U.S. Jan Ciechanowski, a longtime colleague and friend of Biddle remembered:

"Tony Biddle was more than an Ambassador in the eyes of the Polish people. He was Poland's best friend, the most popular of foreign envoys, beloved for his human qualities, his wonderful spirit, and his charming manners...He had a ready smile and a welcome for rich and poor alike and...certainly succeeded in opening all Polish hearts even wider to the United States..."[28]

Biddle was in Poland when the German attack began on September 1st, and by telephoning direct to Ambassador William Bullitt in Paris, he was able to describe first-hand the German terror-bombing of civilians. Roosevelt used Biddle's intelligence in publicly demanding an end to aerial bombing of civilians by any aggressor. When the Polish government withdrew from Warsaw on September 6th ahead of the German attack, Biddle accompanied it, and he and his wife Margaret came under repeated German dive-bomber attacks while traveling with a refugee column to the Romanian border. Although he had fastened a large American flag on the roof of his car to indicate neutrality, the Germans seemed to single him out. In a cable of September 8th, Biddle wrote that his car, "was bombed fifteen times and machine-gunned four, forcing me to take refuge in a roadside ditch." The State Department published this cable to prove that Hitler was lying when he denied that German aircraft were bombing civilians. On September 12th, the Germans bombed the small town of Krzemieniec, killing many civilians, apparently because of the presence of retreating government and diplomatic personnel. Biddle again cabled Secretary of State Cordell Hull: "I find it difficult in many cases to ascribe the wanton aerial bombardment by German planes to anything short of deliberate intention to terrorize the civilian population and to reduce the number of child-producing Poles..."[29]

Ambassador Ciechanowski later wrote of the conduct of the American Ambassador Anthony Biddle in the final days of this unequal battle:

"Throughout the tragic September and October days in 1939, when the Polish Army and the entire nation fought their desperate struggle against Hitler's might, Tony and Margaret Biddle had, with the utmost courage and serenity, faced all the dangers and hardships at the side of the Polish government, whom they never abandoned, and with whom they ultimately crossed into Romania, driven there by the German onslaught from the west and the Soviet invasion from the east."

As the Polish forces were being driven back, they were suddenly attacked in the rear by an invading Soviet army on September 17th. Over 200,000 Polish military POWs were captured by the Soviet forces and these were immediately deported to prison camps in the Soviet Union. The Red Army was accompanied by a cloud of Stalin's secret police. In the ensuing occupation of eastern Poland, Ivan Serov, the NKVD commander appointed by Beria to carry out the operation, deported between one and a half and two million "anticommunist" Poles to forced-labor camps in the arctic north of Russia and in Siberia. In addition to the some 230,000-250,000 Polish soldiers confined in Russian POW camps, at least 1,200,000, and possibly more than 1,500,000 Polish civilians were deported to Soviet gulags. After Poland's collapse, an army of NKVD secret police operatives under command of Ivan Serov (later head of the KGB) descended on the defenseless eastern half of the country and began mass arrests of anti-Soviet elements of the population.[30]

The partition of Poland resulted in the Soviets gaining 77,620 square miles of Polish territory, with the Germans adding 72,866 square miles to their conquests. In November 1939 the Soviets attacked Finland, precipitating a war in which the Red Army was embarrassed before the entire world by the tenacious resistance of the Finns. Although Stalin's forces finally prevailed at enormous cost in officers and men, Hitler took note of the poor performance of the Soviets.

The April-May 1940 Katyn Forest massacre of 5,500 Polish POWs by Stalin's NKVD secret police near Smolensk, Russia, and the disappearance of 10,000 other Polish POW officers by Stalin's NKVD secret police was exposed in 1943. Yet, in order to maintain wartime Allied unity, President Franklin Roosevelt and Prime Minister Winston Churchill publicly blamed the Germans and suppressed the truth, although the secret cables of U.S. Ambassador to Poland, Anthony J. Drexel Biddle Jr., Ambassador Winant, and now-declassified War Department documents, reveal that U.S. and Allied intelligence already knew that the Soviets had murdered the Polish prisoners. Censorship was imposed on Soviet guilt in the Katyn Forest massacre, which inexplicably continued long after the end of WW II.[31]

After an interval of 22 years, the western front erupted in 1940, but this time a Nazi blitzkrieg overwhelmed Belgium, Holland, and France with a speed that astonished and terrified the defeated. In consequence hundreds of thousands of British/Dominion, French, Belgian and Dutch POWs were captured by the Germans and eventually transported to POW camps in Poland, East Prussia and the Baltic Republics, to keep them as far away as possible from possible rescue or sanctuary.

On June 22, 1941, Hitler launched the German attack on the Soviet Union, code named "Operation Barbarrossa," with 175 Divisions of Nazi troops, and massive aerial bombings of Russian cities and military formations. The German invasion should not have surprised Stalin as he had received numerous warnings through both Soviet intelligence and British, American and other government sources. Churchill had even warned Stalin of the impending German attack by the use of Britain's most important state secret of the time, the successful decoding of German orders by a British team that had used a smuggled "Enigma" encoding machine, which had been obtained from Polish intelligence, which became known as the "Ultra" traffic. (By 1942, this cryptanalytic capability would allow the British, and by extension, the Americans with whom the subsequently shared the secret, to know virtually all German plans and military deployments.) Yet, the Soviet dictator remained convinced until the battle erupted that all warnings were part of a plot by the British and their capitalist allies, including the United States, to incite a war between his regime and Hitler's.

In the United States, the left, which up until this moment had been against any aid to the beleaguered western European democracies, and had accused Roosevelt of being a "warmonger," immediately demanded American aid and intervention on behalf of Soviet Russia. Little more than a month after the German attack on Russia, Roosevelt had dispatched his close confidante and administrator of Lend-Lease aid, Harry Hopkins, to Moscow, with a promise to supply American military aid as soon as possible, and a message for Stalin that the dictator should feel "the identical confidence you would feel if you were talking directly to me."

Roosevelt met with Churchill shortly thereafter in Newfoundland, to formulate common war aims, in what came to be called the "Atlantic Charter," and the two leaders agreed that all possible aid should be sent to the Soviets, even if it drew down British and American supplies and equipment. The Soviet armies meanwhile, suffered major defeats in rapid succession, which cost them millions of casualties, including hundreds of thousands of Russians taken prisoner in first German encirclements, and many more afterward. Averell Harriman and Lord Beaverbrook were sent to the USSR to make the Allied intentions toward Russia more plain, and shortly after this the German Army advance had penetrated to within thirty miles of Moscow. Stalin seemed to be paralyzed with fear for a time, and ordered the Soviet government to evacuate Moscow and reassemble in Kuibyshev.

Roosevelt also appointed a new ambassador to the USSR, the former Chief of Naval Operations, Admiral William H. Standley, who had allied himself with the President in the campaign to convince the Congress and the American people of the necessity of supplying aid to the nations of Europe who were fighting the Nazis. Colonel Faymonville remained in Moscow as the principal agent for American aid. Stalin had expressed displeasure with the embittered Ambassador Steinhardt, which Harriman had conveyed to Roosevelt. Harriman himself thereafter continued in his role as a special envoy for Roosevelt and expediter of Lend-Lease aid, headquartered in London.

The German attack resulted in the re-establishment of diplomatic relations between Poland and the USSR on July 30, 1941. This was a small price to pay for Stalin, who was truly desperate and wanted aid from the British, and particularly the Americans. On August 16th, the Polish Ambassador to the U.S. met with the Soviet Ambassador Oumansky in Washington. They discussed formation of a Polish army from among the some 250,000 Polish soldiers held prisoner in Russia since 1939, to fight the Germans alongside the Red Army. Oumansky hinted then that there might be fewer Poles, particularly officers, in Russia then expected: "...The Soviet Government...looks after the Poles in Russia...I am afraid there are fewer officers and soldiers in our camps than your government imagines."

Polish Ambassador Ciechanowski replied: "I think we have good reason to know that there are quite considerable numbers..."

They then discussed Polish civilian deportees in Siberia and the Russian arctic north. Oumansky retorted: "I am not aware that there are any great numbers of Poles now in Russia..."

Ciechanowski replied bluntly to the Russian that there were, "over a million and a half Polish deportees in Russia."[32]

In his later memoirs, Ambassador Ciechanowski remembered:

"In the permanent Soviet blackout it had never been possible to determine the exact number of Poles forcibly deported from Poland in 1939 and 1940, concentrated in prisons and forced-labor camps in Soviet Russia. As far as it was possible to ascertain a conservative estimate fixed their total somewhat above a million and a half men, women and children. On September 17, 1940, the official Soviet Army paper Red Star declared that in addition to civilians 181,000 Polish prisoners of war were held in Soviet camps and that this included twelve generals, fifty-eight colonels, seventy-two Lieutenant Colonels, and 9,227 officers of other ranks.

"The liberation of these Polish people was a slow process... General Wladyslaw Anders, just released from the Lubyanka Prison in Moscow...a personal friend of General Sikorski, was appointed commander in chief of the Polish forces in Russia...(In late October 1941)...The freeing of our deportees was not proceeding so smoothly... The Polish Embassy in Moscow had to intervene repeatedly by means of notes and verbally with the Soviet government, and to point to the desperate situation of Poles still deprived of freedom in the Arctic regions of Russia."[33]

Anders soon found that there were few officers among the 46,000 released Polish soldiers who had reached his training camps in Russia. Sikorski had the Polish Ambassador to Russia confront Stalin on the some 10,000 missing Polish officers on November 14, 1941: ... We have not yet heard from even one of the officers, prisoners of war, detained in the three camps of Starobielsk, Kozielsk, and Ostashkov, who were transferred to an unknown destination in the spring of 1940..."

Stalin was anxious to accommodate his new western allies and ordered the POWs release, telephoning for further information on them. After hearing back from his security organs, the Soviet dictator suddenly changed the subject.

Generals Sikorsky and Anders, and the Polish Ambassador to Russia, met with Stalin in the Kremlin on December 3, 1941. Sikorski confronted the dictator: "...The amnesty granted by you to our people is not being carried out. Many of our most valuable men are still in labor camps or in prisons."

Stalin addressed both Molotov and Sikorsky:

"That cannot be correct, since the amnesty was to embrace all, and all Poles are liberated now."

General Anders then produced depositions from released Polish prisoners testifying to the fact that many Poles were still being confined by the Soviets. Anders attempted to supply an explanation that Stalin and Molotov would understand:

"You see, many camp commanders who have to carry out a production plan do not wish to deprive themselves of workmen, without whom the execution of their plan may be impossible."

Molotov seized on the explanation saying: "I think that may be so."[34]

Anders commented that perhaps the labor camp commanders did not understand the importance of the new Allied alliance against the Nazis. Stalin said: "Such commanders should be punished."

Sikorski then produced a list of the names of about 4,000 Polish officers who had been taken prisoner by the Soviets but had never appeared at Anders' military camps after the amnesty. He made it clear that this was only a partial list of the missing POW officers. According to a report of the meeting personally revised by General Sikorski, he said to Stalin, "Not one of these officers—and their number is probably at least twice as large as the 4,000 shown on my list—has turned up as yet."

"That is impossible," snapped Stalin. "They must have escaped somewhere."

Anders interjected: "...Where could they have escaped?"

Stalin, who already knew his secret police had murdered the officers was uncomfortable: "Well, to Manchuria, maybe."

Anders cut in boldly for a man so recently released from the terrors of the Lubyanka and still in Russia: "It is out of the question that all of them have escaped. Moreover, all correspondence with their families ceased abruptly in April and May 1940 when they were transferred from their former prison camps to an unknown destination."[35]

The Soviet dictator was already trapped in his lies, but there was to be no Soviet admission of the NKVD massacres at Katyn and elsewhere for fifty years. Indeed, throughout WW II and into the Cold War of the early 1950s the U.S. Government was to suppress certain evidence of Soviet guilt in order first to "maintain Allied unity" and later apparently to avoid unnecessary tensions, or perhaps to avoid any unnecessary prying into the entire subject of Soviet-held Allied POWs.

With the Japanese attack on Pearl Harbor on December 7, 1941, the United States entered the Second World War against the Axis powers, and American policy towards the Soviet Union was thereafter greatly affected by the need to maintain wartime "Allied unity." The only discernible price Stalin ever paid for this Allied compliance in covering-up one of the greatest known mass-executions in modern history, was to allow the release of other POWs and civilians still alive in Soviet captivity. Not all were released. Many thousands disappeared forever.[36] Others among them had no doubt been also been executed in unknown massacres, but many had either died in the NKVD's forced-labor camps, or were simply retained by the Soviets to be worked to death during the Second World War. The Polish military POWs who were released later served with distinction in the Polish Corps in the North

African campaign, and later in Italy.

When the Polish ex-prisoners of war left Russia in 1941 and 1942 for service with Allied forces, they staged in Iran. Lt. Colonel Szymanski of the U.S. Army was the American intelligence officer attached to the Poles as a liaison officer. By the middle of 1942 Szymanski and his British counterpart, Colonel Hulls, with free access to the ex-POWs, were amassing data on the missing Polish officers. Szymanski's report to the War Department, together with the similar British report clearly indicated Soviet guilt in the massacre, but both were suppressed by their governments as dangerous to the ongoing wartime alliance with Stalin. (On April 30, 1943, Szymanski sent another report on the Katyn Forest massacre to General George V. Strong, then Chief of U.S. Military Intelligence, which was also suppressed.[37] Lt. Colonel Szymanski was later punished for uncovering the truth of Soviet guilt by the sending of a cable to his superiors censuring him for his "bias in favor of the Polish group which is anti-Soviet." News of this type of action against a field grade officer travels through the Army's "grapevine" and would have had a dampening effect on other officers finding any fault with the Soviet allies. Thus is a "party line" established in the military services by the political powers-that-be.

While the Soviets made a show of cooperating with the Polish government-in-exile in London, to assure that massive American aid would continue to reach the USSR for its fight against Germany, they simultaneously began to prepare for eventual reoccupation of Poland. In 1942 Stalin's regime fostered the founding of the communist "Polish Workers Party," recruiting exiled Polish communists in Russia. Wladyslaw Gomulka, a communist Pole who had spent most of his life in the USSR took over as head of this party in 1943.

REVELATION OF THE KATYN FOREST MASSACRE

In February 1943 the German government announced that mass graves of thousands of Polish officers had been discovered in the Katyn Forest near Smolensk in western Russia. Over two months of exhumations followed, with many independent observers being transported to the scene by the Nazi government, in hopes of turning the entire episode into a stunning propaganda blow at Stalin's regime, at the same time the Germans were themselves massacring Jews, Soviet POWs and others who they considered sub-human. It is necessary to analyze the official U.S. and British response in 1943 to revelation of the Soviet Katyn Forest massacre, in order to understand how the same attitude about the importance of maintaining "allied unity" led to the abandonment of Soviet-held U.S., British and western Allied POWs in 1945.

A purge had occurred in the Department of State in March 1943, carried out at the behest of the Roosevelt White House by Undersecretary of State Sumner Welles, despite opposition from

Cordell Hull. The Soviets had expressed a loss of faith in the Allies over their failure to open a "second front" on the European mainland in 1942, and the charges by Litvinov and other Soviet officials that the U.S. and Britain were letting Russia down had a profound impact on Roosevelt and Churchill, who constantly feared a Soviet-Nazi armistice that would free the Germans to concentrate all their forces against the West. This led to a meeting between Litvinov and Welles during early March 1943, after the Nazi revelation of the Katyn Forest massacre, during which the Soviet presented a list of U.S. officials who were accused of sabotaging Soviet-American relations. Loy Henderson, Chief of the European Affairs Division was one of the primary Soviet targets, and after a public smear campaign in the media, with encouragement from Eleanor Roosevelt, Harry Hopkins and Joseph Davies, he was eventually exiled from Washington to Iraq. The naive Mrs. Roosevelt absurdly believed that Henderson had turned against the Soviet regime, at least partially, "because he missed the atmosphere of high society there."[38] By this time William C. Bullitt, who had so ably defended the Bolsheviks in the past, had turned against the Soviet regime with a vengeance, due to his own bitter personal experiences, and he used the Henderson incident to warn Hull that his political enemy, Welles, was allowing the Soviets to dictate the makeup of the U.S. State Department, thus angering the Secretary of State, who already felt that Welles had undermined his authority by dealing directly with the pro-Soviet New Dealers in the White House. This incident was to lead to Welles' own downfall during the Katyn Forest massacre episode.[39]

In Poland, two U.S. prisoners of war, Colonel John H. Van Vliet and another American officer, Captain Donald Stewart, were brought from German POW camps in Poland to witness some of the disinterments at Katyn Forest. Although the evidence was clear to Van Vliet and his fellow prisoners that the Soviets had done the deed, the POWs made no statements for fear of them being used for German propaganda. (It was not until 1945 that Van Vliet was repatriated and sent to Washington to make his report on Katyn, which was then classified top secret, and disappeared.) In addition, Kathleen Harriman, the daughter of then-U.S. Ambassador to Moscow Averell Harriman, went to witness the site and also wrote a report about Katyn which stated, in part: "...The party was shown the graves in the Katyn Forest and witnessed post mortems of the corpses. As no member was in a position to evaluate the scientific evidence given, it had to be accepted at face value...We were expected to accept the statements of the high ranking Soviet officials as true, because they said it was true."[40]

On April 21 Stalin cabled Churchill:

"...The anti-Soviet slander campaign launched by the German fascists in connection with the Polish officers whom they themselves murdered in the Smolensk area...was immediately seized upon by the Sikorski government and is being fanned in every way by the Polish official press....The Hitler authorities, having perpetrated a monstrous crime against the Polish officers, are now staging a farcical investigation...The Soviet government has decided to interrupt relations with that government."[41]

Information already in the hands of Allied intelligence that contradicted the Soviet version of Katyn was then being suppressed at the highest levels of the American and British governments. Ever fearful of a disruption of Allied unity, Churchill wired back in a conciliatory manner on April 24th:

"...We shall certainly oppose vigorously any 'investigation" by the International Red Cross or any other body in any territory under German authority...I hope therefore that your decision to 'interrupt' relations is to be read in the sense of a final warning rather than a break."[42]

Stalin replied on April 25th: "...The matter of interrupting relations with the Polish government has already been settled and today V.M. Molotov delivered a Note to the Polish Government."[43]

Churchill again tried to mollify Stalin on the 25th: "...Sikorski emphasized that previously he had several times raised the question of the missing officers with the Soviet government and once with you personally...Sikorski has undertaken not to press the request for the Red Cross investigation...He will also restrain the Polish press from polemics...I am examining the possibility of silencing the Polish newspapers in this country which attacked the Soviet government and at the same time attacked Sikorski for trying to work with the Soviet government. In view of Sikorski's undertaking I would now urge you to abandon the idea of any interruption of relations..."[44]

Polish Ambassador to the U.S. Jan Ciechanowski met with Undersecretary of State Welles the next day in Washington: "Since July 30, 1941, when Polish-Soviet relations had been resumed, we had been repeatedly asking the Soviets for news of these Polish officers. We had never obtained any definite or satisfactory reply to our enquiries...On April 26 I called on Mr. Sumner Welles...I sensed for the first time a tense atmosphere in my relations with the Undersecretary. He started by saying that he could not understand how the Polish government could have appealed to the International Red Cross to investigate an accusation made—as he put it—'by the German propaganda machine.'"

Ciechanowski was "taken aback" by this and asked Welles, "Did the American Government think that any self-respecting government could possibly pretend to ignore such an outrage to its fighting men?..."[45]

Clearly, both the United States Government and that of the

British, did desire that the Poles ignore the massacre of their POWs. A few days later Ciechanowski saw the British Ambassador in Washington, Lord Halifax, who asked him if Sikorski might consider removing some of his cabinet ministers considered unfriendly to the Soviets. The Polish Ambassador pointed out that this would be "equivalent to an acceptance of dependence of the Polish government and a recognition of the right of the Soviet government to influence the appointment of Polish Cabinet Ministers...Lord Halifax asked me what I thought would be the next phase in Soviet policy toward Poland. I replied that it would probably be the announcement of the formation of a puppet government for Poland, and of an allegedly Polish armed force...Lord Halifax became thoughtful, but said he still could not believe that the Soviets wanted to dominate Poland or to Sovietize it..."[46]

President Franklin Roosevelt sent a message to Stalin on the Katyn revelations on April 26th, which, while not admitting his personal acceptance of the Soviet version of Katyn, nevertheless hinted at it, offering assistance in quieting the Poles about the massacre in return for Stalin releasing other Polish prisoners:

"...I hope that you can find a way in this present situation to define your action as a suspension of conversation with the Polish government in exile in London rather than label it as a complete severance of diplomatic relations between the soviet Union and Poland...I cannot believe that Sikorski has in any way...collaborated with the Hitler gangsters. In my opinion however, he has erred in taking up this particular question with the International Red Cross. Furthermore, I am inclined to think that Prime Minister Churchill will find a way of prevailing upon the Polish government in London in the future to act with more common sense. I would appreciate it if you would let me know if I can help in any way in respect to this question and particularly in connection with looking after any Poles which you may desire to send out of the Union of Soviet Socialist Republics..."[47]

Stalin replied to Roosevelt on April 29th, that it was too late to alter the break in diplomatic relations and added assurances of his care for the Poles in Russia which appear ludicrous in light of the sufferings of the more than one and a half million Polish prisoners in Soviet forced-labor camps:

"...On April 25 the Soviet Government was compelled to interrupt relations with the Polish Government...As the Polish Government for nearly two weeks, far from ceasing a campaign hostile to the Soviet Union and beneficial to none but Hitler, intensified it in its press and on the radio, Soviet public opinion was deeply outraged by such conduct...As regards Polish subjects in the USSR and their future, I can assure you that Soviet Government agencies have always treated and will continue to treat them as comrades, as people near and dear to us...If they themselves wish to leave the USSR, Soviet Government agencies will not try to hinder them, just as they have never done..."[48]

Winston Churchill sent another personal and secret message to Stalin on April 30th, in which he carefully hinted at the real reason behind the Soviets breaking diplomatic relations with the Poles, and hinted that if it were true that the Soviets were forming a communist puppet government for Poland it would not be recognized by Britain or its allies:

"I cannot refrain from expressing my disappointment that you should have felt it necessary to take action in breaking off relations with the Poles...Mr. Eden and I have pointed out to the Polish Government that no resumption of friendly or working relations with the Soviets is possible while they make charges of an insulting character against the Soviet Government and thus seem to countenance the atrocious Nazi propaganda. Still more would it be impossible for any of us to tolerate inquiries by the International Red Cross held under Nazi auspices...

"I am glad to tell you that they have accepted our view and that they want to work loyally with you. THEIR REQUEST NOW IS TO HAVE DEPENDENTS OF THE POLISH ARMY IN IRAN AND THE FIGHTING POLES IN THE SOVIET UNION SENT TO JOIN THE POLISH FORCES ALREADY ALLOWED TO GO TO IRAN...The cabinet here is determined to have proper discipline in the Polish press in Great Britain...So far this business has been Goebbel's triumph. He is now busy suggesting that the USSR will set up a Polish Government on Russian soil and deal only with them. We should not of course, be able to recognize such a Government..."[49]

On May 5th Roosevelt proposed a curious private meeting between he and Stalin on the Bering Straits between Alaska and Siberia, in which their respective staffs would not be present. Although the meeting never came off, the proposal indicated Roosevelt's propensity to ignore traditional diplomatic channels and rely on his own charisma to conduct delicate affairs of state.

In the meantime, on May 2nd, Churchill, who was fearful of the Soviets making a separate peace with the Germans as they had in 1918, thus leaving the western Allies to bear the brunt of the war, had written another conciliatory message to Stalin: "I have just read with the utmost satisfaction and admiration your splendid speech on May Day, and I particularly appreciate your reference to the united blow of the Allies and you can indeed count on me to do everything in my power 'to break the spine of the Fascist beast...'"[50]

Stalin attempted a practically psychotic explanation of Soviet conduct in a message to Churchill on May 8th, containing promises which were already, and continued to be, broken:

"In sending my message of April 21 on interrupting relations with the Polish Government, I was guided by the fact that the notorious anti-Soviet press campaign, launched by the Poles as early as April 15...had not encountered any opposition in London... It is hard to imagine that the British Government was not informed of the contemplated campaign...Since the Poles continued their anti-Soviet smear campaign without any opposition in London, the patience of the Soviet government could not have been expected to be infinite. You tell me that you will enforce discipline in the Polish press. I thank you for that but I doubt if it will be as easy as all that to impose discipline on the present Polish Government...As to the rumors, circulated by the Hitlerites, that a new Polish Government is being formed in the USSR, there is hardly any need to deny this fabrication. Our Ambassador has already told you so...As regards the Polish citizens in the USSR, whose number is not great, and the families of Polish soldiers evacuated to Iran, the Soviet Government has never raised any obstacles to their departure from the USSR."[51]

The U.S. Minister to the Polish government-in-exile, Ambassador Anthony J. D. Biddle, Jr., was one of the few U.S. officials who would not simply let the Katyn Forest issue die. Biddle had sent a secret cable to the State Department as early as June 2, 1942, which clearly outlined the seriousness to the Polish Allies, of the missing Polish POWs who were already believed to have been murdered by the NKVD at Katyn:

"The absence of these officers is the principle reason for the shortage of officers in the Polish forces in Russia, whither officers from Scotland had to be sent recently. The possible death of these men, most of whom have superior education, would be a severe blow to Polish national life...."[52]

This had of course, been the precise intention of Stalin and his NKVD henchmen, Beria, Merkulov and Serov. Their plans for a postwar Poland did not include a strong national life. A secret report on, and evidence of the Soviet conducted massacre at Katyn Forest, compiled from solid Polish intelligence sources, was forwarded from London to Undersecretary of State Sumner Welles by Ambassador Anthony Biddle on May 20, 1943.[53] This report, which was suppressed by Welles and remained secret until the 1952 Congressional investigation in to the Katyn Forest Massacre, could have, with other secretly suppressed supporting evidence, changed the course of postwar history had it been revealed to the world at the time. The final report of the 1952 Congressional investigation on the Katyn Forest massacre makes this clear:

"It becomes apparent to this Committee that the President and the State Department ignored numerous documents from Ambassador Standly, Ambassador Biddle and Ambassador Winant... who reported information...which strongly pointed to Soviet perfidy..."[54]

According to US documents in the National Archives, top secret intelligence reports of the deportations of over 1,500,000 Poles, and of the Katyn Forest massacre and disappearance of some 15,000 officers and NCO's of the Polish Army by the Soviets were carefully studied during 1942 and 1943 by the American OSS, U.S. Military Intelligence and the British. These reports were apparently used to assess Stalin's future possible conduct with military POWs.[55]

Another important secret report on the Katyn massacre, which also indicated Soviet guilt and was covered up, was written much later by U.S. Army Colonel John H. Van Vliet, who was one of the 2 American POWs in a Nazi camp in Poland brought as a witness in 1943 (with several British POWs and Kathleen Harriman) to the scene of the exhumations by the Germans. Although Van Vliet had no love for the Nazis who then held him prisoner, he came to the "clear and irrevocable" conclusion that the Polish POWs at Katyn had been massacred by the Soviets. He wisely refused to make any statement while a German prisoner, however, for fear that it would be used as Nazi propaganda, and was forced to wait until he was (fortunately) repatriated in 1945. At that time he was immediately ordered to come to Washington, D.C., where he was required to write a report on his conclusions about the mass murder, by Major General Clayton Bissell, then Chief of U.S. Military Intelligence (Assistant Chief of Staff, G-2) in the Pentagon. According to the official record of 1952 Congressional testimony, Bissell then classified the only handwritten copy of this report Top Secret and although the war in Europe was then over and confrontation with the Soviets over missing Allied POWs had already begun, he ordered Colonel Van Vliet to "maintain absolute secrecy concerning his report."[56]

According to records of the Congressional hearings, "This Top Secret document has disappeared from the Army Intelligence (G-2) files, and to this date has not been found. THE SEARCH FOR THIS VAN VLIET REPORT HAS BEEN ONE OF THE MOST IMPORTANT TASKS OF THIS COMMITTEE. An independent investigation conducted by the Army's Inspector General in 1950 concluded the report had been 'compromised' and that there is nothing to indicate it had ever left Army Intelligence (G-2). This finding was in response to General Bissell's allegation that he 'believes' he had forwarded Van Vliet's report to the Department of State."[57]

No evidence could ever be found to substantiate Bissell's half-hearted claim. Clayton Bissell, a former WW I officer and an intriguing character whose name was to surface during Congressional investigations into the communist affiliations of the Institute for Pacific Relations (IPS), was questioned during the Katyn hearings and the final report of the proceedings contains this summary:

"It is this Committee's conclusion that General Bissell is mistaken in his claim that he might have forwarded the Van Vliet report to the State Department. General Bissell himself admitted to the Committee that had the Van Vliet report been publicized in 1945, when agreements for creating a United Nations organization at Yalta were being carried out in San Francisco, Soviet Russia might never have taken a seat in this international organization.

"In justifying his actions for designating the Van Vliet report "Top Secret," General Bissell said he was merely carrying out the spirit of the Yalta Agreement. He admitted the report was explosive and came at a time when the United States was still trying to get a commitment from the Soviets to enter the Japanese war. General Bissell contradicted his own theory when he told the committee that the Van Vliet report couldn't have been sent to the Secretary of the Army 'because it had nothing to do with the prosecution of the war at that time." This committee was dismayed to learn that the United States Assistant Chief of Staff, Army Intelligence, was considering political significance of the Van Vliet document, which should have been treated objectively from a strictly Military Intelligence standpoint.

"In the opinion of this committee, it was the duty and obligation of General Bissell to process this document with care so that it would have reached the Department of State. General Bissell testified":

'"I saw in it (the Van Vliet report) great possibilities of embarrassment; so I classified it the way I have told you, and I think I had no alternative.'

"More amazing to this committee is testimony of three high-ranking American Army officers who were stationed in Army Intelligence (G-2) during General Bissell's command of this agency. TESTIFYING IN EXECUTIVE SESSION ALL THREE AGREED THERE WAS A POOL OF 'PRO-SOVIET CIVILIAN EMPLOYEES AND SOME MILITARY IN ARMY INTELLIGENCE (G-2) WHO FOUND EXPLANATIONS FOR ALMOST EVERYTHING THAT THE SOVIET UNION DID.'

"THESE SAME WITNESSES TOLD OF TREMENDOUS EFFORTS EXERTED BY THIS GROUP TO SUPPRESS ANTI-SOVIET REPORTS. THE COMMITTEE LIKEWISE HEARD TESTIMONY THAT TOP-RANKING ARMY OFFICERS WHO WERE CRITICAL OF THE SOVIETS WERE BYPASSED BY ARMY INTELLIGENCE (G-2)...EVIDENCE UNEARTHED BY THIS COMMITTEE SHOWS THAT...HIGHLY CRITICAL REPORTS OF SOVIET RUSSIA WERE BURIED IN THE BASEMENT OF ARMY INTELLIGENCE (G-2), AND SUBSEQUENTLY MOVED TO THE 'DEAD FILE' OF THAT AGENCY."[58] Bissell and his group of pro-Soviet military intelligence specialists were to remain in authority in that agency until early 1946 and Allied acknowledgment of the Cold War with Russia.

The secrecy imposed by Sumner Welles, General Clayton Bissell and his assistants such as Colonel Alfred McCormack of U.S. Military Intelligence, and others in the U.S. Government, ensured that

the truth about the Katyn Forest massacre would not be allowed
to emerge until it was to late to alter the course of postwar Soviet
domination and partition of Poland, an issue which had brought
on World War II. President Roosevelt himself was determined that
nothing would be permitted to break up the alliance between the
Western democracies and Soviet Russia against Nazi Germany.
Winston Churchill concurred in this view. Implacable Soviet policy
toward Poland was to lead to another NKVD operation that began in
Soviet-occupied Poland and which was suppressed with even greater
secrecy, in which the Russians were to kidnap thousands of American,
British and Allied POWs from Nazi camps in Poland, Germany and
Austria. General Clayton Bissell was to remain in command of
Army Intelligence, assisted by his clique of pro-Soviet colleagues,
throughout this 1945 NKVD atrocity, and he very likely played a role
in suppressing the truth of the matter, "in the spirit of Yalta."[59]

The constant veiling of real Soviet motives and actions by
Allied leaders, in the name of wartime (or postwar) Allied unity, made
the subsequent, sudden onset of the Cold War all the more confusing
to the people of the West.[60]

## BRITISH, FRENCH AND BELGIAN POWS HELD WITH THE POLES IN RUSSIA

The Soviets had already captured several hundred British,
French and Belgian POWs, in 1941, who had escaped from German
captivity in Poland, only to be arrested by the NKVD in Russia.
Among the imprisoned Polish civilians confined by the Soviets, later
fortunate enough to be allowed to join General Anders' Polish Army
in Russia, was a writer named Tadeuz Wittlin, who survived the war
and eventually wrote of his experiences in the Soviet gulags. Of all
his memories, one in particular serves to illustrate both the paranoia
of Stalin and his regime and the complicity of the Allied governments
in keeping the obscene secrets of the Soviet dictator's slave empire,
in the name of allied "unity." Of the hundreds of thousands of British,
French and Belgian prisoners captured by the Germans during the
1940 campaign which ended at Dunkirk, approximately 200 escaped
Nazi captivity and eventually crossed the Soviet-Polish frontier.
Instead of being well treated or repatriated to Britain as hoped, all
were arrested and confined by the NKVD. Tadeuz Wittlin met one of
them in a Soviet prison in March 1941:

"My nearest neighbor not only understood nothing that was
said to him, but also nothing that went on around him. The nightmare
reality of a Soviet prison is not easily grasped by a Briton. This man
was English, born in Manchester, by trade a driver. His name was

Edward Baldwin. As a sergeant in the B.E.F. (British Expeditionary Force), he was captured by the Germans in France and taken to a P.O.W. camp in Poland. It was not easy to escape from a camp like that, but, somehow, Baldwin managed to slip through the barb wire to freedom. By day he lay low in woods and barns. At night he marched East. Always East, making his way by the stars, the bark of trees and finally with the help of a cheap compass which he bought in a village store. The peasants he came across did not understand him, but they gave him what they could. And so at last he reached the River Bug, meaning to cross into the Russian occupation zone, where he hoped to be given asylum by the Russians and eventually sent back to England. The naive Baldwin did not know that the Russians have their own peculiar way of looking at these things. Caught while crossing the border, he was accused of espionage and thrown in prison.

"They flung him into our cell during the night. The black and white room, filled with the bodies of sleeping men, looked like a woodcut out of 'Dante's Inferno.' The soldier's wide eyes gleamed with terror. He did not answer any of the questions thrown at him but simply stood there as though paralyzed.

"'Deaf...' said someone.

"'...and dumb,' another voice added.

"'English,' mumbled the Sergeant, taut with fear.

"'What's that? English? Well, I'll be... How did you get here?'

"I took him by the arm and led him to my corner. He sat down, smiling his gratitude.

"Thereafter, we struck up a friendship and he became my English teacher. Lectures began in the morning after breakfast. The Sergeant used an entirely new version of the direct method, without textbook, pen or paper. We had just finished subjugating the verb 'to have,' when he took off his shoe.

"'What is this?'

"'This is a shoe, I answered.

"'Yes.'

"The teacher then removed a fragrant sock and, waving it under my nose, asked:

"What is this?'

"This is a sock.'

"All right."

"He turned the sock inside out and began removing the troublesome little lice. Seeing my teacher thus engaged, I promptly pulled the shirt off my back and saying, 'What is this? This is a shirt,' following his example.

We were working away like this when the door opened to admit the orderly with a piece of paper in his hand. He was at once surrounded by an uneasy crowd. It could only be one of two things—

trial or transportation. The orderly would shoot a name like a bullet into the crowd and depart with the stricken victim. Like a hawk waiting to sink a few words, talon-like, in the chosen creature and carry him off. The iron doors would no sooner shut than the usual questions and doubts would be voiced. What's he gone for? Release? Siberia? Death? Or maybe he's only been moved to another cell in the very same prison?

The guard drew his eyes over the tense assembly and shouted:

"'Edward Edwardovitch Baldwin.'

I nudged my friend,

"'Sir! shouted the Sergeant in his English parade-ground voice.

"'Pick up your stuff and follow me!.'

"Edward 'Edwardovitch' Baldwin, driver from Manchester and sergeant in the British Army, did not grasp the meaning of the particular Russian words, but he understood their portent. He fastened all his home-made tunic buttons, sewn on by himself, and thrusting his way to the door, threw us a final:

"'Good-bye, boys!'

"The door shut behind him. All this took place in March 1941, in a building which, intended as a monastery, had been adapted by the Soviets for a purpose more in keeping with the times.

"When, a few years later, I became editor of an Army periodical published by the British Ministry of Information, I wrote the above in memory of my first English teacher. Unfortunately, the story was not published in the magazine and the manuscript was returned to me with the stamp: 'Stopped by Censor.'

"'I am extremely sorry, but we could not possibly pass your excellent contribution,' said the excessively suave Intelligence Corps Captain in the British Censorship Department, where I went for an explanation. 'But, you know war entails casualties,' he added mysteriously, as if making a revelation of some sort. 'Let us just think that Sergeant Baldwin died at the front. Just at present we can't upset our great Ally. You do understand?'

I did not."[61]

This tragic incident had underscored the lengths that the British government under Churchill had been willing to go to, in avoiding any disruption of the wartime alliance with Soviet Russia, and makes more understandable the conduct of the British and American governments in covering up the Katyn Forest massacre.

Nearly half a century later, after the collapse of Communism in Eastern Europe, an independent Soviet newspaper published more facts about these British, French and Belgian prisoners of 1941, in Russian prisons. Writer Vladimir Abarinov located references to 195 of these men in the Soviet archives of the NKVD's escort troops. Two days after the Presidium of the Supreme Soviet divided the NKVD under Beria and the NKGB (state security) under Merkulov, an order was issued to transfer the British and French POWs from a Moscow prison to a concentration camp at Kozelsk. The "top secret and very urgent," February 5, 1941, order was addressed to Captain of State Security Soprunenko, Chief of the Main Directorate of Escort Troops of the USSR Directorate for POWs:

"By order of the People's Commissar for State Security Comrade Merkulov, I hereby request that forty-five interned Englishmen and Frenchmen should be urgently transferred from the Butyrskaya Prison to the Kozielsk camp and separated from other inmates."

Archives of the 136th Escort Battalion at Kozielsk revealed that on June 28, 1941, 6 days after Hitler's invasion of Russia began, 195 interned French, British and Belgian prisoners of war were transferred to the Gryazovets concentration camp. In the process of this transfer 11 of the English prisoners were separated for release to the British government, put together with three others and on July 9, 1941 all 14 were turned over to the British Ambassador in Moscow, Sir Stafford Cripps. At least one other, a British Navy petty officer named Maurice Barnes, was reported killed by guards at the Soviet border.

British historian Lord Nicolas Bethell (author of "The Last Secret") assisted the Soviet researcher Abarinov with some British records of the affair and in locating several of the 15 survivors who had been returned to England in 1941. All 15 had been decorated upon return, but also ordered to forever remain silent about their prison experiences in Russia. This was done to avoid negative publicity about Britain's new Soviet ally at the same time the Polish government-in-exile was asking Moscow about the 15,000 missing Polish POWs. The form which this silencing took in attempts to publicize the other British POWs of 1941 who disappeared in the USSR is evidenced by Tadeuz Wittlin's account of Sergeant Baldwin. Hubert Lovegrove, a corporal in the Gordon Highlanders before his capture and eventual release from Russia as one of the 15, told Lord Bethell many years later, "Never in my life have I been so cold and hungry."

Later in the war when American bombers were attacking Japan and Germany and the Romanian oil fields at Ploesti, from 1942-1945, and as a result sometimes crashed or made forced landings on Soviet territory, they too were interned by the NKVD and in some cases sent to gulag forced-labor camps, even though the U.S. and Russia were officially "allies" at the time. (This act of treachery by Stalin and his henchmen was not to be publicly admitted by Moscow until after the collapse of the Soviet Union and the rise of Boris Yeltsin as President of a new, post-communist Russia in 1992.)

Undersecretary of State Sumner Welles had, in the meantime, resigned from the U.S. Government by the end of August 1943.

Accusations of his homosexuality had been publicly revealed after he reportedly propositioned a black porter on a railroad journey, and his political enemy and rival for Roosevelt's attention, Ambassador William C. Bullitt, had chosen to publicize the matter in the newspapers.[62] In addition, Secretary of State Cordell Hull accused Welles of formulating U.S. policy on his own and undermining his authority in the Department. Hull publicly avoided discussing the personal habits of his Undersecretary, however, but made it clear to Roosevelt just before the Quebec Conference, that the President had to remove Welles, or that he, Hull, would have to leave. Welles had been one of Roosevelt's closest collaborators, and Bullitt had engendered the President's displeasure in forcing the issue and he never recovered Roosevelt's favor. In September 1943, Edward Stettinius, the straightforward and popular former president of U.S. Steel who had been one of Roosevelt's early supporters and served as Lend-Lease administrator, replaced Welles as Undersecretary of State.[63]

In September 1943, a new U.S. Ambassador to Russia was also selected to replace Admiral Standley, who had lost favor with Stalin's regime in Moscow by not blindly accepting every Soviet demand. Standley had become furious at Colonel Philip Faymonville for his unswerving support of Stalin and the Soviet system, and his manner of bypassing the Ambassador in dealing directly with the White House. Admiral Standley had finally demanded that Faymonville be subject to the Ambassador's orders and eventually that he be removed from Moscow. By October, however, Standley had been retired himself, another of the casualties of the wartime alliance with Russia. Averell Harriman, a friend of the President and the candidate of the New Dealers closest to Roosevelt, was chosen to replace Standley. According to then-Polish Ambassador to the U.S. Jan Ciechanowski, who maintained relations with several of this group, Roosevelt's closest advisor was behind the appointment:

"It soon became apparent that W. Averell Harriman would be appointed. He was the personal candidate of Harry Hopkins, the power behind the throne, and Hopkins was seldom known to fail in putting through his appointments..."[64]

Under continual pressure since 1941 from Stalin to open a "second front" against Hitler, the western Allies had finally mounted a new offensive against the Germans with the U.S. landings in North Africa in November 1942, to support the British army already fighting there. By January 1943, the American Army had advanced to the Kasserine Pass in Tunisia, where they suddenly were assailed by a massive counterattack by Field Marshall Erwin Rommel's Afrika Corps. The American forces were forced to give ground and many U.S. POWs and MIAs fell into German hands. The names of 3,724 American soldiers who remain missing in action (MIA) from the Tunisian campaign are inscribed on the memorial walls of the Carthage American Cemetery, maintained by the American Battle Monuments Commission, while 2,840 other known American killed in action (KIA) and 234 unknown American dead were buried where the Romans and Carthaginians had once fought. Aside from the POWs captured in this campaign who were eventually released from German POW camps at the end of the war, among the missing in action were many who were actually POWs of the Germans, who would eventually be overrun by the Red Army's 1945 advance into Poland, eastern Germany and Austria. These American prisoners and thousands of others later captured or listed as MIA in Sicily and Italy in the fall and winter of 1943-44, had gradually been shipped to various German POW camps in Italy, Poland, Austria and eastern Germany, where they joined United Kingdom, British Commonwealth, French, Belgian, Dutch, Polish, Norwegian and other Allied prisoners, who had already spent years in captivity. (In the Allied advance against Salerno alone, from September 9th to the 18th, U.S. losses included 1,084 killed in action (KIA), 1,869 Missing in action (MIA) and presumed POWs and 3,252 wounded (WIA). In this same short campaign, which represented only a small share of Allied losses in the Italian campaign, the British lost 1,066 killed, 2,230 missing in action and 2,0002 wounded.)[65]

The British deciphering of German Enigma traffic should have enabled both them and the Americans to record with some certainty the numbers of British, American and other Allied POWs captured by the Germans, and to know how many others initially listed as MIAs had actually been captured by the Nazis and transported to the POW camps in eastern Europe. However, this ULTRA traffic has never been declassified and remains secret to the time of this writing. Aside from President Roosevelt, the ULTRA secrets were shared by the British with a very few other Americans who were initially involved through Sir William Stephenson's British Security Coordination headquarters in New York, and some of these individuals later rose to prominence. As author William Stevenson, a wartime fighter pilot who first met Sir William Stephenson on special assignment to intelligence, was to write decades later:

"A future chief of the Central Intelligence Agency, Allen W. Dulles, moved into Room 3663 at Rockefeller Center a month after Pearl Harbor. American academics began to be seen in country lanes around Bletchley, so that in time ULTRA was served by Telford Taylor, the future distinguished Columbia University professor; William Bundy, future Assistant Secretary of State and editor of the influential Foreign Affairs; Lewis Powell, who would sit on the Supreme Court."[66]

In the War Department in Washington, ULTRA and MAGIC intercepts were controlled by a small group within the U.S. Military

Intelligence "Special Branch," which was headed by a prewar New York lawyer brought into the War Department by Assistant Secretary of War John McCloy, Colonel Alfred McCormack, who served under the Chief of Military Intelligence, General George V. Strong. Another U.S. official with access to ULTRA traffic was James Murphy, head of the OSS X-2 Branch.[67] Later, the OSS chief in Italy James Angleton, a future CIA senior official would also be cleared to handle Enigma traffic originating from Bletchley Park. During the war, Winston Churchill ordered that selected ULTRA intelligence that he knew would be of value to the Soviets in defeating the Nazis be supplied to Stalin, but the Soviets were of course not satisfied with what was doled out to them. They wanted everything, so that they could better formulate their plans for the postwar domination of eastern Europe. In Britain, the code breakers of CCHQ at Bletchley Park's "Station X" included an important Soviet agent, John Cairncross, who had been recruited by Moscow Center while at Trinity College, Cambridge, and was later revealed to have supplied the NKVD with a vast amount of Enigma-generated intelligence. Anther Soviet agent who supplied Ultra traffic to the Russians was Leo Long.[68]

A significant step in Soviet-American relations had meanwhile been taken during the winter of 1942-43, when the Russians succeeded in gaining acceptance of their demand for veto power in future Allied war relief activities. This power was later to be used to keep Allied personnel out of Soviet-occupied areas so that NKVD actions would remain hidden from the free world. These negotiations, between December 21, 1942 and April 1943, were conducted for the Soviets by the old Bolshevik diplomat and revolutionary colleague of Stalin, Maxim Litvinov, with the Soviet Ambassador to the U.S. (and later Foreign Minister) Andrei Gromyko, and for the United States and Britain by Assistant Secretary of State Dean Acheson and Lord Halifax, respectively.

The Russians demanded that in Soviet-occupied territory they would be free to assume total responsibility for "relief and rehabilitation matters." Assigned by Secretary of State Cordell Hull to negotiate for the United States, Assistant Secretary of State Dean Acheson strongly and steadily supported the Soviet demands throughout these negotiations and gradually pushed Lord Halifax to gain British acceptance of this position by the British government. Thus was born the UN principle that all decisions of the Policy Committee must be unanimous or they could not be implemented.

The right of the Soviet Union to veto the other great powers in UNNRA matters was confirmed and expanded on during the conferences in Teheran, Iran, in November 1944, at Yalta in February 1945 and during the founding conference of the UN at San Francisco in April 1945. It was to become a major principle of the postwar functioning of the United Nations. This commitment by the United

States and Britain, to be bound by Soviet vetoes ordered by dictator Josef Stalin, led to innumerable difficulties for the Allies in the immediate postwar period.

Following American and British acceptance of this principle of Soviet veto power, Stalin's government was emboldened to increase its verbal attacks on the Polish government-in-exile, in preparation for forming their own communist puppet regime for postwar Poland. Among other things, the power of veto also was to ensure that the Soviets would be legally permitted to deny entrance into Poland (and other Soviet-occupied areas) for American and British relief officials and military officers attempting to contact and repatriate thousands of Allied POWs overrun by the Red Army in 1945.[69]

General Sikorski, the embodiment of Polish nationalism and military strength, was considered a serious threat to Soviet plans for dominating postwar Poland. Stalin and his cohorts sensed that Sikorsky was the one man capable of creating a strong and unified anti-Soviet force in the country if he ever got the opportunity, especially with the potential nucleus of a Polish Army then fighting under General Anders with the Allies in the Mediterranean Theater. Although Sikorsky was closely guarded by the London Polish government-in-exile, the NKVD still got their chance to kill him. In the summer of 1943, General Sikorski traveled by air to the Mediterranean area to visit the Polish Corps under Anders, which had performed well with the American and British forces in North Africa. A Soviet agent in the British government, reported to have been the later-unmasked spy H.A.R. "Kim" Philby, supplied the NKVD with the top secret itinerary of Sikorski's routes and stopping-points. This allowed the NKVD to mobilize an operation on the island of Malta, where Sikorski's aircraft was to be refueled on the return trip. A bomb was placed on the aircraft by a Soviet agent and it detonated on July 4, 1943, about 80 miles from Gibraltar. Sikorski and his entire staff were killed in the crash, but the Czechoslovakian pilot miraculously survived.[70]

The incident was immediately shrouded in secrecy under wartime censorship regulations, and was announced to the world as an accidental crash by the British and American governments. In an eerie addendum to the Katyn Forest massacre cover-up, the Polish government-in-exile was forced to publicly accept this story as the price for continued Allied "support." Again, the cause of American and British unity with the Soviet allies was the ostensible purpose for this travesty of justice. (As late as 1947, when former Polish Ambassador to the United States Jan Ciechanowski published his comprehensive history of wartime Polish-Allied relations, the death of General Sikorski was still recorded as accidental.)

As the Second World War continued in Europe and western Russia, vast armies of millions of slave laborers toiled to their deaths for Hitler's Nazi war machine and for Stalin's Communist regime. Hitler singled out Jews, Gypsies, Poles and others for extermination in death camps that used poison gas for mass killing and advanced crematoria for disposing of the millions of human corpses.[71] Nazi death camps were known to the Allied governments, but little was done by Allied bombing raids to disrupt the flow of doomed prisoners from all the occupied nations of western and eastern Europe. Likewise, with the wartime alliance between the Western democracies and Russia, the existence of the Soviet Gulag slave empire was either downplayed or denied completely by Roosevelt Administration officials and the mainstream American media, while Nazi atrocities, when uncovered, were widely published.

In January 1943, after nearly five years at forced labor in Stalin's gulags, the American prisoner Thomas Sgovio was still alive, now in the Valley of Death, a slave labor camp near Neksikan, in the Chai-Urrya Administration of Kolyma, Siberia. No better description of conditions there survives than what he later remembered of that winter:

"By mid-December more than half of my comrades from Srednikan had perished. I was amazed—those old-timers, strong in faith and constitution, experienced in the art of survival...I thought surely I would go before them...By midwinter almost all my Srednikan contingent had perished. It was the time of the year when one would normally notice the dwindling of our ranks at line-up. And yet, they did not. New rats were thrown in among us...When we awoke in the morning we glanced at the fellow next to us. Was he alive? If he was dead we hurriedly took his rags and covered the corpse. Since there was no check-up, the barracks orderlies often stuffed the dead Z/K's (Zeks) under the bottom planks for days so as to keep on receiving their bread portions.

"And there was the time when we were astonished to receive a different kind of soup. It tasted like no other food we had received before. It had meat and bones in it! The chunks were large...couldn't be American spam. The meat tasted fresh...Perhaps venison from the Yakut reindeer breeding farms. Two weeks we gobbled it down, then word was spread that the cooks had been arrested! They had stolen and sold the American lend-lease pork to free-citizens (who were also on rations), and substituted the flesh of Z/K corpses. That meant there was no count of the dead! And how did we feel when we had eaten human meat? At the time we were not human beings, incapable of feeling revulsion.

"It is difficult for me to analyze what happened in Satellite Camp GE7 because of the eternal darkness. I believe it would have been even worse had it not been for the blatniye (criminals) who tried to keep a semblance of order among themselves. Being masters at adapting themselves to the worst of camp conditions, they obtained food from the outside, had softer job; hence they did not decline to our depths. However, even their select group was beginning to deteriorate. There were those who, unable to get off general work, violated the blatniye laws, thus becoming Sukas (bitches). Cards were drawn to decide who was to be the executioner. Among those who had their throats slit were the ones caught stealing bread. There were cases where free-citizen blatniye attended special councils outside the camp zone to decide the fate of certain thieves. One such case was that of Vitya Sudarkin, our barrack orderly. It happened two months after I was transferred from Camp 7. After being knifed his corpse was stuffed under the floor..."

"Then there was self-mutilation...I heard of cases where blatniye, during logging operations on the mainland, chopped off their fingers, then printed letters-pleas for help with their bloody finger stumps on the logs which were being shipped to England...In the Chai-Urya Administration, self-mutilation was widespread. Prisoners blew off their fingers in the drying room with stolen American Lend-Lease detonators. During the war USVITL, to counter the rising tide of self-mutilation issued another decree. The guilty ones were to receive ten-year sentences as saboteurs! So the Z/K's devised a new method...To make it appear unintentional, they urinated in their mittens or burkkiis (depending on whether you wanted to lose a hand or a foot)—they urinated because water was scarce. Then when they were certain the member was thoroughly frozen, they notified the brigadier or foreman, who sent them to the infirmary. In the majority of cases there was no investigation. However there were instances when ten-year sentences were handed out. In any case they were sent to the Administration hospital near Neksikan. It meant amputation... in exchange for living on a cot all winter; moreover they had their lives—for the time being at least.

"There were blatniye and non-politicals of the simple, hearty type who chopped off the head of the brigadiers while the work leaders were asleep. The killer then went calmly to the gate-house to notify the guard of his deed. A two-month investigation period in the lock-up (heated for killers) awaited him—and a ten-year sentence. What motivated the prisoner to kill? To escape general work for two months in the wintertime, lie all day in the heated penal hut, and rest! Besides, if say the killer had eight years to go-the new ten years was not added to his original sentence. In reality he received a mere two years—two years for a murder!

"...After the last of the brigades marched through the gates, the round-up of so-called refusers and work-shirkers began. The Camp Elder, work allocator and their aides searched the barracks, ferreting out prisoners from their hiding places—beating while dragging them to the feet of the officials and guards. The Z/K's pleaded and whimpered, 'someone stole my mittens...' another blabbered, I have no burkiis.' Others did not have the strength to get up. They were on their last legs—and the wolf-hounds, obeying their masters chewed up the prostrate figures. And how many others were dragged outside the zone and shot through the head? That happened daily! Despite the dying and the killing, our ranks did not dwindle. Men died—fresh ones replaced them...The Ukazniki kept pouring in! No wonder they called it the valley of death! What about the war? The Russians must be winning—Japanese waters must be safe for navigation again?

"Whereas in the Southern Administration the dead were buried by male nurses, in Camp 7 a whole brigade was required to perform the operation. According to regulations each dead prisoner was to be buried in his underwear in an individual grave. A wooden tag with his name, year of birth and dossier number was to be tied on the ankle...But there is quite a difference between Soviet theory and Soviet practise. Even in the Southern Administration they didn't bury them in their underwear! There was not enough to go around for the living! And in the Chai-Urya Administration...who was to dig individual graves in the permafrost? Imagine the manpower required for such a task.

"In Camp 7 a nondescript brigade of about twenty men, most of them doghodyagas having at most a month or two to live, made at least two trips daily up the slope, two kilometers away from the compound. On their shoulders they carried naked (except for a tag tied to an ankle) frozen corpses to the burial site. The bodies were piled like logs. When three or four hundred accumulated, holes were bored and blasting place. The bodies were thrown into a mass grave, then covered. Occasionally, a dokhogyaga, gasping and panting from the climb up the slope, tossed his burden alongside the snow-beaten path, then kicked snow over it."[72]

Thomas Sgovio's life was saved by a friendly camp brigadier who took him off the death-work of hauling frozen gold-bearing clay by hand carts, in temperatures far below zero, before it was too late. He was eventually transferred to another camp, OLP MESTPROM, where he learned later that his savior had lost his privileged position and had himself died in slave labor in Camp 7. At his new prison he did different work, but this camp provided an example of how nearly costless and self-sufficient the Soviet slave-state system had become:

"Our work assignment was carrying logs on our shoulders from the forest to the camp—a distance of six kilometers. We did this twice daily...The camp was located only six kilometers from Neksikan, the Administration center...The largest structures were the garment and boot-making shops in which clothing and burkiis were manufactured for the prisoners of all the Chai-Urya Administration camps; moreover they sewed quilted coverings for American Lend-Lease Studebaker trucks, and clothing for officials, guards and free-citizens. The Z/K's clothing and burkiis were sewn from United States, Canadian and Argentine flour sacks. American denim and cotton khaki were used for the clothing of officials and guards. Then there were woodworking, machine and other shops. The carpenter brigade...worked on various construction projects...In the woodworking shops, skilled cabinetmakers made fine furniture for very high officials. Most of the prisoners in OLP MESTPROM were invalids—becoming so as a result of self-mutilation. They worked in garment and bootmaking shops. Several brigades were assigned to logging. Those without arms carried logs on their shoulders. The footless hobbled along on wooden stumps, falling trees in the forest..."

"On March 21st (1943) I did not go to work. In the registration office a sonorous voice called out, 'Sgovio, Tomas...born in 1916, has completed his five-year sentence, it has been decreed by the North-East Corrective Labor Camp Administration that the said Z/K is to be a prisoner until the end of the war...However, it should be noted that by working hard and conscientiously he can be freed before then.'

"The work allocator handed me a pen, 'Here, sign that you have been notified of your status.' It became official. I was now a holdover..."[73]

Adolph Hitler had learned much by studying Josef Stalin's long regime of gulag terror, and he had applied those methods, with German modifications and thoroughness, to millions of Jews, Poles, Gypsies and other unwanted minorities in Nazi-occupied territory. During nearly four years of total war on the eastern front which followed the German attack on Russia, millions of Russian and German military prisoners were captured, exploited, and eventually worked to death, by both sides. The Soviet Union had refused to sign the pre-war Geneva accord on the treatment of POWs, and Soviet POWs were considered criminals by Stalin for having surrendered. Nazi military intelligence specialists, led by General Reinhard Gehlen, made enormous gains by extracting information from starving Russian POWs. German prisoners in Soviet hands fared no better. (German treatment of most Western Allied POWs, including thousands of Americans captured in North Africa and Italy, or in bombing raids, was considerably better.)

In the months following the June 1944 Allied invasion of Normandy, U.S. and Allied forces captured many thousands of Russian troops in German uniforms. While some hated Stalin's brutal Communist regime and had served the Germans willingly, hundreds of thousands of others had been forced to, for survival. The embarrassed Soviet government at first denied that some 1,000,000

Soviet prisoners had served Germany but soon began making strident demands for their immediate repatriation, whether voluntary or forced, to USSR control. Even the act of surrender in battle was punishable by death under Soviet law.

Mr. Roy Inman (of Cordova, Alaska) is one of many American soldiers who witnessed the capture of these "Osttruppen" on the western front during the summer of 1944. Inman drove a tank in the first U.S. assault wave at Omaha Beach, during the June 6, 1944 invasion of Normandy, and he later participated in the breakthrough at Saint Lo and the mass roundup of German POWs at Falaise gap:

"They were going by us in long columns as far as you could see, pretty downcast and sorry... There were thousands of them; some were speaking in Russian or other dialects, and many looked part-Asiatic, or almost Oriental. We couldn't understand how they had ended up in the German Army." Later during 1944, in France and Belgium, Inman encountered other Russian ex-POWs and forced-laborers who told him, and any other American who would listen, that they were extremely afraid of being sent back to Stalin's Russia by the western Allies. A U.S. naval officer of the time, Lt. Commander Anthony Drexel Duke, captain of USS LST 530, landed Canadian assault troops at Sword Beach on the first day of the Normandy invasion. In the first return trips of the ship to Britain, Duke remembers carrying thousands of Russian prisoners of war captured in German uniforms on the Normandy battlefields.[74]

At Kolyma, in far eastern Siberia, the American "holdover" prisoner Thomas Sgovio, by now a veteran Zek and a hardened survivor, felt the effect of the simultaneous Soviet offensives.

"By 1944 the flow of prisoners to Kolyma had reached its prewar rate. They started coming in by land by means of a new truck route, still under construction by Z/K's—from Aldan to the Indigirka region. Prisoners transferred from eastern Siberian labor camps began to arrive by the new auto trass, part of the way by truck, the rest by foot. Most of the prisoners however, were transported by ship. As the Soviet armies re-occupied German-held territories, the KGB sent hundreds of thousands to Kolyma. There were many lads16-17 year olds. The stories they told were hair-raising. Whole villages had been arrested. We asked, 'What did you do to be sent here?' (We had been asked the same question so many times in 1938.) 'My father was arrested,-so they arrested me...'" was the answer. ...Most of the boys became dokhodyagas. A few learned to adapt themselves to camp life and became hardened criminals.

"Looking back, I often wonder...What would it be like—were it not for American Lend-Lease products and equipment...Everything around us was American products, machinery, tools, Studebaker trucks, steam shovels, Diamond bull-dozers, ammonal (explosives) in fancy wax paper coatings, detonators, etc. Once the head of the supply Section asked me to...the warehouse...I saw the floor littered with hundreds of sheets of American newspapers. The Russian closed the door. 'Only don't tell anyone I let you see these...tell me what's written here.'...I remember reading front-page accounts of battles in the Pacific. I learned the United States and Great Britain were actually engaged in a war! And here in Kolyma, from bits of talk we heard around us, we thought the Allies were making the Soviet Army bear the brunt of the fighting...That heroic Russians were dying on the battlefield-while America was sitting back—helping only with supplies!"[75]

The hatred of the Germans engendered by the calamitous losses of two world wars led to the presentation of the Morgenthau Plan for the eventual postwar occupation of Germany at the September 1 1-16, 1944, Quebec Conference of wartime leaders Franklin Roosevelt, Winston Churchill and Canada's Mackenzie King. Despite the high ideals for human freedom enunciated in the Atlantic Charter, this plan would have mandated the dismembering of Germany and the conversion of that nation's industrial base into a pastoral economy that would never again threaten it's European neighbors. But Winston Churchill had the foresight to oppose the plan, because he already had envisioned the coming postwar confrontation with the Soviet Union, and he realized that a strong Germany would inevitably be required to act as a buffer between the western democracies and Soviet Russia.

Morgenthau's right hand man at the Treasury Department, Harry Dexter White, had long been a Soviet agent, who had been a leading figure at the July 1944 Bretton Woods Conference, which he took part in with the British economist Lord Maynard Keynes. At this conference, plans for the International Monetary Fund (IMF) and the International Bank for Reconstruction and Development were drawn up. White supplied samples of occupation currency to the Soviets that had been printed by the U.S. Treasury, prompting them to request paper and ink for production, which they ultimately received. The eventual cost to the United States of this generosity was never known. In January 1945 he became an Assistant Secretary of the Treasury, and he proposed in a memorandum to Morgenthau that the USSR should receive a $10 Billion postwar loan at 2% interest, and in another memo proposed that the German industrial base be transferred to Russia as war reparations. Although Churchill opposed the idea, Roosevelt eventually accepted the necessity for war reparations to Russia, although this same principal, adopted after WW I, had helped precipitate the Second World War. White later took part in the Yalta Conference and became the first Director of the IMF. (White was eventually to be unmasked as a Soviet agent with the assistance of GRU defector Igor Gouzenko, and died of a heart attack in 1948.)[76] Although the Morgenthau Plan was not adopted in its entirety, significant parts of it were approved at the 1945 Potsdam Conference, and would cause great human suffering in the postwar era.

In the summer of 1944, Vice President Henry Wallace had traveled to Soviet Russia, and returned to the U.S. belittling reports of widespread slave labor there, in an effort to bolster the wartime alliance. Wallace was a liberal internationalist and a willing apologist for Stalin's brutal regime, and was soon to be dropped as Roosevelt's running mate in the coming 1944 campaign and was replaced with Missouri Senator Harry S. Truman. The American prisoner Thomas Sgovio later wrote of Wallace's trip to far eastern Siberia:

"I was serving my 6th year as a convict slave laborer then. The watchtowers of the Magadan camps were chopped down and prisoners were locked indoors and shown movies for the three day duration of the visit. Wallace and (Owen) Lattimore went to an evening performance at the Gorky Theater. The actors were prisoners who were hurried back to the compound after the performance. Prisoners were made to sign affidavits that, if confronted by the American visitors, not a word that they were prisoners, otherwise a 25-year sentence awaited them. After being wined and dined and duped, Wallace and Lattimore returned to the United States and wrote glowing accounts about Kolyma, thus participating in the totally false picture, painted by the Soviet communists, that the building of Kolyma was being done by volunteer young communists...I can't help remembering the taunts of my fellow-prisoners during the autumn of 1944...'Oh, you stupid, naive Americans!'...'Oh, you foolish men of good will!'"[77]

Meanwhile, by early 1944, Winston Churchill, British Foreign Secretary Anthony Eden and others, were already worried about the repatriation of hundreds of thousands of Allied POWs held by the Germans in the east, who would ultimately be liberated by the Red Army. In August of that year the British War Cabinet adopted a policy of involuntary repatriation of Soviet prisoners to avoid giving Stalin any reason for withholding British and Commonwealth POWs of the Germans who would be overrun by the advance of the Red Army.[78]

Major General John R. Deane, Chief of the US Military Mission of the Moscow Embassy, then under Ambassador Averell Harriman, was one of the Americans who was most concerned about American POWs in the east. General Deane seems to have been an honorable man who had difficulty in suspecting or believing the worst of his Soviet counterparts, some of whom, although they were in Red Army uniforms, were not in fact regular Russian military officers but undercover officials of the NKVD (KGB) or GRU (Military Intelligence) who worked directly for Stalin. Deane later remembered:

"Aside from the few interned at the beginning of the war, there was only one category of Americans in Germany-those captured in military action. In the winter of 1944-45 these totaled about seventy-five thousand...Germany kept her war prisoners as far from the countries of which they were nationals as possible. This meant that most of the Americans were in prisoner-of-war camps in eastern Germany, Poland, or the Balkans. Through the International Red Cross we had a fairly accurate idea of where the American camps were situated and of the numbers in each."[79]

The figure of 75,000 American POWs used by General Deane is, in fact, the number of US prisoners KNOWN BY NAME to have been German prisoners during the winter of 1944-45, and does not include some 25,000 missing-in-action (MIA) Americans who were later in 1945 recovered as prisoners of war. Many of these were late captures from the Battle of the Bulge and elsewhere, who had not been registered by the Germans as prisoners by the time of which Deane was writing, but were carried on the rolls as MIAs. Some 10,000 of the known-by-name US prisoners were to disappear in Russian hands, as were many thousands more Americans still listed as MIA.

General Deane recorded the initiation of the POW repatriation problem with the Soviet General Staff:

"Since the British and the Americans were acting as a combined force on the western front, Lieutenant General Burrows, head of the British Military Mission, and I approached the problem of repatriating liberated prisoners on a joint basis. We made our first approach to the Red Army General Staff on June 11,1944. We called attention to the quick advance of the Red Army into Rumania and presented the Russians with a list of British and American prisoner-of-war camps known to be in its path. We requested the General Staff to inform us without delay of any of our camps which might be taken...At the same time I gave General Slavin, my General Staff contact, A FULL LIST OF THE NAMES OF ALL AMERICANS KNOWN TO BE IN RUMANIA AND HUNGARY."[80]

It may never have occurred to General Deane at the time that the Soviet NKVD might later use a list such as this to enable them to arrest and permanently retain large numbers of imprisoned American Army POWs and downed aircraft crews of the Ploesti oilfield bombing raids who were still carried as MIA's, and therefore not recorded on a known-POW list.

Concerns for the Allied prisoners in the east increased following the Soviet action of allowing merciless German destruction of the non-communist Polish Home Army in the September 1944 Warsaw Uprising, as the Red Army of "liberation" waited nearby. By this time the British government had decided to adopt a policy of forcible repatriation of all "Soviet citizens," to gain the hoped-for Soviet reciprocation of returning all British and Commonwealth prisoners of war overrun by the Red Army. The American Government subsequently adhered to the British policy by late 1944, despite resistance from U.S. Ambassador to Moscow Averell Harriman, Undersecretary of State Joseph Grew and Attorney General Francis Biddle.

The United States was fortunate in successfully evacuating a total of 2,200 of the American POWs held in Romania and Bulgaria, before the final takeover by Soviet forces in September 1944. A reported total of 1,700 of these men were evacuated by air to Italy from Romania in a combined OSS-military operation run by OSS officer (and later CIA official) Frank Wisner. The courageous young King Michael of Romania lent crucial local assistance to this operation, although the United States did little to help him later in confronting the implacable Soviets.[81]

Romania had surrendered to the Allies on the 23rd of August 1944 after a pro-Allied coup in which the young King Michael participated. The Soviet armies continued on conquering the country anyway, treating the people with great brutality and deporting hundreds of thousands to slave labor in Siberia. The Americans and British were concerned for their prisoners held by the Germans there, and reacted quickly to an unexpected opportunity.

Major General John R. "Russ" Deane later wrote:

"The first large-scale release of American prisoners of war came with Rumania's collapse and withdrawal from the Axis toward the end of August 1944. I first heard of their liberation on August 29, in a message from General Ben Giles, who was commanding the American forces in the Middle East. He said his Air Force representative in Rumania had reported that all British and American prisoners of war were being concentrated in Bucharest and that King Michael had personally guaranteed their safety...There were about one thousand officers and enlisted men involved...Giles worked out a plan with General Eaker, commanding our Mediterranean Air Force, whereby American planes were flown to designated fields in Rumania on September 1, 2, and 3 to pick up our ex-prisoners. This mass evacuation by air was accomplished through arrangements made with Rumanian military authorities before the Russians had complete control of the country. It was successful through a fortunate combination of circumstances, which included the presence of a few Americans who seized the initiative and put the project over before the Russian regime had become fully established."[82]

In his memoirs Deane neglected to mention that not all the US and British POWs in Romania were evacuated. Reports of American and British in Soviet hands in Romania were to continue for months, and beyond the end of the European war. Other Americans held by the

Germans remained behind in Hungary and the Balkans (in Bulgaria and Yugoslavia).

In Romania, the Allies also left behind stay-in-place agents of the OSS and the British Special Operations Executive (SOE) to monitor the Soviet takeover, and this angered the Russians, who eventually ordered them to leave. The Russians wanted no witnesses to their merciless methods of mass arrests and executions while establishing total communist control in areas they occupied.[83]

Of some 420,000 Romanian military prisoners taken prisoner by the Russians during and after the war some 240,000 were to disappear forever. Many suffered agonizing deaths in Soviet slave labor camps of the far north or in Siberia. Only 50,000 of these were officially announced as dead by the Soviets. Aside from the military prisoners some 250,000 Romanian civilians were also deported to the Gulag camps, most of whom died in the first few years of forced labor. The Allied actions in Romania may have influenced later Soviet conduct in Poland, while also giving the Soviets a better chance to seize US and British POWs as hostages.

Early in the Romanian operation, while Warsaw was dying, Deane had initiated a 3 point proposal for POW repatriation measures on August 30th, which was forwarded through Averell Harriman to Soviet Foreign Minister Molotov, and to General Antonov of the Soviet General Staff. A similar action was taken by the British Military Mission at this time. This initiative on the part of the Allies may have been an attempt to mollify a Stalin furious over unilateral American actions in Romania. Aside from urging the prompt return of prisoners, and prompt reporting of their existence, Deane proposed the use of American and Soviet officers to contact the prisoners for the purpose of establishing their nationality. A long wait ensued. As Deane later recalled:

"Neither Harriman nor I had replies to our letters for several months although we each kept pressing for one...General Eisenhower's staff became concerned about Russia's unwillingness to make any commitments concerning the treatment of American prisoners liberated by the Red Army. They wrote us a letter in which they stated that there was a group of Russians at General Eisenhower's headquarters in connection with the repatriation of displaced Russian and Red Army prisoners of war which was being afforded every facility for its work. There was the possibility of curtailing their activities if the Russians did not show some disposition to reciprocate. I did not favor doing this because there was no chance of winning in a competition of discourtesy with Soviet officials...Harriman was away, so on November 6 I asked our Charge, George Kennan, to write Molotov again, asking for a Foreign Office reply..."[84]

The Soviet General Slavin initiated the first counterattack in a war of words with an immediate formal complaint about the American handling of Soviet prisoners liberated on the western front. Four months after the invasion of Normandy the American Army alone had captured 28,000 Soviets in German uniforms fighting for the Nazis in France. Deane later wrote:

"The Government-controlled Soviet press, never as concerned with maintaining Allied unity as we were we in the United States, began to voice Soviet displeasure. Slavin's protest to me was expanded upon by Colonel General Filip Golikov in the November 9 issue of PRAVDA."[85]

Soviet General Filip Golikov, who was the senior "Repatriations Commissar," was in reality a high-level intelligence aide to Josef Stalin, hated and feared by most regular Red Army officers. Kennan believed that he was an NKVD officer in a military uniform.[86]

Deane's and Kennan's November 6th request to Molotov on US POWs, and the simultaneous British proposal, had coincided with the first large-scale British repatriation by ship of some 10,200 Russian ex-POWs to Murmansk on the same date. In what could hardly be construed as anything but a clear indication of British (and Allied) intentions, they had been loaded from POW camps in Great Britain under armed guard on October 31, 1944, after months of resisting the propaganda of the Soviet repatriations chief in Britain, General Vasiliev.[87]

On the 14th of November the Soviet news agency TASS aired a purported account of the Soviet repatriates arrival in Murmansk, saying they were warmly welcomed by Golubev's and Golikov's Repatriations Commissariat and that they were being sent in groups "to their native places." TASS said the prisoners "expressed their deep gratitude to the Soviet Government and to Comrade Stalin for their solicitude...On November 6th they heard Stalin's speech."

A different view of this same return was preserved in a report by a British officer on the scene, Major S.J. Cregeen:

"On November 7th, in Murmansk, I was in a car returning from the Naval Mission Headquarters to the War Port. En Route we were passed by a long column of Russian repatriated nationals, who were being marched from their transport, the Scythia, under armed guard to the camp just outside the town. It appeared that they were being treated as having the status of nothing more than enemy prisoners of war. The guards were armed with rifles and were probably allotted at the rate of one per 10/15 Nationals. There was no sign of a welcome reception being arranged for these repatriates, whose demeanor was added proof of their unfortunate status..."

Officials at the British Foreign Office who had long pushed for the forced repatriations of Soviet nationals were satisfied with the first major shipment of Russian POWs. They refused to believe the

Soviet treatment of them would be anything other than correct. Foreign Office official Geoffrey Wilson (a committed Socialist) irritated by Major Cregeen's note of reality, wrote on the report, "I should like to know a good deal more about Major Cregeen."[88]

From the Soviet point of view, the Americans and British always acted in concert, so the ex-POWs arrival in the USSR must have been viewed as a signal that the western Allies were planning to fulfill Stalin's desires. Yet the Soviets still chose to issue their challenge through Kennan and Deane and in PRAVDA. Perhaps Stalin felt that the United States needed to learn more from the British. President Roosevelt's chief of staff, Admiral William Leahy was already advising that the United States follow the British lead and adopt a policy of forced repatriations.:

"Since the British War Office, with Foreign Office concurrence, has agreed that all captured Soviet citizens should be returned to Soviet authorities without exception...it is not advisable for the United States Government to proceed otherwise."[89]

General Deane recalled the next stage in the negotiations on prisoners:

"On November 25, 1944, Molotov, in a letter addressed to George Kennan (the Charge d'affairs at the US Embassy in the absence of Harriman) finally replied to the proposals and looked toward reciprocal treatment of liberated prisoners. Molotov ACCEPTED OUR PROPOSALS 'IN Principle' and said that Soviet representatives would meet with me to work out the details...he also registered a complaint about the way Russian citizens were being detained by the Allies in German prisoner of war camps both in Western Europe and in the United States..."[90]

Kennan's cable to the Secretary of State on November 27th added additional information not recorded by Deane:

"At the request of General Deane I sent a letter to Molotov on November 6...on treatment of prisoners of war. I referred to Mr. Harriman's letter of August 30...to which no answer has been received and stated that since the advance of the Soviet armies had already enveloped the location of one prisoner of war camp KNOWN TO HAVE FORMERLY HELD AMERICAN WAR PRISONERS and since Soviet forces were apparently APPROACHING ANOTHER SUCH CAMP IN THE BUDAPEST AREA it was desirable that we should not delay any longer arriving an understanding...I have now received a reply from Molotov dated November 25..."[91]

Molotov's letter to Kennan of November 25th contained both an acceptance in principle of the Deane-Harriman repatriation proposals and a long series of complaints about the treatment of Russian prisoners. It also stated:

"It goes without saying that those special questions regarding American prisoners in the Budapest and Rumanian areas, brought up by the Ambassador's and Mr. Kennan's letters, may be discussed at the meeting of our representatives..."[92]

Late in 1944, in an attempt to reassure the Soviets of future Allied intentions, first Britain and soon afterward the United States, repatriated to the USSR several thousand Soviet POWs, some of whom did not wish to return to Russia because they had been captured in German uniforms. At the end of November, 1,979 Soviets screened from among more than 2,100 who had been captured in German uniforms and interned in the U.S. were assembled in Portland, Oregon, from POW camps in several areas of America, and repatriated by ship from the Pacific northwest of the U.S. Of this number 74 were forcibly returned and two apparently committed suicide as their bodies were recovered near the mouth of the Columbia River.[93] The tedious POW-exchange negotiations with the Soviets continued long after the Romanian POW operation, but no conclusive agreement was to be reached until the February 1945 Yalta Conference. At the same time, the furious Battle of the Bulge was raging on the western front in Europe from December 1944 into January 1945, and many thousands more U.S. soldiers became prisoners of war or POW/MIAs (those who were not yet known by name to be in German captivity). Some of the American Army units engaged, from the U.S. 28th Infantry Division and the U.S. 106th Infantry Division lost extremely large numbers of U.S. soldiers listed simply as "missing in action," and the ultimate fate of many of these men remained unknown after the war ended that spring.[94]

After weeks of waiting, Deane had Ambassador Harriman write Molotov again on December 28th urging that repatriation discussions be held. Molotov's deputy Andrei Vyshinsky (Stalin's vicious 1930s purge prosecutor) replied the next day. Harriman cabled the Secretary of State on the 29th of December:

"I have received a letter from Vyshinsky...stating that ...Golubev and...Slavin have been appointed...to conduct negotiations...with Deane...on...prisoners of war...in accordance with principles set forth in Molotov's letter to Kennan of November 25..."[95]

The new Secretary of State, Edward Stettinius, replied to Harriman on January 3, 1945, in a message drafted by E. A. Plitt which reflected fear on the part of some old-line State Department officials of linking the return of American POWs to the future forcible repatriations of unwilling "Soviet citizens."

"SECRET...Department is extremely anxious that in any discussion concerning the repatriation of American and Soviet prisoners of war and civilians THAT THERE BE NO CONNECTION BETWEEN THE RETURN OF AMERICANS FOUND IN GERMAN...CAMPS ON THE ONE HAND AND SOVIET NATIONALS FOUND AMONG GERMAN PRISONERS OF WAR TAKEN BY THE AMERICAN FORCES ON THE OTHER

HAND.

Some difficulty has arisen here in the determination of claimants to Soviet nationality whom this government is prepared to turn over to Soviet authorities...OVER 1,100 SOVIET NATIONALS FOUND FIGHTING WITH GERMAN TROOPS WERE TURNED OVER TO SOVIET AUTHORITIES AT A WEST COAST PORT LAST WEEK. A FURTHER REPORT ON THE PROBLEMS WHICH HAVE ARISEN IN THIS CONNECTION WILL BE SENT TO YOU..."[96]

On January 5th, 1945, there was an important meeting of the State-War-Navy Coordinating Committee (SWNCC). On that same day Attorney General Francis Biddle told Secretary of War Henry Stimson that "THE RUSSIANS HAVE ALREADY THREATENED TO REFUSE TO TURN OVER TO US AMERICAN PRISONERS OF WAR WHOM THEY MAY GET POSSESSION OF IN GERMAN INTERNMENT CAMPS."[97]

This would indicate that senior American policymakers had a far clearer understanding of Soviet intentions with US POWs than most declassified official documents or contemporary memoirs indicate. (Author Mark Elliott, who wrote a book on the forced repatriations in 1982 interviewed Elbridge Durbrow on the subject some 30 years later. Durbrow remembered "THE RUSSIANS BLACKMAILED US IN THAT." He did not inform Elliott that the Soviets had eventually withheld some 20,000 American POWs and MIAs)[98]

By the time of the Yalta Conference the British had already sent home 17,500 Soviet citizens.[99] To Stalin, the time was now right, with the approaching conference of the big three war leaders-himself, Roosevelt and Churchill, at Yalta on the Crimean Peninsula of southern Russia's Black Sea coast, for the Soviet Union to insist on its demands for continued forced repatriations. Deane wrote later:

"I had my first meeting with Golubev on January 19, 1945, just a little over six months after my first approach to the (Soviet) General Staff on the subject... Golubev had followed the usual Soviet procedure of avoiding arguments that might arise in working out a joint plan with a foreigner by having one ready to hand to me. It was a reasonable plan and with a few minor amendments was exactly what we wanted."[100]

An undated copy of the Russian redraft of the repatriation agreement was found among the papers attached to a note of February 5, 1945, from British Foreign Secretary Anthony Eden to U.S. Secretary of State Edward Stettinius. According to General Deane it was similar to the Soviet redraft of Soviet-US repatriation agreement presented to him on January 19, 1945.[101] The British version was forwarded by British Ambassador Archibald Clark Kerr and Admiral Archer, the new Chief of the British Military Mission to Moscow who had replaced General Burrows. This was the proposed agreement Deane forwarded to the Joint Chiefs of Staff, Eisenhower and McNarney (US commander in the Mediterranean area).[102]

The key points of Soviet-proposed agreement read:

"1. All Soviet citizens liberated by Allied Armies and British (US) subjects liberated by the Red Army will, without delay after their liberation, be separated from enemy prisoners of war and will be maintained separately from them in camps or points of concentration until repatriation...

"2. Both sides shall ensure that their military authorities shall without delay inform the competent authorities of the other side of Soviet or British (US) citizens...taking the steps which follow from this agreement. SOVIET AND BRITISH (US) REPATRIATION REPRESENTATIVES (contact officers) WILL IMMEDIATELY BE PERMITTED INTO THE CONCENTRATION CAMPS AND POINTS WHERE CITIZENS OF THEIR COUNTRY ARE LOCATED AND THEY WILL HAVE THE RIGHT TO APPOINT THE INTERNAL ADMINISTRATION... The commandant's organization of the camp and civilian protection will be established in accordance with the direction of the military commandant within the zone in which the camp is located. THE REMOVAL OF CAMPS AS WELL AS TRANSFER FROM ONE CAMP TO ANOTHER OF LIBERATED CITIZENS WILL BE ACCOMPLISHED ONLY IN AGREEMENT BETWEEN COMPETENT SOVIET AND BRITISH (US) AUTHORITIES. Enemy propaganda directed against the contracting sides or against the United Nations will not be allowed...

"3. The competent British (US) and Soviet authorities will supply liberated Soviet citizens and British subjects (US citizens) with food, clothing, housing and medical attention both in camps or points of concentration and en route, and with transport UNTIL THEY ARE HANDED OVER TO THE AUTHORITIES AT THE OTHER SIDE AT PLACES AGREED UPON BETWEEN THE SIDES...(There follows a supply proposal)

(Point 4 deals with monetary advances for prisoners.)

"5. Ex-prisoners of war (with the exception of officers) and civilians of each of the parties UNTIL THEIR REPATRIATION MAY BE EMPLOYED ON WORK IN AIDING THE COMMON WAR EFFORT AS TO WHICH COMPETENT SOVIET AND BRITISH (US) AUTHORITIES SHALL AGREE AMONG THEMSELVES...

"6. Both parties shall use all means at their disposal to ensure the evacuation to the rear of the above-mentioned citizens or subjects of the other party IF THIS PROVES NECESSARY AND QUICKEST POSSIBLE REPATRIATION OF THESE PERSONS.

"7. If the British (US) Government agree to these principles the Soviet Government suggests presentation of note and British (US) reply to it constitute an agreement..."[103]

## Notes on Chapter Three

1    *Harvest of Sorrow*, by Robert Conquest, Oxford University Press, 1986, p. 306
2    Dallin and Nikolaevsky, Forced Labor in Soviet Russia; Gulag Archipelago, by Aleksandr Solzhenitsyn; *Harvest of Sorrow*, by Robert Conquest, Oxford University Press, 1986
3    *The New KGB*, p. 95, op. cit.
4    Ibid.; and *Utopia in Power*, op. cit.
5    Ibid.
6    Bohlen, *Witness to History*, pp. 14-54 op. cit.; Kennan, *Memoirs*, Atlantic Monthly Press,1967, pp. 79-86; also: *Russia and the West under Lenin and Stalin*, op. cit.
7    The author's father, John M. G. Brown Sr., then a student at Saint George's School in Rhode Island, who had been shaken by events of the Great Depression, was among the first American visitors to travel in post-recognition Russia at this time, with his mother, brother and sister. His account of this experience in Stalinist Russia made a profound impression which he conveyed to his own family many years later.
8    *So Close to Greatness*, Will Brownell and Richard N. Billings, MacMillan's, NY, 1987
9    *So Close to Greatness*, p. 163,op. cit.
10   *Witness to History*, pp. 44-45 and p. 57, op. cit.
11   *Witness to History*, p. 49, op. cit.
12   *Russia and the West Under Lenin and Stalin*, p. 255
13   *Utopia in Power*, p. 306, op. cit.
14   *Gulag Archipelago*, Volume I, p. 439, op. cit.
15   *The Great Terror and Kolyma*, by Robert Conquest, op. cit.
16   *KGB-The Inside Story*, pp. 163-169, op. cit.
17   *Witness to History*, p. 41; also: *A Pretty Good Club*, op. cit.
18   *A Pretty Good Club*, p. 153, op. cit.
19   Ibid.
20   Sgovio, *Dear America*, op. cit.
21   Ibid.
22   *The New KGB*, p. 447, op. cit.
23   *Utopia in Power*, op. cit.
24   Mission To Moscow, by Joseph Davies, NY, 1941, 1942, 1943, Garden City, pp. 208-209, Publishing edition, with permission of Simon and Schuster
25   *America and Russia*, pp. 248-49, Simon and Schuster, NY, 1962
26   Ibid.; also: *Gulag Archipelago* and others already cited
27   NYT, Nov. 14, 1961
28   *Defeat in Victory*, by Jan Ciechanowski, Doubleday's, NY, 1947
29   Poland And The Coming of the Second World War, The Diplomatic Papers of A.J. Drexel Biddle Jr., Edited by Cannistraro, Wynot and Kovaleff, Ohio State University Press, 1976, p. 30
30   *Defeat in Victory*, op. cit.; also: *Utopia in Power* and others
31   Record of hearings by the House Select Committee on the Katyn Forest massacre, 82nd Congress, 1951-1952; Also: *Death in the Forest*
32   *Defeat in Victory*, op. cit.; and *Death in the Forest*, by J.K. Zawodny, Hippocrene Books, NY, 1962
33   *Defeat in Victory*, op. cit.
34   Ibid.

35    *Defeat in Victory*, op. cit.

36    Ibid.; also: *Utopia in Power; KGB -The Inside Story*, and others

37    *Death in the Forest*, op. cit.; Archival documents and Congressional Record

38    Lash, *Eleanor Roosevelt*, p. 86

39    *A Pretty Good Club*, pp. 140-141 op. cit.

40    Hearings of the House Select Committee to Conduct an Investigation of the facts, Evidence and circumstances of the Katyn Forest Massacre, Part 7, U.S. Government Printing Office, 1952

41    Volume I, Correspondence Between the Chairman of the USSR And the Presidents of the U.S.A. and the Prime Ministers of Great Britain During the Great Patriotic War, Foreign Languages Publishing House, Moscow, 1957, p. 120, reprinted by Dutton's, NY, 1958 as: "Stalin's Correspondence With Churchill, Atlee, Roosevelt and Truman"; Cable # 150

42    Ibid: Cable#151

43    Ibid., Cable # 152

44    Ibid., Cable # 153

45    *Defeat in Victory*, op. cit.

46    Ibid, pg. 159

47    Volume 1, Stalin's Correspondence, p. 61, Cable #81, Personal and Secret From the President to Mr. Stalin"

48    Ibid, p. 62, Cable no. 32

49    Ibid., pp. 124-125, Cable #154

50    Ibid: p. 126, Cable # 155

51    Ibid: pp. 127-128, Cable # 156

52    Hearings before the House Select Committee to investigate the Katyn Forest Massacre, 1952, op. cit.

53    Exhibit 22, Part VII, Hearings held in Washington, D.C. November 11-14, 1952. p.2092

54    Ibid., Final Report, 1952, U.S. Government Printing Office

55    NA, RG 31947/1047 documents recording case studies of Polish evacuees from the Soviet Union to Teheran, Reports of Lt. Colonel H.I. Szymanski, U.S. Army, November 1942 onward

56    House Select Committee Hearings, Part VII, op. cit.

57    Report of House Select Committee Hearings on the Katyn Forest Massacre, op. cit.

58    Report of House Select Committee on Katyn Forest Massacre, op. cit.

59    Final Report of the Select Committee to Conduct an Investigation and Study of the Facts, Evidence and Circumstances of the Katyn Forest Massacre, 82nd Congress, 1952

60    Ambassador Biddle, much loved by the Polish people, became the only senior U.S. WW II official to resign in protest in January 1944 over President Roosevelt's and his old friend's submission to Stalin over the control of Poland's future. Ambassador Biddle subsequently reentered the U.S. Army having served as an infantry officer in France in WW I as a Colonel on General Eisenhower's staff at SHAEF Headquarters, Deputy Chief of the European Allied Contact Section (EACS), involved with underground forces in German and later Soviet - occupied eastern Europe.

61    Personal memoirs of Tadeuz Wittlin

62    *So Close to Greatness*, pp. 294-297, op. cit.

63    Ibid.

64    Jan Ciechanowski, *Defeat in Victory*, Doubleday, 1947, op. cit.

65    According to the research of a West Point researcher and Assistant Professor, Dominic J. Caraccilo, published in the Veterans of Foreign Wars (VFW) Magazine,

September 1993
66  A *Man Called Intrepid- The Secret War*, by William Stevenson, Harcourt, Brace, Jovanovich, NY and London, 1976
67  *Wild Bill Donovan-The Last Hero*, by Anthony Cave Brown, p. 280, op. cit.
68  *Mask of Treachery*, by John Costello, op. cit.
69  Defeat in Victory, Ambassador Jan Ciechanowski, op. cit
70  Corson, *The New KGB*, 230-231, op. cit
71  *The Holocaust*, by Martin Gilbert, Henry Holt & Co., NY, 1985
72  *Dear America- Why I turned Against Communism*, by Thomas Sgovio, Partners' Press Inc., Abgott & Smith Publishing, Kenmore, NY. 1979
73  Ibid.
74  Personal interviews with the author, 1986-1988
75  *Dear America*, op. cit.
76  *Nemesis at Potsdam*, by Alfred M. de Zayas, University of Nebraska Press, 1977; also: *KGB, The Inside Story*, pp. 335-336, 370, op. cit.
77  Tom Sgovio, Letter to the Arizona Republic, 6/12/1989
78  FRUS-1944, Alexander Kirk to Hull, 16 Sept. 1944; Also: Bethell, *The Last Secret*; Elliott, *Pawns of Yalta*
79  *Strange Alliance*, Viking Press, NY, 1947
80  *Strange Alliance*, pp. 183-184, op. cit.
81  *Wild Bill Donovan, The Last Hero*, by Anthony Cave Brown, Times Books, NY, 1982, pp. 670-674
82  *Strange Alliance*, p. 184, Op. Cit.
83  "BishopTraffic", pp. 679-682, *The Last Hero*, op. cit.
84  *Strange Alliance*, p. 185, op. cit
85  Ibid., p. 187
86  Letter from George Kennan to the author cited elsewhere in this book
87  *The Last Secret*, by Nicolas Bethell, Basic Books, NY, 1974; and others
88  Tolstoy, *Secret Betrayal*, p.120, Charles Scribner's Sons, NY, 1977
89  Leahy to Hull, 2 November 1944, FRUS, IV, 1262
90  *Strange Alliance*, p. 188, op. cit
91  p.p. 413-415, Foreign Relations of the United States, The Conference at Malta and Yalta, 1945; also: National Archives, RG 165, entry 421
92  NA, State Department Files
93  *Pawns of Yalta* and NA documents
94  Records of casualties in individual units
95  FRUS, 1944, IV, p. 1272; also: NA, RG 165/Entry 11/Box 473
96  NA, RG 165/11
97  Minutes of SWNCC, 5 Jan., 1945, NA RG 218, IGCCS 334 SANACC, 19 Dec., 1944 and Pawns of Yalta, Mark R. Elliot, pp. 41 & 53, "Russia- Safe File" NA RG 107, Sec. of War, Stimson 1940-45
98  *Pawns of Yalta*, Mark Elliott, University of Illinois Press, 1982, p. 41
99  Ibid.
100  *Strange Alliance*, p. 188, op. cit
101  Ibid.
102  Foreign Relations of the United States, The Conferences at Malta and Yalta, p.416.
103  NA RG 165/421/473

# CHAPTER FOUR

## THE YALTA AGREEMENT AND THE KIDNAPPINGS IN POLAND

At the February 1945 Yalta Conference, on the Crimean Peninsula in Russia's Black Sea region, the repatriation of prisoners of war and displaced persons (DPs) was formalized in agreements between Soviet, British and American military officials, which then became part of the overall Yalta accords signed by Stalin, Churchill and Roosevelt. Among the most important provisions of the military repatriation agreements, were those guaranteeing access for U.S. contact officers to concentrations of American POWs inside Soviet-occupied territory. Roosevelt's advisors on the scene included Harry Hopkins General George C. Marshall, Secretary of State Edward Stettinius, Averell Harriman, Charles Bohlen, H. Freeman Matthews, Alger Hiss and others. The obsessive secrecy of the Soviets, who had no intention of allowing the promised Allied POW contact officers into their occupation zones, and the fact that the repatriations would be under control of Lavrenti Beria's NKVD, increased Allied fears that Stalin might renege on his promise to return all Allied prisoners.

Poland, the final cause of Britain and France going to war with Germany, was finally overrun by the Soviet Army in January and February 1945. The Russians had been long preparing for this by gradually forcing de facto Allied recognition of their communist puppet regime.

During January of 1945 the Soviet Army was already overrunning American POWs in Poland and eastern Germany but the American Embassy in Moscow was not yet aware of the fact, as the Russians were already concealing their possession of them. Constituents of Texas Senator William Connally with connections to intelligence sources inside Poland seem to have been the first to cause an alert at the US Embassy in Moscow. A secret telegram from Acting Secretary of State Joseph Grew was routed to Charge D'Affairs George F. Kennan, in the absence of Averell Harriman, who was then on his way to Yalta:

"PERSONAL FOR THE AMBASSADOR (Harriman) FROM SENATOR CONNALLY. THERE ARE 1,200 AMERICAN PRISONERS IN A GERMAN CAMP CALLED OFLAG 64 SITUATED NEAR A TOWN CALLED SCHUBIN IN THE POSEN AREA, OF WHICH 65 ARE TEXAS SOLDIERS. DID GERMANY MOVE THE CAMP AND PRISONERS TO GERMANY BEFORE THE RUSSIAN ARMY MOVED IN OR HAVE THESE PRISONERS BEEN LIBERATED BY THE RUSSIANS? YOUR FRIEND AMON CARTER'S SON, LT. AMON CARTER, JR., IS ONE OF SUCH PRISONERS. ALL AVAILABLE INFORMATION IS

DESIRED. PLEASE CABLE A REPLY. GREW—ACTING[1]

Later intelligence and events were to confirm that the Russians had indeed overtaken many Americans, including some seven hundred U.S. officer POWs from OFLAG 64, prior to the Yalta Conference. Kennan seems to have been unaware at this time that the Russians had overtaken Americans, basing his belief on Soviet statements. In a January 27 answer to Grew's message Kennan indicated some faith in the Soviets:

"WITH RESPECT TO SENATOR CONNALLY'S PERSONAL TELEGRAM TO THE AMBASSADOR...CONCERNING AMERICAN PRISONERS OF WAR IN POLAND...DEANE HAS BEEN IN CONSTANT TOUCH WITH SOVIET MILITARY AUTHORITIES IN AN EFFORT TO OBTAIN INFORMATION...IN THE PAST HE HAS ALWAYS RECEIVED PROMPT INFORMATION OF AMERICAN PRISONERS OR STRANDED AIRMEN AND THESE HAVE BEEN PROMPTLY EVACUATED. AS YET, NO AMERICAN PRISONERS HAVE BEEN REPORTED AS FOUND BY THE RUSSIANS ON THEIR PRESENT DRIVE...GEORGE F. KENNAN[2]

In light of subsequent Soviet actions with American and Allied prisoners, it seems unlikely that the Soviets had always been as forthright about their possession of Americans in the past as Kennan believed at this date. As was to emerge in this latest case, it is more likely that the Russians had all along only released those Americans who were known by the US to be in their hands. It appears certain that American prisoners taken by the Russians in Hungary and Romania prior to this time had already been interned, and since no contact was had with them later, they remained missing and never reappeared.

Prior the Yalta meetings of the "big three" wartime leaders in early February 1945, as noted, the repatriation of Russian ex-prisoners had already begun; but it was at the Crimea Conference that the repatriations of POWs were formalized. President Franklin D. Roosevelt's advisors on the scene included his influential personal advisor Harry Hopkins, Admiral Leahy, Secretary of State Edward Stettinius, Averell Harriman, Charles Bohlen, H. Freeman Matthews, Alger Hiss and others. Roosevelt refused to permit Assistant Secretary of State for Political Affairs James Clement Dunn, a relative hardliner towards the Soviets, to participate in the Yalta conferences, saying: "He'll sabotage everything."[3] Alger Hiss, who specialized in United Nations matters at the Department's Office of Special Political Affairs (SPA), was thought to be more amendable to the Soviets, and was substituted in Dunn's place. Hiss and Hopkins remained in close proximity to Roosevelt during the conferences. The fact that the Soviets were always secretive in their actions and that the repatriations would be under control of Lavrenti Beria's NKVD increased fears by some that Stalin might renege on his promise to return all Allied prisoners.

On the day the Yalta Conference began, February 5, 1945, Churchill's Foreign Secretary, Anthony Eden, had written a letter urging immediate action on Allied POWs and repatriations of Russians to the American Secretary of State Edward Stettinius (both of whom were present at the conference):

"In present circumstances where the Soviet forces are overrunning the sites of British and United States prisoner-of-war camps very fast, and we know that a number of British prisoners-of-war (though not exactly how many) are in Soviet hands, and no doubt some United States prisoners-of-war also, it is really urgent to reach agreement with the Soviet Government ON THIS DRAFT AGREEMENT during ARGONAUT." (ARGONAUT was the code name for the Yalta Conference.) I intend therefore to ask M. Molotov for discussions to be opened between the experts of the three parties concerned at once, in order to reach agreement on a satisfactory text..."

Eden then tied in the recovery of Allied POWs to the future repatriations of Russians:

"It is clear, as SHAEF have already reported, that the only real solution to the problem of the Soviet citizens who are likely to fall into British and American hands shortly is to repatriate them as soon as possible...we have already sent 10,000 back from the United Kingdom and 7,500 from the Mediterranean. It seems to me it would materially help the proposed negotiations if we could inform the Russians at a suitable moment of our plans to repatriate their citizens... I am however without any information on the United States plan for this. General Eisenhower has recently pressed the Combined Chiefs of Staff once again to provide two ships to take 3,000 each from Marseilles until the present large numbers have been cleared...."[4]

Also on February 5th, Eden wrote Soviet Foreign Minister Molotov:

"You will remember that during the Moscow conversations of last October, I discussed with you and Marshal Stalin the question of caring for and repatriating Soviet citizens and British subjects... Since then our two Governments have exchanged drafts and on 20th January our embassy received from your Government a redraft of a reciprocal agreement on this matter...I understand that a similar draft was put forward to the United States...In view of the integrated character of the Allied Commands in western and southern Europe, it seems to us essential that any agreement should be tripartite and cover British and United States Combined Commands. We have accordingly PREPARED A REDRAFT OF THE TEXT...to cover...certain other points where alterations APPEAR TO US AS NECESSARY. I have brought with me the experts on this matter...I have given a copy of this letter to Mr. Stettinius. I enclose a copy of the British redraft.[5]

Within a short time of the Yalta Agreements the Soviets were to refuse American and British contact officers access to their prisoners in Poland, East Prussia, Germany, Austria, Hungary and elsewhere; often on the basis that their presence was not needed, as there were sufficient Soviet officers present in the Russian-controlled camps to handle the repatriations.

It is not known who in the British delegation added the words: "AND THERE ARE INSUFFICIENT OFFICERS," which were not included in the original Soviet redraft. Aside from Churchill, Eden and Lord Cadogan, those involved in the Yalta negotiations for Britain included Sir Archibald Clark Kerr, Admiral Archer, Geoffrey Wilson of the Northern Department of the British Foreign Office (a long-time radical socialist), Field Marshal Alan Brooke, General Ismay and others.[6]

While the minutes of several important Yalta meetings on prisoners are missing, a Top Secret memorandum of conversation survives of some of the negotiations on the wording of the prisoner repatriation agreement. The notes were made by Edward Page at the time:

"Subject: Examination of the draft relating to prisoners of war and civilians liberated by the Soviet and Allied armies.

Preamble. No Comment.

Article 1. No comment.

Article ii, paragraph I. Mr. Novikov (Soviet) requested that the words "UNDERTAKE TO FOLLOW ALL" be replaced by the words "AT THE SAME TIME TAKE THE NECESSARY STEPS TO IMPLEMENT."

Article ii, paragraph 3. Mr. Novikov requested that the words "NOTIFYING THE COMPETENT SOVIET OR ALLIED AUTHORITIES" be replaced by "EFFECTED AS A RULE BY AGREEMENT OR IN ANY CASE ONLY AFTER NOTIFICATION TO THE COMPETENT SOVIET OR ALLIED AUTHORITIES."

Novikov also stated the Soviets desired their version of article iii dealing with supply of prisoners, and several other changes including two insertions of the words "in agreement with the other party" after the words "liberty to use."

In Article VI the British desired a minor insertion on the right to exercise prisoners who refused to work, and on Articles VII and VIII there was no comment.

According to the Yalta record the British representative also stated that his Government desired to exchange notes with the Soviet Government concerning nationals of other countries, (Belgium, Holland, Poland), IN BRITISH UNIFORM who were liberated by the Russian armies.[7]

The veteran diplomat and Undersecretary of State Joseph Grew, (Acting Secretary in Washington), appears to have been the only senior U.S. official to protest strongly the terms of the POW repatriation agreement, in a series of urgent cables from Washington. The implicit promise to return all Soviet citizens to Stalin, whether or not they were willing to go, was the root of the disagreement. Stettinius replied that the Americans at Yalta had determined to follow the British lead, regardless.

While no one in the American delegation at the conference altered the Soviet and British word changes or additions, Joseph Grew, the old-line diplomat who was one of those in the State Department who would have had some knowledge of earlier Bolshevik hostage-taking and prisoner-blackmail, did not trust Soviet intentions and cabled his reservations on the proposed agreement:

"TOP SECRET, WASHINGTON, February 7(8), 1945,

"War Department has just made available message dated February 7 from Marshall which indicates that JCS (Joint Chiefs of Staff) approved with certain changes British preliminary text of Agreement with Soviet Union for exchange of prisoners of war and APPARENTLY ALSO FOR LIBERATED PERSONS (civilian citizens)... While it is not definitely clear what preliminary British text is referred to, if it is the preliminary text included in JCS 1266, the agreement would not appear to cover the following specific points which were incorporated in the United States COUNTER PROPOSALS FORWARDED TO JCS STAFF WITH YOU:

Protection of the Geneva Convention which we have informed Soviet Government we will accord to Soviet citizens captured in German uniform who demand such protection.

Soviet citizens in the United States not prisoners of war whose cases the attorney General (Francis Biddle) feels should be dealt with ON BASIS OF TRADITIONAL AMERICAN POLICY OF ASYLUM>

Persons liberated by the United States forces no longer in their custody.

QUESTION OF THE LIBERATION AND REPATRIATION OF OTHER UNITED NATIONS CITIZENS.

Persons claimed as citizens by the Soviet authorities who were not Soviet citizens prior to outbreak of war and who do not now claim Soviet citizenship...IT IS FELT THAT THESE QUESTIONS AND OTHERS...SHOULD BE BROUGHT TO YOUR ATTENTION IN ORDER THAT CONSIDERATION MAY BE GIVEN TO THEM BEFORE FINAL AGREEMENT IS REACHED."[8]

In this message Undersecretary Joseph Grew had anticipated many of the problems which would lead to the Soviets withholding of American, British, French, Belgian, Dutch and other Allied prisoners in the coming months, and a crisis which led directly to the Cold War. At this point General George C. Marshall was assisting the President with high level advice on decisions related to military matters, such as this. Admiral Leahy, the President's military assistant and General John Deane were also involved. In addition, Major General Clayton Bissell, the Director of Military Intelligence (G-2), who helped cover up the Soviet massacre of Polish POWs at Katyn Forest, was also with the President's party at Yalta.[9] For many months to come he was to remain in a critical position of control over intelligence about U.S. POWs who fell into Soviet control. Other civilian advisors, particularly Harry Hopkins, Averell Harriman, Charles Bohlen, and to a lesser extent Edward Stettinius, also were in a position to urge alteration of the agreement or accept it as it was. The primary translators and interpreters included Charles Bohlen and Edward Page, who had come from the Moscow Embassy to assist Deane and Harriman.[10]

At this time Eden had information indicating that nine German prison camps containing some 50,000 British and Commonwealth POWs had been, or were in the process of being, overrun by the Soviets at the time of the Yalta conference. Winston Churchill raised this issue on February 9th in a private conference with Stalin.

On February 9, Stettinius sent a telegram in answer to the suspicious acting-Secretary of State, Joseph Grew, which gives indication of the final process of agreement on repatriations at Yalta:

"TOP SECRET, Argonaut 125... The text referred to...is the British redraft of the Soviet redraft submitted to the British and American governments on January 20. IN ORIGIN IT IS A SHAEF PAPER. The British have subsequently made a few changes to it... THE BRITISH ARE MOST ANXIOUS TO PRESENT THIS DRAFT TO THE RUSSIANS TODAY FOR THEIR CONSIDERATION. JCS ARE IN FULL AGREEMENT. I can see no objections to the redraft...It does not cover the numbered points mentioned in your reference...THE CONSENSUS HERE IS THAT IT WOULD BE UNWISE TO INCLUDE QUESTIONS RELATIVE TO THE PROTECTION OF THE GENEVA CONVENTION AND TO SOVIET CITIZENS IN THE US IN AN AGREEMENT WHICH DEALS PRIMARILY WITH THE EXCHANGE OF PRISONERS LIBERATED BY THE ALLIED ARMIES AS THEY MARCH INTO GERMANY. With respect to 'claimants', notwithstanding the danger of German retaliation, we believe there will be serious delays in the release of our prisoners of war unless we reach prompt agreement..."[11]

Stettinius appears to shift the blame for the ill-advised agreement onto the British. It is clear that Anthony Eden had been pushing very hard for a protocol and had long been in favor of forced-repatriations, if necessary, of all Soviet citizens, to obtain reciprocal returns of British and Commonwealth POWs. The American Secretary of State also makes clear in this message that the power of General Marshall and the JCS were behind the agreement as worded. While practically ignoring Grew's concern over the inclusion of all "citizens," the impression he gives is that despite the wording, the agreement is really about prisoners (of war) and that to get the Allied POWs back this version must be signed quickly, as is.

In fact, it was to make no difference at all. The Soviets were already at this very time hiding their possession of American, British and other prisoners overrun on the eastern front in Poland, eastern Germany and Hungary. The other Allied prisoners were practically ignored in the final discussions and agreements. A secret message from the US Military Mission in Moscow to General Eisenhower dated February 9th, 1945, during the Yalta Conference indicates the magnitude of this omission:

"THIS IN REPLY TO YOUR S-78110. FRENCH HERE ESTIMATE ABOUT 300,000 FRENCH POW HAVE BEEN TAKEN BY SOVIET ARMIES, AMONG WHICH ARE 10,000 WOMEN AND CHILDREN RECOVERED FROM CAMP AT KATTOWICE AND 41,000 POW RECOVERED IN EAST PRUSSIA. THEY BELIEVE LARGE PART OF ALL FRENCH POW WILL BE RECOVERED ON EASTERN FRONT. FRENCH HERE SAY THEY MADE REQUEST TO UNNRA FOR ASSISTANCE IN CARE AND EVACUATION OF POW FROM RUSSIA. COMPLIANCE IN THIS REQUEST WAS DEFERRED. THEY THEN MADE REQUEST TO AMERICAN RED CROSS... THEY ARE NOW NEGOTIATING WITH SOVIETS."[12]

Franklin D. Roosevelt (then an ill and exhausted man) signed the Yalta Agreement containing prisoner-repatriation provisions, which although the subject of endless argument as to interpretation, proved disastrous to the United States. The American military officers, including Major General John Deane, Bissell and Marshall, who agreed to the specific wording of Soviet POW repatriation proposals, seemed to have no memory of the Soviet blackmail with American prisoners held as hostages from 1919-21, and into the 1930s. The same can be said of the civilian officials at Yalta who urged the formalizing of the provisions within the Agreement. Perhaps the official secrecy surrounding the matter in 1921 was partially to blame for what was to come. Certainly the wartime years of pro-Soviet propaganda resulting from the alliance against Hitler also had an effect, as did hopes for a post-war United Nations. It is possible that some presidential advisors who should have foreseen dangerous trouble with Stalin on the POW issue were secretly sympathetic to any Soviet positions or demands. Harry Hopkins, who had enormous influence on Roosevelt, is a candidate for this type of influence. It seems questionable to the author, however, in light of subsequent actions by Stalin and Beria's NKVD, that any slight change in the wording of the Yalta Repatriation Agreement would have actually altered Stalin's subsequent conduct on Allied prisoner returns.

One of those who advised the President at Yalta, was Assistant Secretary of State Charles Bohlen, who also served as Roosevelt's Russian translator during conferences with Stalin, Molotov and others. Bohlen was later to maintain silence on the fate of American POWs remaining in Russian hands, but in his memoirs he later made clear that something had gone badly wrong with the post-Yalta prisoner repatriations:

"Another agreement at Yalta that was certainly open to question and PERHAPS CAUSED THE UNITED STATES AND BRITAIN MORE MORAL ANGUISH THAN ANY OTHER was the one made by our military representatives and signed by Roosevelt for the return of the citizens of one country in territories overrun by the armies of another. Our military leaders believed that without some such arrangement, the Soviet Union might find a pretext to retain the thousands of American prisoners who had been sent to camps in parts of Poland or east Germany subsequently captured by the Soviet armies if we did not force reluctant Russian prisoners of war to return to the Soviet Union."

Bohlen seems to over simplify the problems with the terms and wording of the final agreement which had so worried Joseph Grew and others:

"It has always been my feeling that the execution of this agreement went beyond conditions laid down in the text. The controlling paragraph of the agreement said, 'All Soviet citizens liberated by the forces operating under United States command and all the United States citizens liberated by forces operating under Soviet command will without delay after their liberation be separated from enemy prisoners of war and will be maintained separately from them in camps or points of concentration until they have been handed over to the United States or Soviet authorities as the case may be, at places agreed upon between the authorities.' There was nothing in this agreement that required the forcible repatriation of unwilling Soviet citizens to the Soviet Union. Yet this is exactly what happened..."

Bohlen goes on to blame all subsequent troubles on the military: "The execution of the repatriations was entirely determined by the American and British military..."[13]

The Yalta military POW repatriation agreements were of course, subject to final approval by General Marshall and implementation by he and General Eisenhower. Although altered, they were in fact the result of the prior negotiations by British and American military officers, including Major General John Deane of the US Military Mission of the Moscow Embassy, who later wrote:

"My darkest days in Russia were in the winter of 1944-45 when I was trying to arrange for the best possible care and speedy repatriation of American prisoners of war liberated by the advance of the Red Army...None of my negotiations with Soviet officialdom met with less success. The story is not a pleasant one. It is marked by broken agreements, vindictiveness, recriminations, and stupidity on the part of Soviet leaders."[14]

In retrospect, the stalling, the lying and the misinformation of the Russians that was to appear to Deane and others at the time as "stupidity on the part of Soviet leaders," was in fact cold and deliberate scheming by the communists to achieve their goal of dominating postwar eastern Europe and forcing the Allies to return millions of persons who did not wish to live under Stalin's rule.

A different view of the prisoner repatriation agreement worked out by Deane and Archer with Golubev, Slavin and Novikov, and signed at Yalta, is provided by one of Beria's NKVD guards at the conference, Ilya Dzhirkvelov, who after many succeeding years as a KGB officer was eventually to defect to the West. Writing in Britain in 1987 Dzhirkvelov remembered:

"A telephone cable had been laid along the highway from Yalta to Sevastopol to enable our allies to communicate with each other and it was our duty to protect it. In a number of places on my section special equipment was attached to the cable so that KGB specialists from Moscow could listen in to any conversation. Those of us guarding the Yalta conference knew nothing of course, about what was decided there and it was only in 1948 that I learned what had gone on at the conference table:

"At a meeting of operatives, Andrei Vyshinsky, who had been one of Molotov's deputies at Yalta, asked if any of us had taken part in guarding the conference... 'Do you realize what a tremendous significance the conference had for the future foreign policy of the Soviet Union and its security?' Vyshinsky asked, and went on to say that the results of the conference reflected the experience and maturity of Soviet diplomacy and that future generations of Soviet officials handling relations with the capitalist world should learn from the example of Yalta."

"After discussing in general terms the purpose of the Yalta conference, Vyshinsky described the way in which the final document of the conference was arrived at. 'we realized,' he said, 'that the most important thing would be the final document, but the Soviet Union was in a minority because, however good our relations were with America, the views of the American delegation would certainly be closer to the British point of view. So we did not want to be the ones to produce the draft of the final document. Stalin therefore proposed that the American and British delegations should produce it, because, as he said, they were in the majority.'

"According to Vyshinsky, Stalin's proposal was accepted, though not without some opposition from the allies. 'We knew,' Vyshinsky continued, "that Churchill and Roosevelt would try to draw up a document that would be unacceptable to us and then, if we refused to accept it, they would hand the job over to the Soviet delegation. Our supposition proved correct...During our study of the draft (author's note: Deane's first 3-point proposal?) we came to the conclusion that the preliminary talks we had had with the other delegations had not achieved what we wanted, although we appeared to have arrived at common solutions to the problems confronting us. These were basically the post-war frontiers in Europe...THE FATE OF SOVIET PRISONERS OF WAR AND THAT OF NON-_GERMANS WHO HAD FOUGHT AGAINST THE SOVIET ARMY AND WERE IN EUROPEAN COUNTRIES. To have rejected the document would have meant assuming responsibility for drawing up another draft which might not be acceptable to the British and American delegations, and that might not leave us any chance of achieving our aims. There were many suggestions as to what our next step should be. Stalin listened to them all with attention and then said that we should tell our allies that on the whole the Soviet side accepted the draft of the final document but would like at the same time to introduce a number of amendments. He went on to say that since we had accepted the draft, they could not refuse to let us make some amendments. "We will make the document,' Stalin continued, 'what we would like it to be in general terms and we won't bother about the details even if they do not quite suit our book.'"

'Our allies were greatly surprised," Vyshinsky said. "They could not refuse us, and in that way, by introducing amendments, we obtained practically everything we had aimed for. That was one of the most important successes of Soviet diplomacy,' he concluded."[15]

From January through April 1945, the advancing Red Army overran as many as 60,000 U.S. POWs and POW/MIAs, and some 1,500,000 British, French, Dutch, Belgian and other Western Allied POWs and civilian deportees (including ethnic Jews) held by the Germans in Poland, eastern Germany and Austria.[16] Although many Allied POWs taken by the Red Army were relatively well-treated and ultimately released, now-declassified SHAEF documents indicate that an officially-estimated 500,000 of these Allied prisoners, including approximately 20,000 Americans, subsequently disappeared inside Soviet-occupied territory. (As many as 4,000,000 German/Axis and Japanese POWs also fell under Soviet control.) The deportations were controlled by the murderous NKVD chief Lavrenti Beria, and in Poland and Germany by his deputy Ivan Serov (later head of the KGB). On the eastern front, in Poland and East Prussia during the weeks following the Yalta conference, the Germans forced hundreds of thousands of Allied POWs and civilians to retreat westward with them, but many, including thousands of American and British POWs, were left behind and taken by the Red Army. In violation of the Yalta Agreements, the Soviets attempted to hide their possession of many thousands of the American, British and Commonwealth POWs and refused Allied contact officers access to the camps and areas in Poland where they were held.

Among the first post-Yalta reports on Americans in Russian-controlled Poland came from Jan Ciechanowski, the Polish Ambassador to the United States representing the non-communist Polish Government in exile, based in London. Ciechanowski was a particular friend of the former United States Ambassador to Poland, Anthony J. Drexel Biddle Jr., who had been recommissioned as a Colonel in the Army, on Eisenhower's headquarters staff, as chief of the European Allied Contact Section (EACS). Biddle maintained connections to an intelligence network in Poland and other occupied nations. Some members of the Polish Home Army loyal to the London-based Polish government acted as intelligence agents for the Western Allies inside Poland. This report was sent to Perkins at the embassy who forwarded it to George Kennan:

"MR. CIECHANOWSKI, OF THE POLISH EMBASSY HERE, CALLED ME THIS AFTERNOON TO SAY THAT THEY HAVE JUST RECEIVED WORD FROM THEIR(POLISH) GOVERNMENT THAT OVER 1,000 OFFICERS AND ENLISTED MEN OF THE AMERICAN ARMY ESCAPED FROM GERMAN PRISON CAMPS IN THE POSNAN DISTRICT AND ARE NOW SCATTERED IN VARIOUS TOWNS IN POLAND. THE PRISON FROM WHICH THEY ESCAPED WAS SZULIN IN THE POSNAN AREA.[17] TWO AMERICAN OFFICERS: COLONEL CHARLES KOHN AND MAJOR JERRIS EDEN CALLED AT THE FOREIGN OFFICE IN LUBLIN..."[18]

A note sent on February 12 from the British Embassy in Moscow to George Kennan at the US Moscow Embassy calls attention to a February 9th Soviet radio broadcast which announced a list of United Kingdom, Dominion and American prisoners liberated by the Red Army. When pressed for details by Western journalists in Moscow the Soviet and Polish communist authorities denied all knowledge of the broadcast. The message ended on an ominous note: "WHEN PRESSED BY A CORRESPONDENT THE POLISH EMBASSY STATED THAT 1,400 BRITISH AND 600 AMERICAN OFFICER AND OTHER RANK PRISONERS OF WAR HAD BEEN LIBERATED OR HAD ESCAPED FROM THE CAMP AT SHUBIN...[19] ...LATER IN THE DAY THE POLISH EMBASSY TELEPHONED THE SAME INFORMATION TO THIS EMBASSY, WITH THE DIFFERENCE THAT THE FIGURE FOR BRITISH PRISONERS WAS STATED TO BE 500..."[20]

On February 13, 1945 the US State and War Departments made a joint announcement that all German POW camps in Poland, East Prussia and Pomerania were being evacuated to the west. Sir James Grigg, the British Secretary of State for War, made an official statement in the House of Commons on February 13th, 1945 in reply to questions Conservative and Labor questions as to whether any British prisoners had been liberated by the Russians:

"12 CAMPS...HAVE EITHER BEEN OVERRUN BY THE SOVIET FORCES OR ARE IN THEIR DIRECT PATH. THERE WERE ABOUT 60,000 PRISONERS FROM THE BRITISH COMMONWEALTH IN THESE CAMPS...THE GERMANS INTENDED

TO MOVE THE PRISONERS...TO CENTRAL GERMANY...IT IS LIKELY, HOWEVER, THAT MANY HAVE BEEN OVERTAKEN BY SOVIET FORCES... FOLLOWING ARE THE CAMPS: STALAG IIB, STALAG IID, STALAG IIIB, STALAG IIIC, STALAG 344, STALAG VIIIB, STALAG VIIIC, STALAG XXA, STALAG XXB, STALAG LUFT III, STALAG LUFT IV, STALAG LUFT VII..."[21]

This information was transmitted to Secretary of State Stettinius by Carlos Warner, Second Secretary of the American Embassy in London under Ambassador Winant (whose son was a POW of the Germans). These prisoners included British Army, RAF and Merchant Marine, and also thousands of Australians, New Zealanders, Canadians and South Africans who would eventually be declared missing and disappear. OFLAG 64 at Shubin was not mentioned.

The Western Allies knew for certain that up to about mid-February 1945, at least 35,000 Americans and some 73,000 British and Dominion prisoners of war had been confined in Poland, East Prussia and the eastern most part of Germany now overrun by the Soviet Army. The actual numbers may have been higher than this due to as yet unrecorded, late captures. In addition the allies also knew of 170,000 French, 20,000 Belgians and many thousands more of other nationalities who had been held prisoner in Poland and eastern Germany. 100,000 Allied prisoners were marching along the north German coast ahead of the Russian advance through East Prussia and Poland. In the central group some 60,000 POWs were moving through an area bounded by Leipzig, Dresden and Berlin. About 80,000 prisoners, including a reported 25,000 Americans, were on the southern line of march through the Sudetenland. Many tens of thousands of U.S. and Western Allied prisoners of war, however, went missing behind Soviet lines at this time.[22]

In late January and early February 1945, prior to the Yalta Agreement on repatriation of prisoners, the American Embassy in Moscow had begun receiving reports of US prisoners in Soviet hands through US intelligence, Senator Connally and the Polish Ambassador to the United States, Jan Ciechanowski. Ambassador Harriman, his #2 at the US Embassy, George Kennan, and General John Deane, Chief of the US Military Mission, had already begun responding with demands for information from the Russians. Upon receiving evidence that many Americans had been held incommunicado by the Soviets, who hadn't revealed the fact, Kennan, as Charge d'Affairs demanded contact with the prisoners, and subsequently suggested a stiff protest to Foreign Commissar Molotov. Another foretaste of trouble occurred when the Embassy received a note dated February 14th,1945, signed F.H. Brooks, saying:

"...AT THE MOMENT WE ARE WITH THE RUSSIAN ARMY IN NOWY SACZ FOR OVER THREE WEEKS AND NO RESPONSIBLE  PARTIES HAVE BEEN NOTIFIED OF OUR PRESENCE AND WHEREABOUTS.

WE HAVE NO FREEDOM AND HAVE BEEN TOLD WE ARE INTERNEES."[23]
    Contrary to the Yalta Agreements the Soviets attempted to hide their possession of these and thousands of other prisoners until a lucky 3 Americans escaped eastwards, eventually reaching the Military Mission at the US Embassy in Moscow where they were debriefed by General John Deane. More handwritten notes from other groups of prisoners continued reaching the American and British embassies into March 1945.[24]
    Among the camps known to have been overrun by the Soviets at this approximate time were: OFLAG-64, Stalag II B, Stalag II D, Stalag 11E, Stalag III B, Stalag III C, Stalag 344, Stalags VIII A, Stalag VIII B, and Stalag VIII C, Stalag XX A, Stalag XX B, Stalag Luft III, Stalag Luft IV, Stalag Luft VI, Stalag Luft VII.[25]
    Harriman and Deane pressed the Soviets hard to allow U.S. contact teams into Poland and East Prussia to locate American POWs and ensure their safe evacuation, but the Soviets, led by Stalin's cohorts Molotov, Vyshinsky, Dekanozov, Golikov and Golubev continually stalled and blocked such action. There seemed to be little the American Embassy could do. A typical note of this period to Harriman on February 28, from his #2 (and sometime Charge D'Affairs) George Kennan, ended with:
    "...The only thing I could suggest would be a stiff, further note to Molotov."[26]
    A state of continual crisis in the US Moscow Embassy followed for over 3 months as the Soviets under Stalin and Beria, particularly Foreign Minister Molotov, his Deputy (and Stalin's 1930s purge prosecutor) Andrei Vyshinsky, and Repatriations Generals Golubev and Golikov, (a high-level intelligence aide to Stalin) lied, stalled and accused the Americans themselves of perfidy in POW handling. The Russians also simultaneously demanded American and British diplomatic recognition of their Polish-communist regime and indicated their expectations of a Postwar $6 Billion war reconstruction "credit."
    US and British POWs from German camps such as Stalag III-C-Kustrin, Stalag II-B-Hammerstein, Stalags VIII-A, B and C containing many British, New Zealanders and Australians, Stalag XX-A, near Torun on the Vistula, Stalag XX B, near Marienberg, southeast of Danzig, OFLAG-64 northwest of Szubin, and others, had been left behind or escaped during the German retreat and were overtaken by the Red Army.
    Thousands of U.S., British and Allied prisoners were being withheld and confined by the Soviets instead of being repatriated through Odessa, on the Black Sea, as had been promised following the Yalta Agreements. Thousands more American, British/Commonwealth and French ex-prisoners were in hiding from the Russian forces, many of the PWS being assisted by the Polish people and even by the

Polish militia. Only a single, small American contact team of three men, consisting of Lt. Col. James D. Wilmeth, a surgeon, Lt. Col. Curtis Kingsbury and a U.S. interpreter-T-5 Paul Kisil, were finally allowed into Poland, on February 28, and it was confined, for political reasons, to Lublin, where the Russian-controlled Polish puppet government had been headquartered. In time the U.S. contact team would be confined to a guarded hotel in Lublin, and ordered to leave in mid-March.[27]

By the 28th of February Stalag Luft III, at Sagan, on a tributary of the Oder near the Polish border in eastern Germany, had been liberated. Arthur Robinson, a special representative of the American Red Cross, had learned from the German Intercross delegate, Shirmer, in Geneva, that in late January 1945, the Germans had held 6,844 US officers and thousands more Allied personnel in Sagan, and that at least 300 severely wounded American officer-aviators had been abandoned there and overrun by the Russians. This scenario was being repeated across Poland.[28]

On March 1, 1945, a secret message from General Deane to General Clayton Bissell, Chief of US Military Intelligence in the Pentagon, conveyed precise US prisoner intelligence gleaned from 252 Americans who had reached Odessa, on nearly 5,000 US POWs still remaining in Russian-controlled Poland:

"...ODESSA REPORTS ARRIVALS OF LIBERATED U.S. PW THERE TO DATE AMOUNT TO 146 OFFICERS AND 106 ENLISTED MEN... INTELLIGENCE INDICATES THERE ARE 28 SICK AT WOLLSTEIN...700 FROM OFLAG 64, SZUBIN; 1,900 FROM STALAG 3C, KUSTRIN; AND 2000 FROM STALAG 2B, HAMMERSTEIN..."[29]

(This number did not include the 300 wounded U.S. officers from Luft III left behind at Sagan, and other groups of American POWs in Poland.)

In view of Bissell's 1945 actions in suppressing evidence of Soviet complicity in the Katyn Forest massacre to avoid "great possibilities for embarrassment" to the Russians, "in the spirit of Yalta," a question arises as to whether U.S. Military Intelligence under Bissell's command properly forwarded secret reports of American POWs under Soviet control such as this one, to top U.S. policymakers.

On that same 1st of March, President Franklin Roosevelt gave a speech to the US Congress claiming that the decisions reached at Yalta, especially in regards the United Nations, spelled the end of the system of unilateral action, exclusive alliances and spheres of influence. Writing later of this period, Roosevelt's wartime Chief of Staff, Admiral William Leahy, remembered:

Even as he spoke, a crisis was in the making in our relations with Soviet Russia that was to reach dangerous proportions and require his best attention up to the actual day of his death. The swift advance of the Russians through Poland was overrunning many Nazi prison camps where Americans had been incarcerated...We asked Stalin for an explanation."[30]

Now-declassified U.S. documents reveal that the March 1, 1945 U.S. prisoner of war total sent by Deane in Moscow to Clayton Bissell in the War Department was incomplete, as it did not include many more Americans who had been left behind or had escaped from the German retreat, and who had been or were soon to be taken under Soviet control. These POWs were from camps such as: Stalags XX A (Torun) and XX B (Marienburg), Stalag Luft III (Sagan), Stalag Luft IV (Gross tychow), Stalag Luft VI (Heyderug, Lithuania) and others. A total of 2,858 American POWs were to be eventually recovered through Odessa, while an additional 2000-3000 U.S. prisoners were to disappear from Russian-controlled Poland and adjacent east German border regions while being shipped eastwards into the Soviet Union under NKVD authority, ostensibly toward "Odessa."[31]

OFLAG-64 at Szubin, Poland contained approximately 1,557 American officers, and had been overrun by the Red Army on January 21, 1945. 86 disabled Americans were left behind in the camp itself. 1,471 U.S. officers left on foot for an intended 345 mile trek to Brandenburg, Germany, but many of them were subsequently overtaken by the Soviets. The 86 Americans left behind were taken by the Russians in trucks from Szubin to Rembertow on the 28th, where they were confined by Soviet guards with 5,000 others of many nationalities for nearly a month. It was not until February 22, 1945, after repeated complaints from the U.S. Embassy in Moscow that they were loaded into boxcars and shipped to Odessa for repatriation. During this confinement several of the American officers escaped and two of them eventually reached Moscow where they were fortunate enough to elude the NKVD and report to General Deane. Their debriefings provided Deane and Harriman with their first eyewitness reports of Soviet actions with American prisoners of war.[32]

In the March 1st intelligence cable from Deane to Bissell, it was estimated that 700 of these officers from OFLAG-64 came under Soviet control, but this estimate was later raised to 1,000 by Lt. Col. James Wilmeth, from his first-hand debriefings of American ex-POWs in Lublin, Poland, from February 27 to March 28. 186 of the 1,471 retreating American officers escaped from the Germans near Exin and made their way to Rembertow where they joined the 86 disabled men left in the camp under a Colonel Drury, and all were eventually shipped to Odessa, reaching that port in early March. About 375 of the officers from OFLAG-64 were reported to have reached POW camps in Germany including OFLAG 13-B, Stalag III-A at Luckenwalde (which fell under Soviet control) and VIII-A at Moosburg, which was liberated by American forces. Several hundred

American officers from OFLAG-64, Szubin, may thus have disappeared into Soviet captivity.

Stalag III-C at Kustrin, containing 2,053 American NCO's (at last report), was overrun by the Soviets on the 31st of January. Three U.S. prisoners present during the German evacuation, who escaped the Russians and were eventually debriefed by U.S. officers, supplied the following eyewitness testimony of the Soviet takeover, which remained secret for many years thereafter. Their names were Staff Sergeant J.B. Blasko (90th Infantry Division), Sergeant G.G. Gann (101st Airborne Division) and T/4 H.P. Andrews (101st Airborne Division):

"As the column was being marched down the road, a Russian spearhead consisting of one Russian tank, one Sherman tank, one tank destroyer and a small body of troops (appeared). The spearhead fired on the column (of U.S. prisoners) and there were about 15 casualties. The German staff which was riding at the head of the column in automobiles took off, and the column scattered. About 300 of the column of 2000 ran off, the rest being taken back into the camp by the Germans. When the Russian spearhead found out that the column consisted of Americans they apologized for having fired. The Germans tried to take the column out again but ran into Russians and this time 500 POWs got away. The Germans took 200 back to Kustrin and this is the last heard of them. The Soviets soon took over Kustrin.[33]

"Gann and Andrews stayed with the Russians and on the third day of their freedom Blasko joined them. They stayed on the front for a total of five days with the crew of a 37mm anti-tank gun, the gun being American-made. Sergeant Gann made friends with a Russian Sergeant on the gun and the men were treated all right. The Germans counterattacked and the Russians pulled back 12 kilometers. After they had withdrawn, a Russian Lieutenant in charge sent the men (U.S. ex-POWs) back. They caught a ride on a truck and rode to the small town of Witten, and here they caught a train and rode into Warsaw. They had been told there was an American Consul in Warsaw but found none. In Rembertow, outside the Warsaw/Praga area, they met about 300 (American) officers and 300 (American) enlisted men, some of whom described conditions of the camp to them. The set-up didn't appeal to the men and they went on to Lublin. A Colonel in charge of the men at Rembertow had told them they would be shipped to MOSCOW. In Lublin the men placed themselves in the hands of the Polish militia. They were kept in homes or in Barracks, AND ONE AIM OF THE POLES WAS TO KEEP THEM AWAY FROM THE RUSSIANS. From Lublin the men were taken (by the Poles) to Rzeszow where they spent a total of nine days, spending four days in Rzeszow and the balance of the time at a nearby airfield where they caught a plane to Poltava. (The U.S. airbase in the

Ukraine.) They stayed with Poles and received good treatment."[34]

U.S. Major Gerald Rich reported to General Deane on March 14th about the American casualties among the POWs fired on by the Russians at III-C, Kustrin on January 31. His sources were eyewitnesses who had returned to military control: 1st Sergeant Percy Coleman, Sergeant Raymond P. Carraway, T/Sgt. Donald Flaherty and Sergeant George Lukashewitz. They reported 9 American POWs had been killed and 12 wounded, a few of whom may also have died subsequently, and listed all 21 names.

An American ex-POW who was held at Stalag III-C Kustrin from September 1944 to the end of January 1945, Arley L. Goodenkauf of Table Rock, Nebraska, remembered the last days in the camp:

"In January of 1945, we heard the reports of the Russian armies advancing rapidly towards us, and began to consider seriously the possibility that the camp might be evacuated across the Oder River. Three of my roommates joined me in digging out a hole beneath a floor, and, when the camp was assembled for evacuation we moved in. In the confusion we were not missed and the Americans were moved out. The column ran into a Russian patrol, was fired upon, and ended up back at the camp shortly after...minus casualties...Three days later, on the night of February 3rd, the Russians moved us back about three miles, and told us we were on our own. We broke up into small groups and eventually made our way to Wreschen, Poland, where the Russian Army put us on a train for Odessa. While many of the Russian soldiers we encountered were friendly, our treatment was not always what you would expect from our allies. One of my group lost his overcoat to a gun-waving Russky—a serious matter during that winter—and I came close to losing my life at the hands of a drunk Russian soldier on another occasion...I remember watching the Russian Army marching a column of Hungarian prisoners through a Polish city, and hearing someone say that several Americans were included among the prisoners. I never actually saw anyone in that circumstance, however. We arrived in Odessa on March 16 and stayed about one week...In Odessa we were confined to one building...and we had two men (Yates and Timmerman) killed by a falling wall in Odessa. We shipped out to Naples on the British freighter Circassia, arriving on April 2nd. Our ship was the second to leave Odessa, the first one going to Port Said, Egypt... Robert Bradley has done a lot of research on the men from III-C and one of the stories told to him insisted that some of the wounded Americans at III-C were flown out by a bomber which landed on the Oder River ice sometime after February 3, 1945. He is extremely skeptical of this, but has talked to a man who insists he was flown out in this way...I have been corresponding with several of my friends from III-C, and we are in agreement that many of the 2,000 men have not been returned..."[35]

American intelligence officers debriefing escaped American prisoners later made a careful estimate that some 1,700 Americans from Stalag III-C-Kustrin fell into Soviet control. These two camps alone (OFLAG-64 and Stalag III-C) are stated to have accounted for at least 3000 American POWs in Soviet control, in recently-declassified original reports by Lt. Colonel Wilmeth, Major Gerald Rich and Captain P.S. Hall, located in the National Archives. (Only 2,858 American POWs from Poland and eastern Germany were to be eventually repatriated through Odessa.) In addition to these camps however, at least 6,100 American enlisted men were held in Stalag II-B (Hammerstein) in eastern Germany, which was evacuated during this same Russian advance, just prior to the Soviets overrunning the camp on the 26th of February. This camp had included 9 satellite work camps (Kommandos) containing 5,315 Americans, and according to formerly restricted and classified Red Cross records in the National Archives, Stalag II-B and its satellites actually contained a total of 7,200 Americans in January 1945. In addition, some 16,000 French POWs and about 1,000 Belgian POWs and more of other nationalities were in separate enclosures within the same North Compound as the Americans, while some 10,000 Russian POWs were held in a separate East Compound of II-A.

During the evacuation of the main camp some 1,200 Americans were forced-marched out of the main compound while at least 500-700 U.S. prisoners were left behind in the barracks. (Some of these men were shipped west in February while some failed to reappear.) The additional 5,315 American POWs in 9 satellite work camps of II-B were apparently evacuated at the same time, as the Soviets advanced, but few reports of their subsequent movements and ultimate fate are recorded, other than the 1,700 Americans from work-Kommando Lauenberg, who joined the remnant of the main column during late March, in Germany.

Initial U.S. Military Intelligence reports from escaping Americans from II B-Hammerstein, and later reports from those who eventually reached Odessa, indicated that some 1,000 to as many as 2,000 Americans from II-B and it's 9 satellite work camps fell into Soviet hands during early stages of the retreat. (One later report was to reduce this number to an estimated 100 "known to be" in Soviet control, but with no substantiation or explanation of what happened to those missing from II-B, Hammerstein.) U.S. prisoners of war who eventually were liberated by British forces at Westertimke, Germany, in April reported many American POWs from Stalag II-B had escaped, disappeared or been forcibly detached on labor details during the retreat from the Russians. At Westertimke they met some of the Americans who had been left behind at Hammerstein and shipped on February 18th in an easy 3 day journey to Germany. From all available evidence, it appears that several thousand Americans simply vanished inside the Soviet zone of control in late February and early March 1945.

At Torun (Thorn) on the Vistula River, Stalag XX-A, "containing American and British PWs," was reported as evacuated just ahead of the Russian advance. At the end of February the American contact officer in Lublin, Poland, Lt. Colonel Wilmeth, upon being refused by the Russians radio and telephone communication with General Deane in Moscow, sent four POWs (2 Americans and 2 British) with a report for the U.S. Military Mission and the Embassy on conditions in Poland.

One of these messengers who got through was F/Sergeant V.A. Panniers, an RAF ex-POW from Stalag XX-A at Torun. He reported that XX-A was evacuated ahead of the Soviet advance on January 20, 1945 (one day prior to OFLAG-64 which was further west). Panniers also stated that 500 of the British prisoners from XX-A were forced to march for Bromberg, and that large numbers of POWs from XX-A had escaped and were at large in the Russian-controlled area. He further reported that 250 sick British soldiers had been left in the hospital at XX-A, Torun, which had been overrun by the Red Army on the 21st of January. He had heard hearsay reports that a German SS unit had shot up a group of British POWs being marched from XX-A, but he had not confirmed this report. A later 14 March report (see below) on XX-A stated: "Portions will be transported to unknown destination by rail... MARCH 9 PHONE CALL FROM GENEVA (from Shirmer the German Red Cross delegate) APPEARS TO INDICATE THAT A SIZABLE GROUP MAY HAVE ESCAPED, SEE REPORT ON STALAG XX-B, NO NEWS HAS BEEN OBTAINED ON HOSPITAL AT KONRADSTEIN."[36]

Stalag XX-B at Marienburg (north of XX-A at Torun and about 45 miles southeast of Danzig, Poland) also contained British and some American POWs. The messenger sent from Lublin, Sergeant Panniers, also brought a report that some 900 British POWs from XX-B, Marienburg, had been out on working parties when the camp was overrun by the Red Army. These men also no doubt fell into Soviet control. In a recently-declassified report by Major General Ray W, Barker, Assistant Chief of Staff, G-1, at SHAEF Headquarters, dated March 14, 1945, little news of this camp was reported during the Russian advance from late January into February 1945. The only note for XX-B is:

"A SMALL PART CAPTURED WITH XX-A, BUT MOST OF THESE PRISONERS OF WAR ARE IN MARIENBURG AND ARE COMPLETELY SURROUNDED BY THE RUSSIANS."

Both Stalag XX-A and XX-B were reported as containing American and British POWs and as having been overrun by the Russians in a now-declassified report to Secretary of State Edward Stettinius by Carlos Warner, Second Secretary of the American Embassy in London, on February 14, 1945. Warner quotes as his source a statement to this effect made by the British Secretary of

State for War, Sir James Griggs, in the House of Commons on February 13th. An American Journalist traveling in Poland, Anna Louise Strong, reported to Colonel James C. Crockett of the U.S. Military Mission to Russia, on seeing 400 British and 24 American POWs just south of Torun at Ciechoginek, on about February 12, and these prisoners were most likely escapees from XX-A. She reported, "conditions there are poor." She also reported American POWs further west from this point, "at Woolstein, south of Posnan, 23 Americans seriously ill in hospital. Two American doctors are there; Dr. Godfrey and Dr. Monahan. This was the condition on February 13."

Subsequently only a few scattered reports of these camps appear in War Department records in the National Archives, mostly in the form of cables asking questions about where these prisoners had gone.

Thousands of British and Commonwealth POWs who had been held in Stalag III-A, Stalag II-D, Stalag II-E, Stalag VIII-B, Stalag VIII-C, Stalag XXI-A (Schilberg), Stalag XXI-D (near Posen), Stalag XXI-E (east of Torun) and Stalag Luft VII (at Bankau), also disappeared behind the Russian advance during this period, and subsequent messages concerning them were mostly in the form of questions as to where they were.

In early February the site of Stalag Luft III at Sagan, on a tributary of the Oder River near the Polish border in eastern Germany, had been liberated by the Red Army. According to Red Cross documents, Arthur Robinson, a special representative of the American Red Cross, learned from the German Intercross delegate in Geneva, Shirmer, that in late January this camp had held 6,844 U.S. officers and thousands more Allied POWs, and that at least 300 severely wounded American officer aviators had been abandoned in Sagan and overrun by the Russians. The majority of the POWs from Luft III had been evacuated in the middle of the night of January 28/29, 1945, by a forced march in intense cold. The scenario was being repeated across Poland.

Stalag III-B, at Furstenburg on the Oder River, had an officially reported camp strength of 6,600 American POWs 17,000 French POWs, 1,970 Italian POWs, and 840 Yugoslavs, on January 31, 1945. The camp was evacuated ahead of the Soviet advance on February 2nd. According to Red Cross records, of the 6,600 Americans the German Intercross delegate Shirmer reported that 4,600 arrived at Stalag III-A (Luckenwalde, Germany), between February 11 and February 16, in poor condition. After a few days rest these men were reported moved on foot to a new satellite camp of Stalag III-A, which was located 18 kilometers (10 miles) west of Luckenwalde (which was overrun by the Soviets on April 22nd). Many of the 2,000 remaining Americans of Stalag II-B appear to have been left behind or escaped in the path of the Russian advance. An unknown number, presumably many of these 2,000, were taken under control of Soviet authorities in February and March 1945.

At Gross Tychow, Stalag Luft IV was evacuated early in the morning of February 2, 1945, as 6,165 Americans, with 800 British NCO's among them, moved in forced marches ahead of the rapidly advancing Red Army. According to restricted and previously classified Red Cross reports in the National Archives, Luft IV-Gross Tychow held 8,033 Americans and 820 British POWs at it's peak strength on January 31, 1945. No official reference to the fate of these nearly 2,000 unaccounted for American prisoners has been located. On February 25th and 28th Intercross reported that 1,500 American POWs from Luft IV were marching, and 209 were riding by train to Stalag Luft I at Barth in northern Germany, and 1,550 to Stalag Luft III. (Luft III had alreadt, in fact, been evacuated by the Germans and overrun by the Russians.) 3,600 others were being sent to other locations. Luft IV prisoners later reported that many hundreds, perhaps thousands of the American POWs disappeared in the ensuing weeks of what was remembered as a series of "death marches." Many of these missing American POWs fell under Soviet control at this time.

The POW camp at Bankau on the Polish-German border, Stalag Luft VII, was also evacuated by the Germans before the Soviet advance. 1,500 American and British POWs from Bankau arrived at Stalag III-A Luckenwalde on March 10th, but others from the camp had disappeared in Soviet-occupied territory. Stalag VIII-B at Teschen in southern Silesia, was evacuated on January 23rd with the Soviets reported to be only 5 miles away. Some 8,000 POWs were forced-marched from this camp westward through Bohemia, in Czechoslovakia. The Allied prisoners from Stalag 344 at Lamsdorf, near the Polish border in southeastern Germany, also marched into Czechoslovakia. Many prisoners also escaped or were left behind from these groups, directly in the path of the Soviet advance. It appears from available records that at least 5,000, and possibly many thousands more American POWs were taken under Soviet control in Poland and eastern Germany from late January to early March 1945. Due to the fact that some records in the National Archives remain classified, under British authority, the number of British and Commonwealth POWs who fell under Soviet control in Poland and East Prussia is more difficult to compile, but based on available documents would appear to have been some 10,000-15,000, or more.

With information gained from many sources the staff of the U.S. Embassy in Moscow under Averell Harriman became alarmed over the treatment of American POWs by the Soviets, and were now certain that Stalin's senior officials were lying about their evacuation of U.S. POWs to Odessa. From Moscow on March 2nd, Deane cabled General Marshall at the War Department in Top Secret:

"...VYSHINSKI...INFORMED THE AMBASSADOR THAT ALL OUR EX-PRISONERS OF WAR HAD BEEN EVACUATED BY TRAIN FROM POLAND ENROUTE TO ODESSA..."

"...I URGENTLY RECOMMEND THAT YOU ASK THE PRESIDENT TO SEND A MESSAGE TO STALIN...URGENTLY REQUEST...AMERICAN AIRCRAFT WITH AMERICAN CREWS...TO FLY BETWEEN POLTAVA AND POINTS IN POLAND..."[37]

The withholding of American prisoners had become so blatant that Harriman also prevailed upon President Roosevelt to make a personal appeal to Stalin on March 4th (which was the message referred to by Admiral Leahy in his memoirs):

"I HAVE RELIABLE INFORMATION REGARDING THE DIFFICULTIES ENCOUNTERED...EVACUATING AMERICAN EX-PRISONERS OF WAR AND AMERICAN AIRCRAFT CREWS WHO ARE STRANDED EAST OF THE RUSSIAN LINES...IT IS URGENTLY REQUESTED THAT INSTRUCTIONS BE ISSUED AUTHORIZING AMERICAN AIRCRAFT WITH AMERICAN CREWS..."[38]

Stalin refused, and in his reply denied there were any groups of Americans remaining in Poland.

Averell Harriman called on Molotov's deputy, Vladimir Dekanozov, urgently requesting that General Deane of the US Mission be allowed into Poland. Dekanozov, a Georgian who was a close ally of Lavrenti Beria of the NKVD, refused approval, saying Harriman would be given a final answer later.

On March 8th, 1945 Harriman cabled the President (m-23119) in top secret:

"...INFORMATION RECEIVED FROM OUR LIBERATED PRISONERS INDICATES THERE HAVE BEEN FOUR OR FIVE THOUSAND OFFICERS AND ENLISTED MEN FREED. THE RUSSIANS CLAIM TODAY THERE ARE ONLY 2100...I AM OUTRAGED...THAT THE SOVIET GOVERNMENT HAS DECLINED TO CARRY OUT THE AGREEMENT SIGNED AT YALTA IN OTHER ASPECTS, NAMELY, THAT OUR CONTACT OFFICERS BE PERMITTED TO GO IMMEDIATELY TO POINTS WHERE OUR PRISONERS ARE FIRST COLLECTED, TO EVACUATE OUR PRISONERS, PARTICULARLY THE SICK, IN OUR OWN AIRPLANES..."[39]

That night in Moscow the Ambassador's daughter, Kathleen Harriman, wrote to her sister: "Our gallant allies at the moment are being quite bastard-like. Averell is very busy—what with Poland, PWs and I guess, the Balkans. The house is full of running feet, voices and phones ringing all night long—up until dawn."[40]

On this same date, Harriman's number two at the Embassy, the career Soviet specialist George Kennan was disagreeing with the American response to the Russian challenge. He later wrote:

"I opposed the tenor of the notes of protest with which we were assailing Molotov: Notes in which expressions of plaintive surprise

# 194 Moscow Bound

were mingled with empty pleadings to the Soviet authorities to do otherwise than that which they were doing."[41]

Harriman, nevertheless, continued to urge the Soviet Foreign Minister, Molotov, to allow General Deane into Poland to report on the situation and to particularly assist the some 300 additional severely wounded American prisoners at Sagan, but the Soviets refused this and other requests.

Although some Americans were in fact being transported over a thousand miles to Odessa in filthy boxcars, to be held there under armed guard, handwritten notes from desperate American prisoners remaining behind in Poland continued to be smuggled into Lublin and the US Moscow Embassy:

"A NUMBER OF AMERICAN OFFICERS (PWs) ARE IN BROMBERG, WOULD LIKE TRANSPORTATION AND REPATRIATION...REQUEST IMMEDIATE ANSWER. R. W. ANDERSON, CAPT. 01286584." (dated 12 March)

A note attached to this says: "Anderson not on list!!" (Orig. Emphasis.)[42]

After fruitless attempts to persuade the Russians to live up to their agreements on prisoners, the Ambassador took the unusual step of again asking Roosevelt himself to personally intercede with Stalin. In a Personal and Top Secret message to the President on March 12, 1945, Harriman outlined the seriousness of the situation:

"...AFTER 48 HOURS OF CONTINUED PRESSURE ON THE (SOVIET) FOREIGN OFFICE I FINALLY RECEIVED AN ANSWER LAST NIGHT DISAPPROVING GENERAL DEANE'S TRIP TO POLAND ON THE GROUNDS THAT THERE WERE NO LONGER ANY AMERICAN EX-PRISONERS OF WAR IN POLAND...TODAY I AM INFORMED BY GENERAL DEANE THAT THE PERMISSION FOR OUR CONTACT TEAM TO REMAIN AT LUBLIN HAD BEEN WITHDRAWN...IT SEEMS OBVIOUS THAT THE SOVIETS HAVE BEEN ATTEMPTING TO STALL US OFF BY MISINFORMATION..."[43]

The Soviets insisted the Americans deal with the Polish communist puppet regime on the POWs, in a blatant attempt to gain US acceptance of its legitimacy, when in fact the composition of the future Polish government was a major diplomatic difference between the Soviets and their allies at this time. In a secret March 14 cable to Secretary of State Stettinius, also read by President Roosevelt, General Marshall, Secretary of the Navy Forrestal and others, Ambassador Harriman clearly saw the political implications, (#738):

"I ASSUME THE DEPARTMENT HAS BEEN INFORMED BY THE WAR DEPARTMENT OF THE GREAT DIFFICULTIES GENERAL DEANE AND I HAVE BEEN HAVING IN REGARD TO THE CARE AND REPATRIATION OF OUR LIBERATED PRISONERS OF WAR...

"I SAW MOLOTOV TODAY...HE MAINTAINED THAT...BOTH THE RED ARMY AUTHORITIES AND THE POLISH PROVISIONAL GOVERNMENT

OBJECTED TO THE PRESENCE OF OUR OFFICERS IN POLAND...
HE POINTED OUT WE HAD NO AGREEMENT WITH THE POLISH
PROVISIONAL GOVERNMENT...

"...I FEEL THAT THE SOVIET GOVERNMENT IS TRYING TO USE
OUR LIBERATED PRISONERS OF WAR AS A CLUB TO INDUCE US
TO GIVE INCREASED PRESTIGE TO THE PROVISIONAL (communist)
POLISH GOVERNMENT BY DEALING WITH IT IN THIS CONNECTION.
GENERAL DEANE AND I HAVE NOT, REPEAT NOT BEEN ABLE TO FIND
A WAY TO FORCE THE SOVIET AUTHORITIES TO LIVE UP TO OUR
INTERPRETATION OF OUR AGREEMENT...

"IT IS THE OPINION OF GENERAL DEANE AND MYSELF THAT
NO ARGUMENTS WILL INDUCE THE SOVIETS TO LIVE UP TO OUR
INTERPRETATION OF THE AGREEMENT EXCEPT RETALIATORY
MEASURES..."

"CONSIDERATION MIGHT BE GIVEN TO SUCH ACTIONS AS...
THAT GENERAL EISENHOWER ISSUE ORDERS TO RESTRICT THE
MOVEMENTS OF SOVIET CONTACT OFFICERS-THAT LEND-LEASE
REFUSE TO CONSIDER REQUESTS OF SOVIET GOVERNMENT-
ALLOWING OUR PRISONERS...TO GIVE STORIES TO THE
NEWSPAPERS..."[44]

(In his later published memoirs, "Special Envoy," Harriman was
to delete the part of his cable recommending US curtailment of Lend-
Lease to Russia. The eventual cut-off of Lend-lease on VE day was to
infuriate Stalin, and be used by him later as an example of American
hostility.)

During this period the question of American diplomatic
recognition of Soviet puppet governments in eastern Europe was a
major US government concern, but it was in regards to Poland that
the most serious concern was felt. At this time the United States
government was in the middle of negotiations over the establishment
of a Polish Provisional government of National Unity. The Russians,
of course, wanted their totally controlled Lublin-Warsaw communist-
dominated regime to remain predominant. The Americans and British
wanted representatives of the London-based Polish government-in-
exile to at least share power, and free elections to be held as was
agreed to by the Soviets. With this in the background, the real reason
for the Soviet action of retaining and concealing Allied prisoners at
this time, was becoming clear to others besides Harriman, Deane and
Kennan in Moscow.

On the 14th of March, Secretary of State Edward Stettinius
attached a memo to Harriman's telegram # 1445 and sent it to
President Roosevelt:

"I BELIEVE YOU WILL BE INTERESTED IN LOOKING
OVER THE ENCLOSED MESSAGE FROM HARRIMAN IN WHICH
HE DESCRIBES THE DIFFICULTIES WE ARE ENCOUNTERING
IN FACILITATING THE EVACUATION FROM POLAND OF
LIBERATED UNITED STATES PRISONERS OF WAR. IT
WOULD APPEAR THAT THE SOVIET AUTHORITIES MAY BE

ENDEAVORING TO USE OUR DESIRE TO ASSIST OUR PRISONERS AS A MEANS OF OBLIGING US TO DEAL WITH THE WARSAW GOVERNMENT."[45]

The fact that the warnings were coming from Harriman only lent greater weight to them. It was Harriman, after all, whose experiences with Soviet leaders went back to the early 1920s when he had made financial deals with them and obtained a mining concession in Stalin's home province of Georgia. It was Harriman who had often said that he could deal with the Russians, that they were just people like anybody else. It was Harriman who, in January of 1944, (when US Ambassador to Poland, Anthony J. Drexel Biddle Jr., was resigning in protest) had told American reporters in Moscow that he did not feel that Moscow wanted to impose a communist system on Poland.[46] His now-urgent cables on hostile Soviet conduct sent a wave of fear throughout the highest levels of the US government.

The State-War-Navy Coordinating Committee (SWNCC) meetings, functioning as a forerunner to the present-day National Security Council, was the area in which policy decisions were hammered out. These meetings were chaired by the powerful Assistant Secretary of War John (Jack) McCloy, who was Stimson's protege. They were attended by Stettinius, Stimson, Forrestal and Marshall, and also Undersecretary of State Joseph Grew (a pragmatist suspicious of Soviet moves), Charles "Chip" Bohlen, and at various times Assistant Secretary of War Robert Lovett, Assistant Secretary of State Dean Acheson, and others. Following Harriman's and Deane's first warning cables on POW repatriation troubles in Poland and President Roosevelt's first unsuccessful appeal to Stalin on March 4th, the State-War-Navy Coordinating Committee had met and put together a policy memorandum on the subject, specifically to answer a query by Gen. Joseph McNarney, US Mediterranean commander.

This memorandum, dated March 9th, and signed by Assistant Secretary of State for Political Affairs James Clement Dunn (who had married into the Armour family), set forth the United States government interpretation of the Yalta Agreements on repatriation of POWs and captive citizens in Allied and Russian hands. It also clarified the categories of repatriates, the importance of reciprocity between the United States and the USSR in repatriations, and recommended coordination with Britain on the course of action. Subsequently, Harriman's continued urgent cables from Moscow became a major topic of discussion at these policy meetings, and after a meeting of senior officials following his March 14th cable from Moscow, Stettinius sent the following message to Harriman (in part):

"... IT WOULD ALSO BE HELPFUL IN OUR CONSIDERATIONS IF YOU AND GENERAL DEANE COULD GIVE US YOUR VIEWS AS TO THE

PROBABILITY WHICH HAS BEEN RAISED HERE THAT RETALIATORY MEASURES ON THE PART OF THE UNITED STATES GOVERNMENT MIGHT RESULT IN COUNTERMEASURES DETRIMENTAL TO OUR LIBERATED PRISONERS."[47]

Harriman was to answer this concern over his proposed retaliatory measures, including a partial cutoff of US Lend-Lease to Russia on the 17th with the following:

"WITH REGARD TO THE LAST PARAGRAPH OF YOUR MESSAGE, I DO NOT FEEL COUNTERMEASURES DETRIMENTAL TO OUR PRISONERS WILL BE TAKEN SHOULD WE ADOPT THE RETALIATORY MEASURES SUGGESTED IN MY 738 OF MARCH 14. AT THE PRESENT TIME THE SOVIET AUTHORITIES COULD SCARCELY GIVE OUR PRISONERS LESS CONSIDERATE TREATMENT THAN THEY ARE RECEIVING...WE HOLD TEN OF THEIRS FOR EVERY ONE THAT THEY HOLD OF OURS...."[48]

(Through its controlled spies and moles at high levels of the American and British governments, the Russians were to receive copies of messages like this one, giving them unique insight into secret Allied positions in the face off that was coming.)

Meanwhile, the British Ambassador in Moscow, Sir Archibald Clark Kerr, was recommending to Anthony Eden and Churchill a more conciliatory response to the disappearance of thousands of British and Commonwealth prisoners in Poland and East Prussia (15 March):

"THERE ARE OBVIOUS OBJECTIONS TO ASKING FAVORS OF THE POLISH EMBASSY IN PRESENT CIRCUMSTANCES BUT IN VIEW OF HUMANITARIAN CONSIDERATIONS AND OF THE HELP THE POLISH RED CROSS ARE ALREADY GIVING OUR MEN IT OCCURS TO ME THAT YOU MIGHT BE PREPARED TO AUTHORIZE AN INFORMAL APPROACH WITHOUT PREJUDICE TO GENERAL QUESTION OF RECOGNITION."[49]

In another March 15 cable, Clark Kerr (long considered by many of his colleagues an apologist for Stalin and a friend of the Soviet regime) stated:

"...DEKANOZOV'S LETTER ALSO SHOWS THAT IN SPITE OF THE ASSURANCES GIVEN, MOLOTOV HAS EITHER BEEN UNABLE TO OVERCOME THE OBJECTIONS OF THE SOVIET MILITARY AUTHORITIES OR IS USING THIS QUESTION AS BLACKMAIL TO FORCE US INTO SOME RELATIONSHIP WITH THE PROVISIONAL POLISH GOVERNMENT...UNSATISFACTORY THOUGH THE PRESENT POSITION IS, I DOUBT WHETHER THE TIME HAS YET COME FOR A DIRECT APPROACH TO STALIN...I THINK THEREFORE THAT IT WOULD BE A MISTAKE TO TAKE ACTION NOW AT THE HIGHEST LEVEL, WHICH MIGHT BE INTERPRETED AS A CHARGE OF BAD FAITH..."[50]

Despite Clark Kerr's urging for moderation, Churchill was worried, and cabled Roosevelt on the 16th of March:

"AT PRESENT ALL ENTRY INTO POLAND IS BARRED TO OUR REPRESENTATIVES. AN IMPENETRABLE VEIL HAS BEEN DRAWN ACROSS THE SCENE. THIS EXTENDS EVEN TO THE LIAISON OFFICERS, BOTH

BRITISH AND AMERICAN, WHO WERE TO HELP IN BRINGING OUT OUR RESCUED PRISONERS...THERE IS NO DOUBT IN MY MIND THAT THE SOVIETS FEAR VERY MUCH OUR SEEING WHAT IS GOING ON IN POLAND... WHATEVER THE REASON, WE ARE NOT TO BE ALLOWED TO SEE."[51]

The British however, were at this time attempting to rigidly adhere to the terms of the Yalta agreements, hoping that their compliance would eventually result in a reciprocal Soviet response. On the 15th of March a secret directive) titled "Disposal of Soviet Citizens" had been issued by British Army authorities at the behest of Whitehall to Headquarters 8th Army and to 5th Army-British Increment, reiterating the Yalta agreements on access by contact officers to POW camps and collection points, and other articles which the Soviets were in the process of breaking. Also, the British directive stated that these provisions, "apply to all Soviet citizens, civilian or military, who have come under British control, from whatever source, irrespective of whether they have served in enemy forces or not."[52] This is a clear indication of the appeasing direction that British policy was moving in, to achieve an eventual return of missing British and Commonwealth POWs.

The Soviets demonstrated on March 18, 1945, just how violently they would react to attempted violations of their airspace by American or British aircraft. On that day a Soviet fighter airwing attacked a large force of heavy American bombers above the Soviet bridgehead over the Oder River, near the Polish-German border. Whether or not the bombers were bound for a secret POW rescue mission inside Poland is still classified, but they suffered no losses in the ensuing aerial battle—which resulted in six Soviet fighters being shot down. The incident was hushed up, but more aerial combats between the Russians, Americans and British occurred along the front from this time forward.[53]

In Odessa, where American and British prisoners and stranded aircrews were supposed to be well-treated prior to their evacuation by ship, Soviet conduct was also hostile. Although Harriman had cabled Washington that there "was no present reason to complain about the situation in Odessa," an American Red Cross official in Odessa at the time, Mr. George K. Hundley (in a 1945 report still in the National archives) took a different view. He had sailed from Naples to Odessa on March 7th, on the USS Tackle, a relief ship loaded with supplies for liberated POWs. Upon landing, the crew had extreme difficulty in even getting a small amount of relief supplies to the prisoners because the Russians would not cooperate in any way. Then they had great difficulty in even getting in to see the American POWs, let alone try to get some of them out of the camp. Hundley wrote: "Although the men in the camp are "liberated" POWs they are still "prisoners" because the Russians allow them no freedom whatsoever. Men and officers are CONFINED at all times to the camp building. They are always under guard." He continued that not only was it impossible to get any POWs out, it was nearly impossible to even get in to see them, and that at one point after he had managed to get in with some relief supplies, the Russian guards would not let him out again and he became a prisoner himself, for a time.[54]

Prisoners who were allowed to leave Russia reported that American officers were continually being interrogated by Soviet NKVD and SMERSH counterintelligence agents, sometimes in uniform, and sometimes in civilian clothes. Others later reported that they were ordered by American and British officers to remain silent about Russian mistreatment and about the disappearance of many of their fellow prisoners in Poland, and even in the Odessa area. (See below)

Roosevelt had again sent a "Personal and Secret" cable to Stalin at Harriman's and Stettinius' urging on March 18th, stating he knew American POWs were still in Poland and almost beseeching the Russian to allow American contact with them, and their subsequent release:

"In the matter of evacuation of American ex-prisoners of war from Poland I have been informed that the approval for General Deane to survey the United States prisoner of war situation in Poland has been withdrawn. You stated in your last message to me that there was no need to accede to my request that American aircraft be allowed to carry supplies to Poland and to evacuate the sick. I have information that I consider positive and reliable that there are still a considerable number of sick and injured Americans in hospitals in Poland and also that there have been, certainly up to the last few days and possibly still are, large numbers of other liberated American prisoners either at Soviet assembly points awaiting entrainment to Odessa or wandering about in small groups not in contact with Soviet authorities looking for American contact officers.

"I cannot, in all frankness, understand your reluctance to permit American contact officers, with the necessary means, to assist their own people in this matter. This Government has done everything to meet each of your requests. I now request you to meet mine in this particular matter. Please call Ambassador Harriman to explain to you in detail my desires."[55]

Stalin again refused, nearly going so far as to call Roosevelt a liar, and stating in his reply, which was received on March 22nd, 1945, that there were only 17 sick Americans in Poland and accusing the Americans of abusing and beating Soviet POWs in their control:

"I am in receipt of your message about the evacuation of former U.S. prisoners of war from Poland.

"With regard to your information about allegedly large numbers of sick and injured Americans in Poland or awaiting evacuation to Odessa, or who have not contacted the Soviet authorities, I must say that the information is inaccurate. Actually, apart from a certain number who are on their way to Odessa, there were only 17 sick U.S. servicemen on Polish soil as of March 16. I have today received a report which says that the 17 men will be flown to Odessa in a few days."

Stalin then presented a carefully crafted series of excuses for refusing Roosevelt's simple and personal request to allow U.S. contact officers access to American POWs in Poland:

"With reference to the request contained in your message I must say that if it concerned me personally I would be ready to give way even to the detriment of my own interests. But in the given instance the matter concerns the interests of the Soviet armies at the front and of Soviet commanders who do not want to have around odd officers who, while having no relation to the military operations, need looking after, want all kinds of meetings and contacts, protection against possible acts of sabotage by German agents not yet ferreted out, and other things that divert the attention of the commanders and their subordinates from their direct duties. Our commanders bear full responsibility for the state of affairs at the front and in the immediate rear, and I do not see how I can restrict their rights to any extent."

This argument by Stalin was both ludicrous and insulting in view of the swollen Soviet mission of over 40 officers which was then being tolerated at General Eisenhower's headquarters and also in the face of Stalin's well known ruthlessness toward his own military commanders. Characteristically the Soviet dictator ended this, his last message to Franklin D. Roosevelt on U.S. POWs, with more lies and an accusation of his own:

"I must also say that U.S. ex-prisoners of war liberated by the Red Army have been treated to good conditions in Soviet camps-better conditions than those afforded Soviet ex-prisoners of war in U.S. camps, where some of our men were lodged with German war prisoners and were subjected to unfair treatment and unlawful persecutions, including beating, as has been communicated to the U.S. Government on more than one occasion."[56]

Stalin neglected to admit the fact that hundreds of thousands of Soviet citizens had been captured in German uniforms, and that both they and hundreds of thousands of other Soviet citizens, who had not served the Germans, had no desire to return the Soviet Union under Josef Stalin's regime.

Coming less than two years after Stalin's heated denials of the Soviet massacre of Polish POWs at Katyn forest, which Allied leaders knew to be lies, his denials of the existence of thousands of American and Allied prisoners known by Allied intelligence to be remaining in Soviet control in Poland (and already secretly confined in the USSR) must have sent a chill down the spines of many senior policymakers in Washington and London.

The arrival in Moscow of three British prisoners of war, who had gotten through NKVD patrols ringing the city to intercept any unwanted visitors, forced the British Ambassador, Sir Archibald Clark Kerr, to convey a more realistic assessment of the situation in Poland to London on March 19, 1945:

"THREE LIBERATED BRITISH PRISONERS OF WAR ARRIVED RECENTLY IN MOSCOW, BRINGING WITH THEM A REPORT FROM THE MILITARY MISSION REPRESENTATIVES IN LUBLIN WHICH SUGGESTS THEIR ARE THOUSANDS OF OUR LIBERATED PRISONERS STILL IN POLAND...WE HAVE OBTAINED NO SATISFACTION, SINCE THE RUSSIANS CONTINUE TO DENY THAT THERE ARE ANY OF OUR MEN IN POLAND...MR. HARRIMAN TELLS ME THAT HE HAS EXHAUSTED ALL OF HIS ARGUMENTS WITH MR. MOLOTOV... WHEN ALL ELSE FAILED PRESIDENT ROOSEVELT SENT ON THE 17TH OF MARCH WHAT MR. HARRIMAN DESCRIBES AS A "ROUGH" PERSONAL MESSAGE TO STALIN, TO WHICH THERE HAS NOT YET BEEN A REPLY...I FEEL THAT WE CANNOT FALL SHORT OF THE AMERICANS... I THEREFORE SUGGEST THE TIME HAS COME FOR THE PRIME MINISTER TO SEND A PERSONAL MESSAGE TO MARSHAL STALIN...MY DOMINION COLLEAGUES, AND PARTICULARLY THE NEW ZEALAND MINISTER, ARE SERIOUSLY WORRIED ABOUT WHAT THEY FEEL IS THE DELIBERATE FAILURE OF THE RUSSIANS TO FULFILL THEIR AGREEMENT..."[57]

It appears from this message that Clark Kerr was being pushed strongly to act by not only the military realities but also the Australian, Canadian, and especially the New Zealand government representatives in Moscow, who were aware that large numbers of their own countrymen were disappearing in Russian-controlled Poland and East Prussia. Many Dominion or Commonwealth prisoners had been held in camps such as Stalag VIII A, VIII B and VIII C. The Soviets meanwhile retaliated on the British for using first-hand intelligence, with Dekanozov refusing to grant exit visas to the three escaped British prisoners. This gives indication of Vladimir Dekanozov's power and influence in the Soviet hierarchy. Like Vyshinsky, Dekanozov had served Stalin zealously in the 1937 purges. It must also be remembered that he had been appointed by Stalin as the NKVD master of Lithuania in 1940, and was already responsible for the murders of scores of thousands of arrested persons who were shot or deported to the Gulag death camps thereafter. (The cruel and evil Vyshinsky, it will be remembered, had been given Latvia where he ordered the killing or deportation of tens of thousands more innocent persons, within months.) Dekanozov, like each of Stalin's henchmen, was fully experienced at mass-kidnapping by the spring of 1945. The chance to use with practical impunity the finely developed Soviet hostage-taking technique against the envied and hated Americans and British was no doubt a source of great satisfaction to the Stalinist elite. The individual sufferings of the thousands of vanished American POW officers and enlisted men at the hands of experienced NKVD interrogators and torturers, bound to extract "espionage" confessions or intelligence information, can only be imagined.

During the course of the prisoner of war episode in Poland, Odessa and Moscow, the American and British forces were beginning their final campaign in the west against the Germans. On March 7, Patton's 3rd Army had broken through the German front and reached the Rhine near Coblenz. Also, the US First Army had advanced and captured an intact Rhine bridge at Remagen, near Bonn, and established a bridgehead across the river. Patton's army crossed the Rhine south of Mainz starting on the 22nd of March. (That same night, the Soviet Army surrounded and isolated the German pocket of Danzig (Gdansk) in the east.) To the north the British Army under Montgomery, with attached Canadians and the US 9th Army under General Simpson, began crossing the Rhine on the 24th of March. Winston Churchill was at Montgomery's headquarters, and it was here on the 23rd of March that he received important messages from Stalin and Molotov on British and Dominion prisoners in Poland.

British Foreign Secretary had urged Churchill to send a strongly worded telegram to Stalin in coordination with Roosevelt's on the 17th of March. Very likely influenced by Clark Kerr's urging of moderation, Churchill had messaged Eden:

"I DO NOT WISH TO SEND THIS TELEGRAM TO STALIN, AS IT WOULD ONLY MAKE A ROW BETWEEN US AFTER A MONTH'S SILENCE."

Instead he drafted his own, milder wording of a message, telling Eden that it, "GIVES NO EXCUSE FOR A ROUGH ANSWER." (from Stalin)

Churchill, (then at British Army headquarters on the Rhine River in France) cabled Stalin, "personal, private and Top Secret" on the 21st of March 1945 (in part):

"THERE IS NO SUBJECT ON WHICH THE BRITISH NATION IS MORE SENSITIVE THAN ON THE FATE OF OUR PRISONERS...I SHOULD BE VERY MUCH OBLIGED IF YOU WOULD GIVE THE MATTER YOUR PERSONAL ATTENTION...WE SEEM TO HAVE A LOT OF DIFFICULTIES NOW, SINCE WE PARTED AT YALTA, BUT I AM QUITE SURE THAT ALL THESE WOULD SOON BE SWEPT AWAY IF ONLY WE COULD MEET TOGETHER."[58]

Given the relatively subdued tone of his message Churchill seems to have been somewhat startled by the Soviet response. Molotov immediately sent a "venomous" message to Churchill on the Rhine accusing the British (and Americans) of attempting to negotiate a secret, separate surrender of the Germans facing the Western Allies. This referred to the negotiations between SS General Karl Wolff and Alexander and McNarney on the Italian front, being conducted in Berne through OSS Chief, Allen Dulles. The Russians had already linked their extreme displeasure over this matter (which may have been exaggerated or even fabricated) with American and British concern for their POWs in Soviet control, by raising the subject simultaneously with Harriman's strident POW complaints in Moscow.

Stalin's retort to Churchill, which followed Molotov's telegram, was a sharp series of lies and insinuations, and remarkably similar to the previously cited 22nd of March message for Roosevelt, clearly illustrating the orchestrated nature of Soviet actions:

"SO FAR AS CONCERNS BRITISH PRISONERS OF WAR, YOU HAVE NO GROUNDS FOR ANXIETY. THEY ARE LIVING IN BETTER CONDITIONS THAN SOVIET PRISONERS OF WAR IN ENGLISH CAMPS, WHEN THE LATTER IN A NUMBER OF CASES SUFFERED PERSECUTION AND EVEN BLOWS. MOREOVER THERE ARE NO LONGER ANY ENGLISH PRISONERS IN OUR CAMPS, THEY ARE EN ROUTE FOR ODESSA AND THE VOYAGE HOME."[59]

Churchill, by now clearly alarmed, mentions that he showed Molotov's insulting letter to both Montgomery and Eisenhower, and that on March 25, in minutes to Anthony Eden he warned of the importance of coordinating any further response to Soviet challenges over Poland:

"WE SHOULD ASK THE UNITED STATES WHERE THEY STAND AND WHETHER THEY WILL NOW AGREE TO A TELEGRAM FROM THE PRESIDENT AND ME TO STALIN, AND SECONDLY WHETHER THIS SHOULD, AS YOU SAY, COVER OTHER TOPICS-EG, ACCESS TO POLAND, TREATMENT OF OUR PRISONERS, IMPUTATIONS ABOUT OUR GOOD FAITH ABOUT BERNE, RUMANIA, ETC."

"MOLOTOV'S REFUSAL TO GO TO SAN FRANCISCO (UN Conference) IS NO DOUBT THE EXPRESSION OF SOVIET DISPLEASURE. WE SHOULD PUT IT TO ROOSEVELT THAT THE WHOLE QUESTION OF GOING TO SAN FRANCISCO IN THESE CONDITIONS IS CALLED IN QUESTION AND THAT QUITE DEFINITE FORMING UP BY BRITAIN AND THE UNITED STATES AGAINST BREACH OF THE YALTA UNDERSTANDINGS IS NOW NECESSARY IF SUCH A MEETING IS TO HAVE ANY VALUE."

"HOWEVER, I MUST SAY WE CANNOT PRESS THE CASE AGAINST RUSSIA BEYOND WHERE WE CAN CARRY THE UNITED STATES. NOTHING IS MORE LIKELY TO BRING THEM INTO LINE WITH US THAN ANY IDEA OF THE SAN FRANCISCO CONFERENCE BEING IMPERILLED."[60]

For Franklin D. Roosevelt and several of his influential advisors, however, the founding of the United Nations was of the highest importance. Nothing, not even Soviet mass-kidnapping of American prisoners of war, was to be permitted to interrupt progress toward the President's idealistic vision of the future world organization. Another reason for American moderation in the face of such severe provocation was the US policy of attempting to bring Russia into the war against Japan. In his memoirs, General Deane notes: "Harriman and I considered that our primary and ultimate mission in Russia was not only to obtain Russian aid in the defeat of Japan, but also to insure that Russian efforts would be coordinated with ours."[61] However, the Russian announcement that Molotov would not attend the founding UN Conference which Eden and Stettinius were going to, apparently tipped Roosevelt toward appeasement.

In Moscow, Harriman, Deane and Kennan were seething over Stalin's latest message to Roosevelt and the Soviet countercharges of American and British mistreatment of Russian prisoners. On the 24th of March Harriman again cabled the President, (in part):

"MOLOTOV HAS GIVEN ME A COPY OF STALIN'S ANSWER TO YOUR MESSAGE REGARDING AMERICAN LIBERATED PRISONERS OF WAR IN POLAND... WE HAVE HAD CONTINUED DEFINITE STATEMENTS...THAT THERE ARE NO PRISONERS LEFT IN POLAND AND EACH TIME THESE STATEMENTS HAVE BEEN PROVED TO BE WRONG...STALIN'S STATEMENTS THAT THE RED ARMY CANNOT BE BOTHERED WITH A DOZEN AMERICAN (contact) OFFICERS IN POLAND IS PREPOSTEROUS WHEN WE THINK OF WHAT THE AMERICAN PEOPLE HAVE DONE IN SUPPLYING THE RED ARMY... WHEN THE STORY OF THE TREATMENT ACCORDED OUR LIBERATED PRISONERS BY THE RUSSIANS LEAKS OUT I CANNOT HELP BUT FEEL THERE WILL BE A GREAT AND LASTING RESENTMENT ON THE PART OF THE AMERICAN PEOPLE..."[62]

President Roosevelt's concern for the success of the United Nations, and some military leaders—particularly General Marshall's concerns for the entrance of Russia into the Japanese war, led to an ignoring of Harriman's warnings. One sign of the direction that U.S. policy towards American POWs in Soviet control was to take, is contained in a confidential memorandum for Major General Alexander Surles from Major General Clayton Bissell, Assistant Chief of Staff, G-2, the chief of U.S. Military Intelligence at the time who was to play a major role in suppressing evidence of Soviet guilt in the Katyn Forest massacre. The March 23, 1945 memorandum from Bissell to Surles stated:

"The following policy will govern information which may be released by American personnel liberated from German prison camps by the Russians. a. NO CRITICISM OF TREATMENT BY THE RUSSIANS..."[63]

On March 26th, two weeks before his death, Roosevelt made or supported a policy decision to not confront Stalin further on the disappearance of US POWs in Soviet hands. On that day the President cabled a Top Secret answering message to Harriman:

"... IT DOES NOT APPEAR APPROPRIATE FOR ME TO SEND ANOTHER MESSAGE NOW TO STALIN... YOU SHOULD MAKE SUCH APPROACHES TO ENSURE THE BEST POSSIBLE TREATMENT OF

AMERICANS WHO ARE LIBERATED."[64]

On this same date Roosevelt made clear his top priority was for the future of the UN, even above the return of US prisoners in Poland. He cabled Stalin:

"THE STATE DEPARTMENT HAS JUST BEEN INFORMED BY AMBASSADOR GROMYKO CONCERNING SOVIET DELEGATION TO THE SAN FRANCISCO CONFERENCE...I CANNOT HELP BUT BE DEEPLY DISAPPOINTED THAT MR. MOLOTOV APPARENTLY DOES NOT PLAN TO ATTEND...I HOPE YOU WILL FIND IT POSSIBLE TO LET HIM COME..."[65]

From this point on the Soviets had evidence that once pushed hard the American leadership would back down. On this same 26th of March General Marshall in Washington sent a secret cable to the Moscow Embassy and AFHQ in Europe addressing Harriman's 24th of March concerns, and ordering the new policy of a very ill Franklin Roosevelt, in part):

"... REVISED POLICY LIBERATED PRISONERS: ...CENSOR ALL STORIES. DELETE CRITICISM RUSSIAN TREATMENT..."[66]

In Moscow, General Deane made one more attempt to urge the American command in Washington to retaliate on the Soviets, in at least a token manner, on the day that this message reached the Military Mission there. He again reported the totally unsatisfactory Soviet actions with American prisoners in Poland and recommended "that the activities of Soviet representatives in the European and Mediterranean theaters be restricted in the way in which U.S. representatives in Russia (and Poland) were restricted."[67]

Acting on Deane's message, General McNarney, U.S. commander under Field Marshall Alexander, reported on the 30th of March (CM-IN 31431) that he had suspended the authority, which he had already granted, for Soviet (repatriations) representatives to go to Fifth Army Headquarters, "allowing them, however, continued access to the concentration camps near Bari." (In Italy, where Soviet ex-prisoners and Russians captured in German uniforms were being held.) The Supreme Allied Commander in Europe, General Dwight Eisenhower, seems to have made the key decision against retaliation against the Soviets at this time:

"On 1 April 1945, (CM-OUT 64858) GENERAL EISENHOWER RECOMMENDED AGAINST THE IMPOSITION OF RESTRICTIONS ON SOVIET REPRESENTATIVES IN HIS THEATER."

Five days before the death of President Franklin D. Roosevelt, "On 7 April 1945 (CM-OUT 64858) the Joint Chiefs of Staff advised General Deane that "RUSSIAN FAILURE TO COOPERATE IN THESE MATTERS WOULD NOT BE FOLLOWED BY RETALIATORY ACTION ON OUR PART AT THIS TIME."[68]

For the harassed and beleaguered Major General John Deane in Moscow (and for Harriman and Kennan) this must have been a greatly disheartening order to receive from General Marshall.

After months of continual confrontation over the vanishing American POWs with Stalin and his arrogant Soviet officials-Molotov, Vyshinsky, Dekanozov, Golubev and Golikov, the line had been laid down by the highest U.S. authorities that nothing would be done by the United States in retaliation. Since the Soviets, through their many successful espionage sources, had access to practically all U.S. and Allied communications, they must have known of Washington's decision immediately, thereby further encouraging them to seize more U.S. POWs as hostages for achieving future bargaining points.

Churchill later remembered of this critical time: "...The Soviet policy became daily more plain, as also did the use they were making of their unbridled and unobserved control of Poland. They asked that Poland should be represented at San Francisco only by the Lublin (communist) Government. When the Western powers would not agree the Soviets refused to let Molotov attend."[69]

As the Soviets continued their multi-pronged offensive, Churchill knew that the only way to resist was to pull Roosevelt closer to his position. He cabled the American President on the 27th of March, (in part);

"AS YOU KNOW, IF WE FAIL TO ACT TOGETHER TO GET A SATISFACTORY SOLUTION ON POLAND AND ARE IN FACT DEFRAUDED BY RUSSIA BOTH EDEN AND I ARE PLEDGED TO REPORT THE FACT OPENLY TO THE HOUSE OF COMMONS. THERE I ADVISED CRITICS OF YALTA TO TRUST STALIN...SURELY WE MUST NOT BE MANEUVERED INTO BECOMING PARTIES TO IMPOSING ON POLAND—AND HOW MUCH MORE OF EASTERN EUROPE—THE RUSSIAN VERSION OF DEMOCRACY?...IS IT NOT THE MOMENT NOW FOR A MESSAGE FROM BOTH OF US ON POLAND TO STALIN/...IF WE ARE REBUFFED IT WILL BE A VERY SINISTER SIGN, TAKEN TOGETHER WITH...MOLOTOV'S RUDE QUESTIONING OF OUR WORD... THE UNSATISFACTORY PROCEEDINGS OVER OUR LIBERATED... PRISONERS, THE COUP D'ÉTAT IN RUMANIA...WHAT DO YOU MAKE OF MOLOTOV'S WITHDRAWAL FROM SAN FRANCISCO?... DOES IT MEAN THE RUSSIANS ARE TRYING TO RUN OUT, OR ARE THEY TRYING TO BLACKMAIL US?..."[70]

Despite Churchill's warnings, Roosevelt was to stick with his policy decision on not confronting Stalin over the Allied prisoners in Poland until his death two weeks later. Churchill and the British government were, in effect, forced by American preeminence in the overall war effort to acquiesce to a moderated tone on vanishing Allied POWs. Stalin and his cohorts clearly understood the meaning of the revised policy on POWs. On this same 27th of March, Lt. Colonel James Wilmeth, leader of the beleaguered American contact mission at Lublin, was finally ordered out of Poland by the Russians. Wilmeth was later to admit that American POWs were left behind in Soviet-occupied Poland, and one of his subordinates was to state in a final report on the subject that the equivalent of the number of U.S. POWs evacuated from Poland through Odessa came equaled the total number of known American POWs in only two of the camps liberated by the Russians.[71]

The pragmatic and habitually suspicious US Secretary of the Navy James Forrestal recorded in his diary on April 2nd, 1945:

"...THE SECRETARY OF STATE (Stettinius) ADVISED OF SERIOUS DETERIORATION IN OUR RELATIONS WITH RUSSIA. THE PRESIDENT HAS SENT A STRONG MESSAGE TO STALIN DEPLORING THIS CONDITION...brought to a focus by the request to have the Lublin Poles invited to San Francisco. He recites the fact that the ties between Russia and this country, knit together by the necessities of war, are in grave danger of dissolution and asks the most serious consideration by the Marshal of the questions involved..."[72] Although the editors of the published version of the Forrestal Diaries (Walter Millis and E.S. Duffield) here note that, "It is apparent that Forrestal was at once put on the alert," and go on to say that the next 30 or more pages of his (original-unedited) diary, "are mainly occupied by copies of the telegrams in which Averell Harriman, from the Moscow Embassy, analyzed Russian policy..."; there is nothing whatsoever in the published version on the POW crisis in Poland which had so alarmed Harriman, the President, Churchill, Forrestal and the others.

In Moscow, Ambassador Averell Harriman, falling in line with the new presidential policy, sent a distinctly mollifying letter to Molotov on Soviet POW conduct, in which he expressed his government's, "appreciation to the Red Army for the liberation of our POWs and for the many acts of kindness and generosity by individual officers and men of the Soviet armies toward our released soldiers..."[73]

As the Allied armies advanced into Germany on March 26, 1945, the first big POW camp for Americans, Stalag XIII-C, near Hammelberg in Bavaria, was liberated by U.S. forces detached from General George Patton's Third Army. It contained 4,700 POWs, including 1,400 Americans, one of whom, Patton's son-in-law, Colonel John K. Waters, was severely wounded by German SS guards in the process. Prisoners unable to travel, including Waters, were returned to the Germans under a flag of truce while others were recaptured later. Some were again freed on April 6. The US First Army liberated 6,500 Allied prisoners, including 3,200 American POWs, from Stalag IX-B, southeast of Orb, Germany, on April 2nd. Many of the prisoners in this camp had been captured in the Battle of the Bulge and had been mistreated and starved. (Some US POWs captured in the Bulge or later had not been officially reported as POWs by the Germans, and had been kept in an MIA category in US War Department records.)

Oflag IV-C (Colditz) east of Leipzig was liberated on April 16;

and at Stalag XII-A (Limburg) the Americans didn't reach the camp until April 25, at which time the Soviet Army was already there. This camp was hurriedly evacuated by air transports and bombers in the next few days. The advancing American Army continued to liberate large numbers of prisoners, including 15,568 Americans out of 110,000 prisoners freed by elements of Patton's Third Army at Stalag VII-A on April 29, 1945. This German POW camp at Moosburg in southern Germany contained the largest prisoner of war concentration in the Western Allied area.[74]

On the battlefields of the European front at the end of April, from the Baltic Sea to the Adriatic, the Soviet Army was approaching or linking up with the Americans and British, squeezing the remnants of the Wermacht between them and even bypassing some formations. On the western front by the fourth week of April, most large-scale German military resistance had collapsed. The first link-up between Soviet and American forces occurred at Torgau on the Elbe River on the 25th of April. In other areas isolated Nazi units provoked last-ditch firefights and holding actions while American and Russian units made their initial contacts.

The British Army under Montgomery, with the attached U.S. 9th Army under General Simpson, continued their fighting advance along the north German coast capturing Bremen on the 26th of April and moving towards Hamburg, on the lower Elbe. Their objective was the Lubeck-Wismar area on the Baltic Sea east of the strategic Kiel Canal. At Churchill's urging Montgomery not only wanted to cut off the Danish Peninsula from a possible Russian advance, but also to liberate thousands of American, British and other Allied POWs beyond Lubeck, near Barth, at Stalag Luft I, and others in nearby camps and hospitals, before the Russians did.

The Western Allies were already reacting operationally to the Soviet seizing of American and British POWs in Poland and eastern Germany. Transport aircraft and hastily converted bombers were being readied and used to rescue POWs beyond the American and British lines who were in danger of disappearing behind the Soviet front.

At the same time, masses of German soldiers and civilian refugees fleeing ahead of the Soviet advance were attempting to surrender to the western Allies or escape through the lines, all along the front. Some of the German commanders were continuing a fierce fight against the Soviets on the eastern front, while putting up little resistance in the west, or surrendering when confronted with U.S. forces. The cost to the Americans for the last weeks of fighting in Europe was still not cheap. General George Patton's Third Army lost over 8,200 battle casualties from 23 March-22 April 1945, including 1,685 killed in action, 5,815 wounded and 773 missing in action.[75] Thus, the loss of American MIAs continued to the very end of the war in Europe, at a time when the German system was breaking down in the shrinking rear area between the Allied armies, and many such MIA who had actually been captured were not reported as POWs by the Nazis. Some of these men fell into Soviet hands with the rapid advance of the Red Army, but the United States Government was uncertain about their number. Soviet losses in the last battles were indeed much greater, and intelligence reports of German behavior reaching Stalin seemed to confirm the possibility of his worst fear in the later part of the war being realized: That the Americans and British would make a separate peace with the Germans, and then unite against Soviet communism.

East of the Elbe River in the Russian zone of Germany, from Barth on the Baltic coast to Austria in the south, a great mass of Allied POWs remained in Nazi camps such as Stalag III-A, near Luckenwalde, some 30 miles south of Berlin. Here, on April 22nd, 1945, advance armored elements of the Red Army flanking Berlin from the south, entered and "liberated" the camp which had previously contained approximately 8,800 American and British prisoners of war, but now had administrative control over thousands more U.S. and U.K. POWs in satellite camps. As the U.S. 9th Army advance was then in the vicinity of Magdeburg on the Elbe, about 60 miles to the west, the prisoners expected to be conveyed to the American lines, or at least allowed to get there on their own. Instead, the Soviets surrounded the Allied POWs with armed guards and kept them under confinement in the camp.

Patton's Third Army attack down the Danube valley in southern Germany began on April 22nd and soon penetrated into Austria. The American Vth Corps advanced into Czechoslovakia towards the oncoming Soviets who had fought their way through the Czechoslovakian mountains in a brutal all-winter campaign, The U.S. First Army was consolidating and holding along the Elbe river east of Leipzig, as Patch's 7th Army captured Munich on April 30th.

The British 8th Army under Field Marshall Harold Alexander was advancing north of the Adriatic Sea, in the Austrian-Italian-Yugoslav border area. Beyond lay other German POW camps, particularly in Austria, where prisoners herded back from the east by German columns retreating through Czechoslovakia had been assembled in both large temporary camps, and in some cases added already-crowded established camps like Stalag XVII-B, up the Danube from Vienna and west of Bratislava; all being in the Soviet zone of advance.

Marshal V.I. Tolbukhin's 400,000-strong 3rd Ukrainian Front of the Red Army had inflicted a severe defeat on the Germans in the middle of March, at the fierce battle of Lake Balaton in western Hungary.[76] The Germans were totally defeated in Hungary by April 4th, and the Red Army was already advancing into Austria. With the known disappearance of at least 2,000 Americans and thousands more British POWs in Poland, and with evidence of more being taken into Russian control in other areas, there was a deepening concern among western Allied leaders for the American and British POWs in the path of the Red Army in Austria.

Vienna was first assaulted on April 6th, and captured by Tolbukhin's 3rd Ukrainian Front on April 13th, 1945. The 3rd Ukrainian Front (a name originally designating zone of battle rather than ethnic Russian troop make-up), quickly overran a large expanse of Austria, including Allied POW concentrations in several camps and hospitals, and others from retreating German columns. According to some estimates in official U.S. 1945 documents, as many as 24,000 American and British POWs, possibly including many carried in records as MIAs, and many others who had been evacuated through Czechoslovakia from camps in Poland, were estimated to have been taken under Marshall V.I. Tolbukhin's control, along with thousands of other western Allied POWs and hundreds of thousands of German/ Axis prisoners.

The Soviet Ambassador in Washington (later Foreign Minister) Andrei Gromyko, sent a series of messages to Stettinius during April 1945 which perfectly illustrate the Soviet strategy of turning the entire POW blackmail attempt around and putting the American government on the defensive about repatriations of prisoners. In a letter of 10-11 April, Gromyko begins with a long, haranguing explanation as to how it was that hundreds of thousands of captured Russian soldiers had ended up serving in the German Army against Stalin and their own country. Then he rejected the American contention that some Russians serving in the German Army were not liable for repatriation as they were in a similar legal status as foreigners, including Russians, who might be serving in the American forces. Gromyko clearly enunciates the second Soviet demand, after Allied recognition of Soviet hegemony in Poland and eastern Europe, which must be met by the United States before the missing US POWs would be released:

"...CONSEQUENTLY, THE SOVIET GOVERNMENT AGAIN INSISTS THAT ALL SOVIET CITIZENS, LIBERATED AND BEING LIBERATED BY AMERICAN ARMIES IN THE COURSE OF MILITARY OPERATIONS AGAINST GERMANY, ARE SUBJECT TO TRANSFER TO THE SOVIET AUTHORITIES FOR RETURN TO THE SOVIET UNION..."[77]

A similar message was sent to the British government. It was to be only through Russian internment of Allied POWs that their rigid insistence on the terms of Article I of the Yalta Agreements, along with their all-embracing view of what constituted "Soviet citizenship," would be zealously adhered to by British and American officers during the coming mass-repatriations.

In another cable of the 18th/19th of April, as the San

Francisco UN Conference was opening, and Washington officials such as Alger Hiss were assisting in the preparation of U.S. negotiating positions on the Polis-POW crisis, Gromyko again leveled Soviet charges of bad faith in the repatriations at the Americans:

"To the Secretary of State (Stettinius),

"Your Excellency: In accordance with instructions of the Soviet Government, I have the honor to bring to your attention the following: "According to information published in the press within the past few days the Anglo-American command in Europe is taking measures not to return Soviet prisoners of war and Soviet citizens....interned in Germany, to the Soviet Union..."

Gromyko went on to accuse the United States of violating the Yalta Agreement by allowing Soviet prisoners and DP's to work on German farms, (in order to both assist the Allied armies in their massive supply problems for millions of prisoners, and also to insure a crop for the coming winter in the war-ravaged land). He demanded a series of immediate steps toward their repatriation.[78]

In the last two weeks of his life Roosevelt stuck to his decision of not confronting Stalin over American POWs. Stalin knew then that if pushed hard the American leadership would back down. On the day Roosevelt died, April 12th, 1945, the senior American contact officer in Odessa, Lt. Col. Fenell, reported that 2,687 American prisoners had already been shipped from Odessa. This was out of a total of 2,858 Americans who would leave Russia by that route, yet only 3 days before he had conveyed information to Deane in Moscow (at 1720 hours on April 9th) that an American prisoner in Odessa had reported 2,000 more Americans still remaining in Poland under Russian control.[79] These were the men, whose existence had already been acknowledged by Ambassador Averell Harriman, who would disappear in Soviet captivity. Also, on April 13th the US political advisor to SACMED, Alexander Kirk, (later Ambassador to Italy) sent a Top Secret cable (#1505) to Secretary of State Stettinius stating that 2600 (actually 2,687) Americans had been shipped from Odessa to date but that a shipment of 884 Americans who had arrived there on April 2nd were being held up by the Russians.[80] Another top secret U.S. document of this period, for the Mediterranean Theater commander, Field Marshall Alexander, which was sent to the Adjutant General of the War Department in Washington and the SACMED U.S. Political Adviser, Alexander Kirk, confirms this hostage incident involving American prisoners: "...A SHIPMENT OF 884 AMERICANS ARRIVING HERE ON 2 APRIL REPORTED BEING HELD AT ODESSA, AS VIRTUAL POWs, UNDER SOVIET ARMED GUARD..."[81] Since only 171 more American POWs were released from Odessa by the Russians after this date, It appears that some of these U.S. POWs, in addition to the reported 2,000 more Americans then still in Poland, were never released by the Soviets.

Thus, the new President, Harry S. Truman, was being suddenly thrust into a secret crisis with the Soviet Union at a time when military cooperation with the Russians on the European Front, and eventually in finishing the Japanese war, was still considered essential by General Marshall and several other Presidential advisers. General Douglas MacArthur, the U.S. commander in the Pacific, was also anxious for Soviet participation in the war against Japan, in anticipation of a costly American invasion of the Japanese home Islands. Outward appearances of essential harmony were to be kept up, to avoid giving the Germans and Japanese any cause for hope that the alliance might collapse. However, behind the scene the confrontation over Allied prisoners of war held by the Soviets in Poland and the USSR intensified. An April 13th letter from Eisenhower's Chief of the Communications Zone, General John C. H. Lee, to General Courtney Hodges, Commander of the U.S. 1st Army, responded to his query about whether he should forcibly repatriate the Soviet prisoners overrun by his forces:

"Answering your inquiry as to how much force an Army commander should use in the control of displaced Russians, I can only say that the ETO letter of 8 April 1945, subject: Liberated Citizens of the Soviet Union, covers the matter in general...TALKING WITH JUDGE McCLOY TODAY, HE AGREED THAT OF COURSE AN ARMY COMMANDER COULD USE ANY FORCE NECESSARY TO INSURE THE SUCCESS OF HIS OPERATIONS. He expressed the hope, which I share, that most of these people can be kept well forward and thus be enabled to reach their Fatherland over the shortest possible route..."[82]

Harry S. Truman was 60 years old when he succeeded to the presidency. A native of Missouri who had graduated from high school and then gone to work, he helped raise troops at the beginning of World War I, and served in France as a captain in a field artillery battery at St. Mihiel and in the Meuse-Argonne campaign. After returning home in 1919, he had remained in the reserves and become a judge. Truman had been elected to the U.S. Senate in 1934 at the age of 50, and ten years later had become an influential voice on Capitol Hill. The left liberals of the Democratic Party had supported Henry Wallace for Vice President that year, but the majority of the party's leaders wanted more of a centrist candidate in case Roosevelt did not live through his fourth term, and Truman was ultimately chosen to run. In the November 1944 election Roosevelt and Truman had easily defeated their Republican opponents, Governor Thomas Dewey of New York and Governor John Bricker of Ohio. Truman, who was not close to Roosevelt as Wallace had been, subsequently felt a debt to the former Vice President, who would have become President if Truman had not succeeded in capturing the 1944 nomination, and appointed him to a cabinet position.

In his memoirs Truman recorded that on April 13th he received a report titled "Special information for the President" from Secretary of State Stettinius, on the critical areas of concern facing the United States, which stated in regard to the Soviet Union:

"Since the Yalta Conference the Soviet Government has taken a firm and uncompromising position on nearly every major question that has arisen in our relations. The more important of these are the Polish question, the application of the Crimea agreement on liberated areas, the agreement on the exchange of liberated prisoners of war and civilians, and the San Francisco Conference...PERMISSION FOR OUR CONTACT TEAMS TO GO INTO POLAND TO ASSIST IN EVACUATION OF LIBERATED PRISONERS OF WAR HAS BEEN REFUSED ALTHOUGH IN GENERAL OUR PRISONERS HAVE BEEN REASONABLY WELL TREATED...The present situation regarding Poland is highly unsatisfactory with Soviet authorities consistently sabotaging Ambassador Harriman's efforts in the Moscow Commission to hasten implementation of the decisions at the Crimea Conference..."[83]

This report indicates the new President may not have been fully briefed as yet, on the seriousness of the prisoner of war situation in Poland. With the approaching United Nations Conference in San Francisco, Winston Churchill felt that it was essential for his views to be conveyed to the new President. On April 16, 1945, on his way to San Francisco, British Foreign Secretary Anthony Eden, accompanied by British Ambassador Lord Halifax and Secretary of State Stettinius met with President Truman in the White House executive office. In his memoirs Truman mentions that Eden brought "messages from Churchill as well as the Prime Minister's version of the joint communication we were to send to Stalin on the Polish issue." Truman wrote that Eden and he together agreed to a final text, "... and discussed the importance of having Molotov present at the San Francisco conference, and I informed the Foreign Secretary that Stalin had just sent word through Harriman that Molotov would attend."[84]

Truman recorded in his memoirs the text of the joint message which he radioed to Averell Harriman in Moscow for delivery to Stalin. While the message accuses the Soviet Union and its puppet Lublin/Warsaw regime of failing to live up to the Yalta agreements on Poland, and named the Polish leaders Washington proposed to be included in a postwar government, it said nothing specifically (as recorded) about the thousands of missing U.S. and Allied prisoners in Poland and East Prussia.

In Eden's memoirs the Foreign Secretary said considerably less, noting merely that he had telegraphed to Churchill that Truman was "honest and friendly...overwhelmed by his new responsibilities...his references to you could not have been warmer. I believe we shall have in him a loyal collaborator, and I am much heartened by this first conversation."[85]

What was not recorded in the memoirs of either leader about this April 16th White House conversation, was that the subject of the missing U.S. and Allied prisoners of war in Poland was also raised in the meeting. In a formerly top-secret memorandum of a later conversation on that same day (located by the author in the National Archives and declassified in 1989), reference is made to the Allied POWs by Anthony Eden. In this meeting, which occurred later on April 16th, Eden, Stettinius, James Clement Dunn and Alger Hiss were present, and the discussion centered on the Polish crisis and the proposed agenda for Soviet-Allied talks at the upcoming UN conference to be attended by Molotov. According to the official minutes, the British Foreign Secretary raised the issue because of a statement of concern about the missing POWs by Secretary of State Stettinius during the meeting with President Truman:

"MR. EDEN SAID THAT THE SECRETARY HAD MENTIONED TO HIM WHILE THEY WERE WITH THE PRESIDENT TODAY THAT HE WANTED TO TALK ABOUT A 'PRISONER OF WAR' MATTER. THE SECRETARY (Stettinius) SAID WHAT HE HAD IN MIND WAS THE DIFFICULTY WE BOTH WERE HAVING WITH REGARD TO MEMBERS OF THE U.S. AND U.K. ARMED FORCES WHO WERE PRISONERS BEHIND THE RUSSIAN FRONT LINES. HE SAID THE TREATMENT WHICH OUR PRISONERS HAD OBTAINED WERE QUITE UNSATISFACTORY AND THAT WE HAD NOT BEEN ABLE TO HAVE CONTACT TEAMS SENT INTO THE AREA TO CARE FOR AND PROVIDE FOR OUR MEN. THE RUSSIANS HAD PROMISED TO DO SOMETHING ABOUT THIS AT YALTA BUT THEY HAD DONE NOTHING. HE WANTED TO RAISE THE MATTER WITH MR. MOLOTOV WHEN HE ARRIVED."[86]

It thus appears that Edward Stettinius insisted that his objection to Soviet treatment of U.S. and British POWs in both the meeting with Truman and the meetings thereafter, be noted in the record, while none of the other public figures present mentioned the subject later.

That same day on the European war front, as the Soviets continued to take control of more Allied prisoners, Eisenhower cabled the British War Office:

"Request has been received from 21 Army Group for lists of British and U.S. PW in German PW camps. U.S. lists are being forwarded immediately with supplements to follow. Will War Office please deal with 21 Army Group direct concerning British lists."[87]

The next day, April 17th, the situation of American and British prisoners in the path of the Red Army in Austria was reaching the critical point. The following message[88] was marked "Secret and Important":

"STALAG XVII-A KAISERSTEINBRUCH LAST KNOWN STRENGTH

1025 BRITISH, ALREADY OVERRUN IN LATEST SOVIET ADVANCE. FOLLOWING LIKELY TO BE OVERRUN IN NEAR FUTURE: A) STALAG XVII-B GNEIXENDORF (KREMS), STRENGTH 4,280 BRITISH AND AMERICAN. (B) STALAG XV111-A WOLFSBERG, STRENGTH 9,752 BRITISH NO INFORMATION HERE THAT THESE CAMPS WERE EVACUATED BY THE GERMANS. GRATEFUL YOU FORWARD SOONEST, DETAILS, NUMBERS RECOVERED ETC.[89]

In Washington, on the 18th of April, Gromyko had made his charges that it was the United States that was mistreating and refusing to repatriate Soviet prisoners. With the continual stalling and hostility of the Soviets over prisoners of war, the Combined Chiefs of Staff made a decision on this same April 18th, to alter Allied policy in regard to U.S. and British troops entering the future Russian zone of occupation to rescue American and British/Dominion POWs. At the same time it was made clear that military confrontations with the Russians during any such incursions were to be avoided at all costs. A top-secret (book) message is recorded for this date[90] to Field Marshal Alexander for action, and to Deane and Archer in Moscow for information:

"It was earlier interpreted by the Combined Chiefs of Staff that Anglo-American forces should not enter the ultimate Russian zone, even while still held by the enemy, without Soviet permission, and vice versa. Reconsideration of this question now interprets that an 'ultimate zone of occupation' is not a 'zone of operations' limited to any of the allies, and that prior to the arrival of the ultimate occupying forces, any and all portions of enemy held territory are considered as 'no man's land' and, as such, can be entered by any Anglo-American or Soviet forces having military interests in the reign.

"UNDER SUCH CONDITIONS IT IS NOT NECESSARY TO OBTAIN SOVIET APPROVAL FOR SECURITY OF PRISONERS OF WAR (POW). HOWEVER, SPECIAL CARE AND COORDINATION WITH THE SOVIETS WILL BE NECESSARY TO AVOID UNFORTUNATE INCIDENTS. IN CONNECTION WITH THIS TYPE OF OPERATION.

"...In light of the present tactical situation, topographical differences and SACMED's limited resources and facilities, the Combined Chiefs of Staff are now of the opinion that the responsibility for the security of British and U.S. prisoners in Austria should be transferred from SACMED (Alexander) to SCAEF (Eisenhower). It is therefore desired that SCAEF, in cooperation with SACMED, plan for implementing the operations visualized in FACS-145 and FAN-495 in Austria as well as elsewhere..."[91]

2,000 more American prisoners had been reported to the U.S. Military Mission in Moscow to be still in Soviet-controlled Poland by a U.S. POW who reached Odessa on April 9th, 1945. By that time 2,660 Americans had already been repatriated from Odessa according to declassified cables. Although 2,858 Americans and some 4,400 British were eventually repatriated from Poland through Odessa, according to recently declassified government documents, 2,000 or more Americans, and thousands more British and Commonwealth prisoners disappeared in Poland during this first phase of the NKVD kidnapping operation. An estimated 700 more Americans were kept by the Soviets in and around Odessa. In addition, American bomber pilots and aircrewmen who had crash-landed in the USSR after bombing the Romanian oil center at Ploesti had already disappeared in late 1944. A British prisoner of war who escape from Poland that year, Alec Masterton, formerly of the Argyll and Sutherland Highlanders, now of Vancouver, B.C., had been imprisoned by the Soviets in the fall of 1944 in Odessa., with American airmen whose presence was unknown to the U.S. authorities. With the help of an American sea captain in the port, his presence had been made known to a British captain, who successfully obtained Masterson's release, but the American POWs disappeared. On October 29, 1989 in a front-page Toronto Star article, Masterton stated that he knew of American bomber crews in Odessa who never returned from Soviet control. He also stated that upon his return he was sworn never to tell by MI-6, Britain's foreign intelligence agency, and a British Army Captain "on my honor as a British soldier."[92]

As European war drew to a close, the new President, Harry S. Truman, who had been studying Harriman's cables from Moscow on the American POWs in Poland, met with Secretary of State Edward Stettinius and British Foreign Secretary Anthony Eden, prior to Molotov's arrival for the San Francisco U.N. Conference. A now-declassified Department of State memorandum for the use of Secretary of State Stettinius in confronting Molotov, dated 19 April 1945, listed among the most important topics for negotiation: "THE FAILURE OF THE SOVIET GOVERNMENT TO PROVIDE FACILITIES FOR THE ENTRY OF UNRRA AND RED CROSS PERSONNEL INTO POLAND AND THE BALKANS; THE IMPLEMENTATION OF THE CRIMEA AGREEMENT REGARDING THE EXCHANGE OF LIBERATED PRISONERS OF WAR AND CIVILIANS...THE SOVIET REQUEST FOR A SIX BILLION DOLLAR LOAN..."[93]

In the accompanying position paper prepared by the Department of State for Stettinius' meeting with Molotov, the POW problem was dealt with by an American promise of future forced repatriations of Soviet citizens:

"You are familiar with the fact that despite the intervention of the President the Soviet government would not allow our contact teams to proceed to Poland to assist our liberated prisoners...Mr. Molotov may raise the question of our refusal to return to Soviet control Soviet nationals captured in German uniforms who claim the protection of the Geneva Convention as German prisoners of war... YOU MIGHT ASSURE MR. MOLOTOV, HOWEVER, THAT WE HAVE NO INTENTION OF HOLDING SOVIET CITIZENS AFTER THE COLLAPSE OF GERMANY REGARDLESS OF WHETHER THEY DESIRE TO RETURN TO THE SOVIET UNION OR NOT."[94]

The same position paper connected the Soviet request for a six billion dollar loan to the crisis in Poland: "...You may consider it advisable to point out to Mr. Molotov that in considering this matter the Congress will doubtless be influenced by the prospect for full collaboration between the United States and the Soviet Union in the establishment and maintenance of peace and stability. You might further...remind Mr. Molotov that Congress reflects public opinion and that public opinion in this country has been greatly concerned over developments in eastern Europe since the Crimea Conference..."[95]

Following Roosevelt's death Harriman had rushed home, bringing General John Deane and leaving George Kennan in charge of the crisis. Harriman personally briefed the new President in the White House on April 20, and found that Truman had already studied all his cables from Moscow and the Yalta Agreements. Harriman told Truman that, "the Russians needed American help in postwar reconstruction and would not, therefore, wish to break with the United States. America could afford to stand firm on important issues without serious risk."[96]

Harriman was remembering his discussions with Molotov and Mikoyan and their expectations of $6 Billion in US war-reconstruction "credits," and his own recommendation from Moscow in March that Lend-lease be curtailed in retaliation for Soviet conduct with U.S. POWs. Harriman reiterated concern about Russian actions, particularly in Poland and Romania. Truman in turn, made clear to Harriman his overriding concern for Russian participation in the United Nations.

Perhaps because of this, and despite Harriman's warnings, a key policy decision was apparently made by the Truman Administration on April 20th, 1945, to reaffirm Roosevelt's decision to not confront Stalin on the missing American prisoners. A Top Secret summary of the analysis by the Operations Division of the War Department General Staff (WDGS) leading to this policy decision on U.S. POWs in Soviet control, which had actually been formulated at the time of Roosevelt's death on April 12, was approved and distributed after Truman became President, included the explanation:

"...The question of what action, if any, should be taken by the United States as a result of Soviet non-cooperative actions, including the refusal to allow us to aid our own prisoners liberated by the Soviet forces, was referred by the Chief of Staff (Marshall) to the Joint Chiefs of Staff. On 7 April,[97] the Joint Chiefs of Staff advised General Deane that Russian failure to cooperate in these matters would not be followed by retaliatory action on our part at this time.

Harriman had thoroughly briefed top policymakers such as Stimson, Stettinius, Grew, Forrestal, Marshall and others including Leahy and King, as he was going on to San Francisco to confront Molotov during the founding-UN Conference.[98] At this Conference, recently declassified documents prove that Soviet retention of Allied POWs became a secret topic of negotiation, although "Poland" was the only related topic released to the public. "Poland" had thus become a euphemism for the crisis with the Soviets over their actions in eastern Europe, the most serious and threatening of which was the deliberate withholding of American prisoners of war in Soviet control. The senior U.S. State Department team at San Francisco included Alger Hiss, Director of the State Department's Office of Special Political Affairs (SPA), who later to be convicted of perjury for denying he had spied for the Soviets. (Hiss was to be the acting Secretary-General of the UN during the Conference and was later proposed by the Soviets for permanent appointment to this position.) On the secret agenda and in the position-papers for Secretary of State Stettinius, which Hiss helped plan, Item # 6 concerns implementation of the Yalta Agreement on prisoners of war, particularly in Poland. In the attached official US position-paper the expected demand by Molotov for the repatriation of Soviet nationals captured in German uniforms is to be countered with a promise to deliver them after the German collapse, "whether they desire to return to the Soviet Union or not." This was to become a de facto American promise of forced-repatriations of Soviet citizens.

The senior policymakers and advisors in the new Truman Administration gathered in the White House on April 23, 1945, to be briefed further by Ambassador Averell Harriman and Major General John Deane, on their first-hand views about the dangerous Soviet challenge over U.S. POWs in Poland, and for all present to give their advice to the President. Besides Truman, Harriman and Deane, also present were Secretary of State Stettinius, Secretary of War Stimson, Secretary of the Navy Forrestal, Admiral King, General Marshall, the Chief of Staff to the President-Admiral Leahy, Charles Bohlen-State Department liaison to the White House, and Assistant Secretary of State James C. Dunn.

Contemporary U.S. documents declassified thus far, and later published accounts, are deleted of details concerning the prisoner of war aspect of the Polish crisis relayed to the group by Harriman and Deane. The "sanitized" versions of the meeting make it clear that the Soviets non-compliance with the Yalta agreements and their demand for the recognition of their puppet regime in Warsaw, by giving it a seat in the newly established United Nations during the upcoming

San Francisco UN Conference, predominated in the discussion.[99] It is also clear from the Forrestal diaries and other sources that, Harriman's urgent cables from Moscow (on POWs) had been a source of alarm to other American officials besides the President.

Stettinius made it clear in this meeting that the Russians had receded from the agreements made at Yalta (on POWs and Poland). He stated that "a complete deadlock had been reached on the subject of carrying out the Yalta agreement on Poland."[100] It would appear certain, based on Stetinnius' raising the Soviet treatment of U.S. and British POWs a few days before with Truman and Eden, that he would have referred specifically to the POW crisis in this meeting, whether recorded as such, or not. In any case Stettinius must be considered a "hard-liner" towards the Soviets at this time. In his later memoirs Stettinius was to write:

"Shortly after the Yalta Conference had closed the Soviet Union began to hedge on some of its agreements...At Yalta a military agreement called on each nation to allow the others to send missions behind its lines to deal with its own liberated prisoners. The Soviet Union however, would not allow a United States mission to function behind Soviet lines."[101]

Truman then asked the opinions of others present, saying it was now or never, and that the Yalta agreements had so far been a one way street and that could not continue. He vowed to go ahead with the UN plans in San Francisco and if the Russians stalled they could "go to hell." He asked for the opinions of the others present.[102]

Secretary of War Stimson, who could perhaps be considered the senior advisor present, gave a cautious warning on the necessity of finding out "how seriously the Russians take this Polish question." In his diary he wrote that he was very doubtful about the wisdom of too strong a policy; "So I... told the President that I was very much troubled by it....I said that in my opinion we ought to be very careful and see whether we couldn't get ironed out on the situation without getting into a head-on collision..."

Stimson essentially advised Truman to give in to Soviet demands on Poland, in the interest of insuring their participation in the war against Japan. Stimson felt Harriman and Deane were being too strongly influenced by their personal involvement with the Soviets in Moscow and their own previous bad experiences, and that they had in effect had too strong an influence on the President.[103]

Forrestal thought that the Soviet action in Poland, "was not an isolated event but was one of a pattern of unilateral action on the part of Russia...I thought we might as well meet the issue now as later on." He agreed with Harriman's and Deane's assessments on Soviet intentions to dominate eastern Europe, and felt that the Navy (and the Air Force) could finish the war against Japan without the Russians.[104]

General George C. Marshall was categorized by Forrestal as "even more cautious," and brought up what was undoubtedly on the minds of others. He said, "from the military point of view the situation in Europe is secure," and that he hoped for Soviet participation in the war against Japan "at a time when it would be most useful to us." Marshall warned that the Russians could wait until the Americans had "done all the dirty work" before attacking Japan, and that problems with the Soviets in the past had usually straightened out. Marshall, whose opinion must have been important to Truman, therefore was one of those who advised the President to back down from the blatant Soviet aggression. Admiral Leahy waffled between Stimson's and Forestall's positions.

Harriman, the author of the urgent, secret cables on POW blackmail which had brought the crisis home to the policymakers in Washington, clarified for the assembled group that the real issue was whether the United States was to be a party to the Soviet program for dominating Poland. The intimation was, again, that the POWs were being used as hostages in an attempt to force the U.S. to accept total Soviet domination of Poland. By confronting the Russians openly and strenuously over their heavy handed blackmail attempts on behalf of their Polish client regime, he felt the U.S. would benefit in the end, and so would Poland. By not confronting them immediately, the Truman Administration would be sending the Russians a signal that their bluff had won and the U.S. was going to go along with their program for eastern Europe.

Major General John "Russ" Deane agreed with Harriman that the Soviet Union "would enter the Pacific war as soon as it was able, irrespective of what happened in other fields." Deane said "the Russians must do this because they could not afford too long a period of letdown for their people who were tired," and that, "we should be firm when we are right."

Truman had apparently been affected by Stimson's and Marshall's advice to not have a serious confrontation with the Soviets, as he commented that he was not going to deliver an ultimatum to Molotov. And in the end this was to prove the pattern of gradual appeasement to Stalin throughout the prisoner-of-war blackmail episode.

The new President's decision to not confront Stalin with an ultimatum was reiterated later on April 23rd, in a personal meeting between Harry S. Truman and Soviet Foreign Minister Vyacheslav Mikhailovitch Molotov. In addition to the importance of the United Nations issue, Soviet entrance into the Japanese war was still seen by Truman and several of his advisors as an overriding concern prior to the still-uncertain results which might be obtained from use of the secret atomic bomb, which was not to be successfully tested until July. Postwar Allied unity of purpose with Russia was viewed by Harry Truman, as it had been by Franklin Roosevelt, as more important than confronting Stalin over a possibly-temporary Soviet blackmail attempt involving American war prisoners.

Truman met with Molotov for the serious part of their discussions shortly after the meeting with his advisors. Others present included Stettinius, Averell Harriman and Charles "Chip" Bohlen, who translated for the President. The first topic was the lack of progress on the overall "Polish problem," already becoming a euphemism for secrets veiled by the simple question of diplomatic recognition for what remained a Soviet puppet regime. With Harriman's and Deane's (and British) intelligence backing up his irritation, Truman confronted Molotov in a brusque way which surprised the diplomats present. The President was firm with Molotov and stated that the United States had gone as far as it could go to meet Stalin's demands, and it could not recognize a Polish government that failed to represent all democratic elements. Truman announced to Molotov that he was going ahead with his plans for the UN organization, no matter what other issues might have arisen. He reiterated Roosevelt's April 1st message to Stalin on the necessity of public confidence and support for any American Government policy, and warned that any war reconstruction aid must be appropriated through Congress, thus tying all the outstanding issues to future American aid to Russia.

Molotov implied that the Americans and British were trying to impose their will on Russia, and he had the audacity to speak of honor, and of overcoming "difficulties." Truman suddenly became angry, and stated that the problem was the Soviet failure to comply with the Yalta agreement; "...for Marshal Stalin to carry it out in accordance with his word." Molotov quickly tried to switch the subject to the Japanese war (perhaps an indication of how well informed Soviet intelligence was of at least part of the American command's true concerns). At this point Truman said:

"That will be all, Mr. Molotov, I would appreciate it if you would transmit my views to Marshal Stalin."

Molotov blustered: "I've never been talked to like that in my life." (A ludicrous statement from a survivor of Stalin's vicious, self-destroying hierarchy.)

Truman retorted in anger:

"Carry out your agreements and you won't get talked to like that."[105]

Although there is no publicly recorded specific reference to the disappearance of U.S. POWs in Poland by any of the participants in the meeting, Harriman's cables from Moscow along with his and Stettinius' presence, point to the underlying cause of bitterness in the exchange.

It must also be remembered that Gromyko had just sent the

latest Russian demand for the immediate repatriation of all Soviet citizens, with the accusation that the United States was failing to live up to the Yalta agreements on POWs. George Kennan, the Charge D'Affairs at the Moscow embassy while Harriman and Deane were in America, was living with the knowledge of the Soviets true actions and was also being subjected by them to accusations of American perfidy in POW repatriations. As will be seen, he continued to cable his alarm to Washington.

Declassified U.S. documents indicate that the new Truman Administration also considered Soviet participation in the war with Japan and in the United Nations to be the primary policy goals of the United States. On April 20, 1945, the day of Truman's first consultation with Harriman upon the ambassador's return from Moscow, Marshall issued an order forbidding the use of military force to recover American POWs in Soviet control. This policy directive had actually been formulated at the time of Franklin D. Roosevelt's death, on April 12th, and thus its distribution on April 20th amounted to a confirmation that the Truman Administration was not going to confront the Soviets over the disappearing American prisoners of war in the east. Coordination was to be accomplished through Major General Ray Barker, Eisenhower's Assistant Chief of Staff, G-1 (Personnel), at SHAEF Headquarters. General George C. Marshall issued the orders on the day of President Truman's first meeting with Harriman, upon the Ambassador's return from Moscow:

"... TOP SECRET FROM MARSHALL FOR MCNARNEY CMA INFORMATION EISENHOWER...POLICY IS THAT NO REPEAT NO RETALIATORY ACTION WILL BE TAKEN BY US FORCES AT THIS TIME FOR SOVIET REFUSAL TO MEET OUR DESIRES WITH REGARD TO AMERICAN CONTACT TEAMS AND AID FOR AMERICAN PERSONNEL LIBERATED BY RUSSIAN FORCES."[106]

Although it was recorded that Truman had spoken roughly to Molotov at their meeting in Washington, demanding that the Soviets adhere to the Yalta Agreements, there is no record (as yet declassified) of any ultimatum by the President on the Soviet-held U.S. POWs.

Patton's 3rd Army had captured the entire Gestapo files on American and Allied POWs and the German operating staff, intact, in mid-April,1945. The Wehrmachtauskunftstelle fur Kriegsverluste und Kriegsgefangene (Information Center for War Losses), located at Saalfeld and Meiningen, Germany, transmitted casualty information through neutral channels to Allied governments. The German International Red Cross delegate, Shirmer, had transmitted much of this information to Switzerland, where the OSS chief (and future CIA director) Allen Dulles was involved in forwarding the information about American POWs to Washington, D.C. According to declassified U.S. documents, ""The departments concerned with records and personal effects were reopened (on) 23 April 1945, under the supervision of the adjutant General, this Headquarters, EMPLOYING GERMAN PERSONNEL TO COMPLETE THE UNFINISHED WORK. Posting of records of American personnel was approximately six weeks in arrears."[107]

Thereafter, the War Department knew exactly how many American prisoners, known by name up until early March 1945, had been alive in German camps subsequently overrun by the Red Army, but later researchers, including the author and research colleague Thomas V. Ashworth have been repeatedly denied access to this information by the U.S. Government, for years. According to U.S. documents obtained by the author, In the last six weeks of combat in Europe, thousands more American soldiers had been originally listed as MIAs and the fate of many remained unknown thereafter, which resulted in their being declared killed in action, body not recovered. (KIA-BNR). This contributed substantially to the extraordinary number of WW II U.S. MIAs, still officially listed at over 78,750 Americans.

The Russians still refused to allow Americans to send contact officers to visit American POWs in Soviet hands, as noted in a secret cable of April 21, 1945 from Alexander Kirk at SACMED, to Secretary of State Stettinius:

"WE ARE INFORMED BY GENERAL KEY THAT (RE OUR 1011 MARCH 17) SOVIET ACC HUNGARY NOW HAS FORMALLY DISAPPROVED ENTRY INTO HUNGARY OF US REPATRIATION CONTACT TEAM."[108]

During April 1945 and thereafter, US aircraft were involved in an aerial-reconnaissance photography operation over Soviet-occupied Europe named "Operation Casey Jones." This operation was conceived by OSS General William Donovan, and along with its associated operation, "Ground Hog," the Donovan-Sibert plan photographed 2,000,000 square miles of Europe, including Soviet-occupied Germany and Austria, all central Europe and most of the Balkans including parts of Yugoslavia, Bulgaria and Albania. It appears certain that the purpose of this classified operation was revealed to the Soviets by moles and spies within the U.S. and British governments. The Soviets viewed the U.S. aircraft as hostile, and six US-Soviet aerial combats were reported by General Nathan Twining, Commander of the Mediterranean Allied Strategic Air Force during the single day of April 2, 1945: Four Mustangs were attacked by between 15 and 20 Soviet LG-5's over western Austria and one Mustang was shot down; ten Soviet Yaks attacked a group of U.S. Lightnings near Bratislava, Czechoslovakia, but the Americans avoided contact; four Yaks attacked 15 U.S. Lightnings over Neusiedler Lake, near Vienna, but the Americans fled; an hour later in the same area six or eight Soviet fighters attacked four Mustangs, and two LG-5's opened fire on two Mustangs who returned fire, but 'no damage was inflicted or received.' The British used unmarked Mosquito aircraft to obtain photos from 25,000 feet for their part of the operation over Soviet-occupied territory, but the American aircraft carried U.S. markings. More aerial dogfights with the Russians occurred along the front from this time on and they continued after the war's end.[109]

British Foreign Office officials reported identical NKVD action in withholding British and Commonwealth prisoners in Soviet hands during the April 1945 founding-UN Conference in San Francisco, to Churchill's Foreign Secretary, Anthony Eden. In a typical series of messages from this period the British acting-Foreign Minister Sir Orme Sargent, cabled British Ambassador Lord Halifax, for Anthony Eden and Lord Cadogan (in the United States then, also) on April 20th (#3936):

"IT IS CLEAR THAT THE SOVIET GOVERNMENT WILL NOT ALLOW OUR CONTACT TEAMS INTO POLAND. THE RUSSIANS DENY THE EXISTENCE OF ANY BRITISH PRISONERS OF WAR IN POLAND BUT WE HAVE EVIDENCE THAT THERE ARE PRISONERS OF WAR CONCENTRATED AT CRACOW AND CZESTOCHOWA, AND IN HOSPITALS. THIS IS A CLEAR BREACH OF THE YALTA AGREEMENT."[110]

(It must be remembered here that Halifax's First Secretary at the British Embassy in Washington, Donald Maclean, was at this time one of the premier Soviet spies, and forwarded to his Russian control every shred of secret information he could lay his hands on throughout the entire POW kidnapping episode.)

Another cable of April 20th to Lord Halifax for Eden and Cadogan (at the San Francisco UN Conference), from Sir Orme Sargent (#3923) warned:

"...While I agree that we should await reports from our officers at Lwow and Volkoyysk (note: in Russia) I doubt if these will in fact provide much additional information...THERE IS I AM SURE GREAT OPPOSITION TO ALLOWING OUR PEOPLE TO HAVE ACCESS TO POLAND AND ALSO SOME INCLINATION TO BLACKMAIL US INTO DEALING WITH THE WARSAW AUTHORITIES."[111]

In a secret message from the Foreign Office to the British Embassy in Moscow and also "repeated to United Kingdom Delegation San Francisco," (where Foreign Minister Anthony Eden and his Deputy Lord Cadogan were then confronting Molotov, (#2440) on the 5th of May,1945, the War Office reports knowledge that 40,000 or more British prisoners are still in Russian hands.[112] The Soviets went to such lengths as actually transferring several hundred of the Allied prisoners to control of the Polish communist-dominated government, to strengthen the ongoing demand for diplomatic recognition. As General Eisenhower's grandson was to write over four decades later, with access to certain records unavailable to other researchers:

"Simultaneously, Moscow expelled Allied air liaison officers, suspended shuttle bombings, and permitted snags to develop over the release of Allied prisoners of war liberated in the Russian offensive. Several hundred prisoners were transferred to Polish care in an effort to force the Allies to deal with Lublin directly, while others were detained in Odessa on various pretexts while Moscow sought assurances that the Allies would repatriate hundreds of thousands of Russian nationals in Allied hands classified under Soviet law as deserters."[113]

From the U.S. Moscow Embassy, the Charge d'Affairs during Harriman's and Deane's absence, George Kennan (who was later to achieve fame as "X," who was later credited with authoring the "Containment Doctrine," in the publication of the Council on Foreign Relations, Foreign Affairs), had cabled Secretary of State Edward Stettinius on April 30th:

"GOLIKOV MAKES STARTLING ALLEGATIONS REGARDING MISTREATMENT OF SOVIET CITIZENS IN BRITISH AND AMERICAN PRISONER OF WAR CAMPS...HE ALLEGES THAT ALL BRITISH AND AMERICAN PRISONERS HAVE BEEN REPATRIATED EXCEPT FOR SMALL GROUPS AND COMPLAINS OF DELAYS IN REPATRIATION OF LIBERATED RUSSIAN PRISONERS OF WAR."[114]

At midnight on the same day George Kennan sent another message to the Secretary of State that left no doubt about his view of the situation from his position in Moscow:

"Vyshinski called me to the Foreign Office this evening and proposed that we agree, now that are forces have linked up in Germany, to effect repatriation of our respective liberated prisoners of war directly across our line of contact in Germany instead of by sea... IN CASE THE PROPOSAL IS ACCEPTED, I THINK IT LIKELY THAT THE RUSSIANS WILL DO THEIR BEST TO INTERPRET OUR ACCEPTANCE IN SUCH A WAY AS TO OBLIGE US TO HAND OVER AT ONCE ALL THE RUSSIANS WE FIND, REGARDLESS OF THEIR STATUS, BEFORE WE HAVE HAD A CHANCE TO DO ANY SIFTING AMONG THOSE FOUND TO HAVE BEEN FIGHTING WITH THE GERMANS. The Department may wish to phrase its answer in such a way as to anticipate this possibility.

"The British Charge has wired his Government in connection with the Golikov statement recommending that a factual refutation of Golikov's shameless distortions be given to editors in England. I am heartily in sympathy with this recommendation. I CAN SEE NO REASON WHY OUR PUBLIC OPINION SHOULD BE LEFT UNDER ANY MISAPPREHENSION AS TO THE TRUE FACTS OF THIS SITUATION. Kennan"[115]

Thus, one of the foremost Soviet experts in the American Foreign Service, who was on the scene in Moscow, warned the policymakers in Washington that the Soviet line on POWs not only remained unchanged, but had actually hardened as World War II was drawing to a close in tortured Europe.

## Notes on Chapter Four

1    NA RG 334

2    Ibid.

3    Lash, *Eleanor and Franklin*, p. 920; also Washington Post, December 21,1944; and *A Pretty Good Club*, op. cit.

4    NA, RG 218

5    p. 693, FRUS, The Conferences at Yalta and Malta, 1945, also: NA, State Department files

6    pp. 694-95, The Conference at Yalta and Malta, 1945, U.S. Government Printing Office

7    Ibid.

8    p.697, FRUS, The Conferences at Yalta and Malta

9    *Last Hero*, p. 697, op. cit.

10   On above documents: See: Foreign Relations of US, Conferences at Malta and Yalta, pp. 691-7

11   pp. 756-757, FRUS, The Conferences at Yalta and Malta, als: NA

12   A copy of the original document from the National Archives, supplied by the author to the U.S. Senate, appears in Part II, "POW/MIA Policy and Process," Hearings of the Select Committee on POW/MIA Affairs, November 1991, p. 669, U.S. Government Printing Office, 1992, ISBN 0-16-038479-6

13   Bohlen, *Witness to History*, p. 199, op. cit

14   *Strange Alliance*, Op. Cit.

15   *Secret Servant, My Life With the KGB and the Soviet Elite*, by Ilya Dzhirkvelov, Harper and Row, NY, 1987

16   Declassified documents in the National Archives cited hearafter

17   OFLAG 64

18   POW/MIA Policy and Process, P. 670, op. cit.; also: NA RG 84

19   OFLAG 64

20   NA RG 84

21   Ibid., p. 673; also NA State Department files

22   Declassified Red Cross documents, NA

23   SHAEF SGS, RG 331/Entry 1/ Box 87; also: p. 671, Part II, POW/MIA Policy and Process, op. cit.

24   *Strange Alliance*, op. cit and NA

25   Ibid., pp. 673-676

26   NA, RG 84, Moscow Embassy

27   Lublin Trip, NA, RG 334/ Box 22

28   Declassified Red Cross files, National Archives

29   POW/MIA Policy and Process, Part II, Op. Cit. P. 684; also NA RG 334, Box 22

30   *I Was There*, by William D. Leahy, Whittlesey House, NY, 1950

31   NA, RG 84; 165; and 334

32   Red Cross Records, NA; also: RG 334, Evacuation of American Prisoners of War

33   Ibid.

34   NA, Red Cross Files; and RG 334

35   Ibid.

36   Ibid.

37   NA RG 334, Box 24

38   Foreign Relations of the United States, 1945; also: Stalin's Correspondence, p. 194, Op. Cit; and National Archives; this document is reproduced on p. 685, Part II, POW/MIA Policy and Process, Op. Cit.

39   POW/MIA Policy and Process, Volume II, p. 686, Op. Cit, also RG 84, National Archives

40   *Special Envoy to Churchill and Stalin*, by Averell Harriman and Elie Abel, Random House, NY, 1975, p. 419

41   *Memoirs*, 1925-1950, by George Kennan, Atlantic Monthly Press, 1967, pp. 253-254

42   NA RG 334/ Box 22

43   M-23174, RG 84, NA

44   State Department Files, NA; also: The entire cable is reproduced on pp. 692-697, POW/MIA Policy and Process, Volume II

45   NA, RG 334, USMMM-POWs

46   Special Envoy, P. 290, op. cit.

47   NA, RG 334

48   NA, RG 334

49   NA, RG 334

50   POW/MIA Policy and Process, Volume II, p. 698, reproduced from the original document located by the author in the National Archives

51   *Triumph and Tragedy*, by Winston S. Churchill, Houghten-Mifflin, NY, 1953, p. 429

52   IG/3491/2/G1 (br

53   *Wild Bill Donovan- The Last Hero*, Anthony Cave Brown, p. 640-642, Op. Cit.

54   Declassified Red Cross records, NA

55   NA, and Stalin's Correspondence, Op. Cit., p. 195

56   Ibid., pp. 196-97

57   NA, RG 334

58   Stalin's Correspondence With Churchill, Attlee, Roosevelt and Truman, pp. 306-307, op. cit

59   National Archives, also: a slightly different translation appears on p. 308, Stalin's Correspondence, op. cit.

60   Churchill, *Triumph and Tragedy*, p. 444, op. cit.

61   *Strange Alliance*, op. cit

62    NA, RG 334, USMMM-POWs

63    NA, RG 319/47/940, filed under M.I.D.)

64    NA, RG 334

65    Roosevelt Library; also: p. 197, Stalin's Correspondence, op. cit.

66    WARX-58751, NA, RG 334; this document is reproduced on p. 702, POW/MIA Policy and Process, op. cit.

67    CM-IN DB28852, 27 March 1945, NA RG 334

68    Operations Division, WDGS, OPD 383.6 (12th April 1945, NA

69    *Triumph and Tragedy*, p. 431, op. cit.

70    Ibid., pp. 432-433

71    National Archives, Mission to Moscow Files; "The Lublin Trip."

72    *Forrestal Diaries* pp. 38-39

73    NA, RG 334, USMMM files

74    New York Times, and For You the War is Over- American Prisoners of War In Nazi Germany, by David A. Foy, Stein and Day, NY, 1984

75    The Patton papers, in the Library of Congress

76    *The Last Six Months*, by General of the Soviet Army S.M. Shtemenko, translated by Guy Daniels, Doubleday's, NY, 1977

77    NA, RG 59

78    Foreign Relations, 1945, Vol.5, p. 1091.

79    These documents from RG 334, National Archives are reproduced on pp. 704-705, POW/MIA Policy and Process, U.S. Senate, op. cit.

80    NA, State Department files, reproduced on p. 707, POW/MIA Policy and Process, op. cit.

81    NA; AG 383.6/292 A-0

82    Original document from the National Archives reproduced on p. 709, POW/MIA Policy and Process

83    *Memoirs*, by Harry S. Truman, Volume 1, *Year of Decisions*, Doubleday and Co. NY, 1955, p. 15

84    *Year of Decisions*, by Harry S. Truman, Doubleday's, NY, 1955, p. 37

85    Eden, *Memoirs*, Houghten-Mifflin, NY

86    NA, Department of State, Memorandum of Conversation, April 16, 1945; this document is reproduced on p. 712, POW/MIA Policy and Process, op. cit.

87    NA, FWD 19334

88    30 Mission to Troopers 86094 PW5

89    NA, RG 331

90    FACS-183 to Eisenhower, FAN 524, National Archives, RG 331

91    NA, RG 331

92    1989 interview by the author. Mr. Masterton has since died, but he left a manuscript on his experiences as a prisoner of war.

93    NA, State Department files; these documents are reproduced on pp. 713-715, POW/MIA Policy and Process, op. cit.

94    Ibid., p. 715

95   Ibid., p. 716

96   *Special Envoy to Churchill and Stalin*, p. 447, op. cit.

97   CM-OUT 64858, National Archives

98   *Special Envoy, The Forrestal Diaries and The Wise Men*, op. cit.

99   *Forrestal Diaries*, p. 49, and *Special Envoy*, op. cit

100  This all-encompssing statement was quoted by Harriman in *Special Envoy*, p.451

101  Roosevelt and the Russians, The Yalta Conference, by Edward R. Stettinius Jr., Doubleday & Company, NY, 1949, p. 311

102  Special Envoy, p. 452

103  *Special Envoy*, p. 452, and *On Active Service in Peace and War*, by Henry L. Stimson and McGeorge Bundy, Harper and Brothers, NY, 1947,1948, p.608-9

104  *Forrestal Diaries and Special Envoy*, p.452, op. cit

105  Harriman, *Special Envoy*, Bohlen, *Witness to History*, op. cit., and others

106  Record Group 165/15/11, National Archives; also NA RG 165/419/Box 164; this document from the Operations Division of the War Department General Staff, Policy Section, WDOPD, signed by Major General J.E. Hull, Assistant Chief of Staff, OPD, declassified top secret, is reproduced on p. 722, POW/MIA Policy and Process, U.S. Senate, op. cit.

107  NA, RG 407/360/2439, AG 383.6 I, German records of prisoners of war, to: The Adjutant General, War Department, 11 June 1945; also: NYT Article- AP wire April 19th, 1945. Despite Freedom of Information requests by the author, the National Archives and the Department of Defense repeatedly denied to the author that these records could be located, although fragments of them have been recovered scattered among other records in the Archives which indicate that American POWs known by name to have been alive in German captivity in the spring of 1945, never returned from areas of eastern Europe under Soviet control. Likewise, many U.S. Red Cross records and International Red Cross records (in Switzerland) on American and Allied POWs of WW II have not been made available to researchers, and have remained classified to the time of this writing.

108  NA, RG 331

109  *Wild Bill Donovan-The Last Hero*, pp. 641-642, op. cit.

110  NA, Record Group 84/Box73; State Department files; declassified in 1988; this document is reproduced on p. 720, POW/MIA Policy and Process, op. cit.

111  POW/MIA Policy and Process, U.S. Senate, p. 721

112  *Secret* Hiss-Stettinius agenda

113  *Eisenhower At War,* by David Eisenhower, Random House, NY, 1986, p. 697

114  NA, State Department files; this document is reproduced on p. 723, POW/MIA Policy and Process, op. cit.

115  National Archives, Department of State files, Telegram #1426, midnight, April 30, 1945, from Moscow via Army, declassified "Secret."

# CHAPTER FIVE

## FROM VE DAY TO THE POTSDAM CONFERENCE

As the Soviet Army advanced into Germany and Austria, tens of thousands more U.S. and Allied prisoners had fallen into their hands but they would not permit nearby U.S. and British forces to recover these men either by aircraft or overland convoy. Instead they were either held in the stalags in which they had been imprisoned under the Nazis, or in some cases they were moved in large groups to other locations by the Soviets. The Germans surrendered on May 7th, and Truman declared May 8th to be Victory in Europe Day, but the Russians still held the American, British, French, Belgian, Dutch and other Allied POWs. On VE Day Lend-lease aid to Russia was suddenly and drastically curtailed, as had been (secretly) urged by Harriman in mid-March. This had infuriated Stalin, who had already demanded $6 billion dollars in American postwar reconstruction aid. On May 11th, 1945, three days after the European war's end, Eisenhower cabled Harriman's military chief, General John Deane, in Moscow:

"...INFORMATION RECEIVED FROM PRISONER OF WAR CAMPS IN REAR OF RUSSIAN LINES INDICATES THOUSANDS OF UNITED STATES AND BRITISH PRISONERS OF WAR HELD IN CLOSE CONFINEMENT...PLEASE EXPRESS TO RUSSIANS THE URGENCY OF THIS MATTER...UNLESS THIS EVACUATION CAN BE EFFECTED PROMPTLY, THERE MAY WELL ENSUE THE MOST UNDESIRABLE CONSEQUENCES..."[1]

The Soviets, in fact, merely took control of thousands more American prisoners in scattered groups in many locations across eastern Germany, steadfastly refused US contact officers access to them, and accused the United States of retaining and "abusing" Russian prisoners. Many of the Russians were actually refusing to return to Stalin's promised revenge, for having surrendered to the Germans following high-level 1941-42 Soviet blunders. Numerous Allied soldiers present during repatriations witnessed immediate firing squad executions of Russians, while the rest were deported to forced labor camps.

Eisenhower's demand resulted in the Soviets grudgingly allowing a mass air evacuation of 7,700 Americans and some British POWs from Stalag Luft I at Barth, on the north German coastal plain near Rostock, although at the end the Soviets attempted to block the runway to prevent the last POWs from leaving. Luft I was abandoned by its German guards, who fled the approach of the Russians on April 29th and 30th. John Werner, an American POW at Barth remembered:

"The Germans, who had planned to defend the place as they had dug trenches around it, fled quickly to the west as they naturally wanted to be captured by the Americans and British instead of the Russians...We were only alone for a day or so..."[2]

Bryce Robison, an enlisted U.S. POW at Luft I when the Soviets arrived, had been in several other German camps prior to liberation. He had enlisted in the Canadian Army prior to American involvement in the war and fought on the ground in the disastrous 1942 raid on Dieppe. Later he had succeeded in getting transferred to the American forces and was assigned to the 306th Bomb Group (the "First Over Germany") of the 8th U.S. Air Force, only to be shot down and captured by the Germans. He had been held at Memel during the late 1944 Soviet offensive he and many other POWs were moved out of that Baltic port by the Germans, handcuffed to the bottom of a barge. Later confined at Stalag Luft IV-Gross-Tychow near the Polish-East Prussian border. He was considered a recalcitrant by the Germans because of escape attempts and moved around from camp to camp, once as far away as Stalag XVII B near Gneixendorf, Austria, before eventually being sent back to Grosstychow early in 1945. With the renewed Soviet offensive in January 1945, many of the American POWs at Grosstychow were forced-marched west by the retreating Germans but Robison, along with a number of other recalcitrants and some of the sick and wounded, were shipped in boxcars. While the marching POWs took 27 days, those in the boxcars were 9 days enroute to their new destination, Stalag Luft I at Barth. Bryce Robison remembered when the Soviet Army approached Luft I after the German guards had fled the camp on the 29th and 30th of April:

"The first Russian came riding up on a fine white horse. He was wearing a lot of medals and everyone thought he was a General. He said something like: 'You been prisoner for two years, tear down the wire, everybody out!' The American senior officer, Colonel Zemke, was concerned about security, but told an aide to pass the word to the men to, 'tear down the inner fence.' The lower ranks thought this meant to tear down all the wire, so they did it, or most of it...Later we figured out that the first Russian was just a scout, probably had been heavily decorated for valor in the past..."

George R. Simmons, now of Oklahoma city, was a U.S. POW in the camp at Barth and also remembered those first few days:

"I was at Stalag Luft I which was liberated by the Russians on April 30, 1945. Colonel Hubert Zemke was our camp commander and Lt. Colonel (now General) Gabreske was my 'compound commander' (north 4). We were advised that Eisenhower wanted to fly us home from the local airstrip, and he was awaiting approval from Stalin. Stalin advised that we were to be 'repatriated' and that we were to be sent to Odessa on the Black Sea. I remember that the Russians were calling us in on a 24 hour schedule to give our name, rank and serial numbers, as well as our thumb print. I was called in during the middle of the night to do so. I was told that the local Russians were unaware of Stalin's refusal to allow us to fly out."

Eric H. Sherman, a POW at Barth who later retired from the Air Force as a Lieutenant Colonel, remembered long weeks of waiting under Soviet control:

"The Germans pulled out and left the camp in Allied POW control. The Russian Army arrived a day or two later and had us tear down the barbed wire fences. They provided us meat (beef or horse) and Red Cross parcels were located in a warehouse. Things looked great for us. The war was over and were we free. But were we? Days went by and there was no move to get us back to American control. In all it was three weeks before the first B-17's were allowed to land and start transporting the POW's west..."

Ex-POW Bryce Robison also recalled the weeks of waiting under Russian control at Luft 1, and since much of the barbed wire perimeter had been torn down by the prisoners he remembered, "... Many men left the camp, they scattered..."

An estimated 1,000 US Army Air Corps prisoners of war left the unguarded camp at this time, many of them seeking some form of transport to the west where the British and American lines of Field Marshal Montgomery's combined force lay. Hundreds of these prisoners who later fell into Soviet control outside the camp were in many cases arrested by the NKVD, who refused to accept G.I. dog tags as proof of identity in lieu of passports (which none of the prisoners possessed). The camp itself was taken under full control of the Soviet Army on May 2, 1945, and the stalemate continued over whether the American Air Force would be allowed to land and evacuate the POWs.

Finally, for reasons unknown to the POWs, the Russians gave in and allowed some U.S. aircraft to land. It was later rumored that U.S. Air Force General "Hap" Arnold had threatened to load his planes with bombs and drop them on the Soviet positions if he was not allowed to send his bombers in empty to evacuate the U.S. and Allied prisoners. Ex-POW George Simmons remembered the day:

"On May 12th enough planes landed to fly out 2,000 British POWs who had been held the longest. The following day the planes started coming in at about 4:00 A.M. without a break."

The American air force began a frantic POW evacuation from a German airfield two miles away. For ten solid hours 20 bombers landed and took off every hour, without cutting their engines. George Simmons got out on May 13th, 1945:

"While one flight loaded (17 men on a B-17) another was clearing the runway and taxiing around to 'D' shaped airport, then the third flight was landing as close as possible and as quickly as possible on the 4000' strip. 7000 men were taken out on May 13th. I went off about 4:00 P.M. and was flown to Rouen, France..."

Bryce Robison didn't get out on the 13th and was in fact one of the last American POWs to leave the camp. He recalled the date as about the 22nd or 23rd of May. He remembered that on the last day he was there the Soviets were angry with the American air evacuation:

"During the last airlift the Russians landed two aircraft on the field, then they pulled out the starters and walked away. They were attempting to block the airfield...The Americans were hustling around trying to find two starters that would fit those aircraft...I finally got out 14 days after VE day..."[3]

Although it was officially reported that a total of some 7,700 American prisoners were successfully extracted from the Russian zone in this operation at Barth, now-declassified US documents of June 25, 1945 indicate that hundreds of American POWs who had fled from Stalag Luft I soon after liberation or during the weeks of waiting, disappeared into Russian hands at this time. (This matter is dealt with more fully later in this book.)

Other prisoners who had been at Stalag Luft IV-Gross-Tychow had disappeared when the Red Army overran that area; whether temporarily or permanently was as yet unknown. Among them was a young Army Air Force prisoner of war, Staff Sergeant Bernard Solomon, who had gotten to know President Truman when he worked at then-Senator Truman's campaign headquarters in Missouri Sergeant Solomon had been shot down and captured on June 2, 1944, while serving in the 385th Bomb Group-551st Bomb Squadron and had written his parents, Mr. and Mrs. Albert Solomon on November 24, 1944. The appeal of his parents to Presidential aide Colonel Harry S. Vaughn, resulted in a response from Vaughn that, he said, "gives you the War Department's latest information on your son." This consisted of a memorandum from Lt. Colonel B.W. Davenport, of the War Department General Staff that indicates how well informed the Pentagon was of the movements of U.S. POW columns fleeing the Russian advance with their German guards:

"The information I received indicates that Sergeant Solomon was listed as a prisoner at Stalag Luft 4, Gross-Tychow, Germany. The prisoners at that camp were divided into four groups and moved to Stalag Luft I at Barth, Stalag IIB at Fallingbostel, Stalag 10B at Sandbostel, Stalag 357 at Oerbke and Stalag 13D at Nurnberg. It is believed that some of these men also reached Stalag 7A at Moosburg. All of these camps except Stalag Luft I have been overrun by Allied forces but the War Department has not received a report that Sergeant Solomon has been liberated..."[4] (The ultimate fate of this POW is not known by the author since the National Archives and the Department of Defense have refused the author's Freedom of Information requests for the lists of American POWs liberated and returned to the United States.)

According to contemporary accounts and still-living

witnesses, truck convoys attempting to reach U.S. prisoners in the Russian zone of Germany were fired on and driven back empty, U.S.-manned trains loaded with prisoner relief supplies were hijacked in the Russian zone, where the American guards and crews disappeared, and US and British reconnaissance aircraft were attacked and shot down by the Soviets.

A veteran of the U.S. 42nd Infantry Division late in the war and in the post-VE Day period 1945, Val Stiegal (of Bethesda, Maryland), recalled for the author how in the post-VE Day period he had been ordered to accompany a train shipment of relief supplies into the Soviet zone, but the shipment was suddenly halted by U.S. authorities because two previous shipments into Russian-occupied territory had vanished without a trace, together with their crews and American Army guards.[5]

According to declassified documents, U.S. POWs and some British prisoners escaped by "exfiltrating" the Soviet lines. However, thousands of prisoners in Germany, both in large groups from camps such as Stalag II-A Neubrandenburg (with 3,700 Americans reported under Russian control on May 10), Stalag III-A, Luckenwalde (overrun by the Soviets with at least 4,893 Americans, although some reports said more than 6,000 Americans had been held there, and moved to "satellite camps" and "many more expected"), Stalag IV-B-Muhlberg (with 3,000 more U.S. POWs in Russian control), and in smaller groups of 20 to several hundred, scattered across hundreds of locations in eastern Germany, such as several hundred U.S. POWs who were arrested outside Stalag Luft I at Barth, were retained by the Russians and subsequently disappeared. Other U.S. prisoners from camps in Germany such as Stalag IV-F, Hartmannsdorf-Chenitz with 3,000 Americans last reported, and Stalag IV-G-Oschatz (which had held over 7,000 Americans in early April 1945), may have been removed from the area of the camps deeper into Soviet-occupied territory, to await results of negotiations. Declassified confidential, secret and top secret SHAEF cables of the time described conditions in the Soviet-controlled camps, such as Stalag II-A Neubrandenburg on May 10th:

"PRESENT CONDITIONS OF CAMP: NO CAMP CONTROL, RUSSIAN OFFICER IN CHARGE USUALLY DRUNK, DRUNKEN RUSSIAN SOLDIERS MOLESTING AMERICAN SICK, AMERICANS ROBBED BY SOME RUSSIAN SOLDIERS MAKING IT DIFFICULT TO AVOID TROUBLE."[6]

On May 12th the Allied command at the Reims Advance Headquarters released to the media a statement about the American and British POWs remaining under Soviet control, which was reported by Associated Press:

"Nearly half of the estimated 200,000 British and 76,000 American prisoners of war in Germany are believed to be within the Russian zone of occupation and Supreme Headquarters has twice requested a meeting or an arrangement to arrange their return...Up to late today there has been no reply from Marshal Ivan S. Koneff (Koniev), to whom the requests were addressed, one through the Twenty-first Army Group and the other through Moscow..."[7]

American prisoners who escaped from such Soviet-controlled camps reported that the Russians had announced that American and British POWs were still to be shipped east into Russia for repatriation, ostensibly to Odessa, instead of being repatriated across the lines as most POWs and the U.S. command expected and desired. Indeed, General Deane had reported from Moscow that Soviet repatriations commissar Golubev had announced on May 12 that no more U.S. POWs would be repatriated through Odessa. Thus U.S. and British prisoners still being sent east vanished in Russia, then a nation of thousands of vast gulag camps containing millions of prisoners of many nationalities. Some of the U.S. POWs sensed the possible danger and took chances in escaping westward to Allied lines. A MAY 17th cable from General Simpson, Commander of the U.S. 9th Army, to SHAEF Headquarters, revealed the conditions under the Soviets a week after VE Day:

"REPORTS RECEIVED THAT 7,000 UNITED STATES AND BRITISH EX-PWs FORMERLY IN MUHLBURG AND NOE REISA, NEED MEDICAL SUPPLIES AND FOOD. MANY HAVE LEFT BECAUSE OF CONDITIONS. REPORTS INDICATE CAMP LEADER DOING ALL IN HIS POWER TO ENFORCE STAY-PUT ORDER. RUSSIANS ALLEGED TO HAVE THREATENED TO USE FORCE TO PREVENT ESCAPE. SUGGEST IMMEDIATE EVACUATION..."[8]

But the Soviets would not permit the evacuations to occur at several of these camps, and Americans continued to risk being shot in escape attempts.

This short message, which immediately concerned the fate of some 3,000 Americans and thousands more Allied POWs from ten nations at Stalag IV-B Muhlberg, who had by then been under Soviet control for nearly three weeks with no prospect of being freed, has been corroborated by an American prisoner who escaped, Martin Siegel:

"IV-B Muhlberg was indeed a multinational camp, whose prisoner of war population grew at a great rate in late February. As more and more PWs were shipped in, we were forced to not only double up in bunks, but had to sleep in shifts when possible. Rations then were cut accordingly. There were British PWs (mostly infantry but also a high proportion of enlisted airmen), French partisans, Serbs, Russians, some Italians—seemingly representing every faction that opposed the Germans. The Russian prisoners were treated even more harshly than anyone else; they had little or no rations and eagerly scraped the barrels that our soup ration came in. There were many deaths due to starvation among Russian prisoners, and the bodies were carted off each morning in trucks. Their compound was entirely separate from ours...From the time I arrived at IV-B to the time we took-off I never learned of, nor met an American camp leader.

"When we were liberated by a Russian tank battalion (in late April 1945), I did notice there were a number of Sherman tanks, as a matter of fact, these were spotted as the first vanguard of tanks approached the camp that morning. Some PWs were running through the barracks shouting that GI tanks were coming to get us. Conversely, the cheers coming from the Russian compound, the waving of arms and pieces of clothing soon made us realize that it was 'Ivan' that was about to take over...

"A group of us enthusiastically sought out a Russian officer to try to find out what was happening, and when we could expect to be repatriated back to American control. This officer spoke no English, but did have a smattering of German and since my mother had spoken German/Yiddish and my father had come from Russia and taught me some of that language, I became sort of an unofficial translator for some of the men in my section."

The Soviet tank battalion commander's name was Major Vasilli Vershenko, and Siegel's unofficial interpreter position gave him a unique opportunity over the next few days to learn of Russian plans for the Allied POWs as the European War drew to a close in early May:

"Vershenko was a dark, heavy set man, most polite, but aloof and had a commanding presence. In later years I noticed that pictures of (Andrei) Gromyko bore a resemblance to Vershenko. This is exactly what I was told by Major Vasilli Vershenko:

"The Russians were first concerned about the repatriation of the Russian prisoners held in the separate compound at IV-B, and the Major indicated that they had to be interviewed individually since they felt that there were many 'cowards, traitors and deserters among them and they had to be dealt with expeditiously.' Then he told me that the Russians and the Americans had agreed to a pact wherein the Russians would receive 'credits' for each American POW returned...I cannot recall any reference to a credit 'ratio' from Vershenko nor from the three NKVD (we assumed they were NKVD because of their non-combat uniforms) that were present. This, he explained was a complex logistical matter, best handled by sending us to Odessa for treatment and repatriation. The callousness of his response and the officious tone in which this information was given, gave me real pause and I tried to explain to others that I was suspicious of their methods and motivations. Most of the men, however, felt that the Russians were our allies and that we were going to be well treated and returned home shortly. That night, my bunkmate, Corporal William Smith of the 9th Division shared our mutual concerns and decided to take off... Our decision was also based on the fact that by this time, most of the PWs were emaciated, lice-ridden, and stricken with diarrhea. Since our rations were less than minimal during February, March and April, I had lost over 45 pounds, and Smitty almost 65 pounds. The next evening we 'liberated' two Russian bicycles, got through a gap in the wire where a Russian tank was parked and took off toward the west where we believed the American Army would be..."9

Sergeant Les Neal of the U.S. 232nd Infantry Regiment, which had lost a great number of men listed as MIAs initially, was also an American POW at Stalag IV-B Muhlberg, but he was one of the many who remained behind in Soviet control after Siegel and Smith had fled toward the American lines. Neal was part of a group of U.S. POWs taken from Muhlberg by the Russians in early May:

"...They took us across the Elbe River to Mecklenberg. They kept us in some sort of German military academy. There were dormitories, mess hall and other buildings. There was a tall iron fence around it and sentries at the gates. It was here I was assigned as a cook with eight other men. We prepared Russian Musch. A thick stew...this was served with chunks of black bread. Sometimes we served watery macaroni soup. There was no breakfast. You served twice a day only. The cooks worked under Russian women supervision. They took us over to a building the first day. Here we had to disrobe, go into a room, and were given a physical by three Russian women doctors. Later we were issued three surgical gowns which we had to wear on duty in the kitchen. My friend (a cook from Company L) and I shared a room over the mess hall. His name was Henderson. Russians were very hard to get used to. It seemed like you couldn't trust them too well. I was in this place when the war with the Germans ended. (May 8-9 th, 1945) The Russians all got drunk on their vodka (180 proof), and danced and played music all night. We were not allowed to attend but we watched them from the mess hall windows...Later in the night after we had gone to bed they started shooting their rifles off. Some of the slugs came up into our room breaking windows. It was about eleven o'clock in the night when they knocked on our door. When I went to the door two Russian officers came in carrying a large container of vodka. They heard we had a radio that we had found in the building and had fixed it so it would play. We turned on the radio and got BBC. One of the Russians went downstairs and came back with gravy bowls, and that is what we poured the vodka in...it felt like fire going down. They can handle it but my buddy and I got pretty sick."

A few days after VE day, about the middle of May, a somewhat extraordinary thing happened to Les Neal and the other American POWs from Muhlberg who were being confined by the Soviets in Mecklenberg. For some inexplicable reason a lone American officer (with an enlisted aide) was allowed into the Soviet zone, made brief contact with Les Neal and the other Americans and just as inexplicably was then permitted to leave by the Russians. It would almost appear that the Soviets had allowed an eyewitness to view their American hostages to prove their serious intention of demanding the forced-repatriations of Soviet citizens in American and British control:

"One morning an American medical officer (a Captain) and his aide arrived in a jeep. The jeep had white flags on the front fenders. The captain gave us some cigarettes and a few candy bars. HE TOLD US THAT THE RUSSIANS WERE NOT GOING TO RELEASE US. One man asked him what we should do. He replied, 'YOU WILL HAVE TO ESCAPE, IT'S YOUR ONLY CHANCE.'

The Russians were not going to let the Americans come across the Elbe River. The Russians came up and told the captain to leave. I guess our hearts really sank when we saw the jeep leave that day..."[10]

The experiences of the earlier escapees from the Muhlberg camp, Martin Siegel and William Smith, during the first few weeks of May 1945 illustrated the different treatment given two American POWs by front-line Russian troops when not under the watchful gaze of the dreaded NKVD or SMERSH counterintelligence:

"Two days after we left IV-B we did see two American planes, flying low toward the camp and waved and shouted-to know avail. What neither of us knew at the time we headed west...was that, by agreement, the American and British forces were pulling back, leaving east Germany to Soviet occupation, and as a result, whenever we reached a village or town we met up with Russian troops. These field troops were most kind, and provided us with food, shelter and directions. During our travels we confiscated a smashed Volkswagen, got it running, siphoned some gas from an upended Tiger tank and continued looking for Americans. Just before we came to the Elbe River, we were fired upon, surrounded and beaten by a squad of die-hard Hitler Youth, led by a fanatical young officer. Fortunately we had fashioned an American flag, by painting stars and stripes on the backs of our jackets. Since this group was on the run from the Russians, they wanted to execute us, but after lengthy discussions decided they better head for the hills. A few hours later, we came across a Russian platoon that was chasing them and we continued on our way. In Halle, we were accosted by a German woman who said that her husband was a physician who had been assigned to a 'political prisoners' camp and wished to surrender to the Americans rather than the Russians. When we saw that they both had American uniforms, from helmets to combat boots, we took them as far as the next Russian encampment and turned them over...The VW had broken down and we proceeded by foot until we reached the Americans at

Wurzen..."[11]

Martin Siegel, the U.S. 42nd Infantry Division POW who had escaped from Soviet-controlled Stalag IV-B Muhlberg with his 9th Infantry Division bunkmate, Corporal William Smith, was officially recorded as "returned to military control," on May 23rd, the same day that the Halle Conference ended, and a telegram stating this was sent to his family by the War Department. Siegel and Corporal Smith were debriefed by U.S. intelligence officers soon after their return who appeared skeptical and annoyed at their reports that thousands of American and Allied POWs from Stalag IV-B Muhlberg remained behind in Soviet control:

"The officers who debriefed us were young, newly arrived junior officers, who looked like they had graduated from law school that year. I was required to sign a form agreeing to remain silent about our escape and the circumstances, etc. At this point I would have admitted to being Judge Parker if it meant going home sooner..."

When Martin Siegel related to the U.S. intelligence officers exactly what Major Vasilli Vershenko had told him of the Soviet plans to ship the American POWs a thousand miles east into Russia, "to Odessa for treatment and repatriation," instead of releasing them at the American lines a few miles to the west, he was cut off:

"I tried to tell American Intelligence officers about the Major's comments, but was told that we were probably told that version in order to preserve order in the camp."

With the clear Soviet intention of holding the remaining American and British POWs as hostages, which had been revealed both to General Barker at Halle and to prisoners who had succeeded in escaping from Russian-controlled camps, the American command instituted even more precautions to ensure secrecy concerning U.S. POWs still being held captive in the Russian zone of Germany. Martin Siegel remembered what happened many years later, in a letter to the author:

"...MY CONCERNS FOR THE OTHER PRISONERS LEFT BEHIND AT IV B WERE TREATED WITH INITIAL SKEPTICISM, THEN ANNOYANCE AT MY PERSISTENCE, AND FINALLY REASSURANCES THAT THE MATTER, "WOULD BE INVESTIGATED.'"[12]

Sergeant Les Neal of the 232nd Infantry Regiment, the other American POW from Stalag IV-B who had escaped from forced-labor at the Soviet military hospital weeks after Siegel, reported a very different sort of ending to his exfiltration from the Red Army zone of Germany. After the strange mid-May visit of the American captain in the jeep with white flags who had urged them to escape, Neal and two other POWs built up their courage. Around the 20th or the 21st of May they decided to try it:

"It was about a week later my buddy wasn't feeling very well. We had worked in the kitchen and were returning upstairs to our room. As we turned to go down the hall we saw our belongings being thrown out in the hall. There were three large Russian women taking over our room. Across the hall several other American kitchen personnel were being evicted from their quarters by Russian women. This was the last straw. I spoke to the others about escaping immediately. My buddy was too ill to make the escape. Finally three of us agreed to make a try."

Sergeant Les Neal was luckier than many thousands of other American, British and other Allied POWs who were confined inside barbed wire ringed POW camps at this time. At least he and the other two men had a good chance at the beginning. Nearly 44 years later Neal remembered his actual escape:

"I had been walking with a cane at times because of my right foot so the other two (escaping POWs) put fake bandages on their arms. We walked to a side gate and the Russian guard stopped us. We told him we had to go to the hospital. It was on the outside of the gate alongside of a road.

"He let us through and when we got near the road we laid down in a gully. It was getting toward dark when we one by one crossed to the other side of the road. We crawled up a gully to the nearest house. We knocked on the door but were refused. It was at the third house we were admitted. It was an old couple with a daughter living with them. They were very nice to us. These Germans were scared to death of the Russians. We got to take a bath and later they fed us. They drew us a map of the best way to leave there. It was through a swamp, that way we could stay off the roads. We went to bed and they woke us up about two o'clock in the morning. They gave us some black bread, ersatz honey and a bottle of wine. We thanked them and said our goodbyes.

"We entered the swamp and found a path and proceeded on. About twenty minutes later we had a Russian confronting us with an automatic rifle. One of the men with me spoke Polish to the Russian. He turned around and told us if we had enough cigarettes we wouldn't be shot. We gave him all we had. I gave my last four, a small price to pay for my life. He led us on out to the road and pointed the way to Reisa. This was where the Americans were located, but on across the Elbe River. We had to be very careful because of Russian patrols. The road at times was filled with refugees pulling carts with all their belongings. We were about ten kilometers from Reisa when we were spotted by a Russian patrol. They jumped out of a large truck and forced us to get in the back. There were two Russian soldiers guarding us. We were greatly surprised when instead of turning around and taking us back, they went on to Reisa. On the outskirts of town they stopped and a Russian Colonel ordered them to let us go. He told us goodbye and got into the cab of the truck and they left. This is what I mean about the Russians; you never could figure them out.

"In the city of Reisa when we entered, everything was in chaos. Drunken Russian soldiers were urinating in the streets. There were women being molested, cars were overturned, and stores were being looted. We hurriedly made our way toward the Elbe River. The river was up and it was fairly wide but there was a bombed-out bridge. It was bombed-out on the Russian side, but was fairly intact going over to the American side. First we had to cross a footbridge out to an island. This was being guarded by four Russian soldiers. To our surprise we walked right past them, over the bridge before they ordered us to halt. We kept going and they started shooting at us.

"We ran toward the bombed out bridge. It was lucky for us that the Americans had put a makeshift ladder down within eight feet of the ground. We had to jump to catch the last rung. After the second try I made it but my fingers were bleeding. An American 2nd Lieutenant reached down and pulled me on up. The three of us gave him a big hug. We sat down and cried a little and thanked God for helping us.

"We went on across the bridge, we could hardly believe our eyes. On the American side there were no drunken soldiers, there were no women being molested, and no overturned vehicles. Everything was quite normal; people were shopping and American MP's directing traffic. It was the direct opposite of what was going on across the river in the Russian sector..."[13]

Martin Siegel was exhausted and ill following his return, "...by this time I was running a high temperature, the shrapnel wounds in my foot had opened and I don't remember much about the ensuing days. I was then flown from Frankfurt to Le Havre (Camp Lucky Strike) where I boarded ship for home...Since we left IV-B so early-on I did not hear of nor did we meet up with any other PW's who exfiltrated... When I finally came back to the States, recovery from wounds that I had received at the time of my capture, the malnutrition and amoebic dysentery I developed, and sheer exhilaration of surviving took precedence and life soon resumed some semblance of normalcy. When I got back to the States, I do not recall any public references to the repatriation of PW's through the USSR."[14]

Although most of the 10,000 American and British POWs brought up to Reisa by the Soviets for exchange purposes were from Stalag IV-B, Muhlberg, and 7,000 were still there on May 18th, of which 3,000 were eventually exchanged at the end of the Halle Conference, 3,000-4,000 other American and British POWs who had been at Muhlberg disappeared in Russian control.

Siegel's and Smith's experiences were being repeated with numerous other Allied POWs who succeeded in escaping Soviet control, and in all known cases they were ordered to remain silent.

Sergeant Les Neal of the U.S. 232nd Infantry Regiment, the American POW from IV-B who had escaped to the American lines under Soviet gunfire at Reisa, never forgot the other American POWs he had left behind in Russian captivity. He still questions their ultimate fate:

"Is there some way we could find out if the rest of them ever came out of Russian hands? The day I escaped from there there were still lots of them being confined. One of the two who escaped with me was Lang from the 1st Ranger Battalion. He lived somewhere in Michigan..."[15]

President Truman himself received appeals from the families of American POWs who had been at Stalag IV B-Muhlberg, who were constituents of his from Missouri. A frantic letter from George W. Coffin about his son, who had been confirmed as a POW but had disappeared, to Truman's military aide, Colonel Harry Vaughn, read:

"I am very anxious to know more about my son, Bruce E. Coffin, of Independence (Missouri). We have received a card from him dated December 28, 1944, from Stalag IV-B, Germany. His prisoner number is 311,219. Will you please inquire from the Provost Marshal's office, if this Stalag has been liberated—also, if they can confirm this postal card received in Independence, April 6th, 1945. We are sure that the writing on the card is genuine and that the message was written by our son. Anything you can do will be greatly appreciated by this soldier's whole family. Do not hesitate to wire me at my expense if you can get this desired information anywhere in Washington."

Vaughn, who had been the treasurer of Truman's campaign committee and a reserve officer in St. Louis, Missouri, who later accompanied the President to Potsdam, passed the letter to the War Department, which resulted in a memorandum from Lt. Colonel B.W. Davenport, on General Marshall's staff, showing that Coffin and many other American POWs had not been known to have been held by the Germans:

"...I find that a photostat of the postcard received by Mr. Coffin has been used as the basis for adding Private Coffin's name to the official list of prisoners held by Germany at Stalag IV-B. Many other names have been added as the result of this kind of evidence. Since Mr. Coffin wrote, Stalag IV-B at Muhlberg, Germany has been over-run by Allied forces. (Davenport did not say it was the Soviets who had overrun the camp-ed.) Those prisoners not liberated were undoubtedly moved elsewhere by the enemy..."

Vaughn sent this report to Mr. Coffin, which resulted in the distraught father sending a May 18th telegram asking for "any further information..." To this Colonel Davenport replied to a White House secretary, Miss Bonsteil: "May 18th-No further information available at this time." On May 19th, Harry Vaughn again wrote George Coffin: "I am unable at this time to furnish any further information regarding your son. The War Department assured me that you will be the first to be notified of any change in his status or if he has been liberated, Harry H. Vaughn, Military Aide to the President"[16]

An American soldier who didn't escape from the Russians until late June 1945, Staff Sergeant Anthony Sherg, stated in now-declassified U.S. documents dated June 25, that some Americans were "tried and convicted of espionage" after Soviet officials refused to accept GI dog tags as proof of identity, in lieu of passports, which none had.[17] Allied prisoners who were eventually returned by the Russians, such as Australian flying officer Norman Dodgson, a prisoner from Stalag III-A, have reported that the Russians shot and killed POWs who tried to reach American or British lines. Long afterward Dodgson wrote:

"Even today there is cause to believe the Russians are still holding some of our types. My own crew have never been traced, and except for myself and the bombardier, the other 5 have simply vanished."

Retired U.S. Army Captain Sidney C. Miller was a witness to the shoot-to-kill orders by the Russians at Stalag III-A Luckenwalde, and he subsequently escaped to Allied lines, leaving 4,000 to 5,000 other Americans behind at Stalag III-A, under Soviet guard. The indications are they were shipped eastwards in railway boxcars on May 28th, 1945 and thereafter. In a June 1990 letter to the author, Captain Miller remained certain that many of them were never released by the Russians:

"I personally think that there were a great many Allied POWs who came under Russian control and were never heard from again, similar to the Katyn Forest episode...What happened to those POWs in Camp Luckenwalde subsequent to my escape in many cases has been documented by various agencies as well as individuals. The Russian military mind remains a mystery to me except that they were under control of the political branch who always took control over the territory they had won."[18]

General Eisenhower cabled the essential facts to the Adjutant General of the War Department, under General George C. Marshall and Secretary of War Henry Stimson, on May 19th, 1945:

"NUMBER OF U.S. PRISONERS OF WAR RECOVERED FROM GERMAN CUSTODY NOW IN U.S. OR BRITISH HANDS—80,000. THIS FIGURE INCLUDES 27,000 ALREADY EVACUATED TO US. NUMBERS OF U.S. PRISONERS ESTIMATED IN RUSSIAN CONTROL—25,000."[19]

This estimate by Eisenhower's staff of 105,000 known U.S. prisoners of war in Europe, including 25,000 in Russian hands on May 19th, was forwarded to Washington during the tense, May 16th-22nd, 1945, Halle Conference on POWs between Soviet and U.S. generals, near Leipzig, Germany. A subsequent classified "secret" report of the conference revealed that an estimated 90 Soviet personnel came to the meeting:

"The Russian party arrived in requisitioned German vehicles of all makes, an American-type armored car, fully equipped, and a radio truck, which...was in operation during most of the time. All Russian male personnel were heavily armed with pistols, submachine guns and rifles. The first conference was held in a former Luftwaffe officers club, 0900 hours, 16 May..."[20]

Now-declassified secret minutes of this critical meeting reveal that in addition to the American Chief of the delegation, Major General Ray W. Barker, Assistant Chief of Staff G-1 at SHAEF Headquarters, the Allied representatives also included civilian and military political advisers such as the State Department's G. Frederick Reinhardt (later a U.S. Ambassador to South Vietnam), British Assistant Political Adviser F.D.W. Brown, former-U.S. Ambassador to Poland, Anthony J. Drexel Biddle jr.—now a Colonel on Eisenhower's SHAEF staff, U.S. Brigadier General Stanley R. Mickelson, U.S. Major General David Schlatter of the Air Staff, British Brigadiers R.H.S. Venables—who was Barker's deputy, H.R. Carson, A.C. Salisbury-Jones and others. The senior Soviet members included Lt. General Golubev, Major General Ratov and Major General Dragun.[21]

The U.S. Chief of the Allied delegation, General Barker, later reported: "After opening statements...I proposed...prompt release and return to Allied control of all British and American prisoners of war then in Russian captivity, using air and motor transport. This proposal was firmly resisted by General Golubev, who...stated that serviceable air fields did not exist, which was known by myself to be not the case and I so informed him. THE RUSSIAN POSITION WAS VERY CLEAR, THAT NEITHER NOW, NOR AT ANY TIME IN THE FUTURE, WOULD THEY PERMIT ALLIED AIRPLANES TO BE USED FOR THE MOVEMENT INTO OR OUT OF THEIR TERRITORY OF PRISONERS OF WAR OR DISPLACED PERSONS..."[22]

Barker had demanded access to 32,000 U.S. and British POWs then being held by the Soviets at four of the many known POW locations in eastern Germany, named by Barker in the Halle minutes as: "Neubrandenburg, Luchenbach (Luckenwalde), Muhlberg and Muterborg." Golubev responded: "32,000? BUT HUNDREDS OF THOUSANDS OF SOVIET CITIZENS ARE STILL IN FRANCE..."

General Barker had stated that these were only "some" of the American and British POWs behind Soviet lines, and confronted Soviet Deputy Repatriations Commissar with the facts:

"There is a good airfield nearby, very near, and we understand these airfields are good ones. WILL YOU ALLOW US TO MAKE A RECONNAISSANCE OF THESE AIRFIELDS AND COME IN AND GET OUR PRISONERS OF WAR?...I MUST AGAIN INVITE YOUR ATTENTION TO

ARTICLE 4 OF THE YALTA CONVENTION WHEREIN THE RUSSIAN GOVERNMENT AGREED THAT THEY WOULD ALLOW THE ENGLISH AND AMERICANS TO COME IN AND GET THEIR OWN PEOPLE."[23]

The Soviet General, Golubev, responded: "THESE CAMPS EXISTED WHEN THE GERMANS WERE THERE...WE SET UP SOME NEW ASSEMBLY POINTS...LOCATED NOW IN THE REGION OF THE SEVEN POINTS I ALREADY MENTIONED..."[24] These locations included Crivitz, Magdeburg, Dessaw, Reisa, Torgau, Wismar and Parchim. The Halle Conference minutes thus confirm that some U.S. and Allied POWs had already been moved by the Soviets from their original camp locations (including a reported total of 15,000 U.S., British, French and Norwegian POWs at Stalag III-A, Luckenwalde), and that the Russians refused to let U.S. aircraft land and remove the prisoners from nearby airfields. (Excess prisoners in some of the camps may have reflected recent POW arrivals from German camps in the path of the Russian advance in Poland and eastern Germany.) Although the Soviets claimed to have been moving the U.S. and British POWs to points near the front line, the Allied command did not know how many had actually been moved in the opposite direction, towards Poland and Russia. In any case, the Soviets demanded the immediate return of all Soviet POWs and civilian DP's in U.S. and British territory, before the bulk of American and British POWs would be released.

Barker countered with a statement of certainty that American and British POWs also remained under Russian guard in the four camps he had named: "OUR PEOPLE ARE STILL THERE. AN OFFICER RETURNED FROM ONE OF THEM ONLY THREE DAYS AGO."

Golubev responded: "THERE WAS NO CAMP AT MUTERBORG AND THE CAMP AT MUHLBERG WAS TRANSFERRED. THERE ARE ONLY PLAINS THERE NOW."

Barker asked: "DOES THE GENERAL SAY ALL OUR PEOPLE HAVE BEEN MOVED FROM LUCHENBACH (Luckenwalde) NOW? Golubev responded: "YES." Barker: "AND MUHLBERG?" Golubev: "YES." Barker: "There were about 400 sick and wounded at Muhlberg. Are they still there?"

At this, Golubev became cagey, saying he wasn't sure, but added: "IF YOU CAN GIVE US THE NAMES OF THE SICK AND WOUNDED, I CAN TELL YOU WHERE THEY ARE." The Americans did not know all the names of all U.S. POWs in the German camps, which the Soviets were no doubt already aware of, and this was another hint of the future difficulties to be faced by the Allies. Barker then asked: CAN THE GENERAL TELL US WHERE THE PEOPLE ARE WHO WERE AT LUCHENBACH?" (Luckenwalde) Golubev would not reveal their location, promising to do so later.[25]

The conference continued in the same vein for days, with the Soviets eventually promising the Americans that U.S. contact officers would be permitted into the locations where Americans were held, a promise they later reneged on in every case but one. At the planning meeting on May 18th for working out deliveries of POWs of both sides at which Colonel (former Ambassador) Anthony J.D. Biddle Jr., was present, the Soviets complained that the Americans had not informed them of atrocities perpetrated by the Germans on Russian POWs. General Mickelson then stated that it was not the purpose of the conference to deal with Nazi atrocities. British Brigadier Venables also responded: "Every day, for weeks now, SHAEF has been notifying the Russian authorities of atrocities committed by the Germans. They are notified in Paris through Colonel Biddle's office to General Dragoon's (Dragun) office."[26]

After this, Venables, Mickelson and Biddle argued over numerous points and word changes to the agreement, while Soviet General Ratov and his deputies demanded entrance into the SHAEF zone for between 125 and 300 Soviet repatriation officers, through the fourth and fifth sessions of the conference. This was after they had consistently refused entrance for American and British contact officers into the Soviet zone, for months. Barker's deputy, British Venables negotiated an exchange of POWs from Stalag II-A, Neubrandenburg which occurred on the 18th and 19th, but records of returns indicate that over 1,000 Americans who had been held at that camp disappeared. Meanwhile, the Halle Conference continued. In the end an agreement was hammered out in which all POWs and displaced persons of both sides would be delivered through the army lines, and it was signed on May 22nd, but the Soviets failed to honor it in the weeks to come.

Declassified documents reveal that the Russians made a token release of 2,200 Americans across the front lines near the end of the Halle Conference, causing the skeptical U.S. military chief negotiator of the time, General Barker, to write: "There is every indication that the Russians intend to make a big show of rapid repatriation of our men, ALTHOUGH I AM OF THE OPINION THAT WE MAY FIND A RELUCTANCE TO RETURN THEM ALL, FOR AN APPRECIABLE TIME TO COME, SINCE THESE MEN CONSTITUTE A VALUABLE BARGAINING POINT."[27]

On May 23, 1945, more than two weeks after the end of the war in Europe, Supreme Allied Commander Dwight Eisenhower's chief POW negotiator with the Soviets at the Halle Conference, Major General Ray W. Barker, (Who was an old-line U.S. Army officer from the Cavalry branch), stated in his official, secret report, that remained classified for 44 years afterward, to SHAEF Chief of Staff (and later CIA Director) General Walter Beadle Smith:

"THE SHAEF REPRESENTATIVES CAME TO THE FIRM CONVICTION THAT BRITISH AND AMERICAN PRISONERS OF WAR WERE, IN EFFECT, BEING HELD AS HOSTAGES BY THE RUSSIANS... IT WILL BE NECESSARY

FOR US, THEREFORE, TO ARRANGE FOR CONSTANT LIAISON AND VISITS OF INSPECTION TO 'UNCOVER' OUR MEN."[28]

After General Golubev had promised at the Halle Conference to allow U.S. contact officers access to the more than 17,000 U.S. POWs then under Soviet control in eastern Germany, who had originally been held in five nearby German camps including Stalags III-A, IV-B, IV-G, IV-F, and others such as II-D and II-B, the Russians in fact allowed only one American contact officer to visit a single POW camp, after which he was ordered to leave Soviet-occupied territory by written order of General Golubev. In addition to the American and British prisoners within or moved from large camps such as the above, or in the area of Stalag II-A Neubrandenburg, recently declassified U.S. documents in the National Archives list hundreds of other locations overrun by the Soviets in eastern Germany, where small groups of 20-100 or more American and Allied prisoners had been held, and these many small groups could well account for several thousand more of the American POWs and MIAs who ultimately disappeared in Europe.[29] British Brigadier Venables negotiated an exchange of POWs with the Soviets at Stalag II-A Neubrandenburg during the Halle conference, and this exchange occurred on the 18th and 19th of May, but records of return indicate that over 1,000 of the 3,700 U.S. POWs at Neubrandenburg also remained in Soviet captivity.

As the Halle Conference in Germany was drawing to a close the American command appears to have realized at last that in all likelihood the Russians would retain many of their American POW hostages for an indeterminate period, and that the time had come to tone down the entire subject of American POWs in Soviet control. This was done by releasing to the SHAEF press corps a new version of the situation, in which it was now stated that most of the Americans had been recovered from the Russian zone. The New York Times, the most prestigious newspaper in America, accepted the sudden new line without question.

On May 22, 1945, the day the Halle Conference ended, the New York Times reported a May 21st SHAEF announcement that by May 17, nearly 63,000 of 75,850 Americans known to have been in German hands had been evacuated by air, and that the Americans had begun to leave the zone occupied by the Soviet Army. It was now reported that only 7,000 Americans remained in the Soviet zone of Germany, and that approximately 1,500 transports and bombers had been used in the evacuations. The report stated that official figures for the middle of March 1945 included 199,500 British (and Commonwealth) prisoners in German control. This original figure of the number of British POWs was soon to disappear from public view as the censorship tightened around Allied POWs inside the Soviet zone. Likewise, individual Allied POWs escaping or "exfiltrating" from

Soviet control were ignored when they reported many others remaining behind in a captive status.[30]

The statement that only 7,000 Americans remained in the Soviet zone was blatantly untrue. POW camps in Russian-controlled Germany at that time, such as Stalag III-A, Luckenwalde, Stalag IVB, Muhlberg, Stalag IV-G, Oschatz, Stalag II-D and others had contained many more than this number of Americans in Germany alone, who had not yet been returned, and who were in fact being held as hostages by the Soviets.

Stalag III-A, Luckenwalde, had at least 4,894 American POWs, of which the records indicate only some 1,100 were released by the Soviets.

Yet other records located by the author and research-colleague Ashworth in the Red Cross files at the National Archives indicate there may have been many more Americans than this at III-A Luckenwalde. A Red Cross response to a researcher 16 years after the war stated:

"Re: Your file no. G932-Rai, Stalag III A, Luckenwalde, Germany No records of this prisoner of war camp can be located in American Red Cross Archives. However, a review of our Prisoner of War Bulletins published during World War II did produce a few brief references to this camp. In the December 1944 issue of an article by Isabelle Lynn on 'German camp locations' but does list Stalag III A as one of the camps where Americans were being held. The February 1945 issue shows the camp on a map and a typewritten list in Red Cross Archives/ for that same month lists the camp stating it held BRITISH AND IRISH POWs and also included at present Stalag Luft 7.' The March 1945 issue states that late in January 1945 'most of the men/ American officers/ were moved from Szubin (Altburgund to Stalag III A.' The only description of life in Stalag III A found in our records was published in the May 1945 issue of the Prisoner of war Bulletin.

"When visited by a delegate of the International Red Cross on February 8th, Stalag III A at Luckenwalde was being used as a Dulag (transit camp) for prisoners of war evacuated from some of the camps in the 3rs, 8th, and 21st military districts. Of the 38,413 prisoners at Stalag III A on February 8, 5,716 were Americans-including 4,600 non-coms from Stalag III B and 266 officers from OFLAG 64. A later cable gave the American strength at at Stalag IIIA on February 28 as over 6,000. MANY MORE PRISONERS WERE STILL EXPECTED AT THIS CAMP., BUT, THE CABLE STATED, 'SEVERAL IMPORTANT GROUPS WILL BE TRANSFERRED ELSEWHERE.' ON MARCH 28, THE 4,600 NONCOMS FROM STALAG III B WERE REPORTED IN A NEW CAMP AT MARKISCH RIETZ, ABOUT 12 MILES WEST OF LUCKENWALDE. Stalag III A was reported to be overcrowded in February, with some recent arrivals being housed in tents. There were no epidemics..."[31]

Stalag IV-B at Muhlberg held some 10,000 American and British POWs at the time the Soviets took control, and these POWs were used as hostages at the time of the upcoming Halle Conference, but it appears that some 3,000 were never released. Stalag IV-G at Oschatz held 7,087 Americans when the area was overrun by the Red Army but only a few hundred U.S. POWs appear to have been returned to American control from this camp population. Stalag II-D held 6,894 Americans of which it appears that most were not repatriated. Other German POW camps from which few Americans ever reappeared included Stalag IV-A at Hohenstein and Oflag IV-B at Konigstein, both in the Dresden area. (A secret cable to London from Eisenhower of April 17th (FWD-19362) had warned that POWs from other camps such as Colditz had been moved by the Nazis to Stalag IV-B, Konigstein, south of Dresden, "to be held as hostages, on order of Himmler." Among them was Ambassador Winant's son, Lt. John Winant, and Captain Alexander. Movement of U.S., British and French POWs to be held as hostages continued to be reported until the Soviets overran the area, and it is not known how many disappeared forever. A new camp was also reported at Parchim, since the 19th of April, that was later taken over by the Soviets. Several of these camps also held British and French POWs.[32]

How many of these Americans were known by name to the War Department and how many were late-captures listed only as MIAs is as yet unknown. The declassified U.S. documents released thus far give only the total numbers of Americans interned there at given times up until the overrunning of these areas by the Soviet Army. In addition thousands more American POWs and MIAs not listed by name in Poland, Austria, Hungary and the Balkans were still in Soviet control.

Despite the press releases that the crisis over Americans in the Russian zone was almost over, General Marshall was by now seriously concerned about Soviet actions, as is indicated by another secret, priority message from him to Eisenhower dated 21 May 1945, (#W-85496):

"Concerned over report your S-88613 that 25,000 US prisoners still in Russian hands.

"Request completest details and when transfer to US control expected."[33]

From this point on the Allied command received more intelligence from the field that US and British POWs were being seen "moved to the east," or "toward Odessa," when Odessa was already in the process of being closed as a POW repatriation shipment point for US and UK POWs. Another declassified War Department cables of May 23rd stated:

"GENERAL MARSHALL IS PARTICULARLY CONCERNED WITH THE SHAEF REPORT THAT 25,000 US EX-POWS ARE UNDER RUSSIAN

CONTROL"

The reply to this question, a critical one in attempting to understand the confusion over the missing POWs, came from Eisenhower's Adjutant General for the European Theater, Brigadier General Ralph B. Lovett, a 54 year-old regular army staff officer (who, like his commander was a native of Kansas):

"MANY OF THESE PERSONNEL BECAME PRISONERS OF WAR FROM THE NORTH AFRICAN AND MEDITERRANEAN THEATRE. AND THERE HAS NOT THEREFORE BEEN ANY PREVIOUS RECORD OF THEM IN EARLIER RECORDS OF THIS THEATRE...THE ONLY SOURCE OF INFORMATION FOR OUR LEARNING POW STATUS IN PAST HAS BEEN FROM LISTS SENT HERE FROM TIME TO TIME BY CASUALTY BRANCH WAR DEPARTMENT."[34]

This message makes it absolutely clear that SHAEF Headquarters did not have all the names of POWs then in the European or Mediterranean theaters. Apparently full lists of the names of all POWs and MIAs were kept centralized under General Marshall's control in the Pentagon.

Although 2,200 American POWs were reported returned across the lines by the Soviets near the end of the Halle Conference, and an officially-reported total of some 5,241 Americans returned or exfiltrated from the Russian zone BETWEEN MAY 23rd AND MAY 28th, many thousands of other Americans from Soviet-controlled POW camps in Germany were withheld under guard in a number of locations, and eventually shipped east into Russia, where few were ever heard from again.[35] (The official total of some 5,241 Americans returned after the Halle Conference may be exaggerated by as much as 1,000 men, according to daily evacuation cables now declassified in the National Archives.[36] After May 28, 1945, declassified documents, such as the official Repatriation of American Military Personnel (RAMP) study, state that the repatriation of Americans from Soviet-occupied territory was "CONSIDERED COMPLETE."[37] Subsequently, on June 1, 1945, the American people were informed through the media, such as the New York Times of that date, that: "substantially all" U.S. prisoners in Europe had been returned, and that, "arrangements have been made with Soviet authorities for return of Americans liberated by the Red armies. Five exchange points have been established at Wismar, Wustmaru, Ludwiglust, Magdeburg and Leipzig."[38]

The sudden administratively-declared end to the return of American POWs from Soviet control in Europe and the simultaneous public announcement was backed up by a single SHAEF cable (#FWD-23059), also dated June 1, marked as signed by General Eisenhower, located in archival files which contain numerous other documents indicating that many thousands of American prisoners and MIAs were in fact still inside the Russian zone:

"The 25,000 prisoners of war reported in S-88613 as being in

Russian hands was an estimate of 19 May. It has been subsequently determined that of this number several thousands were in transit or already under U.S. control and not yet reported on nominal rolls on 19 May. The numbers reported as returned to U.S. control are as follows:

A. From 1 to 23 May, 23,421. Names of these were either processed and reported to the War Department by Machine Records cards or were included in cable transmissions which began with the large group on 22 May.

B. From 23 to 28 May, 5,241. Names have been cabled to the War Department immediately upon receipt, being included in the more than 12,000 names sent by cable during the period (23-31 May).

"IT IS NOW ESTIMATED THAT ONLY SMALL NUMBERS OF U.S. PRISONERS OF WAR STILL REMAIN IN RUSSIAN HANDS. THESE ARE NO DOUBT SCATTERED SINGLY AND IN SMALL GROUPS AS NO INFORMATION IS NOW AVAILABLE OF ANY LARGE NUMBERS IN SPECIFIC CAMPS. THEY ARE BEING RECEIVED NOW ONLY IN SMALL DRIBLETS...EVERYTHING POSSIBLE IS BEING DONE TO RECOVER U.S. PERSONNEL..."

This document and the RAMP study are presently used by the Pentagon (and the government-contracted RAND Corporation) to disprove the entire mass of U.S. and British documentary evidence and still-living human witnesses to the fact that many thousands of Americans remained in Russian control after June 1, 1945. Such usage, combined with the unrealistically sudden end to the return of Americans from the chaotic Soviet zone so soon after many had been reported held there in numerous locations, raises suspicion that this message was inserted in the SHAEF records at the time (or later) to end the POW/MIA problem bureaucratically, and thus justify the public announcement that all U.S. prisoners in Europe had returned, except a few stragglers. The simple fact that a large percentage of the 78,750 Americans who remain missing from WW II disappeared in the European theaters lends weight to this explanation. (To the time of publication of this book the U.S. Government has refused to release the names of the 78,750 U.S. missing of World War II to the author, so they could be checked against POW lists, in spite of freedom of information act requests and appeals to the Department of Defense and the archives.)

The now-declassified secret minutes of the 16-22 May Halle Conference and other documents in the National Archives listing POW populations and daily evacuations reveal that at least 17,000 Americans were held in nearby locations in Germany alone, of which, by the U.S. Government's own admission only 5,241 Americans, including stragglers and exfiltraters were returned following the Conference. The message could contain a germ of the truth if the Russians had already moved all LARGE GROUPS of U.S. prisoners eastward or even into Russia, as has been reported by eyewitnesses of the time. If such was the case, perhaps only scattered small groups and individuals were being reported through the limited first-hand intelligence sources then available to the U.S. command from inside the closed-off Russian zone. Otherwise, the June 1 message marked as signed by Eisenhower represents a deliberate distortion of the truth inserted in the records to justify the simultaneous public announcement that all American POWs had been returned.

In reality, at this same time in Austria, the Supreme Allied Commander Mediterranean Theater, (SACMED) Field Marshall Harold Alexander, was facing another Soviet POW blackmail threat. The Russians were demanding Allied recognition of their puppet Renner regime in Vienna, refusing entrance to prisoner of war camps and temporary POW concentrations by U.S. and British POW contact officers, and demanding forcible repatriations of anti-communist Cossacks and other "Soviet citizens," who had served under or been captured by the Germans. A now-declassified secret May 16, 1945 cable signed by Field Marshal Alexander (#FX 76272) to London and SHAEF Forward stated:

"Exchange of liberated PWs between 8th Army and Russians agreed...Understand exchange began 14th May at Graz from 2,000 fit 250 sick Russians immediately available from WOLFSBERG...Large numbers surrendered personnel supposedly of Russian origin ex German COSSACK DIVISION were in VOLKERSMARKT area on 12 May. RUMORED THAT 500 TO 1000 BRITISH PWS ARE INSIDE RUSSIAN LINES NEAR GRAZ. IF AND WHEN FREED WILL EVACUATE BY AIR FROM KLIGEFURT (Klagenfurt)."[39]

Meanwhile, in late May 1945, Patton's 3rd Army had located American and British prisoners being withheld after the war by the Soviets in eastern Austria and the American military high command secretly continued to try negotiating their release. Recently declassified secret U.S. documents in the National Archives record that General George Patton's Chief of Staff, Major General Hobart Gay, confronted the Russians at Linz, Austria, on May 23rd, and Soviet General Derevenko (later a commander of postwar Gulag camps) admitted holding the American and British POWs near Weiner Neustadt, farther down the Danube near the Hungarian border. An important meeting was held at Linz, Austria on the 23rd of May, in conjunction with the Halle meeting in Germany of General Barker, and the other similar POW conferences with the Russians in Austria at Wolfsberg and Graz. In the Linz meeting General George Patton's Chief of Staff, Major General Hobart R. Gay, in making a repatriation agreement with Soviet General Derevenko, succeeded in confirming that the Soviets were still holding American and British prisoners in Austria at Weiner-Neustad.[40] This was deep in the Russian zone south of Vienna and west of Bratislava, Czechoslovakia and close also, to the Hungarian border. Here, the Soviets had concentrated

60,000 former prisoners of several nationalities, and this was only one of several such concentrations in Austria. Among those who witnessed the large number of prisoners concentrated by the Russians at Weiner Neustadt, was N.N. Krasnov Jr., a son of one of the Cossack atamans from Wrangel's Don Army in the Civil War, who had fought on the German side in WW II, and who was later to be forcibly repatriated by the British in Austria into the hands of the NKVD.[41] This was at a time when Alexander had intelligence of the continued shipment of Allied POWs to the East or to "Odessa." As has been noted, the Soviets decision to close Odessa had been made on May 12th, and announced to General Deane in Moscow, who had already begun steps to remove the harassed American contact team there.

Prior to the Linz, Austria meeting American intelligence had received information that there were US POWs concentrated at Melk, Austria which the Soviet General Derevenko, of the 1st Russian Guards Army denied, but in so doing he admitted to another concentration:

General Gay asked: "DO YOU HAVE BRITISH AND AMERICAN PRISONERS OF WAR ON HAND AT MELK NOW?"

General Derevenko answered: "THE BRITISH AND AMERICAN PRISONERS OF WAR ARE NOT IN THE VICINITY OF MELK. THEY ARE FURTHER DOWN. THEY ARE IN THE VICINITY OF WIENER-NEUSTAD..."[42]

To the Soviet demands in Austria were added Tito's demands on Trieste and for the forcible repatriations of hundreds of thousands of anticommunist "Yugoslav citizens," many of them anti-communist Croatians and Royalists of Mikhailovich. Patton's 3rd Army was used in a threatening demonstration against Yugoslav forces in Austria during late May 1945 and as a warning to their Red Army supporters but there is no evidence that the majority of the Allied prisoners held by the Russians in Austria or those still in Yugoslavia were ever returned. A "Top Secret and Personal" message from President Harry Truman released from the White House Map Room, for Prime Minister Winston Churchill on May 20th, 1945, states:

"I agree we cannot leave matters in the present state. It seems our immediate action should be to reject Tito's answer as unsatisfactory and urge him to reconsider his decision. At the same time, I suggest we have Field Marshal Alexander, with assistance from General Eisenhower, immediately reinforce his front line troops to such an extent that the preponderance of force in the disputed area and the firmness of our intentions will be clearly apparent to the Yugoslavs...I SUGGEST THAT WE NOW DIRECT GENERAL EISENHOWER AND FIELD MARSHAL ALEXANDER TO PROCEED WITH THE IMPLEMENTATION OF A SHOW OF FORCE, BOTH AIR AND GROUND, and that the presentation in Belgrade of our rejection of Tito's stand be timed, if practicable, so that our commanders' troop movements will already be evident to Tito...IT MAY BE THAT A HEAVY SHOW OF FORCE WILL BRING TITO TO HIS SENSES...I MUST NOT HAVE ANY UNAVOIDABLE INTERFERANCE WITH THE REDEPLOYMENT OF AMERICAN FORCES TO THE PACIFIC. TRUMAN"

The primary policy consideration of finishing the war with Japan was thus still in the forefront of Truman's mind even as a confrontation between Yugoslav (and Soviet back-up) forces and Patton's Army in Austria ensued.[43] Tito backed down but his Soviet sponsors remained inflexible in their demands.

Field Marshal Harold Alexander (SACMED) cabled London (and Eisenhower at SHAEF) on May 22, 1945, after forced repatriations of Soviet citizens had already commenced in Austria (#FX-80335):

"AGREEMENT REACHED WITH...TOLBUKHIN'S HQ. AND GENERAL GRAZAKIN 57 ARMY GRAZ FOR HAND OVER TO THEM OF ALL SOVIET EX PW IN BRITISH ZONE. 2,874 TRANSFERRED 17/18 MAY. PERMISSION OBTAINED FROM LOCAL SOVIET COMMANDER FOR REPATRIATION DETACHMENT TO ENTER RUSSIAN ZONE BUT DIFFICULTY EXPERIENCED IN TRACING BR/US PW's. APPROX. 300 HAVE EXFILTRATED (escaped) INTO KLAGENFURT. UNCONFIRMED REPORTS SUGGEST BR/US PW STILL BEING EVACUATED ODESSA BY RAIL IN BOXCARS WITH GERMAN PWs. MANY INSTANCES OF THEFT OF CLOTHING AND PERSONAL EFFECTS."[44]

It seems clear in this message that the "unconfirmed reports" of the shipment of American and British prisoners east into Russia from Austria, with German POWs, after the Moscow-announced May 12 closure of Odessa as a repatriation point for Americans, came from the 300 Allied prisoners who had just escaped from the Russians into Klagenfurt, south of Judenburg where the Russian forced-repatriations took place.

An important cable from Field Marshal Alexander, sent in the midst of the forced-repatriations of Soviets to obtain release of American and British POWs in Soviet-controlled Austria, was widely distributed, to the U.S. and British embassies in Moscow, to Eisenhower at SHAEF and significantly, to the governments of Australia, Canada and New Zealand, on May 26, 1945 (#FX82606):

"AGREEMENT WITH RUSSIANS AT GRAZ ONLY APPLIES TO HANDING OVER SOVIET CITIZENS IN BRITISH ZONE, AUSTRIA. NO REPEAT NO RECIPROCAL GUARANTEES IN RESPECT OF BRITISH PW's OBTAINED EXCEPT HALF-HEARTED PROMISES WHICH HAVE SO FAR NOT BEEN HONORED. EVACUATION TO ODESSA STILL CONTINUING FROM THIS AREA."

This meant that Alexander had accurate intelligence that American and British prisoners were still being shipped east into Russia, after the Soviets themselves had announced the port of Odessa was closed for repatriation of U.S. and British POWs. He added:

"PREMATURE TO PLAN ON OVERLAND EXCHANGE ON A LOCAL

CONTACT BASIS TILL MOSCOW ISSUES DIRECTIVE TO GRAZ COMMANDER." (Meaning the Soviets were refusing to repatriate Americans and British POWs in Austria.)

"CONSIDERED ESSENTIAL MOSCOW BE ASKED TO ANNOUNCE THEIR AGREEMENT TO LOCAL OVERLAND EXCHANGE AS THERE ARE 15,597 U.S.A. ACCOUNT, 8,462 BRITISH ACCOUNT AWAITING REPATRIATION IN THIS THEATRE."[45]

On this same day, former US Ambassador to Moscow Joseph Davies (long sympathetic to the Soviet system and Stalin) spent the night with Winston Churchill as a personal emissary of Truman. This happened to be the same day that SACMED Political Adviser Harold Macmillan was ordered back to London on May 26, from the crisis atmosphere of Field Marshal Alexander's headquarters. (Churchill would have, of course, been receiving copies of the alarming telegrams from Alexander, concerning Soviet conduct in Austria.)

According to his own written report, Davies told Churchill that the British Prime Minister's recent warnings against the spread of Communism in Europe, and the threat of Soviet domination, put him in the same camp as Hitler and Goebbels. Churchill was furious and rejected indignantly the suggestion relayed by Davies that the President meet alone with Stalin somewhere in Europe before being joined by the Prime minister. Stalin meanwhile, had notified Churchill that he would be pleased to meet with him and Truman in the vicinity of Berlin in the "very near future." The original suggestion to send Davies came from Steve Early who saw it as a way of balancing Harry Hopkins' visit to Stalin.[46]

As a result of the violent hand-over of about 50,000 Cossacks in Austria who had fought against Stalin, in late May and early June 1945, Alexander received some 2,000 British POWs from the Russians in Austria, according to SACMED Political Adviser (and later British Prime Minister), Harold Macmillan; indicating an exchange ratio of 25-1. Macmillan remembered:

"The Yugoslavs were claiming part if not all of the Province of Carinthia...In this conflict for the possession of Klagenfurt and its surrounding area we had a few brigades to match 30.000-40,000 Yugoslavs...(General Keightey has to deal with nearly 400,000 surrendered or surrendering Germans, not yet disarmed...On his right flank Marshal Tolbukhin's armies have spread into what is supposed to be the British Zone of Austria, including the important city and road center of Graz. With the Russians there were considerable Bulgarian forces...Among the surrendering Germans there were about 40,000 (sic) Cossacks and White Russians with there wives and children. These were naturally claimed by the Russian commander, AND WE HAD NO ALTERNATIVE BUT TO SURRENDER THEM...AT LEAST WE OBTAINED IN EXCHANGE SOME 2,000 BRITISH PRISONERS AND WOUNDED..."WE ALSO PERSUADED THE RUSSIANS TO DELIVER

THEM IMMEDIATELY INSTEAD OF SENDING THEM, AS WAS THE ROUTINE PRACTISE, THROUGH ODESSA..."[47]

After June 7th there were still up to 6,000 British POWs inside the Russian Zone of Austria, but less than 1,000 were repatriated after that date, according to a June 7th SHAEF report.[48]

The Times of London reported on May 30th that, according to the British Secretary of State for War Sir James Grigg, "One hundred and fifty-six thousand British Commonwealth prisoners have now been repatriated, over 140,000 of them by air. About 10,000 are awaiting repatriation either in General Eisenhower's or Field-Marshal Alexander's zone and about 400 in Odessa. ABOUT 8,500 ARE IN THE PART OF AUSTRIA CONTROLLED BY THE RED ARMY, AND IT IS HOPED THAT ARRANGEMENTS WILL SOON BE MADE FOR THEIR TRANSFER TO THE BRITISH OR AMERICAN FORCES. The number of stragglers on the Continent is not likely to be large."[49]

In addition to this public announcement in the Times of London on May 30, 1945, General George C. Marshall cabled a top secret message (#WX-90429) to the senior American General in Alexander's command, Joseph McNarney, and also to Deane in Moscow and Eisenhower at SHAEF Headquarters:

"INFORMATION RECEIVED FROM BRITISH INDICATES 15,597 U.S. AND 8,462 BRITISH NOW IN MARSHAL TOLBUKHIN'S HANDS. UNDERSTOOD FURTHER THAT SOVIET COMMANDER PROPOSES TO CONTINUE THE EVACUATION OF THESE POWs TO ODESSA RATHER THAN REPATRIATE THEM OVERLAND AS HAD BEEN PROPOSED BY SOVIET GOVERNMENT AND ACCEPTED BY UNITED STATES AND BRITISH. REQUEST FULLEST INFORMATION URGENTLY."[50]

Whether these figures represent an accurate estimate or not, the fact remains that substantial numbers of American and British POWs were behind Soviet lines in Austria. Aside from some U.S. prisoners who disappeared after being held in regular German camps such as Stalag XVII-A Kaisersteinbruch, Stalag XVII-B Gneixendorf and Stalag XVIII-A at Wolfsberg, thousands of other American and British POWs who had been forced-marched from Poland through the Czechoslovakian mountains ahead of the January 1945 Russian offensive, some of whom had escaped, or were housed in temporary camps and bivouacs when overtaken by Soviet forces, also vanished. While the Soviets temporarily admitted the existence of U.S. and British POWs under their control at various locations in Austria, including the concentration at Weiner Neustadt, they refused US and British contact officers entrance into the camps holding these men (which resulted in Churchill's complaint to Truman), as they had previously in Poland and Germany.

General Deane wrote to Soviet General N.V. Slavin, Assistant Chief of Staff of the Red Army, about the reported 15,597 Americans in Russian controlled Austria, on May 31, emphasizing that SACMED intelligence was also reporting U.S. and British POWs were still being shipped east into Russia instead of being exchanged across the western lines, as had been agreed to:

"FIELD MARSHAL ALEXANDER HAS ALSO CABLED TO ME STATING THAT THE REPATRIATION OF PRISONERS OF WAR IN MARSHAL TOLBUKHIN'S AREA CONTINUES TO BE THROUGH ODESSA. WE HAVE BEEN REQUESTED BY THE SOVIET AUTHORITIES TO ABANDON OUR PRISONER OF WAR CONTACT TEAM ACTIVITIES IN ODESSA AND I HAVE ALREADY TAKEN STEPS TO ACCOMPLISH THIS..."[51]

The top secret messages about 24,000 American and British POWs in Soviet-controlled Austria caused consternation in the War Department, at a time when Undersecretary of War Patterson was publicly announcing the return of all U.S. POWs in Europe. General George C. Marshall demanded more information from SHAEF and from Deane in Moscow.

A cable of May 30th from SHAEF Headquarters (FWD 22790) again reiterates the connection between the repatriation of Soviet citizens in western Allied hands and the release of U.S. and Allied POWs in Soviet control:

"The airlift for Soviet citizens from France and Belgium is dependent on there being sufficient U.S./British PW's and Western European repatriates available at the Russian border zone for (the) return life (lift)."[52]

Unfortunately, the Soviet repatriations authorities and Beria's NKVD had already instituted their plans for retaining this U.S., British and western Allied "return life" for far more lucrative exchange rates.

During this same period of frantic American efforts to extract U.S. and British POWs from the Russian zone by air or on the ground, declassified documents indicate the operational officers were urgently asking General Eisenhower not to withdraw the Special Allied Airborne Reconnaissance Force (SAARF) Teams from active duty along the front, as had been suddenly and inexplicably ordered. On May 20th, even before the Halle Conference was concluded, General Eisenhower cabled the War Office in London: "Request you issue disbandment orders for Special Allied Airborne Recce Force...Disbandment to take place at Headquarters SAARF in United Kingdom commencing 20 June...Some personnel will be available for disposal earlier..."[53] These teams had been used in aerial insertions to assist in contacting and extracting POWs from what was now the Soviet zone, beyond the reach of U.S. and Allied ground forces. In a typical operation, "Six teams were dropped near Altengrabow after an escaped prisoner had reported that it was being used as a center for prisoners from five other camps..."[54] Eisenhower's order for disbanding SAARF resulted in an urgent plea from operational officers that the decision be reversed, at a time when many

American POWs were still held behind Russian lines:

"STRONGLY REQUEST PLEASE ARRANGE POSTPONEMENT WITHDRAWAL TEAMS FOR 1 MONTH. INVALUABLE MEANS COMN FOR PWX/DP MOVEMENT AND EXCHANGES..."[55]

This resulted in a message back from SHAEF Forward that, "G-3 state that disbandment of SAARF unlikely to be postponed beyond 1 July, but possible that position of SAARF might be transferred to 21 Army Group for retention as long as required."[56]

But, at SHAEF Headquarters plans were being implemented to officially announce that POW repatriations from the Soviets were complete. In a classified memo from the Prisoner of War Branch at SHAEF dated May 31, 1945 British Brigadier R.H.S. Venables, U.S. General Ray W. Barker's deputy, issued the orders:

"IT IS CONSIDERED THAT THE TIME HAS NOW ARRIVED WHEN CERTAIN DOMINION LIAISON OFFICERS MAY BE WITHDRAWN FROM THIS THEATRE IN THE IMMEDIATE FUTURE. PRISONERS OF WAR NOT YET EVACUATED AND IN RUSSIAN HANDS WILL BE DEALT WITH THROUGH EXCHANGE POINTS ALREADY AGREED UPON... THERE MUST INEVITABLY BE A NUMBER OF PRISONERS OF WAR STILL 'MISSING' THOUGH THE TRACING OF THESE WILL BE A LONG TERM PROJECT. IT IS INTENDED THAT THIS SHALL BE THE SUBJECT OF A FURTHER COMMUNICATION IN DUE COURSE, AND WHEN THE RELEVANT DATA OF THIS CATEGORY IS AVAILABLE FROM YOUR CASUALTY BRANCH..."[57]

In a top secret cipher telegram dated May 31, 1945 from the Joint Staff Mission in Washington to the British General Staff via AMSSO, the U.S. War Department questioned the reports from Field Marshal Alexander on the numbers of U.S. and British POWs remaining in Soviet-controlled Austria:

"Arguments in your telegrams were communicated to appropriate American authorities who;

"(A) Expressed incredulity at the statements in paragraph one that 15,597 Americans and 8,462 British PW's released by Marshal Tolbukhin were still in Soviet hands and that Marshal Tolbukhin intended to continue to evacuate British and Americans to Odessa..."

"With regard to paragraph one (A) above, the War Department inquired position from AFHQ who have today replied that the figures quoted by you refer to Russian Nationals held by U.S. and British forces in Italy and not to American and British liberated PW's in Russian hands. They also showed our representatives a message from Major General John Deane to General Marshall of which the following are extracts:-

"Deputy Head-Soviet Repatriation Committee informed me on 12 May that henceforth all repatriation of American PW's would be westward and overland. Admiral Archer was informed by Soviet Repatriation Committee on 27 May that instructions would be issued to stop sending British Officers, soldiers and civilians to Odessa, also that henceforth those Allied PWs liberated in Southern Regions would be despatched to Graz to be handed over to Eighth Army."⁵⁸

General Deane, head of the U.S. Military Mission in Moscow, does not seem to have been informed that the U.S. and British POWs reported in Soviet-controlled Austria were now officially considered to have been Russians. On the night of June 1st, Deane had an opportunity to confront General Golikov, Stalin's Repatriations Commissar, during a dinner given by the Soviet dictator for Truman's emissary, Harry Hopkins. Deane cabled General Marshall from Moscow on June 2nd (M-24524) in top secret:

"General Golikov, Head of the Soviet Repatriation Commission, assured me last night that they would make every effort to locate individual Americans who might be stranded in Soviet-occupied territory and when located, they would be evacuated by any route that I requested...CONCERNING THOSE LIBERATED PRISONERS OF WAR IN MARSHAL TOLBUKHIN'S AREA, ESTIMATED IN EXCESS OF 15,000, GOLIKOV ASSURED ME THAT THEY WOULD BE EVACUATED WESTWARD IN ACCORDANCE WITH THE HALLE AGREEMENT. HE CONFIRMED MY PREVIOUS BELIEF THAT ODESSA WAS TO BE ABANDONED AS A TRANSIENT CAMP FOR THE REPATRIATION. OF AMERICAN PRISONERS OF WAR."⁵⁹

Meanwhile, according to a declassified cable dated June 2, 1945 (released by the Pentagon in 1989 handouts), the War Office asked Alexander's Headquarters for confirmation on the figures for American and British POWs reported in Soviet-controlled Austria:

"Do figures 15,597 USA accounts, 8,462 British accounts refer to numbers American and British PW in Soviet hands or Soviet nationals held by U.S. and British forces in Italy awaiting repatriation. If latter request (a) numbers of British Commonwealth PW hitherto transferred to you by Marshal Tolbukhin (b) your estimate British Commonwealth PW still in Tolbukhin's zone."

The response by AFHQ in Alexanders theater to this message was also released by the Pentagon in 1989, which reported that the document was obtained from the British Public Record office. It is a secret cipher telegram dated June 3rd:

"Figures did refer Soviet citizens then held by U.S. and British. NO BRITISH COMMONWEALTH PW TRANSFERRED TO US BY MARSHAL TOLBUKHIN THOUGH SMALL PARTIES HAVE COME THROUGH RUSSIAN LINES. NO KNOWN ESTIMATE OF BRITISH COMMONWEALTH PW STILL IN TOLBUKHIN'S ZONE."⁶⁰

While this message would appear to set the record straight, other declassified U.S. and British documents indicate that at his time some 100,000 Russians were in fact under control of the Americans and British in AFHQ area of Alexander's command. Likewise the actual number of British and American prisoners in Tolbukhin's zone is not revealed within, but if this cable is correct, they were definitely not being released by Tolbukhin as of June 3rd. Alexander's political adviser, Harold Macmillan, reported in his memoirs that Tolbukhin eventually released 2,000 British POWs in Austria for the 40,000-50,000 forcibly-repatriated Cossacks. Without saying outright that British POWs had been withheld by Tolbukhin's forces, General of the Soviet Army S.M. Shtemenko, of the Soviet Stavka or Soviet Army General Staff, later recorded the Russian explanation of how Stalin's government forced the British to forcibly repatriate the Cossacks, and others:

"In the foothills of the Alps some ancient enemies of the Soviet regime were discovered: Generals P.N. Krasnov, A.G. Shkuro, K. Sultan-Girei, and others...Krasnov, former commander in chief of all the armed forces of Kerensky's provisional government and ataman of the 'Don Host'...Shkuro, former commander of the 3rd Cavalry Corps of Denikin's army...Sultan, former prince, strangler of the 1905 Revolution and commander of the 'Savage Division'...THE SOVIET GOVERNMENT MADE A FIRM DEMARCHE TO OUR ALLIES IN THE MATTER OF KRASNOV, SHKURO, SULTAN-GIREI AND OTHER WAR CRIMINALS. THE BRITISH STALLED BRIEFLY; BUT SINCE NEITHER OLD WHITE GUARD GENERALS NOR THEIR TROOPS WERE WORTH MUCH, THEY PUT THEM INTO TRUCKS AND DELIVERED THEM TO THE SOVIET AUTHORITIES...ALL THE GENERALS IN THIS FOUL-SMELLING 'BOUQUET' WERE TRIED BY A SOVIET COURT AND SENTENCED TO DEATH."[61] Despite the forcible repatriations of Krasnov and the other generals and an estimated 50,000 of their troops and dependents, the Soviets failed to reciprocate by releasing all British POWs in their zone of Austria, as will be seen.

The U.S. Defense Department has claimed, without substantiation, that reports of American POWs disappearing in Soviet-occupied Austria are false. Yet, in a June 6, 1945 meeting in Moscow, General John Deane was still requesting information from the Soviets about a by-then-reduced estimate of 5,500 U.S. POWs in the Russian zone of Austria, as will be seen. Although some 4,000 known-U.S. POWs were returned from Soviet-occupied Austria, it appears that thousands more Americans, who may have been carried as MIAs or POW/MIAs disappeared in Austria.

That same day, the New York Herald Tribune featured a report by it's correspondent, Carl Levin at SHAEF Headquarters entitled, "25,000 Missing U.S. Soldiers Turn Up Alive." The report quoted SHAEF Lt. Col. W.P. Schweitzer (a New York paper manufacturer in civilian life) as stating that of nearly 90,000 U.S. POWs recovered from German POW camps by that time, some 25,000 had been listed as MIAs (not known by name to have been prisoners of war). 77,500 U.S. POWs had been listed by name as held by the Germans in the European Theater in March 1945, which would indicate that on June 6th, approximately 12,500 U.S. POWs, known by name to have been prisoners, had failed to return from the Soviet occupation zone of Germany. Five months later the number of American POWs still carried as "not yet returned to military control" from the Soviet-occupied zone of Germany had been reduced by the return of some stragglers, and by the Presumptive Finding of Death process (PFOD) to 6,125 prisoners.[62] In addition, 20,351 Americans were listed at the end of 1945 as having been captured in the Mediterranean Theater, of whom 19,753 were returned, but many had been carried as MIAs and on October 31, 1945, only 470 of these were still carried as "POWs not yet returned to military control."[63] Contemporary news accounts often use the figure 76,000 known U.S. POWs in German hands. Lt. Col. Schweitzer was quoted as saying,

"IT IS BELIEVED THAT THE 89,776 AMERICAN PRISONERS OF WAR RECOVERED FROM CAMPS IN GERMANY CONSTITUTE THE ENTIRE AMERICAN PRISONER GROUP WHICH WAS IN NAZI HANDS DURING THE FINAL STAGES OF THE WAR."[64] A total of 78,750 Americans remain missing in action from WW II, most from the European theaters. The publication of this story served to deflect questions about American prisoners remaining in Soviet territory after this date. Other U.S. officers were not permitted to tell the truth because of security regulations. (Levin was to achieve fame later in 1945, for contributing to the removal of General George Patton as Commander of 3rd Army and military governor of Bavaria, through his published charges that Patton was sympathetic to Nazis and hostile to Russia.)

Although 5,241 Americans had been officially reported by the U.S. as returned by the Soviets in Germany after the 16-22 May Halle Conference, between May 23rd and May 28th, available documents declassified thus far indicate that approximately 14,000 or more U.S. POWs and POW/MIAs disappeared in Soviet-occupied Germany, along with the 2,000 or more Americans who had vanished previously in Poland and an estimated 5,500 (to as many as 15,000) U.S. POWs and POW/MIAs who never returned from Soviet-occupied Austria, Hungary, Czechoslovakia, Romania, Bulgaria and Yugoslavia.

All U.S. Ex-POWs were screened by the intelligence service MI S-X and other intelligence branches who saw to it that each ex-prisoner signed a pledge not to disclose any information on methods or routes of escape or any other facts concerning experiences as a prisoner. ALSO, EACH PRISONER WAS WARNED IN THE PLEDGE TO BE PARTICULARLY ON GUARD AGAINST SPEAKING TO ANYONE REPRESENTING THE PRESS AND FORBIDDEN TO GIVE ANY ACCOUNT OF PRISONER EXPERIENCES IN BOOKS, NEWSPAPERS, PERIODICALS, BROADCASTS OR LECTURES. The pledge threatened disciplinary action against ex-prisoners for any disclosure of information.[65] The author has learned from ex-POW informants who escaped from Soviet control-leaving other Americans behind in captivity, that threats like this, and genuine concern for future possible American escapees ensured their continued silence; in the belief also, that higher authorities were secretly negotiating for the POWs release.

The crisis over Soviet-withheld U.S. and Allied POWs in Poland, Germany and Austria was thus kept secret from the American people, who had been assured on May 31st, 1945, by Henry Stimson's Undersecretary of War, Robert Patterson, that all U.S. prisoners of the Germans had been returned, except for a few stragglers. Yet a now-declassified SHAEF memo of the day before, to Eisenhower's chief surgeon at SHAEF, General Kenner, reported that "20,000 American prisoners" and "20,000 British prisoners" still remained in Russian control, along with hundreds of thousands of French, Dutch and Belgians and other Allied prisoners then in the closed-off Russian zones.

Clues to a possible second Truman policy decision concerning POWs in Russian control oh or about May 30-31, that General Eisenhower and Field Marshal Alexander would have had to adhere to, may be revealed by the events of those two days, when it had become clear in Washington, D.C., through urgent cables from U.S. commanders in Europe, that hundreds of thousands of U.S. and Allied prisoners were being secretly and illegally withheld by Soviet authorities.

Harry Hopkins, an intimate adviser to Roosevelt and negotiator for Truman who expressed admiration for Stalin, exhibited an ardent desire to support Roosevelt's dictate of maintaining Allied unity with Russia at all cost, and sometimes appeared to act more on behalf of the Soviet Union than for the United States. His secret late-May/early-June 1945 meetings with Stalin in Moscow at the behest of President Truman, as US POWs were disappearing in Russian-occupied Austria and Germany, may have been decisive to the ultimate outcome of the Soviet POW blackmail operation. Declassified minutes of these meetings do not record any forceful request or demand by Hopkins on behalf of the United States Government that Soviet-held American POWs be repatriated. In a top secret, eyes-only message to President Truman from Moscow on May 31, Hopkins merely wrote:

"YOU CAN BE SURE THAT AT YOUR NEXT MEETING (Potsdam) STALIN WILL HAVE SOME PRETTY SPECIFIC PROPOSALS TO MAKE ABOUT PRISONERS OF WAR, AND MORE PARTICULARLY, I BELIEVE ABOUT WAR CRIMINALS. HE DID NOT AS WE ANTICIPATED, EXPRESS ANY CRITICISM OF OUR HANDLING OF WAR PRISONERS."[66]

In a previously unknown two and a half-hour conference on June 10th, between Harry Hopkins and former U.S. Ambassador to Poland, Colonel Anthony J. Drexel Biddle Jr., (who had been at the Halle Conference with General Barker), Truman's personal envoy to Stalin referred in couched diplomatic terms to the Polish crisis and the American and allied POWs who had disappeared in Soviet-occupied Poland:

"In a two and one half hour conversation with Mr. Harry Hopkins, this date, the following were the main points we discussed: Russian-American relations in terms of American public opinion. Mr. Hopkins said that President Truman had called him in to go immediately to Moscow to have a frank talk with Marshal Stalin. In requesting him to go, the President said he had no definite directive to give him, OTHER THAN TO ASK HIM TO STATE THE CASE CLEARLY AND FRANKLY TO MARSHAL STALIN. The President said he wished to carry on President Roosevelt's policy of friendship with Russia, and of a common search for a friendly and practical modus vivendi. He was not willing, however, to stand idly by with his hands in his pockets while the Russians wrote their own ticket contrary to American traditional principal. He would like to know what, in the back of Marshal Stalin's mind, were his actual aims; what was he shooting at; was he willing to impart this frankly and was he just as desirous as the President to find a modus vivendi. DID HE REALIZE THAT PUBLIC OPINION IN THE UNITED STATES WAS REACTING VERY UNFAVORABLY THIS TIME TOWARDS CERTAIN ACTIONS OF THE RUSSIANS; that this only served to torpedo the administration's efforts to bring about a better understanding between the two countries...Poland...was a matter of urgent consideration as it had become an outstanding test case in the minds of the public...WHAT WAS THE MARSHAL'S ATTITUDE TOWARD POLAND AND THE POLISH PEOPLE AND THE POLISH PRISONERS OF WAR GOING TO BE. Could he and would he define those points. The President thought that if Mr. Hopkins shared the above reactions, he might care to emphasize to Marshal Stalin that he personally associated himself with this trend of public opinion."

"In his several conversations with Marshal Stalin, Mr. Hopkins had followed to the letter the President's views, as above expressed, and had furthermore given his own observations on the trend of American opinion, associating himself definitely with those views. In reciting the above, Mr. Hopkins added that the President had made it clear to him that he was not completely familiar with the continuity of the late President's contacts with Marshal Stalin, and that both he and the former Secretary of State, Mr. Hull, had agreed that Mr. Hopkins was perhaps the only person who was entirely familiar therewith. The President furthermore felt that if Mr. Roosevelt were alive at this time, he would share his views, i.e., that notwithstanding whatever Marshal Stalin had previously agreed on, the impression was rapidly gaining ground in America that Moscow was letting Washington down badly, and was acting contrary to the spirit of whatsoever agreement might possibly have been made..."[67]

According to the 1945 U.S. Political Adviser to General Eisenhower in Europe, Robert Murphy, Hopkins was "bubbling with enthusiasm about his meetings with Stalin," and he remembered that Hopkins had "changed our perspective at SHAEF." Hopkins told Murphy (and Eisenhower): "We can do business with Stalin! He will Cooperate!"[68] Yet, few of the 20,000 POWs and MIAs who disappeared in Soviet control after May 30, 1945, were ever returned, and on June 1, despite massive evidence to the contrary, General Eisenhower, Undersecretary of War Robert Patterson and the major American media were all suddenly saying, in concert, that nearly all U.S. POWs in Soviet control, except for small, scattered groups or individuals had been returned. The fact that there were small, scattered groups in hundreds of locations in Poland, eastern Germany, Austria and elsewhere in southeastern Europe was downplayed or totally ignored, and not revealed to the American people.

The full role of Harry Hopkins in the postwar POW/MIA crisis with the Soviets may never be known. Hopkins was terminally ill and died in 1946. A later Soviet defector, Colonel Oleg Gordievsky revealed that early in his KGB career he had attended a lecture given in Moscow's Lubyanka Prison which was given by the former top NKVD illegal in the United States during the war years, Iskhak Abdulovich Akhmerov, who was reported to have been the wartime controller of Alger Hiss, Elizabeth Bentley and other Soviet agents. Akhmerov used the aliases Michael Adamec, Michael Green and Bill Greinke. As early as 1938 he had met in the U.S. with Michael Straight, the owner and editor of the New Republic whom Soviet agent Anthony Blunt had already tried to recruit for the NKVD at Cambridge. In his postwar lecture attended by Gordievsky, Akhmerov stated that the most important wartime agent in the United States for the Soviet Union was Harry Hopkins, although Gordievsky concluded from what he learned that Hopkins had been an unconscious rather than a conscious agent, whose devotion to Stalin may have swayed his sensibilities. In his contacts with Hopkins from 1941-1945, Akhmerov usually said he brought personal and confidential messages from Stalin.[69]

This accusation has been disputed by members of the Hopkins and Harriman families, who contend that Hopkins was merely keeping Roosevelt's policy of getting along with the Russians for the sake of the wartime alliance, foremost in his mind. It must also be remembered that Hopkins had a son killed in action fighting the Japanese in the South Pacific, and as a lifelong liberal, he had a deep and abiding fear of the Fascist and Nazi dictators, while he had long sympathized with the ideals espoused by the first "workers state."

Meanwhile, the hundreds of thousands of missing French, Belgian and Dutch POWs were also vanishing inside Russian-occupied territory. While Hopkins was meeting with Stalin, French General Cherrière reported to SHAEF headquarters on May 30th, 1945, nearly a month after the fighting in Europe had ended:

"...ACCORDING TO CONFIRMED REPORTS, RUSSIANS STILL DO NOT RELEASE THOUSANDS OF FRENCH EX-POWs AND CIVILIANS, FORCING THEM TO WORK. MANY TRANSFERRED EASTWARDS TO UNKNOWN DESTINATION. PLEASE INFORM HIGH AUTHORITY."[70]

The Soviets were now in the process of denying the existence of over one million French, Belgian, Dutch and other Allied POWs and civilian political deportees, known to have been held in German camps now inside the Russian zone. On May 31st, 1945 a secret message to Harriman's assistant in Moscow, General Deane from the Supreme Commander Allied Expeditionary Force (SCAEF) General Eisenhower, admitted the disappearance of over a million Western Europeans in the Soviet zone and added:

"...THIS DISCREPANCY OF OVER A MILLION WESTERN EUROPEANS IS CAUSING THE DUTCH AND FRENCH GOVERNMENTS CONSIDERABLE ANXIETY. CAN THE SOVIETS CLARIFY SITUATION...?"[71]

While some may have died in German captivity, subsequent investigation revealed that an estimated 500,000 of these western European prisoners who were believed by their governments to be alive after the dead were accounted for, simply vanished behind the Russian lines. The Soviets stonewalled for years on the return of these prisoners and an unknown number of them were eventually to disappear into Stalin's secret gulags. Many probably died at forced-labor on starvation rations during the first few postwar winters among the millions of German, Austrian, Japanese and other foreign POWs in Soviet POW camps. A secret, June 1st, 1945 report by Donald Heath, Deputy to US Political Advisor Robert Murphy, issued the same day the American people were told that all U.S. POWs in Europe had been returned, illustrated the process of disappearance:

"...ALTHOUGH GENERAL GOLUBEV WOULD NOT AGREE TO FIRST PRIORITY DELIVERY OF US AND UK EX-PRISONERS OF WAR, HE GAVE HIS MOST SOLEMN PERSONAL ASSURANCES THAT ALL US AND UK EX-PRISONERS WOULD IN FACT BE GIVEN PREFERENTIAL TREATMENT. A REQUEST FOR SECOND PRIORITY FOR WESTERN EUROPEAN EX-POLITICAL DEPORTEES...WAS COUNTERED BY THE FLAT ASSERTION THAT 'ALL' POLITICAL PRISONERS HELD IN GERMAN CONCENTRATION CAMPS HAD BEEN RELEASED AND THAT THERE WERE, ACCORDINGLY, NO MORE POLITICAL PRISONERS IN SOVIET-OCCUPIED TERRITORY. WITH RESPECT TO THIS CATEGORY OF DISPLACED PERSONS, NOT EVEN VERBAL ASSURANCES WERE TO BE HAD."[72]

Thus, Eisenhower's June 1 report of "scattered small groups," of American prisoners in the Russian zone could represent an attempt to conform to another as-yet-undocumented Presidential policy decision made by Harry Truman at the end of May, in response to Soviet intransigence on returning prisoners, that there would be no military confrontation with Stalin over the hundreds of thousands of missing U.S. and Allied POWs in Soviet-occupied Europe. A top secret priority message from General George C. Marshall in Washington to General Deane in Moscow with copies to Eisenhower and McNarney, sent on May 31 and received on June 1st, has the appearance of a policy directive:

"Request report of interviews with Russians include information concerning INDIVIDUAL U.S. STAGGLERS WHO MAY BE IN RUSSIA and arrangements to be made for return such personnel to U.S. control."[73]

A message such as this from the Chief of Staff to Deane at the time of Harry Hopkins' meetings with Stalin, would have had the effect of limiting the seriousness of the prisoner of war problem, for the record, and would have instructed Deane on how Washington viewed the matter, notwithstanding what he had learned about the missing U.S. POWs since the time of the Polish crisis.

POWs and repatriations were also a secret topic of the 1945 founding UN Conference in San Francisco, while "Poland" was the topic most often released to the public and where the Soviets raised their demand for a $6 Billion U.S. war reconstruction loan, diplomatic recognition of the Polish regime and forcible repatriations of Soviet citizens. Allied leaders continued to confront the Soviets with evidence that large numbers of Western Allied POWs were in their control, which the Soviets denied and countered with demands for postwar reconstruction aid, or U.S. diplomatic recognition of the Soviet-puppet regime in Poland, and forcible repatriations of Soviet "citizens."

By early June 1945, 25,000 late-captured U.S. MIAs whose names had not been carried in a known-POW status, had been recovered in Europe. But later declassified documents indicate that at least 12,500 of the 77,500 German-held U.S. POWs known by name to have been held in the German Theater, and 10,000 or more U.S. POW/MIAs known by number to be in captivity (but possibly not by name) in Germany and Austria, in the Mediterranean Theater, disappeared inside Soviet-occupied territory.

On June 2nd, while Truman envoy Harry Hopkins was departing Russia, Stalin's long-time intelligence aide, General Filip Golikov, an NKVD man hated and feared by regular Soviet military officers and according to Nikita Khrushchev, cashiered from combat command for cowardice at Stalingrad, had again charged in a Moscow meeting with General Deane that the Allies were "abusing" liberated Soviet POWs, contrary to what Hopkins had just cabled the President. On June 7, Allied officials stated they were "puzzled" by Golikov's charges, and on June 11, Britain denied Golikov's charges and hinted that they reflected Soviet resentment about the western Allies failure to repatriate Polish people and citizens of the Baltic republics of Estonia, Latvia and Lithuania, now part of the USSR, but not recognized as such.[74]

The June 2nd and June 3rd secret cipher telegrams between the War Office, AFHQ-SACMED and Washington cited by the Pentagon and earlier in this article indicate that while the numbers of POWs in Tolbukhin's control in Soviet-occupied Austria were not officially admitted to be 15,597 Americans and 8,462 British, there was also, as yet, "no known estimate of British Commonwealth PW still in Tolbukhin's Zone." Alexander's command was also reporting on that date that, "No British Commonwealth PW transferred to US by Marshal Tolbukhin, though small parties have come through Russian lines." These ex-prisoners however, were largely escapees, known as "exfiltraters," and did not represent Soviet reciprocation on POW exchanges.

One of the few reports that passed the censors and was published in the American press at this time that depicted Soviet conduct in their zone of Austria as thousands of American, British and allied prisoners disappeared behind Russian lines, appeared in the Chicago Daily Tribune on June 5th, and revealed the U.S. POW concentration at Melk which had been confirmed a fortnight earlier by Third Army's General Hobart Gay:

"Americans trying to enter the part of Austria occupied by Russia are meeting with as much frustration as those trying to get home without those 85 points. Unsmiling Russian soldiers, with tommy guns ready, guard every bridge, every highway, and every byway. It takes a special pass from Marshal Tolbukhin to get by them and such passes are hard, if not impossible, to obtain. THE ONLY AMERICANS KNOWN TO BE IN THE RUSSIAN ZONE OF AUSTRIA ARE SOME OFFICERS AND MEN AT MELK, MIDWAY BETWEEN THE AMERICAN LINES ON THE ENNS RIVER AND VIENNA. Heads are counted for convoys enter there almost every day, and every driver must be accounted for when they return."[75]

During a meeting in Moscow on June 6th, in response to Major General John Deane's remonstrance that an estimated 5,500 U.S. POWs (reduced from the earlier reported 15,597 estimate) remained under Marshal Tolbukhin's control in Soviet-occupied Austria, the Soviet Deputy Repatriations Commissar, General Golubev, off-handedly denied that Tolbukhin was holding any American prisoners in Austria, or that any Americans had been shipped east into Russia from Austria, which was contrary to the Allied intelligence repeatedly reported by Field Marshal Alexander from late May into early June 1945:

"GENERAL GOLUBEV SAID THAT HE WAS SURPRISED TO HEAR THE CLAIM THAT THERE MIGHT BE 5,500 U.S. PRISONERS STILL IN MARSHAL TOLBUKHIN'S AREA, AND THAT THESE WERE SUPPOSEDLY BEING EVACUATED THROUGH ODESSA. GENERAL GOLUBEV SAID THIS WAS NOT SO."

Deane retorted that he, "had an opportunity to speak with General Golikov (Stalin's chief Repatriations Commissar) the other night at the reception which was given for Mr. Hopkins, and that General Golikov said the remaining U.S. stragglers and prisoners of war being released from hospitals may be evacuated westward by any means which General Deane might desire."[76]

The head of the U.S. Military Mission in Moscow also raised the issue on June 6th, of hundreds of thousands of western Allied POWs and civilian deportees who had disappeared in Soviet-occupied territory:

"General Deane indicated that the representatives of Holland and Belgium believed that they had more war prisoners to be repatriated than the combined sum indicated by both the Allies and the Russians."

Here Golubev showed his teeth, claiming that the Soviets had overrun only 600,000 of these prisoners, total (whereas over 1,000,000 were then missing in eastern Europe), of which he claimed 300,000 had been returned, and revealed one of the Soviet motives for retaining U.S. and Allied POWs:

"GENERAL GOLUBEV SAID BELGIUM AND NORWAY SHOULD NOT WORRY ABOUT REPATRIATION OF THEIR CITIZENS, BUT THAT IT WOULD BE WELL FOR THEM TO CONCERN THEMSELVES WITH SPEEDING UP REPATRIATION OF SOVIET CITIZENS FROM THEIR COUNTRIES."[77]

The Russian Commissar's apparently unrelated mention of Norway is significant when viewed with a declassified secret cable (#M-24293) of May 13, from General Deane in Moscow to Eisenhower at SHAEF Headquarters:

"Norwegian General officer, who has been a prisoner of war at Luckenwalde (Stalag III-A), south of Berlin, has arrived in Moscow. He states that conditions in Luckenwalde are extremely bad. Camp Commandant is a major who is usually intoxicated and treatment of British and American prisoners is not good. He states that recently a great many British and American prisoners were flown out in American transports against the Russians wishes and that the Russians have vented their resentment on the remainder.

"The Norwegian does not want his name used for fear of the effect it might have on Russian treatment of other Norwegian prisoners; however, if you have any supporting data that you can furnish, I would appreciate your sending it to me so that I can put pressure on the Russian repatriation committee."[78]

(Stalag III-A, Luckenwalde was a Soviet-controlled German POW camp from the administrative area of which thousands of American, British and other western Allied POWs disappeared.)

Were the Soviet Generals Golubev and Golikov, who actually answered to NKVD Chief Lavrenti Beria and Josef Stalin, thus sending a message through the American General Deane on June 6th, for the Norwegians, and by extension all western allies, that not only were their secret cables and confidential dealings with the Americans compromised, but also that they must look to themselves, not the Americans, for return of their own missing prisoners?

This kind of pressure resulted in total Norwegian compliance later in 1945, in the repatriations of over 80,000 Soviet "citizens," including Baltic people and Poles, who had been transported by the Germans to Norway.[79] Some were shot by NKVD firing squads near the docks of Murmansk, while most were reported by Harriman and others to have been shipped east to the forced-labor camps of the Gulag. As an apparent result, relatively few Norwegians vanished inside Soviet-occupied territory.

General Filip Golikov, who, according to Deane and others, was a despotic physical giant, and Stalin's right-hand man on U.S. and British POW repatriation matters, had repeatedly blocked the return of many thousands of POWs then in Soviet control. When questioned about this episode decades later, Harriman's former deputy in Moscow (and later Ambassador to Russia), George Kennan, wrote to the author:

"I thought it highly probable that General Golikov, while wearing a regular military uniform, was actually a high official of the NKVD (KGB), and not of the regular Russian military authorities. I believe I made my view plain to General (John) Deane, but that he felt he had no choice but to deal with the man...I resented this form of obfuscation and saw it as an attempt to prevent contact between our Military Mission and the regular Russian military establishment."[80]

Golikov had also been hated and feared by other Red Army officers as Stalin's Chief of Army Intelligence, and was responsible for past battlefield blunders which had cost the Red Army dearly. He was himself responsible for many Soviet POWs being taken prisoner, yet it was they who later paid a terrible price for having been captured. In this regard it is interesting to note former Soviet Premier Nikita Khrushchev's personal attack on Golikov in his later memoirs, wherein he accuses Golikov of supreme cowardice and panic at Stalingrad, and of doing the Red Army no good, whereupon he was relieved of command on the battlefield and ended up in Stalin's Moscow clique.[81]

Meanwhile, in the United States, J. Edgar Hoover's FBI had arrested six persons connected to a pro-Communist magazine called "Amerasia," and had charged them with spying for the Soviet Union: An Office of Naval Intelligence (ONI) officer, Lt. Andrew Roth; a State department Foreign Service Officer, John Stewart Service; another State department officer who served as liaison to ONI through Roth, named Emmanuel Sigurd Larsen; "Amerasia" editor Philip Jacob Jaffe and "journalists" Mark Julius Gayn and Kate

Louise Mitchell. The OSS had suspected leakage of official secrets to the magazine from government sources due to the content of some of the articles, and a clandestine OSS team entered the offices and confirmed the presence of classified documents, in March 1945, when the case had been turned over to the FBI for investigation. A pro-Chinese Communist and pro-Soviet publication, Amerasia's offices were found to contain over 1,000 classified documents from the State Department, ONI, OSS and British intelligence.[82] Publicity about this case revealed the depth of Soviet intelligence penetration inside the U.S. Government, at a time when Stalin needed to know what American policymakers were planning to do about the Soviet challenge. The day before these arrests were announced, Soviet repatriations General Golikov again charged that the western Allies were abusing Russian POWs.[83]

On June 8th, General Omar N. Bradley said at a news conference that from 100,000 to 150,000 Russian "traitors" had been captured by American forces in Czechoslovakia and were, according to the New York Times, "being held as prisoners of war and presumably will ultimately face Soviet justice...The question of the capture of Russian traitors came up when General Bradley was asked to comment on reports that Moscow was angry over the alleged treatment of some Russians we have captured...He added dryly: 'I don't believe those people have much future!'"[84]

During the first fortnight of June 1945, British Prime Minister Winston Churchill had been urging President Harry Truman to display strength to Stalin, by not withdrawing the western Allied armies out of the previously agreed to Soviet zone. He sent a message to Truman on June 9, on the critical situation in Austria:

"Our missions have been ordered by Marshal Tolbukhin to leave by the 10th or 11th of June. THEY HAVE NOT BEEN ALLOWED TO SEE ANYTHING OUTSIDE THE STRICT CITY LIMITS...On the other hand the Russians demand the withdrawal of the American and British forces in Germany...Would it not be better to refuse to withdraw on the main European front until a settlement has been reached on Austria?"[85]

Two days later, contrary to the public assurances that all American POWs had been returned by June 1, a now-declassified U.S. document dated June 11, 1945, entitled "Repatriation of British, U.S. and other United Nations Prisoners of War as of 7 June 1945," states:

"IT IS BELIEVED THAT THERE ARE MORE BRITISH AND U.S. PWs TO BE RECOVERED AND THE WAR OFFICE/WAR DEPARTMENT ARE ENDEAVORING TO ASCERTAIN THE NUMBER AND IF POSSIBLE THE LOCATION OF SUCH PERSONNEL. AS REGARDS THE NUMBER OF U.S. PWsTHEFIGUREISSOMEWHATCOMPLICATEDBYTHELARGENUMBER OF MISSING IN ACTION PERSONNEL RECOVERED AS PW BUT NEVER

REPORTED AS SUCH DUE TO LACK OF TIME BETWEEN THEIR CAPTURE AND RECOVERY."[86]

In a message from Moscow to Washington, also on June 11th, U.S. Ambassador Averell Harriman reminded Secretary of State Edward Stettinius that the Russians had never signed the Geneva Convention accords, and that during the war, "they refused all attempts of enemy governments to come to an agreement regarding the treatment of POWs." Harriman further stated that repatriates to Russia were met by Soviet guards and marched to unknown destinations and that, "TRAINLOADS OF REPATRIATES HAVE BEEN GOING TO MOSCOW ENROUTE EAST. NO CONTACT MAY BE HAD WITH THE PRISONERS WHILE THE TRAIN REMAINS IN THE MOSCOW RAIL YARDS."[87] This part of the cable underscores the difficulty facing Allied intelligence in attempting to trace American and British prisoners, known to have been shipped east with German (or Soviet) POWs. Harriman added that those repatriates accused of anti-state activity were probably being shot (and mentions reports of executions at Murmansk), while the largest number were destined for forced-labor camps.

When a United States aircraft was shot down near Cape Lopatka, Kamchatka, Siberia, on June 10th, the Soviets buried the six crewmembers on the spot, rather than return their ostensible allies' remains (or any survivors) to the United States authorities for identification. For three years to come this incident was to be a test case concerning the refusal of the Soviets to allow U.S. search and recovery teams into their territory.[88] This was the same area of Siberia about which General Deane in Moscow had been negotiating for months, trying unsuccessfully to obtain Soviet permission to set up an American airbase for bombing missions over Japan. Deane believed that Soviet intransigence in the Kamchatka matter was directly related to their demands about Poland.[89]

On June 12th, over Churchill's objections, Truman cabled that he had decided to complete the withdrawal, regardless.[90] This greatly weakened the Allied bargaining position on their forcibly-withheld POWs in the Soviet zone. In the House of Commons, also on June 12th, the British Secretary of State for War, Sir James Grigg, was asked whether all British/Commonwealth prisoners of war liberated by the Red Army had been returned. Grigg, who was well aware of the Times of London reporting on May 30th that 8,500 British POWs were still in the Soviet zone of Austria, answered:

"THERE ARE STILL SOME PRISONERS IN PARTS OF CENTRAL AND SOUTHERN EUROPE OCCUPIED BY THE SOVIET FORCES BUT FIGURES ARE NOT AVAILABLE. Arrangements are being made for their transfer westwards to areas in British and American occupation."[91]

The War Secretary said that Britain had tried for a list of names of those held by the Russians, but he felt that it was, "a hopeless task." Declassified U.S. and British documents state that in March 1945 the British knew that 199,500 British and Commonwealth POWs had been held by the Germans, but after the final repatriations from the Soviet zone, only 168,476 returned. From the Yalta Conference in February 1945 and into March, British Foreign Secretary (later Prime Minister) Anthony Eden estimated the Soviets had overrun 50,983 British and Dominion POWs, and declassified British of May 1945 inform the British Ambassador in Washington, Lord Halifax, and Eden that an estimated 40,000 British POWs are still in Soviet hands; yet by September 1945, after the exchanges, Soviet repatriations General Golikov stated officially that only 23,744 British POWs had been freed by the Russians. It appears that nearly all the 31,000 missing British POWs in Europe had been in the Soviet occupied zone.

A formerly-secret British document of the DIRECTORATE OF PRISONERS OF WAR (Curzon Street House, London) dated June 9, 1945 is entitled: "Tracing Personnel Missing or Otherwise Unaccounted For." Recently released by the Canadian government, this report sheds light on the possible moving of British personnel from a prisoner of war category to another category of: "POSSIBLE PRISONERS OF WAR AWAITING REPATRIATION." According to this document the representatives of the British and Commonwealth services present, "WERE ONLY CONCERNED WITH PERSONNEL WHO HAD BEEN PRISONERS OF WAR AND WERE LATER UNACCOUNTED FOR." The War Office representative, Mr. A.S. Weston, "...EXPLAINED THAT THE WAR OFFICE EQUALLY DID NOT CONCERN ITSELF WITH MEN PRESUMED DEAD."

"The Chairman...pointed out that...to recover all liberated prisoners of war it was essential to discover so far as possible how many had still to be located and brought home...Presumably cooperation could be looked for in the U.S. and French zones...In the Russian zone, however, things would be different. The Russians kept no records of their service personnel (?-author) and made no notification of missing or casualties to next of kin...

"MR. (W. St. C. H.) ROBERTS (of the Foreign Office) EMPHASIZED THAT THE RUSSIANS HAD ALWAYS BEEN SLOW TO GIVE US WHAT WE WANTED, AND IN FUTURE BRITISH REQUIREMENTS WOULD BE RELATED TO WHAT THE RUSSIANS WANTED FROM US. IN THIS CONNECTION THE CHIEFS OF STAFF HAD CONSIDERED RECOMMENDATIONS FOR SETTING UP A CO-ORDINATED BODY TO WHICH WOULD BE REFERRED ALL REQUIREMENTS FROM THE RUSSIANS SO AS TO RELATE THEM TO RUSSIAN COMPLIANCE WITH OUR REQUIREMENTS...DETAILED INSTRUCTIONS ON THE RESPONSIBILITY FOR TRACING PERSONNEL MISSING OR OTHERWISE UNACCOUNTED FOR IN GERMANY SHOULD BE ISSUED TO FIELD MARSHAL MONTGOMERY BY ARMY COUNCIL LETTER."

This statement of the situation on June 9th clearly indicates

British determination to continue forced-repatriations of Russians in order to gain Soviet reciprocation in returning British POWs in their control. The figures given in this document showed the condition of British and Commonwealth POW repatriations as of 0600 hours, June 8, 1945:

"Repatriated from April 1st to June 8th, 164,002," including United Kingdom (Royal Army, Navy and Air Force), Canadians, Australians, New Zealanders, South Africans and Indian Army POWs.

"BALANCE OF RECORDED POWs BELIEVED AWAITING REPATRIATION: 3,136, including both British and Commonwealth POWs.

"POSSIBLE PRISONERS OF WAR AWAITING REPATRIATION, INCLUDING MISSING: 31, 809, including: 20,517 United Kingdom Forces, 3,292 Canadians, 2,360 Australians, 2,573 New Zealanders, 1,243 South Africans, 1,627 Indian Army and 857 Colonial POWs."

These last listed soldiers, in a newly created and uncertain category, were the British and Dominion prisoners who disappeared in the Soviet zones of Europe, and who largely remain missing to the present day.[92]

Despite such measures already being taken in London and Washington to maintain secrecy about the missing POWs, Churchill understood the consequences of showing weakness to Stalin by withdrawing from the Soviet zone in June 1945, and replied to Truman's June 12th message announcing that U.S. forces were withdrawing from the Russian zone, on June 14:

"Obviously we are obliged to conform to your decision, and the necessary instructions will be issued...As to Austria, I do not think we can make the commanders on the spot responsible for settling the outstanding questions...I consider the settlement of the Austrian problem is of equal urgency to the German matter..."[93]

Stalin cabled Churchill on June 17, apparently stalling for time on the return of the Allied POWs in Austria:

"WITH REGARD TO AUSTRIA, I HAVE TO REPEAT WHAT I HAVE ALREADY TOLD YOU ABOUT THE SUMMONS OF SOVIET COMMANDERS (Tolbukhin, et al) TO MOSCOW AND THE DATE OF THEIR RETURN TO VIENNA."[94]

Meanwhile, the disappearance of an estimated 1,000,000 French, Dutch and Belgian POWs and civilian deportees in the Russian zone was the subject of urgent and high level communications. On June 19th, in a declassified secret cable signed by General Eisenhower (#S-91662) to General Deane in Moscow, the SHAEF commander stated:

"A FURTHER APPROACH TO THE SOVIETS REGARDING NUMBERS OF WESTERN EUROPEANS IN SOVIET-OCCUPIED AREA OF EASTERN EUROPE IS URGENTLY NECESSARY. ABOUT 1,200,000 FRENCH HAVE BEEN REPATRIATED. LESS THAN 100,000 REMAIN IN SHAEF-OCCUPIED

AREA. FRENCH INSIST TOTAL POW AND DISPLACED PERSONS IS 2,300,000. EVEN ALLOWING FOR SEVERAL HUNDRED THOUSAND UNACCOUNTED FOR TREKKERS, DISCREPANCY IS STILL VERY GREAT. ABOUT 170,000 DUTCH HAVE BEEN REPATRIATED, WITH LESS THAN 25,000 IN SHAEF AREA. TOTAL DUTCH ESTIMATE OF DEPORTEES IS 340,000. BELGIAN DISCREPANCY IS COMPARATIVELY SMALLER. Have Soviets any further evidence of Western Europeans who are accessible to Allied lines? Is it possible to obtain figures for those not so accessible—that is, Western Europeans in the USSR, Poland, Austria, Czechoslovakia and Balkan countries?"[95]

General Deane repeated the substance of Eisenhower's message to the Soviet repatriations commissar, General K.D. Golubev, in a letter dated June 20th (No. 1100), and added, "APPARENTLY THE MATTER IS CAUSING THE GOVERNMENTS CONCERNED GREAT ANXIETY," and asked for an early response.[96] Deane was unable to get a reply from the Soviets for Eisenhower until June 25:

"Upon receipt of S-91662 dated 19 June, we presented the queries contained therein to GOLUBEV and have received the following reply:

"In answer to your letter of 20 June, I do not have the exact data on the moving around of persons from Western Europe and therefore cannot say much about them. I know there have been freed by the Red Army:

"French, about 250,000 of which 202,456 persons have already been sent home and about 50,000 who are getting ready to be sent home... Belgians: 27,980 persons freed of which 25,920 have been sent home..."[97]

The cable continued with Soviet figures on other European nationalities which in no way corresponded with the information available to U.S. and Western Allied officials. In this manner approximately 500,000 western European prisoners disappeared inside Soviet-occupied territory, among millions of German/Axis and Soviet POWs. As France and the USSR announced a repatriation agreement had been signed on June 29th, a report that appeared in the New York Times the next day gave the Allies' estimate of French prisoners who had died in German control: "Approximate figures made available today indicated that the number of French war prisoners and political and racial deportees who died or were killed in Germany without possibility of identification might reach 320,000."[98] This figure does not include many French POWs and civilians still carried as missing. According to the declassified SHAEF documents cited previously, 2,300,000 French had been prisoners or missing, of which 1,300,000 were returned from the western Allied zones, and 1,000,000 or more French remained missing in the Soviet zones, on June 20, 1945.

A delayed dispatch of June 30th from Paris by Dana Adams

Schmidt which was printed in the New York Times on July 2nd reported the signing of a repatriation agreement between France and Russia which would regulate the return of up to 600,000 French POWs and civilian deportees still in Soviet-occupied territory. French prisoners were reported to have been stripped of their personal belongings by the Russians and only two members of the French repatriation mission had been allowed to enter Russian territory. It was reported that the western allies had repatriated almost all of the 1,500.000 French captives in their zones, but that only 90,000 had so far been returned by the Soviets. The rise to prominence of a future French Socialist head of state can be traced to this period and to the disappearance of the French prisoners in Soviet control. As the New York Times reported:

"Under the title of "Silence in the East," Francois Mitterand, vice president of the Prisoners' Federation, declared in a newspaper article that political and security considerations should not be allowed to weigh in the balance with the accumulated anguish of thousands of prisoners. 'Excessive precautions hardly favor sincere and ever so necessary friendship,' he said. Phillippe Viannay, in another article, recalled the cries of indignation in the French press when the repatriation from the American and British zones seemed slow. He added: 'Very few voices, by comparison, are raised in alarm about the frightening mystery of our nationals east of the Oder...We have scarcely any news and no letters at all...I am far more disturbed that no mission has reached them and that, far from speeding up, the returns are slowing down.'"[99]

In the next year and a half the Soviets were to claim that they returned some 300,000 French prisoners of war and political prisoners from the eastern zones under their control, while an estimated 320,000, including French Jews were listed as dead, but some 350,000 French believed to be alive at the time of the collapse of Germany, thus remained missing in the east after the dead were accounted for and the Russians had released all they claimed to have held. The records reproduced by SHAEF repatriation official and author Malcolm Proudfoot in The Military and UNRRA chapter of his 1956 book, European Refugees, indicate that the figures for Western European prisoners who remained missing in the Soviet-occupied zones at the end of 1946 were: Belgium, 40,000; France, 350,000; and the Netherlands, 150,000.[100]

While a relatively small percentage of these missing western Allied prisoners of war and civilians had served the Germans willingly (including SOME of the 50,000 or more Alsace-Lorrainers then still missing in Russia, or some of the missing Dutch and Belgians who had also served in the German Army), hundreds of thousands of the French, Belgian and Dutch missing from WW II had simply been captured by the Germans in 1940 blitzkrieg and were interned in the east after Hitler's attack on Russia, or sent as forced laborers from German-occupied France during the war years that followed. (Some of the British, French and other POWs captured in 1940 had escaped from the Germans eastward, into Soviet territory, where according to recently released Soviet documents they were reimprisoned, and many died in Stalin's gulag camps during the war. Some of the survivors who were secretly returned to Britain after Hitler's 1941 attack on Russia, were sworn to secrecy by the British intelligence service, to protect wartime unity between Russia and the West against Nazi Germany.

In the case of the 350,000 French POWs and civilian deportees who were believed to be alive but had disappeared in that part of eastern Europe occupied by the Red Army in the 1945-46 period, available evidence indicates that these were not prisoners murdered by the Germans, and in a most careful recapitulation of all wartime shipments of French Jews to Nazi extermination camps, a noted Israeli researcher and author on the subject of the Nazi holocaust, Leni Yahil, using French and Israeli sources, documented a total of 85,000 French Jews sent to the gas chambers or otherwise murdered in the east by the Germans[101] This was somewhat less than the earliest postwar estimates of French Jewish dead, of 120,000-130,000, published in the New York Times and elsewhere. Other estimates report some 40,000 non-Jewish French POWs and civilian deportees were killed or died in German control. By late 1947, figures from official sources listed 500,000 western Europeans as "missing," when the murdered French Jews and other French killed by the Nazis were largely known and accounted for as dead. French (and Belgian) WW II POWs were seen alive by eyewitnesses, and reported by Israeli Soviet expert and Gulag survivor Avraham Shifrin, to a U.S. Senate Committee in 1973, and to other Federal agencies. (In 1962 the Queen of Belgium traveled to the USSR and obtained release of a token number of Belgian WW II POWs who had survived in forced labor camps along a railroad line in Siberia.)

A month and a half after VE Day, and three weeks after the American people had been told that all U.S. POWs in Europe had been returned, classified intelligence reports revealed the disappearance of many American POWs in Soviet captivity. A now-declassified secret cable of June 25th (#S-92930), signed by Eisenhower and sent to General Deane in Moscow, 21 Army Group, 12 Army Group and 6 Army Group concerned Americans in Soviet control near the Baltic coast:

"The following communication has been received from General Conrad, Acting Deputy Chief of Staff, G-2, ETOUSA:

"POSSIBILITY THAT SEVERAL HUNDRED AMERICAN PRISONERS OF WAR LIBERATED FROM STALAG LUFT 1, BARTH, ARE NOW CONFINED BY THE RUSSIAN ARMY IN THE

ROSTOCK AREA PENDING IDENTIFICATION AS AMERICANS IS REPORTED BY AN AMERICAN WHO RECENTLY RETURNED FROM SUCH CONFINEMENT.

"Staff Sergeant Anthony Sherg was one of 1,000 Air Force officers and non-commissioned officers who left Stalag Luft 1 immediately prior to assumption of control in Barth by the Red Army in order to obtain rumored air transport from Wismar. The group of ten in which Sgt. Sherg travelled was arrested by Russian soldiers and held in jails in Bad Doberan, then Rostock. Ten other Americans were seen in similar circumstances in Rostock.

"RUSSIAN AUTHORITIES DEMANDED IDENTIFICATION PAPERS, WHICH NO PRISONER POSSESSED, AND REFUSED TO CONSIDER DOG TAGS PROOF OF THE AMERICANS' STATUS...AFTER 25 DAYS HE ESCAPED FROM JAIL AND MADE HIS WAY TO BRITISH FORCES. FROM HIS OWN OBSERVATIONS AND CONVERSATIONS WITH OTHER FORMER PRISONERS HE BELIEVES SEVERAL HUNDRED AMERICANS MAY BE HELD IN LIKE CIRCUMSTANCES IN THE WISMAR-BAD DOBERAN-ROSTOCK AREA.

"Will you please expedite any action that can be taken by you regarding the above AND ALSO POSSIBILITY THAT SAME SITUATION MAY EXIST ELSEWHERE IN RUSSIAN CONTROLLED TERRITORY. FOR 21st, 12th, 6th ARMY GROUPS. IT IS ENTIRELY POSSIBLE THAT YOUR CONTACTS WITH RUSSIANS MAY ENABLE YOU TO UNCOVER SIMILAR CASES AS ABOVE. IMMEDIATE REPRESENTATION SHOULD BE MADE BY YOU TO RUSSIANS WITH A VIEW TO RECOVERY OF ANY BRITISH/UNITED STATES PRISONERS OF WAR WHO MAY AT PRESENT BE HELD UNDER SIMILAR CIRCUMSTANCES BY THE RUSSIANS."[102]

Weeks later, on July 12, and prior to the Potsdam Conference of the wartime leaders, Stalin, Truman and Churchill, Deane finally got an answer on this case from the Soviet Deputy Repatriations Commissar:

"GOLUBEV TODAY INFORMED ME THAT HE IS STILL CONDUCTING AN INVESTIGATION CONCERNING THE POSSIBILITY OF AMERICAN PRISONERS OF WAR BEING IN THE WISMAR-BAD DOBERAN-ROSTOCK AREA, AS SUGGESTED IN YOUR S-92930. HE HAS PROMISED TO LET ME KNOW THE RESULTS OF THE INVESTIGATION AND I WILL INFORM YOU."[103]

In this subtle manner, behind hundreds of miles of the Red Army front from the Baltic to the Adriatic, American and British POWs disappeared into Stalin's NKVD gulags. U.S. aerial reconnaissance missions over Soviet territory like "Casey Jones," conceived by OSS General William Donovan, or the operations of secret agents, could not be relied on completely to pinpoint locations of prisoners, if the Russians continued to conceal the prisoners and move them eastwards into Russia in sealed boxcars.

Formerly-secret U.S. 1945 SHAEF documents in fact record the continued Soviet refusal to allow Allied contact officers access to nearly all the U.S. and British POWs in the Russian zones, as had been agreed to by Stalin's government at Yalta. A secret June 29, 1945 cable from British Air Marshal Tedder, Eisenhower's SHAEF deputy commander in Europe, to General Deane in Moscow stated:

"...BEFORE THE HALLE CONFERENCE (18-22 May 1945) WE HAD MADE NUMEROUS ATTEMPTS TO VISIT PW CAMPS IN THE RUSSIAN ZONE AND ALWAYS MET A FIRM REFUSAL. AFTER THE HALLE CONFERENCE...GOLUBEV... AGREED TO ALLOW ONE OF OUR OFFICERS TO VISIT FIVE CAMPS...AFTER VISITING THE FIRST AND NEAREST CAMP THE RUSSIAN OFFICER PRODUCED ORDERS SIGNED BY GENERAL GOLUBEV RESTRICTING OUR OFFICERS VISIT TO THE ONE CAMP. THIS IS THE ONLY INSTANCE OF SOVIET AUTHORITIES PERMITTING US OR BRITISH OFFICERS TO VISIT CAMPS IN THEIR AREA, WHICH IS IN SHARP CONTRAST TO THE LIBERAL POLICY PURSUED BY US."[104]

Despite Stalin's refusal to return many U.S. and Allied prisoners, the American and British armies in Europe completed their final withdrawal from the permanent Soviet occupation zone on July 1, 1945, leaving behind for Stalin's vengeance hundreds of thousands of Soviet ex-prisoners. As Churchill wrote later: "Soviet Russia was established in the heart of Europe. This was a fateful milestone for mankind."[105] General Eisenhower later defended this act in his memoirs:

"The policy of firm adherence to the pledged word of our government was challenged shortly after the close of hostilities. Some of my associates suddenly proposed that when so requested by the Russians I should refuse to withdraw American troops from the line of the Elbe...The argument was that if we kept troops on the Elbe the Russians would be more likely to agree to some of our proposals, particularly as to a reasonable division of Austria. To me, such an attitude seemed indefensible. I was certain, and was always supported in this attitude by the War Department, that to start off our first direct association with Russia on the basis of refusing to carry out an arrangement in which the good faith of our government was involved would wreck the whole co-operative attempt at the very beginning."[106] Thus Eisenhower apparently hoped that by sticking to its agreements and continuing to forcibly repatriate all Soviet citizens, the Soviets would be shamed into reciprocating by releasing the American POWs and MIAs they were holding.

In a nearly simultaneous attempt to mollify the Soviet dictator, the Truman Administration extended U.S. diplomatic recognition to the communist-dominated Soviet puppet regime in Poland, on July 5, 1945, just prior to the Potsdam Conference between Stalin, Truman and Churchill. The fact that England and France had initially gone to war with Germany in 1939 to protect Poland's independence was officially ignored by the United States, although Truman recorded in his memoirs that Churchill expressed, "real surprise that he should have only a few hours' notice of my decision to recognize the new Polish government. He pointed out that the old Polish government was located in London and that it had under its strength a Polish Army of 170,000 men, whose attitude had to be considered...He asked me to postpone my announcement until July 4."[107]

One reason why this unfortunate event did not occur on July 4th, American Independence Day, was because the Polish Ambassador to the United States (of the London exile Polish government), Jan Ciechanowski, lodged an urgent warning with U.S. State Department officials, of the potential national disgrace in the proposed timing of the announcement, causing them to advance the date by one day.[108] In his memoirs Truman reports he went on to reply to Churchill: "The twenty-four hour delay suggested by you would mean that we would accord recognition on Independence Day. I, therefore suggest and hope you will concur that we postpone recognition...until...July 5..."[109] And so it was done, thus accepting the status quo in Soviet-occupied Poland, where 2,000 or more American POWs had vanished in March and April.

In the border zone between the Soviet and Anglo-American armies, the Soviets suddenly stopped receiving their own POWs and civilian deportees from the Americans and British. This had followed a dispute between the French and the Soviets over the repatriation of French POWs from Russia through Odessa and Murmansk. A July 2nd cable from General Deane in Moscow to Eisenhower's headquarters explained:

"I have again requested General Keller, Head of French Repatriation Mission in Moscow for information requested... At prisoner of war camp in Odessa, there are now 9,000 French nationals and at NA, Mission to Moscow Filesa camp near Murmansk, there are 3,000. THE MOVEMENT OF FRENCH PRISONERS OF WAR TO ODESSA HAS CEASED> The French are making efforts to obtain water transportation to France for both the liberated prisoners in the Odessa area and those in the Murmansk area. REPATRIATION OVERLAND WILL ONLY BE UNDERTAKEN AS A LAST RESORT..."

Yet this was what was ultimately attempted by desperate French prisoners in Russia when the Soviet authorities cut off the sea route.

Shortly after this and a July 2nd New York Times article about the disappearance of French POWs and civilians in Soviet territory and the signing of a Soviet-French repatriation agreement, on July 8th the Soviets halted all repatriations by blocking off their zone and refusing to receive their own Soviet nationals whom they had been demanding the return of, for months. A July 8th cable to Deane and Gammell in Moscow and also sent to 12th Army Group stated:

"Acute congestion in U.S. and British zones results owing to sudden refusal of Soviet authorities to accept Soviet citizens across the border zone. Situation aggravated by unauthorized transfer of Italian citizens across our line..."[110]

Meanwhile, the American and British armies had turned back to the Soviets large numbers of German soldiers who had put up ferocious resistance to the Russians on the eastern front, and since the end of the war had continued to try and reach western Allied lines to surrender. An estimated 4,000,000 German POWs captured during the war, or surrendered after it's end, were taken into the USSR, where vast numbers faced death at forced labor in the coming decade.

Even Jewish survivors of Nazi death camps overrun by the Red Army in the east disappeared in the gulags. In July 1945, an Italian Jew named Primo Levy, who had survived Auschwitz only to be shipped east into Russia with thousands of others, witnessed massive numbers of former prisoners of the Germans held haphazardly in many Soviet camps, including, "Negroes in American uniforms, Germans, Poles, French, Greeks, Dutch, Italians and others," at Slutsk, south of Minsk, Byelorussia.[111] Due to the kindness of an individual Soviet officer, Marshal S.K. Timoshenko (who as a young Red Army colonel had negotiated over U.S. POWs on the Archangel front in 1919), Levy and some of his comrades were lucky enough to be repatriated from another camp later, but many thousands who were less fortunate remained behind to die in Soviet concentration camps. A German Jew named Eleanor Lipper, imprisoned in Stalin's gulags since 1937, witnessed Jewish survivors of the Nazi holocaust, of many nationalities, sent east to Soviet prison camps in 1945, where, ironically, they were confined with Germans and Austrians at forced labor.[112] Israeli Soviet expert Avraham Shifrin, a former Red Army officer and Gulag inmate who learned his English from kidnapped American Army officers in the camps, has subsequently recorded more details of this subject.

Thomas Sgovio, now of Mesa, Arizona, the civilian American political prisoner, was still confined in the Soviet gulags at this time. Sgovio had already survived 10 years in the forced-labor camps with Ukrainians, Russians, Poles, Latvians, Estonians, Lithuanians, Romanians, Hungarians, Germans, Japanese and others. He had witnessed and survived massive death tolls among entire prisoner transports in far eastern Siberia. Sgovio was not confined with American military POWs, but in July 1989 he wrote to the author, "...From time to time I encountered those who had either seen or heard of American G.I.s who had been arrested in Germany."

Thomas Sgovio was also present in labor camps that received Soviet POWs who had been forcibly-repatriated by the American and

British armies. He remembered the NKVD method of emptying a slave camp for handling special contingents of POWs:

"It was in late June or July (1945) when they began to herd them in—the SPETZ-special contingent...Soviet prisoners of war. Russians, Ukrainians, Cossacks... All the z/k's ("zeks" or prisoners) of OLB Pobyeya, excepting the camp personnel, were transported to other camps...A squad of special KGB case-officers and prosecutors arrived from Moscow to investigate the cases of the prisoners of war. They passed verdicts on the spot. Some were shot, others were given twenty-five-year sentences, the rest six years exile in Kolyma...They were robbed of their possessions, clothing, boots, etc. We were not allowed to come in contact with any of the Spetz-contingent while they were under investigation. They were sent to work in the gold-fields under heavy guard and locked up at night...A revolt occurred at OLB Pobydeda-rather I would call it a group suicide. The men knew they did not have a chance. They preferred to die fighting... Those of the Spetz-contingent who received 25 years were sent to OLP Neriga. They were specified as KATORJHINKI-The Shackled Ones-no names, only numbers on their backs! Of the thousands sent to the Katorga camps in the Indigirka Administration, I can say with absolute certainty that not one of them survived! In 1947 as a free-citizen in Ust-Nera I learned there were no more Katorga camps...all had died!"[113]

One Russian former-POW repatriated by the Americans in the summer of 1945 was sent to a huge labor camp complex some 40 kilometers from Omsk, Siberia. He survived, and remembered the vast numbers of prisoners:

"I do not know the total number of camp inmates, but I estimate that THERE MUST HAVE BEEN SEVERAL HUNDREDS OF THOUSANDS IN THE HUTS, PEOPLE OF VARIOUS NATIONALITIES AND FROM MANY COUNTRIES. They were all engaged in building vast industrial plants for which the equipment was brought from Germany. We christened the project "New Germany"...The mortality was high. This was due to the lack of proper clothes, the terrific cold, the speed up and the competition, the bad food and the shortage of medical supplies...No distinction was made between Germans and Russians like myself. One had the impression that the destruction of the human beings was being carried out at the direction of Moscow. Suicides were frequent..."[114]

On July 21, 1945, Soviet Deputy Repatriations Commissar Golubev wrote a secret memo (#04997) to General Deane concerning the numbers of Allied prisoners returned by the Russians:

"DURING THE ENTIRE TIME OF REPATRIATION, THE FOLLOWING NUMBERS OF FOREIGNERS WERE HANDED OVER TO THE ALLIED COMMAND: AMERICANS, 22,010; ENGLISH, 20,483; FRENCH, 225,111; BELGIAN, 31,530; DUTCH, 29,558...INASMUCH AS A CHECK HAS NOT

YET BEEN FINISHED, I AM UNABLE TO TELL YOU THE PRECISE
NUMBER OF WESTERN EUROPEANS WHO HAVE REMAINED IN THE
SOVIET ZONE OF OCCUPATION...AFTER ALLIED ANGLO-AMERICAN
TROOPS WITHDREW. WE KNOW EXACTLY THAT IN THIS ZONE
AFTER THE WITHDRAWAL OF TROOPS THERE WERE MORE THAN
100,000 FOREIGNERS OF WHICH MORE THAN 70,000 WERE POLES,
80 AMERICANS AND 15 ENGLISH."[115]

With such outright lies by Soviet authorities, over 500,000
Allied prisoners disappeared in the Soviet zones, many into secret
Soviet POW camps and prisons in both eastern Europe and the USSR
proper.

In Dresden, Germany, on July 6th, the Soviets arrested an
American born civilian named John Noble, along with his father,
who was German-born but also an American citizen. They had been
interned in their home in Germany during the war, where they had
a camera factory but the Soviets were suspicious of their lenient
treatment by the Nazis. Since May 13th when eight U.S. POWs had
come to them for help, the Nobles had been taking groups of up
to 30 liberated American and British POWs stranded inside Soviet
territory into their home and assisting them in reaching the American
lines to the west. The U.S. 76th Division had succeeded in getting
through the Russian lines with a few vehicles on several occasions,
bringing back small groups of American ex-prisoners assembled by
the Noble family. The border was already blocked with barbed-wire
entanglements, floodlights and machine-gun nests manned by Red
Army soldiers.

The day before confining the Nobles the Soviets had already
arrested a number of American soldiers at their house who had come
from the American lines to pick up U.S. prisoners of war; they were
held for five days and released, probably because the Russians were
certain the American command knew they were there, and all of their
names. The fate of the American POWs still at large in the Dresden
area and trying to reach the western lines at this time was undoubtedly
the same as that which befell the Nobles. John Noble and his father
Charles were held at the NKVD headquarters in Dresden at first, and
then transferred to the central prison there, Munchenerplatz, which
had previously been used by the Nazis. Each was placed in a solitary
confinement cell. John Noble later remembered:

"In the door was a spy hole...through the hole I could see my
father being shoved through a door directly opposite. I stood for half
an hour at the hole, staring at the faceless rows of doors. I wondered
how many eyes were similarly peering through spy hole. Then I heard
the screams. Someone was being whipped. The screams were clear
but directionless...Later I was able to place the direction from which
they came; the whippings took place in the 'questioning room...By
looking sharply upward from the bottom edge of the spy hole I could

see the hallway that ran past the fourth-floor cells. As I looked, cell door opened. Guards dragged a struggling prisoner out and threw him on the floor. He tried to fight his way up, and they pounced on him and pinned him to the floor with their knees. Then they stripped him, tearing his shirt away and pulling his trousers off in violent tug that left the prisoner tumbled head down in a heap against the wall.

"One guard had a short leather whip. The other hastily pulled off his belt. Then they began beating the man, not slowly and methodically but rapidly and in semifrenzy. They kicked him and shoved him along the floor while they tore his skin with their cutting lengths of leather. His screams were terror-filled and anguished. I couldn't keep from watching; the horror of it was hypnotic. Long after the guards had finished, panting, and had flung the bloody whimpering man back into his cell, the scene and sounds stayed on in my mind, even into sleep. And added to them were new screams from the questioning room."

In late July the German Army rations ran out and faced with having to feed their prisoners from Red Army stocks the Soviets chose to let them starve. Noble later wrote in his memoirs of Gulag life:

"When it became apparent on the first of those days that there was to be no food, loud protests, uncontrolled curses and screaming were let loose. They became louder as the second, third and fourth days went by. Men went out of their minds, women prisoners became hysterical. Some Moslem prisoners chanted their prayers. Then death struck, right and left. Cell doors were opened and dead bodies pulled out by an arm or a leg. I wondered when it would be my father's time and mine. When I no longer was strong enough to lift my feet off the floor, I put myself into the hands of God. Some seven hundred prisoners had entered that starvation period. I was one of twenty-two or twenty-three, that survived, along with my father."[116] John Noble was eventually assigned to reorganizing and managing the records of 21,000 prisoners of the Russians, which may have ensured his continued imprisonment for nine more years to come.

General George Patton had meanwhile been ordered back to the US until the crisis was secretly contained in July 1945. After his return to Europe he continued to urge confrontation with the Russians until his death following a truck-car collision in December 1945. One of the reporters whose biased charges caused the removal of General George Patton as Commander of Third Army and as Military Governor of Germany was Carl Levin of the New York Herald Tribune. It was also Levin who had written the cover story on POWs on June 5th, 1945, under the heading: 25,000 US PRISONERS TURN UP ALIVE. In what appears to be the primary explanation concocted by the US command and deliberately released to a journalist, Levin quoted U.S. Lt. Col. W.P. Schweitzer as saying, "That it was unlikely that any others would be found, except STRAGGLERS who may come in from Russian-occupied Germany." This article by Levin helped seal the fate of thousands of American POWs in Russian hands. Due to security regulations, other US military officers were not permitted to tell the truth about the POWs to the media.

The U.S. and its western allies were unsure of the exact numbers and movements of many POWs and MIAs known or believed to be in Soviet control, which quite naturally led to Allied intelligence and aerial reconnaissance missions, such as "Operation Casey Jones" (conceived by OSS General William Donovan), which were immediately countered by the Soviets, in the beginnings of what became the Cold War. During the summer of 1945, American and British aerial reconnaissance aircraft were attacked over Soviet-occupied areas where Allied prisoners were known to have disappeared. In July 1945 two British Anson bombers engaged in aerial photography over the Russian zone were shot down near Klagenfurt, Austria, where the POW exfiltraters from the Soviet zone had reached allied lines. Other incidents followed in August 1945, when the Soviets filed more than 300 complaints of airspace violations of their zones by U.S. and British aircraft.[117]

---

## Notes on Chapter Five

1    National Archives, RG 331 SGS, Box 88; This document, supplied by the author to the U.S. Senate Foreign Relations Committee Republican staff in 1990, and to the Senate Select Committee on POW/MIA Affairs in 1992, is reproduced on p. 727, POW/MIA Policy and Process, Hearings before the Senate Select Committee on POW/MIA Affairs, Part II of II, November 5, 6, 7, 1991, U.S. Government Printing Office, 1992, Washington, D.C., ISBN 0-16-038479-6, and National Archives, Washington, D.C.

2    Letter to the author from John Werner

3    Interview by the author, 1990.

4    Personal Papers of Harry S. Truman, Truman Library, Independence, Missouri, Box 675

5    Letters from and interview with Val Spiegal, a 42nd Infantry Division veteran of Bethesda, Maryland, 1990.

6    POW/MIA Policy and Process and National Archives, RG 331, p. 726

7    Associated Press and New York Times, May 12 & 13, 1945

8    POW/MIA Policy and Process and National Archives, RG 331, p.729

9    POW/MIA Policy and Process, letter to the author turned over to the U.S. Senate Foreign Relations Committee in 1991 and to the Senate Select Committee on POW/MIA Affairs in 1991, pp. 732-733

10    Ibid.

11    Personal statement of Dr. Martin Siegel in correspondence with the author, 1990.

12    Letter to the author, op. cit.

13    Personal statement by Mr. Les Neal, of Bradenton, Florida, 1988.

14   Personal Correspondence and interview by the author with Dr. Martin Siegel, Washington, DC, 1990.

15   Letter to the author

16   Letter from George Coffin, Independence Missouri, April 13, 1945; Memorandum for Colonel Vaughan, May 4, 1945, from Lt. Col. Davenport, Office of the Chief of Staff-War Department; May 10, 1945, letter from Colonel Harry Vaughn to George Coffin; Telegram from George Coffin to Col Harry Vaughn, May 18, 1945; and May 19, 1945 letter from Col. Vaughn to Geoge Coffin. Box # 675 Papers of Harry S. Truman Official File, Harry S. Truman Library, Independence, Missouri

17   POW/MIA Policy and Process, pp. 784-785; and National Archives, RG 331

18   Personal Statement of Captain Sidney Miller

19   POW/MIA Policy and Process, p. 738; and National Archives, RG 331-SGS files

20   Report of Major General Ray W. Barker, 23 May 1945, Records of the Halle Conference, NA, SHAEF G-1

21   Report To The Secretary of State from the Deputy U.S. Political Adviser, Donald R. Heath, June 1, 1945, No. 449- Distribution list, 711.4/840.la; and the Halle Conference minutes, National Archives, SHAEF, G-1, 337/2 Halle Conference, 17W4 13/18/C, Box 1&2, obtained by the author through Freedom Of Information request and made available to the U.S. Senate in 1991-92

22   23 May Report of General Barker, op. cit.

23   Halle Conference minutes, p. 14, op. cit.

24   Halle Minutes, p. 14, op. cit.

25   Halle Conference minutes, p. 15, op. cit.

26   Fourth Session, Halle Conference minutes, p. 9, op. cit.

27   Report of Maj. Gen. Barker, 23 May, op. cit.

28   General Barker's 23 May 1945 report on the Halle Conference, located in the National Archives, as previously cited, was provided by the author to the Senate Foreign Relations Committee and to the Senate Select Committee on POW/MIA Affairs, and a portion of it is reproduced on pp. 744-745, POW/MIA Policy and Process and National Archives

29   RG 331, SHAEF AG Records, 383.6-7, "Location of known and Possible Prisoner of War Installations in Germany and Occupied Countries, as known to PWX-G-1 Division, SHAEF," 18 March 1945, Map Reference G.S.G.S., 4346 unless otherwise specified; marked "Screened, 26 September 1945."

30   NYT, May 22, 1945

31   National Archives, Red Cross Files, Letter to Mr. Fred Clarke, Research Editor, Information Service, 575 Lexington Ave., NY, NY, May 1, 1961, from Clyde E. Buckingham, Acting Director, Office of Research Information, American Red Cross

32   NA, 383.6-7, #2 and # 3, Box 187, RG 331, Stack area 174W, Allied Operational and Occupation Headquarters, WW II, SHAEF Special Staff, Adjutant General Division, Executive Section, Decimal File 1945, 383.6-11 thru 12 to 383.7-1, Box #188; 383.6-15; Also: 383.6-17, "Repatriation of Allied Nationals From Russia", A.G. Records/SHAEF-Main and others, courtesy of research assistant Marion P. Lelong, of Falls Church, Virginia, and supporting documents located in the Archives by the author.

33   POW/MIA Policy and Process, p. 741; and National Archives

34   POW/MIA Policy and Process, p. 742-743; and National Archives

35   NA RG 331; also: 383.6-7 op. cit.; also Official RAMPS Study.

36   NA, RG 331/6 Box 22

37   Official RAMPs study, National Archives

38   New York Times, June 1, 1945, see also: NYT, June 2nd: "99% of U.S. Captives in Reich Survived, Red Cross Reports."

39   POW/MIA Policy and Process, p. 737; and National Archives

40   National Archives, RG 331/6/Box 22

41   *The Hidden Russia*, by N. N. Krasnov, NY, 1960, p. 47

42   POW/MIA Policy and Process, pp. 752-753; and National Archives, RG 331/6/22

43   NA, Top Secret Priority message from Opnav To: London, No. 44, "Top Secret and Personal From the President For the Prime Minister, 20 May 1945, stamped: 7332; also: H.W. Putnam, Major, A.C.

44   POW/MIA Policy and Process, p. 740; and National Archives, RG 84, Moscow Embassy files

45   POW/MIA Policy and Process, p. 754; and National Archives, RG 84, Moscow Embassy

46   *Special Envoy*, Harriman, op. cit.

47   MacMillan's memoirs, "Tides of Fortune, 1945-1955, Harper and Row, NY and Evanston, 1969, pp. 16-22

48   Repatriation of British, U.S. and other United Nations Prisoners of War as of June 7, 1945, "#X-38316, SHAEF G-1, PWX Branch, and RAMPS official report on repatriation, op. cit. A portion of the June 7th document cited appears on p. 768, POW/MIA Policy and Process, op. cit.

49   Times of London, May 30, 1945

50   POW/MIA Policy and Process, p. 755; and National Archives, RG 84, Moscow Embassy

51   National Archives, RG 84 Moscow Embassy; also: p. 758, POW/MIA Policy and Process

52   NA, RG 331

53   NA, FWD 22001 "secret" from Eisenhower to War Office,20 May 1945

54   Stars and Stripes, May 21-22, 1945, from the records of Colonel A.J.D. Biddle Jr. It appears that a policy decision had indeed been made to officially end the POW repatriations in the European theaters, for political purposes and for reasons of policy relating to the highest stated U.S. priority, of winning the war against Japan—with Soviet assistance.

55   NA, From EXFOR Rear to SHAEF Forward G-3, 24 May 1945, Ref. No. AG-3685

56   NA, SGS Files, FWD-22406, 27 May 1945 SAARF records are still "classified" and now in the custody of the CIA, which has refused to release them to the author until the time of this writing.

57   POW/MIA Policy and Process, p. 762; and National Archives, 383.6, SHAEF G-1, PWX Branch

58   Documents supplied by the Pentagon to the author in response to a Freedom of Information request

59   POW/MIA Policy and Process, p. 765; and National Archives

60   This document was supplied to the author by the Pentagon following a Freedom of Information request for records pertaining to American prisoners remaining in Soviet control.

61   The Last Six Months—Russia's Final Battles With Hitler's Armies in World War II, by General S.M. Shtemenko (translated by Guy Daniels), pp. 419-420, Doubleday and Company, New York, 1977

62   POW/MIA Policy and Process, p. 802

63   POW/MIA Policy and Process, p. 802 and pp. 742 & 743, ibid.

64   NY Herald Tribune, June 6, 1945; also: p. 763, POW/MIA Policy and Process

65   Occupation in Europe Series, Repatriation of American Military Personnel (RAMPs study in the National Archives, pp. 24 & 125.

66   A Copy of the original document in the National Archives is reproduced on pp. 756-757, POW/MIA Policy and Process

67   National Archives, Records of the European Allied Contact Section-SHAEF, Main SH/13/EAC/F; 911091-458, Declassified through Freedom of Information request by the. author, February 27, 1992

68   *Diplomat Among Warriors*, by Robert Murphy, Doubleday, NY, 1964

69   KGB-The Inside Story, by Andrew and Gordievsky, Harper/Collins, NY,1990, pp. 286-290

70   A copy of the original document MF-14427, 30 May 1945, declassified secret, was found in the National Archives, RG 331/6/334 and RG 84, Moscow Embassy, and is reproduced on p. 760, POW/MIA Policy and Process

71   POW/MIA Policy and Process, p. 759, declassified "Confidential," found in National Archives RG 331/1/88; also RG 84 Moscow Embassy

72   A copy of the original document located in the National Archives, SHAEF, G-1, RG 337/2-Halle Conference Records, is reproduced on p. 761, POW/MIA Policy and Process. One copy of this document in the author's possession, is marked with an official stamp as are many other POW repatriation documents of this period: "This copy only MAY BE SHOWN to duly accredited unofficial researchers Order Secretary of the Army by the Adjutant General 29 December 1949." This is believed to be related to the writing of secret but "unofficial" histories of the events of 1945 pertaining to U.S.-Russian prisoner of war exchanges, and other matters, which were not made available to the public, and remained "classified" at the time this book was written.

73   NA RG 331/1/88, Ref. No. WX-91036, 31 May 1945, declassified Top Secret

74   NYT, June 7-12, 1945

75   Chicago Daily Tribune, Associated Press report of June 4, printed on June 5, 1945

76   Minutes of June 6, 1945 Deane-Golubev meeting in Moscow, NA, RG 84, Moscow Embassy

77   Ibid.

78   The original document from RG 331, NA, is reproduced on p. 734, POW/MIA Policy and Process, op. cit.

79   NYT reports and declassified documents in National Archives

80   Letter to author from former-Ambassador George Kennan, dated December 13, 1988

81   *Khrushchev Remembers*, pp. 194-95, Little, Brown, 1970.

82   New York Times, June 7, 1945; and p. 483, *Mask of Treachery*, by John Costello, William Morrow, NY, 1988

83   Ibid.

84   NYT, June 9, 1945

85   *Triumph and Tragedy* by Winston Churchill, p. 63, op. cit.

86   National Archives, RG 331/6/29, also: Reproduced (in part on p. 768, POW/MIA Policy and Process

87   POW/MIA Policy and Process, pp. 769-771; also: *Special Envoy*, op. cit.

88   NA, AG 150, 22 December 1949, QM-M, "Soviet Claim For Reimbursement For Recovery of Remains," From General Headquarters Far East Command to the Quartermaster General, Department of the Army, declassified 17 November 1988

89   Strange Alliance, op. cit., 252-265

90   FRUS

91   New York Times, p.6, June 13, 1945

92   War Office File No: 0103/6929, RG 23 Canadian Archives. This document was first obtained by Canadian researcher James Bacque, who kindly forwarded it to the author, who turned it over to investigators of the Senate Select Committee on POW/MIA Affairs; whereupon it was ignored. Bacque is an authority on German POWs held by the Americans and French after VE Day, who perished in large numbers under U.S. control, which he described in his book: Other Losses, Stoddart Publishing, Toronto, Canada, 1989

93   *Triumph and Tragedy*, Winston Churchill, p. 605, op. cit.

94   Ibid., p. 608

95    NA, RG 84, Moscow Embassy; also RG 331 US Military Mission in Moscow files

96    POW/MIA Policy and Process, pp. 780-781; also RG 84 Moscow Embassy and RG 165 Military Mission

97    POW/MIA Policy and Process, pp. 782-783

98    NYT, June 30, 1945. Time Magazine published on June 11, 1945, p.23, U.S. casualty figures released by the Pentagon at that time, which were admitted to be several weeks behind. At that time the most influential U.S. newsmagazine reported: "...60,000 are missing. More than 75,000 are still listed as prisoners of war. Almost a quarter of a million are dead...Costliest week (Dec. 19-25) was during Runstedt's counterattack at Ardennes (Battle of the Bulge) when 27,194 U.S. soldiers were lost to combat. (1,721 killed). Bloodiest week: The Normandy landing (June 6-12) when 2,880 men were killed." (30, 338 U.S. military personnel were also listed as missing in the Pacific Theaters in this article along with 67,733 dead in the Pacific.) In this report Time Magazine failed to inform its readers that in addition to those Americans reported killed in action, some had been killed or drowned in a still-secret German attack on a preinvasion Allied exercise flotilla off the British coast, and many thousands more U.S. soldiers were still listed as missing in action from the Normandy landings and the Battle of the Bulge; that thousands of the Americans officially listed as POWs remained unaccounted for in the Soviet-occupied zones at this time, or that the majority of the Americans listed as missing in action were lost in the European theaters and many were believed to be alive in areas now overrun by the Red Army. This was known to correspondents in Europe at the time, and to military officers at SHAEF headquarters. Later the number of U.S. MIAs was to be raised to 78,750, most of whom were lost in the European theaters, but Time and other major media publications in America were to ignore these facts for more than four decades to come.

99    New York Times, July 2, 1945, p.5

100    European Refugees, 1939-1952, by Malcolm Proudfoot, Northwestern University Press, Evanston, Illinois, 1956, p. 270, POW/MIA Policy and Process, p. 798

101    The Holocaust, The Fate of European Jewry, 1932-1945, p. 434, by Leni Yahil, Oxford University Press, 1990.

102    POW/MIA Policy and Process, pp. 784-785; NA RG 331

103    POW/MIA Policy and Process, p. 785; NA RG 331

104    POW/MIA Policy and Process, pp. 786-787; National Archives RG 331/1/88

105    Triumph and Tragedy, p. 609, op. cit.

106    Crusade in Europe, p. 474, op. cit.

107    Volume I, Year of Decisions, Memoirs of Harry S. Truman, p. 322, op. cit.

108    Defeat in Victory, by Jan Ciechanowski. op. cit.

109    Volume I, Year of Decisions, Memoirs of Harry 5. Truman, p. 322

110    National Archives, Mission to Moscow files

111    The Awakening, Primo Levy, Collier Books, MacMillan, 1965

112    11 Years in Soviet Prison Camps, Elinor Lipper, Henry Regnery Co., Chicago, 1951

113    Letter to the author from Thomas Sgovio, dated July 17, 1989, and "Dear America!" by Thomas Sgovio, Partners Press, Inc., Abgott & Smith Printing, Kemore, NY, 1979

114    Forced Labor in Soviet Russia, p.288, op. cit.

115    POW/MIA Policy and Process, p. 788, NA, RG 331 and 84 Moscow Embassy

116    I Was A Slave In Russia, John Noble, The Devin-Adair Company, New York, 1958, pp. 16-21

117    Wild Bill Donovan—The Last Hero, Anthony Cave Brown, Times Books, NY, 1982, p. 642

# CHAPTER SIX

## THE POTSDAM CONFERENCE AND THE ATOMIC BOMB

At the Potsdam Conference near Berlin in late July 1945, Stalin attempted to curry favor with Truman by revealing that Japanese Prince Konoye had brought a peace offer to Moscow, which he desired the President's thoughts on. Truman may have been too anxious for some sign that the Soviets were finally going to cease their confrontational actions, and he was committed, as Roosevelt had been, to a new world order represented by the United Nations. His advisers at Potsdam included his new Secretary of State, James Byrnes, along with Averell Harriman, Robert Murphy, "Chip" Bohlen, Byrnes' principal adviser James C. Dunn, Benjamin V. Cohen, H. Freeman Matthews, Admiral Leahy, and others from State and War departments.

Truman recorded in his memoirs: "I was greeted at the airfield by a large delegation including Secretary of War Stimson, Assistant Secretary McCloy, Assistant Secretaries (Will) Clayton and Dunn, Ambassadors Harriman, Pauley, and Murphy, Fleet Admiral King, Minister Lubin, Lieutenant General Clay, Major General Floyd Parks, and Soviet ambassadors Gromyko and Gousev."

At the first meeting the Soviets present were recorded by Truman as: Premier Stalin, Molotov, Vyshinsky, Ambassador Gromyko, Gousev, Novikov, Sobolev and the translator, Pavlov. The British present were Churchill, Eden, Attlee, Cadogan, Clark Kerr and others. With Truman were Byrnes, Leahy, Harriman, Pauley, Davies, Dunn, Clayton, Cohen, Matthews and Bohlen.

The Potsdam Conference resulted in creation of a Council of Foreign Ministers, which would ostensibly negotiate for peaceful solutions to confrontations over such matters as prisoner repatriations and war reparations. Last-minute Soviet participation in the Japanese war was welcomed, but somewhat halfheartedly, although Churchill concurred. Actually, the Soviets wanted to enter the war to share in the spoils of victory in the Far East, although the Americans did not need them, now.[1] Churchill complained about Soviet conduct in Austria, saying, according to Truman, "that the British had not even been allowed to enter the sector assigned to them because of intervention by Russian troops." According to Truman, Churchill also restated "his reasons for refusing to accept Stalin's proposal to cede the eastern territory of Germany to Poland. Stalin, in turn, challenged the Prime Minister's reasoning..."[2]

Truman had himself objected to his own Secretary of the Treasury Henry Morgenthau's continual efforts to single-handedly make U.S. foreign policy with his pro-Soviet plan to "pastoralize" and partition Germany after the war, and did not wish him to accompany the U.S. delegation to Potsdam, whereupon Morgenthau rendered his resignation, which the President accepted. Churchill wanted Germany to remain strong enough to act as a buffer to Soviet power facing western Europe. Nevertheless, important provisions of the Morgenthau Plan were adopted at Potsdam, over Churchill's earlier objections, which resulted in Soviet Russia annexing a large portion of eastern Poland and Poland's borders being moved far to the west, at or beyond the old Curzon line at the expense of much German territory east of the Oder and the Neisse rivers. As a result some 15,000,000 German refugees were subsequently removed or fled from the east, resulting in a great human tragedy that cost millions of dead. The brutality and mass-murders of the Nazis had caused a savagery in the avenging Russian armies that continued during the postwar German exodus. In addition, the U.S. approved the payment of war reparations in kind from Germany to Russia at Potsdam, which resulted in the forced-removal of many of the German industrial plants to the USSR. The Russians also deported many Germans to forced labor camps in the Soviet Union at this time, to reconstruct the German industrial base in the east, and to obtain labor to replace their own war losses.[3]

It must be noted, however, that the Americans, British and French also made use of hundreds of thousands of German POWs for forced-labor in the postwar period, and recent research, together with the eyewitness accounts of German ex-POWs who survived, indicates that thousands of other German POWs in American and French control in the western Occupation Zones, died of malnutrition and disease in overcrowded POW camps, lacking adequate food and water. British treatment of the German POWs, who in most cases were merely soldiers rather than Nazis or SS men, was far more humane.[4]

In his memoirs, Truman recounted that Churchill complained about Soviet conduct elsewhere in eastern Europe: "With regard to Rumania, and particularly to Bulgaria, he added, the British knew nothing. Their mission in Bucharest, he asserted, had been penned up with a closeness approaching internment...An Iron fence, he charged, had come down around them. Stalin interrupted to exclaim, "ALL FAIRY TALES." Truman went on to write: "I stated that, in the case of the United States, we had been much concerned about the many difficulties encountered by our missions in Rumania and Bulgaria. The exchange continued sharp and lengthy..." During the conference Truman also had preliminary talks about adjusting the enormous eleven billion dollar Lend-Lease account with the Soviets, which was the total of U.S. aid supplied to Russia during the war, but he wrote, "there was no opportunity," to go further into it with Stalin.[5]

Truman had been informed through secret channels at the beginning of the Potsdam conference, about the successful detonation of the first atomic bomb at the Los Alamos test site. At the Conference Truman had informed Stalin officially about the awesome new bomb, but the Soviet dictator had acted unimpressed. (Bohlen and others have reported Molotov gave the impression he already knew) Truman had said then that the atomic bomb was, "Too dangerous to let loose in a lawless world...that is why we who have the secret of production do not intend to reveal the secret until means have been found to control the bomb (and) to protect the world from total destruction." Through later-revealed spies within the U.S.-British-Canadian atomic development project, Stalin no doubt already knew that the Americans were very close to success in this endeavor, and that it would change history dramatically. The Soviets were already working on developing their own atomic bomb, using information obtained from their extensive spy network, and within six weeks after the Potsdam Conference, the defection of a Soviet GRU (military intelligence) cipher expert named Igor Gouzenko, from the Soviet Embassy in Ottawa, Canada, was to reveal a massive penetration of U.S. and Allied government agencies, and even the most-secret nuclear program.[6]

After losing a British general election to a Labor-Socialist coalition on July 26th, Winston Churchill and Anthony Eden had suddenly been replaced at the Potsdam meetings by Clement Attlee and his new Foreign Secretary, Ernest Bevin. The Soviets immediately perceived a weakened front at the Potsdam conference, but not before Churchill had already made certain promises to Stalin for future forced-repatriations of Soviet POWs still in British control. Stalin particularly wanted the Ukrainian POWs from a Waffen-SS Division who were held in camps under Field Marshall Harold Alexander's command), apparently in the hope that more British POWs would thereby be released from Soviet control.[7]

At least some of the POW/forced repatriations discussions at the late-July 1945 Potsdam Conference between Stalin, Truman and Churchill (and later Attlee), have remained secret. U.S. Political Adviser Robert Murphy states in his memoirs, "The American Government made no official transcript of the proceedings because overzealous security officers ruled against admitting stenographers to the plenary sessions."[8] The British Prime Minister said that on July 25 (the last meeting he attended before being voted out of office), in a disagreement over Poland's borders and population shifts, Stalin complained of being short of labor for food, metal and coal production because of war casualties. When Churchill also complained of coal shortages Stalin said:

"Then use German prisoners in the mines, that is what I am doing." And a little later he added: "Our position is even more difficult than yours, we lost over five million men in the war, and we are desperately short of labor."[9]

Whether these recorded words veil a more secret exchange about British and American prisoners shipped east into Russia with German POWs, may never be known, but an August 10 cable from SACMED Political Adviser Alexander Kirk to the new Secretary of State James Byrnes confirms that the subject of POWs/forced repatriations was discussed in more detail at Potsdam:

"We have learned from our British colleague (Harold Macmillan) that in accordance with Churchill's promise to Stalin at Potsdam, General Basilov will arrive at AFHQ in next few days to arrange visit to British POW camp near Cisenatico. Russians claim that some ten thousand occupants of this camp are Soviet citizens whereas British claim they are Poles from east of Curzon line. Field Marshal Alexander was not enthusiastic about receiving this mission but decided to do so when he was informed that Churchill had given his promise at Potsdam."[10]

A deputy POW repatriations commissar, with an arrogant manner that demonstrated how certain the Soviets were of British compliance in order to recover their own POWs in Russian hands, Basilov would continue to demand the forcible repatriation of 'approximately 30,000 Soviet citizens' in the Polish Corps in Italy, who had fought heroically on the Allied side under General Anders, which Alexander continued to refuse. In closing his narrative of Potsdam and his wartime actions, Churchill says (in part):

"I take no responsibility beyond what is here set forth for any of the conclusions reached at Potsdam...There were many other matters on which it was right to confront the Soviet government, and also the Poles...I had in view, namely, a 'showdown' at the end of the Conference."[11]

Truman was silent about the POW-Repatriations matter at Potsdam, and thereafter said nothing about the American prisoners of WW II who had disappeared while under Soviet control. The policy decision initially made by Roosevelt in March 1945, confirmed by Truman in late April, reaffirmed by U.S. actions at the end of May and adhered to by Churchill in June, to avoid military confrontation with Stalin over some 20,000 vanished American POWs and MIAs, and 20,000 or more British and Commonwealth POWs and MIAs who had disappeared, was allowed to stand at Potsdam. United States commitment to coerced and forcible repatriation of all "Soviet citizens" was still seen as the only possible way to recover U.S. and Allied prisoners being withheld by the Russians at secret locations. Implementation of this policy was required of the departments and agencies. It appears from available declassified evidence that Field Marshal Alexander, General George Patton, and to a lesser extent others, including General Eisenhower, resisted the course of events prior to and following the unsuccessful September 1945, London Council of Foreign Ministers meetings that followed, involving Molotov, Eden's successor, Ernest Bevin and James Byrnes (who had replaced Stettinius).

George Kennan, then #2 in the U.S. Moscow Embassy under Averell Harriman, had to fill in for the Ambassador in meetings with Stalin's officials, when Harriman was away attending important conferences such as the one at Potsdam. When questioned by the author on the disappearance of U.S. and Allied POWs in the Soviet zones during the summer of 1945, Kennan wrote in a letter to the author, long afterward:

"I was vaguely aware of the fact that the Russians were dragging their feet on the repatriation of these American prisoners...They were doing this as a means of forcing us to return the Soviet POWs previously in German hands, whom we had found in the area overrun by our forces in Germany. I had, I may say, no illusions about the fate of these men if they should be repatriated to the Soviet Union; and I deplored the necessity of repatriating them; but I understood the dilemma with which this confronted our military authorities."[12]

The forced repatriations of Soviet POWs and citizens by rail, truck and on foot, were only accomplished after mass-suicides, shootings and other violence. In addition to more than 2,000,000 Soviets repatriated directly across the lines in Germany, many others among the three million displaced Soviets recovered in the east zone had been turned back by U.S. and allied armies or left in Soviet-occupied territory when the allies withdrew. In addition, some Russian prisoners who had been transported to America were also being forcibly repatriated back to the USSR in the summer of 1945. 118 prisoners screened from POW camps in many parts of the U.S. had been assembled at Fort Dix, New Jersey by late June. The legal counsel of the Department of State, R.W. Flornoy, objected strenuously to the proposed forced-repatriation from Fort Dix, arguing that the "spirit and intent" of the Geneva accord on POWs was being violated. E.A. Plitt of the State Department's Special War Problems Division (SWP) countered with the suggestion that by repatriating the Russians first to Germany where the U.S. authorities could turn the prisoners over to Soviet officials, to satisfy the terms of the Yalta agreement on prisoners. Flournoy was overruled and U.S. guards began preparations to transfer 154 Russians to Germany. The desperate Russians had attempted to force the American military guards to fire on them and then barricaded themselves into a barracks which they tried to set on fire. Camp authorities used tear gas to drive the prisoners from the building, who attacked and wounded three guards with knives, and were then fired upon, resulting in seven wounded POWs. Inside the barracks, three of the prisoners were found dead after hanging themselves and 15 other ropes were already prepared for more suicides.[13] The surviving prisoners were moved under heavy guard to New York for shipment to Europe, At the last moment Secretary of War Stimson ordered them to be taken back to Fort Dix until a policy decision could be made concerning their fate.

The policy decision had been made on July 11, 1945, just prior to Truman's departure for the Potsdam Conference with Stalin and Churchill, that the "Vlasov" prisoners must be repatriated in order to conform to the repatriation agreement signed at Yalta; and that force must be used if necessary. Because so much publicity had surrounded this particular group of Soviet POWs, the government was still reticent to repatriate them. On July 23rd, Undersecretary of State Joseph Grew, who had long opposed the forced-repatriations, and Undersecretary of War Robert Patterson, who had declared the missing American POWs in Europe dead at the end of May and was to succeed Stimson, as Secretary, ordered another investigation, perhaps hoping to delay the inevitable until after the Potsdam Conference. There the new American Secretary of State, James Byrnes, was in attendance with Truman, Stalin and Churchill, who was to be succeeded by Clement Attlee after the British elections. The 146 remaining Fort Dix Russians were ultimately transferred under heavy guard to Soviet control on August 31st.[14]

In the Potsdam Conference also, lay some of the roots of the Korean War, as Stalin agreed at that meeting to a U.N.-supervised election there, but later the Soviet-backed Communist regime of Kim Il Sung refused to let UN observers cross the 38th Parallel to witness North Korea's compliance.[15]

The first atomic bomb to be used in war was dropped from an American B-29 aircraft upon the city of Hiroshima, Japan on August 6th, as President Harry Truman was returning from the Potsdam conference. Its use was a signal of American military strength that could hardly have failed to impress Stalin. The single nuclear blast resulted in 92,000 dead or missing Japanese, and an equal number of others crippled or sickened by radiation. The Soviet Union then declared war on Japan at midnight on August 8th, 1945. On August 9th, the date that the second (and sole remaining) U.S. atomic bomb was dropped on Nagasaki, resulting in 40,000 more Japanese deaths, massive Soviet forces under the command of Marshal Vasilievsky, augmented since the defeat of Germany, attacked across the Manchurian frontier. While the Japanese fought back against the Russians, they almost immediately stated that they were ready to accept the terms laid out in the Potsdam declaration, subject only to assurances that the Emperor could continue as ruler of Japan. To this the Americans stated that the surrender must be unconditional. Molotov, meanwhile, advanced a Soviet demand, after Russia had been in the war all of two days, that their own Marshal Vasilievsky should share joint-command with General Douglas MacArthur, in the postwar control of Japan.

In a secret message to the British government on August 11th, Secretary of State James Byrnes laid out the proposed US conditions for a Japanese surrender, which included a key sentence on Allied prisoners:

"IMMEDIATELY UPON THE SURRENDER, THE JAPANESE GOVERNMENT SHALL TRANSPORT PRISONERS OF WAR AND CIVILIAN INTERNEES TO PLACES OF SAFETY, AS DIRECTED, WHERE THEY CAN QUICKLY BE PLACED ABOARD ALLIED TRANSPORTS."[16]

Japan accepted the terms of surrender on August 14, 1945. As it turned out however, the Japanese were so fearful of the Soviet Union that they were to refuse to evacuate American and Allied prisoners to places where they could be rescued by U.S. or British forces, without first obtaining permission from the Red Army. The Soviet army in Manchuria, under the immediate command of Marshal Malinovsky, continued the offensive. (Malinovsky, one of Stalin's key commanders, had commanded the 2nd Ukrainian Front Army in the conquest of Romania, Hungary, Austria and Czechoslovakia, operating on Tolbukhin's flank.)

At Spaso House, the residence of U.S. Ambassador to Moscow, Averell Harriman, a celebration of the end of the war took place on August 15. General Dwight D. Eisenhower was there, on his Moscow visit, and Harriman recalled that he was treated to a genuine hero's welcome in the Soviet capital. Marshal Zhukov and many other Soviet officers and officials attended, and the Americans present recorded having difficulties getting rid of the drunken Russian guests.[17]

During the Moscow visit, Eisenhower, Harriman and General Deane were invited to stand with Stalin, Molotov, Kalinin, Zhukov and other Soviet leaders on top of Lenin's tomb and review a huge victory parade, which was viewed as a great honor for foreigners. In his memoirs, Eisenhower records an interesting fact: "Throughout our stay Marshal Zhukov and other Russian officials pressed me to designate the spots I should like to visit. They said there was no place, even if it took us as far as Vladivostok, that I could not see..."[18] It would appear that Zhukov may have been trying to use what influence he and the other regular Red Army officers had to initiate a diplomatic move on behalf of their esteemed American Army allies, perhaps in hope of countering the omnipotent NKVD/MGB. In light of the events with American prisoners and missing over the previous six months it is strange that Eisenhower did not seize upon Marshal Zhukov's suggestion and demand to go to locations where American prisoners had been reported by intelligence sources. Eisenhower said that his time was short and decided on a visit to a military museum in the Kremlin. Harriman recorded that Eisenhower, "had come to Moscow with some definite ideas of his own about the postwar future." The SHAEF Commander was sure that Zhukov, whom he considered a friend was going to succeed Stalin, and that after that the United States would cease to have major difficulties with Russia.

Both Harriman and General John Deane told him he was wrong, based no doubt on their years of dealing with Stalin, Molotov and Vyshinsky in Moscow, and a better understanding of the Soviet Communist system. Perhaps Eisenhower thought that the POW matter could be worked out peaceably with the Soviets given some more time, and continued expressions of America's peaceful intentions. Eisenhower may not yet have seen Gehlen's report on the power of the political Commissars with direct links to Stalin, over officers of the regular Red Army like Zhukov, regardless of their personal popularity. In regards to Zhukov, Eisenhower evidently didn't fathom the Marxist-Stalinist dread of military takeover, known among communists as "Bonapartism," or the power and deviousness of Stalin himself. By the time he did it was too late, and the American and British people had already been told by the War Department that all their prisoners were home, thus automatically initiating a coverup. (Zhukov was eventually downgraded to a lesser command at Odessa, by Stalin personally.)

Eisenhower had made at least one last personal and roundabout effort to secure permission from Marshall Zhukov to obtain entrance for American contact officers, in Berlin during August 1945, by offering the Soviets liberal access to the western zones. He was unsuccessful, as Stalin was the real authority.[19] At this same time General Edwin Sibert, Eisenhower's Chief of intelligence for the Communications Zone (later Deputy Director of CIA), had begun developing Nazi east-front intelligence expert Rheinhard Gehlen. Gehlen's organization included a stay-behind network inside Soviet-occupied territory. Gehlen's organization later worked for US Military Intelligence and CIA and became the foundation of the West German Intelligence. Evidence has surfaced of secret-returns of small numbers of Allied prisoners for the forcibly-repatriated Soviet citizens of whom Solzhenitsyn, Lord Bethell, Julius Epstein, Nikolai Tolstoy and Mark Elliott have written so eloquently. The 1945-46 repatriations to Stalin's gulags were marred by mass suicides, shootings by Allied soldiers and other violence. Stalin reneged on full reciprocation and most of the Allied POWs disappeared into secret, special camps. Now-declassified U.S. and British documents state that some 199,500 British and Commonwealth prisoners, including Canadians, Australians, New Zealanders, Indians and South Africans, had been held by the Germans up to mid-March 1945, yet only 168,746 were eventually reported repatriated. Most of the 30,000 or more missing were in prison camps in the Soviet occupation zone.

In combining reports from various formerly-classified US documents, the total of US prisoners in German hands in early 1945, including late-captured MIAs not known individually by name to be in a POW status, appears to have been some 110.000-112,000, of which 90,937 (including 25,000 MIAs) had been officially reported repatriated by January 1946, although the actual number returned may have been lower, due to double-counting of some. In retaining such numbers of prisoners, the Russians gained the additional benefit of free labor from thousands of technologically-advanced (by Soviet standards) U.S. and Allied prisoners. Others were no doubt made useful for NKVD/KGB and military intelligence, and in training KGB and GRU deep-penetration agents and spies.

While several hundred American POWs were reported returned to military control, "...after being released or having escaped from enemy territory," during July, August and September 1945, and were subsequently shipped back to the United States, sometimes under "secret" orders, it appears that in some cases these POWs had actually been released or escaped earlier. In addition, the numbers contained in these declassified records, are often in small groups of 3 to ten, occasionally 20 to as many as 43 in one case, but the total, in the hundreds, in no way approximates the vast number of American missing and POWs who had disappeared in the Soviet occupied zones.[20]

Meanwhile, President Truman had sent a Top Secret cable to Stalin on August 15th in which he clarified the U.S. position on surrender of Japanese forces, and among the conditions was included:

"...B. THE SENIOR JAPANESE COMMANDERS AND ALL GROUND, SEA, AIR AND AUXILIARY FORCES WITHIN MANCHURIA, KOREA NORTH OF 38 DEGREES NORTH LATITUDE AND KARAFUTO SHALL SURRENDER TO THE COMMANDER IN CHIEF OF SOVIET FORCES IN THE FAR EAST."[21]

(Stalin's reply merely added a further demand that the Soviets be allowed to occupy the northern part of the Island of Hokkaido in addition to what areas had already been agreed to at Yalta.)

In an urgent, top secret message from General Douglas MacArthur of August 17, 1945 (C-34359), the US Commander in the Pacific informed the War Department that the Russians were still attacking the Japanese and advancing in Manchuria:

"THE FOLLOWING MESSAGE HAS JUST BEEN RECEIVED FROM THE JAPANESE GHQ:

'While on our side the Imperial order has already been given to cease hostilities, the Soviet forces are still positively carrying on the offensive and their spearhead is reaching a point west to Mukden early this morning...'

MacArthur continued the message with his own analyses:

"ON ALL FRONTS BUT THE RUSSIAN, OFFENSIVE ACTION ON THE PART OF THE ALLIES HAS CEASED. IT WOULD UNQUESTIONABLY IMPROVE THE OVERALL PROSPECTS OF A BLOODLESS SURRENDER IF THE RUSSIAN GOVERNMENT SHOULD CONFORM TO THE ACTION TAKEN ON THE OTHER FRONTS..."

At this same time the Chinese Communists were putting forth claims to the US Commander in China, General Wedemeyer, about their conquests of large expanses of territory from the Japanese, and demands that they be allowed to occupy any city or area which they desired. They also demanded an immediate end of American Lend-Lease shipments to the Kuomintang (KMT) Nationalist Chinese Army under Chiang Kai-shek, on the 17th of August.

Meanwhile, on August 18th, Lord Mountbatten, the Commander in Burma (who was a member of the Royal family and would ultimately be murdered in an IRA terrorist attack) raised the issue of "Disarmed enemy forces" versus those carried as POWs entitled to protection under the Geneva Convention, that had caused widespread suffering and death among German prisoners of the Allies along the Rhine, and which the Russians would later exploit to the maximum with both German and Japanese whom they retained as permanent forced-labor. MacArthur transmitted the message (C-34637) to Washington:

"Following message received from Mountbatten, 'ascertain from General MacArthur if we may use Japan's soldiers as labor to repair the ravages of the war, WHETHER THEY ARE TREATED AS PRISONERS OF WAR OR MERELY DISARMED. What is your policy with reference to this matter."

On the same day, August 18, Truman cabled Stalin refusing the Soviets American permission to occupy northern Hokkaido or any of the main Japanese islands, and requesting American air base and landing rights on the Kurile Islands, now being occupied by the Soviets.

Stalin replied sharply to Truman on the 22nd:

"I UNDERSTAND THE CONTENTS OF YOUR MESSAGE IN THE SENSE THAT YOU REFUSE TO SATISFY THE REQUEST OF THE SOVIET UNION FOR THE INCLUSION OF THE NORTHERN PART OF THE ISLAND OF HOKKAIDO IN THE REGION OF THE SURRENDER OF THE JAPANESE ARMED FORCES TO SOVIET TROOPS. I HAVE TO SAY THAT I AND MY COLLEAGUES DID NOT EXPECT SUCH AN ANSWER FROM YOU..."[22]

Stalin then went on to refuse the American request for airbases in the Kurile Islands.

On the 19th of August an American OSS/Army POW contact team succeeded in dropping into Mukden, Manchuria, where nearby Japanese POW camps held some 1,500 US prisoners of war, along with a few other Allied POWs. Mukden was still under Japanese control, and the Americans had made a desperate effort to contact

US POWs before advancing Soviet forces overran the area. General Wainwright, who had stayed behind until Bataan and Corregidor fell to the Japanese in 1942, after MacArthur's withdrawal to Australia, was a prisoner in the area. The American high command had made a priority of getting Wainwright back as soon as possible, and requested the Japanese to allow American aircraft into Mukden to evacuate him and the other American POWs in the area, before the Soviets arrived.

The Japanese were afraid to authorize this American action without prior approval from Stalin's government. They were no doubt already concerned about the treatment they themselves would receive when they became prisoners of the Red Army. General Deane, in Moscow went to see General Antonov of the Soviet General Staff with a request that American aircraft be allowed into Mukden to evacuate Wainwright and the other American captives, explaining that, "...Wainwright and all other Americans who had fought on Bataan and Corregidor were national heroes, and I emphasized the gratitude the American people would have for Soviet assistance in effecting their immediate release. A few hours later Antonov informed me that instructions had been sent to Vasilievsky to do everything he could to further the prompt release of General Wainwright and other American prisoners of war in the Mukden area. He was also authorized to allow our planes to bring supplies to the Americans until all had been evacuated from Manchuria,"

The way things actually worked out in Mukden were not quite so simple. Vasilievsky sent a group of Soviet officers to the Mukden area to locate General Wainwright and found that Wainwright was with a smaller group of POWs at a camp at Hsian about 100 miles north of Mukden. The Soviets brought Wainwright and some other American officers to Mukden by road. Meanwhile the American attempts to gain entrance for a number of US aircraft sufficient to evacuate all the American POWs in Manchuria ran into typical Soviet obstinacy. General Wedemeyer cabled the War Department and MacArthur on the 19th of August (CFBX-5225):

"...First report from our Mukden contact team indicated definite possibility early evacuation General Wainwright from Mukden area...FURTHER REPORT FROM MUKDEN CONTACT PARTY GIVES ADDITIONAL INFORMATION RECEIVED BY THAT PARTY FROM JAPANESE AT 1000 HOURS 19 AUGUST. SUBSTANTIALLY JAPANESE STATED IT IS VIOLATION OF JAPANESE-RUSSIAN 'PEACE TERMS' (SICO TO ALLOW US PLANE TO LAND AT MUKDEN AND UNLOAD SUPPLIES. JAPANESE WILL WAIT RECEIPT OF WORD FROM RUSSIANS BEFORE SUPPLIES CAN BE UNLOADED. EVACUATION OF ANY PRISONERS IS TO BE HELD UP BY JAPANESE UNTIL THEY RECEIVE DEFINITE WORD FROM SOVIET RUSSIA...DUTCH AMBASSADOR HAS MADE STRENUOUS REPRESENTATION TO ME SEVERAL TIMES DURING THE PAST WEEK,

CITING HIS DIRECTIVE FROM...QUEEN WILHELMINA...TO EVACUATE THE FORMER GOVERNOR GENERAL OF N.E.T., BELIEVED TO BE IN THE SAME AREA AS GENERAL WAINWRIGHT..."

Wedemeyer cabled further information on Mukden to the War Department and to MacArthur, with the addition, "Pass to Deane (Moscow)," on the 21st (CFBX-5396):

"...B-24 was reported to have left Hsian China 20 August at 1100 for Mukden with ETA at Mukden 1700. Landing was to be made if considered advisable from air observation. Major Watson Agas (M.I.S. X) and Major Helm OSS plus 2 photographers were aboard. No later information on this...It is probable that all American personnel who land will be interned until word is received by Japs from Soviet Russia that Americans can unload supplies at Mukden for prisoners of war. Urgent that immediate contact be made between American Headquarters and Soviet Russia...Request Soviet authorities inform Vasilievsky..."

This message was relayed to Moscow and immediate copies were sent to General Hull, Admiral Leahy, General Arnold, General Somervell, OPD-State Department, General Henry, General Clayton Bissell at G-2, War Department, Admiral King and to Mr. McCloy, as were most of the cables in the Manchurian episode.

General Deane cabled back to the War Department on the same day from the US Military Mission in Moscow that he had passed the request to the Soviet General Staff and they had verbally assented to the American request for US aircraft to be allowed into Mukden to evacuate POWs. MacArthur then informed the War Department that a US aircraft was taking off from Kadena to Mukden that same day.

Also on the 21st of August, the Joint Chiefs of Staff replied to Lord Mountbatten's request on how Japanese POWs were to be treated as regards to their POW status and their use as labor. The message was sent to MacArthur, Nimitz, Wedemeyer and other commanders (WARX 52674):

"United States policy is as follows: THE FOUR ALLIED POWERS ARE BOUND BY THE POTSDAM ULTIMATUM TO PERMIT THE RETURN OF JAPANESE MILITARY FORCES, AFTER THEY ARE COMPLETELY DISARMED, TO THEIR HOMES. PENDING SUCH RETURN, WHICH OF NECESSITY DEPENDS ON THE AVAILABILITY OF TRANSPORTATION, THESE PERSONNEL MAY BE USED FOR SUCH PURPOSES AND SUBJECT TO SUCH CONDITIONS AND DIRECTIVES AS MAY BE PRESCRIBED BY THE NATIONAL COMMANDERS AUTHORIZED TO RECEIVE THE SURRENDER. SURRENDERED JAPANESE SOLDIERS SHOULD BE CONSIDERED AS DISARMED PERSONNEL AND NOT NECESSARILY AS PRISONERS OF WAR...DESIRE YOU INFORM MOUNTBATTEN OF ABOVE..."[23]

The acute need for intelligence inside Soviet-occupied territory, to (among other major concerns) locate hundreds of thousands of missing Allied prisoners for possible secret rescues or negotiations, had caused covert US-Soviet air-battles in Europe and now led to postwar US naval threats against the Russians at Dairen and Port Arthur, in Manchuria. In Moscow, General John Deane was informed that Stalin had told President Truman that Lt. General Kuzma Derevyanko would represent the Soviet Union during the surrender negotiations at General MacArthur's headquarters. Derevyanko (spelled Derevenko by Harriman, in his memoirs) was one of Stalin's trusted subordinates who had been chosen to confront Patton's Third Army negotiators over U.S. and British POWs under Soviet control in Austria in May and June 1945. That he was chosen by Stalin to represent him in the surrender of Japan is a clear indication of the favor in which he was held by the dictator.

Deane was instructed to assist Derevenko in getting to Tokyo in a long-range American aircraft as the Soviets did not have one capable of making such a flight at the time. When a great typhoon in the north Pacific intervened and prevented the arrival of the U.S. aircraft for ten days, Deane was accused of deliberately dragging his feet so that Derevenko would not be able to attend the surrender ceremony. This absurd charge was alleviated when the storm passed and MacArthur's aircraft arrived in Siberia in time to transport Derevenko to Japan. Derevenko then became the Soviet diplomatic representative in Tokyo during the occupation. The formal surrender ceremonies took place on board the battleship USS Missouri on September 2, 1945. The enormous size of the battleship, photographed from an angle which further magnified it's proportions and published in the Soviet Union, impressed many Russians with the military might of the United States. A subsequent overflight by 400 B-29 bombers and 1,500 carrier aircraft must have further impressed Derevenko and the other Russians present, although no record of Soviet impressions of U.S. power has so far been revealed.

(Later, after leaving Japan, Derevenko would be placed in charge of an entire region of Gulag camps containing millions of former POWs, of many nationalities, engaged in forced labor projects.)[24]

It is possible that the temporary halting of the forced-repatriations during the last three months of 1945 which followed the final Russian repatriations of Japanese-held Americans from Manchuria through Mukden and Dairen, and other areas, was related to an unfulfilled American expectation that the U.S. would receive thousands of missing U.S. POWs from the European Theater who had been shipped to Siberia with the German, Austrian, Italian, French and other European prisoners of war. Whether or not this was the case, the United States government was already seriously concerned about the Japanese-held prisoners in Manchuria who would soon fall into Russian control.

After the atomic bomb had been dropped on Japan, while the

Soviets were still capturing prisoners for slave labor in Manchuria, another violent forced-repatriation of Russians had been attempted at Kempten, Germany, on August 12th. 410 Russians were supposed to be returned, but all were terrified of the fate that awaited them at the hands of the Soviet NKVD and SMERSH operatives. The original guards had been friendly and let some escape, but the rest were later surrounded by military police. 100 were seized during a church service in a violent operation, with U.S. soldiers dragging some by the hair and eventually resorting to shooting. Some refugees claimed that thirty-five Russians died in the repatriation, but most said that only wounded resulted:

"On Sunday, 12 August, Russian DPs were in their (Orthodox) camp church. The religious services were interrupted by the arrival of an armed detachment of American troops. An American Military Government officer told the Russians to leave the church and to board the trucks for repatriation. The Russians did not comply as they had already declared that they did not wish to repatriate. Shortly thereafter—the open windows of the church faced the Baltic camp—we saw armed Military Government troops enter the church, upset the altar, drag the aged priest by the beard, push and drag men, women and children out of the church and bodily lift about 70 persons on the waiting trucks. The people offered no resistance-they just wept and begged for mercy and help...Russian mothers attempted to save their infants by tossing them through the church windows into the Baltic camp yard. One mother succeeded in tossing two children through the window. Some adults jumped into the Baltic section and ran into the milling crowd of Balts who stood trembling as helpless onlookers of the shocking scene. Some soldiers fired several shots into the crowd. Two Ukrainians were seriously wounded and were operated on in the Kreigslazaret II DP hospital. Ten persons were contused...Russian "repatriation officers" observed the scene with evident amusement and repeatedly took pictures of the clubbing of Russian DP's by American soldiers...Where are the rights of man, the principals of humanity... the Atlantic Charter, and, above all, Christian principals of American democracy?"[25]

Meanwhile, at Fort Dix, New Jersey, Soviet repatriation officers took custody of the 146 Russian 'Vlasov Army' prisoners of war who had rioted at the end of June, when three had committed suicide rather than to return to Stalin's Russia. These were the last of 4,000 Russians who would be forcibly repatriated from the continental United States to the Soviet Union. On September 6th, U.S. authorities failed in an attempt to move 600 Ukrainians and 96 Armenians from Mannheim to the Stutgart DP center for eventual repatriation to the USSR, and other, similar incidents occurred elsewhere in Europe.

General Walter Bedell Smith, Eisenhower's Chief of Staff in

Europe (later U.S. Ambassador to Moscow and CIA Director), briefly mentioned the 1945 forced-repatriations of Soviet citizens in his later memoirs. Considering that by mid-August 1945 some 2,000,000 Soviet citizens had been repatriated from the U.S. zones, forcibly or by coercion, or in other cases, voluntarily, through fear of appearing to be unwilling to onlooking Soviet authorities, Smith greatly understated the scale and pathos of these operations. He also gives the apparently innocent impression that he and other American leaders were surprised and perplexed at Soviet demands, when in fact Soviet threats to retain U.S. and British POWs unless their own repatriation demands were met, had been well known at the highest civilian and military levels since before the Yalta agreements:

"Our most serious problem with the Soviet Government while I was in Germany concerned the implementation of an Allied agreement made at Yalta and reaffirmed at Potsdam, looking toward the return of displaced persons—prisoners of war, slave laborers and others—to their country of origin...the Soviet interpretation and our interpretation differed considerably. We believed that we were talking about facilitating the return to Russia of those Soviet nationals who desired to return plus, of course, such individuals as might be charged specifically with war crimes. We certainly did not intend to violate the traditional American attitude toward giving sanctuary to political refugees, of whom we found thousands in Germany."

Bedell Smith even blandly alluded to the specific August 1945 forced-repatriation by American authorities at Kempton, although his account and those of others differ considerably:

"I found on one occasion that an American unit, somewhat overzealous in carrying out its responsibilities, actually had begun to load forcibly on a train some of these Russian displaced persons who had refused to return voluntarily to the Soviet Union. When their pleas seemed unavailing, one or two actually committed suicide...We immediately issued instructions that we would not forcibly repatriate anybody except actual war criminals, since it entirely contrary to our principles to force the return to the Soviet Union of any of these people who did not wish to go..."[26]

Smith's sketchy, inaccurate and piously self-serving account of the forced repatriations belies the fact that he had full access to all the gruesome details of these operations, some of which are still secret at this writing. From his memoirs it would appear that one isolated incident of violence caused a U.S. policy change which he himself was instrumental in. He says nothing about the continuation of forced repatriations until 1948, long after the incident mentioned. Nowhere does he mention the disappearance of many thousands of American and other Allied POWs in Soviet control, or the fact that the forced repatriations were a brutal attempt to ransom those prisoners from Stalin.[27]

The Supreme Allied Commander in the Mediterranean area, Field Marshal Harold Alexander, was by now openly refusing to submit to Soviet demands of repatriating to Soviet-occupied Poland the Polish soldiers of General Anders corps in the AFHQ area. A cable of August 21st to the Secretary of State from U.S. Political Adviser Alexander Kirk states:

"At a meeting held at AFHQ on August 18 presided over by Field Marshal Alexander stated flatly that according to his information there were approximately 30,000 Soviet citizens in Polish Corps in Italy and asked permission for Russian officers to visit Polish Corps units in order to seek Soviet citizens. Supreme Allied Commander states he assumed Basiliov in referring to 30,000 Soviet citizens in the Polish Corps was adopting the Soviet definition of citizens as all persons residing east of Curzon Line. This was confirmed by the Russian delegation and Alexander replied he did not concur in this view. He added that while he would be very glad to investigate all allegations made by Soviet delegation, he could not permit Russian officers to visit Polish units. He insisted that it must be clearly understood that Poles were Allies whose forces formed a corps which had fought exceedingly well...

"Field Marshal Alexander has informed British Chiefs of Staff that he will not, repeat not guarantee the personal safety of any Polish Provisional Government representative who may be allowed access to Polish troops. He stated also that he is not prepared to answer for discipline or morale of Polish troops in such an eventuality. During recent interview with Sir Orme Sergeant at Foreign Office, General Morgan, (the) Supreme Allied Commander's Chief of Staff warned him in this matter. SAC added in his message to British Chief of Staff, however, that he would be prepared to accept visit at AFHQ of small delegation from Polish Provisional government."[28]. (Other documents in this series remain "security classified" under authority of OSS/CIA and of Great Britain, according to 'security-classified' withdrawal cards placed in the State Department records in the U.S. National Archives.)

At least superficially appalled by the violent conduct of the forced-repatriations of Russians in Europe, with little apparent gain, General Eisenhower had ordered a temporary halt to them in late August, and was urgently requesting during September 1945, a review of the policy by Washington. The real reason for this concern, however was probably the publicity which occurred when the secret began to leak out. Robert Murphy, the U.S. Political Advisor in Germany had asked Washington for clarification on August 27th:

"...IN APPLYING THE POLICY OF FORCIBLE REPATRIATION THERE HAS BEEN A NUMBER OF UNPLEASANT INCIDENTS INVOLVING VIOLENCE...A CONSIDERABLE NUMBER OF SUICIDES BY RUSSIANS IN

THIS CATEGORY APPARENTLY ARE ALSO TAKING PLACE...DID WE AT YALTA ASSUME THE SPECIFIC OBLIGATION TO RETURN THESE RUSSIANS BY FORCE IF NECESSARY?...GRAY HAS SUGGESTED THAT WHERE FORCE IS NECESSARY WE MIGHT WISH TO PERMIT RUSSIAN TROOPS TO ENTER OUR ZONE FOR THE PURPOSE OF REMOVING THESE INDIVIDUALS..." (Cecil W. Gray, author of the final, disgraceful suggestion in this message, was Counselor of Mission, Office of the US Political Advisor for Austrian affairs.)[29]

The answer for Murphy and Eisenhower came back on August 27th signed by Byrnes but written by Assistant Secretary of State H. Freeman ("Doc") Matthews, who, with Hopkins, Bohlen and Alger Hiss, had advised Roosevelt at Yalta:

"...WHILE AGREEMENT MAKES NO MENTION OF SOVIET CITIZENS CAPTURED IN GERMAN UNIFORMS NOR THE USE OF FORCE, SOVIETS HAVE CONSISTENTLY CLAIMED IT COVERS ALL THEIR CITIZENS AND DEPARTMENT HAS INTERPRETED IT AS COVERING POWS OF SOVIET NATIONALITY PRIOR TO 1939 AND IN CONCURRENCE WITH WAR DEPARTMENT ORDERED RETURN TO EUROPE AND TURNED OVER TO SOVIET AUTHORITIES A NUMBER OF POWS BROUGHT TO THIS COUNTRY...INCIDENTS INVOLVING RESISTANCE REQUIRING USE OF FORCE...OCCURRED...

"FOR YOUR CONFIDENTIAL INFORMATION, DEPARTMENT HAS BEEN ANXIOUS IN HANDLING THESE CASES TO AVOID GIVING SOVIET AUTHORITIES ANY PRETEXT FOR DELAYING RETURN OF AMERICAN POWs OF JAPANESE NOW IN SOVIET OCCUPIED ZONE, PARTICULARLY IN MANCHURIA"[30]

Thus, it was clearly reiterated to Eisenhower by the policymakers in Washington, that the forcible repatriations still had as their object the recovery of US prisoners in Soviet hands. At this time Eisenhower also received as visitors, Averell Harriman and General John Deane, who briefed him on the dismal situation in Moscow.

It is also certain, as evidenced by this Matthews/Byrnes cable of August 27, that the United States was extremely concerned about the American POWs in Manchuria who were in the process of being overrun by the Red Army.

Instead, American OSS teams searching for US POWs in August and September 1945, met with hostile and sometimes violent reactions from the Russians and communist Chinese. (In one instance, OSS Captain John Birch was killed.) Also, on August 29th, 1945, the Soviets shot down a US B-29 bomber on a POW rescue mission over Korea.[31] The number of US and Allied prisoners of war in Manchuria given in a recently declassified US 1945 document, under the heading, "Latest reported population and nationality breakdown, as of 7 September 1945," was given as: In Mukden, 1709 total Allied POWs, including 1139 U.S., 274 British and 158 other nationalities. In (Wei)hsien, 1463 total Allied POWs, including 198 U.S., 976 British and 289 other nationalities. The total for the two camps therefore, was given in this report as 1,337 U.S., 1,250 British, and 447 other nationalities, or 3,034 Allied POWs in Manchuria. Other U.S. and Allied POWs reported held in northern China included 985 (including 128 Americans) in Peking and 1182(?) in Tienstin.

It is by no means certain that this is a complete accounting of all US prisoners in Manchuria and northern China in 1945. It is the only document the author located in the National Archives containing the numbers of POWS. Archivists approached by the author declined to assist the author's search for more records on the subject of American POWs in Manchuria. The United States Department of Defense wrote to the author on 11 July 1990 in response to a Freedom of Information Act request for other official records on this subject: "The Department of Defense does not have any of the lists and records for the time frame you requested. If they exist they are under the cognizance of the National archives..."

One other document, located in the same archival area and dated, "As of Sept. 20, 1945," is titled: "The Total of P.O.W. Death And The Number Officially Reported, From Japanese Sources." In this document, reporting on US deaths in 17 areas, the Manchurian prison camp at Weishien is not listed at all, and under Mukden 222 Americans (and 2 British) are listed as dead, according to Japanese sources. In parentheses the "officially reported" (according to U.S. government figures) deaths at Mukden are listed as 254, and also again listed, are 2 British. No explanation is given for the extra 32 Americans reported as dead, although on a later list of "prisoners not yet returned to military control in the China/Burma/India Theater," it is interesting to note that 33 Americans are listed as such.

In the National Archives there exists an accounting of "Captured or Interned United States Army Personnel," accurate up to October 31, 1945. In these statistics, under the heading of "Japanese Theaters," a total of 23,460 Army-only prisoners were listed as having been taken. Of these 6,941 are listed as "Prisoners died while in captivity," 15,296 are listed as "Returned to Military Control," and 1,223 American POWs are listed as "PRISONERS NOT YET RETURNED TO MILITARY CONTROL." These were not MIAs, but known US POWs.[32] By the last day of 1945, U.S. Army Strength Accounting Office reported that only 86 American POWs remained in this status as still-POWs, while 1,742 more U.S. Army personnel in the Japanese Theater had been recently "Declared Dead, and 9,685 remained in a current status of MIA. Of the 1,742 MIAs in the Japanese theaters that had just been "declared dead," 1,202 are listed as "reported dead from MIA" status. At this time 8,436 Americans are listed as dead during imprisonment in the Japanese theaters and 16,961 as returned to U.S. military control.[33] Yet a much later, and presumably more "accurate" accounting lists only 16,358 U.S. POWs returned to American control in the Pacific theater, while 8,452 are reported dead of non-battle causes and 182 dead of wounds and injuries.[34] It is not as yet known how many of these men were taken as captives by the Soviets in Manchuria or the Chinese Communists in northern China, whose subsequent disappearance was kept classified by the U.S. government. U.S. Marine Corps losses in missing in action and POWs in the Pacific theater at the end of World War II could not be located by the author in the Archives and NARA personnel would not assist in finding them.

The Department of Defense and the National Archives have refused to make any other, more detailed statistics concerning U.S. POWs who fell under Russian or Chinese communist control in Manchuria or north China, available to the author or to ex-POW veterans. In 1945, U.S. officers on the scene were dependant upon the War Department in Washington for precise information on how many American, British and other POWs were to be expected in the Mukden area who should have been recovered. Likewise, to this day no information or full accounting has been made available on the fate of the U.S. and British prisoners at Weishien, who were not evacuated with General Wainwright. In short, an aura of mystery and incompleteness hangs over the repatriation of U.S. and Allied POWs from Manchuria and north China. This is illustrated by a strange series of incidents which occurred at the time of the evacuation of the last POWs, known by officers on the scene, to leave Manchuria.

Both before and after the last U.S. POWs delivered by the Russians in Manchuria had sailed, the United States naval forces made several unexplained military demonstrations of power against the Soviet forces in Manchuria. The order for a US military "SHOW OF FORCE" to the Russians came from such a high level that senior US officers at the scene were not informed of the reasons for these threats. These actions were recorded in the secret report of OSS Lt. Colonel James Donovan:

"ON ONE DAY BETWEEN THE 2nd AND 8th OF SEPTEMBER (before the U.S. POWs had sailed), AMERICAN NAVAL CARRIER AIRCRAFT MADE A 'SHOW OF FORCE' OVER THE HARBOR OF DAIREN...THIS EXCITED THE RUSSIAN AUTHORITIES IN DAIREN A GREAT DEAL AND THEY ASKED COMMODORE C.C. WOOD WHAT IT WAS ALL ABOUT. AS HE HAD NOT BEEN GIVEN ANY ADVANCE INFORMATION FROM THE NAVY, HE WAS JUST AS PUZZLED AS THE RUSSIANS AND COULD OFFER ONLY A WEAK EXPLANATION. ON THE AFTERNOON OF THE 13th OF SEPTEMBER, AFTER THE EX-POWs HAD SAILED, ADMIRAL SETTLE WAS INFORMED BY RADIO THAT ANOTHER 'SHOW OF FORCE' WOULD BE MADE THAT AFTERNOON, THIS TIME BY CRUISERS AND DESTROYERS OF THE SEVENTH FLEET.

THIS WORRIED ADMIRAL SETTLE CONSIDERABLY AND HE SENT A STRONG PROTEST TO HIS HEADQUARTERS. HE WAS VERY PUZZLED BY THE SHOW OF FORCE WHICH WAS DULY MADE OUTSIDE THE HARBOR ABOUT 1500 HOURS, 13 SEPTEMBER."

"ALTHOUGH SOME RUSSIAN OFFICERS CAME TO SEE HIM AFTER THE CRUISERS AND DESTROYERS HAD DEPARTED ADMIRAL SETTLE...DID NOT TALK WITH THEM...I DO NOT KNOW IF THEY WERE LODGING A PROTEST AGAINST THE SECOND NAVAL DEMONSTRATION...THE ADMIRAL EXPECTED A PROTEST, HOWEVER, AND WAS AT A LOSS FOR A REASONABLE EXPLANATION. HIS INFORMATION WAS THAT THE SHOW OF FORCE WOULD BE REPEATED OFF PORT ARTHUR...THE ADMIRAL HAD TOLD ME THAT A 'SHOW OF FORCE' IS TECHNICALLY RATHER A BELLIGERENT ACTION..."

"IT IS PERHAPS WORTH MENTIONING THAT THE RUSSIANS WERE EITHER NOTABLY LACKING IN FRIENDLY FEELINGS TOWARD THE BRITISH OR OPENLY CONTEMPTUOUS OF THEM. SOME MADE VERY UNFRIENDLY REMARKS, SUCH AS DESCRIBING ENGLAND AS 'THE PROSTITUTE OF THE WORLD'. I COULDN'T QUITE GET THE APPLICATION OF THIS INVECTIVE BUT DIDN'T ASK FOR AN ELABORATION AND REMAINED NONCOMMITTAL, IN SPITE OF SOME EFFORT TO GET ME TO MAKE SIMILAR EXPRESSIONS OF OPINION..."[35]

The series of high-level American threats to Soviet forces in Manchuria in September 1945, reported in secret by Donovan, but unrecorded in any official history or personal biography of the period, clearly represented serious U.S. displeasure with Soviet actions in the area. Whether policymakers at the highest level in Washington had information that some number of American, British or other prisoners were being secretly retained by Soviet authorities in Manchuria or nearby in Siberia, without the knowledge of U.S. officers on the scene, is still unknown, as the U.S. government has chosen to never reveal the full and complete records of the POW subject to researchers, veterans or POW/MIA family members. Some of the American POWs returning from Manchuria were reported killed when a U.S. transport (Colbert) struck a mine and sank.

As recently as late 1986, in response to disability claims by American ex-POW veterans of Manchuria, the Defense Department refused to make all records of the Mukden area POW camps available to substantiate the veterans' claims against the Japanese government for their use in illegal medical experiments. The American prisoners at Mukden had been injected and sprayed by Japanese doctors with deadly viruses such as plague bacillus, cholera, dysentery, typhoid, and anthrax. The prisoners reported that many had died of these causes or deliberate freezing experiments. Subsequently Japanese doctors performed body dissections and administered medical and psychological tests on survivors. At a House Veterans Affairs subcommittee hearing, John H. Hatcher, Chief of Army Records Management testified that the service "has no records from the Mukden camp," and "only about 200 pages of secondary documents, such as interrogations of Japanese officials." U.S. Representative Pat Williams of Montana said it was not known how many survivors of Mukden were still alive or how many had been denied compensation. According to official records revealed during the testimony 1,318 U.S. servicemen were liberated from captivity in this area at the end of the war. This figure must represent prisoners liberated from both the Mukden camp and from Weishien. According to the official records located in the National Archives and cited previously in this book 1,337 Americans had been in these camps on September 7, 1945, and apparently, at the time of the Russian takeover. It is interesting to note that both the Pentagon and the National Archives have refused this author access to any records of the Mukden or Weishien POW camps and the 200 pages of secondary material referred to by Mr. Hatcher in 1986.

In the hearings it was reported that some 300 Americans had died during these Japanese medical experiments with POWs, although official U.S. records in the National Archives record that 254 Americans had died while in captivity at Mukden, of which the Japanese had reported 222. It is not as yet known whether the U.S. government used the Japanese-caused deaths of many prisoners at Mukden to cover for the disappearance while under Russian control of some relatively small number of U.S. POWs in August and September 1945.[36]

Colonel James Donovan's mention of the Russians denouncing England as, "the prostitute of the world," is likely a reference to the forcible repatriations of the some 50,000 Cossacks, old anti-Bolshevik leaders and other anti-communist Soviet citizens in Austria and elsewhere, in the attempt to ransom British POWs remaining in Soviet control. Colonel Donovan had mentioned in his report that many of the Red Army soldiers in Manchuria were veterans of the German war, transferred to the far east in the summer of 1945. Stories of the British betrayal of their old anti-Bolshevik Cossack and Russian allies must have spread rapidly throughout the Red Army "grapevine." It is not known whether the contemptuous remarks about the British recorded here by Lt. Colonel Donovan reflected some additional British action or lack of action on behalf of the reported 1,250 British POWs in Manchuria, or many others held in northern China.

General William Donovan, Director of the OSS, sent a secret, though comparatively incomplete memorandum on Soviet actions in Manchuria to President Truman on September 17, 1945:

"...For the President: The following information, obtained by OSS representatives in Mukden and transmitted by the OSS representative in Chungking, is dated 13 September: Americans are very unpopular with the Soviets in Mukden, PROBABLY BECAUSE THE SOVIETS DO NOT WISH THE AMERICANS TO OBSERVE THEIR ACTIONS...They are carrying out a program of scientific looting...The Soviets are indiscriminately killing Chinese and Japanese...Three OSS men have been held up by Red Army soldiers armed with submachine guns and their watches, wallets and pistols taken from them..."[37]

It is not known what other intelligence concerning the treatment and American, British and Allied prisoners in Manchuria was conveyed to the President by Donovan as such OSS records and documents have been denied to the author by the National Archives for national security reasons and because of objections to their release by the Central Intelligence Agency.

In any case, the OSS was about to be terminated and dismembered by Truman, who distrusted Donovan. An Executive order signed by the President on September 20, 1945 disbanded the wartime intelligence agency. It was to officially go out of existence on October 1st, and its components—with a portion of the staff parceled out to the War Department and State Department.

According to declassified US documents and other sources, some 800,000 to 1,200,000 or more Japanese prisoners of war were also taken by the Soviets from Manchuria, and according to formerly classified documents and then-Secretary of State James Byrnes, some half a million eventually remained in Siberia as slave-laborers in Stalin's Gulag. A declassified Secret OSS cable after the surrender of Japan reported to Washington:

"RUSSIANS SENDING TRAINLOADS OF JAPANESE TO SIBERIA FROM HARBIN AND MUKDEN. A RUSSIAN SOLDIER SAYS THE JAPANESE TO WORK THERE...ALL WHITE RUSSIANS ARE BEING SENT TO SIBERIA UNLESS THEY ARE CITIZENS OF AN ALLIED POWER..."

Another Strategic Services Unit (SSU) cable a few months later stated: "...NEARLY 200,000 JAPANESE SOLDIERS FROM THE KWANGTUNG ARMY WERE TRANSFERRED BY THE RUSSIANS TO CHITA, SIBERIA...APPARENTLY THE SOLDIERS WERE USED AS LABORERS..."[38]

## THE CONTINUING SOVIET INTELLIGENCE CONNECTION TO POW REPATRIATIONS

The actions of highly-placed Soviet moles and spies in the 1945 British and American governments severely compromised the Allies, and greatly assisted the Soviets throughout the entire 1945 POW-repatriations crisis. One example of this may be found in the published Diary of Harold Nicolson (former Parliamentary Secretary and Ministry of Information official), for February 26, 1945, concerning a long-time Soviet spy, Guy Burgess, who passed every secret document he could to the Soviets, and who eventually

defected to Moscow:
"I DINE WITH GUY BURGESS WHO SHOWS ME THE TELEGRAMS EXCHANGED WITH MOSCOW."

Harold Adrian R. "Kim" Philby was chief of all Soviet counterintelligence for the British Secret Intelligence Service, MI-6 at the time of the 1945 POW crisis. He also was to cultivate relationships with American intelligence officers of the OSS such as Allen Dulles, James Angleton and Frank Wisner, who later became senior CIA officials. Donald Maclean was First Secretary of the British Embassy in Washington under Lord Halifax, with access to all high level plans and secret documents on POWs. Unknown to the West for decades, Anthony Blunt, later knighted for his services to the Crown, had penetrated MI-5 and had already compromised the secret communications of the Polish government-in-exile, and of the other occupied nations. British Ambassador to Moscow, Archibald Clark Kerr had long and close relationships with later-revealed Soviet spies such as Guy Burgess, Stig Wennestrom, Richard Sorge and Gunther Stein. He had long been known to his colleagues as a supporter and admirer of the Soviet system and Stalin. Documentary evidence now indicates that in mid-March 1945, when Harriman was urging direct Presidential action with Stalin over Allied POWs in Poland, Clark-Kerr was belittling their numbers and urging Churchill not to confront Stalin, for the sake of Allied unity. Alger Hiss was also an advisor to Roosevelt at Yalta, where the repatriation of prisoners was settled, and subsequently had access in the Department of State to most critical and secret US documents and strategies during the POW kidnapping crisis.[39]

Intelligence reports compiled by OSS and it's successor the Central Intelligence Group (CIG) on the thousands of missing American, British and allied POWs known to have come under Soviet control are still secret, and under CIA jurisdiction. (Requests for certain U.S. POW-related records by this writer through Freedom of Information Act (FOIA) certain requests have been denied or ignored, and appeals rejected by the CIA, Department of State, U.S. Army Intelligence and Security Command, Department of Defense, FBI and other federal agencies.) It has been revealed that at the time of the secret POW crisis OSS had a working relationship with NKVD (KGB) to share critical intelligence and willingly identified many U.S. agents in eastern Europe and that the NKVD exploited this to gain intelligence which ultimately proved harmful to American goals and fatal to some western sympathizers and agents behind Soviet lines. In a February 1945 appeasement attempt, by direct orders of Roosevelt and Stettinius, the OSS returned to the Soviets the secret Russian diplomatic and military codes which U.S. intelligence had just acquired clandestinely from Finland.[40] The subsequent BRIDE-VENONA decryptions of late 1945 Soviet intelligence traffic revealed, aside continued high Soviet interest in POWs, that OSS was penetrated by some 14 Soviet agents (one of whom turned out to be Allen Dulles' assistant, James Speyer Kronthal)[41]

Declassified U.S. documents cited elsewhere within this book reveal that U.S. Military Intelligence was reluctant to supply OSS with all the sensitive POW data requested of them. Yet the Army's 1945 Chief of Military Intelligence, a prewar New York lawyer named Clayton Bissell, who had been promoted to General over professional intelligence officers through the influence of Assistant Secretary of War John McCloy, and who had had access to all the most-secret Ultra traffic during the war, and all U.S. and British decryptions in the immediate postwar confrontation with Stalin over U.S. and Allied POWs, was later to be censured by a 1952 Congressional investigation, for withholding and destroying U.S. intelligence in 1945 that was detrimental to the Soviet Union. The Congressional Committee headed by Representative Madden (D-Ind), recommended that court-martial proceedings be instituted against Major General Bissell, and in a statement accompanying their final report said,

"Testimony recorded from numerous witnesses,... governmental, military and civilian revealed a deplorable tendency to conceal facts and reports concerning Russia's flagrant violation of international laws."[42]

Until all CIA-held records relating to POWs are declassified, it will be difficult to assess the role played in this matter by OSS (and later CIA) officials then in Europe, such as Allen Dulles, William Colby, Richard Helms, Frank Wisner, William Casey and many others. Despite Soviet expertise in accomplishing the mass-disappearance of prisoners of war, secret reports were continuing to reach Averell Harriman, George Kennan and General Deane at the US Embassy in Moscow during September 1945 and thereafter, of American prisoners in NKVD captivity and under torture to extract "espionage" confessions. When questioned by the author in 1988 about the disappearance of American POWs in Soviet control during 1945, the retired State Department Russian expert indicated that POWs were a military responsibility and he had not been greatly involved. In a December 13, 1988 letter to the author, Kennan alluded to the classified nature of such information at the time: "I may have from time to time heard gossip about these matters, but I cannot recall doing so." In the same letter Kennan wrote of the author's POW research:

"It would seem to me that the sort of documentation you have sent to me should permit you to form a pretty fair picture of the course of these exchanges with the Russians on the part of both the British and ourselves, and of the results obtained."

Following this, the author sent Ambassador Kennan a copy of a declassified secret September 26, 1945 cable from the United States Embassy in Moscow drafted by Edward Page (a close friend of 'Chip' Bohlen since they had been in college, and had later lived together in Paris), who was in 1945 an interpreter and personal aide to Averell Harriman. This cable passes intelligence (stated within to be FROM KENNAN), to Secretary of State James Byrnes, of the recent release by the Soviets of a British POW who had been held incommunicado as a prisoner of the Soviets at KRASNOGORSK, where German Marshal Paulus and General Seydlitz were also held, and in a Moscow prison where the Britisher reported seeing an American prisoner, Stanley Young.[43] (The author had wondered if the Soviets had purposely segregated this British prisoner for later use as a 'messenger.')

In a letter of February 9, 1989, regarding the author's specific question about this 1945 case, former Ambassador Kennan replied from Princeton's School of Historical Studies at the Institute For Advanced Study: "I am afraid I can, again, be of no help to you. I had no recollection of the episode in question or of the message, a copy of which you sent me." The author's research indicates that Mr. Kennan was involved in high level intelligence-gathering and analysis before, during and after 1945, that some details of the Soviet withholding of American and Allied POWs in 1945 still remain "classified," and that knowledgeable 1945 participants are apparently still sworn to secrecy.)

At the time that this incident involving Kennan had occurred, in September 1945, an important Soviet GRU defector named Igor Gouzenko was seeking political asylum in Canada because he feared for his family's future under Stalin. From the moment he defected Gouzenko began to reveal high-grade intelligence on Soviet espionage operations in North America. While Canadian prime Minister McKenzie King vacillated when first informed of Gouzenko's initial revelations to Canadian intelligence, and nearly let the Soviet code officer be retaken by the NKVD in Canada, he soon understood why Sir William Stephenson (known as "Intrepid") the chief of British Security Coordination (BSC), had protected Gouzenko from the NKVD. Mackenzie King then took Gouzenko's shocking information on Soviet moles and spies to President Truman in Washington. They met on September 30th, and King delivered a briefing to the President on Gouzenko's initial information, that had been sent over from the Canadian Embassy by Ambassador Lester Pearson (later Prime Minister of Canada). The brief summarized "what we have learned about espionage in the United States (and) Soviet intelligence requests for information about U.S. military matters....the case of a courier to the U.S. who turns out to be an inspector for the Red Army, sizing up chances for further espionage in the United States." This same brief had already been delivered to the Office of Special Political Affairs (SPA) under Alger Hiss in Department of State. At the meeting, Truman proposed that the U.S., Canada and Britain all act together in using the information to track down the spies whom Gouzenko knew by code-names or by their positions. Ironically, Truman then urged King to go to the British Embassy and brief Lord Halifax on the revelations, not knowing that the First Secretary of the British Embassy at the time, Donald Maclean, was also a Soviet spy, and would sit in on the meeting. As author William Stevenson wrote decades later:

"It was Donald Maclean who later sent an inside account to Moscow of the consequences of Gouzenko's defection, and the estimates of what the Soviets had better do about it...And Burgess? Professor Wilfred Basil Mann, a distinguished British nuclear physicist, shared offices with (Guy) Burgess in the cramped main chancery of the British Embassy in Washington...Working in Russia's favor was the West's obsession with keeping on the right side of Stalin. This had been illustrated for Stephenson by the inability of the OSS to take legal action in the case of a magazine now defunct, in whose offices three hundred stolen secret documents had been found, including reports from Stephenson's BSC headquarters and labeled A-BOMB...The Soviet dictator's reputation for being difficult had the effect of 'making any concession to Stalin's demands seem necessary,' wrote a BSC political analyst. 'THE FURTHERANCE OF GOOD RELATIONS WITH THE SOVIET UNION HAS BECOME ALL-CONSUMING. ANY CONCESSION IS POSSIBLE, NO MATTER HOW BASE OR CRUEL, IF IT DOESN'T DRAMATICALLY AFFECT STRATEGIC OR POLITICAL INTERESTS.' There was the political tendency in London as in Washington and Ottawa, to brush under the carpet anything embarrassing to Stalin. But exposure of spy rings run by Moscow throughout the Atlantic alliance was not going to be so easily concealed...The real enlightenment came for the FBI with the first captured Gestapo documents to reveal the existence of what the Nazis had dubbed die Rote Kapelle, the Red Orchestra, their own cryptonym for the networks discovered in Europe and extending into Britain. German interrogators had pieced together the history from Russian intelligence officers who belonged to the Red Orchestra and were caught after the 1941 invasion of Russia. It was clear now that the first nets were created as early as 1935; that the main targets had been Britain and the United States; and that the highest priority was given to the penetration of U.S. and British intelligence."[44]

Gouzenko's September 1945 revelations and his subsequent work with the FBI and other intelligence agencies, led to a chain-reaction of exposures of Soviet agents within the U.S., British and Canadian governments, including a British nuclear spy, Dr. Allan Nunn May, who had worked on the atomic bomb project, and had known

Donald Maclean at Cambridge in the 1930s, when Maclean had begun serving Stalin. Also exposed was John Cockcroft, director of the Canadian Atomic Energy Division, which had shared secrets with the Los Alamos project. The coded GRU traffic revealed that the Canadian passport office had been a source for documenting many Soviet agents, who then traveled with impunity in the Western Hemisphere. Some identities were also discovered to have been taken from deceased Canadians, including Canadian volunteers killed in the Spanish Civil War. The use of false identities taken from the dead or from prisoners was an NKVD/KGB trademark that was to continue. An extensive network of GRU (Soviet military intelligence) spies were tracked down after Gouzenko's defection. Gouzenko's defection confirmed, through the use of stolen Soviet documents, that one liaison between the Soviet intelligence and a network of Soviet agents inside the U.S. Government was Elizabeth Bentley, a Vassar graduate who was ideologically committed to Communism. Bently revealed to FBI interrogators how for five years she had acted as a Soviet courier between Washington and New York, carrying secret U.S. documents obtained from 30 government officials including Lauchlin Currie, a Special Assistant to President Roosevelt throughout the war and Harry Dexter White, formerly Assistant Secretary of the Treasury, and then executive director of the International Monetary Fund (IMF), and also Alger Hiss, who held a key position at the State Department. However, it subsequently became clear that the Soviets had made great gains in developing their own atomic weapons through their spies in the West.[45]

The subsequent BRIDE-VENONA decryptions of secret 1945 Soviet NKVD radio traffic, confirmed Gouzenko's statements and revealed that additional Soviet spies and agents-of-influence existed in the American White House, the OSS, the Pentagon and the State Department; and in Britain's Foreign Office, signals intelligence and MI-5, giving the Soviets precise information on every American and British move and countermove, before they were made. One top Soviet spy was identified in the VENONA intercepts as having flown back from Moscow to the United States with Harriman in his private aircraft. Alger Hiss has been mentioned as the candidate, but other senior U.S. officials, including Harry Hopkins, had also traveled with Harriman on occasion. In a February 1945 appeasement attempt, by direct orders of Roosevelt and Stettinius, the OSS had returned to the Soviets the secret Russian military and diplomatic codes which US intelligence under General Donovan had just acquired clandestinely, thus warning Stalin that his secrets had been compromised up until then. (However, a copy had been kept by U.S. intelligence and decryption work secretly continued on Soviet radio traffic.) It has also been revealed that at the time of the POW crisis, the OSS had a working relationship with NKVD (KGB), to share critical intelligence and to willingly identified many US agents in eastern Europe, and that the NKVD exploited this to gain intelligence which ultimately proved harmful to American goals and fatal to some western sympathizers and agents behind Soviet lines.[46] Not surprisingly, Whereabouts and Welfare (of US POWs) records of American POWs in the Soviet zones originating from U.S. agents there, have largely remained secret, as have many other pertinent G-2 and G-5 POW records, although the author has filed Freedom of Information requests.

The Operation BRIDE-VENONA decryptions also revealed that most of the decrypted September 1945 week's NKVD-KGB British traffic was taken up with numerous messages from Moscow, according to MI-5 counterintelligence officer Peter Wright, "detailing arrangements for the return of Allied prisoners to the Soviet authorities, groups like the Cossacks and others who had fought against the Soviet Union." Wright, who read them, said that, "many of the messages were just long lists of names and instructions that they should be apprehended as soon as possible." This clearly indicates that the forced-repatriations were still at the highest priority level for Stalin.[47]

With the surrender of Japan and the rising confrontation with Stalin, Secretary of War Henry Stimson chose to retire on his 78th birthday, September 21, 1945. Robert Patterson, the former judge who had declared the missing Americans as dead at the beginning of June, succeeded him as Secretary of War. One era was closing while another, even more perilous, was unfolding. On the day of his retirement, at the final Cabinet meeting he attended, Stimson strongly advocated sharing the control of the atomic bomb. He was supported in this by former Vice-President, now Secretary of Commerce Henry Wallace. Stimson's strongest opponent was Secretary of the Navy James Forrestal, who said that the Russians were a different kind of people, adding: "It seems doubtful that we should endeavor to buy their understanding and sympathy. We tried that once with Hitler. There are no returns for appeasement."[48] In his later memoirs, written with assistance from a future national security adviser, McGeorge Bundy, Stimson glossed over the lost American POWs of the war that had just ended:

"A...sharpness developed in negotiations over prisoners, both Americans in the Soviet lines and persons from Russian or Russian-occupied territory in American hands. Although there seemed to be little doubt that the ordinary Russian soldier and officer were friendly to liberated Americans, official obstructionism to American efforts to care for Americans was extremely irritating and finally led to a sharp telegram from the President to Stalin At the same time the Russians indicated a keen interest in the 'repatriation' of many men in American hands who showed no desire whatever to be handed over to Russian control, and the Americans were faced with the unpleasant alternative of offending a great ally or abandoning the great principal of political asylum."[49]

Such a mild and incomplete statement on a great human tragedy involving millions of human beings, may be partially explained by Stimson's co-author, McGeorge Bundy, who would serve as national security adviser under a President beleaguered by a future Asian war, two decades later. In an acknowledgment at the close of Stimson's 1947 memoirs, Bundy wrote: "We owe a particular debt to the Department of the Army, whose officers have read and cleared Part III as free from violations of military security."[50] Thus did Bundy absolve Stimson and himself from responsibility for the great secrets they had kept together, about the onset of the Cold War with Soviet Russia.

Stalin's failure to reciprocate in the repatriations of 1945-1946, resulted in an official tendency to minimize the extent of Allied POW losses. Still-living U.S. and British prisoners who returned from Soviet control have stated that they were ordered by intelligence officers to remain silent about their knowledge of other Allied POWs remaining in Soviet captivity. In classified US and British documents the totals of prisoners known to have fallen under Soviet control in the German theaters were gradually scaled down to some 5,400 American POWs and 7000-8,500 British and Commonwealth prisoners. Continuing research indicates that these figures reflect a shifting of missing, known prisoners into other categories, such as MIA and presumed killed. According to the 1953 Senate testimony of Charles Bohlen, an adviser and translator for Roosevelt and Truman, some 60,000 Americans may have come under Soviet control, of whom only some 28,662 were repatriated or exfiltrated from the Soviet Zone. Facing tough confirmation hearings, Bohlen may have been threatening possible consequences for his defeat. After this revelation Bohlen's nomination as Ambassador to the USSR was swiftly approved. Some 78,750 Americans remain missing-in-action from WW II, mostly from the European theaters.

The military was forced to implement what was in fact a decision made by the supreme civilian authority. Top Secret lists of known US prisoners in Soviet hands were classified and buried in the Pentagon under the command of General (later Secretary of State) George C. Marshall. The present writer has interviewed retired US Army Major Carl Heinmiller, of Haines, Alaska, who served under General Marshall in the Pentagon in 1945-46 (after prior combat-infantry service). Major Heinmiller stated 1n 1987 and 1989 interviews with this writer that in early 1946 he had handled classified US documents which listed figures, by prison camp numbers, totaling thousands of Americans remaining in Soviet captivity at that time. He further stated he had long remained silent for security reasons. Major Heinmiller also revealed to the author, that he later worked with a U.S. intelligence agency and a Senate aide in resettling, inside the United States, and particularly in Alaska, escapees of Baltic, Polish and other nationalities who came out through the "ratline," from the Soviet Zone of eastern Europe, nearly all of whom stated they had seen American prisoners in Soviet captivity. These eyewitnesses also stated to Major Heinmiller, who had won their confidence, that they had been ordered to never reveal this information if they wished to remain in America.

Former U.S. Ambassador to Poland and the Occupied Nations, Anthony J. Drexel Biddle Jr., now a Colonel on Eisenhower's staff at European Command Headquarters, who had taken part in the Halle Conference with Soviet General Golubev and his staff, was another of many U.S. officers forced by strict security requirements to remain silent about American and other Allied prisoners who had disappeared behind Soviet lines in 1945. He was instrumental in the formation of the Polish Guard" units recruited from among Polish displaced persons and used in policing the DP camps and assisting U.S. troops in repatriations.[51] Later appointed head of the Allied Contact Section (ACS) of U.S. Forces European Theater, Biddle remained in liaison and intelligence work, utilizing his many contacts in eastern Europe, and was promoted to Brigadier General before his retirement. Eventually he was to be appointed U.S. Ambassador to Spain by President John F. Kennedy, prior to his untimely death in that position. In a subsequent letter to the author, Ambassador Angier Biddle Duke, who had served in the Army Air Corps during the war and was among the first Americans to enter the Nazi concentration camp at Buchenwald, remembered his uncle Tony Biddle speaking of American and Allied prisoners remaining in Soviet control after 1945:

"I do recall Uncle Tony's concern for the non-return of the POWs taken by the USSR...Your documentation is first-rate, and reveals a chapter or chapters of our history of which we are so ashamed that we seem to prefer a curtain of silence and forgetfulness over all..."[52]

A September 19, 1945, "Memorandum for the files," from the personal papers of President Truman, was given to Colonel B. W. Davenport, who had been designated by Presidential aide Colonel Harry Vaughn to respond to numerous requests from distraught U.S. POW family members whose loved ones had disappeared when the Soviets overran German stalags in eastern Europe: "Telegram of 9/19/45 to the President from Reverend H. Dewitt Smith, of Holly Springs, Mississippi; ASKS IF THE PRESIDENT COULD WIRE CONCERNING POLICY USED IN NOTIFYING FAMILIES OF PRISONERS

OF WAR; PARENTS ARE FRANTIC AND DESPERATE..." There is no record of the response to this message among the personal papers of President Truman.[53]

In September 1945 General Eisenhower had tried briefly to stop future forced-repatriations at a time when some 26,400 Soviet citizens were still in American custody. (A number which is interestingly close to the estimated number of Americans withheld by the Russians.) An Associated Press report from Frankfort, Germany of October 4th reported that the United States had ceased complying with the Yalta agreement on the repatriation of prisoners:

"General Dwight D. Eisenhower, in an order amounting to a temporary abrogation of one phase of the Yalta agreement, has instructed that American troops discontinue forcing Russian nationals to return home unless the United States Government rules otherwise. The existence of the order, affecting 26,400 Russians still in the American zone of Germany, was disclosed today by commanders of displaced persons camps. Questioned on reports that troops had fired over the heads or near the feet of some Russians to compel them to board Soviet-bound trains, one officer said: 'Possibly for a time some of them were being pushed onto trains...but that's all stopped now.' Many of those left in Germany have threatened to commit suicide if forced to go home..."

It appears that by this time Eisenhower feared it was hopeless to expect Soviet reciprocation in returning the disappeared American and British prisoners, and sought to put an end to the cruel charade, which continued to involve self-hangings, throat-cuttings and shootings of repatriates. His action caused a reassessment of the policy of forced-repatriation in Washington, D.C. By January 1946 Eisenhower's decision was clearly overruled by higher authority in Washington, which ordered further screenings and forced-repatriations of Soviets, indicating continued official reluctance to declare the missing American POWs lost, or to confront the Russians. According to now-declassified documents in the National Archives, Assistant Secretary of War John McCloy and Undersecretary of State Dean Acheson were among those who concurred in the necessity for continuing, in 1946, the forcible return of Soviet citizens, (although the numbers would be much smaller).[54]

By the end of October 1945, the British Labor government was also publicly expressing dissatisfaction with Soviet reciprocation in the repatriations of British and Commonwealth POWs, who had been alive in German camps when the Red Army overran eastern Europe. On the 30th of October, Great Britain asserted that the Soviet Union was barring British POW teams from entering Russian-occupied territory to search for the thousands of missing British prisoners. In a most significant account that appeared in the American press at

the time, The New York Times reported on this refusal the next day:
"WAR SECRETARY JAMES LAWSON TOLD THE HOUSE OF COMMONS TODAY THAT THE SOVIET GOVERNMENT HAD NOT YET GIVEN PERMISSION FOR BRITISH SEARCH TEAMS TO ENTER RUSSIAN-OCCUPIED GERMANY TO LOOK FOR MISSING BRITISH AND DOMINIONS PRISONERS OF WAR. ASKED WHETHER THE SOVIET GOVERNMENT HAD GIVEN ANY REASON 'FOR THIS EXTRAORDINARY REFUSAL,' THE SECRETARY SAID THAT IT WAS A MATTER FOR THE FOREIGN OFFICE. HE REFUSED TO ANSWER ADDITIONAL QUESTIONS."[55]

Yet, the United States government failed to inform the American people that the same situation existed with regard to the American POWs who had vanished inside Soviet-occupied territory. Truman recorded in his memoirs in this period, that, "...Our own demobilization program was reviewed at a Cabinet meeting on October 26. Secretary of the Navy Forrestal and Secretary of War Patterson outlined the program and expressed the warning that its acceleration threatened to jeopardize our strategic position in the midst of the postwar tensions that were building up around the world. I agreed entirely with this view..."[56]

After months of speaking out on the dangerous threat posed by the Soviets, General George Patton had been relieved of his command of 3rd Army by General Eisenhower on October 8, 1945, and also of his position as Military Governor of Bavaria. The reason given was his "failure" in the denazification of Bavaria, as reported by a small group of sensation-seeking, pro-Russian journalists, who were critical of Patton's warnings about the Soviet threat.

One of these reporters, Carl Levin of the New York Herald Tribune, had written the key report of June 5, 1945, entitled "25,000 US Prisoners Turn Up Alive," which appears to have been one of the primary cover-up stories concocted by the U.S. command and foisted on a chosen and willing journalist. General Patton had been praised by Secretary of War Henry Stimson in September 1945, as having, "the best governed area in the European Theater of Operations," and Assistant Secretary Jack McCloy relayed a similar report to Beatrice Patton; but no high U.S. official came forward to aid him now. On September 29, he had written: "I believe Germany should not be destroyed but rebuilt as a buffer state against the real danger which is Bolshevism from Russia." On October 1st he had written his wife Beatrice:

"I relinquish command of the 3rd Army and assume command of the 15th Army on the 8th. If by some miracle this should reach you keep it dark and of course keep the inclosure secret...In a sense I am glad to get out as I hate the role we are forced to play and the unethical means we are required to use...Later if the pendulum swings I can be of use. At the moment I feel pretty mad. Be sure to keep all the extracts from my diary secret. WARN SHERMAN MILES TO

EXPECT IMP. SHIPMENTS. Lucian Truscott gets my job." (It is not known what the inclosure was that General Patton wished his wife to keep secret as it is not included in the Patton Papers Collection in the Library of Congress.).

On October 15th General Patton again wrote to Beatrice: "I will resign when I have finished this job which will be not later than Dec.26. I hate to do it but I have been gagged all my life and whether they are appreciated or not America needs more honest men who dare to say what they think, not what they think people want them to think." On the 20th he wrote her: "I know I am right and the rest can go to hell or I hope they can but it is going to be very crowded...;" and on the 22nd: "By now you will have gotten quite a few letters from me telling why now is not the time to begin the offensive. It could only be a limited objective attack..."

In a letter of October 30, to Beatrice, General Patton again questioned his wife about much of his mail to her disappearing: "As I keep saying the whole damned world is going communist. Russia is back of it... It is just possible that this last attack on me is another act of God. Definitely know we are up against the Reds. I think I wrote you that the auto accident was nothing, so your dream didn't work...I certainly wrote you several times between Sept. 27 and Oct. 5 let me know if they ever show up..."

On November 17th, General Patton was sending diaries and reports of German generals analysis of the Soviets to Eisenhower's intelligence chief, General Edwin Sibert, Gehlen's U.S. controller who had been discussing the German intelligence network since August. It is noteworthy that Patton believed that the U.S. was already at war with the Soviets, indicating that the General knew the Soviets had already perpetrated an act of war. His comment:

"It is pointed out that the establishment of a powerful German-Anglo-American-French army might induce the communists to desist from their war of aggression, which will not be the result if the English and American statesmen try to satisfy the communists by continued concessions, which is what they are doing now."

On Nov. 18th he wrote to his wife Beatrice..."Darling B, I hope to leave here by boat around the 1st of the year. I AM GOING BY BOAT SO I CAN KEEP MY NUMEROUS BOXES OF PAPERS WITH ME IN MY STATEROOM AND NOT HAVE THEM LOST IN TRANSIT...I really shudder for the future of our country. I love you, George."

On November 23rd when Eisenhower was back in the United States and McNarney was coming to take over the European theater, Patton was temporarily the American commander in Europe. He noted:

"When General Eisenhower spent the night with me on 16 Sept. he told me that McNarney would replace him. It was on this same night that he directed me to remove the guards from all D.P. camps...I ARGUED WITH HIM AGAINST REMOVING THE GUARDS AT ALL AND PARTICULARLY AGAINST REMOVING THEM FROM THE POLISH AND RUSSIAN CAMPS. He acquiesced in the retention of guards in the last named camps. EVIDENTLY SINCE MY DEPARTURE FROM THIRD ARMY, THE GUARDS HAVE ALSO BEEN REMOVED FROM THESE CAMPS...I saw a telegram sent by (General) Beedle Smith under my authority as Theatre Commander, recommending that all repatriated persons who fail to request repatriation by 31 December be removed from DP camps and have their rations stopped..."

In a previously unpublished Diary entry of November 24, 1945 General Patton wrote:

"Admiral Lowry, the Navy member of the General Board, returned from Vienna. He was appalled at the utter destruction of the city and stated that in that portion of it under Russian control, the Russians are removing every movable article...large numbers of Viennese from the Russian sector moving into the American and English sectors with nothing left... I ADVISED HIM TO SPREAD THE INFORMATION WHEN HE GETS HOME, as it is simply another evidence of the inevitable war with Russia and another evidence of our criminal folly in letting them take over any part of Western Europe."

General Patton's last (surviving) diary entry was on December 3, 1945: "General (Walter Bedell) Smith gave a luncheon for General McNarney, the new Theater Commander...I HAVE RARELY SEEN ASSEMBLED A GREATER BUNCH OF SONS-OF-BITCHES...I had a good deal of fun at luncheon, quoting from recent articles on the military Government of Germany...which removed the appetite of Bob (US Political Advisor, Robert) Murphy...Their answer is they could not do anything about it as THEY WERE CARRYING OUT ORDERS. My answer is that a man who receives a foolish order should not carry it out-but such is not the breed of cats now in authority. It is certainly quite a criticism...that the deputy Theater Commander General Lucius Clay, and the Theater Commander General McNarney, have never commanded anything, including their own self respect, or if that, certainly not the respect of anyone else."

On December 5th, General Patton wrote his last (surviving) letter to Beatrice:"...I will radio you where and when I land and will stay a day or two in Wash to see how the hand lays and to insure jobs for Hap Gay and. I hate to think of leaving the Army but what's there?"[57]

Following General Patton's removal, his old 3rd Army had forcibly-repatriated a further 243 Soviets who had reportedly fought for the Germans. Whether any American POW-hostages were secretly returned for these doomed Russians is still a classified matter in the United States, although the Soviets admitted in 1990 and 1991 that American POWs were released from their control until the end of 1945, with an additional handful of 25-30 released in 1946 and 1947.

As General George Patton started his fateful journey homeward that would end with his death following a truck collision with his staff car, General George C. Marshall set out for China on December 5, 1945, with the stated intention of averting, "the tragic consequences of a divided China." Marshall was to demand a coalition government in which the Nationalists under Chiang Kai-Shek would be forced to share power with all "liberal" groups, including Mao Tse Tung's communists. When the Nationalists resisted sharing power with the Communists, Marshall was to order their military aid from the U.S. to be cut off until they complied.

At the end of December 1945, Averell Harriman was in Romania and later made special note in his memoirs of a long conversation at a reception there with Soviet Marshal Tolbukhin, (commander of the Soviet Third Ukrainian Front in Austria during the disappearance of the Allied POWs.) Tolbukhin spoke to Harriman of his meeting with General Patton at Linz in May 1945, and said he had a photograph of he and General Patton which he wished to send via Harriman to Mrs. Patton. Harriman remembered:

"...That was the last I heard of the matter, although Tolbukhin had spoken of Patton with such intensity of emotion that I felt he fully intended to send it. Perhaps the NKVD intervened. It was profoundly depressing to me that a man of Tolbukhin's importance should have been debarred, presumably for political reasons from doing this simple, human thing."[58]

Harriman might just as well have been speaking of the power of the NKVD and SMERSH counterintelligence during May and June of 1945 to kidnap thousands of the Allied POW's under Tolbukhin's control, whether the Red Army Marshal personally approved of such perfidy towards his wartime allies or not.

In December 1945, Congressional hearings resulted in revelations that British, French, Belgian and other Allied POWs had been mistreated and in some cases killed by Soviet troops during the Red Army advance through Poland and eastern Germany. Author Alfred de Zayas cites the German Federal Archives (Ost-Dokumente) in Koblenz "and personal interviews with ten French and Belgian prisoners of war who were liberated by the Red Army in 1944-45: "M. Georges Hautecler of the Centre de Recherches et d'Études Historiques de la Seconde Guerre Mondiale of the Belgian Ministry of Education informed me that 209 Belgian prisoners of war died or disappeared in the months between January and May 1945."[59]

Intelligence received by the U.S. and Allied governments revealed the fate of some of the prisoners who had disappeared in Russian-occupied territory. One example is a declassified December 20, 1945 livesighting report from OSS/CIG files, of a Polish ex-prisoner who escaped from the Soviet Union. He reported being held in a forced labor camp at Tambov with 20,000 prisoners of war including French, American, British, Belgian and other POWs including "even a few citizens of Luxembourg." He reported tens of thousands of prisoners had already died in this camp:

"ALL PRISONERS WERE FORCED TO WORK AND THE FOOD THEY WERE GIVEN WAS VERY BAD AND MONOTONOUS. THEY WERE HOUSED NOT IN HUTS BUT IN DUG-OUTS. THE MONOTONOUS FOOD CAUSED SOME STRANGE DISEASE WHICH MADE THE LEGS AND ARMS SWELL; THE SWELLINGS SPREAD AND THE SWOLLEN PLACES DEVELOPED SORES AND WOUNDS. AFTER A TIME MEN AFFLICTED WITH THIS DISEASE DIED...MORE THAN 23,000 ITALIANS, MORE THAN 2,500 FRENCH AND APPROXIMATELY 10,000 ROMANIANS AND HUNGARIANS HAD DIED IN THIS MANNER. THERE WERE ALSO MANY CASUALTIES AMONG THE POLES AND THE OTHER NATIONALITIES.

"WHEN INFORMANT LEFT THE CAMP THERE WERE STILL NUMEROUS FRENCHMEN...ALSO SOME BELGIANS AND DUTCH AND OTHERS, INCLUDING SOME ENGLISHMEN AND SEVERAL SCORE AMERICANS, THE PRESENCE OF WHOM IN THIS CAMP IS PROBABLY UNKNOWN TO THE BRITISH AND USA AUTHORITIES. WHEN HE WAS LEAVING, THESE ENGLISHMEN AND AMERICANS ASKED HIM URGENTLY (AS DID THE FRENCH OFFICERS AND MEN) TO NOTIFY THE ALLIED AUTHORITIES OF THEIR PLIGHT. INFORMANT SUCCEEDED IN REACHING FRANCE WITH A CONVOY OF ALLIED NATIONALS."[60]

Declassified U.S. documents obtained by this writer from CIA through FOIA request indicate that Soviet intelligence adopted a cover story for the Tambov livesighting, that all the Americans, British, French and other Allied POWs held there had been "captured in German uniforms," thus in effect throwing back in the faces of the Americans and others, their own charges about Soviet POWs captured in German uniforms. The declassified records of this case indicate a tacit acceptance of this explanation by U.S. intelligence as early as 1946, an early example of what came to be known as "wink and nod diplomacy." (That is, you take ours and we'll take yours.) This could well explain why the Soviets released new information on the Tambov camp on the eve of the February 1992 visit by the Chairman and Co-Chairman of the Senate Select Committee on POW/MIA Affairs, and in June announced the location of the grave of a U.S. POW in the Tambov area (noted elsewhere). While it is true that many of the thousands of French, Belgians and Italians, and even a few hundred Americans and British held there had served the Germans in one way or another, the author does not accept as valid the statements by the Soviet NKVD (KGB) that ALL the American, British and other prisoners in the Tambov area had served Germany. The ruthless 74-year record of the Chekists precludes such faith in their pronouncements.

U.S. casualties of WW II released by the Pentagon for publication long afterward state that there were 292,131 American battle deaths in all theaters, including 78,750 U.S. missing in action (and non-repatriated prisoners of war), mostly in the European theaters, and 115,185 U.S. dead from other, non-combat causes such as non-hostile aircraft losses, vessel losses, accidental deaths and deaths by disease. (Despite Freedom of Information Act (FOIA) requests by the author, the Department of Defense has refused to release a further statistical breakdown of the 78,750 U.S. MIAs of WW II.)

After the October 1945, temporary halt in forced-repatriations of Soviet citizens, a declassified U.S. Government document titled: "Captured or Interned United States Army Personnel," indicates that on October 31, 1945, 6,595 U.S. Army POWs from the German and Mediterranean theaters were still carried as currently-held "prisoners not yet returned to military control."[61] This was after all the U.S. POWs who died in German camps or were killed in the retreat had been accounted for as killed, body not recovered (KIA-BNR), and long after all stragglers had returned to U.S. lines. According to this document, of 97, 879 total American prisoners taken in Europe, 90,552 had been returned to military control (including as many as 25,000 initially carried as MIAs upon their return). Of these 97,879 prisoners taken in Europe, 77,528 had been captured in the European Theater and 20,351 in the Mediterranean Theater. 6,125 of the American prisoners of the Germans who had disappeared and remained in a "current POW status" at this time were lost in the European theater in eastern Germany and Poland, while only 770 Americans listed as known-POWs remained unreturned from the Mediterranean Theater. The missing POWs in Austria were apparently listed in the MIA category, as had, initially, many of the POWs recovered in May who had been captured in Alexander's SACMED area. According to this report, 11,753 POW/MIAs had been declared dead by the end of October, and 2,997 Americans, who may have originally been listed as POWs, and then been moved to the missing category, remained in an MIA status on November 1st. This information remained classified for over forty years to come.

An original hand-written note on this document says, "I phoned Colonel Ballard and suggested he do some more work on this."[62] At some point following the final 1945 returns of American and British POWs from Russian territory, and the failure of the September 1945 London Council of Foreign Ministers meeting attended by Molotov, Byrnes and Bevin, a high-level decision was apparently made to maintain the secrecy surrounding the Soviet retention of U.S. POWs, while continuing to push for a Russian reciprocation through more forced repatriations of Soviet citizens in early 1946. By the end of 1945, 8,532 "unidentified U.S. remains" were recovered by U.S. Graves Registration and as many as 17,000 Presumptive Findings of Death (PFOD's) of Americans, without remains, had already been made.

On January 1, 1946, 5,414 U.S. Army prisoners of war from the German theaters, were still carried as live POWs—"current status," in a now-declassified document of the U.S. Army's Strength Accounting and Statistics Office, entitled: "Missing in Action U.S. Army Personnel." 90,937 U.S. POWs had been "returned to Military Control," of the 97,209 reported as "Total Prisoners Taken," (which included some 25,000 initially recovered as MIAs). These figures indicate that 142 American POWs were returned to U.S. military control during the last three months of 1945, presumably from the Soviet-occupied zone. (Yet the total of "prisoners taken" had been inexplicably reduced from October through December from 97,879 to 97,209.)

Thus, it would appear that the reduction in "POWs—Current Status" of 1039 others between late October and the end of December 1945 was achieved by the Presumptive Finding of Death (PFOD) process, whereby U.S. POWs were declared dead one year after being taken prisoner if no new information was received about them to indicate they were still alive.[63] It was essentially the same process that had been followed with the missing U.S. POWs of north Russia from 1919 to 1921. These 5,414 American POWs were known by name to have been held in Nazi POW camps, and the U.S. MIAs were suspected of being late-captured prisoners of the Germans, overrun by the Russians before their identities had been forwarded to the American and International Red Cross in Switzerland, through the German Intercross delegate (Shirmer), all of whom had disappeared when the Soviet Army overran the German stalags in eastern Germany, Austria and Poland.

A subsequent 25 February 1946 letter from Lt. Col. J.L. Ballard, Chief of the Army's Strength Accounting and Statistics Office, GSC, the source of these POW-MIA statistics, explained the meaning of the work done there to Mr. Maurice Pate, Director, Relief to Prisoners of War, American Red Cross:

"It will be noted that the item 'prisoners of war (current Status)' and 'missing in action (current status)' are still large. The reason of course is that as of 31 December 1945, THESE CATEGORIES REFLECTED LATEST DEFINITE REPORTS AVAILABLE for statistical compilation, and the situation to date has not materially changed. YOU WILL APPRECIATE THAT FOR STATISTICAL PURPOSES THESE CASUALTIES CANNOT BE MOVED TO OTHER CATEGORIES UNTIL DETAILED DISPOSITION RECORDS HAVE BEEN PROCESSED. IN MANY CASES, FINAL DISPOSITION MUST AWAIT A LEGAL DETERMINATION OF DEATH UNDER FEDERAL LAW 490 WHICH MAY TAKE UP TO NEXT

SEPTEMBER..."[64]

A much later compilation of the Department of the Army (in 1953) was to claim that 92,820 Army and Air Force POWs of World War II had been returned to military control (with no substantiation or any previous announcements about when the additional 1,883 men carried as returned since January 1946 had been "returned") while only 257 were reported dead from wounds and injuries and 575 dead from other, nonbattle causes. Another important and unexplained alteration in statistics in this document is that the total of prisoners taken in the European theaters had been reduced to "93,653," from a stated total of "97,879" in October 1945 (of which 25,000 men had been reported by the government as having been listed as MIAs, not previously known as POWs). 4,226 more American POWs had apparently disappeared from the official records.[65] Thus, in a simple and passive bureaucratic manner, the living American POWs secretly held in Stalin's forced labor camps were declared dead by their own government, and the records of this deed remained classified for over 40 years.

For thousands of American, British, French and other allied POWs shipped east by the Soviets with German/Axis prisoners, the winter of 1945-1946 may have been their last on earth. The conditions they faced in the NKVD's Siberian gulags were best described by a surviving American civilian. In January 1946, after nearly eight years at forced labor in Stalin's labor camps, the American political prisoner Thomas Sgovio was still alive, in the Magadan area of far eastern Siberia, witnessing the arrival of new convoys of Japanese war prisoners from Manchuria. During the last war years he had nearly perished with millions of others in the camps.

For those unacknowledged and doomed American and Allied prisoners of war who were held in large groups of hundreds or thousands, in forced labor camps engaged in mining, logging or road and railroad building, such as those that Sgovio, John Noble and others witnessed, with various nationalities, the initial die-off during the first arctic winters must have been very great. Based on what is known of the Gulag Administration, however, it is likely that still-secret KGB archives would have records of all such deaths. Other American POWs were held in smaller groups of up to 200, working in industrial plants or construction projects, with perhaps better rations and living conditions, which resulted in their survival for years to come. This is attested to by now-declassified eyewitness livesighting reports from the late 1940s through the 1950s. Others, such as British Private Frank Kelly, from Stalag IV-B Muhlberg, were held in solitary or segregated from any other Americans or British for their entire imprisonment, to be used eventually for a show release which could not lead to proving the existence of other POWs in Soviet captivity. For the NKVD and its successor organization, the KGB, deniability about potentially embarrassing revelations, was a constant goal.

## Notes on Chapter Six

1    *Witness to History*, Charles Bohlen; *Present at the Creation*, Dean Acheson
     *Witness to History*, Charles Bohlen; *Present at the Creation*, Dean Acheson

2    Volume I, Year of Decisions, Harry S. Truman, p. 365-372, op. cit.

3    *Nemesis at Potsdam*, Alfred M. de Zayas, University of Nebraska Press, 1977

4    *Other Losses*, James Bacque, Stoddart Publishing, Toronto, Canada, 1989

5    *Year of Decisions, Memoirs of Harry S. Truman*, Vol. 1, pp. 384-85, p. 477, op. cit.

6    *Intrepid's Last Case*, William Stevenson, Ballantine, NY, 1983

7    Foreign Relations of the United States, Vol. 5, 1945

8    *Diplomat Among Warriors*, p. 276, op. cit.

9    *Triumph and Tragedy*, p. 671,op. cit.

10   State Department Records, National Archives #3250 August 10, 1945

11   *Triumph and Tragedy*, pp. 671-672, op. cit.

12   December 1, 1988, Letter to the author from former Ambassador George Kennan

13   NYT, June 30, 1945

14   *Pawns of Yalta*, Mark Elliot; *Operation Keelhaul*, Julius Epstein, Devin-Adair; and *The Last Secret*, Nicolas Bethell, Basic Books

15   Foreign Relations of the United States; Intrepid's last Case and other sources

16   Archival documents & FRUS

17   *Special Envoy*, p. 501

18   *Crusade In Europe*, Dwight D. Eisenhower, Doubleday, NY, 1948, p. 464

19   *Diplomat Among Warriors*, Robert Murphy, pp. 287-88, op. cit.

20   NA, Records of the Adjutant General's Office, War Department, "Secret." Examples are: AGPC-R 383.6 (216076), 6 August 1945; AGPC-R 383.6 (219102, 8 August 1945; AGPC-R (235063, 23 August 1945, signed by the Adjutant General, "ETA was 22 August 1945 and Port of Debarkation was New York."

21   FRUS

22   FRUS

23   National Archives: FRUS

24   Interviews with John Noble by the author and Thomas V. Ashworth

25   Lithuanian Bulletin, January-February 1947, published by the Lithuanian-American Council, New York, Volume V, No 1-2, Summary of the petition addressed to the American Military Government by the leaders of the Baltic Camp Administration in Kempton-Allgau, from the personal papers of Colonel A.J. Drexel Biddle Jr.; also: Pawns of Yalta, by Mark Elliott, University of Illinois Press, 1982

26   *My Three Years in Moscow*, Walter Bedell Smith, Lippincott, Philadelphia, 1950

27   *My Three Years in Moscow*, pp. 24-25, op. cit.

28   National Archives, State Department files, classified secret telegram #3327, dated August 21,1945 to the Secretary of State

29   Foreign Relations of the United States

30   Foreign Relations of the United States...1945, V, pp 1104-5.0; also: p. 789, POW/MIA Policy and Process

31   New York Times

32   A copy of the original document, supplied by the author to a Senate Select Committee, is reproduced on p. 802, POW/MIA Policy and Process

33   A copy of the original document "Missing in Action U.S. Army Personnel, 1945, is on p. 801, POW/MIA Policy and Process

34   Army Battle Casualties and Nonbattle Deaths in World War II, Final Report, Department of the Army, 1 June 1953

35   Report of Lt.Col. James F. Donovan, 0338695, 22 September 1945. AG 1945, RG 407, Entry 360, Box 2437; also reproduced in part on p. 791&792, POW/MIA Policy and Process

36   *The Quican*, an U.S. ex-POW publication, Nov., 1986

37   POW/MIA Policy and Process, p. 790, Also: Harry S. Truman Library, Independence, Missouri

38   Copies of the originals of these documents are reproduced on pages 793 and 794, POW/MIA Policy and Process

39   *Spycatcher*, Peter Wright; *The Philby Conspiracy*, Page, Leitch and Knightly, Doubleday, NY, 1968; *The Fourth Man*, Andrew Boyle, Dial Press, NY, 1979; and *Mask of Treachery*, John Costello, Morrow, NY, 1988

40   *The Last Hero*, Anthony Cave Brown, op. cit.

41   Widows, by William Corson, Susan Trento and Joseph Trento, pp.7-16, Crown Publishers, NY, 1989

42   December 22, 1952 INS news service report of the Madden Committee hearings in the House of Representatives, by Sam Fogg, which appeared in the New York Daily Mirror; also: Polish-American Journal, December 6/December 12, 1952, article on General Bissell by Julius Epstein, from the files of Dr. Stephen H. Johnsson, a friend and former Hoover Institution research colleague of Dr. Julius Epstein.

43   National Archives, Record Group 84/Box 73, Code MCA Secret via Army to Secretary of State, No. 3381, September 26, 1945, from E.P. (Edward Page) at the Moscow Embassy

44   *Intrepid's Last Case*, op. cit.

45   Ibid.

46   *Shadow Warriors*, Bradley Smith, Basic Books, NY,1983.

47   *Spycatcher*, Peter Wright, p.185, Viking, NY, 1987

48   *Wise Men* and the *Forrestal Diaries*, op. cit.

49   *On Active Service in Peace and* War, Henry L. Stimson and McGeorge Bundy,

Harper Brothers, N.Y., 1947, p. 608

50    Ibid., p. 675

51    Personal files of Ambassador, A. J. D. Biddle Jr.

52    1988 letter to the author. Ambassador Duke was later instrumental in seeing that the essential facts of the author's investigation were publshed in the October 1991 American Foreign Policy Newsletter.

53    Box 675, Papers of Harry S. Truman, Truman Library, Independence, Missouri

54    Foreign Relations of the United States, Volume V, 1945

55    NYT, October 31, 1945

56    *Year of Decisions, Memoirs of Harry S. Truman*, Vol. 1, p. 509, op. cit.

57    All of the above letters and Diary notes of General George S. Patton Jr. are to be found in The Patton Papers Collection at the Library of Congress; consulted and copied by the author with permission of Major General George S. Patton III, U.S. Army—Ret'd.

58    *Special Envoy*, op. cit.

59    Congressional Record, Senate, 4 December 1945, p. 1374; also: p. 67 & p. 202, *Nemesis at Potsdam*, op. cit.

60    This document from the National Archives is reproduced on p. 797, POW/MIA Policy and Process, Volume II

61    National Archives, Red Cross Files (formerly classified)

62    A copy of this document, "Captured or Interned United States Army Personnel," which was supplied to the U.S. Senate Foreign Relations Committee in 1990 and the senate Select Committee on POW/MIA Affairs in 1991, by the author, is reproduced on p. 802, POW/MIA Policy and Process. The original document is located in the formerly classified files of the Red Cross in the National Archives.

63    POW/MIA Policy and Process, p. 801, a document of the Army Strength Accounting & Statistics Office entitled: "Missing in Action U.S. Army Personnel," which the author turned over to Senate Foreign Relations Committee investigators in 1990. This document was located by the author among formerly classified records of the Red Cross in the National Archives in 1988

64    A copy of the original document from the National Archives is reproduced on p. 800, POW/MIA Policy and Process. This document was located by the author and research colleague Ashworth in the formerly classified files of the Red Cross, at the National Archives, in 1988.

65    Army Battle Casualties and Nonbattle Deaths in World War II, Final Report, Table 2, Department of the Army1 June 1953.

# CHAPTER SEVEN

## THE EARLY COLD WAR PERIOD

President Harry S. Truman addressed a joint session of the U.S. Congress on January 21, 1946, in a State of the Union address in which he defined, "The fundamental foreign policy of the United States," at a time of military confrontation with Soviet dictator Josef Stalin:

"It will be the continuing policy of the United States to use all its influence to foster, support, and develop the United Nations Organization in its purposes of preventing international war. If peace is to endure it must rest upon justice no less than on power....We may not always succeed in our objectives. There may be instances where the attainment of those objectives is delayed, but we will not give our full sanction and approval to actions which fly in the face of these ideals...We seek no territorial expansion or selfish advantage. We have no plans for aggression against any other state, large or small...We believe in the eventual return of sovereign rights and self-government to all peoples who have been deprived of them by force. We shall approve no territorial changes in any friendly part of the world unless they accord with the freely expressed wishes of the people concerned...We are convinced that the preservation of peace between nations requires a United Nations Organization composed of all the peace-loving nations of the world who are willing jointly to use force, if necessary, to insure peace..."[1]

That same day in Germany, a Soviet prisoner of war died while under American control, after a suicidal outbreak of what were termed by the New York Times: "renegade Russians," at the Dachau prison camp, when twenty other Soviet POWs were injured while resisting forcible repatriation. The Dachau operation represented the renewed American policy of forcibly repatriating Soviet prisoners, apparently to gain reciprocal action from the Russians with American prisoners still under Soviet control. Truman had said nothing specifically about the thousands of Americans missing inside Soviet-occupied Europe, but in January 1946 France raised the subject of her own MIAs and POWs, who had vanished after the Red Army overran the German stalags in the east. The French chief of Russian repatriation in Moscow, Lt. Colonel Raymond Marquie, who was later to be unmasked as a Soviet agent, had attracted some suspicion in France and America when he stated on January 14th, that the controversy with the Soviet Union over the where abouts of the French Alsatians who had been drafted into the German Army and since held prisoner in the USSR, was nearing a solution. He claimed that only 28,000 out of 130,000 French missing still remained unaccounted for, and that the difficulty in tracing missing persons in Russia stemmed from German POWs who hoped for repatriation by pretending to be French Alsatians. Colonel Marquie barely mentioned the more than 300,000 other missing French POWs and displaced persons who had been captured in 1940 or deported east, and were in German control when overrun by the Red Army in 1945. The New York Times reported that Marquie had said, "There was not the slightest evidence of 'secret' camps in Russia or that the Russians desired to hold French or Alsatian prisoners."[2]

At the time of Marquie's remarks, the French government knew his statements were untrue, but pro-Russian sentiment in the postwar French National Assembly delayed his exposure and recall from Moscow until late in 1947. The day after Marquie's statement about French POWs in Russia, perhaps not coincidentally, the official press agency of the Soviet-controlled Polish Communist regime announced the discovery of the bodies of 40,000 murdered Allied prisoners who had been held by the Germans. The bodies were said to have been unearthed from a mass grave in Niemodlin, Lower Silesia, and were reportedly American, British, Russian, French, Polish, New Zealand and Greek POWs, soldiers who had been identified as from the Lambinowice concentration camp. A search was being conducted for the bodies of thousands of other prisoners believed to have been executed.[3] If true, this would have represented a far larger massacre of POWs than the 1940 Katyn Forest killings by the NKVD, but the Soviets were in no haste to invite Western observers to inspect Russian-occupied territory such as this. Rumors of mass-shootings by NKVD and SMERSH operatives in the eastern zones were already reaching the West through refugees. The coincidence in the two announcements was not remarked upon in the American press, and no mention was made of the 78,750 missing Americans of World War II.

Meanwhile, the Department of State continued to receive inquiries about the fate of American prisoners detained by the Soviet Union, but they were kept classified, and in many cases remain classified to the time of this writing. One example in the files of the National Archives is a card from the Special Projects Division dated January 23, 1946, referring to a question sent from Harrisburg, Pa (with the name blacked-out) saying: "Regarding Doolittle fliers interned by the Soviet Union."[4] Another related withdrawal card from the State Department's Special Projects Division signed by one "Bailey," is a still-classified memo dated February 6, 1946:

"Memorandum of conversation between Colonel Kavanaugh, from (the) War Department, and Captain George and Mr. Bailey, regarding Doolittle fliers interned by the Soviet Union."[5]

These documents refer to U.S. airmen still missing after Doolittle's bombing raid over Japan early in the war, when some of the aircraft landed in Soviet and Chinese territory. While some had eventually been repatriated, others had disappeared.

In a speech of February 9, 1946, Stalin reiterated the fundamentals of Marxism-Leninism, predicted continuing sacrifices by the Soviet people to attain the goals of his five-year plans, and warned of the possibility of a "violent disturbance" involving the capitalist nations. He hailed the Soviet system, and hinted at Soviet efforts to develop an atomic bomb:

"...The point is that the Soviet social system has proved to be more capable of life and more stable than a non-Soviet social system, that the Soviet social system is a better form of organization of society than any non-Soviet social system...Our Cheka is not so powerful that it could abolish the laws of social development...I have no doubt that if we render the necessary assistance to our scientists they will be able not only to overtake but also in the very near future to surpass the achievements of science outside the boundaries of our country."

Stalin's speech was interpreted by many Western leaders as a public declaration of hostilities with his former allies. American aircraft were already at peril over Soviet airspace, but the incidents of American aircraft being fired upon or downed had been downplayed or classified altogether. Shortly after Stalin's speech, on February 20th, Soviet aircraft attacked a U.S. Navy Mariner seaplane, but no American casualties were reported. On that same day the Soviet-surrogate Albanian Communist regime gave notice to the United States that it would fire upon any American aircraft overflying Albanian airspace without previous authorization.[6]

On February 23rd a delayed report of Pope Pius XII's condemnation of forced repatriations in an allocution on Wednesday, February 20th, reached the United States. The New York Times headlined a story with: "Rebuke By the Pope to U.S., Britain Seen," and reported that the Pope's important allocution Wednesday was aimed at the United States and Britain even more than at the Soviet Union and specifically at the Yalta Agreement..." Correspondent Herbert L. Matthews wrote from Rome:

"One high prelate gave me the typical feeling here when he said that the Yalta Agreement 'went against humanity and justice.' What did you gain by it,' he asked. 'The Russians violated their engagements, yet you are still keeping to clauses that are immoral and unjust. This prelate incidentally reproachfully complained that the American press had been afraid to publish facts like these. I assured him that the American press was not afraid to publish anything and that much of this nature had been published." The fact was however, that the major American media had published very little about the repatriations of millions of Soviets and other eastern Europeans who had been sent back to live under Communist regimes, and nothing at all since June about the thousands of American POWs who had vanished behind Soviet lines.

In the United States, meanwhile, the PFOD process of presuming missing in action Americans dead, went on. In a typical example of February 25th in the New York Times, a Brooklyn officer missing in action since March 1944 was now listed as killed over Germany on the date he went missing:

"First Lieutenant Robert A. Meyer, son of Mrs. Edna B. Meyer... of Brooklyn, was killed in action over Germany on March 16, 1944. while serving as a fighter pilot with the Army Air Forces, according to word received by his family from the War Department. He previously had been reported missing..."[7]

This process was occurring all over the United States in this period, and would continue until all the MIAs and those 5,414 Americans still carried as "POWs-Current Status," in the German Theater were all declared dead, on the basis that nothing further had been heard from them in captivity after the surrender of Germany and the occupation of eastern Europe by the Red Army. At this same time, General Albert C. Wedemeyer announced that there were 3,000,000 Japanese in China, 2,000,000 of whom were in Soviet-occupied Manchuria and northern Korea. While repatriations to Japan supervised by U.S. Marines were reported to be proceeding at a rate of 3,000 per day, the fact that hundreds of thousands of Japanese POWs held by the Soviets had already disappeared into Siberia was all but ignored in the American media.[8]

Churchill's "Iron Curtain" speech in Fulton, Missouri, on March 5th, which followed an introduction by President Truman, who had encouraged the former Prime Minister to speak candidly, was interpreted as the western Allied response to Stalin. Churchill reminded the Soviets that the United States alone possessed the atomic bomb, which gave the West a "breathing space" and that American power must be used to bring order to the world through the principles enunciated by the United Nations. In this important policy address, Churchill reiterated his own 1945 complaints about Soviet refusals to allow Allied contact officers access to prisoner concentrations in the Soviet zones, charging that eastern Europe had been totally cut off from the West, and called for a measured response to Soviet threats against world order, through the United Nations:

"We understand the Russian need to be secure on her western frontiers by the removal of all possibility of German aggression. We welcome Russia to her rightful place among the leading nations of the world. We welcome her flag upon the seas...It is my duty however...to place before you certain facts about the present position in Europe. FROM STETTIN IN THE BALTIC TO TRIESTE IN THE ADRIATIC, AN IRON CURTAIN HAS DESCENDED ACROSS THE CONTINENT. BEHIND THAT LINE LIE ALL THE CAPITALS OF THE ANCIENT STATES OF CENTRAL AND EASTERN EUROPE. WARSAW, BERLIN, PRAGUE, VIENNA, BUDAPEST, BELGRADE, BUCHAREST AND SOFIA, ALL THESE FAMOUS CITIES...LIE IN WHAT I MIGHT CALL THE SOVIET SPHERE... All are subject...to Soviet influence...and in some cases increasing measure of control from Moscow. Police governments are pervading from Moscow. But Athens alone, with its immortal glories, is free to decide its future at an election under British, American and French observation. The Russian-dominated Polish government has been encouraged to make enormous inroads upon Germany, and mass expulsions of Germans on a scale grievous and undreamed-of are now taking place. The Communist parties, which were very small in all these eastern states of Europe, have been raised to pre-eminence and power far beyond their numbers and are seeking everywhere to obtain totalitarian control. Police governments are prevailing in nearly every case, and so far, except in Czechoslovakia, there is no true democracy. Turkey and Persia are both profoundly alarmed and disturbed at the claims which are being made upon them and at the pressure being exerted by the Moscow government.

"An attempt is being made by the Russians, in Berlin, to build up a quasi-Communist party...At the end of the fighting last June the American and British armies withdrew westward, in accordance with an earlier agreement, to a depth at some points of 150 miles upon a front of nearly 400 miles, in order to allow our Russian allies to occupy this vast expanse of territory which the western democracies had conquered...THE AGREEMENT WHICH WAS MADE AT YALTA, TO WHICH I WAS A PARTY, WAS EXTREMELY FAVORABLE TO SOVIET RUSSIA, BUT IT WAS MADE AT A TIME WHEN NO ONE COULD SAY THAT THE GERMAN WAR MIGHT NOT EXTEND ALL THROUGH THE SUMMER AND FALL OF 1945...I was a Minister at the time of the Versailles treaty and a close friend of Mr. Lloyd George...In those days their were high hopes and unbounded confidence that the wars were over, and that the League of Nations would be all-powerful. I do not see or feel the same confidence...at the present time...I DO NOT BELIEVE THAT SOVIET RUSSIA DESIRES WAR. WHAT THEY DESIRE IS THE FRUITS OF WAR AND THE INDEFINITE EXPANSION OF THEIR POWER AND DOCTRINES...Our difficulties and dangers will not be removed by closing our eyes to them. They will not be removed by mere waiting to see what happens; nor will they be removed by a policy of appeasement...What is needed is a settlement and the longer this is delayed the more difficult it will be... If we adhere faithfully to the Charter of the United Nations and walk forward in sedate and somber strength, seeking no one's land and treasure, seeking to lay no arbitrary control upon the thoughts of men...the high roads of the future will be clear, not only for us but for all..."

Stalin counterattacked in the war of words a week later, during an interview that was published in Pravda, charging that Churchill and his allies were attempting to rule the world, and saying this was a, "set-up for war, a call to war with the Soviet Union."[9]

Churchill's "Iron Curtain" speech was followed by a toughening attitude towards Russia on the part of the Truman Administration in Iran, Turkey, Greece and elsewhere. Shortly after Churchill's speech, on March 15th, 1946, Stalin also rejected American terms for a $1 Billion loan which had been previously requested by the Russians. This request had been scaled down in August 1945, from the much larger loan request of $6 Billion, first sought by Stalin in January 1945, through Averell Harriman, in Moscow. The $6 billion loan request had been shelved by the United States until Stalin acceded to American demands in Western Europe, which included, among other conditions, implementation of the Yalta accords on repatriation of prisoners of war. It was for this reason that the two issues of war prisoners and the $6 billion war reconstruction loan, had been linked in the secret agenda for discussions at the April and May 1945, San Francisco UN Conference. The scaled-down Russian request of August 1945 had been ignored by the United States, for obvious reasons, until the collapse of the Foreign Ministers meetings in Moscow, in December, when it was suddenly remembered by the Americans, apparently in a belated attempt to resurrect the diplomatic discussions.

The particular 1946 American terms which appeared to offend the Soviets most were those that would mandate that the U.S.S.R. must join the World Bank and the International Monetary Fund, now headed by the Soviet agent Harry Dexter White, who had moved to the IMF from the U.S. Treasury Department. In what appears to have been yet another American effort to gain entrance into the Soviet Union for observers who might obtain information on the hundreds of thousands of missing American and Allied prisoners, the Russians would have been required under these terms to open their territory to observers from international agencies. This they were, of course, unwilling to do, as such observers would have quickly stumbled upon the existence of 20,000,000 or more prisoners of war of all nationalities and Soviet "enemies of the state," in the Gulag slave camps.[10] The Soviet Union was not to accede to these American and allied requests until 1991, following the failed coup against Mikhail Gorbachev, and the collapse of Communism in eastern Europe and in the USSR.

In March 1946, the world focus was on Soviet Russia and its

moves in the Middle East. The Soviets were exerting pressure on Turkey for concessions on the Dardenelles region, which controlled the entrance to Russia's warm water ports on the Black Sea, and at Batum. Also, Soviet troops remained in occupation of northern Iran, which reflected Stalin's interest in Middle East oil reserves and in access to the Persian Gulf for the USSR. But the real reason for hostility between Russia and the West, the hundreds of thousands of missing American and Allied prisoners, remained veiled in the American press, and classified by the U.S. government.

In addition to the thousands of American POWs in German hands who had been secretly withheld by the Russians, other mysterious cases of U.S. personnel captured after the end of the war by the Soviets were reported by eyewitnesses returning to Europe from the USSR. In some of these reports, the existence of records of U.S. prisoners, identified by name as being alive in Soviet captivity, is denied by the American command in Europe. In some cases the missions from which these Americans were captured are also denied. One example is a March 21, 1946, report to U.S. Air Force headquarters in Europe of a German ex-POW's debriefing, revealing the existence of Americans captured from a secret flight to Leningrad, which was shot down over Romania:

"A Secret report received here from interrogation of a German prisoner of war repatriated to his home in Hamburg, states (that) 5 American fliers, including one Jerry Helsmayer, a Chicago bombardier, being held since January 1946 in the camp at Maramar Sebeth(?), near the Hungarian-Romanian-Jugoslav border. (The) Men flew a transport mission to Leningrad in September 1945, were downed in Romania on the return trip, and were taken into custody by Russians. Above forwarded for your action. Suggest information distributes forward USAFE available data immediately. Request instructions if further action desired this headquarters."

The ultimate fate of these reported Americans is unknown. All along the border of the Soviet occupation zone, American military personnel and civilians were also kidnapped by Soviet authorities during this period. Many formerly-secret U.S. documents of these incidents have recently been declassified, during 1990-1991, by a component of the National Security Agency (USAINSCOM).

A refugee was permitted to respond to a report of a CIO investigation in the New York Times on March 21st, on conditions inside the Gulag camps:

"Having spent almost two years in eight prisons and a labor camp in Soviet Russia (1939-41) and having, after my release, worked for some months as assistant accountant in a village in Siberia, I know that the number of prisoners, political and otherwise, in Soviet Russia is estimated at 20,000,000 by the Russians themselves. I also know that prisoners food and accommodations are much worse than in pre-revolutionary times. In fact all of us suffered from scabies and scurvy.

"As for the nonexistence of unemployment in Soviet Russia, which was casually mentioned in the CIO report, may I observe that if everybody in this world were forced to work under the aegis of his government for two platefuls of cabbage soup...and 500-900 grams of black bread a day and a few potatoes, there certainly would be no unemployment even in the poorest and most backward countries of the world..."

Pyotr Deriabin, a Red Army combat veteran who had been sent as a student to the SMERSH Counter-intelligence School in 1944, was subsequently involved in "screening" returned Soviet war prisoners. He later defected to the West and revealed the fate of those forcibly repatriated by the Americans and British from late 1944 to early 1946, in the hope of gaining release of U.S. and Allied POWs in return:

"There were...at least five million Soviet soldiers taken prisoner by the Germans during World War II. The survivors, on their return from captivity, were individually judged, tried, and, almost automatically, convicted of 'anti-Soviet' activity. They received sentences varying from immediate execution to imprisonment or deportation to Siberia... The single fact that a man's brother was captured by the Germans is evidence enough for the KGB to begin a searching investigation of his position and 'character.' The actual fate of the returned war prisoners was regulated by authority according to a rough sliding scale of their imagined complicity in 'ant-Soviet activities.' Almost twenty percent of them were either imprisoned for 25 years or shot. Another fifteen to twenty percent received jail sentences of from five to ten years. About ten percent were sent to frontier territories in Siberia for periods of not less than six years. Another fifteen percent went to industrial areas in need of work conscripts, like the Donbas or the Kuzbas. They could not return home after their assignment."[11] There was little wonder then, that the estimated 25,000 Soviet ex-POWs who remained in American control at the end of 1945, strongly resisted repatriation to the USSR.

By early 1946 forced-repatriations of Soviet citizens were tried again and then halted, as Stalin once more reneged on ordering reciprocal action by releasing comparable numbers of U.S. and Allied prisoners. The 1946 forced-repatriations were far smaller in scale than the mass-movements of the summer of 1945. In the old Nazi concentration camp of Dachau, 135 out of 271 Russians were repatriated in a violent incident that left 11 of them dead of suicide, and 21 severely injured in suicide attempts. Although the secret was beginning to reach the public and Pope Pius XII had protested the forced-repatriations in his allocution of February

20th, 1946, the operations continued. Of some 3000 Russians concentrated at Plattling, 1,590 were forcibly returned to the Russians on February 24th,1946, in another violent operation in which 5 men killed themselves rather than face the NKVD, and many others attempted suicide. On May 13th, 222 more Russians from the Plattling camp were returned, and one more man succeeded in committing suicide despite the most elaborate precautions. The Russians insisted on receiving the dead bodies of suicides during the repatriations. How many secretly-returned American POWs were exchanged for these unfortunates is still highly classified. However, the exchange rates between Americans and Russians in 1945 had often been 10, 20 or many more Russians, for each American or British prisoner.[12]

The former U.S. Ambassador to Poland, Colonel Anthony J. D. Biddle Jr., who had served on General Eisenhower's SHAEF staff, was at the U.S. Frankfurt Headquarters in December 1945. He met his second wife, Margaret Biddle (Robbins), when she was involved with DP and POW repatriations teams in 1945 and 1946, which infiltrated the Soviet border zone to recover American POWs and civilians caught behind the Soviet lines. (It was, in her words: "A section under G-5 (Military Government) with a G-2 (intelligence) atmosphere.") A typical repatriation team included: one American (a U.S. paratrooper), one English agent and one Belgian. In the Soviet border zone they met with members of the civilian underground in Soviet-occupied Germany (and Austria) and escorted back small groups or individual American and western Allied civilians and ex-POWs, who had been captured first by the Nazis and later overrun by the Red Army. Cover stories may have been used for some of these missions to identify as civilians some ex-U.S. military POWs. She also served at the old Nazi concentration camp at Dachau at the time of the forced-repatriations of both Soviet citizens and east Germans bound for the forced-labor camps and remembered with despair: "The Russians even insisted we send children; I will always remember one particular train load of people going into the Soviet zone and the children were all crying...I was madder than I have ever been in all my life." She already considered the Russians untrustworthy and remembered that Colonel Biddle, who at that time was the Chief of the Allied Contact Section, acting as Eisenhower's military buffer and go-between with his many contacts in Allied forces in eastern Europe, now behind Soviet lines, also did not trust the Soviets to reciprocate. It was her understanding also that in Austria the POW/Repatriations matter was handled by General Lyman Lemnitzer, who had served under General Mark Clark and later as a high-level aide to Field Marshal Harold Alexander in the SACMED Theater during 1945-46, and who much later became Chairman of the Joint Chiefs of Staff. Anthony Biddle remained in the Army in Europe and later in the Pentagon until he retired as a Brigadier General in the 1950s.[13]

Yet, at the very time of the last major forced repatriations in early 1946, the Russians were still shipping American and British POWs east into Russia. Following publication by the Toronto Star of essential details of this writer's (and Thomas V. Ashworth's) 1989 published report, A Secret That Shames Humanity, a Downsview, Ontario resident confirmed this fact to that newspaper, which published his account. During the January-February period of 1946, Mr. Ben Kovacs was an 18 year-old in Novi Mesto, Hungary, about 22 miles from the Soviet border. He remembered seeing Allied POWs being shipped into the Soviet Union at this time. He remembered passing within 15 meters of the train that was under Soviet military guard:

"THERE WERE ABOUT 25 RAILWAY WAGONS FILLED WITH TROOPS...I KNEW THEY WERE SPEAKING ENGLISH."

Mr. Kovacs is certain that the Allied POWs had to be heading into the Soviet Union. While the usual load for a boxcar would be 40 men, indicating this shipment was at least 1,000 POWs, in reality the Soviet NKVD/KGB secret police were notorious for packing far more men into a "Stolypin" or boxcar than the standard number. These POWs were very likely from among the thousands withheld by Marshal Tolbukhin's Third Ukrainian Front Army in Austria and Hungary during June 1945, on Stalin's orders. The relatively small numbers of Russian forced-repatriates from Plattling apparently caused Stalin to keep several thousand American prisoners who had been slated for secret trades.

Charles Bohlen, the State Department Soviet expert who had served as an adviser and translator for Roosevelt at Yalta (and later Ambassador to the USSR), inaccurately stated in his memoirs that all American POWs were out of Soviet control by 1946, and blamed the military for the fiasco:

"THE EXECUTION OF THE REPATRIATIONS WAS ENTIRELY DETERMINED BY THE AMERICAN AND BRITISH MILITARY. IN THE SPRING OF 1946, THE FORCIBLE REPATRIATION WAS STOPPED. BY THAT TIME, ALL AMERICAN POWs WERE OUT OF SOVIET-HELD AREAS, AND THERE WAS LITTLE MOLOTOV COULD DO ABOUT IT."[14]

Soviet officials admitted in mid-1990, at a Moscow meeting with U.S. war veterans Michael Van Atta, J. Thomas Burch and William T. Bennett, that 13 American POWs of WW II had been returned in 1946-47, something the United States Government had not and still has never officially admitted to. In the September 1991 Russian publication Nezavisimaya Gazeta, the Soviets revised this figure to 22 American POWs released in 1946 and 10 more in 1947. (Burch was then Chairman of the National Vietnam Veterans Coalition, Van Atta had been a long-time POW researcher, and Bennett was the Secretary of the N.V.V.C., whose father, former

Assistant Secretary of State William Tapley Bennett, was a career diplomat who had been George Bush's Deputy when he was U.S. Ambassador to the United Nations.)

It is uncertain whether German prisoners were actually traded with the Russians for secretly-returned Americans. Some 180,000 or more Germans were returned to the Russian zone by the American command. Many were immediately arrested by NKVD or puppet secret police and disappeared.

Meanwhile, in January 1946, Averell Harriman had resigned his position as US Ambassador in Moscow. He did not get along well with the new Secretary of State, James Byrnes, whom Harriman felt did not use his lengthy experience in dealing with the Soviets. There were also other powerful figures in Washington who he didn't see eye to eye with. Harriman later wrote:

"General Marshall, in particular, refused to concede that the United States was facing a time of trouble with the Russians until he had to deal with Molotov, face to face, at the final Council of Foreign Ministers meeting in Moscow during the spring of 1947."

When Eisenhower's former chief of staff, General Walter Bedell Smith, was being briefed before taking over the job of US Ambassador to Russia in January 1946, George C. Marshall said to him: "You must talk to Harriman of course, in view of his long experience. But you ought to discount what he says. The Russians gave him a hard time and that made him too pessimistic."[15]

Considering what General George C. Marshall must have known about the disappearance some 20,000 US prisoners and MIAs in Soviet-occupied territory, and the fact that under his direction, and that of his Chief of Intelligence, Major General Clayton Bissell, this fact had become highly classified and hidden from the American public, Marshall's judgment and actions during this period are inexplicable. However, he continued to remain at the center of power in Washington, DC. Given Marshall's apparent reluctance to recognize the already-demonstrated Soviet threat at this time, the possibility exists that some of the conclusive intelligence concerning U.S. POW/MIAs disappearing in Soviet control may have been withheld from Marshall, by the former New York lawyer and Soviet apologist Clayton Bissell, who headed Army Intelligence until early 1946. Evidence later revealed during 1952 Congressional hearings regarding Bissell's deliberate suppression of evidence of the Soviet Katyn Forest massacre of Polish POWs by the NKVD, supports this possibility. It can only be conjectured upon at the present time whether such action by Bissell was of his own volition or at the behest of higher authorities, such as Presidents Roosevelt and later Truman, or of his patron, Assistant Secretary of War John McCloy, to follow a policy of deliberately avoiding confrontation with Stalin at this time.

Some perfunctory efforts were made by the U.S. Government to "ascertain the whereabouts and welfare of American prisoners," who had vanished in the Russian zone, but these efforts were doomed to fail in areas such as Poland, where 300,000 Soviet troops under Marshal Rokossovsky still held that country in late 1945 and early 1946, despite the promises by Stalin at Potsdam that they would be withdrawn. Such a powerful force, watching every move of the Polish Communists, precluded any chance for an actual investigation or search for the thousands of missing American and other allied POWs. The new U.S. envoy to Poland, who had finally been appointed by the President to succeed Anthony Biddle, was a professional diplomat, Ambassador Arthur Bliss Lane, who had served in the U.S. Legation at Warsaw in 1919. Lane was prevented from traveling to Poland until the end of July 1945, after he had stopped off in Germany to be briefed by General Eisenhower, Averell Harriman and others, during the Potsdam Conference. On August 1, 1945, while first viewing postwar Warsaw, Lane witnessed part of the mass deportations then occurring in Poland under Soviet NKVD and Polish Communist supervision:

"We observed thousands of Soviet troops marching east. Many of them were without uniforms, and these we presumed to be liberated prisoners of war from Germany...In addition we noticed, as we did on many future occasions, groups of men, women, and children being marched eastward to the railroad station at Praga, accompanied by Soviet and Polish armed guards. Evidently they were being deported to the Soviet Union. It was prohibited for anyone to question these people who, from their dejected expressions, were not proceeding on their journey voluntarily."

Whether or not it occurred to Lane that American or Allied prisoners might be among the masses of "liberated" Russian prisoners or Polish deportees being moved east into the USSR, went unrecorded. Since October 1944 and through the spring 1945 Soviet offensive, the Russians and the Polish Communists had blocked entry into Poland of UNRRA and Red Cross contingents to Poland, for fear that they would witness the mass-murders and deportations of Polish anti-Communists, and members of the Home Army who had fought the Germans but were now treated as enemies of the State, in violation of the agreements at Yalta and with the United Nations. In addition to the 2,000 or more U.S. POWs who had vanished after the Soviet conquest, thousands of other native-born and naturalized American citizens then in Poland, were living in daily fear of arrest and disappearance, while the woefully understaffed American Embassy was unable to help most of them.

Despite the proven perfidy of the Polish Communist regime, Ambassador Arthur Bliss Lane received news from Washington in early March 1946 that the United States was granting fifty-million dollars in credits to the Warsaw government for obtaining U.S. war surplus material. (It was later revealed that this was the result of lobbying by undersecretary of State Dean Acheson, and negotiations by Donald Hiss, brother of Alger Hiss.) The Ambassador protested this decision to Washington on the basis of Warsaw's repeated violations of the human rights of its own citizens, but on April 21, Lane was informed that the U.S. Department of State intended to grant not only the $50 million credit for war surplus, but also $40 million more from the Export-Import Bank to buy locomotives and railroad cars. Nothing was said about the possibility that such equipment would be used to transport even more Polish prisoners to the USSR. In his protest at this time to the State Department Ambassador Lane noted:

"...I beg the Department not to approve the extension of any credits at this time. When the terroristic activities of the Security Police come to an end, when freedom of the press is restored, and when American citizens are released from Polish prisons—not until then should United States public funds be used to assist the Polish Provisional Government of National Unity."

Neither President Truman or the State Department followed Lane's advice, perhaps because of a misguided concept that granting economic aid to Stalin's surrogates would somehow wean them away from reliance on the Soviet Union. In the coming months, anti-Jewish pogroms and massacres erupted in Poland, with the Communist authorities either aiding and abetting the murders or standing idly by as others did the killing; yet the U.S. aided the Warsaw Communist regime.

Nearly a year after Ambassador Lane's arrival in Poland, on July 18, 1946, the by now well-established American Embassy in Warsaw, forwarded a response from the Polish Red Cross to a request for information about U.S. prisoners of war who had disappeared in Poland. Edmund J. Dorsz, who had accompanied Lane to Potsdam and Poland and was Second Secretary in an Embassy which was then under constant surveillance and harassment by the Polish Communist UB (Urzad Bezpieczenstwa) secret police under Stanislaw Radkiewicz, and their Soviet NKGB mentors, wrote on that day to the Secretary of State in Washington, confirming that the War Department was still trying to find out what had happened to 2,000 or more American POWs who had vanished in Poland:

"I have the honor to inform the Department that the Polish Red Cross at Warsaw, Poland, has provided the Embassy with the names of 230 persons who were registered with the Polish Red Cross at Radom, Poland, Between February 1, 1945 and May 28, 1945, as American war prisoners. While the Polish Red Cross has stated that for several reasons beyond its control, some of the data set forth in the list may not be entirely accurate, the list as it was received from the Polish Red Cross is transmitted herewith in original and hectograph.

"It is suggested that the Department may wish to refer copies of the enclosed list to the office of the Provost Marshal General, War Department, for such action as that office may deem appropriate IN THE EFFORTS IT IS EXERTING TO ASCERTAIN THE WHEREABOUTS AND WELFARE OF FORMER AMERICAN PRISONERS OF WAR.

"The Embassy has sent an appropriate note of appreciation to the Polish Red Cross for its action in sending the Embassy the list of names under discussion."[16]

Although the Polish Communist regime was extremely unlikely to willingly reveal any substantial lists of names of U.S. POWs still held by the Soviets or their own UB, and many of the names on the list of 230 prisoners were in fact subsequently marked by the War Department as returned to military control, "RMC," some of them were not. And their fate remained unresolved. Conditions at the Warsaw Embassy continued to deteriorate over the next three weeks, to the point where the UB secret police kidnapped, interrogated and sentenced a U.S. Embassy employee, and there was nothing that the Ambassador or the Government in Washington could, or would do about it. On August 23, 1946, Mrs. Irena A. Dmochowska, an American citizen who served as an Embassy translator, was arrested by the UB security police on trumped-up charges and sentenced in a public show trial four months later to five years imprisonment. As the Ambassador himself was later to write:

"The helplessness of the American Embassy in coming to the assistance of one of its own employees who, according to our laws, was considered an American citizen, was clearly shown in this case. We were at the mercy of the Polish authorities who ignored precedent and international comity...Most dreaded of all the military was the UB. At the time of my departure from Poland in February 1947, it too was estimated to have at least a hundred thousand agents..."[17, 18] One native-born American prisoner who had been held prisoner by the Soviets for seven years, both in Russian-occupied Poland and in the USSR, was released by the NKVD in April 1946, and allowed to board an American Liberty ship in Odessa. He was John Czechel, aged 26, who had been born in Providence, Rhode Island, but traveled to Poland with is parents in 1925, and had been wounded by the Germans while fighting with the Polish resistance during the war. After the Soviet takeover of eastern Poland he had attempted to flee to the West through Lithuania on skis. Captured and interrogated by the NKVD, Czechel had been beaten. He recalled: "I told them I was an American but they just laughed, although they examined my documents that showed I was an American...The NKVD chief said: 'After a couple of years we will be in America. Why do you want to go to America now?'"[19] The release of the American civilian John Czechel may have been a deliberate gesture by the Communists to deflect attention from the 2,000 or more U.S. military prisoners of war who had disappeared in Soviet-occupied Poland in 1945.

The author located declassified U.S. documents in the National Archives indicating that in 1946 and 1947, US Fifth Army intelligence and the F.B.I. were also aware of, and investigating the questioning of US POWs in postwar, Russian-controlled Poland, by at least 2 American-born Soviet Intelligence officers. One of these Soviet intelligence officers was known by name. The source of the information was a former escaped-POW, a United States Army officer who had been in Soviet-occupied Poland as American POWs were being taken captive. Military Intelligence was trying to stop the FBI's investigation of this source and J. Edgar Hoover's FBI was attempting to halt Army G-2 Intelligence involvement in the case, in a dispute over jurisdiction.[20] The document states:

"In making a contact in this matter at the Ford Motor Company the FBI learned that one Major Prentice G. Morgan, Headquarters, Fifth Army, had been to Detroit during October 1946 investigating this matter. The FBI has complained to the Intelligence Division regarding this apparent violation of the Delimitations Agreement by the Fifth Army." Whether this interference represented high-level direction to stop the investigation, remains unknown.[21] It is possible that the Army was anxious that the source not discuss with the FBI the disappearance of other American POWs in Poland. The Army subsequently claimed that "Since the two persons referred to...are not civilians but commissioned officers in the Russian Army and so far as is known, are not in the United States nor engaged in espionage, counterespionage, subversion or sabotage, it is difficult to see how any violation of the Delimitation Agreement, real or imagined, can be involved."[22]

Archival documents also indicate that the Soviets used bilingual German POWs to translate into Russian the German POW camp records, and identified Americans in this manner, whom they continued to hold prisoner. In many government documents of the late 1940s and early 1950s, there is an apparent pattern of subtle debunking of witnesses offering information on Americans in Soviet captivity.

The United States forces still in China at this time, mostly Marines, were meanwhile caught between the Communists and Nationalists in their struggle for supremacy. Although they avoided conflict, the Americans lost some casualties in killed and missing in Manchuria, although U.S. officials denied it. On May 28th, Secretary of State James Byrnes, "denied today that American Army personnel was taking part in the fighting between factions in Manchuria. His denial was along the lines of one made last week by the War

Department after reports that American aviation had been shot down in that region."[23] Despite the aggressive moves of the Soviets and their allies, many American leaders blamed the U.S. itself for the standoff.

Former Vice President Henry Wallace, the apologist for Stalin whose blindness to the Soviet slave labor empire during his 1944 trip to Siberia had aroused such controversy, submitted to Truman a lengthy written criticism of the administration's policy toward the Soviet Union, in July 1946. He accused the Truman Administration of attempting to build overwhelming force with which to, "intimidate the rest of mankind." He also wrote that the United States was working to prevent Russia from building it's own nuclear weapons. (This development had, in fact, already been ensured by Soviet atomic spies such as Julius and Ethel Rosenberg in the United States and Dr. Allan Nunn May in Britain). In what could have been construed as a warning to Stalin, the United States tested two atomic bombs of 20-kiloton force during July 1946 in the south Pacific Ocean, at Bikini Atoll.

Wallace and other influential liberals in government, as well as media pundits such as Walter Lippman had been greatly disappointed in the failure of a proposal to share America's nuclear secrets with Soviet Russia, through a proposed International Atomic Development Authority. This idea dated back to tripartite discussions at Potsdam, when Truman had formally informed Stalin about the atomic bomb and had indicated America's willingness to share nuclear technology for peaceful use. Since that time John McCloy and Dean Acheson had been among the strongest proponents of this idea, which both the New York Times and the Washington Post endorsed when it was made public by Acheson. Bernard Baruch, a liberal political contributor and manipulator whose money had gained him some influence with Truman, was selected by that president to present the plan to the United Nations. It contained however, provisions that would permit Americans to inspect mines and production centers in the USSR; yet another of the repeated American attempts to gain access to Soviet territory for on-site inspections that had continued since Yalta and the Halle Conference in 1945. The possibilities for intelligence-gathering that might lead to, among other things, the discovery of American prisoners, or information about them among the forced-laborers used to construct and operate such industrial sites in the USSR, was more than Stalin, Beria and Molotov would ever agree to. The Soviets instead called for a complete ban on nuclear weapons, which they were themselves secretly working on with critical data supplied by spies in America, Canada and Britain. In September, however, Henry Wallace further defended Soviet domination of eastern Europe at a leftist political rally in New York. Shortly thereafter, Truman demanded and got Wallace's resignation from his position of Secretary of Commerce. Also during the summer of 1946, General George C. Marshall was asked by Truman whether he would serve as Secretary of State to succeed James Byrnes, who had infuriated the President by his conduct of the Foreign Ministers conferences, particularly in Moscow. Marshall agreed, but Byrnes remained for the time being. Marshall spent much of 1946 in China, where he continued to force the Nationalists to share power with Mao Tse Tung's Communists. U.S. Marines had been caught in the middle of a renewed civil war, and a Marine convoy from Tientsin to Peking had been ambushed by the Communists. In his later memoirs, then Undersecretary of State Dean Acheson remembered:

"By the end of July fighting had broken out again and was spreading rapidly...General Marshall believed that military aid to the Nationalists, originally justified under the plan for a united Chinese army, had become wholly inconsistent with impartial efforts to suppress incipient civil war. He therefore called for and obtained immediate embargo of arms and munitions from the United States to China. This was to continue until may 1947."

Congressional conservatives were later to charge that this cut-off in U.S. aid to the Nationalists gave Mao Tse Tung the chance he needed to rebuild his forces with Soviet aid and captured Japanese weapons and munitions, turned over to him by the Russians in Manchuria, for a final campaign that ultimately defeated Chiang Kai-shek and established the Chinese Communist regime. Later Congressional investigations were to uncover a network of pro-Communist American officials in China whose influence upon American policy also contributed to this unwelcome outcome as was charged by U.S. the U.S. Commander in China, General Albert C. Wedemeyer.[24]

A member of the U.S. Naval Port Facilities in Tsingtao, China, at this time, Donald Blair (now of Simi Valley, California), was wounded in action during combat action by the Chinese Communists against U.S. forces, on December 3, 1946. He was captured by the Communists and held prisoner from that date until January 15, 1947, making him the last U.S. POW of the official World War II era (which ended on December 31, 1946). Blair was not officially awarded the Purple Heart Medal for this action until September 12, 1990. The ammo dump at Tsingtao, China was subsequently blown up by suspected Communist guerillas, killing hundreds of innocent Civilians.[25]

Meanwhile, the confrontation with Tito in northern Yugoslavia over control of the city of Trieste had continued since the end of the war. American and Allied troops stationed in the enclave had been fired upon several times by Yugoslav Communist troops and threatened by Soviet forces. On August 9th an American plane was forced down near Kernj, Yugoslavia, and the crew reported missing. A far more serious "incident" occurred on August 19, 1946, when a U.S. aircraft shot down over Yugoslavia resulted in 5 American airmen being killed.[26] Undersecretary of State Acheson was involved at this time in preparing a note relating to the ongoing Soviet threat to Turkey when the second attack occurred, and he later recalled:

"On the day we delivered our note Tito's planes shot down the second American transport. We did not know at the time that the crew had been killed, and, therefore, demanded the release of BOTH crews within forty-eight hours. The President ordered our troops along the Morgan Line augmented and the reinforcement of our air forces in northern Italy. As soon as the necessary naval forces could be assembled, they were moved into Greek waters. Tito quickly got the point and backed down. The offensive against the Straits quieted. The Russian offensive moved to the northern border of Greece, the eastern provinces of Turkey, and northern Iran. The autumn would witness Soviet fire increasingly concentrated there."[27]

Other incidents involving U.S. aircraft occurred along the border of the Soviet occupation zone in Austria and Germany, but the details have largely remained classified.

The secret POW crisis led inevitably to the first major American covert actions inside Soviet-occupied territory, executed by agents of the OSS, Army G-2 intelligence and SSU, the State Department's later Office of Policy Coordination and the CIA, using among other assets, existing German intelligence networks under Nazi General Reinhard Gehlen, and others within Soviet republics. As late as December 10, 1945, the War Department had refused official permission to Gehlen's controller, General Edwin Sibert, for initiating covert activity against the Soviets. In light of the February 1946 date of the final large-scale forced repatriation of 1,590 Russians from Plattling, it is significant that such official approval was not forthcoming until March 1946, although some intelligence activities had been underway unofficially for months. This date is also pertinent to the January 1946 reduction of (classified) numbers of "currently held" U.S. POWs in the European theaters subsequently overrun by the Soviets, to 5,414, as noted in documents previously. The further shipment of U.S./British POWs east into the USSR in boxcars from Hungary, indicated a Soviet decision to not fully reciprocate in the Plattling repatriations. (It is now known that the Soviets, by their own admission, returned only 22 U.S. POWs in 1946.) This decision had coincided with Stalin's warning to the Soviet people of a coming confrontation with the U.S. and Britain, Churchill's "Iron Curtain speech" at Fulton, Missouri, and a toughening attitude toward the Soviet Union on the part of the Truman Administration.

It is clear, however, that whatever intelligence successes may have occurred, by mid-1946 most of the U.S. prisoners in Soviet hands had been given up as lost. A September 16, 1946, State Department document of the Special Projects Division (SPD), located by the author in the National Archives, indicates something about the methods used in reaching the conclusion that the Russian-held U.S. POWs were deceased. Edward McLaughlin, Assistant Chief of the Division wrote:

"The War Department has informed this Department that its overseas commanders were recently queried as to the probability of personnel now carried in a missing status being located alive, and that the answers from all commanders reveal that no information had been brought to light by intensive investigations which would indicate the possibility that any will be found alive..."

Yet, an August 17, 1946, French livesighting report, made available to the U.S. Department of State, reveals that some information on the fate of missing U.S. prisoners of war was indeed available, even if some of it was not specific enough to identify individual Americans. Henri Meck, a Deputy of the French National Constituent Assembly, wrote:

"I have brought to the attention of the Minister for ex-prisoners of war the testimony of Mr. Joseph Bogenshutz, who was repatriated last July 7 from Russia, from Camp 199-6 at Inskaya, which is 75 kilometers from NOVOSOBIRSK...BOGENSHUTZ...ALLEGES THAT THERE STILL REMAIN AMERICAN, BRITISH, BELGIAN, POLISH, ROMANIAN, LUXEMBOURG, ETC. NATIONALS IN THE CAMP..."

Despite such information reaching the West, millions of tons of American food, medical supplies and other war relief was pouring into Russia at this time, under the direction of the United Nations Relief and Rehabilitation Administration (UNRRA), largely through the port of Odessa on the Black Sea.[28]

Another State Department memorandum dated October 30th, 1946, for use in answering POW family members who reached high government levels with their concerns, explained why the U.S. Government was unable to look for Americans held by the Soviets:

"THE QUESTION OF PERMISSION FOR UNITED STATES MILITARY TEAMS TO ENTER RUSSIAN TERRITORY IS ONE WHICH HAS BEEN DISCUSSED BETWEEN UNITED STATES AND SOVIET MILITARY AUTHORITIES...THE RUSSIANS HAVE ADOPTED A POSITION THAT THEY ARE THEMSELVES MAKING ANY NECESSARY SEARCHES..."[29]

With few exceptions, the senior officials and officers who were in decision-making positions involving the POW/Russian forced-repatriations affair, remained prominent in the upper echelons of the political or military establishments. Churchill, Truman, Eisenhower, Eden and others long retained and protected the secret from their heights of power. Field Marshal Sir Harold

Alexander was made Governor-General of Canada after the 1945 repatriations, presumably to get him out of Europe. Averell Harriman would later play a key role in negotiations over the "demilitarization" of Laos that would affect the outcome of the Vietnam War, and in the post Tet-1968 peace negotiations with the North Vietnamese in Paris. George Kennan went on to achieve fame as the author of the "Containment Doctrine," and an honored position as an elder statesman of the American foreign policy establishment. OSS figures in Europe, Allen Dulles, Richard Helms, William Colby and William Casey became Directors of the CIA. The future American Chairman of the Joint Chiefs of Staff, General Lyman Lemnitzer, served on Alexander's staff, General Mark Clark, later commander in Korea (who would admit in his memoirs that Americans were left in captivity there), also served under Alexander at this time and both played roles in the 1945 POW-repatriations. At this time future US Commanders in Vietnam were also involved in the repatriations of POWs; Paul Harkins as Deputy Chief of Staff to General Patton at 3rd Army, and William Westmoreland as Chief of Staff of the U.S. 9th Infantry Division on the Elbe River, and later on the Danube. Vernon Walters, an Army translator for Roosevelt and Truman who would eventually become a General, a Deputy Director of the CIA, UN Ambassador under Reagan, and U.S. Ambassador to West Germany under President George Bush, was to play a crucial role in the Indochina POW/MIA matter, following the Vietnam War.

Meanwhile, in secret interviews with U.S. intelligence agents in Europe during 1946, escapees from NKVD-run POW labor camps in Soviet-occupied territory sometimes revealed information on the fate of specific U.S. and Allied POWs, including some known by full names, or by partial and disguised names, as had been attempted by Lindsay Retherford of the 339th U.S. Infantry, in 1927. In addition to large groups of hundreds or more unidentified American prisoners reported held in various locations, American POWs were also confined in smaller groups among German POWs in Soviet-occupied Europe and in many camps across the USSR, as a now-declassified secret August 20, 1946, U.S. Army Counter Intelligence Corps report of CIC Detachment B-225 indicates:

"On 16 August 1946, Friedrich Grzybeck, (address blacked out by U.S. Government censors), crossed the border into American occupied territory from the Russian occupied zone and was arrested for illegal border crossing. During routine questioning Grzybeck told the following story:

"During the war, Grzybeck was a member of the Luftwaffe Ground Force Division Number 21, and was taken prisoner by the Russians on or about 10 April 1945. From 10 April 1945 to 11 August 1946, Grzybeck was a Russian prisoner of war and was held at the former German prisoner of war camp, Dresden/Ost number 11.

This same had been used for Russian civilian workers. There were 1,800 men in that camp and they were transported daily to an airfield near Dresden to work.

"Grzybeck states that in May 1945, three United States soldiers were brought to the camp as prisoners. He gave their names as follows: PFC Olen Taylor of (address blacked out), Corporal Bucki(y) Okhane, (O'Kane, O'Cain?) of (address blacked out), and Private Billy Hafers of (address blacked out). These names were written by Grzybeck and may not be spelled correctly, except for Olen Taylor, as Grzybeck has in his possession a picture of Taylor in a United States Army uniform, signed by Taylor. (See attached photograph.)

"The three United States soldiers and Grzybeck were confined in the camp near Dresden from May 1945 to August 1946. Before being confined to this camp near Dresden, the three United States Soldiers were German prisoners of war, having been captured by the Germans shortly after the Normandy Invasion near Cherbourg, France. According to Grzybeck, the three American soldiers were in an American Airborne Division which called itself "King of the Air.

"During the morning of 13 August 1946, three airplanes crashed at the airfield near Dresden and the Russians, suspecting sabotage, loaded eight hundred prisoners in railroad cars in preparation of shipping them east. Grzybeck and seven others escaped from the train and Grzybeck made his way into the United States Zone of Occupation. PFC Olen Taylor did not attempt to escape because his two buddies were sick..."

This report was signed by CIC Special Agent Manfred Maier, and forwarded to CIC headquarters by Harold L. Kirkpatrick, the Special Agent Commanding. A special report on this incident was forwarded to the Director of Intelligence for the U. S. Military Government of Germany on August twenty-sixth which supplied a few more details:

"Grzybeck claims that Americans were treated the same way as the Germans, most likely because they attempted to escape once but the Russians succeeded in recapturing them. Although subject never saw the uniforms of the three soldiers, he claims they belonged to the 101st Airborne Division.

"On an airport near the camp on which the prisoners worked, there were a total of about two hundred planes, of which 125 were four-motored. There were some U.S. fortresses too, and Russian B-34's. On 12 August 1946, three Russian planes crashed. The same day eight hundred prisoners were supposed to be shipped to Varsovia, as the Russians suspected sabotage. At 2300 (11:00 PM) on the twelfth of August, Grzybeck and seven other prisoners (three Austrians and four from the British Zone) escaped. The train was then still at the railway station at Dresden, and Grzybeck told the three U.S. soldiers about his plans. O'Kane and Hafers were sick, but Olen Taylor gave Grzybeck his picture and asked him to try to help all three of them as soon as he could. TAYLOR DID NOT WANT TO ESCAPE BECAUSE HE DID NOT WANT TO DESERT HIS TWO COMRADES."

The picture smuggled out of the Russian zone of PFC Olen Taylor, an unknown American hero, was attached to the report. He stands there in a GI uniform and an overseas cap, in front of a modest frame home with an older sedan parked out front, perhaps on his last pre-overseas leave at home. He has a huge grin on his face that reveals a warm and friendly character, just the kind of U.S. soldier who wouldn't desert two sick and disabled buddies at the most fateful moment in all three of their lives. A common American, but nevertheless a man of the highest honor.[30]

Another American prisoner confined by the Russians during this period, in a different Dresden-area prison, Munchenerplatz, was the U.S.-born civilian John Noble, who had been held since the summer of 1945, and eventually been placed in charge of the records for over 21,000 prisoners of many nationalities. Noble was able to keep track of hundreds of executions by the Soviets there, and after being freed nine years later he wrote:

"Death was a day-by-day statistic in my bookkeeping at Munchenerplatz Prison. Regularly I received a list of prisoners who were being ordered to appear before the prison court. The lists were written in red ink, in keeping with the official color motif of the Soviet court system. Everything about the system was red. Red cloths cover the court tables. Red hangings muffle the rooms...The MVD court members were a lieutenant-colonel, a major and a captain...Trials were not held anywhere. There was only a reading of the charges and a sentencing...The entire paperwork of the sentencing was prepared in advance...I came to realize that among the court members there was a sincere indifference to the sentences. With some, the indifference was that of tribal Asia, where death is no more meaningful than a leak in a sod roof. Others had the trained attitude that death simply was a Soviet instrument of correction. It was the highest form of social hygiene, not unlike burning away a slum area or cauterizing a wound. These men felt no passion in sentencing people to death. They were merely snapping off a light switch.

On two days in June 1946 the sentencing was particularly memorable. The guard brought me the lists, on which as usual I wrote the cell numbers of the prisoners. Each day there were between twenty and thirty names in red; all, whatever the charges, were sentenced to death. The punishments would ordinarily have been varied to fit the crimes; but on those two days the members of the court were hopelessly drunk....A prisoner sentenced to death was put on half-rations; since humans are animals, there is little sense in giving full rations to one that is about to be destroyed. The squad assigned to bring the prisoners to the place of execution consisted of a junior-grade MVD lieutenant and two enlisted guards...When the guards came to a cell where a sentenced prisoner was, they ordered him to remove his clothes...While the undressing went on, guards and officer would joke and laugh...

"In that joking was summed up a startling difference between these guards and the Nazi death squads abut which those prisoners who had known both sometimes spoke. The Nazis, they said, killed viciously, because they were convinced that the people being killed were actually their enemies. The Russians killed because, almost literally, a number had been drawn from a hat, because some meaningless document in some meaningless proceedings had said to snuff out the candle. No ferocity attended the executions. The reasons for the killings were as remote and irrelevant to the Russian guards as was the concept of death itself. Their joking, then, was not forced. When they patted a prisoner's shoulder, the action came easily. Life had to end for certain integers in the state table of statistics. That's all, comrade. Nothing personal, comrade.

"Horribly, the laughter of the guards marked those days more than did the sounds of the killings themselves. The process of execution, about which the guards sometimes boasted because it was so 'humane,' was simplicity itself. After a condemned prisoner had undressed, he was led to a partly shattered wing of the prison. As he rounded the corner into a corridor of the wing, a guard shot him in the back of the head. It was 'humane,' because it came without warning.

"As each prisoner was shot, his body was dragged to the end of the corridor. By the end of a clay's killing, a stack of sprawling bodies, naked or in undershirts, stood in the dark and dirty hall. A guard doused the bodies with gasoline and tossed on a match. The flames from the pyre made a light that often was seen by prisoners in other parts of the building. A gurad, if questioned, would explain that trash was being burned. As smoke from the burning bodies drifted from the execution corridor into other parts of the prison, it was difficult to make anyone believe that it was anything but exactly what it was. On execution days, in many cells not even the pitiful scraps that passed for rations were eaten."

Noble later remembered his last sight of this execution area:

"The guard led me along the hall and down the steps to the first floor. As I passed a certain door, I stopped for a second to look at it once again. Hundreds of men had walked through it naked, or nearly so, in their last minutes of life. A few others, including myself, also had passed through that door occasionally, accompanied by guards. We were looking for prisoners old clothes, blankets, sheets, aprons, and other articles. That part of the building was quite destroyed. Once, I came a little closer to the execution department than prisoners were permitted to do, but the guard was friendly and he let me look. I saw only a smoke-stained end of a corridor. I acted as if I were not interested, and I searched from cell to cell for clothing. BLOODSTAINED RAGS I FOUND WHICH, VARIOUSLY, WERE ONCE THE UNIFORMS OF PRISONERS OF WAR, DANISH STREETCAR CONDUCTORS, AND DUTCH POST OFFICE CLERKS."[31]

On August 31, 1946, John Noble was transferred from Munchenerplatz Prison in Dresden to Muhlberg Prison in Germany, site of the old German Stalag IV-B, from which some 1,500 American POWs had disappeared after the Soviet takeover. When Noble reached the Muhlberg camp in September 1946, the American POWs had long disappeared into the east, probably in special, secret train shipments., bound for secret camps housing only American, British, French and other allied prisoners of war. At Muhlberg Noble remembered seeing a British woman who was also being held captive by the Soviets. Her history or fate were never determined. He rembered the camp a decade later:

"Muhlberg was like a vast sewer, with rotten things, the prisoners, floating in it. Rottenness seemed to touch almost everybody."

From Muhlberg John Noble was transferred to Buchenwald, another old Nazi camp still being used by the Soviets, and wrote in his memoirs later: "I heard there were Americans, British and Frenchmen here."[32] Eventually, John Noble, like thousands of other Americans before him, was to be shipped to the Soviet Union, to become part of the vast, unacknowledged army of millions of slave laborers in Stalin's gulags.

In an effort to clarify American-Soviet policy in the fall of 1946, for the record, President Truman had directed his naval aide and troubleshooter, Clark Clifford (with George Elsey), to prepare a summary of American relations with the Soviet Union. Clifford, a pre-war lawyer whose rank of Captain in the U.S. Navy stemmed from political influence rather than active sea duty, was to become a political adviser to Truman and later an influential Washington lawyer, lobbyist and power-broker. He would ultimately be appointed Secretary of Defense by President Lyndon B. Johnson, in 1968, at the height of the Vietnam War. (Near the end of his long life in lofty circles he would be accused, in 1991, of serving as a front-man for a subsidiary of the scandal-ridden bank known as BCCI.)

Clark Clifford submitted to Truman what can only be categorized now as a sanitized germ of the truth, in that part of the report which dealt with American POWs of 1945 who had been overrun by the Red Army in eastern Europe. But by the mere mention of existing difficulties over repatriation of Soviet-held Americans, the report was destined to remain secret for over 20 years. In the now-declassified study (located in the Harry S. Truman Library), Clifford reported to Truman that he had, "consulted the Secretary of State, the Secretary of War, Fleet Admiral Leahy, the Joint Chiefs of Staff, Ambassador Pauley, the Director of Central Intelligence and other persons who have special knowledge in this field." In the introduction to this document, dated September 24, 1946, which Clark Clifford entitled "AMERICAN RELATIONS WITH THE SOVIET UNION," he noted: "The gravest problem facing the United States today is that of American relations with the Soviet Union." Clifford also identified the crux of America's problem with Josef Stalin's regime:

"...Peace seems far away and American disillusionment over the achievements of peace conferences increases as the Soviet Government continues to break the agreements which were made at Teheran, Yalta and Berlin, or "interprets those agreements to suit its own purposes."

Since one of the major violations of the Yalta Agreement concerned the Soviet failure to repatriate all American and Western Allied POWs and displaced persons, this bland paragraph served to refer to the missing POWs of 1945. The disappearance of American POWs in Soviet control is hinted at in the Clifford-Elsey Report, but the subject was dealt with summarily and (in light of recently-declassified documents) incompletely:

"AN AGREEMENT RELATING TO PRISONERS OF WAR AND CIVILIANS LIBERATED BY ALLIED MILITARY FORCES WAS ALSO SIGNED AT YALTA ON FEBRUARY 11,1945. THE RECORD OF THE SOVIET UNION IN CARRYING OUT THIS AGREEMENT FOR THE CARE AND REPATRIATION OF AMERICAN AND SOVIET CITIZENS HAS NOT BEEN SATISFACTORY. IN GENERAL, LIBERATED AMERICAN PRISONERS OF WAR IN SOVIET OCCUPIED AREAS OF GERMANY WERE FORCED TO MAKE THEIR WAY AS BEST THEY COULD ACROSS POLAND TO SOVIET TERRITORY. DURING THEIR JOURNEY ACROSS POLAND THEY WERE FORCED TO RELY FOR FOOD AND NECESSITIES ON THE GENEROSITY OF THE POLISH PEOPLE WHO THEMSELVES HAD VERY LITTLE. WHEN THEY ENTERED THE USSR, THEY WERE GATHERED TOGETHER, PUT IN BOXCARS, AND SENT TO ODESSA. THE SOVIETS REFUSED PERMISSION FOR AMERICAN AIRCRAFT TO BRING IN SUPPLIES TO LIBERATED U.S. PRISONERS OF WAR BEHIND SOVIET LINES OR TO EVACUATE THE SICK AND WOUNDED BY AIR. THE ONLY UNITED STATES CONTACT TEAM ALLOWED IN SOVIET TERRITORY WAS THE ONE IN ODESSA, THE TRAFFIC POINT WHERE THE AMERICANS WERE ASSEMBLED PRIOR TO BEING SHIPPED TO THE UNITED STATES. EVACUATION OF U.S. LIBERATED PRISONERS OF WAR WAS ACCOMPLISHED UNDER THE MOST DIFFICULT CONDITIONS."

Although nothing was said anywhere within Clifford's 26,000-word report about the actual disappearance of many thousands of

American and Allied POWs and displaced persons inside Soviet-occupied territory, nevertheless the report stated that Soviet repatriation of U.S. POWs had "not been satisfactory," which was a strong hint of serious difficulties experienced in the liberation of American prisoners in Soviet control. In the context of world events in the fall of 1946, Clifford's report, if publicly released, might have caused serious questions to be raised in the American media about the ultimate fate of Soviet-held U.S. prisoners of war, and also about conduct of the still-ongoing forced-repatriations of Soviet ex-prisoners. Clifford's official summation to Truman of the repatriation of "Soviet Citizens" had also obscured the truth of those often-horrifying events:

"With respect to the repatriation of liberated Soviet prisoners of war in U.S. hands, the Soviet interpretation of the Yalta Agreement was that the United States would forcibly repatriate all persons claimed by the Soviet Union to be Soviet citizens. The United States interpretation was that assistance would be given for the repatriation of those who wished to return to the U.S.S.R., while forced repatriation would be limited to those war criminals demanded by the Soviets. THE UNITED STATES HAS NOT MET MANY SOVIET DEMANDS FOR REPATRIATION OF UNWILLING U.S.S.R. CITIZENS NOT CLEARLY SHOWN TO BE WAR CRIMINALS."[33]

Declassified documents and eyewitness accounts indicate that many Soviet citizens had, in fact, been forcibly repatriated in 1945 and in 1946, who had not been shown by rigid U.S. screening to have been "war criminals." The testimonies of U.S. witnesses involved in the forced-repatriations bears this out.[34] Yet Clifford's report, if publicly revealed in late 1946 would have proven both that the United States had not fully complied with Stalin's "demands for repatriation of unwilling USSR citizens" in U.S. control, and that Soviet repatriation of American prisoners in their control had "not been satisfactory." This was apparently too close to the truth for President Harry Truman in September 1946, when that president was facing a growing Soviet threat in Europe, the Middle East and Korea; together with declining domestic support for large overseas troop deployments, and serious labor and other domestic problems at home in America. Also, the fact that the U.S. Government had written off thousands of American POWs and MIAs then in Soviet control by declaring them "presumed Dead," when intelligence to the contrary existed, would have required an explanation to the American people that might have cost Truman the Presidency.

Truman decided that the cost of even the slightest candor about the 1945 prisoner-of-war debacle was too high. While other parts of the top-secret report contained passages which, if publicly revealed could be damaging, only the truth about Soviet conduct with U.S. prisoners of war posed a grave political danger to Truman and those advisers in his Administration who, unlike James Forrestal, had urged moderation and accommodation with Stalin.

In his memoirs, published many years later, Clifford emphasized the sensitivity of the "Clifford-Elsey Report." He recorded that after receiving the first copy on September 24, 1946, President Truman had asked him to bring all 20 copies of the report to the Oval Office at once, and when Clifford had done so, Truman said to him:

"I read your report with care last night. It is very valuable to me- but if it leaked it would blow the roof off the White House, it would blow the roof off the Kremlin. We'd have the most serious situation on our hands that has yet occurred in my Administration."

According to Clifford, Truman felt it was too early to tell the public the seriousness of the Soviet threat, but he also claims that his report had a significant impact on the President in that shortly after its completion Truman's policy toward the USSR hardened.[35]

In spite of the presidential policy of official secrecy, at least one accurate report on the Soviet mass-kidnapping of U.S. prisoners of war leaked into the American press in 1946, largely through pressure from POW/MIA family members who had refused to give up hope. In the WISCONSIN STATE JOURNAL of December 1, 1946, the essential details of the truth were published under the headline: "Iron Curtain shrouds Lee's Fate, Parents Believe Russians Hold Him With 20,000 Other Yanks."

"...Frank Lee...being held prisoner with a reported 20,000-25,000 other Americans...For almost two years we have been trying to get information but we are balked at every turn. The Lees got a letter through to the military governor of Linz (Austria) asking for information...but he answered he was unable to check in the Russian zone...Mrs. Lee has been corresponding with Kin Seeking Missing Military Personnel, Inc... getting signatures to petitions...to bring 100,000 signatures to President Truman asking him to terminate all UNNRA shipments to Russia and to postpone loans to that country until she gives us the right to search for missing personnel in zones controlled by the Russians."

"The Lees...have talked with Madison men, such as Tom Tierny, 2525 Sherman Avenue, who have been held by the Russians. Tierny managed to escape before the governments of the United States and Russia settled up his board bill, BUT HE DOES REMEMBER THAT THOSE WHO TRIED TO ESCAPE AND WERE CAUGHT OR INVOLVED IN SOME OFFENSE AGAINST RUSSIA WERE SHIPPED TO OTHER CAMPS INSIDE RUSSIA."

"...In May 1945, several months after their son had been listed as missing, Mr. and Mrs. James Thomas, of Scarsdale, N.Y., began to get the first of a mysterious series of messages from sources abroad, saying that their son, pilot of the B-24, was among a group of prisoners in Austria...In December, the Thomas' got a letter from their sources abroad saying that their son and others had been moved from Linz, Austria, Hospital, on April 25, 1945, farther north in Austria, and a friend connected with the Federal Bureau of Investigation (FBI) later claimed that records in Munich bore this out

An interview with the Russian consul in New York was obtained by James Thomas, but, "It resulted in nothing. They just didn't have anything to tell me and indicated that if they did know anything, they wouldn't tell me anyway."[36]

Bound stacks of tens of thousands of the petition signatures referred to in this Wisconsin State Journal article were delivered at the White House for the President in 1947, urging Mr. Truman to demand entrance for US search teams to Russian-controlled territory to locate the missing American prisoners and remains Of the dead, and were located in the National Archives in the late 1980s. This was the work of "Kin Seeking Military Personnel, Inc," the organization of POW and MIA family members referred to by Mrs. Lee in the December 1946 article. Within some of them are original notes from American families who believed their loved ones were still in Soviet captivity, and which apparently went unread. A brief attached note by a Pentagon colonel reads: "file, no action necessary." (This episode is a sad preview of similar actions by Vietnam veterans and Indochina-POW/MIA family members, delivering petitions to the White House in 1986-87.)

Despite such courageous journalism on the part of the Madison, Wisconsin newspaper, most of the American media failed to investigate the fate of the 78,750 Americans who remained missing in action from WW II, the majority of whom had been held in German camps in the part of eastern Europe overrun by the Red Army. Criticisms of the Soviet system were largely confined to a more general complaint about the lack of freedom there, or about Soviet preparations for a conflict with the West. On December 3rd, for example, the New York Times reported that the Vice President of the American Federation of Labor (AFL), Matthew Wo II, had charged that "Billions of American Lend-Lease dollars were being used by Russia 'in building new vast arsenals of aggression for Soviet imperialism.' and that the arsenals were being created by 'millions of enslaved laborers."[37] Although the New York Times and other major media in America had access to far more precise facts about U.S. war prisoners among the millions of enslaved Gulag workers, they chose not to cover the subject, no doubt with the active encouragement of government officials who had access to the classified information on such American POWs in the USSR.

A most influential American, Mrs. Franklin D. Roosevelt, spoke out on December 6th against a Soviet proposal that a United Nations commission investigate conditions in European refugee camps, some of which still held non-repatriated Soviet and east-bloc citizens who did not wish to return to their homelands. The New York Times reported:

"Valentin I. Tepliakov of the Soviet Union and Leo Mattes of Yugoslavia charged that organized propaganda in the camps was preventing repatriation and therefore should be suppressed. They repeated accusations by Andrei Y. Vyshinsky, Soviet Vice Foreign Minister, THAT CIRCUMSTANCES IN THE CAMPS WERE A THREAT TO PEACE...THE FRENCH, FOR EXAMPLE, 'REFUSE TO REPATRIATE CITIZENS OF THE BALTIC STATES, WHITE RUSSIA, THE UKRAINE AND THE SOVIET UNION.' HE SAID."

The widow of the late President spoke for the United States in defending the occupation forces of America, Britain and France in western Europe:

"'This charge is made against authorities which have already repatriated over 2,000,000 Soviet nationals, some 600,000 Poles and nearly 100,000 Yugoslavs.'"

Mrs. Roosevelt said that anti-repatriation literature judged unfair was immediately destroyed and suppressed, and that the Soviets should improve the 'caliber' of their repatriation liaison officers and provide them with specific information on alleged war criminals among east Europeans who wished to remain in the West. In a significant development, the French turned the tables on the Soviets by supporting them in their proposal to inspect European camps, but requested also that Russia allow observers into their camps holding French POWs. According to the New York Times:

"Her delegate, Leon de Rosen, asked Mr. Teplkiakov to request his government to open camps that the Soviet Union maintains for prisoners of war and refugees to French officials tracing 300,000 Frenchman and women, many of whom were in German Army service."[38]

The United States government remained publicly silent about the 78,750 Americans still missing from World War II, but various American citizens, often relatives of missing Americans, would not let the matter drop. An October 25, 1946 telegram to Secretary of State James Byrnes, read:

"The undersigned Chicagoans, relatives of American citizens now held illegally as slaves on farms in Yugoslavia and in mines in Russia, ask that you present to the United Nations Assembly without delay a formal demand that the Yugoslav delegate to the meeting be disqualified until proper action to free those prisoners has been vouched. We demand that you take a firm stand against this monstrous action on the part of representatives of Yugoslavia and its Chief of Staff, Marshal Tito and that you call for an end of the mockery of a representative of such a sovereign nation sitting in deliberation with unclean hands. We, as freedom-loving Americans and believers in human liberty and dignity, demand that the U.S. also take immediate action to lay before the UN assembly the case of these illegally held American citizens..."[39]

A letter to U.S. Ambassador Patterson dated January 19, 1947, is from the mother of an American MIA, Mrs. Lynn Paul of Crandon, Wisconsin, and mentions not only the disappearance of U.S. airmen who parachuted into Yugoslavia near the end of the war, but also the December 1946 Wisconsin State Journal article on American prisoners held by the Soviets, quoted earlier in this book:

"I have read with interest the item in the Chicago Tribune of January 9, 'Envoy pledges help if it's proved Slavs hold any Americans.' My son was a gunner on a B-24 which flew from lower Italy to Vienna, Austria, during the last months of the war. On February 19, 1945, they bombed Vienna and were hit over the target, killing one member of the crew. But they flew back over Yugoslavia and radioed their intention to bail out. Last radio coordinates (were given) The entire crew is still missing and presumed dead by our government. Since Yugoslavia is Russian dominated, we understand free travel is forbidden and we have heard and still feel that American flyers are being held in that country. In a Madison, Wisconsin paper, December 1946, an item said the Russians are holding 20,000 Americans. If our searching teams are allowed in Yugoslavia it seems to us that some evidence of the plane or the crew should have been found by now..."[40]

U.S. State Department documents of the 1946-47 period which were declassified in the late 1980s, indicate official American suspicion that U.S. prisoners held in Yugoslavia and other Balkan countries were being transferred to Soviet control, prior to Tito's "break" with Stalin. One example is a December 10, 1946, Department of State memorandum from the Special Projects Division (SPD) to the European Division's Llewelyn Thompson (later U.S. Ambassador and Assistant Secretary of State):

"When Ambassador (to the USSR, Walter Bedell) Smith was here he stated that the attitude of the Soviet government in regard to AMERICAN DEPORTEES FROM THE BALKANS was entirely unsatisfactory but that we would not be in a position to protest until they flatly refused to release a person proven to be an American citizen and proven to be in the Soviet Union. That has not occurred and that is the important point in my opinion to stress to the public. The fact that we are presenting these cases to the Soviet Union is not in my opinion any reason why we should make a Pollyanna-ish presentation of the matter to the public."[41]

Seven years later, families of POWs and MIAs missing in Yugoslavia were still writing the President:

"...I do feel that I have a son in prison in Russia, His name is Victor Norman Heinselman, 35742032, Air Corps...He was aboard a B-

24 aircraft bomber, which departed Manduria, Italy, on March 4, 1944, on a combat mission to Steyr, Austria., and was seen to sustain damage from enemy aircraft and fall into the Venetian Gulf of the Adriatic Sea. There is no further record of him, whether he was drowned or whether he could have found a haven in the Yugoslavia mountains, AS SIX MEN WERE SEEN TO HAVE BAILED OUT OF THE PLANE. AS GENERAL MICKHILOVICH STATED BEFORE HIS DEATH THAT HE HAD SOME AMERICAN FLIERS HIDDEN IN THE MOUNTAINS, SUCH MEN COULD BE WORKING IN RUSSIAN FACTORIES AT THE PRESENT TIME. As some men have been located and returned by the Russians, I am anxious to have an effort made to see if my said son is still in Russia or her satellites; and if so, to have him returned to the United States authorities in Germany..."[42] But the Pentagon claimed all these men had gone down with their plane, although no evidence was offered to substantiate the claim, and the memory of the American POWs and MIAs who disappeared in Yugoslavia gradually faded away.

Despite this officially-expressed desire to keep the matter of the vanished World War II prisoners out of the public eye, the issue would not die, as millions of families in western Europe and the United States were missing family members, whom they refused to forget about. The American media however, stressed the enormous number of German POWs who remained in Soviet labor camps. A United Press report of December 16th was titled: "3,000,000 Axis POWs on Siberian Projects," and repeated official Soviet figures from "Irkutsk Pravda" that 2,000,000 German POWs and 800,000 Japanese POWs, together with Italians, Finns and Hungarian POWs, were building a new railway and connected road system extending thousands of miles from the Volga River to TAISHET (Siberia), some 400 miles northwest of Lake Baikal. The Soviet newspaper editorialized with none-to-subtle irony that, "'Nazi and Japanese dreams of meeting in Central Asia are realized as Japanese and German prisoners of war are working on railway and highway construction.'"[43]) Nothing was said in the report of the horrifying conditions in the Gulag prisons, many of which were, in reality, death camps.

American, British, French, Dutch and Belgian prisoners were sometimes held in the same camps as the Germans, but the most influential U.S. media declined to cover that subject, while the government chose to keep its evidence of such prisoners classified. On December 19th, the Soviet Union and Japan concluded an agreement whereby 50,000 Japanese captives per month would be returned from the Soviet Union. The agreement was signed by General Kuzman Derevyanko, the Soviet member of the Allied Council for Japan, and by General Paul Mueller, Chief of Staff for General Douglas MacArthur. The agreement not only covered Japanese prisoners of war, of whom 900,000 were estimated to still be in

Soviet control, but also "an undetermined number" of Japanese civilians. Some of these returned Japanese later reported seeing American military POWs in Siberia and Kazakhstan. The winter of 1946-1947 was the worst on record for decades, and with the new year blizzards and below-zero weather swept out of Siberia and across northern Europe. The death rate mounted among the millions of prisoners still confined under the harshest conditions in Soviet gulags nearly two years after the end of the war, and concern for them mounted in the West, despite preoccupations with devastation and unrest in Western Europe.

Truman finally replaced James Byrnes with General George C. Marshall as Secretary of State in January 1947. Upon his retirement, the embittered Byrnes began to speak publicly of "secret agreements" with the Soviets being the cause of problems with Stalin, but he never revealed that the Soviet violation of one of those secret agreements, at Yalta, concerning the repatriation of war prisoners, was at the root of the serious deterioration of relations between the two countries. In his 1947 memoirs Byrnes mentioned the disappearance of French prisoners of war from World War II, inside the Soviet Union, which had been discussed at a Council of Foreign Ministers meeting with Molotov in London:

"The Council had discussed a complaint from the Soviet Union that France was failing to repatriate Soviet citizens. FRANCE IN TURN COMPLAINED THAT THE SOVIET UNION WAS FAILING TO REPATRIATE NATIONALS OF FRANCE. That evening the communique contained no information other than this simple statement of subjects discussed. It was unanimously agreed to by members of the Press Committee. But later that night...a Soviet representative told Mr. Walter Brown, who represented the United States on the Press Committee, THAT MR. MOLOTOV OBJECTED TO THE COMMUNIQUE AND WANTED IT HELD UP. Mr. Brown replied that it had already been given out by the representative of the United Kingdom..."[44]

The prisoner of war matter was to become a serious international issue in 1947, largely due to a public rupture between the French and Soviet governments over the non-return of many thousands of French POWs and displaced persons who had disappeared in Soviet-occupied territory. Although the total number of French missing from World War II at this time exceeded 350,000 persons, the American media concentrated on the relatively small percentage of missing Frenchman who came from Alsace-Lorraine, estimated at 40,000, many of whom had been drafted into the German Army during the Nazi occupation. Unlike the United States Congress, the French parliament was engaged in a lively debate about the prisoners in Russia, and the Foreign Minister, Georges Bidault, had earlier asked Soviet Ambassador Bogomolov about the missing prisoners. Bogomolov had claimed then that only a handful of

'Frenchmen' remained in Russian control, and he said that these were satisfied to remain in the USSR.[45] No contact with the French prisoners in the USSR was permitted by Soviet authorities, however.

France's concern for her missing prisoners was mirrored by Belgium in March 1946, as public opinion heated up there over the fate of Belgian POWs and displaced persons who remained missing in Soviet-occupied eastern Europe. This was after a fairly complete accounting had been made of Belgian Jews sent to the Nazi gas chambers, and of other of Belgians killed by the Germans. Many of the Belgian missing were prisoners of war who had been captured in 1940 or deported later during the war and overrun by the Red Army in 1945. Although Allied repatriation officials knew that more than 40,000 Belgians remained missing in the Soviet-occupied areas, The New York Times reported on March 17th that no one knew whether hundreds or thousands of Belgians remained in Soviet control, and added:

"The Russians claim that 35,000 came back, but the point now is that many others have not returned. Of those who have, some were ordinary POWs, some were simply displaced persons. Few, such as Alois Damma, in Flanders, returned with shocking tales of cruel treatment in slave camps, as related in these dispatches yesterday. There are four types of Belgians who may still be found alive in the Russian zone of Germany or the Soviet zone itself: prisoners of war, deported workers and political prisoners, men from Eupen and Malmedy conscripted by Hitler into the German Army, and finally, Belgian traitors...A STALEMATE HAS PERSISTED SINCE MOSCOW IN 1945 CLAIMED THE REPATRIATION OF SOME 400 FORMER CITIZENS OF THE BALTIC STATES RESIDENT IN BELGIUM...Brussels refused to surrender anyone who did not wish to leave, and as a retaliatory measure the Russians simply sealed the door on Belgians caught east of the Oder. THE PROBLEM HAS BEEN COMPLICATED IN RECENT MONTHS BECAUSE REPATRIATION IS HANDLED HERE BY A YOUNG COMMUNIST MINISTER, JEAN TERFVE...WHAT THE BELGIANS ARE DEMANDING IS A MISSION QUALIFIED TO TRACK DOWN TIPS ABOUT MEN IN RUSSIA. There are at least 40,000 missing persons causing grief in this country, and a most unfortunate impression is spreading that the men in question languish in Russia."[46]

The newspaper said nothing about the American missing and known-prisoners who had vanished under similar circumstances, at the same time, but speculated without any evidence that most of the Belgian missing were "unquestionably" dead, and quoted a pro-Soviet Belgian publication as saying there were "few, very few," Belgians in the Soviet zone. The Belgian government was doubtlessly most concerned about the growing public belief that the missing prisoners were held in Russia. Little more than two weeks before, a mob of over 50,000 disaffected war veterans had stormed the

Parliament in Brussels, after their demands had been ignored.[47]

In the United States the issue of the missing Americans remained muted, although a few devoted family-member groups, such as "Kin Seeking Missing Military Personnel, Inc., and individual relatives of missing POWs, relentlessly pressed for answers about the fate of their loved ones. Within the secret confines of the government some effort was made to discover their fate, but all substantial information was kept "classified," and the most influential U.S. media showed no interest in fate of over 78,000 missing Americans.

To gain proof of American POWs in Soviet gulag camps and general intelligence on the Soviet Union, U.S. intelligence agents, working in conjunction with German police and intelligence, interviewed thousands of returned German and German-Axis POWs in Europe, and this contributed to increased U.S. pressure for release of more Germans still held in Russia. A now-declassified secret memorandum for the Chief of the Collection Branch of the Central Intelligence Group (a forerunner of CIA), raised the issue of millions of missing German POWs still remaining in Russian control:

"In official communiques from June 1941 through 14 May 1945, published in the New York Times, the Soviets claim a minimum of 4,240,000 German prisoners of war, of which number at least 1,230,000 were rounded up after 8 May 1945. These figures do not include satellite prisoners...The total figure given above may have been underestimated by as much as one million men."[48]

A few days later, on April 7th, 1947, Newsweek Magazine reported the disappearance of two million or more German POWs, in an article entitled: "Where Are The POWs?," which stated, in part:

"This week Secretary (of State George C.) Marshall planned a sharp and detailed challenge to the Soviets on the whole question of German prisoners. He was armed with a fresh intelligence report that estimated that the Russians had captured about 4,000,000 Germans. Foreign Minister Molotov had previously asserted 1,003,924 had been returned and 890,582 remained in Russia. That left 2,000,000 unaccounted for."

Newsweek went on to report that, while many of those missing might be dead already, as many as 2,000,000 "may have been deliberately misclassified as civilian workers instead of POWs."

In mid-August the summer before, western Allied troops had carried out a transfer of 1,163 Soviet prisoners and their dependents, called "Operation Keelhaul," in preparation for forcibly returning them to the Russian zone. The Soviet prisoners included Crimean Tatars, Azerbaidzhani, Ukrainians, Russians and Don and Kuban Cossacks, most of whom had fought for the Germans because of their hatred for Stalin. After a final screening of the ex-Vlasov soldiers and Osttruppen, the United States and Britain carried out another forcible-repatriation of Soviet citizens, deemed to be 'war criminals' or 'traitors' between May 8th and 10th, 1947., when 256 Soviet prisoners were moved from Riccioni and Pisa to San Valentino, southeast of Linz, Austria. The Soviet prisoners were also formally transferred from UNRRA control to military control. A draft of the "Operation Eastwind," plan to return these prisoners uses the phrase "dead or alive," and AFHQ requested assurance from the Russians that dead bodies would be acceptable in the event of suicides or death from use of lethal force by U.S. troops. On the 8th and 9th of May the British repatriated 171 prisoners; nine more married men, whose wives declined to accompany them into MVD control, were forced over the border on the 9th and 10th. U.S. forces forcibly repatriated 76 Soviet prisoners on the 9th and 10th of May.[49]

Yet, according to official Soviet statistics given to three Vietnam veterans in Moscow more than forty years later, only a handful of American POWs, less than ten, were returned from Soviet control in 1947. (This subject is covered in a later chapter of this book.) Through the rest of 1947 and into 1948 Walter Bedell Smith the former Chief of Staff to Eisenhower who was now the U.S. Ambassador in Moscow, was confronted with the issue of claimants to American citizenship among dual-citizens of eastern European or Russian descent who were being sent to the gulag The American embassy officials were refused permission to visit or contact these Americans, wherever they were held, which led to a July 31, 1947 State Department consideration to bar Soviet liaison officers from U.S. and western Allied DP camps, or by considering the Yalta Agreement to be no longer in effect.[50]

The Soviets were stung by Western criticism about the missing prisoners of war inside their territory of eastern Europe, and in their usual form turned the charges around by bitterly denouncing the French government, on May 27th, for allegedly refusing to repatriate 12,000 Russians said to be remaining in French-run DP camps in Austria. They claimed that the French were permitting anti-repatriation propaganda into the camps and were also secretly negotiating with the United States about a "naval base at Marseille." The French government swiftly denied the charges and said they had strictly followed the June 1945 repatriation agreement they had signed with Russia.[51] On June 4th, French officials admitted publicly, for the first time, that Soviet military and special police had been in charge of the Russian repatriation camp in France known as Camp Beauregard. The French government had delivered to the Soviet agents there any Soviet citizens who might be "persuaded" to go back to Russia. Stories of prisoners who had escaped clearly indicated that prisoners from Beauregard were being forcibly returned to the USSR. France also released figures showing that 7,504 Soviets had been sent back from Austria, while the Russians claimed only 632 had been received. French officials stated that nearly 5,000 DPs remaining under their control in Austria were former Polish citizens of Ukrainian descent, 1,244 were Baltic people and only 296 were Russian citizens.[52] Little was said in the American press about the thousands of French and other allied prisoners still missing in Soviet-occupied territory.

In the United States on July 2nd, 1947, Lt. Colonel Jerry M. Sage, a former POW of the Germans who was now involved in repatriation matters with the Soviets for the U.S. command, testified before the House Subcommittee on Immigration and Naturalization:

"I am a Lieutenant-Colonel in the United States Army and have been called here from my station, the Headquarters of European Command in Frankfurt, Germany. Here I work with displaced persons as Chief of the Field Contact Section. In that capacity, I have been in most of the 300 DP Assembly Center groups which contain nearly 500 installations in the United States Zone of Germany...IT HAS BEEN ERRONEOUSLY STATED THAT 80% OF THE DPS ENTERED THE OCCUPATIED ZONES AFTER THE END OF HOSTILITIES...THE TRUE SITUATION IS EXACTLY THE REVERSE...OF THE MILLIONS DISPLACED PERSONS REMAINING, LITHUANIANS, LATVIANS, ESTONIANS, POLES, JEWS, YUGOSLAVS, UKRAINIAN AND STATELESS PERSONS, OF WHOM WE ARE TALKING HERE NOW, THE UNITED STATES HAS CONTROL OF ABOUT 600,000 IN GERMANY, AUSTRIA AND ITALY... ON SEVERAL OCCASIONS IT HAS BEEN PART OF MY JOB TO VISIT SUCH INSTALLATIONS TO QUIET THE PANIC AMONG THE PEOPLE BY GIVING THEM THE TRUE FACTS ABOUT THE MOVEMENTS AND REITERATING THAT IT HAS NOT BEEN AND IS NOT THE POLICY OF THE UNITED STATES GOVERNMENT TO FORCE DISPLACED PERSONS TO RETURN TO THE AREA FROM WHICH THEY CAME..."[53]

Sage thus lied to the U.S. Congress about forcible repatriations that had been going on in Europe for more than two years, on a massive scale. Sage's performance helped set the stage for later instances of military officers giving false testimony to Congress about war prisoners, to cover for policy decisions by senior government officials. Nothing was said publicly about the missing American POWs of World War II, whose disappearance in Russian control had caused the forcible repatriations of Soviets which were, in turn, also denied by the U.S. Government.

During September, the "National Association of Barbed Wire Clubs," made up of 1,000 former prisoners of war living in thirty-five cities across the country, adopted a resolution appealing to the United Nations to call for the release of all former enemy prisoners of war, whether held by Britain, France or the Soviet Union. The resolution, which also called for a limit of two years imprisonment for all war captives, was sent to Trygve Lie, the secretary-general of the UN.[54] But the American public remained completely uninformed about U.S. prisoners in Russia, as such information remained "classified" by the government.

U.S. intelligence agencies of this period worked in cooperation with German (and Austrian) authorities and with the newly resurrected West German intelligence service under east-front expert General Reinhard Gehlen, to locate German and other western European POWs returned from the Soviet Union who had been confined in gulag camps with American military POWs of WW II. Senior U.S. Counter Intelligence specialists of the 1945-47 period, such as CIAs future Counterintelligence chief, James Jesus Angleton (who worked primarily in Italy and Austria), had access to such debriefings as part of their everyday work.

A Counter Intelligence Corps report of June 23, 1947 from C.I.C. Headquarters U.S. Forces in Austria, and a follow up C.I.C report of 10 July, concerned the eyewitness report of ex-POW Ernst Schmidt to the Austrian Veterans Aid Society, which was passed to U.S. intelligence:

"...Informant escaped from Subject camp in May 1947... Secret Intelligence from CCIC USFA...reports that in Camp 4108/3 STARNOVKY, near Budapest, Hungary, which is under direct administration of the Russian MVD, there are approximately 35,000 prisoners, the majority of which are German and Austrian. There are however, a considerable number of American, British, French, Italian, Dutch and Belgian nationals in confinement at the camp. THE AMERICAN AND BRITISH PERSONNEL REMAIN IN THE CAMP FOR APPROXIMATELY EIGHT DAYS AND ARE THEN DISPATCHED TO OTHER, MORE PERMANENT PLACES OF DETENTION."

A U.S. C.I.C. agent subsequently interviewed Ex-POW Ernst Schmidt and recorded more details of his experiences at Soviet Camp # 4108 Starnovky, in a top secret August 1947 report which was not declassified (by the NSA at Fort Meade, Maryland) until 1991:

"WORKING HOURS ARE 16 HOURS DAILY...THE GUARD CONSISTS OF GPU PERSONNEL AND BLOODHOUNDS. POOR WORK OUTPUT BY A PRISONER IS PUNISHED WITH 25 STROKES OF A LASH, AND IF HE SHOULD COLLAPSE FROM THIS, THE VICTIM IS KILLED...VERY FEW AND POOR RATIONS...NO HOSPITAL FACILITIES, NO CAMP DOCTOR, NO MEDICINES. NEITHER SWISS RELIEF PACKAGES OR CARE PACKAGES ARE RECEIVED BY THE POWS. ALL CHARITY PACKAGES ARE CONFISCATED BY THE GPU PERSONNEL...ALL QUALITY CLOTHING IS TAKEN FROM THE POWS AND DIVIDED AMONG THE GUARD...JEWELRY AND SOUVENIRS ARE CONFISCATED BY THOSE IN CHARGE OF THE CAMP AND MAILED IN PACKAGES TO MOSCOW...ALLIED NATIONALS ARE TREATED AS POWS AND RECEIVE FAR HARDER TREATMENT THAN THE

AUSTRIANS OR GERMANS. THEY ARE SEPARATED FROM THE OTHERS WHILE WORKING AND MORE IS DEMANDED OF THEM; THEIR RATIONS ARE THE SAME AS THE OTHERS...AMERICANS AND ENGLISHMEN ARE USUALLY TRANSFERRED FROM THE CAMP AFTER 8 DAYS...DOCUMENTS ARE FORWARDED TO USSR-LENINGRAD. IN THE CAMP THERE IS A LARGE SPY NET COMPOSED OF POWS OF ALL NATIONS. IF A PRISONER OF WAR CONFIDES IN ANY OF THESE SPIES, HE IS LIQUIDATED. THIS IS DONE USUALLY IN FRONT OF THE ASSEMBLED MEMBERS OF THE CAMP, ORDINARILY BY MEANS OF A SHOT...OR HE IS USED AS A TARGET, BEING SUSPENDED BY HIS FEET FROM (the gun barrel of) A TANK WHILE STILL ALIVE."

"No mail arrives in the camp. Appeals made by the prisoners of war to family members or to the International Red Cross are destroyed by members of the camp staff. Prisoners of war have been used in experiments in the search of a cure for tuberculosis and other diseases, causing the death of 7 or 8 of them per 90 cases."

Ernst Schmidt also added that the Soviet intelligence network for recapturing escaped prisoners extended to the Allied zone: "I suspect that escaped POWs are recaptured even in Graz, through the medium of GPU (MVD) spies, because three of my comrades, who intended to return to Salzburg and upper Austria, vanished there, after having been taken on a truck in Graz, by civilians."

This top secret report is significant in that it reveals the process by which American and British POWs of 1945-46 held captive in Soviet-occupied Hungary, Austria and Germany were separated from other Allied prisoners by the MVD and shipped to the USSR, as Mr. Ben Kovaks witnessed at Novi Mesto, Hungary, in early 1946. It also bespeaks the shockingly cruel deaths faced by many American WW II POWs who, in the early years, must have believed that their own government would exert every effort to rescue them, if only their existence was made known. Sadly, their existence in Soviet captivity was to remain a state secret for 45 years to come.

A clue to the ultimate fate of thousands of such U.S. and British POWs is contained in the eyewitness account of German ex-POW Hans Joachim Hofmann, who stated to a Counter Intelligence Corps agent that he was held at forced labor in 1947 with 1,100 American, British and French POWs in Novosibirsk camp #311, near Krasnoe, Siberia. This was recorded in a Top Secret U.S. CIC report dated August 4, 1947[55] declassified through FOIA request to USAINSCOM-NSA in 1991, 44 years later:

"Reason For Investigation: Reference is made to Orientation and Guidance Report #2 dated 30 April 1947 re: Project 154032. A Casual informant, Hans Joachim Hofmann, recently released prisoner of war from Russia contacted CIC Ulm thru the local American Red Cross on 2 August 1947, to give information about American and Allied personnel held as prisoners in Siberia, USSR. Informant

Hofmann was released from the Dachau Prisoner of War Processing Center on or about 17 July. 1947. The informant did not contact American authorities there, but German officials in Dachau told informant to contact UNNRA in Ulm. The informant did this and was referred to the American Red Cross in Ulm.

"The informant gave this agent three (3) pictures of American soldiers attached hereto as exhibit A, B, and C. The informant claims to have talked with the soldier shown on the picture marked exhibit "A," whose name is Viktor Boehm (long blacked-out area follows, which probably gave the American POW's rank and U.S. military unit.)" ...emigrated to the United States of America in 1928, lived in (blanked-out area where the POW's address had been).

"According to the informant Boehm served in the Army of the United States, was taken prisoner by the Germans and then taken over by the Russians on or about 5 May 1945, in Oelmuetz, Silesia, Germany and transported to Siberia on 17 July 1945, to the Nor(v) osibirsk Camp #311, near Krasnoe, Siberia, where Boehm is presently working in a tank factory. The informant worked with Boehm, who speaks German. Boehm gave Hofmann the attached pictures.

"THE INFORMANT CLAIMS THERE ARE TWO HUNDRED (200) AMERICAN SOLDIERS WORKING IN THIS PLANT AND ABOUT NINE HUNDRED (900) ALLIED SOLDIERS, MOSTLY ENGLISH AND FRENCH SOLDIERS."

"On the back of the picture attached as exhibit "B" is the address of Mr. James E. Green (address blacked out by U.S. Government censors) who is working in said tank factory in Siberia. The picture attached as exhibit "C" could not be identified by the informant. The informant knows the names of the following Allied personnel: 1. Sergeant Riedel 2. Schanno Ambrosini, also working in said tank factory...It is further recommended above information be made available to the Adjutant General, Washington, D.C. the F.B.I., and the American Red Cross. If the F.B.I. confirms informants statements this matter should be of interest to the State Department."[56]

The photographs attached to this CIC report, which had been smuggled out of the Soviet Union in 1947 by Hans Joachim Hofmann, and remained for 44 years in the classified files of U.S. Army Intelligence, clearly show three U.S. soldiers in American military uniforms, wearing overseas caps. The Counter Intelligence Corps Special Agent who wrote and forwarded this report, John J. Menken, evaluated this intelligence as being of unknown reliability, noted that the informant claimed to be the son of a German Colonel Hofmann, killed by the Gestapo in Tunisia, and recommended that CIC Headquarters in Wiesbaden recheck and investigate the informant in a "discreet" manner. While Special Agent Menken had recommended that the information on the 1,100 U.S., British and French POWs in Soviet camp #311, Novosibirsk, be made available to the Adjutant General of the Army, the FBI and the American Red Cross, this and much other, similar information was to remain secretly classified within these agencies, and would not be revealed to the American people for 44 more years to come.

A follow-up communication of CIC Headquarters, European Command, regarding this livesighting by Mr. Hofmann serves as a classic example of the official tendency to obscure the meaning and significance of such eyewitness reports, which over the years became what can only be termed as an automatic debunking process. The Goeppingen CIC agent was, ludicrously, required to determine if the 1,100 American, British and French soldiers working at forced labor in the Siberian tank plant were held against their will!:

"This headquarters is contacting Sub-Region Goeppingen in an attempt to clarify the following: a. WHETHER ALLEGED AMERICAN AND ALLIED PERSONNEL ARE HELD UNDER DURESS. b. THE PURPOSE FOR WHICH INFORMANT WAS ALLEGEDLY GIVEN ENCLOSED PHOTOGRAPHS. This office (Headquarters CIC EUCOM) is withholding action on Agent's recommendations pending clarification of points listed in paragraph above. Signed, William Wood for Herman M. Krom, Special Agent CIC, Operations Officer."[57]

One can imagine the verbal comments of the astonished Goeppingen CIC agents to such a reply from Headquarters, but in the hard-copy response of August 22nd, the commanding officer of the Region 1 CIC Detachment at Esslingen, Agent Quinn, answered:

"FIELD AGENT INITIATING REPORT STATES (the) SUBJECTS HELD IN DURESS; PHOTOS GIVEN WITH EXPRESS PURPOSE THAT THEY BE PLACED IN PROPER HANDS. INFORMANT LEFT AREA, PRESUMABLY FOR WIESBADEN."

Such a reply left no room for doubt as to how the field agents felt about the report, and therefore it must have been forwarded to higher military and civilian authorities, although which of the highest authorities actually saw the seven copies of this report remains unknown. If the system was functioning properly, this information would have been read by then-senior counterintelligence officers of the time, such as James Angleton, who had been transferred from OSS to the Strategic Services Unit (SSU) and had remained in Europe when the SSU became part of the Central Intelligence Group, (CIG), which received copies of such CIC livesightings. General Edwin Sibert would have had access to it, as would have the Chief of U.S. Military Intelligence, the Director of Central Intelligence, the Director of the FBI, J. Edgar Hoover. The Chief of Staff, General Eisenhower, and the Secretary of State also should have seen it. A copy should also have been given to President Truman, but whether the intelligence agencies of the time, or the

White House Staff, screened such information from the President, to "protect" him, remains unknown. (The FBI FOIA Chief, J. Kevin O'Brien, has denied receiving the author's Freedom of Information requests for declassification of FBI's voluminous holdings of intelligence reports such as this one, on U.S. POWs held by Soviet and Soviet-surrogate regime after WW I, WW II, Korea and Vietnam.)

There were hundreds of such forced labor camps in Stalin's gulag system of the 1940s and early '50s, including many in secret locations inside restricted areas, and hundreds of them must have held large and small groups of American, British and French POWs either separately, or among the aggregate of millions of Soviet, German and other foreign prisoners.

The War Department's Strategic Services Unit (SSU) had established intelligence operations in areas such as Poland, the Ukraine, Czechoslovakia and elsewhere in Soviet-occupied territory, seeking to gain more information about Soviet military strength and intentions, and also about related subjects like Ernst Schmidt's eyewitness account of American POWs shipped to the USSR from Soviet Camp 4108/3 Starnovky, or Hans Joachim Hoffmann's report of 200 U.S. POWs near Novosibirsk. Despite the intelligence acquired by the Army's Counter Intelligence Corps and G-2 /SSU, including that from sources identified by the Gehlen Organization's debriefings of ex-POWs, more indigenous human intelligence sources inside the Soviet Union were required, to further confirm such accounts.

Another source of reports on American prisoners in Soviet gulags were the "Wringer Reports" compiled by Air Force Intelligence in the late 1940s and early 1950s. Although most of the Wringer reports declassified for scholars in 1990 were about the observations of German and Austrian POWs in Russia, which might relate to U.S. national security concerns, and had been screened for almost every reference to American prisoners by U.S. government employees, one example concerning a prison labor location at Krasnokamsk, Borovsk, USSR, which escaped the censors, carries a date of information of mid-1945 to November 1947. It concerns an Austrian civilian taken prisoner by the Soviets because of his blue train conductors uniform, in May 1945, and transported to the USSR for labor purposes. The report said

"Source was a well mannered, fairly polite and neatly dressed elderly man. He was rather hard of hearing and had difficulty expressing himself clearly...He stated that during his stay in the Soviet Union his only goal was to survive the captivity and to return healthy to Vienna, Austria." The report further stated that the source had been captured in Krems, and first interned by the Soviets at Doellersheim Prisoner of War Camp in lower Austria Transported to Russia, along with so many others in July 1945, he was assigned to do manual labor at a forest Kolkhoz and paper mill. the report described the 'Paper Mill Main Building':

"Source recalled that it was a 4 storied brick structure, about 200m by 200m in size with a flat roof. He stated that the factory was engaged in the production of all kinds of paper goods. He further stated that the factory was operating in three shifts. THE TOTAL LABOR FORCE WAS ABOUT 20,000 TO 25,000 WORKERS, 60% MEN AND 40% WOMEN. MOST OF THE WORKERS WERE NATIVES FROM KRASNOKAMSK, BUT SOURCE LEARNED FROM EXPERIANCE THAT ALSO SEVERAL GERMANS, AMERICANS, JAPANESE, POLES, CZECHS, AND HUNGARIANS WERE EMPLOYED THERE."[58] Such information was ignored by the U.S. Government as being not precise enough to establish American identities and thus have specific names of Americans to demand the return of from the Russians.

In mid-November 1947, on the same day that Josef Stalin and Molotov "consented" to run in the upcoming Soviet elections, the prisoner of war repatriation issue, which had been simmering for months between France and the Soviet Union, suddenly erupted. Pressured by French Deputies and citizens who believed the Russians were being given a free reign in France to search for and forcibly return Soviet citizens from Camp Beauregard, while deliberately stalling on returning thousands of French POWs and civilians held in the Soviet Union, the French Socialist Ramadier government ordered a raid on the repatriation camp near Paris. Ostensibly acting to gain custody of three small children whose citizenship was contested, 2,000 heavily-armed French police, supported by four tanks, surrounded Camp Beauregard and conducted a search. On November 16th the New York Times also reported:

"The French Foreign Office delivered a note to the Soviet Ambassador Saturday night informing him of the government's intention to wind up the Russian repatriation camp within five days the camp will be taken over by French authorities and disbanded... French officials said that only this one Russian camp existed in France, that it worked for Russians in transit, and that there was no extra territoriality as the Soviet officers controlled only the interior of the camp."

Yet, a French newspaper had reported that kidnappings of Russians in France had been conducted by these Soviet officers, including one incident in which a Russian named Nicolas Lapchinski had been kidnapped from the house of his friend Jean Tolstoy, a grandson of the great Russian novelist. The three children were discovered in the camp with 58 other persons who had not yet been sent back to Russia, and a cache of small arms was also uncovered at Beauregard. The Soviet guards had been unarmed or absent at the time of the surprise raid. The New York Times reported that under the terms of the June 1945 French-Soviet repatriation agreement,

102,481 Russians had been repatriated, and that this accord had been negotiated to facilitate the process of repatriating thousands of French prisoners from Russia, where they had been taken from Germany.

In a nearly simultaneous action, French troops in Marseille raided several Communist-dominated union centers, arresting 82 persons, among whom were six held for trial on charges of having led a riot on November 12th, and the strikes that had paralyzed that port. Dock and port workers at Nice and on the island of Corsica had also gone on strike to support Marseille workers.[59] The Soviet-French clash over repatriations of war prisoners was part of a larger confrontation over Russia's support for western European Communist parties that threatened to undermine the strength of America's allies there. In postwar France, and elsewhere in western Europe, there was widespread poverty among the working class and millions of war veterans, which caused deep disaffection and a swing of voters towards the Socialists and Communists.

The United States was fearful of the rapid postwar growth in western European Communist parties, with the menace of the Red Army so close by. In response, the Marshall plan for rebuilding Europe's economy had been approved in May, and with passage of the National Security Act, the CIA had been established in September 1947, from already-functioning components of U.S. intelligence. In France, Italy and elsewhere, instead of supporting the right-wing anti-Communists (represented in France by General Charles DeGaulle), as some might have expected, the CIA chose to use other leftists to fight the Communists, by backing the Socialists with covert financial aid that allowed them to grow stronger. Socialist leaders in Paris and Marseille ordered police attacks on striking Communist-led workers, but CIA operatives supplied weapons and financial aid to Corsican mobsters in Marseille, led by the Guerini brothers, for attacks on picket lines and anti-Communist propaganda campaigns. In response, the leftist labor federation in Marseille, made up of both Socialists and Communists, was forced by the rank and file members to call for a general strike on November 14, which soon spread to other parts of France. As a major port of entry for U.S. relief supplies and as a harbinger of the political future in France, Marseille was important to both the Americans and the Soviets in their Cold War confrontation. In Marseille, the Corsican gangsters, led by the Guerini brothers, controlled the underworld and the waterfront, and engaged in international smuggling operations which expanded to include trafficking in heroin, refined from opium originating in Turkey and Indochina, that had previously been controlled by the Italian underworld. The Guerinis acted as the Corsican liaison with American CIA agents and a psychological warfare team dispatched to Marseille to break the November 1947 strike.

This unholy alliance between the CIA and the criminal underworld, considered absolutely necessary at a time of confrontation with Stalinist Communism, was to have far-reaching implications for American society in the decades to come. It was, however, merely a continuation of a practise adopted by U.S. military intelligence and the OSS during World War II. In combatting German sabotage on the New York waterfront, the Office of Naval Intelligence (ONI) had worked with Italian-born mafia capo "Lucky" Luciano and his deputy, Meyer Lansky. This relationship was expanded upon by military intelligence during the invasion of Sicily, when contacts between the American and Sicilian mafia resulted in subversion of an Italian army facing General Patton in 1943. After the invasion of Italy the OSS, the predecessor of the CIA, had worked directly with Don Calogero Vizzini, chief of the Sicilian Mafia, through one of Luciano's lieutenants, Vito Genovese, in combatting Communist agitators and strikers from 1944 onward. Lucky Luciano had been rewarded for his wartime services to the United States by being released from prison and deported to Italy. There he proceeded to build a world-wide heroin empire based first on Turkish-produced and Lebanese-refined opium, with later expansion into Cuba and Indochina, to supply the illicit drug trade in the United States.[60]

During the fall of 1947, the issue of repatriation of French war prisoners held by the Soviet Union, and Moscow's demand for the return of Soviet citizens from France, was thus part of a larger struggle over public opinion and ultimately about control of the population. The French Socialist government was forced by the domestic upheaval to reveal Soviet crimes to the French people in order to sway public opinion against Communist-led strikers. (The CIA was to repeat this tactic in Japan during the Korean War.) The Soviets aggravated the crisis by making their own countercharges. Associated Press reported from Moscow on November 16th, on the Soviet Tass news agency's shrill charge that France had broken the POW repatriation agreement (signed in June 1945), by its raid on Camp Beauregard, which was termed an "illegal search" on Moscow radio. In France, with wildcat strikes spreading from Marseille across the country, a Communist deputy of the National Assembly questioned the government on the use of such a large force in the raid, while the Soviet publications Pravda and Trud, simultaneously, "...pictured France and Italy as threatened by both their own Fascist forces and "enslavement" under the Marshall plan," which had been proposed by the United States in May to aid in rebuilding war-torn western Europe.[61]

Stalin and Molotov seethed over France's action for three weeks before deciding how they would counterattack. In the meantime, the whole issue of war prisoners was addressed by the American Red Cross and the International Red Cross, whose classified files contained information on the American and Allied POWs who had disappeared in the Soviet zone. The International Committee of the Red Cross made a formal appeal on November 28th, to all governments, for the release of 2,500,000 World War II prisoners on humanitarian grounds, without naming the countries where they were held. The appeal was endorsed by the American Red Cross on December 8th, and a resolution by the ARC'S Board of Governors was presented by chairman Basil O'Conner which stated, in part:

"The United States Government and the American Red Cross have consistently urged prompt repatriation of prisoners of war on the grounds of humanity and in recognition of the fact that long detention of prisoners destroys the prospects of resuming normal family life for large groups of people, as well as retarding the recovery of national economies, and stores up hatred and a spirit of revenge that breeds future wars."

In presenting this story, the New York Times supplied out-of-date statistics of the International Red Cross in Geneva to indicate that the Russians were holding only 800,000 war prisoners when the number was in fact several times that figure, if millions of Germans and hundreds of thousands of Japanese were included. France was reported to be holding 600,000 Germans, Britain 250,000; The French zone in Germany, 20,000; British zone in Germany 70,000; Yugoslavia, 90,000; Poland, 50,000; Belgium, 20,000 and Austria, mostly in British hands, 13,000. The United States officially no longer held any prisoners of war, but about two-thirds of the prisoners held in France had been captured by American armies, and the United States was said to retain "an active interest in their fate."[62] In the case of Belgium, for example, Jan-Albert Goris, Commissioner of Information, subsequently wrote a correction to the New York Times about the IRC report that Belgium was holding 20,000 prisoners. He said that under an agreement with the American military authorities in Europe signed in April, 6,000 prisoners had been released by Belgium in May; 6,000 in June; 5,000 in July; and in August and September 2,000 more left each month. Only 3,000 to 4,000 German ex-POWs remained in Belgium, who had a special work status and wanted to stay.[63]

In France at the end of November, three million French workers were out on a general strike that had spread from the initial riot and strike in Marseille, and the nation was practically immobilized. The new Socialist government of Robert Schuman was forced to call out a quarter of a million regular troops and reservists to oppose the strikers and regain control of the factories, mines and port facilities.

Stalin struck back on December 9th, when the Soviet government suddenly expelled the French repatriation mission in Moscow and broke off trade negotiations with France. The Russians charged the French government with 'unilaterally annulling' the Soviet-French treaty on repatriations by conducting the raid and denounced a French charge that two members of the Soviet mission had taken part in 'subversive activities against France,' as 'foul slander,' fabricated by France to justify its actions. At this same time, Truman's new Secretary of State, George C. Marshall, had made it clear that the United States was actively opposing Soviet claims for $10 billion in war reparations from future German industrial production, which Stalin had demanded since the time of Yalta and before, indicating that an open break with Russia over a broad range of issues, had occurred.[64]

Under the heading, "The New Slavery," The New York Times editorialized about the POWs of World War II still in captivity on December 11th:

"The expulsion of the French repatriation mission from Russia, with the similar action taken by France against the corresponding Russian mission, calls attention again to a situation that has become a dark blot on our victory and our civilization. Behind the innocuous and even appealing term of 'repatriation' there lurk the horror and the agony of a modern slave system and of the most extensive and ruthless dislodgment of populations known to history. These practises were first applied on a vast scale by the Axis Powers in a frank effort to rebarbarize' the human race for purposes of conquest. Now, barring the last development, represented by the Nazi crematories, these same practises are being continued by the victors on an even vaster scale. They threaten to make a mockery not only of Allied wartime professions but also of the war crimes trials of the Nazis and of the Japanese accused of similar offenses.

"The Russo-French moves concern only a tiny fraction of the problem, but they illustrate the issue. The French mission was in Russia to repatriate an undetermined number of French citizens, including Alsatians and Lorrainers, either incorporated in the German armies and captured by the Russians or transplanted from Nazi slave labor camps to like slave labor camps in Russia...Nor are the Russians the only ones to use these methods. Two and a half years after the end of the war there are still four million prisoners of war being used as slave laborers—Germans, Japanese, Hungarians, Poles and many other nationalities. Far more than half of them are in Russia, where they join millions of Russians in the same camps. But hundreds of thousands of war prisoners are employed in a like capacity throughout eastern Europe, in Britain, France and Belgium—even in the Middle East and Africa. Their numbers have been swelled by hundreds of thousands of civilian captives deported, expelled, imprisoned or kidnapped by Russia and her satellites, or turned over to them by the Western Allies either in the name of 'repatriation' or under some other guise...The question of slave labor has already been brought before the United Nations by the American Federation of Labor (AFL), and is due for discussion in February."

The American POWs and MIAs in Soviet prisons and labor camps were not mentioned, apparently because detailed reports of their existence remained "classified" by the newly-named Department of Defense, of which James Forrestal, the Navy chief in Roosevelt's and Truman's administrations, had become the first Secretary of Defense.

In France, meanwhile, the nationwide general strike had begun to collapse under pressure from the government's troop mobilization, and the 80,000 workers of Marseille abandoned the strike on December 9th, the same day that the Stalin regime expelled the French POW repatriation mission from Moscow. By December 10th, the New York Times reported, workers all over France were "streaming back" to their jobs. As a none-to-subtle reward to the Corsicans and their Socialist allies, 87 boxcar loads of desperately needed American food supplies rolled into Marseille, for distribution to the people. The CIA and its Socialist allies had won an early victory in the Cold War.[65] In Italy however, the struggle continued, as Communist and Socialist labor union leaders called a strike of 1,500.000 workers in Rome on December 10th, despite the fact that the government had given in to nearly all their demands.[66] Here too, the CIA was active in covert operations aimed at breaking the Communist hold on Italy's labor unions.

That same day, as France's government under Premier Robert Schuman turned back the diplomatic note from the Soviet foreign ministry that announced the expulsion of the French mission in Moscow, as "unacceptable," the chief of that French POW repatriation mission in Moscow during the critical past two years from late 1945 to 1947, was exposed as a Soviet agent. He was Lt. Colonel Raymond Marquie, a former prisoner of the Germans who claimed to have escaped in the east, and to have subsequently served with Soviet partisans in the Ukraine and Poland. (French officials, however, later said this was untrue.) In the midst of the French-Soviet crisis over prisoner repatriations, Raymond Marquie now suddenly announced at an unauthorized Moscow news conference that was not held at the French Embassy in Moscow, that the French government had "plotted" for the last three months to expel the Soviet repatriation mission in France and accused his own government of a "systematically malicious attitude" towards the USSR. Although he had been in Moscow for several years, Marquie stated that French charges against the two Soviet officers arrested at Camp

Beauregard were "without proof and false," accused his own government of being fully responsible for "events of the last few day,." and said "none of these events can be blamed on the Soviet Union." An underling of Marquie at the mission in Moscow, one Henrietta Dumas, also signed a statement released to the Soviet and Western press which said that the French repatriation mission had received every cooperation from Soviet authorities. The New York Times also reported:

"The French mission head said that the Russians were not to blame for a recent raid on a Soviet repatriation camp near Paris, nor for the expulsion by the French of nineteen Russian nationals last month when the French strike crisis was at its peak." Associated Press reported that Marquie had "denounced as a 'lie and propaganda' statements he said had been made in France that thousands of Frenchmen remained in Russia, declaring that only 'several tens' still had been unrepatriated." The traitor parroted the postwar Soviet line on the missing Allied POWs by adding that such propaganda, 'is entirely false and represents the efforts of certain groups to play on the hearts of Frenchmen and Frenchwomen whose sons, brothers and husbands fell, the unknown victims of German slave labor, or on the battlefield against the fascists.'[67] Thus the Soviets were handed a world-wide propaganda coup by Marquie, their sleeper agent within the French mission. The New York Times reported from France that the POW repatriations crisis was part of a much larger conflict:

"Many quarters in France see the dispute less a quarrel over technical missions in these two countries than a pretext seized upon by the Soviet government to express its resentment first with the course of recent events in France, and second with the growing anti-Russian sentiment in this country. They link it with the trend in the Council of Foreign Ministers in London, where the French have sided with the Western powers against the Soviet Union more wholeheartedly than ever before..."[68]

On December 11th, France countered with public charges that the Soviets intelligence operatives had kidnapped or lured into Camp Beauregard at least 60 Frenchmen and secretly transported them to Moscow.[69] One of them may have been the furtive Frenchman who was imprisoned in the Gulag with former Red Army officer Aleksandr Solzhenitsyn, who remembered in his later book on the Soviet prison camp system that the Frenchman said he had been warned about going near the Russian repatriation mission, but paid no heed, until he was suddenly kidnapped and put on a plane to Moscow.[70] (Solzhenitsyn also wrote of another fellow prisoner: "...they say he was an American Colonel.")

France also revealed that since liberation, over 15,000 Soviet citizens had been repatriated to Russia from Camp Beauregard.

Francois Mitterand, Minister of Veterans Affairs (and later French leader), announced that Lt. Colonel Marquie had been suspended from his position as head of the French repatriation mission in Moscow, and it was announced also that he had been ordered to return to France. On December 12th, the government announced that Marquie had been dismissed as head of the repatriation mission "for his lack of discipline in having made statements attacking his own country."[71] The new French government of Premier Robert Schuman also contended that 23,000 French Alsatians and Lorrainers originally deported by the Germans who had been forced to serve in the Wermacht were still missing and might be alive in Soviet prison camps. In a dispatch from Strasbourg to the French newspaper Figaro, it was stated that 90,000 of these men had been repatriated since the war's end; 16,500 others were listed as dead, but the fate of the missing 23,000 remained uncertain. The French Charge D'Affairs in Moscow, Pierre Carpentier, was instructed by the Schuman government to return the Soviet note to the Kremlin as "unacceptable."

Associated Press reported from Paris on December 11th that a French Veterans Ministry spokesman had stated Raymond Marquie was appointed to his repatriation position in Moscow when Laurent Casanova, a Communist, was head of the Ministry. Marquie's records revealed that he had been a foreman in a chemical factory before his mobilization and later capture by the Germans in 1940. Officials said that he had been repatriated to France in 1943 because of illness, and had not, as he claimed, escaped and joined the partisans. The New York Times reported that Marquie "was believed to have Communist affiliations," and that ALL the Communist Deputies in the French National Assembly took sides with the Soviet Union throughout a debate on the repatriations crisis, some shouting that Marquie "was right," for having attacked the French government. One of these Communist Deputies was the same Laurent Casanova who had appointed Marquie to his position in 1945. Casanova denounced the French government for its allegedly deliberate anti-Soviet policy, which he claimed was taking France towards "another Munich."

In countering the Communists, Pierre Closterman, a Deputy from the lower Rhine, "declared that numerous families in Alsace had received postcards from released German prisoners, who, he said, maintained they had seen missing sons from Alsace in Russian camps. By this means only, the Deputy said, many families now had learned that their sons were living." Francois Mitterand, chief of the French Veterans ministry, said that since the expulsion of the French mission, "the French Embassy and consulates in the USSR had been urged to use every means to obtain the release of French citizens in Russia." That same day the League of the Rights of Man passed a resolution condemned Lt. Colonel Marquie for his remarks concerning France. The damage, however, had already been done.[72]

The questions surrounding the fate of the 350,000 missing French POWs and deported civilians who had disappeared when eastern Europe was overrun by the Red Army, of which the 23,000 missing Alsatians and Lorrainers were only a portion, largely remained unanswered after 1947. This number had been determined by Allied repatriation officials after deducting the number of known-dead French Jews, prisoners of war, deported civilians, and Frenchman impressed into the German Army who had been either killed or worked to death by the Nazis. French battle deaths of World War II are listed as 201,568., while French missing and prisoners of war numbered 1,798, 543.[73]

SHAEF records reveal that over a million of these people had been expected to return from the Soviet zone after the war ended in 1945 (although these documents, cited in the previous chapter, remained classified for more than 40 years to come). Yet, even the Soviets, and their now-unmasked agent Raymond Marquie claimed that from the end of the war until the fall of 1947 no more than 315,000 Frenchman had been returned from Soviet-occupied territory. Hundreds of thousands of other French prisoners and missing had thus disappeared in the east. As researcher and author Malcolm Proudfoot wrote in 1947, "During 1946 European tracing bureaux prepared what were then considered to be careful estimates of the number of persons who had lost their lives as a result of the war, and in addition compiled records pertaining to individual missing persons." The numbers of missing reported then included 350,000 French, 150,000 Dutch and 40,000 Belgians.[74] While it is true that some of these may have been unknown wartime dead, at the very least it can be said that hundreds of thousands of missing French, Dutch and Belgians, known to have been in German captivity late in the war, according to declassified documents noted in the previous chapter, must have inexplicably died after the war ended, inside Soviet-occupied territory, while thousands had been reported moved east into Russia where they were later seen alive in POW labor camps and prisons, from which most never returned.

George Kennan, Harriman's number two in Moscow throughout the 1945 prisoner of war crisis, had been recalled to Washington after the success of his February 1946 "long telegram" analysis of Soviet intentions, which had been published in the New York Times through the efforts of Arthur Krock, Washington Bureau Chief for the New York Times. Kennan was assigned to brief senior civilian officials and military officers on the background and expected results of Soviet policy. In 1948, Kennan was appointed chief of the State Department's Policy and Planning Staff which functioned as an operational arm of the National Security Council. Kennan produced a plan for what was to become a covert action agency: The Office of Policy Coordination, or OPC, which would in turn become a major component of the CIA. The POW hostage-crisis had been contained by the secrecy surrounding the disappearance of thousands of American and Allied prisoners into the Soviet Union, and Kennan's assignment was to institute a crash program for obtaining intelligence inside Soviet-occupied territory. A key deputy of Kennan was Robert Joyce, a former OSS agent in Yugoslavia, and specialist on eastern Europe. In putting this plan into practice, Kennan formulated an Office of Policy Coordination within the State Department, a bland name for what was to become America's premier covert action agency during the Cold War. President Truman did not consult Congress when he authorized the OPC with a secret presidential directive known as NSC 10/2. Early CIA covert actions in France, Italy and Germany had been a prelude to implementation of a general U.S. policy of combating Communism by any means necessary, including the use of criminal underworld assets or known war criminals in intelligence operations.

At the suggestion of Allen Dulles, ex-OSS agent Frank G. Wisner Jr. was put in charge of the OPC as a Deputy Assistant Secretary of State. Wisner had been named to his State Department post at the suggestion of Undersecretary Dean Acheson (who was himself later to succeed George C. Marshall as Secretary of State). Resistance by U.S. Military Intelligence to Wisner's appointment was overcome by support for him from Reinhardt Gehlen's controller, General Edwin Sibert, who was now a Deputy Director of the CIA, with the help also of Secretary of the Navy (later Secretary of Defense) James Forrestal, and General Lucius Clay (all of whom knew the magnitude of the Allied POW loss to the Russians). Wisner's service during the 1944 U.S. POW rescue operation in Romania and Yugoslavia, where a number of American prisoners held in small, isolated groups had also disappeared, made him a prime candidate for the job.)

Frank Wisner recruited Tracy Barnes, Richard Bissell, Desmond Fitzgerald, and later others, such as Cord Meyer, and they in turn became senior CIA officials in covert operations and keepers of the secrets. By early 1949 Frank Wisner had forged a solid link with Gehlen at Pullach, Germany, by providing the former Nazi general's organization with desperately-needed extra financing. The U.S. Army had previously kept Gehlen on a short leash by curtailing his resources.[75] The greatest single source of intelligence from within the Soviet Union for Gehlen and for Wisner's OPC was the extensive and systematic debriefings by the U.S. Counter Intelligence Corps (CIC), of German POWs returning from Soviet Gulag labor camps, many of whom, like Ernst Schmidt or Hans Joachim Hoffmann, gave detailed descriptions of U.S. and Allied prisoners of war still being held alive in many parts of eastern Europe and the USSR. American intelligence activities had resulted in a number of former-Nazi war criminals being brought to America to assist in anti-Soviet operations, but also produced more livesightings of American POWs in Stalin's Gulag camps. These remained classified until the 1990-91 period, however, possibly to avoid a potentially dangerous public reaction, which was finally alleviated by the collapse of Communism in eastern Europe and the USSR.

KGB defector Peter Deriabin later revealed the contemporary Soviet view of U.S. intelligence during this period, when the CIA was in its infancy:

"The official view of American intelligence was interesting, although confused in the details of its announced 'orientation.' The most complete organizational description, in fact was given not about CIA, but about a unit of U.S. military intelligence, the Counter Intelligence Corps, with which the Russians had grown familiar through its free-wheeling operations in Europe. There was some scanty information about...the Office of Naval Intelligence, Air Force Intelligence and the Federal Bureau of Investigation. All were regarded as divisions of the same office..."[76] The Russians apparently had little inkling of the deep divisions that existed within the U.S. intelligence community, or indeed among the senior officials who had an impact on American policy towards the USSR.

The former Moscow Embassy official and OSS Chief in Austria, Charles Thayer, had been named to be the Director of Voice of America (VOA) in 1947, and he thereafter largely controlled official American propaganda directed at the USSR and the Soviet-occupied nations of eastern Europe. Under Thayer's direction, as late as 1950, with U.S. troops fighting Soviet-backed Communist forces in Korea, VOA refused to air the truth about the Soviet NKVD's massacre of Polish POWs at Katyn Forest. It is thus not surprising that the truth about the 20,000 U.S. prisoners in Soviet captivity continued to be kept from the American public.

King Michael of Romania had finally been forced by the Communist Groza government to abdicate, on the last day of December 1947. The "Romanian People's Republic" was announced, with the real power consolidated in the hands of Gheorghiu-Dej and the secret police, known as People's Security, or DGSP (Directoratul General al Sigurantei Poporolui). Gheorghiu-Dej's successor, Nicolae Ceausescu, was to maintain a dictatorial Stalinist regime in Romania until he was killed, along with thousands of others, in a bloody 1989 anti-communist revolution.

In the 1947-1950 period there also occurred a sordid trade for small numbers of American skeletal remains in Soviet hands, undoubtedly in response to the continued pressure from POW/MIA family groups such as "Kin Seeking Missing American Military

Personnel," then still demanding information on the fate of their loved ones. The declassified documentary evidence in the National Archives indicates that this was a minor and perfunctory effort on the part of the American government, in order to be able to report to concerned POW/MIA family members that 'everything possible was being done' to ascertain the fate of their loved ones.

An April 23, 1947 memorandum for Major General S. J. Chamberlin from Lt. General LeRoy Lutes of the War Department General Staff Procurement Division requests, "the status of arrangements made...relative to search and recovery of American dead in Russian occupied...or dominated territory."[77] Later declassified documents refer to Public Law 368, of the 80th Congress, approved August 5, 1947, which resulted in American authorities attempting to negotiate the return of American war dead from Soviet territory:

"American authorities undertook to negotiate with Soviet authorities as early as 1945 to secure the right of Graves Registration personnel to operate within Soviet territory for the purpose of locating and recovering the remains of American World War II dead. Early negotiations were handled by the Commanding General, U.S. Forces, European Theater (USFET). However, because of many restrictions of the Soviet authorities placed on the Search and Recovery Teams, normal search and recovery operations were not possible. For this reason negotiations are presently being handled by the Department of State. Subsequently Soviet authorities advised our Department of State officials of their readiness to disinter the remains of American military personnel of World War II now buried on the territory of the USSR, PROVIDED THEY WERE FURNISHED WITH THE NAMES AND EXACT BURIAL LOCATIONS OF DECEDANTS AND THAT ALL EXPENSES CONNECTED WITH THE EXHUMATION AND DELIVERY OF SUCH REMAINS TO BERLIN WERE BORNE BY THE UNITED STATES GOVERNMENT...."

"Based on a survey of casualty reports...there is a total of 646 (U.S.) remains not yet recovered from Russian occupied and controlled territory. Because of the circumstances under which these casualties occurred, a specific place of burial is not known...It has been possible to effect the recovery of certain remains from Soviet controlled territory when specific information was available. On 27 October 1948, the Soviet authorities were furnished a list of names and burial locations of all...for whom...burial location was known...(twenty-four)...Soviet authorities have advised they have been able to locate the graves of sixteen of the twenty-four listed and that the cost for disinterring and delivering these sixteen remains would be 204,000 rubles or at the present diplomatic rate of exchange, $25,500.00. These remains are interred in Archangel, Molotovsk, Murmansk, Odessa and Poltava...By separate action Russian authorities have also submitted a claim to the Far East Command, seeking reimbursement in the amount of 122,977 rubles ($15,372.00) for exhumation, preparation, casketing and transportation of the remains of six (6) U.S. airmen recovered by them within Soviet territory near Cape Lopatka (Kamchatka, Siberia) and delivered to the U.S. Far East Command in September 1948."[78]

They actually ended up costing more than that, as the Soviets first raised the price for 16 remains to $34,000.[79] By the time delivery was made, the first 16 remains cost $51,000, or the equivalent of $500,000. in the present inflated currency. Considering the massive amount of wartime American aid to Russia, this is a revealing indication of the state of Soviet-American relations at the time.[80]

Among those remains recovered in this pro-forma exchange with Stalin's government were seamen from the convoy runs to Murmansk and Archangel, an airman from the old U.S. shuttle-bombing base in the Ukraine at Poltava, and Corporal Lyle Timmerman and Sergeant Ted Yates, two U.S. POWs from Poland who had been killed by a falling wall in Odessa and buried there. The Soviet Foreign Office subsequently demanded an extra fee of $1,000. more to "open unidentified graves at Matzicken POW camp in an effort to locate the missing bodies of Nies, Walker and Teaff."[81] These three POW dead, Staff Sergeant Walter Nies, Tech.Sergeant George Walker, and Tech. Sergeant William Teaff, had been known by others held at "POW Camp 6. (Stalag Luft VI) Heydekrug, Lithuania."

One important case which tested Soviet willingness to cooperate in what amounted to a charade in searching for remains while living American prisoners were being held in Soviet POW and labor camps, involved six American airmen reportedly buried at Cape Lopatka, Kamchatka, Siberia. A letter from the Far East Command to the Quartermaster General of the U.S. Army stated:

"...The Soviet government is seeking reimbursement in the amount of 122,977 rubles for...the exhumation, preparation, casketing and transportation of the bodies of six (6) United States airmen whose remains were interred within Soviet territory...Casualty data received by this Headquarters indicated that a United States aircraft had been shot down near Cape Lopatka on 10 June 1945 and further indicated that the crew had been buried in the Cape Lopatka area."

The Soviets denied permission for an American Graves Registration Search and Recovery team to enter the Cape Lopatka area, saying they would recover the remains themselves.

An October 5, 1949, memorandum from the Allied Council for Japan, member for the USSR, to General MacArthur's Chief of Staff, General E. Almond, records the delivery of the purported remains of the 6 American fliers from Kamchatka and says that they were turned over to a Japanese vessel at the Port of Nahodka, which thousands of Japanese POW repatriates from Siberia were then passing through. At that time the cost of these skeletal remains to the U.S. was $15,372.00, which at the present writing would equal some $150,000, in relation to the increase in the value of gold. The money was to be deposited, interestingly, "in The First National City Bank of New York in Tokyo." The American airmen were: Captain Edward Irving, Captain J. Nathan Eiser, 2nd Lt. Orvil Lord, S/Sgt. Roland Ernser, S/Sgt. Frederick Lang, and Corporal Leslie Denton.[82]

It is unknown whether any of these American airmen survived the crash and were taken into custody by the Soviet secret police, with remains of other prisoners substituted later. At that time there was a central identification system for remains brought back from Europe, similar to today's controversial Central Identification Laboratory.

The 646 cases of known American dead in Russian-occupied territory are broken down into categories, in a classified memorandum attached to the above noted documents, under the heading "Remains to be recovered": "Soviet Zone-419; Russia,-15; Austria-122; Poland-20; Czechoslovakia-0; Hungary-28; Roumania-42." A note included on the back of this file orders: "Do not include Bulgaria, Yugoslavia and Albania." Another "restricted" official U.S. table gives the same total of 646 "remains to be recovered," but also lists 148 others as "Unknowns declared unidentifiable;" 794 more "casualties for which association with recovered remains or non-recoverability has not been established; 309 "total unknowns generally associated"; 173 "casualties definitely associated with unknowns;" 1,276 "Casualties in a non-finalized status;" 84 "Casualties to be declared non-recoverable;" and 1,360 "Total Unresolved U.S. casualties" in the Soviet Zone. This total of 4,790 U.S. servicemen obviously does not include most of the 78,750 Americans listed as "missing in action" initially, or the thousands of "prisoners of war, current status" who vanished inside Soviet-occupied Poland, Germany, Austria and elsewhere, all of whom were subsequently declared dead, by "presumptive findings of death (PFOD). The U.S. Government refused to make further information on the missing in action of World War II available to the author up until the time of this writing.

Records of many internments in National Cemeteries, such as the one at St. Augustine, Florida and at Arlington, Va., of American soldiers who were in combat units whose dates of death are listed as throughout 1946, have also been located in the National Archives.[83]

By early 1948 the threat of war with Russia loomed in both Europe and America. On February 25th, the Soviet-backed Prime Minister of Czechoslovakia, Klement Gottwald, purged all those in the government who were perceived as anti-Communist. The STB secret police and their Russian intelligence "advisers" carried out a new wave of arrests, deportations and executions. On March 10th, Czech Foreign Minister Jan Masaryk, who had nobly tried to maintain relations with the West since the end of the war, fell to his death from an office window, and was widely believed to have been murdered. The "Coup d'état" in Czechoslovakia was seen in the West as a Soviet response to the actions against Communist parties in France and Italy and U.S. proposals to massively aid Europe's recovery.

Meanwhile, with increased tension on the border, the U.S. Military Governor of Germany, General Lucius Clay, reported to Washington that war with the Soviet Union could erupt "with dramatic suddenness." A Time Magazine report of March 21st noted somberly that, "All last week in the halls of Congress, on the street corners, U.S. citizens had begun to talk of the possibility of war between the U.S. and the USSR." Truman had addressed a joint session of Congress on the 17th, calling for enactment of the European Recovery Program (known as the Marshall Plan) and a reinstatement of the military draft. To emphasize American deterrent force during the spring of 1948, at Enewetak Atoll in the south Pacific, the United States exploded three atomic bombs of far greater power and efficiency, including one with six times the destructive force of the bomb used on Nagasaki. On June 11th the Soviets stopped all rail traffic into Berlin, closed the roads the next day, and withdrew the Soviet representative from the Kommandatura, the four-power military control of Berlin, thus precipitating another crisis. On June 18th the Soviets imposed a full blockade on Berlin, thus forcing the United States to conduct a massive an expensive airlift to supply the city and the U.S. and allied garrison there.

In Washington, D.C., the newly-formed National Security Council (NSC) had authorised the covert action agency proposed by George Kennan, which was to be run by Frank Wisner, that became known as the Office of Policy Coordination; first connected to the State Department and later incorporated into the CIA. The OPC was to assume responsibility for penetrations of Soviet territory and for obtaining information about American prisoners of war in Communist captivity.[84] Another hint about the fate of some of the still-living American prisoners of World War II at this time was contained in a classified Counter Intelligence Corps report sent to the Director of Intelligence, Army General Staff, from Headquarters, U.S. Forces in Austria on July 13, 1948, as the Berlin Blockade continued:

"The following information was obtained from a former forced laborer who claimed to have been confined in an UNREGISTERED

LAGER with subject personnel. Approximately 60 km from Moscow in the direction of Kaline, there is an unregistered labor camp. The confinees, 150 men and 50 women, work in coal mines in the vicinity of the camp. Among those confined are 3 American Air Force soldiers who were captured by the German Wermacht, Czechoslovakia, during April 1945. The men are: Charlie, 21 years, 170 cm., blond, blue eyes, has paralyzed right shoulder; Joe, 25 years, 165 cm., dark blond, dark eyes, has stomach wound and is confined in lager infirmary; Albert, 27 years, 170 cm., black hair, brown eyes, has stiff left hip and burn scar on left side of face, is from Texas. THE LAGER CONFINEES WILL NEVER BE REPATRIATED AND ARE NOT PERMITTED TO WRITE LETTERS."[85] (Most documents of this nature have remained classified to the time of this writing, despite repeated Freedom of Information requests and appeals by the author to relevant agencies.)

It has been estimated that in the Soviet-Japanese war in Manchuria, which had begun the day following Hiroshima's destruction by an atomic bomb and lasted only a fortnight, the Russians captured some 2,600,00 Japanese. Of an estimated 800,000-1,200,000 Japanese military POWs from Manchuria taken into Siberia for forced labor, hundreds of thousands vanished in gulag labor camps along the trans-Siberian railroad route, which Stalin had ordered constructed near the Chinese border. Others were held in camps in the Ukraine, Kazakhstan and arctic north Russia. The American political prisoner Thomas Sgovio, who still survived in the Gulag in 1947 and 1948, remembered when great masses of Japanese POWs from Manchuria were moved by sea transport into far eastern Siberia, also: "They flooded Kolyma with Japanese prisoners-of-war, they were kept mostly near and in Magadan, the capital... In Magadan construction was going on. The manpower was all Japanese POW's-and here it was 1947...At first they had refused to accept the black (rye) bread portions, demanding rice instead. But after two weeks they begged for the black bread because it filled their stomachs..."[86]

Former Secretary of State James Byrnes indicated in his memoirs, "Speaking Frankly," that at least 500,000 Japanese POWs remained in Soviet control in 1947, while 900,000 Japanese military POWs had been admitted to be in Soviet captivity during the prisoner exchange agreement signed by General Derevyanko in December 1946. A declassified report from General Douglas MacArthur's Far East Headquarters (dated December 12, 1950) indicates that 316,339 of these Japanese held by the Russians were known by name or had been heard from in captivity. It has been recently reported that 254,000 Japanese POWs of WW II died in the slave camps, and that some 93,000 others still remain missing. (The author believes the number of missing Japanese may actually be far greater.) The missing Japanese POWs (together with the Russian occupation of the Kurile Islands) continues to affect Japanese-Russian relations to the present time. For more than a decade after WW II, Moscow periodically released groups of prisoners which eventually totalled hundreds of thousands of the Japanese, although many of these men had been starved and worked so ruthlessly in the camps that their health and spirits were broken. Many of those who were returned sailed in ships from the Siberian port of Nakhodka, near Vladivostok, to the port of Maizuru, north of Tokyo, where they were screened by U.S. intelligence.

A former U.S. Counter Intelligence Corps agent named Wymo Takaki, a Japanese-American who retired in 1972 after 28 years of U.S. Army intelligence service, revealed the existence of an American woman prisoner once held in far eastern Siberia. Takaki was assigned to a C.I.C. detachment at Maizuru with another Japanese American C.I.C. agent named Yoshinobu Oshiro, who remembered that, "The Japanese were sent to camps located all along the Trans-Siberian railway...They stretched from the Maritime Province all the way to the Ukraine, to the Caspian Sea, Moscow and up north." Several hundred Japanese infantrymen who were repatriated by the Soviets in 1947 from Nakhodka to Maizuru, who had been forced laborers in the gold mines near Magadan, where Thomas Sgovio had been confined, were screened by the C.I.C. in an attempt to identify those recruited to serve as spies for the NKGB. Every one of the Japanese POWs whom C.I.C. agent Takaki interrogated mentioned the same American woman prisoner they had seen, who had been working as a railroad crossing guard at Magadan, Siberia:

"SHE WAS THERE...AND WHEN THE JAPANESE POWS RECEIVED THEIR ORDERS FOR TRANSFER TO NAKHODKA, SHE BROKE DOWN AND CRIED AND TOLD THE JAPANESE SOLDIERS THAT SHE WOULD NEVER HAVE THE CHANCE TO GO BACK AND STEP ON AMERICAN SOIL AGAIN...BUT, FOR THOSE WHO WERE RETURNING TO JAPAN, SHE WISHED THEM GOOD LUCK. BUT THEY WERE GOING BACK UNDER AMERICAN COMMAND, SO THEY HAD NOTHING TO WORRY ABOUT."[87]

Some of the Japanese POWs taken to Siberia from Manchuria, who were fortunate enough to return home alive years after the war, had also been imprisoned with individual American military POWs of WW II, who for various reasons had been separated from other Americans and confined with foreign prisoners. A June 1948 livesighting report of one American POW by Japanese POWs in Kazakhstan, recorded in a later Foreign Service report, stated:

"AT 99-13 PW CAMP IN KARAGANDA...25-26 YEARS OF AGE, 5'6" TALL, WEIGHT 110 POUNDS, BLOND HAIR, BLUE EYES, LEAN FACE, THE AMERICAN WAS AN 'EX-GI' WHO WAS CAPTURED IN EUROPE NEAR A LARGE RIVER (POSSIBLY THE ELBE) AFTER BEING LOST FROM HIS UNIT IN THE LAST WAR. HE WAS REPORTEDLY TRIED AT MOSCOW AND

KARAGANDA BY A MILITARY COURT ON CHARGES OF ESPIONAGE AND RECEIVED A 15-YEAR SENTENCE."

A similar livesighting of October 1949 in Karaganda, Kazakh SSR, gave evidence of the treatment accorded American POWs by the Soviet secret police in the gulag camps:

"ON HIS WAY TO WORK WITH HIS GROUP...HE SAW SOVIET MILITIA AND MVD (KGB) SOLDIERS START BEATING A MAN WHO HAD BEEN WALKING IN THE AREA. THE MAN CRIED OUT THAT HE WOULD NOT STAND FOR THAT SORT OF TREATMENT AND THAT HE WAS AN AMERICAN CITIZEN. THEN THEY BEAT HIM ALL THE MORE AND CALLED HIM A SPY. A CAR OF THE MVD POLICE ARRIVED...THE MAN WAS DRAGGED INTO IT AND TAKEN TO MVD HEADQUARTERS...SOURCE NEVER SAW THE MAN AGAIN NOR DID ANY OF HIS COMRADES."

Forty-three years later, retired U.S. Counter Intelligence Corps agent Takaki revealed more details about these two American POWs of WW II held with the Japanese POWs at Karaganda Camp 99. Returning Japanese POWs in 1948 and 1949 stated that Camp 99, Karaganda also held 335 Spanish Fascists who had fought for the Germans, 900 Japanese officers, an unknown number of Soviet prisoners in a separate compound, and the two American POWs reported in the above Foreign Service dispatches at the time:

"They were captured by the Russians in Germany on May the 8th, 1945, VE Day...These two GIs had crossed this huge river...and were sent to Karaganda POW Camp 99...One of the Americans went by the name of Steve...the Americans were allowed to mingle freely with the Japanese...And when the Japanese were sent to Nakhodka for repatriation, they were allowed to go with them to the railway siding and see their friends off...I think this was (the Soviet intelligence) idea of breaking the willpower (of the two Americans).

Takaki said in the news report that this conduct of the Soviets with these two particular Americans was unusual, and may have been related to the Soviets attempts to recruit spies among the soon to be released POWs. Takaki believes that the two GIs, "may have entered the United States years later, after agreeing to work for the Soviets, perhaps using papers the Russians obtained from Abraham Lincoln Brigade members in the Spanish Civil War, or assuming the identities of Americans captured in Korea or Vietnam." Takaki gave no substantiating details as to how he arrived at the conclusion that these two U.S. POWs may have been returned many years later.[88]

The Karaganda Camps (south of Omsk and northwest of Lake Balkhash in the Kazakh Soviet Socialist Republic) were known to contain at least 120,000 prisoners of many nationalities, who were used as forced labor in coal, iron, copper and silver mines. This number of prisoners was constantly being replaced due to the high death rate in the camps, executions and deaths resulting from prison uprisings. How many Americans or other Western Allied prisoners of

WW II were secretly imprisoned within the huge Karaganda complex will remain a secret until the KGB chooses to open its files on the prison transports and death records of the inmates. The surviving American POWs may have later taken part in the desperate Gulag prisoner revolts which occurred in 1952 at Karaganda, and also in May 1954, when many hundreds of "Zek" prisoners were ruthlessly killed by KGB and military guards to restore the utter submission of inmates.

Another American prisoner, Alexander Dolgun, an employee of the United States Embassy in Moscow who had been kidnapped by the MGB in 1948, was brought through Karaganda during this period, on his way to a slave camp at Dzhezkazgan, southwest of Karaganda. He described his arrival in a convoy of prisoners, and revealed the conditions faced by American WW II POWs in Kazakhstan:

"Not long after Karaganda, the grasses began to thin out and there was nothing to see but flat expanses of rock and sand. The guards told us maliciously that it was called Bet Pak Dala, the Dead Steppe, and that we'd all be part of it soon enough...At three o'clock on the third morning, the train stopped at a station that seemed to be in the middle of empty space. The first sound I heard was a continual barking of dogs, as if the train had run into a huge kennel. When we came out, there was nothing but flat rock reaching off into the darkness, and milling dogs pulling at their leashes, and dozens of guards in tropical uniforms, shivering a lot in the cold because it was not dawn yet, and holding back their German shepherds.

"They sat us on the ground and came around to us with file folders and heard our prayers. Several of us were judged too weak to walk the eleven kilometers to camp and were put into a truck with several guards. The sky was brilliant with stars. By the time we got to the camp it was beginning to get light in the east. We were unloaded beside a huge stone wall that seemed to stretch for a half a mile in each direction. There were watchtowers and barbed wire on the top, and a great gate a hundred feet or so from where we were told to sit and wait for the rest of the prisoners, who were marching in convoy from the station...After the sun appeared, there was some noise from inside the gates, and in a moment they swung open. A thin, tired horse appeared, drawing a flat farm wagon with wooden wheels. Ten or twelve corpses were stretched out on the wagon. Somehow I found this normal. I was watching indifferently until the wagon stopped and two guards appeared with axes. Then I felt quite sick. The guards methodically walked from corpse to corpse and swung the axes up and down. Soon each skull was split wide open. The man leading the horse tugged the reins and led it off. Each corpse had a small metal tag wired to the big toe and the metal tags waved back and forth as the wagon moved away across the steppe."

"I began to feel as though I was hallucinating again because I could hear music, a band, playing some kind of bravura march. It sounded weak and the instruments were not well tuned, but the rhythm was fast and I was sure it was coming from inside the gate. I had a deep sense of cosmic horror that made me dizzy. In the distance I could see the silhouette of the corpses on the wagon. The band seemed to be playing some kind of grotesque farewell. Then it got worse. Out of the gate came, in lines of five abreast, a column of walking corpses in black cotton jackets with white number patches. They could hardly drag themselves. Their faces were pale and drawn and expressionless and they stared straight ahead. Somehow I learned that they were from a hard punishment barracks called BUR. The band is for them I thought. They marched or shuffled, away into the distance surrounded by guards and dogs...The band kept playing. Now in one of the distances I saw a black line approaching and heard more dogs. It was the rest of our etap. When it arrived, we were formed into columns of five and marched through the gate..."

By the end of 1948, James Forrestal was the last surviving senior member of the wartime policymakers in the Truman Administration who knew, from the beginning, what the Soviets had done with American and Allied prisoners in their control, and who also continued to take a hard line toward Soviet aggression. Hostile news reports during the election of 1948 had hinted that he had made a deal with the Dewey campaign, in expectations of a Republican victory. Truman asked for the Defense Secretary's proforma resignation, as was normal following a Presidential election, yet the President made it clear to Forrestal that he wanted him to stay on, at least for a while. But, the intense and devoted Forrestal was worn out and emotionally disturbed from the great and sad burden of knowledge he carried with him, which could not be revealed to the American public, or to his detractors and enemies who accused him of unwarranted anti-Soviet hatred. Truman had apparently decided that the great secret could never be revealed without causing his own political demise.

As a result of a high-level policy decision by the Truman administration, on February 16th, 1949, General Lucius Clay finally ordered the Soviet DP repatriation mission out of Germany. The charade of assisting the Soviets in returning their own citizens to a probably dismal fate, when they continued to kidnap American military personnel and civilians in addition to holding thousands of known and suspected World War II prisoners of war in secret camps, had apparently worn too thin. In a front-page report, The New York Times quoted a communication from General Clay to Soviet Marshal Vassily D. Sokolovsky as saying: "'It is apparent that sufficient time has elapsed since the surrender of Germany for voluntary repatriation to be completed...I must therefore advise you that effective March 1st the repatriation mission will no longer be accredited and request you withdraw its personnel by that date.'" The newspaper also reported:

"...There are about 12,600 Russian displaced persons left in the United States zone. At least 11,000 of them live outside the camp at Ansbach, chiefly because they do not want to be subjected to Soviet attempts to get them back to Russia. Many persons here feel that the Russians have used their missions in Frankfort to conduct 'spying' activities, either on United States occupation forces or on the displaced persons. Elimination of the Russian repatriation mission would leave only the Polish repatriation office..."

An American official stated, however that the Soviet repatriation mission in Austria would be allowed to remain, thus leaving a window open for the unlikely event of future reciprocity. The unnamed official was quoted by Associated Press as saying the Soviets there were not accomplishing much but that once in a while the Soviets came up with somebody who had decided to go back to Russia.

On that same day, Secretary of State Dean Acheson had briefed Senate leaders, including Democrat Tom Connally of Texas and Republican Arthur Vandenberg of Michigan, on both the proposed NATO defense pact between America and its western European allies, and a U.S. policy of massive retaliation in the event of any armed attack by the Soviets. In an oblique reference to America's nuclear bomb and delivery capability he had said: "'If we can make it clear, in advance, that any armed attack affecting our national security would be met with overwhelming force, the armed attack might never occur.'" It had also been reported from Lake Success, New York on February 17, 1949, that no action had been taken the day before by the United Nations Economic and Social Council "on the controversial issue of slave labor conditions in the Soviet Union." Associated Press reported that the entry into the United States of strategic manganese and chrome ore mined in the Soviet Union, had doubled during December 1948. The irony of the United States importing vital metals known to be produced by the forced labor of prisoners of war and other "zeks" in Soviet gulags would not have been lost on Forrestal.[89]

Several friends and associates of James Forrestal have noted that by this period the Defense Secretary was so overcome with his burdens and worries that he never smiled, that his mouth was almost always thinly pursed in almost a grimace, and that during NSC meetings he sometimes scratched his head in the same place so continuously that his scalp bled. By the spring of 1949 Truman had become aware that Forrestal was deeply upset and he decided to replace him as Secretary of Defense with his deputy, Louis Johnson, a large and bluff political operator, who possessed none of Forrestal's greatness of character. Forrestal had become extremely paranoid by this point and confided with several close friends his fear of the Soviet threat and their domestic allies in the United States, and that unidentified enemies were conspiring to destroy him.

At last he was flown to Hobe Sound, Florida, to be cared for by his old friends. Exactly what occurred thereafter is not fully known, but a senior Navy psychiatrist was flown to Florida to examine him, and recommended that Forrestal be returned to Washington, D.C. for immediate hospitalization. His wife Josephine was later to be quoted as saying that members of the Truman Administration feared he might reveal dangerous state secrets. Forrestal was confined to a room on the 16th floor of Bethesda Naval Hospital, guarded 24 hours a day, and visitors were tightly screened by U.S. officials. Although Forrestal had already hinted at suicidal tendencies, he was kept in a location which in fact facilitated his eventual death, apparently by suicide. Forrestal was a bitter foe of Stalin, and was castigated by Soviet apologists in the American media, such as columnist Drew Pearson. Physically and emotionally exhausted from leading the resistance to Soviet expansionism, while unable to counter his detractors by revealing classified information that would have justified his stance, Forrestal chose to die. His passing removed another threat to those who wished to forget the American POWs then dying in Siberia. The last words Forrestal wrote, in transcribing Sophocles' Chorus From Ajax, were:

> Fair Salamis, the billows roar
>   Wanders around thee yet,
> And sailors gaze upon thy shore
>   Firm in the Ocean set
> Thy son is in a foreign clime
>   Where Ida feeds her countless flocks,
>   Far from thy dear, remembered rocks,
> Worn by the waste of time-
> Comfortless, nameless, hopeless save
> In the dark prospect of the yawning grave...
>
> Woe to the mother in her close of day,
> Woe to her desolate heart and temples gray,
>   When she shall hear
> Her loved one's story whispered in her ear!
>   "Woe, Woe!" will be the cry-
> No quiet murmur like the tremulous wail
> Of the lone bird, the querulous nightingale...

Forrestal's successor as Secretary of Defense, Louis Johnson, did nothing about the POWs in Russia, but succeeded in reducing

American defense-preparedness. Smaller U.S. forced-repatriations of Soviets until well into 1948 only confirmed Stalin's intransigence, and now-declassified documents reveal that the Soviets continued to kidnap American soldiers and confine captured U.S. airmen from downed reconnaissance flights through the early Cold War period. The explosion of the first Soviet atomic bomb, at Semipalatinsk in northern Kazakhstan, in September 1949, signaled a new peril facing U.S. forces confronting the Red Army in western Europe, which would eventually threaten the existence of the United States itself in the post-Korean War era.

The Soviets had just released what they claimed to be the "last" Austrian war prisoners held in the Soviet Union, during July and August, in the hope of gaining a favorable propaganda advantage prior to their first successful atomic detonation, which had been achieved with the help of prisoners of war of many nationalities. In the United States, Life Magazine minimized the number of POWs remaining in Russia: "Although Austria claims that some 6,400 prisoners are still to be accounted for, the grim presumption is that they are dead or lost in the void behind the Iron Curtain." A photograph of a group of returned POWs and an elderly man in the Vienna train station was captioned: "A father learns from two prisoners that his missing son died in prison. While they watch in silence, he slowly puts the snapshot that identified his boy back in his wallet."[90]

West German Chancellor Konrad Adenauer, who walked a tightrope between the Soviets and the Americans, was tireless in his efforts to recover German and Austrian POWs from the Soviet prison camps, offering aid in the form of trucks, tractors, farm equipment and other industrial goods in return for minimal cooperation from Stalin's government. After gaining agreements from the Soviets that resulted in the repatriation of thousands of surviving POWs, Adenauer was cheered by adoring Germans, and the mothers of POWs still in Russia would kiss his hands, when he greeted the public.

The missing American prisoners of World War II had not been forgotten by all Americans, but the Department of State was so concerned about maintaining secrecy around the issue, that even inquiries from concerned citizens about them, were classified. One example found in the National Archives dated November 10, 1949, is a withdrawal card representing a letter from a Blanche L. Moore, "Regarding a demand for the immediate release of 400 American veterans that are now being held by the Russians."[91] It is unknown whether this figure represents the American prisoners from one camp overrun by the Russians, or a compilation of other information discovered by Blanche Moore, as the document still remains classified at the time of this writing.

Others who were aware of the secrets that burdened Forrestal were able to live with the knowledge, sometimes by turning their hatred of the Soviets into covert actions against Stalin's regime. One of these was Frank Wisner. According to former Justice department investigator and author John Loftus, Wisner's agents rarely shared files with anyone else at CIA, so that to this day CIA has apparently been unable to locate all OPC files.[92]

Wisner also created false file for Heinz Felfe, Gehlen's liaison to NATO Headquarters, who turned out to have been a Soviet mole from the beginning. It is obvious that all NATO's secrets were thus compromised because of this betrayal. Operationally, Wisner's program was supposed to rapidly develop a network of U.S.-controlled agents inside Soviet-occupied territory. A large number of agents were recruited, including many who had come to the attention of the OSS during and after the war in Romania, Czechoslovakia, Poland, Yugoslavia, the Baltic republics, Byelorussia, the Ukraine and elsewhere. Cover and support was provided for the OPC by CIA, but Wisner was responsible to, and reported directly to the Secretary of State and the Secretary of Defense.[93] Among those supplying agents for the OPC was the former U.S. Ambassador to Poland and SHAEF staff officer, Anthony J. Drexel Biddle Jr., who had become the Chief of the Allied Contact Section at European Theater Command, in Frankfurt. An example from his personal files is the recommendation that a Czechoslovakian officer who had acted as a U.S. intelligence agent in Soviet-occupied Prague and reported to him, was to be "sent to Frank Wisner."

Some (but by no means all) of these agents, including a number brought to the United States for intelligence purposes, were in fact war criminals who had served the Nazis in their occupied homelands, and were protected from prosecution by the intelligence agencies because of their ostensible value as sources. Active operations inside Soviet occupied territory were initiated. Intelligence on (among other things) numbers of American, British, French and other war prisoners in Soviet captivity was brought out of the Soviet Union by released German and Axis POWs interrogated by Gehlen's organization and revealed to U.S. CIC agents, American Consular Services and embassy officials and other intelligence personnel. The information, however, remained classified and was not revealed to the public, apparently because of the overriding policy interest of avoiding armed conflict with the Soviet Union. People who were brought out via the "ratline" from behind the Iron Curtain, with certain knowledge of American prisoners in the Soviet gulag camp system, were ordered to remain silent.[94] The livesighting intelligence on American prisoners was kept secret and strictly compartmentalized, and was not generally known with certainty even within governmental circles except through what has been termed by those interviewed by the author as: "gossip"[95]

H.R. "Kim" Philby, the KGB mole who was head of Soviet counterintelligence for the British secret intelligence service, MI-6, and other Soviet spies within Allied intelligence, betrayed Frank Wisner's networks as they had betrayed many other agents and thousands of British and American war-prisoners held in the NKVD/KGB's labor camp system. In addition, many of Wisner's recruits turned out to be double and triple agents actually working for the KGB, and Wisner eventually committed suicide in 1965. It was later revealed that controlled Soviet intelligence agents, such as James Speyer Kronthal, a former OSS officer and a deputy of Allen Dulles in Switzerland during WW II, who the NKVD-KGB had blackmailed by homosexual entrapment and who committed suicide in 1953, were operating at the highest level of the early CIA.[96] James Angleton, the head of OSS counterintelligence in Italy during the war, who later became head of CIA counterintelligence, had a long and close relationship with Kim Philby. For a time Philby was Angleton's only source on what was revealed by the BRIDE-VENONA decryptions, as CIA didn't yet have access, but the British SIS did. Whether high-level Soviet moles inside US intelligence affected any actions or non-actions in regards to captive US prisoners is not yet fully known. It is known that KGB defectors in Britain and the United States revealed details on western prisoners in Soviet prison camps and information from Soviet intelligence archives, which facts were not revealed to the American public. (The author has unsuccessfully sought to obtain these debriefings from CIA through Freedom of Information request up until the time of publication.)

Some intelligence received on American military prisoners in Soviet camps, of the type acquired by Wisner's OPC from German ex-POWs, which were not very useful for identification purposes in several cases, did indicate the widespread locations of interned U.S. personnel in the USSR:

"Set out below is a series of reports OBTAINED FROM ANOTHER US INTELLIGENCE AGENCY. This information, pertaining to several Americans allegedly held in Soviet prisons, was obtained by the US intelligence agency through interviews with German prisoners recently returned from the Soviet Union.

"Name, unknown, rank, Colonel, U.S. Army or Air Force, age, 45-48 years, height, 5'11", hair: dark blond, balding. The source met the Colonel in VORKUTA Transit Camp #58 during the week of 21-25 August 1949. The Colonel, speaking perfect German, claimed he had been dropped behind German lines during WW II for the purpose of espionage. Neither the Colonel or his parents were born in Europe. The Colonel was captured in the East Zone, possibly East Berlin. The Colonel arrived in transit camp #58 in early August 1949."

"Name, unknown, rank, Major...From August 1949 to July 1949, source was at VORKUTA in Distribution Camp #61, where he met a man who claimed to be an American Major. The Major claims he was kidnapped by the Russians in 1945 while the American troops were still at the Elbe River. (May-June 1945). He claims the Russians sentenced him to twenty-five (25) years for espionage. The Major wore an American uniform and was treated in accordance with his rank. The Major claimed to be a staff officer and was constantly questioned by the Russians about German and American industrial installations. Sepp Wresshwoik, another prisoner, also spoke to the Major. Wresshwoik is expected to arrive in a prisoner transport from STALING in the near future.

"Name, unknown, rank, Captain U.S. Air Force, age 28-33 years, hair reddish. Sometime between May and July 1950, source met a U.S. Air Force Captain in a prisoner of war camp in PETCHORA-KOSCWA. Source had heard that the Captain had been sentenced to ten (10) years for his part in a gun battle which took place in a Moscow restaurant at the end of WW II. It was rumored that the Captain had killed a Soviet officer. The Captain spoke broken German which he had probably learned from his German girlfriend who was also a prisoner in the camp. The Captain worked in a lumber yard. Major Kurt Broikman, a German prisoner of war, was a very close friend of the Captain. Broikman, who speaks fluent English, is expected to return to West Germany in the near future."

It is possible that the names of some of the Germans due to be repatriated from the USSR who had close personal relations with the American POWs, and knew their names, may have been compromised to Soviet intelligence through insecure U.S. communications or by Soviet agents like Heinz Felfe within Gehlen's command, or others within U.S. intelligence. It would thus be interesting to know if the German prisoners named in these reports ever did return from Russia, and if so, whether they were ever interrogated by American agents.

Another case from this file involves a named American civilian, a WW II spy who was reported captured by the Germans and then reimprisoned by the Soviets in 1945:

"Source was imprisoned in UCHTA prison camp, at KOMI, USSR, along with an American newspaperman named James Stafford... WHO IS USING THE COVER NAME OF KURT NISLONE. During World War II, Stafford was working as an American spy in the German Army. During the last few days of the war he was discovered and sentenced to death by the Germans. In 1944, Stafford was imprisoned in Reval, Estonia, where he was liberated by the Russians. Stafford has an Estonian girlfriend who has borne his child. In 1945, Stafford tried to escape with a group of Estonians to Finland where he had previously worked as an American newspaperman in Helsinki. His I.D. card number as a newspaperman was K-226, which is well known in

Helsinki newspaper circles. While escaping, Stafford was caught by the Soviets and sentenced to ten (ten) years in UCHTA prison camp. Stafford killed a Russian in an escape attempt and went as far as KIROV before he was recaptured and brought back to UCHTA. This is when the source first met Stafford. In September 1949, Source and Stafford failed in an escape attempt. In February 1950, source was transferred to MINSK while Stafford remained in UCHTA. THE RUSSIANS CLAIM THAT STAFFORD IS A GERMAN CITIZEN AND NOT AMERICAN. Stafford has a wife and child and other relatives in San Francisco."

The name of the source of this report has been blacked-out by U.S. Government censors. A note below states that, "On 3 December 1953, a check of the files of Regional Registry, Region III, 66th CIC Group, Offenbach, revealed no information concerning any of the personalities mentioned within."[97]

George Kennan was appointed U.S. Ambassador to the Soviet Union during the Korean War, but was unable to maintain his composure in the midst of a police state. He was declared persona-non-grata by the Soviet government after he compared Stalin's communist regime to Hitler's Nazi dictatorship during a trip to London, and barred from returning. Although he later served briefly as U.S. Ambassador to Yugoslavia, Kennan largely devoted himself to academic pursuits, particularly the study of Russian-American relations, although he avoided the subject of Soviet-held American and allied prisoners of war. Within the highest levels of the U.S. Government, however, the knowledge of the missing American WW II POWs kept surfacing in government memos and reports in times of crisis (which were to remain classified for over 30 years). In one declassified secret State Department cable from Australia dated August 16th, 1950, with the communist attack in Korea only two months old, Secretary of State Dean Acheson raised the possibility of revealing the truth:

"Department has proposed to UK, France and Australia, through Embassy here, that they join US in submitting to next General Assembly (of UN) question of failure of USSR to repatriate or otherwise account for POWs and civilian internees detained in Soviet territory. Case would consist of factual record of Soviet failure to either repatriate or account for vast number of German, Japanese AND OTHER POWs, as well as civilians deported to USSR, despite repeated Allied appeals and in clear violation of Moscow Agreement, April,1947; Potsdam Proclamation, July,1945; SCAP-USSR Agreement 1946 and recognized standards international conduct."—ACHESON

Some insight into the actual handling within the State Department, during the Acheson era and thereafter, of any information about the American prisoners who survived the first postwar winters with the "German, Japanese and other POWs" has been provided by former inmates of the gulags who survived the camps and were eventually allowed to leave the Soviet Union. Some of these living witnesses, including John Noble, the American civilian imprisoned in July 1945, have revealed that they were instructed by U.S. State Department officials, "to not speak of, or write about, large numbers of U.S. prisoners of war in Soviet captivity."[98]

In the February of 1953, while thousands more American war prisoners in Korea were in the process of disappearing into Chinese and Soviet captivity, Charles Bohlen was nominated as US Ambassador to Russia by the new President, Dwight D. Eisenhower. Then an Assistant Secretary of State, Charles "Chip" Bohlen was the former Russian translator and advisor to Roosevelt and Truman who had played such an important role in the events of 1945-46. Although Secretary of State John Foster Dulles opposed the nomination, Eisenhower was adamant, considering Bohlen a friend since their time in France together. The State Department's internal security chief opposed Bohlen's nomination on the basis of Bohlen's actions at Yalta and Potsdam, and for other security reasons. The Congressional opposition was led by the right-wing anticommunist Senator Joseph McCarthy, who had already exposed Bohlen's brother-in-law Charles Thayer, one-time head of VOA, as a security risk. McCarthy considered Bohlen, "so blind that he cannot recognize the enemy," and enlisted the support of Everett Dirksen, Homer Capehart and other conservatives to stop Bohlen's confirmation, despite the fact that he was going against his own party's president.

McCarthy introduced as part of his evidence against Bohlen the sworn affidavit of a former Soviet Foreign Ministry official (and of Soviet intelligence according to one informant, see Henry Regnery note below), Igor Bogolepov, who had worked for the State Department in the early 1950s.

Bogolepov had defected to the Germans because of his opposition to Stalin, and later in WW 11 had worked on General Vlasov's political warfare project. He had been saved from the 1945-46 forcible-repatriation of Vlasov men by the efforts of a U.S. Office of Naval Intelligence (ONI.) agent named Herschell Williams. Williams was involved in the forced-repatriations, and had been given a list of names of those who were not to be sent back to Stalin. Bogolepov was on that list, as having been part of the Vlasov Army.[99] Bogolepov was part of the NTS, a Russian nationalist exile organization, which had links to U.S. intelligence and to anticommunist agents remaining inside Soviet territory. Later in the 1940s he was brought to the United States by U.S. intelligence. In his affidavit he stated he had worked for CIA for several years, but became dissatisfied with that agency. In his affidavit for the Bohlen confirmation hearing, Bogolepov claimed that he had learned from contacts in the Soviet secret police (NKVD) when he was in the Soviet Foreign Ministry in Moscow, that Bohlen was regarded by the NKVD as a "friendly diplomat" and as a "possible source of information," when he was with the US Embassy in Moscow.

Bogolepov's affidavit was treated with contempt and derision by the American left, but his charges were taken seriously by the anti-communists in Congress. Other allegations concerning alleged "immoral" conduct and hints of homosexuality surrounded Bohlen during the hearings.

The most important subject of all was touched on however, when Bohlen was asked about the bloody continuation of the forced-repatriations to Russia in 1945-47, at the Senate Foreign Relations Committee hearing. Facing what had now become a difficult confirmation, concerning his role at the Yalta Conference and other summits, and McCarthy's and Bogolepov's allegations about his loyalty and personal conduct, Bohlen fell back on his knowledge of the most damaging secret of the Second World War. He said:

"The purpose of it was there were 60,000 American prisoners in Poland and Germany under the control of or about to be under the control of the Red Army, and the purpose was to get those prisoners back."

After what thus appears to have been a threatened revelation of the fate of US prisoners in Russian hands, Bohlen's nomination as Ambassador to the USSR became a major issue to President Eisenhower who exerted all his influence to see that it was swiftly approved, which it was soon afterward. Bohlen was to be in Moscow at a critical time in the history of Soviet-held American POWs, at the end of the Korean War.

Following the descent of the Iron Curtain along the Soviet front from the Baltic Sea to the Adriatic in 1945, a concerted effort was made by the United States and its Allies to penetrate this invisible wall, both to gain warning of Soviet intentions and troop movements and also to discover if possible by aerial reconnaissance the fate of hundreds of thousands of Allied prisoners and civilian political deportees who had disappeared. While many of the aerial attacks by the Soviets on American and Allied reconnaissance aircraft remained secret for years, some were revealed, and declassified documents clearly indicate that many more American prisoners disappeared inside Soviet territory during these incidents. American POWs from WW II and the Cold War were reported alive in Stalin's gulags into the 1960s, in now-declassified livesightings in the National Archives. Such livesightings caused the United States government to periodically request information about, and the return of all U.S. prisoners captured during Cold War intelligence gathering missions. One such request from the

Department of State concerned shoot-downs of a U.S. Navy intelligence aircraft in the Baltic Sea area in 1950 and a B-29 over the Sea of Japan, in 1952. These cases and other Cold War "incidents" are noted in chapter nine of this book.

Nine years after the Second World War, in April 1954, during the liberalization period following Joseph Stalin's death, the Soviet Union suddenly released 248 Spanish survivors of the famous "Blue Division," which had been ordered by Franco to fight for the Germans on the Russian front. After landing in Barcelona from a Liberian steamer chartered by the Red Cross, the returning Spanish POWs were mobbed by hundreds of hysterical relatives and friends. A news report of the time said:

"Glad to be back in Spain, the returnees were gladder still to be out of Russia. They painted a chilling picture of their years as prisoners, but insisted that the lot of the Russian peasant is but little better than that of the labor-camp inmate. Said Telesforo Moreno: 'Communism? Communism is cabbage soup, hard work and every man for himself.' While the Spanish press made little mention of the fact, not all the Semiramis' passengers were Blue Division P.W.s. With them were four bitter young men with pinched faces and premature wrinkles who had been sent to Russia by the Republican government in the 1930s..."[100] Thousands more Spanish citizens, who had been sent away to Russia by the Republican government of Spain during the war with Franco, vanished forever.

Most Cold War-era gulag prisoners, such as the Russian Jew Avraham Shifrin, who testified before the U.S. Senate in 1973 on American and European prisoners and conditions in the gulags, were not held in camps with large numbers of American or British prisoners, but witnessed other WW II Allied and Axis prisoners still remaining alive in the camps in the 1950s, and even into the 1960s. Shifrin was a former Soviet Army officer who had served with distinction in WW II. He was born in Minsk, Byelorussia in 1923 and lived in Moscow until 1941, when he was inducted into the Red Army. His father, a Russian Jew, had been arrested during one of the earlier purges for telling an anti-Stalin joke, and he had died in the Gulag camps after ten years at forced labor.

As the son of a convicted Zek, Shifrin was assigned to a delinquent, or penal battalion of the Soviet Army in 1941. Of 500 men in the battalion only 100 were issued rifles. They were ordered to capture weapons from the Germans for the rest. He was wounded in action twice and was promoted to First Lieutenant on the battlefield, after which he served through the remainder of the war, fighting all the way to Konigsburg, East Prussia. Shifrin was demobilized as a Major in the Soviet Army in September 1945, and after demobilization, he was employed by the "Department to Combat Banditry," as an investigator in the Krasnodr region. Shifrin subsequently became chief of the investigative unit and in 1946 he was transferred to the headquarters of the Department in Moscow. In this period he studied law and became a legal advisor and attorney. In 1947 he was sent as a chief investigator to Tula, a major military center. Here he became Chief Legal Advisor in secret war plant No. 535. This plant made automatic guns for fighter aircraft, Anti Aircraft guns for the Navy and mine launchers. He served here from 1948-1952. In that year he moved to Moscow where he became Chief Legal Advisor in the Contract Division of the Soviet Ministry of Defense. Shifrin was arrested on June 6, 1953 during the peak of the anti-Jewish terror, which had begun as the "Doctors Plot" imagined by the paranoiac Stalin, and was continued by Beria and his MGB successors for months after the death of the dictator. Shifrin's boss, Aleksandr Lieberman, another Jew who was head of the Legal Division for the entire Soviet Ministry of Defense, committed suicide by jumping out of a 6th story window rather than submit to arrest by the secret police. Shifrin was confined in the Lubyanka and Lefortovo prisons and sentenced to death for what amounted to fabricated anti-Soviet activities, but his sentence was subsequently commuted to 25 years imprisonment. Among his KGB interrogators was Merkulov, the deputy to Beria who had interrogated and tortured the forced-repatriates in 1945, including the Cossack leader, Krasnov.

Later, in the Butyrka Prison he met former Soviet POWs of WW II who were still confined, and also Vlassov men who had served in the German Army. Here also Shifrin met Prince Konojo, a member of Emperor Hirohito's family and a former General in the Japanese Kwangtung Army. There were also still many German prisoners from WW II.

After this he was deported to prisons at Novosibirsk and then to Omsk, where he remembered the death rate among prisoners was staggering. He was held here in a camp which was part of the Kamyshlag complex, containing some 100,000 to 150,000 people. At the Omsk camp where he was confined, out of 5,000 Zeks, every day some 30 to 40 naked corpses of prisoners, who had died of starvation or been murdered, were loaded on horse-drawn sleds for disposal. Here the KGB used killings to incite prisoners of different nationalities against each other. After a 1953 melee between the 2000 Russians 2000 Ukrainians and 1000 prisoners of other nationalities, the KGB guards opened fire killing 300 of the prisoners in Shifrin's camp. Afterward the KGB executed 30 more prisoners who had been nearest to the gate at the time of the uprising.

After this Shifrin was transported to Semipalatinsk the nuclear testing site in Kazakhstan. (A location that was later to be the destination of some U.S. Vietnam War prisoners.) Many prisoners were being transferred here from other camps after prison revolts. In this camp Shifrin met a former American officer named Alexander Shornik who had come to Russia to get his father out in 1948. Shornik had tried to smuggle him out through Ukrainian Nationalist channels (the Bandera Underground) but both had been caught and sentenced to ten years in the slave camps. In November 1955, Shifrin was sent to Tayshet on the Baikal-Amur railroad in Siberia. Along the way he was held in the Novosibirsk transit prison. Here Shifrin met people who had come from the Gulag of Kingir, where there had been a major camp uprising. Tanks had crushed to death 500 women who had tried to protect their men by standing in front of them. (The fences separating the people had been broken down.) Also in Novosibirsk, he met people who had taken part in major uprisings in Norilsk and other areas.

Shifrin was released after ten years in the Gulag and spent four more years in exile. He became an advocate for Soviet Jews wishing to emigrate to Israel, and in 1970 he was finally permitted to emigrate to Israel himself, where he now resides. Today, he is the director of The Research Center for Prisons, Psych-Prisons and Forced Labor Concentration Camps of the U.S.S.R., in Israel. As a result of his studies Shifrin estimates that some 25,000,000 persons, including five million prisoners of war, died during the peak years only, in Stalin's slave labor camps. He believes the figure to actually be higher than this. On February 1, 1973, Shifrin testified before the Internal Security Subcommittee of the Judiciary Committee of the U.S. Senate, under the Chairmanship of Senator James Eastland. Other Senators of the subcommittee were: Sam J. Ervin, Birch Bayh, Hugh Scott, John J. McClellan, Edward Gurney, Strom Thurmond, and Marlow W. Cook. Many of the questions were asked of Shifrin by Senate Chief Counsel J. G. Sourwine. Avraham Shifrin testified:

"FIRST I MUST ASK YOU TO EXCUSE MY ENGLISH, BECAUSE I CANNOT SPEAK LIKE YOU. I LEARNED MY ENGLISH IN CONCENTRATION CAMPS AND FIRST MY TEACHERS WERE KIDNAPPED AMERICAN OFFICERS...I ALSO WANT TO TELL YOU I AM VERY GLAD TO BE HERE...TO GIVE TESTIMONY IN SENATE OF UNITED STATES ABOUT THESE QUESTIONS. ALSO I MUST TELL TO MY FRIENDS IN CONCENTRATION CAMPS FROM THIS ROOM THEY MUST EXCUSE ME, BECAUSE I CANNOT MENTION ALL THESE HUNDREDS, HUNDREDS AND HUNDREDS OF NAMES WHICH I KNOW SIT NOW IN PRISONS..."

Shifrin told the Senators of the day he received his sentence: after 14 days, came an officer and they opened the little door in my cell and he read to me that they had changed my sentence to 25 years in the concentration camps, 5 years in exile, and 5 years of deprivation of rights of a citizen. And in the same minute, they brought me from this death cell to another cell, cell number 58. I remember this because I was sentenced under article 58 in code and I came to cell 58. In this cell was—it was a transit cell—sometimes 60, sometimes 75 people together. All of them have a 25-year sentence. And it was corridor with cells and each cell with prisoners with 25 year-sentence."

"IN THIS CELL, I FIRST MET SOME AMERICAN OFFICERS WHICH WERE KIDNAPPED FROM THE BORDER AND ALSO JAPANESE PRINCE KONOYA."

Senator Gurney: "Kidnapped from where?"

Avraham Shifrin: "FROM OSTERREICH (Austria), FROM VIENNA. AND I WOULD GIVE YOU THEIR PHOTOGRAPHS AND NAMES. I HAVE THIS."

Senator Gurney: "What were their names, the American officers?"

Mr. Shifrin: "IT WAS ONE OF THEM, ALEXANDER SHORNIK, ONE OF YOUR OFFICERS. HE WAS NOT KIDNAPPED. THEY ARRESTED HIM WHEN HE CAME TO VISIT FATHER IN U.S.S.R....ANOTHER OFFICER WITH THE NAME STANLEY WAS KIDNAPPED FROM VIENNA."

Senator Gurney: "That is the one they kidnapped?"

Mr. Shifrin: "YES."

Senator Gurney: "Stanley who?"

Mr. Shifrin: "I HAVE A LITTLE PROBLEM WITH THIS OFFICER, BECAUSE HIS NAME IS A LITTLE SECRET HERE. MAYBE I WILL TELL YOU PRIVATELY ABOUT THIS? IT WAS AN IMPORTANT AMERICAN OFFICER WHICH THEY KIDNAPPED FROM VIENNA. HE WAS ONE OF MY FIRST TEACHERS OF THE ENGLISH LANGUAGE."

At this point in the testimony no more questions were asked of Avraham Shifrin by the Senators about these kidnapped American officers, including Stanley_____?, from Vienna. Shifrin's testimony however, raises a serious question about the identity and military rank of one American, STANLEY YOUNG, who had been reported by Edward Page of the U.S. Moscow Embassy in a secret cable of September 26, 1945 to Secretary of State Byrnes as being: "HELD BY THE SOVIETS IN ROOM (cell) 92 AT BUTIERSKAYA PRISON IN MOSCOW..."

No more questions about Americans in Soviet captivity were asked of Avraham Shifrin during his public Senate testimony, but later in the day on February 1st, 1973, after lengthy testimony on conditions in the camps, he again brought up the subject of POWs:

Mr. Shifrin: "NOW, I SHOULD TELL YOU ABOUT POWs."

Senator Gurney: "I wonder if this is not a good time for us to recess and then we can start on that afterward. So, the subcommittee will recess until 2:30." (Whereupon at 12:50 p.m., the hearing was recessed, to reconvene at 2:30 this same day.) It is not known what sort of caution may have been urged upon Mr. Shifrin during this recess, on his coming statements on any aspect of the subject of POWs in Soviet prison camps. According to the Senate record, the subject of POWs did not come up again until the afternoon of February 2nd, 1973:

Senator Gurney: "Now Mr. Shifrin, would you like to proceed with where we left off this morning? I think you were on the prisoners of war as I recall it...And if you can summarize as concisely as possible, I would appreciate it. Proceed."

Avraham Shifrin: "Excuse me, please. Sometimes my answer is not short but you must understand me that I have too many things what I must tell you."

Senator Gurney: "I understand."

Avraham Shifrin: "BECAUSE I AM RESPONSIBLE FOR MY FRIENDS, WHICH ARE NOW IN MY PLACE IN CONCENTRATION CAMPS. I WILL BE SHORT. NOW ABOUT THESE PRISONERS OF WAR. WHEN I CAME TO TAYSHET (near Krasnoyarsk, Siberia) I MET WITH THOUSANDS OF PRISONERS OF WAR: GERMAN, ITALIANS, SPANISH, AND ALSO JAPANESE, BELGIAN..."

Senator Gurney: "These were mostly prisoners from World War II?"

Avraham Shifrin: "WORLD WAR II; YES. THESE PEOPLE EXPLAINED TO ME THAT THEY WAS BROUGHT TO THIS TAYSHET FOREST IN 1945. IT WAS NOT HERE ANY RAILWAY THEN AND IT WAS NOT HERE ALSO HIGHWAY. THEY PUT THEM IN THE FOREST AND THEY BUILT, AT FIRST, HIGHWAY FROM CAMP TO CAMP. EXCUSE ME, BEFORE THIS THEY BUILD CONCENTRATION CAMP FOR THEMSELVES. THEY CUT THIS FOREST AND BUILT HIGHWAY, AND THEN NEAR THIS HIGHWAY THEY BUILT RAILWAY. AND THEY TOLD ME THAT NOW IN MY TIME-IN 1954—IT IS HERE LIKE RESORT. BUT I HEARD THIS FROM LIVING CORPSES..."

"IN 1945, 1946, 1947, 1948 IT—THEY TOLD ME—WAS DAYS WHEN PEOPLE DIED IN HUNDREDS EACH DAY FROM STARVATION. THEY SHOWED ME KILOMETERS OF CEMETERIES ON THE SIDE, AND THEN, I TOLD YOU, I HAVE SEEN THESE CEMETERIES WHICH GO NEAR THE RAILWAY. THEY EXPLAINED TO ME THAT ALONG THIS RAILWAY WAS THE KWANGTUNG ARMY AND HUNDREDS OF THOUSANDS OF OTHER POWs. AND IT WAS THE TRUTH. I DON'T KNOW HOW MANY DIED THERE-MAYBE MILLION AND MAYBE MORE BECAUSE IT IS A 600 KILOMETERS RAILWAY AND HIGHWAY, AND ON BOTH SIDES CEMETERIES WITHOUT INTERVALS."

"WHEN I CAME I MET NOT MILLIONS OF POWs; I MET ONLY THOUSANDS. BUT WE KNOW THAT IN KWANGTUNG ARMY WAS MILLION SOLDIERS. IN 1956 WHEN THEY RELEASED THESE PRISONERS, I HAVE SEEN HOW THEY PUT THEM IN TRAINS NEAR OUR CONCENTRATION CAMP. THEY BROUGHT THEM FROM ALL ALONG THE RAILWAY; FROM ALL THE CONCENTRATION CAMPS., AND I HAVE SEEN THAT IT WAS ONLY SOME 10,000 PEOPLE; GERMANS, FRENCH, SPANISH, JAPANESE."

"NOW TRY TO UNDERSTAND HOW MANY DIED THERE. I HAVE SEEN ONLY MAYBE HUNDREDS OF THESE  GERMAN OFFICERS, AND

THEY TOLD ME THAT MOST OF THESE PRISONERS OF WAR DIED THERE. AFTER 1956 KGB RELEASED OTHER FOREIGNERS FROM TAYSHET, IT WAS CITIZENS FROM CZECHOSLOVAKIA, POLAND, IRAN, AND OTHER COUNTRIES."

"IT WAS ALSO BELGIAN OFFICERS. WAS MAYBE 12 TO 15 OF THESE OFFICERS. I DON'T KNOW WHY THEY DIDN'T RELEASE THEM IN 1956. I DON'T KNOW. THEY WERE THERE IN 1962. AT THE END OF 1962 THEY RELEASED THEM. WHEN QUEEN OF BELGIANS CAME TO U.S.S.R., SHE ASKED TO SEE THESE OFFICERS AND IN BIG HURRY CAME, KGB TO OUR CONCENTRATION CAMP. THE BELGIANS WAS IN CONCENTRATION CAMP NUMBER 7 IN POTMA BECAUSE IN THIS DAYS WE ARE NOT IN TAYSHET; WE WAS IN POTMA. I TOLD YOU YESTERDAY THEY BROUGHT US FROM TAYSHET TO POTMA IN 1960, AND THESE BELGIANS CAME TOGETHER WITH US. ONLY IN 1962 KGB SENT THEM HOME.

"ALSO IN 1963 ONCE I HAVE MET WITH GROUP OF CRIMINALS WHICH CAME TO OUR CONCENTRATION CAMP, NUMBER 7, AND WHEN I ASKED THEM IN WHICH PLACE THEY WERE, THEY TOLD ME WE WERE IN ISLAND OF WRANGEL. I ASKED, 'ISLAND OF WRANGEL?' 'KGB HAS CONCENTRATION CAMP THERE?' THEY ANSWERED ME, 'YES; THERE IS THREE CONCENTRATION CAMPS FOR PRISONERS OF WAR AND THERE SITS ONLY BIG OFFICERS FROM GERMANY, ITALY, FRANCE AND SPAIN.'"

"I DIDN'T BELIEVE THEM, BUT NOW IN ISRAEL I MET WITH ONE MAN IN 1971—WE HAVE HIS LETTER; I SHOW YOU HIS LETTER—AND HE EXPLAINED TO ME THAT HE WAS ALSO IN ISLAND OF WRANGEL IN 1962 AND HE HAVE SEEN THERE THREE CONCENTRATION CAMPS WITH THOUSANDS OF PRISONERS OF WAR, AND HE EXPLAINED TO ME THAT THEY HAVE IN ONE CONCENTRATION CAMP ATOMIC REACTOR, AND THEY MAKE EXPERIMENTS WITH LIVE PEOPLE WITH RADIATION. IN ANOTHER CONCENTRATION CAMP, THEY HAVE EXPERIMENTS WITH PHYSICIANS ON THE PEOPLE, AND IN THE THIRD THEY HAVE SUBMARINES AND THEY HAVE EXPERIMENTS WITH LIVE PEOPLE UNDER WATER. AND IT IS BIG SECRET, AND HE WAS THERE IN THE GROUP OF PRISONERS WHICH GAVE TO THE PRISONERS FOOD.; HE WAS NOT INSIDE THE SECRET CAMPS. HE HAVE HEARD FROM OTHER PRISONERS ABOUT THIS WORK, AND HE SENT TO YOU THIS LETTER...IN WHICH HE WROTE REAL NAMES OF THESE PRISONERS IN WRANGEL."

"ALSO HE TOLD ME ABOUT TWO BIG PEOPLE WHICH HE MET ON WRANGEL. ONE OF THEM, ALEXANDER TRUSHNOVICH, HEAD OF ANTI-COMMUNIST RUSSIAN ORGANIZATION, N.T.S.; HE WAS KIDNAPPED IN 1953 FROM WEST BERLIN. AND ANOTHER MAN WHICH HE HAVE SEEN THERE AND ALL THE WORLD DON'T KNOW... ABOUT HIS FATE; THIS MAN IS RAOUL WALLENBERG, DIPLOMAT OF SWEDEN, CONSUL IN BUDAPEST, WHO SAVE MANY THOUSANDS OF HUNGARIAN JEWS FROM NAZIS. THEY KIDNAP HIM IN 1945, AND THEY ALWAYS ANSWER TO FREE WORLD THAT THEY DON'T KNOW NOTHING ABOUT THIS MAN..."

And thus, while hundreds of living American POWs were

disappearing in Communist control in Vietnam during early 1973, did a foreign ex-prisoner of Stalin's gulags expose to the U.S. Senate a secret world, that it apparently did not wish to see. Nine years later, Edward Hurwitz of the Department of State's Office of Soviet Union Affairs released an official statement about the reported prison camps on Wrangel Island, which had originally been settled by Americans and later forcibly occupied by Soviet forces in 1924:

"I am replying to your letter of April 10 to Deputy Secretary of State Stoessal concerning the status of Wrangel Island and the existence of Soviet prison camps there...Although the United States has never had occasion formally to recognize Soviet sovereignty over the Island, our government has long adopted the practise of dealing with authorities in De Facto control of a given piece of territory. Careful review of more recent information has produced no evidence of the existence at this time of prison camps on the Island. The above mentioned Soviet settlements do not appear to have the characteristics of penal facilities. Previous studies also were unable to confirm reports of prison camps alleged to exist after World War II..."[101]

---

## Notes on Chapter Seven

1   NYT, January 22, 1946

2   NYT, January 15, 1946, p.8

3   New York Times, January 16, 1946

4   NA, Department of State 740.00114 PW/ 1 -2346

5   NA, Department of State, 740.00114 PW/1-2346

6   Department of State Memorandum: "Inquiry Concerning Incidents Involving U.S. aircraft," dated July 25, 1952, originally declassified through Freedom of Information request by Theodore Grevers

7   NYT, February 25, 1945

8   Ibid.

9   NYT, March 6, 1946

10   NYT, April 17, 1946.

11   *The Secret World*, Peter Deriabin, Doubleday, NY, 1959

12   *The Last Secret*, by Nicolas Bethell; also original records on microfilm in the National Archives

13   After his death, which occurred while he was serving as U.S. Ambassador to Spain under President Kennedy in 1961, his widow Margaret eventually remarried, to Colonel Edwinston Robbins, General Biddle's former deputy.

14   *Witness To History*, Charles Bohlen, Norton's, NY, 1973.

15   *Special Envoy*, op. cit.

16   State Department Files, 740.00112A EW/7-1846 CS/JEC; also reproduced on p. 810, POW Policy and Process, Volume II

17   *I Saw Poland Betrayed*, Arthur Bliss Lane, NY, Bobbs-Merrill, 1948.

18   Ambassador Malcomb Toon, who in 1992 was appointed by George Bush to head the American delegation in the U.S.-Russian Commission on POWs, began his State Department career in 1946 as a junior foreign service officer in Soviet-occupied Poland, later rising to Director for Soviet Affairs in he Department of State and U.S. Ambassador to Moscow.

19   NYT, June 4, 1946

20   Military Intelligence Service Memorandum, May 20, 1947, Fifth Army Intelligence Report No. 2715, pp. 821-22, Volume II, Hearings before the Select Committee on POW/MIA Affairs, November 1991, U.S. Government Printing Office, 1993, ISBN 0-16-038479-6

21   NA, RG 319 Box 284, Army Intelligence Project Decimal File, 1946-1948, From 381 USSR to 385.2 USSR, Military Intelligence Division memorandum # 918, 21 May 1947

22   Ibid, ALGF 091(240) SN 2853, HQ 5th Army, 27 May 1947, To Director of Intelligence, War Department

23   NYT, May 29, 1946

24   Wedemeyer Reports!, Holt, NY, 1958

25   Veterans of Foreign Wars Magazine, December 1991

26   State Department memorandum, "Inquiry Concerning Incidents Involving U.S. Aircraft," July 25, 1952

27   *Present At The Creation*, p. 196, Dean Acheson, Norton, NY, 1969

28   NYT, SEPTEMBER 28, 1946, P. 5

29   On the above three 1946 U.S. documents located in the National Archives, pp. 809, 810 and 813, POW/MIA Policy and Process, Volume II, Hearings of Select Committee on POW/MIA Affairs, November 1991, U.S. Government Printing Office, 1992, ISBN 0-16-038479-6

30   Case # 0507, CIC Region VI, Bamberg, 26 August 1946, Declassified by Commander, U.S. Army Intelligence and Security Command (INSCOM, 20 September 1991, through FOIA request.

31   *I Was A Slave In Russia-An* American Tells His Story, by John Noble, Devin-Adair Company, NY, 1958, pp. 28-40

32   I Was A Slave in Russia, op. cit. p. 67

33   A copy of this report is in the Harry S. Truman Library, Independence, Missouri

34   The Last Secret; Pawns of Yalta and Operation Keelhaul, op. cit.

35   The neglected topics within the Clifford-Elsey Report of 1946 prompted the author to produce a report condensed from this book for the U.S. Senate Select Committee on POW/MIA Affairs, and for President George Bush, in April 1992.

36   Ibid., this article is reproduced in full on page 81 5, Volume II

37   NYT, December 4, 1946

38   p. 9, NYT, December 7, 1946

39   NA, Department of State, October 25, 1946, 740.00114 AEW/10-2546

40   NA, Department of State, January 19, 1947, 740.00114A EW/1-1947. This document, supplied by the author, is reproduced on p. 817, POW/MIA Policy

and Process, U.S. Senate, op. cit. It was this document that led the author and research colleague Ashworth to search the archives of the United States for the December 1946, Madison, Wisconsin newspaper (The Wisconsin State Journal) quoted previously in this book. It was located by Ashworth in the Library of Congress.

41    p. 816, POW/MIA Policy and Process, Volume II

42    NA, Department of State records, Letter to President Dwight D. Eisenhower, received at Department of State, June 16, 1953, from Noah Heinselman, Parkersburg, West Virginia

43    p. 18, NYT, December 17, 1946

44    Speaking Frankly, by James F. Byrnes, Harper & Brothers, NY, 1947, pp. 249-250

45    p. 4, NYT, February 7, 1947

46    p. 6, NYT, March 17, 1947

47    NYT, February 27, 1947

48    p. 818, "POW/MIA Policy and Process," Volume II

49    Pawns of Yalta, pp. 115-119, Pawns of Yalta, by Mark R. Elliott, University of Illinois Press, Urbana, Chicago and London, 1982; also original Operation "Keelhaul" and "Eastwind" documents on microfilm in the National Archives. consulted by the author in 1988.

50    Pawns of Yalta, by Mark R. Elliott, op. cit., pp. 121-122; Smith to State Department 16 January 1947, NA RG 740.162114/1-1647, 5 February 1947, FRUS 1947, Vol. I V, p. 720,State Department to Soviet Embassy, 28 May 1947, Ibid, pp. 708-29

51    p. 5, NYT, May 28, 1947

52    p. 19, NYT, June 5, 1947

53    Statement of Lt. Colonel Jerry M. Sage, United States Army, to the House Subcommittee on Immigration and Naturalization, July 2, 1947, #543, Department of State; from the personal papers of Ambassador (Colonel) Anthony J. Drexel Biddle Jr., at the Historical Society of Pennsylvania, Philadelphia, Pennsylvania.

54    NYT, September 8, 1947

55    File # I-G-820

56    Headquarters Subregion Goeppingen, Counter Intelligence Corps Region I, File Number I-G-820, Project 154032, "Copy #1 of 7 copies." Declassified on September 20, 1991 by order of the Commander U.S. Army Intelligence and Security Command (INSCOM Freedom of Information and Privacy Office.

57    File No. I-3.32, I-G-820, to Chief Counter Intelligence Corps, Region I Stuttgart, Charles T. Herring, Special Agent, CIC, "1 of four copies."

58    USAF "Wringer" Report # IR-545-A-51, 7 September 1951, prepared by Edmund, Captain, USAF, Country: Austria, in the National Archives, Secretary of the Air Force Files

59    NYT, November 16, 1947

60    Politics of Heroin, pp. 20-45, Alfred W. McCoy, Harper and Rowe, NY, 1972

61    p. 3&10, NYT, November 17, 1947

62    p. 4, NYT, December 9, 1947

63    NYT, December 17, 1947

64    NYT, page 1, December 10, 1947

65    Politics of Heroin, p. 45, op. cit.

66    NYT, December 11, 1947

67    NYT, December 11,1947

68    Ibid.

69    NYT, December 12, 1947

70    *Gulag Archipelago*, Aleksandr I. Solzhenitsyn, Harper and Rowe, NY, 1973

71    NYT, December 13, 1947

72    NYT and Associated Press, December 12&13, 1947

73    1991 Almanac, Houghten-Mifflin and World Book Encyclopedia

74    European Refugees, The Military and UNRRA, 1947

75    *The Belarus Secret*, John Loftus, Knoph, NY, op. cit.

76    *The Secret World*, Peter Deriabin, Doubleday, NY, 1959

77    NA,WDGSP/C3 17689

78    National Archives, 293 Russia, QMGMP, Removal of Remains of American World War II dead, 17 January 1950, Lt. Colonel Johnson, 3184, "Restricted", signed by Major General H. Feldman, the Quartermaster General, 2 pages, declassified 17 November,

79    NA, Department of State Records, April 18, 1950, Letter to Secretary of the Army Frank C. ace, Jr. from Deputy Undersecretary of State John E. Peurifoy

80    National Archives, examples are on pp. 825 & 826, POW/MIA Policy and Process, Volume II

81    State Department Classified message No. 1041, dated 17 November 1950, from the American Embassy in Moscow to the Secretary of State, and following memorandum for the record(Berger/53328)

82    Letter of Soviet Member, Allied Council of Japan to Major General Almond, Chief of Staff, Supreme Commander For the Allied Powers, with enclosures, 5 October 1949, No. 1304; also: 22 December 1949, AG 150, General Headquarters Far East Command, For the Commander in Chief, "Soviet Claim for Reimbursement For Recovery of Remains, to the Quartermaster General Department of the Army."

83    See for example: NA, SPQYC, 314.6, To the Superintendent, St. Augustine National Cemetery, St. Augustine, Florida, "Correction of Records, 5 March 1946, 29 March 1946, etc.30 April 1946

84    Testimony of Lt. Colonel Phillip Corso, NSC, November 10, 1992, Senate Select Committee on POW/MIA Affairs

85    This document from the National Archives is reproduced on p. 828, POW/MIA Policy and Process, Volume II Many other CIG documents of this era, now under CIA control, remain classified despite repeated Freedom of Information requests and appeals by this writer.

86    In a letter to the author of July 1989.

87    July 19, 1992 report in the Honolulu Sunday Star & Advertiser, written by Night City Editor editor Mark Matsunaga

88    July 19, 1992 Honolulu newpaper report noted previously

89    NYT, February 17, 1949

90    Life Magazine, August 8, 1949, 27:30-31

91    NA, Department of State, Nov. 10, 1949, 740.00114 a EW/11-10491

92    *Belarus Secret*, p. 123, Knoph, NY, op. cit.

93    *The CIA and the Cult of Intelligence*, Marchetti p. 45, op. cit.

94    1987 and 1989 interviews with Major Carl Heinmiller, AUS-Ret'd.

95    Letter quoted previously from former Ambassador to the USSR, George F. Kennan.

96    *Widows*, Corson, Trento and Trento, Crown Publishers, NY, NY, 1989, pp. 7-16

97    Report 111-33330, Region III, 66th CIC Group, 11 December 1953.Declassified 20 September 1991, by U.S. Army Intelligence and Security Command.

98    Confirmed in interviews of John Noble by Thomas V. Ashworth and John M.G. Brown

99    Personal interview with Henry Regnery by the author, 1988.

100    Time, April 12, 1954; see also: Life, April 19, 1954

101    Eur/Sov: E. Hurwitz/bd/4-23-1982, 22248, declassified in 1992

# CHAPTER EIGHT

## THE KOREAN WAR

The mountainous Korean Peninsula of Asia has been inhabited for more than 30,000 years. In 108 B.C. China conquered the tribes of northern Korea and introduced Chinese sciences, art and government to that region, but in 313 AD the Koreans succeeded in driving the Chinese out. By the late 600s the southern Korean kingdom of Silla had succeeded in uniting what they called "Koryo" into one country, which continued to have close cultural ties with China. In 1259 the Mongol armies conquered Koryo, but by 1368 the Koreans had driven them out, too. The Japanese invaded Korea in the 1590s and were also driven out, but this was followed by a Manchu invasion from China in the 1630s, and Korea continued to pay tribute to China until the late 1800s. Korea became known as the "Hermit kingdom" in the 1800s, because of its policy of excluding foreigners and persecuting Christian missionaries and converts. After the American Civil War the U.S. Navy and Marines intervened in Korea to protect American citizens and interests, fighting one severe battle there, and by 1876 Japan had forced Korea to open some of its ports for trade. After Japan defeated China in the Chinese-Japanese War, and then Russia in the 1904-05 war, the Japanese moved to take over the Korean Peninsula, which was completely accomplished by 1910. Japan ruled Korea brutally and expropriated land for Japanese settlers until 1945 and the end of the Second World War.

The origins of the U.S.-Korean War can be traced to the same World War II Yalta Conference that had resulted in such serious problems with the repatriations of American and western Allied prisoners of war. Earlier, in the wartime Cairo Declaration, it was stated that China, the United States and Britain, "are determined that in due course Korea shall become free and independent."[1] At the Teheran Conference in 1943, President Roosevelt had told Stalin that he didn't believe Korea was capable of governing itself, and that the country should be put under a 40-year tutelage.[2] But it was at Yalta where formal plans for the postwar occupation of Korea were discussed in an inter-Allied consultation, and on February 8, 1945, between President Roosevelt and Marshal Josef Stalin.

The Department of State's position papers on the postwar status of Korea for the use of the President and other top U.S. officials at Yalta, indicate that at the Yalta Conference the Soviet Union was to be offered a share in the postwar occupation and administration of Korea, in hope of gaining Soviet participation in the war against Japan, but even if Russia didn't attack Japan, the same offer was to be made:

"It is the Department's tentative opinion that an interim international administration or trusteeship should be established for Korea...AND THE SOVIET UNION SHOULD BE INCLUDED IN ANY SUCH ADMINISTRATION... The entrance of the Soviet Union in the war against Japan would result in the presence of Soviet forces in Korea which would be an important factor in determining the composition of the occupational forces...the traditional interest of the Soviet Union in Korea raises the possibility that it will wish to participate in the military occupation of Korea even though the Soviet Union may not enter the war in the Pacific...The position of the Soviet Union in the Far East is such that IT WOULD SEEM ADVISABLE TO HAVE SOVIET REPRESENTATION ON AN INTERIM INTERNATIONAL ADMINISTRATION REGARDLESS OF WHETHER OR NOT THE SOVIET UNION ENTERS THE WAR IN THE PACIFIC."[3]

In the spring of 1945, General Douglas MacArthur was one of those urging that Russia be encouraged to enter the Pacific war, after many senior policymakers in Washington had ceased regarding Russia's participation as necessary or even desirable.[4] The published official record of the Yalta Conference indicates that President Roosevelt proposed to Stalin in a meeting on February 8th that Great Britain be excluded from participating in the administration of Korea, but Stalin replied that the British should be invited. Otherwise, Stalin said, Prime Minister Winston Churchill might, "Kill us."[5] The United States Government officially denied, as early as June 8, 1945, that any secret pledge or division had been made at Yalta by the Roosevelt Administration on Korea. The New York Times repeated the government's position: "Reports of a secret understanding made at Yalta with the Russians whereby they would obtain Korea and other concessions for their entry into the Pacific war were denied today by Joseph C. Grew, Acting Secretary of State."[6]

However, Stalin's forces were allowed to occupy the northern portion of the Korean Peninsula with a wing of the Red Army that had attacked the Japanese in Manchuria, following the atomic bombing of Hiroshima in August 1945. The record of the State-War-Navy Coordinating Committee (SWNCC) reveals that on the night of August 10/11, 1945, in an emergency session, Assistant Secretary of War John McCloy directed an aide of General George C. Marshall, former Army Colonel Dean Rusk (later Secretary of State) to decide with another colonel, C.H. Bonesteel, on the demarcation line that would divide the Soviet and American occupation zones in Korea. Rusk and Bonesteel decided on the 38th parallel, which the Soviets accepted immediately.[7] Thousands of Communist-indoctrinated north Koreans who had served with Kim IL Sung in the Soviet Union, or in Manchuria, during the Second World War, now returned to set up a Communist police state on the Soviet model. On the 29th of

August, one of the earliest violent incidents between the U.S. and the Soviets in the Far East occurred, with the Russians shooting down an American bomber over Korea which had been involved in a POW rescue mission.[8] U.S. documents declassified by the U.S. Army Intelligence and Security Command (INSCOM/NSA) in the early 1990s indicate that Americans were also kidnapped by Soviet authorities in Korea during the immediate postwar period, as had occurred in Germany, Austria, Poland, Romania, Manchuria and elsewhere, and some were later reported alive in Soviet captivity.

The Soviets, thus installed their own Stalinist regime in North Korea headed by Kim Il Sung, complete with GRU and NKVD-trained secret police and summary executions and imprisonment of dissidents, along with mass-propaganda indoctrination of the population. This was to prepare the North Koreans for an eventual war of conquest against non-communist South Korea, which was then occupied by U.S. forces. Secretary of War Robert Patterson subsequently recommended withdrawing American troops from Korea as early as 1947, while the civil war in China continued, and in September of that year the Joint Chiefs of Staff stated that the United States had "little strategic interest" in keeping troops and bases there. General Eisenhower, then Chairman of the Joint Chiefs of Staff, was among those concurring, in a September 25, 1947 memorandum to Secretary of Defense James Forrestal, which also had the concurrence of the National Security Council and President Harry Truman. In October 1947, the United States asked the United Nations to conduct elections in Korea by the spring of 1948, and the General Assembly agreed to do so, but the Soviet Union refused to participate, saying the U.N. had no jurisdiction there. The Communists launched terrorist attacks against polling places during the subsequent elections in South Korea, killing more than one hundred persons, but a U.N. commission certified the election of President Syngman Ree in a heavy turnout of voters in the south, as valid. Later in 1948, the Soviets officially announced the formation of their Communist puppet government under Kim Il Sung, and that Soviet forces would be withdrawn from the peninsula by the beginning of 1949.

American intelligence estimates on Korea varied widely. In a secret CIA "Review of the World Situation," produced for the President of the United States on June 15th, 1949, less than one year prior to the Communist attack on South Korea, the following analysis was presented to Harry Truman:

"Korea: The scheduled start of US troop withdrawals produced so much official apprehension, publicly communicated, that symptoms of hysteria appeared. Unless the Republic assumes an air of confidence—justifiable at least for the short run—hysteria can easily grow into panic. Actually, since the Republic's armed forces are at least equal in number and superior in equipment to those of North Korea, an immediate test of strength is not likely. Popular panic, stimulated by hysterical government publicity, has recently done more to prepare the ground for the destruction of the Republic than have direct acts of the Communists."

The United States had withdrawn most of the remaining of American troops by the end of June 1949, leaving some advisers. Although the State Department (and specifically, Secretary of State Dean Acheson) was later to be blamed by General MacArthur and others for bringing on the Korean War, in reality that department had, at least superficially, counseled caution in withdrawing American forces from the peninsula. Acheson had omitted South Korea and the nationalist-controlled Island of Formosa, from his description of the U.S. defense perimeter in the Pacific, in a speech before the National Press Club on January 12, 1950. But MacArthur himself had done the same thing in an interview with a British journalist a short time before.[9] Most other CIA intelligence predictions, as well as those of Army Intelligence (G-2) of MacArthur's Far East Command, under Major General Charles Willoughby, failed to note the significance of a buildup of Soviet-supplied tanks on the 38th parallel, the evacuation of villagers in the area and other signs of impending aggression by North Korea.[10]

After consultation with Stalin, whose boldness had increased following the Berlin Blockade and the successful 1949 Soviet atomic bomb test, on June 25, 1950, Kim II Sung unleashed his Soviet-supplied North Korean army on the Republic of South Korea. The United Nations immediately called for North Korea to halt its aggression and the ten-member UN Security Council passed a resolution demanding that the communists cease their attacks and withdraw to the 38th Parallel. Fortunately, the Soviet Union's delegation was boycotting the Council meetings as a protest against Nationalist China's membership, so the Soviet delegate (Mikoyan) was absent, and therefore the USSR did not veto the resolution. North Korea continued its devastating attack.

As North Korean forces approached Seoul, the South Korean capital, on June 27, President Truman ordered U.S. air and naval forces to help defend South Korea. On the same day the United Nations asked UN member nations to aid South Korea and on the 30th of June, Truman ordered U.S. ground forces into action in South Korea. U.S. troops of the Army's 24th Infantry Division were flown from soft occupation duty in Japan to Pusan in southern Korea on July 1st, and began taking up positions near Taejon the next day. Other UN troops soon began arriving, and American soldiers first went into combat at Osan, south of Seoul, on July 5th, 1950. General Douglas MacArthur was named Commander-in-Chief of the UN Command by President Truman on July 9th, with the compliance of the United Nations.

The American Congress did not declare war but supported the policy of President Truman and the United Nations. Averell Harriman and others urged that the President obtain a resolution of support from the Congress for prosecuting the war in Korea, but Secretary of State Dean Acheson, an icy and formal anglophile who was already under attack in the Congress for having appeased the Communists, scoffed at the idea, claiming it was unnecessary. Truman agreed with Acheson and both began to refer to the war in Korea as a "police action," a term that was to infuriate veterans of the war for decades to come. Harriman had been administering the Marshall Plan in western Europe, until he was asked by Truman to serve as the "President's Special Assistant for National Security Affairs" (The position of national security adviser that would later rival the Secretary of State's role when it was held by McGeorge Bundy, Henry Kissinger, Zbigniew Brzezinski and Brent Scowcroft). Clark Clifford had acted in this role earlier, before a formal position had been created. Harriman and Acheson got along well most of the time, as they had been friends since Harriman had coached Acheson on the Yale crew decades before. Forrestal's successor, Louis Johnson, was still Secretary of Defense when the Korean War broke out, but he had angered Truman with his political attacks on others in the Administration and Truman was to replace him with General George C. Marshall three months later.[11]

General MacArthur directed the war from his Far East Command headquarters in Tokyo, Japan, and Lt. General Walton Walker of the U.S. 8th Army was named field commander in Korea on the 13th of July. The heavily outnumbered U.S. 24th Division was reinforced by elements of the American 1st Cavalry Division and 25th Infantry Division on July 19th, but the UN forces continued a fighting retreat. On July 21 the city of Taejon, south of Seoul, was captured by North Korean forces, and the American and allied forces continued to fall back. During the month-long, July 4th to August 3, 1950, delay of the North Korean invasion campaign, U.S. forces lost 1,991 killed in action and 2,588 wounded.[12] Of those nearly two thousand Americans listed as killed, many had been last reported as presumed killed, body not recovered, or as missing in action. Nearly a thousand Americans were reported killed in the Battle of Taejon, from a 3,933-man U.S. force of the 24th Infantry Division on July 19th and 20th. In a July 25-26th ambush at Hadong, the 757-man 3rd Battalion of the U.S. 29th Regimental Combat team lost 313 killed in action, of whom 100 were reported captured.[13] Many hundreds, and perhaps thousands of American prisoners of war were captured by the Communists during the early weeks of the war, and hastily force-marched north by their captors. Those unable to march were executed in cold blood on the battlefield by their Communist captors.

One of the Americans taken prisoner at this early stage of the war was Captain Alexander M. Boyson, a medical officer, whose experiences typify those of many other POWs:

"I was captured in an area eight miles north of the city of Chochiwon, South Korea, on 12 July 1950. At this time I was attached to the 3rd Battalion, 21st Regiment, 24th Infantry Division. We were operating a battalion aid station on the 11th of July 1950 when we were overrun by North Korean troops. After the fire-fight was over, I managed to escape from the area and was at large in the surrounding mountains for approximately 24 hours. On 12 July 1950 I inadvertently met an enemy patrol of North Korean soldiers and was captured. I was completely stripped of all belongings except a pair of socks, fatigue pants and fatigue shirt. The next 24 hours were spent marching across the mountains...I was taken to a command post area of the North Korean Army, where I joined four other American captives. The next 24 hours was spent marching northward to Pyon Taek, guarded by six North Korean Army personnel.

"Approximately 200 24th Division men had been collected at Pyon Taek. They were quartered on the second floor of the city hall and were sleeping on the floor. Ten or fifteen wounded men had also been collected here. All of these men had been able to walk. Wounded men unable to walk had been shot on the field of battle...During the sojourn at Pyon Taek, three or four meals of rice and soup were given to us...All the prisoners had had all personal articles taken away, and were dressed in summer fatigue uniforms, many of them without boots or any footwear whatsoever. Their food had consisted of at the most two or three rice balls since their capture. We were then forced to march to Seoul, Korea. This trip took us four days. We were guarded by a constantly changing number of guards. They refused to allow the men to get water when it was available, and insisted on keeping up a rapid marching pace...The march to Seoul was grueling, and many men had their feet blistered and cut...We remained in Seoul until 21 July...(then) we were placed on two dilapidated railroad cars and taken to Pyonyang...We arrived in Pyonyang on 25 July 1950 and were taken into the courtyard of the local railroad station, which was surrounded by a mounted guard. We began our march through the center of Pyonyang, paced by a North Korean major who later became known as the 'bullfrog'...A group of 124 American prisoners, also from the 24th Division, most of them being from the 34th Regiment, had already arrived, before us. Approximately one week later another group of prisoners arrived, who were also from the 24th Division and associated units, giving us a total of 724 prisoners...."[14]

Although further American reinforcements from the U.S. 2nd Infantry Division and the 1st Marine Provisional Brigade arrived in late July, the UN forces were driven back to a battle line across the southeastern tip of the Korean peninsula, which became known as the Pusan Perimeter. Despite huge losses in killed and wounded the North Koreans continued attacking, but were finally stalled by American airpower, tanks and artillery fire. During the August 4th to September 16, 1950, defense of the Pusan/Naktong Perimeter, U.S. forces suffered severe losses of 4,599 killed in action, 12,058 wounded, 2,107 missing in action and 401 others subsequently found to have been taken as POWs.[15]

During the fighting retreat, thousands of wounded and unwounded Americans had actually been captured by the Communists but were simply listed as "presumed killed" and "missing in action." Many of the known-prisoners and these unknown POW/MIAs were forced-marched by their brutal North Korean captors for hundreds of miles through rugged mountains in a starving condition, and hundreds were shot along the way or massacred in groups. Thousands of others subsequently died of starvation, disease and later from the cold, in makeshift POW camps. Almost from the beginning of the Korean War American POWs were also being reported by U.S. intelligence sources as transported across the Manchurian border into Communist China, and also into the Soviet Union (which also bordered North Korea near Posyet and Vladivostok).

It was subsequently learned by U.S. investigators that in 1950 the MVD (KGB) had produced a thousand-page study on the expolitation of foreign POWs based on the experiances with World War II POWs held by the USSR. This was a top secret document entitled: "About Spies, Operatives Work with POWs and Internees taken Prisoner During the Great Patriotic War of the Soviet People, 1941-1945." This was no doubt used as a manual by Soviet interrogators of U.S. POWs during the Korean War, but to the time of this writing it has remained secret, and no copy has been located.[16]

In Washington, meanwhile, a debate was raging over the what the war aims of the United States should be, as outnumbered American soldiers fought for their lives in southern Korea. The State Department's top Soviet expert, Charles Bohlen, along with George Kennan, and a former German diplomat in Moscow who was now a consultant to the State Department, Gustav Hilger, argued against a proposal by Acheson and Assistant Secretary of State for Far Eastern Affairs, Dean Rusk, that U.S. forces should counterinvade North Korea and overthrow the Communist regime there. Acheson and Rusk believed that the invasion signaled a change in Soviet policy towards further expansion of the Communist system into new territory by conquest, indicating that they might also attack in Germany. Germany was particularly critical to U.S. national security at this time because the American delivery system for atomic bombs targeted against Russia in the event of world war was still the B-29 bomber, with a range of about 2,000 miles without refueling. Those based in Germany thus threatened the Soviet capital and major population centers. Kennan and Bohlen believed that a U.S. force approaching the Manchurian and Siberian borders in northern Korea would be perceived as a direct threat to the preservation of the Communist system, and would provoke China and possibly the USSR into a major war.[17] Another policymaker who supported Kennan's and Bohlen's view on the danger of a general war was Paul Nitze, the principal drafter of NSC-68, a highly-classified plan for massively increasing American defense capabilities to meet the Soviet threat. Truman ended up supporting Acheson and the Pentagon in deciding that MacArthur must be given latitude to conduct the war into North Korea. After the outbreak of the Korean War, initial Congressional opposition to the cost of implementing NSC-68 had crumbled, and greatly increased defense appropriations were approved. Acheson's judgment may have been affected by charges that he was "soft on Communism and that he had helped bring on the Korean War by excluding the Korean peninsula from the American defense perimeter in the Pacific.

By early September the Communist advance had been halted and MacArthur personally planned and directed a surprise amphibious attack in their rear at Inchon, on September 15, 1950. From Inchon the U.S. Xth Corps, composed of Army and Marine units, advanced on Seoul and recaptured the capital on September 26 after heavy fighting. Simultaneously, Walker's 8th Army troops counterattacked north from the Pusan Perimeter and linked up with the Xth Corps on September 28. MacArthur then, rather prematurely, demanded the surrender of North Korea, which was ignored. The Communists began to take the precaution of moving the captured Americans and other UN POWs in their control farther north. The 24th Division medical officer, Captain Alexander Boyson, was one of those men:

"We left Pyonyang on 6 September 1950 and proceeded by train to Mom Po on the Yalu River. The men were placed in one coach and two open gondola cars. Another coach was used for the sick and wounded...A third coach contained 70 foreign nationals, including American and British missionaries, as well as some White Russian and Turkish families...We traveled at night...three men died on this trip...We arrived in Mom Po on 11 September (1950)...Our camp in Mom Po was organized on an American military basis by a very competent senior officer, Lt. Col. John J. Dunn. He, with other American officers approached the North Korean Camp Commander with the demand that the camp be allowed to be organized in this method....During our stay in Mom Po, we lost 28 men. The cause of death was malnutrition which was complicated by dysentery and pneumonia... Approximately 50% of the men received either a heavy padded Korean jacket or a pair of pants on October 1, 1950. The rest of the men continued to be clothed in summer fatigue uniforms, some of these becoming ragged and inadequate in every form. Many of the men continued to remain barefoot..."

South Korean forces invaded North Korea on October 1, followed by U.S. Eighth Army troops on the 8th. The North Korean capital at Pyonyang was captured on October 19th, and the communists retreated north, forcing thousands of U.S. and U.N. POWs to accompany them. UN forces, led by the 8th U.S. Army, attacked northwest from Pyonyang toward the Yalu River, which formed the border with the Chinese province of Manchuria. The Xth Corps simultaneously attacked northeastward, although Communist China warned that it would not tolerate further advances toward it's border. Far ranging American aircraft attacked targets of opportunity ahead of the advancing armies, bombing targets near or on the Manchurian border, and they sometimes strayed into Chinese airspace. On October 9th, as a result of reported navigational errors, several U.S. aircraft crossed into Siberia and attacked an airfield some 60 miles inside Soviet territory. (The U.S. government apologized for the incident and offered to pay damages.)

The North Koreans were furious at the turning of the tide of battle and increased their brutality toward the American POWs already in their control. Major Jesse V. Booker, a Marine Corps pilot who was shot down near Chorwon, in central Korea, on August 7th, had been harshly treated from the beginning:

"Upon capture I was manhandled incessantly, also, a short time thereafter, bayonets were used to inflict at least six wounds in the lower part of my back and buttocks. In one small village I was forced into a small wooden cage and placed in the center of the village as an exhibition...also in this same village, I was tried by a 'peoples court.' This court found me guilty and sentenced to be stoned to death with the entire populous joining in the execution..."[18]

Booker was rescued by a group of North Korean soldiers and marched north to Wonson and then Pyonyang, where he joined the same group of 726 prisoners as Alexander Boyson. His group of POWs, minus some who had died of wounds and malnutrition, was moved to Manpo on the Yalu River by the 11th of September, where they remained until the American/UN invasion of North Korea began in early October. Booker remembered:

"On the 9th of October, the prisoner group departed Manpo on a series of relatively short forced marches...These were miserable moves due to cold weather, lack of proper clothing and very little food. Due to these conditions and the fact that dysentery ran rampant among the prisoners, at least 50 men were killed or else died during these moves..."[19]

The U.S. medical officer, Captain Alexander Boyson, also vividly recalled these same forced marches:

"On 9 October we were moved to the village of Ko San,

approximately 14 kilometers west from Mom Po...and left there on the 19th of October. During this time six men died from starvation and dysentery...On the 19th we moved across the mountains...two men died while we were in Konagon...On 25 October we moved...to a corn field on the west side of Mong Po. We were forced to leave seven men in a police station in Goson who were subsequently shot. These men refused to continue the march, believing that they would soon be relieved by the advancing American troops..." Captain Stanley G. Zimmerman, another 24th Division officer who had also been captured in July remembered the same incident:

"...We moved out and left seven men in the police station in K(G)osan and were told by a North Korean interpreter that they were shot. We moved back by foot toward Manpojin. While we were enroute, two men were left by the roadside and were shot..."[20]

President Harry S. Truman and his advisors, including Assistant Secretary of State Dean Rusk (later Secretary of State during the Vietnam war), flew to Wake Island for a conference with General MacArthur. Truman was worried about the potential of a sudden Chinese Communist and Soviet participation in Korea, and the possibility of a general world war ensuing. In fact, the Soviets, Chinese and North Koreans were being kept fully informed of American and allied war plans and intentions by highly placed Soviet agents in Washington and London such as Kim Philby, Guy Burgess and Donald Maclean. Secretary of State Dean Acheson had been opposed to the President traveling to meet his Theater Commander on the grounds that it was a demeaning gesture to a mere military commander, popular though he might be.

Chinese Communist troops attacked American forces for the first time on October 25th, near the Chongjin Reservoir and Onsong, North Korea. In another ominous development, Russian MiG-15 jets began to appear over Korea on October 31st, some of them flown by Soviet pilots. In the Battle of Unsan on November 1st and 2nd, the 8th Cavalry Regiment of the U.S. 1st Cavalry Division lost heavy casualties in the first major Chinese attack, with 600 American soldiers listed as killed in action or taken as POWs.[21] On November 8th a U.S. F-80 Shooting Star shot down a MiG-1 5 in the first all-Jet aerial combat in history. American airmen subsequently reported that some of the enemy MiG pilots spoke Russian to each other. After two weeks of heavy fighting, the Chinese suddenly pulled back on November 6th, and American and UN troops also withdrew to regroup and count their losses. Many more American POWs (many of them actually unknown MIAs) marched north, this time as captives of the Chinese communists.

For the American POWs already in North Korean captivity, the Chinese intervention seemed to cause even more vicious treatment then they had been previously subjected to. Captain Stanley

Zimmerman, one of the 24th Division officers captured in July, was in the cornfield with the medical officer Captain Alexander Boyson and the Marine pilot Major Jesse Booker, in late October. Zimmerman recalled:

"On 31 October 1950, there was a change of command in the Korean officials over the camp, and a North Korean Major from the Security Police took charge of our camp. Prior to this time we had been under the command of the North Korean Army. The North Korean Major, hereafter referred to as the 'Tiger', told us that we were going on a march that would last nine or ten days, and we were going to a town called Chungon. He wanted a headcount made of the prisoners, and there were 650 or 651 prisoners who started this march. When we left the cornfield, we left 19 still alive, but very weak. They were unable to make the march. We were promised that they would be taken care of by the North Korean people. From later reports, it is surmised that these men were killed..."[22]

Master Sergeant Calvin C. Creeson, a 24th Division soldier who had been captured in July, was on the same death march under the command of the murderous North Korean major known as the "Tiger." He remembered speaking to two of the American civilian missionaries at the start of the march:

"Miss Dyer and Miss Rossier...both about 50 years old and both of whom understood Korean, told several of us that they had overheard the 'Tiger' talking with other North Korean People's Army officers of his staff about the feasibility of simply shooting all of us then and there..."[23]

Captain Stanley Zimmerman recorded the early phase of the death march that followed:

"On the night of 31 October 1950, we moved first to the edge of Mampo-jin and stayed there until 2200 hours; then we moved on through the city...and spent the rest of the night in a rice paddy. The next morning we were given some boiled, whole-kernelled corn; then all the officers were called together and told that we were about to begin the march. He gave us the instructions—the rules for the march...which included the fact that no men were to fall out on the march unless they had his permission...We marched for about 15 minutes, and then the column was stopped. All of the section leaders, the officers, were told to make a headcount of their sections and report the number of men missing. There were no men missing from my section; however there were about five sections that did have men missing. The section leaders were Lieutenant Culbertson, Lieutenant Rountree, Lieutenant Thornton, Lieutenant Adams, and one other officer whose name I do not remember. Thornton had two men missing from his section; Adams had eight men missing from his section; and the others I do not remember. These officers, Major (Lt.Col.) Dunn, and the Commissioner Lord, a member of the British Salvation Army who was acting as interpreter for the group, were taken to a little knoll beside the road and approximately 15 to 20 feet from my position.

"The 'Tiger,' through the interpreter asked these officers why the men had fallen out of their sections. He was told that they didn't know that the men had fallen out of their sections because they had been marching at the head...He told them they had disobeyed the rules... and they must be punished. He told them they were on trial by the Korean people and asked what they thought their punishment should be. He received different answers...generally that they should be made to fall out themselves and help these men along the road. He said that he was going to shoot all of these men on this knoll, then turned to the column and asked us what we thought-should they be shot?

"He received, of course, an answer of no. So... he picked Lieutenant Cordius Thornton, of L Company, 34th (U.S.) Infantry, 24th (Infantry) Division, and told the other section leaders that they could go back to their sections. He asked Lieutenant Thornton, 'why these men had fallen out of his section,' and Lieutenant Thornton said, 'that they were too weak to carry on.' The "Tiger" said that he had, 'disobeyed rules and that he must be shot.' He turned to the column and asked it what they thought, and of course, he received the answer, 'no, that he shouldn't be shot.' He turned to the guards who were, most of them, sitting on the hill on the opposite side of the road...he asked them should he, 'shoot this officer,' and I think the answer was, 'yes, very definitely so.' So he turned to Lieutenant Thornton; he knocked his hat off; he took a towel around Lieutenant Thornton's neck and told Lieutenant Thornton to, 'turn around.' He did so, and he placed the towel around his eyes, turned to the column and told the column to 'turn around.' We all turned around, and then, 'some of us turned back;' and he took out a pistol, placed it to Lieutenant Thornton's head and shot him...This was 1 November 1950."[24]

A successful UN counteroffensive drove the North Koreans almost to the Manchurian border, but caused more U.S. and UN POWs to be moved across the Yalu River. Communist China then intervened in the war with about 300,000 Chinese "volunteer" troops, who had entered North Korea in October and November 1950. MacArthur's foreign-born Chief of Intelligence, General Willoughby, had misread and underestimated Chinese intentions and strength. MacArthur believed that the U.S./UN forces still outnumbered the Chinese Communists, that the Chinese would only be used for defense, and that U.S. air supremacy would be decisive in defeating them. Soviet spies in Washington and London again ensured that the Chinese were fully informed on U.S. and U.N. troop strength and deployments. MacArthur had misjudged Chinese intentions, and couldn't realize the impact of communist spies who had revealed Allied intentions and plans to the enemy, when he ordered a renewed advance north on November 24, 1950.

On November 26th and 27th the Chinese replied with massive attacks on U.S. and allied U.N. positions, sometimes overwhelming the defenses with sheer weight of numbers, despite massive losses. In one of the most savage battles in American military history, the U.S. 7th Infantry Division's Task Force Maclean/Faith, consisting of elements of the 31st and 32nd U.S. Infantry regiments with supporting artillery and tank units, was annihilated by the Chinese Communists east of the Chosin Reservoir. While under operational command of the hard-pressed Marine Colonel Lewis "Chesty" Puller, whose larger force was west of the reservoir, this small U.S. Army task force was told to fend for itself. The Army infantrymen were already encumbered by over 600 wounded, and they continually lost more killed, wounded and missing during their fighting retreat. Surrounded repeatedly by masses of Chinese and unable to obtain aid from the American Marine force west of the reservoir, the task force was virtually annihilated. Colonel Allan MacLean was later reported to have died in Communist captivity, while Colonel Donald Faith was killed in action, along with most of the wounded Americans, at one of a series of Chinese Communist roadblocks and ambushes. Of the 3,200 soldiers engaged (including 700 ROK troops), only 385 survived.[25] In this one action, over 2,000 U.S. Army soldiers were listed as presumed killed or missing in action, but there can be no doubt that many were captured alive and transported north to Manchuria, where they subsequently disappeared.

The Chinese attacks drove a wedge between the Xth Corps in the east and the 8th Army in the west by the end of November, and by December 4th, the UN forces began retreating from Pyongyang.

In the region west of the Chosin Reservoir (Changgjin in Korean), 20,000 other U.S. Marine and Army troops, who were also surrounded by Chinese forces, began a fighting retreat to the coast of northeastern Korea. Despite the below-zero cold, and being under constant ground attack and heavy shelling by enemy forces, the Marines and Army troops brought out all of the wounded that could be found, and even some of their dead comrades, but there can be little doubt that more Americans, both wounded and unwounded, were captured during this withdrawal. From late October to mid-December, including the November 27-December 9 Battle of the Chosin Reservoir, the 1st Marine Division lost 718 killed in action or died of wounds,192 listed as MIA and 3,500 wounded.

A day after a unit of the 7th Infantry Division's 32nd Infantry reached the Yalu River on the Manchurian border, the Chinese attacked the 2nd U.S. Infantry Division in the Battle of Kunu-Ri, in northwestern Korea. This resulted in a battle from November 29th to

December 1st, in which 4,940 more Americans, most from the 9th Infantry Regiment and 38th Infantry Regiment, became casualties; almost 1/3 of the 2nd Infantry Division. In northeastern Korea, meanwhile, by December 24th, when MacArthur had hoped the war would be over, 105,000 U.S. and Korean troops had been evacuated by ship from the port of Hungnam, along with over 17,000 vehicles, tanks and artillery pieces.

The Joint Chiefs of Staff were demoralized by the rout and informed Acheson that a cease-fire must be obtained to permit the withdrawal of American troops. The Chairman of the Joint Chiefs, General Omar Bradley, speculated about whether the U.S. should pull out and forget about Korea. Secretary of Defense George C. Marshall, who had been sworn in in September, and his Deputy Secretary, Robert Lovett, had permitted MacArthur to do as he wished in the attack toward the Yalu River border with Manchuria, and now were practically paralyzed by the disaster that was unfolding. Truman wrote in his diary: "It looks very bad." Acheson however, remained firm and argued that America and the UN must fight back, even if another Dunkirk occurred in Korea. He was strongly supported by Assistant Secretary of State Dean Rusk, who also recalled the British defiance in what Churchill had called their darkest hour. After a meeting in Washington with George Kennan at the Secretary's home which had been suggested by Bohlen in Paris, Kennan wrote to Acheson: "... If we try to conceal from our own people or from our allies the full measure of our misfortune, or permit ourselves to seek relief in any reactions of bluster or petulance or hysteria, we can easily find this crisis resolving itself into an irreparable deterioration of our world position-and of our confidence in ourselves."[26] The Republicans in Congress blamed Acheson for the disaster and on December 15th House Republicans voted unanimously that the country had lost confidence in Acheson and he should resign. The vote on a similar measure among Republicans in the Senate passed by a margin of 20-5.

And in England, a writer named Erich Arthur Blair, better known by his pen name of George Orwell, had died that year of a neglected lung ailment, at the age of 46. His masterpiece, 1984, published the year before, predicted many dreadful aspects of a totalitarian future faced by advanced societies that distorted or altered the truth in the name of security and peace, while engaged in perpetual warfare. His earlier book, Animal Farm, had been a satire of a Communist society, but 1984 was a warning to Britain, America and other industrial democracies, about the danger of adopting totalitarian methods in the name of defense, that in the end would destroy privacy and punish love. The secret outcome of the prisoner of war repatriations of World War II and Korea conformed to Orwell's mythical party slogans of the future: "War is Peace;

Freedom is Slavery; Ignorance is Strength."

In Korea, on the 8th Army front, the U.S. and UN forces now retreated back into South Korea. General Walker was killed in a jeep accident during the retreat and General Matthew Ridgway had taken over command of the 8th Army on December 27th. By December 31st, Communist forces were again attacking Seoul and they recaptured the capital on January 4, 1951. The Allies established a defensive battle line 25 miles south of Seoul and halted the communist attack, inflicting severe losses on the North Koreans and Chinese.

General Ridgeway altered U.S. and allied tactics by ordering slower, more thorough advances beginning on January 16, 1951. All enemy forces were to be wiped out rather than bypassed, so that rear areas would be secure from guerilla attacks. In a counteroffensive from January 25-February 20, 1951, the 1st Cavalry Division and the 2nd, 3rd, 7th, 24th and 25th U.S. Infantry Divisions lost 667 more killed and 3,570 wounded. During the February 11-13th Battle of Hoengsong, the 2nd and 7th U.S. divisions, and the 187th Airborne Regimental Combat Team lost over 2,000 casualties, including 530 Americans who were killed in a single ambush, the largest concentrated loss of American lives in the war. Between February 12th and 21st, the Chinese counterattacked in the Wonju/Chipyong area and the 1st Cavalry, 2nd Infantry and 7th Infantry divisions lost 651 more U.S. soldiers listed as killed in action.[27] By March 14th, however, the Communists had retreated from Seoul, and the South Korean capital was liberated by U.N. forces once again. The two opposing armies dug in along a front just north of the 38th Parallel across the Korean Peninsula, and the character of the war changed. Instead of massive thrusts and counterthrusts, fighting raged around individual positions in what became known as the "Battle for the Hills."

MacArthur was all for widening the war by blockading the coast of China and involving Nationalist China in attacks on the mainland, to divert Chinese Communist forces from the Korean battlefront. The Soviets referred to the General as a maniac and the principal American threat to world peace. MacArthur became increasingly outspoken about the constraints placed upon him and had continued to leak his blunt criticisms of U.S. policy to the media and sympathetic Congressmen, a situation the President found intolerable. General Marshall, Bradley, Harriman, Lovett, Rusk and others of Truman's advisers urged him to recall MacArthur, but the President was concerned about the impact of such an action on his critics in the Congress. Then Truman received copies of secret communications of Portuguese and Spanish diplomats intercepted by the NSA monitoring station at Atsugi Airbase in Japan, indicating that MacArthur had said he was confident of being able to turn the Korean War into a major conflict that would defeat Communist

China, and if Russia entered the war it would be destroyed also.[28] President Truman considered this to be evidence of dangerous disloyalty and removed General MacArthur from command on April 11, 1951, replacing him with General Matthew Ridgway. At home in the United States, the removal of MacArthur caused a nationwide outrage, and resulted in calls for the impeachment of Truman in the Congress. MacArthur came home to a hero's welcome, and Acheson faced a severe grilling by his enemies in the Congress, but in the end, the decision stood, since Truman, as the Commander-in-Chief, had been left no choice but to exert his authority over a military commander whose actions were counter to the Administration's policies. The incident further divided the country at a time when their was widespread public mistrust of the motives and loyalty of many senior officials in Washington.

Alger Hiss had been convicted of perjury, in 1950, after denying he had supplied classified documents to Soviet agents and Acheson's subsequent statement of continued support for Hiss had contributed to the Secretary of State's unpopularity among conservatives. Other State Department officials, including John Carter Vincent, John Stewart Service and John Paton Davies were also to be dismissed because of questions about their loyalty. In the Senate, Joseph McCarthy and others continued to denounce the government for having brought about the crisis in Korea through weakness of will and traitorous acts. From the highly placed Soviet spies and agents of influence in Washington and London, the Communists had in fact gained solid intelligence of the Truman Administration's self-limited plans for Korea and fears of a wider war.

Donald Maclean was head of the American desk in Whitehall while Guy Burgess was second secretary of the British Embassy in Washington prior to their defection to the Soviet Union in May 1951 after years of spying for the Russians. The Soviet master spy H.A.R. "Kim" Philby was First Secretary of the British Embassy in Washington and the chief liaison between MI-6 and the CIA and FBI. Philby had access to basically all U.S. secret intelligence By this time, the Office of Policy Coordination (OPC) under Frank Wisner had been brought into the CIA by DCI Walter Bedell Smith and combined with the Office of Special Operations (OSO) to form the Directorate of Plans under Allen Dulles, which came to be known inside CIA as Clandestine Services.[29] Frank Wisner remained as Dulles' deputy and both of them, along with CIA counterintelligence expert James Angleton, were on close terms with Philby, sharing virtually everything with the Soviet spy in MI-6. After the defections of Maclean and Burgess, Walter Bedell Smith consulted with the White House and the FBI and then demanded that the British government recall Philby to London. But the damage was already done. The

Assistant Secretary of State for Far Eastern Affairs (FE) at the time, Dean Rusk said of these penetrations: "It can be assumed that anything we in our government knew about Korea would have been known at the British Embassy and that officers of the rank of these three would have known what the British embassy knew."[30]

In addition to these three, other spies were later discovered to have been operating in the American and British governments and in the United Nations. The subsequent 1952 Congressional hearings on the Katyn Forest massacre of Polish POWs by the Soviets revealed the pro-Soviet bias of the former chief of U.S. Military Intelligence, Major General Clayton Bissell, and his suppression of intelligence information damaging to the Soviets in 1945-46, as noted previously in this book. The Bride-Venona decryptions of Soviet intelligence traffic had revealed that many other Soviet agents of influence or spies were operating within the U.S. and allied governments. To capitalize on these and other disclosures, some of which were downplayed or remained classified, conservative Senator Joseph R. McCarthy (R-Wisc.) attempted to monopolize a subsequent series of 1953-54 public hearings on the penetration of the U.S. Government by Soviet spies. He was to claim that Secretary of State Dean Acheson was "soft on Communism," that concessions to the Soviets at Yalta and Potsdam had undermined American security, that the State Department, CIA and even the Army contained many disloyal senior officials and that the loss of China to the Communists and the stalemate in Korea were the results of a conspiracy. But McCarthy's methods of personal character assassination, combined with insufficient evidence or unsubstantiated charges, in many cases, caused a backlash in the Congress and in the American media which nullified any good that was done by calling attention to the original charges, some of which were based on fact.

Through their secret agents the Communists were certain that the war would eventually reach a stalemate and that the Americans would seek a negotiated settlement, because of their fear of Korea escalating into a world war. Knowing this, during the winter campaign of 1950-51 they took precautions to insure that the major bargaining chip they possessed, the American and allied UN POWs, would stay securely in their control. In addition, the Soviets had already been obtaining numbers of the American POWs since early in the war, to exploit their technological expertise as forced labor, and for espionage purposes. Since some Americans would have to be returned following any negotiated settlement of the war that would prove acceptable to the United States, it was necessary for the Communists to separate, as soon as possible, those who would be repatriated from those who would be retained as captives by the Soviet Union, Communist China and North Korea.

One destination of these vanished POWs is revealed in a March

12, 1951, declassified State Department cable to the Political Adviser of the U.S. Far East Command, which relayed information from French Military Intelligence that 3000 American prisoners captured in Korea were concentrated at Chi An and Sin Ni Chow near the Chinese-North Korean border, and that: "...1,200 LIGHTLY WOUNDED AMERICANS WERE AT THE SAME TIME PLACED IN 3RD GENERAL AIR DEFENSE HOSPITAL AT ANTUNG MANCHURIA."

Since only a handful of U.S. POWs who were returned at the end of the war had been held captive by the Chinese and Soviets in Manchuria, these 1,200 Americans are among those who were reported subsequently as transported to USSR control in Siberia, or who remained in ever-diminishing numbers, for the next four decades, in Communist Chinese labor camps and restricted areas. The Soviet MGB (KGB) had been sending large numbers of military and intelligence advisers under an agreement reached with Mao tse Tung following establishment of the Chinese People's Republic in October 1949, but they were never permitted to gain control of the local intelligence organs as they had done in eastern Europe. Yet, according to Colonel Oleg Gordievsky, a later Soviet defector, "China did, however, provide much intelligence on U.S. military technology obtained during the Korean War, and gave the MGB (KGB) a base on Chinese territory, where it could train ethnic Chinese illegals for work against the Main Adversary and other Western States. The MGB (Ministerstvo Gosudarstvenoi Bezopasnosti) was created in March 1946 by the merging of the NKVD and the MVD (Ministry of Internal Security). This was altered back into its original two organizations after Stalin's death in March 1953. What had once been the NKVD became the KGB. The MGB was also given unrestricted access to Western POWs held by the Chinese and North Koreans. Among them was George Blake."[31]

Blake was originally a Dutch national, the son of a naturalized Briton of Egyptian-Jewish origin. After service in the Dutch resistance he entered the Royal Navy during WW II and became a naval intelligence officer by 1945. He was recruited by MI-6 in 1948 after Russian studies at Cambridge, and in 1949 was posted to Seoul, Korea under cover as the Vice Consul, where he was captured by the Communists in June 1950. He later agreed to become a Soviet spy while in Chinese captivity, and was infiltrated back with returning POWs at the end of the war.

A CIA intelligence source (quoted in declassified CIA report SO-61735) revealed: "Officers captured in North Korea by the Chinese Communists are now interned in a former Army prison in Mukden (Shenyang, Manchuria). Enlisted men are confined in concentration camps in T'unghua. The daily routine includes physical exercise, political training in Marxism and Leninism..." Another CIA report (SO-65823), which was dated August 11, 1951, stated:

"AMERICAN PRISONERS OF WAR WERE SENT TO CAMPS IN MUKDEN, TUNGHUA AND ANTUNG PROVINCES OF MANCHURIA, WHERE THEY WERE PUT TO HARD LABOR IN MINES OR FACTORIES."[32]

For those U.S. POWs who didn't disappear into Manchuria or the Soviet Union, but remained in North Korea, conditions were also brutal, and some 3,000 or more U.S. POWs froze or starved to death in Communist captivity. The last enemy symbol that many American prisoners of war saw on this earth was the Communist star, which had so little meaning to the civilian population in their faraway homelands. Other POWs continued to be murdered by their guards. Survivors reported many interrogations of Americans by Chinese intelligence officers, and the presence of Soviet intelligence officers in the camps. Prisoners were subjected to a steady stream of communist propaganda, and some of them actively participated in the communist indoctrinations of fellow POWs. A few of these Americans, perhaps out of fear of being prosecuted, eventually refused repatriation to the United States. Corporal Walter R. Williams, a soldier of the 24th Infantry Division, considered himself lucky to have survived:

"I was one of a group of nineteen from the 24th Infantry who became lost approximately 26 November 1950, from our organization somewhere north of Kuneri. We joined the 503rd Artillery Battalion. On 30 November 1950, with two men whose names were unknown to me, I was captured and marched by three armed Chinese guards north about six miles to a cave. We were not allowed to talk to each other. There were two prisoners in the cave. After dark we again proceeded north and had been given no food nor water. We were joined by other prisoners at relay points and continued to march twelve hours at a time until we reached a mining camp in Death Valley, toward the end of December.

"We remained there 8 to 10 days. The weather was very cold and shelters were only sufficient for camouflage. Our food consisted of cracked corn. The buildings at Death Valley were open, and over 100 prisoners died of cold and hunger. Edward Baylor of the 24th Infantry was one of them. There were besides the Americans, Turks, Filipinos, Puerto Ricans, a few British, French, South Koreans and two or three Japanese."

"About the end of December 1950, I was included in a group of from 1300 to 1600 prisoners (who were) marched 75 or 80 miles north to Camp #5, Pyoktong, on a branch of the Yalu River, arriving on 14 January 1951. Our group, joined with men already in the camp, made a total of approximately 2,500. North Koreans were in charge of the camp at that time. On this march many of the prisoners were beaten and kicked. I was hit and kicked three or four times. There was no medical treatment, and no clothing nor covering...Prisoners were spaced and not permitted to talk to each other. The guards were very young, had not been in combat and were very rough; only one of ten would let men stop for a drink of water, etc. I remained at Camp #5 from 14 January 1951 to 10 August 1953. On the latter date approximately 1,100 prisoners remained, since approximately 2/3 of all those who arrived had died.

"In the latter part of 1951, Chinese troops took over operation of the camp, dividing the prisoners into six companies according to nationalities. Besides two companies of Chinese guards, (approximately 100 men), 25 to 30 English speaking Chinese...who stated they were from Chinese Peoples Volunteers-Intelligence, were attached to each company of prisoners...During 1951 I was assigned to grave detail nearly every day. The greatest number of prisoners to be buried in one day was 85 and there were never fewer than 12 a day in burial detail...Men died from the intense cold and lack of food and medical care. The majority of deaths was among the younger men..."

"Among Americans who worked with the Chinese assisting the 'study program' were Lawrence Sullivan, W.C. White, Clarence Adams, Rogers Herndon, Fred W. Porter, Roscoe Perry, Roy Atkins and Leroy Carter. They lectured, presented study schedules and attempted to propagandize fellow prisoners...The British prisoners appeared to be more influenced by the communist program than any of the other nationalities. British newspapermen visited the camp to lecture every two or three months...In the 1st Company three Americans chose to remain (refused repatriation). They were Lawrence Sullivan, W.C. White, and Clarence Adams.

"During the winter of 1951 some Russian military men were seen at the camp. They carried arms but I did not know exactly what their duties were. We thought they were administrative. One Russian instructor remained during the entire period in charge of the Turkish company..."[33]

Soviet intelligence officers reported at various camps in North Korea by American POWs were responsible for the disappearance of some of the U.S. soldiers, who were segregated or called in for interrogations and failed to reappear. Many American POWs who were eventually repatriated had had little contact with the Soviets, however, other than being aware of their presence at the camp headquarters. Sergeant Ellis J. Yancey, a medic of the 35th U.S. Infantry Regiment, was captured in the great Chinese offensive of November 27th, 1950, at Yongsandong. After forced marches with a group of 300 American and South Korean prisoners, and temporary incarceration in a small POW camp, he arrived at Camp #5, Pyoktong on 18 January 1951. He later remembered:

"At Camp Number 5 we had about two thousand deaths from malnutrition, dysentery, diarrhea, pneumonia, wounds received in combat and freezing. I was moved to Camp Number 3, which is approximately eight miles north of Changsong. There were about forty or fifty Chinese guards for the one hundred and sixty-two men who were there. These one hundred and sixty-two prisoners were pulling hard labor as punishment for not having accepted Communist indoctrination. The camp population was increased to about eight hundred and forty including thirteen British; the other prisoners were Americans. For the increased population there were about three hundred guards. At Camp Number 3 we had approximately one hundred men to die from dysentery, diarrhea and malnutrition. Since there were so many who died...we never tried to get to know each other...

"In order to receive our daily food, all of us were forced to sing progressive type songs. We were forced to attend lectures and discussions on Marx, Lenin and Stalin, and various Communist indoctrinations...Prior to cease fire we had many men put in solitary confinement, some staying about sixty-five days with handcuffs on. The men were forced to sign confessions that they had planned an escape or were the leader of a gang. Others were beaten because of their opinions on communism and their failure to submit to the germ warfare indoctrination. We were told by the Chinese that Lieutenant Quinn, Lieutenant Enock, Major Schwabble, and some other air force fliers had confessed to dropping germ bombs or spraying germs from planes. While at Camp Number 3, I saw around headquarters, and while on detail, Caucasians with fur caps and fur collared coats. These persons drove Russian-made jeeps, but their uniforms could not be seen..."[34]

Meanwhile, the fighting continued in the hills of Korea, costing thousands more U.S. dead and missing. After an April 1951 Chinese Communist "spring offensive," which had cost 314 American dead, between May 16th and 21st, in the battle of the Soyang River, the 2nd U.S. Infantry Division reported 190 more American MIAs. During an August 18-September 5 battle at "Bloody Ridge," the Americans lost 326 killed and 414 more U.S. MIAs. From September 13-October 15, the 2nd Infantry Division suffered 597 killed in action and 84 more Americans were listed as missing in action. How many of these nearly 700 missing U.S. soldiers were actually captured by the enemy and transported north, may never be known.

The "Germ Warfare" propaganda campaign was begun in earnest by the Communists with publication of "confessions" by American airmen in the Soviet newspaper Pravda on May 6, 1952. This was orchestrated into a world-wide propaganda campaign by the Communists and willing accomplices in the western media. American, British and other pro-Communists traveled to North Korea and China to assist in interrogating American POWs and obtaining more "evidence." One of the most infamous was the Australian "journalist" Wilfred Burchett, who personally took part in interrogations of British, American and other UN POWs, and in threatening the uncooperative with death. (Burchett was later to continue his activities during the Vietnam War, with great effect.) Other Korean War prisoners later reported that U.S. Airmen had been subjected to intensive interrogation and torture to obtain confessions. The Communists attempted to turn this into a massive, world-wide propaganda offensive against the United States and the UN forces in Korea. A typical CIA report about such efforts, dated June 1952 (SO 91011), conveyed information about Chinese germ warfare propaganda acquired on May 23rd:

"A letter from Peiping to (censored) containing a New China News Agency English language summary dated 12 May stated that two captured U.S. airmen whose plane was shot down in North Korea on 13 January had admitted participation in germ warfare.

"The airmen were identified as follows:

"Pilot John Quinn...age 29, white, 1st Lt., 3rd Bomb Wing, 3rd Bomb Group, 8th Squadron, captured 10 miles southwest of Sunan, North Korea. Navigator Kenneth L. Enoch, 1st Lt., age 27, same unit.

"Enclosed photographs of the men carried the stamp of the New China News Agency on the reverse side in both English and Cyrillic."[35]

Now-declassified CIA (and other) intelligence reports reveal that an aggregate total of 4,000 or more American POWs from Korea were REPORTED as seen by eyewitnesses at many locations in China including Mukden, Tunghua, Canton, Kweilin, Hankow, Liuchow, Shanghai, Chekiang, Peking, Tientsin, Manchouli and other cities, where many were forced to participate in communist propaganda activities or worked as forced-laborers.[36] CIA report SO-66740-51 (dated June 27, 1951), was titled: AMERICAN PRISONERS OF WAR IN SOUTH CHINA, and revealed:

"In early April American prisoners of war from Korea began arriving in Hankow, where they were turned over to the Chinese Communist Central and South China Military Command. By 15 April approximately 500 had arrived in Hankow, and on 18 April some of these prisoners were paraded through the streets of Hankow under heavy guard.

"In mid-April 60 prisoners of war, most of whom were American and the rest British, arrived in Canton via the Canton-Hankow railway. In early May they were being detained in a foreign-style compound at the corner of Tunghua Road...There were barbed wire barricades around the compounds, and a Public Security division mounted a heavy guard around the area. No one was allowed to enter without permission from high Communist authorities. The prisoners were treated fairly well, and were given food and billets.

"In mid-June 52 American prisoners of war from Korea were incarcerated in the Baptist church on Tunghua Road, Canton. These prisoners were sent to Canton because the Chinese Communist authorities hoped to obtain military and medical supplies from the United States Government in return for their release. They planned to demand U.S. $100,000. worth of supplies for each prisoner released. The British and Indian governments were to be used as intermediaries."[37]

A large volume of intelligence reports from military intelligence and CIA sources of mid-1950 to early 1951 also reported the shipment of many American POWs captured in Korea to the Soviet Union. (Many of these intelligence reports have remained classified by authority of the National Security Council, Department of State, Central Intelligence Agency, U.S. Army Intelligence and Security Command/NSA, FBI and other Federal agencies, despite repeated Freedom of Information requests and appeals, by the author and two other published researchers who testified before the U.S. Senate in 1992.)

American POWs held in camps on the Manchurian-North Korean border also later reported the disappearance of American prisoners who were taken away for interrogation. U.S. Army Major Fred A. Smith, a POW at Pyoktang in early 1951, was told by the Chinese Communist Vice-Commander of the camp:

"You are here to learn. This may take one year, 10 years, 20, 30 or even 40 years and some of you may die here. But if you die don't worry, we will bury you deep so that you won't stink."[38]

Corporal James R. Young, a soldier of the U.S. Army's 1st Cavalry Division who was captured in December 1950, witnessed the deaths of many American POWs at Kanggye, Anjum and Chongsong that winter. Later he reported what he personally knew of Americans who disappeared: "Fifteen prisoners disappeared from Camp #1 during 'midnight repatriations'...These men would be seen one day and then be gone the next day and it is not known if they were repatriated." James Young also remembered what happened to attempted escapees: "I know of seven or eight attempted escapes from Camp #1. When the prisoners were recaptured they were sent to the prison for six months to a year..."[39]

According to Captain Homer A. Curtis, a POW in an enclosure for officers of Camp #2, about 10 kilometers northeast of Pyoktang on the Yalu, "some of our officers were isolated and interrogated for many months on this subject (germ warfare). Lt. Joe Green, formerly a West Point graduate, was in isolation for four months that I know of." Some of these isolated prisoners never reappeared. Others were convicted of "crimes" and sentenced to prison terms by the Chinese communists. PFC James J. Coogan, captured in the Chorwan area by Chinese Communist forces on September 6th, 1951, and imprisoned at Camp #1, Changson, and later at Camp #2, was questioned by U.S. intelligence officers after his repatriation. He remembered prisoners tried for escape attempts: "...There were three fellows I knew who were sentenced. William Carter... for 5 years, Corporal Doc Fraser...for 5 years, and Wilford Ruff...sentenced for 10 years."[40]

By September of 1951, U.S. intelligence was reporting the arrival of thousands of fresh Communist troops on the battlefronts, which were referred to in classified briefings of U.S. ambassadors at the time as "Soviet and/or other Caucasian troops."[41] Soviet anti-aircraft and coastal defense units, security forces, signal and radar units and counterespionage personnel were identified.[42] According to a U.S. Government report of Russian participation in the Korean War, written from declassified Soviet sources four decades later:

"The Soviet Union initiated its battlefield testing in the Korean War with the activation of the 64th Fighter Aviation Corps Headquarters in Antung (now Dandong) Manchuria, in November 1950, just as North Korea teetered on the edge of destruction. The Corps was charged with...air defense of the area north of the 38th Parallel; protection of the trans-Yalu bridges; and training of North Korean and Chinese pilots....The 64th had yet another mission: the management of the overt and covert human intelligence (Humint) effort targeted against the U.S. air forces. A review of the documents provided by the Russians reveals regular and intense coordination between Moscow, the senior advisers to the Korean General Staff, and the commander of the 64th Aviation Corps (General Georgii A. Lobov) on a variety of topics related to prisoner of war interrogation and control..."[43] Thus the Soviets were in Korea actively seeking to obtain U.S. POWs in the fall of 1950.

In addition, Soviet pilots who had engaged American aircraft in aerial combat were reported, and CIA or military intelligence were not certain that Soviet troops had not been engaged on the front lines, posing as Chinese or North Koreans. In a U.S. News and World report story published nearly a year after the war ended, Pentagon sources were quoted as saying that Soviet advisers and ground troops had reached some 20,000 in Korea by the middle of 1951.[44] The actual number was higher. (Four decades later, with the collapse of Communism in the USSR, the Russian government admitted to a substantial Soviet military presence in Korea during the war, and the loss of significant Soviet casualties there.)

When former SHAEF Chief of Staff (and former Ambassador-to-Moscow) General Walter Bedell Smith had become Director of Central Intelligence in 1950, he had obtained Truman's agreement to the merging of Frank Wisner's Office of Policy Coordination (OPC) with the Office of Special Operations (OSO), to become the Directorate of Plans, under overall CIA control. This was accomplished on January 4, 1951. The Directorate of Plans became known within CIA as "Clandestine Services." Frank Wisner now worked as a deputy directly under Allen Dulles, who was the head of Clandestine Services (and future CIA Director). With large-scale CIA operations during the Korean War and expanded anti-Soviet Cold War operations in eastern Europe, the Agency grew rapidly at this time. According to author (and former CIA official) Victor Marchetti, the CIA grew from 5,000 employees in 1950 to 15,000 in 1955.

Among the many new CIA officers serving under Frank Wisner during the Korean War was Cord Meyer Jr., at work at CIA by October 1951, and assigned to the OPC, still headed by Wisner. A Yale man and World War II veteran who had been wounded at Guam in the Pacific campaign, Meyer had become involved in the postwar world peace movement with a group of anti-war veterans, but was eventually convinced that the Soviets were bent on conflict with the West and joined the CIA. He arrived at the same time that NSC directive 10/5 was issued, on October 21, 1951, which assigned to CIA responsibility "for mounting expanded covert action program with policy guidance to be supplied by an NSC subcommittee."[45] Cord Meyer and another CIA employee who would later rise to prominence in the American national security establishment, William Bundy, were among those singled out by Senator Joseph McCarthy as being of questionable loyalty, and both nearly had their careers destroyed during the Korean War anti-Communist "witchhunt."

The CIA's covert action chief during the Korean War was a Danish-American named Hans V. Tofte. Tofte had been recruited by Frank Wisner, and his old friends Desmond Fitzgerald and Richard Stilwell, who had known him during the Second World War. Tofte had lived for eight years in Manchuria before WW II, and had become familiar with northern Korea at that time. During the war he had worked for William Stephenson's British Security Coordination (BSC) in Burma and Malaya, and later was detailed to Donovan's OSS. A reserve Lieutenant colonel in the U.S. Army by the time of the 1950 North Korean attack, he agreed to run the Agency's covert Korean War operations, through an OPC unit he set up in Japan, holding a rank equivalent to a major general. He established an Escape and Evasion (E&E) operation across Korea to rescue downed fliers, which included two islands off the east and west coasts of the peninsula above the 38th parallel held by CIA agents and communications specialists, a "belt" across the peninsula with guerillas and guides in place, covert agents operating in "MIG Alley" around the Yalu River and two "fishing fleets" controlled by CIA to rescue downed fliers on either coast. The CIA's air assets included Civil Air Transport (CAT), which had originally been General Claire Chennault's "Flying Tiger" air force in WW II, that had been taken over and renamed by CIA. Approximately 1,200 Korean guerillas were trained by a CIA detachment on Yong-do Island off the southern Korean coast near Pusan, and were inserted incrementally into North Korea by boat and aircraft.[46]

Tofte's CIA agents were also inserted inside the Soviet Union's Maritime Province, in the Vladivostok area, while other CIA agent teams penetrated eastern Siberia and Manchuria, and also the Shantung Peninsula and Tientsin areas of China. Another successful operation mounted by Tofte was the cutting of an underwater cable across the Yellow Sea used by the Chinese and North Koreans for their most secret communication, which had caused difficulties for NSA technicians attempting to intercept enemy messages. Once the cable had been cut by Tofte's "fishing fleet," the Chinese and North Koreans were forced to revert to radio communications, which the NSA listeners were able to intercept and decipher, thus providing the U.S. command with the best quality intelligence. (These intercepted Chinese, North Korean and Soviet communications, which would contain much detailed and comprehensive information on the fate of thousands of American and UN POWs and MIAs, have never been declassified by CIA or NSA.) Among the CIA's covert programs run by Hans Tofte during the Korean War was a publicity campaign about the Japanese POWs in Siberia, who were still being released in groups of hundreds at a time in late 1950 and 1951, as noted in the previous chapter. The Soviets were gaining a propaganda advantage by the releases, and left-wing organizations in Japan were extoling the Soviets humanity and good treatment of the POWs in Siberia. Tofte made a counter-propaganda film in Japan about the actual treatment of the Japanese POWs in Siberia, of whom untold thousands had died in unimaginably harsh conditions of the forced labor camps.[47] Although the film was a big success in Japan, the CIA of course said nothing about the American POWs of World War II, who had been reported alive in Soviet camps by some of the returning Japanese.

In a repeat performance of 1945 Soviet policy, the Chinese and North Koreans demanded forcible repatriations by the US of many thousands of their nationals in U.N. captivity, who were then refusing to return voluntarily to Communist control. Forced repatriations of Communist prisoners unwilling to return to their homelands became the central issue of the later Korean War period, and all attempts to achieve a negotiated settlement from 1951-1953. Dean Acheson, the American Secretary of State at the time, somewhat piously made mention of this dilemma in his memoirs, while carefully covering past U.S. government injustices by labeling the 1945-46 Soviet forced repatriates as all being deserters:

"On our side, we saw ourselves confronted again with the horrors encountered in Europe in 1945 when large numbers of Soviet civilians and Soviet soldiers who had DESERTED TO THE GERMANS (sic) and been recaptured by the allies committed suicide as they were being forcibly repatriated. The communists were determined never to open wide the invitation to desertion and escape that voluntary repatriation presented.

"Not only did the matter precipitate a deep issue between the two sides, but also one between the State and Defense departments. The military were, understandably enough, primarily concerned with getting back our own men (a much smaller number) at the end of the fighting. They had been properly interested in separating out of the prisoner pens those of our own Korean allies who had been swept up in the confusion of the war. BUT TO INSURE THE RETURN OF OUR ENEMY-HELD PRISONERS, THE PENTAGON FAVORED THE RETURN OF NORTH KOREAN AND CHINESE PRISONERS AND CIVILIAN INTERNEES REGARDLESS OF THEIR WISHES...."

"MY COLLEAGUES AND I WERE MOVED BY HUMANITARIAN REASONS AND BY THE EFFECT UPON OUR OWN ASIAN PEOPLES OF THE FORCIBLE REPATRIATION OF PRISONERS WHOSE LIVES WOULD BE JEOPARDIZED. WE WERE ALSO AWARE OF THE DETERRENT EFFECT UPON THE COMMUNISTS OF THE ESCAPE DEFERRED TO THEIR SOLDIERS BY FALLING INTO OUR HANDS."[48]

Recently declassified top-secret Truman Administration, U.S. Psychological Strategy Board reports (a subcommittee of the National Security Council), in the Harry S. Truman Library (located for the author by researcher Michael Caron), involving such officials as President Dwight D. Eisenhower's future National Security Advisor, Gordon Gray, the Army's Special Warfare expert General Robert A. McClure and CIA's Frank Wisner and Richard Bissell, indicate that senior U.S. officials opposed a "repetition of our previous mistake," of forcibly repatriating prisoners to Communist control-after Stalin's previous reneging on reciprocal returns of prisoners.[49]

The Psychological Strategy Board was a subcommittee of the National Security Council established on April 4, 1951, which was, according to author William Corson, "charged with determining the 'desirability and feasibility' of proposed covert programs and major covert projects."[50]

The actual top secret policy discussions in Washington which lay behind Acheson's apparently reasoned, 'humanitarian' position, revealed a good many more details about the decision-making process which ultimately resulted in the American government's stand against forced-repatriations of Communist POWs in Korea, and the subsequent abandonment of thousands of American (and UN) Korean war prisoners in communist captivity. Acheson failed for instance, to make it clear to the nation that he and other policymakers in Washington were well aware that U.S. refusal to repatriate Chinese and North Korean prisoners would result in the Communists keeping American POWs. This veiling of the reality led to a subsequent downplaying, by U.S. government officials, of the numbers of American prisoners withheld by the communists after the war, which took the form of lying to the public and to the families of POWs about their existence as prisoners, and a purposeful government campaign to keep from the public the fact that thousands of Americans, listed in the MIA (or another) category and later declared dead by the PFOD process, were in fact alive in enemy captivity. In this regard the failure of Truman and Acheson to allow the military commanders in Korea the means and freedom to achieve a total military victory must be considered as part of the overall POW problem.

The recently declassified reports found among the private papers of President Harry S. Truman, document official US reluctance to again forcibly repatriate prisoners of war to the communists. The Psychological Strategy Board contained representatives of the NSC, CIA, State Department and military Psychological Warfare branches. On October 22, 1951 a key, Top Secret, "Memorandum to the Executive Secretary of the National Security Council," entitled, "Report on the Situation With Respect to the Repatriation of Prisoners of War," had been prepared and sent to Secretary of State Dean Acheson, Secretary of Defense Louis Johnson, his successors George C. Marshall and Robert Lovett, and to the Joint Chiefs of Staff, National Security Council and Central Intelligence Agency. According to recently declassified U.S. documents, Frank Wisner, by now a Deputy Director of the CIA, also received a copy. He and Tracy Barnes, as anti-Soviet specialists for Allen Dulles, were involved in formulating and supporting this policy in 1951 and 1952. This report and subsequent US documents warned the highest U.S. policy makers that the potential loss of known American prisoners remaining in communist captivity after cessation of the fighting, must be weighed against the propaganda damage, within the communist countries, of continued forced-repatriations of Communist POWs.[51]

The decision that obtaining the freedom of the "lost" American prisoners of Korea could not be permitted to outweigh other important U.S. policy considerations was an agonizing process, requiring a high-level review of the earlier, failed policy of forced-repatriation, as a declassified top secret PSB/NSC document for Gordon Gray, dated 28 December 1951, indicates:

"Our treatment of Soviet and satellite expatriates has an unfortunate history...As a result of an agreement at Yalta, the United States in the years immediately after World War II assisted the Soviet Union in the repatriation of various categories of Soviet bloc persons-chiefly prisoners of war, escapees, and displaced persons. The result of our cooperation was that MORE THAN FOUR MILLION SOVIET CITIZENS WERE RETURNED TO THE SOVIET UNION and that thousands were executed or punished in other ways...In addition, persons escaping from the Soviet area after World War II were forcibly returned to Soviet control as a matter of U.S. policy until well into 1948. This treatment of Soviet expatriates became well known to the populations within the Soviet area and, as has been well documented, became the cause of widespread despair. It practically stopped the flow of defectors...

"THIS IS THE BACKGROUND WITHIN WHICH THE QUESTION OF FORCIBLE REPATRIATION OF CHINESE AND NORTH KOREAN PRISONERS OF WAR MUST BE EXAMINED. REPETITION OF OUR PREVIOUS MISTAKE WOULD DISCOURAGE DEFECTION BY CHINESE COMMUNIST FORCES IN ANY FUTURE CONFLICT. IT WOULD THEREFORE IN THE LONG RUN COST US MORE AMERICAN LIVES THAN ARE INVOLVED IN THE EXCHANGE OF PRISONERS PROBLEM."[52]

The principles outlined in the above document were in fact those adopted as United States policy toward enemy and American POWs for the remainder of the Korean War, and they were supported by President Truman and Secretary of State Dean Acheson, and eventually by President Eisenhower. Although this report makes it clear that the policy had already been communicated to General Ridgway in Korea in late December 1951, Dean Acheson noted in his memoirs that this policy was put forward as a proposal to the Communist negotiators at Panmunjom on January 2, 1952, by a U.S. negotiator, Admiral Ruthven Libby.

The Truman Administration thus officially adopted the policy of "Voluntary Repatriations" in January 1952. Among its strongest proponents was Secretary of State Dean Acheson, but, according to declassified top-secret U.S. documents, President Harry Truman was known to be opposed to any further forced repatriations of POWs during the formative process of the U.S. policy change. As Truman wrote later in his memoirs:

"Just as I had always insisted that we could not abandon the South Koreans who stood by us and freedom, so I now refused to agree to any solution that provided for the return against their will of prisoners of war to Communist domination."[53]

Charles Bohlen later wrote about his role in the Korean War POW/repatriations matter:

"In the Department of State, I became directly involved in the trickiest and most involved question connected with the Korean armistice negotiations—what to do about the prisoners of war... the major sticking point was the fact that the majority of prisoners captured by the United Nations forces did not wish to return to North Korea or China. They were definitely anti-Communist. The enemy refused to repatriate the prisoners they held unless we turned all in our hands over to them. The argument in Washington split along classic lines. The military, whose primary interest was the safety of our men, was inclined to be rather callous about forcing the enemy prisoners to return. The State Department argued that the humanitarian reputation of the United States would be damaged by forced repatriation...the State Department warned that dissidents in Communist-controlled countries would be less likely to defect in the future if they thought they were going to be turned back like cattle. I was in the State Department group that tried to come up with a formula to break the impasse..."[54] Neither Acheson or Bohlen commented in their memoirs on the subsequent abandonment of thousands of American prisoners of war in Communist control.

Many other PSB/NSC-prisoner of war documents, which are among the Presidential papers at the Harry S. Truman Library, remain classified and not subject to FOIA request. They involve U.S. officials consulted on POW/MIA policy, such as Eisenhower's former SHAEF Chief of Staff (and CIA Director) Walter Bedell Smith, General Omar N. Bradley, Allen Dulles' pet assistant at CIA-C. Tracy Barnes, Robert A. Lovett, Robert Cutler, Gordon Gray, Robert Potter, Palmer Putnam and others. (The location and identification of these documents was given to the Senate Select Committee on POW/MIA Affairs by the author, to assure that they would be declassified through the Mandatory Review Process affecting Presidential papers. Other classified documents on Korean War POWs taken to China and the USSR are in the Eisenhower Library. (Despite several written requests, neither the Select Committee nor the Truman Library ever informed the author as to the declassification of these documents up until the time of publication of this book.)

The North Koreans, Soviets and Communist Chinese had, in any case, already decided to keep many American prisoners after any future POW exchanges. This was clear to the policymakers who had formulated the plan for voluntary repatriations, not only from secret intelligence reports of Americans being moved into China and the U.S.S.R., but also from the recent prior experiences with WW II and Cold War POWs (and with kidnappings) and from analysis of the strategic concerns of the Soviet Union. The Soviets and the Chinese were simply expected to retain some of the American POWs because of their technological expertise in areas of primary concern to Russian military and intelligence planners, and in some cases for skilled industrial work. However, with the U.N. proposals on voluntary repatriation at the Peace Conference, the Chinese appear to have gathered up a number of the American and UN POWs in China and moved them toward the northeast, to at least put on an appearance of possibly exchanging them. CIA report SO-79282, of the 5th of January 1952, is titled "Preparation for exchange of United Nations prisoners in central and southern China."

"Chinese Communist authorities ordered all United Nations prisoners of war in Central and South China sent to Hankow prior to 23 December 1951 for subsequent transfer to the Northeast to await exchange.

"On 13 November, 105 Republic of Korea prisoners of war were transferred to the Northeast via the Hankow-Canton railway. On 18 December, 13 American and 8 British prisoners of war who were formerly in a building at 52 Fu Hsing Road, Shameen, were transferred by rail to Hankow. The Americans and British were escorted to Hankow by a company of Public Security troops under the command of Li Kuo-Liang.

In the comments on this report it was also noted that another source had previously reported that, "US prisoners were performing hard labor on airfields in the Canton area."

On March 9, 1952, the South China Morning Post, a respected newspaper published in Hong Kong, reported that South Korean and United Nations prisoners of war were being moved into Manchuria and from there to other areas of China. A Captured Chinese lieutenant had reported details of a POW processing center in Harbin, Manchuria, holding about 1,000 prisoners, to which he had escorted captured U.N. POWs, including Americans.

In South Korea, meanwhile, Communist North Korean and Chinese prisoners revolted on Koje Island, off the southern Coast of Korea. The Island had served not only as a prison camp but as a haven for many thousands of refugees from battle areas on the mainland during Communist offensives. During the revolt, heavily armed U.S. guards (including the 20th Battalion of the Philippine Expeditionary Force, U.S. Military Police and the 187th Airborne Regimental Combat team) had to enter the prison compound and suppress the uprising in May and June 1952, an action which caused the loss of many human lives. 55 POWs were killed, 22 died of wounds and 140 were wounded or injured. One U.S. soldier was speared to death and 14 were wounded.[55] The Chinese Communist and North Korean prisoners had armed themselves with a few handmade, smuggled and captured firearms along with knives, spears and barb wire flails. On May 8th, while the Commander of Koje Island, Brigadier General Francis Dodd, was at the wire outside Compond 76, a work detail let out of the gate suddenly seized him and dragged him back inside the gate where he was held as a hostage.

Dodd's successor in command at Koje Island, Brigadier General Charles Colson, made concessions to the Communist POWs so that they would not harm General Dodd. After negotiations Dodd was released on May 11th, and on the 13th General Mark Clark removed Colson and repudiated the concessions he had made to the Communist POWs. In December 1952, civilian internees confined by UN forces at Pongam-do (an island near Koje-do attempted a mass-breakout, in which 85 POWs were killed and over 100 wounded.

At this time, the Chinese Communists were segregating U.S. and UN POWs in Manchuria into separate categories; those who would be released and those who would be held in captivity forever. A now-declassified CIA report (#SO 91634), dated July 17, 1952, with the date of information being January-May 1952, revealed that a Soviet Far Eastern Military District officer: "...Controlled prisoner of war camps in Manchuria and North Korea," and further stated:

"THE OFFICE, FORMERLY IN MUKDEN, EMPLOYED 30 PERSONS, SEVERAL OF WHOM WERE ENGLISH-SPEAKING SOVIETS... The office had developed three types of prisoner of war camps. Camps termed 'Peace camps,' detaining persons who exhibited pro-communist leanings, were characterized by considerate treatment of the prisoners and the staging within the camps of communist rallies and meetings. The largest peace camp, which held two thousand prisoners, was at Chugchun...Reform camps, all of which were in Manchuria, detained anti-communist prisoners possessing certain technical skills. Emphasis at these camps was on reindoctrination of the prisoners. Normal prisoner of war camps, all of which were in North Korea, detained prisoners whom the communists will exchange. PRISONERS IN PEACE AND REFORM CAMPS WILL NOT BE EXCHANGED."

CIA report SO-91634 continued with descriptions of various camps containing South Korean and UN prisoners:

"Officials of North Korean prisoner of war camps sent reports on individual prisoners to the (Soviet-run) War Prisoner Administrative Office. Cooperative prisoners were being transferred to Peace Camps. ROK army officers were being shot; ROK army soldiers were being reindoctrinated and assimilated into the North Korean army.

"Kangdong Camp: In May (1952) the largest North Korean prisoner of war camp, detaining twelve hundred prisoners, was near T'ai Ling (1132/1545) mountain, six miles southeast of the Kangdong railroad station. (126-05, 39-09) (BU-4837) The compound, divided with barbed wire and mud embankments into four partitions for American, English, and Turkish prisoners and prisoners of other nationalities, held 840 Americans, 100 English, 60 Turkish, and 200 French, Dutch and Canadian troops. Most of the United States prisoners were members of the 1st Cavalry Division and the 24th Infantry Division. General William Dean (captured commander of the 24th Infantry Division) was moved from Harbin and Mukden (Manchuria) to this camp in 1951.

"The Kangdong camp, organized into study, management, sanitation and finance sections, compelled the prisoners to study for three hours, to labor for four hours, and to discuss political problems for two hours.

"On 1 May (1952) nine thousand (sic) ROK army prisoners and fifty United Nations prisoners were in caves at the Kangdong camp, extending from approximately BU-492363 to BU-494368 in a valley at Adal-ni, Kangdong-myon. Of the ROK army prisoners 10 percent were officers, 50 percent non-commissioned officers, and 40 percent privates. Of the United Nations prisoners, 10 percent were negroes. The prisoners, who received 600 grams of cereal and salt each day, were not required to work and spent only two hours of each day out of the caves. An average of two prisoners were dying daily from malnutrition and eruptive typhus. The majority of prisoners at this camp were extremely anti-Communist in thinking. Three North Korean Army guards, armed with PPsh's and rifles, were at the entrance of each cave.

"Camp Number 106, Mirim: On 1 May approximately sixteen hundred ROK army prisoners of war, including one hundred officers and five hundred non-commissioned officers, were at North Korean prisoner of war camp Number 106 at approximately YD-472214, 1.6 kilometers southwest of the Mirim railroad station (125-51, 39-01) (YD-4722). Prisoners held here, having been processed through five ideological screenings, were believed to be potential converts to Communism. The prisoners believed that they were to be assimilated into the North Korean Army. Members of political and security detachments maintained strict surveillance of the prisoners. The surveillance often was carried out by members of these bureaus who entered the camps disguised as prisoners..."

"Each prisoner received 50 won monthly, 1 kilogram of grain and 45 grams of soy bean oil, vegetables, salt and soy bean paste daily. The prisoners were wearing North Korean army uniform. The prisoners were constructing air raid shelters near the Mirim-ni airfield ten hours a day. Two hours of indoctrination lectures were also held daily. The prisoners had been organized into squads of ten men. Each of the camp's four battalions had three platoons and each platoon, four squads. A guard platoon, armed with M-1's, carbines, and PPsh's, was at the camp.

"On 5 May (1952) 200 ROK army prisoners and 110 prisoners from other United Nations armies, including 80 negroes, were at the North Korean prisoner of war camp at the site of the former Suan mine. (126-23, 38-47), ten kilometers north of Suan. Lieutenant Colonel Kim Kyu-hwan commanded the camp. Although other United Nations prisoners were not required to work, the ROK army prisoners constructed shelters and trenches throughout the entire day. The prisoners received only rice balls for food. Several of the other United Nations prisoners had obtained wheat paste from villagers in exchange for watches and other personal possessions. Thereafter the villagers were prohibited from entering the camp area. No sanitary facilities were offered to the prisoners. Approximately fifteen prisoners, including both ROK and United Nations personnel, were too ill to stand.

"Sariwon: In late April (1952) approximately eight hundred

United Nations prisoners were in a series of underground shelters at approximately YC-436673 in a valley between two hills four kilometers northeast of the Sariwon railroad station. (125-46, 38-30). On each side of the valley at the base of the two hills were 25 shelters. One shelter in every five accommodated the camp guards. Eight hundred Chinese Communist soldiers, armed with PPsh's and rifles, guarded the area. Dummy guards were also used at night. The majority of the guards were billeted in a nearby village of fifteen homes. The guard billets were easily visible from the air. Each prisoner received pork soup and 600 grams of cereal three times daily and a package of cigarettes each day. The prisoners were required to work for one hour and attend indoctrination meetings for two hours daily.

"Mukden (Manchuria): On 6 January 1952, 400 United States prisoners, including three hundred Negroes, were being detained in two buildings at Hsiao Nan Kuan Chiak, at the southeast corner of the intersection, in Mukden (Manchuria). One building, used as the police headquarters in Hsiao Nan Kuan during the Japanese occupation, was a two-story concrete structure, 30 meters long and 20 meters wide. The other building, one story high and constructed of grey brick, was behind the two-story building. Both buildings had tile roofs. All prisoners held here, with the exception of three second lieutenants, were enlisted military personnel. The prisoners, dressed in Chinese Communist army uniforms, with a red armband on the left arm, were not required to work. Two hours of indoctrination were conducted daily by staff members of the Northeast Army Command. Prisoners were permitted to play basketball in the courtyard. The attempt of three white prisoners to escape caused the withdrawal of permission for white prisoners to walk alone through the streets in the vicinity of the camp. Two Chinese Communist soldiers guarded groups of white prisoners when such prisoners left the buildings. Negroes, however, could move outside the compound area freely and individually. Rice, noodles, and one vegetable were served daily to the prisoners in groups of 10 to 15 men. One platoon of Chinese Communist soldiers guarded the compound."[56]

Decades later, a special U.S. Army investigative unit called "Task Force Russia" obtained further confirmation of the transfer of American POWs in Chinese Communist camps to Soviet control during this period in the winter of 1951-52, from a ranking participant in the program:

"An interview with Shu Ping Wa, a former head of a division-level POW collection team (164th Division) in the so-called Chinese People's Volunteers (CPV) serving in Korea, showed that a policy existed to turn over (U.S.) pilots to the Soviets. As he testified in the video recording shown at the April 1993 Commission meeting in Moscow,[57] HE HIMSELF TURNED OVER THREE AMERICAN PILOTS TO

THE SOVIETS JUST NORTH OF THE FRONT LINES SOME TIME IN THE WINTER MONTHS BETWEEN NOVEMBER 1951 AND MARCH 1952. HE STATED THAT HIS SUPERIOR TOLD HIM THAT THE 'RUSSIANS WANTED THE PILOTS.'"[58]

In a meeting between Josef Stalin and Chinese Foreign Minister Chou En-lai, on September 19, 1952, the Soviet dictator had referred to the Communist practise of withholding thousands of UN war prisoners, saying:

"Concerning the proposal that both sides temporarily withhold twenty percent of the prisoners of war and that they return all the remaining prisoners of war-the Soviet delegation will not touch this proposal, and it remains in reserve for Mao Tse-tung."[59]

Further confirmation that thousands of the South Korean POWs noted in CIA reports of the time had in fact been moved to Communist China and later to the Soviet Union, also resulted from the "Task Force Russia" investigation of Korean War, four decades later, when a report of that agency stated:

"The essence of the Stalin-Chou En-lai meeting was corroborated by a senior retired Soviet officer, Kahn San Kho, who had been seconded to the North Korean People's Army, promoted to the rank of lieutenant general, and who eventually served as the deputy chief of the North Korean MVD. He stated in November 1992 that HE ASSISTED IN THE TRANSFER OF THOUSANDS OF SOUTH KOREAN POWs INTO 300-400 CAMPS IN THE SOVIET UNION, most in the taiga (forest) but some in Central Asia as well. Lt. General Kahn's testimony shows the POW element of the Gulag was operating efficiently at this time in absorbing large numbers of UN POWs..."[60]

On the Korean War battlefront meanwhile, more Americans had disappeared in combat and their fate remained unknown. In one case, during the Battle for old Baldy, from July 17-August 4, 1952, the 23rd U.S. Infantry Regiment of the 2nd (Indian Head) Division, lost 39 killed and 234 wounded, and 84 more Americans were listed as missing in action. In the ensuing period, the U.S, carried out the largest air raids of the war, including a 1,403-plane assault on the North Korean capital at Pyongyang on August 29th.[61] In these air attacks U.S. pilots and crews continued to disappear, and an unknown number listed as MIA by the Air Force, Navy and Marine Corps were actually captured.

Since early in the Korean War U.S. POWs had been reported by intelligence sources as transferred to Soviet control and shipped into Siberia. Individual American prisoners returning from captivity at the end of the war also reported U.S. POWs being shipped to the Soviet Union, and sometimes this knowledge was gained in curious ways. Marine Master Sergeant John T. Cain, an observer in a reconnaissance aircraft was shot down during his 129th mission over western Korea on 18 July 1952 and captured by Chinese troops.

While confined in a hut with 5 other American prisoners in the village of Obul one of the Chinese guards turned out to have been a Captain in Chiang Kai-shek's Nationalist Chinese Army, who had been captured and given the choice of fighting for Mao Tse Tung or dying. He had chosen to live. Now while guarding Cain and the others he was friendly, giving them cigarettes and matches and what news he could of several other recently shot down U.S. aircraft. In broken English he told Cain and the other Americans of one captured American helicopter pilot with the rank of 2nd lieutenant, "being taken to Russia in March 1952." He did not know the American's branch of service.[62] The Soviets were most anxious to equal American technological success at mass-producing efficient helicopters for military and internal security use. Having captured American helicopters to study was enhanced by the presence of well-trained American pilots of the aircraft.

The superb American F-86, fighter, together with all its high-tech equipment, was a sought-after prize for the Soviets who were assisting the North Koreans and Chinese in the war effort, particularly with anti-aircraft artillery units engaged in active combat, whose orders were to capture as many American pilots as possible, alive.

Retired KGB Lt. Colonel Yuriy Lukianovich Klimovich, who served in Korea, recounted, four decades later, that the Soviets made an effort to capture intact F-86s, and that one F-86 Sabrejet had been forced down on a beach and transported to the Sukhoi Design Bureau in Moscow for exploitation. He also said on a Russian television program that he knew of two F-86s transported to Moscow in 1951/52. Klimovich told investigators of the U.S. Army's "Task Force Russia" about a "close friend, now deceased," who had confided to him that a U.S. F-86 and an American pilot had been brought to Moscow:

"...One of the aircraft was in excellent condition and was disassembled at the Sukhoi Design Bureau in an attempt to copy it. Klimovich said that neither his friend nor he knew what happened to the alleged American pilot since he fell immediately into KGB hands... Klimovich then escorted Task Force Russia interviewers to the Sukhoi Design Bureau where they met designers who clearly remembered that an F-86 had been brought to the Bureau during the Korean War. These designers confirmed Klimovich's assertion that two F-86s had been brought to Moscow...stripped of markings and serial numbers. None of them had spoken to an American pilot...They did, however, receive information from a member of the project that appeared to be from a pilot. One of the designers remembered that this individual had once told him he was participating in the interrogation of the aircraft's pilot. The designers also stated that the aircraft had been at the Mikoyan-Gurevich (MiG) Design Bureau."

American investigators then visited the former MiG design Bureau, now the Zhukovsky Central Aerohydrodynamics Institute and spoke to the chief historian, Yevgeniy I. Rushitskiy, who confirmed that an F-86 had been delivered to the institute which no longer had markings or identification numbers of any kind on it. A subsequent report of the visit to this institute sent by cable from the American Embassy in Moscow four decades later stated:

"One of the designers distinctly remembered the study and disassembly of...an F-86 at the design bureau. This source also remembers an American pilot having been available at another location for follow-up questions. This story was repeated by other personnel from the Design Bureau."[63]

During the tedious negotiations over POWs at Panmumjom, the conflict in Korea had evolved into a stalemated form of trench-warfare, which continued during 1952. Through the spring and summer the Communists steadily increased the number of their forces in Korea and strengthened their artillery and armor...The UN waged its heaviest offensive attacks on the north from the air. Major bombing attacks occurred on the Yalu dams and Hydroelectric plants. The U.S. also carried out secret bombing raids inside Manchuria, against targets which would conserve the lives of U.S. and UN soldiers fighting in Korea. Many U.S. missing in action and prisoners of war resulted from such covert activities, which may not be reflected in reports of combat casualties, but may be carried under another, non-combat category of loss.

Admiral C. Turner Joy was replaced as chief U.N. negotiator by Major General William K. Harrison Jr. on May 22nd, 1952. Secretary of State Dean Acheson, desperate to get the U.S. out of Korea at almost any cost before he left office, proposed a weakening of the U.S. bargaining position to keep the talks from being recessed. Truman supported the Pentagon against Acheson, instructing the military to keep up the pressure on the battle front and the negotiators to put forth a strong position against forced repatriations to the Communists. If they walked out of the peace talks, Truman was willing to live with it.

The Communists chief negotiator, Nam II, rejected yet another UN proposal on October 8, 1952, calling it "unacceptable." In Washington D.C., Dean Acheson reaffirmed the American position against forced repatriations in announcing the break in the peace talks somewhat rhetorically: "We shall not trade in the lives of men... We shall not forcibly deliver human beings into Communist hands."[64]

In the November 1952 presidential election campaign, the former U.S. commander in Europe and Chairman of the Joint Chiefs of Staff, General Dwight D. Eisenhower, who had promised to end the war in Korea, defeated Harry Truman's chosen candidate, Adlai

Stevenson. During the campaign Stevenson and his advisers had come under attack by Senator Joseph McCarthy and his supporters. Harry S. Truman and Dean Acheson had sought for a way to bring the war in Korea to an end, but Truman refused to defer the POW matter in order to achieve a quick peace, and supported the Defense Department position. The Democrats twenty-year-long hold on the White House was thus finally broken, amidst the clamor and recriminations of the McCarthy era. The incoming president designated John Foster Dulles as Secretary of State on December 20, 1952, and his younger brother, Allen Welsh Dulles as Director of Central Intelligence on February 11, 1953.[65] John Foster Dulles, educated at Princeton and the Sorbonne, was an influential New York lawyer who had been brought into the State Department to act as a buffer between the Department and the Republican right wing. He was an ideological anti-Communist who wished to liberate eastern Europe from the Soviets.[66] Allen Dulles was the former OSS station chief in Berne, Switzerland during WW II, who had received all the classified Red Cross and military intelligence information about the American POWs in German hands being overrun by the Soviet armies in 1945. Dulles had known ever since that the Soviet threat was real and presumably had conveyed this information prior to his becoming Secretary of State. Lt. General Charles P. Cabell, USAF, was selected by Eisenhower in 1953, upon DCI Walter Bedell Smith's recommendation, to serve as Deputy Director of Central intelligence (DDCI) and Deputy Director of CIA. Involved with Air Force Intelligence, Cabell was in a position to know a great deal about the American prisoners held by the Soviets and he subsequently had charge of the CIA's clandestine services. (He and Dulles were to be dismissed a decade later by President Kennedy for their role in the disastrous Bay of Pigs invasion of Cuba.)

Eisenhower announced that Robert Cutler would be his Special Assistant for National Security Affairs, a position that Averell Harriman had held under Truman which would come to be called "national security adviser." Cutler, who had been involved in POW repatriations matter, recommended that the NSC senior staff be renamed the Planning Board, the members of which would include representatives of the Joint Chiefs of Staff, CIA, and the Psychological Strategy Board (PSB). (The Psychological Strategy Board of the NSC was abolished on September 2, 1953 and replaced by the Operations Coordinating Board, OCB, which controlled U.S. covert operations, and a future Vice President, Nelson Rockefeller was later to be designated by Eisenhower as the president's representative on the OCB from December 1954-December 1955.)[67] Gordon Gray was to become the President's national security adviser from July 1958 to January 1961 and C.D. Jackson had charge of the POW/MIA matter on the Committee in 1953-54.

Subsequent to Eisenhower's inauguration, U.S. intelligence received an alarming report on February 24, 1953, concerning the shipment of a group of 68 UN POWs from China to the Soviet Union late in 1952, solely for espionage usage. This information was forwarded by the Ministry of Foreign affairs of South Korea from Nationalist Chinese sources. According to the report of Combined Command for Reconnaissance Activities, Korea (CCRAK), "THE CCF HAS TRANSFERRED UN POWs TO RUSSIA IN VIOLATION OF THE GENEVA CONFERENCE. THESE PWs WILL BE SPECIALLY TRAINED AT MOSCOW FOR ESPIONAGE WORK. PWs transferred to Moscow are grouped as follows: British 5, Americans 10, Canadians 3, and 50 more from various countries."[68]

It must be assumed that among the 50 others were South Koreans and likely also, Turks, British, Belgians, Dutch, French, Canadian, Australian and other UN prisoners, who, if cooperative, would be invaluable to the Soviet MGB secret police, for training deep penetration agents and spies, or perhaps to be used in an operational manner themselves. (This theme was later popularized in the best-selling novel: "The Manchurian Candidate," which was also made into a film.) The conversion of British MI-6 POW George Blake into a Soviet spy after his capture in Korea and brainwashing by the KGB while in Chinese Communist captivity, was a real-life result of this long-established Communist policy for Western POWs. The CCRAK comment on this intelligence was as follows:

"THIS OFFICE HAS RECEIVED SPORADIC REPORTS OF PRISONERS OF WAR BEING MOVED TO THE USSR SINCE THE VERY INCEPTION OF THE HOSTILITIES IN KOREA. THESE REPORTS CAME IN GREAT VOLUME THROUGH THE EARLIER MONTHS OF THE WAR, THEN TAPERED OFF TO A STANDSTILL IN EARLY 1951, BEING REVIVED BY A REPORT FROM JANUARY OF THIS YEAR (1953). IT IS DEFINITELY POSSIBLE THAT SUCH ACTION IS BEING TAKEN AS EVIDENCED BY PAST EXPERIENCE WITH SOVIET AUTHORITIES. ALL PREVIOUS REPORTS STATE PRISONERS OF WAR WHO ARE MOVED TO THE USSR ARE TECHNICAL SPECIALISTS WHO ARE EMPLOYED IN MINES, FACTORIES, ETC. THIS IS THE FIRST REPORT THAT THEY ARE BEING USED AS ESPIONAGE AGENTS THAT IS CARRIED BY THIS OFFICE."[69]

In the Soviet Union, meanwhile, dictator Josef Stalin had finally died, apparently of natural causes, after nearly 30 years of supreme power. His death from "a stroke" was announced to the world on March 5, 1953. Lavrenti Beria was then conducting Stalin's last purge, which involved the arrest and execution, or imprisonment of thousands more imagined enemies, many of them Soviet Jews in influential positions, who had supposedly been involved in what came to be known as the "doctors plot." Stalin had come to believe that a cabal of Jewish doctors in the Kremlin were planning to murder him and other Soviet leaders, and that these plotters were connected to a vast conspiracy. (One of those later sent to a forced labor camp as a result of this purge was Avraham Shifrin, whose subsequent testimony on the Gulag was quoted in the previous chapter.)

The Soviet anti-Semitic purge had begun in the satellite nation of Czechoslovakia with a show trial (conducted behind the scene by the MGB) in 1952 in which 11 of the 14 defendants were identified as being "of Jewish origin." 11 of the defendants were sentenced to death, and the purge of Jews then spread to the Soviet Union, with the removal of all Jews from the MGB (KGB) and gradually from other Soviet government departments. The government-controlled Soviet newspaper Pravda, signaled the start of the campaign on January 13, 1953, with a report that "monsters and murderers," who were agents of American and British intelligence acting through a "corrupt Jewish bourgeois nationalist organization," were "hiding behind the honored and noble calling of physicians."

Stalin's successor to supreme power, Nikita Khrushchev, later wrote in his memoirs that "Stalin was crazy with rage, yelling at Ignatyev (head of the MGB) and threatening him, demanding that he throw the doctors in chains, beat them to a pulp and grind them into powder." The interrogation of the doctors was turned over to M.D. Ryumin, deputy chief of the MGB. As Khrushchev wrote, "It was no surprise, when almost all the doctors confessed to their crimes."[70] Stalin became increasingly suspicious of his old henchmen in the last months of his life ordering the arrest of his chief bodyguard Vlasik, and suspecting Molotov (whose Jewish wife was sent to a labor camp) of being a CIA agent.[71] The purge continued after Stalin's death, under Beria, who ordered it halted a month later, when he was certain that sufficient terror had been instilled and he could pretend to be a saviour of those who had so far escaped arrest. Soviet belief in a vast Zionist conspiracy continued, however, for decades to come, resulting in persecution of Russian Jews and Soviet support for the Muslim enemies of the newly founded state of Israel.

With the death of Stalin, a triumvirate, composed of G. M. Malenkov, Nikita Khrushchev and Nikolai A. Bulganin took over the Soviet government, from which Khrushchev eventually emerged as the sole ruler of the USSR.[72] Lavrenti Beria, however, remained a threat to the triumvirate's survival for a few more months. Intent upon consolidating their rule over Stalin's vast empire, the new Soviet leaders decided that better relations with America were necessary, and signaled their interest in seeking a settlement of the Korean War.

Shortly after this, on March 28, 1953, the Communists accepted a U.N. proposal to exchange a small number of wounded and sick prisoners of war, and on April 11th the liaison officers at Panmunjom reached agreement on what came to be called by UN negotiators, the "Little Switch." 5,100 Korean and 700 Chinese prisoners were to be exchanged for 450 South Koreans and 150 U.N. prisoners. The Communists agreed to release the most sick and worst wounded POWs first in "Little Switch," but reneged when the operation got underway. They decided instead, for propaganda reasons, to withhold the most severe cases and instead to release some POWs in relatively good condition, including a number who had succumbed to communist propaganda and who could be counted on to discount the tales of communist atrocities which had leaked out or would inevitably result from any prisoner release. Still, some of the Americans and South Koreans who got out revealed that the Communists had reneged on the agreement and had secretly withheld at least 500 of the sick and wounded American and Korean POWs. Despite the Communist attempt at achieving a propaganda victory, some of those who were released were in tragic physical condition, starved and emaciated, and some of them had gone insane from months and years of brutality and torture in captivity.

One of the American POWs released in Little Switch told Army authorities on April 28th that 50 of the U.S. prisoners listed as missing or dead were actually alive in Communist prison camps. This soldier was able to memorize all their names and addresses. Two of the POWs whose names he had memorized were from Hawaii, and when the prisoner reached Honolulu he contacted them. According to a member of one family quoted in an Associated Press story of April 29, "the man memorized the names so perfectly that he even pronounced Hawaiian street names correctly."[73] From returning prisoners like this one, U.S. Army investigators learned the names of many other Americans who were known to be alive in captivity but who were not on Communist lists, or were claimed by them to be dead.

An authoritative eyewitness to the negotiations leading to Little Switch and the Communist retention of many of the sick and wounded American prisoners at that time, Lt. Colonel Philip J. Corso, testified under oath about these events before the U.S. Senate four decades later. Corso reported to C. D. Jackson, a member of the President's Committee on International Information activities of the National Security Council. Senate investigators verified in 1992 that Colonel Corso was head of the Special Projects Branch of the Far East Command during the Korean War, serving in the G-2 (military intelligence) Division, and was responsible for tracking the movement and treatment of American and other UN POWs, verifying locations of POW camps and monitoring the locations and conditions of the POW camps and the numbers of the prisoners held there. Corso revealed:

"I also served as a military intelligence officer with the

United Nations Truce Delegation at Panmunjom in the closing days of the war in April 1953. There I assisted with discussions dealing with the exchange of sick and wounded prisoners, known as 'Operation Little Switch.' Both sides eventually reached complete agreement on exchanging sick and wounded POWs. Soon after that eventful day, however, I caused an incident that almost brought Little Switch to a halt. I had prepared a detailed intelligence report showing that the Communists had violated the agreement by not giving up all the sick and wounded they were holding. We estimated they still held 500 (sick and wounded) POWs. Furthermore, the condition of these POWs was such that if they did not receive adequate treatment immediately, they would surely die.

"Rear Admiral John C. Daniel, Deputy Chief of the UN delegation, read my intelligence report at a meeting at which Chinese Communist General Bien Zhang-Wu presided. He became so angry at hearing my facts that he snapped in half a pencil he was holding. He got up from the table and turned to leave. Then, he caught himself, came back, sat down and glared at me in anger. ALTHOUGH ADMIRAL DANIEL MADE THIS CHARGE IN OPEN SESSION ATTENDED BY NEWSPAPER REPORTERS FROM MANY NATIONS, THE AMERICAN NEWS MEDIA NEVER REPORTED THIS DAMAGING INFORMATION TO THE AMERICAN PUBLIC. THE COMMUNISTS NEVER RETURNED THOSE 500 SICK AND WOUNDED POWs. AND THIS FACT TOO, WENT UNREPORTED BY THE AMERICAN PRESS. NOR DID THE POLICY MAKERS AND HIGH GOVERNMENT OFFICIALS IN WASHINGTON EVER MAKE AN ISSUE OF THESE CHARGES."[74] Corso was later to have an extraordinary meeting with President Eisenhower on the subject of American POWs from Korea transferred to the Soviet Union.

The official secrecy and major media silence surrounding the Communist withholding of many U.S. prisoners and the shipment of large numbers of American and UN POWs to Manchuria and Soviet Russia made all the more extraordinary an article written by Zygmunt Nagorski Jr., which appeared in the May 1953 edition of Esquire Magazine, entitled: "Unreported G.I.'s In Siberia." The article cited a secret source who traveled widely in Soviet-occupied territory and had been tested by Nagorski on the accuracy of other intelligence supplied to that writer's "Foreign News Service," which regularly reported on events inside the USSR, despite Soviet censorship and security. U.S. POWs from Korea were also reported by intelligence sources in KGB-run Siberian camps at Khabarovsk, Omsk, Chita, Shivanda, Kudymkar, Chermos, Gubakha, and in Yakutsk and Irkutsk and other locations in the USSR, for exploitation of their technological or linguistic skills, or in the cases of recalcitrant prisoners, as forced laborers.

The American media had been deeply divided and politicized by the on going anti-Communist "witch hunt" being conducted by

Senator Joseph R. McCarthy and other extreme conservatives, both Republicans and Democrats. The shock of Maclean's and Burgess's treason and defection to Moscow had still not subsided, and J. Edgar Hoover's FBI was pursuing many other possible traitors, while reportedly providing the McCarthy camp with classified files as ammunition in their public campaign. Julius and Ethel Rosenberg, the relatively low-level atomic spies for the Soviets who were exposed by the arrest of Klaus Fuchs in 1949 and of David Greenglass in 1950, had been and found guilty of spying in a trial beginning in 1951, and were executed by electric chair on June 19, 1953. The ending of the stalemated Korean War by an armistice, without a victory for the U.S., contributed to a growing sentiment among a substantial portion of the public and in the Congress that American interests had been subverted by disloyal government officials who were actually serving the Communists. The wild and often unproven charges by McCarthy and some of his allies in the Senate had put much of the media on the defensive, resulting in a tendency to avoid publication of truly damaging facts about the Soviets and their Asian surrogates which should have been revealed to the American people, but which would have justified some of the claims of the anti-Communist crusaders. McCarthy worked closely with John V. Grombach who had helped depose Dean Acheson's ally, the former military intelligence Colonel Alfred McCormack, from the State Department in 1946. Their targets in 1953 included Charles Bohlen, his brother-in-law Charles Thayer, William Bundy, Carmel Offie (who had been an aide to William C. Bullitt and Anthony J. Drexel Biddle Jr.), Cord Meyer Jr., John Paton Davies and others.[75] In 1954 McCarthy overstepped in his insinuations against the U.S. Army during televised hearings. His public questioning of the actual loyalty, rather than the judgment of revered American leaders including George C. Marshall, President Dwight Eisenhower and other former senior commanders ultimately brought about his downfall and vilification.[76] One of the issues which McCarthy embraced and which much of the major the media avoided, was the disappearance of thousands of American prisoners of war in Communist control.

In the Soviet Union meanwhile, the arrest of KGB chief Lavrenti Beria was announced on July 10th, along with the arrest of the former head of the NKGB, Merkulov, and Vladimir Dekanozov, who had helped in the kidnapping of American POWs in 1945. In a ludicrous but typical Soviet manner, they were charged with, and found guilty of plotting "to revive capitalism and to restore the rule of the bourgeoisie." They were shot. Khrushchev, Malenkov and Bulganin subsequently amnestied and released some Gulag prisoners, but did not alter the Soviet policy of seizing American prisoners wherever they could, in accordance with Soviet national security requirements, and continued to hold U.S. POWs who had been transported to Siberia from the Korean War theater, and from Manchuria.

Secret intelligence reports from CIA sources (whose reliability was admittedly uncertain) indicated that large numbers of American and other UN prisoners were still being held inside Communist China, near Peking. A July 15, 1953 CIA report (CS-14835) was distributed to the Army, Navy and the Department of State:

"IN LATE MAY 1953, APPROXIMATELY 1,500 UNITED NATIONS PRISONERS OF WAR WERE CONFINED IN A CAMP AT TUNGCHUTUA, TIENTSIN. THE MAJORITY OF THESE PRISONERS OF WAR WERE AMERICAN MARINE OFFICERS AND MEN WHO WERE SENT TO THIS CAMP AFTER RECOVERING FROM WOUNDS."[77]

These reported U.S. POWs were never repatriated from Communist China. (The most conclusive evidence of the fate of U.S. POWs in Korea retained by the Chinese and Soviets would be the records of "special intelligence," obtained by the NSA's (and others) intercepting and decoding of enemy orders and communications, but this documentary evidence on Korean war POWs has remained highly classified and has been unavailable to researchers or U.S. Senate Select Committee on POW/MIAs investigators whom the author worked with in 1991-1992)

On the battle lines in Korea, a June 10-18 attack on the U.S. 3rd Infantry Division's Outpost Harry had cost the Chinese Communists an estimated 4,200 casualties, but killed 174 Americans and wounded 824 others. From July 6-10, renewed attacks on Pork Chop Hill caused the 7th Infantry Division to evacuate defensive positions, and in the last major Communist offensive, from July 13-20, against the U.S. IXth Corps, nearly a thousand more Americans were killed and wounded. On July 14, 1953, the U.S. 555th Artillery was overrun and 300 Americans were listed as KIA or MIA. On the 19th and 20th of the month, a small force of 81 U.S. Marines was attacked at Outposts Berlin-East Berlin and reinforcements were brought in. Of the combined U.S. force, 6 were killed, 86 wounded, 12 were known to have been taken as POWs and 44 more Americans were listed as MIA. The last significant ground fighting of the war, between the 24th and 26th of July, cost some 350 more American casualties in the "Boulder City" area, as the Communists tried to gain important hill positions. Other action occurred on "Sniper Ridge." The last official U.S. air kill of the war was an enemy transport shot down on July 27th "near the Manchurian border."[78]

On July 29th, 1953, Soviet MiG-15 jet fighters attacked and shot down a U.S. Air Force RB-50 reconnaissance aircraft near Cape Povoratay, off the coast of Siberia, south of Vladivostok. It was subsequently revealed that the Soviets were retaliating for the downing of a Soviet aircraft over Manchuria by U.S. aircraft. The RB-50 aircraft had seventeen American crewmen, one of whom, the copilot, was rescued by the U.S. Navy about 40 miles off the cape. In a note delivered in Moscow by U.S. Ambassador Charles Bohlen on July 31st, the United States said that other crewman had been rescued at sea by the Russians: "Information has also been received that other survivors have been picked up by Soviet vessels in the vicinity of the crash. My government has instructed me to request an immediate report from the Soviet authorities regarding the condition of these survivors and what arrangements are being made for their early repatriation." The Soviets refused to return these U.S. POWs.

One of the crewmen on this aircraft was 1st Lieutenant Warren Sanderson, and nearly 40 years later his son Bruce Sanderson, of Fargo, North Dakota, testified before the Senate Select Committee on POW/MIA Affairs about his efforts to obtain classified records from the U.S. Government pertaining to his father's case. Senate investigators noted some of the important facts in this RB-50 case: 1. The Soviets admitted they shot it out of the sky. 2. Survivors, beyond the sole individual who was rescued by a U.S. Navy ship, were seen in the water by search and rescue aircraft. 3. NKVD patrol boats were seen in the area, moving to and away from the crash site. 4. The co-pilot was rescued 22 hours after the crash, 17 miles from the coast.

The younger Sanderson was assisted by Task Force Russia (under authority of U.S. Army Major General Bernard Loeffke), during a 1992 visit to the former Soviet Union, in gaining access to Soviet archives and sources. One Russian citizen who Sanderson located, admitted to personal involvement with the interrogation of U.S. military prisoners from 1950-1954, and according to Sanderson's subsequent testimony, this same Soviet, "Also reaffirmed the information from the first meeting that ALL U.S. PERSONNEL UNDER SOVIET CONTROL WERE PHOTOGRAPHED, FINGER-PRINTED AND GIVEN RUSSIAN NAMES, and that these men were then moved frequently from camp to camp. IT WAS COMMON PRACTISE TO CREATE A FALSE DEATH CERTIFICATE OR RECORD WHEN A PRISONER WAS MOVED." Thus was obtained further confirmation that the Soviets hid their possession of American POWs by giving them new, Soviet identities, so that even their own prison records were covered, insuring that the U.S. prisoners would vanish forever.

On the sour note of the Soviet shootdown of the RB-50, Operation Big Switch, the final POW exchange in Korea, began at Panmunjom on August 5, 1953. By early September, at the conclusion of Big Switch, 3,597 American POWs had been returned, along with 7,862 South Koreans and 1,314 from other nations. U.N. forces returned 75,823 Communist prisoners, including 70,183 North Koreans and 5,640 Chinese. During this second and major return of American POWs in Korea, almost none of the thousands of American POW/MIAs reported being transferred to China, and none of those U.S. POWs interrogated by Soviet intelligence and shipped to Siberia, were reported as returned. In another case, hundreds of U.S. NCO's who were known to be prisoners in Camp #4 at Wiwan, yet the North Koreans released only 25 American POWs from that camp. The Pentagon was soon to admit that approximately 1,000 U.S. prisoners, known by name to have been alive in Communist captivity, were not released in Operation Big Switch, although now-declassified documents indicate that this was an artificially low number, intended to admit the problem but not reveal the full scale of it, to avoid a further prolonging of the fighting.

General Mark Clark, former U.S. commander in Korea, who had served under Field Marshall Alexander in Austria in 1945, repeatedly stated that the Communists had kept many US and allied UN POWs from Korea. General Clark gave a press conference in Washington DC on August 6, 1953, following his return from Korea and his report to President Eisenhower. Quoted in a New York Times report of that day, General Clark announced his sudden decision to retire from the Army, accused the Communists of withholding many U.S. POWs (alluded to by the New York Times on that day as "several thousand"), and proposed use of the atomic bomb if the Communists broke the truce agreement. Beyond the 3,313 American prisoners the Communists had admitted to holding, General Clark was quoted as saying there remained:

"...Other information that leads us to believe they have more. We do have certain evidence that there are additional prisoners alive who should be returned, and I assure you as Commander in Chief, while I am there, I will press that in the military and political conferences."

According to this New York Times report, "...General Clark pledged to press the Communists for further information on the additional troops he believed they held and for a possible exchange. He said that he had pointed out the wide discrepancy between his information and that supplied by the Communists on prisoners during the truce negotiations. He had been advised by his superiors in the Pentagon, he said, not to delay the armistice negotiations over the discrepancy but to reserve the privilege of later protest..."

In a public statement the same day President Eisenhower rhetorically announced that, "The Korean truce demonstrated the determination of the free world to resist aggression anywhere." The Chinese Communists meanwhile, announced publicly that they and the North Koreans might not return all the American and U.N. prisoners they had admitted to holding. The communists claimed they had not screened all the U.N. prisoners and that they reserved the right to not return some who they said might not wish to be repatriated. In his memoirs, published in Great Britain son after his retirement, General Clark wrote:

"Through Big Switch we learned that the Chinese and North Koreans, like the Communists in Russia, had refused to return all the prisoners they captured. Why the Reds refused to return all our captured personnel we could only guess. I think one reason was that they wanted to hold the prisoners as hostages for future bargaining with us, possibly for some concession such as a seat for Red China in the United Nations..."[79]

The New York Times published an Associated Press report about the lost American POW/MIAs of Korea on August 7th, 1953. General James A. Van Fleet, retired commander of the U.S. 8th Army in Korea, who had had access to the most secret intelligence reports (some of which still remain classified at the time of this writing), made the most truthful and important statement about the 8654 U.S. MIAs (then officially listed) of any senior U.S. Commander of the Korean War, saying:

"A LARGE PERCENTAGE OF THE 8,000 AMERICAN SOLDIERS LISTED AS MISSING IN ACTION IN KOREA ARE STILL ALIVE." He was further quoted in the New York Times of August 8, 1953, as stating, in an interview in Belmont, Massachusetts, that General Mark Clark's estimate of at least 3,000 G.I.'s still held captive by the communists was "CONSERVATIVE."[80]

One of those missing was General Van Fleet's son, U.S. Air Force Lieutenant James A. Van Fleet Jr. (promoted to Captain while missing), who was carried on the POW/MIA list of 5,000 as the husband of Mrs. Yvonne L. Van Fleet of Seattle, Washington. This helps explain the General's strong public stand on the American MIAs of Korea, which he continued to express for years to come. Captain Van Fleet was number 3,031 on a list of nearly 5,000 U.S. POW/MIAs who were to be officially presumed dead by early 1954, even though his name was kept on a shorter list of 944 Americans who were known to have been alive in captivity. (The date of his presumptive finding of death was officially recorded as March 31, 1954.)

In another New York Times report of the same day it was noted:

"...Evidence that not all the Allied prisoners held by the Communists would be returned came from a number of sources. PFC. Thomas R. Murray of Baltimore told reporters that several Americans in his camp were held in 'jail' and at least one of his friends had been sentenced to a year's imprisonment..."

This and other isolated cases reported in the press at the time were secretly confirmed by classified CIA, CKRAK and other reports of Americans in Communist China and the Soviet Union, and backed up by hundreds of other eyewitness reports by returning prisoners who related details and names of many cases of American prisoners of war who had disappeared during interrogations or 'midnight repatriations,' or had been illegally "sentenced" to prison terms and withheld by the Communists after Big Switch. Many of these long-classified reports have now been uncovered in the National Archives, and serve to illustrate the magnitude of such Communist actions in the Korean War.

Prisoners refusing repatriation in Big Switch totaled 22,604 Communist POWs, including 14,704 Chinese, and 359 U.N. POWs, including 325 Koreans. They were turned over to the Neutral Nations Repatriation Commission inside the Korean DMZ, and representatives of their own nations were given access to them in attempts to win their agreement to voluntary repatriation. Of the 22,604 Communist POWs who refused repatriation only 137 changed their minds and returned north.

Of the 359 U.N. non-repatriates, only 10 changed their minds and decided to return home, including 8 Koreans and 2 Americans. 325 Koreans, 21 Americans and 1 British POW decided to remain with the communists. The 21 Americans who had refused repatriation were branded traitors and turncoats in the press, but over the next 20 years about 12 of them changed their minds about life under Communism and were allowed to return home.

In the end, the North Koreans, Chinese and Soviets had their revenge for the American refusal to carry out forced repatriations of communist POWs. They had secretly withheld an estimated 4,000-5,000 or more American POWs who had been hidden in special, secret prison camps in North Korea, China and the Soviet Union. By early 1954 over 4,900 of these Americans, many of whom were listed in the MIA category by the United States, were declared dead by their own government through the Presumptive Finding of Death (PFOD) process, in use since the missing POWs of the American Intervention in Russia had been declared dead in 1920.

Many of the prisoners who were recovered by the United States had reported extensive interrogation of US prisoners by Communist Chinese and some reported interrogations Soviet intelligence officers. Few of the thousands of Americans held in the special, secret prison camps on the Manchurian border ever reappeared.

A now-declassified CCRAK report (#66-53), dated August 10th, 1953, during the period of the postwar prisoner exchanges, states: "A compilation of reports indicate that during the past two years several PSW have been transferred from PW camps in North Korea to points in Manchuria, China and Siberia...MANY PRISONERS OF WAR TRANSFERRED HAVE BEEN TECHNICIANS AND FACTORY WORKERS. Other prisoners transferred have had a knowledge of Cantonese and are reportedly used for propaganda purposes...Figures show that the total number of MIA plus known captured less those to be U.S.

repatriated leaves a balance of over 8,000 unaccounted for."[81]

Lt. Colonel Phillip Corso, who served on President Eisenhower's NSC staff from 1953 to 1957, later testified under oath before a Senate Select Committee that KGB officer Yuri Rastvorov, who defected to CIA from the Tokyo Residency in January 1954, revealed details of the movement of Korean War POWs into the Soviet Union and their subsequent use as forced labor. (The full debriefings of Rastvorov and of several other NKVD/KGB defectors of the Cold War period have long been unsuccessfully requested through Freedom of Information requests to CIA by the author, despite the CIA's recent publicity campaign that it was making such information available to the public.)

After the author had printed and distributed the first edition of this book in 1992, the U.S. Army's "Task Force Russia" obtained more information from Lt. Colonel Corso about his sources of intelligence:

"I secured this information from I'd say, hundreds of prisoner of war reports, from Chinese and North Korea(ns), who actually saw these prisoners being transported and later I talked to a few high-level Soviet defectors who confirmed it—that this transfer was going on... and that they were being taken to the Soviet Union....The operation, as far as we were concerned, was a GRU/NKVD operation in those days. And it was mostly to elicit information from them, possibly take over their identities or use them as agents...we had information...that this was being done....The source of this information, as I said, was hundreds of prisoner reports...AND OTHER INTELLIGENCE THAT I CAN'T DESCRIBE FOR CERTAIN REASONS."

Corso, who may have been referring to NSA decryptions of Soviet, Chinese and North Korean encrypted communications at this point, believed that many of the American prisoners from Korea in the Soviet Union were "eliminated" after they were no longer useful.

Task Force Russia subsequently reported to the U.S.-Russian Joint Commission on POW/MIAs in November 1992:

"LTC Corso's single most dramatic source was North Korean Lt. General Pak San Yong. Pak was a Soviet colonel of Korean ethnicity who had been seconded to the North Korean People's Army and promoted to lieutenant general. He was also a member of the North Korean Communist Central Committee. Pak had been captured and disguised himself as a private but had been denounced by anti-Communist fellow prisoners. UNDER INTERROGATION HE REVEALED THAT U.S. POWs HAD BEEN SENT TO THE SOVIET UNION AND THAT THEY HAD BEEN PRIORITIZED BY SPECIALTY AND THAT HE HAD A LIST OF THOSE SPECIALTIES."[82]

Lt. Colonel Corso also had participated in U.S.-Korean and Chinese prisoner exchange negotiations prior to Little Switch and as noted previously in this book, he witnessed that release, during which the Communists had withheld 500 wounded and sick American POWs, as well as witnessing the later Big Switch exchange. In a prepared statement for his November 10, 1992 Senate testimony, Lt. Colonel Corso recalled the circumstances leading to silence of the American media about the non-returned prisoners:

"Later, during my tour as a military assignee to President Eisenhower's National Security Council-Operations Coordinating Board staff, from 1953 to 1957, I discovered that U.S. policy contributed in no small way to this strange silence. One policy document, NSC-135/3, contained such statements as these: 'Begin talks at Panmunjom in the interests of peace...The United States should not refer to the USSR as the aggressor in Korea...Agree to the term that Chinese Communists in Korea were volunteers...Agree to a Neutral Nations Repatriation Commission which includes Czechs and Poles.'

"Upon completion of 'Operation Big Switch,' the final exchange of POWs in September 1953, we concluded—based on reliable intelligence—that more than 8,000 American POW/MIAs were not returned or accounted for by the Communists. Again, I feel that our national policy contributed significantly to the silence on this grave issue. For example, NSC-68/2 said, 'The USSR, Chinese Communists and North Koreans are not co-conspirators.' And NSC-135/3 stated, 'Korea is an inconclusive operation, and continued maintenance of military operations would create the grave danger of general war.'"

Policymakers in Washington were well aware that the Communists were withholding American POWs. An August 26, 1953 State Department memorandum stated:

"The U.S. Government believes that some American and other United Nations military personnel held prisoner by the Chinese and North Korean Communist forces in Korea will not be returned to friendly control."[83]

Official U.S. statistics on American losses in the Korean War which have been released for publication report 54,246 American DEAD OF ALL CAUSES in Korea, of which 33,629 Americans were listed as battle dead (including over 8,000 non-returned U.S. POWs and POW/MIAs who were later declared dead by the PFOD process.), and 20,617 OTHER DEATHS from non-battle causes in Korea (accidents, non-combat aircraft crashes, vessel losses and disease; a suspiciously high percentage in comparison to such losses in WW II or in the later Vietnam War.[84] A statistical analysis of these figures reveals that military personnel in Korea were almost 3 times as likely to have died or disappeared from non-hostile causes as those who served in Vietnam. It may be that these statistics mask another group of American MIAs who were later administratively moved to other categories of death.

Following the Korean War (as in 1920 and 1946), and in

accordance with U.S. laws and regulations, the status of over 5,000 of the more than 8,000 officially-listed American POW/MIAs was changed to "Killed In Action," through the administrative process of "presumptive findings of death." An official U.S. Progress Reports and Statistics document of the Office of the Secretary of Defense (OSD), dated November 4, 1954, lists a total of 33,629 U.S. dead of all causes in Korea. Of this number 7,140 were carried as captured or interned POWs, of which 4,418 were returned in Operations Little and Big Switch during the spring and summer of 1953, and 2,701 were carried as died while interned as POWs (or murdered in camps and on forced-marches). Beyond these, 5,866 other Americans were carried as MIA, of whom 715 were returned to U.S. military control. Thus, by November 1954, a total of 5,233 U.S. POWs and MIAs of Korea had been returned, but 5,127 other U.S. POW/MIAs were PRESUMED by the U.S. Government to have died while missing. Many of these approximately 5,000 Americans were those who disappeared in the Manchurian Peace and Reform camps, or at other locations in China listed by CIA sources, and across the Siberian border into the Soviet Union. (Other declassified U.S. documents give slightly varying, but substantially similar figures to the above losses.) The writer provided approximately 5,000 names of these U.S. POW/MIAs of Korea, together with their home addresses of 1953, ranks and PFOD dates, to the Senate Select Committee on POW/MIA Affairs through staffer Colonel William LeGro. Many others were declared dead on the basis of reported later evidence (such as human skeletal remains recovered by U.S. Graves Registration on South Korean battlefields, and assigned to the names of missing Americans).

By a similar process, the numbers of American known-POWs, listed among the 5,000 lost POW/MIAs of Korea and publicly admitted by the U.S. to have been in Communist control but never repatriated by the Communists, was gradually reduced in succeeding years from just under 1,000 in early 1954, to 389. (5 additional sets of human skeletal remains were returned by North Korea in 1990.) In January 1954, Assistant Secretary of the Army Hugh Milton II noted one of the reasons for such reductions in the 954 U.S. POWs then admitted by the U.S. Government to have been alive in captivity but never repatriated, in a secret memorandum for the Secretary of the Army:

"...A FURTHER COMPLICATING FACTOR IN THE SITUATION IS THAT TO CONTINUE TO CARRY THIS PERSONNEL IN A 'MISSING' STATUS IS COSTING OVER ONE MILLION DOLLARS ANNUALLY. IT MAY BECOME NECESSARY AT SOME FUTURE DATE TO DROP THEM FROM OUR RECORDS AS 'MISSING AND PRESUMED DEAD.'"

(This memo and hundreds of other recently-declassified U.S. POW/MIA documents of the Korean War era in the possession of the author were turned over to the Senate Select Committee on POW/MIA Affairs, which declined to use them in it's final report. The author had previously given them to the Senate Foreign Relations Committee Republican staff in 1990-91, which used them in their May 23, 1991 final report on U.S. POW/MIA policy.)[85]

The relatively few American non-repatriates who had chosen at first to live under Communism, gradually opted to return to the United States. A former U.S. Counter Intelligence Corps agent assigned to debrief such later returned Americans, Mr. Mike McGowan, has informed the author that he was briefed by his CIC superior at the time, on the disappearance of 2,000 or more U.S. prisoners of war under Chinese Communist control in Manchuria after Operation Big Switch. This was done to assist the agents in knowing what information was critical to obtain from any Americans who had been held in China.[86] It appears that the thousands of secretly-held American POWs in China were largely kept in different locations than the "progressives," who had chosen to remain after the war.

The list of 954 Americans was subsequently reduced administratively to 944 and then lesser numbers until it reached 452. A Department of the Army cable from the JCS to Seoul, Korea, dated July 28, 1960 (declassified through FOIA request by the author in 1990), illustrates how the names of 61 more of these POW/MIAs were deleted from the then already reduced list of 452 Americans admitted to have been in communist control:

"THE DEPARTMENT OF DEFENSE, IN REDUCING THE FIGURE TO 391 EMPHASIZES THAT WHILE ALL OF THE ORIGINAL LIST OF 944 EXCEPT FOR THESE 391 HAVE NOW BEEN ACCOUNTED FOR... LARGELY...THROUGH THE EFFORTS OF U.S. GRAVES REGISTRATION UNITS AND THE U.S. INTELLIGENCE AGENCIES..."[87]

That is, largely by matching human skeletal remains, mostly located on South Korean battlefields, with the names of those American prisoners known to have been alive in North Korean or Chinese control, but not repatriated in 1953. Such action was merely a continuation of U.S. policy since 1919 toward secretly-held prisoners of the Communists.

The U.S. government however, in official statements, chose to minimize the problem and to secretly classify the number of American prisoners known to have been alive as captives, but never repatriated. Also, the military services and government officials lied to the family members of U.S. prisoners known to be held alive, telling their wives, children and parents instead that they were dead, that they had been "killed in action" and that their, "bodies had not been recovered."

A typical example of the duplicity involved in the U.S. government's handling of the POW/MIA matter during and after the

Korean War can be seen in the case of U.S. Air Force Captain Arthur Heise. Captain Heise was a World War II veteran of the China-India-Burma theater, originally from Ohio. He was married and the father of two sons, and had just turned 35 years old when he became missing in action in Korea on the 10th of January 1953. An "incident report" (AFM 200-25) from an Air Force manual containing the names of 187 Air Force men listed as "still missing in action" (which his wife was not informed of for over 30 years), and dated 16 January 1961, contains some of the details on Heise's disappearance. (This was out of a total of 1,303 U.S. Air Force personnel declared missing for all reasons from June 1950-July 1953.[88] In addition, over 5,500 U.S. Air Force non-combat deaths were reported for the Korean War period.)

Captain Arthur Heise was the Aircraft Commander of a B-29 bomber of the 372nd Bomb Squadron of the 307th Bomb Wing. The aircraft, with a crew of 13 men took off from the base at 1554 hours on the 10th of January 1953, for a bombing mission on the Anju Marshalling Yards. After a 4 1/2 hour flight and just before bombs away over the target at 7:38 PM, his B-29 was hit by fire from enemy aircraft. Heise's plane dropped their bombload on the target while still under enemy attack and then started losing altitude. They managed a "Mayday" call which was received at 7:48, and five minutes later Heise's aircraft reported they were "hit bad, have wounded men, will try to get to K-13." At 8:06 PM Heise himself radioed "We are bailing out, number-three engine on fire."

According to the Air Force report, no parachutes were observed due to limited visibility, but the aircraft was reported to have hit the ground and exploded. It was also recorded by the Air Force that all search efforts "reported negative results," with the exception of "flares and fires sighted by B-26 the night the aircraft crashed."

Seven months later, in the post-Armistice August 1953 prisoner exchange known as Operation Big Switch, the entire crew of the B-29 12 men) was repatriated by the Communists except for Captain Arthur Heise. In the official account, all of the survivors were reported by the Air Force to have "advised that they never saw or heard of subject (Captain Heise) after they bailed out of their disabled aircraft."

In the debriefings of Captain Heise's crewmembers (which were not revealed to Mrs. Jean Heise for over 30 years), several contradictory stories of his fate were reported. 1st Lt. Frederick W. Forsythe, Jr., the radar observer, stated that Captain Heise was still in the aircraft when he bailed out. 1st Lt. George F. Barmes, student radar observer and 1st Lt. Albert L. Seavers, the navigator, stated upon repatriation that a fellow prisoner in the camp, one, "Major David F. MacGhee, HAD ASKED A CHINESE INTERROGATOR ABOUT SUBJECT (Heise's) STATUS. INTERROGATOR ADVISED THAT HEISE WAS

DEAD AND THAT HIS BODY HAD BEEN FOUND IN THE AIRCRAFT WRECKAGE."

The Air Force report said further, "According to 1st Lt. Samuel E. Massenberg, pilot, subject stayed with the aircraft in attempt to reach K-13 Air Base, believing that the injured crewman, A1/C Robert K. Starkson, gunner, could not withstand bailout. Massenberg believes subject stayed with aircraft too long and went down with it. These negative sightings however, would appear to be totally nullified by the report of the flight Engineer, Master Sergeant Richard M. Kaufman who, "STATED THAT HE RECALLS SEEING THREE PARACHUTES OPENING AFTER HE CLEARED THE AIRCRAFT, WHICH WOULD INCLUDE SUBJECT (Heise).

Despite this apparently certain evidence from an eyewitness member of the crew that Captain Heise had indeed parachuted from the aircraft and not gone down with it, the Air Force chose to accept the hearsay second hand report from a Chinese communist POW interrogator, that Captain Arthur Heise was killed, and to relay this information from communist sources to his wife and children. The Air Force conclusion was:

"IT IS POSSIBLE THAT CAPTAIN HEISE MAY HAVE STAYED WITH THE AIRCRAFT TOO LONG AND THAT THE INFORMATION REPORTED BY MAJOR DAVID F. MacGEE IS CORRECT."

According to the Air Force none of the 12 surviving crewman from the plane reported that they had seen or heard of Captain Heise alive in captivity, although no interviews are recorded with 7 of the 11 crewmen, in the report.

Captain Heise's wife, Jean, wrote to the author that her husband Arthur was declared "Killed in Action" in January 1954, in accordance with regulations—because of lack of evidence to the contrary, and she was so informed. Jean Heise was not informed that the U.S. Air Force was actually carrying her husband in a category of "still missing in action." In official letters from the Department of Defense she was told that the Chinese said her husband, "was dead," and that he "died the day the plane crashed, and that, "no body had been recovered." She also recalled that each of these pieces of information came to her, "in separate letters."

Mrs. Heise, believing the U.S. government and the Air Force, that her husband had in fact been killed in action, subsequently remarried and became Mrs. Jean Heise Earl. Years later she discovered that her declared-dead husband Arthur Heise's name was on a long-classified list of 389 men known to have been alive as prisoners in enemy hands in Korea. (The list was first published in the "Spotlight" in 1979.) When brought to her attention she, "Dismissed it at the time, because I had no information indicating that he was ever a prisoner, and actually assumed he was on the list because he was 'unaccounted for,' not because he was thought to have been a prisoner."

In actuality however, the list of 389 was a scaled-down version (of originally much longer lists) of those men whom the U.S. government had definite knowledge of their having been live prisoners, in enemy captivity. That is, that they had been seen and reported alive as POWs, usually by other U.S. prisoners held with them who were fortunate enough to have later been repatriated. In Captain Heise's case, as in many others, he had obviously been held separately from his own crewmembers. The list of 389 American POWs (out of over 8,000 MIAs), on which Arthur Heise's name appears are those men of which the United States Government had no doubt that they had been alive as prisoners. Yet Mrs. Jean Heise (Ray) had never been told, or given the slightest reason to suspect that her husband was a live POW.

Thus, in a totally roundabout way, and 36 years after the fact, Mrs. Jean Heise Earl discovered that her husband had been a live POW at the end of the Korean War. Yet when she attempted to receive a POW Medal due her husband and authorized by Congress, she received a letter dated 9 May 1989 from the Department of the Air Force, signed by a Colonel Joseph P. Tencza, which stated:

"This is in response to your recent letter...concerning your deceased husband's eligibility for the Prisoner of War (POW) Medal. We regret the delay in responding to your original request but had to obtain your husband's military personnel records from the National Personnel Records Center, in Saint Louis, Missouri. WE HAVE THOROUGHLY REVIEWED YOUR HUSBAND'S RECORDS BUT COULD NOT FIND ANY DOCUMENTATION THAT VERIFIED HE WAS HELD AS A POW IN KOREA. YOUR HUSBAND'S RECORDS INDICATE THAT HE WAS KILLED IN ACTION...Under the provisions of Public Law 99-145, the POW Medal is only awarded to individuals who were taken prisoner and held captive...Individuals who were killed in action are not eligible for the award. Our determination in this matter is based on the provisions of public law and Department of Defense policy and does not negate any hardship or suffering you and your family experienced as a result of your husbands service..."

Despite this categorical Air Force denial, 9 months later Mrs. Jean Heise Ray finally got a copy of a letter dated February 8, 1990, signed by Defense Intelligence Agency Colonel Joseph A. Schlatter (Chief, Special Office For Prisoners of War and Missing in Action), with an attached list containing her husband Arthur Heise's name as one of the 389 prisoners of war. Schlatter wrote of these names:

"THE LIST INCLUDES ONLY THOSE 389 MEN THAT WERE AT ONE TIME LISTED AS PRISONERS AND FOR WHOM THE NORTH KOREAN GOVERNMENT SHOULD REASONABLY BE ABLE TO PROVIDE AN ACCOUNTING."

Mrs. Jean Heise Earl wrote to the author that "it took me 1 1/2 years to find an office which contained the information of the crash site. This finally came from the Mortuary and Casualty Affairs office in Alexandria, Virginia, complete with detailed maps, etc. Also included were on-site reports from Marine fighter pilots who flew alongside the plane and saw it crash. This was the sort of information I had been wanting for so long, and was told in one instance that records of this nature 'had been destroyed, ten years after the war.'

"One of the of the crew kept a diary. A part of it was gotten out through another crewmember during repatriation. But the rest of it was taken away from this crew member (by US officers) and never returned to him. He has been trying to recover it...It has been implied that it is still government property and perhaps still classified!" It seems to me that this is a violation of his civil rights. We are planning to have a crew reunion—those who are willing and interested, so I am in close contact with some of them. However, they are not all willing to come. Understandable, I think...I have received copies of my husbands 201 file (personnel), and Form 5, neither of these indicated the crash site, nor that he might be a POW, just MIA. I do not feel his 201 is complete, as others I have seen contain much more information, but here again, I am stopped..."

Mrs. Jean Heise (Earl) and her two sons had clearly been misled and repeatedly lied to in 1953 and 1954 and thereafter, by her husband's military superiors and the government he had been fighting to uphold, in Korea. In reality, for an as yet unknown number of years, he had been another of the thousands of abandoned living-dead Americans, in hundreds of secret gulags inside North Korea, Communist China or the Soviet Union.[89]

One report of American POWs in the USSR during this period came from a Japanese World War II POW who was repatriated from POW Camp No. 21 at Khabarovsk in December 1953. He had learned from a prison camp guard that two Americans had been brought to Khabarovsk Prison charged with being "spies." He had also been told by Soviet guards, prisoners and free citizens about 12 or 13 American prisoners from a U.S. aircraft shot down by the Soviets were in a Khabarovsk prison.; and that in 1951 an American fisherman, captured in the Gulf of Alaska, was brought to the Magadan region. He had also learned from a guard on a Soviet prisoner train at No. 2 Station in Khabarovsk that there was a special prison camp in the USSR for Americans only.[90] Yet this information and much more remained classified for decades to come.

As had happened to New York Representative Hamilton Fish a quarter-century before, members of the U.S. Congress who pressed the Department of State or the Pentagon on the "954," or later the "451" or later still the "389" American POWs admitted by the U.S. to have been held in captivity, were responded to with deliberately vague statements, intended to mislead them into thinking that no certain evidence of any known number of secretly-held POWs existed. Thus, in a series of proposed changes to official State Department answers to Senate Majority leader Lyndon B. Johnson's December 21, 1953 request for information on the "944" Americans then reported as still in Communist control, the final January 20, 1954 draft changes the official "line" from there being a precise number of 944 U.S. POWs known to be in Communist captivity to: "AMERICAN PRISONERS OF WAR WHO MIGHT STILL BE IN COMMUNIST CUSTODY."[91] With each successive draft the Department of State clerks were ordered to "retype" the words about American POWs in Communist control until a suitably vague and noncommittal response to Senator Johnson was finally achieved, by stapling on the desired words so that they covered-up the more precise wording and number.

At this same time Assistant Secretary of State Walter Stoessal attempted to deflect any interest LBJ might develop regarding Soviet-held U.S. prisoners from Korea or U.S. POW survivors of WW II:

"IT IS SUGGESTED THAT THE FOLLOWING BE INCLUDED IN A LETTER WHICH N.A. DRAFTS IN REPLY TO SENATOR JOHNSON: THE DEPARTMENT OF STATE HAS NO INFORMATION TO THE EFFECT THAT THERE ARE APPROXIMATELY SIX OR EIGHT HUNDRED AMERICAN SOLDIERS IN THE CUSTODY OF THE SOVIET GOVERNMENT. A FEW OF THE PRISONERS OF WAR OF OTHER NATIONALITIES RECENTLY RELEASED BY THE SOVIET GOVERNMENT HAVE MADE REPORTS ALLEGING THAT AMERICAN CITIZENS ARE IMPRISONED IN THE SOVIET UNION. ALL OF THESE REPORTS ARE BEING INVESTIGATED BY THIS DEPARTMENT WITH THE COOPERATION OF OTHER AGENCIES OF THE GOVERNMENT."[92]

The State Department did not possess ALL the facts, which were buried in top-secret files of the Pentagon, CIA, FBI and National Security Council. The classification and compartmentalization of the facts permitted the answering department to be evasive or lie to influential members of Congress, as had been done to Hamilton Fish of New York in 1930. Yet the Department of State constantly received livesighting reports from eyewitness escapees from the Soviet gulags who reported American POWs, sometimes by name held in captivity.

Despite such duplicity, the United States Government simultaneously continued to press the Communists for an accounting of a selected number of U.S. prisoners, gradually reduced over the years from some 5,000 to 954 and then to 389, who had been known to be alive in captivity but never released. As a signal of official American concern, an unsigned column appeared in the New York Times on January 5, 1954, entitled: THE OTHER PRISONERS, which dealt primarily with the experiences of two Americans recently released from Soviet prison camps and one from China. But buried within the article was a reference to the U.S. POWs of WW II who had disappeared in the USSR:

"All three confirm that the Soviet bloc and the Chinese Communists are holding in their jails and slave camps many foreigners, including soldiers, and civilians, women and children...according to State Department figures, the total number of Americans held by the Soviets and their European satellites exceeds 5,000...Many of these Americans, like many Europeans, were residents in the iron curtain countries caught by the Communist tide; OTHERS WERE DEPORTED FROM GERMAN WAR PRISONER CAMPS; some like (Homer) Cox were simply kidnapped."[93]

At this time the U.S. Embassy in Moscow was thoroughly penetrated by both Soviet agents working on the non-American portion of the staff and by hidden microphones in the most secure rooms of the building. After George Kennan's disastrous period as U.S. Ambassador in Moscow during the Korean War, it was decided to find another envoy whom the Soviets had long known. The new ambassador, Charles Bohlen had come to Moscow in 1953 after enduring acrimonious hearings in the Congress over his appointment by the new President, Dwight D. Eisenhower. Bohlen had been challenged by several Senators, including Joseph McCarthy, who questioned Bohlen's loyalty over his role as Roosevelt's and Truman's translator and advisor in important wartime conferences, including the crucial meetings at Yalta where the fate of POWs was settled. McCarthy called Bohlen a "security risk." Chip Bohlen's brother-in-law, Charles Thayer, the former Moscow embassy staffer, OSS officer and Voice of America director, had been one of the victims of the McCarthyites, "separated on the basis of morals charges," from the State Department, which at that time meant homosexuality.[94] McCarthy taunted Bohlen about his brother-in-law, and used the affidavit of former Soviet official Igor Bogolepov, who claimed that the NKVD regarded Bohlen as a "possible source of information" and as a "friendly diplomat" in the 1930s.[95] Bogelepov defected from the Soviets to the Nazis during the war and had later been allowed to enter America where he was employed by the CIA. and came to know John V. Grombach, the former military intelligence officer who had engineered the dismissal of Colonel Alfred McCormack and his team from the State Department in 1946-47.[96] Yet, the very nature of diplomacy and intelligence-gathering made a charge like this very difficult to prove disloyalty in the State Department's resident Soviet expert, and Bohlen had the personal support of President Eisenhower. In a tough confirmation hearing Bohlen had been confirmed, but not until he had revealed significant information about the U.S. POWs and forced repatriations of World War II, as noted earlier in this book. Bohlen went to Moscow on April 11, 1953.[97]

During construction work in 1953 on the new U.S. Embassy in Tchaikovsky Street, Bohlen admits in his memoirs that American security men were withdrawn from the site at the end of each workday "to save money" and through "carelessness." Although Bohlen attempted to belittle the importance of this incident later, the fact is that all secret discussions and ciphered messages in the new Embassy were monitored from this point on by over 40 hidden microphones or "bugs" in the most important rooms in the building, including the CIA station.[98]

Declassified US government documents clearly indicate that an eyewitness reported in early 1954 that large numbers of US prisoners of war had been transferred through Manchuria to Soviet control in Siberia during 1951. A March 23rd, 1954 US Foreign Service despatch (#1716) to the U.S. Department of State from the American Consulate General in Hong Kong states (in part):

"A RECENTLY ARRIVED GREEK (actually Polish-Russian) REFUGEE FROM MANCHURIA HAS REPORTED SEEING SEVERAL HUNDRED AMERICAN PRISONERS OF WAR BEING TRANSFERRED FROM CHINESE TRAINS TO RUSSIAN TRAINS AT MANCHOULI NEAR THE BORDER OF MANCHURIA AND SIBERIA...THE POWS WERE SEEN LATE IN 1951 AND IN THE SPRING OF 1952 BY THE INFORMANT AND A RUSSIAN FRIEND...THE INFORMANT WAS INTERROGATED ON TWO OCCASIONS BY THE ASSISTANT AIR LIAISON OFFICER AND THE CONSULATE GENERAL AGREES WITH EVALUATION OF THE INFORMATION AS PROBABLY TRUE..."

This 1954 report was generated by Lt. Colonel O'Wighten Delk Simpson, now of North Palm Beach, Florida, who was at the time the Assistant U.S. Air Force Attaché to the American Consulate General in Hong Kong. Colonel Simpson interviewed the refugee source, who was actually a Polish-Russian emigre from Communist China. The refugee was a young man who was in the process of emigrating to Australia, and asked for nothing in return for the critical intelligence he had supplied. The refugee stated to Colonel Simpson that while he was working at the Manchouli station on the Siberian border he had two times personally witnessed hundreds of American POWs being taken from trains while railroad crews prepared to change the undercarriages of the railroad cars to conform to the different Russian track gauge. He also reported that his best friend in Manchuria, a Russian railroad worker at Manchouli had seen another shipment of several hundred more American POWs transferred to Soviet control a few months later, and that he had subsequently witnessed several more shipments of Americans. (This would indicate a total of thousands of Americans transported to Siberia.)

In the March 23rd, 1954 document located in the National Archives by the author and other researchers, the above quoted first paragraph is followed by the explanation:

"The full text of the initial Air Liaison Office report follows: First report dated March 16, 1954, from Air Liaison Office Hong Kong, C2.

"This office has interviewed refugee source, who states that he observed hundreds of prisoners of war in American uniforms being sent into Siberia in late 1951 and 1952. Observations were made at Manchouli (Lupin), 49 degrees 50 '-117 degrees 30' Manchuria Road Map, AMSD 201 First Edition, on USSR-Manchurian border. Source observed POWs on railway station platform loading into trains for movement into Siberia. In railway restaurant source closely observed three POWs who were under guard and were conversing in English. (These three) POWs wore sleeve insignia which indicated POWs were Air Force noncommissioned officers. Source states that there were a great number of Negroes among POWs shipments and also states that at no time later were any POWs observed returning from Siberia. Source does not wish to be identified for fear of reprisals against friends in Manchuria, however is willing to cooperate in answering further questions and will be available Hong Kong for questioning for the next four days."

The March 23, 1954 despatch also recorded the official response from Washington to the first March 16 cable:

"Upon receipt of this information, USAF, Washington requested elaboration of the following points:

1. Description of Uniforms and clothing worn by POWs including ornaments.

2. Physical condition of POWs.

3. Nationality of guards.

4. Specific dates of observations.

5. Presence of Russians in uniform or civilian clothing accompanying movement of POWs.

6. Complete description of three POWs specifically mentioned.

"The Air Liaison Office complied by submitting the telegram quoted below.

"FROM USAIRLO SGN. LACKEY. CITE C 4. REUR 53737 following answers submitted to seven questions.

1. POW wore OD (olive drab) clothing described as not heavy inasmuch as weather considered early spring. Source identified from pictures service jacket, field, M1943. No belongings except canteen. No ornaments observed.

2. Condition good, no wounded all ambulatory.

3. Station divided into two sections with tracks on each side of loading platform. On Chinese side POWs accompanied by Chinese guards. POWs passed through gate bisecting platform to Russian train manned and operated by Russians. Russian trainmen wore dark blue or black tunic with silver colored shoulder boards. Source says

this regular train uniform but he knows the trainmen are military and wearing regular train uniforms.

4. Interrogation with aid of more fluent interpreter reveals source first observed POWs in railroad station in spring 1951. Second observation was outside city of Manchouli about three months later with POW train headed towards station where he observed POW transfer.

SOURCE WAS IMPRESSED WITH SECOND OBSERVATION BECAUSE OF LARGE NUMBER OF NEGROES AMONG POWs. SOURCE STATES THAT HE WAS TOLD BY A VERY CLOSE RUSSIAN FRIEND WHOSE JOB WAS NUMBERING RAILROAD CARS AT MANCHOULI EVERY TIME SUBSEQUENT POW SHIPMENTS PASSED THROUGH MANCHOULI. SOURCE SAYS THESE SHIPMENTS WERE REPORTED OFTEN AND OCCURRED WHEN UNITED NATION FORCES IN KOREA WERE ON THE OFFENSIVE."

5. Unknown.

6. Only Russians accompanying POWs were those who manned train.

7. Three POWs observed in station restaurant appeared to be 30 or 35. Source identified Air Force noncommissioned officer sleeve insignia of Staff Sergeant rank, stated that several inches above insignia there was a propeller but says that all three did not have propeller. Three POWs accompanied by Chinese guard. POWs appeared thin but in good health and spirits, were being given what source described as good food. POWs were talking in English but did not converse with guard. FURTHER INFORMATION AS TO NUMBER OF POWs OBSERVED STATES THAT FIRST OBSERVATION FILLED A SEVEN PASSENGER CAR TRAIN AND SECOND OBSERVATION ABOUT THE SAME. Source continues to emphasize the number of Negro troops, which evidently impressed him because he had seen so few Negroes before. Source further states that his Russian railroad worker friend was attempting to obtain a visa to Canada and that he could furnish more information. The railroad worker's name is Leon Strelnikov whose mother's sister lives in Canada and is applying for a visa for Strelnikov (phonetic). Comment reporting officer: Source is very careful not to exaggerate information and is positive of identification of American POWs. In view of information contained in Charity Interrogation Report No. 619 dated 5 February 1954, Reporting Officer gives above information rating of F-2. Source departing Hong Kong today by ship. Future address on file this office."

"In this connection the Department's attention is called to Charity Interrogation Report No. 619, forwarded to the Department under cover of a letter dated March 1, 1954, to Mr. A. Sabin Chase, DRF. Section 6 of this report states:

"ON ANOTHER OCCASION SOURCE SAW SEVERAL COACHES FULL OF EUROPEANS WHO WERE ALSO TAKEN TO USSR. THEY WERE NOT

RUSSIAN. SOURCE PASSED THE COACHES SEVERAL TIMES AND HEARD THEM TALK IN A LANGUAGE UNKNOWN TO HIM."          (signed)
>    Julian F. Harrington
>    American Consul General
cc:    Taipei
>    Moscow
>    London
>    Paris[99]

Two shipments of seven carloads or coaches each of American POWs are definitely recorded in this sighting report, along with at least several more such shipments about which less definite information is known. Considering that railroad coaches could carry from 50 to 90 passengers each, depending on how jammed in the prisoners were, 14 carloads represents anywhere from 700 to 1,260 American prisoners. The other, less precisely reported shipments represent at least several hundred to possibly several thousand more American POWs. The other report of Europeans being shipped into the USSR during this same period may have referred to other United Nations POWs or to captive European civilians, or both. It is therefore clear from this combined report that the U.S. government had gained additional knowledge by early 1954 that thousands of Americans and other U.N. personnel were being transported from Communist China into Russia in 1950-1951, and that none were ever seen returning. This report merely substantiated earlier indications which had already been received by U.S. intelligence, that such actions were being initiated by the CCF and the Soviets. "Charity report 619" for instance, referred to another eyewitness account by a Turkish traveler who had been in the same area where he:

"...saw several coaches full of Europeans who were also taken to the USSR. They were not Russians Source passed the coaches several times and heard them talk in a language unknown to him." The source stated that one of the coaches was full of wounded Caucasians who were not speaking at all."[100]

The wide distribution accorded the March 1954 State Department cable, by sending it to the U.S. Embassies in the Soviet Union, Great Britain, France and Nationalist China ensured that some sort of follow-up would occur. In addition, on the original document found in the National Archives it is recorded that after March 30, 1954, 7 more copies went to the Central Intelligence Agency and 6 were distributed within the State Department. Accordingly, this particular despatch proves that it was widely known at the highest official levels within the U.S. government that large numbers of U.S. Korean War POWs were being shipped into the Soviet Union as captives.

It is apparent that this particular report proved to be one of the final pieces of evidence required by the U.S. government and caused an official demand for the return of Soviet-held American POWs. Lt. Colonel Philip Corso, on the national security Council staff at the time, had estimated from many other intelligence reports of eyewitnesses that at least two trainloads of U.S. POWs and possibly three, totaling 900-1350 POWs "were transferred from Chinese to Soviet custody at the rail transshipment point of Manchouli on the Manchurian-Chita Oblast border of China and the Soviet Union." He estimated that each trainload of a maximum of 450 POWs, for a total of 900-1350 U.S. POWs transported into Siberia from Manchouli. These were not necessarily the same shipments reported by Lt. Col. Simpson's source, and they did not include at least 300 and possibly more than 600 other U.S. POWs shipped into the USSR through Posyet near Vladivostok, and from thence to Molotov (now Perm) in February 1952 under heavy MVD guard, or those transported by aircraft and by sea. An unknown number of American prisoners had been moved by ship to Siberian ports in the summer months.

According to Zygmunt Nagorski, cited previously in this book, who obtained information from two members of the MVD and an employee of the Trans Siberian Railroad, in addition to an extensive 'source network' of truck drivers and other working class Soviets employed at or near prisons in Molotov, Khabarovsk, Chita, Omsk, Chermoz and elsewhere, Pos'yet was also a U.S. POW transit point, when ice closed the Pacific coast and the Tatar Straits between November 1951 and April 1952. Prior to this, from August to November 1951 other U.S. POWs had been moved from Chita, Siberia to Arkhangelsk (Archangel) Oblast to camps at Kotlas on the northern Dvina River (where American POWs had been captured in 1919), and to Lalsk. In March of 1952, POWs passed through Khabarovsk and Chita about every two weeks in small groups of up to 50 men...From December 1951 through the end of April 1952, trains of U.S. and European (probably British) POWs passed at intervals into Komi-Permysk National District to Molotov, Gubakha, Kudymkar and Chermoz. In April 1952 a number of U.S. officer POWs, referred to informally as the 'American General Staff, were kept under strict isolation in Molotov (Perm). In the town of Gubakha and in the industrial regions of Kudymkar and Chermoz, there were three isolated camps and one interrogation prison for U.S. POWs. At a camp called Gaysk about 200 (American) POWs were kept and forced to work in workshops assembling rails and doing various technical jobs. These camps were completely isolated."[101]

Faced with the overwhelming intelligence of many U.S. POWs transported to the USSR from Korea, Secretary of State John Foster Dulles, on April 19th, 1954, instructed the United States Embassy in Moscow, under Ambassador Charles Bohlen: "In your discussions with (Soviet) Foreign Office you may desire inform Soviets without revealing sources that we have reliable accounts transfers of POWs Manchouli." This was followed by a note from the American Embassy (found in the National Archives with the March 23 despatch from Hong Kong) which formally demanded the return of Korean War prisoners from the USSR on May 5th, 1954:

"The Embassy of the United States of America presents its compliments to the Ministry of Foreign Affairs of the Union of Soviet Socialist Republics and has the honor to request the Ministry's assistance in the following matter:"

"THE UNITED STATES GOVERNMENT HAS RECENTLY RECEIVED REPORTS WHICH SUPPORT EARLIER INDICATIONS THAT AMERICAN PRISONERS OF WAR WHO HAD SEEN ACTION IN KOREA HAVE BEEN TRANSPORTED TO THE UNION OF SOVIET SOCIALIST REPUBLICS AND THAT THEY ARE NOW IN SOVIET CUSTODY. THE UNITED STATES GOVERNMENT DESIRES TO RECEIVE URGENTLY ALL INFORMATION AVAILABLE TO THE SOVIET GOVERNMENT CONCERNING THESE AMERICAN PERSONNEL AND TO ARRANGE THEIR REPATRIATION AT THE EARLIEST POSSIBLE TIME."[102]

On May 12th, 1954 the Soviet Foreign Ministry, still under Molotov despite the post-Stalin "thaw," denied holding American prisoners from Korea and the message reached the United States on the 13th:

"In connection with the note of the Embassy of the United States of America, received by the Ministry of Foreign Affairs of the Union of Soviet Socialist Republics on May 5, 1954, the Ministry has the honor to state the following:

"THE UNITED STATES GOVERNMENT'S ASSERTION CONTAINED IN THE INDICATED NOTE THAT AMERICAN PRISONERS OF WAR WHO PARTICIPATED IN MILITARY ACTIONS IN KOREA HAVE ALLEGEDLY BEEN TRANSFERRED TO THE SOVIET UNION AND AT THE PRESENT TIME ARE BEING KEPT UNDER SOVIET GUARD IS DEVOID OF ANY FOUNDATION WHATSOEVER AND IS CLEARLY FAR-FETCHED, SINCE THERE ARE NOT AND HAVE NOT BEEN ANY SUCH PERSONS IN THE SOVIET UNION."[103]

A US State Department spokesman that day in Washington confirmed the exchange but said he did not know who the prisoners were or how many there were. Thus, the new post-Stalin Soviet government, led by Malenkov, Nikita Khrushchev and others who had deposed and had then killed the NKVD/MVD Chief Lavrenti Beria, still denied holding "American" POWs, "under Soviet guard." The power of the KGB overrode any desire by the new Soviet leaders to fundamentally improve relations with the United States.

In view of this declassified report of large numbers of Air Force enlisted personnel being shipped to the USSR from China, the high ratio of 20,617 reported U.S. non-battle deaths in Korea to 33,629 U.S. combat dead (including the 8,000 officially-listed missing), may be significant. For example, non-combat Air force dead in Korea (5,884) were approximately 5 times greater than the combat dead (1,200), while in Vietnam non-combat Air Force dead (766) were less than 1/2 the total of combat dead (1,738). Thus, the possibility exists that many Americans listed as dead of non-combat causes may have originally been missing in action when U.S. rear areas were overrun several times in 1950 and their names later moved to other categories. Others remained missing from shootdowns of U.S. aircraft engaged in secret bombing missions over Manchuria. It is also possible that when secret Soviet (or other Communist) files become fully accessible to U.S. researchers, that many of the American names in KGB dossiers on Soviet-held U.S. prisoners may not be on the official POW, POW/MIA, or presumed dead-in-captivity/ or while missing lists. When researchers pointed out the extremely large number of reported non-battle dead for the Korean War, the Pentagon announced that "recent research" had determined that previous totals of some 20,000 "other deaths," beyond the 33,629 battle dead and 2,789 "other dead," which had previously been listed for Korea had included 4,043 Navy, 5,884 Air Force, 9,429 Army and 1,261 Marine personnel who had died "WORLDWIDE" during the Korean War era, and not exclusively in Korea. No explanation is given as to where else in the world, and under what circumstances, nearly 20,000 Americans could have died during the Korean War era.[104]

A March 1954 letter concerning news reports of Communist-held American POWs, written to Massachusetts Senator Leverett Saltonstall by Assistant Secretary of State Thruston Morton, at the direction of John Foster Dulles, contrasts sharply with the definite information on Korean War POWs in Soviet control contained in the secret U.S. diplomatic communications noted previously:

"Secretary Dulles has asked me to reply to your communication of February 19, acknowledged by telephone on February 23, enclosing a letter from Mrs. Ruth King of Windsor, Massachusetts. Mrs. King writes concerning an article in "U.S. News and World Report" regarding American civilians and military personnel who MAY BE in Communist custody. The Department does not overlook any reports that American citizens abroad are in difficulties of any kind. In each case it makes a careful investigation into the circumstances in order to determine what action can and should be taken. We share your constituent's concern for those Americans who MAY BE imprisoned in Communist countries, and for those individuals with legitimate claims to American citizenship who, while not actually imprisoned, are not permitted to leave the countries in which they presently reside. It should be pointed out however, that most of the possible American citizens in Eastern Europe are dual nationals whose claims to American citizenship are in many cases not recognized by the governments of the countries in which they are now residing..."[105]

The soothing tone of this deliberately misleading official document, which diverted attention away from the U.S. military POWs and towards American dual citizens, made it seem as if the thousands of kidnapped American prisoners from Korea then still being held in Chinese and Soviet prisons had disappeared on some holiday excursion while "abroad." An explanation of why Senator Saltonstall and many other members of Congress over the years accepted such obviously untrue and incomplete answers from federal agencies responding to their questions about such classified "national security issues" was recorded by a former CIA official-turned author, Victor Marchetti, two decades later:

"Senator Leverett Saltonstall, who served for many years as ranking Republican on the (intelligence) oversight subcommittee, expressed the same view publicly in 1966: "IT IS NOT A QUESTION OF RELUCTANCE ON THE PART OF CIA OFFICIALS TO SPEAK TO US. INSTEAD IT IS A QUESTION OF OUR RELUCTANCE, IF YOU WILL, TO SEEK INFORMATION AND KNOWLEDGE ON SUBJECTS WHICH I PERSONALLY, AS A MEMBER OF CONGRESS AND AS A CITIZEN, WOULD RATHER NOT HAVE."[106]

In sworn testimony before the Senate Select Committee on POW/MIA Affairs nearly four decades later, retired U.S. Army Lt. Colonel Phillip J. Corso shed much light on the making of a policy decision by President Dwight D. Eisenhower in early 1954 to abandon the American POWs from Korea who had been transported into the USSR from Manchuria:

"Upon my return to the United States, at the request of Mr. C. D. Jackson, Special Assistant to the President, I was assigned to the Operations Coordinating Board (OCB) (of the National Security Council-NSC). My initial duties were to do the basic research and prepare data for Dr. Charles Mayo in Committee I at the United Nations on the subject of 'Bacteriological Warfare' and 'Questions of Atrocities Committed by the North Korean and Chinese Communist Forces Against United Nations Prisoners of War in Korea,' delivered by Ambassador Henry Cabot Lodge. "My findings revealed that the Chinese Communists with their allies, the North Koreans and Soviet Union, carried on a detailed scientific process aimed at molding war prisoners into forms in which they could be exploited. Prisoners who underwent the experiences and later returned to their own countries reported that the experts assigned to wearing down prisoners were highly trained, efficient and well educated. They were, in short, specialists in applying psychological rather than physical torment. These experts were trained in highly specialized schools in the Soviet Union and Communist China. Their methods of eliciting individual compliance were the same in eastern Europe and the Soviet Union as they were in China and Korea. During my tour of duty with the Operations Coordinating Board, virtually all projects pertaining to prisoners of war and the Soviet ideological threat were assigned to me. I did not agree with the items in our present Code of Conduct (for military personnel), especially those portions which instructed a soldier on how to act when he became a prisoner of war, mainly because they were defensive and did not fully consider the nature of the Communist captors and the methods they used to extract confessions or elicit individual compliance."

"Some of the returned U.S. prisoners of war that I interviewed stated 'they had been taken to Mukden, Manchuria.' This was in itself a violation of the so-called neutral status of Communist China. MORE REVEALING WAS THE INFORMATION THAT THE U.S. PRISONERS OF WAR HAD BEEN SENT TO THE SOVIET UNION; NONE OF WHICH EVER RETURNED. This information came to me from various sources. It was declassified and placed in one of Ambassador Lodge's speeches. My information was derived from two main sources: Chinese prisoners of war and a high-level Soviet defector. The Chinese prisoners stated that they had seen and talked to U.S. prisoners of war at Manchou-li, Manchuria. Here the passengers had to be transferred from Chinese to Soviet trains because of the difference in the gage of the tracks. The (Soviet) defector stated to me that Soviet Embassy personnel coming from the Soviet Union had told him there were Korean War U.S. prisoners of war in the Soviet Union...The defector who confirmed that American POWs had been taken to the Soviet Union...was Yuri Alexandrovitch Rastvorov, a Russian KGB officer who defected to the United States from the Soviet Embassy in Japan in January 1954.

"My superior, C.D. Jackson, prevailed upon CIA Director Allen Dulles to deliver Rastvorov to me. U.S. Air Force Brigadier General Dale O. Smith, on a one-year assignment to the NSC, also sat in on the interview. RASTVOROV REVEALED HE HAD PERSONALLY SEEN U.S. POWs HEADING INTO SIBERIA AFTER CHANGING TRAINS AT THE MANCHURIA-RUSSIA BORDER. Rastvorov told me he believed that selected U.S. soldiers—especially those with no roots or family ties—were taken to Russia as part of a joint KGB-GRU operation to be debriefed on details of their backgrounds. Then, Russian nationals would be able to assume the identities of these POWs and be 'played back' at appropriate times to carry out espionage activities in the U.S. and Canada.

"I submitted my report on my interrogation of Rastvorov to C.D. Jackson and President Eisenhower. I recommended that the report not be made public because the POWs should be given up for being dead since we knew the Soviets would never relinquish them. Out of concern for the POWs families, the President agreed."[107]

The author and research colleague Ashworth met with Lt. Colonel Corso in Washington prior to his testifying on the Soviet-held American POWs, who confirmed his sworn statement on the numbers American POWs he knew had been shipped into the USSR:

"I knew about 800 or 900 for sure, but the number might have been as many as 1,200. My intelligence reports had referred to three trainloads, and each train could hold up to 450 people." In his prepared testimony Corso added:

"The chief reason I ultimately was forced to rely on Rastvorov's knowledge to confirm my intelligence reports about American POWs being taken into the USSR is that we were hard pressed during the Korean war to be able to confirm exactly what was happening inside Manchuria or Communist China. THE RESPONSIBILITY AND JURISDICTION FOR RUNNING AGENTS INSIDE COMMUNIST TERRITORIES BELONGED TO A UNIT CALLED THE OFFICE OF POLICY COORDINATION." (then under Frank Wisner). The OPC, a covert action organization reporting to the U.S. State and Defense Departments but funded and staffed by CIA, was created in 1948. (The CIA now claims that most of the records of OPC cannot be found.) Once again, U.S. policy hindered such intelligence-gathering activities. The State Department, for instance, had put pressure on our intelligence sources in Taipei and Hong Kong to refrain from reporting anything that might show Peking's preparedness to enter the Korean War and thus support General MacArthur's position. Policy makers in Washington so strongly pressed the notion of the likelihood of a general war that they inserted in NSC-11335/3 the declaration, 'Aid to anti-communist forces in China would increase the danger of general war.' To be productive, American intelligence efforts on the Chinese mainland would have had to elicit the cooperation of the anti-communists. But policy forbade any such working relationship with our friends there. Prohibited from contacting Chinese anticommunist, no American agents were able to operate on the mainland. Policy was deliberately designed to dry up any intelligence and curtail or agents' activities on the Chinese mainland."[108]

Testifying under oath on national television before the Senate Select Committee on POW/MIA Affairs on November 10, 1992, Lt. Colonel Corso, was then 77 years of age and alert and dignified in his bearing, but he expressed great sadness about the secret burden he had carried for forty years. Colonel Corso described his meeting with President Dwight D. Eisenhower in early 1954 while he was serving on that President's National Security Council Staff:

"I had a call from my principal, C.D. Jackson, one day, who was Special Assistant to the President. He said, get over, we have to go see the President. Bring your prisoner of war report. My prisoner of war report that I handed him was one page. I walked in the office. The President was in the Oval Office, the three of us, and I saw him, and he said, I UNDERSTAND YOU HAVE A REPORT ON PRISONERS OF

WAR GOING TO THE SOVIET UNION? I told him, yes, that's what I'm here for.

"Now I compiled this report not only here but information in Korea, that close to 1,200 that we suspect, but about 900 certainly, did go there. Our information is solid, as solid as intelligence information can be, because that's the nature of intelligence.

"I handed (President Eisenhower) the report, and he read it. And he had a very serious look on his face...This was not a pleasant meeting. It did not last long....He said, 'COLONEL, DO YOU HAVE ANY RECOMMENDATIONS?'...Because in the military, generally the writer of the report has to make a recommendation to his superior who then decides what to do with it. I said, yes. The nature of this report-these men will never come back alive because they will get in the hands of the KGB who will use them for their purposes. Espionage, play-backs or whatever. This is not uncommon in the intelligence business. Once they fall in their hands, there's little hope of them coming back.

"And I told him, Mr. President, you are aware of the system of the KGB, how they use prisoners of war and defectors? And he said, YES, I AM. He said, IS YOUR RECOMMENDATION NOT TO MAKE IT PUBLIC? I said, my recommendation is not to make public the part-the KGB operation. It's difficult to understand at its best. It hasn't been revealed. The part on prisoners, that I don't know. So the President said, WELL, I ACCEPT YOUR RECOMMENDATION. He said, WELL, I AGREE, WE CANNOT GIVE IT TO THE FAMILIES. Then I said, Mr. President, though, May I send a copy of this report to the Department of Defense? He said, YES."

Later interviewed and videotaped about these experiences by the U.S. Army's "Task Force Russia," Lt. Colonel Corso revealed that fear of a "general war" with the Soviet Union that would have caused millions of casualties, was foremost in President Eisenhower's mind. A later U.S. Government report of the interviews with Lt. Colonel Corso stated:

"According to Colonel Corso, President Eisenhower did not press the POW issue to the hilt because he feared that it could have precipitated general war. Eisenhower feared 8,000,000 American dead if war occurred at this time."[109]

Faced with Soviet, Chinese Communist and North Korean refusal to return thousands of American prisoners, and having decided to shroud this enemy intransigence in official secrecy, the Eisenhower Administration chose to publicly concentrate on the conduct of a relatively small percentage of U.S. POWs who had totally succumbed to Communist "brainwashing" in the camps, including those 21 Americans who refused repatriation. Between 1950 and 1955, nearly 4,000 case histories on returned American POWs had been assembled by U.S. Army Intelligence, some of the individual files being as thick as a city telephone book.[110] After much soul searching and public lamentation about the behavior of American POWs in Korea, on August 17, 1955, Eisenhower authorized a new Code of Conduct for members of the armed forces of the United States, which every service person was sworn to uphold from then on, containing six articles:

"I am an American fighting man. I serve in the forces which guard my country and our way of life. I am prepared to give my life in their defense.

II I will never surrender of my own free will. If in command I will never surrender my men while they have the will to resist.

III If I am captured I will continue to resist by all means available. I will make every effort to escape and aid others to escape. I will accept neither parole nor special favors from the enemy.

IV If I become a prisoner of war I will keep faith with my fellow prisoners. I will give no information or take part in any action which might be harmful to my comrades. If I am senior, I will take command. If not, I will obey the lawful orders of those appointed over me and will back them up in every way.

V When questioned, should I become a prisoner of war, I am bound to give only name, rank, service number, and date of birth. I will evade answering further questions to the utmost of my ability. I will make no oral or written statements disloyal to my country and its allies or harmful to their cause.

VI I will never forget that I am an American fighting man, responsible for my actions, and dedicated to the principals which made my country free. I will trust in my God and in the United States of America."

One of those officials involved in promoting this code of conduct for military personnel was President Eisenhower's Assistant Secretary of the Army, Hugh Milton II.[111] Milton was well aware that U.S. POWs had been abandoned alive by the U.S. Government in Communist control, as was the President. Thus, while the highest officials of the U.S. Government departments and agencies did virtually nothing about thousands more Americans held secretly in Korea, China and the USSR, they demanded irreproachable conduct from U.S. servicemen, on pain of imprisonment and disgrace.

Declassified U.S. 1954 documents claimed that, "A world-wide collection effort" for information on the missing U.S. POWs of Korea was being initiated, and that the British were asked by the U.S. to help negotiate with Chinese communists; but a U.S. 1956 document declassified by INSCOM (a component of NSA) states that, "No real effort has been made toward the collection of information of an intelligence nature regarding prisoners of war or missing personnel." A declassified secret 1957 U.S. Air Force document however, reveals the July 1954 classified collection effort titled "Project American," which conducted secret research into the fate of non-returned U.S. prisoners, and concluded:

"Through information collected from repatriated U.S. and U.N. prisoners of war, Japanese repatriates (from Russia), foreign refugees, and numerous intelligence reports, A STRONG POSSIBILITY EMERGED THAT A LARGE NUMBER OF 'MISSING IN ACTION' MAY STILL BE ALIVE, AND INTERNED IN COMMUNIST PRISON CAMPS." But the U.S. Air Force has told researchers in recent years that the classified files on "Project American" cannot be located.[112]

A now-declassified Department of State memo dated July 26, 1956, obtained by the author through FOIA request, states that the Department of Defense's 'Prisoner of War Officer,' a Colonel Monroe:

"...Is leaving for a three weeks trip to Japan and Korea... TAKING WITH HIM SOME NEW MATERIAL DEVELOPED ON MISSING AIR FORCE PERSONNEL AND WOULD CONSULT WITH THE UNITED STATES ELEMENT...CONCERNING THE PRESENTATION THEY HAVE BEEN INSTRUCTED TO MAKE TO THE COMMUNIST SIDE."[113]

A British ex-POW, Private Frank Kelly, who had been a British airborne infantry medic captured at Arnhem, Holland in 1944 and interned at Stalag IV-B Muhlberg, was suddenly returned by the Soviets in November 1953. He had left the Soviet-controlled camp in May 1945 and after months of hiding in the Soviet zone, afraid of trying to escape to Allied lines, he had been arrested by the NKVD and held incognito in Soviet-run prisons in Germany, in Moscow's Lubyanka, and in Vladimir Prison, tortured and charged with espionage, but almost always without contact with any other British or American POWs. It appears the Soviets had kept him segregated for deniability purposes and for the eventual use he was put to. On one occasion in Vladimir he was able to communicate with a U.S. POW named Sergeant Robert Cummings, who had been arrested in Czechoslovakia, and like Kelly, was subsequently transported to the USSR but held entirely alone. Upon Kelly's release by the Soviets he was court-martialed by the British government for alleged desertion in 1945, but allowed to return to his home.

Between late 1953 and 1957 a handful of American and British prisoners were released by the Chinese Communists and the Soviets. In June 1954 the Chinese Communists had suddenly admitted they were still holding 15 American POW fliers and one Canadian, who Peking Radio said later in November had been sentenced for "spying." After months of negotiations, the 15 token U.S. POWs were released by the Chinese by August 1955. The Chinese Communists had thus forced the United States to deal with them on a diplomatic level. Several of these ex-POWs, including Korean War Canadian flier Andrew Mackenzie and U.S. ex-POW Steve Kiba revealed upon their release that other American POWs were still being held captive. Kiba later testified before a Senate committee that he had seen approximately 15 other Caucasian prisoners he was certain were Americans who did not return from captivity, including one, Lieutenant Paul Van Voorhis, who had been a member of Kiba's own B-29 crew. The Chinese Communists claimed that Van Voorhis had been killed in the crash of the aircraft, but Kiba had seen him alive in captivity in China. Upon his return home to the United States, Kiba was told by U.S. Air Force Intelligence, the CIA and the Department of State to forget about the POWs he had seen, that he 'hadn't seen them' and he should not discuss the subject further, in public.[114] Another of the returned POWs from China, Captain Elmer Llewelyn, had seen the names of U.S. prisoners on walls, who were still listed as missing. The Chinese admitted holding other Caucasian prisoners whom they labeled as spies, and a few more were released, but the vast majority of the thousands of U.S. and other U.N. prisoners of war who had vanished into China and the Soviet Union were to disappear.

After the mid-1950s a few of the American civilian prisoners released by the Soviets, such as John Noble, revealed to U.S. officials that other gulag inmates of the Ukrainian Bandera underground had reported up to 3,000 American POWs of Korea, the Cold War and WW II still surviving in special camps.[115] Following his release in Moscow on January 5, 1955, John Noble had explained to U.S. News and World Report magazine, in a report published on January 21, the reason given by the communists for his (and his father's) 1945 arrest in Germany:

"The reason for the arrest of my father and I was for allowing American soldiers to come over to our property in east Germany and having a stock of American food products in our house."

When asked by the reporter where he had gotten the American food, Noble replied:

"American officers and soldiers brought it along when they came over to support American prisoners-of-war who had come to us... I was arrested in July of 1945. After 14 months of being under arrest I had my first interrogation...I was sentenced to 15 years...but the actual reason for sentencing was not given to me...on the 11th of August 1950..."

(U.S. News and World report printed nothing about these American prisoners of war inside the Soviet zone of Germany in July 1945, or their subsequent fate.)

Other excerpts from Noble's interview revealed conditions, numbers and populations of Soviet forced labor camps in the Vorkuta area of the Arctic north, and his memory of the 1953 prisoners revolt at Vorkuta. He also mentioned he had information on other American prisoners:

"In Vorkuta there are approximately one half million prisoners, divided into camps...200 to 250 camps...Most of them were Soviets, but not Russians. Most of them, I'd say, were Ukrainians...Quite a few Poles and Czechs...There was quite some trouble up there. Also quite a lot of blood was shed...the end of July, beginning of August 1953. The workers had gone on strike, not at all the camps but at a great many of the camps approximately 80,000 to 100,000 prisoners were on strike. In tent number 29 the guards opened fire immediately. After the fire raid, 100 to 110 men were killed at the place. Later on, 50 or 60 died and 400 to 500 were wounded. All kinds (of weapons) were used...Also heavy machine guns. There were about 1,200 guards around the camp. They were special guards sent up there to break up the strike. We were not sure about the reasons (for the strike). It was the result of Stalin's death and the disappearance of Beria. (Lavrenti Beria, former head of the Soviet secret police, who was arrested and executed.) It was organized..."

The American POWs of WW II and Korea held by the Soviets were largely held in secret camps outside the system of constant prisoner transfers of the regular Gulag Administration. Only rarely did American civilian prisoners such as John Noble come into contact with other American civilians or hear of the secretly segregated U.S. military POWs taken in wartime.

"I only saw two (Americans), one, William Warchuk (Marchuk) who returned with me and the other, William Verdine. Verdine was not in the best of shape. He is slim built. Not that he has any sickness of any kind, but he's very weak. (Another) one was Chapman. I don't know his first name. He was up in Camp Department Number 10...In Vorkuta.

(He was there) approximately three years, three and one half years, something like that. His age was about 35 years."

When asked if Chapman was a soldier Noble replied:

"I cannot say...he was physically in fair shape. I only saw him several times because I was in his tent very little.

Noble was then asked if there were many Germans in these camps, to which he replied:

"Yes, there are a great deal. It's hard to say—several thousand." When asked if these camps were mainly for foreigners, he replied:

"No. They were Russian camps. At first when we came, they were partly for political prisoners and partly for criminals; but, later on, more and more the criminals were taken out of the camps...The living conditions were very poor...there was one Frenchman. Rene Feret was his name, I guess."

Noble was then asked whether he had heard of other Americans who were interned and he answered:

"I have heard of some but not there names and not the exact place they stayed. I heard of some pilots who were shot down over the Baltic Sea as well as some Koreans or some American soldiers from Korea."[116]

In his subsequent memoirs, *I Was A Slave In Russia*, published in 1958, Noble again revealed his knowledge of American prisoners from reconnaissance aircraft shoot-downs and POWs from Korea transferred to the Soviet Union:

"'I spoke with eight of your countrymen,' the Yugoslav told me. 'They said they were American fliers who had been shot down by the Russians over the Baltic Sea...One of them told me he was afraid they would never get back to America. The Russians had reported them dead, saying there were no survivors of the crash.'

"Prisoners being funneled into Vorkuta from camps in Tadzhik(istan) and Irutsk in Soviet Asia, Omsk in Siberia, and Magadan in the Far East said that there were many Americans, including veterans of the Korean War, both GIs and officers, and South Korean soldiers, working as slave laborers in their camps. From what I heard, they were PWs captured by the Communist Chinese and North Koreans who had been shipped to the Soviet Union for safekeeping."[117]

Years later John Noble reported that the U.S. State Department had requested that he not discuss publicly the large numbers of American POWs of the Korean War being held in the Soviet Union, reported to him by other Gulag inmates. Mr. Noble confirmed this fact in a conversation with the author in June 1990.

U.S. documents recently declassified (in 1991) by an Army intelligence component of the National Security Agency, confirm that Mr. Noble did indeed report the presence of the American POWs in the USSR. Other examples of recently declassified (1990-1991) U.S. documents released by USAINSCOM-NSA, reveal first-hand reports of Korean war POWs alive in USSR labor camps, such as those American POWs at KIROV-(9 U.S. pilots from Korea in January 1955); in another camp at NAIZURA-(American POWs from Korea just arrived in USSR); or at KHABAROVSK-(12-13 U.S. intelligence aircraft POWs, 1953-54). Additional locations reported by first-hand sources of American POWs of Korea (and earlier) in Soviet camps in the 1950s include: OMSK, CHELYABINSK, VERCHNI-URALSK, TAISHET, VORKUTA, TOMAR(gishi)-SAKHALIN, GORKI, KARAGANDA, NARINSK-(lower Yenisei), and NORILSK.[118]

Other declassified livesightings by returning European prisoners, of U.S. POWs from WW II, the Cold War and Korea held in Soviet gulag prison camps, exist in the American National Archives-and many more remain classified, but declassified U.S. State Department documents indicate an official U.S. position that American POWs could not be proven to be still alive from reports by such one-time sources, who were of unknown reliability. Gulag survivors have stated that Soviet citizenship and Russian names were assigned to some American and Allied prisoners held in secret Soviet POW camps. The existence of at least six secret POW camps in the USSR was revealed by Israeli Soviet expert Avraham Shifrin, 3 of them near the Chinese-Soviet border east of Chita, Siberia, one of which was at Nerchinsky Zavod, and 3 more camps on Wrangel Island, where secret experiments were carried out on POWs; and an escaped Romanian POW (since 1945), George Risiou, has reported 900 U.S. military POWs held in another secret Soviet camp in 1975. Russian names were reportedly assigned to American, British, French and other POWs held there, for deniability. William Robertson, a U.S. WW II POW reported in a declassified State Department document as taken (with all the other occupants) from a German POW camp near Dresden to the USSR in 1945 and held in International Camp 6062/X111-Kiev, stated in 1955 to a German POW who later reached the west that, "NO ONE WOULD BELIEVE THAT HE WAS AN AMERICAN."

As noted previously in this book, Avraham Shifrin supplied much information to the U.S. Senate (in 1973) about European and American prisoners in secret Gulag camps, held for their technological expertise, general intelligence purposes and as forced laborers. Following the printing of the first edition of this book by the author in 1992, and delivery of a copy to Colonel Stuart Herrington of the U.S. Army's Task Force Russia, that agency interviewed the former Red Army Major Avraham Shifrin in Jerusalem about the secret shipment of American Air Force POWs from Korea to the USSR, for technological exploitation. A subsequent report of that interview stated:

"The most specific comments by former Soviet officers concerning the transfer of F-86s and their pilots to the USSR were those made by former Major Avraham Shifrin, at that time a lawyer in the Ministry for Military Production. Shifrin discussed his relationship with renowned aircraft cannon designer A. Nudelmann and General (NFI) Dzhakhadze, commander of Vasilii Stalin's support regiment at Bykova, near Moscow. Shifrin recalls that Nudelmann expressed regular concern about the F-86, and about recurring jamming problems with the cannon he designed for the MiG-15. He also recalled that Dzhakhadze related having to fly to Korea in his 'Douglas, to pick up crash parts of MiGs and F-86s.' DZHAKHADZE HAD RELATED TO SHIFRIN THAT WHILE HE WAS IN KOREA ON SUCH A MISSION, THE 'SECURITY ORGANS' HAD ASKED HIM TO TRANSPORT A GROUP OF F-86 PILOTS TO KANSK IN WESTERN SIBERIA. THE MOVE HAD BEEN DONE CLANDESTINELY, WITH THE PILOTS TRAVELLING IN CIVILIAN CLOTHES UNDER SECURITY ESCORT."[119]

Another section of this report identified Sary Shagan (Saryshagan) in Kazakhstan as a center for exploitation of captured American technology, in a facility controlled by the MGB (KGB), which was developing radar systems and missile systems and antiaircraft weaponry.[120]

Four decades after these events in Korea and the USSR,

evidence that Shifrin and the Romanian POW George Risiou had also been correct about the Soviet policy of name changes for American and other foreign prisoners in the Gulag was also confirmed by investigators of the then newly-formed U.S. Army "Task Force Russia" (TFR). The U.S. investigators had located a retired KGB Lieutenent, Colonel Valerii Lavrentsov, at Khabarovsk Krai, who confirmed much evidence already supplied to TFR by Soviet military intelligence officers of the GRU:

"Lavrentsov stated that during his research on Japanese and Korean POWs he ran across some interesting information that suggests that some Americans may have been held in Khabarovsk in 'Special Houses" until they were able to recover from their wounds and were sent on to Moscow and other places...He speculated that the Americans could have been moved by either train, ship or air to the USSR, and that when they were in Soviet custody, their names would most certainly have been changed to Slavic ones. Larentsov suggested that an entire false background would have been concocted for each prisoner[121]

Further confirmation of the practise of name-changes for POWs came later from retired Soviet Lt. General Yuriy Filippovich Yeserskiy, of the Ministry of Internal Affairs (MVD):

"Yezerskiy stated that tracking down specific foreigner prisoners in the former Soviet prison system would be difficult because the names of foreigners were routinely changed, usually to foreign rather than to Russian names He suggested that the best source for the real names of prisoners would likely be other prisoners who knew them. He suspected that records of name changes may exist, most likely somewhere in Moscow."[122]

In addition to the Americans in Russia, many U.S. POWs remaining in Chinese Communist and North Korean captivity were in effect being held as hostages to a demand for eventual U.S. diplomatic recognition of those regimes, and fulfillment of China's desire for membership in the United Nations. A classified June 1955 report by James Kelleher, of the Office of Special Operations (OSO) of the Department of Defense, had admitted that American POWs from Korea were being held as hostages for political purposes:

"At the time of the official repatriation, some of our repatriates stated that they had been informed by the Communists that they were holding 'some' U.S. fliers as 'Political Prisoners,' rather than prisoners of war and that these people would have to be 'negotiated for' through political or diplomatic channels."[123] By forcing the U.S. to deal for the prisoners through diplomatic channels, the Communists expected to achieve a de facto recognition of their legitimacy and eventually full diplomatic recognition. The Soviets may also have viewed their own holding of American prisoners in this light, to support the demands of their Asian allies.

The previously-cited U.S. Government report entitled "The Transfer of U.S. POWs To the Soviet Union" stated: "It is possible that Stalin, given his positive experience with Axis POWs, viewed U.S. POWs as potentially lucrative hostages."

Formerly secret U.S. documents declassified by a component of the National Security Agency (USAINSCOM-NSA) in May 1991, reveal WW II, Cold War and Korean war U.S. POW locations in the USSR, including one at MAGNITOGORSK (South Russia), where in 1952 there was a special camp, "...EXCLUSIVELY FOR WESTERN ALLIED OFFICERS..." (including American), and another at TSCHURBAI-NURA, which was reported to be 40 kilometers southwest of KARAGANDA (Kazakhstan) which contained a, "SPECIAL CAMP FOR FOREIGNERS," including Americans.[124]

Individual American POWs from Korea were also reported by name, sometimes spelled phonetically, or by partial name, to be alive gulag labor camps by returning European ex-prisoners, such as Adalbert Skala, an Austrian who stated he was held in Prison #2 Irkutsk, and in Moscow's Lubyanka Prison with an "American Lieutenant RACEK," of armored troops, who had been captured in Korea and was originally from New York. The Austrian said he had been threatened with reprisals upon his release by the Soviets, "if he should give any information regarding his imprisonment." An August 21, 1956 State Department Foreign Service report distributed to CIA, Army and other agencies quoted the Austrian as saying: "Racek is not in particularly good condition, having had a number of teeth knocked out, having lost his hair, and generally having suffered the effects of mistreatment."

Another Foreign Service report from the following year, dated October 21, 1957 concerned a livesighting of three American soldiers who had been captured in the Korean War, by a Polish ex-prisoner:

"He said he had been until July 1957 a prisoner in the Russian concentration camp of Bulun, in the Province of Yakutsk. AMERICAN FELLOW PRISONERS IN THE CAMP, HE SAID, HAD URGED HIM TO REPORT THEIR PRESENCE THERE AT THE FIRST AMERICAN CONSULATE HE COULD CONTACT AFTER HIS RELEASE."[125]

The Department of State repeatedly requested information on suspected and reported American military prisoners in the 1950s, 60s and 70s, but the Soviet Foreign Ministry usually denied their existence. Many livesightings like these were rejected as invalid by U.S. officials because the names of Americans reported were misspelled, partial or disguised names, and the POWs were routinely denied by the Pentagon as ever having served in U.S. forces in the same manner as had been done since 1930 or before. A few accurately identified Americans were secretly returned from Communist control, but these later repatriations of WW II and Korea

POWs were not publicly announced by the U.S. Government.

At the 20th Congress of the Soviet Communist Party in 1956, Nikita Khrushchev startled those assembled with his secret report on Stalin's crimes. Although Khrushchev and his cohorts (and most others present) had taken part in, or had knowledge of these crimes, his open admissions presaged a short era of liberalization. Hundreds of thousands of political prisoners and former prisoners of war who still survived in the Gulag camps where so many millions had died, were actually released. These were to include several thousand Germans and other Western Europeans, and even a handful of Americans civilian prisoners. However, according to U.S. Government statements and available declassified records in the National Archives, no U.S. military WW II POWs were among those freed, although a very small number may have been secretly returned.

Throughout the period since the end of the Korean War, groups of POW/MIA family members had confronted the U.S. government, demanding information and action on recovering their loved ones, who had in many cases, been reported alive in enemy control. One such organization was "Fighting Homefolks of Fighting Men," a national organization founded in 1954, which grew from about 350 members of POW/MIA families to over a thousand by the late 1950s. It was headquartered in Glenwood Springs, Colorado and was largely the result of the devoted leadership of retired U.S. Army Captain Eugene Guild.

Guild later described an April 1954 demonstration by 83 mothers of American MIAs and POWs who had vanished in North Korea and Manchuria, which he and other war veterans witnessed:

"Some 83 mothers of the missing appealed to the UN in New York for help. Neither Dag Hammarskold (Secretary General of the UN) nor Henry Cabot Lodge (U.S. Ambassador to the UN) would see them. They went on to Washington and on two days stood heartsick and weary on the sidewalk in front of the White House waiting in vain for a word from the President, who passed by them on social engagements without a glance. I know; I was there.[126]

In a subsequent letter to Senator Richard Russell of the Senate Armed Services Committee, Captain Guild added that "when these same mothers asked the courts of justice to enforce equal loyalty between the Government and the soldier, they were told in effect, 'the Commander in Chief is above the law.'"[127]

Hearings on the return of unaccounted-for U.S. POWs chaired by U.S. Representative Clement J. Zablocki of Michigan, were held in May 1957, by a subcommittee of the House Foreign Affairs Committee. During the hearings it was publicly demonstrated that some American prisoners had been known to be alive in captivity but were not returned by the Communists. In one case, President Eisenhower had personally written to Representative Page Belcher of Oklahoma about the POW son of Harry D. Moreland of Tulsa, who had been seen alive in enemy captivity in 1952, and that since that time, "Nothing further, official or unofficial, has been obtainable since then, despite our ceaseless efforts."[128] Yet, in the Zablocki hearings it came out, quite by accident, THAT IN MARCH 1956, THE COMMUNISTS OFFICIALLY ADMITTED HAVING HELD MORELAND'S SON, BUT THEY CLAIMED HE HAD "ESCAPED." For over a year after that, however, THE U.S. GOVERNMENT CONTINUED TO TELL MR. MORELAND THAT THE COMMUNISTS DENIED KNOWING ANYTHING ABOUT HIS SON. This had included the President of the United States. As Captain Eugene Guild wrote to other members of the Fighting Homefolks organization: "According to experiences of mothers of other missing, if you go to Washington inquiring about your son and are properly docile, you will be met with a shower of sympathy and courtesy. But if you persist you will be silenced with the brutal and unjustifiable question, "DO YOU WANT AMERICA TO GO TO WAR OVER YOUR SON?"[129]

One of the most active and forceful POW mothers in the United States involved with Fighting Homefolks of Fighting Men was Mrs. Rita Van Wees, whose son, Ronald Van Wees had been captured in Korea and was known to have been alive in Communist captivity but was not released in Operation Big Switch. His name was on the 944-man list, and the later-reduced 450-man list of Americans known to have been alive in captivity. It was she who had attempted to sue President Eisenhower in a U.S. District Court in Washington, D.C. during January 1957, for abandoning military personnel in enemy captivity as the Commander-in-Chief; but Assistant U.S. attorney E. Riley Casey had moved for dismissal, claiming the court had no jurisdiction. Judge Matthew McGuire, reluctantly agreed, and the case was dismissed. In an October 25, 1957 report to the members of the Fighting Homefolks organization about her meeting with Assistant Secretary of State for Far Eastern Affairs Walter S. Robertson, a political appointee who had been a Virginia banker, Mrs. Van Wees recalled:

"I began by asking why he had withheld from the Zablocki Committee and the public the existence of the 2,691 (POWs) outside of the 450 of our missing boys who may yet be alive...He then said: "You know as much about those boys as I do, and there was no evidence that they were alive.' I asked, 'Have you evidence that they are dead? He said, 'No,' and went on to tell how the government has made demands for an accounting of all the missing, but always receive the same answer, that they don't have them, or (they) died, or escaped, and that is the answer we get, and it is not true that the boys have been forgotten. I looked him in the eye, smiled and said, 'If you will permit me, Sir, I have here with me official documentary evidence, that no matter what you are telling me, you did withhold the evidence from the Zablocki Committee hearings, the public and the press, the fact that there are 2,691 (U.S. POWs) plus the 450. May I be allowed to read the presentation and the questions which is very important for the liberty of our sons? He turned red and said he would be glad to answer the questions but there was a delegation waiting and then said, 'What kind of questions.' I picked question eight in a hurry; he then said: 'THIS QUESTION CAN BE ANSWERED BY THE DEFENSE DEPARTMENT,' and not all questions could be answered by him. I smiled again and went on to tell how the President passed the buck to the UN and the UN forgot...and allowed our sons to be taken prisoner for life...'and now you sir, are passing the buck to the Defense Department...' He began to talk again, that it was impossible to deal with the Reds, and asked 'What do you think we can do? I said, 'Why not throw Russia out of the UN, AND WHY DON'T YOU DEMAND OF RUSSIA THE RELEASE OF ALL OUR MEN, AND THAT THIS GOVERNMENT SHOULD CUT OFF ALL AID TO THE REDS...I told him this administration is soft and cowardly in dealing with the Red murderers.

"He then threw up his hands and said, 'I have no argument with you...but the Congress passes the laws...What more can I do?' I told him, to begin with, the truth should be told to the American people. 'IF OUR SONS RECEIVE NO PROTECTION, WILL THE SONS OF THE FUTURE RECEIVE THE PROTECTION DUE TO EVERY AMERICAN WHO WEARS THE UNITED STATES UNIFORM?' I told him the mothers were brushed off, some threatened. At this he said, over and over, 'Who is threatening the mothers? I want to know who would do this.' I said I could provide some evidence of this. He said the talks in Geneva still go on for the release of the boys, and that a list of all the names, including Ronald's, was handed over to the Reds. I told him then, perfumed notes, and blasts from some of our officials for 'trade,' and a seat in the UN for the Red Chinese will do our sons no good; and inviting the Red Chinese athletes to play in the next Winter Olympics is a shameful outrage, and this we protest. The Japan delegation was waiting for Mr. Robertson. He got up...I kept insisting I read the presentation and questions. He then told his secretary to make the appointment to the Pentagon, Defense Department I told him, 'I will not see some office boy, I want to see someone who can really be of help...'"

"At the Pentagon we were introduced to a Captain Allen. I thought he would take us to the proper office, but it was his office. I felt a polite brush-off and told Captain Allen so. He had all of Ronald's record on his desk. I told him I knew what was in the record and for an hour he talked...He said all the names were brought back by the returned POWs, and it was impossible for our sons to be alive; the ones who were in jails had a way of keeping the others on a list, and above all, the Government is doing everything possible, and that in view of the fact that the Reds may only be holding a few of our men, WE SHOULD BELIEVE OUR SONS DEAD; OTHERWISE THEIR NAMES WOULD FOR SURE HAVE BEEN BROUGHT BACK. The talk went along this line, mostly that we should BELIEVE our sons to be dead. Then I blew my top, saying, "don't you dare try brainwashing me to believe this lie...it would be impossible for the returned POWs to have brought back all the names. MANY OF OUR BOYS WERE THROWN IN DIFFERENT JAILS, EVEN SENT TO SIBERIA. You tell your State Department was very unsuccessful. I didn't come to see you anyway; I will not waste your time. What you are telling me we have heard before, and this is the well known brush-off given to the parents. I handed him the other copies of the papers and said to give them to the proper people, and walked out on him.[130]

Assistant Secretary of State Walter S. Robertson subsequently wrote to Mrs. Van Wees, clearly enunciating the Department of State's policy that they could not ask for the release of specific personnel unless the Defense Department supplied their names towards prisoners of war who had not been released.:

"Having met you personally I was surprised at the offensive tone of the statement and questions which you left with me. I would judge that those papers were prepared not by you but by others with some ulterior motive beyond the recovery of military personnel...I EXPLAINED TO YOU WHEN YOU WERE HERE THAT THE RESPONSIBILITY FOR DETERMINING THE STATUS OF AMERICAN MILITARY PERSONNEL MISSING IN ACTION RESTS WITH THE DEPARTMENT OF DEFENSE. THE DEPARTMENT OF STATE MUST RELY UPON THE DEPARTMENT OF DEFENSE FOR THE ESSENTIAL INFORMATION UPON WHICH TO BASE ITS REPRESENTATIONS TO FOREIGN AUTHORITIES CONCERNING SUCH PERSONNEL..."

"I can only repeat here what I related to you in person, namely, that the United States Government, through the Military Armistice Commission in Korea and through Ambassador U. Alexis Johnson in Geneva, has continuously pressed the Communists for the release of all American personnel which they were known to be holding or who might be in their hands...IN OUR DISCUSSION HERE YOU AGREED THAT WHILE FORCE MIGHT PUNISH THE COMMUNISTS FOR THEIR PERFIDY, SUCH ACTION WOULD ALMOST CERTAINLY INSURE THE MURDER OF ANY PRISONERS WHICH THEY MIGHT BE HOLDING...I yield to no one in my condemnation of the ruthless, heartless regime in Peiping nor in my desire to do everything possible to achieve the release of any of our citizens whom they MAY be holding. On September 10, 1955, they made an unequivocal public commitment to release 19 (American) civilians which they had illegally jailed. Today, over two years later, they have released only 13, and in callous disregard of this commitment are still holding 6 as political hostages seeking to obtain political concessions...I deeply resent the insinuations of the memorandum you submitted questioning my sincerity in a problem which lies very close to my heart...Walter S. Robertson" (Assistant Secretary of State).[131]

Since the Pentagon maintained a position that it did not know of any American military POWs still alive in enemy captivity, as the War Department had done after WW I and WW II, the State Department could not ask for their release by the Soviets, or the Chinese, or the North Koreans. It was essentially the same position that had been taken by the War Department and the State Department from 1921-1933, as noted in the first part of this book.

Ambassador U. Alexis Johnson was the American representative who had met with the Chinese Communist representatives in Geneva, Switzerland, numerous times (77 meetings occurred, in all), on the subject of the missing American POWs and MIAs of Korea, but the Chinese repeatedly stated that the POW issue must be discussed at Panmonjum. In his memoirs, Johnson recalled the perfunctory nature of his requests: "As far as I was concerned, I raised it in Geneva only for the record."[132]

An Associated Press report of January 22, 1959, from Panmunjom, Korea, however, illustrated that some of the U.S. officials were willing to say publicly that they knew some of the American POWs of Korea were still alive in Communist captivity:

"The UN Command Thursday demanded an accounting of 2,147 Allied prisoners still missing from the Korean War. U.S. Rear Admiral Ira H. Nunn of the UN Command said he was "sure some of these prisoners are still alive. The Communists promptly rejected the demand and countercharged that the allies are detaining 98,742 war prisoners."[133] Captain Guild pointed out a discrepancy in the AP report: "The story carefully refrained from saying that, among the 2,147 prisoners are 450 named Americans as listed in the hearings of the Zablocki congressional committee. It did not mention that in addition to the 450 Americans there are still 2,691 other American servicemen still missing and unaccounted for, many of whom must reasonably been captured without eyewitnesses or other evidence. The 2,691 are in addition to the 832 buried in Hawaii as unidentified."[134] These were "American remains" recovered after the war and accepted as U.S. servicemen by the Department of Defense.[135]

In October 1959, the national magazine "Argosy," popular among many veterans and working-class Americans, published a major article on the POWs of Korea by Edward J. Mowery which revealed substantial information about over 3,000 American POWs and MIAs left behind there, including 450 admitted by the U.S. Government to have been held as prisoners but not released or otherwise accounted for. Captain Guild urged the some 1,000 members of Fighting Homefolks of Fighting Men to obtain it, and to spread it around to others in the states where they lived.[136] Guild also informed the members on his view of U.S. policy towards POWs from Korea and from the U.S. reconnaissance aircraft shoot-downs over the USSR:

"The official policy is to take the Soviets' word for it when they deny our men are alive. On July 16, 1956, the Secretary of State surprisingly notified the Kremlin that the U.S. was 'compelled to believe' 22 named fliers were being held alive at certain named places. But when the Communists denied it, their word was officially believed by the U.S. Government and the matter was dropped-except for perfumed notes requesting an 'accounting.' If we know a man is alive on Monday, but on Tuesday the Reds deny it, then he is dead in the Administration's book. The Congress tried to do something by means of the Zablocki House Committee, but the President's supporters sabotaged the resolution by having it urge the President to 'continue' to do everything possible-which of course meant to continue doing nothing. Many (Congressmen) are protesting right now, but as soon as those citizens discover that they are at odds with the political gods of both parties, they usually subside for reasons of self-protection. Few will risk their own security for the soldiers who sacrificed everything for them..."[137] Mrs. Van Wees had learned in 1959, meanwhile, from a Russian source in Europe in contact with underground sources in the USSR, that her son was alive in the Soviet Union, with other American prisoners, but when she took this information to Washington the White House denied any American prisoners had been taken there from Korea.

In 1960, as the Presidential contest between John F. Kennedy and Vice President Richard Nixon began, yet another example of a livesighting of Korean war POWs held captive in Soviet concentration camps, was received by the Department of State, but remained "classified" for 30 years to come. State Department officials such as Assistant Secretary of State Robertson were well aware of such reports when telling POW mothers like Rita Van Wees that he knew nothing about the American POWs from Korea held by the Soviets. This document was declassified in 1990, with the names of the POWs blacked out, ostensibly "to protect their privacy," according to U.S. Government officials. This concern for "privacy" was used as justification for such censorship despite the fact that these men had been abandoned by their government for 13 years after the Korean Armistice. A classified Foreign Service Despatch to the U.S. State Department in Washington from the American Embassy in Brussels, Belgium on September 5, 1960, under the heading "Korean War Prisoners in the Soviet Union," stated the following:

"On Labor Day, September 3, the Embassy duty officer, Joseph W. White, met with a "walk-in" Polish refugee, (name blacked out) said that he was released on May 1, 1960, after seven and one half years detention, from Soviet prison camp near Bulun, supposedly about ?0 kilometers from Yakutsk. He stated that he later escaped from Poland via East and West Germany and entered Belgium on August 9, 1960. He is now a political refugee in Belgium and he is attempting to immigrate to Australia."

"(name blacked-out) said that he became acquainted in the Soviet camp with two American Army prisoners who were captured in Korea in 1951: (First and last names blacked-out) an infantry lieutenant, and (First and last names blacked-out) a commando or paratrooper sergeant. They had asked him to report their presence to the American Embassy in Warsaw, but (name blacked-out) indicated he was afraid to do so. Consequently, he was taking this opportunity to report the facts to the American authorities. However, he asked that his name be kept confidential as its revelation might jeopardize the safety of his relatives in Poland. A full memorandum of conversation is enclosed. The office of the Army attaché is separately reporting on this matter through military channels, which is also attempting to check further on (name blacked-out) story and bona-fides. For the Ambassador:

Stanley M. Cleveland, First Secretary of the Embassy."

In a U.S. State Department "confidential memorandum of conversation," dated September 3, 1960, the duty officer of the American Embassy in Brussels, Belgium, James W. White, recorded the following details:

"Subject: American military personnel in Soviet concentration camp. Background:

"Between 1800 and 1900 on September 3 (5?), the Marine guard called Mr. White, the Embassy duty officer, to inform him that a Polish refugee had telephoned the Embassy and asked to speak with the "consul" in order to give him information on Americans held in a Soviet camp. The Marine guard suggested that the refugee come to the Embassy. When White arrived at the Chancery, the refugee (name blacked-out) was already there. Before interviewing him Mr. White telephoned Mr. Cleveland at the Political Section to ask if there were any special instructions concerning a case like this. Mr. Cleveland suggested that White interview (name blacked-out) obtaining as much information as possible, particularly as to where he might be reached in the future. (Name blacked-out freely gave data concerning his background. He repeatedly stressed however, that his name should be kept confidential and should in no way be associated with the American Embassy. His relatives in Poland might be jeopardized."

(A large blacked-out area follows on this page of the report, presumably containing more information, or perhaps a photograph. The report then continued on p. 2 of the classified "confidential"

memorandum:
"He is living off public aid given at the rate of BF 50 per day, an amount which he says is insufficient for him to pay room rent and eat properly. He has a temporary Belgian identity card which does not permit him to work. (Several lines of blacked-out area follow.)
"At present he is trying to arrange emigration to Australia. "Information on American prisoners: (Name blacked-out) states that the two Americans whose names are given below were captured in Korea in 1951. These men, who think that their families believe them dead, asked (name blacked-out) to inform the Embassy in Poland that they were prisoners in the USSR. (Name blacked-out) said that he feared to go to the Embassy in Warsaw because all Poles visiting it are checked. The information on the Americans is as follows:
"1. (First and last names blacked out)-infantry lieutenant; 32 years old (birthday April 12 or 13), not married, has mother, father, and two sisters; comes from Buffalo (presumably New York, but (name blacked-out) was not sure).
"2. (First and last names blacked-out)-Sergeant; a commando or paratrooper; grandparents were Polish emigres; father is dead, has two sisters, from Chicago, liberty Street.
"The health of both men is poor as a result of working in the phosphorous mine associated with the camp. (Name blacked-out) said the chemical attacks the head and liver. He showed me how his own head is constantly scaling.
"Comment: On appearance (name blacked-out) seemed sincere. When leaving, he indicated that he did not have sufficient funds to return to liege. White gave him BF 100 which he appeared to accept reluctantly. He more readily accepted some American cigarettes and pipe tobacco. It was at this point that he said he was living off BF 50 per day, 30 of which goes for room rent, leaving him only 20 for daily food. It would not seem that he came to the Embassy for help, but it is not impossible that he might request some assistance later on. (One Belgian franc equals $.32 U.S.)"
This livesighting, as usual, was widely distributed within the State Department, with multiple copies sent to the Army, Navy and Air Force, and also to U.S. Information Agency (USIA), OCB, Central Intelligence Agency and the National Security Agency (NSA). It is obvious from cited examples that the existence of American POWs from Korea in Soviet forced labor camps was widely, if secretly known to U.S. policymakers as the 1960s brought a new President, John F. Kennedy, to the White House, and America began a military commitment to defend Vietnam from a Communist threat.
(After this document was first declassified by the Department of State and brought to the attention of U.S. Representative John Miller (R-WA), that Congressman wrote to Secretary of State James

Baker III in 1991, protesting the continued classification of the two U.S. servicemen's names in this report: "While I understand the need at that time to protect the confidentiality of the refugee, and the servicemen, it is now 31 years later...Meanwhile, families of POW/MIAs continue in their search for answers to the fate of their loved ones." The State Department subsequently relented, in this case, and revealed to researchers with outstanding Freedom of Information requests (including the author) that the two Korean war American POWs seen by the Pole at Bulun in the Yakutsk region up until 1960 were: U.S. infantry Lieutenant Ted Watson and commando or paratrooper sergeant Fred Rosbiki.)

One of several stock replies were used by the Department of State to answer questions by members of the U.S. Congress about rumors of such reports of Americans in Soviet captivity. An example used about eight weeks after the previously quoted livesighting, to answer such questions by Senator Wayne Morse of Oregon, during the administration of President John F. Kennedy, was dated April 21, 1962, and signed by Assistant Secretary of State William Macomber:

"This Department has over a period of years received a number of reports alleging that American citizens have been seen in Soviet prisons and labor camps. Each of these reports is investigated with a view of establishing the identity and the American citizenship of the person concerned. If sufficient evidence is obtained to establish to a reasonable degree the American citizenship of the subject, representations are made to the Soviet government.

"Because of numerous RUMORS that had come to our attention concerning Americans detained in the Soviet Union, on July 16, 1956, we sent a note to the Soviet government requesting information concerning American airmen who had been lost under circumstances leading some to believe that the airmen might have been detained by the Soviet government. The Soviet government replied on August 13, 1956, that a careful investigation by appropriate Soviet authorities had established that no AMERICAN CITIZENS from the personnel of the U.S. Navy and Air Force were in the territory of the Soviet Union."

Very few books on the Korean War which achieved mass-circulation included the information that an officially-admitted 944 American prisoners of war were still held by the Communists in North Korea after Operation Big Switch. One which did, entitled: "This Kind of War," by T.R. Fehrenbach, was published by the Macmillan Company in New York in 1963.

The State Department was consistent over many years in avoiding as much as possible, any public statement or reference to U.S. military prisoners of war from WW II in Soviet captivity. When forced to address the subject of American prisoners in Russia at all, the Department used terms such as "American citizens," or gave examples of discrepancies over the fate of crewmen in reconnaissance aircraft shootdowns. Yet the questions over the fate of the huge number of American POWs missing in eastern Europe, never faded away entirely. On April 17, 1967, as yet another American expeditionary force fought an anti-Communist war in Asia, the distinguished U.S. Senator Karl Mundt wrote to President Lyndon B. Johnson's Secretary of State, Dean Rusk (who was then involved in a secret Vietnam POW exchange negotiation with the Soviets, through the East Germans):

"It is my understanding that on December 10, 1951, the Department of State sent a mission behind the Iron Curtain in Russia, seeking to establish the number of Americans still held in Russia. It is my further understanding these Americans were not the type of tourists we were talking about at the time of the ratification of the Consular Treaty, BUT RATHER THEY ARE MEMBERS OF THE AMERICAN ARMED FORCES ORIGINALLY CAPTURED IN WW II AND LATER TRANSFERRED TO RUSSIAN PRISON CAMPS. I HAVE HAD REPORTS FROM VARIOUS SOURCES THAT THIS CONDITION ACTUALLY EXISTS, and would appreciate a letter from you providing as much authentic information as our Department of State has on this particular situation...I would appreciate hearing from you in response to this inquiry."[138]

Senator Mundt's letter was hard-hitting and revealed that he had what he considered to be good sources of information on not only the Soviet-held WW II U.S. POWs, but also on what may have been a highly secret mission, by the covert action arm of the State Department, the Office of Policy Coordination, to determine their number in the USSR, perhaps achieving deniability by using eastern European agents such as Frank Wisner routinely employed. The reply to Senator Mundt on April 25, 1967, was worded very carefully by the State Department:

"WE TOO OCCASIONALLY RECEIVE REPORTS OF THE SORT YOU DESCRIBE. ANY SUCH REPORT COULD CONCEIVABLY BE BASED ON FIRST-HAND KNOWLEDGE IS EXHAUSTIVELY INVESTIGATED IN COORDINATION WITH THE DEPARTMENT OF DEFENSE.

"Our Posts abroad and agencies of the U.S. Government have CAREFULLY interviewed individuals who have been imprisoned in the USSR and have since come to the West. In these interviews we seek to establish whether the individual saw or heard of any Americans under detention. It is our conclusion, shared by the Department of Defense, that there is no RELIABLE information indicating that American servicemen captured during WW II are NOW being held in the USSR."

"It is likewise not true that a U.S. GOVERNMENT MISSION went to the Soviet Union in 1951 to investigate the problem of Americans detained there."

The Department thus attempted to disarm the Senator by

admitting that it too, received such reports, but that the sources, who were almost always one-time informants, with no record of providing intelligence later proven correct, therefore could not categorized as RELIABLE, in the strict intelligence-usage meaning of the word. And even if some, like Senator Mundt in 1967, believed such sources were reliable, the Department of State, in conjunction with the Department of Defense (then under Robert S. McNamara), maintained that they did not prove that American servicemen captured during WW II were "NOW" being held in the USSR. This undeniable assertion reveals the methodology that is at the very crux of public U.S. policy toward missing American prisoners of war, reported at any time by eyewitnesses, to have been alive in Soviet (and also Korean, Chinese and Vietnamese) captivity. Just because an American military prisoner of war may have been reported alive in a Soviet prison by an eyewitness yesterday, that did not mean he was alive today.

Many years later (after yet another U.S.-Asian war in Vietnam), a Romanian engineer who had immigrated to the United States named Serban Oprika, revealed to the State Department that he was an eyewitness to American POWs being held in North Korea in 1979, twenty-six years after the end of the Korean War:

"I thank you for inviting me to this meeting. My name is Serban Oprica. I am from Romania. My parents suffered loss of jobs and private property while the communists were taking power, the only source of happiness remaining their children's education.

"In the year 1967, I received my master's Degree in Construction(s), and I started working at (the) Institute of Designing in Bucharest city. In September 1979, I was sent to North Korea to help built(d) the first television factory in that country. In the month of October in a visit to a museum I saw the event for which I attended this meeting.

"It was a beautiful day. The Korean bus with twenty-five Romanian people and North Korean guides drove along (a) winding rural road. After a turn a large field with tiny houses appeared before my eyes. From those houses a man that was taller than the Koreans approached the bus. Behind him, near a house, a Korean woman could be seen. As the bus came closer, everyone in the bus was shocked because the man was not Korean. He was in his 50s and had light-colored eyes. To his right, farther away, we saw more people bent toward the ground. The ones that were closer raised their heads and stared toward the bus. None of these people were Korean. They were tall, in their 50s, and curiously looked at the bus. The rest, at least 50, continued working. An atmosphere of silence filled the bus. We didn't exchange any gestures, and we saw no guards.

"After this incident, the colleagues that had worked longer in

Korea said that they knew something about the existence of Americans and that this is something normal in a communist country. After returning from Korea, I completed the documents for emigration with my family in the USA. After 5 years my dream came true. In (at) present, I work as an engineer for Savage Engineering in Bloomfield, Connecticut. I thank the American people for their hospitality. I read in the newspaper about POWs, and that many people don't believe their existence. I decided to tell what I saw. Now, I feel better and I hope that in a good day they (POWs) will come back home."[139]

Mr. Serban Oprika, a cultured and sensitive man with a sense of honor, later testified before the Senate Select Committee on POW/MIA Affairs in November 1992, about this same incident. In a lengthy interview with the author and research colleague Ashworth in Washington, D.C., prior to his testimony, Oprika also remembered that most of the American prisoners he saw looked older but one of those in the group of 7-10 standing behind the closest American was another who also had blue eyes but had bright red hair and looked younger:

"We see a land like a camp, where vegetables, and my attention was to— because I saw a person with a European face, with blue eyes, very close (to) the bus. And I was very shocked. And everyone on the bus was shocked. And I was looking behind him, I saw seven or ten peoples with Caucasian face. And behind them, I saw more people working the camp...They were dressed with North Korean dress, like Chinese..."

During the interview with the author and research colleague Ashworth in Washington D.C., Mr. Oprika added that behind the closest group there were up to 50 others visible of the same appearance and in the same uniform, working in the camp. He said that one of the Caucasians in the group of seven to ten American prisoners of war behind the closest man looked younger, with bright red hair and blue eyes. Mr. Oprika also recounted to the author and testified to being guided through a museum by North Koreans which contained many relics of the Korean war, such as weapons, uniforms and flags, but most particularly body parts of American soldiers preserved in containers of alcohol, including hands, feet and whole heads. In the interview Mr. Oprika stated that the North Koreans kept these human heads on display both to frighten people and to demonstrate their superiority over the Americans.

Other livesightings of American POWs from Korea remaining in Communist China and the Soviet Union also occurred. In 1990, Zygmunt Nagorski, who had written the powerful article for Esquire Magazine in 1953, was quoted by the Los Angeles Times as saying his sources had informed him that there were still up to 1,000 American POWs in Siberia from the Korean War when he last had contact with them in the late 1950s.[140] The survivability demonstrated by prisoners of war of various nationalities released after decades of Soviet captivity, indicates that many of the former U.S. prisoners of war captured in Korea or earlier, should in fact still be living at the time of this writing, in remote or restricted areas of the former USSR, and in North Korean or Chinese captivity. Indeed, there have been persistent reports from intelligence sources in South Korea that the North Korean Communist regime of Kim Il Sung and his son and heir-apparent Kim Jong-Il maintains at least eight known forced labor camps, which together contain a minimum of 105,000 prisoners. The extensive camps are reported to be surrounded by barbed wire and minefields, and to have been discovered by "highly sensitive and scientifically reliable information" (U.S. aerial reconnaissance) and to have been confirmed by North Korean defectors. Two camps in North Pyongyang Province containing some 20,000 captives, and another in adjacent province, are in areas where the North Koreans held many U.S. POWs during the war, and may hold hundreds of surviving American prisoners to this day.[141]

More than a year after the first edition of this book was printed and distributed by the author, the U.S. Government finally admitted, in September 1993, that American prisoners of war from Korea had been taken to the Soviet Union as part of a KGB intelligence-gathering program. A 77-page report entitled: "The Transfer of U.S. Korean War POWs To the Soviet Union," supported by new information obtained by the U.S. Army's Task Force Russia, charged that certainly hundreds and probably over 1,000 U.S. prisoners, including a small number of F-86 Sabrejet pilots and many technical personnel, were moved secretly to the Soviet Union for interrogation by the KGB and disappeared in Soviet prisons. This report did not depend on the intelligence of shipments of trainloads of other U.S. POWs to Siberia to Manchouli, as noted earlier in this book and in other still-classified State Department files, but revealed many new sources of confirmation. The American press then began, once again, to use the figure of "54,246" total American dead in Korea, including 8,140 "unaccounted for," and now reported that the "missing for which there is no evidence of death is set at 2,195"[142]

This however, was an unrealistically low figure.

This report particularly identified 9 U.S. POWs about whom the Russian Archives should have information. They were: Captain Albert G. Tenney, an Air Force F-86 pilot; Major George V. Wendling, also an F-86 pilot; 1st Lieutenant Charles A. Harker Jr., an F-84 pilot; 1st Lieutenant Robert F. Niemann, an F-86 pilot; Major Charles E. McDonough, an RB-45C pilot; Captain Halbert C. Unruh, a B-26 pilot; Captain John W. Shewmaker, an F-80 pilot; Staff Sergeant Elbert J. Reid, a B-29 gunner; and Staff Sergeant Louis H. Bergmann, a B-29 radio operator.

For example, a 1993 U.S. Government report stated: "Captain Tenney's name appears on the "List of 59" entitled 'A list of United States Air Force Personnel Shot Down in Aerial Combat and by Anti-Aircraft Artillery During Military Operations in Korea, Who Transited Through an Interrogation Point.' Additionally, the Joint Commission Support Branch believes that further information on Captain Tenney (shot down 3 May 1952 and listed as MIA) exists in the Russian archives ..."[143] The U.S. had detailed information on a total of 31 such F-86 pilots whose loss indicated possible capture. (Author's note: Captain Tenney's name had first been obtained from the Russian archives by clandestine effort at the time of the collapse of the Soviet Union, and was turned over to U.S. documentary film producer Ted Landreth, who passed his name and others to the author, who in turn passed them to retired U.S. Army Colonel William LeGro, then on the staff of a Senate Select Committee on POW/MIA Affairs. LeGro later was assigned to the staff of the U.S. Army's Task Force Russia's Washington, D.C. headquarters.)

In a nearly simultaneous development, yet another livesighting of an American Korean War POW in the Soviet Union had been reported by a former MVD Colonel named Vladimir Malinin, who had seen U.S. Marine Corps Sergeant Philip Vincent Mandra alive in the USSR in 1966. Sergeant Mandra had been missing in action, and was later declared killed, after a battle with Chinese infantry on August 7, 1951. He was a native of Farmingdale, N.Y., and was 21 years old at the time of his capture. Malinin had identified Sergeant Mandra from among 121 photographs of other missing U.S. servicemen whose fate was under investigation. The former MVD colonel said he had first seen Sergeant Mandra walking alone in a Magadan, Siberia, prison in 1963 and saw him again at the same prison in 1966. Malinin was told that Sergeant Mandra was segregated from other prisoners because he was "an American spy." After receiving the new information obtained by U.S. Government POW/MIA investigators, his sister, Irene Mandra, reported that she had kept a 1954 letter on her wall from President Eisenhower that expressed sympathy for the death of her brother, and said that she was "praying and hoping the Pentagon will follow up."[144]

# Notes on Chapter Eight

1    NYT, June 9, 1945

2    *Korea, The Untold Story*, by Joseph C. Goulden, McGraw-Hill, 1982

3    Executive Secretariat Files, Briefing Book Paper, Inter Allied Consultation Regarding Korea, pp. 358-361, 770, 952, Foreign Relations of the United States, The Conferences at Yalta and Malta, 1945

4    *Korea*, by Goulden; also: The Forrestal Diaries and others)

5    FRUS, Ibid., p. 770

6    New York Times, June 9, 1945, pg. 1

7    *Korea, The Untold Story* of the War, op. cit., p. 19

8    NYT, op. cit.

9    *Korea*, by Goulden, op. cit.

10   *Korea*, Goulden, pp. 39-41, op. cit.

11   Wise *Men*, op. cit.; *Witness to History*, op. cit., *Korea, The Untold Story of the War*, op. cit.

12   Official battle statistics of the Defense Department which were published in the Veterans of Foreign Wars, VFW Magazine, in July 1993

13   Official statistics reprinted in the VFW Magazine, op. cit.

14   For Boyson's debriefing and those of other U.S. POWs, see: National Archives, Record Group 153, Entry 183, boxes 1-4

15   Official statistics reprinted by VFW Magazine, op. cit.

16   "The Transfer of U.S. Korean War POWs To the Soviet Union," by the Joint Commission Support Branch, 26 August 1993, p. 26

17   *Witness to History*;, pp 292-293; also: *The Wise Men*, op. cit.

18   NA RG 153 Entry 183.

19   Ibid.

20   Ibid.

21   Official statistics reprinted in the VFW Magazine, Op. Cit.

22   NA, RG 153, Entry 183

23   RG 153, Entry 183

24   RG 153 Entry 183

25   Official statistics reprinted in the VFW Magazine, June/July 1993; See also: "Korean War Almanac, by Harry G. Summers Jr., Facts on File, Inc., NY, 1990

26   Wise *Men*, p. 543

27   Pentagon statistics reprinted by VFW Magazine, op. cit.

28   *Korea, The Untold Story*, op. cit., p. 477

29   *CIA and the Cult of Intelligence*, p.45

30   *American Caesar*, William Manchester, 596-97, Little, Brown, 1978

31   *KGB-The Inside Story*, Andrew and Gordievsky, p. 404

32   This document and other declassified CIA reports, obtained through Freedom of Information request and supplied by the author to the Senate Foreign Relations Committee in 1991, and to the Senate Select Committee on POW/MIA Affairs in 1992, is reproduced in full in "POW/MIA Policy and Process," Part II, Hearings before the Select Committee on POW/MIA Affairs, United States Senate,

November 5, 6, 7, and 15, 1991, U.S. Government Printing Office, 1992, ISBN 0-16-038479-6

33　Debriefing of Corporal Walter Williams, NA RG 153 Entry 183, Op Cit.

34　National Archives, RG 153, Entry 183, box 6

35　Declassified CIA reports made available to researchers through Freedom of Information request for all declassified CIA reports pertaining to American prisoners of war in Communist control

36　Declassified CIA documents supplied by the author to the Senate Foreign Relations Committee in 1990-1991 and reproduced on pp. 841-867, POW/MIA Policy and Process, Op. Cit.

37　POW/MIA Policy and Process, Op. Cit., pp/ 847-848

38　NA RG 153 Entry 183 Box 7

39　NA RG 153 Entry 183 Box 6

40　NA RG 153 Entry 183

41　Briefing of Ambassadors, Department of State records, September 18, 1951, Harry S. Truman Library

42　*Korea*, by Goulden, p. 573, op. cit.

43　"The Transfer of U.S. Korean War POWs To the Soviet Union," p. 3, op. cit.

44　U.S. News and World Report, May 28, 1954

45　*Facing Reality—From World Federalism to the CIA*, by Cord Meyer, Harper and Rowe 1980, University Press of America, 1982, p. 65

46　*Korea, The Untold Story*, pp. 464-475, op. cit.

47　Ibid.

48　*Present At The Creation*, Dean Acheson, W.W. Norton, NY, 1969, p. 653

49　Declassified records of the Psychological Strategy Board, in the personal papers of President Harry S. Truman, Truman Library

50　*The Armies of Ignorance*, p. 341, op cit.

51　Box 32, Papers of Harry S. Truman, Psychological Strategy Board

52　The papers of of Harry S. Truman, Psychological Strategy Board, Box 22, Truman Library, Independence, Missouri This document and many others from the PSB were supplied by the author to the Senate Foreign Relations Committee in 1990, which used them in a May 1991 Report on U.S. POW/MIA Policy, and to the Senate Select Committee on POW/MIA Affairs in 1991, which declined to use them in it's final report.

53　*Memoirs*, Harry S. Truman, op. cit.

54　*Witness to History*, p. 300

55　*Korean War Almanac*, by Harry G. Summers Jr., Facts on File, NY, 1990, p. 213

56　The entire July 17, 1952 CIA report SO 9163 is reproduced on pp.868-870, POW/MIA Policy and Process, U.S. Senate, Op. Cit.

57　Produced by Ted Landreth and Australian TV

58　"Korean War POW Transfers to the Soviet Union: Eyewitnesses "(RT:18:35), prepared by Task Force Russia, April 1993

59　Minutes of the Meeting Between Comrade Stalin with Chou en-lai, 19 September 1952, translated in Draft Task Force Russia 37-11

60　American Embassy Moscow Message 271140Z, quoted in "The Transfer of U.S. Korean War POWs To the Soviet Union", 26 August 1993, by the Joint Commission Support Branch (successor to Task Force Russia, U.S. Army

61　Pentagon statistics reprinted in the VFW Magazine, op. cit.

62　Debriefing of Master Sergeant John T. Cain, National Archives, RG 153 Entry 183 box 1&2

63   American Embassy Moscow Message, 1411521Z April 1993; also pp.14-15, "The Transfer of U.S. Korean War POWs To the Soviet Union," 26 August 1993, by the Joint Commission Support Branch

64   *Korea, The Untold Story of the War*, by Joseph Goulden, p. 622, op cit.

65   *The Armies of Ignorance*, by William R. Corson, Dial Press, NY, 1977, pp. 336-343

66   *Wise Men*, p. 561, op. cit.

67   *Armies of Ignorance*, p. 341,op cit. Cutler was a Harvard graduate of 1916 who had served in France during WW I and in the office of Secretary of War Henry Stimson from 1943-1945, and thus was also aware of the earlier prisoner of war repatriation problem with the Soviets. To coordinate security problems within the NSC Cutler had Eisenhower establish the President's Committee on International Information Activities. The Committee was composed of Cutler, Gordon Gray, Barklie Henry, John Hughes, C.D. Jackson, Roger Kyes, Sigurd Larmon and Chairman William H. Jackson. p. 341, *Armies of Ignorance*

68   CCRAK documents, National Archives, RG 349, Entry 95C, Box 36-40, CCRAK, 101

69   CCRAK # M-101, 24 February 1953, declassified in 1990, National Archives, RG 349, Entry 95C, Box 38

70   *Khrushchev Remembers*, Strobe Talbott translation, pp. 282-287, Volume I; also: KGB, The Inside Story

71   Ibid.

72   *KGB, the Inside Story*, op. cit.; also, The New KGB, op. cit., and others

73   New York Times report

74   Testimony of Lt. Colonel Philip J. Corso, Hearings of the Senate Select Committee on POW/MIA Affairs, November 10, 1992.

75   *Blowback*, p. 237, op. cit.

76   *McCarthy and his Enemies*, William F. Buckley Jr. and L. Brent Bozell, Henry Regnery Company, 1954; also: *McCarthy*, Jack Anderson and Ronald May; and *Facing Reality*, Cord Meyer Jr., University Press of America, 1980

77   POW/MIA Policy and Process, Part II, p. 871, op cit.

78   Defense Department records reprinted in the VFW Magazine, June/July 1993

79   *From the Danube to the Yalu*, by General Mark Clark, George Harrap & Co. London, 1954, p. 298.

80   NYT, August 7, 1953

81   National Archives, RG 349, Entry 95-C

82   Annex B to Task Force Russia Biweekly report, 13 November 1992, Subject: Interview with LTC (Retired) Philip Corso

83   National Archives, Department of State Files, Box 144, August 26, 1953, "Information on UN Prisoners Retained Under Communist Control."

84   *The World Almanac* 1987-1992

85   National Archives, Records of the Office of the Secretary of Defense

86   Interview with the author, 1992

87   Declassified through FOIA request of the autho and turned over to the Senate Foreign Relations Committee in 1990 and the Select Committee on POW/MIA Affairs in 1991

88   "The Transfer of U.S. Korean War POWs To the Soviet Union," p. 5, op. cit.

89   Interviews and correspondence with Mrs. Jean Heise Earl, by the author

90   information Report, 29 December 1953, referred to in "The Transfer of U.S. Korean War POWs To the Soviet Union," p. 44, op. cit.

91   National Archives, Department of State Records, 611.61241/12-2153, Office of

the Secretary of State (John Foster Dulles) from Senator Lyndon B. Johnson, December 21, 1953 (#215486), and enclosures. The 14-page file of State Department documents related to this matter was given by the author to investigators for the Senate Foreign Relations Committee staff in 1991, through Senator Charles Grassley of Iowa, and were utilized in the May 1991 Final Report on U.S. POW/MIA policy issued by the minority staff of that Committee. The same documents were supplied to the Senate Select Committee on POW/MIA Affairs in 1992, but were not utilized in that Committee's final report, apparently because of objections by the Chairman, Senator John Kerry of Massachusetts.

92   National Archives, Records of the Department of State. The above cited document was among those given by the author to the Senate Foreign Relations Committee Republican Staff in 1990, and in 1992 the Senate Select Committee on POW/MIA Affairs also received it from the author, but declined to use it or hundreds of other historically-significant documents cited throughout this book in their January 1993 final report on U.S. POW/MIAs.

93   NYT, January 5, 1954

94   NYT, April 21, 1953; April 28, 1953; also: pp. 238-241, *Blowback*, by Christopher Simpson, Weidenfeld and Nicolson, NY, 1988

95   NYT, March 28, 1953; also: *Blowback*

96   Ibid.

97   *Witness to History*, op. cit.

98   *KGB*, Gordievsky & Andrew, pp. 452-453, op. cit.

99   National Archives, RG 87, State Department Records, "American POWs Reported en Route to Siberia," March 23, 1954

100   Charity Interrogation Report No. 619 referenced in declassified State Department cables dated 23 March 1954, "The Transfer of U.S. Korean War POWs To the Soviet Union," by the Joint Comission Support Branch, 26 August 1993, p. 38, op. cit.

101   "The Transfer of U.S. Korean War POWs To the Soviet Union," pp. 37-38, op. cit.

102   NA, RG 87, State Department files

103   NYT

104   Veterans of Foreign Wars, VFW Magazine, December 1991

105   National Archives, RG 87, State Department files: SEV 611.61241/2-1954)

106   *The CIA and The Cult of Intelligence*, p.324, Victor Marchetti and John D. Marks, Dell Publishing, NY, 1974

107   Sworn statements and prepared testimony of Lt. Colonel Philip Corso before the Senate Select Committee on POW/MIA Affairs, November 11,1992

108   Corso testimony

109   "The Transfer of Korean War POWs To the Soviet Union," Joint Commission Support Branch, pp. 51-52, op. cit.

110   *In Every War But One*, by Eugene Kinkead, W.W. Norton's, NY, 1959

111   Ibid.

112   6004th Air Intelligence Service Squadron, Semi-annual history of 1 July-31 December 1957)

113   Obtained through 1990 FOIA request of the author

114   Statement and testimony of Steve Kiba, before the Select Committee on POW/MIA Affairs, November 11-12, 1992, and interview by the author, same date

115   Interviews of Noble by the author and research colleague Thomas V. Ashworth

116   January 21, 1955, U.S. News and World Report

117   *I Was A Slave In Russia*, by John Noble, Devin-Adair, NY, 1958, p. 119

118   Obtained by the author in 1991 through FOIA request to USAINSCOM

119    "The Transfer of U.S. Korean War POWs To the Soviet Union," p. 12 of a classified report dated 26 August 1993 by the Joint Commission Support Branch, Research and Analysis Division, (Formerly Task Force Russia, U.S. Army)

120    Ibid., pp. 16-17

121    American Embassy Moscow Message, 311004Z December (31), 1992, POW/MIA, TFR-M Members visit to Irkutsk and Khabarovsk; p. 18, "The Transfer of U.S. Korean War POWs To the Soviet Union, op. cit.

122    American embassy Moscow Message 2711132Z May 1993; also: ibid, p. 40.

123    17 June 1955 report, "Recovery of unrepatriated Prisoners of War," Records of the Office of the Secretary of Defense

124    Declassified by USAINSCOM-NSA, in 1991, after FOIA requests by the author and research colleague Thomas V. Ashworth

125    Microfilmed, declassified Department of State records obtained by the author through FOIA request

126    February 14, 1959 letter from Captain Eugene Guild, Headquarters, Fighting Homefolks of Fighting Men, to Mrs. Cecil E. Moore, in Box 75, Hoover Institution on War, Revolution and Peace, courtesy Dr. Stephen H. Johnsson, a research colleague of Dr. Julius Epstein

127    Letter dated March 3, 1959, same source as cited previously

128    Report of the Far East and Pacific Subcommittee, House Foreign Affairs Committee hearings on prisoners of war, p. 22

129    February 14, 1959 letter, op. cit.

130    Bulletin # 383, October 25, 1957 (The Rita Van Wees Report)

131    Bulletin #382, same source as above, Assistant Secretary of State Walter S. Robinson, November 20, 1957 letter to Rita Van Wees

132    Microfilmed State Department Files on American citizens reported detained in the USSR, obtained by FOIA request of the author

133    Associated Press report, January 22, 1959

134    Bulletin #402, February 3, 1959, Fighting Homefolks of Fighting Men, op. cit.

135    #402, February 2, 1959

136    The author was shown this article by a Korean War veteran, Francis Homsher of Ardmore, Pennsylvania, in the early 1960s

137    Bulletin #439, Fighting Homefolks of Fighting Men, same archival source

138    Ibid.

139    Original statement of Serban Oprika to the Defense Intelligence Agency, "Sighting of Possible U.S. POWs (Possibly Korean War Eraln North Korea."

140    Senator Robert Smith (R-NH) citing the Los Angeles Times, 8 July 1990

141    U.S. Veteran Special Report by Dak Rhee, 1992)

142    News reports in USA Today, September 27, 1993; also: New York Times and others

143    "The Transfer of U.S. Korean War POWs To the Soviet Union," by the Joint Commission Support Branch, p.7& p. 62, op. cit.

144    Associated Press, published in the Santa Rosa, California, Press Democrat, Thursday, September 16, 1993

# PART III

# FROM THE COLD WAR TO THE COLLAPSE OF THE SOVIET UNION

## CHAPTER NINE

### THE VIETNAM-INDOCHINA WAR

The scene of the last major struggle in the decades-long global conflict between the Soviet Communist empire and the Western Capitalist democracies, led by the United States, was to be in Indochina. This great peninsula of southern Asia, of vast tropical jungles and rugged mountains, was to be an unfortunate choice of terrain upon which to fight an extended land battle. First the French and then the Americans engaged the formidable forces of an implacable Communist regime headed by Ho Chi Minh, supplied with modern Soviet and Chinese weapons, combat advisers and eventually with Soviet armor, anti-aircraft missiles and crews.

The prehistoric Vietnamese had pushed into the Indochinese Peninsula from the north and conquered the indigenous people, who for centuries to come were driven farther back into the mountains from the coastal plains, where they remained traditional enemies of the Viet people and would one day be referred to as Mois, or savages. A Chinese general named Chao T'o united the Vietnamese tribes and part of adjoining south China into a kingdom called Nam Viet in the 2nd century B.C. China conquered Nam Viet by 111 B.C. and by 679 A.D. had changed the name to Annam. Meanwhile, in southern Vietnam the Kingdom of Funan had developed in the Mekong River delta by 200 A.D., while another kingdom was growing in central Vietnam called Champa.

In the 600s A.D., Khymer people from the west conquered Funan in the south and built a great empire centered at Ankor Wat (Cambodia). The Chinese left Annam in 939 and for 500 years to come the Vietnamese maintained a large degree of independence and developed their own small empire. Under the Ly and Tran families from 1009-1400, Annam was called Dai Viet and defeated its enemies to the west and north. The Vietnamese defeated invading Mongol armies several times between 1225 and 1400, but the country was reconquered by the Chinese in 1407. They were driven out again in 1427. This fight was led by the Le family who ruled Dai Viet until 1827.

By the 1600s and 1700s European traders and Catholic

missionaries had begun to penetrate the country and they were to have a lasting impact on Vietnamese culture. After 1802 the southern part of Dai Viet was ruled by Nguyen Anh, who defeated the Tay Son ruler of the north in 1802 and declared himself emperor Gia Long, of all Vietnam. The Nguyen family remained emperors of Vietnam until 1945 and the end of the Second World War, but real power was in the hands of the French who had begun to conquer Vietnam in 1858. They seized Saigon in 1861 and the rest of southern Vietnam by 1867, using naval and ground forces. By 1883 the French controlled northern Vietnam and they forced the Nguyen ruler to sign a treaty that year recognizing French control over all Vietnam. The French called their new colony Indochina and divided it into three parts: Tonkin in the north, Annam in central Vietnam and Cochin China in the south around the Mekong delta. France also gained control of Cambodia and Laos. Several uprisings and revolutions against French rule occurred and in 1919, a young Vietnamese nationalist named Nguyen Ai Quoc traveled to Paris at the time of the Versailles Peace Conference to plead for independence for his people. He was ignored and after later adopting the name Ho Chi Minh, he was to lead his people against the French occupiers in a war.

After Nazi Germany defeated France in 1940, its ally Japan took control of French Indochina. The French colonial administration was permitted to function until March 1945, when the Japanese suddenly arrested all French officials and declared that Annam and Tonkin were independent of France. Near the end of the Second World War, the United States position concerning Indochina was enunciated by President Franklin D. Roosevelt, in a meeting with Soviet dictator Joseph Stalin at Yalta, on February 8, 1945. Roosevelt then told Stalin that what he had in mind for Indochina was a trusteeship, but added that the British did not approve of this idea, as they wished to give it back to the French, once the Japanese occupation had been overthrown. Stalin agreed with Roosevelt in opposing the British view. According to the official record of Yalta, Roosevelt had little appreciation of the fighting qualities of the Indochinese, who had once defeated the Mongol hordes:

"The President said that the Indochinese were people of small stature, like the Javanese and Burmese, and were not warlike. He added that France had done nothing to improve the natives since she had the colony. He said that General de Gaulle had asked for ships to transport French Forces to Indochina. Marshal Stalin inquired where de Gaulle was going to get the troops. The President replied that de Gaulle said he was going to find the troops when the President could find the ships, but the President added that up to the present he had been unable to find the ships."[1]

A few days after the Japanese defeat in August 1945, Ho Chi

Minh arrived in Hanoi from China, as head of the Revolutionary League for the Independence of Vietnam, which came to be called the "Vietminh." After the Japanese surrender, in September 1945 the French made immediate moves to recover their colonial empire in Indochina. Formerly interned French troops were rearmed and, assisted by the British, began reasserting authority in Saigon and later in Hanoi. Ho Chi Minh and Vo Nguyen Giap, who were by now Communists and leaders of a small guerilla resistance force supplied by the American OSS out of China, had begun making a few attacks on the Japanese late in the war. After years of Marxist indoctrination and preparation they felt ready to make the bid for Vietnamese independence which had long before taken Ho Chi Minh to the Versailles Conference, out of which so much evil had come to the world. By 1945 and the reestablishment of European colonialism that had been temporarily interrupted by the Japanese, Ho Chi Minh and Vo Nguyen Giap believed themselves to be the only legitimate leadership for an independent Vietnam.

The American government's policy toward Indochina at this time still favored nationalist, anti-colonialist movements opposed to a reassertion of French control. This policy had been enunciated by Franklin D. Roosevelt during the war and at the time of the Yalta Conference. General Wedemeyer's headquarters in China had sent instructions to General Claire Chennault that "no arms and ammunition would be provided to French troops under any circumstances." Wedemeyer, according to his memoirs, received his instructions directly from the President. OSS agents were reported at Ho Chi Minh's headquarters from May 1945 on, and OSS medic Paul Hoagland is credited with saving Ho Chi Minh's life by administering sulfa drugs and quinine to counter the effects of malaria, dysentery and other tropical diseases, in August 1945. American intelligence agents encouraged and supported Ho Chi Minh at a time when France was an important ally of the United States in the European confrontation with Stalin over the closure of eastern Europe to UN observers attempting to locate hundreds of thousands of missing allied POWs and displaced persons inside the Soviet occupation zones. OSS officer Archimedes Patti, chief of the Indochina desk of the OSS Secret Intelligence Branch in Kunming China organized a combined Vietnamese-French force for operations against the Japanese in Tonkin, North Vietnam. A seven-man OSS team under Lieutenant Colonel A. Peter Dewey was also dispatched from Ceylon to Saigon in July 1945 to assist in liberating 200 American POWs who had been captured by the Japanese. These survivors were subsequently flown out in seven DC-3's. Dewey's team was in Saigon on September 2, 1945, as Ho Chi Minh declared Vietnam's first independence day in Hanoi, and he remained there when the first British soldiers, in the form of Indian Gurkhas from

Nepal, arrived at Tanh Son Nhut airport on September 12th, joining forces with a company of French paratroopers from Calcutta.

Dewey was on an intelligence-gathering mission in addition to his POW contact work, as hostilities broke out between the Viet Minh and the French. The first American casualty in Vietnam occurred when one of Dewey's officers, Captain Joseph Coolidge, was shot and wounded in a Communist Viet Minh ambush on September 26, 1945, while returning from Dalat with some allied ex-POWs. Colonel Dewey himself became the second U.S. casualty in Vietnam (and the first American "MIA" in Indochina) when he was reported killed by machine-gun fire in another Viet Minh ambush in the Saigon-Tan Son Nhut area on the same day; his body was never recovered. Ho Chi Minh sent a note of condolence to Colonel Archimedes Patti, who was by then the OSS Chief of Mission in Hanoi. Publication of the Pentagon Papers later revealed that Ho Chi Minh had also written to the American presidents Roosevelt and Truman on at least eight occasions in 1945-46, hoping to reach an accommodation with the United States on the future of Indochina, but he had been ignored. Another OSS officer in Indochina at this time who was to influence future events, Captain Lucien Conein, believed that Ho Chi Minh was actually a great statesman, but Conein also assisted in rescuing French officials from the Viet Minh. Conein was later reported in the Pentagon Papers to be the CIA officer in charge of paramilitary operations and sabotage in North Vietnam at the time of the French withdrawal in 1954.

The Vietminh seized control of a large part of norther Vietnam and also enclaves in Annam and Cochin China. Ho Chi Minh traveled to France in an effort to negotiate a solution to the crisis but failed in achieving his goal. The Vietminh attacked French forces throughout Indochina on December 19, 1946, and the First Indochina War had begun.

Substantial quantities of small arms, machine-guns and heavier weapons, together with ammunition and other military supplies, had been transferred to Viet Minh control by the OSS, which would be later used against the army of France. Subsequently, American policy gradually began to tilt away from the Viet Minh, as evidenced by a $160 million arms sale to the French, administered by the OSS. This was followed by more American weapons sales and increasing American involvement on the ground as the Cold War intensified in the late 1940s. With the U.S. war in Korea against Soviet proxy forces in 1950, American support for the French war effort in Indochina became more massive and overt. The French however, were hampered not only by much of the Indochinese population's hostility towards the reassertion of their colonial rule, but also by the proximity of a vast sanctuary in Communist China after 1949. The same situation which faced U.S. and U.N. forces in

Korea, of a large bordering enemy sanctuary, used for marshalling armies and supplies, was later to be duplicated for the American forces in the Second Indochina War.

In the latter part of the French Indochina War, from 1950-1954, the United States massively assisted the French against the Viet Minh, hundreds of American advisers served with French forces and American aircrews and intelligence provided direct combat support, in which some minor American losses occurred. In addition, with the collapse of the Nationalist Chinese or Kuomintang government of Chiang Kai-shek in 1949 and its escape to Formosa, the American CIA aided several defeated remnants of the Kuomintang Army which had retreated from China's Yunnan Province into the territory of the Shan states in Burma.

To maintain these thousands of soldiers and their camp followers, the CIA-supported Kuomintang generals in Burma rapidly assumed control of a large portion of the Shan states opium trade, expanding production by 1,000 percent in the ensuing ten years. During the French Indochina War, with Mao Tse-tung (Mao Zedong) supporting the Vietminh, these Kuomintang forces engaged in harassing attacks on China, and in other covert operations directed by the CIA and OPC. Shadowy American front companies supplied arms and ammunition to them, airlifted by Civil Air Transport (CAT) a CIA front airline which later became Air America. Simultaneously, in prosecuting their war against the Vietminh, the French military and intelligence services were supporting the Laotian mountain tribesmen known as the Meos, against their traditional Vietnamese enemies. Much of the Meo opium crop was transported to Vietnam and beyond, by French aircraft and stored in warehouses of the French Montagnard advisory group, MACG (Mixed Airborne Commando Group). The limestone mountain opium-growing region straddling the borders of Burma, Laos and Thailand was known as the "Golden Triangle" and eventually supplied more than half the word's supply of illegal opium, from which the stronger drug heroin is refined. Thus did the Cold War ideological confrontation lead to involvement between Western intelligence agencies engaged in covert operations and criminal elements controlling the illicit drug trade of Southeast Asia, and elsewhere, which was to have far-reaching implications in the coming decades.[2]

Despite the massive American aid to France, when faced with the annihilation of an entire French army at Dien Bien Phu between March and May 1954, President Dwight D. Eisenhower refused to permit U.S. combat forces to intervene on the side of the French with U.S. tactical aircraft strikes. The anti-colonialist legacy of the Roosevelt and Truman administrations was still strong in some Washington circles, and the bitter American experience in Korea had made Eisenhower and some American military leaders leery of escalating America's commitment to the unpopular French occupation. Negotiations for an Indochinese peace settlement involving the French, Vietnamese, Americans, Russians and British in Geneva resulted in an armistice signed on July 20, 1954.

A statistical summary compiled by the French Army's Historical Service which was published in Paris during 1991, indicates that in 9 years of bitter guerilla warfare from 1945-1954, a total of 39,888 French Union prisoners and MIAs were estimated to have fallen into Vietminh control, and of this number 29,954 were never repatriated. 2,350 of those not repatriated were French nationals and 2,867 were French Foreign Legionnaires of various European nationalities. The rest were North Africans, Africans, Vietnamese, Laotians and Cambodians who had been fighting on the French side.[3]

Official French figures published 30 years before stated that 2,587 French Mainland POWs had been returned from July-October 1954, in the prisoner exchanges following the end of the war, as were 2,567 Foreign Legionnaires, 3,369 North Africans, 796 Africans and 1,435 Vietnamese, totaling 10,754 prisoners returned by the communists in 1954. As the French Union missing in action during the war had totaled 36,987, including 6,449 French Mainland, 6,328 Foreign Legion, 6,695 North African and 1,748 Africans, other missing soldiers had been presumed dead by the French government. Of those who disappeared in Viet Minh control, thousands are known to have died, sooner or later, while in captivity.

Upon capture by the Vietnam People's Army (VPA), or "Vietminh," both wounded and unwounded French Union prisoners were often forced-marched many hundreds of kilometers and denied the most rudimentary medical care or sufficient rations for survival. While some French POWs were summarily executed, most of the deaths resulted from deliberate maltreatment at the hands of Viet Minh guards and communist political cadre. Death rates given by a careful French researcher indicate that conditions in Viet Minh camps were similar to the worst of Stalin's Gulag slave camps: "Camp 5-E counted 201 deaths out of a total of 272 inmates between March and September 1952. Camp 70 lost 120 out of a total of 250 men in July-August 1954. Camp 123 lost 350 men (one-half of its population) between June and December 1953. Camp 114 maintained an average of two deaths a day throughout 1952...Of the 10,754 PWs returned after the cease-fire, 6,132 required immediate hospitalization. Of these, 61 died within the next three months."[4]

Following the 1954 Geneva accords which ended the French Indochina war, the Soviet Union sent a delegation to North Vietnam to accomplish the forced repatriation of former Soviet-bloc nationals serving in the French Foreign Legion. Indochina expert and journalist/author Bernard Fall wrote of the shipment from Vietnam to other Communist countries, of eastern European former D.P.'s captured in Indochina while serving with the French Foreign Legion: "Some Eastern European Legionnaires were repatriated to their countries of origin against their own will only to be tried there as 'Fascists' by People's Courts."

One of Fall's sources was a book entitled Legion Étrangère, by Gunter Halle, published in the Soviet Zone of Germany in 1953, which cited, "A SOVIET AIRCRAFT (ILYUSHIN-12, REGISTRATION NUMBER SSSR-P.1783, AIRCRAFT COMMANDER GREGORY IVANOV, FLIGHT ENGINEER PETROV) AS HAVING TRANSPORTED SUCH PRISONERS FROM INDOCHINA TO EASTERN EUROPE. NEITHER FRANCE NOR THE U.S. RAISED A PROTEST AGAINST SUCH AN OPEN VIOLATION OF THE RULES OF WAR."[5]

The French Expeditionary force was withdrawn to southern Vietnam after the Geneva accords, and by 1956 only 5,000 French troops remained there. French advisory teams cut off in remote locations of North Vietnam were ordered to fight their way out to Laos and Thailand, but many French MACG advisors were gradually overrun and their desperate radio signals faded out year by year. With France's withdrawal the American advisory support for the Diem regime in South Vietnam rapidly assumed greater significance and the CIA engaged in a struggle for control of the Saigon area with the remaining French Deuxième Bureau intelligence operatives.[6] Secret negotiations and exchanges for the non-returned French POWs of the Indochina war continued from 1954 into the 1960s. Journalist and author William Stevenson has written of his being deliberately introduced to brainwashed French POWs in Indochina by Communist authorities during the interwar phase, after all French POWs were supposed to have been returned. Stevenson has recounted to the author how the secretly-withheld French prisoners appeared too terrified to even raise their eyes from the ground in the presence of their North Vietnamese security police captors, and how when they did speak it was only to parrot the Communist line that they had been 'reeducated' voluntarily to see the error of having served against the Vietnamese captors in an illegal and immoral cause. In recounting the difficulty facing American searchers, decades later, in attempting to locate similar secretly-held U.S. POWs, Stevenson said:

"IF THEY HADN'T TURNED OFF A TRAIL IN THE JUNGLE TO DELIBERATELY EXPOSE THE FRENCH POWs TO ME I WOULD NEVER HAVE KNOWN OR SUSPECTED FOR A MOMENT THAT THEY WERE THERE. THE VIETMINH WERE SO BRAINWASHED WITH COMMUNIST RHETORIC THEMSELVES, THEY THOUGHT THEY COULD USE ME AS A PATSY, TO TELL THE WESTERN WORLD HOW WELL THEY HAD TREATED THE FRENCH POWs, WHO THEY CLAIMED HAD REMAINED BEHIND VOLUNTARILY AND WHOM THEY SAID DID NOT WISH TO RETURN TO FRANCE."[7]

About 40 "Metropolitan" French POWs were returned to France from Vietnam in 1962. They had been accused of being "progressives" in the prison camps, as they were court-martialed upon return for various offenses and some received prison sentences of up to five years. The Hanoi regime admitted the existence of about twenty other French POWs who declined to return to France. This does not mean that there are not other French POWs still alive in communist Vietnamese captivity, only that they do not admit to holding them.[8]

One case in point is that of Private First Class Yves Le Bray, a radioman of a French artillery battalion captured in a night ambush near Haiphong and interned for six months in 1954 until after the fall of Dien Bien Phu and the subsequent exchange of prisoners with the Vietminh. A January 4, 1965 Newsweek Magazine article revealed his fate thereafter:

"After the fall of Dien Bien Phu in May 1954, a fresh inscription was chiseled in the gray stone war memorial in the little Breton town of Pleudihen. Lettered in gold, it read: 'Yves Le Bray, mort pour La France.' (died for France) And on All Saints Day every year thereafter, someone from the Le Bray family joined in placing a wreath beside the plaque honoring Yves and other heroes of Pleudihen who had died for France. Le Bray's lost decade began...when he was a 21-year-old PFC..."

"...AT THE WAR'S END, INSTEAD OF BEING RETURNED TO FRANCE LIKE MOST OF HIS FELLOW PRISONERS, HE WAS PACKED OFF BY THE NORTH VIETNAMESE GOVERNMENT TO LANGSON NEAR THE CHINESE BORDER, TO BECOME A SLAVE LABORER...HAVING LOST ALL TRACE OF HIM, FRENCH AUTHORITIES PRESUMED THAT LE BRAY HAD BEEN KILLED IN ACTION AND THUS IT WAS THAT HIS NAME WAS ADDED TO THE 'MONUMENT AUX MORTS', BACK IN PLEUDIHEN...EVENTUALLY, MORE THAN TEN YEARS AFTER HIS CAPTURE, THE FRENCH LEGATION IN HANOI FOUND OUT ABOUT LE BRAY, OBTAINED HIS FREEDOM, AND SENT HIM WINGING HOMEWARD ON AN AIR FRANCE JET."

Livesightings of French POWs are possibly still occurring, as such reports continued in Vietnam through the 1970s and into the 80s. Many declassified CIA and other intelligence reports from the Vietnam war years and later in the 1970s reported the continued existence of French POWs, sometimes held in the same camps as American POWs captured 10 or 15 years later. While the French government has maintained a public policy of not paying any form of ransom for the return of war prisoners, in reality the French have secretly paid for the return of some 1,000 or more French Union POWs since the end of the first Indochina war. This diametrically opposite public and clandestine policy towards secretly-held POWs is in keeping with a long tradition of French (and European) diplomacy, which resulted from many centuries more experience in the realities of world power-politics than the United States of

America possessed.

## THE COLD WAR TURNS HOT

The development of an actual Soviet nuclear threat to the population of the United States in the post-Korean war period, with the successful 1953-54 Soviet testing of a hydrogen bomb and the 1957 launching by rocket of the Soviet "Sputnik" satellite, made any future direct conflict between the superpowers more perilous, and secondary wars between proxy forces of both sides continued on battlefields in remote corners of the world. At this time the Soviets also faced serious revolts in East Germany and Poland, and in the remote Gulag camps of Siberia, which were mercilessly crushed with great loss of life. By March 1954 the premier Soviet intelligence agency and successor of the NKVD, then known as the MGB, had again been renamed, to become: Komitet Gosudarstvennoy Bezopasnosti, the "KGB." The personnel remained the same and the terror instilled in the population by the "Chekists" continued unabated.

The American intelligence services continued their war against the KGB throughout the world. In addition to CIA covert operations in eastern Europe and Asia, the agency's clandestine services at this time were active in Central America, where a leftist government had taken power in Guatemala. Arbenz Guzman initiated land reforms for the rural poor Maya Indians and mestizos which angered the Guatemalan elite and the American United Fruit Company, resulting in a CIA-run coup d'état against the democratically-elected government there, in 1954. CIA operatives C. Tracy Barnes and E. Howard Hunt, who were involved in the Guatemalan coup in 1954, would later attempt a similar operation in Cuba. The CIA coup in Guatemala delivered that country into the control of a series of military dictators, and caused widespread bitterness among the poor and disaffected in Latin America for years to come.

Hungary revolted against its Soviet masters in 1956, after hearing encouragement from the Voice of America, and thousands of Hungarians died fighting against Russian tanks; yet the United States did nothing but complain to the U.N. and offer sanctuary for Hungarian refugees. Thousands of other Hungarians, imprisoned by authority of then-KGB Chief Ivan Serov, who personally took command in Budapest at the time, were shipped east to KGB forced labor camps in Siberia, where many of them vanished, as had thousands of East Germans and Poles arrested in earlier 1950s rebellions against Communist authority in their countries. During the Hungarian Revolution, Vladimir Kryuchkov, who was to become KGB Chairman under Mikhail Gorbachev, was active in Budapest under cover as a third secretary to Ambassador Yuri Andropov, another future KGB chairman who later became the supreme Soviet leader after Leonid Breshnev. Andropov was the Soviet Ambassador in Budapest during the 1956 revolution and served as a proconsul for Khrushchev and the Central Committee, in his ruthless suppression the revolt.[9] The Soviets made gains in their influence in the Middle East following an unsuccessful 1956 war in Suez by the British and French against Egypt's Gamel Abdul Nasser. Within the Soviet Union the repressive policies of the Communist regime continued. Soviet writer Boris Pasternak was persecuted for his 1957 novel "Doctor Zhivago," until his death in 1960, and was repeatedly vilified and disgraced for his portrayal of the death of the Russian intelligentsia after the Bolshevik Revolution. Although the population of the Gulag labor camps had been somewhat reduced by the liberalization under Khrushchev, millions of prisoners still toiled for the state in thousands of camps and prisons. Those who were released were often kept in a form of internal exile or not allowed to travel within the Soviet Union from their appointed place of residence. Thus, foreign POWs and civilian prisoners allowed out of the camps were rarely able to make their presence known to carefully watched Western embassies.[10]

The thousands of American prisoners of the Korean War and World War II held captive in Gulag camps were periodically augmented by U.S. airmen shot down and captured from Cold War reconnaissance missions; and 1956 reports that some were alive in Soviet captivity prompted the U.S. Government to request their return from the USSR. On April 8, 1950, Soviet MiG-15 jet fighters had shot down a U.S. Navy PF4Y-2 Privateer reconnaissance aircraft, approximately 20 miles off the Soviet Baltic Sea coast. The aircraft, which contained a highly trained crew of 10 men, was a modification of the B-24 Liberator Bomber of WW II, and contained sophisticated electronic systems and eavesdropping equipment. The Soviets claimed that the unarmed Privateer had attacked their aircraft and that there were no survivors. The US Navy stated at the time that the Privateer was on a routine flight from Germany to Greenland. In 1951 the U.S. Government had announced that all ten crewmen of the Privateer were "presumed dead." This was an early example of what became known in the intelligence community as "Wink and Nod Diplomacy," in which both antagonists know full-well that the intercepted aircraft, or agents, or captured military personnel, are involved in a secret intelligence gathering operation, and if captured, their actual mission and even their existence will be denied. Major diplomatic incidents between the powers would thus be avoided by declaring the missing personnel dead and ignoring the disappearance of survivors.

However, in this particular case, William Marchuk, an

American who was later released from the Soviet forced-labor camps after the Korean War, revealed that he had spoken with a Russian named Rusin, in a slave camp at Vorkuta, who stated that he had personally helped to rescue 8 of the 10 crewman of the Privateer from the Baltic Sea, in 1950, and that they had been retained as captives by Soviet authorities. John Noble, another American who was released from the Gulags in 1955, also revealed that he had spoken to Russians who knew that the crewmen of the Privateer had been taken prisoner.[11]

The publicity which grew out of the statements by Marchuk, Noble and others subsequently forced US officials to revise the previous public statements. In 1956, according to declassified official documents, U.S. Secretary of State John Foster Dulles directed that the U.S. Embassy in Moscow deliver a three-page request to the Soviet Foreign Ministry that Soviet authorities provide an accounting of the missing Americans from the Privateer and from the subsequent B-29 incident, involving a total of 22 Americans. The State Department request was dated July 16, 1956 (#388):

"The Embassy of the United States of America presents its compliments to the Ministry of Foreign Affairs of the Union of Soviet Socialist Republics and has the honor to refer to the question of the detention of United States military personnel in the Soviet Union. THE UNITED STATES GOVERNMENT HAS FOR SOME TIME RECEIVED, FROM PERSONS OF VARIOUS NATIONALITIES FREED FROM SOVIET GOVERNMENT IMPRISONMENT DURING THE LAST SEVERAL YEARS, REPORTS THAT THEY HAVE CONVERSED WITH, SEEN OR HEARD REPORTS CONCERNING UNITED STATES MILITARY AVIATION PERSONNEL, BELONGING EITHER TO THE UNITED STATES AIR FORCE OR TO THE UNITED STATES NAVY AIR ARM IN ACTUAL DETENTION IN THE SOVIET UNION."

"The United States Government has sought in all such cases to obtain...precise identification of American nationals detained by the Soviet Government...THE REPORTS CONCERNING SUCH PERSONNEL HAVE NOW BECOME SO PERSISTENT AND DETAILED, AND SO CREDIBLE, THAT...IT REQUESTS THE SOVIET GOVERNMENT TO INFORM THE UNITED STATES GOVERNMENT IN DETAIL CONCERNING EACH AMERICAN MILITARY PERSON WHO HAS BEEN DETAINED...SPECIFICALLY ONE OR MORE MEMBERS OF THE CREW OF A UNITED STATES NAVY PRIVATEER-TYPE AIRCRAFT WHICH CAME DOWN IN THE BALTIC SEA AREA ON APRIL 8, 1950. THE UNITED STATES GOVERNMENT HAS SINCE THAT TIME RECEIVED REPORTS THAT VARIOUS MEMBERS OF THE CREW OF THIS AIRCRAFT WERE, AND ARE DETAINED IN SOVIET DETENTION PLACES IN THE FAR EASTERN AREA OF THE SOVIET UNION. In particular, in 1950 and in October 1953, at least one American military aviation person, believed to be a member of the crew of this...Privateer, was held at Camp No. 20 allegedly near Taishet, and

Collective Farm No. 25, approximately 54 kilometers from Taishet, said to be under sentence for alleged espionage. This American national was described as having suffered burns on the face and legs during the crash and using crutches or a cane."

"Reports have been received from former prisoners...at Vorkuta that in September 1950 as many as eight American nationals, believed to be members of the crew of the...Privateer...had been seen in the area of Vorkuta...One or more members of the crew of a United States Air Force B-29 which came down on June 13, 1952, either over the Sea of Japan or near the Kamchatka area of the Soviet Union. AN OFFICER, BELIEVED BY THE UNITED STATES GOVERNMENT TO HAVE BEEN A MEMBER OF THIS CREW, WAS OBSERVED IN OCTOBER 1953 IN A SOVIET HOSPITAL NORTH OF MAGADAN NEAR THE CROSSING OF THE KOLYMA RIVER BETWEEN ELGEN AND DEBIN AT A PLACE CALLED NARIONBURG. THIS OFFICER STATED THAT HE HAD BEEN CONVICTED, WRONGFULLY, UNDER ARTICLE 58 OF THE SOVIET PENAL CODE."[12]

This July 16, 1956 document listed by name the 22 Americans missing from these two crews and further requested information from the Soviets on any other Americans captured from shot down aircraft "engaged on behalf of the United Nations Command side of the military action in Korea..." The Soviet government subsequently denied knowledge of any of these prisoners remaining in captivity.

These reports of the Americans seen by Soviet prisoners may indicate that they were "recalcitrant," who had resisted cooperating with Soviet authorities, and as punishment were separated from their fellow U.S. prisoners and thereby exposed to the hazards of the general Gulag population. It is interesting to note that then-Undersecretary of State Robert Murphy, who had long concealed his knowledge of the US POWs of WW II remaining in Soviet captivity after 1945, also publicly downplayed these postwar livesightings of Americans in the gulag camps. Subsequent to Dulles's message to the Soviet government, and despite accurate livesighting reports of the locations of these American prisoners, Murphy was quoted in a United Press report of July 19, 1956 as stating:

"...THIS COUNTRY'S PROTEST TO THE KREMLIN THAT RUSSIA IS HOLDING SOME US AIRMEN ILLEGALLY WAS SOMETHING OF A STAB IN THE DARK."[13]

Incidents like these two shoot-downs occurred throughout the 1950s and 1960s, but the Soviets usually withheld all survivors as captives unless the political situation warranted a publicized release, as in the Gary Powers-U2 incident.

It was within this Cold War context, in 1957, that an intense and bespectacled 18-year-old U.S. Air Force volunteer, from Watertown, South Dakota, named Jerry James Mooney, began a brilliant 20-year intelligence career as a U.S. cryptanalyst, that was to span the entire Vietnam-Indochina War era. After two decades of distinguished service, Mooney's conscience was to lead him into a direct confrontation with the high and mighty leaders of America from 1986-1992, over the fate of more than 1,000 U.S. prisoners of war who had vanished and were eventually declared to be dead by their own government.

Following 22 weeks of intensive training in Air Force Communications School, which led to the award of an initial security clearance, Mooney was also indoctrinated for the first time in the secret and unacknowledged "Wink and Nod Diplomacy," which had been practised by leaders of the United States and the Soviet Union, to prevent the inevitable violent Cold War 'incidents' from escalating into a devastating nuclear exchange. Mooney's first teacher, veteran Air Force Technical Sergeant Saint George gave a final and most valuable 1957 lesson to those young communications specialists who would eventually replace him as America's invaluable Cold War codebreakers, that consisted of a few well-chosen words which covered the entire history of secret Soviet-American relations since 1919:

"IF YOU'RE CAPTURED, DON'T EXPECT TO BE RETURNED. THEY TAKE OURS AND WE TAKE THEIRS, AND YOU'RE GOING TO MOSCOW, BUDDY, YOU'RE MOSCOW-BOUND."

It was this early introduction to an already-old intelligence term, "Moscow-Bound," (albeit then-unknown to the American public), by intelligence veterans of WW II the Cold War and Korea, which intentionally inoculated Mooney and other young communications intelligence trainees of the late 1950s from succumbing to any later shocks at the sight of irrefutable evidence in decrypted messages on the disappearance of American prisoners in Soviet or other hostile control, throughout their subsequent careers. This was intended to avoid future security violations resulting from emotional human reactions to handling highly-secret decrypted enemy communications which would time and again prove, beyond a shadow of doubt, to Mooney and others, that American prisoners captured alive were taken into the Soviet's or Soviet-surrogates prison camp systems, where they subsequently vanished.

Jerry J. Mooney also felt privileged later to be among the last of a younger generation who received guidance and instruction from the "grandfather of American communications intelligence" at the National Security Agency (NSA), Dr. William Friedman. Friedman delighted more than anything else in teaching young cryptanalysts the methodology and tricks-of-the-trade he had developed for the United States since the 1918-1919 era of Soviet-American relations, in which he and Herbert Yardley of MI-8 had played such a prominent role. Mooney later remembered: "He loved showing us how easy to solve some problem really was, and when you had finally figured it out he made you feel like it was as simple as a child's picture puzzle."

After more training in cryptanalytic, during his first overseas duty assignment gathering intelligence on the Turkish-Soviet border on September 2, 1958, Mooney received his first personal confirmation of 'wink and nod diplomacy,' from Air Force Technical Sergeants Sanchez and Dapalito, following the Soviet shoot-down of a U.S. EC-130 aircraft with 17 Americans aboard, over Kinegi near the Armenian SSR-Turkish border. He was informed by these two Surveillance and Warning supervisors, who were on duty when the incident occurred, that the American intelligence aircraft had been, "lured into a trap with Soviet fighters inside Soviet airspace, with a Soviet beacon, while the U.S. beacon was either being jammed by the Russians, or after the American beacon was overshadowed by a much louder Soviet beacon, and that after the Russian attack there were 'PROBABLE SURVIVORS' in Soviet control, who then disappeared inside the USSR. (Other evidence indicated that the American aircraft may have purposely violated Soviet territory. U.S. intercepts of Soviet radio traffic from the fighter that intercepted the EC 130 were eventually revealed: "218 are you attacking?... Yes, Yes... the target is burning. The tail assembly is falling off the target...82, do you see me? I am in front of the target...Look! Oh!...Look at him, he will not get away, he is already falling, I will finish him off, boys, I will finish him off on the run, 62...the target has lost control and is going down."[14]

Although six bodies were returned by the Soviets on September 24th, 1958, only four of them could be identified and the other 13 Americans were listed as missing. Premier Khrushchev denied Russia was holding the survivors as prisoners, but the U.S. Government under President Eisenhower, pressed the matter. Llewelyn Thompson, the U.S. Ambassador in Moscow, cabled the Secretary of State on May 4, 1959: "...PROBABLE SOVIETS HAVE ALL OR SOME OF MISSING AIRMEN BUT SEE NO WAY OUT OF IT OTHER THAN TO STICK TO THEIR STORY."[15]. Although Vice President Richard Nixon later confronted Nikita Khrushchev on these MIAs, nothing came of it, or of livesightings of members of the crew as actually captured in the Yerevan (Erevan) area of Armenia, reported by East German writer Wolfgang Schreir in the Communist magazine OGONEK.

In a contemporary cable dated September 20, 1958, from Davis at the U.S. Embassy in Moscow, to the Secretary of State, is an official allusion to the "wink and nod diplomacy" then in practise:

"Now that Soviets have denied twice in writing that they have any information about 11 missing crewmen it appears unlikely that in future they will admit to having captured them or that they have in mind any sensational trial charging espionage...FROM KNOWLEDGE AVAILABLE TO U.S. ABOUT THIS INCIDENT IT SEEMS SOVIETS HAVE DETERMINED TO MAKE 11 CREWMEN 'UNPERSONS' WITH OBJECTIVE

IMPRESS ON U.S. AS WELL AS OTHER COUNTRIES DANGERS FLYING
CLOSE TO SOVIET BORDERS. I discussed...with Swedish Ambassador,
who has been here 10 years...He immediately recalled incidents
back in 1951-1952 when two Swedish planes shot down over Baltic
Sea. Crew of first plane disappeared without trace and only bullet-
punctured life raft found later. In second instance Swedish crew
picked up by passing ship. He recalled these incidents led to heated
exchange of notes between Swedish and Soviet governments but
Soviets consistently denied any knowledge missing crew first plane.
Despite rumors crew might be in Soviet hands, Swedish authorities
have never been able discover any evidence..."[16]

U.S. Undersecretary of State Robert Murphy, who had
witnessed many such incidents since the Soviet mass-kidnapping
of POWs in 1945, met with the Soviet Ambassador, Menshikov,
on October 10th. A secret, limited distribution State Department
memorandum of conversation recorded his demand that the U.S. be
permitted access to the crash site, and the Soviet response:

"Mr. Murphy raised with Ambassador Menshikov, who called on
another matter, the question of the eleven airmen still unaccounted
for and missing...Mr. Murphy stressed that not only have we been
unable to get any information regarding these airmen but that we
have also not been given access to the site near Yerevan where the
plane crashed...Ambassador Menshikov then observed that the site
of the crash was probably in a restricted frontier area and stated
that we, in effect, were asking permission to visit a closed area in
the Soviet Union when we would not permit members of the Soviet
Embassy, including himself, permission to travel to closed areas in
the United States... Mr. Murphy emphasized that these restrictions
had been initiated by the Soviet Union and that the United States
had repeatedly proposed their abolition on a reciprocal basis...
Ambassador Menshikov observed that when more 'trust' has been
developed between our two countries, these restrictions might then
be eliminated... Mr. Murphy added that we had evidence indicating
that the C-130 was shot down by Soviet fighters but wanted to avoid
publicity on this and other aspects of the case if at all possible."

Thus, a public silence gradually ensued on the fate of these 11
American airmen.[17]

A recently declassified letter of August 1, 1959, from Vice
President Richard Nixon to Nikita Khrushchev admitted the aircraft
was downed over Soviet territory and reveals that another effort
was made by the White House to recover the living survivors of this
incident who remained in Soviet captivity:

"During our conversation on July 26, I mentioned to you that
several problems of a bilateral nature continue to be a source of
friction between our two countries...One problem is the fate of the

eleven missing men who were aboard the C-130 U.S. Air Force transport aircraft that was shot down over Soviet Armenia on September 2, 1958...IT IS IMPOSSIBLE FOR THE AMERICAN PEOPLE TO BELIEVE THAT THESE ELEVEN CREW MEMBERS DISAPPEARED WITHOUT A TRACE AND THAT NOTHING IS KNOWN ABOUT THEM BY THE SOVIET AUTHORITIES..."

Khrushchev responded on August 22nd:

"I have taken note of your letter of August 1, 1959., in which you again raise the question of the United States Air Force plane which crashed September 2, 1958 on Soviet territory, 55 km. northwest of the city of Erivan. Frankly, I must say that your raising this question again after the Soviet Government has done everything in its power to bring to light the circumstances of this crash and has given full and complete information to the American side regarding the results of the investigation, can only puzzle us to say the least. This cannot be regarded as other than an attempt to set up artificial barriers in the way of an improvement in American-Soviet relations...An American military plane crashed...near the city of Erivan, where the remains of military personnel were discovered. From these remains, which were handed over to the American side on September 24, 1958, it might be supposed that six crew members had perished. No other American airmen from the aforementioned plane were found, and consequently there are no such airmen in the Soviet Union...similar flights of American planes and violations by them of Soviet state boundaries are still continuing. In particular, such flights were noted in the eastern part of our country literally on the eve of your visit to the Soviet Union..."

The personal intervention of President Eisenhower in the matter of these captive Americans was to no avail either. They had become "non-persons."[18]

Long after this shootdown, while visiting the U.S. intelligence facility at Trabzon, Turkey in 1961, Mooney saw the actual decoded Soviet communications on this 1958 'incident' in which Soviet pilots first reported there were 'good parachutes' emerging from the aircraft after it had been crippled by the Soviet fighter, and the subsequent decrypted traffic stating variously that there were 'FIVE TO EIGHT SURVIVORS IN SOVIET CONTROL.'[19]

This was cryptanalyst Jerry J. Mooney's first personal experience with U.S. POWs who were 'Moscow-bound,' and 33 years later, in an interview with the author in his Montana home, he recalled: "IT WAS THIS INCIDENT THAT ISOLATED ME FROM EITHER SURPRISE OR ANY GRAVE CONCERN OVER THE U.S. POWs OF VIETNAM WHICH I CARRIED ON LISTS AT NSA IN 1972 and 1973 AS 'MOSCOW-BOUND,' IT WAS JUST BUSINESS AS USUAL..."

During 1959, while analyzing decrypted Soviet radio communications, Mooney got his first personal confirmation of the

Soviet use of slave-labor in the crash construction of Russian missile and space testing facilities in places such as Saryshagan, on Lake Balkash in Kazakhstan.

"THE TRAFFIC WE WERE MONITORING INDICATED THAT THEY WERE BRINGING IN THOUSANDS OF FORCED-LABORERS, IN FLIGHT AFTER FLIGHT OF JET TRANSPORT AIRCRAFT AFTER THEY FINISHED CONSTRUCTION OF THE AIRFIELD DURING WHICH THEY HAD ALSO USED PRISONERS, FROM ALL OVER THE SOVIET UNION. YOU CANNOT IMAGINE THE MANPOWER IT TOOK TO CONSTRUCT THESE ENORMOUS, SOPHISTICATED FACILITIES, USING HUGE AMOUNTS OF CONCRETE AND STEEL PILING, IN LESS TIME THAN IT TOOK AMERICA TO CONSTRUCT OUR OWN MISSILE AND SPACE TEST CENTERS WITH OUR ACCESS TO UNLIMITED HI-TECH EQUIPMENT OF A TYPE NOT AVAILABLE TO THE SOVIETS FOR DECADES, IF EVER. THEY USED UP THE LIVES THOUSANDS OF PRISONERS IN HANDLING THE MASSIVE AMOUNTS OF STEEL AND CONCRETE REQUIRED, AND I HAVE NO DOUBT THAT SOME OF THE PRISONERS USED WERE SURVIVING U.S. AND ALLIED PRISONERS FROM THE KOREAN WAR AND EARLIER SEIZURES OF PRISONERS FOLLOWING WORLD WAR TWO, AND THEIR OWN GULAG INMATES OF ALL SOVIET NATIONALITIES DURING THE COLD WAR PERIOD."

On May 1st, 1960, a favorite day of celebration for loyal Communists all over the world, the American U-2 pilot Gary Powers was shot down over the Soviet Union and captured alive, whereupon he, like so many American military and civilian prisoners before him back to 1918, was transported to Moscow. He eventually spent time in the Lubyanka Prison and other places of confinement. The Soviets did not reveal that they had captured Powers alive when they announced the shootdown of the U-2, and upon the recommendation of Richard Bissell, the CIA denied that Powers could have been captured alive, whereupon Khrushchev embarrassed President Eisenhower by proving to the world that Powers was a prisoner. When the incident occurred, Mooney, then a Soviet decryption specialist under authority of the 'A-Group' of NSA, concerned with countering the primary nuclear threat against the United States, of the Soviet Union, was serving under Master Sergeants DeLauis and "Jug" Hatfield. Almost immediately after the shootdown incident occurred, according to Mooney during a 1992 interview by the author in his home, their Air Force Security Service commander, a Major Mazarowski, came running into Mooney's work-area shouting: "DESTROY EVERYTHING, DESTROY ALL THE DATA ON THE U-2." Mooney vividly remembers that his immediate supervisor, Staff Sergeant DeLauis, refused this direct order of his commander, and based upon this commitment of the senior NCOIC, Mooney himself also refused the Major's subsequent panicked orders to HIM to destroy all intercepts on the U-2 shootdown. A few minutes later they were glad to have resisted such a questionable order, as a call suddenly came from the White House saying that President Eisenhower was demanding that everything be reviewed and packed up for shipment to the President's oval office. The President's orders were: "I WANT EVERYTHING TO DO WITH POWERS AND THE U-2 SHOOTDOWN ON MY DESK, NOW!"

The captured U-2 pilot Gary Powers was subsequently freed in a trade for Soviet spy Rudolph Abel, after a confrontation between Eisenhower and Soviet leader Nikita Khrushchev, which had caused the cancellation of a Summit conference on the German problem. The release of Powers illustrated to Air Force/NSA analyst Jerry Mooney, in remembering the incident in an interview 30 years later, "that the political system could work if the policymakers did their job." Another shootdown of a U.S. RB-47 intelligence aircraft over the Kola Peninsula area of arctic Russia on July 1st, 1960, also resulted confrontation with the Soviets and the eventual release of two airmen, Captain John McKone and Captain Bruce Olmstead, in early 1961, after the inauguration of John F. Kennedy as President. However, three of the "Ravens," the technicians most valuable to the Soviets, who manned the sophisticated electronic intercept gear in the rear of Mckone's and Olmstead's RB-47 aircraft, still remain missing. They are: Major Eugene E. Posa, Captain Dean B. Phillips and Captain Oscar L. Goforth.[20]

In a subsequent confidential State Department memorandum dated October 10, 1961 (no. 1767), the views of Soviet premier Alexi Kosygin and other Kremlin officials on the return of the American prisoners McKone and Olmstead from the USSR were recorded by U.S. Ambassador Llewelyn Thompson. The Soviets clearly expected both a change in U.S. policy and increased access to American technology in return:

"At Indian reception last night Kozlov, Kosygin, Kuznetov and other Soviet officials as well as diplomatic colleagues I encountered expressed great satisfaction release American airmen...Marshal Bagramyan brought up subject and said it was NOW up to us which direction world would go. In long talk with Kosygin he repeatedly emphasized heavy responsibility I would have in expressing opinions about Soviet Union to new administration...During conversation he stressed importance expanding our technical exchanges. Kozlov expressed some concern as to how our press would handle returned American fliers."[21]

But many other American airmen and earlier U.S. prisoners of war remained in captivity or unaccounted for by the Russians, and the U.S. intelligence flights around the Soviet periphery did not cease. Indeed, neither did reports of American POWs held inside the Soviet Union. One example is a confidential September 15, 1962 telegram from McSweeney at the U.S. Embassy in Moscow, to Secretary of State Dean Rusk:

"Rose Wood, member of delegation of American women visiting USSR, passed report from American Armenian community in Erevan concerning American servicemen allegedly imprisoned in Siberia after World War II. Source in Erevan claims to have seen American soldier Gunner Zigerman of Philadelphia and 'MANY OTHER AMERICANS' in Siberia camp in 1955. Other members delegation and Time correspondent Stevens aware this report. Embassy has no record. (In) View public emotions which can be aroused by this sort report, Embassy would appreciate receiving any information Department or other government agencies can develop on Zigerman."

A subsequent airgram to Moscow from Washington signed by Secretary of State Rusk, claimed that no records existed in the Army, Navy, Air Force or Marine Corps which could be associated with this name, but then the federal departments and agencies controlled all records, so no independent check could be made.[22]

A third incident which underlined the unstated policy of 'wink and nod diplomacy' to the NSA's Mooney and other intelligence analysts with whom he served, occurred when he was assigned to the Turkish-U.S. Logistics area at Samsun, Turkey in 1961. When the Berlin Wall was suddenly erected during the Presidency of John F. Kennedy, resulting in a further rapid deterioration of U.S.-Soviet relations, the personnel at Samsun became aware that a Soviet naval task force had approached their area and the Russians were pointing their heavy guns right at the American intelligence-gathering facility. It was the first moment in Mooney's career when he and the other personnel with him who were monitoring the Soviet Union realized that they too, might suddenly be carried as "gone forever' and "Moscow-bound.' Nothing happened in the end, as the crisis eventually dissipated, but Mooney remembers writing fearful letters home to his family and seriously thinking for the first time, about the deadly serious 'game' that he and his comrades were engaged in, on the flanks of the hostile and dangerous Soviet empire where so many American prisoners had vanished with barely a trace, only to become statistics of 'classified information' in the secret recesses of Washington, D.C. Every time a U.S. intelligence aircraft was lost in the constant secret war between the U.S. and Russia in the 'peripheral Reconnaissance' shootdowns of the Cold War, Mooney and his fellows realized what could happen to any of them, at any time, and that their own Government, to avoid a dangerous confrontation leading to a nuclear alert, would be disowned, ignored and abandoned. In 31 known Cold War intelligence aircraft shootdown incidents over, or near Soviet, Chinese and North Korean territory, which occurred between 1950 and 1970, at least 24 Americans were killed and 138 pilots or crewmembers disappeared and still remain missing. 90 other Americans were recovered by the United States.[23]

Meanwhile, throughout the early years of cryptanalyst Jerry J.

Mooney's Air Force Security Service and NSA career, the business of government went on, as did the never-ending process of grooming replacements for the now-aging Cold Warriors, who must one day, of necessity, retire. In 1956, a Harvard University government professor, Henry A. Kissinger, had been picked by former Assistant Secretary of War John J. McCloy (who after WW II had become the Rockefellers' banker) to serve as chairman of a special study of Soviet-American relations. Kissinger had been born in Germany to an Orthodox Jewish family that escaped to America before Hitler's holocaust. He had served as a translator in the Army Counter Intelligence Corps during WW II, where he was reputed to have attached himself to a senior U.S. officer so that his own future career would be enhanced.[24]

It was the influential McCloy who thus gave Kissinger his first chance to serve with the American policymaking elite, which had collectively demonstrated three times since 1919 that it had the determination to ignore American prisoners who had vanished into the hells of Leninist-Stalinist concentration camps, in adhering to its own vision of correct American foreign policy. Kissinger was to listen and learn in the years to come, while connected to the most powerful vested interests in the nation, but he still maintained close ties with the liberal academia of America. Thus he began a meteoric rise to eventually become President Richard M. Nixon's national security advisor, Secretary of State under President Gerald Ford, and a household name representing American diplomacy. Yet one of his unwanted and unstated legacies was to be the disappearance of approximately 1,000 living American POWs and MIAs of the next major war, who were to be secretly held in Vietnam, Laos, Cambodia, the Soviet Union and China, and to subsequently deny for many years that such was the case. This was the man whom Newsweek Magazine was to characterize, three decades later as: "Our greatest diplomatic technician.[25] First, however, the United States was to engage in a dangerous nuclear showdown with the Soviet Union over the installation of Soviet missiles on an island in the Caribbean which was only 90 miles from the Florida coast.

As a result of the successful guerilla campaign waged by Cuban revolutionary Fidel Castro, who had been lionized by the New York Times and other influential segments of the American media, dictator Fulgencio Batista fled the island republic on January 1st, 1959.[26] That same month, Fidel Castro stated in a speech that only 400 enemy prisoners from the former regime would be shot, but in fact mass-executions, totaling thousands of prisoners, took place under his authority in La Cabana Prison in Havana, and under his brother Raoul's authority in Oriente Province and in many other areas, from 1959-1961.[27] Although Castro had denied in April 1959 interviews in the United States that he was a Communist, he had long been planning—with his brother Raoul and the Argentine-born revolutionary, Che Guevara—to adopt that system. Major Ramiro Valdes, Castro's intelligence chief, began a series of secret meetings in Mexico City with the Soviet ambassador and the KGB. Among the more than one hundred Soviet advisers subsequently sent to the island to train Castro's intelligence service were many of the "Los Niños," children of Spanish Communists brought to the USSR during and after the Spanish civil war; those who had not died in the gulags. The Soviets cautiously provided the first military assistance through their Czechoslovakian Warsaw-Pact allies, after a visit to Prague by Raoul Castro.

At this same time, in October 1959, and prior to establishment of formal diplomatic recognition by Moscow, a Soviet "cultural" delegation went to Havana which was led by KGB official Aleksandr Shitov, using the alias "Alekseev," who later became the Soviet ambassador to Cuba. In a speech attacking America on July 9, 1960, Soviet leader Nikita Khrushchev stated: "We shall do everything to support Cuba in her struggle...now the United States is not so unreachable."[28]

More mass arrests in Cuba followed a December 28, 1960 order by the Castro regime, with thousands of political prisoners being confined in the Islas de Pinos gulag, and elsewhere. A disorganized anti-Communist revolt grew, which took the form of bombings, burnings and armed attacks by small guerilla units on Castro forces during early 1961, aided by CIA-backed landings of small groups of anti-Communist rebels along with arms and ammunition. Raul Castro later admitted that these numerous attacks had cost the Cuban Communist troops some 500 dead. In the prisons, common criminals were separated and treated more leniently than anti-Communist political, prisoners, in the long-established Soviet tradition, and forced-relocation of rural poor, "campesinos," to a centralized villages also occurred, particularly to the notorious "Sandino" camps in the westernmost part of the island. Faced with such a situation only 90 miles from the Florida coast at the height of the Cold War, the outgoing Eisenhower Administration had already put into action a covert plan to train an exile Cuban force for an invasion of the island leading to the installation of a government friendly to the United States. Vice President Richard Nixon was involved in the initial planning of this operation, which was taken over by the incoming Kennedy Administration with the approval of the new President, early in 1961.

John F. Kennedy, a wealthy young naval hero of World War II, who had earlier served in the Senate as a Democrat from Massachusetts, had defeated Richard Nixon in the 1960 election largely because of his personal charm. Despite his liberal views on domestic issues, Kennedy was still a willing Cold Warrior. Yet, in the 1961 Bay of Pigs fiasco in Cuba, the American CIA trained and fielded a small army of exile Cubans, only to have the United States abandon them on the battlefield, when President Kennedy considered the possibility of a Cuban-U.S. war which might conceivably involve the Soviets. Kennedy refused to permit the use of overwhelming U.S. airpower to relieve the surrounded anti-communist Cubans. A hostile, Marxist Cuba was not deemed to be a sufficient threat to America's vital interests to risk confrontation with the nuclear-armed Soviet Union, by an American invasion of the island.

Senior CIA officials who had been involved in formulating the National Security Council's policy decision that thousands of U.S. POWs withheld by the Communists after the Korean Armistice could not be recovered, such as Richard Bissell and C. Tracy Barnes, were responsible for planning and executing the Bay of Pigs operation. Bissell, a Yale graduate with no prior military or intelligence service in WW II, had thrived at CIA since the Korean War period, under the patronage of Allen Dulles.[29] He was credited with pushing the development of the successful U-2 spy planes in the late 1950s, but when U-2 pilot Gary Powers was shot down over Sverdlovsk, USSR, and captured on May 1, 1960,

Richard Bissell was the CIA official who claimed that it was impossible for Powers to have survived the crash, when in fact such was the case.[30] Bissell's assurances prompted Eisenhower to make a public denial that Powers' U-2 had penetrated Soviet air space which was publicly refuted by Khrushchev when he revealed that Powers was a prisoner in Moscow. In a later Senate Intelligence Committee report on Foreign Assassinations, it was to be revealed that Bissell also took part in meetings with CIA Director Allen Dulles and DDCIA General Charles Cabell, in which plans to assassinate Fidel Castro, with assistance from underworld elements, were confirmed.[31] Tracy Barnes, a CIA veteran of the 1954 Guatemalan coup, was later accused of not properly briefing the President on the inadequacies of the Bay of Pigs invasion force, by Brigadier General David W. Gray, the Joint Chiefs of Staff liaison to CIA for the operation, who said a 'full scale military briefing' for the President was never held.[32] Another CIA veteran of the Bay of Pigs operation, E. Howard Hunt, the future Watergate burglar who had been a political officer for the Cuban exile Brigade, continued in the covert operations business through the Vietnam War, and at that time became significantly involved with the families of the missing American prisoners of that war.[33]

The invasion had been preceded by bombings of selected targets from CIA-owned B-26s on April 15, 1961, followed by the exile Cuban infantry and armor landing at the Bay of Pigs on April 17th. Upon hearing of the invasion in Moscow, Soviet leader Nikita

Khrushchev immediately sent a threatening communication to the United States promising "all necessary assistance" to Castro if the Americans didn't halt the invasion. At the critical moment of the battle, after consultation with Secretary of State Dean Rusk and National Security advisor McGeorge Bundy, President Kennedy refused to let U.S. air power—from the Navy fleet off Cuba—intervene on the side of the outnumbered invaders. Kennedy believed that a full-scale American commitment to invade Cuba would result in a Soviet move against Berlin, possibly precipitating a nuclear confrontation.[34]

Poorly equipped and lacking decisive air support, then pinned down within their beachhead by superior forces under personal direction of Fidel Castro (which lost over a hundred killed), 114 men of the Cuban exile Brigade (and a few Americans) were killed in action or died in the savage fighting. According to Theodore Sorenson, an aide to President Kennedy, the CIA cancelled an ammunition resupply convoy "without consulting the President." The CIA had also failed to destroy Castro's air force on the ground in an operation that was controlled by Deputy Director Charles Cabell.[35] A reported total of 1,189 Brigade members, including many wounded (and a few more Americans) were captured and became the latest prisoners of war of a Communist regime. Some died in captivity in the first few days as they were transported to Havana. Others died later during interrogations and under confinement in inhuman conditions. In the aftermath of the abortive invasion, many more mass-executions by Castro security forces occurred in Pinar del Rio, El Moro, and La Cabana prisons, as well as in Matanzas, Camaguay and Oriente provinces.[36]

The Chairman of the Joint Chiefs of Staff at this time was General Lyman Lemnitzer, who in 1945 had been involved with the prisoner of war hostage crisis in Soviet-occupied Austria, and who had faulted the CIA-backed invasion of Cuba as poorly planned and supplied. The failed invasion led directly to the Cuban missile crisis of October 1962, when the United States and the Soviet Union went to the brink of nuclear warfare over Cuba. Nikita Khrushchev had decided that Kennedy would back down in the face of a military bluff, and in the summer of 1962 ordered the shipment and installation of Soviet missiles in Cuba for the defense of Castro's regime. In addition to 54 intermediate and medium range ballistic missiles, the Soviets transported advanced MiG fighters, surface to air (SAM) missiles, and coastal patrol vessels as well as more than 20,000 Soviet troops, all unloaded in secrecy, at night. The United States had recently installed nuclear missiles in Turkey which threatened the USSR and Khrushchev's countermove in Cuba was viewed as a direct threat to most major American cities. In his memoirs Khrushchev later recalled: "We had installed enough missiles already to destroy New York, Chicago and other huge industrial cities, not to mention a little village like Washington. I don't think America had ever faced such a real threat of destruction."[37]

Upon proof of the construction of nine Soviet ballistic missile sites in Cuba being presented to Kennedy in the form of aerial reconnaissance photographs from U-2 overflights, the President ordered an American blockade of the island and Soviet ships were ordered to be stopped and searched. Oleg Penkovsky, a senior Soviet GRU officer recruited initially by British MI-6 to spy for the west, had provided photographs of Soviet documents detailing the various phases of Soviet missile site construction, which enabled the CIA to confirm the U-2 photos. Between October 26 and 28, 1962, with the shooting down of an American U-2 reconnaissance aircraft over Cuba with a Soviet missile, and the death of its pilot, Major Rudolf Anderson Jr., the U.S. and Russia were at the brink of war. Four U.S. reconnaissance aircraft had been fired-on that day. Kennedy administration official Roger Hilsman called this "the blackest hour of the crisis."[38] Another U-2 incident in the northern polar region was then caused by the American aircraft straying into Soviet airspace, which resulted in Soviet fighters scrambling, and the launching of U.S. fighters from Alaska. The missile crisis caused a wave of nuclear war fear to spread across the United States, and although Khrushchev's October 28th agreement to dismantle the missile bases in Cuba was portrayed as a victory for the young American president, the United States also secretly agreed to withdraw it's own offensive missiles from Turkey, thus indicating a mutual retreat from an impending nuclear Armageddon. Penkovsky was soon afterwards caught by the KGB, tortured and shot. His friend, Ivan Serov, former head of the KGB, was disgraced by the revelations of Penkovsky's interrogation, removed from his command of the GRU, and committed suicide.

True to Marxist-Leninist form, the Exile Brigade prisoners captured at the Bay of Pigs were subjected to a publicized mass-trial and Cuban dictator Fidel Castro demanded a $62 million ransom for the POWs release. Since established U.S. policy towards communist-held war prisoners forbade a direct U.S. Government payment of ransom for their release, U.S. pharmaceutical corporations and baby food manufacturers were encouraged by Attorney General Robert Kennedy to contribute to a $53 million shipment of drugs and food to Cuba, which Castro accepted, at least superficially. Cuban gulag survivor Armando Valladares remembered when 214 of the Bay of Pigs prisoners held at Islas de Pinos were chosen by the Castro regime to be the first ransomed, at a value of $100,000 each, after the American negotiator Donovan had reached an agreement: "We were all overjoyed for them...We were saying goodbye and thanking those who had tried to liberate Cuba from Communism. And too, a little window opened for us. The precedent had been set for the Cuban government to enter into that type of negotiation. We might be ransomed too. Twenty years later, that hope of being exchanged is still clutched at by imprisoned patriots."

With public fanfare it was announced that the first Bay of Pigs prisoners had been released on December 23, 1962, and the exile-Brigade survivors were addressed by President Kennedy on the 29th of December in the Orange Bowl before 40,000 spectators.[39] Among the American POWs of the Bay of Pigs still held in Cuba at this time were three U.S. CIA agents who were subsequently released after renewed secret negotiations by Donovan. Armando Valladares later remembered: "The whole negotiation had been carried out for their sake, to keep them from falling into the hands of the Soviets. Carswell, Dunbrunt and Karansky were their names, at least at that point. They were agents of the CIA whom the KGB had been looking for with special zeal for years and who had slipped through their fingers again."[40]

Yet, there were Bay of Pigs prisoners who were not released by Castro's regime. Instead, some were secretly withheld in prisons and forced labor camps in Cuba for decades, in the same manner as had occurred under every Communist regime since the time of Lenin and Trotsky; and some few were still being released after secret negotiations in the late 1980s.[41]

Later Congressional hearings revealed that the CIA was also funding and planning the assassination of Cuban revolutionary leader Fidel Castro, because of his hostility to the United States and his adoption of a Communist system aligned with the Soviet bloc, and that Mob boss Santos Trafficante and other underworld figures were involved with the CIA in the plot. Castro had angered American organized crime bosses Carlos Marcello and Santos Trafficante, whose profitable operations in Cuba had been halted by Castro, and who wished to return to the island after his overthrow. Kennedy's conduct during the Bay of Pigs operation infuriated some CIA and Pentagon officials and he had also angered the syndicate figures and others such as Teamsters Union leader Jimmy Hoffa, with his brother Robert Kennedy's assault on organized crime as Attorney General of the United States. Subsequent Congressional hearings revealed links between organized crime figures and the CIA.[42]

Following the debacle at the Bay of Pigs, President Kennedy reportedly threatened to "splinter the CIA in a thousand pieces and scatter it to the winds,"[43] CIA Director Allen Dulles was forced by Kennedy to resign because of the Bay of Pigs disaster, ending the public career of a man who had been involved in covert operations and intelligence-gathering since his WW I State Department service and the Versailles peace conference of 1919. John McKone, a careerist, replaced him as DCI. Another CIA official whom Kennedy fired was the Agency's Deputy Director for clandestine operations, General Charles Cabell, who had been involved in the Korean War POW debacle, and in covert operations with E. Howard Hunt, and whose name would repeatedly surface in the investigation of Kennedy's subsequent assassination.[44] General Cabell's brother, Earle Cabell, was the Mayor of Dallas, Texas, who reportedly helped organize President Kennedy's motorcade through the city on the day of the assassination.[45] Richard Bissell was also forced to leave the CIA. Richard Helms, a future director, became DDCI in Cabell's place.

President John F Kennedy was assassinated on November 22, 1963. The subsequent Warren Commission Report on the assassination stated that he was killed by a lone gunman, Lee Harvey Oswald. Prior to his own murder shortly afterward by the Mob-connected Jack Ruby, Oswald claimed to have been set up as a scapegoat for the assassination. Oswald was a former U.S. Marine who had once been assigned to a secret airbase in Japan (Atsugi) from which U-2 reconnaissance aircraft like that of Gary Powers, departed on flights over the USSR. Oswald became a Russian speaker in the Marines and was tested for his proficiency in the language by the U.S. Government. Following his discharge Oswald had gone to live in the Soviet Union for three years; after his return to America he associated not only with Communist-front groups but also with CIA-front organizations and CIA-connected individuals. His relationship to U.S. intelligence was never adequately explained. For example, a witness before the later House Committee on Assassinations, formerly employed by CIA at the Tokyo station, named James Wilcott, stated under oath that he had been convinced from conversations with colleagues that Oswald "had been recruited by the agency to infiltrate the Soviet Union."[46]

Two months after Kennedy's assassination a high-level KGB officer named Yuriy Ivanovich Nosenko defected to the CIA. In establishing his bona fides, Nosenko revealed information which led to the arrest of a Soviet spy in the British Admiralty, William John Vassal, and of an American Army sergeant named Robert Lee Jackson, who was also a deep-penetration agent of the KGB. Nosenko also provided locations of secret microphones being used to bug the American embassy in Moscow. Nosenko claimed that the KGB had shown no interest in Oswald, and had no direct contact with him in the USSR; which seemed unlikely to some. Through the influence of earlier defector Anatoli Golytsin, CIA counterintelligence chief James Angleton became convinced that Nosenko was a double agent still working for the KGB, and amounted to a "sent messenger." The Golytsin-Nosenko controversy was to fuel a runaway "molehunt" within the Agency which would eventually come close to paralyzing the CIAs Soviet Division during the coming Vietnam War years. The location and condition of Kennedy's wounds and the testimony of eyewitnesses indicated to many Americans that the assassination was carried out by multiple gunmen firing from different directions. A subsequent Congressional investigation of Kennedy's murder was to find that many of Oswald's military records had disappeared, and revealed substantial evidence that a conspiracy had existed. The controversy about the Kennedy assassination continues to the present time.[47]

## THE U.S.-VIETNAM WAR

While monitoring Soviet radio traffic during the administration of President John F. Kennedy in 1961, at the 6922nd Security Wing, based on the island of Okinawa, in the western Pacific ocean, Jerry J. Mooney became aware through intercepted communications of a massive Soviet airlift of war material to Vietnam, in preparation for an obvious war buildup there. His superior officer, Colonel John Kennedy, called him in to his office at that time and asked Mooney what he knew "about Vietnam." In an interview with the author 31 years later Mooney recalled:

"I told him of the huge airlift to Hanoi and he sent me immediately to the 6925th Security Group in the Philippines, with instructions to find out everything they knew about the airlift that we had been following through the decryptions of radio traffic concerning this war buildup. This particularly involved radar locations, capability area and coverage and interpreting. They knew hardly anything, according to Mooney:

"When Colonel Kennedy debriefed me he said, "WE'RE GOING TO WAR, AND WE HAVE TO BEGIN BRIEFING THOSE RESPONSIBLE TO GET THE INTELLIGENCE TO THE TACTICAL COMMANDERS, TO PREVENT LOSS OF AIR-CREWS AND AIRCRAFT TO THE VIETNAMESE AND TO THE SOVIETS. WE HAVE TO TIE IN THE SIGINT (signals intelligence) TO THE TACTICAL COMMANDERS TO PREVENT LOSSES."

But, in fact, according to Mooney, "THIS WASN'T REALLY DONE UNTIL 1974, LONG AFTER THE WAR WAS OVER."[48]

In Vietnam, which was apparently deemed by American leaders to be more "vital to the interests of the United States" than Hungary or Cuba, the U.S. policymaking establishment was determined to oppose the further spread of Communism. When President Dwight Eisenhower was about to leave office in January 1961, during the Communist takeover in Cuba, he had told his young successor, John F. Kennedy, that Laos was "the key to the entire region of Southeast Asia," and U.S. forces might be required there. Kennedy, who later withheld American military power in Cuba during the Bay of Pigs invasion, committed more U.S. troops to Southeast Asia and decided that South Vietnam was more vital than Laos. (In this he may have been influenced by his naval training, as he had relied on a maritime blockade during the 1962 Cuban missile crisis.) The evidence from decryptions of Soviet communications revealed by retired U.S. intelligence analyst, Jerry J. Mooney, indicates that the Soviets had begun a major buildup of war material in North Vietnam during 1961, which may indicate a strategic plan to draw American interest away from Cuba and encourage a deeper American commitment to the defense of Indochina.

Whatever the reasons for such policy decisions in Washington, the Communist government in Hanoi under Ho Chi Minh was aware that Laos was of the utmost strategic importance in flanking the long border of South Vietnam, the non-Communist southern part of a partitioned country created from the old French Indochina, which Ho Chi Minh was still determined to conquer.[49] Communist-initiated warfare against an American-supported Royalist regime in Laos during the late 1950s and early 1960s created a crisis for President Kennedy which he asked Assistant Secretary of State Averell Harriman to resolve for him. Harriman's closest personal aide during these negotiations was a 40-year-old career foreign service officer named William Sullivan, who had attracted some attention with policy papers he had written a year before recommending a neutral Laos and better relations with the People's Republic of China. According to Sullivan, Harriman viewed the Communists as "human beings, not automatons." In Geneva, Harriman dealt directly with Soviet Vice Foreign Minister G.M. Pushkin, who convinced the old American diplomat that the Soviets and Americans shared a parallel interest in keeping peace in Southeast Asia and controlling the Chinese.[50] In light of what is now known of the ongoing Soviet war material buildup in North Vietnam, it appears that Harriman was deluded by Pushkin's words. Nevertheless, Kennedy had told Harriman to get a peace settlement in Laos, and Harriman did as he was told.

The agreement which had been gained by Averell Harriman and his aide, William Sullivan, from the willing Communists during the 1962 Geneva Conference on Laos, which was signed by the United States and the Soviet Union, led to the withdrawal of U.S. Special Forces "Green Beret" troops from Laos by October, although the Vietminh forces supporting the Pathet Lao Communists remained in place, and continued consolidating their hold on the strategic border areas. The CIA-run military units which had been withdrawn to Thailand formed the nucleus of forces clandestinely redeployed in Laos to counter the Communists. As ex-CIA official Victor Marchetti was to write later, "The Laotian operation became one of the largest and most expensive in the agency's history: more than 35,000 opium-growing Meo and other Lao mountain tribesmen were recruited into the CIA's private army, L'Armée Clandestine; CIA hired pilots flew bombing and supply missions in the agency's own planes; and...the agency recruited and financed over 17,000 Thai mercenaries for its war of attrition against the Communists."[51] The Pentagon prepared a plan for the insertion of U.S. combat troops from South Vietnam's northern DMZ border, across the narrow panhandle of Laos to the Mekong River and the secure border of Thailand. This plan, if it had been carried out, would have been the only effective U.S. method of halting the infiltration of soldiers and war material down through Laos and into South Vietnam, on a road system that became known as the "Ho Chi Minh Trail," without which the Viet Cong Communists of the south would not have been able to carry on their costly guerilla war.

Under President Eisenhower and his successor John F. Kennedy, the United States advisory role and military assistance to the non-Communist Ngo Dinh Diem regime in South Vietnam rose steadily, but combat losses among U.S. advisers were light and the American people were hardly aware that the United States was gradually committing itself to another major Cold War conflict with a Soviet-surrogate power on the Asian mainland. Most Americans were also unaware that there was a new group of U.S. prisoners of war in communist control, in Vietnam and Laos. Some were merely listed as missing and others were presumed dead. One who was captured on October 29, 1963, Captain James "Nick" Rowe, was also to be one of the few Americans to later escape successfully from Communist captivity in Vietnam. The captured Americans, known and unknown, were kept in small groups surrounded by many guards in primitive jungle prison camps, and some of them were to disappear forever.

In Laos during the early 1960s years of the Kennedy Administration, the Communist Pathet Lao forces, which had remained in place since the Vietminh war and the 1962 Geneva agreement negotiated for Kennedy by Averell Harriman, were rearmed and supplied from the USSR and China through Hanoi, and increased their attacks on positions of the Royal Lao Army, which was supported by the United States and its Pacific allies. The Communist campaign was clearly intended to flank the defensible Demilitarized Zone, which had been arbitrarily drawn across the middle of Vietnam during the 1954 settlement of the French-Indochina war. A U.S. Army officer serving in DCSOPS in the Pentagon during this period, Major William LeGro, was tasked by his superiors to develop secret U.S. military plans which would effectively counter the North Vietnamese war buildup in Laos on South Vietnam's western flank, which had been mandated by Hanoi's Communist leader Ho Chi Minh and no doubt encouraged by his protectors in Moscow. LeGro, who had been commissioned from the ranks after WW II, in which he had served as an infantry NCO in the

Biak/New Guinea campaign and in the Philippines, and other Defense Department planners, developed a viable military strategy to isolate the panhandle of Laos from the northern bulk of that country and the major supply lines originating in North Vietnam. The plan recommended insertion of a force of American ground troops at the narrowest part of northern South Vietnam, from the "DMZ" across to the Mekong River on the border of the Kingdom of Thailand, which could in turn have protected its own frontiers from North Vietnamese invasion, with American and other allied assistance. Institution of this plan, which was delivered during the administration of the new President, Lyndon B. Johnson, in 1964, would have cut off the route of access of Communist troops and supplies, and made their conduct of a full-scale war inside South Vietnam virtually impossible.

This entirely practical and workable plan was not approved by U.S. policymakers however, due to White House political decisions resulting from the Geneva agreement gained by President Kennedy's special Ambassador in Laos, Averell Harriman. Harriman had thus played an important part in international negotiations with the Communists again, seventeen years after his key role at Yalta and in Moscow. Harriman was to be involved in later unsuccessful negotiations on Soviet-held American POWs in East Berlin during 1967, and in the first negotiations to end the war in Vietnam, in 1968. LeGro went on to serve in Vietnam as the 1st Infantry Division's G-2 (Chief of intelligence) during 1966-67, and still later as the last chief of U.S. Military Intelligence in Vietnam from 1973-1975, but as with many other professionals, it was to be a heartbreaking end to a lifelong military career. As with President George Bush, and retired NSA analyst Jerry Mooney, the thoughtful and compassionate William LeGro was to have his later life become intertwined with resolving the fate of thousands of American POWs and MIAs who had disappeared in Communist control following WW II, the Cold War, Korea and Vietnam, long after he had retired from active duty and had believed his career was finished.

Following Harriman's negotiations in Geneva and after subsequent Pentagon plans such as LeGro's were permanently shelved by the Joint Chiefs of Staff, due to official American unwillingness to repudiate what was in fact a flawed agreement, the covert war continued in Laos unabated. The de facto enemy "sanctuary" of Laos was to cause tens of thousands of additional American casualties during the agonizing and costly years of war that came now to neighboring Vietnam. This prompted a vast increase in the level of the "secret war" conducted by the United States in Laos, run, naturally, by the CIA, in which "sheep-dipped" military personnel, officially separated from the military services and serving as civilian mercenaries, together with secret U.S. military forces whose existence inside Laos was denied, joined together in a vicious clandestine conflict with the Pathet Lao and their North Vietnamese allies. The CIA also took over the local Lao and tribal Meo and Hmong anti-communist allies of the former French regime, many of whom had been encouraged to finance their operations through the growing and sale of opium. This practise of encouraging and even supporting the illicit drug traffic by providing supply and transport services through CIA, was to ultimately cause far more harm to that agency and to the United States than was ever done by those secret proxy forces to the Communists in Laos. American organized crime figures like Santos Trafficante, who traveled to Vietnam in 1968, took advantage of the situation, and the importation of opium and particularly its derivative, "China white" heroin, to the United States in the late 1960s, increased enormously. The use of such hard drugs by thousands of young American servicemen in Indochina also increased dramatically at the same time, causing serious morale and disciplinary problems within the U.S. expeditionary force, which resulted in the return of many addicts to middle American communities not heretofore involved in the epidemic.[52] And in the end, this secret war in Laos was also to contribute another group of lost U.S. POWs, those hundreds of Americans who were abandoned there because their government denied both the war they fought in, and their very existence, for foreign policy and national security considerations.

Influential senior advisers under Kennedy who had pushed for the disastrous Bay of Pigs invasion of Cuba, including Secretary of Defense Robert S. McNamara (a former automobile industry executive and future President of the World Bank), National Security Advisor McGeorge Bundy (who, as an aide to Secretary of War Henry Stimson had kept the secret of the Soviet-held American POWs of WW II), his brother, William Bundy of CIA, and the Pentagon's General Maxwell Taylor, favored increased American military intervention in Vietnam. Others, including Undersecretary of State George Ball, and National Security Council staffer Michael V. Forrestal (the son of Roosevelt's Navy Secretary and Truman's first Defense Secretary), strongly opposed the escalation. Shortly after John F. Kennedy's inauguration, the Department of Defense announced the creation of a U.S. Military Assistance and advisory Command in South Vietnam (MACV) in February 1961, under General Paul Harkins.[53]

Kennedy had at first embraced the defense of South Vietnam, and permitted an increase of the American advisory role there, which included covert U.S. air attacks on Vietcong positions. A young West Point-educated Air Force officer, Captain Richard Secord (later Major General, of Iran-Contra repute), was assigned to covert combat service in Vietnam under President Kennedy with the 4400th

'Combat Crew Training Squadron,' in March 1962. Deployed to Bien Hoa Airbase, just north of Saigon, and to other airfields, this squadron actually used American pilots to conduct continuous close air support fighter-bomber missions against Vietcong positions, in support of South Vietnamese infantry operations, from the Mekong Delta and Plain of Reeds area in the south, to the Da Nang region, near the DMZ border with North Vietnam.

As Secord was to write later in his memoirs: "Officially we weren't anywhere near Vietnam. Our cover, should we be rooted out by reporters, was that we were 'instructors' for Vietnamese pilots... our 'copilot' was often a crew chief or maintenance man or some poor soldier who happened to be standing around...We Americans wore no name tags or unit patches or even rank on our flight suits; just dog tags so that our bodies could be identified by friendly forces in a wreck. We also cleansed ourselves of all 'pocket litter,' wallets or religious tokens or photos...that would help the enemy prove Americans were participating in the war. We even gave up our Geneva Convention cards, which every serviceman carries to show he's entitled to humane treatment as a POW." But Secord also records the visit of Air Force Chief of Staff General Curtis LeMay to their base, indicating strong JCS support for the covert air operations occurring in Vietnam.[54] Kennedy had also supported expansion of elite guerilla warfare units such as the Army Special Forces, known as the Green Berets, and the Navy's SEAL (Sea/Air/Land) commandos and their use in low intensity warfare with Soviet proxy forces in Vietnam, Laos and Cambodia and other third world nations deemed vital to American and free world interests.

Yet, despite this earlier commitment, declassified U.S. documentary evidence indicates that by the fall of 1963, just prior to his assassination on November 22nd, President Kennedy had ordered a withdrawal of 1,000 men from the U.S. advisory force in Vietnam, in an "Accerated Withdrawal Program." This indicated the implementation of a policy change by Kennedy, who was becoming wary of any deeper involvement there, and who was influenced by a highly optimistic report on Vietnam submitted by Secretary of Defense Robert S. McNamara and General Maxwell Taylor, Kennedy's military favorite who had commanded U.S. paratroopers in Europe, in 1945. Retired U.S. Air Force Colonel Fletcher Prouty recalled later: "I was one of the primary writers of NSAM 263... When I was working on NSAM 263, '(the) Taylor-McNamara' Report, I was very well briefed on this plan of Kennedy's. He was getting out..."[55] This report recommended the withdrawal of the first 1,000 U.S. combat advisors by the end of 1963. Kennedy adopted this as his new policy on Vietnam in October 1963.

Kennedy aide Kenneth O'Donnell, in JOHNNY WE HARDLY KNEW

YE, wrote: 'The President's order to reduce American military personnel in Vietnam by one thousand before the end of 1963 was still in effect on the day he went to Texas.'[56] Whether this was merely political posturing by Kennedy is uncertain, but Colonel Fletcher Prouty, who was Chief of Special Operations in the Office of the Joint Chiefs of Staff in 1962 and 1963, later recorded:

"This report was discussed with the President and the full National Security Council on October 3, 1963. It contained two important recommendations that were basic to Kennedy's plans for Vietnam. They were: a) 'The Department of Defense should announce plans...to withdraw 1,000 U.S. personnel by the end of December 1963...We believe the U.S. part of the task can be completed by the end of 1965.'" A Kennedy decision to decrease rather than escalate American involvement in Vietnam would have angered hardliners in the CIA and Pentagon, and those leaders in private industry who backed a U.S. commitment to defend South Vietnam against a Communist takeover, as part of an overall policy to protect American markets in Asia. In 1963, Lucien Conein, the former OSS officer who in 1945 had believed Ho Chi Minh to be a great statesman, was the CIA's liaison with the cabal of Vietnamese generals who, after initial American encouragement for a coup, overthrew and then murdered President Ngo Dinh Diem in South Vietnam, three weeks before the assassination of President John F. Kennedy. Ironically, some U.S. analysts and policymakers believed that Diem was the only South Vietnamese leader with sufficient stature as a nationalist to maintain South Vietnam's independence.[57]

Soon after taking the oath of office following John F. Kennedy's assassination in Dallas, Texas on November 22nd, 1963, President Lyndon B. Johnson reversed this decision, halting the withdrawal of U.S. forces in Vietnam. On November 26, 1963 Johnson approved National Security Council Action Memorandum (NSCAM) #273, reversing Kennedy's decision, pledging total commitment to defending South Vietnam and approving U.S. covert operations there.[58] Johnson was a conservative southern Democrat who although personally liked by FDR, had been part of the bloc that opposed the more liberal policies of Roosevelt's New Deal administration in the late 1930s and 1940s. "L.B.J." was the son of a Texas Hill country ranch family that had served the Confederacy, fought the Comanches to hold their land, and driven cattle up the Chisholm trail to the Kansas railroads in the 1800s. While in Congress he had fulfilled the New Deal dream by obtaining construction of dams that electrified vast rural areas and brought twentieth century life to many of his constituents. Yet, Johnson was a complex man, whose loyalties to political allies who believed in the Cold War confrontation, and to early political supporters like Brown & Root Inc., the Texas contractors, would become apparent during the Vietnam War.[59] Johnson identified with the figures in the Pentagon and CIA who favored a military confrontation with the Communists in Vietnam. By the end of 1963 some 16,300 U.S. troops were in South Vietnam.

Johnson also initiated bombing strikes against North Vietnam in response to highly questionable incidents of "provocation" in alleged North Vietnamese torpedo-boat attacks on U.S. ships in the Tonkin Gulf in 1964, which, according to retired U.S. Navy Admiral James Stockdale who was on the scene at the time, never occurred. (Later a prisoner of war in Hanoi for seven years, Stockdale had to keep this secret through severe torture and privation, knowing that if he revealed it, the Communists would make world-wide propaganda against his country with the information.)[60] In the August 2nd 5th airstrikes against North Vietnam, two U.S. planes were shot down by anti-aircraft fire and Navy Lt. Everett Alvarez became the first known U.S. POW to be held in North Vietnam. He was to remain in captivity until the Americans who were to be released in Operation Homecoming, nearly nine years later, came home. In December 1964, the U.S. Air Force began attacking Pathet Lao and North Vietnamese targets in "neutral" Laos. By the end of 1964 some 246 Americans had been killed by hostile action in Vietnam and 1,641 others had been wounded.

In March 1965, the Johnson administration escalated the war sharply in response to Viet Cong attacks at Binh Gia, Pleiku and Qui Nhon that killed 34 Americans, and wounded 150 more. This marked the end of the American advisory phase of the war. Operation Rolling Thunder, a U.S. bombing campaign against North Vietnam, was initiated on March 2nd, and by the end of that year, 171 American aircraft had been lost in combat, resulting in many more U.S. POWs being captured. Elements of the 9th Marines, 3rd Marine Division and the Army's 173rd Airborne Brigade arrived in Vietnam between March and May, and began combat operations. They were augmented by the 1st Cavalry Division and 1st Infantry Division later in the year. The Strategic Air Command began conducting bombing raids on Viet Cong sanctuaries in South Vietnam by June, using B-52's based on the island of Guam. By the fall of 1965, President Lyndon Johnson had committed division-size U.S. combat units to battle with local mainforce Vietcong (VC) battalions, and with North Vietnamese Army (NVA) regiments that had infiltrated into South Vietnam via the Ho Chi Minh Trail through "neutral" Laos and Cambodia to war fronts from the Central Highlands to Saigon and the Mekong River Delta.

In the Marine Operation "Starlite," during August, U.S. losses were 250 killed, wounded and missing, while in the October 23rd to November 20th Ia Drang Valley battle, U.S. Army losses were approximately 1,000 killed wounded and missing. With such a vast sanctuary nearby, shielded from easy attack by jungle and mountains, and by international agreements on Laos negotiated under Kennedy by Averell Harriman, the Communist forces could afford to lose enormous casualties while wearing out the patience of the Americans.

After the executions by the South Vietnamese of two Viet Cong terrorists who were accused of attempting to assassinate Secretary of Defense Robert McNamara on May 12, 1964, the Viet Cong announced over "Liberation Radio" on September 26, 1965, that they had executed two American prisoners of war. These turned out to have been Army Special Forces Captain Humberto "Rocky" Versace and Sergeant First Class Kenneth M. Roraback. Prior to Captain Verace's public execution, he was forced to kneel and apologize for his "war crimes." A few months earlier the Viet Cong had executed Army E4 Harold Bennett of Perryville, Arkansas. The U.S. State Department protested these violations of the Geneva Convention on prisoners of war and called the acts "wanton murder."[61] Other executions and murders of American war prisoners occurred later.

Substantial numbers of U.S. infantry soldiers were listed as MIA or "presumed killed," but were believed by veterans to have been actually taken prisoner in the first large-scale fighting between American and North Vietnamese units in the Ia Drang Valley battle. Numerous other Army troops and Marines disappeared during later large-scale search and destroy operations during 1966, code named "Masher," "Utah," "Prairie" and "Attleboro," but many of those initially reported missing were later declared dead on the dates they disappeared, according to declassified unit records and after-action reports in the U.S. National Archives consulted by the author in 1984.

Losses in downed aircraft over North Vietnam and the South also rapidly increased the number of U.S. POWs in 1965 and 1966. On the single day of December 2nd, 1966, eight U.S. aircraft were shot down over Hanoi. HANOI, MEANWHILE, VIEWED THE AMERICAN MILITARY OPERATIONS AGAINST THEIR COUNTRY AS ILLEGAL, BASED UPON ACTS OF PROVOCATION WHICH THEY MAINTAINED HAD NOT OCCURRED, AND THUS STATED THAT ALL U.S. PERSONNEL SUBSEQUENTLY CAPTURED WERE 'WAR CRIMINALS' AND 'AIR PIRATES', WHO DID NOT DESERVE LEGAL STATUS AS 'PRISONERS OF WAR.' And indeed, if Navy POW (and later Vice Presidential candidate) James Stockdale had revealed his secret while in captivity, this may have become the position of much of the world, given the envy and malice felt for the United States in this period.

During the worst seven war years for America in Indochina, from 1965 to 1972, U.S. infantry and armored forces, backed up by often-devastating air power, fought hundreds of battles and thousands of skirmishes, over much the same terrain, year after year, against a highly organized and tenacious enemy, well-armed with Soviet and Chinese Communist weapons, whose leaders, unlike America's, were dedicated to total victory. From 1968 on, Soviet-supplied 122 mm rockets and artillery, surface to air (SAM) missiles (with Russian combat crews), tanks and Soviet ground combat advisers made the forces of the Vietnamese Communists more dangerous and effective. Thus, during the total of ten years of war from 1963-1973, American pilots, navigators, advisers, infantrymen, reconnaissance teams and others vanished in a continuous series of missing in action or "presumed" killed, "Loss Incidents"—as they were officially referred to by the military services.

Throughout the Vietnam War, precise intelligence obtained from intercepted and decoded Vietnamese communications about U.S. POW/MIAs remaining in enemy control, was again classified and much of this "special intelligence" remains classified at the time this book is being written.

But from the 1965 war escalation until late 1968, Air Force Security and NSA analyst Jerry J. Mooney was deeply concerned, after realizing that tracking of U.S. POWs through intercepts of Vietnamese communications was only an "idle analytic problem" developed by analysts like himself, and not a fixed policy and priority of U.S. intelligence. Until the early 1970s such information on the capture and movements of American prisoners was only collateral data of secondary importance to the primary mission of countering the North Vietnamese air defense buildup. The priceless POW data was virtually ignored and largely went unreported, although volumes of this raw intelligence were filed for storage.

As with the earlier French prisoners, many American POWs in Vietnam, Laos and Cambodia were confined in remote jungle prison camps and sometimes in caves where the death rate was high. Of the more than 58,000 officially reported U.S. dead of all causes in the Indochina war, 47,356 were reported as killed by enemy action and of these over 3,500 Americans were actually reported as missing in action or "presumed" killed, body not recovered. One group of U.S. POWs in North Vietnam, which grew to number nearly 600, were eventually acknowledged and held openly in Hanoi for eventual repatriation at the war's end in what became known as "The System," which included four prisons in Hanoi: The Hanoi Hilton (Hoa Lo Prison), the Plantation, the Zoo and Alcatraz, and 6 other prisons within 60 miles of Hanoi: Son Tay, D-1, Briarpatch, Faith, Skidrow and the Rockpile. One other camp in the "System," called Dogpatch by the POWs, was located only five miles from the Chinese border. Most returned U.S. POWs indicated that any American prisoners within this "System" (whose names became known) were among the 591 repatriated, or those who died in captivity. But some returning POWs, such as Colonel Laird Gutterson, revealed that not all U.S.

prisoners had been tracked in the System, and some Americans may have been held in other locations. As Gutterson recalled later:

"Our system for keeping track of men inside the prison was a good one, but it didn't go far enough. When we returnees refer to 'The System,' we're talking about the POW camps that we knew about."[62]

Approximately 1,000 other American MIAs and POWs had indeed been held in a secret, parallel prison camp system from which none were to be returned after the war. In several of these secret Vietnamese prison camps, selections were made among POWs for shipment to the Soviet Union or China, in much the same manner as had previously been done in Communist China during the Korean war. Any of the prisoners held in these secret selection camps, staffed by Soviet intelligence officers, who did not break under interrogation and agree to shipment to the USSR voluntarily, faced a slow death at forced labor on North Vietnamese war-related projects. The North Vietnamese could not afford to release any of these specially-interrogated POWs because of their dangerous knowledge of the Soviet intelligence connection, and because their observations would impact on North Vietnamese national security interests, which remained in effect after the American withdrawal and indeed until the final collapse of South Vietnam in 1975. All of this information on U.S. POWs was to be revealed to U.S. intelligence during the later Vietnam war years by NSA decryptions of North Vietnamese radio communications, copied and translated by NSA cryptanalysts and cryptolinguists, and made available to other intelligence services such as DIA and CIA. This information was to remain highly classified and compartmentalized for decades after the war, while these same U.S. intelligence agencies consistently debunked the eyewitness testimony of Indochinese human sources who had actually seen such prisoners alive in captivity.

While the North Vietnamese and their Soviet allies were secretly assembling and interrogating American POWs prior to shipping them to the USSR, a second overt propaganda campaign was orchestrated by the communists around the POWs, to portray to the world the humanity of Hanoi's leaders towards prisoners. In order to carry this off successfully, the Vietnamese communists depended upon their American "antiwar" allies to consistently reiterate Hanoi's kind and leniently treatment of American POWs in the American media, to which they were being increasingly granted access. As U.S. prisoners were being tortured, shot or beaten to death in captivity, American leftists defended Hanoi's definition and treatment of the American 'air pirates' and 'war criminals' who fell into Communist control, but they also encouraged the Viet Cong to release a token number of them as a propaganda ploy to lull the American public into believing that this Communist regime was actually humane.

To release prisoners long held in inaccessible regions of South Vietnam was in itself a propaganda coup in that it demonstrated the ineffectuality of American and Army of the Republic of Vietnam (ARVN) forces at even being able to liberate U.S. POWs in the South, within a relatively short distance from Saigon. This adhered to the communist line that the war in South Vietnam was a revolution by the people of the South, without interference by the regular North Vietnamese Army (NVA). (Then, in fact actually in the South in large numbers, but which fact was denied by the Vietnamese communists and their American supporters.)

Hanoi's leaders and intelligence officials decided to use the most pro-communist members of the American antiwar movement to accomplish their object of portraying their kind and humane policy towards American POWs. In choosing those to be released they naturally picked enlisted men—as proper representatives of the American working-class proletariat—forced to fight an "imperialist war for their capitalist masters." In addition, they made sure that two of the three were black American soldiers, to show "solidarity" with the civil rights movement then occurring in the United States. And in choosing the American liaison, they had to pick an American of proven loyalty to Hanoi, who could be depended upon to adhere to the correct line on POWs when the actual release was accomplished, and forever afterward. Their choice was Thomas Hayden, the American Marxist radical who had helped found SDS (Students for a Democratic Society) at Port Huron, Michigan, and who had been one of the first American visitors to Hanoi used to condemn the U.S. bombing campaign.

One of the American POWs who, unbeknownst to him, was about to be released, was a Special Forces (Green Beret) NCO adviser named Dan Pitzer, a medic who was captured during the advisory war under Kennedy in an October 1963 firefight in An Xuyen Province in South Vietnam's Mekong Delta. (A lieutenant in his Special Forces A Team, Nick Rowe, who was also captured at this time, later became one of only 31 American POWs to successfully escape the Vietcong or North Vietnamese during the entire war.) Pitzer was taken by his captors deep into the U Minh Forest, in the southernmost area of Vietnam, where, throughout the decade-long war U.S. and South Vietnamese military operations rarely ventured. He was held with a total of only 8 other American POWs for four years. Of these, Captain Rocky Versace was executed by the communists, three U.S. POWs died of starvation, one escaped (Lt. Rowe), and two others, Jackson and Johnson, both black American soldiers, were later released with Pitzer. The POWs were beaten and tortured by their captors and subjected to indoctrination and brainwashing techniques. According to Pitzer, the two blacks were singled out because of their race for special attempts to propagandize them.

In an interview for a 1985 book, Pitzer remembered the circumstances of his release:

"By September 1967, I had been a prisoner of the Vietcong for four years. Nick Rowe and I were held in isolation in the U Minh Forest the entire time. We did not get a true picture of what was going on in the world, only the propaganda the VC gave us. After the war escalated, we kept seeing the aircraft change. We often wondered how many Americans were in Vietnam...I picked up a piece of this paper one day (Stars and Stripes). It had the total amount and name of every American unit in Vietnam. We couldn't believe it. We counted five hundred thousand people. We said, 'What the hell are we still doing in this prison camp? This war should be over with.'"

"...They tried to get us to say something like: 'The war is bad. American soldiers go home. Join the National Liberation Front and they'll have a lenient policy toward you.' The interrogators told us that if we did not repent we would be tried as war criminals and treated accordingly. But the most frightening thing was how they tried to manipulate our behavior through Thought Reform, which they did to their own people to maintain political control. One way they conducted psychological warfare was to show us reports from the New York Times, the Washington Post, news magazines, and the Congressional Record, displaying antiwar sentiments at home. It was hard to drive ourselves on day after day when the guards showed us newspaper stories of protestors flying the VC and North Vietnamese flags around the Washington Monument, while students burned American flags.

"The interrogator, Major Bai, would say, 'Look, this is your capitol. Look where our flag is, look where your flag is. Why do you resist? Your own country looks at the people who support our cause as heroes. Why do you stay here and suffer?' What was being published in the U.S. was much better than anything the Viet Cong could write. And while it was disheartening for us, it was a real boost to their self-confidence.

"In October 1967 there were four of us in the prison camp: Nick Rowe, myself, and Jackson and Johnson, who were black. The VC based a lot of their propaganda on what the American people were doing. So when Martin Luther King marched on Washington, D.C., the Viet Cong made an announcement. To show appreciation to the black Americans for protesting the war, they would release some black prisoners. And though this black march on Washington was just a civil rights demonstration, the Communists saw it as an opportune time. So they prepared these two blacks and myself, because I was a medic...I was told, 'You keep Johnson alive until we reach a neutral country.' Well, that neutral country turned out to be Cambodia."

"It took about thirty days to move us. Walking, but mostly on the canals at night by sampan. I sat there with a coolie hat on watching thousands and thousands of sampans moving south during the night from the VC sanctuaries in Cambodia. Once we got into Cambodia we saw cache after cache of weapons, huge stockpiles. They kept us for about two weeks on the border. They fed us—I was down to about ninety-five pounds...Jackson and I literally dragged Johnson home. I was told that this man had to live until we got to Phnom Penh, 'or else'..."

"In Phnom Penh the only people we saw were Communists. No free world representatives or press people there. They tried to put us in front of the cameras for the propaganda thing...And Tom Hayden was there waiting for us in Phnom Penh. We immediately told him what we thought of him...We were put on a Czechoslovakian airliner...And Tom Hayden sat with me..."[63]

Soviet and Cuban personnel were also on the aircraft, which was supposed to take the American POWs to Prague, Czechoslovakia, for a Communist demonstration showing international support for the Viet Cong and North Vietnam. There were even rumors that they would then be taken on to Moscow for a similar exhibition. It was intended to be a repetition of similar propaganda ploys with prisoners of war dating back to the December 1918 Comintern meeting in Petrograd, where American and British POWs had been used in the same manner.

The aircraft made stops in Bombay and Kuwait and at one of them Pitzer succeeded in smuggling out a note to the American Embassy through an aircraft maintenance worker. At Beirut, Lebanon, a U.S. official named Art Beaton, of the FAA came aboard the plane and interposed himself between the Soviets and Cubans and the American POWs. He escorted Pitzer, Jackson and Johnson off the aircraft. They were soon flown to New York and freedom, along with Hayden, and the planned Communist propaganda show had failed.

Hayden later testified to a Congressional Committee on how the POW release had been planned with the Vietcong and North Vietnamese representatives at a conference between American radicals and the Vietnamese Communists at Bratislava, Czechoslovakia in September 1967. At the hearings, the following quotation from a document Written by American supporters of the Vietcong was read into the Congressional Record:

"The Prague conference is intended to create solidarity and mutual understanding between revolutionaries from Vietnam and their American supporters who are trying to change the United States." The document was signed by Dave Dellinger, and Hayden was asked by the Committee whether he had seen it. Hayden responded:

"I had a hand in writing this document. I was one of the people who helped to write the agenda and work out an agenda for the conference."

Hayden was asked by the committee whether he was one of the

American supporters of the "revolutionaries" in Vietnam, to which he answered, "I am." He was then asked about how he came to be involved with the propagandized release of three U.S. prisoners of war in 1967. (Pitzer, Jackson and Johnson) Although Hayden may not have realized at the time the secret chain of prisoner repatriation extending back to the time of the Bolshevik Revolution, he was a willing participant. Hayden answered:

"Propaganda works both ways. Every time the Vietnamese have released prisoners, the United States releases prisoners and announces it although the actions are not reciprocal...It was at this conference in Bratislava during the discussions with the Vietnamese about the state of American prisoners that some Vietnamese approached myself and said they were contemplating the possible release of some prisoners from South Vietnam. They were not sure how to do it technically. They had a lot of problems."

Sticking rigidly to Hanoi's line that there were no North Vietnamese troops in South Vietnam, Hayden went on:

"Contrary to public opinion, people do not run up and down from North to South Vietnam on the Ho Chi Minh trail. These prisoners were deep in South Vietnam. A way had to be found for them to be released without the National Liberation Front having to hand them directly to the Americans because they did not recognize each other..."[64]

Hayden had been deliberately and carefully chosen by Hanoi's communist leaders as their tool, and in return for carrying out their propaganda mission he would gain press coverage and acclaim which he could then use to further promote Hanoi's (and Moscow's) aims in the United States and Vietnam, and his own political aspirations. (20 years later Hayden would continue to boast about his part in this POW ploy in expensively produced campaign literature.) Tom Hayden never indicated whether it was he or other American radicals who recommended to the Vietnamese Communists that two of the three released prisoners should be black, in order to increase the propaganda effect in the United States.

In an interview many years later Pitzer mentioned Hayden's continued efforts to turn the 1967 POW release to his own political advantage:

"During the last election time, Hayden called me up, wrote me letters wanting me to help him on his election. He said, 'I did have something to do with the release of you prisoners.' I said, 'No you didn't.' He said, 'But I was there when you guys were released.' And he said, 'All I'm asking you for is a statement that I had something to do with your release.' And at that time I found an old picture of Jane Fonda (Hayden's then-wife) sitting at the antiaircraft guns in Hanoi, wearing a North Vietnamese Army helmet. I bundled up two of those, mailed them to him, and put a note on it: 'My grandmother has an old saying, 'Sleep with dogs, you wake up with fleas.' Sorry, I can't help you.'"[65]

In the fall of 1967, as a North Vietnamese offensive raged on the battlefields of Dak To, Con Thien and Loc Ninh in South Vietnam, and American infantry soldiers and airmen suffered thousands of casualties in killed, wounded and missing in action, Hayden recorded propaganda statements over Radio Hanoi (on October 28, 1967 and November 4, 1967). Mrs. Dagmar Wilson, Mrs. Ruth Krause and Mrs. Mary Clark, who represented "Women Strike For Peace," also interviewed a U.S. Air Force POW, Lt. Larry Carrigan, on Sept 11, 1967. Mrs. Krause spoke once over Radio Hanoi and Mrs. Wilson made two broadcasts over Radio Hanoi. After the interview with Lt. Harrigan, one of the women told the North Vietnamese that Carrigan was a "wayward individual." As a result Carrigan was beaten by the Vietnamese so severely that he suffered pain in his shoulders for six months afterwards.

Some American POWs were transported to North Vietnam in times of crisis, such as the February-March 1968 Tet Offensive, when, during eight continual weeks of the worst ground-combat of the war, approximately 4,000 U.S. troops were killed in action, more than 19,000 were wounded, and 604 more Americans became missing in action. For the U.S. combatants it was a battle equal to any in American history, and it had dismal implications for American POWs held in secret camps in South Vietnam, Laos and Cambodia.

A Declassified April 23, 1969 set of CORDS documents released by CIA through FOIA request by the author years after the war, revealed that many of these positively-identified U.S. POWs subsequently disappeared and were later presumed dead on the date they went missing, or are still carried as MIAs on the official lists of the U.S. Government. (This document first came into the author's possession in the late 1986-early 1987 period, but was obtained again through FOIA request to CIA in 1989.) The report number is blacked-out by CIA censors in the author's copy, but the document carries the heading: "REGIONAL OFFICE DANANG, DISTRIBUTION-CONSULATE, CORDS, XXIV, III MAF, 525th MILITARY INTELLIGENCE, 1st CORPS G-2 ADVISOR. INFORMATION PASSED TO 525th MILITARY INTELLIGENCE, PHU BAI FOR BRIGHT LIGHT PROJECT. (the Codename for POW/MIA intelligence project) ATTACHED ARE LIST OF U.S. PRISONERS POSITIVELY AND TENTATIVELY IDENTIFIED BY (blacked-out) FROM PHOTOGRAPHS, LIST OF VIETCONG HUONG THUY DISTRICT COMMITTEE AND SKETCHES OF THE COMMITTEE'S HEADQUARTERS."

The documents reveal the location and many details about a, "VIET CONG HUONG THUY DISTRICT COMMITTEE HEADQUARTERS AND PRISON CAMP IN THUA THIEN PROVINCE, VIETNAM." In this camp it is reported that "FOLLOWING TET 1968, THE SECURITY SECTION

HANDLED 43 AMERICAN AND 326 VIETNAMESE PRISONERS. THE AMERICANS LATER WERE TRANSFERRED TO NORTH VIETNAM, AND THE VIETNAMESE WERE TRANSFERRED TO NORTH VIETNAM OR TO AN AGRICULTURAL CAMP AT AN UNKNOWN LOCATION NEAR THE BORDER OF LAOS."[66]

Attachment #1 of this CORDS document, which is dated 11 April 1969, is titled "List of U.S. Personnel IDENTIFIED FROM PHOTOS, Photo list: AVGJ-HCIS (10 November 1968) 525th Military Intelligence Group." It listed 53 U.S. POWs, of whom 22 were rated unequivocally as POSITIVE IDENTIFICATION, while the remaining 31 were carried as POSSIBLE IDENTIFICATION. Of those positively identified, many were never repatriated at the end of the war prisoner exchange in early 1973, or thereafter.

Among those positively identified who subsequently vanished forever in Communist captivity were: Lt. Colonel Ronald E. Storz, officially listed as "killed in action" on 28 April 1965; David R. Devers, listed as "killed in action" on 13 August 1966; Major Richard J. Schell, listed as "missing in action" on 24 August 1967; Lt. Colonel Albert Pitt, listed as "missing in action on 24 January 1966; Captain Ronald E. Pfeifer, listed as "killed in action" on 6 October 1966; Staff Sergeant Fred T. Schreckengost, "missing in action" on 7 June 1964; 1st Lieutenant Robert Lewis, listed as "killed in action" on 10 January 1968 (or possibly Specialist 4th Class Robert Lee Lewis, listed as "killed in action" on 15 November 1965; both vanished forever); Staff Sergeant Harry Brown, listed as "missing in action" on 12 February 1968; Staff Sergeant John C. Stulier (Stuller), originally carried as "MIA" but listed on the Vietnam Memorial Wall as "killed in action" on 12 May 1968; Sergeant First Class Lyle E. Mackedanz, listed as "missing in action" on 21 April 1968; Sergeant First Class Earl E. Shark, listed as "missing in action, 12 September 1968; and Sergeant Steven M. Hastings, listed as "missing in action" on 1 August 1968.

In addition, many of the POWs in this camp listed as "possibly identified from photographs," also failed to return in the postwar 1973 prisoner exchange, or thereafter. They are: Sergeant First Class Joe Parks, listed as "KIA" on 22 December 1964; Staff Sergeant David F. Demmon, listed as "KIA" on 9 June 1965; Major John C. Jacobs, listed as "KIA" on 7 June 1966; Sergeant Major Edward R. Dodge, listed as "MIA" on 31 December 1964; Major Richard D. Smith, listed as "MIA" on 11 March 1965; Navy EN2 Richard D. Musetti, listed as "MIA" on 28 September 1967; Sergeant First Class Donald S. Newton, listed as "MIA" on 22 February 1966; Major Gerald E. Olson, listed as "MIA" on 13 March 1966; Major Orien J. Walker, listed as "MIA" on 23 May 1965; Robert Monahan, listed as a civilian, but could be Lance Corporal Robert E. Monahan listed as "KIA" on 28 May 1967; Captain William E. Johnson, listed as "KIA" on

6 October 1966; Captain Morvan D. Turley, listed as "KIA" on 13 January 1967; Master Sergeant William B. Hunt, listed as "MIA" on 4 November 1966; Staff Sergeant Daniel R. Nidds, listed as "MIA" on 21 April 1967; Sergeant Major Ronald J. Dexter, listed as "MIA" on 3 June 1967; Sergeant Major Raymond L. Echevarria, listed as "MIA" on 3 October 1966; Colonel Donald G. Cook, listed as "MIA" on 31 December 1964; Major Wilbur R. Brown, listed as "MIA" on 3 February 1966; Colonel William R. Cook, listed as "MIA" on 28 April 1968; Master Sergeant Michael R. Werdehoff, listed as "MIA" on 19 April 1968; Lt. Colonel Donald Burnham, listed as "MIA" on 2 February 1968; Captain Jerry Lee Roe, listed as "MIA" on 12 February 1968; Specialist 5 Buford Johnson, listed as "KIA" on 24 April 1968; Specialist 4 Daniel M. Kelly, listed as "KIA" on 25 April 1968; and Sergeant First Class Roy C. Williams, listed as "MIA" on 12 May 1968.

It is most significant that one of the five U.S. civilians also on the "positively identified" list in this document was Mr. Eugene A. Weaver, and that according to a statement read publicly 23 years later by Senator John Kerry, Chairman of the Senate Select Committee on POW/MIA Affairs on January 20, 1992, Weaver was subsequently "INTERROGATED BY THE SOVIETS," according to Hanoi's Communist Government, but was eventually repatriated to the United States. Such a statement by Hanoi in 1992, may have been a clear indication of the ultimate destination of some of the positively and tentatively identified American POWs on the two lists who were never released. In various cases and for various purposes, some of them may, in fact, have also been among those Americans who were "Moscow Bound," and ended up in the USSR. Five civilians were listed as tentatively identified, including a Canadian prisoner named Marc O. Cayer, and on the bottom of the document is an additional note:

"(Source blacked out) STATED THAT TWO (2) AMERICAN WOMEN WERE ALSO PRISONERS. (NO PHOTO WAS IDENTIFIED) HE RELATED THESE WOMEN WERE SCHOOLTEACHERS FROM HUE. HE STATED THEY WERE YOUNG, ONE WAS TALL AND THIN AND ONE WAS SHORT AND HEAVY. THE WOMEN TOLD (BLACKED-OUT) THAT THEY WERE NEW IN VIETNAM."

Such lists as these and many others like them in the possession of U.S. intelligence throughout the war, assembled from captured enemy documents or from photo identifications by captured or defected enemy personnel, which correlate with missing in action or "presumed" killed in action Americans whose bodies were never recovered, give a clear indication of how many Americans, known to have been alive in captivity, were deliberately withheld as prisoners at the end of the war; whether to remain in Vietnamese custody for future bargaining purposes, or to be shipped to other Communist-bloc nations for intelligence use.

Meanwhile, from the period of the January-March 1968 Tet Offensive and counteroffensive, and on through the five years of American military involvement that followed in Vietnam, a ring of Soviet spies, led by Warrant Officer John Walker, were betraying the U.S. Navy's most secret ciphers to the KGB. Walker had been a walk-in to the Soviet Embassy in January 1968; a man with no ideological motive, but simply greedy for money to support a better lifestyle. In February 1968, as the Tet Offensive raged around hundreds of military bases, provincial capitals and villages of Vietnam, John Walker was delivering highly secret Navy cipher key cards to a Soviet intelligence officer in the Washington, D.C. area. Such information in the hands of the Soviets was immediately put to use in deciphering the U.S. Navy's secure communications between the U.S. command, the fleet, the fighter and bomber aircraft, and by extension the American Marine and Army ground troops in combat with the enemy. Coordination necessary for safety in combat operations would have also betrayed Army, Air Force and Marine Corps positions and plans, in being kept informed of Navy operations. The communications secrets passed to the KGB by the Walker ring, of ever increasing quality from the late 1960s through the early 1970s, assisted the Soviets and North Vietnamese in selecting and targeting U.S. pilots and other desired air and ground personnel for capture, and no doubt also caused thousands of additional American and allied war casualties.

An agent of such value was handled most carefully by Moscow Center, and by Walker's Soviet controller in the critical Vietnam war years, Oleg Kalugin, who was then head of the KGB's counterintelligence in Washington, D.C. (Over 20 years later, in the fall of 1991, KGB General Kalugin, former Chief of KGB Military Counterintelligence, became the first senior Soviet official to admit that U.S. prisoners of war in Vietnam had been interrogated by Soviet intelligence (although he claimed that the KGB had not transferred any to Russia, he did not rule out the possibility that the GRU may have done so). Other traitors had existed in the ranks of CIA and NSA who betrayed U.S. secrets to Moscow, and this also affected American security during the Vietnam War period.[67]

Only 591 American POWs were released by Hanoi following the January 1973 Paris peace agreement, while some 2,000-3,500 Americans (including covert U.S. POWs captured during clandestine operations in China, North Vietnam, Laos and Cambodia) may actually have been taken as prisoners in Southeast Asia, according to formerly-secret U.S. documents and retired U.S. intelligence sources interviewed by this author. A declassified CIA report dated 15 May 1968, identified two of the many secret Vietnamese POW camps, one west of Hanoi (perhaps the Lam Thao superphosphate plant) and another north of VINH, and further stated:

"...THERE WERE SOME 2,500 ALLIED PRISONERS IN NORTH VIETNAM, MOST OF THEM AMERICANS. (Blacked out)...AMERICANS AT TWO WELL GUARDED CAMPS, ONE ABOUT 40 MILES WEST OF HANOI, ANOTHER SOMEWHAT NORTH OF VINH. THE CAMPS (blacked out) HAVE FACILITIES FOR ABOUT 100 POWS AND WERE LOCATED IN RURAL AREAS."[68]

Some of the missing American POWs are known to have died at forced labor in a secret and unacknowledged parallel prison system, and others were executed, but many were held captive for years to come in Indochina, while still others, those with "special talents," were secretly transported to the Soviet Union and to Communist China. (A small number of those not returned, no more than 20, were possibly genuine defectors, a few of which had actually fought against their own countrymen.) It had long been known within the American intelligence community that the Soviet KGB (and the GRU) had orchestrated the shipment of captured American military equipment to the USSR, in the same manner as during the Korean War. By 1967 the second largest element of the KGB's First Chief Directorate, known as Department T, and the GRU (Soviet Military Intelligence) were ordered to conduct operations in Vietnam to achieve this end. According to respected researcher and author William Corson:

"One phase was limited to the theft of high-tech military equipment deployed by the United States throughout Southeast Asia. This work was carried out by North Vietnamese agents and with the aid of communist agents in the government and the army of South Vietnam. Captured and stolen material made its way to Department T's research institute, located near the Belyourski Railroad Station in Moscow, by a variety of routes and means. Some went overland from South Vietnam via the Ho Chi Minh Trail to Hanoi and thence to Moscow by air or by sea to Vladivostok. These 'exports' were dispatched under many guises, ranging from military equipment concealed as 'construction equipment in need of repair' to 'household goods."[69]

What was never admitted by the U.S. government is that the Soviets, just as naturally, demanded and obtained American prisoners of war from Vietnam to further their operations in studying, testing, modifying or reproducing American high-tech weaponry, war material and equipment. This was in accordance with Soviet-Russian national interests and long-time Soviet military intelligence policy (since 1918), as has been outlined throughout this book. At this same time, in 1967, Senator Karl Mundt was confronting the State Department with information he had received on American POWs from WW II who had been taken into the Soviet Union 22 years before, as noted previously in this book. The Department's response had been that they too, sometimes received such reports, but that the source's reliability was questionable and even if true did not prove that such prisoners still remained alive. It was known by U.S. intelligence analysts of the time, through decoded enemy communications, that the Soviets were requesting the Vietnamese to target certain categories of U.S. prisoners for capture and interrogation, and for eventual transfer to the Soviet Union, along with the sophisticated war equipment that was shipped to Moscow. But throughout the late 1960s, at the height of the Vietnam war, collection of this information was not specifically tasked, requested or even deliberately collected. It was only after Mooney had himself, on his own initiative, made a study of the targeting, fate and destination of prisoners, reported as captured alive in decoded enemy communications, that he was instructed to continue to compile such data on American POWs beginning in 1971. At that point he also began to go back through old reports dating to 1965, correlating them with new information obtained in decoded Vietnamese traffic, and building up a list of earlier, certain and possible, Moscow-Bound and China-Bound U.S. POWs.

The process by which lists of "Moscow-bound" American prisoners were compiled from intercepted and decrypted enemy communications has been revealed to the author by a retired Air Force and NSA intelligence analyst, Mr. Jerry J. Mooney, who in 1986 and thereafter until 1992 provided three Congressional investigations with this same information:

"Special POW camps in Laos and Vietnam (with a Soviet presence) were used for POW debrief and interrogation. The intent to break was tempered with a dedication to convince, for a willing/converted POW was more useful than one cooperating in fear and pain. And in our government's stupidity, it collaborated by declaring them dead—or if some were alive—by branding them as traitors, a most powerful tool for interrogators. When, not if, broken or convinced, several USSR locations were probable, depending on the POW's high tech value. I had many conversations with my DIA counterparts on this for specific POWs and my comment at the time was 'well, they are gone forever,' and the response was 'you betcha!'"[70]

The Soviet-staffed POW camps monitored through decrypted enemy communications by Mooney and other NSA analysts included one at Ron Ron near the Demilitarized Zone, others at Thach Ban, Long Dai, Cua Luoi near Vinh (code-named Son Tay and distinct from another camp of the same name), the Lam Thao superphosphate plant in Vinh Phu Province west of Hanoi, Lang Son near the Chinese border and Sam Neua, Laos, known to a few intelligence personnel as 'Tentacle MB." Soviet advisers also operated inside South Vietnam and were no doubt useful in making preliminary selections of some of the prisoners sent to the special interrogation camps.

Further south in the Cambodian border region, where the author served as a rifleman in the 1st and 9th Infantry Divisions throughout 1968, it was common knowledge among combat soldiers that U.S. forces operating clandestinely on raids and reconnaissance missions inside the "Fishhook" and "Parrot's Beak" areas of Cambodia were constantly losing casualties, and that missing in action personnel had disappeared in these North Vietnamese/Viet Cong sanctuaries for years. In addition, Soviet military advisers were known to be present in the Cambodian border regions, and there is little doubt that interrogation camps with a Soviet presence, similar to those far to the north reported by Mooney at Ron Ron, Cua Luoi and Tentacle MB, also existed in Cambodia during the war years. The situation was also to be reversed on rare occasions. In one late-July 1968 incident, during a combat infantry reconnaissance-in-force operation of the 1st Infantry Division and 11th Armored Cavalry along the Song Be River, south of the "Fishhook" region of the Cambodian border in Vietnam, the author's platoon of Company A, 1st Battalion, 16th U.S. Infantry (Rangers), captured a large, non-Vietnamese, but Eurasian-appearing POW in a tunnel complex of the VC/NVA Dong Nai Regiment, who, upon interrogation in the field claimed in broken Vietnamese to a South Vietnamese interpreter, in the author's presence, to be a "Soviet adviser," who had recently infiltrated into South Vietnam from Cambodia with 300 North Vietnamese regulars. (This POW, slightly wounded by grenade fragments, was evacuated to the rear for further interrogation by a U.S. LOH helicopter (which also brought in ammunition), and he was not heard from again. A record of his capture exists, but does not note his nationality.[71]

U.S. POWs were also reported held in several locations inside Cambodia, where the North Vietnamese controlled all the border regions, and from which they operated against the American Infantry divisions defending South Vietnam on the southern war front. After the "border battles" near Cambodia in the fall of 1967, South Vietnamese agents reported to U.S. military intelligence that American POWs (from the 173rd Airborne Brigade and 4th Infantry Division) captured in the heavy fighting around Dak To and Hill 875, ended up in two groups, one of 36 U.S. POWs held in the Pro Hut Woods and the other totaling some 100 U.S. POWs held at Ta Mo, Cambodia. Most of these men disappeared forever.[72] Although later official records were to list only 18 U.S, MIAs in the Dak To battle, which had cost 285 Americans their lives and wounds to some 1,000 more, many of the POWs reported by intelligence sources inside Cambodia from this fight, initially reported by their units as missing in action, had subsequently been listed as Killed, Body Not Recovered (KBNR). Families of these soldiers may have received "unidentifiable remains" from the U.S. Government. Other American POWs and MIAs were taken into Cambodia during and after the vicious fighting of the 1968 Tet Offensive, and several American counteroffensives which followed that year and into 1969, when more severe fighting occurred along the Cambodian border.

The author consulted records of several U.S. combat units in Vietnam at the National Archives in 1984-85, and found that in many cases U.S. Army soldiers initially listed as MIA were later carried in a category of Killed in Action, Body Not Recovered (KIA-BNR).[73] It appears that the same policy was applied to many Marine MIAs of Vietnam. Edward "Ned" Tuthill, a U.S. Marine rifleman who served in combat in Vietnam during 1968, was reassigned out of the field to serve in the Marine "War Room" in Danang, where he was an orderly and assisted officers who were posting casualties for daily combat actions of Marine units. Tuthill repeatedly witnessed the radio reporting of Marine MIAs from skirmishes and battles in the I Corps zone, sometimes in groups as large as 10-20, or more, who were in nearly all cases moved from the MIA category to the Killed, body not recovered category, usually within 24-36 hours. In these cases the government may also have eventually returned "unidentifiable remains" to family members, with a recommendation that they not be viewed.

Soviet intelligence officers were reported to be present inside the Cambodian border where the Communist headquarters for the South (COSVN) was often located, and in the context of the Cold War of the time, it must be assumed that at least some of the American POWs held in Cambodia were interrogated by them. It is also likely that any individual American POWs held in Cambodia possessing "special talents" would have been moved on towards the Soviet Union, with North Vietnamese permission and in return for Soviet assistance, in the same manner as occurred in North Vietnam and Laos.

## THE SOVIET INTELLIGENCE CONNECTION
## TO AMERICAN POWS OF VIETNAM

Official U.S. documentary confirmation of Soviet-held American POWs from Vietnam largely remains classified, despite FOIA requests and appeals pursued by the author for years with the relevant agencies, such as CIA, DIA, FBI, Pentagon and State Department. Yet, a set of 1967 U.S. State Department documents, declassified in 1992 and turned over to a Senate Select Committee late that year, proves that the Johnson Administration was aware the Russians had obtained at least some American POWs from Hanoi by the beginning of that year and were transporting them to Soviet-bloc nations. Such information would have been passed on to the succeeding Nixon Administration. These documents and cover sheets are stamped 'Secret-Exclusive Distribution (EXDIS), Special

Handling, Transfer by hand delivery and receipt only. Not to be discussed with or communicated to anyone except on a strict 'need to know' basis, or discussed with anyone outside of recipient's organization who has not received copies distributed by S/S." They contain signature record sheets for those handling them and are reproduced here in their entirety, due to their historic significance. A secret January 26, 1967 telegram from the U.S. Mission in Berlin (#956), to Secretary of State Dean Rusk, who must have had painful memories about the fate of America's earlier POWs, stated:

"West Berlin lawyer, Stange, who has long been active in East-West exchanges, told a mission officer January 25 that the GDR was making a new offer to get back the Soviet spies Peter and Helen Kroger. THE GDR IS WILLING, IN EXCHANGE, TO TURN OVER TO US TWO WOUNDED U.S. FLIERS CAPTURED IN VIETNAM AND CURRENTLY IN AN EAST GERMAN HOSPITAL. Stange said that he had been informed of this offer January 20 by his East Berlin partner, Vogel and a representative of the GDR State Prosecutor's office on January 21 and 22. STANGE HAD LEARNED THAT SEVERAL—PERHAPS FIVE OR SIX—WOUNDED U.S. FLIERS WERE NOW IN THE HOSPITAL JUST OUTSIDE EAST BERLIN. The two offered in exchange for the Krogers were both severely wounded (one of them had lost both legs), Stange said). Stange thought that the East Germans would be willing in the end to exchange more than two for such valuable prizes as the Krogers. HE ASSUMED THAT THE OFFER HAD BEEN CLEARED WITH BOTH THE NORTH VIETNAMESE AND THE SOVIETS."

"Mission officer replied that the U.S. of course wished to explore any reasonable possibility that might get our captured servicemen released. He reminded Stange however, that the Krogers are United Kingdom, not U.S. prisoners. Mission officer also requested the names and serial numbers of the U.S. servicemen now in the East German hospital. Stange was not sure he could obtain these data until there had been 'reaction in principle' from the Western side. HE ADDED THAT THERE WAS NO DOUBT IN HIS MIND THAT THE EAST GERMANS DID IN FACT HAVE THE WOUNDED U.S. FLIERS, HOWEVER ADDRESSEES WILL RECALL (BERLINTEL 790, MARCH 7, 1966, AND PREVIOUS) THAT THE EAST HAS USED THE VOGEL-STANGE CHANNEL BEFORE IN ITS PERSISTENT EFFORTS TO OBTAIN THE KROGERS. Wyman" (This document also contains a note, "Passed S/S-O, White House, 1/26/67, 11:10 AM," indicating that President Lyndon B. Johnson was informed.)[74]

In order to understand the significance of this message, it is necessary to review some contemporary Cold War spy episodes that followed the defections of Donald Maclean and Guy Burgess to Moscow during the Korean War, and which preceded the defection to Moscow of Kim Philby of British Secret Intelligence Service (MI-6), in 1963.

Peter and Helen Kroger, actually Morris and Lona Cohen—who were originally Americans involved with the Rosenberg ring, were now New Zealand nationals living in Britain, and were important sub-agents of a KGB illegal using the name Gordon Lonsdale, who was actually a deep-cover Soviet national named Konon Trofimovich Molody.[75] Lonsdale-Molody was a high-level KGB officer and son of a Soviet scientist, who, in 1955, had taken on the identity of a long-deceased Finnish-Canadian, Gordon Lonsdale, and in Britain ran a group of Soviet agents known as the "Portland Spy Ring." According to retired British counterintelligence officer Peter Wright, through an unidentified Soviet mole in MI-5, Moscow center already feared that Lonsdale-Molody was suspected of being a Soviet agent by the British, but after a recall to Moscow he was sent back to London by the KGB, because to have not done so would have exposed the unknown Soviet agent in MI-5 to a molehunt, leading to worse consequences for the Russians. Thus, the precaution was taken by the KGB to have Lonsdale turn over some of his best agents at the main British antisubmarine warfare base, at Portland, Dorset to his sub-agents, the Krogers, for their secret communications with Moscow. As the advanced British and American nuclear-armed submarine force, operating in conjunction, was critical to the free world's deterrent strength against Soviet aggression, the case was of great importance.[76]

In January 1961, an important CIA agent inside the Polish intelligence service (UB), code-named "Sniper," had decided he must defect to the United States. He was Michal Goleniewski, who, while working for CIA had revealed in 1959 that the Soviets had two important spies in British intelligence that he knew of, one in the British intelligence service and the other in Naval intelligence. "Sniper" also uncovered George Blake, a senior MI-6 officer, who was another deep-cover Soviet spy, having been captured during the Korean War while serving as the British Consul General at Seoul, Korea in 1950. During three years in North Korean and Chinese Communist captivity, Blake was turned by the Soviets into a KGB agent prior to his return from captivity. George Blake served as a Soviet spy from 1953 to 1961 and has been ranked with H.R. "Kim" Philby, in terms of the damage he did to the west during his career. He identified to the Soviets most MI-6 plans, operations and personnel and betrayed numerous British spies in Europe.[77] After his return from the Korean War as a secret Soviet agent, Blake had played a key role in the "Berlin tunnel" episode. Known as "Operation Gold," this was a 1,500-foot tunnel under the border from West to East Berlin, built at a cost of $25 million, which became operational in 1955. After nearly a year of use in providing vast quantities of intercepted GRU and KGB radio traffic, which was then laboriously deciphered, it was "exposed," although the exposure was likely staged by the Soviets. George Blake's arrest in 1961 raised serious questions as to how much of the Berlin Tunnel's product had been Soviet disinformation all along. According to retired MI-5 officer Peter Wright, MI-6's transcribing of the great volume of intelligence from the tunnel took over seven years, and it finally revealed at that time that George Blake had betrayed the Berlin tunnel to the Soviets from the beginning, as he was Secretary of the Planning Committee for MI-6. Blake was sentenced to 42 years imprisonment in Old Bailey Prison, London, but, incredibly, he was reported to have escaped after six years and subsequently lived in the USSR.[78] (One of the CIA officers involved in the Berlin tunnel debacle was Theodore Shackley, who would later figure prominently in the Indochina War.)[79]

MI-5 realized that Gordon Lonsdale and the Krogers would be exposed by "Sniper's" defection to CIA, and in January 1961 arrested all three of them, and also Harry Houghten of British Naval intelligence (identified first by "Sniper"-Goleniewski), who was a Soviet spy for Lonsdale at the British Underwater Weapons Establishment at Portland, Dorset. All were sentenced to long prison terms, and the Soviets had been making persistent efforts to get them back ever since.

Stange and Vogel, referred to in the January 1967 State Department cables, were then already well-known as, respectively, the West German and East German conduits to their intelligence services; acting as brokers in deals between the Soviets and East Germans and the Americans and British. As researcher and author Tom Mangold later wrote: "East German lawyer Wolfgang Vogel, (was) the man who had previously arranged several major spy swaps and whose career paralleled the history of Cold War espionage. Vogel, and Juergen Stange, his opposite number in West Germany, had pioneered and monopolized what came to be known as Freikauf—the commercialized and controversial business of exchanging prisoners of the East German regime for hard western cash."[80] Thus, when the Soviets, through their East German surrogates and the Stange-Vogel team, offered American POWs from Vietnam held in East Berlin to the U.S. Government in exchange for Peter and Helen Kroger in January 1967, American and British intelligence would have considered it a serious offer indeed, and probably as a feeler for a possible Lonsdale exchange, too. Recovering exposed deep-cover agents such as Lonsdale and Blake would have been a high priority for Soviet intelligence, indicating that the offer to supply U.S. prisoners of war from Vietnam must have come from the highest level in the Kremlin.

Secretary of State Dean Rusk replied to the initial message on the POW exchange offer with a secret cable to Berlin of January 26th headed, "Subject: Captured U.S. Fliers in GDR," with an action copy to the American Embassy in London and information copies to the U.S. embassies in Bonn, Moscow and Saigon. The message was drafted by F. Sieverts, evidently Mr. Frank Sieverts, who was to be involved with the POW/MIA matter through two more presidencies:

"Department wishes follow-up Stange's offer concerning release of wounded U.S. fliers reported in RETEL. Mission Berlin should inform Stange that we are favorably disposed to discuss his offer and that as useful first step we would be interested in obtaining names and serial numbers of the U.S. servicemen now in East Germany of whom he spoke. Mission should point out that, as Stange-Vogel aware, Krogers present difficulties, and in any event are not in U.S. custody. Mission should sound out Stange to see if there is any other quid pro quo he or his contacts might have in mind, Mission should not however, take initiative in suggesting what this might be. For London: Please give us your assessment of probable British attitude towards release or exchange of Krogers. Do not discuss with UK authorities for time being. POSTS SHOULD NOTE SENSITIVITY OF THIS MATTER AND NEED FOR UTMOST SECRECY. (DEAN) RUSK."[81]

It is noted on this document that it was supplied to W. Averell Harriman, the former U.S. Ambassador to Moscow in 1945, who still remained involved with the matter of Communist-held American prisoners of war at this time, with the rank of Assistant Secretary of State.(Later sworn testimony before a Senate Select Committee by a succeeding Secretary of State, Melvin Laird, was to confirm this fact.)

Berlin cabled Secretary of State Dean Rusk on February 9th:

"VOGEL TOLD A MISSION OFFICER FEBRUARY 8 THAT HE HAD DISCUSSED THE DEPARTMENT'S REACTION WITH GERMAN DEMOCRATIC REPUBLIC STATE PROSECUTOR STREIT ON FEBRUARY 3. TWO SOVIETS WERE ALSO PRESENT DURING THE DISCUSSION, VOGEL SAID. HE ASSUMES THAT THEY WERE SECURITY MEN. THE EASTERN STAND WAS THAT THE WEST MUST COMMIT ITSELF IN PRINCIPLE TO RETURNING THE KROGERS BEFORE PROGRESS CAN BE MADE TOWARD FREEING THE CAPTURED U.S. FLIERS, ACCORDING TO VOGEL. He pointed out to his interlocutors that the Krogers were not U.S. prisoners, but to no avail. He also inquired whether the East would be interested in getting something else, for example Thompson, in exchange for the U.S. flyers. He detected little, if any Eastern interest in Thompson. THE EAST ALSO REFUSED TO PROVIDE NAMES AND SERIAL NUMBERS, MAINTAINING THAT THIS INFORMATION WOULD REPRESENT A CONCESSION IN ADVANCE TO THE U.S., WHICH DID NOT NOW KNOW FOR SURE WHICH OF ITS MEN MISSING IN VIETNAM WERE ACTUALLY DEAD AND WHICH WERE STILL ALIVE. Stange, Vogel's West Berlin partner who was also present during the conversation with the Mission officer, interrupted to say that the U.S. might make a counteroffer for the fliers. He, like Vogel, recognized our difficulties in bringing the British to release their prisoners for our servicemen. Stange also speculated that, when it got down to hard bargaining, the East would be willing to give up more than two fliers for the Krogers. Vogel neither contradicted nor agreed with Stange on these two points. Vogel plans to be away...through early March. He said he would be reachable, if needed, however. Departments instructions are requested. Wyman."[82]

On February 15th the Department of State cabled the U.S. London embassy (with copies to the Bonn, Moscow and Saigon U.S. embassies, to the U.S. mission in Berlin, and to Averell Harriman:

"Embassy should inform Fonoff of Stange-Vogel offer to swap Krogers for injured U.S. pilots, noting that offer, while unofficial, APPEARS TO HAVE EAST GERMAN AND SOVIET APPROVAL. Should point out that we have no SUBSTANTIAL evidence that captured U.S. pilots anywhere but North Vietnam, THOUGH PRESENT PROPOSAL DIFFERS FROM PREVIOUS STANGE-VOGEL INITIATIVE IN THAT VOGEL STATES SUCH PILOTS ARE NOW PHYSICALLY IN EAST GERMANY. As minimum we have obvious interest in keeping proposal alive long enough to find out if this is true. AS REFTELS INDICATE, OTHER SIDE'S CHIEF INTEREST APPEARS TO BE KROGERS. ENTIRE 'OFFER' MAY BE NO MORE THAN NEW PLOY BY SOVIETS TO ACHIEVE THEIR RELEASE, WHILE IRRITATING U.S.-U.K. RELATIONS IN BARGAIN. HOWEVER WE WOULD APPRECIATE HER MAJESTY'S GOVERNMENT'S COMMENTS ON PROPOSAL, AND THEIR SUGGESTIONS ON HOW TO PROCEED."[83]

The American Embassy in London replied on February 16th (#6625), "Two key people in HMG competent to deal with this sensitive subject are both out of London...We will bring up matter with them when they return first of next week."[84]

The U.S. mission in Berlin sent a cable to the Secretary of State on February 20th, which revealed a few more details about the original Soviet involvement in the offer:

"Mission officer met with Vogel, Stange's East Berlin partner, on January 31 and informed him of U.S. reaction as instructed...Vogel undertook to pass the message to his principal in the GDR government, Chief State Prosecutor Streit. No new factors emerged from the conversation. VOGEL DID MENTION THAT THE U.S. FLIERS-KROGERS OFFER HAD FIRST BEEN BROACHED TO HIM BY SOVIET SECURITY OFFICERS. VOGEL SAID THIS HAD OCCURRED DURING A BRIEF TRIP WHICH HE PAID TO MOSCOW RECENTLY, IN CONNECTION WITH THE RELEASE OF THE WEST GERMAN NEWSPAPERMAN KISCHE. After his return to the GDR, Vogel raised the matter with Streit. Wyman."[85]

It was not until March 14, 1967, that a decision of sorts was reached by Dean Rusk and the Johnson Administration on the proposed trade for the American POWs, and this proved to be a rejection of an exchange for the Cohen-Krogers:

"Mission should inform Stange and Vogel that Krogers not available for exchange. Krogers not in U.S. Government control, and not up to us to release them. You should express concern over fate of captured pilots, and request information about their health and welfare. You should further ask Vogel to try to arrange visit to prisoners by representative of International Committee of the Red Cross (ICRC) or other neutral organization or country. We would hope this would be permitted as act of basic humanity. YOU SHOULD ASK VOGEL IF THERE IS ANYTHING OTHER THAN THE KROGERS THAT MIGHT BE EXCHANGED FOR THE PILOTS, SUCH AS VC OR NORTH VIETNAMESE MILITARY PERSONNEL HELD IN SOUTH VIETNAM. For Your Information (20 spaces blacked-out by U.S. Government censors upon declassification) casting doubt on report that pilots held in Bestensee. It is reasonable presumption that information in Bonn's 9865 comes from Vogel or connected source, and thus provides no independent verification. (Dean) RUSK."[86]

On May 4th, Dean Rusk sent another cable to Berlin—drafted by Sieverts—with copies to Assistant Secretary of State for European Affairs Walter Stoessel (who had been involved with POWs of earlier wars), and to Averell Harriman, who also had much prior experience with American prisoners in Soviet control. This cable effectively killed any chances for a secret trade for the American POWs by casting doubt on the whole offer, and urging that the matter be pushed into the legal realm involving treaty rights of POWs. The message also revealed that the East German Communist government had already been put on the spot about the POWs by a U.S. request for verification about their presence and for their repatriation. This was a clear signal that Lyndon B. Johnson's Administration did not want to deal with the Soviet bloc on U.S. POW exchanges for Soviet spies:

"We are disinclined to rise to bait suggested by Stange by discussing Thompson, who in any case, not repeat not available for exchange. Seems apparent that other side's interest is in Krogers... Ref. B and previous indications lead us to doubt any U.S. pilots actually in East Germany. STANGE-VOGEL APPEAR TO US TO BE TRYING US OUT ON A KROGERS FOR PILOTS EXCHANGE. IF WE PRODUCED KROGERS, THEY WOULD PRODUCE PILOTS. Under those circumstances we see no value in responding favorably to suggestion which by its own terms comes not from OTHER SIDE but as personal thought of Stange.

"We of course remain deeply interested in obtaining the release of captured U.S. pilots wherever they may be held, and would appreciate Mission thoughts on how best to proceed. As we see it we have little to go on, since Stange-Vogel channel is weak basis for approaching East German or Soviet authorities directly. IF PRESENCE OF PILOTS IN EAST GERMANY COULD BE ESTABLISHED, MIGHT BE

POSSIBLE TO DEMAND THEIR RETURN IN ACCORDANCE WITH HUEBNER-MALININ AGREEMENT. ASSUME THERE HAS BEEN NO RESPONSE TO INQUIREY SET FORTH IN PARAGRAPH TWO, REFERENCE C, IF SICK AND WOUNDED PILOTS ARE IN EAST GERMANY THEIR REPATRIATION TO U.S. IS OBLIGATION UNDER GENEVA CONVENTION ON PRISONERS OF WAR. REQUEST MISSION COMMENTS. (DEAN) RUSK."[87]

In reality, the Soviets and East Germans could have given the U.S. Government the names of any U.S. POWs then held in Vietnam, if that was the game they were intending to play, knowing that they could have prevailed upon the North Vietnamese to deliver those identified if that was all it took to accomplish the trade. The fact that they refused to do so indicates that the "five or six" American POWs were in East Berlin, but the East Germans and Soviets were not so stupid as to identify those particular POWs, who had already been segregated from the publicly-acknowledged POW system in North Vietnam and processed through the secret parallel prison system which was staffed by Soviet intelligence officers, then already functioning in Vietnam, when the American Government was obviously bent on an expose and public demand for repatriation, rather than a secret trade. Cables such as these must be assumed to be compromised to the main adversary, through decryptions or spies, thus even when "classified" they represent the public position of the government which sends them.

The U.S. Government ploy to ascertain the location of the American POWs in East Germany was rejected in a meeting with Wolfgang Vogel, the East German connection to the Soviets, which is discussed in a cable of June 2nd, 1967, from the U.S. Mission in West Berlin:

"At postponed meeting (Ref A) Vogel said flatly that he had been authorized to negotiate on the pilots only in connection with the Krogers, VOGEL WAS TOTALLY NEGATIVE ON ARRANGING FOR PARCELS OR VISITS TO THE PRISONERS...HE ADDED THAT HE HAD BEEN UNABLE TO OBTAIN ANY ADDITIONAL INFORMATION ABOUT THEIR WELFARE. HE UNDERSTOOD THAT THEY HAD BEEN UNDER THE CONTROL OF THE EAST GERMAN ARMY, PROBABLY IN AN INSTALLATION WITHIN A RESTRICTED AREA, BUT WAS NOT EVEN SURE THAT THEY WERE STILL IN EAST GERMANY. VOGEL KNEW THE SOVIETS HAD HAD DIFFICULTY IN CONVINCING THE NORTH VIETNAMESE TO PERMIT THE PILOTS TO BE OFFERED IN EXCHANGE FOR THE KROGERS."[88]

The Krogers (or Cohens) story does not end there. In late 1962, Oleg Penkovsky, a senior Soviet GRU (military intelligence) officer who spied for MI-6 and for CIA in 1961 and 1962 and provided much important intelligence about the Soviet military to the West, had been arrested by the KGB in late 1962 with his British MI-6 cut-out, Grenville Wynne, and put on trial. Penkovsky had served GRU on the Intelligence Directive of the Soviet General Staff in Moscow and had passed some of the "highest grade Soviet military intelligence ever received in London and Washington," including some of the minutes from secret Central Committee meetings of the Soviet Communist Party. His information was sent directly to President Kennedy during the Cuban missile crisis, allowing the White House to monitor Soviet missile emplacements that targeted and threatened most major cities in the U.S.[89] While Penkovsky was reportedly tortured and shot, WYNNE WAS IMPRISONED IN THE USSR AND EVENTUALLY EXCHANGED FOR GORDON LONSDALE AND THE KROGERS.[90]

It is as yet unknown whether any American POWs from Vietnam were secretly traded in this or other Cold War spy exchanges, but the CIA has always classed spies as a different category of prisoners than military POWs.

James Angleton, Chief of the CIA's Counterintelligence Staff, who appears to have been completely under the spell of KGB defector Anatoli Golytsin, promoted a theory that Penkovsky was a deception agent; a false defector. No other Soviet defector ever supported this theory, and Colonel Oleg Gordievsky, an important KGB defector who was consulted by Presidents Bush and Reagan, and by British Prime Minister Margaret Thatcher, confirmed that Penkovsky was a genuine 'defector-in-place.'

In 1965 Angleton had attempted to place a CIA counterintelligence unit in Saigon under his personal control. This was opposed by William Colby, then Chief of the CIA's Far Eastern Division, by CIA-DDO Desmond Fitzgerald, and by Major General Joseph McChristian of U.S. Army Intelligence, on General William Westmoreland's staff. Angleton later accused Colby of complicity in the deaths of American fighting men by his opposition, and he raised doubts about Colby's loyalty because of that official's unreported meetings with a French medical doctor who later turned out to have been a Soviet GRU agent, who was in radio communication with Moscow from Saigon.[91] By the 1968 period, Angleton maintained that Averell Harriman, the former ambassador in Moscow who now advised President Johnson on Indochina and prisoners of war, was a Soviet agent, and had been one since the 1930s.

Angleton opened a file on Harriman under the code name "Project Dinosaur," as a result of allegations of the paranoic Soviet defector Anatoli Golytsin, whom the CIA counterintelligence chief believed in implicitly, put together with intelligence gleaned from the Bride-Venona decryptions of 1945 Soviet radio traffic and noted earlier in this book. He had already accused British Prime Minister Harold Wilson and Canadian leader Lester Pearson of being Soviet agents. By the mid to late 1960s, the CIA's Soviet Division had been nearly paralyzed, at the height of the Vietnam war, by a runaway "molehunt" instigated by James Angleton and his Soviet source and confidante, Golytsin, which ruined the careers of many CIA Soviet specialists and endangered the CIAs overseas secret agents. Angleton was given complete freedom and authority by his old friend Richard Helms, as DDO from 1962-1965 and as Director of Central Intelligence from 1966-1973. Such was the situation at CIA's Soviet Division when the offer from the East to trade American POWs for Soviet spies came through Wolfgang Vogel. It is not as yet known what role Angleton or Golytsin played in this case.

The State Department's apparent questioning of Juergen Stange's (or Vogel's) competency and motives within the cited documents on the 1967 East German/Soviet-U.S. POW offer does not square with the fact that he and Wolfgang Vogel continued to play an important role in East-West prisoner exchanges and both were directly involved when KGB defector Yuri Loginov was forcibly repatriated to Soviet control in East Germany during 1969, in one of the most atrocious acts linked to CIA counterintelligence Chief James Angleton's obsessive mole hunt which was abetted by Golytsin's unsubstantiated accusations. Yuri Loginov had been recruited as a double agent by CIA in 1959 and run by CIA inside the KGB for six years under the code name "Aegusto," and as "Eyeball" to the FBI. Angleton had betrayed the CIA's own agent Loginov to the South African intelligence service, fingering him as a Soviet spy, when he was actually a bonafide defector and a loyal CIA agent. Loginov was arrested by South African intelligence in July 1967 and held in solitary confinement for two years, while under interrogation. Peter Kapusta, Chief of the Illegals Section in the Counterintelligence Branch of the Soviet Division had been Loginov's case officer and interrogator, also supported Angleton's view that Loginov was still loyal to the KGB, but other case officers who had handled Loginov such as Richard Kovich, were certain he was a genuine defector. Loginov's defenders came under immediate suspicion themselves, and CIA careers were later destroyed by Angleton, secretly. Angleton and South African General Van den Bergh decided to trade Loginov for a reported "eleven West German spies," held by the Communists in East Germany. Based on Angleton's information, the CIA's Deputy Director for Operations, Desmond Fitzgerald, had ordered the betrayal. (Fitzgerald's daughter, Frances Fitzgerald, had achieved national prominence with the 1966 publication of her book, "Fire in the Lake," which decried American involvement in Vietnam.)

On July 12, 1969, Yuri Loginov was secretly flown from South Africa to Berlin for the "swap," which was in fact a forcible repatriation, in which Stange and Vogel were involved. Researcher and author Tom Mangold reported in his study of Angleton: "Juergen Stange is the West German lawyer who represented the BND (West German Intelligence) during the Loginov exchange."

Loginov was delivered against his will to Soviet KGB officers and East German guards in the presence of Wolfgang Vogel. Loginov tried desperately to avoid being forced across the border, appealing to the West German intelligence (BND) agents present, and a South African intelligence service officer, to be kept as an agent in West Germany. Nonetheless he was forced across into KGB control by CIA officers. Juergan Stange refused to discuss the Loginov exchange as recently as 1989, saying, "I have been forbidden." Loginov's fate was not revealed by the Soviet government.[92]

Following the July 4th, 1991 publication by the Washington Post of an editorial on the history Soviet-held American POWs of four wars, written by the author, Loginov's CIA interrogator, Peter Kapusta, contacted the author. He indicated that he had more information on the author's report of American POWs transported to the Soviet Union from Vietnam, and earlier from Korea and China, but he did not reveal whether he had questioned Yuri Loginov on the subject. Both before and after being subpoenaed to testify before a Senate Select Committee on POW/MIAs in mid-1992, the author gave Colonel Kapusta's name to Senate investigators and urged that he be deposed, with unknown results. The author is uncertain whether any American military prisoners of war from Vietnam or other conflicts, were returned by the Soviets or their east-bloc surrogates in Europe, in the spy exchanges which occurred subsequently.

In Laos, meanwhile, where the CIA-run secret war involved both U.S. military units operating clandestinely and American contract civilians, some of whom were "sheep-dipped" military personnel working for CIA, U.S. POWs were held in many locations from the 1960s on. A declassified CIA report states: "In early May 1965, six Americans were imprisoned in a compound in the vicinity of the Pathet Lao/North Vietnamese Army National military Headquarters in Houa Phan Province, ABOUT 18 KILOMETERS EAST OF SAM NEUA. THE AMERICAN PRISONERS WERE A COLONEL AND FIVE CAPTAINS... DOWNED BY ANTIAIRCRAFT FIRE...THE COMPOUND WAS A TEMPORARY HOLDING FACILITY AND FOREIGN PRISONERS WERE LATER TRANSPORTED TO HANOI...A PORTION OF THE COMPOUND WAS A CAVE...IN EARLY MAY 1965 A CHINESE MISSION, REFERRED TO AS "THE EMBASSY" WAS IN A CAVE...TOONG SING SAID THAT THERE WERE APPROXIMATELY 80 CHINESE STATIONED AT THE PATHET LAO NATIONAL MILITARY HEADQUARTERS..."[93]

There can be little doubt that the Chinese interrogated and took control of some of the POWs thus held in the caves of Ban Nakay and elsewhere in the Sam Neua area, particularly in the earlier stage of the war when Chinese-North Vietnamese relations had not yet deteriorated.

One of the U.S. POWs held in the Ban Nakay caves was Air Force Lt. Colonel Charles E. Shelton, who had been shot down over the Sam

Neua area of Laos on April 29, 1965. Shelton was reported alive in captivity by CIA and Air Force intelligence sources for years to come, and due to this he remained the only officially-listed U.S. prisoner of war, after all others had been presumed dead. He was the POW reported in a secret CIA cable as having killed one of his North Vietnamese interrogators in 1968, and was not repatriated at the war's end. His wife, Marian Shelton became a nationally known spokesperson for the POW/MIAs who had been left behind and for their families and her lifelong fight with Pentagon and CIA officials to uncover more of the classified evidence about her husband and the other POWs who disappeared. A small but tough woman with a big heart, Marian Shelton quickly befriended the author and other Vietnam veterans who joined her cause, perhaps too late. (Worn out and anguished by the struggle, Marian Shelton apparently took her own life, by a gunshot wound in the fall of 1990. Across the United States her many friends mourned her death, but only increased their determination to uncover the truth.)[94]

Another of many CIA documents on American POWs held during the war in Laos which have been declassified, dated December 6, 1968, reveals a different fate for other American POWs held in nearby caves in Laos at this time:

"As of late 1965, 15 American, four Thai and one Philippino prisoners of war were incarcerated in a prison cave complex near Ban Long Kou...26 KILOMETERS EAST OF SAM NEUA IN HOUA PHAN PROVINCE. Two of the Americans were a major and a Lieutenant Colonel who were shot down with their F-105 jet near Ban Long Kou in late March-early May 1965...ALL FLIERS SHOT DOWN IN LAOS ARE KEPT IN LAOS AND NOT SENT TO NORTH VIETNAM."[95]

Nothing better illustrates the secret communist "parallel prison system" than these two CIA cables on two groups of U.S. POWs held in two different cave-prisons east of the Pathet Lao headquarters Sam Neua. One group which is withheld in Laos by the Pathet Lao for future use, and another, constantly changing group held separately but accessible to a known Chinese Communist mission and a later-discovered Soviet military and intelligence mission, which came to be titled within the confines of the NSA as: "Tentacle MB," where "Moscow-bound" American POWs assembled in the Sam Neua area, were interrogated and moved on through Hanoi to the Soviet Union.

A 1967 CIA report lists Pathet Lao POW camps at Muong Sing (reference to Americans blacked-out), Nam Tha, Muong Sai, Muong Liet ("pilots, caves"); Ban Nakay (caves) holding "American POWs;" Ban Na Viet Neua, Xieng Mene, "Cave with American;" Dane Phao, "American prisoners in cave;" Sam Neua, "American prisoners;" Ban Nhoummarth; Mahaxay; Ban Tha Pha Chon, "Cave with American;" Ban Na Nhom, "American and Thai prisoners;" and more locations too numerous to list here.[96]

In the years which followed, from the late 1960s until the 1973 U.S. military withdrawal, other CIA reports list many locations where numerous American POWs were held in Laos. Some locations, such as the Ban Nakay Caves in the Sam Neua area, were continually reported to contain U.S. POWs during and long after the war. One CIA source in Laos reported that a Pathet Lao official had claimed that there were up to 300 U.S. POWs in one location in Laos, named the "Supreme Command Training Center." (CIA believed this to be an exaggeration.)[97]

One of the U.S. POWs who may have been secretly captured in Laos was 32-year-old U.S. Air Force Technical Sergeant named Melvin A. Holland, who was from Toledo, Washington, and was still carried as missing in action on the date he disappeared in combat, March 11th, 1968, during the extended Tet Offensive. Sergeant Holland was one of the "sheep dipped" Americans of the secret war in Laos. Officially discharged from the military, their records were transferred to secret files and ostensibly they were hired by civilian companies such as, in Holland's case, Hughes Aircraft, but which employment was, in reality, a cover for a CIA-controlled secret military intelligence operation at "Lima Site 85," only some 25 miles west of Sam Neua, where the Soviet-run POW selection camp known to NSA analysts as "Tentacle M-B" was then in operation. Site 85 was one of a chain of secret CIA "Lima Site" bases in Laos, which were used to guide bombers over Hanoi and monitor communications in North Vietnam.

Richard Secord, who retired as a Major General, was a Major in 1968, assigned to air operations for the CIA in Laos under the Vientiane Chief of Station, Theodore Shackley, and working closely with the CIA chief-of-ground in Laos, Thomas Clines. Shackley was an old CIA hand who had been involved in the controversial Berlin Tunnel incident, in which a supposed conduit for Soviet-bloc intelligence eventually turned out to have been thoroughly compromised by the KGB from the beginning. Shackley was also involved in the subsequent Bay of Pigs fiasco and became the CIA's Chief of Station in Miami, Florida following the abortive invasion of Cuba, which had ended with hundreds of anti-communist Cuban POWs (and some Americans) being held in Fidel Castro's prisons and forced labor camps. He later was involved in obtaining intelligence on Soviet deployments that led to the Cuban missile crisis. In Vientiane, Shackley reported to Ambassador William Sullivan, Harriman's deputy in the 1962 Laos peace negotiations, who ruled Laos as a proconsul.

According to Secord, Sullivan was referred to by CIA operatives as the 'field marshal', who often interfered with purely military decisions to the detriment of the Allied war effort, for political or personal reasons. William Sullivan later became a key assistant to Henry Kissinger at the Paris peace talks. Tom Clines, whom, Secord relates in his memoirs, he became a life-long friend to, had served as an enlisted man in the Army in Korea and later joined the CIA, where he trained Cuban exiles for the Bay of Pigs, and took part in other Latin American covert operations prior to his transfer to Laos. Secord later wrote that the Strategic Air Command's TSQ-81 ground based radar bombing system, which had been installed at Site 85 in 1967, was "used to radically improve all-weather bombing accuracy in tactical fighter' route packages' in North Vietnam," and, "...represented a major escalation in a war that technically didn't exist. It also had to be manned around the clock by about 15 technicians who were rotated weekly from a cadre of 40 workers based at Udorn (Thailand)."[98] Site 85 also contained a TACAN station (Tactical Air Navigation system and was guarded by indigenous forces under CIA control.

Melvin Holland was one of a team of sixteen such U.S. technicians at the secret Site 85 base, perched atop the 5,600-foot Phou Pha Thi mountain, which was attacked and overrun by the North Vietnamese Army as the air war over Hanoi intensified during the 1968 Tet offensive. In what may have been evidence of a coordinated Communist strategy, the U.S. intelligence-gathering ship Pueblo had also been seized by North Korean forces in January 1968, just prior to the nationwide Tet attacks. Secord claims that the fall of the Site 85 base to North Vietnamese forces, after weeks of siege warfare and a sudden assault, could have been prevented, but that Ambassador Sullivan refused to allow American reinforcements, despite an order from Washington to U.S. officials in Laos to 'hold the site at all costs.' Only 6 Americans survived the assault, including the CIA case officer in command of Site 85 at the time, Evan Washburn. Of the five sheep-dipped technicians who escaped with Washburn, only one, Sergeant Gary, was unwounded. One of the wounded technicians was killed by NVA antiaircraft fire during the extraction. Secord later wrote: "Of the ten left on the hill, most were confirmed dead by Evan, but a number were inexplicably MIA, even after a quick but thorough search of the site..."

Aerial photographs taken by a CIA aircraft after the battle showed what appeared to be some of the MIA technicians, apparently dead on a ledge below the summit on a cliff side of the mountain, just below the summit, where they had attempted to escape by lowering themselves down on aircraft tie-down straps. Still, several of the Americans at Site 85 had vanished and their bodies were never recovered. In his memoirs Secord remembered: Ambassador Sullivan met with Laotian Prime Minister Souvanna Phouma on March 15 (1968) to give him the bad news: The sensitive facilities on Site 85 had not been totally destroyed prior to evacuation, and some Americans had been left behind; presumably killed, but possibly captured, although this was unlikely." Secord later told an investigator for Secretary of Defense McNamara, that the loss of Site 85, which had directed over 25% of the bombing missions in North Vietnam, had cost not only the loss of most of the U.S. team there, but also "the compromise of our TSQ technology and a variety of top-secret encryption systems that we could only assume were now safely on the way to Moscow."[99]

Sergeant Melvin Holland was one of the high-tech "special talents" U.S. personnel at Site 85 whose capture was a top-priority for the North Vietnamese and Pathet Lao and their Soviet and Chinese allies. The U.S. Ambassador in Laos at the time, William Sullivan, signed the death certificate for Holland as an alleged American civilian, even though the U.S. Embassy in Vientiane later denied to his wife that Holland was missing in Laos or anywhere in Southeast Asia. His loyal wife, Ann Holland, who was ordered by the government to say nothing about his disappearance, never believed Melvin was killed in 1968 based upon evidence she herself uncovered. After the war a high-ranking officer told her that POWs had been captured by the North Vietnamese at Site 85, and she filed suit to demand information on her husband and other MIAs. She later reported: "I found the man who the Air Force claimed to have seen my husband dead...He told me that he last saw Mel alive and running."[100]

Ann Holland kept fighting for his release and for the freedom of all secretly-held American POWs for more than two decades to come, and she has reported receiving threats against her life for these activities. Her courage and resourcefulness at confronting the military and governmental bureaucracy and gaining access to classified information was an inspiration to many other POW/MIA family members, and to many veterans of the war (including the author), who later joined her struggle and became her friends. Ann Holland obtained top-secret U.S. documents on the capture of Site 85 which indicated that American POWs were captured alive there by the Communists and taken away for interrogation. These documents were later reclassified by the U.S. Government. A Thai prisoner of war who escaped from the Communists reported that several (at least three) of the Americans at Site 85 had been led away into captivity by the North Vietnamese forces after the battle. It appears likely that one of them was Mel Holland, and that he and the other high-tech U.S. POWs captured at Site 85 ended up under interrogation by Soviet, or possibly Chinese intelligence, and may have joined the secret convoys of Moscow-bound POWs.[101]

In his memoirs, Richard Secord clarified the structure for rescuing American POWs at the time Site 85 fell to the North Vietnamese Army:

"During my stay in Laos, MACV's (Military Assistance Command

Vietnam) Studies and Observation Group ((SOG) was led by Colonel (later General) John Singlaub. SOG was really a cover for special operations...in North and South Vietnam all during the war. Singlaub and his staff met with us periodically at Udorn...to coordinate covert activities, usually clandestine helicopter infiltrations into Laos and Vietnam...Singlaub was responsible for several POW rescue attempts between 1966 and 1968, leading eventually to the much-publicized Son Tay attempt ...in which Singlaub was not involved. Sadly, not a single prisoner was sprung by these efforts. The reason they all failed, I believe, was lack of real-time intelligence, poor security, and to much outside 'hep,' as we called it: 'help from heaven or help from higher up-' from bureaucrats and pencil pushers. In Laos, we had plenty of intelligence, but lacked high quality ground forces and tactical leadership necessary to bring off such raids."[102] When questioned under oath long after the war about American POWs and MIAs in Laos, by a Senate Select Committee, on POW/MIA Affairs, Secord admitted that American POWs had been left behind under enemy control in Laos, and that, "None of them, that I know of, have been located or even heard of since the Paris Accords. But we did know to, I think, a reasonable degree of certitude, that there were more..."[103]

Back in the United States, a powerful homegrown antiwar movement, made up of some extreme radicals, but also of millions of ordinary Americans who opposed the apparently endless war in Asia, and who disagreed with its conduct for widely varying reasons, had grown up in the 1965-1968 period. A large and vocal march on the Pentagon by war protestors in the fall of 1967 had shocked many Washington insiders into realizing how strong the antiwar movement had become, and this was reinforced by the hostility on many college campuses of the time, towards anything to do with the U.S. Government and its unpopular war. American Marxists and assorted radicals traveled to Cuba to pay homage to Castro's successful revolutionary experiment, and there met with the Vietcong and North Vietnamese in December 1967. One of these was Todd Gitlin, who wrote later of a meeting he and SDS leader Carl Davidson had with them:

"...The North Vietnamese delegates we met with one day told the Americans, 'We have faith in the conscience of the American people.' 'Sometimes I wish we had as much faith,' Davidson told them. The Vietnamese laughed: 'We have faith that you can AWAKE the conscience of the American people.'"[104]

War protesters began to surface among returning veterans of the fighting, in the Congress, within the Department of State, the Pentagon, and even inside the CIA. Senior officials of the Johnson Administration, such as McGeorge Bundy and Robert McNamara, who had engineered the American war in Indochina, were now turning against its prosecution, and urging that the United States begin to withdraw from what they now saw as a losing proposition. When the Tet Offensive exploded throughout Vietnam at the end of January 1968, costing approximately 4,000 U.S. battle dead in the seven weeks of heavy fighting that followed, some 19,000 wounded and 600 more MIAs, many saw it as proof that America was indeed losing the war or, as news commentator Walter Cronkite said at the time, that it had become a "stalemate."

Johnson appointed Truman's long-ago naval aide, Washington lawyer and lobbyist Clark Clifford to replace McNamara as Secretary of Defense in February, and Clifford, who had also turned against the war, immediately began assembling a group of like-minded past and present policymakers who would convey the view to Johnson that America must disengage. They came to be known as the "Wise Old Men," and included such leaders as Dean Acheson, Averell Harriman, John J. McCloy, Omar Bradley, Henry Cabot Lodge, and Matthew Ridgway, all of whom had been involved with U.S. POWs in the past. Other "Wise Men" consulted on the future of the war included McGeorge Bundy, George Ball, Maxwell Taylor and others. The consensus of this group, which represented the establishment that had committed over 500,000 troops to Vietnam by that year, was that the war had gone wrong and the time had come for the U.S. to de-escalate. This had led to President Lyndon B. Johnson's surprise March 31, 1968, announcement, during the Tet counter-offensive by U.S. forces, that he would not run for reelection, and also his ordering a halt of U.S. bombing of North Vietnam above the 17th parallel, across North Vietnam's narrow southern panhandle.

The effect on the American fighting men in Indochina of this announcement was profound, as the author himself witnessed at the time. They and their successors would be required to fight and die in the jungles of Vietnam, Laos and Cambodia, for four more years, while the enemy's heartland went largely unscathed. This unilateral action was, at best, detrimental to those daily risking their lives in combat with a tenacious enemy. Meanwhile, the American negotiators in Paris, Averell Harriman and Cyrus Vance, argued with North Vietnamese Communists about the proposed shape of the "peace conference" table. Severe fighting continued for months in Vietnam, resulting in the heaviest combat casualties of the war for the United States, with 14,595 Americans killed in action during that year, and some five thousand more U.S. personnel listed as dead of non-hostile causes." Nearly 100,000 U.S. personnel were wounded, which brought the war home to that many more families in America.

The American POWs in Vietnam were being subjected to brutal treatment by their guards and interrogators, with torture and solitary confinement being commonplace. But the North Vietnamese and Vietcong convinced most of the world that the American "air pirates" and "war criminals" were being leniently treated by the generous Vietnamese people. One of the American POWs held in Hanoi, Navy flier James Stockdale, who had succeeded in getting coded messages back to the U.S. in letters to his wife Sybil that revealed the prisoners were being tortured, blamed Averell Harriman for not protesting their harsh treatment:

"By the time I and my fellow prisoners were returned to Hoa Lo Prison on the night of January 26, 1967, my first two coded letters had completed their processing in the Washington chemical labs and their messages had been read. But Uncle Sam failed to do his part for two more years and four months, during which time torture was rampant and Americans were killed in the North Vietnamese camps. It was Averell Harriman who insisted on keeping the torture evidence under wraps in the interest of furthering his 'delicate' negotiations on the prisoners' behalf, all of which ultimately came to nothing. Even after the Johnson Administration was out of office, Harriman tried to prevail upon the new secretary of defense, Melvin Laird, to continue the 'keep quiet' policy."[105]

The Soviets took advantage of America's painful preoccupation with Indochina, and crushed a democratic liberalization movement in Czechoslovakia that had become known as the "Prague Spring," which had captured the hearts of idealists throughout the West and inside the Communist world. An invasion by the Red Army, supported by troops from other Warsaw Pact nations, began on the night of August 20-21, 1968. Czech leader Alexander Dubcek, and many others, were arrested by the KGB and the Czech STB secret police and taken as captives into the Soviet Union, as had so many others before them.

One unforeseen, but valuable result of the brief Czech liberalization period for U.S. intelligence, was the early-1968 defection of Major General Jan Sejna, a senior Czech security official in the Communist administration of First Secretary Antonin Novotny, whom the young Alexander Dubcek had replaced in January 1968. In the course of his duties with the Czech defense ministry, General Sejna had witnessed up to 90 American prisoners of war who had been captured in Southeast Asia, transiting Prague as they were being transported to the USSR. Sejna also revealed that American prisoners of war may have been used in secret biological warfare experiments, in which some were reported to have died. U.S. intelligence was later to rate General Sejna as highly credible, and after being used by the CIA for a time he was employed by the Department of Defense for many years to come; but for security reasons he was to remain silent about the "Moscow-Bound" American POWs from Vietnam until 1992.

Thus, what had been expected to happen with U.S. POWs in Vietnam by Mooney and other U.S. intelligence analysts, had already happened, both in East Germany and Czechoslovakia by the 1967-1968 period, but the knowledge of such "Moscow-bound" Americans remained highly secret and compartmentalized, even from others within U.S. intelligence community, who should have been informed. Compartmentalization of intelligence for security reasons had resulted in restricting such cross-fertilization by establishing "need to know" orders for sharing intelligence information with others. Sejna's information would have been made known to high-level policy makers such as Dean Rusk and Averell Harriman, however, and should have been passed on to President Nixon and senior officials of his administration.

As with the U.S. POWs brought to East Berlin by the Soviets in 1967, the identities of these American POWs in Prague remains unknown, as does their ultimate fate. While some may have indeed died in biological warfare or medical experiments, others would have been to valuable as technical eperts to be wasted so casually by the KGB or GRU. Sejna was extensively debriefed in the United States and the CIA later sent him to Europe to share his knowledge about Warsaw Pact military and intelligence matters. For twenty-five years to come he would maintain secrecy about the American POWs he had seen being shipped through Czechoslovakia from Vietnam, on their way to the Soviet Union.

The Chinese Communists also took advantage of the American preoccupation with Vietnam by expanding their control over the nation of Tibet. After invading that country during the Korean War they had attempted to rule that country through a puppet regime using the Dalai Lama as a figurehead. But after large-scale fighting broke out in 1959 that religious leader fled his country and in 1967 the Chinese People's Army had taken over the security bureau, newspapers, radio stations and banks, while attacking Tibetan dissidents and rebels in remote areas during 1967 and 1968 and destroying many ancient monasteries.

Meanwhile, at home in America following the Tet Offensive, 1968 had turned into a year of violence and upheaval that saw the assassinations, first, of Martin Luther King, the apostle of nonviolent protest against racial discrimination, and then of Senator Robert F. Kennedy, during the Democratic primaries. Race riots broke out in several major cities during the year resulting in the use of military forces to assist police in restoring order. In Chicago, on August 28th and 29th, a major riot erupted from antiwar protests led by Tom Hayden, David Dellinger and other radicals, during the Democratic National Convention. Youthful demonstrators waving Vietcong flags and hurling rocks and bottles were savagely attacked by police, and hundreds were injured and arrested. News media coverage brought the violence into most homes in America, millions of which had family members then serving in the military. The

Democrats chose Hubert Humphrey as their candidate, but he inspired little faith in a war-weary nation that wanted badly to be led out of the quagmire of Vietnam. In western Europe antiwar protests against the American war in Indochina erupted in France, Holland, Belgium, Sweden, Italy and in England. For years to come, demonstrations in Britain and other foreign nations were sometimes led by young American students, including one who attended Oxford University named William Jefferson Clinton, who would one day become President of the United States.

On November 1st, 1968, President Lyndon B. Johnson halted all U.S. bombing of the remainder of North Vietnam, as American soldiers continued to fight and die in the south for a cause already seen as 'lost' by millions of Americans, and by many of their leaders at home. In that same month Republican candidate Richard M. Nixon defeated the contender Hubert Humphrey, and was elected President, having promised to "end the war with honor" for the United States.

---

## Notes on Chapter Nine

1   The Conferences at Malta and Yalta, p. 770, 1945, Foreign Relations of the United States

2   *Politics of Heroin, op. cit.,;* see also: *Armies Of Ignorance,* by William Corson, pp. 321-323; and the *Pentagon Papers,* Bantam Books and the New York Times, N.Y., N.Y., July, '71

3   Le Figaro, Paris, France, May 6, 1991.

4   *Street Without Joy,* Stackpole, 1961, pg. 278

5   *Street Without Joy,* p.276, op. cit.

6   *Politics Of Heroin,* op. cit

7   Interview by the author, Seattle, Washington, 1992)

8   Senate Foreign Relations Committee Republican Staff Report, "An Examination of U.S.Policy towards POW/MIAs, pp.96-97.

9   *KGB,* Andrew and Gordievsky, op. cit. Also Aleksandr Solzhenitsyn and Avraham Shifrin

10   *The Gulag Archipelago,* Aleksandr Solzhenitsyn; also, Avraham Shifrin's *Directory of Gulag* camps for locations and numbers of prisoners.

11   Reiterated in interviews of Noble by Brown and Ashworth.

12   Department of State #388, July 16, 1956

13   UPI report, July 19, 1956

14   *Puzzle Palace,* James Bamford

15    RG 761.5411/5-459

16    Incoming Telegram, Department of State, Control 13192, Sept.20, 1958, 11:22 A.M., distributed to Pentagon, CIA, etc., declassified through Freedom of Information request by the author in 1993.

17    Department of State Memorandum, October 6, 1958, 761.5411/10-658, declassified by FOIA request of Theodore Grevers, Case # 730163

18    These two Nixon-Khrushchev documents were declassified by the Department of State for the author on September 8, 1993, through Freedom of Information request No. 9104470

19    T.U.S.LOG 3-1, Turkish-U.S. Logistics

20    In 1992 the former wife of Captain Oscar Goforth, Maria McAttee, hand-delivered a letter to Russian President Boris Yeltsin requesting to know her husband's fate. Yeltsins intelligence aide General Dimitri Volkogonov, demonstrated continued Russian intransigence and prevarication in the case by claiming her husband was dead, "in the depths of the Barents Sea." USN&WR, 1993

21    Incoming Telegram, Department of State, January 27, 1961, Control 14723, No: 1767, declassified in 1993 by the Department of State through FOIA request by the author

22    Moscow No. 691, September 15, 1961 and Department of State Airgram No. A-77, October 16, 1962

23    U.S. News And World Report, March 15, 1993

24    *Kissinger: A Biography*, Simon & Shuster, 1992

25    On September 7, 1992

26    Matthews in: *Bad News, The Foreign Policy of the New York Times*, Russ Braley, Regnery-Gateway, Chicago, 1984

27    Valledares, op. cit.

28    Andrew and Gordievsky, *KGB—The Inside Story*, op. cit

29    Marchetti, *Cult Of Intelligence*, Knopf, N.Y., 1974, pp. 32-49

30    *Bay of Pigs*, Peter Wyden, Simon and Schuster, 1979, p. 11

31    Senate Intelligence Committee Report on Foreign Assassination Plots, p. 74; HSCA Report on CIA-Mafia plots

32    *Bay of Pigs*, Peter Wyden, Simon and Schuster, 1979, p. 163

33    Marchetti, *Cult of Intelligence*, pp. 163, 226, 249-250, 306, 346, op. cit.; also: p, 270, *First Heroes*, by Rod Colvin, Irvington Publishers, NY, 1987

34    Bay *of Pigs*, op. cit.

35    *High Treason*, p. 313 & 434, op. cit.; also: Sorensen, p. 334

36    Valledares, *Against All Hope*, Alfred A Knopf, NY., 1986, op. cit.

37    *Khrushchev Remembers, The Last Testament*, Strobe Talbott, translator, Little, Brown & Co., Boston, 1974

38    Hilsman, *To Move a Nation*, p.220

39    Ibid.

40    Armando Valladares, *Against All Hope*, p. 174, Alfred A. Knopf, NY, 1986

41    NCAFP-December 1986, also interviews with Brigade survivors by Luis Alcebo

42    *Contract on America, Crossfire* and *High Treason*, op. cit.

43    *Cult of Intelligence*, pp. 31-32, op. cit.

44   Washington Post, September 16, 1973

45   The Washington Post reported on September 16, 1973, that: "New Orleans District Attorney Jim Garrison as late as March 1971, was preparing to accuse another person of conspiring to assassinate President John F. Kennedy. Garrison's intended defendent this time was the late Air Force General Charles Cabell, former Deputy Director of the Central Intelligence Agency." High Treason, by Robert J. Groden and Harrison E. Livingston, Berkley Books, NY, 1990. General Cabell's brother, Earle Cabell, was the Mayor of Dallas at the time of the assassination, and was reported to have helped organize the President's motorcade of November 22nd, 1963.

46   High Treason, p.384, op. cit.; also: Report, 198-200

47   See: Crossfire, by Jim Marrs; High Treason, op. cit.; Cold Warrior, pp.167-175; also Contract on America, op.cit.

48   Interview of Jerry J. Mooney by the author, also in a letter to the author on file

49   One American official serving under Ambassador Charles Yost in Laos during the 1955-56 transition period, was a World War II veteran named Fitzhugh Green, who would later also serve under Ambassador Yost in the United Nations during 1964-65, and who ultimately was to write a book about an old friend and WW II friend comrade-in-arms, President George H.W. Bush, who was still a private oil business executive in the early 1960s. It was in turn the fate of the later Ambassador, CIA Director and President, George Bush, in the 1990-1992 with the unresolved and long-classified problem of American prisoners of war of World War II, Korea, the Cold war and the Vietnam war who had been secretly withheld in Communist control for decades.

50   The Wise Men, Isaacson and Thomas, p. 616, op. cit.

51   THE CIA AND The Cult Of Intelligence, pp. 31-32, op. cit.

52   Politics of Heroin and Kiss the Boys Goodbye, op. cit.

53   The Pentagon Papers, op. cit.

54   Honored and Betrayed, op. cit.

55   Kennedy and the Vietnam Conflict., op. cit.

56   Crossfire, Jim Marrs, Carroll & Graf, NY, NY

57   OSS, Ch. 10, R. Harris Smith, University of California Press, 1972, and the Pentagon Papers,.

58   Contract on America, David E. Scheim, Zebra Books, 19889, N.Y., N.Y.

59   The Years of Lyndon Johnson, The Path To Power, Robert A. Caro, Vintige Books-Random House, NY, 1981

60   In Love and War, James and Sybil Stockdale, Harper and Row, N.Y., N.Y., 1984

61   Newsweek Magazine, October 1, 1965

62   First Heroes, p. 163, 164, op. cit.

63   Interview with Pitzer: To Bear Any Burden, Al Santoli, Ballantine Books, NY, 1985.

64   Congressional Record of Hearings

65   To Bear Any Burden, p. 163, op. cit.

66   Declassified by CIA and obtained by Freedom of Information Act (FOIA) of the author

67   The New KGB, op. cit. and KGB, The Inside Story, op. cit., also: Breaking the Ring, by John Barron, Houghten Mifflin, Boston, 1987

68   Declassified CIA Report TDCS-314/08006-68, May 15, 1968, obtained through Freedom of Information request by the author

69    Corson & Crowley, *The New KGB*, William Morrow, 1985.

70    Interviews of Mooney by the author, also: 1990-1991 affidavit and letter

71    *Rice Paddy Grunt*, John M.G. Brown, Regnery-Gateway, Chicago, Illinois, 1986

72    Volume 7, p. 654, Uncorrellated Reports, op. cit.

73    Individual Unit records, Operational Reports and Lessons Learned, Quarterly Reports, After-action Reports, Daily Journals and Logbooks reporting radio Transmissions of: 1st Infantry Division; 1st Cavalry Division; 4th Infantry Division; 5th Infantry Division (Mechanized); 9th Infantry Division; 25th Infantry Division; Americal Division; 82nd Airborne Division; 101st Airborne Division; 173rd Airborne Brigade; 199th Light Infantry Brigade; and other Army Units declassified and in the National Archives.

74    Cable # 956, Berlin to Secretary of State, January 26, 1967

75    *KGB:The Inside Story*, p. 440-444, op. cit)

76    *Mask of Treachery, Spycatcher* and *Cold Warrior*, op. cit.

77    *Cold Warrior*

78    *Mask of Treachery, Cold Warrior*, p. 90 and *Spycatcher*, p. 47, op. cit.

79    Secord, *Honored and Betrayed*, op. cit.

80    *Cold Warrior*, p.26, op. cit.

81    Cable # 127418, Department of State to Berlin, Jan. 26, 1967

82    Cable #1037, Berlin to Secretary of State, February 9, 1967

83    Department of State to U.S. Embassy, London, February 15, 1967

84    # 6625, London to Department of State, Feb. 16, 1967

85    Berlin to Department of State, Feb.20, 1967

86    Cable # 188110, Secretary of State to Berlin, March 14, 1967

87    Secretary of State to Berlin, May 4, 1967

88    Berlin to Department of State, June 2, 1967

89    *Cold Warrior*, op. cit.

90    *Spycatcher*, p. 204, op. cit.

91    *Cold Warrior*, p. 311, op. cit.)

92    *Cold Warrior*, Tom Mangold, Simon and Schuster, NY, 1991

93    CIA Report# CS-311/03670/ 69, obtained through Freedom of Information request by the author

94    Interviews of Marian Shelton, by the author, 1986-87)

95    CIA report, TDCS DB-315/04550, 6 December 1968, obtained through FOIA request)

96    CIA Report CS-311/10503-67, obtained through FOIA request) Another declassified CIA report of November 1968 listed six different locations where American POWs were held in Laos; CIA Report CS-311/10502-68, November 1968

97    Uncorrelated Information Reports, Volume 14 p. 347

98    *Honored and Betrayed*, op. cit.

99    *Honored and Betrayed*, by Richard Secord, John Wiley and Sons, NY, 1992

100   Interviews with the author, and "MIA Families Won't Give Up Hope", by Al Santoli, Parade Magazine, May 30, 1993

101   Interviews with Ann Holland by the author, *Kiss The Boys Goodbye and the Bamboo Cage.*

102   *Honored and Betrayed*, p. 65

103   Secord, 09/24/92, Hearings of the Senate Select Committee on POW/MIA Affairs

104   *The Sixties, Years of Hope, Days of Rage*, by Todd Gitlin, Bantam Books, NY, 1987

105   *In Love And War*, op. cit.

# CHAPTER TEN

## NIXON'S WAR

Born in California to a Quaker family in 1913, and educated at Whittier College, Richard Milhous Nixon had served in the Pacific during WW II as an officer in a Naval air transport unit, and in 1946 he was elected to Congress as a virtual unknown. He worked on a committee that laid out the framework for the Marshall Plan, and on the House Committee on Un-American Activities. Nixon had achieved prominence when he insisted that the charges against an accused Soviet spy in the State Department, Alger Hiss, should be either proven or disproven. Nixon was elected to the U.S. Senate from California in 1950, by portraying his opponent, Helen Gahagan Douglas as an ultra-liberal who was soft on Communism. Nixon became a popular campaigner for Republican Party functions. Nixon joined in the attacks on Dean Acheson that characterized the time, for which he was detested by many powerful establishment figures, including Averell Harriman, who once refused to eat dinner with the young California Senator, at the Stewart Alsops' house.[1]

Nominated by the Republican National Convention to run as Vice-President with General Dwight D. Eisenhower in 1952, Nixon survived a political attack by the Democrats over his campaign finances, and served as Vice-President until January 1961. He had been involved in the planning of the Bay of Pigs invasion of Cuba, and long before that he must have been fully aware of the fate of the American POWs who had disappeared in Korea and Manchuria. With U.N. Ambassador Henry Cabot Lodge as his running mate, Nixon had run against John F. Kennedy and Lyndon Johnson in 1960, and had lost. He subsequently was defeated by "Pat" Brown in the 1962 California gubernatorial campaign, and many thought his political career was finished, but by 1968, public dissatisfaction with the conduct of the Vietnam War gave him another chance, which he took, and won. He appointed William P. Rogers to be his Secretary of State, Melvin R. Laird as Secretary of Defense, and Henry Kissinger to be his Special Assistant for National Security Affairs. Spiro T. Agnew had been elected Vice President and, like Nixon, he was destined to resign in disgrace.

By the time of Nixon's inauguration in January 1969, some 40,000 Americans had been killed in Vietnam, and within days of taking office the new president was greeted with a renewed Communist offensive, that cost thousands more American casualties in killed, wounded and missing. (Also, the North Koreans downed a U.S. EC-121 aircraft on April 15th, and 28 of the 30 crewmen disappeared, remaining MIA and unaccounted for to the present time.)

Yet, instead of using the mandate to finish the war that had been given to him by the American people, Nixon continued the bombing halt, over North Vietnam. He may have thus encouraged a ruthless enemy, who had no intention of seeking a 'negotiated settlement' of the war, but instead had rededicated themselves to victory over the 'foreign imperialists.' The Vietnamese Communists were led still by Ho Chi Minh, although he was to die in September of that year. Ho had instilled a dedication to achieving victory in the subordinates who would succeed him, such as Army commander General Vo Nguyen Giap, the victor over the French at Dien Bien Phu, Pham Van Dong, who inherited Ho's rule over North Vietnam, and Le Duc Tho, who was to negotiate in Paris for years over an eventual American withdrawal, with Nixon's national security advisor, Henry Kissinger, until January 1973. The American prisoners of war held in Vietnam, Laos and Cambodia were to be the most important bargaining tool that the Communists possessed, and shortly after the Nixon Administration took office in January 1969, a secret RAND Corporation study for the U.S. Government predicted that the Vietnamese would retain American POWs and the U.S. would have to pay in some manner, to get them back when the war ended. This study would remain classified however, through the years of war and negotiations that followed. In the meantime, the Vietnamese Communists repaid their Soviet and Chinese supporters by giving them access to high-tech American military equipment and U.S. POWs who knew how to operate it in the same way the North Koreans had from 1950-1953. This should not have surprised anyone in the U.S. Government who knew of it. In fact it should have been expected and predicted by U.S. intelligence, based on historical precedents established in America's earlier 20th century wars, although any documentary evidence of such an official expectation has remained classified to the time of this writing.

While the North Vietnamese and their Soviet and Chinese allies were secretly assembling and interrogating American POWs prior to shipping them to the USSR and Communist China, and before the analysts at NSA (who had not been briefed on the U.S. POWs under Soviet control in East Germany and Czechoslovakia) were even certain that this was occurring, a second overt propaganda campaign was orchestrated by the Politburo around the POWs, to portray to the world the "humanity" of Hanoi's leaders toward all war prisoners. They encouraged their American antiwar allies in the United States to consistently reiterate Hanoi's kind and lenient treatment of U.S. POWs in the American media, to which they were increasingly being granted access. As U.S. prisoners were being beaten and tortured in captivity, American apologists for the Hanoi regime used the mass media to deny charges of such treatment, and to label the POWs of their own nation as "war criminals."

The most radical pro-Hanoi faction of the American anti-war movement had been greatly encouraged by the heavy U.S. losses in the 1968 Tet Offensive, and the change in American public sentiment brought on by that surprise attack and the violent follow up known as the "May Offensive," which had cost over 2,000 more U.S. military personal killed in action. These casualties had included a substantial number originally listed as "missing" or "presumed killed, body not recovered." Tom Hayden of SDS, the seasoned Hanoi traveler who had brought out three American POWs late in 1967 as part of a communist propaganda ploy, wrote to one of his North Vietnamese contacts on June 4, 1968: and later admitted to a Congressional investigating committee that he had in fact written the letter:

"Dear Col. Lao (Lau),

"This note is to introduce you to Mr. Robert Greenblatt, the coordinator of the National Mobilization to end the war in Vietnam. He works closely with myself and Dave Dellinger, and has just returned from Hanoi.

"If there are any pressing questions you wish to discuss, Mr. Greenblatt will be in Paris for a few days.

"We hope that the current Paris discussions go well for you. The news from South Vietnam seems very good indeed.

"We hope to see you this summer in Paris or at a later time.

"Good fortune!

"Victory!

Tom Hayden[2]

It is unlikely that Hayden, or any of his fellow Marxist-Leninist and left-leaning American or foreign compatriots, with the exception of a few, like the veteran Australian Communist POW interrogator and "journalist," Wilfred Burchett, had any inkling of the long and still-secret history of Soviet and Soviet-surrogate kidnappings of American war prisoners, dating back to 1919. The official secrecy still surrounding those events had prevented both American leftists and conservatives from making accurate judgments on the actual reasons for the Soviet-American hostility which had become apparent since the 1945 onset of the Cold War, and indeed, since the 1917 Bolshevik Revolution and the Soviet Civil War which had followed. Without knowledge of this most fundamental, but secret, cause of hostility between the United States and Russia, many interpretations could be given for each stage in the development of the Cold War, and to some, America's role and actions could be seen as hostile to the USSR, for insufficient cause, or because of a failure by the American establishment to correctly analyze Soviet national interests, both after the Bolshevik Revolution, and following the 1945 Yalta Agreements.

The American left of the late 1960s had largely succumbed to an orchestrated, world-wide communist propaganda campaign about the injustice of the American attacks on North Vietnam and the definition of American POWs as "war criminals" and "air pirates" held in Southeast Asia, as they were described by Hanoi and the Vietcong of the south, and also by Radio Moscow and the organs of all other communist regimes. In the Communist view, any release of U.S. prisoners at the end of the war would only be for propaganda purposes, not for humanitarian reasons, or because of the demands of international law. (The only hope for obtaining release of all American prisoners would be if the United States had admitted to war crimes against the Vietnamese—thus admitting ideological defeat, and thereby perhaps obtaining the release of the majority of the POWs through a pardon and parole procedure.) Many among the American opposition to the war ascribed to the view that the U.S. POWs in Hanoi were "dupes of big business and the U.S. 'military-industrial complex', who had done criminal things to the Vietnamese, but were merely following the orders of the real criminals, who were in the White House, the Pentagon, the State Department, the CIA and on Wall Street.

An American in Hanoi at the time, U.S. Navy POW James Stockdale, later wrote about the reality of treatment accorded U.S. prisoners who tried to escape from North Vietnamese captivity, long after the cessation of American bombing over North Vietnam:

"In 1968, USAF Captains Ed Atterbury and John Dramesi had planned an escape from a cell they shared with seven other prisoners in an area of the Zoo after I left it. Americans came to know it as the Zoo Annex. The two officers' preparations were elaborate...Dramesi and Atterbury had hoped to go earlier, but were delayed. Finally when one of their sought-after dark rainy nights occurred on Saturday, May 10, 1969, they got all their gear together and went up through the roof. Their disguises worked well and saved them in many close calls with unexpected confrontations, both in getting down from the roof and across the prison yard and later. As they suspected, the wires atop the wall were electrified. They shorted them together, blew the fuses, and got away. Feeling that search parties were forming in the suburbs, they were trying to work their way through the open countryside. Before daylight they hid in a thicket in a churchyard. After sunrise they were discovered when, as an afterthought, after his party had conducted a futile search of the churchyard, a young soldier decided to crawl into their thicket. They were taken back to the Zoo, separated into two different locations, whipped unmercifully, manacled, and given the ropes until late afternoon on May 15..."[3]

Dramesi, who fortunately survived, remembered what happened after that:

"On the night of the eighteenth of May, I could hear them

beating Ed. Suddenly the hush of death seemed to fall over the whole prison. I could not hear any noise anywhere. I held my breath, afraid I would miss a sound. I leaned toward the door, and I heard them patting Ed's face and talking to him in low voices as though they were trying to revive him. They knocked him out, I thought."[4]

Atterbury had been beaten to death by his North Vietnamese guards, which became apparent to the other POWs when they never heard from him again. His body was eventually returned to the U.S. after the POWs were released. In the meantime, Averell Harriman's long-time policy of keeping official silence on the torture and mistreatment of American POWs in North Vietnam was finally ended by the new Nixon Administration, largely because of the tenacity of the new secretary of defense, Melvin Laird. Ex-POW James Stockdale later wrote:

"On May 19, 1969, Laird, convinced of the stupidity of keeping quiet about the mistreatment of American POWs, lifted the wraps Harriman had imposed. American headlines then delivered the message, and within six months a politically sensitive North Vietnam had made a scapegoat of their overall prison commissar, Major Bui (the Cat), and relieved him of command. Five months after that he was permanently sent packing, away from all prison camps. Soon the prisons of North Vietnam were operating more or less in accordance with the dictates of international law."[5] Stockdale was referring to those prisoners who were held openly in prisons accessible to American and other western visitors, albeit of the antiwar stripe. Other Americans, held in secret interrogation camps with Soviet and Chinese intelligence present, did not benefit from the new regime. Defense secretary Laird was to authorize a study of all letters received from POWs in Vietnam which resulted in discovery of the names of 474 to 478 additional U.S. POWs known to be alive in captivity who were never released (noted elsewhere in this book). Those Americans still being captured in combat in South Vietnam were also subject to the same harsh treatment. The fate of many of these POWs remains unknown.

At Ap Bia Mountain in 1969, U.S. paratrooper battalions of the 101st Airborne Division repeatedly assaulted dug-in North Vietnamese Army positions near the Laotian border, in a vicious battle which became known to the American people, who were already sick of the endless war in Indochina, as "Hamburger Hill." Hundreds of American soldiers were killed, wounded or missing in this extended action, but the American high command downplayed it's significance, in order to appease public concern about Vietnam War losses, and even in histories written long after the war the casualties often given for the battle were in fact only those U.S. soldiers officially listed as killed in action (or MIA), thus reducing the significance of a heroic but wasted action by American soldiers.

One of the soldiers present, Sergeant Michael Booth of Washington State, who was serving as a pointman for his platoon of the "Screaming Eagles" at the time, recalled for the author why he had become involved in the MIA/POW issue more than a decade after the war ended. He and the exhausted men of his squad were pinned down by NVA automatic rifle and machine gun fire in the blasted and burnt-out stumps of a jungle on the slope of Ap Bia Mountain, having pulled back under heavy enemy fire from a forward position which their company had momentarily attained after three previous assaults against dug-in NVA regulars.

Amid the cracking and popping of enemy AK-47 bullets and the explosions of Chinese 82mm mortars and Soviet RPG antitank rocket shells, while U.S. helicopter gunships attacked overhead, Booth and the other men in his platoon heard terrifying screams coming from the steep slopes above them, where the green AK-47 tracers of the enemy spat out death to the careless ones, and then a clear voice: "HEEELP! HEEEELP MEEE! THEY'VE GOT ME! THEY'VE GOT ME, YOU GUYS! THEY'VE GOT ME PRISONER...!" It was a voice that Booth and the other men of his platoon recognized, a black American soldier of their own unit who had disappeared after they had fallen back before the last NVA counterattack. They never knew what became of the man, who did not return in the POW repatriations of 1973, but it was his voice, this incident, which forced Booth, 17 years later, to become involved in the MIA issue with a former Marine named Gino Casanova.[6]

By 1969, the "Stockholm Conference," organized by the Soviet-front World Peace Council under the Indian Communist Romesh Chandra, planned activities against the U.S. Government around the world, along with assistance to U.S. draft evaders and military deserters and other antiwar actions. The KGB was actively involved in mobilizing world opinion against the U.S. participation in the Vietnam War, but the effect of graphic American reporting on the bombing in battles, shown in many parts of the world, was far more important.

In the People's Republic of China the tortures and murders of the "Cultural Revolution," which had begun in earnest in 1967, still continued. Mao Tse tung (Mao Zedong) had encouraged this great purge to consolidate his power and to root out "reactionaries" throughout Chinese society; it is estimated that some one million Chinese were thus killed. The violent outburst in China led to serious clashes between the Chinese and Soviet armies in Manchuria during March and August of 1969. The growing Chinese threat to the USSR was dramatized by Chinese nuclear tests which took place in Xianjiang during September 1969. The Soviet KGB-sponsored journalist Victor Louis (Vitali Lui), usually a reliable indicator of Soviet thinking, hinted that the USSR was considering a preemptive nuclear strike against China, before it could develop nuclear missiles capable of threatening Russia.[7] These events led in turn to a gradual American rapprochement with Communist China, which was conducted by Richard Nixon and Henry Kissinger over the next three years, in the interest of 'U.S. national security.'

At home in the United States a novelist named Kurt Vonnegut Jr. wrote from his own memories about the savage bombing of Dresden and the tense exchange of U.S. POWs with the Soviets after the May 1945 Halle Conference, in his 1969 book: Slaughterhouse-Five. Its publication must have caused some tremors among those who knew the truth about Stalin's reneging on the prisoner of war exchanges, and who dreaded the day when the truth would be exposed.

The U.S.-Soviet space race, begun after the 1957 Sputnik I satellite launch and accelerated by cosmonaut Yuri Gagarin's orbit of the earth in 1961, reached a crescendo in the summer of 1969 with the first landing of human beings on the moon. They were U.S. astronauts Neil Armstrong and Edwin Aldrin, who manned the Apollo 11 lunar module as it landed in the Sea of Tranquility. The American-Russian competition in space had resulted in the rapid development and use of ever more powerful rocket boosters and other sophisticated technology in an attempt to gain the edge in space, which was now perceived as a potential battlefield, in the struggle for control of the earth. Thus, American prisoners of war captured in Vietnam who had had any experience during their prior service with the Mercury, Gemini and Apollo space programs were prime candidates for transport to the USSR, if their past experience was revealed. Although the U.S. had gained an edge in space it was unable to extricate itself from the agonizing guerilla war in the jungles of Vietnam.

President Richard Nixon authorized a U.S.-South Vietnamese invasion of the VC/NVA Cambodian border sanctuaries in the spring of 1970, which caused an already powerful anti-war movement in the United States to explode anew, in protests against the incursion that involved hundreds of thousands of students and other citizens. At Kent State University in Ohio, four students were shot dead in one anti-war demonstration by undisciplined National Guard troops that had been called up to enforce order. Meanwhile, in Cambodia, American forces fought a series of running battles with retreating North Vietnamese/ Viet Cong forces, that cost the American Army and supporting air units approximately 2,000 more casualties in killed, wounded and missing. The Communists removed most of their prisoners from the border areas at this time.

Two U.S. intelligence reports of 1971 indicate that 64 American POWs were confined in a facility identified as a 'museum' located near the city of Kratie, Cambodia, which is situated on the

Mekong River northwest of the "Fishhook" region.[8] American journalists had also disappeared inside Cambodia during the incursion, including Sean Flynn, son of the famous actor Errol Flynn, together with his friend Dana Stone, a cameraman working for Time Magazine. Intercepted radio communications from the Communist COSVN Headquarters indicated that they had been captured alive, and Cambodian peasants reported they had witnessed their capture by Vietnamese soldiers. A declassified CIA report dated October 5, 1971, reports two other American journalists in captivity in Cambodia:

"As of 27 August 1971, two American journalists were being held prisoner by the VC/NVA in the Tapao plantation. Both had been held in the area since their capture...in October or November of 1970. Both were captured near the Svay Antor airstrip after having become separated from government units to which they had been attached... Until the capture of the two Americans (Khmer Communist) Region 304 Headquarters had been in a fixed location. It is not clear whether the capture of the Americans was a factor in the new policy of continual relocation of this element."[9]

Meanwhile, the U.S. received information that the North Vietnamese were holding many more U.S. POWs than they publicly admitted. This is recorded in an October 1970 classified-confidential cable (#05276 090939Z, from the American Embassy in London to the Secretary of State which was not released to the American public for twenty-three years to come:

"British labor leader Clive Jenkins, who returned from Hanoi last week, inquired about POWs. HE WAS GIVEN USUAL FIGURE of 'about 350' WHEN HE PRESSED, INFORMATION AND PROPAGANDA OFFICER HOANG TUNG, IT WAS INDICATED THERE MAY BE MORE THAN THAT FIGURE, WORD "NUMEROUS" USED. JENKINS SAID IMPRESSION AMONG HIS GROUP IS THAT FIGURE CLOSE TO 900."

"When Jenkins asked Hoang Tung Why 'they' could not give out more information on 'humanitarian' basis. He was told they are 'PAWNS,' can't publish names, etc./ because of military reasons. Jenkins was told there is 'FAIR AMOUNT OF DISATISFACTION' among POWs. Some want to make political statements favorable to Hanoi, however, captors not encouraging this because it suggests 'BRAINWASHING' which of course is not the case, according to Hoang Tung. Jenkins did not ask to see the prisoners, nor was this offered."

The intelligence gained by Jenkins was to be confirmed nearly twenty-three years later, with the release of two Soviet documents which were translations of North Vietnamese Politburo documents made available to Leonid Breshnev and the Soviet politburo by the Hanoi government at the time. These documents together confirmed that between 1970 and in 1972 the North Vietnamese were holding at least 700 more American POWs than they had admitted to publicly. The two documents are discussed in more detail in the final chapter of this book. These POWs never returned after the war. Hundreds remained secretly confined in North Vietnam, while others were transported to the USSR.

Retired NSA analyst Jerry J. Mooney (and another retired intelligence professional who spoke to the author) believes that the transfer of U.S. POWs who were not used for labor by the North Vietnamese Army from Soviet interrogation points in North Vietnam at Ron Ron, Cua Luoi/Vinh, Thach Ban/ Long Dai, Lam Thao-Vinh Phu Province, and other locations, through Tentacle MB at Sam Neua to Hanoi's Gia Lam Airport or to Haiphong, and on to the USSR, went forward from at least 1967. He thought it highly probable that it had really begun in 1965, based on his later working backwards through earlier decryptions of NVA radio traffic. Actual monitoring of Russian aircraft movements from Vietnam back to the USSR was accomplished by NSA's A3 Division, and in an interview with the author Mooney remembered, 'THEY WOULDN'T SHARE THAT DATA WITH US IN THE B GROUP, BECAUSE WE DIDN'T HAVE A DEMONSTRATED NEED-TO-KNOW IT."[10]

Most Soviet aircraft transport between Tentacle MB and Hanoi's Gia Lam international airport consisted of light single engine luxury AN-2 or AN-1 2 aircraft, on "VIP flights" occurring about 3 or 4 times a week, for years, which could only carry a few passengers, but would have been sufficient to transport a small but steady stream of U.S. POWs on their way through Hanoi, and on to the Soviet Union in larger jet aircraft of Aeroflot or of the Soviet military. Also servicing Tentacle MB at Sam Neua was an occasional Soviet IL-14, a twin-engine propeller aircraft roughly equivalent to a DC-3, which had the passenger capability to substantially increase the flow of Moscow-bound American POWs, who had undergone extensive interrogation, been psychologically broken and had agreed to cooperate in order to survive, by willingly going on to the USSR under Soviet control. To carry Americans, any of these aircraft were expected by the U.S. cryptanalysts to have special GRU or KGB flight crews, which would bring the aircraft in and out on such missions, for obvious security reasons.

Once transported to Hanoi's Gia Lam airport, Moscow-bound American POWs were transferred to larger multi-engine Soviet jet aircraft, such as AN-14 transports or IL-62s, which regularly flew to the Soviet Union from Hanoi, or in some cases they were sent by sea from the port of Haiphong on Russian freighters bound for Vladivostok, Siberia. The advantage of the sea route, was that there was always a chance that a Soviet aircraft might be forced down over a neutral country and its cargo of illicit U.S. POWs possibly exposed to the world, but if a ship went down for any reason, there would be no physical evidence of the deed being perpetrated. The

Soviet jet aircraft normally followed a route from Hanoi over northern Laos and Burma, just below the border of the Chinese province of Yunnan; thence across northern India south of the Himalayas and the borders of Bhutan, Nepal, with fueling stops in Delhi/New Delhi, capital of the "neutral," but friendly-to-the-Soviets nation of India, and from there northwestward above Kashmir and Afghanistan, east of the border of unfriendly Pakistan and into TASHKENT. Here some of the Americans bound for the high-tech missile and space facilities of Kazakhstan and elsewhere may have been dropped off, while others, bound for high-level intelligence interrogations, went on to Moscow's Lubyanka Prison, and likely to several GRU-run prisons in Russia.

On November 21, 1970, a U.S. helicopter-borne assault force of U.S. Army Rangers and Green Berets carried out a raid on the North Vietnamese prison at Son Tay, some 20 miles west of Hanoi. Although many enemy soldiers identified as non-Vietnamese foreigners, apparently Russian or Chinese, were killed in the ensuing attack (and none of the Americans), all the U.S. POWs had been moved. This may have reflected forewarning to the North Vietnamese, through the Soviet possession of U.S. code secrets that had been passed by the Walkers and others. An estimated 100-200 hostile troops were killed by the raiders at a compound 400 meters south of the prison camp. A later book on the incident, "The Raid," by Benjamin Schemmer, reported that, "Simons men noted as they took the others under fire that they were much taller, 5 feet 10 inches to 6 feet, oriental, not wearing the normal NVA dress." Former U.S. Army intelligence Captain Larry J. O'Daniel learned from intelligence sources that, '"some' of the persons killed in this incident could fairly be described as Caucasian." O'Daniel later obtained through FOIA request from DIA in 1978 a description of these soldiers: 'Even more unique is the description of most of the troops manning the military facility 400 meters south of the objective. They were described as 'LARGE ORIENTALS.'[11]

Intelligence from human sources in Southeast Asia which was eventually declassified a decade later, indicated that a large movement of American POWs to different camps occurred after this unsuccessful raid, as the Vietnamese were now fully alert to the value of the POWs and the lengths that the U.S. would go to in recovering them. These included both prisoners from camps reported by intelligence sources who never returned home and others who did. Many American POWs within "the System," including James Stockdale, later expressed the belief that all U.S. POWs had been concentrated in locations from which 591 were later repatriated in 1973. Others, such as Air Force POW Laird Gutterson, suspected that a parallel prison system existed, and believed that the POWs within "the System" did not know about all of the U.S. prisoners in

Vietnamese captivity, and thus could not know what percentage were actually allowed to return home at the end of the war.

The planner and overall commander of the raid was Air Force Brigadier General Leroy Manor, while the operational commander on the scene was U.S. Army Colonel Arthur "Bull" Simons. Future Deputy Director of Defense Intelligence, Admiral Jerry Tuttle, who helped plan the raid, was to remain involved in the POW/MIA issue long after the war ended for the United States. The Son Tay rescue POW mission had also been a personal project of Secretary of Defense Melvin Laird, who had decided that the POWs were being neglected, and had single-handedly pushed the new Nixon Administration into publicizing their plight. Among those Americans who assisted Laird in this mission was a Texas businessman named H. Ross Perot, whose involvement with missing U.S. POWs was to continue for more than two decades to come. In significant testimony before the Senate (in 1992), Laird stated that at the suggestion of Henry Kissinger, he had discussed the prisoner of war and MIA matter with former Ambassador to Russia, Averell Harriman, in 1969: "I remember Ambassador Harriman coming to see me and urging me not to go public on this particular issue." Laird persisted, however, and prisoners returning home years later stated that their conditions had improved from this period on, and the guards became more respectful.[12]

Meanwhile, the United States was still following a policy of forcibly-repatriating Soviet citizens in an arbitrary manner during this period, although publicly maintaining that America stood by the 1951 Geneva Convention protocol on political asylum. A Lithuanian citizen named Simas Kudirka, who leapt from a fishing trawler off the Massachusetts coast on November 23, 1970, and sought political asylum on the U.S. Coast Guard vessel Vigilant, was forcibly returned to Soviet control, despite the fact that the United States had never recognized Soviet rule of the Baltic Republics. The decision was made by senior Coast Guard officers, in what appeared to be a routine conformance to a U.S. policy, which was little known to the public. Russian officials were allowed to board the U.S. vessel at sea and forcibly remove Kudirka, who, according to U.S. coastguardsmen present, was savagely beaten unconscious during the struggle.[13]

The Lithuanian was returned to the Soviet Union where he was sentenced to ten years in a prison camp, but the incident sparked widespread protests in the United States, particularly among Americans of Baltic descent, who had first-hand or family knowledge of Soviet methods. The protests continued until the Soviets were ultimately forced to release Simas Kudirka, but the damage that had been done to the reputation and prestige of the United States was immense. It is not known whether the forcible return of the Lithuanian was related in any way to American prisoners then under Soviet control.

The American prisoners of war in Southeast Asia were finally being tracked by intercepted enemy communications at this time, after years of ignoring this open window into the Communists own system. According to retired NSA analyst Jerry J. Mooney, by 1970, the climate had indeed changed within the intelligence community, but the results were disappointing:

"MY SECTION CHIEFS AT NSA (B34) IN 1970, RAY RILLEY AND SKIP PERKY, KNEW THAT THE NORTH VIETNAM SAM MISSILE-ANTI AIRCRAFT ARTILLERY AIR DEFENSE CRYPTOSYSTEMS WERE UNDERDEVELOPED, AND GAVE ME FULL RESOURCES TO BUST IT WIDE OPEN, AND IT WAS DONE!" But Mooney stated in a 1992 interview with the author that the NSA analysts to whom he and his colleagues were sending the information they gleaned from decrypted NVA radio transmissions in 1970, Chuck Semich, Bill Service and William Melton, "Were doing a fair job on the Air Defense analysis, BUT NOT MUCH REPORTING. EVEN WHEN WE STARTED FEEDING THEM HIGH VOLUMES OF DECODED AND TRANSLATED MESSAGES, THEY DIDN'T HANDLE THEM PROPERLY...[14]

Mooney's cryptanalytic supervisors at NSA, Dick Chun and Joe Burgess, "knew the important data wasn't being reported, that Semich was the problem, and they raised hell, after which I was transferred to the REPORTING SHOP, B-31, in the summer of 1970... Higher up, at the NSA's B office level staff, Jean Frankenburger handled the EYES ONLY and LIMITED DISTRIBUTION messages on Americans executed by the North Vietnamese as reprisals for U.S. actions, which information was withheld from the public, and particularly from American aircraft crews, for fear that it might make U.S. fliers reluctant to go on certain missions. Mooney remembered that NSA's Richard Chun was particularly incensed at the way important POW data wasn't being properly reported. Chun was a Korean war veteran who had made it clear to Mooney that many American POWs had been left behind in Chinese, Soviet and Korean captivity after that war ended. Chun was a strong supporter of Mooney's effort to collect all significant information possible through decryptions on the capture of U.S. POWs who might later be identified through corollary data as prisoners in future negotiations.

According to Mr. Mooney, other key personalities in the NSA information chain included two "super cryptolinguists," Jerry Kelly and Berkeley C," who were in the NSA's B64 from 1971 to 1973. In an interview with the author, Mooney recalled: "They supervised the cryptolinguistic Air/Air Defense effort. KELLY AND "C' PERFORMED ALL THE BOOKBREAKING AND TRANSLATION OF MESSAGES AND PASSED THEM ON TO ME FOR MY SECTION'S REPORTING. Berkeley "C." IS STILL WITH THE NSA AND IS INVOLVED WITH THE CURRENT NSA

'REVIEW OF MATTERS' PERTAINING TO CRYPTOANALYTIC RECORDS OF KNOWN U.S. POWs WHO WERE NOT REPATRIATED IN OPERATION HOMECOMING IN 1973... My Division Chief, who signed off on all my electrical messages was Wally Marshall AND HE KNEW OF MY EXTENSIVE POW LISTS, INCLUDING THOSE CARRIED AS MOSCOW-BOUND...MY BRANCH CHIEF AT NSA WAS MARVIN CONNERS, WHO WAS AN EXCELLENT ANALYST; IT WAS HE WHO TOLD ME THAT HE BELIEVED THE POW DATA I HAD PRODUCED HAD BEEN SENT TO THE DEFENSE INTELLIGENCE AGENCY (DIA). Frank Buck was another analyst who was mad that we couldn't get prior knowledge of POW executions in product and he was the person who helped me sneak it into the DIA's 'Blue Book.'"

The NSA, which has been called America's most "supersecret" intelligence agency, was then under the directorship of Admiral Noel Gayler, but the real power in the agency for 13 years was in the hands of Deputy Director Louis Tordella. Tordella was the keeper of the agency's secrets and he would have been knowledgeable about the intercepted communications concerning American POWs in Vietnam, Laos and Cambodia, and those transported to the Soviet Union, Communist China, North Korea or Cuba. As author James Bamford wrote later in his definitive study of the agency, "The Puzzle Palace":

"If NSA was the darkest part of the government, Tordella was the darkest part of the NSA. Even General Carter, Director of NSA for the longest period of time, could never be totally sure he knew ALL of the Palace's secrets; he believed there may have been times when, 'because of the sensitivity they felt, well, why burden me with it?'"[15]

But the NSA, or it's past directors and senior officials, were never to be seriously challenged by the legislative branch of government about what was revealed by intelligence intercepts about American POWs in Indochina. When, years later, the subject was finally raised by a Congressional investigation, the NSA was to claim that it could not locate the intercepted communications which were the subject of sworn testimony and had been stockpiled for years at Fort Meade and Fort Holabird, Maryland.

Sometimes the "operational intelligence" gained by CIA or military intelligence in Southeast Asia from "human sources" bore out the information obtained through the "special intelligence" collected and reported by Mooney, Minarcin and others involved in electronic eavesdropping, that there were many more American POWs being held in North Vietnam, South Vietnam, Laos and Cambodia, then those listed officially by the U.S. Joint Prisoner Recovery Center. As one example, at a time when the Center listed only 496 U.S. POWs in all of Southeast Asia, in 1971, A CIA INFORMANT CLAIMED TO HAVE SEEN A PRISON CAMP IN HA TAY

PROVINCE, NORTH VIETNAM, CONTAINING MORE THAN 1,000 AMERICAN (AND AUSTRALIAN) PRISONERS OF WAR.[16] This sighting occurred after the North Vietnamese had moved many POWs following the failed U.S. raid on Son Tay Prison, and had also closed Ben Pha Den Prison as a result of the raid.

An August 1969 CIA report states that 300 American POWs, being used as forced-laborers, were being held with 700 South Vietnamese POWs in a prison containing bamboo and thatched roof buildings five kilometers south of Tuyen Quang, North Vietnam.[17] In another CIA report, 200 Americans were reported moved from the YEN BAI area and being held in Tuyen Quang, North Vietnam. Many other declassified CIA reports from human sources report hundreds of Americans held at various locations in Vietnam which indicate probable total numbers of U.S. POWs there in the 1,000-2,000 range, as reported in a 1968 CIA document quoted earlier in this book. And many of these reports concerned prisoners and prisons WHICH WERE NOT PART OF THE REGULAR NORTH VIETNAMESE PRISON SYSTEM, FROM WHICH U.S. POWs WERE TO BE REPATRIATED, FOLLOWING THE AMERICAN MILITARY WITHDRAWAL FROM THE WAR IN 1973.

A now-declassified 1971 CIA intelligence report, obtained by the author from CIA through Freedom of Information request, confirms that a number of U.S. POWs in North Vietnam were interrogated by Soviet and Chinese Communist personnel over a period of several years, at a secret site in a Vinh Phu Province superphosphate plant (west of Hanoi), and changed clothing for transport to unknown destinations, accompanied by different guards:

"THE GUARDS LINED UP, FORMING A CORRIDOR THROUGH WHICH THE PILOTS ENTERED THE BUILDING. AT THIS POINT A SOVIET, A CHINESE AND A VIETNAMESE GREETED THE PILOTS AND LED THEM INTO THE BUILDING. THE PILOTS USUALLY REMAINED IN THE BUILDING FOR SEVERAL HOURS. WHEN THEY EMERGED THEY HAD CHANGED FROM UNIFORMS INTO CIVILIAN CLOTHING...AFTER SHAKING HANDS WITH THE SOVIET AND CHINESE, THE PRISONERS WERE LED TO A DIFFERENT VEHICLE FROM THE ONE THAT BROUGHT THEM TO THE SITE. THEY WERE ESCORTED FROM THE PLANT BY A DIFFERENT SET OF GUARDS WHO WORE YELLOW AND WHITE UNIFORMS AND WERE ARMED WITH RIFLES AND PISTOLS. (Blacked-out) DID NOT KNOW THE DESTINATION OF THE PRISONERS.[18]

Retired NSA analyst Jerry Mooney believes that this interrogation camp was similar in function to the one at Cua Luoi, near Vinh, which was code-named Son Tay, and that declassification of more CIA documents could reveal other, similar sites. The U.S. Government maintains that no U.S. POWs interrogated by the Soviets in Vietnam were returned after the war.

As their price for assisting North Vietnam in its war with the U.S., the Soviets had presented Hanoi with "shopping lists" of particular aircraft and technological equipment which Moscow needed, together with profiles of American pilots and specialized equipment operators to be targeted for capture and transfer to Soviet control. Through their intercepted and decoded orders the North Vietnamese were known by Mooney and other NSA analysts to have assembled pre-selected U.S. prisoners from North and South Vietnam at the secret POW camp at Cua Luoi near Vinh, NVN, which was code-named Son Tay (but was distinct from the other, more famous camp of that name unsuccessfully raided by U.S. forces). At Cua Luoi/Vinh the Americans were interrogated and screened by Soviet intelligence. These POWs were categorized within NSA as "MB" or "Moscow-Bound." The same situation existed in South Vietnam, although the numbers reported were often smaller.

Decoded enemy communications and subsequent U.S. tracking of Soviet aircraft indicated that from Cua Luoi/Vinh the screened, Moscow-Bound U.S. prisoners were shipped by truck to the Bai Thuong airfield, and from there to another special POW camp near the Pathet Lao headquarters at Sam Neua Laos, which was known to select NSA intelligence personnel as "Tentacle M-B." Here the final interrogations by Soviet intelligence officers (probably of the GRU) determined which Americans with technological expertise or intelligence value would cooperate. POWs in this category were referred to by NSA analysts and their Defense Intelligence Agency (DIA) counterparts as "GONE FOREVER." Those who willingly or unwillingly submitted, in order to survive, were then transported by air and by sea to the Soviet Union, while those who refused and were thus deemed unreturnable, such as Navy Commander Larry Van Renselaar, faced death at forced labor on the Ho Chi Minh Trail system. Sam Neua, inside Laos near the North Vietnamese border, was to be the site of 91 livesighting reports over the years from 1973-1991, involving U.S. POWs seen in small groups and in groups as large as over 100.

According to a DIA counterpart of Mooney's, with whom he discussed the Tentacle MB situation at Sam Neua, China-Bound U.S. POWs were reportedly assembled at other locations, including a special, secret camp at LANG SON, between Hanoi and the Chinese border (where Russians were also present). There, many seriously wounded and burned U.S. prisoners may have been treated by Chinese doctors prior to being moved across the border. In Laos, numerous reports indicate that American POWs were seen in captivity in the Pathet Lao Headquarters area of Sam Neua, where NSA analysts had also discovered the presence of many "FRIENDS" (Soviets) and the existence of the Tentacle M-B connection for U.S. POWs being interrogated for selection to go either on to the Soviet Union, or to enter the 'pick and shovel' brigades as forced-labor supporting the North Vietnamese supply system down the Ho Chi Minh trail down the length of Laos.

Among those Air Force/NSA officers and officials who had knowledge of the Moscow-bound U.S. POWs as co-workers with NSA analyst Jerry J. Mooney, were: Master Sergeant Greenman, Senior Master Sergeant Groves and CMSGT Wilson, of the Joint Sobe Processing Center, Okinawa, 1968, where secret, encrypted North Vietnamese communications were intercepted and decrypted by a U.S. 'listening post.' Many years later Mooney recounted to the author that at this time, following the 1968 Tet Offensive, "They knew the North Vietnamese SAM missile/AAA problem was being ignored, and thus assigned me with Jerry Wolton and Terry Minarcin to develop it and the Air Defense Cryptosystems."[19]

From these decryption analyses the first evidence of the existence of a 'Tentacle M-B" interrogation center at Sam Neua, Laos, staffed by Soviet intelligence officers, fed with prisoners captured in North Vietnam and processed through a secret parallel prison camp system with locations such as Ron Ron, near the DMZ, or Cua Luoi, near Vinh, was to be revealed to Mooney and his coworkers. But such information was only a secondary by-product of accomplishing their primary mission of analyzing NVA air defense capabilities, and the Moscow-bound U.S. POW/MIA information thus painstakingly assembled—largely because of NVA security breaches in decrypted NVA radio traffic they were monitoring, was to be officially ignored by the NSA hierarchy and their White House masters until the 1971-1972 period.

Yet, Mooney immediately understood the significance of discovering the secret parallel prison system and it's TENTACLE-MB connection at Sam Neua, and on his own time, without extra pay, he began to go backwards through other formerly-decrypted NVA communications to 1965, to find confirming data on other POW captures which gave solid indication of being earlier cases of an as yet undiscovered link to American POWs who had never shown up in the regular North Vietnamese prison system, and thus were earlier candidates for the Moscow-bound category. This was how he began to develop his own lists of U.S. POWs, based upon all available information from decryptions, and on their "special talents" which would make them obvious priority targets of the Soviets to service their High-Tech needs relating to Russian national security requirements, who had undoubtedly also been shipped to the USSR, by air or by sea, after processing through Tentacle MB at Sam Neua, and other locations such as Hanoi's Gia Lam international airport, and the North Vietnamese port of Haiphong. (From collateral data, such as NVA "rallier" or defector reports, which he became aware of through intelligence colleagues at the time, Mooney also learned that substantiating data existed from human intelligence sources inside North Vietnam and Laos indicating that the Soviet presence at Sam

Neua was variously estimated at between 75 and 500 personnel.)

An example of the form which the information took from decryptions of NVA traffic was revealed by Mooney in a 1991 interview with the author:

"A NORTH VIETNAMESE OBSERVER ON THE ROAD FROM CUA LUOI-COVER NAME SON TAY-NEAR VINH, TO BAI THUONG AIRFIELD, WOULD REPORT THE PASSING OF A SOVIET-BUILT TRUCK CARRYING TWO AMERICANS WITH THEIR GUARDS TO OBSERVERS AT ANOTHER NORTH VIETNAMESE POSITION: 'TWO BANDITS (Americans) JUST PASSED HERE ON THE WAY TO BAI THUONG AIRFIELD.'"

Such a report as this, and many others intercepted over the radio from 1968-1973, actually constituted a breach of security by the North Vietnamese observers, who were low-level personnel, and probably not conscious of the fact that they were being constantly monitored, or uncaring that their breach of security, possibly uttered to another North Vietnamese out of boredom, or envy that the Americans were riding in a truck, actually gave solid confirmation to NSA cryptanalysts listening in to details of an otherwise highly-secret program of interrogating American POWs prior to moving them on to Bai Thuong where most would be transported by either a Soviet IL 14 or a smaller AN2 or 12 to Sam Neua and thence to the Hanoi Haiphong area for shipment out of country on regular Soviet AN14 or IL62 flights, or by Russian ship to Vladivostok. (Others may have been transported overland to Sam Neua or other secret camps from the Cua Luoi/Bai Thuong/Vinh area.)

In later sworn testimony before Senators Grassley and Kerry of the Senate Select Committee on POW/MIA Affairs, and in earlier affidavits and interviews with the author over a three-year period, retired NSA analyst Jerry J. Mooney revealed the substance of what the NSA analyses of decrypted North Vietnamese radio communications had taught him and other intelligence specialists about the fate of hundreds of OTHER U.S. prisoners, who were not moved on to Tentacle MB, or to the regular prison system in the Hanoi area for eventual exchange at the end of the war. Instead they became forced-laborers on a project which Hanoi considered to be in its highest national security interest:

"OUR AMERICAN POWs, PARTICULARLY IN SOUTH AND NORTH VIETNAM WERE (ALSO) USED IN AN AMERICAN PICK AND SHOVEL BRIGADE. FOLLOWING THE TET OFFENSIVE OF 1968, THE VIETNAMESE REALIZED THAT IF THEY WERE EVER GOING TO WIN A WAR IN THE SOUTH, THEY WOULD HAVE TO HAVE SUFFICIENT AIR DEFENSE POWER TO COMBAT AIR POWER, AND THEY HAD TO SHORTEN UP THEIR LOGISTICS LINES. IN THE SUMMER OF 1968, UNDER THE COMMAND OF THE 367TH (NVA) DIVISION... OPERATING IN THE HA TIWA AREA OF SOUTHERN NORTH VIETNAM, THEY BEGAN A MASSIVE UNDERGROUND CONSTRUCTION COMPLEX IN THE THACH BAN-HONG DAI AREA NEAR

THE DMZ. THIS CONSTRUCTION TOOK OVER TWO AND ONE HALF YEARS.

"IN THE CONSTRUCTION FORCE, HANDLED OUT OF RON RON, WERE AMERICAN POWS SHOT DOWN IN THIS AREA—NOT ALL, BECAUSE SOME WERE SENT NORTH—BUT MEN LIKE (Air Force Major) SAN D. FRANCISCO (MIA, shot down on 25 November 1968), (Air Force Colonel) JOSEPH C. MORRISON (MIA, also on 25 November 1968), (Navy Lt. Commander) LARRY J. VAN RENSELAAR (MIA, 30 September 1968), and (Navy Lt. Commander) DOMENICK SPINELLI (MIA, also on 30 September 1968)...THESE MEN WOUND UP IN THE AMERICAN PICK AND SHOVEL BRIGADE. THEY WERE USED AS LABOR AND ALSO PROBABLY AND MOST LIKELY, THEIR TRAINING WOULD GIVE THEM SOME EXPERTISE IN CONSTRUCTING THESE MASSIVE UNDERGROUND FACILITIES...COMPLETED IN EARLY 1972. (This information was obtained by us from)...ENCRYPTED MESSAGES...THAT WE READ...THE SPECIFIC TITLES (Cryptographic titles at NSA) ARE IN THE VCZD SERIES, THEY START IN THE VCZF SERIES. IN THE VCZD SERIES THEY RUN FROM VCZD001 THROUGH AT LEAST VCZD300, THAT REPRESENTS 300 CODE SYSTEMS. THESE ARE IN THE POSSESSION OF THE NATIONAL SECURITY AGENCY AT FORT MEADE, (also)...THERE WERE VCZF'S, AND THIS RAN FROM AT LEAST 001 THROUGH 008 WAS THE LAST ONES I SAW..."[20]

Other, luckier American prisoners captured in the South were sent to Hanoi to boost the numbers of those POWs who would ultimately be permitted to return to their homes at the end of the war. David Harker, a Marine rifleman captured in January 1968 and held in a POW camp in South Vietnam, was being marched up the Ho Chi Minh Trail to North Vietnam with 5 other American POWs when the 1971 South Vietnamese invasion of Laos occurred. He remembered: "The Saigon government launched Operation Lam Son 719...as we started up the Ho Chi Minh trail...We saw no ARVN soldiers fighting, but we did see whole companies of captured troops, many of them badly wounded...The ARVN were badly beaten, and they soon retreated...The sky was lit nightly with fiery tracers and exploding puffballs of flak. Literally hundreds of antiaircraft positions were firing around us in the distance...We saw no jets shot down...Each day on the trail we passed battalion after battalion of fresh North Vietnamese troops heading south...We passed Highway 9, which cut across Laos into Khe Sanh...We heard a dull rumble like thunder in the distance. The rumble grew louder. As the last bit of twilight disappeared a seemingly endless line of trucks began to roll past...all of them axle-heavy with supplies, heading south...A jet made a run on our area, flak guns boomed, and we tumbled into another bunker. Everything around us came to a standstill. When it was quiet once more we were told to get into a truck. There were ARVN POWs from the Laos invasion, ourselves and the guards crammed into the truck's rear—maybe 20 or more men...As day broke we finished the mountains and hit level ground. Off to the sides, in the paddies, we could see mist-shrouded silhouettes of SAM missiles. North Vietnam...We were put in irons for the first time since our capture, cuffed by twos with ancient French shackles. Next day we resumed our truck journey to Vinh. We were blindfolded but I managed to pull mine down...At 10:00 that night we were taken to Vinh's train station and put in a boxcar with seventy ARVN POWs...We reached Hanoi at 5:00 in the afternoon the next day, April 1, 1971..."[21]

In the United States, the nationwide anti-war sentiment was causing serious problems for security officials. With the June 1971 leak of the Pentagon Papers (a classified official study of the origins of the war in Vietnam), to the New York Times and others, by Daniel Ellsberg, a former Pentagon official, one of the most secret and valuable intelligence sources of the NSA and the White House, had been compromised. This was an eavesdropping operation in Moscow that intercepted the scrambled radio-telephone communications of top Soviet leaders including Leonid Breshnev and Alexei Kosygin, in and around the Kremlin, that was code-named "Gamma Guppy." In August 1972 a former Air Force Security Service Staff Sergeant who served with NSA named Perry Fellwock, under the pseudonym Winslow Peck, revealed NSA secrets in the antiwar Ramparts Magazine. Fellwock had served as a traffic analyst at NSA listening posts in Vietnam, Turkey and West Germany, and stated that the NSA was able to decipher all secret Soviet communications: 'We're able to break every code they've got.'[22] These type of security breaches infuriated intelligence bureaucrats and the Nixon White House, causing a particular focus on leaks and security. Whether the intercepted Kremlin traffic ever included references to American prisoners in the Soviet Union or Southeast Asia remains unknown, or, classified.

At the CIA, James Angleton's molehunt continued to disrupt the Soviet division. The CIAs counterintelligence chief was still under the spell of Soviet defector Anatoli Golytsin. Angleton's staff had even begun an investigation of Henry Kissinger, the President's national security adviser; during 1971. This was initially based on accusations made in 1970 by the Polish intelligence defector, Michal Goleniewski, who claimed that he had seen a Soviet colleagues file on one of his agents, who was identified as Henry Kissinger, recruited while the Soviet officer had been operating from East Germany, and remaining in contact with the Russians after returning to the U.S. Goleniewski said he had learned from the Soviet file that at Harvard Kissinger was working on a CIA project, which was classified information and raised suspicions at CIA. Counterintelligence officer Clare Edward Petty gave the Kissinger file to Angleton, who did nothing with it, and declined to share it with the FBI which is mandated to investigate domestic espionage.[23]

Kissinger was known for his lengthy meetings with Soviet Ambassador Anatoli Dobrynin, and declined to be debriefed by CIA about them. Despite the fact that Golenieski had produced valuable intelligence in the past that had led to the arrest of KGB mole George Blake inside MI-6, and exposure of Gordon Lonsdale/Konon Molody and Harry Houghton of the Portland spy ring, he was discredited by Angleton from this point on. It was said he was bitter about being ignored by CIA after being of such value before and now had delusions of being descended from the Czar. Angleton now contended that Goleniewski was a Soviet plant, who had fallen back under Soviet control before he defected in 1961.[24] Yet, Angleton and Golytsin also reportedly suspected Kissinger of being a possible KGB mole. Years later, after he left the CIA, Angleton was to claim that Kissinger was "objectively a Soviet agent."[25] Kissinger had reportedly undergone an extensive security check by the FBI prior to being confirmed to high position in the U.S. Government. He later told author Tom Mangold in an interview that he hadn't known Angleton was smearing him, and that he normally dealt directly with the DCIs, with no involvement by Angleton.[26]

By 1972 the treatment of those American POWs held openly in known prisons in North Vietnam had improved considerably, largely due to the public campaign being waged by family members in America, which was encouraged by Richard Nixon's Secretary of Defense, Melvin Laird. Colonel Larry Guarino, the 50-year-old American veteran of WW II and Korea who was a POW of the North Vietnamese for eight years, remembered this period:

"...The Vietnamese informed us that China had been admitted to the United Nations General Assembly! THAT news was earthshaking enough, but they added that President Nixon had recently visited Peking, and had met with Chairman Mao and Premier Chou En-Lai! Some of us were very disturbed at those revelations, because our view was that our president was meeting with one of the parties who was exerting a major effort to keep North Vietnam in the war. However, our trust in President Nixon was immense... From the sixteenth of April on, our planes came North regularly. In early May the (liars) Box put up a helluva squawk about the U.S. Navy mining the harbor at Haiphong...We were finally putting some REAL pressure on North Vietnam again...Shortly afterward the gooks suddenly moved over two hundred POWs out of Unity to another camp (or camps) unknown to the rest of us...In the fall of 1972, the Box hailed the arrival of a famous American movie actress who had 'come to demonstrate her friendship with the Vietnamese people and her opposition to the illegal, immoral and unjust war.' The box told us that 'their friend Jane Fonda' had met with some American prisoners, to talk about the war and their treatment. We guessed that the

Americans had to be the same two who had refused to join us...
After we were released, she called us 'liars and hypocrites.' Fonda's
comments are on record...At Duke University, December 11, 1970,
she said, 'I am a socialist, therefore I think we should strive toward
a socialistic society all the way to communism.' ...Right after Fonda's
visit, Hanoi had another American visitor, the former attorney general
of the United States, Ramsey Clark! This one really threw us for a
loop—our own former attorney general What the hell is going on back
there? we wondered!"[27]

Throughout the interminable years of peace negotiations in
Paris conducted by Henry Kissinger for the Nixon administration, the
Vietnamese Communists had always demanded war "reparations" as
a precondition to a negotiated settlement, and they had long tied
the reparations issue to the release of American POWs held by them.
Philip Habib, a member of Kissinger's negotiating team in Paris later
reported that the North Vietnamese had bracketed the two issues
together within a single numbered point in their demands. The North
Vietnamese demanded $9 billion in reparations divided between
North and South Vietnam.[28]

Kissinger had made clear the United States' willingness to pay
billions of dollars in war reconstruction aid, but for political reasons
he refused to allow such an agreement on what amounted to US
payment of war reparations to appear in the final, publicly released
Paris peace agreement. (Indeed the idea of postwar reconstruction
aid to the North had first been proposed by President Johnson, but
it had been United States policy long before and since then to never
tie such aid directly to a release of prisoners of war, to avoid setting
a precedent of paying ransom.) The Vietnamese in turn, had refused
to sign the Agreement without the provision for US postwar aid
and further refused to provide the list of American prisoners held in
Laos, until the reparations were promised in writing. This obstacle
in reaching a peace agreement was overcome in principle prior
to the concurrence on the October 1972 agreement, which Thieu
had refused to sign. The US would promise unconditionally to pay
reparations as "postwar reconstruction aid" to assist in "healing the
wounds of war," but the U.S. would not link payment of the aid to any
other provision of the Agreement or specify the amount or method
of payment. (These specific details were to be promised only in a
separate "Unilateral Document." This document took the form of a
highly secret letter from President Richard Nixon dated four days
after the 27 January 1973 Paris Peace Agreement.)

The breakdown in negotiations ultimately resulted in a
sudden week-long renewal of intensive U.S. airstrikes against North
Vietnamese targets in December 1972, code-named "Linebacker II."
Within the OSD of the Pentagon, Lieutenant Colonel Richard Secord,

veteran of the covert war in Vietnam under Kennedy and the secret CIA war in Laos under Johnson, was ordered to dust off a previous plan for unlimited air war against North Vietnam. In his memoirs Secord records that the major opposition to the plan in the Pentagon came from the Joint Chiefs of Staff. According to Secord, aside from projected possible losses of U.S. aircraft in such an operation, the final protest from an officer representing the JCS staff was: "'BUT IT'LL SCREW UP SIOP!' referring to the Strategic Integrated Operations Plan (SIOP), the plan that allocated nuclear resources to World War III."[29]

This reaction gives a clear indication as to how seriously the Soviet nuclear threat was perceived to be at the Joint Chiefs of Staff level, even in the midst of an ongoing major war that had already cost some 58,000 U.S. dead or missing, while many American POWs yet remained in Communist control. The aerial campaign subsequently launched on December 18, 1972, was labeled in the American press as the "Christmas bombing" and was almost universally condemned, with the influential Washington Post claiming that President Nixon had "taken leave of his senses." Talk of impeachment proceedings spread in the Congress and the specter of the nationwide anti-war movement arose again. An American POW in Hanoi, Colonel Larry Guarino, remembered:

"Just after dark, the air raid sirens went off, and we thought we were in for another airstrike or reconnaissance flight by tactical aircraft. But no!...The antiaircraft fire and the SAM launches were deafening! Then came the bombs, and by the long volleys of explosives, we knew immediately the birds were B-52s! Our side meant business..."[30]

Another POW in Hanoi, the Marine rifleman David Harker, who had been held since 1968, also remembered the "Christmas bombings," and the resulting move of 108 U.S. POWs from the "Plantation" to the "Little Vegas" section of the "Hanoi Hilton.":

"A week before Christmas the bombing started again. This time it was B-52s. We heard the SAMs whooshing off and saw a flash of orange as they exploded. The ground rumbled, our cells shook...I had confidence in dear old Uncle Sam and aerial reconnaissance. Every day a high-flying SR-71 came down the middle of North Vietnam and broke the sound barrier over Hanoi...I was very happy to see the picks and shovels the Vietnamese brought us on December 20. We dug a five-by-fifteen bomb shelter in our cell and slept in it at night. We finished the bomb shelter in one day through thick cement... Two days after Christmas we were told to get our gear together. Our spirits hit rock bottom. We thought we were being moved to China. The trucks arrived and all one hundred eight prisoners were herded aboard. We backed up turned around, traveled two blocks north, four east, and in ten minutes we were there. We got out.

It was a big prison. The walls were much higher than at the Plantation Gardens and broken glass was imbedded in the cement...The prison was massive and divided into various cell blocks, but as we walked through it we heard no sounds and saw no one. It could have been deserted...We were put into a section of the Hilton that American pilots almost a decade before had named Little Vegas...The ventilation was much better at the Hilton, at least in our cell; the windows were open and we could see the sky and didn't feel so closed in. That night the B-52s came over. We watched as one went down in flames. There were no parachutes..."[31]

In the twelve days of bombings from December 18th to December 30th, sixteen U.S. Air Force B-52's were shot down or lost, in addition to many other tactical aircraft. B-52s and fighter-bomber support crews suffered losses in killed and missing, some of which constituted the last wave of American POWs to reach the prisons and later be released. (This was at a time when NSA analysts like Jerry J. Mooney were monitoring enemy communications indicating that the powerful political commissars in North Vietnamese AAA units were ordering most new prisoners to be processed for release, to raise those numbers in anticipation of a peace agreement.) Part of Gia Lam International Airport, long protected from American attack because of the presence of Soviet aircraft, was partially destroyed in the Christmas bombings, as were Hanoi's rail yards and the two main rail lines into Communist China, used to transport war material. The North Vietnamese Air Defense Ministry was practically destroyed and port of Haiphong was also severely damaged after years of sanctuary due to the presence of Soviet shipping and U.S. fears of possible escalation to a global confrontation. For the first time in the long agonizing war the North Vietnamese realized how restrained in the use of air power the Americans had been.

The December 1972 bombing campaign brought North Vietnam back to the conference table quickly, and restored hope in many of the POWs. The agreements finally worked out and signed on January 27, 1973 by Henry Kissinger and Le Duc Tho set a timetable for the withdrawal of the last 5,000 U.S. troops remaining then in Vietnam and provided for the simultaneous return of all POWs in Vietnam and adjacent Laos and Cambodia, within 60 days of the signing of the accords.

Yet, right up to the end of official American involvement in the war, U.S. pilots continued to disappear who should have been released, or accounted-for shortly afterward. One example is U.S. Navy Commander Harley Hall, a "top gun" pilot of an F4-J and former commander of the Navy's "Blue Angels," who was shot down on January 27, 1973, just a few hours before the signing of the Paris peace agreement reached by Henry Kissinger and Le Duc Tho. Harley

Hall was last seen alive on the ground, disengaging from his parachute and running, but he subsequently vanished. He may have seemed a worthwhile prize to Vietnam's Soviet allies.

Another U.S. aircraft which had been shot down on December 21, 1972, near Pakse, Laos, was an Air Force AC 130A Spectre gunship containing 15 crewmembers, one of whom was Airforce Captain Thomas Hart III. Five parachutes were seen and several beeper signals were monitored. Two men were rescued but the other 13 became "category 2" MIAs. They were not to be returned in the POW exchange that was being negotiated at the time and was accomplished in the spring of 1973. Later they were all declared dead by the U.S. Government, presumed to have been killed in the crash. Months after the prisoner returns were completed in the coming year, aerial reconnaissance photography was to record an apparent message stamped out in long grass, in the area, a "T.H." and the date: "1973." A note from an intelligence analyst concluded "This was thought to be Captain Hart." But this information was to remain secret and unknown by Ann Hart until she found it in another MIA case file of Hart's fellow crewman, Master Sergeant James Ray Fuller. When a picture of another crewman of this plane-in captivity, appeared in the Bangkok Post, the U.S. Embassy in Bangkok promptly "classified" the picture. The AC 130A case was to surface again years later, due to the persistence of POW/MIA family members like Ann Hart, who refused to give up hope that their loved ones might someday be released.[32]

Indochinese prisoners of the Hanoi regime reported seeing Americans in captivity during the war who could not have been released at the war's end. British researcher and author Nigel Cawthorne later interviewed a Laotian named Le Thom, living in London, who had been held next to two Americans who were being interrogated by Soviets. A guard officer named Lieutenant Nguyen Tan Hat informed him that the American POWs were being prepared for transport to the Soviet Union. The guard later informed Le Thom that the Americans he had seen 'were only two of a number of American prisoners who were sent to the USSR.'[33] No U.S. prisoners of war interrogated by the Soviets were released at the war's end.

The Washington Special Actions Group (WSAG) had met on January 29, 1973, at the White House, two days after North Vietnam released their list of U.S. POWs in Vietnam. Admiral Thomas Moorer, Chairman of the Joint Chiefs of Staff in 1973, testified two decades later on the certain cases of POWs in Laos which in reality masked some 300 more U.S. prisoners believed to be held there. Senator John Kerry questioned Moorer:

"Admiral, you attended a WASAG meeting, as they're called, the Washington Special Actions Group, on January 29th at the White House. And as you recall, that meeting took place two days after

Vietnam released the list of U.S. POWs in Vietnam...it was three days before Laos produced its list...And Dr. Kissinger turned to you for your thoughts on the situation in Laos. And you said to him, quote, we're hoping for 40 more on the list of those in Laos. So you had an expectation of some 40 people coming out of Laos...before they gave you the list."

Moorer responded: "Well, I don't recall saying specifically that, and this is the first time I've seen those minutes...IF I had a number in writing it came from DIA..."

Kerry then said: "Now, the reason obviously I'm asking this is, we've had people making allegations of hundreds of prisoners being held in Laos..."

Moorer replied: "Well...that paragraph says, from the Defense Department, not from me."

Kerry said: "It says Defense. And Defense was represented by Admiral Murphy. But Admiral Murphy says he doesn't recall making that statement..."

Moorer denied any memory of the meeting: The Deputy Secretary of Defense (William Clements) was at all WASG meetings... could have been (him)...but Defense is not me..."

Kerry replied: "The two spokespersons for Defense were Secretary Eagleburger...(and Clements)."

Admiral Murphy (later assistant to Vice President George Bush) then interrupted: "I have no knowledge."

Admiral Moorer ended on a challenging note: "I'm not stonewalling you. I don't recollect anything about how that came up... We would have had to go to Laos and go to the caves and go all over the place, and we could start tomorrow and put 500,000 troops back in there in Indochina...That's the only way we're ever going to get an answer to the question..."[34]

At another point in the hearings Senator Robert Smith addressed Admiral Moorer: "For several weeks before the signing of the accords in January of 1973, General Eugene Tighe, who was then head of intelligence for the Pacific command, was asked by the Joint Chiefs to make a list of American POWs that we could reasonably expect to be repatriated both from Laos and Vietnam. The list contained some 900-1,000 names, yet, as we know, only 577 showed up on the Vietnamese list. Now, Tighe has testified in depositions that he was 'shocked' by the difference in those two lists...Were you shocked at that? Admiral Moorer answered: "Well, at least I was surprised..."[35]

The Vietnamese Communists had long demanded billions of dollars in U.S. war reconstruction aid, and while only mentioned in the agreement specifically, this was finally promised by President Nixon in a secret February 1, 1973 letter to Pham Van Dong which was later delivered by Henry Kissinger, and which remained "classified" until 1977. Vietnam clearly expected this aid before they would return certainly hundreds more, and probably over 1,000 American POW/MIAs confined in the secret parallel prison system, whom they publicly denied holding prisoner. At least 80 irrefutable test cases of known-POWs from among 1,277 suspected U.S. POWs not on the Vietnamese list of POWs they had admitted to holding, were taken by Kissinger to Hanoi on February 11th, but the Vietnamese subsequently refused to release any of these men. The draft of Nixon's promise to Hanoi was kept SECRET from Congress under authority of President Ford's national security adviser (and former military assistant to Henry Kissinger in January 1973), General Brent Scowcroft, until 1977.

On February 1, 1973 five days after the signing of the Accords, both sides met in Paris again to settle the fate of the missing POWs. The Americans were expecting a list of many more prisoners still held in Laos and came prepared with a promise to pay for them in the form of the letter from President Richard Nixon. The Vietnamese presented the Laos list, which contained the names of only ten men, nine Americans and one Canadian (all of whom had actually been held by the North Vietnamese, not the Pathet Lao).

It has not as yet been revealed, by any of the participants at the conference, whether the United States delegation made any request for information from the Vietnamese on the U.S. prisoners who had been reported by NSA analysts to have been shipped to the Soviet Union or to Communist China. Nor has there ever been a satisfactory explanation by any of the participants as to why more questions were not asked publicly about the large number of nonreturned Americans who had been used as forced labor by the North Vietnamese Army on the Ho Chi Minh Trail system, or held in secret Vietcong camps as far south as the U Minh Forest, at the southern end of the Mekong Delta.

Yet, the American delegation under Henry Kissinger handed the Vietnamese Communists a letter from President Nixon promising between $3.25 and $4.75 billion in war reconstruction aid. It appears that Kissinger and the American negotiators were unprepared for such an obvious lie on the number of prisoners remaining, and apparently had no previously prepared fallback position for such an eventuality. Yet they should have been ready. Kissinger and most of his negotiating team would have known something of the difficulties over POW repatriations following WW II and Korea, aside from the continual State Department requests to Moscow for over two decades about Cold War prisoners who had disappeared in the USSR. With the massive intelligence on the existence of some 250-350 US prisoners in Laos, and hundreds more held in South and North Vietnam, China and the USSR, which was available to Kissinger and his staff upon request, it remains incomprehensible why the United

States offered so much in return for such a blatant communist lie. The failure to reveal to the public the real numbers of both known and suspected U.S. prisoners remaining in Vietnam, or that hundreds of others had been shipped to the USSR, China and elsewhere must, however, also be viewed in the context of the Cold War mentality of the time.

The only possible justification for such secrecy by Richard Nixon and his representative, Henry Kissinger, or of his personal staff, including Lawrence Eagleburger, Winston Lord, John Negroponte, Brent Scowcroft and others who acceded to it, would be a fear that revealing the truth about secretly-held American POWs could ultimately cost far more lives than would be saved. Those in the NSC, JCS and CIA who had knowledge of similar historic precedents with POW/MIAs, and access to classified material on U.S. prisoners at the time, would have had an impact upon such a decision. The domestic situation in America in 1973 was volatile and unpredictable, which gave the policymakers all the more reason to say to each other: There is nothing we can do.

This "Unilateral document," which was to be kept secret from the American people until the Carter Administration, was the basis of all subsequent demands by Hanoi for American war reconstruction aid. This was the actual substance of Nixon's and Kissinger's years of diplomacy, paid for by the deaths of 20,000 additional American fighting men since 1969, in achieving what was to be called "peace" and the alleged repatriation of "all American POWs." The Nixon letter, which was "classified" in the United States until 1977, read as follows:

"THE GOVERNMENT OF THE UNITED STATES OF AMERICA WILL CONTRIBUTE TO POSTWAR RECONSTRUCTION IN NORTH VIETNAM WITHOUT ANY PRECONDITIONS...PRELIMINARY UNITED STATES STUDIES INDICATE THAT THE APPROPRIATE PROGRAMS FOR THE UNITED STATES CONTRIBUTION TO POSTWAR RECONSTRUCTION WILL FALL IN THE RANGE OF $3.25 BILLION IN GRANT AID OVER FIVE YEARS. OTHER FORMS OF AID WILL BE AGREED UPON BETWEEN THE TWO PARTIES."

Under a separate heading of "Note Regarding Other Forms of Aid," later in the letter is the following clarification:

"IN REGARD TO OTHER FORMS OF AID, UNITED STATES STUDIES INDICATE THAT THE APPROPRIATE PROGRAMS COULD FALL IN THE RANGE OF 1 TO 1.5 BILLION DOLLARS DEPENDING ON FOOD AND OTHER COMMODITY NEEDS OF THE DEMOCRATIC REPUBLIC OF VIETNAM."[36]

A total of 113 American personnel were officially listed as "U.S. Military Personnel Who Died While Captured in the Vietnam War, 1957-1985," in a September 5, 1985, U.S. document, yet the North Vietnamese list of 'Died in Captivity' produced at the Paris peace talks listed only 55 Americans in that category. The 58 other names on this list are among those American losses which are still considered discrepancy cases. There were, however, many other discrepancy cases.

From the 1967 East German/Soviet offer, Sejna's intelligence from Czechoslovakia, intercepted communications and analysis, it was known within a small circle of the U.S. intelligence community at the time that a substantial number of missing American POWs, reported as 200-300 or more, had already been (and were no doubt still being) shipped to the Soviet Union as partial Vietnamese payment for the massive war aid provided by Moscow. But the specific details obtained from intercepts on this aspect of the POW matter were highly classified and compartmentalized; eventually to be filed away and all but forgotten, except by those who had produced the intelligence.

It was a secret which could conceivably threaten the existence of the U.S. through a possible escalation of tensions to a nuclear confrontation with the Soviet Union, if revealed. This existence of the Soviet "threat" was one of the underpinnings of U.S. policy toward secretly-withheld military POWs; by now a decades-old policy which dictated that a military confrontation with the Soviet Union and its allies over POWs was not worth the potential cost to the people of the United States. The government officials charged with the responsibility to recover all POWs were also faced with the responsibility to protect the lives of all Americans, by keeping the nuclear peace intact.

Known American prisoners carried on secret government lists as "M-B," or "MOSCOW BOUND," were referred to within the confines of the U.S. intelligence community as "manna from Lenin," according to retired USAF and NSA intelligence analyst Jerry J. Mooney. High level officials of the time, such as Richard Nixon and Henry Kissinger do not speak of the MOSCOW-BOUND or CHINA-BOUND American prisoners in their memoirs, in the same manner that American officials of earlier wars had also largely avoided this sensitive subject.

While the North Vietnamese admitted to having only nine Americans in Laos on their 1973 list, they denied any knowledge of, or control over hundreds more Americans who had to be held either by Hanoi's Pathet Lao allies or by the North Vietnamese Army in Laos. The Pentagon admitted to having expected at least 53 names on the list of Americans who, out of the nearly 600 Americans missing in Laos were considered as held alive there. None of these 53 names appeared on the list of ten names given to the US. The report of a later Senate Select Committee on POW/MIAs stated:

"The DIA talking points prepared for Dr. Kissinger stressed the fact that the prisoners on the DRV/Laos list (of 9 Americans) had been captured not by the Pathet Lao, but by the North Vietnamese. The DIA also stated that approximately 215 men from the 350 U.S.

personnel missing in Laos 'were lost under circumstances that the enemy probably has information about their fate.[37] This was a euphemism for believed to have been captured alive, and did not include many more Americans "presumed killed," "body not recovered" in Laos. A June 1972 interagency report completed under Harriman's former deputy and Kissinger's assistant William Sullivan had concluded that American POWs were being used as leverage, and speculated on the possibility that U.S. POWs would be retained as bargaining chips. The report cited Laos as a particular area of concern because of the more than 300 Americans listed as "missing" in that country. Despite indications that some U.S. POWs captured in Laos had been moved to North Vietnam, there was reason to believe that a number of Americans could still be held as prisoners in Laos.[38] In return for the deliberately insulting 1973 list of 9 American POWs from Laos, the secret letter, hand-delivered from the American President by Henry Kissinger, had committed the US to paying between $3.25 Billion and $4.75 billion in postwar reconstruction aid to North Vietnam.

On February 2, 1973 President Nixon cabled North Vietnamese Prime Minister Pham Van Dong:

"The list of American prisoners held in Laos which was presented in Paris on February 1, 1973 is unsatisfactory. U.S. records show that there are 317 American military men unaccounted for in Laos and it is inconceivable that only ten of these men would be held prisoner in Laos. The United States side has on innumerable occasions made clear its extreme concern with the prisoner issue... It should also be pointed out that failure to provide a complete list of prisoners in Laos or a satisfactory explanation of the low number thus far presented would seriously impair the mission of Dr. Kissinger to Hanoi.[39]

No record exists in the National Security Council or White House files of a response from the North Vietnamese government to this message or of further threats about Kissinger's upcoming visit. A military aide, Colonel Guay, who delivered Nixon's cable to the DRV officials, said the Vietnamese subsequently complained of the "difficulty involved in finding pilots."[40]

Meanwhile, the secret air war in Laos continued, and on February 6, 1973, an EC 47Q US Air Force reconnaissance aircraft, codenamed "Baron 52," with seven crewmen aboard, was shot down over Laos. Four of the crewman bailed out and parachuted to safety but were soon captured by North Vietnamese regular troops. The other three members of the crew were killed in the crash. Yet, the only real concern in the Pentagon about the 4 American captives was whether they were to be listed as held by Hanoi and covered under the provisions of the Paris agreements, or if they were to be ignored and their existence kept secret. The Acting Secretary of

Defense at the time, William Clements (later Governor of Texas), decided to cover up the existence of these prisoners. First the Pentagon declared all seven men as killed in action, although intelligence clearly indicated that four were alive. Secretary William Clements then ordered all persons with knowledge of the incident to remain silent. It was only because Pentagon official Roger Shields refused to comply with this decision, that all seven men were finally relisted as missing in action.[41]

What had actually happened to Joseph Matejov, Peter Cressman, Dale Brandenburg and Todd Melton, the survivors of the EC 47Q "Baron 52" shootdown, was that they were captured by the North Vietnamese Army, and according to analysis and reporting of decrypted enemy communications by the NSA intelligence professional Jerry J. Mooney, they were captured alive and processed by the North Vietnamese army as "Moscow-Bound" U.S. POWs. This information was transmitted through channels from NSA to DIA for the use of American negotiators in Paris The four EC 47Q crewman had in fact already been added to a list of 17 other names of Americans carried by NSA as "Moscow-bound" and "China-bound," and this 21-name list, which was openly referred to within NSA as the "Kissinger list," was forwarded through proper channels for the use of Henry Kissinger in his negotiations for the unaccounted-for American POWs and MIAs. In sworn testimony before a 1992 Senate Select Committee, indicated that he had not seen received solid intelligence on Americans remaining alive in captivity after the 1973 peace accords, or reported as transferred to the USSR or China.

The CIA or DIA did not then disagree with Mooney's or NSA's analysis of the intelligence, yet 19 years later, a DIA analyst named Robert DeStatte, who for years had debunked many eyewitness POW livesighting reports in Southeast Asia, would state under oath to the U.S. Senate, as he had said in previous sworn testimony, that he, (alone in the entire U.S. intelligence community), didn't believe what he referred to as "the musings of Jerry Mooney" of NSA in regard to the four "Moscow-bound" survivors of the Baron 52 incident. Thus, he justified his previous sworn testimony to Congress that "no evidence" existed within the U.S. intelligence community that American POWs had remained alive in enemy control after Operation Homecoming in 1973.[42]

As a producer rather than a consumer of intelligence, NSA is sometimes asked for intelligence that is never used. In early 1973, before the actual signing of the Paris Peace Accords, the analyst concerned was ordered by his chief to produce a list of MB's and CB's" with a memorandum for the U.S. negotiating team in Paris. It was to be a short list of "headliners," from among the some 310 U.S. POWs on Mooney's list only, who were known through intercepted and decoded enemy communications to have been captured alive for their hi-tech intelligence value and processed by the North Vietnamese Army as Moscow-Bound and China-Bound, or withheld in secret camps in Indochina. Four more names were added to a list of 17 in early February, after the signing of the Paris accords by Henry Kissinger and Le Duc Tho. Of the 21 names on the final list assembled by Mooney for NSA, 14 were carried as Moscow-Bound and 7 as China-Bound; also, the first four had been captured from an EC-47Q aircraft on February 5, 1973, while the remaining 16 were earlier F-111, and other high-tech captures. Within Mooney's area of NSA this came to be called the "Kissinger list." It was either Frank Buck, Mooney's Branch Chief at NSA's B-64-1, or Wally Marshall, his Division Chief at B-64 who had first said to him: "Since Kissinger's running the show, we've got to get his attention; because this POW thing won't mean much when whatever number they decide to release come home. So let's send him a Russia and China list. If we go to Kissinger with MB's and CB's and say Morrissey and Brown are in the hands of the Soviets; he'll have to ask about them, and remember they've got the SALT and ABM negotiations to think about."

Mooney subsequently compiled and passed this list up to branch level, and he remembered in an interview with the author years later that, "NSA's B-6-1 Laos were also doing POW names, different ones than mine, from other large areas where many U.S. aircraft were also shot down."[43]

The 21 U.S. prisoners on Mooney's portion of this list, who were not repatriated during Operation Homecoming later in 1973, were: Joseph Matejov, Peter Cressman, Dale Brandenburg, Todd Melton, Robert Brett (or William Coltman, one of these two died), Robert Brown, Robert Morrissey, David Cooley, Edwin Palmgren, Allen Graham, (or James Hockridge), Dennis Graham, Henry MacCann, James Barr, Ralph Disz, Willie Cartwright, James Dooley, Hugh Fanning, Stephen Kott and Charles Lee. While some of these names were long kept secret to protect the POWs, the announced end of the Cold War dispelled such a need and these names, supplied by this writer, had been passed to Soviet military and intelligence officers when the existence of an American POW from Vietnam at Saryshagan, Kazakh SSR was reported by Kommersant and Reuters in November 1991. (This list was also forwarded to the Department of State by the author in December 1991, with a request that Moscow be asked to ascertain the present whereabouts and welfare of these men.)

The February 1973 memorandum listing these 21 U.S. prisoners was not a regular report intended for distribution to other agencies such as DIA or CIA, but was intended as a "teaser" to ultimately invite a request from the consumer (Henry Kissinger and his negotiating team) for a much longer list of U.S. prisoners carried as M-B or C-B, which would be sent to the U.S. negotiators once the issue was raised with the Vietnamese. Insofar as Mooney is aware, no such request was ever received by NSA from the 1973 U.S. negotiating team in Paris, although the longer list and much other intelligence on Moscow-Bound and China-Bound U.S. POWS was made available to national intelligence officers and to DIA for the use of U.S. policymakers and negotiators. Kissinger later testified under oath to the U.S. Senate that "NO CONFIRMED REPORT OF LIVING AMERICAN PRISONERS EVER CROSSED MY DESK."[44]

Due to Hanoi's refusal to return all American prisoners, a new U.S. policy was to take effect in April 1973 that non-returned U.S. POWs were to be considered "presumed dead," following an official announcement by Deputy Assistant Secretary of Defense Roger Shields. But, in the meantime, Henry Kissinger did make an effort to demand an accounting for 80 American test-case POW/MIAs, out of the some 1,200 suspected by the U.S. to have been captured alive.

On February 10, 1973, Kissinger flew to Hanoi from Vientiane, Laos, where he had "publicly called for strict implementation of the Paris Agreement for an early ceasefire in Laos." In his memoirs Kissinger recalled: "The purpose of my journey to Hanoi in February 1973 was to encourage any tendencies that existed to favor peaceful reconstruction over continued warfare, to stabilize the peace insofar as prospects of American goodwill could do so." American prisoners were still being held in Hanoi during Kissinger's visit, as the Operation Homecoming releases of the token 591 American POWs were not to be completed until late March.

A later Senate reports stated:

"In preparation for Dr. Kissinger's trip to Hanoi, the DIA prepared a list of 80 individuals, many of whom the agency listed as POW but who were not on the January 27 DRV or Viet Cong lists. In some cases, these were individuals who had been photographed or interviewed while in North Vietnamese custody...The DIA talking points prepared for Dr. Kissinger stressed the fact that the prisoners on the DRV/Laos list had been captured not by the Pathet Lao but by the North Vietnamese. THE DIA ALSO STATED THAT APPROXIMATELY 215 MEN FROM THE 350 U.S. PERSONNEL MISSING IN LAOS 'WERE LOST UNDER CIRCUMSTANCES THAT THE ENEMY PROBABLY HAS INFORMATION REGARDING THEIR FATE." This again was a euphemism for known to have been captured alive.[45] It appears that in giving Kissinger the "headlined" 80 test cases out of some 1,200 Americans believed to have been captured alive but not repatriated, they hoped to get North Vietnam to admit to the PRINCIPLE that some Americans held prisoner would not be repatriated until the agreements reached in Paris by the United States were honored.

In later Senate testimony Kissinger stated under oath:

"I took the 80 names that we had that we considered discrepancies to Hanoi on February 11th, which was ten days after we had received the Laotian list. AND I BROUGHT WITH ME THE FOLDERS OF 19 ON WHOM WE THOUGHT WE HAD PARTICULARLY GOOD INFORMATION."[46]

After an initial meeting with Le Duc Tho, his long-time adversary in Paris, Kissinger later met several times with the North Vietnamese Premier, Pham Van Dong, Ho Chi Minh's successor since 1969. What he was there to talk about in actuality was POWs and money; the terms and conditions of American postwar aid and the connection of this aid to the release of the remaining US prisoners known by American intelligence to have been alive in captivity, but who had not returned or appeared on any list. In reality there were many hundreds of Americans in this category, probably over a thousand all told, but Kissinger was to argue with the communists over only a relatively few test cases. In his published account Kissinger admitted failure in one of the major reasons for his February 1973 trip to Hanoi:

"Equally frustrating were our discussions of the American soldiers and airmen who were prisoners of war or missing in action. WE KNEW OF AT LEAST EIGHTY INSTANCES IN WHICH AN AMERICAN SERVICEMAN HAD BEEN CAPTURED ALIVE AND HAD SUBSEQUENTLY DISAPPEARED. The evidence consisted of either voice communications from the ground in advance of capture or photographs and names published by the communists. Yet none of these men was on the list of POWs handed over after the Agreement. Why? Were they dead? How did they die? Were they missing? How was that possible after capture? I called special attention to the nineteen cases where pictures of the captured had been published in the Communist press. Pham Van Dong replied non-comittally that the lists handed over to us were complete. He made no attempt to explain discrepancies. Experience had shown, he said, that owing to the nature of the terrain in Indochina it would take a long time, perhaps a year, to come up with additional information, though he did not amplify what the terrain had to do with the disappearing prisoners. We have never received an explanation of what could possibly have happened to prisoners whose pictures had appeared in Communist newspapers, much less the airmen who we knew from voice communications had safely reached the ground."[47]

Although this statement by Kissinger contains elements of the truth, as he saw it, the fact was that he and other White House officials must have been informed about the 474-478 non-returned American POW names in the letters collected by Melvin Laird, and that the number of missing US prisoners was far greater than 80, not only in North Vietnam but in other countries.[48] As national security adviser to the President (and later as Secretary of State), Henry Kissinger should have had access to American intelligence reports by DIA and CIA analysts which clearly indicated that from

200 to as many as 350 American prisoners had been captured in Laos by the North Vietnamese (and some by the Pathet Lao), who were not on any POW lists released by the communists. He should have been informed also, of the 1967 East German/Soviet offer to trade U.S. POWs in their control for Soviet spies, and about the information on American POWs transiting Prague on their way to the Soviet Union revealed by Major General Jan Sejna in 1968.

In Senate hearings conducted nearly two decades later, Senator Robert Smith questioned former Secretary of State Henry Kissinger about the larger number of known and suspected U.S. prisoners at this time:

"The DIA analysis of this list (turned over by NVN on January 27) two days after the lists were turned over to you, the U.S. personnel listed as MIA by DIA who were not either alive or dead on the list, total 1,277. Now did our government take up the question of the 1,277 where there was no proof that they were dead...Because I want to point out that these 1,277 are another group...beyond the 1,100 and some who were listed by our side, who we believed were KIA, based on our information...This is another 1,277, who at that time... we had nothing to indicate they were dead...Now what about those 1,277 people?"

Kissinger replied, under oath:

"I don't remember...The way it generally worked was this...I had enormous confidence in...Admiral Moorer, who was Chairman of the Joint Chiefs of Staff at the time...If any concrete issue came up on numbers, I would turn to the Chairman and the Defense Department, which was primarily responsible for the details."[49]

Henry Kissinger should also have been receiving NSA intelligence on many known and suspected American POWs who were never to reappear, and who had been carried in the category of "Moscow-bound," in the most secret intelligence compilations which were forwarded from NSA, through DIA, to intelligence consumers which included the White House, where Kissinger was the national security adviser to the President and in line to receive such information through his military assistants or directly from the Pentagon and CIA. Kissinger's military assistants in this period were General Brent Scowcroft (later NSC adviser to Presidents Ford and Bush), and General Vernon Walters (later Deputy Director of CIA and Ambassador to the United Nations). In later Senate testimony Vernon Walters stated, "I was translating what he (Kissinger) was saying into French...," but admitted little about POW intelligence.[50]

NSA analyst Jerry Mooney alone had carried some 60-75 U.S. POWs as Moscow-bound and states that his 'share of the analytical pie' on the M-B's was approximately 20%-25%. The fact is that the Soviet-held and Chinese-held U.S. POWs were a high-level "national security" concern which Kissinger or those reporting to him, such as

Admiral Moorer or General Scowcroft, would never be authorized to discuss or write about. The existence those prisoners who remained in Vietnam, Laos and Cambodia had become a political embarrassment, at a time when the Watergate scandal was threatening the Presidency of Richard Nixon.

The burglary of the Democratic Party's national headquarters at the Watergate apartments in Washington, D.C., had been carried out by former-employees of the CIA who were working for a secret White House intelligence unit known as the "plumbers," who were connected to the Nixon White House's Committee to Reelect the President (CRP). Among them was E. Howard Hunt, who had been involved in CIA covert operations since the Bay of Pigs operation, and earlier. From this point on, the emergence of the full details of the Watergate break-in and the subsequent cover-up of that act by officials of the Nixon White House, aided by the President himself, as it turned out later, was to have a profound effect on the final, inglorious chapter of the American intervention in Vietnam.

In his February 1973 meetings in Hanoi with Pham Van Dong, Henry Kissinger should have been aware of the NSA's decryption intelligence on POWs when he talked about a relatively small number of American POW "discrepancy cases." Since he had raised a number of these 80 cases, Kissinger obviously knew, but could never bring himself subsequently to admit, that the North Vietnamese had retained some number of US POWs for bargaining power. Mooney believes that a substantial percentage of these 80 cases were from his NSA lists, including some of those he carried as sent to the Soviet Union or the People's Republic of China. Pham Van Dong's reply was recorded by Kissinger in his memoirs:

"First of all, I would like to express my suspicion... I will speak very frankly and straight-forwardly to you. It is known to everyone that the US had spent a great amount of money in regard to the war in Vietnam. It is said about $200 billion, and in conditions that one would say that the Congress was not fully agreeable to this war. When the war was going on then the appropriation was so easy (laughs), and when we have to solve now a problem which is very legitimate... then you find it difficult."[51]

Kissinger later told a Senate Committee that his hope at the time was that:

"After all this anguish of war...there might be a period in which they would turn to the reconstruction of their country...And in fact when I went to Hanoi in February, that was one of my hopes. I remember one of the newsmen accompanying me on the plane said, what you're really hoping for is that Pham Van Dong, who was then Prime Minister in Hanoi, would turn out to (be) a Chou En-Lai, and I said that's right, that's what I would like to see happen."[52]

But it was not to be.

A spokesman for the Communist Pathet Lao, Soth Petrasy, announced their POW policy on February 19, 1973: "If they were captured in Laos they will be released in Laos." On February 21st, at a press conference, Petrasy further stated: "WHATEVER THE U.S. AND THE NORTH VIETNAMESE AGREED TO REGARDING PRISONERS CAPTURED IN LAOS IS NOT MY CONCERN. THE QUESTION OF PRISONERS TAKEN IN LAOS IS TO BE RESOLVED BY THE LAO THEMSELVES AND CANNOT BE NEGOTIATED BY OUTSIDE PARTIES OVER THE HEADS OF THE LAO."[53]

Petrasy further stated that the Pathet Lao had a detailed accounting of the American POWs and where they were held and that their release was conditional on negotiations between Laos and the US. The United States Government up until the 1990s never subsequently negotiated with Laos for living American POWs and none of the Americans held by the Pathet Lao were ever subsequently released (unless there were secret returns of POWs from Laos). Their number has been estimated on the basis of Pathet Lao statements, later indigenous refugee livesightings and highly trained and experienced US intelligence analysts, as 290-340 or more.

James Schlesinger, who was CIA Director for a short time in early 1973, and subsequently Secretary of Defense by the middle of that year, testified nearly two decades later to a Senate committee investigating POW/MIAs of this period:

"I had a long relationship with Henry Kissinger. We were classmates at Harvard in the same year. He was...tempestuous. But for the most part he regarded me as a bureaucratic ally...But even his bureaucratic allies he tended to regard with a degree of suspicion. He was not a man who reposed enormous trust in anybody else... He was very fond of Elliott Richardson, but he tended to erratically, or seemingly erratically play favorites and move against one element of the bureaucracy using another element of the bureaucracy... By February or March he had, really deeply felt he had gotten these (Vietnam) issues behind him, and he could tend to things that were perhaps more pressing: the Soviet Union, Middle East and so on.

"The monitoring of the agreement was the responsibility primarily of others, including the Department of Defense, to see that it was carried out. This moved to a lower level of the bureaucracy than the President and his chief foreign policy advisor (Four-Party Joint Military Commission)...but the CIA was to monitor this. The State Department was to make representations if they were not living up to the agreement...If there were violations they were to come to the White House."[54]

When asked about POWs left in Vietnam, Schlesinger later testified: "I had other things I was focused on in that period and Vietnam was not the most prominent question...In Southeast Asia, Laos, where we were conducting our own operations was as significant and probably more pressing than was South Vietnam. We were supporting...Vang Pao. And we were conducting airstrikes... the Military assets were...calling on the Department of Defense for strikes out of Thailand..."[55]

In answer to a question about his being briefed as DCI on CIA intelligence relating to American POWs in Laos, and who at CIA would have been in charge of keeping track of POW/MIA intelligence in the spring of 1973 Schlesinger testified: "I don't recall...I may have discussed it to some extent...I don't know...There was the Far Eastern Division, that was under (blacked-out)...(and) my assistant or... deputy for Vietnam Affairs." When asked if there had been concern expressed at CIA about the obviously incomplete list of POWs for Laos, Schlesinger said: "I don't recall it, but that does not mean that it was not brought to my attention..."[56]

Following the signing of the Paris Agreement on ending the war in Southeast Asia on January 27, 1973, and the February 1st "Unilateral Document," the North Vietnamese had released a total of 591 US prisoners of war who had been held in Vietnam up to March 28, 1973. However, the American government knew early-on, from the grossly incomplete lists of prisoners scheduled for release by Hanoi, that many US POWs known to have been captives in Vietnam and Laos were doomed to disappear unless Hanoi adhered to provisions in the agreement for accounting for the missing. Later Senate testimony by former Secretary of Defense Melvin Laird revealed that the Pentagon had collected over 5,000 letters from American POWs in Indochina all through the war years, and the names of other American POWs revealed in these letters were used to compile lists of U.S. POWs expected to be returned. According to Secretary Laird's later sworn testimony, 474 or 478 American prisoners of war known from these letters to have been in captivity, were never released by the North Vietnamese.[57]

One of the U.S. POWs released on March 14, 1973, in the ninth group of Americans to be freed during Operation Homecoming, was Air Force Major Laird Gutterson, who had been shot down over North Vietnam during the Tet Offensive of 1968. Gutterson's biggest concern was for those Americans known to have been alive in captivity, who later vanished:

"Some guys were brought into New Guy Village for torture and we would never see them again. They either died or were taken elsewhere. Several who became mental cases were removed from the prison. Guards told us they had been taken away and that they had died, but you can't believe that since some of those they said had died, showed up later in the system. Other prisoners might have been isolated for other, unknown reasons."[58]

Major Gutterson had personally caused the release of one secretly-held American prisoner. He was a civilian named Bobby Joe

Kieze who had been captured at Hue in 1968 and had been held for five years, whom Gutterson discovered in his last week of captivity, during a short period of relaxed control, when he saw a guard taking food to an isolated cell and followed him. When Gutterson crept up to the cell and asked the prisoner within his identity he found that Kieze had never been held with other Americans since 1968. He had escaped once, but had been recaptured and tortured. Bobby Joe Kieze was one of many Americans who never made it into the "System," that group of prisoners held in known prison locations, for ultimate release at the war's end. Kieze knew nothing about the ongoing POW repatriations. Gutterson and Colonel Jack Flynn immediately confronted the North Vietnamese prison administration about Bobby Kieze and forced his release, but how many others were secretly withheld, perhaps even in close proximity to those who were freed, may never be known.

Upon being freed, Colonel Gutterson and the other ex-POWs were given a direct order in writing to have nothing to do with POW/MIA organizations such as the National League of Families of POWs and MIAs in Southeast Asia, founded by Sybil Stockdale and others, or the west coast group VIVA (Voices In Vital America), which had originated the popular practise of wearing metal bracelets engraved with names of POWs and MIAs. Gutterson was also denied permission to speak out about the American POWs he and others believed had been left behind. Gutterson and all other returning prisoners were assigned military intelligence debriefing officers, who built up files on all experiences of the POW, and names of other Americans they had known in captivity. These files remained classified.

The reality on the ground in Southeast Asia was portrayed officially, but secretly, in a Pentagon-ISA memorandum entitled: US POW/MIA PERSONNEL IN LAOS, for the Secretary of Defense, that was originally prepared by Lieutenant Colonel (later Major General) Richard Secord, on March 23, 1973, for the approval and signature of then-Assistant Secretary of Defense Lawrence Eagleburger. Eagleburger, who had served Kissinger so loyally, was later to become the Deputy Secretary of State throughout the Bush administration and would ultimately ascend to the position of Acting-Secretary of State in August 1992, when James Baker resigned as Secretary to serve on President George Bush's reelection campaign.

Upon Richard Secord's return from conducting air operations for the CIA's secret war in Laos in late 1968, he had trained new pilots for the war and taught at the Naval War College before being assigned to the Pentagon in the Office of the Secretary of Defense in June 1972, in time to participate in planning for the December bombing of Hanoi. Secord became Chief of the Laos desk in the

Southeast Asia Branch, East Asia and Pacific Region, in International Security Affairs (ISA). In his memoirs Secord wrote of this assignment: "This meant advocacy as well as analysis, and, at my level, 'devil's advocacy' to make sure our bureaucratic flanks weren't exposed to unanticipated criticism."[59]

In the spring of 1973 Secord took over as head of the entire Southeast Asia Branch of ISA, and on his staff at this time was a civilian exchange officer from the State Department, named John Kelly (who, 13 years later would achieve some prominence as the U.S. Ambassador to Lebanon, also involved with the Iran-Contra scandal). Secord later wrote in his memoirs of this period, "OUR MAIN FUNCTION WAS TO PREPARE POLICY PAPERS FOR DEFENSE SECRETARY MELVIN LAIRD." (Laird was soon succeeded by James Schlesinger in this period, and during the interim Eagleburger and Deputy Secretary of Defense Clements held great authority.) The March 23, 1973 document on U.S. prisoners in Laos written by Richard Secord for the use of Lawrence Eagleburger and the Secretary of State, which remained classified from the American public for "national security" reasons until the summer of 1992, indicates how approximately 350 American POW/ MIAs in Laos officially disappeared. Because of its historic importance the full text of this document is recorded herein:

"On 1 February 1973, North Vietnam (NVN) released the names of ten POWs who were captured in Laos. North Vietnam claimed that these ten people were prisoners of the Pathet Lao, but DIA analysts indicate these individuals were actually captured by the North Vietnamese in Laos. The ten POWs identified consist of nine Americans (six USAF, one USN, two civilians) and one Canadian. The evidence indicates that most, if not all, of the ten are currently being held in Hanoi.

"DIA lists approximately 350 U.S. military personnel and civilians as missing or captured in Laos. Therefore the Pathet Lao Patriotic Front (LPF) list of ten POWs constitutes only a 2.5% accounting. IN CONTRAST, THE NORTH VIETNAM LIST REPRESENTS 45% AND THE PROVISIONAL REVOLUTIONARY GOVERNMENT (Vietcong) LIST REPRESENTS 20% OF THE POW/MIA PERSONNEL CARRIED ON OUR LISTS IN THESE RESPECTIVE AREAS. There is an obvious lack of reporting on the part of the Lao...Because of the foregoing statistics and analysis of the conditions under which our people have been lost, DIA concludes that the Lao may hold a number of unidentified U.S. POWs ALTHOUGH WE CANNOT ACCURATELY JUDGE HOW MANY. THE AMERICAN EMBASSY, VIENTIANE, AGREES WITH THIS JUDGMENT."[60]

"Several diplomatic moves have been made recently in an attempt to get an accounting and release of American prisoners being held in Laos. On 10 March 1973, the Lao delegation chairman in Vientiane informed us that recent U.S. demarches regarding prisoners in Laos had been conveyed to Souphannouvong, the Lao Communist Chief, personally. No other information was made available. On 15 March the Secretary of State instructed Ambassador (G. McMurtie) Godley (in Laos) to '...SINGLE OUT SOVIET AMBASSADOR FOR THE FULL TREATMENT...' regarding the lack of progress in Vientiane on political and POW matters. Godley was also instructed to inform the Soviet Ambassador in Laos that we '... CONTINUE TO HOLD NORTH VIETNAM TO ITS COMMITMENTS ON RELEASING ALL U.S. POWs IN LAOS BY 28 MARCH AND WILL NOT TOLERATE ANY FURTHER DELAYS...'' No Communist response to this line of action has been noted as yet.''[61]

"On 22 March 1973, the United States informed North Vietnam and the PRG (Provisional Revolutionary government of the Vietcong in South Vietnam) that the U.S. would '...COMPLETE WITHDRAWAL OF ITS MILITARY FORCES FROM SOUTH VIETNAM IN ACCORDANCE WITH THE TERMS OF THE AGREEMENT AND COINCIDENT WITH THE RELEASE OF ALL REPEAT ALL AMERICAN PRISONERS HELD THROUGHOUT INDOCHINA.' This statement was aimed directly at securing release of all U.S. prisoners held by the Pathet Lao. On 22 March, Ambassador Godley addressed the Laos POW/MIA question at length in Vientiane 2139 (attached). Godley separates the 1 February list of ten prisoners from the issue of accounting for the remaining POW/MIAs in Laos. The Ambassador states that the Lao Peoples Front (LPF) '...Just has not focused on the PW repatriation and accounting problem until very recently...'[62] "Godley concludes by recommending that we concentrate on helping the Royal Lao Government get an acceptable military protocol to the Laos cease-fire agreement approved by the Lao Patriotic Front.

"On 23 March, the U.S. delegation to the four-party Joint Military Commission (FPJMC) was instructed by Washington to reaffirm our negotiating stance. General Woodward was instructed to seek a private meeting with the North Vietnam representative and inform him that the U.S. must have assurances that the prisoners on the 1 February list will be released by 28 March. Given these assurances, private or otherwise, we will complete our troop withdrawals. This 23 March guidance to the field also indicates that '...WE INTEND TO PURSUE THE QUESTION OF OTHER U.S. PERSONNEL MISSING OR CAPTURED IN LAOS FOLLOWING THE RELEASE OF THE MEN ON THE 1 FEBRUARY LIST.'

"To review the bidding to date, THE U.S. IS PREPARED TO ACCEPT RELEASE OF THE TEN MEN ON THE 1 FEBRUARY LIST ALONG WITH OTHER U.S. PERSONNEL BEING HELD IN NORTH VIETNAM AS THE FINAL CONDITION FOR COMPLETE U.S. TROOP WITHDRAWAL. However, there has been no accounting of U.S. personnel MIA in Laos other than the 1 February list of ten who were probably all captured in Laos by the North Vietnamese Army rather than the Pathet Lao. Hence

, assuming all the prisoners currently being held in North Vietnam are released by 28 March, we still have the Laos MIA question remaining unresolved. Additionally, AMBASSADOR GODLEY PROPOSES THAT WE RELY UPON THE YET-TO-BE DEVELOPED AND APPROVED LAO MILITARY PROTOCOL AS A MEANS OF GAINING SATISFACTION ON THIS ISSUE. AND FINALLY, AMBASSADOR GODLEY DOES NOT DISCUSS Hanoi's INFLUENCE OR CONTROL OF THE (communist) LAO ON THIS ISSUE.

"...There appears to be need for a well-orchestrated plan for solving the problem of our Lao POWs and MIAs. Therefore, I am recommending below a series of diplomatic moves aimed at gaining a proper accounting of our men lost in Laos. YOU MAY WISH TO PASS ALONG TO THE PRESIDENT PART OR ALL OF THE FOLLOWING DIPLOMATIC TRACK:

A. After recovery of the last prisoners from North Vietnam, Hanoi should be advised unequivocally that we still hold them responsible for the further minesweeping activity as well as all future U.S. reconstruction assistance should be described as wholly dependent upon the accounting for and or release of U.S. prisoners being held in Laos. Once again, North Vietnam should be clearly informed that an accounting for ten men out of a total of more than 350 is considered unacceptable.

B. In the meantime (just after 28 March), a strong demarche should be made to the ranking Lao PF representative in Vientiane by the U.S. Ambassador personally. This initiative should plainly and forcefully that the U.S. will no longer play games with the POW issue in Laos. The Lao PF should be told that we have reason to believe that they hold additional U.S. prisoners, and we demand their immediate release as well as an accounting and information on all those who may have died. Finally, the Lao PF should be advised that failure to provide a satisfactory answer could result in appropriate United States actions.

C. Simultaneous with our representations to the Lao PF, THE U.S. AMBASSADOR SHOULD ALSO ASK THE USSR, PEOPLES REPUBLIC OF CHINA, NORTH VIETNAM, FRENCH, BRITISH AND ICC SENIOR REPRESENTATIVES IN VIENTIANE TO USE THEIR GOOD OFFICES WITH THE LAO IN ORDER TO AVOID A SERIOUS SITUATION.

D. Shortly after 28 March, assuming the Lao have not responded favorably, intensive and obvious tactical air reconnaissance of North And South Laos should commence. Additionally, the movement of a new carrier task force into the waters off Vietnam should be publicly announced.

E. Concomitant with the foregoing, the Lao PF and North Vietnam should be privately advised that Thai Volunteer Forces now in Laos will not be removed until there is a satisfactory resolution of the POW issue.

Other moves that may be worthy of consideration are listed

below...THEY SHOULD ONLY BE CONTEMPLATED IF WE ARE REASONABLY STRONGLY CONVINCED THAT THE PATHET LAO HOLD POWs:

"Ambassador Godley could be instructed to 'lean hard' on Souvanna Phouma and tell him to let the Lao PF know that political concessions in the new Provisional Government of National Union.[63] AS A LAST STEP, U.S. airstrikes and Lao and Thai irregular offensive operations could be resumed in Laos...to force the release of our prisoners in Laos."

Secord showed his skill at providing suggestions leading to deniability for senior U.S. officials involved with the POWs, by indicating that U.S. intelligence on large numbers of Americans held by the Pathet Lao was not adequate, when in fact the special and operational intelligence on the POWs in Laos, taken together, was massive and conclusive, but highly classified and compartmentalized by CIA and NSA. Thus did military officers in key roles gain promotion and influence, instead of resigning or otherwise facing the consequences of stating the truth about their fellow servicemen remaining in captivity:

"The  ...recommended diplomatic/military moves would represent a considerable toughening of the U.S. stance REGARDING OUR POWS BEING HELD IN LAOS. Such a line is even harder to take without a clear picture as to HOW MANY U.S. personnel are ACTUALLY being held in Laos. The intelligence data available is voluminous but IMPRECISE. However, the evidence indicates that the NVN/Pathet Lao forces have captured U.S. personnel since 1964, and the Lao PF have provided no prisoner or casualty data at all other than the ten names listed on 1 February. Therefore, the hard negotiating track outlined in steps A through E above, and POSSIBLY EVEN THE OPTIONAL STEPS A AND B, SEEMS CLEARLY JUSTIFIED. FINALLY, IT IS RECOGNIZED THAT THIS IS A VERY DELICATE SITUATION, AND THE APPLICATION OF ANY OR ALL OF THE ABOVE ACTIONS CANNOT ASSURE SUCCESS.-THERE IS, OF COURSE, LITTLE PHYSICAL RISK ASSOCIATED WITH OPTIONS A THROUGH E. IF YOU APPROVE THE ABOVE LISTED COURSES OF ACTION, I RECOMMEND YOU SIGN THE ATTACHED MEMORANDUM FOR Dr. KISSINGER. "(Signed, Lawrence S. Eagleburger.)[64]

Attachments to this document included a query by the Secretary of Defense, message #2139 from Vientiane, and a "proposed memo to Dr. Kissinger." Also noted on the document is the fact that Colonel Secord and Rear Admiral Bigley prepared this historically significant memorandum, which clearly illustrates how from 250 to as many as 350 living American prisoners of war in Laos were abandoned by officials of their own government in Washington, D.C., due largely to the dictates of politics and policy. Scarcely more than a week after writing this memorandum for Eagleburger, on April Fools' Day, 1973, Richard Secord pinned on the silver eagles of a full colonel, a promotion which even he was surprised by, according to his memoirs, claiming he did not have "the time in grade or sufficient 'friends at court.'"[65]

Secord makes no mention of the March 23rd memorandum or of his role in formulating U.S. policy towards the hundreds of American POW/MIAs in Laos who were not released by the Communists in the postwar prisoner exchange. (After later service as an adviser to the Shah of Iran's security forces in a war against the Kurdish people in the Iran-Iraq-Turkish border area in the late 1970s and early promotion to general officer rank, Secord would achieve national prominence as a Pentagon associate of rogue CIA agent Edwin Wilson and still later as a principal figure in the Iran-Contra scandal that plagued the Reagan and Bush administrations.)

Ambassador Godley and the CIA station Chief in Vientiane once more weakly demanded the return of the American POWs in Laos on March 24, 1973, but the Pathet Lao again ignored the ultimatum which had been presented on the 21st. In testimony years later before the same Committee, former Secretary of Defense James Schlesinger revealed the process by which the 350 known and suspected American POWs in Laos were written off as dead, while many actually remained alive in Communist captivity. The author learned this through staff investigators of the Committee at the time, but the information still remained classified. Portions of this testimony were reported in the Washington Post however, by columnists Evans and Novak, although the former Defense Secretary's name was not used:

"Three days after Admiral Thomas H. Moorer, then Chairman of the Joint Chiefs of Staff, sent a secret directive in late March 1973 to his field commanders to stop the withdrawal of U.S. forces from Vietnam, he reversed himself and ordered the war-ending troop withdrawal to continue. 'WHEN I LEARNED ABOUT IT,' a former secretary of defense told us this week, recalling the sequence of events, 'IT MADE ME FEEL A LITTLE SICK...IT WAS ASTONISHING. ONLY THE PRESIDENT COULD HAVE GOTTEN TOM TO DO THAT.'"

Later Senate testimony revealed, "... a Memorandum dated March 21, 1973, from General John R. Deane Jr. who was acting Director of DIA (and the son of Major General John "Russ" Deane who had served Harriman in Moscow in 1945—author) ...to the Chairman of the Joint Chiefs, Admiral Moorer (which) begins on page 779 of the Joint Chief of Staff files here at the office of Senate Security (with Shields name on it as Chairman of POW/MIA Task Force) within office of the Assistant Secretary of Defense for International Security Affairs...The day after General Deane's memo...Admiral Moorer's message of March 22 orders that the troop withdrawal be stopped unless and until the Pathet Lao provide the U.S. with a list of all U.S. prisoners held in Laos and set the time, date and place for their release."

"Admiral Moorer has testified ...that this was an order that was made by President Nixon and that Admiral Moorer was simply the transmitter...was made to mean that there was a very strong feeling that the list of nine prisoners for Laos weren't really from Laos, and that there were live POWs in Laos whose names had not appeared on any list...A message again (came) from Admiral Moorer the following day, March 23, 1973, this time to the Chief of the U.S. Delegation of the Four Party Joint Military Commission, General Woodward and others, essentially stepping back from the March 22 memo and indicating that the troop withdrawal would proceed on schedule, as long as the nine prisoners on the so-called Laos list were released, but not insisting on any information or on the release of additional U.S. prisoners believed to be in Laos."[66]

To this Schlesinger responded: "I did not know about either of these messages...One must only infer that there was a change of heart on the part of the White House..."[67]

Admiral Moorer and other senior Defense Department officials were aware that Hanoi had agreed to release all U.S. POWs and MIAs by March 28th, but were failing to comply. The American people were tired of the war and the President was already embroiled in the Watergate scandal which would ultimately cause his downfall. The final decision that the POWs were not worth the effort required to gain their release then, was made by President Nixon, as it had ultimately been made by previous presidents after Korea, WW II and WW I. President Richard M. Nixon, in a televised address to the American people on March 29, 1973, claimed that, as a result of the Paris peace accords negotiated by Henry Kissinger and North Vietnam's Le Duc Tho, "ALL OF OUR AMERICAN POWs ARE ON THEIR WAY HOME," but he also alluded afterward to the need for the communists to account for the missing:

"THERE ARE STILL SOME PROBLEM AREAS THE PROVISIONS OF THE AGREEMENT REQUIRING AN ACCOUNTING FOR ALL MISSING IN ACTION ...HAVE NOT BEEN COMPLIED WITH."

However, the impression given to the American people, despite much classified evidence to the contrary, was that all American prisoners of war had been released. To the public, the "missing" were in a nebulous category. This impression was reinforced by subsequent Nixon speeches on May 24th and June 15, 1973, in which he stated that "all our prisoners of war" had been returned home to America.[68]

On March 31, 1973, two days after Operation Homecoming was completed, the Pentagon's weekly status report, from the Comptroller's office of the Department of Defense, which (by his own subsequent testimony) reached Secretary of Defense Elliot Richardson, listed over 1,200 Americans as unaccounted for, including 81 listed as POWs under the category "Current Captured."

(This was the same category that had held 5,414 American POWs in the German Theater, not yet returned to U.S. control on January 1, 1946.) In reality nearly all of these nearly 1,300 Americans were probable, or suspected POWs, and not in a category of "presumed killed, body not recovered," but the 81 in particular were definitely known-POWs. The Nixon administration knew that at least 81 POWs had not been returned, and that actually these represented only test-cases from among 474-478 non-returned U.S. POWs known from other POWs letters, and hundreds more non-returned prisoners identified by correlating evidence from decryptions of enemy radio communications to 'loss incidents.'[69] Vice Admiral Jerry Tuttle, later a deputy director of the Defense Intelligence Agency and director of the Navy's Space and Electronic Warfare, later confirmed that many seriously wounded American POWs disappeared in Vietnam, and that the severely wounded, burned or those missing limbs failed to return in Operation Homecoming.

Kissinger blamed the emerging Watergate scandal, which eventually drove the president from office, for paralyzing Nixon at this crucial time when Vietnam and Laos were clearly planning to violate the provisions of the Paris accord concerning the return of all POWs and an accounting of the missing in action:

"It was a different Nixon in March 1973. He approached the problem of violations in a curiously desultory fashion. He drifted... Nixon clearly did not want to add turmoil over Indochina to his mounting domestic perplexities. The normal Nixon would have been enraged beyond containment at being strung along like this, but Watergate Nixon continued to dither....Nixon was simply unable to concentrate his energies and mind on Vietnam."[70]

In the end, how much information on the non-returned POWs and the MIAs believed to have been captured but not repatriated was actually given to Richard Nixon and Henry Kissinger, and how definitive it was, on the actual status of POWs and POW/MIAs can only be answered by them and by the military and CIA officers who gave it to them.

Sarah Frances Shay, the mother of U.S. MIA Major Donald E. Shay, lost over Laos on October 8th, 1970, took part in the last League of Families meeting with Henry Kissinger, in the period after the Paris accords had been signed and Operation Homecoming. She remembered later:

"I asked him what was being done about those still missing... He thought for a few seconds and said, 'The North Vietnamese have agreed to assume responsibility for accounting for the missing.'

"It hit me like a ton of bricks. Here he had been sitting, meeting with us for months, telling us not to believe anything the North Vietnamese told us. Now, all of a sudden they were wearing white hats and account for our missing men? We questioned, 'What about the men in Laos.' I don't remember his exact response but it was evasive. Again, he said something about the North Vietnamese taking responsibility... It was the last time we met with Kissinger. He would not agree to see us again....Kissinger's cutting us off was typical of what happened after 1973 through about 1978. After Operation Homecoming, the government went out of its way to shut us down... Only a few in Congress were interested in our cause...we couldn't get any kind of public support."[71]

In James Schlesinger's September closed-door sworn testimony before the Senate, nearly two decades later, he indicated that after becoming Secretary of Defense in May 1973, he was cut out of the loop in the later decision-making process, of writing-off the missing American POWs, which type of politically sensitive activity was usually handled by a few people from a Washington "Special Action Group" (WSAG), but in this case apparently included only President Richard Nixon, NSC adviser Henry Kissinger and Deputy Secretary of Defense William Clements. In his public testimony on September 21, 1992, Schlesinger was asked for the record whether he knew that American POWs had been left behind alive in Communist control in Indochina after the 1973 POW repatriations were completed. He answered: "AS OF NOW, I CAN COME TO NO OTHER CONCLUSION. THAT DOES NOT MEAN THAT THERE ARE ANY ALIVE TODAY."[72]

When asked at the same Senate hearings to remember his personal reaction to President Nixon's March 29th speech to the nation that all American POWs throughout Indochina had been released by the Communists, former Secretary of Defense Melvin Laird said: "I MUST SURMISE THAT I WAS SOMEWHAT SURPRISED THAT WE COULD ASSERT THAT ALL OF THEM WERE OUT...I THINK IT WAS UNFORTUNATE TO BE THAT POSITIVE. YOU CAN'T BE THAT POSITIVE WITH THE KIND OF INTELLIGENCE WE HAD WHEN I LEFT."[73]

Senate investigators also learned in 1992 that details of the process of declaring the secretly-held U.S. POWs in Laos dead were recorded on Richard Nixon's White House Oval Office "Watergate" tapes between March 21 and April 11, 1973, which are among those that he had been so reluctant to release to the public in subsequent years. Yet, Laird also testified that the Department of Defense had "letters, eyewitness reports or radio communications" with only 20 U.S. airmen in Laos, and since 10 Americans (held by the North Vietnamese) had later been released after having been imprisoned in Laos, Laird's implication was that only 10 Americans were known for certain to be in Pathet Lao captivity after the U.S. withdrawal.

An indication of acceptance of the official White House line that "all" of the hundreds of U.S. POW/MIAs actually captured in Laos "were dead," by American diplomats who obviously knew better, but showed no character or courage, is reflected in an April 5, 1973 cable to the Department of State from Ambassador G. McMurtie Godley in Laos, who had just been demanding the release of the Americans by the Pathet Lao:

"Although...U.S. PWs may be held in remote areas of PL (Pathet Lao) zone of control, we...received negative response. EMBASSY ACTIVITY IS CURRENTLY BEING DIRECTED TOWARD PROGRAM OF ACCOUNTING FOR MIAs..."[74]

Godley cabled Washington again the next day, April 6th: "WE HAVE TO ASSUME THAT ALL THE AMERICAN POWs LEFT IN LAOS ARE DEAD."[75]

Two decades later, former Ambassador G. McMurtie Godley, the "Proconsul" William Sullivan's successor in Laos in 1973, testified under oath to the Senate select Committee investigating POW/MIAs:

"During my ambassadorship, there were no regular non-American sources we could use concerning POWs and MIAs in Laos... The only reliable sources we had about MIAs or POWs were of course the Air Force reports as to losses over Laos and...Air America, which lost several men in northern Laos. Yet, Godley evidenced knowledge of many POWs at the "Tentacle MB" location, of Sam Neua, although possibly always in motion: "My recollection was that there were never any more than 25 to 27 POWs in Sam Neua. It was a fluid situation..."[76]

The same Committee (in September 1992) revealed in recorded and sworn testimony the contents of a still-secret document (from the Archives, the Nixon Library Project) indicating that on April 11, 1973, one day before Dr. Shields' press conference, he had a meeting in the Oval Office with President Nixon:

"The participants were scheduled to be Scowcroft, Shields and Nixon, and the document I saw was a set of talking points written by General Scowcroft for President Nixon...One of the talking points was a question for President Nixon to ask Roger Shields: Are there any indications that any of our MIAs are still alive...We didn't have that document when we deposed Shields." (This document was denied to the Committee on grounds of executive privilege, by none other than Brent Scowcroft, later national security adviser to President George Bush.)

Schlesinger's response to this information was: "A visit with the President tends to be helpful in persuading somebody to do what he is reluctant to do...It would be very flattering to a deputy Assistant Secretary of Defense to visit alone rather than in the entourage of the Secretary or the Deputy Secretary. This is...an exclusive meeting."[77]

At a news conference the next day, on April 12, 1973, Deputy Assistant Secretary of Defense Roger Shields carried out the unpleasant task of announcing that all the missing POWs were "dead." Deputy Secretary of Defense William Clements had instructed

Shields on what to tell the American press, prior to the conference, by saying to him with great emphasis: "ALL THE AMERICAN POWs ARE DEAD." Shields had then said: "WE CAN'T SAY THAT." Clements responded with the Nixon Administration's line: 'YOU DIDN'T HEAR ME. THEY'RE ALL DEAD." Shields dutifully followed orders, and thus inaugurated the new U.S. policy that all the known POWs were dead.[78]

Such a statement by the U.S. Government's official spokesman on POW/MIA matters had a profound impact on the Congress, which had been asked by Richard Nixon to approve $1.3 billion in postwar U.S. aid for Vietnam; what in effect amounted to the President's attempt to secure funds for a ransom of American POWs, whom the U.S. and Vietnam both publicly denied even existed. An Associated Press dispatch of April 13, 1973, which was carried in many American newspapers of the time said:

"Armed Services Chairman F. Edward Hebert...served notice he will introduce a proposal to prohibit any U.S. aid for Hanoi. The Louisiana Democrat also said JUSTIFICATION FOR PRESIDENT NIXON'S REQUEST FOR $1.3 BILLION AID TO SOUTHEAST ASIA SO FAR IS EITHER NEBULOUS OR NONEXISTENT."[79] Simultaneous tales of privation and torture of U.S. prisoners in Vietnamese prisons further discouraged such politicians, most of whose constituents were sick of the whole subject of Vietnam, from approving any U.S. aid that might reach Hanoi.

After the final releases, ex-POW Colonel Laird Gutterson was certain the North Vietnamese were still holding more Americans, but the Defense Department and the Air Force refused to allow him to speak out publicly, claiming that "negotiations are ongoing for any men left behind." At a White House banquet for the returned POWs and their families attended by 1300 people, including 587 ex-POWs, which was hosted by President Richard Nixon on May 24th, 1973, Colonel Laird Gutterson was seated near Henry Kissinger and chose to act on his concerns. He accosted the man who had negotiated in Paris for the prisoner exchange:

"Mr. Secretary, as a returned POW I've been sternly instructed not to speak out in public about the POW issue or about the possibility of other prisoners still being held. I'm wondering if there are, indeed, any talks underway that could be jeopardized if I or others speak out on the subject."

Kissinger refused to answer the question, saying only: "I do not choose to discuss that issue now," and abruptly turned away.

Colonel Gutterson later requested permission to speak about POWs left behind at the reserve officers annual meeting, a Pentagon Lieutenant General named Chappie James, who had long held responsibility in POW/MIA affairs, used foul and abusive language on Gutterson and warned him to, "Stop being a professional POW and get back on the team. Start following orders for a change."⁸⁰ Instead, following his retirement, Colonel Gutterson continued to speak out fearlessly, for two decades to come.

It was not until September 22, 1992, that Henry Kissinger's role in assuring that some American POWs were written-off as "presumed killed in action," would be revealed to the American people through his testimony under oath before the Senate Select Committee on POW/MIA Affairs. The former Secretary of State came to the hearings flanked by a team of lawyers which included, significantly, a former Carter Administration White House Counsel, Lloyd Cutler. Kissinger attempted to distance himself from the testimony of former secretaries of defense James Schlesinger and Melvin Laird. During the hearing, a formerly Top-Secret memorandum dated May 23, 1973 was revealed, in which it was clearly recorded that during the week that the Watergate cover-up started to unravel because of the "guilty" plea by Nixon aide Jeb Stuart Magruder, Kissinger had pressured North Vietnamese negotiator Le Duc Tho, in Paris to announce to the world that all the American POWs in Laos had been released.

Under cross-examination by the Select Committee Chairman, Senator John Kerry, Kissinger reluctantly admitted the top-secret document had been accurate in recording: THAT HE HAD REPEATEDLY REQUESTED NORTH VIETNAMESE NEGOTIATOR LE DUC THO "HELP US," BY ISSUING A ONE-SENTENCE STATEMENT DECLARING "THAT ALL THE PRISONERS IN LAOS HAVE BEEN RELEASED," and that "IT WOULD BE VERY IMPORTANT TO US," if the North Vietnamese did so for the Nixon Administration, then embroiled in the Watergate scandal. An indignant Kissinger rejected the suggestion by the Senate that he had not bargained hard enough for these and other POWs in Communist control after the repatriation of prisoners, accusing Senator Kerry of not displaying "GOOD GRACE" in questioning Kissinger's negotiating ability, for which he had ironically been awarded the Nobel Peace Prize. He accused Kerry of bias because of the Senator's anti-war activities as a 1973 leader of "Vietnam Veterans Against the War," ignoring the Senator's combat record as a navy gunboat commander, when he had been decorated for valor and twice awarded the Purple Heart for wounds received in action. Kerry thereupon responded: "Sir, this is a filibuster."⁸¹

When Kerry attempted to give Kissinger an easy way out, by calling attention to the weakening of the Nixon Administration's bargaining position on non-repatriated POW/MIAs due to the effects of the Watergate scandal, Kissinger responded by attacking the Congress of 1973, and it's then-prevalent anti-war sentiment, for preventing the Nixon administration from using military force to back up the U.S. negotiators in Paris under his direction. Given the quality and volume of intelligence information on American POWs made available to the White House-NSC, and indeed, information recorded in his own memoirs, Kissinger's sworn testimony on his knowledge of POWs remaining behind in captivity in Laos was difficult to reconcile, as he merely attempted to shift all blame to Hanoi's negotiators, who were, in fact, expecting war reparations that he and President Nixon had secretly promised, but failed to deliver:

"PERSONALLY, I HAVE NO PROOF WHETHER AMERICANS WERE KEPT BEHIND BY HANOI...IF ANY PRISONERS WERE HELD BACK THERE CAN ONLY BE ONE GUILTY PARTY—THE VIETNAMESE COMMUNISTS WHO FLAGRANTLY AND UNFORGIVABLY WOULD HAVE VIOLATED SOLEMN AND CLEAR-CUT AGREEMENTS."

This was a misleading statement, as witnessed by Kissinger's own acknowledgment that he had begged Le Duc Tho to get the North Vietnamese government to "help us" by falsely announcing that all U.S. POWs in Laos had been released. Kissinger also claimed to have no knowledge of any American POWs being held in the Soviet Union.[82]

Of the 591 POWs released, the southern Viet Cong—under the Provisional Revolutionary Government of Vietnam (PRG)—returned only 122 American prisoners of war, and despite the fact that most U.S. war losses were in South Vietnam, claimed that they had held only 40 more, all of whom had supposedly died in captivity. Yet 456 Americans were still listed officially as missing in South Vietnam by the Defense Intelligence Agency (DIA), of whom 22 were carried as known-POWs, even after the 122 POWs had come home from Viet Cong captivity in the South. (At least 500 other missing Americans (and possibly more than 1,000) from the war in South Vietnam had been declared "killed in action, body not recovered (KBNR)," although the proof of such a finding at the time was in many cases, lacking, and many of these men could also have been captured after being wounded and left behind on the assumption they were dead, and due to heavy combat occurring in their immediate vicinity.) Their families may later have received "unidentifiable remains," in caskets, and then been advised to keep caskets closed. Defense Department official Roger Shields later testified under oath to Senate Select Committee on POW/MIA Affairs investigators (in 1992) that he sent a memo to Assistant Secretary of Defense for ISA Robert Hill (Eagleburger's successor in control of POW/MIA matters) dated May 24th, 1973, the same day as Richard Nixon's POW banquet in the White House. In that memo, Shields complained that U.S. Department of Defense policy, that "ALL U.S. POWs WERE DEAD," had come from his statement at the April 12th press conference, and he told Hill, 'I don't think that's accurate.' Shields then brought up two examples the February 5, 1973 EC47-Q shootdown (Baron 52) and the Air America (CIA) pilot Emmett Kay, who had gone down in Laos later.

Former CIA Director and Secretary of Defense James Schlesinger later testified before the same Senate Committee:

"I do not recall Bob Hill ever bringing that up with me and I don't think that any of these subjects were brought to me until I was sworn in. (in July 1973) I THINK THAT THE BUREAUCRACY HAD PROBABLY CONCLUDED AND MAY HAVE BEEN GIVEN REASON TO CONCLUDE THAT THIS IS BILL CLEMENTS' BABY, BRING IT TO BILL CLEMENTS, OR DON'T DISTURB ANYONE ELSE, MORE PRECISELY." (Clements was then acting-Secretary of Defense) "Clements basically was running all the details in the Department....Bob Hill was an old Hill hand...a Republican politician...and I suspect that he kept back from me those things that he thought I should not be interested in...Clements wanted normally, regularly...to ingratiate himself with the White House, and Henry Kissinger could effectively use such an inclination, and he did...So I do not think that when you have a statement as hard and clear as Clements seems to have made that he had not been urged to make that statement by somebody at the White House."... "My recollection is that he tended to be called into Kissinger's office after the WSAG meetings or what have you, for pow-wows."[83]

In South Vietnam, on June 8th, 1973, a North Vietnamese Army "Hoi Chanh" (defector) revealed to the American press that he had seen six American prisoners of war in Communist control who had been secretly held back after the Operation Homecoming POW repatriations. In an interview of the North Vietnamese rallier which was attended by American reporters, a U.S. officer who was present urged the newsmen to downplay the story or not use it at all in their reports, What followed was a blatant example of government-suggested, self-censorship by the American mass media which, like similar actions in past wars, was to have a profound effect on history. A declassified U.S. Embassy cable from Saigon to the Secretary of State illustrates how the U.S. Government insured a news blackout for the American people on any eyewitness reports of U.S. POWs remaining in enemy captivity in Vietnam:

"NVA Rallier/Defector Nguyen Thanh Son was surfaced by Government of Vietnam to press June 8, Saigon. In follow on interview with AP, UPI and NBC American correspondents, questions elicited information that he had seen six prisoners whom he believed were Americans who had not yet been released. American officer present at interview requested news services to play down details: AP MENTION WAS CONSISTENT WITH EMBARGO REQUEST, WHILE UPI AND NBC, AFTER TALK WITH EMBASSY PRESS OFFICER, OMITTED ITEM ENTIRELY FROM THEIR STORIES. Details on rallier's account being reported (in) separate telegram through military channels by BRIGHT LIGHT message today." (Distribution of this message included a copy for Richard Nixon's White House.)[84]

By June 1973 the official list of 81 American POWs carried by the Comptroller of the Department of Defense as "Current Captured" out of more than 1,200 U.S. MIAs who were suspected POWs, had fallen to 67. This was the result of information received from POWs returning home who had identified some 15 as "died in captivity." A July 17, 1973 memorandum from Deputy Secretary of Defense William Clements to President Richard Nixon, revealed how the administration defined the "current captured" category:

"Presently, there are 1,278 military personnel who are unaccounted for in Southeast Asia. Of this number, 67 are officially listed as prisoner of war based on information that they reached the ground safely and were captured."[85]

Following this the military services came to Clements and requested that additional cases of POW/MIAs known to have been captured alive be moved to the POW category. (This amounted to a test of U.S. policy towards known POWs remaining in captivity) A Senate investigator later asked Schlesinger: "Mr. Clements told us that...beginning in July 1973, approximately 50-75 cases were brought to his attention by representatives of various service secretaries WITH PROPOSALS THAT THE STATUS BE CHANGED FROM MIA/KIA TO PRISONER OF WAR. AND ACCORDING TO MR. CLEMENT'S TESTIMONY, NOT A SINGLE ONE OF THOSE CASES SATISFIED HIM THAT THE STATUS CHANGE TO POW WAS APPROPRIATE."[86]

Asked about Clements' control of status changes, former defense secretary Schlesinger answered: "I had no knowledge this was going on....the denial of the 50 to 75 requests, in which all recommendations of the services to the contrary were rebuffed is certainly unusual...and is likely to have not been based on careful and meticulous examination of the evidence in the individual cases. It reflects sort of a broader policy....Clements tended to communicate to me as little as he possibly could, just the bare minimum. It was also brought to my attention that he would have these long sessions at the White House...I presume they started when Bill arrived in Washington, which was the beginning of '73."

A Senate investigator then commented: "Scowcroft was at the White House by that time. He became Kissinger's deputy in January 1973, when Haig went over to the Army...He would probably know a good deal about it..." But Scowcroft and Haig were to keep their secrets.[87]

Despite the evidence that the Nixon Administration had left American prisoners behind in captivity, former President Richard Nixon later protested to a Senate Select Committee, about its handling of an investigation of the POW matter:

"...To convey the impression to the hundreds of families of MIAs that an American President deliberately left behind their loved ones and that some of them might still be alive can only be described as obscene. The Committee owes to the MIA families and to history an honest statement of the facts with regard to POWs and MIAs. Throughout America's military history, casualties are divided into three categories-those known to be killed in action; those known to be and acknowledged by the enemy to be prisoners of war; and all others who are classified as missing in action. My statement on March 29 (1973) was true to my knowledge then and, in view of what I have seen of the Committee's work to date, is true now. Further, the fact that I was not satisfied with the accounting we received for MIAs was true then and is true now."[88]

Yet, by the summer of 1973, American POWs and MIAs, once known through intelligence sources to have been captured alive and never returned in Laos, including some on Mooney's 21-name 'Kissinger" or "Teaser" list of CERTAIN Moscow-bound American POWs, or on his "WHO TO ASK" 1,000-man list, and on his shorter 400+ list of POW/MIAs, 70% of which were believed to be in captivity in North Vietnam, and his 300+ list of KNOWN POWs, also 70% in North Vietnam; were artificially maintained by the U.S. Government as MIAs or "presumed killed," to allow deniability for political leaders. This constituted an arbitrary placement of them in a different category than they should have rightfully been listed under, that clouded for the future any real accuracy on how many Americans actually remained behind alive in captivity. Such was the legacy of official Washington's dishonesty about the U.S. POWs who had been abandoned. This was particularly true in cases where the Pentagon or DIA later conveniently 'lost' or even deliberately destroyed documentary evidence supporting a deserved change in their status to that of known prisoners of war. Therefore, figures used by the Pentagon in all succeeding years, for U.S. MIAs, POWs and 'presumed killed,' must be viewed by researchers and by the American people with extreme suspicion.[89]

The prisoners left behind alive in Laos were thus not the only U.S. POWs who were illegally withheld by the Communists in Indochina. There were also the names of 474-478 known-POWs of the Vietnam war who were not returned, that had been collected from letters assembled under authority of Secretary of Defense Melvin Laird, who later testified under oath before the U.S. Senate about them, and who thus emerges as one of the most honorable men to have served in the Nixon Administration. In addition, of the more than four hundred American MIAs known through decrypted enemy communications to have been captured alive by the NSA's Jerry Mooney and other intelligence analysts (of which some 305 were carried in a definite category of KNOWN POWs), ONLY 15% HAD BEEN CAPTURED IN LAOS, AND SOME 15% MORE IN THE NORTHERNMOST PART OF SOUTH VIETNAM. Thus, some 215 or more U.S. POWs captured in North Vietnam and known with certainty through 'special intelligence' to have been held in North Vietnam, were never repatriated.

Of the estimated 350 U.S. POWs believed to have been captured and held alive in Laos at the time of the 1973 Operation Homecoming, only some 45 were carried on Mooney's lists of known-POWs which had been gleaned from NSA-decrypted NVA Anti Aircraft Artillery unit interceptions; the rest were being handled and carried on lists by other NSA, DIA and CIA intelligence analysts. When Mooney occasionally asked some of his counterparts for the numbers of known-POWs they were carrying on their lists, he was told by several colleagues that he did not have 'a need to know' such information. Mooney's decryption analysis and reporting of his share of NVA radio traffic indicates approximately 600-700 or more American POWs in North Vietnam and Laos were not repatriated, and MOONEY HAS REPEATEDLY EMPHASIZED TO THE AUTHOR IN INTERVIEWS FROM 1990-1992 THAT HIS ANALYTICAL AND REPORTING WORK FOR NSA COMPRISED ONLY 20%-25% OF THE "ANALYTICAL PIE" ON U.S. POWs in INDOCHINA, although he admits there could have been some double-counting by other analysts who did not have knowledge of his work.

Thus, from 1,200 to as many as 2,000 or more American POWs may have been secretly withheld in North Vietnam and Laos at the end of the war, some of whom were known to have been executed or died in captivity, but most of whom were still held alive in many separate groups, from the Chinese border on south to the Panhandle region. This figure squares with numbers quoted in now-declassified CIA documents from CIA human sources inside North Vietnam during the later war years, which are quoted within this book. The approximately 45 U.S. POWs from Mooney's AAA communications decryptions of NVA units in northern South Vietnam's Quang Tri and Quang Nam provinces, on his list of 305 certain POWs out of the 400 or more on the list of suspected U.S. POWs minus those executed or died in captivity, which in turn had been reduced from a list of over 1,000 U.S. POWs on the "WHO TO ASK LIST." This was the actual number of SUSPECTED U.S. POWs Mooney had assembled, and they were only a portion of the American prisoners held in South Vietnam who failed to return from captivity.

The war in South Vietnam had involved an average of 10 or more U.S. Army and Marine Divisions, with all their combined air and ground support, in combat for seven years from 1965-1972, along hundreds of miles of the Cambodian and Laotian border areas, and throughout South Vietnam. This resulted in the disappearance of many hundreds more U.S. POW/MIAs who were suspected or known by U.S. intelligence to be alive in Viet Cong or NVA captivity in the South, but who were not repatriated during the 1973 Operation Homecoming, or thereafter. Some of these POWs were reported by later refugees to have been moved to North Vietnam after the war. In addition, many more American prisoners of war were taken in clandestine U.S. cross-border raids, or "Black Operations" inside Cambodia and Laos, who vanished forever. In some reported cases their names were not even acknowledged, or known to be men serving in U.S. forces, as they were part of a vast 'covert operations' network, which required no public accountability for losses. In some cases families may have later received what were termed "unidentified remains." Others missing in Vietnam were later accounted for on lists of "other losses" of U.S. military personnel lost 'world-wide' during the war years.[90] Much information on these U.S. POWs still remains classified at the time this book goes to the press.

American POWs also continued to vanish in Cambodia (and in Laos) up to and after the signing of the Paris peace accords in early 1973. One example of the use of American POWs to further the Communist war effort is related in CIA report dated February 14, 1973:

"ON 6 FEBRUARY 1973, FIVE U.S. POWS WERE SIGHTED AT A NORTH VIETNAMESE BASE CAMP LOCATED SOUTHEAST OF PHUM CHAN MUL, TAY NINH PROVINCE...ACCORDING TO VIET CONG KATUM DISTRICT CADRE MEMBER THE FIVE U.S. PRISONERS HAD BEEN BROUGHT TO THE AREA ON 3 FEBRUARY FROM RATTANAXIRI PROVINCE, CAMBODIA...THE U.S. PRISONERS WERE BEING USED BY THEIR CAPTORS TO MONITOR RADIO BROADCASTS TO DETERMINE U.S. FREQUENCIES..."[91]

These American POWs never returned from captivity after the end of the war, and other CIA reports of American POWs, both Caucasian and Negro, held inside Cambodia in Kompong Speu Province, Prey Veng Province and elsewhere, in the period following the 1973 Operation Homecoming return of "all" U.S. POWs, have been declassified. They also, like so many other American POWs, never returned.

In analyzing all available documentary and human sources, the author has estimated that 1,000 or more American POWs remained in enemy control after the 1973 prisoner returns, including some 700 or more withheld in North Vietnam, South Vietnam, Laos and Cambodia, and approximately 300 other Americans who were interrogated by Soviet military intelligence agents, processed through the Sam Neua "Tentacle M-B" camp and other centers for shipment by air and sea to the Soviet Union, and an undetermined number, estimated at up to 100, who were carried as "China-bound."[92] An estimated 600-700 or more remained in North Vietnam's secret parallel prison system cited in this book, some 100 to 200 other Americans remained as prisoners in Laos and were periodically moved both into and out of North Vietnam, while some 100-200 U.S. MIAs and "presumed killed" actually remained as prisoners in South Vietnam under Vietcong and North Vietnamese control (most later moved to the north), and up to 100 Americans reported as POWs in Cambodia were never repatriated, although some were moved into Vietnam during the rise of the Khymer Rouge.

Nearly two decades after the end of the war, a U.S. intelligence officer named John McCreary, assigned to a select committee of the U.S. Senate to investigate all recoverable classified records on American POWs in enemy control in 1973, and sources of U.S. intelligence concerning them, who was a special assistant to the Director of the Defense Intelligence Agency, confirmed to the author that after the return of 591 prisoners in Operation Homecoming, 1,300 more U.S. prisoners of war were still under enemy control.[93]

When questioned by the author, ex-NSA analyst Mooney stated that he had carried approximately 60-75 of the 305 known-POWs on his list as "Moscow-bound," and emphasized that his compilation and reporting comprised only 20%-25% of the "analytical pie" on U.S. POW/MIAs. More of these 305 men may have actually been Moscow-bound, but this list of about 60-75 MB's was comprised of those which the best analysis of all information available to NSA indicated were certain to have been processed for shipment to the USSR, no doubt after agreeing to go willingly in order to survive captivity. This gave the Soviet Politburo, then under control of dictator Leonid Breshnev, deniability in the case the shipments were ever discovered and publicly revealed, from charges that the Soviet GRU or KGB were kidnapping U.S. POWs. If they were broken psychologically at Sam Neua's Tentacle MB facility, than they could be labeled as "collaborators" or even as "deserters" by the Russians, and even by the U.S. Government, in the event that such shipments became public knowledge, then or at any time in the future. This would of course fit into the pattern established long before, in the 1919-1921 U.S.-Russia experience, of labeling U.S. POWs returning from Bolshevik control as "deserters."

THESE 60 to 75 "MOSCOW-BOUND" AMERICAN POWs ON MOONEY'S LIST AT NSA, WERE ALL FROM SHOOTDOWNS WHICH OCCURRED ONLY IN THE SOUTHERN PART OF THE PANHANDLE OF NORTH VIETNAM, BELOW THE CITY OF VINH, AND FROM THERE SOUTH TO THE NORTHERNMOST PART OF SOUTH VIETNAM, JUST BELOW THE CITY OF HUE, AND IN THE ADJACENT BORDER REGIONS OF LAOS FROM THE PAK SANE-MUONG CAO AREA TO THE SAVANNAKHET-SEPONE AREA DUE WEST OF HUE CITY, SOUTH VIETNAM. This analytically-determined number of "Moscow-bound" Americans does not include hundreds of other U.S. POWs captured in other areas, including most of North Vietnam, most of South Vietnam, most of Laos and all of Cambodia, who also were held and processed in a secret parallel prison system feeding Sam Neua and other Tentacle MB locations. One of the other such Tenacle-MB's was at Lang Son, near the Chinese border, which

Mooney learned from a DIA counterpart, "was a similar situation as Sam Neua," and which earlier in the war, during the 1964-1967 period when Chinese-Viet relations and Soviet-Chinese relations were better, apparently also served as a Tentacle CB, or "China-bound" connection, for possibly 100 or more U.S. POWs transferred to Chinese Communist control, to serve that country's high-tech developmental needs, in accordance with Chinese national security requirements.

The estimated minimum number of 200-300 "Moscow-bound" American POWs is further substantiated by the statements and sworn testimony before the U.S. Senate of another retired Air Force/NSA analyst (who had for two years or so worked under Mooney), Terrell Minarcin, who indicated that to his certain knowledge, 200-300 U.S. POWs were carried as Moscow-bound and shipped to the Soviet Union during and after the Vietnam war. American POWs left behind in Southeast Asia after 1975 were also sent to the USSR in later screenings of American POWs by Soviet intelligence, according to decrypted enemy communications monitored by Minarcin. Although a later Senate Select Committee stated that no other NSA personnel could be located who would back up Minarcin's assertions, other human intelligence sources were to provide information that supported much of what he had said about hundreds of Americans left behind in captivity after the war, and the shipment of some undetermined number of U.S. POWs to the USSR.

Since these men had already been publicly declared dead by their own government, and given the appalling condition of surviving U.S. POWs as reported in livesightings by Indochinese eyewitnesses of the late 1970s and 1980s, such transport to the USSR may have saved many of their lives, if they indeed agreed to cooperate in order to survive, as is suspected by Mooney, the author and others. This, after all, would have given the Soviet government a certain deniability in regard to any future accusations that they had kidnapped the American prisoners, should the M-B process be revealed, or should Russian-American relations ever change dramatically for the better. Surviving Americans could thus be conveniently labeled as "collaborators," by both governments. Whether many of the Moscow-bound Americans were later killed by Soviet security agencies when they were no longer useful, or, as some have charged, subsequently died in secret Soviet medical experiments, remains to be seen.

# Notes on Chapter Ten

1     *The Wise Men*, by Isaacson and Thomas, op. cit. pp. 546-547. In this incident Harriman threatened to leave the Alsops' home rather than be in the same room with Nixon, and after being persuaded to remain, turned his plate upside down, refusing to eat, switched off his hearing aid, and sat in silence.

2     Congressional Record of Hearings

3     *In Love and War*, Appendix 5

4     *Code of Honor*, John Dramesi, Norton, NY, 1975

5     In Love and War, Appendix 3, op. cit.

6     As an officer of the National Vietnam Veterans Coalition in 1986 and 1987, Booth accompanied the author to meetings with senior officials of the U.S. Department of State in Washington, D.C., such as Shephard Lowman, Director of Vietnam, Laos and Cambodia Affairs, who was to say then that the United States, "HAD NO RELIABLE INFORMATION OF ANY KIND THAT U.S. POW/MIAs REMAINED BEHIND ALIVE IN ENEMY CONTROL AFTER OPERATION HOMECOMING IN 1973."

7     *KGB: The Inside Story*, op. cit. pp. 492-494

8     Volume 6, p. 151.

9     CIA Report # CS 317/09048/71, October 5, 1971, obtained by FOIA request

10     Interview of Mooney by the author

11     *Missing in Action-Trail of Deceit* sequel, 1986, by Larry J. O'Daniel, published by the National Forget-Me-Not Association For POW/MIAs Inc.

12     Senate Select Committee POW/MIA hearings, September 22, 1992

13     NYT, November 29, 1970; Hearings of the Subcommittee on State Department Organization and Foreign Operation of the Committee on Foreign Affairs, House of Representatives, December 1970 and February 1971; also: Operation Keelhaul, pp. 3-11, op. cit.

14     In a 1992 interview with the author in his Montana home.

15     The Puzzle Palace, by James Bamford, op. cit., p. 261

16     Volume 6, p. 259, Uncorrelated Information Relating to Missing Americans in Southeast Asia, Declassified December 1978, but subsequently RECLASSIFIED by the Department of Defense.

17     Volume 13, p. 592, Uncorrelated Reports

18     CIA Report CS/311/04439-71, June 10, 1971. This document, obtained from CIA by FOIA request, was supplied by the author to the Senate Foreign Relations Committee Republican staff in 1990 and used in their May 1991 final report on U.S. POW/MIA policy, op. cit. It was also given by the author to the Senate Select Committee on POW/MIA Affairs in 1991, and is reproduced on pages 896-898, POW/MIA Policy and Process, U.S. Senate, op. cit.

19     1992 interview of retired AF/NSA analyst Jerry J. Mooney, in his Montana home, by the author.

20     Testimony of Jerry J. Mooney before the Select Committee on POW/MIA Affairs, January 22, 1992

21     *Survivors*, by Zalin Grant, W.W. Norton, NY, 1975

22     *Puzzle Palace*, op. cit., p. 262 & 283

23     Widows, by Corson, Trento & Trento, op. cit. pp.54-56

24     *Spycatcher*, by Peter Wright, op. cit., pp. 294 & 303

25     *Cold Warrior*, by Tom Mangold, op. cit, p. 306; also: Daniel Schorr, Washington Post, October 12, 1980; CBS News, January 8, 1975; Clearing The Air, pp. 134-35

26     *Cold Warrior*, op. cit., p. 306

27     *A POW's Story: 2801 Days in Hanoi*, by Col. Larry Guarino, Ivy Books, NY, 1990)

28    Senate Testimony, Sept. 22-24, 1992

29    *Honored and Betrayed*, p. 107, op. cit.

30    *A POWs Story: 2801 Days in Hanoi*, Ballantine Books, 1990

31    *Survivors*, by Zalin Grant, Norton's, NY, 1975

32    *The Bamboo Cage*, pp. 229-230; also interviews of Mrs. Hart by Thomas V. Ashworth

33    *The Bamboo Cage*, by Nigel Cawthorne, Leo Cooper, London, 1991, p. 147

34    Hearings of the Select Committee on POW/MIA Affairs, U.S. Senate, 1992, op. cit.

35    Ibid.

36    Department of State Bulletin, June 27, 1977

37    p. 85, Report of the Select Committee on POW/MIA Affairs, op. cit.

38    Report of the Select Committee on POW/MIA Affairs, op. cit. p. 143

39    Cable from President Nixon to Pham Van Dong, February 2, 1973; Report of the Select Committee on POW/MIA Affairs op.cit., p. 84

40    Ibid

41    Hearings of the Select Committee on POW/MIA Affairs, op. cit.; and interviews with Jerry J. Mooney

42    Hearings of the Select Committee on POW/MIA Affairs, op. cit.

43    Interview with Jerry J. Mooney, Billings, Montana, 199)

44    September 1992 hearings on POW/MIAs

45    Report of the Select Committee on POW/MIA Affairs, op. cit., p. 85

46    Senate hearings on POW/MIAs, Sep. 22, 1992

47    *Kissinger, Years of Upheaval*, p.33-34, op. cit.

48    Testimony of former Secretary of Defense Melvin Laird before the Select Committee on POW/MIA Affairs, U.S. Senate, 1992, op. cit.

49    Hearings of the Select Committee on POW/MIA Affairs, 1992, op. cit.

50    Ibid.

51    Henry Kissinger, The White House Years, p. 41, Little, Brown & Co., Boston, 1979

52    Report of the Select Committee on POW/MIA Affairs, op. cit., p. 86

53    Correlation and Evaluation of Selected Intelligence Reports Concerning the Alleged Sighting of U.S. POWs in Laos, April 1973-September 1975, p. 4

54    Hearings of the Select Committee on POW/MIA Affairs, U.S. Senate, 1992, op. cit.

55    Ibid.

56    Testimony of James Schlesinger, Senate, Select Committee on POW/MIA Affairs, 1992

57    Senate Select Committee on POW/MIAs, hearings of September 1992

58    *First Heroes*, Rod Colvin, Irvington, NY, 1987

59    Secord, *Memoirs*, op. cit.

60    Decoded enemy radio communications actually gave NSA monitors a much more presise estimate than this statement indicates.

61    This paragraph could be construed also as holding the Soviet Union partially responsible ultimately for the non-returned American POWs in Laos, many of whom had in fact been carried as "Moscow-bound."

62    This last comment appears inaccurate in view of Lao actions with American POWs.

63    As if the Lao Communists or their Vietnamese and Soviet backers would care about such a weak and carefully-couched response as Colonel Secord's posturing for Eagleburger!

64  Department of Defense/ISA Memorandum, 1-35174/73, U.S. POW/MIA Personnel In Laos, March 23, 1973; declassified in 1992

65  *Honored and Betrayed*, p. 102

66  P.37, of Box 3 of the State Department, 1973 files.

67  Testimony of James Schlesinger, op. cit

68  Report of the Select Committee on POW/MIA Affairs, op. cit., p. 94

69  Congressional Record—Senate, October 5, 1992, Statement of Senator Grassley, S16601

70  *Memoirs of Henry Kissinger, Volume II, Years of Upheaval*, op. cit., p. 318-328

71  *First Heroes*, op. cit.

72  Deposition of James Schlesinger, former Secretary of Defense, Select Committee on POW/MIA Affairs, U.S. Senate

73  Testimony of Melvin Laird, Select Committee on POW/MIA Affairs, U.S. Senate, 1992

74  American Embassy, Vientiane, message 13647, 5 April 1973, p. 111, Report of the Select Committee on POW/MIA Affairs, op. cit.

75  Vientiane to Department of State, April 6, 1973

76  Hearings of the Select Committee on POW/MIA Affairs, U.S. Senate, 1992

77  Deposition and Hearings before the Select Committee on POW/MIA Affairs, op. cit.

78  This conversation was confirmed in sworn testimony by Mr. Shields before the Senate Select Committee on POW/MIA Affairs in 1992.

79  Associated Press report, April 13, 1973

80  Last *Heroes*, op. cit., pp. 10 & 11

81  Hearings before the Select Committee on POW/MIA Affairs, U.S. Senate, September 22, 1992

82  Deposition of Henry Kissinger to the Select Committee on POW/MIA Affairs, U.S. Senate, 1992

83  Deposition and testimony of James Schlesinger, before the Select Committee on POW/MIA Affairs, U.S. Senate, 1992)

84  State Department No. 112133; p. 5-11 & 5-12, An Examination of U.S. Policy Towards POW/MIAs, Final Report of the U.S. Senate Foreign Relations Committee Republican Staff, May 23, 1991

85  Congressional Record- Senate October 5, 1992, S16602

86  Senator Charles Grassley read this information into the Congressional record, October 5,' 92

87  Depositions and Hearings, Select Committee on POW/MIA Affairs, op. cit.

88  Letter of December 30, 1992, from Richard Nixon to the Select Committee on POW/MIA Affairs, Report of the Select Committee, op. cit., p. 95

89  See final chapter

90  Record of Hearings, Select Committee on POW/MIA Affairs, 1992, op. cit.

91  CIA Report #CS/317/09017-73, February 14, 1973, obtained by FOIA request of the author.

92  Washington Post, July 4, 1991, "A 70-Year-Old Hostage Crisis, by John M. G. Brown

93  Interviews with John McCreary, Defense Intelligence Agency, by the author in 1992-

# CHAPTER ELEVEN

## THE POSTWAR POW/MIA DILEMMA
## The Ford and Carter Administrations, 1974-1981

The unclassified report of the Comptroller in the Office of the Assistant Secretary of Defense, dated December 5, 1973, lists an official total of 3,734 U.S. missing from the Vietnam War. Of these, 2,521 are reported as died while missing in action, 1007 were still carried as missing in action, 50 were reported as died while captured, 59 were still carried as "current captured," and 97 were missing from non-hostile causes, mostly aircraft crashes. 650 U.S., prisoners were reported as returned to military control.[1] The 59 "current captured" involved cases of Americans reported as POWs in Vietnam and Laos who had not been repatriated, although these were only selected test cases, from among those supplied by Mooney and other intelligence analysts, part of the list of 80 POWs Kissinger referred to, but not yet moved to the category of "MIA" or KBNR, "killed, body not recovered." (In a 1991 interview, analyst Mooney told the author that the highest, unofficial, in-house DIA estimate of late-1972 had placed the total of all U.S. prisoners believed to have once been captured alive throughout Indochina, at 3500. Even the most optimistic source, however, admitted that many of these may have subsequently died of wounds or maltreatment in Viet Cong and NVA POW camps, or been executed in reprisals for the deaths of NVA officers, which were detected in signals intercepted by NSA but have remained "classified.") Mooney believed his lists represented only 20% to 25% of the total analytical picture on U.S. POWs, but he still does not know much of his analytical data reached the highest authorities.

On April 17, 1974, Rear Admiral Donald Whitmire of the Defense Intelligence Agency signed a classified agency report which referred to the probability that 300 American POWs remained in Laos and Cambodia, but this report was kept secret from the American people. Also in April, Henry Kissinger wrote a secret memorandum to the beleaguered President, Richard Nixon, advising him not to meet with the National League of POW/MIA Families:

"There is also some danger that taking a strong position (on the POW/MIA issue) in such a forum might encourage the North Vietnamese to think they could use the MIA issue as a lever to press us for concessions on other matters." In a letter for the League of Families at this time Kissinger wrote: None of us can be optimistic about our missing...THERE IS NO RELIABLE EVIDENCE THAT ANY OF THEM ARE ALIVE."

Hundreds of Indochinese eyewitnesses, from this period on into the 1990s would report seeing U.S. prisoners of war still held in captivity in large and small groups in Vietnam and Laos, some of them in locations later reported by a U.S. Marine POW, Robert Garwood, who was to be released in 1979, but all would be judged as fabrications or debunked in some other way by DIA analysts.[2]

In addition to these U.S. prisoners secretly withheld by the Communists in Indochina, some 300 other American POWs carried as "Moscow-Bound" by U.S. intelligence analysts, had been, or were to be shipped to the Soviet Union after screening and interrogation by GRU and KGB officers, in accordance with long-established Soviet policy. NSA analysis of the time indicated locations for U.S. POWs in the USSR at Tyuratam, Kapostin Yan, Sary Shagan, Shuli, Semipalatinsk, Ural'sk and Novosibirsk; this last having also been a certain location for U.S. POWs from the Korean War. Other locations included Tashkent, Krasnodar (where ICBM missiles, telemetric work and a Soviet military-industrial think-tank existed, Baku (where it was expected some Korea and WW II POWs would also be present), and Alma Ata and its environs, particularly in nearby sophisticated radar and tracking installations. Ex-Soviet intelligence personnel have stated to the author's sources that U.S. POWs once under GRU supervision were eventually turned over to the KGB.

According to a former U.S. Army Intelligence specialist, Barry M. Toll (later a private businessman in Florida), senior U.S. policy makers were briefed about American prisoners of war being transported to the Soviet Union during the 1973-74 period. Investigators for a later Senate Select Committee on POW/MIA Affairs verified that Toll was assigned to the unit which he identified in a statement to Senator John Kerry (in 1992), in which Toll revealed he was:

"...Variously assigned within the World Wide Military Command and Control System, or World Wide Airborne Command Post system as an intelligence specialist and Operations Assistant on Battle Staff Teams formulated specifically to implement the nation's highest strategic nuclear policies and plans, and directly assist the President or designated National Command Authority in so doing. I was assigned by Department of Defense, by order of the Secretary of Defense, in Billet J3A1E1A, with duties primarily at the Commander-in-Chief' Atlantic's Airborne Command Post, with duty station Langley Air Force Base, Virginia, or at times temporarily dutied at National Emergency Airborne Command Post, Andrews Air Force Base, or other remote alert stations, to perform our mission...In short, we were there to lead the President through a nuclear execution..."

"To accomplish this mission, members of National Command Authority SIOP Execution Teams held an unusually wide and diversified range of extremely sensitive Top Secret and above security clearances, as we were privy to the combined input of the entire U.S. intelligence community to the President. In order to be capable of assuming responsibilities to brief and assist him in a SIOP environment on a moment's notice, 24 hours a day, 365 days a year."

In a June 14, 1992 statement to Senator John Kerry, Chairman of the Senate Select Committee on POW/MIA affairs, Mr. Toll revealed that the shipment of U.S. POWs from Vietnam and Laos to the Soviet Union and other Soviet Bloc nations had been made known to President Richard Nixon during the 1973-1974 period in his daily intelligence briefing agenda, and it would appear that the National Security Adviser at the time, Henry Kissinger would also have been informed. Toll wrote to the Senate:

I PERSONALLY SAW, DISTRIBUTED AND BRIEFED HIGH RANKING OFFICERS OF THE JOINT STAFF, ON INTELLIGENCE REPORTS, ANALYSES AND OPERATIONS REGARDING THE TRANSFER OF U.S. POWS AND/OR MIAS FROM THE CUSTODY OF NORTH VIETNAMESE OR LAOTIAN AUTHORITIES THROUGH SOVIET BLOC NATIONS, OR DIRECTLY INTO THE USSR. Further, it was the considered opinion of the Joint Chiefs of Staff, and the entire U.S. Intelligence community that at the conclusion of Operation Homecoming in 1973, that there were an estimated 290 to 340 POWs and MIAs alive, and held captive in Laos, and that analyses and reporting was disseminated to the President of the United States, Richard Nixon. I SPECIFICALLY RECALL THAT INFORMATION BEING INCLUDED ON A LIST OF THE PRESIDENT'S DAILY INTELLIGENCE BRIEFING AGENDA MORE THAN ONCE, AND REPORTS ON REAL-TIME TRANSFERS IN PROGRESS (AIRCRAFT BEARING U.S. POWS IN THE AIR ENROUTE TO USSR AND SOVIET BLOC COUNTRIES) BEING PASSED TO THE WHITE HOUSE FOR THE PRESIDENT SEVERAL TIMES IN THIS PERIOD.[3]

Toll's information, revealed years later to the Senate Select Committee (together with the substance of Mooney's testimony), indicates that President Gerald Ford and later President Jimmy Carter would have received similar intelligence on subsequent Moscow-bound U.S. POWs, delivered in a similar manner. The continuation of such shipments of POWS into the period of Jimmy Carter's administration was also eventually revealed by another former NSA analyst and cryptolinguist, Terrell Minarcin, who had once worked under Jerry Mooney at NSA.

In September of 1973, Mooney was overseas again, this time in a supporting role for what was now becoming the greatest defeat in American history, in the 6908th Security Squadron, USAF, at Nakhon Phanom (NLP), Thailand. His squadron flight Commander, Alex Znoj described Mooney in this period:

"Mooney is a professional in every sense of the word. As

Surveillance and Warning supervisor on the flight, he was directly responsible for the issuance of several hundred formal reports and several thousand technical formatted reports. Many of these reports were cited by higher levels in extremely laudatory comments about their contents and timeliness." The unit's Chief of Airborne Operations, Captain Herbert L. Young, wrote of Mooney in early 1974: "Mooney has been an invaluable asset to this unit...HIS PERCEPTION OF REPORTABLE SITUATIONS HAS ENABLED THIS UNIT TO PROVIDE REAL-TIME REPORTING TO LOCAL THEATER AND NATIONAL CONSUMERS." (Meaning the White House, State Department, CIA and JCS)

Reassigned in April 1974 to assist in managing the Operations Exploitation Section and other duties in the 6908th Security Squadron, USAF at Nakon Phanom, he already had seen first-hand how priceless intelligence had been wasted that could have saved the lives of aircrews killed on combat missions or held as prisoners. U.S. fliers were sent into dangerous areas with no tactical warnings that were easily available from NSA intelligence sources which Mooney had worked on, and chances to rescue downed aircrews were also thrown away.

Suddenly, in the summer of 1974, while the final agony of the Watergate scandal was wracking the Nixon administration and leading to impeachment proceedings in the Congress, Mooney learned at Nakhon Phanom that two American POWS had escaped from their Communist captors in Laos and were heading for Thailand. In monitoring the Vietnamese Command unit R-109/VCD through intercepts obtained by Olympic Torch, the U2 spyplane covering Laos, Mooney discovered that all military units in Laos had been put on alert to recapture two prisoners, 'one white' and 'one black,' who had escaped from a North Vietnamese air defense force camp, probably between the Mu Gia and Ben Karal passes near Nhommarath, in the Laotian panhandle. Since these men were regarded as possible escaped U.S. POWs, in accordance with the highest national requirements Mooney reported this incident at NKP and it was sent immediately to all national levels. Mooney estimates that Nixon's White House would have known and all others including JCS, CIA, and State Department minutes later, and he confirmed that a Delta Force special operations unit went on alert. But the incident was to subside as quickly as it arose. The U.S. Ambassador in Thailand reported that the two prisoners were actually a "very dark skinned Laotian and a very light skinned Laotian;" that Thai border police had intercepted them and they were not escaped U.S. prisoners. Mooney is almost certain that the two prisoners did escape but he does not believe to this day that the North Vietnamese Army would have gone on a national alert in Laos over two escaped Laotians.

Shortly after what was jokingly called the "salt and pepper

incident" at NKP, Mooney was contacted directly from NSA Headquarters at Fort Meade by his old Branch Chief there, Marvin Plummer:

"How's it going over there Jerry? By the way, WHAT DID YOU DO WITH THOSE POW LISTS YOU PRODUCED...?"

This communication to Mooney from Plummer at Fort Meade was not a formal message, but came through Ops-Comm, point-to-point teletype (which Mooney believes was witnessed by a Staff Sergeant serving under him at the time):

Mooney had a strong hunch that the sudden interest in the lists was related to the two U.S POWs who had escaped from Laos in the recent "salt and Pepper" incident. He answered, "They're in the upper right hand drawer of my desk... in a multi-packet folder." (Seventeen years later, during an interview with the author, Mooney was to remember that the folder was light green or blue, with speckles ingrained in the surface.) In the folder were all the results of Mooney's years of analytical work, the "Who to Ask List" containing over 1,000 names of American MIA/POWs believed to have been captured alive, whose fate would be known to the enemy, the Warm Body Count list of over 400 POW/MIAs whom Mooney believed were held in captivity, but which contained also some names of executed U.S. prisoners, the "POW List" of 305 names of definitely-known U.S. prisoners, 70% of whom were captured in North Vietnam (of which 305 POWs only about 13 or 14 were released), and the "Kissinger list" or "teaser" of 21 names of POWs believed to have been transferred to the USSR and China, which had in turn been compiled from a longer list of 75 Americans carried by Mooney as "Moscow Bound." And Mooney believed his lists represented only 20% to 25% of the total analytical picture on U.S. POWs, but he still does not know much of his analytical data reached the highest authorities.

Jerry J. Mooney was praised by his superior officers upon his transfer from Thailand back to the U.S. on September 3, 1974: "His performance has been professional and outstanding in every respect. His vast experience was brought to bear in a variety of challenging situations...productivity increased significantly as a result of these actions...The recent timely direct tactical support Msgt. Mooney provided to local air commanders PACA, and a national level agency-concerning a Southeast Asia threat assessment situation, provided a classic example of professional mission accomplishment. His strengths are, technical expertise, acute mission awareness, superior management ability and a conscientious approach to all duties." Mooney's Squadron Commander in Thailand, Lt. Colonel Nicholas Obzut, later wrote of him in a fitness report:

"I am well aware of Master Sergeant Mooney's expertise as an analyst and supervisor. His high degree of professionalism has contributed greatly to the accomplishment of this unit's mission. I recognize him to be an expert in his field and one of the finest NCO's I have served with in 18 years of service."[4]

Mooney never learned what became of his hard-won U.S. prisoner of war lists. The analyst who took over his slot while he was in Thailand would have accepted immediate responsibility for them to familiarize himself with what his predecessor knew about the job. If that individual decided that he could not make use of such information, Mooney says regulations required that, "He pack it up and send it to Fort Holabird or Indiantown Gap, Pennsylvania, for storage, if it wasn't kept at Fort Meade for future reference," as he suspects. Later in 1974, on a return duty assignment to NSA with the 6970th Base Group at Ford Meade after Nixon's resignation and Gerald Ford had become President, Mooney saw his old NSA boss Berkeley "C" and asked him: "Hey...did you ever find my POW lists that Plummer asked me for when I was over in Thailand? "C" wouldn't answer, but gave Mooney a severe look that said: 'You have no need to know the answer to that.' Mooney didn't press "C," but just shrugged his shoulders in resignation and walked on. He had spent years of his life immersed in the "need to know" world of secret intelligence, and he was too duty-conscious to take the issue any further at that time. Besides, Mooney then still believed that the U.S. Government was involved in secret diplomacy to gain release of the prisoners. However, nothing indicating this ever happened after the disappearance of the "salt and pepper POWs," was to come across Mooney's desk in the three more years until his retirement in 1977, after 20 years' service. (Years later he was forced by his own conscience to recall from memory only, hundreds of these names for the use of the U.S. Senate.)[5]

On August 29, 1974, Gerald Rudolph Ford had assumed the presidency of the United States upon Richard Nixon's resignation, and amidst a national furor over public disclosures in the Watergate scandal. "Gerry" Ford had been born in Nebraska in 1913 and attended school in Michigan, where he played football and was an Eagle Scout, in Boy Scouts of America. Ford graduated from the University of Michigan where he also played football, and later attended Yale Law School. During WW II, Ford served as a naval officer on the aircraft carrier USS Monterey, in the major campaigns of the war in the Pacific. Elected to Congress in 1948, Ford became popular because of his ability to get along with others, eventually becoming Minority Leader. When Nixon chose him to replace Vice President Spiro Agnew, who had resigned because of involvement in a bribery scandal, his appointment was approved by a vast majority of both houses of Congress. With the threat of Nixon's impeachment over illegalities of the Watergate break-in, and that President's subsequent resignation, Ford had become President. He retained

Henry Kissinger as Secretary of State, James Schlesinger as Secretary of Defense, William Simon as Secretary of the Treasury and Kissinger's military assistant, General Brent Scowcroft, became Ford's national security adviser.

After 10 years of continual American casualties in a questionable foreign war, the people of the United States were thoroughly tired of Vietnam and all that went with it. The North Vietnamese, by now fairly certain that the American government was no longer able to intervene militarily, made plans for their final offensive to overwhelm South Vietnam. While most of the American people cared little for the fate of their allies, there was substantial public interest in the fate of the prisoners of war and missing in action who had not returned from Indochina, due to efforts by POW/MIA family members and veterans' groups.

Gerald Ford appears to have been dominated by the presence of Henry Kissinger in matters concerning Vietnam. The Secretary of State, then at his apex as the man who had negotiated "Peace" for America, was the primary policymaker in foreign affairs of the United States. Kissinger argued that Hanoi's violations of the Peace Accords invalidated their claim to the promised $3 to $4 Billion in war reconstruction aid. With the most damning evidence of non-returned U.S. POWs and MIAs deeply and safely "classified," Kissinger, Scowcroft and others in the Administration were aware that time alone would ensure that the still-captive Americans would be forgotten by all but a few people. Kissinger's stature as the great peacemaker would thus remain untarnished. (Although Kissinger accepted a Nobel Peace Prize for accomplishing the Paris peace accords, North Vietnam's Le Duc Tho refused it, and turned his attention to winning the war in the south.) Kissinger denied that the United States had ever made a "firm commitment" to extend war reconstruction aid to Vietnam. Only by going against his own Secretary of State, who had gained international fame in his role as an expert on the Vietnamese problem, could Ford have altered the progression of events. The prisoners of war, unable to speak from their places of confinement in Asian camps or in Moscow's Lubyanka prison, went almost unheeded.

The National League of Families of American prisoners and missing in Southeast Asia had been a supporter of U.S. government efforts to return and account for U.S. prisoners during the war years, but after Operation Homecoming and the disappointing number of Americans released in 1973, the League membership became more radical. As the League's legal, Counsel Dermot Foley, whose brother, Colonel Brendan Foley was missing on the Laos-Vietnam border since 1967, remembered later:

"During a 1974 League meeting in Omaha, Nebraska, with vigorous support of over ninety percent of the League membership, the pro-government leadership was replaced. Oddly enough, almost overnight we began to pay the price. Our volunteer lawyer quit. Our headquarters free WATS lines were cut off. ONE GOOD DEVELOPMENT WAS THE LOSS OF OUR ADVISORY COMMITTEE, MADE UP OF FOLKS SUPPLIED BY THE GOVERNMENT, WHO WERE SUPPOSED TO KNOW HOW TO GET THINGS DONE. THESE INCLUDED SUCH ILLUSTRIOUS FIGURES AS E. HOWARD HUNT..."[6] The fact that a veteran CIA covert operator like E. Howard Hunt, one of the Watergate burglars, had been assigned as an adviser to the League of Families of missing U.S. servicemen, indicated how seriously the government took the possibility that the POW/MIA families might get out of control. Hunt was by this time a specialist in domestic intelligence and security for the Committee to Reelect the President and the White House "plumbers," using CIA equipment and safehouses to conduct illegal operations.

Hunt had been involved in CIA covert operations in Central America and Cuba since the 1950s, including the plots to assassinate Fidel Castro, and his name had repeatedly surfaced during the extended investigation President Kennedy's assassination. He had been questioned in 1973 about his forging of a State Department cable directly linking the Kennedy administration to the assassination of South Vietnamese President Ngo Dinh Diem.[7] Following the leak of the Pentagon Papers to the New York Times, and others in 1971, by former Defense Department official Daniel Ellsberg, President Nixon had instructed his aide John Ehrlichman to form a White House Special Investigations Unit (SIU), to enforce security and stop leaks. Ehrlichman appointed Egil "Bud" Krogh, and a former Kissinger aide, David Young to the unit, and Krogh hired G. Gordon Liddy and E. Howard Hunt. Hunt had been involved with covert activities run by Nixon assistant Charles Colson, aimed at discrediting the Democratic Party opposition since early 1970. The group operated out of a room in the White House and became known as the "Plumbers," and the actions of E. Howard Hunt, Gordon Liddy and others, together with the subsequent coverup of illegal acts, were to cause the downfall of President Nixon at a crucial time, which in turn was to encourage North Vietnam's refusal to adhere to provisions of the Paris peace accords. Five of the seven Watergate burglars were ex-CIA employees and one was still an employee, while CIA director Richard Helms had assisted them by helping to organize the White House's top-secret plan for domestic surveillance and intelligence collection., while the CIA's Clandestine Services had provided technical assistance to the "plumbers."[8]

Nixon apparently understood better than his aides the nature of E. Howard Hunt's clandestine activities in the past, and how an investigation of his part in the Watergate burglary of the Democratic Party headquarters could expose earlier secrets. A transcript of a tape recording of a meeting between H.R. Haldeman in the Oval Office on June 23, 1972 revealed the President's concern in his own words:

"...This fellow Hunt...he knows too damn much and he was involved, we happen to know that. And that it gets out that the whole, this is all involved in the Cuban thing, that it's a fiasco, and it's going to make...the CIA look bad, it's going to make Hunt look bad, and it's likely to blow the whole, uh, Bay of Pigs thing which we think would be very unfortunate for CIA and for the country at this time, and for American foreign policy..."[9]

The fact that E. Howard Hunt was assigned to "advise" the League of Families of American POWs and MIAs in Southeast Asia, indicates that the prisoners' families were a major domestic intelligence target at a time when the nation was learning about the involvement of the NSA, FBI, Secret Service, CIA and military intelligence in domestic eavesdropping, spying, interception of telephone communications and other abuses of power. A "Watch List" program for electronic eavesdropping surveillance had begun at the NSA in 1962, and was enlarged after 1967 following an October 20, 1967 'Top Secret Comint Channels only' message to NSA director Marshall Carter from Army Assistant Chief of Staff for Intelligence (G-2) General William Yarborough, "requesting that NSA provide any available information about possible foreign influence on civil disturbances in the United States." The Secret Service also contributed to these "Watch Lists," that included anti-war and civil rights leaders such as Joan Baez, Jane Fonda, Dr. Benjamin Spock, Dr. Martin Luther King, Jr., the Reverend Ralph Abernathy, Black Panther Eldridge Cleaver and Chicago Seven defendants Abbie Hoffman and David Dellinger, among the many thousands of Americans whose communications and activities were monitored by the NSA and other government agencies.[10]

The results of the "Watch list" program was not sufficient for the White House zealots who were intent on crushing opposition to Nixon's policies and ensuring his reelection as president in the fall of 1972. The rise of the antiwar movement and bombings by radical Marxist groups such as the "Weathermen," in 1970, had alarmed White House officials of the Nixon administration, who were dissatisfied with the intelligence gained by the FBI and other federal agencies. With the encouragement of NSA Deputy Director Louis Tordella, the new director (since 1969) Admiral Noel Gayler, and others, a former Army intelligence officer named Tom Huston, who was now a White House staffer, had drawn up a "Domestic Intelligence Gathering Plan: Analysis and Strategy." This became known as the "Huston Plan," and it was to remain secret until revealed by the New York Times on June 7, 1973. In December 1974, reporter Seymour Hersh was to reveal the details of a highly secret and illegal CIA spying program aimed at Americans named Operation

Chaos. The subsequent public furor resulted in Gerald Ford appointing a Commission on CIA Activities Within the United States directed by Nelson Rockefeller, which turned out to be a whitewash, and eventually to Senate Intelligence Committee hearings chaired by Senator Frank Church, which investigated illegal activities of the NSA and CIA.[11]

For years after E. Howard Hunt's involvement with the League of Families, POW/MIA family members would continue to charge that they had been subjected to intimidation, threats, surveillance, illegal wiretaps and other actions, which indicated that the government surveillance continued long after the "Huston Plan" had ostensibly been disapproved and it's author demoted. The frustration felt by thousands of families of MIAs across the nation about the government's apparent inaction in seeking out and returning their loved ones led to their being considered potential security risks by senior officials who did not wish the truth about the abandoned POWs and MIAs to reach the public.

In September 1974, as if to break a logjam, a CIA pilot captured after the 1973 cease-fire, named Emmett Kay, was released by the Pathet Lao, shortly after President Ford's inauguration, and nearly three weeks after the August 29, 1974 signing of an exchange-of-prisoners agreement between the U.S. and the Laotian Communists. In a later interview for a documentary film[12], Emmet Kay revealed that he was moved back and forth across the Lao/NVN border after his capture, and he remembered the Pathet Lao had claimed to him that just before he was captured, many Americans had been moved from Laos into Vietnam, AFTER THE 1973 POW REPATRIATIONS, and that there were still many, many more U.S. prisoners alive in captivity. After his return debriefing Emmett Kay moved to the remote island of Tinian in the Pacific Ocean.

On December 3, 1974, North Vietnam announced that "The U.S. has not carried out and has no intention to carry out in near future, its obligations regarding the healing of the wounds of war in the DRV." In later interviews with the author, Colonel William LeGro, chief of the U.S. Military Intelligence branch at the DAO-Saigon from late 1973 until April 1975, remembered receiving livesighting reports about US POWs still being held in the U Minh Forest, in the southernmost area of Vietnam, below the Mekong Delta and near the Cambodian border sanctuary. This was an area in which the Viet Cong had held substantial numbers of American POWs all through the war. (It was also the area where Green Beret POW Dan Pitzer had been released from (with two others), prior to their return in a communist propaganda ploy aided by American antiwar activists, in 1967.) In a 1986 interview, Colonel LeGro stated that all such livesighting reports of US POWs after the 1973 "Homecoming" of 591 US prisoners were immediately forwarded to the CIA Station Chief in Saigon, and through Army Intelligence channels to Washington, D.C. for DIA and CIA. The last American POW livesightings that LeGro recalled were sent to CIA Station Chief Thomas Polgar in the 1974-75 period, and as far as LeGro could ascertain, nothing was done about them. If Polgar took proper action with livesighting reports such as these they would have ended up in the hands of the CIA's Far East Division chief; at this time, Theodore Shackley.[13]

The Uncorrelated Reports relating to missing Americans in Southeast Asia that were declassified by the Pentagon on December 15, 1978 and then RECLASSIFIED in 1979, contain many examples of such sightings of American prisoners still held in southern Vietnam and the Cambodian border area, during the late-1973 and 1974 period. In the remote Seven Mountains area of Chau Duc Province, in southernmost Vietnam, 80 prisoners of war, with their hands bound, were seen being marched by ten armed Communist guards, on December 23, 1973. Between five and ten of the prisoners were taller men, whom the informant said didn't speak Vietnamese, French or Cambodian but were heard to utter American phrases like, 'take it easy,' 'can do,' and 'all right.' The Seven Mountains area was rarely penetrated by U.S. forces in all the years of war. This information had been brought to a Colonel Wallace in Saigon by a representative of a U.S. "civilian organization" whose identity was not recorded, and was transmitted to the acting commander of the Joint Casualty Resolution Center in NKP, Thailand. Wallace rated the U.S. representative as reliable and honest, who considered his own sources highly credible. Colonel Wallace's assessment of this information was: "Of all the Bright Lights reports and others I have seen this is the most promising. It is a fresh report. It is plausible and if true and properly exploited, it could be a devastating blow to the aspirations of the PRG for diplomatic recognition as a responsible government." Another report records that two black American POWs were seen being moved under guard in the Kompong Cham area of the Cambodian-Vietnamese border during December 1973, by a V.C. defector.[14]

Two Caucasian prisoners, tied together, were seen being moved across the Cambodian border into South Vietnam in January 1974, escorted by 100 Communist troops. On January 15, 1974, a prisoner of the Communists escaped from a camp at Da Dac, in a mangrove swamp near the sea, where he was held with 300-400 ARVN prisoners and six American POWs. Five of the Americans were Caucasian and one was black, with a large heart tattooed on his chest. The U.S. POWs revealed they had been captured in a battle in Ca Mau Province and held prisoner for nearly nine years. This report was remarkable, in that another prisoner of the V.C. who had been held previously in that area, had also seen the black American POW with a heart tattooed on his chest, along with five white POWs;

these included one named Captain Watkins or Walker and another called Sergeant Rockson or Jackson. The ex-prisoner thought they had been captured about 1964. After this informant was no longer in the camp where he had seen these 6 U.S. POWs, he received later information about them until 1974, from a female friend who was able to enter the camp to deliver supplies to the Communist cadre.

Inside Cambodia, 40-50 U.S. prisoners of war were seen by a female Chinese merchant in May 1974, but the DIA considered this report a "fabrication." Yet, almost simultaneously, a Lao who traded with the Communists met a Thai prisoner of war at Stueng Treng, Cambodia, who had become a collaborator and had helped guard 300 fellow-Thai prisoners and 100 American POWs. This informant was repeatedly sent back by the U.S.A.F. Air intelligence group, to confirm aspects of what he had learned. He subsequently reported seeing work details of white and black American POWs in the summer of 1974, being escorted by NVA guards in groups of eight, at the market place in Stueng Treng. The same source reported that this group of prisoners had been moved and 10 Thai and 8 American POWs had escaped on July 13, 1974.[15]

American POWs were also reported by Vietnamese sources as still being held in captivity in the Tay Ninh area and near the "Black Virgin Mountain" (Nui Ba Den). In the central highlands province of Darlac, 20 kilometers north of Ban Me Thuot, six Caucasian men and one Caucasian woman were reported held as prisoners in bamboo cages, reported by an ex-POW. Michael Benge. This mirrors a similar pattern for livesightings of POWs taken in Laos from February 1973, onward.

By the end of 1974 little progress had been made on discovering the fate of these reported U.S. POWs and MIAs by the American officers on the Joint Military Team who had been traveling back and forth between Saigon and Hanoi since the Paris accords and POW repatriations of early 1973.[16] The North Vietnamese had meanwhile been preparing, throughout the cease-fire period, for an overwhelming military conquest of South Vietnam. Former CIA officer Frank Snepp revealed that In December 1974, prior to the initial attack on Phuoc Long which initiated the final campaign, a Soviet general named Viktor Kulikov traveled by aircraft to Hanoi to meet with the North Vietnamese Politburo, and that he immediately alerted CIA headquarters. Snepp had remembered that the last time a Soviet general of this rank had traveled to North Vietnam was late in 1971, just prior to the 1972 "Easter Offensive." Snepp records that the CIA station chief in Saigon, Thomas Polgar sent a message to Washington about this intelligence warning of a possible repetition of that disaster.[17]

Following a successful North Vietnamese Army assault on the provincial capital of Song Be (Phuoc Binh) in January 1975, which was spearheaded by Soviet-supplied tanks, there was a temporary lull in the final campaign of conquest of South Vietnam. The offensive was renewed on March 1st, with a major attack on Ban Me Thuot, considered a key position for controlling the Central Highlands. On the northern coast, Hue and Danang also fell to Communist assaults and the South Vietnamese army in the northern I Corps region collapsed. Great masses of refugees fled south ahead of the victorious North Vietnamese Army.

President Gerald Ford had appealed to a joint session of Congress on the night of April 10th for $700 million in military aid, and $250 million in humanitarian aid for South Vietnam, as had been urgently recommended by General Weyand and others, but the Democratic side of the house began to hiss at him and several walked out. Among the most outspoken opponents of aid who had traveled to South Vietnam to assess the situation was U.S. representative Tom Harkin of Iowa, known as the discoverer of the South Vietnamese-controlled "tiger cages" at the old French prison on Con Son Island. Others such as Bella Abzug, Paul "Pete" McCloskey and Millicent Fenwick arrived and took up valuable time in their predetermined "investigations." In the meantime, the situation in Cambodia had become critical, as the Khymer Rouge Communist forces shelled the capital with rockets, while Henry Kissinger still attempted to make contact with Norodom Sihanouk, hoping to arrange a coalition government with the very leader who had been deposed years before, with American connivance. The U.S. evacuation of Phnom Penh was effected by 36 huge CH-53 helicopters which carried 82 Americans, including Ambassador John Gunther Dean, along with 35 allied nationals and 159 Cambodians, to safety in Thailand. Some Cambodians who could have fled with the Americans chose to remain behind. One was the former Premier Sisowath Sirik Matak, who had taken part in the American-sponsored coup against Sihanouk that had put Lon Nol in power, that had led to increased North Vietnamese intervention in Cambodia, the 1970 U.S. invasion and ultimately, the victory of the Khymer Rouge. He wrote to Ambassador Dean at the end:

"I thank you sincerely for your letter and your offer to transport me to freedom. I cannot, alas, leave in such a cowardly fashion. As for you, and in particular your great country, I never believed for a moment that you would have the sentiment of abandoning people who have chosen liberty. You have refused us protection, and we can do nothing about it. You leave, and it is my wish that you and your country will find happiness under the sky...I have only committed the mistake of believing in the Americans."

To President Gerald Ford, Sirik Matak cabled:

"We will struggle now alone without your support. The Khymer people have already paid a very heavy sacrifice in human lives for you Americans to enable you to disengage from South Vietnam. Your policy of abandoning a poor country, decided brutally without warning or preparation, puts us in a position of heartbreaking betrayal...There will be in this capital of two million a terrible carnage...I lay on the American conscience all Khymer deaths present and future."[18]

After the final conquest of Phnom Penh, the Khymer Rouge began emptying the city of its two million of inhabitants in forced marches to remote jungle locations, where those who hadn't been murdered along the way became forced laborers for the new Communist regime. It was the beginning of the Khymer Rouge genocide that was to kill between 1,000,000 and 2,000,000 Cambodian people by the late 1970s.

In South Vietnam, Da Nang, Pleiku, Kontum, Ban Me Thuot and Nha Trang had already fallen to the communists, and on April 7th, Le Duc Tho, once offered the Nobel peace prize with Henry Kissinger, had arrived at Loc Ninh, in South Vietnam, to oversee the final conquest, called the "Ho Chi Minh campaign." Under command of General Van Tien Dung and his deputies, Tran Van Tra and the Politburo's Pham Hung, the North Vietnamese Army attacked towards Saigon from Ban Me Thuot and Nha Trang, and along the traditional routes south and east towards Bien Hoa from the "Fishhook" and "Parrots beak" regions of the Cambodian border, that had been defended for years by the U.S. 1st, 9th and 25th Infantry divisions. From his forward headquarters near Ben Cat, in Binh Duong Province, Dung then massed 40,000 NVA troops for a cataclysmic two-week battle against remnants of the Army of the Republic of Vietnam. The battle was fought 25 miles northeast of Saigon, at Xuan Loc; the old headquarters of the U.S. 11th Armored Cavalry Regiment during the American war. The South Vietnamese Air Force had used cluster bombs and a giant 15,000-pound American bomb intended for clearing landing zones to deadly effect on massed North Vietnamese regulars equipped with Soviet tanks and armored vehicles, while the ARVN infantry also fought tenaciously; NVA dead were numbered in the thousands. Elsewhere, the offensive maintained its momentum.[19]

As South Vietnam was collapsing before the massive North Vietnamese assault, spearheaded by Soviet supplied tanks and surface-to-air missiles, extremists of the American antiwar movement and far-left lobby felt victorious. Although the mass peace protests had subsided rapidly with the 1972-1973 withdrawal of American troops from Vietnam, the most radical pro-Hanoi American activists had formed new organizations to ensure that the Communist regime retained a strong lobby in the United States until the South Vietnamese regime could be defeated. By this time much of the major media had joined the chorus against further aid to the South Vietnamese government and gave wide coverage to the spokesmen for this diehard core group of "peace activists." Tom Hayden, the long-time supporter of the Hanoi regime who had married the antiwar actress Jane Fonda following her 1972 trips to Hanoi, now worked with his wife through the "Indochina Peace Campaign," to halt all military aid to the embattled South Vietnamese forces. In an interview published on April 19, 1975 in the New York Times, Hayden spoke of the rapidly approaching conquest of South Vietnam by the Communist forces:

'"I see this as a result of something we've been working toward for a long time,' Mr. Hayden said. 'Indochina has not fallen-it has risen. What has fallen is the whole cold war establishment. We now have the opportunity to define a new policy.'" "He said he hoped America would abandon 'knee jerk acceptance of right-wing dictatorships' and recognize that 'Communism is one of the options that can improve people's lives.' But he maintained that there would not be a 'Communist takeover' in Vietnam, saying that a coalition government would be formed under the terms of the original peace agreement. 'The policy of the other side is reconciliation,' he said of the advancing North Vietnamese..."

The tragic fate awaiting the millions of 'anti-communist reactionaries' who were subsequently massacred or died in miserable conditions of slavery at forced labor camps in Cambodia, Laos and Vietnam, was to demonstrate the fallacy of this view.

The United States was now in the midst of evacuating 6,000 Americans, mostly civilians, from Saigon, in addition to some of the 100,000 Vietnamese who had worked directly for the U.S. government and who were therefore dangerously compromised. Day after day, flights of the largest U.S. transport aircraft left Tan Son Nhut Airbase for the Philippines, Guam and Thailand as the NVA forces pushed closer; others left by ship. Unduly optimistic and misled by his Hungarian-born CIA station chief, Thomas Polgar, who may have been affected by disinformation from a Hungarian on the ICCS delegation in Saigon, U.S. Ambassador Graham Martin was convinced that the Communists would permit a last-minute negotiated settlement. (Martin's deputy Chief of Mission, Wolfgang Lehmann, later a senior CIA official, also strongly opposed the "defeatism" of the pragmatists.) In this Ambassador Martin adhered to Kissinger's wishful thinking, although the State Department's intelligence chief at the time, William Hyland (later head of the Council on Foreign Relations), disagreed.

Opposing Kissinger's and Martin's view was Polgar's military intelligence counterpart, Colonel William LeGro of the Defense Attaché's Office, whose intelligence estimates on the NVA military threat had proven consistently accurate during the campaign. Bill LeGro had been ably assisted by dedicated young professionals like Doug Dearth. In briefing General Fred Weyand on March 28th, 1975,

LeGro had stated that only immediate U.S. B-52 bomber strikes would be able to halt the NVA advance and buy time for Saigon to reorganize its defense. The Ford Administration had been unwilling to take the political risk in such a move. When Ambassador Martin had refused permission for initiation of a mass-evacuation, LeGro and others had used their authority to evacuate thousands of refugees secretly. This effort was assisted by a group of mid-level State Department officials in Washington including Lionel Rosenblatt, Frank Wisner Jr. and Kenneth Quinn (of the NSC staff), who also felt the end of South Vietnam was at hand, and clandestinely did everything possible to assist State Department officers in Saigon who were accomplishing the evacuation, such as Lacey Wright and Shephard Lowman.

Among the many Americans assisting in the evacuation of Saigon were members of the American delegation to the Joint Military Team, under Colonel John Madison. These included Lt. Colonel Harry Summers and Captain Stuart Herrington, who had been unsuccessfully negotiating, for 19 months with the North Vietnamese for information on the some 3,500 U.S. MIAs who had disappeared during the war. Herrington later recorded how for a year and a half before the fall of Saigon the North Vietnamese, and Vietcong had refused to cooperate with information about documented cases of American servicemen who were believed captured:

"We confronted the Communists and showed them documents but they refused to answer—or rather they gave us a diatribe about our country's war crimes on the peace-loving Vietnamese people."[20] Now Herrington and the others on his team worked frantically to make sure that a whole new group of American prisoners did not end up in Communist control.

The Americans and their South Vietnamese allies could be ruthless with prisoners too. In late April as South Vietnam collapsed some top allied officials decided that the highest-ranking North Vietnamese agent ever held, Nguyen Van Tai, should die before the Communist takeover. Hanoi had tried to trade for him in 1971, offering a captured State Department officer named Douglas Ramsey, but the CIA was opposed to turning over such a high-level intelligence operative for an unimportant American. Ramsey was later returned in Operation Homecoming in 1973, after brutal captivity in a bamboo cage, but Nguyen Van Tai remained a captive of the Saigon government, held in solitary confinement for over four years. According to former CIA officer Frank Snepp:

"Just before North Vietnamese tanks rolled into Saigon a senior CIA official suggested to South Vietnamese authorities that it would be useful if he disappeared. Since Tai was a trained terrorist, he could hardly expect to be a magnanimous victor. The

South Vietnamese agreed. Tai was loaded onto an airplane and thrown out at ten thousand feet over the South China Sea... This story is not unique. On several occasions during the last years of the Vietnam War, the Communists secretly offered to release various American prisoners, only to be held off by the U.S. government as it angled for 'better terms.' The CIA was particularly inflexible, usually insisting on strict reciprocity, an intelligence operative for an intelligence operative, as if agency personnel deserved first consideration over any other Americans..."[21] Snepp's words recall the U.S. government's highly-secret rejection of the 1967 East German/ Soviet offer to trade wounded U.S. POWs from Vietnam for two convicted Soviet spies held by the British.

When the military position in both South Vietnam and Cambodia had begun to deteriorate rapidly in late March, Henry Kissinger had called a White House meeting of key foreign policy, military and intelligence officials to establish a plan for dealing with the crisis. Kissinger, who had already written off the hundreds of missing American prisoners still remaining in Vietnam and Laos, was infuriated about the unraveling of his orchestrated "peace." General Fred Weyand, commander of MACV, suggested the immediate dispatch of a high-level mission to Vietnam to "develop solutions, or options," and to "assess the needs of the South Vietnamese." The two officials picked by Weyand to carry out this mission were the Pentagon's logistics expert, Erich Von Marbod, and the CIA's Deputy Director for the Far East Division, Theodore Shackley.

Known to some as the "blonde ghost," Ted Shackley was a CIA veteran of the "Berlin Tunnel" fiasco, of the mid-1950s, who had taken over the huge Miami station of the CIA after the Bay of Pigs fiasco in 1961. He later was appointed the Chief of Station in Vientiane, Laos, during the CIAs covert war buildup there, and in 1968 he took over as CIA station chief in Saigon. In the post-Tet Offensive and "Vietnamization" period; perhaps the most prestigious CIA job in the world. His reign in Vietnam coincided with the expansion of the "Phoenix Program" of assassinations of Communist cadre, resulting in the deaths of 20,000 or more suspected Vietcong throughout the country's villages. During Shackley's tenure in Vientiane and Saigon, the involvement of CIA assets and surrogate forces with narcotics trafficking in South East Asia increased greatly, and simultaneously, so did its use by American troops in Vietnam. Shackley was brought back from Saigon in 1972, reportedly in reaction to renegade CIA officer Philip Agee's revelations of CIA operations and agents in that area. In 1973, apparently with assistance from the influential Erich Von Marbod, Shackley got the job he really wanted, as head of the Far East Division of the CIA. From 1973 until after the fall of South Vietnam in 1975, Theodore Shackley was thus one of those responsible for the resolution of, or the ultimate disappearance of classified intelligence reports about live American prisoners of war still being held in Indochina, or about those reported as transferred to the Soviet Union, Communist China or Cuba.

Von Marbod had gained Schlesinger's trust while serving as a logistician with the Defense Department's Security Assistance agency, and was then appointed by Schlesinger as Controller of the Department of Defense. He previously had spent time in Southeast Asia, where he had become friends with Theodore Shackley, and with Thomas Clines. Shackley, Clines and Von Marbod were at this time friends with, and frequent guests of later-imprisoned ex-CIA officer Edwin Wilson, then engaged in expanding his circle of allies who would later attempt to protect him from exposure. It was through Theodore Shackley and his erstwhile deputy, Thomas Clines, that Von Marbod had met Edwin Wilson. Many other connections were also being made at this time that would later play crucial roles in covert operations. According to author Peter Maas:

"Shackley and Clines also introduced Wilson to Richard Secord, back from flying missions during the CIA's secret Laotian war. Now an Air Force colonel in Washington, he first served as the Pentagon's desk officer for Laos, Thailand and Vietnam, and then as executive assistant to the head of the Defense Security Assistance Agency."[22] By 1975 Erich Von Marbod was an influential Pentagon official who had control of all military assistance to Vietnam. This control included the power to authorize US aircraft and ship usage for any type of cargo he might order. Such a position is of great interest to CIA covert operators.

In embattled South Vietnam, President Nguyen Van Thieu resigned on April 21st and four days later fled to Taiwan, accusing the United States of deserting South Vietnam in, "an inhumane act by an inhumane ally." Thieu's Vice President soon handed over his authority to General Duong Van "Big" Minh. The North Vietnamese Army had by now overrun Xuan Loc and had swung around towards Bien Hoa and Saigon; its advance armored elements attacking the retreating ARVN.

Erich Von Marbod's aide in Saigon was the former naval officer Richard Armitage, who had resigned his military commission and was now reportedly working for CIA. Von Marbod assigned to Armitage the task of removing "military stockpiles" from Bien Hoa as South Vietnam was collapsing at the end of April 1975. Armitage was quoted by former CIA officer Frank Snepp on his actions there:

"There were fifty Vietnamese maintenance types at the airfield when I got to Bien Hoa. I promised to take them to Saigon if they'd help me with loading, and told them to shoot anyone who tried to come in over the fence and interfere. During the next half-hour or so a few stragglers did try to climb over the wire. We held them off, though, and in the meantime packed up fifteen pallets."

Armitage did not say what was so valuable as to be loaded on pallets for removal from South Vietnam during the collapse of the country which had already cost more than 58,000 American lives to defend. He was suddenly called back to Saigon by Von Marbod, who met him at Tan Son Nhut airbase. From here Armitage and Von Marbod went immediately to the Vietnamese naval headquarters in Saigon where they met personally with the Vietnamese Navy chief of operations, Admiral Chung Tan Cang, a close advisor to the corrupt General Dang Van Quang President Nguyen Van Thieu's security assistant. Von Marbod, "an old friend" of Admiral Cang, and Armitage who had been a senior naval advisor to the Vietnamese Navy, informed Cang that the time had come to organize an evacuation of all vessels. Von Marbod and Armitage informed Admiral Cang that the rendezvous point for all Vietnamese naval vessels would be at Con Son Island.[23]

Although the United States left behind in Vietnam thousands of modern vehicles, airplanes, helicopters, tanks, artillery pieces, boats and much other high-tech war material, worth billions of dollars, Armitage and Von Marbod had engaged in a risky operation to remove these particular pallets. It could thus be concluded that the contents of the pallets were extremely valuable, but the cover story established by the government early on, indicates otherwise.

In a formerly "secret" report of the Defense Attaché, Saigon, for the months of January to April 1975, is reference to an "element which deserves special mention," concerning: "reactivation of the Bird Contract," which noted:

"The Ambassador objected to removing anything out of the country which was serviceable...We encountered no problem with the removal of unserviceable assets. As a matter of fact, earlier we had loaded out 30 to 40 mixed armored personnel carriers and tanks as well as a large quantity of unserviceable fire control instruments. The bulk of the readily available unserviceable assets belonging to VNAF were at Bien Hoa. At this point in time, on or about 25 April, Bien Hoa was Coming under increasing enemy fire, both rockets and artillery. Nevertheless the Von Marbod team felt they would be able to operate out of Bien Hoa despite the enemy fire, so the Bird Contract was re-initiated. The first C-1 30 arrived at Bien Hoa on 27 April. During the past several days, the DAO transportation element had built up a backlog of from 85-100 pallets of package mail and household goods at Tan Son Nhut for outshipment...On 27 April, one Bird aircraft loaded out pallets. Mr. Von Marbod had agreed to provide us with additional aircraft over the next several days. As it turned out, the shelling of Bien Hoa reached such an intensity on 28 April that Bird was not able to operate there past mid-morning. The aircraft already inbound were made available to DAO to move our palletized mail and household goods to Thailand. In addition a palletized computer was moved by air that date. Had the Bird Contract not been reactivated, all this material would have remained at Tan Son Nhut."[24]

Bird Air was an Asian airline run by a shadowy American named William Bird, with a long history of links to both overt and secret American military operations in the Far East. Although no direct links between CIA or U.S. military intelligence and the Bird Air company have been proven, such affiliation has been suspected for decades. General Herman "Heinie" Aderholt, who had run the CIA's secret air war in Laos until 1973, once recalled meeting with William Bird together with Bernie Houghten, who was associated with the Nugan-Hand Bank (according to a later Australian government investigation). He revealed to author Jonathan Kwitney: "Mr. Bird and I had run an airlift out of Bombay...Houghten was with a company in Bangkok..." (Kwitney was unable to locate the company, named by Aderholt as 'Alkemal.' The CIA-connected Nugan Hand bank, which collapsed in 1980 causing an international scandal, had been chartered after the fall of Saigon, in 1976.)

At a time when South Vietnam was falling to the Communists, Richard Armitage, who ultimately was to be responsible throughout most of the 1980s for the American prisoners left behind in Southeast Asian captivity after 1973-75, was engaged in a series of covert actions that would eventually lead him to a position of high authority in the Department of Defense. Armitage was already closely connected to what has since been labeled a "secret team" of US officials, including Von Marbod, Ted Shackley, Richard Secord and Frank Carlucci, who like Armitage, were known as men who could get anything done, and who were to promote and protect each other for years to come.

Tan Son Nhut had come under increasing North Vietnamese 122 mm rocket fire, making the use of fixed-wing aircraft more dangerous. Under intense political pressure in Washington, Henry Kissinger finally demanded that an emergency withdrawal to the U.S. fleet offshore under overall command of Admiral Noel Gaylor (until recently, Director of the National Security Agency), must begin on April 29th. This became the largest helicopter evacuation ever attempted, as, in eighteen hours, 70 Marine helicopters carried some 1,000 Americans and 6,000 Vietnamese out of Saigon, including 2,000 airlifted from the U.S. embassy compound, in the face of frantic mobs of Vietnamese begging to be saved.[25] Among the last Americans to leave was Captain Stuart Herrington, who had spent so many fruitless months trying to negotiate with Hanoi officials for the return of all American MIAs and POWs still in their control. He later remembered: "When I was airlifted by helicopter from the embassy roof in Saigon, it was a sad, angering experience...our mission was not accomplished."[26]

Other Americans had different priorities. During the helicopter evacuation of Saigon, on April 29, 1975, Erich Von Marbod had flown in an Air America helicopter to the US fleet offshore, under the immediate command of Admiral Donald Whitmire (who had earlier authored a secret 1974 report on the missing American POWs). There he sent a message to Washington asking for assistance and authority in carrying out his assigned mission of evacuating critical material from the collapsing South Vietnam. Secretary of Defense James Schlesinger had replied with a personally signed authorization for Von Marbod, a civilian, to assume command of the USS Dubuque, a helicopter carrier ship. Although Admiral Whitmire protested, the Secretary of Defense's order was clear. Von Marbod was given authority and assigned his aide Richard Armitage to the control of another US navy ship which he then directed to the rendezvous point off the Mekong Delta at Con Son Island.[27]

It is not known what items of such great importance were removed from Vietnam at this time, but both Armitage and Von Marbod also had control of Vietnamese aircraft which were being used to transport cargo and personnel to secure US bases in Thailand. It is, however, likely that what began as a conceivably legitimate order by the Secretary of Defense to ensure the retrieval of important defense material, would be used by covert operators as a carte-blanche for facilitating their own emergency agenda, whatever it may have been. For one thing, Armitage is credited with having used his authority (as did LeGro and others) to help evacuate thousands of compromised Vietnamese, who might have otherwise perished at the hands of the Communists. Some responsible CIA officials in Vietnam at the time apparently did not consider their own numerous assets among the Vietnamese population as a high priority, as most were abandoned to the communists, along with extensive CIA files. This insured that tens of thousands of Vietnamese who had assisted the Americans would be easily rounded up for imprisonment in forced-labor "reeducation camps," or for execution.

In Washington, D.C., Henry Kissinger characteristically announced prematurely at a press conference that the evacuation of all Americans from Saigon was over, before the last U.S. Marines, backing away from a screaming mob, had boarded the final U.S. helicopter to leave the American embassy's roof, at 7:53 A.M. on April 30th. The North Vietnamese had already fought their way into Saigon and soon occupied key government, buildings, establishing authority and searching for Americans left behind.

CIA officer Frank Snepp, who was an assistant to CIA's Saigon Chief of Station, Thomas Polgar, subsequently obtained classified information revealing that only 537 of the Saigon station's 1,900 "indigenous employees," little more than 25%, were evacuated. In addition, thousands more "direct" and "contract" CIA employees in Da Nang, Nha Trang and other South Vietnamese cities had been abandoned. Also left behind intact for the Communists were extensive files of CIA and other U.S. government employees and agents in Vietnam. In addition, more than 5,000 direct-hire employees of the U.S. State Department, SAFFO, formerly CORDS, U.S. Information Agency, USAID and the Mission Wardens' Office were left behind when Saigon fell to the North Vietnamese. At least two American CIA officers were captured and interrogated by the North Vietnamese in Saigon, one of them identified by Snepp only as "Lew James," and the other being Tucker Gouglemann, a retired CIA officer who had returned to Saigon on his own. (Gouglemann died in captivity after being interrogated by the Soviet KGB.) The bodies of two young U.S. Marines, who had been killed by North Vietnamese rockets while guarding Tan Son Nhut airbase during the evacuation, were also left behind. (Nearly a year passed before Senator Edward Kennedy asked for the return of their remains.) The number of Americans left behind in Saigon may never be fully known, but the North Vietnamese later reported they released 71 of the Americans captured there. Hundreds of Korean and Philippine citizens who had been working for the allies in one way or another, also were captured, and some remained prisoners for decades.

By all accounts Major General Homer Smith's Defense Attaché's Office had the best record in evacuating its Vietnamese employees; according to Snepp, 1,500 out of 3,800 direct employees of the DAO were transported to safety. The U.S. chief of the Military Intelligence Branch at the DAO, Colonel William LeGro, aided by members of his immediate staff like John Berwind and Al Hodges, had secretly evacuated 2,000 Vietnamese from the country before April 30, but later admitted that only 20% of them were "high risk;" those who would be in real danger after a Communist takeover.

North Vietnamese Army Colonel Bui Tinh, a Communist Party official, an expert on American POWs, and now editor of the Army newspaper, was the ranking officer on the armored column that captured Saigon's presidential palace, along with Thieu's successor, General Minh and his staff. Bui Tin's speech to the defeated at the time was later to have a hollow ring: "Between Vietnamese, there are no victors and no vanquished..."[28] The North Vietnamese soon began mass arrests of their recent enemies, using an army of secret police brought from North Vietnam along with 30,000 of Hanoi's Communist Party cadre. A North Vietnamese army of some 200,000 men remained in the South to control the country and to hunt down still-resisting partisan units of ex-ARVN soldiers. The North Vietnamese had captured hundreds of modern U.S. helicopters and airplanes, hundreds more tanks and thousands of other armored vehicles, artillery pieces, trucks and naval vessels of all sizes. Millions of U.S. M-16 automatic rifles, machine guns, grenade launchers and other weapons also fell into Communist hands. Some of this captured equipment was later shipped to revolutionary movements on other continents. The secretly-withheld American prisoners in Vietnam were useful for maintaining captured radio equipment, aircraft and vehicles.

Since the fall of South Vietnam to the Communists, the U.S. Government position has remained that Hanoi's 1975 violation of the 1973 Paris peace accords nullified any U.S. promises for postwar American assistance made by Richard Nixon and Henry Kissinger. The Vietnamese Communist regime later claimed to have released a total of 687 American military and civilian prisoners after the 1973 Paris agreement (including the 591 military POWs), and to have released the 71 Americans held after the fall of Saigon.

Meanwhile, two weeks after the collapse of South Vietnam, the Ford Administration had suddenly faced another crisis in Cambodia. This was brought on by the Khymer Rouge seizure of the American merchant ship SS Mayaguez, and removal of the American captain and crew to hold them as hostages. In planning a response to the challenge President Ford convened a National Security Council meeting which involved his deputy national security advisor, Lt. General Brent Scowcroft, Secretary of State Kissinger, Secretary of Defense Schlesinger, CIA director William Colby and others. Ford instructed Kissinger to seek diplomatic assistance from the People's Republic of China (as the primary sponsors of the Khymer Rouge) in freeing the American crew, and instructions were transmitted to George Bush, chief of the U.S. liaison mission in Peking. At the same time, in case of failure in a negotiated release, Ford also initiated military action, by moving U.S. Navy forces and Marines of the 3rd Marine Division into position for a helicopter-borne assault. Although the Khymer Rouge did release the 40 crewmen of the ship, eventually, it was only after Ford's patience had run out and the U.S. Marines had already landed on Koh Tang island, near Kompong Som, Cambodia, where they came under hostile fire.

The fierce May 15, 1975 battle that ensued involved airstrikes by U.S. F-105s and F-111s and even a detonation of the largest non-nuclear bomb the U.S. possessed, a 15,000-pound BLU82 bomb which was dropped from a C-130 Hercules, as had been used in the battle for Xuan Loc. On the Cambodian island of Koh Tang, by now seeking only to follow Ford's order to withdraw because of the crew's release, a reinforced Marine company suffered 68 casualties while fighting at point-blank range against a larger force of Khymer Rouge. 23 other American troops were killed in a helicopter crash while on their way from Thailand to take part in the Mayaguez operation.

President Gerald Ford was preparing to greet the Shah of Iran, on his state visit to Washington D.C., when he was informed by General Brent Scowcroft: "Mr. President, We are reasonably sure that all the Marines got out." Scowcroft also told Ford that the U.S. destroyers Wilson and Holt were cruising the beaches of Koh Tang, using loudspeakers to try and contact any Marines who had failed to return from the island. 15 U.S. Marines had been killed in action and 50 wounded, while three more Marines who had vanished were carried as "missing in action," and later as "presumed killed." Their actual fate remains unknown.[29] The message of the Mayaguez incident to North Vietnam's rulers must have been that they should ensure the utmost secrecy about the locations of various concentrations of captive Americans, then still in Vietnam.

Only five weeks after the fall of Saigon, North Vietnamese Foreign Minister Trinh reported to the National Assembly of the Democratic Republic of Vietnam on June 4, 1975, that: "The DRV government is ready to discuss with the U.S. Government such questions as U.S. help in healing the wounds of war in both zones of Vietnam, the search for Americans missing in action..." It was clearly a message from the Hanoi politburo that if America was interested in its missing servicemen, it must be prepared to pay for them. But the POW/MIA family members were exhausted and their organizations no longer received much publicity, because the U.S. government had said officially that all POWs had returned and the missing were dead.

In Washington, D.C., Russian expatriate historian and writer Aleksandr Solzhenitsyn, author of the Gulag Archipelago, which in two years had altered the world's consciousness about the enormity of the Soviet slave empire, addressed the subject of the missing American POWs of Vietnam on June 30, 1975. His words confirmed the existence of the secret world of "wink and nod diplomacy" that Jerry Mooney of the NSA and Henry Kissinger of the NSC had inhabited for years:

"If the government of North Vietnam has difficulty explaining to you what happened with your brothers, with your American POWs who have not yet returned, I, on the basis of my experience in the Archipelago, can explain this quite clearly. THERE IS A LAW IN THE ARCHIPELAGO THAT THOSE WHO HAVE BEEN TREATED THE MOST HARSHLY AND WHO HAVE WITHSTOOD THE MOST BRAVELY, THE MOST HONEST, THE MOST COURAGEOUS, THE MOST UNBENDING, NEVER AGAIN COME OUT INTO THE WORLD. THEY ARE NEVER AGAIN SHOWN TO THE WORLD BECAUSE THEY WILL TELL SUCH TALES AS THE HUMAN MIND CANNOT ACCEPT.

"A part of your returned POWs told you that they were tortured. This means that those who have remained were tortured even more, but did not yield an inch. These are your best people. These are your first heroes, who, in solitary combat, have stood the test. And today, unfortunately, they cannot take courage from your applause. They can't hear it from their solitary cells where they may either die or sit for thirty years..." The American mass media, however, chose to ignore Solzhenitsyn's warning about the fate of the missing POWs, and to largely ignore the subject as much as possible.

Yet, also in June 1975, North Vietnamese Premier Pham Van Dong wrote a letter to 27 members of the U.S. House of Representatives, stating that the U.S. contribution to healing the wounds of war in both zones of Vietnam was to be linked to information on the missing in action Americans.[30] Ten weeks later, a special Congressional committee to investigate the missing POWs and MIAs of the Vietnam War, was commissioned by the House of Representatives on September 11, 1975. It was headed by Gillespie "Sonny" Montgomery, a Democrat from Mississippi (who, in Washington had once befriended a freshman representative from Texas named George Bush, later to be appointed U.S. Ambassador to the UN and then to China). Other members of the committee were John Moakley of Massachusetts, the outspoken Patricia Schroeder of Colorado, Henry Gonzalez of Texas, Richard Ottinger of New York, a future Presidential hopeful, Tom Harkin of Iowa, Jim Lloyd of California, Paul McCloskey of California, Benjamin Gilman of New York and Tennyson Guyer of Ohio.

During the Committee's time in Vietnam investigating the fate of MIAs and others last reported as POWs, observers noted that Chairman Montgomery seemed unduly anxious to accept the repeated Vietnamese statements that they were not holding any Americans prisoner. And when the U.S. delegation asked for a 'full accounting' of POWs and MIAs that Hanoi would have knowledge of, the Vietnamese Communist replied that they were still expecting the promised U.S. aid that had never been delivered, 'to heal the wounds of war.' Although the Committee had already been told by State Department official and former Kissinger aide Lawrence Eagleburger, speaking under oath, that no secret agreements had been made between Nixon and Kissinger and the North Vietnamese in 1973, in fact when the Committee was in Hanoi asking about American MIAs, they were told by the Vietnam's deputy foreign minister, Phan Hien, that the Vietnamese had the letter written by Nixon to Pham Van Dong on February 1, 1973, promising $3.2 billion in U.S. aid, and they threatened to release it. The astonished Congressmen then returned to Washington and spent months more attempting to get the document declassified and released, which did not occur until the new administration of Jimmy Carter in 1977. Kissinger continued to deny that the document was a secret agreement, and claimed it had no validity anyway because the Congress had not approved and because North Vietnam had broken the peace accords by conquering

South Vietnam. After being threatened with a subpoena by Chairman Lester Wolf, of the House Subcommittee on Asian and Pacific Affairs, Nixon finally admitted on the telephone that he had promised the aid.

In August 1976 the Communist government in Hanoi released 38 Americans and their families, who had been captured during the fall of South Vietnam, but denied holding other Americans. President Ford threatened to block Vietnam's membership in the United Nations unless the MIA issue was addressed, and the U.S. government opposed any aid to Vietnam until 800 American MIA cases were resolved. In September 1976 the Vietnamese regime suddenly released an American captured in the Mekong Delta of South Vietnam in April 1975, named Arlo Gay. He had attended lectures on Communism which helped him survive and become friendly with his guards, but was long held in solitary confinement (his captors even tried to contact him after his release). Arlo Gay's long being held alone in a Laotian cave for his eventual use as a "messenger," to be sent at a significant time, fitted a pattern of similar Communist actions (dating back to 1920, and the release of Private Sydney Vikoren). It was later learned that Mr. Gay had also been held in Son Tay Prison, North Vietnam at the very time the Vietnamese were telling the Montgomery Committee that they were holding no American prisoners. Hanoi's action appears to have been a deliberate diplomatic signal that no captive Americans military or civilian, were considered legitimate prisoners of war by Vietnam, but as criminals deprived of their national status, thus allowing them to say: 'We hold no live Americans.' The indication being that to save face on both sides, the United States must bargain for a group of stateless criminals, and thereby condemn the legitimacy of a ten-year-war which had cost far more casualties than the Korean conflict had.

Yet, the testimony by a U.S. official which carried perhaps the greatest weight in the ultimate findings of this Congressional investigation, was the answer to a question from Congressman Jim Lloyd, by Lt. General Vernon Walters, Deputy Director of the CIA under George Bush, who long before had served as a military translator for Presidents Truman and Roosevelt, and later for Henry Kissinger:

"ALL I CAN SAY, CONGRESSMAN, IS THAT WE DO NOT HAVE HARD EVIDENCE OF THE EXISTENCE OF ANY LIVE MISSING IN ACTION. NOW, I CAN'T TELL YOU THERE ISN'T ANY, ALL I CAN TELL YOU IS THERE IS NO EVIDENCE TO US AT THIS TIME."

Given the CIA's preeminence in the intelligence community, such a statement must carry great weight to members of Congress, if it was not later exposed as yet another lie by a CIA official. The Committee was in the same position as Congressman Hamilton Fish had been in with Colonel Robert Foy of U.S. Military Intelligence during 1930, of having to accept the word of senior U.S. officials, in the latter case from CIA and the Defense Department, that they had no knowledge of American POWs being held in captivity by the Communists. Although the nearly simultaneous, Congressional Church Committee hearings on the CIA established many precedents of CIA lying to Congress about covert actions, the lesson was not noted by Chairman Montgomery, whose protege in the Congress, George Bush, was appointed CIA Director by President Ford during the Committee's life.[31]

The Committee chose to not fully pursue the exposure of a truth already revealed by eyewitnesses, and by NSA-decrypted enemy transmissions, which would have proven conclusively in 1976 that many POWs had been held alive since the war as hostages in yet another Communist demand for diplomatic recognition and financial aid, to thus guarantee the legitimacy of Hanoi's regime. CIA official Vernon Walters' assessment thus became, in effect, the line of the Committee and the future line of the U.S. Government, for years to come.

In sworn affidavits for Congress, retired NSA analyst Jerry J. Mooney later named seven American POWs captured during the Vietnam war and listed as "China-bound" as a result of NSA analysis, who were included on the list also containing 14 "Moscow-bound" POWs, which had been forwarded from the B-Group of NSA for the use of U.S. negotiator Henry Kissinger. The "C-B's" listed in early February 1973 were: James Barr, Navy Lt. Commander Ralph C. Bisz, Navy Captain Billy Cartwright, Navy Lt. Commander James E. Dooley, Marine Corps Captain Hugh Fanning, Marine Corps Major Stephen Kott and Charles Lee. Fanning's remains were reported as identified by the Central Identification Laboratory in Hawaii and returned by the Pentagon in the 1980s, but when his wife Kathryn Fanning had them analyzed by an outside expert, she maintained publicly that they were unidentifiable and, "not the remains of my husband, Hugh Fanning."

During that war at least ten other known American aviators who had been shot down over or near Chinese territory, had disappeared. They included: Navy officers Terence Murphy and Ronald Fegan on April 9, 1965, Lt. Colonel Dean Pogreba on October 5, 1965, reported by radio Peking as captured (whose family claims has been seen in captivity), Navy lieutenants William Albert Jr. and Larry Jordan, with crewmen Reuben Morris and Kenneth Pugh on April 12, 1966, Navy Lt. Commander John Ellison and Lieutenant James Plowman, who was photographed in captivity; and Navy Lieutenant Joseph Dunn, who was seen under a good parachute after being shot down on February 14, 1968, off Hainan Island, south China. All had disappeared, although the Chinese returned what they claimed to be the ashes of navy crewman Kenneth Pugh on December 16, 1975, as the Montgomery Committee was deliberating.

A longtime purveyor of Soviet policy, "journalist" Vitaly Levin (Vitali Lui), who wrote under the pseudonym "Victor Lewis" and was considered an agent of the KGB by western newsmen, published an article in the London Evening News of February 1, 1973 claiming that the Chinese held American POWs from Vietnam in adjacent Yunan Province: "Hanoi has moved American prisoners to China, out of reach of any U.S. rescue attempts." (And to protect them from bombing) "...The camp exactly duplicates Vietnamese conditions so the prisoners remain unaware of their new location in China." (Given the deterioration of Chinese-Vietnamese relations after 1967 it is possible that these POWs were thus held from earlier transfers, or perhaps Victor Lewis was diverting attention from, or providing a veiled explanation for, the Soviet practise of removing American POWs from Vietnam to the USSR.)

George Bush had been asked by Montgomery's House Select Committee on POW/MIAs on September 17, 1975, if there was: "any data on POWs being held in China or elsewhere outside Indochina?" The Records of the Hearings on Americans missing in Southeast Asia quoted Bush's reply:

"Ambassador George Bush then Chief of the U.S. Liaison Office to the People's Republic of China, stated to the Select Committee that the DEPARTMENT OF STATE WAS SATISFIED THAT NO AMERICANS WERE BEING HELD AS PRISONERS OF WAR IN CHINA. Ambassador Bush said that reports that American prisoners of war had been moved from North Vietnam to China were, in his view, false, AND THAT NO SUCH MOVEMENT WAS POSSIBLE WITHOUT THE CHINESE KNOWING OF IT. There has been no change in intelligence to warrant a change in these conclusions. NOR IS THEIR ANY INFORMATION OTHER THAN SPECULATION AND RUMOR TO INDICATE THAT AMERICANS CAPTURED IN INDOCHINA ARE BEING HELD OUTSIDE INDOCHINA IN SOME OTHER COUNTRY."[32]

This statement by the future President may have reflected either an inadequate briefing by Bush's Beijing CIA station chief and military attaché during his years in China, about Chinese-held American POWs from the Vietnam or Korean wars, or possibly a negative Chinese response to a question from Bush or the Department of State on the subject. It may also have reflected his obligation to keep classified "national security" matters from becoming public knowledge, including knowledge of a 1973 Soviet disinformation campaign to shift attention from the "Moscow-bound" American POWs. Bush has never elaborated on the subject.

Shortly after this, as CIA Director from 1976 to 1977, George Bush should have been fully informed by the Agency's deputy directors, NIO's and archivists about the secretly-withheld

American POWs of Vietnam transferred to China and Russia, and those withheld as captives after three earlier wars. Retired NSA analyst Jerry J. Mooney informed this writer (in 1990) that in the spring of 1976 he had gone over to CIA Headquarters at Langley from Fort Meade with an NSA supervisor, "Whitey" Machelski, the A GROUP-NSA Senior Airborne Staff Officer-Programs (COMFY COT). Machelski was a civilian NSA official who had retired from the Air Force as a Colonel after service as a pilot and in intelligence. In a 1991 interview with the author, Mooney remembered: "Whitey Machelski knew about the MB's, the Moscow-bound POWs from Vietnam, we had already discussed them together. We each agreed at that time: 'I wish we could have had control over airborne collection for the Soviet Union, because then we could do the same thing as with China.'

Machelski was one of the senior NSA officials in the COMFY COAT program, which was a planning and program title name for worldwide airborne intelligence collection schedule routes. Mooney was along as a senior NCO cryptanalytic and reporting expert to provide specific data from his specialty areas. In requesting approval for some specific coverage Machelski and Mooney briefed the CIA National Intelligence officer (NIO) for China/Korea area (Korea to Hainan Island), on certain aspects related to the U.S. China-Bound POWs of the Vietnam War. The information on the China-bound U.S. POWs was collected from intercepts by different sections of NSA, the same as the Moscow-bound, but Mooney had become fully aware of these prisoners-without-sightings through his own analytical work and that of several counterparts, both at NSA and DIA.

"We were talking about moving the U2 "Olympic Torch" from Southeast Asia to Korea, which picked up almost everything, to become the "Olympic Game," over Korea and China. At CIA Machelski and I tried to talk the NIO for the China-Korea area, from Korea to Hainan Island off the coast of North Vietnam, into extending the flight path of the U2 out over the Yellow Sea, to pick up Chinese Communist communications from SHEN-YANG (Mukden, Manchuria), where they were doing their CCF-3 fighter plane testing; which was just one area they needed American POWs in. One reason we gave was the existence of the 'China-bound' POWs of Vietnam, who were invaluable to the Chinese in overcoming technological problems, and also to pick up Chinese north Yellow Sea missile testing modes, which NSA (and DIA) felt might also have a China-bound U.S. POW connection. It was really just business as usual, and the NIO didn't resist or disagree at all, he just said: "OK, no problem, that's really logical and it's a good idea, and we will definitely take this into consideration on the area of coverage..."

In Mooney's subsequent fitness report for his September 1975 to June 1976 service, this briefing was referred to by his immediate Chief at NSA's B-209, in a notation signed on June 14, 1976 in which the NSA's Mr. Ross M. Cory wrote:

"SMSGT MOONEY REPRESENTED HIS ORGANIZATION IN DIRECT DEALINGS WITH A U.S. NATIONAL INTELLIGENCE OFFICER, WHO PLACED HIGH VALUE ON SMSGT. MOONEY'S BRIEFING..."

As a staff officer to the Director of Central Intelligence, the NIO concerned would have been remiss in his duty if he had not reported to then-Director of Central Intelligence George Bush the suspicions that U.S. POWs were sent to China from Vietnam. In addition the CIA had its own intelligence sources inside Communist China and North Vietnam. It therefore seems reasonable to assume that as CIA Director, George Bush knew, from this time forward, that the Chinese Communists were holding American prisoners of war. A simple and necessary briefing by the CIA's NIO for the Soviet Union would have provided Director George Bush with information from decoded enemy communications, aircraft tracking and CIA intelligence reports, on the Moscow-bound POWs of Vietnam in addition to the 1967 and 1968 East German and Czech reports of Americans in Soviet control, and CIA archivists had the documented background of Communist-held U.S. POWs from WW I, WW II, the Cold War and Korea. Thus, in his later roles as Vice President and President, it must be considered certain that George Bush knew from 1976 onward, that U.S. POWs had been held in China and were probably still alive, and that despite this, established U.S. security policy dictated that countering the Soviet nuclear threat to America, by overlooking of China's human-rights shortcomings, was the safest policy for the U.S. government to pursue.

The record does reveal that during George Bush's short, 1976-77 tenure as CIA Director, he particularly lobbied for, and obtained a $500 million CIA budget increase for two high-tech photo-reconnaissance satellite systems that could penetrate cloud cover around the world, and also four ground stations and other satellite and navy systems to intercept communications of different countries. His interest in obtaining this additional intelligence-gathering capacity could well have been related to his gaining certain knowledge as CIA Director of classified information about Communist-held U.S. POWs in the USSR, China and Southeast Asia. Bush also permitted a critical, independent analysis of CIA performance by the President's Foreign Intelligence advisory board (PFIAB) when he was director, something that his predecessor William Colby had refused.[33] The new satellite and crypto equipment is reported to have served the CIA well, but nearly all the photographic information and special crypto-intelligence thus obtained on surviving U.S. POWs has remained classified to the present writing. The POW/MIA matter is considered a "national security issue" by the National Security Council and much of this information has even been denied to the 1991-1993 Senate Select Committee on POW/MIA Affairs.

A few months after the 1976 meeting with the NIO for China, Jerry J. Mooney was thinking about retiring from the military and the intelligence community. He and his wife Barbara were a small-town couple at heart, who had taken in foster children in addition to their own, during the years of duty at Fort Meade. Now, they thought about going back to the west to finish raising their daughters in a more natural environment. Mooney could have stayed on with the Air Force, or with the NSA, for years more. At the time of his retirement, when he received an unusual personal letter of commendation from then-Secretary of Defense Donald Rumsfeld, he was offered the job as NSA station chief in Japan (which didn't officially exist), but he turned it down, even though he might have risen high as a civilian NSA official. The things he already knew from 20 years of cryptanalytic and reporting service were a heavy burden to him. He wanted to go back to live in Wolf Point, Montana, a windswept place of the Assiniboine Indians, just up the Missouri River from Fort Buford, North Dakota, where Sitting Bull's last Sioux hostiles had come in from Canada to surrender, in 1881. He was convinced until his retirement in early 1977 that the United States Government would continue to exert every effort, secretly, to recover the American prisoners of war whom he knew for certain had not come home[34][35]

Secretary of State Henry Kissinger had testified before the Senate Appropriations Subcommittee on Foreign Assistance on April 14, 1976, saying:

"We have indicated that we are prepared to have discussions with Vietnam about our relationship. We have a reply from Vietnam that economics is their interest...We cannot accept the proposition that we have an obligation to give aid, which we have not. We believe that the Paris accords have been breached so completely that it would be completely absurd to let only one article survive when all the other obligations have been totally abridged by North Vietnam..."

On July 21st, the Undersecretary of State for Political Affairs, Philip Habib, testified before the Montgomery Committee that there had been no accounting for 320 U.S. MIAs in Laos and said also: "IN GENERAL TERMS, THE QUESTION OF OUR OVERALL RELATIONSHIP WITH THEM HAS BEEN LINKED TO THE SO-CALLED OBLIGATION THAT WE HAVE WITH RESPECT TO BINDING UP THE WOUNDS OF WAR."

President Gerald Ford later addressed the National League of Families on July 25th, telling the members that Vietnam must resolve the POW/MIA issue if it wished to have normalized relations with the United States. In September, President Ford met with the U.S. Ambassador to the UN, William Scranton, and informed him that the United States would veto Vietnam's application for membership in the United Nations. On November 12, the U.S. and Vietnam held talks in Paris that had been announced earlier by Secretary of State Henry Kissinger. Both sides remained intransigent on their respective issues of an MIA accounting and war reconstruction aid.

The December 13, 1976, "Final Report" of Montgomery's House Select Committee on Missing Persons, stated as its major conclusion: "No Americans are still being held alive as prisoners in Indochina, or elsewhere, as a result of the war in Indochina...The U.S. wishes and deserves an accounting for the missing. The Indochinese demand reconstruction of their war-torn countries..." The report also contained a Perspective section with the following explanation: "The purpose of this section is to compare and contrast statistics of previous military experiences with that of the recent Indochina war." This section was compiled by Pentagon officers working for the Committee to illustrate the magnitude of the POW/MIA problem in the past, with particular emphasis on the losses of World War II. As a WW II veteran, this putting of Vietnam into perspective may have appealed to Montgomery, who was well known for his contemptuous attitude toward Vietnam veterans. It is not presently known whether Mr. Montgomery was aware all along of the Soviet retention of US POWs from WW II and Korea, or if he learned anything of the true story of the Moscow-bound POWs from Vietnam during the life of his committee. If he knew, he succeeded well in keeping the secret from the Congress and the American people, to include all of the MIA families of three wars who resided in the State of Mississippi. Representative Paul McCloskey and Representatives Tom Harkin and Patricia Shroeder, later both senators, all supported the Chairman Sonny Montgomery's view, that there was no evidence that live Americans were held in Southeast Asia

Not all members of the Committee agreed with the final report, and Benjamin Gilman and Tennyson Guyer became an opposing minority, against considering the POW/MIAs of Vietnam as 'all dead.' Congressman Moakley also disagreed with continuing the 'case reviews' to presume the MIAs as all dead, and urged the incoming Democratic President, Jimmy Carter, to discuss the issue with the POW/MIA families before any final decisions were made. The National League of Families rejected the results of the Montgomery Committee investigation," and League Director Carol Bates stated: "The Committee staff relied primarily on Department of Defense and State Department Liaison personnel to do the investigation. As investigative journalist and author Rod Colvin noted in his 1987 study of Vietnam POW/MIAs, "First Heros," the media generally lauded the work of the Montgomery Committee, with the WASHINGTON POST editorializing: "...So exceptionally well done...so comprehensive, incisive and compassionate a study that it ought to be treated as a textbook of sorts for further consideration of this excruciating issue left over from the Indochina war..."

Despite the findings of the Montgomery Committee that no Americans remained in captivity in Vietnam, intelligence reports continued to reach CIA Headquarters in Washington, which indicated otherwise. These reports were routinely distributed to the Defense Intelligence Agency's POW/MIA Branch Chief, Commander Charles F. Trowbridge Jr., and to other areas of Government such as the Department of State and the White House-National Security Council (NSC). One such report (FIR-317/09153-76) from an intelligence source in Vietnam is titled "AMERICANS REMAINING IN SOUTH VIETNAM AND DETENTION OF AMERICANS IN NORTH VIETNAM," with the date of information reported as September 1975-January 1976, has many lines blacked-out by U.S. Government censors, but still tells a far different story than the American people heard from Mr. Montgomery and other members of his Committee:

"...(Blacked-out) about seeing a Caucasian male who may be an American, in a 'new economic area' in Dinh Quan District, Long Khan Province (blacked-out) said that the 'American' wore tattered clothes, had no weapon, and was 'very miserable.' The 'American' obtained his food from a settler in the area. No further details on this Caucasian, whose name is unknown. The Communist authorities have opened several rural areas for settlement and cultivation by residents from Saigon and other urban areas. (Blacked-out)...saw Caucasians, who he assumed were Americans, in Saigon...described them as without funds and working as pedicab drivers. (blacked-out)...did not provide the number, names, or physical description of these pedicab drivers. According to FIR-317/09152-76, one American who spoke Vietnamese was seen driving a pedicab in late 1975 in Saigon. There are approximately 50 Americans and their dependents known to be living in Saigon at this time.

"(Blacked-out)...The instructor, Ran Y (phonetic), at a reeducation course held....in Saigon, told.... and the attendees that 'A NUMBER OF AMERICANS WERE BEING HELD IN HANOI PENDING THE SETTLEMENT OF AN AGREEMENT WITH THE U.S. ON AMERICAN RECONSTRUCTION AID TO VIETNAM. THE INSTRUCTOR SAID THAT THE VIETNAMESE WERE 'NOT STUPID' AND WOULD HOLD THESE AMERICANS UNTIL THE THREE BILLION DOLLARS IN PROMISED AID WAS RECEIVED."

The CIA then created it's own solution to this report by adding information to the source's statement, which concerned nine American missionaries who had been released, in order to explain away the reports of American POWs held in Hanoi:

"(Blacked-out)...The purpose of the course instructor was to indicate that the Vietnamese know how to deal with Americans. At the time this course was given, nine Americans, who had been captured in Ban Me Thuot in April 1975, were being held in Hanoi.

These Americans were subsequently released."

Declassified report FIR-317/09155-76, on Vietnam, has an information date of late March-July 1976, and concerns two Americans named "Johnny" and "Bill," formerly employed at the Long Binh military base, who were captured by the Communist authorities with 20 Vietnamese near the seacoast town of Vung Tau, South Vietnam. A mock trial was held in the open by Communist authorities during which all these prisoners were sentenced to death for having been involved in "resistance activities," although the two Americans were reported by the source to be merely in hiding. The report also states that, IN MAY 1976 (blacked-out) A SAIGON RESIDENT...SAID A FRIEND OF HIS TOLD OF SEEING 'MANY' AMERICANS IN THE HANOI AREA. THE FRIEND ADDED THAT ALL OF THE PRISONERS WERE EITHER SERIOUSLY INJURED, I.E. CRIPPLED BY WAR WOUNDS OR SUFFERING FROM MENTAL DISORDERS RELATING TO THEIR LONG IMPRISONMENT OR HARSH TREATMENT...(THE) FRIEND, WITH WHOM HE WAS IN FREQUENT CONTACT, WAS A NORTH VIETNAMESE (Ten lines blacked-out by censors).

"IN THE FIRST HALF OF 1976, (two lines blacked-out) SAID THAT ON SEVERAL OCCASIONS NVA OFFICERS...SAID THAT AMERICAN PRISONERS WERE STILL BEING HELD IN NORTH VIETNAM. THEY ALSO TOLD...THAT SINCE THESE PRISONERS WERE ALL SICK OR BADLY INJURED, THE DEMOCRATIC REPUBLIC OF VIETNAM (DRV) AUTHORITIES WERE AFRAID OF A POSSIBLE UNFAVORABLE IMPACT ON PUBLIC OPINION SHOULD THEY BE RELEASED..."

The U.S. government, however, did not really want to know about such American prisoners, because it had already told the American public that they did not exist. Thus, intelligence about these living-dead Americans was "classified" by the government away from the public eye, as it had been after each succeeding war since 1918, and U.S. officials denied any knowledge about them. In America life went on without the missing prisoners of war as the nation celebrated the Bicentennial of the founding of the United States. Tall sailing ships crossed New York harbor to commemorate the event and veterans groups marched once again in communities across the country, remembering their comrades who had died or disappeared in America's foreign wars. In spite of continuing Cold War "incidents," relations with the USSR were relatively calm, and the year before much propaganda was made about the U.S. Apollo and Soviet Soyuz spacecraft link-up in space. In July 1976, two hundred years after the declaration of independence, an American Viking 1 spacecraft landed on the planet Mars.

The nation was in the midst of a post-war recession, which contributed to a general discontent, manifested in part by a still-strong "counterculture" movement among the younger generation. Many other Americans were more concerned with the outcome of the

World Series or the fall football season than foreign affairs, which were largely conducted in secrecy. That same year in the People's Republic of China, Mao Tse-tung died, while civil wars raged in Lebanon and in Angola, where the CIA was once again engaged in covert operations. Gerald Ford's popularity had suffered from his 1974 pardoning of Richard Nixon for all crimes the former President might have committed in the Watergate affair, and his subsequent performance had apparently failed to impress the voters.

In November 1976, the electorate chose a relatively unknown liberal Democrat from Georgia, named Jimmy Carter, to be the 39th President of the United States, and Walter Mondale as Vice-President. Carter was from Plains, Georgia, and grew up on a farm. In 1942, he had entered the U.S. Naval Academy at Annapolis, where he was assigned during the remainder of the Second World War. From 1946-1953 he had served as a naval officer in surface ships and later in submarines. He was involved in development of the first U.S. nuclear-powered submarines under Admiral Hiram Rickover, and later served on one of them: the Sea Wolf, before resigning his commission and returning to farming in Georgia. Carter had considerable knowledge of the security establishment, but he was essentially a southern agriculturist and politician who in 1970 became a moderately successful governor of a southern state, and had opposed the Vietnam war during that period. The powerful interests who controlled the American mass media had apparently considered him a viable alternative to Ford; but one who could be counted on to maintain the status quo.

In January 1977, James Earl Carter was sworn in as President of the United States. He appointed Cyrus Vance to be his Secretary of State, and Harold Brown as Secretary of Defense. Admiral.

Stansfield Turner was appointed to replace George Bush as Director of Central Intelligence. He announced a human rights theme for his administration, and attracted considerable attention in 1977 with his strong support for human rights in the Soviet Union and elsewhere. Carter banned U.S. aid to a number of right-wing dictatorships in Latin America, Asia and Africa, that were identified as human-rights violators. Communist Vietnam and North Korea also continued to be isolated and denied U.S. aid under Carter. He also achieved ratification of a treaty that would return the Panama Canal to the Republic of Panama, and later in his administration, established full diplomatic relations with the People's Republic of China.

The National League of Families of American Prisoners and Missing in Southeast Asia had disagreed strongly with the findings of the Montgomery Committee's report, which was presented to the new President on January 30, 1977. The League immediately requested a halt in any further "Case Reviews" and "Status Changes" for POWs and MIAs, in which their loved ones were being arbitrarily declared "Presumed Killed" by the Pentagon, as had been done after every war in which Americans disappeared in Communist control since 1919. When the POW/MIA families demanded a meeting with President Carter to argue their points, National Security Council Staffer Michel Oksenberg, wrote a series of "Talking Points" for the President to address the family members, which he forwarded in a February 9, 1977 memorandum to NSC Adviser Zbigniew Brzezinski. This included the following planned-out phrases that the President was to utter for the benefit of POW/MIA family members:

"I am very concerned about this issue. I pledged to make this a priority concern, and I am moving to redeem that pledge...I am trying to improve relations with Vietnam, and have indicated that an accounting of MIAs is essential to move the relationship forward...I place great emphasis, as you know, in healing war wounds. The families of those who were lost in action have suffered great pain and anguish; YOUR WOUNDS WILL NEVER BE HEALED...I met with the Montgomery Committee, as you know, on January 30. I thought that report was useful, but we still have to try to find the MIAs.

"I understand your desire for me to appoint a select Presidential committee. I recognize the merit of that proposal, and I am considering that. But whether through this route or another, you may rest assured I will not act on this issue until I am fully convinced we have done all within my capacity to acquire a full accounting."[36]

Meanwhile, thousands of Vietnamese, Cambodian and Laotian refugees fleeing Indochina to escape the oppression of the Communist masters who were now taking vengeance on those who had opposed them, were never interviewed by American intelligence on the subject of U.S. POWs left behind in captivity after the fall of Saigon, Pnomh Penh and Vientiane. Likewise, the great volume of wartime and post-war intelligence information on U.S. POWs who remained in captivity after 1973-75, from both human and SIGINT sources, was practically ignored in negotiations with the implacable Hanoi regime.

The meeting between Jimmy Carter and the POW/MIA family leaders, which amounted to mere 'window-dressing,' came off as planned, and on February 14, 1977, Secretary of Defense Harold Brown, wrote the following memorandum and sent it to President Carter:

"I understand that at your meeting on February 11 with leaders of the National league of families, you indicated that the moratorium on unsolicited status changes for MIAs would continue. FROM OUR CONVERSATION BEFORE THAT MEETING, MY UNDERSTANDING IS THAT THE DEPARTMENT OF DEFENSE SHOULD GO THROUGH ALL THE FILES, GETTING READY TO MOVE ON A PROGRAM OF UNSOLICITED

STATUS CHANGES LATER THIS YEAR DEPENDING UPON THE OUTCOME OF NEGOTIATIONS WITH THE VIETNAMESE. DO I CORRECTLY UNDERSTAND YOUR WISHES? (signed) Harold Brown."

President Carter was clearly unhappy with the POW families disagreement with the Montgomery report and the overall handling of the MIA situation, and he responded to Secretary of Defense Harold Brown in a hand-written note on White House stationary, also dated February 14, 1977:

"I DID NOT LIKE THE PRESS BRIEFING GIVEN PAST WEEK ON THE MIA MEETING. ALSO, WE ARE GIVING CONFLICTING SIGNALS ON CAPITOL HILL WITH SCHECTER AND BENNETT BOTH REPRESENTING MY, YOUR AND CY'S (Secretary of State Cyrus Vance) POSITION. I REALLY SEE NO REASON WHY NSC NEEDS A CONGRESSIONAL RELATIONS PERSON."[37]

The Woodcock Commission on POW/MIAs was duly appointed by President Jimmy Carter on February 25, 1977. The Commission was chaired by Leonard Woodcock, President of the United Auto Workers, and a major political supporter of Carter; it also included the Senate Majority leader, Mike Mansfield, U.S. Representative "Sonny" Montgomery of Mississippi (to insure, no doubt, that the Commission's findings would not differ too greatly with the report of that Congressman's POW/MIA committee), former Ambassador Charles Yost and Marion Wright Edelman, of the Children's Defense Fund. The Woodcock Commission members were briefed by the Department of State on March 7, 1977 on the Vietnamese position of linking Article 8(b) on POW/MIAs from the Paris peace accords to Article 21 involving $4.25 billion in promised U.S. assistance from the same accords. They were also shown a still-classified copy of the 1973 Nixon letter to Pham Van Dong promising the aid.

While in Hanoi for all of one week after March 16, 1977, Leonard Woodcock seemed anxious to accept the word of the Vietnamese that they were holding no American prisoners, according to then-Defense Department POW/MIA Chief, Roger Shields (who had himself unwillingly helped declare the missing POWs dead on April 12, 1973). Hanoi claimed that they had: 'Asked all Americans who wanted to leave to register and that all who registered wanted to leave...We asked them to register and, if they didn't how would we know?'

Other U.S. officials present included Kenneth Quinn and James D. Rosenthal of the State Department. On March 17th the Commission members met with Vietnamese Prime Minister Pham Van Dong and Vice Foreign Minister Phan Hien. Woodcock was told by the Vietnamese that the issue of POW/MIAs is 'inter-related' with the U.S. obligation to provide reconstruction aid to North Vietnam. Woodcock said to Phan Hien: "Let me stress we are not unrealistic in our expectations. We recognize that no information may ever be recovered on many of the missing." Montgomery informed Phan Hien that he didn't expect Congress to provide war reconstruction aid for Vietnam.

The Defense Department files on POW/MIA cases considered discrepancy cases' of which Hanoi would have knowledge, were not given to the North Vietnamese by Woodcock, who later told President Carter that there, '...IS NO EVIDENCE THAT ANY AMERICAN PRISONER OF WAR FROM THE INDOCHINA CONFLICT REMAINS ALIVE... AMERICANS...WHO REGISTERED WITH THE FOREIGN MINISTRY AND WHO WISHED TO LEAVE, HAVE PROBABLY ALL BEEN ALLOWED TO DEPART THE COUNTRY...WE BELIEVE THEY HAVE ACTED IN GOOD FAITH..."[38]

At this time the Vietnamese media stated that Pham Van Dong had "in particular, underscored" to Woodcock the U.S. obligation to provide aid for Vietnam. Members of the Woodcock Commission had met in Vientiane, Laos with President Souphanouvong on March 20th, and Woodcock reported that: "The Lao made clear to the Commission that they connected the MIA problem with that of U.S. assistance to heal the wounds of war and rebuild their country. They expressed the belief that the two problems should be resolved together."

President Carter approved of the rapid and superficial findings of the Woodcock Commission on March 24th, that no Americans remained alive in Indochina, saying in a press conference on March 24th: "I don't feel we should be forced to pay reparations at all." Later in March he said: "Every hope that we had for the mission has been realized...," and adding that the Vietnamese had not insisted on linking, "economic allocations of American funds with the MIA question. We believe they have acted in good faith."[39] Thus was the POW/MIA issue put to rest, for a time.

Yet during this same period, William W. Wells, Deputy Director of Operations for the Central Intelligence Agency, sent a March 8, 1977, CIA memorandum to James D. Rosenthal, Director of the Vietnam, Laos and Cambodia Desk of the Bureau of East Asian Affairs at the State Department, and to Frank A. Sieverts, also at State Department, and to Commander Bruce L. Heller of the POW/MIA Branch at DIA. It was titled: "Possibility of U.S. Prisoners Being Held by the Socialist Republic of Vietnam for Use in Negotiations on Aid from the U.S. for the SRV." This document carried a "date of information" of February 1977, and was declassified in May 1978:

"IN A PRIVATE CONVERSATION (two lines blacked out by CIA censors) COMMENTED ON THE VISIT OF A U.S. DELEGATION TO HANOI IN MID-MARCH 1977. IN RESPONSE TO STATEMENT ON THE U.S. INTEREST IN THE U.S. PERSONNEL MISSING IN ACTION IN VIETNAM...(blacked-out) SAID THAT HANOI NATURALLY KNOWS WHERE U.S. PILOTS SHOT DOWN OVER NORTH VIETNAM ARE BURIED. IN RESPONSE TO A STATEMENT THAT THE SRV KNOW LONGER HAS ANY AMERICANS STILL UNDER

DETENTION...SAID THAT THERE ARE AMERICAN PRISONERS OF WAR, SOME OF WHOM ARE MEMBERS OF WEALTHY FAMILIES, STILL IN THE SRV AND THEY WOULD NOT BE RELEASED UNTIL U.S. FINANCIAL AID FOR THE SRV WAS FORTHCOMING."[40]

Here, CIA Deputy Director Wells inserted this "comment":

"Other than one American civilian who was arrested in Saigon after its fall in 1975, the SRV is not known to have any live American prisoners." This was a totally untrue statement, given the volume of information from U.S. decryptions of North Vietnamese communications about live U.S. prisoners in Communist control, and other intelligence from human sources which remained classified for years to come. This note constitutes an adherence to a "party line" by the CIA with other branches of the U.S. Government, that there were no other U.S. prisoners. The report continued with more of the source's information in paragraph two:

"(Source blacked-out)...INDICATED THAT THE INFORMATION ON MIAs WAS THE ONE TRUMP CARD THE SRV HAD IN THE FORTHCOMING NEGOTIATIONS WITH THE U.S. AND EXPLAINED THAT THE SRV, AS THE VICTOR IN THE VIETNAM WAR, WOULD TAKE ADVANTAGE OF THIS SITUATION IN NEGOTIATING WITH THE U.S. FOR AID IN ECONOMIC RECONSTRUCTION...SOME WEALTHY PROMINENT FAMILIES HAD BEEN IN DIRECT CONTACT WITH HANOI IN AN EFFORT TO ACQUIRE INFORMATION ABOUT FAMILY MEMBERS WHO WERE MIAs. HOWEVER, THE SRV HAD NOT RELEASED SUCH INFORMATION ON THE ASSUMPTION THAT MORE COULD BE GAINED BY THE SRV ECONOMICALLY BY AWAITING THE OPENING OF FORMAL NEGOTIATIONS WITH THE U.S. GOVERNMENT."[41]

The CIA Deputy Director's "comment" on this further information followed, and consisted of a debunking of the witness and what amounted to another lie in regard to the U.S. Government's inability to determine whether this information was reliable, since the U.S. already knew the prisoners were still in Vietnam, but such information was highly classified and not to be released to the public, as this document was to be in mid-1978:

"(Blacked-out)...APPEARED NERVOUS IN MAKING THESE COMMENTS ON U.S. POWS AND MIAS (TWO LINES BLACKED-OUT BY CIA)...DID NOT INDICATE THE BASIS FOR THE INFORMATION ON U.S. POW'S AND MIA'S, THUS IT CANNOT BE DETERMINED WHETHER...STATEMENTS REFLECT RELIABLE INFORMATION OR UNSUBSTANTIATED HEARSAY."[42]

It was later revealed that March 8, 11, and 16, 1977 reports from the CIA forwarded to James O. Rosenthal, Director of Vietnam, Laos and Cambodia Affairs at the Department of State referred to information collected concerning private comments by Vietnamese Vice-Foreign Minister Phan Hien to a third country diplomat. Phan Hien is reported to have stated privately that his country is still holding American POWs to use as leverage in discussions with the United States on economic aid. The CIA referred to the source of the reports as "a usually reliable source with excellent access."[43]

Despite receiving intelligence like this and additional, far more precise and substantive information from decrypted foreign radio communications on U.S. POWs remaining in captivity (which remained classified for many years to come), on March 16th, 1977, National Security Adviser Zbigniew Brzezinski wrote to Carol Bates of the National League of Families of POWs and MIAs:

"I have learned that some MIA families are distressed by press reports indicating that the Presidential Commission to Hanoi and Vientiane will not press for an accounting of Americans missing in Southeast Asia. Apparently, some fear that the President's mind is already made up, and that he will be content with almost any report that the Commission brings back. May I assure you that this is a misrepresentation. Let me quote to you the mandate which the President gave to the Commission:

"THE COMMISSION'S PRIMARY PURPOSE SHOULD BE TO OBTAIN THE BEST POSSIBLE ACCOUNTING FOR MIAs AND THE RETURN OF THE REMAINS OF OUR DEAD. IN ADDITION, IT SHOULD BE AUTHORIZED TO OBTAIN VIETNAMESE AND LAOS VIEWS ON OTHER ISSUES AND TO REPORT THESE VIEWS BACK TO THE PRESIDENT. BUT ITS MANDATE DOES NOT INCLUDE AUTHORIZATION TO ENGAGE IN NEGOTIATIONS ON THE SUBSTANCE OF THE ISSUES.

"The President reiterated this mandate in his meeting with the Commission on Saturday morning, March 12. I was present at that meeting, and Mr. Woodcock and the other members of the delegation clearly understood that they were embarking on a difficult task. Let us hope that the Commission will bring back an adequate accounting of our fellow countrymen who were lost in the war. I THINK IT WOULD BE UNFORTUNATE FOR THE COMMISSION TO BE CRITICIZED IN THE MIDST OF ITS DELICATE NEGOTIATIONS WITH HANOI. RATHER I WOULD HOPE THAT WE WOULD RALLY BEHIND THESE REPRESENTATIVES OF OUR COUNTRY, in order to demonstrate to the Vietnamese that we are prepared to set aside our bitter memories in hopes that they will do the same...I wish to assure you that the President feels very deeply about this matter and has instructed me to keep you fully informed of all developments. Sincerely, Zbigniew Brzezinski."[44]

A Central Intelligence Agency Memorandum of this same period, dated 11 March 1977, for Rosenthal and Sieverts at the State Department, and Commander Bruce Heller of the POW/MIA Branch at DIA, with distribution to the CIA Deputy Director for Intelligence and the National Intelligence Officer (NIO) of the area concerned, from CIA Deputy Director for Operations, William W. Wells, revealed the following information from a confidential

Vietnamese source, which related to the ongoing U.S.-Vietnamese diplomatic negotiations:

"The attached report, which is of possible interest to you was obtained (blacked-out) March 1977. We are giving this information no further dissemination:

"(Nine lines blacked out by CIA censors prior to public release in May 1978.)...COMMENTS REGARDING SRV RELATIONS WITH THE UNITED STATES AND PRESENCE OF U.S. PERSONNEL IN VIETNAM. (Blacked-out)...FULL DIPLOMATIC RELATIONS BETWEEN THE U.S., AND SRV WILL BE ESTABLISHED IN THE NEAR FUTURE...THE MAJOR STUMBLING BLOCK IN THE PAST HAD BEEN THE U.S. VETO OF VIETNAM'S ENTRY INTO THE UNITED NATIONS BUT THAT NOW... FEELS THE U.S. WILL DROP THESE EFFORTS. REGARDING U.S. INSISTENCE THAT THERE ARE STILL SOME AMERICANS IN VIETNAM WHO HAVE NOT BEEN ACCOUNTED FOR. (Blacked-out)...THE ONLY AMERICANS REMAINING IN VIETNAM ARE THOSE WHO MARRIED VIETNAMESE WOMEN AND HAVE GIVEN UP THEIR U.S. CITIZENSHIP AND HAVE BECOME NATURALIZED VIETNAMESE. THERE ARE ALSO A SMALL NUMBER OF AMERICANS WHO WERE IN VIETNAM AT THE TIME OF LIBERATION AND WHO WERE JUDGED TO HAVE COMMITTED CRIMES AGAINST THE VIETNAMESE PEOPLE. THESE AMERICANS RENOUNCED THEIR CITIZENSHIP (SIC) AND WERE PLACED IN REEDUCATION CAMPS FOR A PERIOD OF THREE YEARS. AT THE END OF THIS THREE-YEAR PERIOD THEY WILL BE EXPECTED TO ADMIT TO THEIR CRIMES AND MAKE A 'SELF-CRITICISM,' AT WHICH TIME THEY WILL BE JUDGED AND EITHER SET FREE AND PERMITTED TO REMAIN IN VIETNAM AS FULL CITIZENS, OR, IN THE CASE OF A LACK OF ADMISSION OF GUILT AND REFUSAL OF SELF-CRITICISM, THEY WILL BE 'SENTENCED.'"(Six more lines blacked-out.)[45]

This intelligence information (and more like it) clearly constituted a Vietnamese explanation and excuse for Hanoi's conduct with captive U.S. POWs since 1975, and also contained a promise that such American POWs would be released from confinement in 1978, following the expected U.S. diplomatic recognition of the Communist regime in 1977 or 1978, by President Carter's Administration. It was no doubt a deliberate leak of information to set the stage for the negotiations of Mr. Woodcock and his Commission, whose members reportedly never pressed the Vietnamese for the release of live American POWs held in captivity, and who were in any case, specifically prohibited by Presidential directive, from negotiating with Hanoi's Communist regime over the Vietnamese expectation of $3 to $4 Billion dollars in U.S. postwar reconstruction aid which had been promised in February 1973 by President Nixon, in the letter delivered to Hanoi by Henry Kissinger.

Another now-declassified CIA report, with an attachment, of March 16, 1977, with date of information given as January 1977, was distributed to the same areas as above with the addition of Captain Raymond Vohden (USN), who was the Principal Advisor for POWs and MIA Affairs in the Pentagon's International Security Affairs (ISA) section of the Office of the Secretary of Defense, Harold Brown:

"(Seven lines blacked-out by CIA censors)...THERE WERE AMERICAN POW'S STILL REMAINING IN VIETNAM AND THAT THESE POW'S POSED A PROBLEM TO THE OPENING OF DIPLOMATIC RELATIONS BETWEEN THE SOCIALIST REPUBLIC OF VIETNAM AND THE UNITED STATES....THESE POW'S INCLUDED SOME WHO DID NOT WANT TO RETURN TO THE U.S."[46]

The official CIA "comment" on this report repeated an already established "party-line' of the CIA and other U.S. Government officials charged with resolving the POW/MIA matter, which was in fact a lie, in stating:

"Other than one American civilian who was arrested in Saigon after its fall in 1975, the SRV is not known to have any live American military or civilian prisoners. Nevertheless, several UNCONFIRMED REPORTS have appeared since May 1975 on the POSSIBILITY of some U.S. prisoners still remaining in Vietnam." (Four more lines blacked-out.)

Upon the return of the Woodcock Commission from meetings with Communist officials in Hanoi, Leonard Woodcock and "Sonny" Montgomery testified before the House Subcommittee on Asian and Pacific Affairs, chaired by Lester Wolff.

Montgomery's testimony included the following statement:

"...Probably some of the committee members will disagree with this statement I make now, but I think President Carter should move ahead as recommended by the House Select Committee and have a case by case review of these 800 Americans that are still missing somewhere in Southeast Asia. They should be reclassified as killed in action. I think this action by the president will bring the close to the final sad chapter in Vietnam."

On May 3rd and 4th the U.S. and Vietnam held two days of talks at the former Viet Cong mission in Paris. Assistant secretaries of State Richard Holbrooke and Kenneth Quinn (later Chairman of the Bush administration Inter Agency Group-IAG on POW/MIAs) met with Vietnamese Vice-Foreign Minister Phan Hien. According to later statements of Kenneth Quinn before a Senate Committee, when the Vietnamese were offered U.S. diplomatic recognition by the Carter administration "without preconditions...Hien said no, THERE MUST BE A U.S. CONTRIBUTION FOR HEALING THE WOUNDS OF WAR." Quinn further said: "It was offered full normalization of relations and lifting of the trade embargo...I was in the room when it happened. They turned it down. They demanded money from us..." The talks subsequently broke down with no agreements reached, but at a May

4, 1977 press conference, Secretary of State Cyrus Vance said:
  "We are pleased with the progress which is being made in the MIA area."[47]
  Yet a declassified CIA document dated May 20, 1977, distributed to Rosenthal and Sieverts at the State Department, Heller at DIA and Vohden at the Defense Department's ISA, by William Wells, CIA's Deputy Director for Operations, was also sent to two DDO's for intelligence and the CIA's area-NIO. It stated:
  "In Spring 1975 while he was on a work detail, (FNU-blacked-out) a former Army of the Republic of Vietnam (ARVN)...who was captured by People's Army of Vietnam (PAVN) forces in the Central Highlands in 1972 and imprisoned in the Hanoi area until May 1975, SAW A GROUP OF CAUCASIAN PRISONERS BEING MARCHED DOWN A RURAL ROAD NEAR HANOI. (Two lines blacked-out) SAID RELATED THIS INFORMATION TO HIM AFTER RELEASE FROM PRISON IN MAY 1975.
  "AN UNIDENTIFIED PAVN ARMY CAPTAIN TOLD A (blacked-out) THAT IN AUGUST 1976 HE SAW A TRUCK CARRYING AN UNDETERMINED NUMBER OF AMERICAN PRISONERS WHO WERE BEING HELD IN A MOUNTAINOUS AREA AT BA VI VILLAGE. THE PRISON CAMP WAS NEAR A TEA PLANTATION ABOUT 12 KILOMETERS FROM SON TAY. (Three lines blacked-out) IN OCTOBER 1976 HE HAD SEEN ABOUT 15 to 20 AMERICAN PRISONERS OF WAR ON A TRUCK NEAR HAI DUONG WHILE HE WAS TRAVELING BETWEEN HANOI AND HAIPHONG. (Blacked-out) SAID HE WAS CLOSE ENOUGH TO OBSERVE THE MEN, DESCRIBED THEM AS CHAINED TOGETHER IN PAIRS, WEARING STRIPED PW UNIFORMS, AND ALL APPEARING SICK OR INJURED. (Blacked-out) DURING HIS YEARS IN VIETNAM, HE HAD PERSONALLY SEEN MANY AMERICAN PW'S, WHICH MADE HIM FEEL 100-PERCENT SURE THAT THE PRISONERS ON THE TRUCK WERE AMERICANS. (Blacked-out) HAD SEEN AN OFFICIAL GOVERNMENT MEMORANDUM THAT CONTAINED A REFERENCE TO AMERICAN PW'S STILL BEING HELD IN VIETNAM. (Blacked-out) AN UNIDENTIFIED PAVN OFFICER HAD ONCE TOLD HIM THAT 'HUNDREDS' OF AMERICAN PRISONERS HAD BEEN WITHHELD, BUT MOST HAD BEEN DISPERSED TO CAMPS NEAR THE BORDER WITH THE PEOPLE'S REPUBLIC OF CHINA (PRC). (Blacked-out) SAID THAT (Blacked-out) A PAVN LIEUTENANT COLONEL, HAD TOLD HIM THAT AMERICANS WERE STILL BEING HELD AS LATE AS MID-1976..."[48]
  These documents and others cited previously were only a few of the many hundreds of intelligence reports from indigenous sources in Southeast Asia then available to the CIA, Pentagon, Department of State and the White House, through the National Security Council, which were declassified and made available to POW/MIA family members and Vietnam war veterans such as the author, through the efforts of a few individuals in the late 1970s, after the Congressional hearings of Montgomery's Committee, and it was this type of information which kept the POW/MIA issue alive

into the 1980s.

Many other first-hand livesighting reports were NOT declassified as these were, but remained secret for years to come under control of the CIA, DIA-Pentagon and NSC, while hundreds (and perhaps over 1,000) surviving American POWs held in many locations throughout Vietnam, Laos, Cambodia and other Communist nations, including the Soviet Union and China, died, one by one, in miserable and forlorn captivity, abandoned by the highest level policymakers of their own government. Because of the dictates of a policy that all U.S. POWs were dead after April 1, 1973, the intelligence bureaucrats debunked eyewitness accounts of the prisoners existence and ridiculed POW/MIA family members and many fellow war veterans who fought for their repatriation. The POWs were no doubt shown American-published evidence by their captors that their own country's leaders had declared them dead, thus adding to their despair and encouraging abject cooperation with their captors.

In August 1977, President Carter recommended that the Defense Department continue "status hearings" on the individual cases of MIAs, the process leading to "presumptive findings of death," in the same manner as had occurred in 1919, 1946 and 1954. A later Senate report by the Foreign Relations Committee Republican staff outlined this process:

"During the Carter Administration...a Department of Defense commission...was established to review the status of individual MIA cases. In these cases, for the purposes of compensation to the next-of-kin, the commission issued the following directive: 'The Commission has used the date of April 1, 1973 as the last date of entitlement to prisoner of war compensation in cases where the actual date of death is not known and where a finding of death has been issued after that date...(because)...the last known prisoner of war was returned to the control of the United States.' The Commission further stated: 'There have been reports of sightings of Americans in Southeast Asia after that date (April 1, 1973), but neither the identities or status of these persons nor the reliability of the reports are known to be established....Therefore, the Commission finds that, in the absence of evidence to the contrary, April 1, 1973 is the last date when members of the U.S. Armed Forces were held prisoners of war by a hostile force in Southeast Asia.

"After a presumptive finding of death has been issued, surviving spouses, next of kin or children are entitled to government-sponsored death benefits, e.g., six-months pay for spouses of deceased military members, government life insurance, etc. The individual is then removed from the active rolls of the military service or agency responsible for him/her."[49]

In its formal report released by the White House on March 23,

1977, the Woodcock Commission noted that the number of MIAs in Vietnam totaled only 4% of those killed in action, while in World War II the number of MIAs (78,794) had totaled 22% of the total reported killed. These figures, which had been provided by the Pentagon staff assistants who did so much of the work of the Montgomery Committee and of the Woodcock Commission, no doubt reflected questions on the part of Commission members as to past US government handling of the POW/MIA problem. Instead of raising doubts in the minds of the investigators about the fate of American POWs and MIAs of previous wars, they were used as justification of the Commission's perfunctory and poor performance. In their report the commission noted:

"This impressed upon the Commission the need to be REALISTIC in its expectations for a future Indochina accounting."[50] It could also have been taken as a signal that the members of the Commission now understood something of the past history of Communist-held U.S. prisoners of war. How much they knew of the truth about WW II and Korean POWs in communist captivity is not presently known.

President Jimmy Carter rewarded Woodcock for his service by appointing him United States Ambassador to the Peoples Republic of China, and Mike Mansfield was appointed US Ambassador to Japan, where he remained comfortably ensconced for over 10 years.

After the formal report of the Woodcock Commission, President Carter pushed harder for normalization of relations with Hanoi, as he simultaneously ordered case reviews of the remaining MIAs, to reclassify them as "killed in action, body not recovered." Thus the U.S. Government policy which had been implemented after the 1918 Intervention in Russia was continued. Although the family members of the missing men and some of the war's nearly three million veterans protested, through the rest of 1977 and much of 1978 the case reviews continued in the Pentagon, and the missing POWs, many hundreds of whom were known to have been alive in the 1973-75 period were declared to have been "killed in action" on the dates they had first disappeared.

The U.S. Air Force had meanwhile been conducting a study of POW debriefings looking for missing Americans witnessed by other U.S. POWs. A May 5, 1978 U.S. Airforce report stated:

"...(W)e extracted an additional possible 51 names of unaccounted for individuals who returnees saw, or heard the name of, in captivity." One of the U.S. POWs, Bobby M. Jones, was seen by returned POW William Metzger, and was also personally observed by another returned POW, Orson Swindle (later an aide to Ross Perot). At the Zoo and at Heartbreak in 1972 the names of Hubert C. Nichols who was a U.S. Air Force Colonel (from Pensacola, Florida, missing in action since September 1, 1966) and Joseph Ross were seen on the walls by returned POWs Myron Young, Rudolph Zuberbuhler and Cecil Bronson.[51][52]

A later Senate Select Committee on POW/MIA Affairs, "attempted to recover/reconstruct the names on each of the lists... including Operation Homecoming debriefings and Vietnam Prisoner of War Analyses." U.S. eyewitnesses who returned from captivity saw: R. L. Bowers, J.A. Champion, J.G. Gardner, Roosevelt Hestle, Larry M. Jordan (and Long). Other names of POWs seen scratched on walls and floors included: Buriss N. Begley (and Ross), and two names on the not on the unaccounted-for list: Crain Nichols and Walter Kosko.[53]

One American intelligence officer who had information on Americans remaining in captivity, particularly in Laos, was Lt. Colonel Albert Shinkle, a recently-retired U.S. Air Force intelligence officer who had served under General "Heinie" Aderholt in the CIA's secret war there. A veteran of WW II and Korea, Shinkle had served in Laos (and neighboring Thailand) from before the big war buildup until the American withdrawal. He was involved in military intelligence and espionage work and developed extensive contacts among indigenous sources while stationed in Laos, from whom he learned the numbers and locations of American POWs in Laos, whether held by the Pathet Lao or more likely, by the North Vietnamese Army. Shinkle's sources combined with other Air Force electronic intelligence, convinced him that of more than 500 American pilots shot down over Laos, 300 or more were still alive in captivity in Laos and North Vietnam. Shinkle later testified that when he refused to ignore the American POWs, "The CIA ordered me out of Laos because I got in the way, and because I pinpointed the Communist camps where Americans were then held."

Shinkle continued his intelligence-gathering after the Communist victory, and testified before a July 1978 House Subcommittee hearing that the CIA's Domestic Contact Service interviewed him in Denver, Colorado in 1975 and told him: "We know what you're doing, Shinkle. We don't like it. It's counter to U.S. policy, and if you don't get out of it, you're going to be in a lot of trouble." U.S. Representative Lester Wolff of New York, Chairman of the Asian and Pacific Affairs subcommittee, then suggested to Shinkle: "Some people are going to have in their mind, this fellow looks like he was a competitor of the CIA." Shinkle replied: If I stayed on active duty, and we had another brushfire war, how am I going to motivate young pilots who go out and risk their lives and may get shot down, if we're not going to come after them? The CIA does not understand military intelligence in time of combat. Every war the CIA had a finger in, from the Bay of Pigs to Laos, they lost...we can't be all under one cap called CIA-They do not understand." Shinkle later learned through continued intelligence gathering in Laos into the 1980s that many of the POWs were moved from Laos into Vietnam, and sometimes groups of POWs would suddenly reappear in Laos, under NVA control.[54]

Fourteen years after Shinkle testified, an Air Force and NSA-B Group cryptolinguist and communications analyst named Terrell Minarcin testified to the U.S. Senate (in 1992) that he had monitored decoded enemy communications during the Carter Administration, from December 1977 to January 1978, which revealed that 22 more U.S. POWs were flown from Vietnam to the USSR in that period, as a result of further screening by the Soviets of American prisoners left behind in North Vietnamese control after the 1975 American withdrawal. Minarcin also revealed, in a 1991 affidavit for the Senate Select Committee on POW/MIAs, that more such shipments occurred until at least the spring of 1983, when 200-300 American POWs were transported from Vietnam to the Soviet Union, after which signals analysis indicated hundreds more American prisoners still remained in Vietnam. Postwar shipments such as these would explain at least some of the Southeast Asian livesightings of the late 1970s, while intelligence sources and continuing eyewitness refugee reports indicated that other U.S. POWs did remain alive in Indochina. (Although Minarcin was later accused of exaggerating both the numbers of POWs in Vietnam and the amount of intelligence available to him at NSA, other sources of information tended to support his shocking statements.)

Human intelligence obtained in Southeast Asia backed up some of Minarcin's conclusions which he claimed to have obtained from intercepted Communist communications. Another of many declassified CIA intelligence reports of this period, dated June 1979, had revealed that, "IN MID-1976 (source's name blacked-out) SAID THAT EARLY IN THAT YEAR THEY HAD SEEN ABOUT 230 U.S. POWS WHO WERE BEING HELD AT BEN BAT, IN THE SOCIALIST REPUBLIC OF VIETNAM, HA SON BINH PROVINCE..."[55]

But Jimmy Carter had brought into his administration many young representatives of the anti-war segment of the Vietnam generation, such as Sam Brown and James Fallows, who pushed enthusiastically, from inside the administration, for diplomatic recognition and U.S. aid for the Vietnamese Communist regime, with little thought for the American prisoners who had been abandoned. The possibility that there were a few, or several hundred of those men whom Hanoi had branded as 'war criminals,' still alive in captivity, could not be permitted to keep America from assuaging its guilt by "normalizing relations."

In addition, and more importantly, the 1967 Berlin-State Department cables, along with the testimony and statements of Mooney, Minarcin, General Sejna, various Indochinese refugees and other sources cited herein about American prisoners from Vietnam transferred to Soviet control, indicate that the POW/MIA matter was indeed a high-level national security concern, directly related to the Soviet nuclear threat. Public disclosure that American war prisoners had been, and still were being shipped to the USSR could conceivably have caused a rapid escalation of public anger and increased tensions between America and Russia, that could have led in turn to a dangerous nuclear confrontation between the superpowers. The actual level of the threat against the existence of the United States was the subject of debate, but the prevailing belief in Washington in the 1978-1979 period was that, rightly or wrongly, the Soviets believed they had a first strike capability.

To his everlasting credit, Lt. General Eugene Tighe, who served as chief of Air Force Intelligence and later as director of DIA, was to be the most forthright and truthful senior U.S. military officer on the subject of U.S. POWs and MIAs left behind as captives in Vietnam and Laos, and on the indications that some of them had been shipped to the Soviet Union. In 1979, General Tighe testified before a Congressional hearing on the ability of the Soviets to launch an unexpected nuclear missile strike that would preclude attempted retaliation by the United States. Tighe said:

"It is the perception that they have, more than anything else, that is important. I think you will find a great deal of disagreement over whether or not they actually have a first strike capability."

Senate aide Richard Kaufman asked:

"Your assessment that they have a first strike capability depends on your perception of what the Soviet perception is of the United States' will to retaliate?

General Tighe answered:

"If the Soviets think they have a first strike capability, it is my judgement that they have it. I believe they believe they have that first strike capability."

Kaufman asked:

"That depends on perception...and not only on objective capability?

General Tighe replied: "That is right."[56]

Since the President and his assistants are charged mainly with protecting the security of the people of the United States as a whole, keeping such a potentially volatile issue as secret shipments of Americans to Russia highly classified, could be justified in the name of the greater good. The Soviet nuclear arsenal could legitimately be said to threaten the national security of the United States and official American policy dictated that any sudden escalation of tensions with Soviet Russia was to be avoided. By keeping the real truth of the POW/MIA matter classified for reasons of "national security," any presidential administration was free to move toward its foreign policy goals, largely unimpeded by any necessity to be truthful with the American people. The national security aspects could appear to justify evasion and secrecy. In this vein, certain unpleasant facts or details are sometimes kept from presidents by their civilian or military assistants, for deniability purposes and sometimes to avoid placing an undue burden on one man.

With the POW/MIA matter thus publicly out of the way for the Carter Administration, and the secret "special" and human intelligence of the existence of non-returned POWs highly classified and compartmentalized by different intelligence agencies, the normalization of relations with Vietnam appeared imminent. However, events leading to United States diplomatic recognition of Communist China suddenly altered the situation. Fearful of Chinese-American collaboration, the Vietnamese decided that they could not trust or expect assistance from the United States, and in October 1978 they turned to the Soviet Union for support.

Historically the Vietnamese had been afraid of China since the time of the Mongol and Chinese invasions centuries before, and the rapprochement between Vietnam's most ancient and most modern enemies was viewed with the greatest alarm in Hanoi. The Vietnamese invasion of Cambodia in December 1978 practically coincided with US recognition of the Peking government in China and led to a short but bloody border war between the former Communist allies, Vietnam and China. The indications from human intelligence sources were that the Chinese invasion caused a sudden movement of American POWs from prison camps in the Sino-Vietnamese border region to the south.

## PRIVATE GARWOOD AND THE NSC

In March 1979, with or without some instigation from Moscow or Peking, the Vietnamese had played a hole card. They had allowed U.S. Marine Private Robert Garwood who had been secretly held since the end of the war in 1973, to leave Vietnam. After 14 years of captivity, Garwood had finally succeeded in smuggling a note out of Vietnam with a foreign visitor, after previous failures. But instead of killing him, the Vietnamese released him, with threats to remain silent about other Americans he had seen long after the war. As the last publicly-acknowledged American POW to depart Vietnam, "Bobby" Garwood finally revealed to Wall Street Journal investigative reporter Bill Paul in 1984, that he had personally seen 70 or more, other secretly-held American POWs in Vietnam years, after the final American withdrawal in 1975. He maintained that he had long remained silent out of fear for the remaining prisoners, after his conviction (perhaps unjustly), for collaborating with the enemy. Garwood subsequently revealed to several researchers that he had seen secretly-withheld American POWs at many locations in

Vietnam following the final American withdrawal from Indochina in 1975.

Aside from Hanoi, among those areas Garwood was held in were: Camp Number 5, which had contained 60,000 South Vietnamese prisoners, the Yen Bai Camps-where 20,000 South Vietnamese remained, and the Camp 776 complex, consisting of twenty-two separate camps, each containing 2,000-6,000 prisoners. Garwood was also in a Hanoi hospital with 22 Palestinians in 1977, who informed him that they were using some of the American POWs as guinea pigs in psychological warfare training, involving films of POWs during capture and then interviews with still-captive POW survivors. One of the Palestinians was PLO terrorist leader Mohammed "Abu' Abbas, later involved in the Achille Lauro terrorist attack.

Garwood was repairing a truck for the camp authorities one night nearby Camp 776, where Highway One crossed the railroad tracks, while a train of twenty boxcars was halted at the crossing a few yards away from him. The North Vietnamese Army guards suddenly opened the doors and prisoners, who had been packed inside like sardines, just fell out. Some of the prisoners had suffocated or died in some other manner while confined, and their comrades began laying their dead bodies alongside the tracks. After one of the boxcar doors had flown open, Garwood, who was thin and bent from 12 years of captivity, and was dressed like a Vietnamese peasant, saw a startling thing. As he later related in a 1985 sworn affidavit which was entered into the record as evidence in a 1986 Senate hearing on POW/MIAs, Garwood was watching approximately 30-40 U.S. POWs unloading from a one suffocating boxcar. He later remembered:

"I FIT IN THE CROWD, WATCHING...OUT FELL CAUCASIANS. THIN, BEARDED, BLUE JACKETS AND PANTS, WITH FLIP FLOPS. (Sandals, often made from the rubber treads of American tires.) "I HEARD THINGS LIKE 'WHERE THE FUCK ARE WE, CHINA?...THIS WAS STREET LANGUAGE, NOT BRITISH ENGLISH. EVERYONE AROUND ME JUST GAWKED. THIS HAD BEEN KEPT A SECRET FROM THE PEOPLE BY THE COMMUNIST ADMINISTRATION. SO EVERYONE'S TAKEN BY SURPRISE. THESE WEREN'T POWS FROM SAIGON. THEY WERE LONG-TIME POWS."[57] The above details were confirmed in an interview of Robert Garwood by the author at National Vietnam Veterans Coalition Headquarters, Washington D.C., 1986.

Garwood had been falsely accused by the U.S. Government of desertion, but it was subsequently proven at his court martial that he'd been captured by the Viet Cong in a road ambush near the end of his tour of duty as a Marine despatcher, in September 1965. The government prosecutors had to be content with gaining his conviction for one count of "aiding the enemy," in order to destroy his credibility as a witness.

A series of 1979-80 National Security Council documents (first uncovered in the Carter Library by investigative journalist Ron Martz) illustrates the policy-making process in response to this latest U.S. prisoner livesighting. It follows a familiar pattern of obscuring the significance of such an event, which in this case centered around the sudden "discovery" of the U.S. POW in Hanoi, Marine Corps Private 1st class Robert Garwood, in 1979. (In reality, however, subsequent Congressional testimony has revealed that U.S. intelligence had been tracking Garwood's movements, and those of many other POWs, for years, as later disclosures of DIA and other agencies information has demonstrated.)

As noted previously, Garwood had actually been captured by the Vietcong in 1965 and held for 14 years. While during his long confinement, it would later be revealed, he had at times "collaborated" with his captors in order to survive, he did not willingly remain in Vietnam as a prisoner. He was simply not released in 1973 along with perhaps 1,000 other American prisoners of that war, held in several countries. In March 1979 when Garwood had smuggled a note proving his existence out of Hanoi, the Carter Administration had long been in the process of conducting "case reviews" of all missing and POWs from the Indochina war, leading to a "Presumed Finding of Death" (PFOD) of all POWs and MIAs. Garwood's appearance alive was a horrifying interruption of this orderly process that had been conducted in the same manner after WW I, WW II and Korea.

NSC official David Aaron, a high-level assistant to Carter's national security advisor Zbigniew Brzezinski, was alerted to the political realities of Garwood's existence in a memorandum dated March 7th, 1979:

"For well over a year, the National League of Families of American Prisoners and Missing in Southeast Asia have sought a meeting with the President. The NSC has consistently turned down these requests. Two reasons now exist for altering our recommendation and for responding favorably to the most recent League request:

"A live American DEFECTOR has been sighted in Hanoi and has indicated that he wishes to return to the US. The Vietnamese had previously given no indication that there were any live Americans in Vietnam—although they clearly knew about this case. The DEFECTOR has also claimed that he knows of other Americans, apparently, who are alive in Vietnam. IT IS POLITICALLY WISE PERHAPS FOR THE PRESIDENT TO PROTECT HIMSELF ON THIS ISSUE BY REASSERTING HIS CONTINUED INTEREST IN A FULL ACCOUNTING."

"In light of the most recent turn of events in Indochina, in my opinion, the time is particularly propitious for a carefully considered Presidential statement on our policy toward Vietnam—

WHETHER WE WISH TO PROCEED TOWARD NORMALIZATION, TO RESUME OUR TALKS WITH HANOI, AND IF SO UNDER WHAT CIRCUMSTANCES."

A hand-written note on the bottom of this memo initialed by David Aaron reads:

"I'VE A LOT OF DOUBT ON THIS. WHY CAN'T STATE PURSUE THE AMERICAN DEFECTOR?"[58]

Who was this Mr. David Aaron who thought it wrong that President Carter meet with the families of missing US prisoners, whose existence in captivity had just been confirmed by an American eyewitness? Mr. Aaron served on the National Security Council under President Nixon as a senior staff member from 1972 to 1974, when hundreds of American POWs were abandoned alive in Vietnam, Laos, China and the Soviet Union. From 1974 to 1976 he "headed an investigative task force for the Senate Intelligence Committee, was a member of the first SALT (Strategic Arms Limitation) negotiations, directed the work of the SALT II negotiations as Chairman of the NSC Mini-Special Coordinating Committee, and participated in the Vienna Summit of 1979. From 1977 to 1981 he was Deputy Assistant to the President for National Security Affairs."

By immediately branding Garwood a defector, (although he had actually been captured by the enemy) the NSC officials managed from the beginning to detract from his future credibility as a witness to other American POWS still held in captivity, a classic method of damage-control. The branding of late-returned U.S. POWs from Communist control as defectors dated to 1921, and the case of U.S. Army Private Sidney Vikoren and others, as has been related earlier in this book.

On March 12, 1979, Oksenberg replied to David Aaron's handwritten note:

"Your response to my memorandum of March 7 asks why State cannot pursue the issue of the American DEFECTOR. State has pursued this issue, and the DEFECTOR is now on his way home. The point is that his return will generate new stories about the MIA issue and PARTICULARLY ABOUT THE POSSIBILITY THAT ADDITIONAL AMERICANS REMAIN IN HANOI....

"THIS IS A "RIGHT-WING" (note, original sub-quotes) ISSUE, AND I THINK IT GAINS THE PRESIDENT SOME POLITICALLY TO INDICATE HIS CONTINUED INTEREST IN THE ISSUE—particularly since he may be moving on the Vietnam front in the months ahead and since the Administration had implied earlier THAT IT BELIEVED VIETNAMESE ASSURANCES THAT THERE WERE NO LIVE AMERICANS LEFT IN HANOI."

David Aaron, who would later achieve some prominence following the publication of his novel STATE SCARLET (in which he extoled the courage and resourcefulness of NSC staff officers in times of crisis), again answered this real-life memo with a handwritten note: "I THINK IT'S A BUM IDEA—David Aaron."[59]

Brzezinski forwarded an action memorandum to President Jimmy Carter on April 18, 1979:

"The National League of Families remains convinced that live American POWs remain in Vietnam. They also believe you are not being adequately informed and that the bureaucracy is not pursuing the matter aggressively. THEY REMAIN OPPOSED TO THE RENEWAL OF CASE REVIEWS WHICH COMMENCED LAST YEAR."

"I attach a letter from them to you (Tab B). THIS CASE HAS LITTLE MERIT...I will answer them myself (Tab A), but wish to be able to say you saw their letter and asked me to reply."[60]

In a later memorandum, National Security Council staffer Michel Oksenberg makes recommendations to Zbigniew Brzezinski on dealing with the League of Families after the Garwood revelations, and numerous other refugee livesightings of Americans in captivity. Mr. Oksenberg, who in February 1988 would be a University of Michigan professor honored by Newsweek Magazine as "one of the top 25 Asia players in the US," wrote in a subsequent January 21, 1980, NSC Action Memo (#386) from Oksenberg to NSC Adviser Zbigniew Brzezinski, revealed the secret position taken by the U.S. Government on first-hand livesightings of U.S. POWs remaining in captivity, to a group of POW/MIA family members:

"...A LETTER FROM YOU IS IMPORTANT TO INDICATE THAT YOU TAKE RECENT REFUGEE REPORTS OF SIGHTING OF LIVE AMERICANS 'SERIOUSLY.' THIS IS SIMPLY GOOD POLITICS; DIA AND STATE ARE PLAYING THIS GAME, AND YOU SHOULD NOT BE THE WHISTLE BLOWER. THE IDEA IS TO SAY THAT THE PRESIDENT IS DETERMINED TO PURSUE ANY LEAD CONCERNING POSSIBLE LIVE MIAs. DO NOT OFFER AN OPINION AS TO WHETHER THESE LEADS ARE REALISTIC..."[61]

It was to be many years before Robert Garwood was properly debriefed by U.S. intelligence, long after his sightings of other American POWs were published in the Wall Street Journal in 1984. In April 1980 Vietnam's Ministry of Foreign Affairs released a 'White Paper' on the 'Question of Americans Missing in the Vietnam War.' While rejecting the legality of the 1973 Paris Agreement regarding it as 'dead' in order to evade its obligations to contribute to postwar reconstruction in Vietnam, the United States demands that Vietnam implement those provisions benefitting the U.S. side, imposing on Vietnam 'obligations' to carry out Article 8(b)..."

Although the Carter Administration did proceed with case reviews and presumptive findings of death for the non-returned U.S. POWs and MIAs of Indochina, some of whom were still being transferred to Soviet control, President Carter himself took a step to assure that the full truth about U.S. POWs in Communist control would eventually emerge. Perhaps mindful of intelligence he received as President on the Soviet connection to the missing prisoners of Vietnam he continued a 30-year declassification order which had instituted a procedure whereby classified U.S. Government documents would automatically be declassified, and strengthened the Freedom of Information Act (FOIA) process (dating from 1968), for requesting other classified records and for appealing decisions that official records must remain classified in the name of "national security." Although these regulations have often not been fully, or in many cases, even partially complied with in regard to U.S. POW/MIA documents refused to the author by FOIA officials of U.S. Departments and intelligence agencies, nevertheless, without such regulations it is doubtful whether anything on the subject of Communist-held American POWs would have ever been declassified by Federal authorities.

Vietnam had become a land of prisons and forced labor camps by the end of the 1970s, but most Americans and much of the rest of the world had little interest in the mass oppression and suffering which was reported by refugees who escaped in boats or on foot to Thailand. From Dien Bien Phu and the Chinese border in the north to the Mekong delta and Phu Quoc island off the southern coast, a vast Vietnamese gulag empire had been created by the North Vietnamese victors to house hundreds of thousands of former Army of the Republic of Vietnam soldiers and officers, Thieu government supporters, religious leaders, liberals and even thousands of former Viet Cong and National Liberation Front officials and supporters, who were no longer needed or trusted by the northern conquerors. Among these was Doan Van Toai, a former student leader who had supported the Viet Cong and been imprisoned several times by the Thieu regime for his antiwar activities. Arrested by the Communist government for criticizing the new regime, Doan Van Toai was held in a relatively small prison in Ho Chi Minh City called Le Van Duyet, where he watched others who had fought or demonstrated against the Americans starve or be beaten to death alongside those who had supported the Thieu government.

Doan Van Toai fortunately succeeded in gaining an exit visa after his release from prison, leaving Vietnam in May 1978, for France, and eventually reaching the United States, where he contacted former antiwar leaders he had known before the fall of Saigon. In Berkeley, California, where his reception was "a good deal cooler" than it had been in 1970 during a visit with activists, he chanced to encounter a friend of Joan Baez, the nationally-known folksinger who had been a leader in the struggle to get American forces out of Vietnam. Baez listened to all that Doan Van Toai told her of the Vietnamese gulag prisons and forced labor camps, and conducted her own further investigation into the human rights violations of the victorious communist regime in Hanoi. He later wrote:

"Finally, Baez wrote an open letter to the New York Times (May 30, 1979) addressed to the Socialist Republic of Vietnam and signed by seventy-eight other prominent individuals, including Cesar Chavez, Daniel Berrigan, I. F. Stone, Jerome Weisner, and Nat Hentoff. It read in part:

'We appeal to you to end the imprisonment and torture—to allow an international team of neutral observers to inspect your prisons and reeducation centers. We urge you to follow the tenets of the Universal Declaration of Human Rights and the International Covenant for Civil and Political Rights which, as a member of the United Nations, your country is pledged to uphold. We urge you to reaffirm your stated commitment to the principals of freedom and human rights...to establish real peace in Vietnam.

"For her efforts, Baez was attacked by Jane Fonda in a letter circulated among a large number of former antiwar movement people and excoriated in an open letter entitled "The Truth About Vietnam"[62] and signed by another group of antiwar people who 'recognize and acknowledge the remarkable spirit of moderation, restraint and clemency with which the reeducation program was conducted.' This letter continued:

"'Vietnam now enjoys human rights as it has never known in history as described in the International Covenant on Human Rights: the right to a job and safe, healthy working conditions, the right to be free from hunger, from colonialism and racism. Moreover, they receive—without cost—education, medicine and health care, human rights we in the United States have yet to achieve.'"[63] During the war Fonda had made pro-enemy broadcasts over Radio Hanoi, referred to American POWs as war criminals and had been photographed in a military helmet manning an anti-aircraft gun in North Vietnam aimed at American aircraft, thus earning the name which she was known by among U.S. war veterans, "Hanoi Jane." She had used her influence and wealth to support the antiwar movement and had subsequently married SDS leader Tom Hayden. It was with such ludicrously false statements that the radical American left of 1979 dismissed the eyewitness accounts of those survivors who had lived to bear witness to the deaths of hundreds of thousands of Vietnamese, Laotians, Montagnard minority people, and over a million Cambodians.

Meanwhile, the livesightings of American POWs still being held in Indochina continued, and many were recorded for posterity through the efforts of Mrs. Le Thi Anh, a Buddhist and former antiwar activist who had opposed the Thieu government in Saigon. At this time the National League of Families was led by George Brooks, and was actively seeking information which would prove the existence of captive Americans in Southeast Asia. Although the U.S.

government officially said they questioned all refugees about American prisoners, the pathetically small staff at JCRC Bangkok under Colonel Paul Mather (who had been assigned to the POW/MIAs since 1973 service in Saigon), made this impossible, and U.S. policy of declaring all POWs provided little encouragement to such bureaucrats to actively seek information that would disprove the legitimacy of this action. (After the initial release of documents following the Montgomery Committee report, all DIA and most other POW/MIA livesighting reports remained "classified" from 1979 until the 1992 period.)

As a result, many hundreds of thousands of Southeast Asian refugees in the late 1970s were never questioned about U.S. POWs, and some made public what they had seen because of an advertisement approved by the National League of Families and written by Le Thi Anh, which appeared in many Vietnamese language publications. Between 1978 and 1983 she received some sixty references to livesightings of Americans in Southeast Asia, which were recorded. One of the first reports Le Thi Anh uncovered concerned a young boat-refugee named Trinh Hung, a former University student who escaped from Vietnam in the fall of 1975, and lived in Pennsylvania. He related to Brooks and others how he had himself seen two thin American prisoners with shaven heads, tied hand and foot in the bottom of a motorized sampan, in the Mekong Delta village of Xeo Ro near Rach Gia, South Vietnam in October 1975. The guards of the American POWs encouraged the people to beat the Americans, and they were later transported to the U-Minh Forest, where many other intelligence reports indicated many other American POWs were held long after 1975.

Dermot Foley, Counsel for the League of Families and a New York lawyer chose this case as an example to argue before a Congressional committee that DIA's subsequent clumsy debunking of Trinh Hung's report was incorrect, and pointed out that "shaving the head is a traditional punishment for persons whom the Vietnamese wish to humiliate and identify as evil."[64]

Le The Anh continued her invaluable human-rights work into the 1980s by locating more eyewitnesses who had seen American POWs, and several subsequent livesightings reported by her are cited later in this book.

Some American POWs from Vietnam have also been reported as turned over to both Cuban and North Korean control. Declassified CIA documents reveal that two of the Cuban interrogators who achieved notoriety for torturing American POWs were officials of the Cuban Embassy in Hanoi, Eduardo Morjan Esteves, known as 'Fidel,' and Luis Perez Jaen, known as 'Chico'. A Cuban informant to the author, 51-year old Ramon C., charged with "sabotage and inciting revolution" has reported to the author and Luis Alcebo, in two interviews, that he was held in a secret G-2 prison near Havana, called Los Maristas, and that in 1966 and 1967 a few American POWs shipped from North Vietnam to Cuba were held there before being moved to other locations in Cuba, or possibly to other Communist-bloc nations. He felt the reason they were there was because they were Communist allies and therefore supported each other's cause: "...It was common knowledge these prisoners were American soldiers from Vietnam." It was impossible to see the prisoners, but he had first been told who they were by a not-unfriendly guard. He believed that:

"...every month or so (from late 1965-1967) they were brought in...but never more than 10-12 Americans held prisoner...The cells they were kept in were called Tapilla cells. They were always covered and the prisoners kept unseen. The prisoners were known to be American because they would yell out during feeding time when the guards would open the small window in their cells. They would yell in English at their captors and soon learned Spanish from overhearing other prisoners. Their Spanish was distinguished by their accent. They yelled: "Hijo Puta!" at their captors...(son of a whore)...Also held with them were Mexicans. These were thought to be anti-Communist Mexicans...The Mexicans received very ruthless treatment, but still... were very tough towards their captors. Maybe some still are there and alive. The prison they were held in was called Los Maristas, near Havana...in the La Vivura section in the subdivision of Sevillano. The location is four to five kilometers south of the city of Los Pinos. Los Maristas used to be one of the biggest Catholic private schools in Cuba...This prison was one of the most top secret where no one saw other prisoners...and was known for people never coming out of there..."

This informant had been interviewed by the FBI upon entering the United States and volunteered the information about the American prisoners, but the FBI agents evidenced no particular interest, nor did they follow up the interview. He had been held later in Castillo Del Principe Prison in Havana with 5,000 other prisoners, including 600 or more confined for political reasons; Granja Fajardo Prison in Pinar Del Rio with 2,000 prisoners; Taco Taco prison, also in Pinar del Rio, with some 2,000 prisoners; and lastly, in the "Sandino" Prison, located "in a new city beyond Mendoza which was estimated to hold as many as 15,000 prisoners." Los Maristas was the only prison where he knew American military prisoners from Vietnam had been held. Neither the CIA or U.S. military intelligence ever interviewed this informant, who has requested that his full name not be used, to avoid harassment.[65] The author received other information about American prisoners still held in Cuba into the 1990s but none were specifically reported as known to be U.S. military personnel.

Much of the intelligence on American POWs shipped from Vietnam to North Korea, allegedly for 'safekeeping,' remains classified. In response to June 26 and July 21, 1991 (and earlier) Freedom of Information requests to the Department of State by the author for release of classified information concerning Soviet/ surrogate-held U.S. POWs, the U.S. Army Intelligence and Security Command, a component of NSA based at Fort Meade, Maryland, released portions of 3 pages of classified Confidential and secret U.S. documents concerning reports of American POWs in North Korean control, stated the following:

"WE HAVE COMPLETED A MANDATORY DECLASSIFICATION REVIEW IN ACCORDANCE WITH EXECUTIVE ORDER 12356. AS A RESULT OF THIS REVIEW, IT HAS BEEN DETERMINED THAT INFORMATION, HIGHLIGHTED IN BLACK, IS CURRENTLY AND PROPERLY CLASSIFIED SECRET AND CONFIDENTIAL ACCORDING TO SECTIONS 1.1 (A)(2):THAT INFORMATION SHALL BE CLASSIFIED SECRET IF ITS UNAUTHORIZED DISCLOSURE REASONABLY COULD BE EXPECTED TO CAUSE SERIOUS DAMAGE TO THE NATIONAL SECURITY, 1.1(A)(4): THAT INFORMATION PERTAINING TO INTELLIGENCE ACTIVITIES, SOURCES OR METHODS SHALL BE CONSIDERED FOR CLASSIFICATION PROTECTION. AND 1,3(A)((9):... PROVIDES THAT CONFIDENTIAL SOURCES SHALL BE CONSIDERED FOR CLASSIFICATION PROTECTION,... OF EXECUTIVE ORDER 12356..."

The attached "released" documents were almost entirely blacked-out from beginning to end. Enough words survived to indicate that the document concerned U.S. POWs, since it was sent to among others, "Stony Beach" (a code-name for a U.S. POW resolution center) to a Defense Intelligence Agency POW/MIA analyst, Mr. Warren Gray, and also to the Secretary of State, CIA, Joint Chiefs of Staff, U.S. Military Attaches in the Defense Attaché Offices in Beijing, Hong Kong, Singapore, Bangkok and Japan; and to many other U.S. Government agencies. Handwritten on one blacked-out 13 November 1989 Confidential Department of State document (#130457Z-Action INR) are the words "OL MOPS LIVESIGHTINGS Refer: DIA." On the non-blacked out portion of the document are the words: "COUNTRY: (U) LAOS (LA); THAILAND (TH)...This is an information report, not fully evaluated intelligence...this is a Stony Beach report...not have full faith in source of his information (Author's note: automatic 'debunking')...Not releasable to foreign nationals—warning notice— Intelligence sources or methods involved."

Yet, absurd as it may seem with such precautions as these by the Department of Defense against unauthorized American researchers seeing the full text of such documents, the DIA and other agencies had made such intelligence material available to the very same Communist governments which had secretly withheld the

American prisoners in the first place, according to later testimony of DIAs one-time POW/MIA Chief, Colonel Millard Peck.

---

## Notes on Chapter Eleven

1    Document, obtained from a confidential source at Bolling AFB, by Thomas V. Ashworth, was turned over to the Senate Foreign Relations Committee Republican staff in 1991, and to the Select Committee on POW/MIA Affairs, U.S. Senate, in 1992, which ignored it, as witnessed by the testimony of the author and research-colleague Ashworth before Senator John Kerry, Chairman, on November 11, 1992.

2    (For a more detailed summary of livesightings reports of American POWs in Indochina, see: *The Bamboo Cage*, op. cit.)

3    Statements of Barry M. Toll, June 14-June 19, 1992

4    Service Records of Jerry J. Mooney, in the author's files.

5    I Interviews with Jerry J. Mooney by the author in Billings, Montana in 1992,

6    *First Heroes*, p. 270, op. cit.; also confirmed in an interview of Dermot Foley by the author in New York, during 1986

7    The CIA and the Cult of Intelligence, by Victor Marchetti and John Marks, op. cit. p.163

8    *The CIA And the Cult of Intelligence*, op. cit. p. 226 &346

9    From the President-Richard Nixon's Secret Files, edited by Bruce Oudes, Harper and Row, NY, 1989, p. 507

10   The Puzzle Palace, by James Bamford, Houghton-Mifflin, Boston, 1982, p. 248-261

11   Ibid.

12   "We Can Keep You Forever", 1987, BBC-Lionheart, producer, Ted Landreth

13   Interviews with Colonel William LeGro, by the author, 1986-87

14   Uncorrelated Reports, Volume 10- p. 374; Volume 7- p. 210

15   Uncorrelated Reports, Volume 10, p. 546; Volume 9- p. 636 and Volume 8- p. 230; Volume 10, p. 191 and Volume 10 p. 186)

16   *They Called it Peace With Honor*, by Stuart Herrington, Presidio Press)

17   *Decent Interval*, by Frank Snepp, Random House, NY, 1977)

18   *Bad News, The Foreign Policy of the New York Times*, by Russ Braley, Regnery Gateway, Chicago, 1984

19   *Vietnam, A History, and Vietnam From Ceasefire To Capitulation*, by Colonel William LeGro, Center of Military History, Washington, D.C.

20   Manchester Union Leader, August 1992

21   *Decent Interval*, p. 37, op. cit.

22   *Manhunt*, by Peter Maas, Random House, 1986

23 DECENT INTERVAL, Snepp, Frank, p. 459; also: The Politics of Heroin, Alfred McCoy, 1973

24 Pp. 16-B-16 and 16-B-17, Defense Attache Saigon, RVNAF Final Assessment, Colonel W. E. LeGro, Secret-Noforn Dissem.

25 Vietnam, A History and Decent Interval

26 Manchester (NH) Union Leader, August 1992

27 Decent Interval, p. 492

28 *Vietnam, A History*, by Stanley Karnow, Viking, NY, 1983

29 *The Four Days Of Mayaguez*, by Roy Rowan, Norton, NY, 1975

30 Montgomery Committee Report

31 Victor Marchetti, *The CIA and the Cult of Intelligence, op. cit.*

32 Item 4, Part II, p. 208, Records of Americans Missing in Southeast Asia, September 17, 1975.

33 *George Bush, An Intimate Portrait*, Fitzhugh Green, Hippocrene Books, NY, 1989, confirmed in an October 1988 meeting between Fitzhugh Green and the author, Washington, D.C.

34 Extensive interviews of Mr. Mooney by the writer and letters over a period of years; also: Affidavit of Mr. Jerry J. Mooney to Senate Select Committee on POW/MIA Affairs.

35 In Mooney's personnel records jacket are the orders assigning him to the 6970th Air Base Group "with duty at the National Security Agency from September 4, 1975 to June 13, 1976,stating that he was the section NCOIC and office level staff action officer for airborne operations at this time, responsible for developing plans and programs pertaining to airborne operations, assisting in developing manpower and equipment resource requirements in accordance with mission objectives, PROVIDED REPORTING GUIDANCE FOR AIRBORNE OPERATIONS, and, PREPARES AND DELIVERS BRIEFINGS AS REQUIRED. "Michie F. Tilley, the Chief of the B2 group at NSA was later to record his views of Mooney in a fitness report dated June 14, 1976: "Senior Master Sergeant Mooney richly deserves the recognition cited by his reporting and endorsing officials. His performance and dedication to duty in an assignment demanding rapid response to both cryptologic sites and command-level users serves well the interest of B Group, NSA and the entire cryptologic community."

36 National Security Council Memorandum, February 9, 1977

37 Memorandum of February 14, 1977, from Secretary of Defense Harold Brown to President Carter.

38 Final Report of Woodcock Commission, 1977

39 *First Heroes, p.* 136-139, op. cit.

40 March 8, 1977, CIA Memorandum, obtained by the author by FOIA request

41 Ibid.

42 Ibid.

43 Policy Related Events/Decisions/Comments From 1973 Onward Concerning Unaccounted For American POWs and MIAs In Southeast Asia. Office of Senator Robert Smith, Vice Chairman, Select Committee on POW/MIA Affairs. September 21, 1992, p. 40

44 March 16, 1977 letter, from Zbigniew Brzezinski to Carol Bates, National League of Families, supplied to the author by Colonel Earl Hopper

45 CIA Memorandum, 11 March 1977, obtained by FOIA request

46    CIA Report #317/091-77, March 16, 1977, obtained through FOIA request

47    Policy Related Events, Office of U.S. Senator Robert Smith, Select Committee on POW/MIA Affairs, 1992, p. 43-44, September 21, 1992.

48    CIA report of May 20, 1977, obtained through FOIA request

49    "An Examination of U.S. Policy Toward POW/MIAs, by the U.S. Senate Committee on Foreign Relations Republican Staff, May 23, 1991, p. 8-6.

50    Final Report of Woodcock Commission, op. cit.

51    Senate Select Committee on POW/MIA Affairs, John Holstine Box 2. Memorandum from Tom Lang to Frances Zwenig. Re: 1978 U.S. Air Force Document containing Information From Returnee Debriefs on Casualty Status of Unaccounted for Personnel." June 24, 1992, courtesy of researcher William T. Bennett

52    Senate Select Committee on POW/MIA Affairs, Tom Lang, Box 1 "51" from Air Force printout, "Air Force Personnel Not Returned From Southeast Asia."

53    Senate Select Committee on POW/MIA Affairs, Gekoski Box 1

54    Testimony of Albert Shinkle before the House Subcommittee on Asian and Pacific Affairs, July 4th, 1978, and Colonel Shinkle's affidavit in the Smith vs. President lawsuit argued by North Carolina attorney Mark Waple; also: "Kiss The Boys Goodbye"

55    CIA Report #FIR-371/09169-79, June 1979, obtained through FOIA request of the author

56    *The Threat:Inside the Soviet Military Machine*, by Andrew Cockbum, Random House NY, 1983, pp. 222-223

57    *Kiss The Boys Goodbye, How The United States Betrayed Its Own POWs In Vietnam*, by Monika Jensen-Stevenson and William Stevenson, Dutton, NY, 1990.

58    National Security Council Memorandum, March 7, 1979; Carter Library, originally located by researcher Ron Martz

59    NSC Memorandum, March 12, 1979; Carter Library

60    NSC Memorandum of April 18, 1979, Brzezinski to President Carter; Carter Library

61    NSC Action Memorandum #386, January 21, 1980; Carter Library. All of these memorandums were turned over to the Senate Foreign Relations Committee Republican staff in 1990 and to the Select Committee on POW/MIA Affairs, U.S. Senate, in 1991. #386 of 21 January 1980 is reproduced on p. 899, POW/MIA Policy and Process, Hearings of the Select Committee on POW/MIA Affairs, November 1991, op. cit.

62    New York Times, June 24, 1979

63    *The Vietnamese Gulag*, Simon and Shuster, NY, 1986

64    *First Heros* by Rod Colvin, Irvington Publishers, NY, 1987.

65    Interview with confidential source Ramon C., in Miami, Florida, conducted with assistance of Luis Alcebo, January 1990.

# CHAPTER TWELVE

## THE POW/MIAS AND THE REAGAN ADMINISTRATION, FROM 1981 TO 1989

The Carter Administration's major foreign policy accomplishment was to bring Egypt and Israel to the conference table, which led to the Menachem Begin-Anwar Sadat accords, and unfortunately, to Sadat's subsequent assassination in 1981. Although President Carter had publicly backed the socialist Sandinistas in Nicaragua, to demonstrate his commitment to human rights, in 1979 he issued a secret authorization for funding aid to the Contra rebels. Carter had halted work on the B-1 bomber and MX missile, yet in the latter period of his Presidency he faced increased tension with the USSR under Leonid Breshnev following the Christmas, 1979 Soviet invasion of Afghanistan, which became a Soviet military occupation by early 1980. In retaliation Carter had called for a U.S. boycott of the Olympic Games, and he ordered a halt to U.S. grain sales to Russia, but thereby caused serious economic hardship for American farmers and political damage to himself. In Iran, an insurrection by Islamic fundamentalists against a government installed with CIA assistance and long supported by U.S. military aid, led to the Shah's departure in January 1979. A new regime under the Ayatollah Khomeini, which was hostile to the United States, took power soon afterward. A mob of Iranian radicals captured the U.S. Embassy in Teheran twice; seizing 52 American hostages the second time.

Thus, the Carter Administration, having already faced a secret hostage crisis from 1977-1979 with the American prisoners still held in captivity in Indochina, now faced a public humiliation over another group of American hostages, which dragged on month after month for the remainder of Carter's presidency. A failed rescue attempt by U.S. helicopter-borne forces on April 24, 1980 (Desert 1) caused the death of eight American servicemen, and the resignation of Secretary of State Cyrus Vance, who had disapproved of the raid. (Warren Christopher became Acting Secretary of State) Carter's subsequent inability to gain the release of the American hostages in Iran caused a dwindling of public confidence in him and contributed to the victory of the Republican candidates, Ronald Reagan and George Bush, in the November 1980 election.

Ronald Reagan was born in Illinois in 1911 and graduated from Eureka College in 1932, where he had been an athlete and studied economics and sociology. He became a radio sports announcer and later a successful film actor. During WW II, Reagan was commissioned as an Army officer, but spent his service time making training films in Hollywood. He served as president of the Screen Actors Guild from 1947-1952, when he was instrumental in removing suspected Communists from the filmmaking industry during the "red scare" of the McCarthy era. Originally a Democrat, who had campaigned for Truman in 1948, Reagan became a supporter of Eisenhower in 1952 and 1956, and became a Republican in 1962. In 1966 he was elected Governor of California, and was noted during the Vietnam War years for his hard-line views against the massive antiwar protests of the time. He was reelected in 1970 and governed California until 1975. Reagan briefly campaigned for the presidential nomination in 1968, and again, with more success in 1976, although the Republican nomination went to Gerald Ford. After the public disappointment with the Carter Administration became apparent, Reagan launched another campaign for the presidential nomination in 1980, and this time won easily. The stalemate over the American hostages held in Iran had subsequently assisted Reagan in winning the November 1980 presidential election.

For years to come charges were to be leveled that Reagan, Bush and William Casey, Reagan's campaign director and security advisor who had served in western Europe with the OSS during WW II, had negotiated secretly with the Iranians to ensure that the American hostages held in Teheran would not be released until after the election.[1] A subsequent Congressional investigation revealed some troubling facts, but failed to prove that such a plot had existed. The American hostages in Iran were, in fact, released as Ronald Reagan assumed office, which indicated, if not a secret deal, then the contempt felt for President Jimmy Carter by the Iranian mullahs. American anger at Iran led to U.S. support and encouragement to Iraq, despite that nation's abysmal human rights record towards its Kurdish and Jewish minorities; a tilt that was to bear bitter fruit in the following decade.

In early 1981, just after the inauguration of Ronald Reagan as President of the United States and the release of the American hostages in Iran, the Vietnamese once again made a secret effort to obtain a ransom for the surviving American POWs who still remained captive in Indochina, six years after the fall of Saigon. A secret communication had been passed from Hanoi to Washington, reportedly through China and with assistance from the Canadian government, informing the new Presidential administration that Vietnam still held American POWs. A U.S. Secret Service agent assigned to the White House had overheard a meeting in the Roosevelt Room about a cable from Hanoi offering to exchange a group of surviving American prisoners of war in Vietnam for the amount of U.S. war reconstruction aid that had been promised by President Nixon eight years before, in 1973. Interviewed by a U.S.

Senate investigator eleven years after this 1981 meeting, the source who was present at the time recalled the conversation he had overheard:

"Casey came into the Roosevelt Room from the Oval Office with President Reagan and Vice President Bush. National security advisor Richard Allen joined them, as they stopped for a moment to talk. They were headed toward another, larger meeting, and Chief of Staff (James) Baker and (his deputy Michael) Deaver stood a few feet away, at the doorway, waiting for the group to enter the meeting. Casey said to the President, apparently referring to a discussion begun in the Oval Office:

'"WHAT DO YOU WANT TO DO ABOUT THE MESSAGE?'

President: 'WHAT MESSAGE?'

Casey: 'THE MESSAGE FROM THE VIETNAMESE, THROUGH THE CANADIANS AND CHINA.' Holding a cable in his hand.

President: (to group) 'WHAT DO YOU THINK.'

Casey: 'I THINK IT'S JUST CHINA RUNNING INTERFERENCE ON VIETNAM.'

Vice President Bush: 'I AGREE.'

Casey: 'WE CAN'T GIVE $4.5 BILLION TO THE VIETNAMESE, IT WOULD BE PAYING BLACKMAIL.'

Vice President Bush: 'YEAH, I AGREE.'

Allen: 'IF THESE ARE LIVE POWs, WE SHOULD DO SOMETHING ABOUT IT.'"

"Baker and Deaver come up. Baker (said):

"'It's time for the meeting.'

"President: 'OK...(to Casey)...DO SOMETHING ABOUT IT. CHECK IT OUT.'

"Group (then) departs room for (the next) meeting."[2]

Five years later the bare facts of this exchange offer were finally revealed to the public by Wall Street Journal investigative reporter Bill Paul, who had also written the first full account of Robert Garwood's revelations about POWs remaining in captivity in North Vietnam. The author confirmed the essential details of this report in 1986 interviews with the attorney for the U.S. Secret Service agent, later identified as John Syphrit, who was a witness to the 1981 meeting in the Roosevelt Room on the POW offer by Hanoi, and with Richard Allen, William Casey, Bill Paul and other, confidential sources.

Long afterward, U.S. Senator Robert Smith of New Hampshire stated that he discussed the subject of the 1981 POW offer by Hanoi with Max Hugel, the CIAs 1981 Deputy Director for Operations, and a personal friend of then-CIA Director William Casey. Senator Smith maintained that DDO Hugel said he was indeed aware of the 1981 offer by Hanoi to trade American POWs for U.S. aid, that he associated the date January 26, 1981, with the offer, and that it was common knowledge at the higher levels in the intelligence community. Hugel stated to Senator Smith that he did not wish to testify about the meeting before a Congressional committee, but would do so if he was subpoenaed. When this was done, Hugel then testified that he had no knowledge of a POW offer as described and denied any memory of having told Senator Smith that he did.[3]

But, in early 1981, photographs taken by an RD-77 satellite and from an SR-71 reconnaissance aircraft showed a stockaded prison camp in the Nhomorroth area of Laos, which clearly showed a number "52" stamped out in the tall elephant grass nearby. This area had been reported by CIA as containing a Communist prison as far back as 1965. Although the exact meaning of the signal was unclear (possibly a member of a B-52 crew), more spy satellite pictures revealed two types of men inside the clearing, one group much taller than the other, and this confirmed in the mind of Admiral Jerry Tuttle, then Deputy Director of DIA, that American prisoners were confined within the barbed wire enclosed camp, now being called "Fort Apache." A Laotian source indicated that 40-50 U.S. POWs were held there, and some of the best intelligence on 25 or more American prisoners at this site had been assembled by retired U.S. Special Forces Lt. Colonel James "Bo" Gritz, according to a Pentagon source of the author's. A Delta Force mission backed by the Army's Intelligence Support Activity (ISA) was planned and deployed while a reconnaissance patrol of Hmong guerillas was sent to photograph the site. They reportedly failed to bring back pictures of American POWs, prompting the Washington Post to publish a story on May 13, 1981, on an unsuccessful U.S.-backed operation to confirm the existence of American POWs in Laos. There can be little doubt that the April 1981 assassination attempt against Ronald Reagan, in which the President, his press secretary and two others had been wounded by a lone gunman, John Hinkley, had diverted the President's personal attention from many matters, including his longtime interest in POW intelligence from Asia.

Gritz had at first been encouraged by the Intelligence Support Activity, a secret Army intelligence group known as the "little CIA" in the Pentagon, under command of Colonel Jerry King, but as word of the clandestine effort to locate American POWs leaked out, the secret Army operatives cut their support by October 1981, and told Gritz that his search effort no longer had official sanction. A CIA agent operating in Thailand under cover as a refugee relief official, Jerry Barker Daniels, aka Michael J. Baldwin, reportedly took over the official part of the operation for CIA, but was later found dead of gas poisoning in his Bangkok apartment. An American citizen recruited by Lt. Colonel Bo Gritz in the U.S. for the team named Scott Barnes, who was transported to Thailand in 1981, later claimed under oath at a 1986 Senate Veterans Affairs Committee hearing that he had been ordered to accompany Daniels/Baldwin on a secret reconnaissance mission into Laos during Gritz's absence from Thailand, and that the two of them had photographed and recorded the voices of two American POWs inside a guarded prison camp in the Mahaxy region of Laos, and that subsequently coded orders were issued from Washington, D.C., to return and kill the two U.S. POWs, to "liquidate the merchandise." Bo Gritz later stated that he did not believe Barnes, although the latter voluntarily submitted to several lie detector and a truth serum session with nationally known experts, passing all of them, and later challenged the 1986 Senate Veterans Affairs POW/MIA subcommittee investigators to charge him with perjury if they did not believe him. The flamboyant retired Special Forces officer continued his intelligence-gathering with private support, and entered Laotian territory several more times on reconnaissance missions, one of which resulted in casualties, when Gritz's patrol made contact with Pathet Lao forces.[4]

The Socialist Republic of Vietnam's government kept up interest in the POW/MIA issue by such statements as an August 1981 communique, referring to Americans who were, "reportedly captured but not registered," and who died or became "missing" on their way to detention centers, because of "war circumstances."[5] At the same time, due to the President's personal interest, U.S. military intelligence, CIA, and special operations personnel were all involved in a general upgrading of POW/MIA intelligence on live American POW survivors in the 1981-1983 period. This was testified to later by U.S. Army Special Forces officers operating at this time on classified intelligence-gathering missions in Thailand, Major Mark Smith and Sergeant First Class Melvin McIntire, and discussed later in this book.

Ronald Reagan's first Secretary of State during this period, former Army General Alexander Haig, had been Nixon's closest security advisor in the waning days of that president's term, after the American prisoners had been left behind in Southeast Asia. Haig vigorously promoted Reagan's combative policy towards further Soviet expansionism in the third world under Leonid Breshnev's leadership, yet in April 1981 the Reagan administration lifted the embargo on the sale of U.S. grain to Russia which had been imposed by the Carter administration after the invasion of Afghanistan. Observers in 1981, such as Richard Secord, recorded that Haig was often highly agitated by the weight of his concerns and chain-smoked cigarettes constantly. Following the 1981 assassination attempt against President Reagan, Haig had exhibited some instability in his immediate claim to be "in control" of the U.S. Government.

The Soviet KGB under Yuri Andropov chose the confused period after the shooting of Ronald Reagan by Hinckley to give the go-ahead for an attack on Pope John Paul II by a Turkish terrorist, apparently recruited by Bulgarian intelligence agents who were doing the bidding of Moscow center. The Soviets were determined to keep control of Poland, and the Polish Pope's relations with Solidarity Union leader Lech Walesa threatened the continued Soviet occupation of Poland, and possibly all of eastern Europe. East German advisers were being brought to Lebanon to train Palestinian Liberation Organization commandos for attacks on Israel, under an agreement made by PLO Chairman Yasser Arafat with East German President Erich Honecker. The Vietnamese, meanwhile, were in the Middle East also, meeting with Arafat in Damascus to negotiate arms shipments of captured U.S. war material left behind in Vietnam.[6]

President Reagan's public categorization of the Soviet Union as the "evil empire" set the tone for his administration. The Soviets were now bogged down in a costly guerilla war of conquest in Afghanistan, involving hundreds of thousands of Soviet troops and thousands of casualties, at a time when their Warsaw Pact client states were chafing under Communist rule. Martial law was declared in Poland in December 1981 and a Communist crackdown on the independent Solidarity labor union and mass arrests of its members began. By June 1982, Alexander Haig was forced to resign and was replaced as Secretary of State by a former Marine officer, George Shultz, as the world's attention was focused on the Falkland Islands where the British fought and won a brief but furious war against Argentina. Leonid Breshnev died in November 1982, to be replaced by Yuri Andropov, the former Soviet master of Hungary during the vicious repression of the 1956 Revolution, who had become head of the KGB from 1967-1982, when he rose to become General Secretary of the Central Committee. Viktor Chebrikov, a protege of Breshnev and of Andropov, took over the KGB in December 1982.

The Defense Intelligence Agency was at this time keeping all of its livesightings of American prisoners in Southeast Asia secret, while publicly debunking them to the American people, yet, in July 1982, Mrs. Le The Anh was still hearing from Vietnamese refugees around the world who had seen American POWs held alive in captivity long after the war. One of many examples is the eyewitness account of a bus driver on the Saigon-Hanoi route during a four-year period. In December 1978, after transporting the possessions of a North Vietnamese Army Colonel named Nguyen Canh, who was traveling from Saigon through Hanoi to a prison he commanded in Son-La Province (NVN), the Colonel had befriended the bus driver:

"After we arrived in his place, my two aides helped unload... we took a bath and ate dinner in his home. After dinner, the prison commander told one of his guards, go and get for me...a uniform...He told me to change...that we would take a tour of his prison camp. DURING THE TOUR, I SAW THAT THE PRISONERS WERE AMERICANS... THEY LOOKED AT ME COLDLY...AND WERE APPREHENSIVE...THEY WERE IN A SORRY STATE, VERY THIN AND PALE, THEIR CLOTHES WERE IN A FADED BROWN COLOR...THEY ATE PART RICE, PART SORGHUM, THE KIND PREVIOUSLY USED TO FEED HOGS. THE PRISON COMPOUNDS WERE COVERED WITH ELEPHANT GRASS... THE PRISON CAMP WAS LOCATED IN A THICK JUNGLE AREA, QUITE LARGE, SURROUNDED ON ALL SIDES BY MILITARY CAMPS, THE ARMIES OF THE SOCIALIST REPUBLIC OF VIETNAM."

"After my question, the camp commander gave me the following explanation: 'You see our Party and our State are not stupid...we want to use them to bargain and set a price for their release with the U.S. imperialists. The reason we need to keep those flying bandits is because these bandits have killed many of our comrades. Our people, our Party, and our State will make the claims, in order that our children will enjoy those.' I asked: Thus how many of those headmen, how many of those 'flying bandits' is our Party and our State keeping. His answer was: 'ABOUT 300 OF THEM, THE HEADMEN AND THE FLYING BANDITS. WE KEEP THEM HERE (IN SON LA PROVINCE) AND WE KEEP THEM IN NHO QUAH, IN FORMER NINH BINH PROVINCE.'" (The Communists now include Son La in a larger Cao Bac Lang Province and Ninh Binh inside Ha Nam Ninh province)

This and much other first-hand intelligence corroborated the earlier closed-door Congressional testimony of Nguyen Cong Hoan, a member of the post-1975 National Assembly of Communist Vietnam, who had defected with the boat-people in March 1977. It appears that Nguyen Cong Hoan may have been deliberately allowed to escape as a "messenger," after the Communist fashion, as his wife and family were later allowed to leave Vietnam on a regular airplane flight. Hoan later testified under oath that he had learned from higher officials that the North Vietnamese government had held back many U.S. POWs to gain U.S. diplomatic recognition and aid and Vietnamese membership in the United Nations. He also testified that Anh Ba, personal assistant to Le Xuan, the Communist party First Secretary, that U.S. POWs HAD BEEN TAKEN TO THE SOVIET UNION.[7]

But the U.S. Government was determined after 1982 to control the POW/MIA issue through maintaining the utmost secrecy on the actual existence of living American POWs in Asia into the 1980s, while simultaneously denying to the public that such certain evidence indeed existed in the form of decryptions of enemy signals and eyewitness accounts. The unspoken hope was that the public would ultimately tire of the subject and forget it, as had happened after WW I, WW II and Korea.

With the past policy guidance indicated by the Reagan Administration's rejection of Hanoi's secret 1981 offer to barter for the surviving prisoners, by January 1982 the POW/MIA matter had been placed in the hands of the NSC's Asian Affairs Director at the time, Colonel Richard Childress, who worked closely with Ann Mills Griffiths, Executive Director of the National League of Families, to ensure that publicity on the subject favored the Administration's current line. Formerly an outspoken critic of the government 's efforts on behalf of the abandoned prisoners, Griffiths was given a security clearance enabling her to gain access to livesightings and other current intelligence, but also requiring that she not divulge classified information. This apparently had the effect of co-opting Ms Griffiths, as in the succeeding years she gradually became more strident in condemning POW/MIA advocates or anyone else who opposed the government's policy. Le The Anh eventually stopped giving the livesightings of Americans by Vietnamese refugees that she acquired to DIA. She suspected the DIA was passing some of this information to the Vietnamese Communist government, who could use it against family members remaining behind in Asia and she was informed that DIA had been harassing the people who wrote to her about the American prisoners remaining in captivity.[8]

In this same period, President Ronald Reagan, still attempting to do something about the POWs in Indochina, also encouraged an effort by two young Republican congressmen, John LeBoutillier of New York and William Hendon of North Carolina, to meet secretly with Laotian officials including Foreign Minister Soubanh Srithirath and Ponmek Dalalov in 1981-82, and offer medical aid in return for information on U.S. POWs remaining in Laos. One delivery worth $275,000 was made to Southeast Asia, but then the story was leaked to the press and the Reagan administration cut off its support. Both of these congressmen had been given briefings by DIA officials which indicated official knowledge that MIAs were alive, and although neither were veterans, this together with their subsequent investigations caused both Hendon and LeBoutillier to become full-time crusaders on behalf of the missing POWs of Vietnam. LeBoutillier began raising money for private intelligence gathering missions inside Laos, by indigenous agents reporting to Colonel Al Shinkle, the retired U.S. Air Force Intelligence officer who had lived in Bangkok, Thailand since the end of the war. Shinkle had been involved in military espionage for the last nine years of his career and had developed a number of human intelligence sources among the Lao people.

According to a subsequent Senate report: "...In August 1982, a tax-exempt POW/MIA organization known as Support Our POW/MIAs (SOP)...began receiving tax-deductible donations, which were then transferred to bank accounts in Southeast Asia and elsewhere for the Skyhook II Project...Approximately $156,000. of the donations were wire transferred to a Bangkok Bank in the name of Mustaq

Ahmed Diwan upon instructions communicated to SOP by (National League of Families Executive Director Ann Mills) Griffiths. Information provided to the Committee indicates that Diwan is a friend or associate of Colonel Al Shinkle (USAF-Ret'd) and that the $156,000.00 transferred to the Diwan account was subsequently provided to Lao resistance forces, presumably to fund efforts on their part to locate, identify and repatriate American POWs..."

Commencing on August 2, 1982 and continuing through 1985, SOP received checks totaling approximately $200,000. Participants in the fundraising for POW/MIA intelligence gathering such as Texas oilman Bert Hurlbut, were informed that NSC staffer Richard Childress and Reagan's then-NSC Advisor, Judge William Clark, had approved of the operation. Some of the donors, including Nelson Bunker Hunt and Ellen St. John Garwood, later helped fund the Contras in Nicaragua, through Spitz Channel. The money raised for Laos was given to Colonel Al Shinkle and Patrick Khamvongsa, a former member of the Royal Lao Air Force with ties to Phoumi Nosavan and other members of the Lao resistance and some of the money was used to purchase guns. Shinkle stated to the Senate committee that, "...the money was obtained through donations by former Congressman John LeBoutillier. The arrangement for the monies to be received in the Diwan account was done by Ann Mills Griffiths and Richard Childress." Khamvongsa began working for Brigadier General Heinie Aderholt in 1984. Because of a 1984 gun purchase by LeBoutillier in Virginia for this project, the U.S. Bureau of Alcohol Tobacco and Firearms (ATF) recommended he be prosecuted, but, "In 1986 the U.S. Attorney for the Northern District of Virginia declined to prosecute, based in part upon the fact that there was evidence that LeBoutillier's activities had been sanctioned by the U.S. Government."[9]

The National Security Council's Asian Affairs Director, Richard Childress, who had control of the POW/MIA issue in the NSC after January 1982, later denied to a Senate committee that he had anything to do with supporting the Lao resistance or facilitating gunrunning: "I personally intervened with Congressman Hendon and LeBoutillier in 1982 and indicated the White House could not deal with them if such activities were contemplated. We learned of this through a State Department cable that alleged Congressman LeBoutillier's representatives intended on raising a private army."

In the Pentagon the POW/MIA matter was under the control of Assistant Secretary of Defense Richard Armitage. After the fall of Saigon, when Armitage had been engaged in removing war material and refugees from Vietnam, he next appeared in Iran in late 1975, in a liasion capacity for supporting the Shah's security forces.[10] Armitage was in Thailand again for an extended time in 1978, involved in a "private" business venture with retired General Herman

"Heinie" Aderholt, called SEATHAI (Ltd.), according to both Aderholt and the U.S. Ambassador to Thailand, Morton Abramovitz. Armitage was picked as an administrative assistant by Republican Senator Robert Dole of Kansas in 1979, which gave him access to the high and mighty of the capitol. By 1980, the veteran covert-warrior Richard Armitage was known on the Washington stage, and was soon tapped by future CIA Director William Casey, to serve on the Reagan election campaign staff as a security affairs adviser. After the Reagan victory William Casey helped Armitage gain an appointment as Deputy Assistant Secretary of Defense, in 1981. At this post in the Pentagon, Armitage soon superseded his own boss, Noel Koch, as Assistant Secretary of Defense for International Security Affairs (ISA), a prestigious policy-level job once held by such renowned Cold Warriors as Paul Nitze. From this influential position, Armitage attended Inter Agency Group (IAG) meetings which formulated high-level national security policy decisions involving problems as far-flung as U.S. counter-terrorist bombing missions against Kaddafi's regime in Libya, missile shipments to Iran during the Iran/ Contra hostage-for-weapons exchanges, Southeast Asian and POW/MIA -related security matters, and all Pentagon covert operations.

Armitage was also a member of the "Operations Sub-Group" (OSG) for review and coordination of U.S. counterterrorist activities. The initial Cochairs of this group were Colonel Oliver North of the National Security Council and Robert Oakley of the State Department. Representatives of the Joint Chiefs of Staff, Pentagon, CIA and FBI were members of this group counterterrorist activities. Armitage represented the Defense Department, and executive assistant director Oliver "Buck" Revell was the FBI's member. "Buck" Revell was not unacquainted with covert operations, having been censured for leaking confidential FBI data to a journalist in Oklahoma, after reportedly failing a lie detector test, yet continuing to rise within the Bureau's hierarchy to the position of associate deputy director for operations. Revell came under suspicion during the 1986-87 Iran-contra investigations because of telephone calls made to him by Colonel Oliver North during the effort to keep federal investigators from seriously pursuing the case, but the public was informed that no proof could be found that Revell had obstructed the investigative process. Revell's name was to continue emerging from shadowy cases until the presidential contest of 1992 between George Bush, Bill Clinton and Ross Perot.[11]

The Operations Sub-Group's activities are classified but reportedly involve the escalating response to Libyan terrorism culminating in U.S. bombing missions and activities of the Intelligence Support Activity (ISA), which had been created in 1980 to deal with the Iranian hostage issue; later involvement with covert POW/MIA reconnaissance missions in Asia, and in 1986 was still functioning, despite public reports of its nefarious activities involving misuse of funds. The Libyan raid of April 15, 1986 was planned by an ad hoc Crisis Pre-planning Group of the National Security Council which included Armitage, Undersecretary of State Michael Armacost, Oliver North and others.[12]

Armitage thus had consolidated his influence within the Reagan-Bush administration and formed alliances with other powerful intelligence bureaucrats. Although he had made repeated trips to Hanoi, often in company with Richard Childress and Ann Mills Griffiths, to discuss the American MIAs while being feted by the North Vietnamese, his "diplomacy" involved only the recovery of alleged skeletal remains of U.S. MIAs, in compliance with the decades-old official U.S. policy that the vanished American POW/MIAs were dead, and the many livesightings of American POWs still held in captivity long after the war were therefore fabrications. Thus Armitage was able to stall any real progress-such as initiating serious diplomatic negotiations for the survivors, year after year, throughout the 1980s. In his official capacity Armitage also suppressed information and evidence on the American prisoners of war by intervening to prevent television exposes of the POW/MIA subject (such as was experienced by CBS 60 Minutes producer Monika Jensen-Stevenson and independent documentary film producer Ted Landreth).

A Pentagon officer who had served in a liaison capacity with Colonel William LeGro in Saigon and had worked under Armitage in ISA, later informed the author that Richard Armitage had been one of those responsible for hindering and cancelling POW/MIA reconnaissance/ rescue missions in Southeast Asia between 1981 and 1985, that his military record had been altered to aid in establishing his legend after he had transferred to CIA, and that he used his influential position, in coordination with the NSC's Richard Childress and others in the State Department to deceive and divert the leadership of the most powerful US veterans groups on the POW/ MIA matter, while using the cloak of national security secrecy to hide these and other covert actions. History has shown that it has ever been the way of powerful leaders to use the most ambitious men to do the worst jobs.

Although President Reagan was personally concerned about the non-returned prisoners, and in a February 1982 policy decision had elevated the POW/MIA matter to the "highest national priority," the national security apparatus was still opposed to any action which could lead an increase in tensions with the USSR, and to a possible unravelling of the whole melancholy story of the American POWs and MIAs left behind in Indochina, the USSR, Korea and China, for decades. Thus, as in earlier wars, it was into the hands of a small clique of U.S. intelligence bureaucrats that the lost American POWs, whom Aleksandr Solzhenitsyn had identified as our "foremost heroes" in 1973, were to be delivered for 8 long years of the Reagan Administration, and 4 more years of the Bush Administration. How many died in misery and pain while Defense Intelligence Agency officials denied their existence, may never be admitted or known. William Casey, Richard Armitage, Richard Childress, Robert McFarlane, Donald Gregg, Gaston Sigur, Paul Wolfowitz, Morton Abramovitz and other U.S. officials were still certain that their actions or nonactions in gaining the freedom of the POWs would remain completely shrouded in national security secrecy and protected by this, they would never be called to task by the American people to account for their conduct.

Through two subsequent investigations by Senate committees, this has proven to be correct. It has been demonstrated that the government is virtually powerless to censure or punish its own employees for their actions in this particular matter. But as has been demonstrated in this book, these men were merely the latest in a 70-year-long series of US officials who had implemented similar harsh policy, or orders, to prove their loyalty. They would now make their own reputations by their willingness to officially perform this latest act in the most tragic and secret drama in American history, which was finally to be irrevocably exposed to the American people in the early 1990s.

Only the unexpected rise of a nationwide, grass-roots movement among Vietnam veterans in the 1984-1987 period, lending powerful support to the exhausted families of the missing who had for too long borne the burden of proof, was to ensure that the POW/ MIA issue did not die with the end of the decade, and of the Cold War. The publicised, courageous actions of Colonel "Bo" Gritz had inspired hundreds of thousands of war veterans across the country, who had been trying to forget the painful memories of the Vietnam era, to do something about their fellow servicemen who had been left behind in captivity in Indochina. It was to be this ever-present pressure from hundreds of thousands of active U.S. war veterans in all 50 states, which never died, but instead ultimately forced the major U.S. veterans organizations to address the issue, that in turn influenced segments of the American media enough to allow a handful of serious researchers to publish the declassified documentary evidence on U.S. POWs in the late 1980s and '90s.

The author had learned long before, during the 1964-66 period, and prior to combat service in Vietnam, of U.S. Korean war POWs secretly withheld by the Chinese Communists, from a Korean war infantry veteran, Francis Homsher, and from others. The author had suspected that Hanoi was holding other U.S. POWs after the 1979 release of Marine Private Robert Garwood, which had followed years of North Vietnamese denials that they had withheld ANY American

POWs after the 1973 ceasefire. In addition, while working as a hunting outfitter and commercial fisherman in the late 1970s and early 1980s on the Kenai Peninsula and Kodiak Island, Alaska, the author had heard national and international news reports from Scandinavian sources in Vietnam of live American POWs asking foreigners they encountered: 'Don't forget us, we're Americans, tell the world about us.' These reports appear to stem from an amalgam of several such eyewitness accounts which were based upon fact. One was the account of a Swedish engineer named Lars Arvling at a foreign aid project paper mill at Bai Bang, NVN, who saw with his own eyes a prison camp nearby with two tall Caucasian prisoners under guard, standing by the gate. Arvling was taken in custody by North Vietnamese police and was lucky to get out of Vietnam. When he informed the U.S. embassy in Stockholm of what he had seen DIA only responded months later, through the Swedish government, and nothing substantive came of it.[13]

The foreign ministers of Vietnam, Laos and Cambodia met in Vientiane, Laos on January 28-29, 1984, and afterwards issued a communique, which mentioned the MIA problem. The three Communist nations agreed to exchange information among each other about missing Americans, but their cooperation with the U.S. would be conditional: "The Indochinese states are ready to 'cooperate' with the United States on the MIA question...if the U.S. government has a cooperative attitude and renounces its hostile policy toward Indochina."

During the 1983-1984 period the author had joined with former U.S. Congressman John LeBoutillier (R-NY) in supporting and funding an intelligence-gathering network inside Laos run by the retired U.S. Air Force intelligence officer, Lt. Colonel Albert Shinkle, then operating from Bangkok, Thailand. Shinkle was the former combat pilot and intelligence officer who had served under Brigadier General Herman "Heinie" Aderholt during the years of the CIAs secret war in Laos. He had already testified about American POWs remaining in Laos before Congress, in the 1978 Pacific and Asian affairs hearings held in Hawaii, by Chairman Lester Wolff, of New York. Shinkle stated at that time that CIA officials based in Denver, Colorado had warned him that his intelligence gathering activities on U.S. POWs in Laos were counter to U.S. policy.

LeBoutillier was a friend of the author's father-in-law, Anthony Drexel Duke of New York, a St. Paul's graduate and former U.S. Navy officer who had commanded a LST 530 at Normandy Beach in 1944, was a trustee of Duke University, and founder of the Harbor for Girls and Boys in New York. The purpose of this "Skyhook II Project," was to fill a gap left by the years of politically-motivated inaction of official U.S. intelligence agencies, in gaining intelligence on the continued existence of surviving U.S. POWs in Laos, many years after the end of the war. The intelligence gathered and confirmed by Colonel Shinkle, together with declassified documents from the 1978 Uncorrelated Reports cited within this book, was sufficient to convince the author that such was indeed the case. Colonel Shinkle had retained his extensive intelligence contacts after retiring from the service and from the late 1970s to the mid-1980s had assembled a volume of intelligence on as many as 300 U.S. POWs still being held inside Laos in a number of separate groups, which were sometimes moved in and out of North Vietnam. At the same time Shinkle was gathering intelligence on hundreds of Americans still reported by Laotian sources to be in captivity, three highly decorated Green Beret officers who were veterans of Vietnam, still on active duty gathering intelligence for the U.S. in Asia, Colonel Robert Howard, Major Mark Smith and Sergeant First Class Melvin McIntire, had come to a remarkably similar conclusion, and they were about to enter into the worst fight of their lives in trying to expose the truth to the American people.

Also in 1984, while the author was researching an earlier non-fiction book on the Vietnam war in the National Archives[14], some U.S. POW/MIA documents relating to individual unit losses during the Vietnam war years came to light, indicating to the author that the U.S. Government's effort to determine the fate of American POWs and MIAs should be fully investigated by an independent researcher. This initial investigation led the author into taking part in a 1985-1987 effort with former Congressman LeBoutillier and Anthony Drexel Duke (aided by his brother, former U.S. Ambassador Angier Biddle Duke), to alter U.S. POW/MIA policy in favor of direct State Department negotiations for all surviving Communist-held U.S. POWs, including those held in the USSR and China, rather than for human skeletal remains from crash sites in Southeast Asia.

In 1984, a secret contact between North Vietnamese officials and the U.S. National Security Council by way of a "cutout" occurred, which went unpublicised until nearly a decade later, when it became known among veterans as the "POW/MIA-Mob connection." At that time a Senate investigator named Tracy Usry revealed that police had seized a diary belonging to I. Irving Davidson during a criminal investigation in the late 1980s, and had turned it over to Senate Foreign Relations Committee investigators because of the references within it to POW/MIA matters and meetings. Mr. Davidson was a Washington, D.C. lobbyist who was acquitted of 12 counts of conspiracy to bribe, racketeering and fraud during the '81 Brilab trial, in which New Orleans crime boss Carlos Marcello was convicted of one count of conspiracy. The trial resulted from a 1979 FBI undercover operation. The diary entries showed Davidson had many meetings scheduled with Richard Childress of the National Security Council and Ann Mills Griffiths, Executive Director of what was now the government-sponsored National League of POW/MIA Families. In a deposition given to the Senate Committee on October 1st and 6th, 1992, Mrs. Griffiths declined to answer questions about her meetings with Mr. Davidson. She said the matter was 'secret' and that she would 'coordinate her story with Mr. Childress,' when he was questioned about the matter. Senator Robert Smith, Vice Chairman of the Senate Select Committee on POW/MIA Affairs called for an "investigation of reported POW/MIA-related meetings at the White House and nearby hotels between Mafia-connected businessman I. Irving Davidson, NSC staffer Richard Childress and Ann Griffiths..." In the subsequent report of this select committee, the matter was only partially dealt with:

"I. Irving Davidson (a civilian with NSC contacts) reported in 1984, that according to his contacts with highly placed officials of an ASEAN nation, it appeared that individuals in the government of Vietnam had indicated that the Vietnamese would welcome an approach by the U.S. to discuss the POW issue. The early reports relating to this subject indicated that the discussions were to cover the sale of both warehoused remains and live POWs ('breathers'). In late 1984, a high-ranking retired general, who was a member of the National Security Council of the ASEAN nation, discussed this matter with Richard Childress of the National Security Council who, with the concurrence of Robert McFarlane (National Security Advisor to President Reagan), traveled to Vietnam to investigate this report. Declassified documents indicate that Assistant Secretary of State Paul Wolfowitz informed Secretary of State George Shultz of a plan to pay for remains and 'POSSIBLE LIVE POWs' in a January 1985 memorandum marked 'super-sensitive.' The memo stated that Mr. Childress intended to fund the initiative with either CIA or private funds. Mr. Childress later reported that he had followed up the possible offer, but that it had led to a discussion only of remains. The Committee did not consider the matter satisfactorily resolved by the reports filed and viewed that open questions remained as to what had actually occurred.

"In 1992, the Chief Counsel to the Select Committee (J. William Codinha) and a Committee investigator travelled to the ASEAN nation to investigate the alleged 1984 live American offer... The general's brother remembered offers for live POWs having been made, the general said that the Vietnamese wanted several hundred million dollars in return for the remains of 50 Americans. The general also said that at some point, Mr. Davidson called him to say that the 'deal was off because of leaks.' Both men indicated that if the Committee desired, the North Vietnamese channel could be reopened for the continued discussion of purchasing remains."[15] The full implications of this 1984-85 matter were not revealed by the Committee's investigation, and remain a mystery at the present

writing. The author met with Irving Davidson in company with former
Congressman LeBoutillier in Washington, D.C. during 1987, and
discussed the POW/MIA matter, but no further information on the
above matter was obtained.

The mercurial former Congressman John LeBoutillier
embarrassed U.S. security officials of the Reagan Administration
in the fall of 1985, when he tape-recorded a response by Reagan's
National Security Advisor, Robert McFarlane, on the subject of POW/
MIAs, during a foreign policy discussion in New York. McFarlane had
been involved with the POW/MIA matter as far back as the Paris
peace negotiations, after the conclusion of which he had attempted
a secret negotiation to ransom those left behind in 1973 with
$100,000,000. in contingency funds, not accountable to Congress
by the Nixon Administration. The Vietnamese had refused, holding
out for the full $3.2 billion which had been promised by Nixon and
Kissinger. In his 1985 response to a question by New York banker John
Thornton about whether any of the missing Americans were still alive
in Southeast Asia, McFarlane had replied with his own assessment,
based upon the number and quality of (still classified) livesighting
reports by eyewitnesses:

"I THINK THERE HAVE TO BE LIVE AMERICANS THERE...
THERE IS QUITE A LOT OF EVIDENCE GIVEN BY PEOPLE WHO HAVE
NO ULTERIOR MOTIVES AND NO REASON TO LIE, AND THEY'RE
TELLING THINGS THAT THEY HAVE SEEN...WHAT WE NEED TO DO
IS HAVE BETTER HUMAN INTELLIGENCE. NOW WE DON'T. IT TAKES
TIME TO GET IT. BUT I WOULDN'T PRETEND TO YOU THAT WE HAVE
DONE ENOUGH EVEN TO START. AND THAT'S BAD. AND THAT'S A
FAILURE."

McFarlane had assumed that he was speaking to a private
group and not to the press. LeBoutillier had committed the
unpardonable political sin of turning over the tapes of McFarlane's
remarks to the Wall Street Journal, where the story appeared in print
at a time when the Reagan Administration was saying publicly that
no intelligence existed that Americans were still held in Southeast
Asia, and as DIA was denigrating the hundreds of livesightings and
attacking the refugee sources as having ulterior motives for stating
they had seen Americans in captivity long after the war ended.[16]
Although LeBoutillier thereafter attempted to gain an appointment
through the Department of State as a POW/MIA negotiator with the
Vietnamese, his action in clandestinely taping McFarlane and leaking
it to the press ensured that this was not to be.

The POW/MIA issue had already heated up after syndicated
columnist Jack Anderson published charges by three U.S. Army
Special Forces officers that evidence of live POWs in Asia was being
covered-up in the Pentagon. A class-action lawsuit had been filed
by a former prisoner of war, Major Mark Smith, and a Special Forces
sergeant named Melvin C. McIntire.[17] The suit was conducted by a

North Carolina attorney and West Point graduate named Mark Waple, and soon attracted other affidavits from Vietnam War veterans who had knowledge of POWs left behind in enemy control. Several of these men, such as the now-retired, former-NSA analyst Jerry J. Mooney, ex-POW Robert Garwood and ex-Marine Captain Thomas V. Ashworth, would devote the next decade of their lives to revealing the truth about the abandoned American prisoners of the Indochina war.

At the same time, retired Air Force Lt. Colonel Delk Simpson, the U.S. air attaché in Hong Kong during 1954 who had reported an eyewitness account of American POWs from Korea being shipped into the USSR, was attempting to find out what had happened to this intelligence. On the last day of August in 1985, the Washington Post had reported on Lt. Colonel Simpson's meeting with officials of the Defense Intelligence Agency, who claimed they hadn't found the "1955" cable and deviated the media's attention from the truth by positing that the prisoners seen being shipped into Siberia had been "French troops being repatriated after the French negotiated an end to the Indochina war in May 1954." The news report said the Pole told Simpson that he was, "standing within 10 feet of about 700 soldiers, many of them black, who had been stopped in Manchouli while the train undercarriages were modified...the Pole drew pictures of the way the soldiers were dressed, including sketches of chevrons that Simpson said strengthened his belief that the men were Americans captured by Chinese forces during the Korean War and sent to Siberian labor camps." (American chevrons for enlisted men and NCO's are distinctly different than French and British chevrons.)

To counter Lt. Colonel Simpson, the Pentagon and DIA advanced the theory that the blacks seen by Simpon's eyewitness source were possibly Senegalese troops fighting for the French. In view of the dates and details in Simpson's original report and the resulting Eisenhower Administration demand to the Soviet government discussed in Chapter 8 of this book, the Defense Department's preposterous response to Simpson's 1985 quest can only be described as deliberate disinformation by the Pentagon to deceive the media and the public, and to deviate attention from the known historic facts of the prisoner of war matter.[18] (This report, combined with information supplied by Korean War veteran Francis Homsher earlier, and the work of John Noble, Solzhenitsyn and others, convinced the author to research the historic roots of the Vietnam POW/MIA issue, with the view of writing a book.)

In January and February 1986, during an extended research period in Washington D.C., author John M.G. Brown became involved with several other concerned Vietnam veterans who either testified, or attended in support of the January 1986 Senate Veterans Affairs

Committee hearings on POW/MIAs. The hearings were chaired by the freshman Republican Senator from Alaska, Frank Murkowski, while a few other Senators, including Dennis DeConcini a Democrat from Arizona, took an active part. These included J. Thomas Burch, Chairman of the National Vietnam Veterans Coalition, and NVVC Secretary William T. Bennett, both Vietnam veterans and Washington D.C. attorneys, and also Vietnam veteran researcher Michael Van Atta, editor of a factual POW/MIA newsletter, "The Insider." Another researcher was Thomas V. Ashworth, who had been a Marine helicopter pilot in Vietnam, and who testified at the 1986 hearings on POW intelligence about 200-300 American POWs remaining captive in Laos, that he had gained from the CIAs Laotian General, Vang Pao, and several of his former staff officers, as well as to being informed that secret returns of U.S. POWs from Indochina had occurred. Other active veteran leaders the author was involved with included John Molloy of New York's Release Foundation and Richard Keeton of the Veterans of Foreign Wars.

At this time the author also met two young Republican Congressman who were to remain in the forefront of the POW/MIA matter seven years later. They were William "Billy" Hendon of North Carolina, a large, forceful man, totally dedicated to uncovering the truth about the POWs left behind; and Robert Smith, a Navy fleet veteran of Vietnam from New Hampshire, who investigated methodically and spoke slowly, but in the end exhibited the most staying-power on the POW issue of any member of the U.S. Congress. Valuable information was gained by the author from POW/MIA wives and daughters who had fought for years to get their loved ones returned, Marcia Welch, Ann Holland, Marian Shelton, Patty Skelly, Dolores Alfond, Diane Van Renselaar, the Standerwick sisters and many others. Among the most knowledgeable POW family members whom the author met was Colonel Earl Hopper, a retired U.S. officer whose son, Earl Hopper Jr., was one of the MIAs in Southeast Asia. Hopper had been head of the League of Families during the Carter Administration, and had sued the government over the issue of mandatory case reviews, that presumed the missing prisoners dead without any real evidence of death. He maintained extensive files of declassified documents which he sometimes shared with other researchers, and provided leadership to a younger generation of war veterans who had become involved in the issue.

The author had first encountered Frank Murkowski when the latter was an Anchorage banker requesting the votes of the people of Kodiak Island, Alaska. As an Alaskan Vietnam veteran from the Kodiak post of the American Legion, the author later met with Murkowski and his counsel during the January 1986 hearings, Anthony Principi (who was later rewarded with an appointment as

Assistant Secretary for Veterans Affairs.) Another aide at Murkowski's side in these hearings was a CIA official named Alan Ptak (who was later to reappear in the POW spotlight in 1991). The author attended these hearings, and during one session, became embroiled in a confrontation with Assistant Secretary of Defense Richard Armitage, who was then in charge of U.S. POW/MIA policy in the Pentagon. Armitage, a massive and pugnacious man, was reiterating the official U.S. position that no evidence existed of live American POWs held in captivity, that any live Americans still there were stay-behinds or deserters, remaining of their own free will, and that there was no government coverup of the matter.

U.S. Marine Corps Vietnam veteran Ed "Gino" Casanova, with the author and former paratrooper sergeant Michael Booth on either side intending to prevent violence, stepped forward and forced a large, black cowboy hat down over Armitage's head and face, to, as the ex-Marine said it: "demonstrate the meaning of a coverup." Casanova then snarled: "You get the black hat award, Armitage..." Armed guards moved forward to defend the glowering Assistant Secretary and an uproar ensued that was filmed by the news cameras, but was subsequently buried. Armitage had recently returned from yet another meeting in Hanoi with the Vietnamese Communists, on January 5th and 6th, during which the U.S. and Vietnam, "pledged to create a favorable atmosphere."

This incident, witnessed by POW/MIA activist Ted Sampley and many others present, including senior DIA officials, indicated the level of frustration which had been reached by veterans at the time. Casanova had been called on the telephone by President Reagan to end his lengthy, late-1985 protest-fast over continued government stalling on the POW/MIA issue, after the charges of a cover up on POWs in Laos by the Green Berets Major Mark Smith and Sergeant Melvin McIntire had achieved wide exposure in the media. Reagan promised to meet with Casanova on the subject of the prisoners of war and Casanova had come to Washington to redeem his pledge, accompanied by the author and another Pacific northwest veteran, of the 101st Airborne Division in 1968-69, Michael Booth. U.S. Secret Service agent James P. Bour, and his partner S.A. Solterer, subsequently questioned all three in Washington D.C., in early February 1986, to determine if a "threat to the President or the President's children existed," and warned all three to leave the Washington D.C. area.

In light of public statements of the time by Richard Armitage and other U.S. officials, that no evidence existed of U.S. POWs being withheld in captivity after the American withdrawal and into the 1980s, it is interesting to note that on January 29, 1986, during the "Murkowski hearings," DIA had sent the Department of State forty-three first-hand, eyewitness livesightings of American POWs held in

North and South Vietnam and in Laos, dating from the late 1970s to the early 1980s. Later obtained by the author through repeated Freedom of Information requests to the Department of State, a letter from the Chief, Special Office for Prisoners and Missing in Action-DIA (name blacked-out by censors but probably Colonel Joseph Schlatter) announced that the information had been supplied to Congress in July 1985, and that since that time "seventeen of the cases have been resolved."[19] "Resolving cases," in DIA terminology translated into debunking the Indochinese (or American) eyewitnesses, by one means or another, in order to conform to U.S. policy since April 1973, that all U.S. POWs and MIAs who had not been released during the 1973 Operation Homecoming, were dead. Thus, important evidence of U.S. POWs sighted by an eyewitness in the VINH area of North Vietnam, which would have assisted in confirming information obtained from decryptions of enemy radio traffic by the NSA analysts Mooney and Minarcin about a collection and preliminary interrogation camp in the VINH area which provided U.S. POWs to the "Tentacle M-B" camp at Sam Neua, Laos, was discounted:

"Source 2623, an ethnic Hmong, recalled that on a visit to Vinh, North Vietnam, at some time in 1972 or 1973, he briefly observed a group of seven men, whom he believed to be American POWs, being escorted by North Vietnamese cadre. At least one member of the group was a black male. Source heard local residents say the group was from a camp in the city."[20] Yet, DIA maintained, "there is no evidence that any group was held in Vinh for a prolonged period."

Another source reported seeing a larger group of U.S. POWs in North Vietnam four years after Operation Homecoming, and two years after the fall of South Vietnam:

"SOURCE 2638 REPORTED THAT IN 1977 HE OBSERVED 20 OR MORE AMERICANS AND AUSTRALIANS CONFINED AT A PRISON CAMP SOUTH OF YEN BAI TOWN, NORTHERN VIETNAM, ABOUT 130 KILOMETERS NORTHWEST OF HANOI. SOURCE...PROVIDED A DETAILED SKETCH DEPICTING THE CAMP AND ITS LOCATION. THE LOCATION WAS A GROUP OF REDUCATION CAMPS FOR FORMER MEMBERS OF THE REPUBLIC OF VIETNAM ARMED FORCES. This was one of six groups of camps established in this region in 1975-76, and was administered by Group 776 of the People's Army of Vietnam (PAVN)..."

DIA maintained that in this case, "Scores of former inmates have provided proven accurate information about this camp system. There is no evidence that Americans or Australians were detained in this system...Although the polygraph might provide insights regarding this source's motives, DIA has insufficient information with which to make a definitive assessment of the source's information without the use of the instrument. In any event, the use of a polygraph is a moot question in that the source was recently jailed for assaulting and seriously injuring several neighbors....Source 2638's account of 20 or more Americans and Australians confined at this camp south of Yen Bai town is a fabrication."[21]

Why the fact that because "scores of former inmates," out of the thousands who were interned in these camps, had reportedly (according to DIA) not seen these 20 or more American POWs at Yen Bai in 1977, should discount the sighting by an eyewitness who HAD seen them, is not explained by the agency. From analysis of many other livesighting reports, it is clear that the North Vietnamese often confined postwar American POWs in secure, or hidden areas at many locations, and often discouraged others from seeing them. Likewise, the alleged assault by this eyewitness (again according to DIA analysts intent on debunking the witness) would appear to have little to do with what had been witnessed years before, and offered voluntarily, probably with the knowledge that rather than be rewarded, the witness could expect harassment and hostility from the DIA analysts, based on years of their performance with other would-be informants on American POWs seen in captivity after the war. In short, there is no adequate explanation of why this source's sighting of 20 or more American POWs was a "fabrication."

Another sighting of a larger group of American prisoners, two years later, proved more difficult for DIA to discredit, at least for the time being:

"SOURCE 1270 REPORTED THAT IN 1979 HE OBSERVED 50-70 U.S. PRISONERS OF WAR AT WHAT HE CALLED CENTRAL PRISON NUMBER 32, IN HA SON BINH PROVINCE, NORTH VIETNAM. ON ONE OCCASION...HE HELPED BURY ONE AMERICAN WHO DIED AT THE PRISON...The source said that a member of the prison cadre told him the group of U.S. PWs...had been held at Son Tay before the unsuccessful U.S. rescue attempt in 1970. All the U.S. prisoners held at Son Tay prior to the 1970 rescue attempt have been accounted for; therefore, this aspect of Source 1270s account is not accurate."

Yet, how could DIA be so certain that some number of American POWs at Son Tay had not been segregated and totally hidden from other U.S. POWs who reported being held at there? How could anyone be certain about such a thing, when throughout the war the Vietnamese Communists had successfully hidden whole companies and battalions of troops underground or in jungle areas in relatively close proximity to American forces. The North Vietnamese had in fact made a practise of segregating groups of U.S. POWs and individual Americans for years, and as this book reveals, they had long run a parallel prison system for U.S. POWs, which makes DIAs analysis of this source flawed from the beginning. DIA goes on to say: Source 1270 is the first source to report the existence of a prison 32 or any prison other than prison 52 (Ha Tay), in Ha Son Binh Province. Source 1270 prepared an annotated sketch of the PURPORTED prison 32. The sketch does not resemble in any way prison 52 (Ha Tay) ...Current HUMINT reporting indicates that there is only one central prison in Ha Son Binh Province and that is number 52 (Ha Tay). As noted earlier, some former inmates of Central Prison number i were moved briefly to Ha Tay before being transferred further south. The case will be kept open until imagery (aerial photography) and HUMINT taskings are completed."

But DIAs record indicates that this tasking would only result in adoption of the foregone conclusion that this, like every other eyewitness livesighting of American military POWs held in captivity long after the war, was, according to DIAs "analysts" such as Robert Destatte, Warren Gray, Charles Trowbridge and Sedgewick Tourison, "a fabrication."

Another example is a much more recent sighting of an American POW held in solitary confinement in the Vientiane area of Laos, at a time when the Pentagon's Richard Armitage, the NSC's Richard Childress, and the Defense Intelligence Agency were all saying in concert that no evidence of still-living American POWs in Laos (or Vietnam) existed:

"In December 1984, a person who described himself as a member of the Lao resistance, contacted the American Consulate in Udorn, Thailand, and offered information about possible American PWs in Laos. This person, who has been designated Source 2676, said that in October 1984, while on an intelligence collection mission in Laos...a friend introduced him to a Lao People's Army soldier assigned to a PW camp north of Vientiane. ACCORDING TO SOURCE 2676, HE WAS INVITED TO THE CAMP AND WAS ABLE TO VISIT THE FACILITY AND SPEND THE EVENING. WITHIN THE CAMP HE OBSERVED A MANACLED AMERICAN PRISONER OF WAR, WHOM HE DESCRIBED AS 65 YEARS OLD, WITH A FULL BEARD. SOURCE...WAS TOLD THAT FIVE OTHER AMERICAN PWS ALSO WERE BEING DETAINED THERE..."

DIA analysts reported: "Extensive imagery analysis of the area north and northeast of Vientiane has failed to locate a camp or compound of the size and description as provided by Source 2676... The vagueness of his story, particularly with regard to the location of the PURPORTED PW camp conflicts with his claim to have been on an intelligence collection mission..."[22] But how could DIA really be so certain that no "compound of 80x200 meters," enclosed by a wooden fence, existed to the north of Vientiane, Phanthaboun District, Vientiane Province? The fact is that for years to come DIA would claim that this or that facility "did not exist" to disprove various livesightings of Americans, only to be later disproven by others.

The author has obtained dozens of similar eyewitness accounts of American POWs held in North Vietnam, South Vietnam and Laos in the late 1970s and early 1980s, but the limitations of a single volume unfortunately precludes their inclusion herein. In every case, it appears to the author that DIA analysts have proceeded with faulty reasoning and haphazard methodology, with a deliberate intention to discredit any source who had the courage to come forth in the hostile and threatening environment provided by the DIA's "Special Office for Prisoners and Missing in Action," which has in reality disgraced both the United States military services and the entire U.S. intelligence community.[23]

Sworn testimony and an affidavit for the 1986 Murkowski hearings of the Senate Veterans Affairs committee, from the last publicly-acknowledged U.S. prisoner of war to be returned from North Vietnam, Marine PFC Robert Garwood, in 1979, substantiated the revelations he had made to Bill Paul of the Wall Street Journal which were published in December 1984, that some 70 to 100 other American prisoners of war, that he himself had seen, were still being held, captive in North Vietnam until the end of the 1970s:

"As a prisoner of war ...in Vietnam between the approximate time frame of 1973 and 1979, I personally saw United States prisoners and heard about others from Vietnamese prison guards. I learned that in the late 1970s some American prisoners of war were being held at four places: prison camps at BAT BAT and YEN BAY, a military complex at LE NAM DE STREET IN HANOI, and at another location referred to as GIA LAM, which was a suburb east of Hanoi...

"TO THE BEST OF MY RECOLLECTION I BELIEVE THERE WERE FORTY TO SIXTY AMERICAN PRISONERS AT YEN BAY, APPROXIMATELY TWENTY IN THE BAT BAT DISTRICT OF SANTE PROVINCE, SIX AMERICAN PRISONERS OF WAR AT GIA LAM AND APPROXIMATELY SIX OR SEVEN AT LY NAM DE. ALSO, IN APPROXIMATELY THE SUMMER OF 1977 I PERSONALLY OBSERVED...THIRTY TO FORTY AMERICANS CLIMB DOWN FROM A BOX CAR DIRECTLY IN FRONT OF ME. THERE IS NO DOUBT IN MY MIND THAT THESE INDIVIDUALS WERE UNITED STATES CITIZENS BECAUSE OF THEIR PHYSICAL APPEEARENCE AND THEIR LANGUAGE...IN LATE 1978 NEAR THE LY NAM DE MILITARY COMPLEX, AFTER SEEING A PRISONER WHO DEFINITLY LOOKED AMERICAN TO ME, I WAS TOLD THAT THERE WERE PRISONERS WHO WERE AMERICAN...APPROXIMATELY SEVEN IN NUMBER AND THAT THEY HAD RECENTLY BEEN BROUGHT FROM A PRISON CAMP AT CAO BANG, NEAR THE CHINA-VIETNAM BORDER..."

Garwood's testimony was convincing evidence by itself that many Americans had indeed been held captive in North Vietnam through the 1970s, but the probability of hundreds of other U.S. POWs still being alive inside Laos until the mid-1980s was testified to by three other witnesses, two of whom had recently been forcibly-retired from the Army for refusing to remain silent about what they knew of U.S. POWs alive in Laos during 1984 and into 1985.

Due to their status as certifiable heroes and widespread media exposure previously, the testimony of these three Army Special Forces (Green Beret) officers, Major Mark Smith, Sergeant First Class Melvin McIntire and Colonel Robert Howard, was viewed as the most sensational information revealed in the 1986 Murkowski hearings. Major Smith had received his commission on the battlefield during the 1968 Tet Offensive, from General Westmoreland personally. In the 1972 NVA Easter offensive, he was an adviser to the ARVN airborne and ranger battalions and ground commander at the battle of Loc Ninh, where he was wounded and captured. Held as a POW inside Cambodia until being repatriated during Operation Homecoming in 1973, he was decorated with the Distinguished Service Cross for heroism. At the Murkowski hearings Major Smith testified under oath and by sworn affidavit:

"In April 1984 while assigned to the Special Forces Detachment Korea (SFD-K)...I was required to turn over certain information to an Army Major General (later identified as Major General Kenneth Leuer). This information was generated because back in 1981, as the detachment commander. I had been given a mission to gather intelligence on the Korea Special Forces because of their ties to Korea's President, and on Thai Special Forces and any other special operations forces in Asia with whom my unit worked. THE SUSPECTED AND KNOWN LOCATIONS OF POWs AS REPORTED TO ME WERE ALL IN LAOS. I LEARNED THAT THE INFORMATION HAD BEEN DEVELOPED THROUGH THAI RECONNAISSANCE TEAMS..." They stated that there was a possibility of obtaining evidence of certainty of identification such as photographs and fingerprints...This first meeting in July 1981 where these conversations took place was at Lopburi (Thailand) After the briefing...I decided that there probably was some truth to reports that there were living Americans in Southeast Asia I WAS ASKED IF I HAD A CHANNEL OF COMMUNICATION THAT WOULD BYPASS (THE) U.S. EMBASSY-BANGKOK, COMMANDER-IN-CHIEF PACIFIC AND JOINT CASUALTY RESOLUTION CENTER (JCRC, also in Bangkok) BACK TO WASHINGTON, D.C.

"I returned to my home station in Korea at Seoul and went to the military intelligence contact from DIA with whom SFD-K worked and drafted a message to DIA headquarters and...INSCOM (United States Army Intelligence and Security Command, a component of the NSA, headquartered at Fort Meade, Maryland.) Five days later I was told that the sources would be closely guarded and no one in the U.S. Embassy, Bangkok would be told of their existence. This message was sent within DIA channels in approximately August 1981. For the next year and a half I went to Thailand every 60 to 90 days and got information from these sources and others. After being debriefed... I operated under the general instructions from DIA to seek additional information about prisoners of war and MIA's. Until approximately April 1984 I...established an agent net among Laos, free Vietnamese and within the Thai military..."

The testimony of Smith's career NCO assistant in Special Forces Detachment-Korea, on these intelligence-gathering missions in Thailand, Sergeant First Class Melvin McIntire, substantiated that of his commander. McIntire was a decorated combat veteran of Vietnam, but he was also a professional Asian languages specialist, whose knowledge of Laotian was complemented by his fluency in two dialects of Thai and two dialects of Korean, which he also read and wrote. McIntire stated under oath:

"...THESE SOURCES WITH WHOM I WAS DEALING TOLD ME THEY DID NOT TRUST THE DEFENSE INTELLIGENCE AGENCY NOR THE CENTRAL INTELLIGENCE AGENCY IN MATTERS RELATING TO AMERICAN PRISONERS OF WAR AND MISSING IN ACTION. THEY ALSO COMPLAINED OF CORRUPTION IN THE REFUGEE PROGRAM AND THE JOINT CASUALTY RESOLUTION CENTER.

"I developed approximately 10 sources that I considered to be credible and reliable. These sources revealed to me that there were living Americans in Southeast Asia and that there were intelligence reports other than those they were giving me which had all been discredited by the agencies of the U.S. Government mentioned above. These sources had agents in the field of their own who were reporting the general grid coordinates or locations of where American POWs were being held. I LEARNED OF APPROXIMATELY 200 LIVING AMERICANS IN LAOS WHO WERE PRISONERS OF WAR, THROUGH THESE SOURCES. I WAS BEING PROVIDED INFORMATION IN DETAIL SUFFICIENT TO IDENTIFY THE NUMBER OF AMERICAN PRISONERS OF WAR BEING HELD IN THE GENERAL VICINITY AND I WAS ALSO BEING TOLD OF THE CONDITIONS UNDER WHICH THEY WERE BEING HELD.

This information was being relayed by me the entire time I was assigned to the SFD-K from February 1982 through August 1984. My continual instructions as this information was being forwarded were, "seek additional information."

"BY JANUARY 1984 MY SOURCES TOLD ME THEY WERE GOING TO BRING OUT TWO AMERICAN PRISONERS OF WAR IN MAY OF 1984...AFTER RETURNING TO KOREA TO REPORT THIS INFORMATION I WAS PERSONALLY DEBRIEFED BY MEMBERS OF THE 501st MILITARY INTELLIGENCE GROUP...AT WHICH TIME (I) SFD-K WAS PREVENTED FROM LEAVING THE REPUBLIC OF KOREA OR TAKING ANY FURTHER TRIPS TO THAILAND. I WAS CURTAILED IN MY OVERSEAS TOUR AND IN AUGUST OF 1984 I WAS SENT BACK TO THE UNITED STATES SIX MONTHS EARLY..."

Major Smith recalled what happened then, in his later sworn

affidavit and testimony before the U.S. Senate:

"IN 1984 THERE WAS A MAJOR COMPROMISE OF ONE LAOS AGENT WHO WAS REPORTED TO HAVE BEEN SHOT BY THE VIETNAMESE AND COMPROMISED MY CONTACT WITH SENIOR THAI OFFICERS ...IN APRIL 1984...I RECEIVED THE CODE WORD FROM A GENERAL OFFICER IN THAILAND THAT THERE WERE THREE AMERICAN PRISONERS OF WAR AVAILABLE TO BE TAKEN OUT OF LAOS IN MAY 1984. WHEN THIS INFORMATION WAS PASSED TO THIS AMERICAN ARMY MAJOR GENERAL (Kenneth Leuer) AND TO THE 501st MILITARY INTELLIGENCE GROUP AND TO (the) CIA STATION IN SEOUL, KOREA, ALL SFD-K OPERATIONS TO THAILAND OR TO SOUTHEAST ASIA WERE DECLARED UNAUTHORIZED AND TERMINATED...I WAS TOLD THAT IF I WANTED TO BE A Lt. COLONEL IN THE ARMY THAT I SHOULD FORGET ABOUT THE POW/MIA INFORMATION WHICH HAD BEEN REPORTED TO INTELLIGENCE CHANNELS FOR THE PAST THREE YEARS...THE BRIEFING PACKAGE RELATING TO THE QUESTION OF POWS IN SOUTHEAST ASIA WAS PREPARED BY MYSELF AND PERSONALLY DELIVERED BY MYSELF TO THIS AMERICAN GENERAL...AFTER READING THE TWO COVER LETTERS THIS GENERAL TURNED WHITE, HANDED THE BRIEFING BACK TO ME AND SAID, 'THIS IS TOO HOT FOR ME TO HANDLE, BIG GUY'... I WAS TOLD THAT IF I WERE SMART, WHAT I WOULD DO WAS TO PUT THE BRIEFING THROUGH A SHREDDER AND FORGET THE ENTIRE ISSUE. I DEMANDED AUTHORIZATION TO GO TO WASHINGTON, D.C., AND SEE ANOTHER GENERAL IN THE OFFICE OF THE DEPUTY CHIEF OF STAFF FOR OPERATIONS, UNITED STATES ARMY. I WAS GIVEN A DIRECT ORDER NOT TO HAVE ANY MORE CONTACT WITH THAT OFFICER IN WASHINGTON, D.C..."

Colonel Robert L. Howard, a career officer who had been awarded the nations highest decoration for valor in Vietnam, the Congressional Medal of Honor, who had operational control over Smith's and McIntire's SFD-K unit, testified that he had become familiar with their reports on definite living POWs in Laos and some of their best human sources which included several high-ranking Thai general officers. Colonel Howard in an affidavit submitted to the Senate stated:

"I witnessed the compromise of a source of information by Lt. Colonel Mather (JCRC) and Colonel Alpern concerning the working relationship which had been developed...It was a blatant security violation by a senior U.S. military officer and it was an effort to undermine the successful intelligence gathering activity of SFD-K on the subject of living Americans in Southeast Asia...I spoke personally with one of these senior Thai officials...and was told that there were living Americans in Southeast Asia and that they were willing to assist and had been trying to provide this information to the U.S. Government through SFD-K. I remained in Thailand on this single occasion for approximately eight days and participated in...a training mission which if completed as planned, could have resulted in the killing or capture of myself, Major Mark Smith, SFC McIntire and others by a hostile force... This mission...was terminated due to the suspicions of SFC McIntire that it would have constituted an illegal mission into a Communist country (Laos).

During 1986 Smith and McIntire also revealed that one of their sources on POW/MIA information in Laos was a British businessman who lived in Vientiane and went by the alias "John Obassy," but whose real name, reportedly, was Robin Gregson. Major Mark Smith had met him in Thailand in 1980, and as an ex-POW he gained the trust of Gregson-Obassy, who subsequently revealed detailed firsthand information on American prisoners in Laos gained during his travels there. Gregson-Obassy was also the agent of McIntire and Smith who had reported through their Thai general officer contact that three U.S. POWs were ready to be extracted in 1984, the incident which had precipitated their eventual recall. Prior to Major Smith's and Sergeant McIntire's appearance before Murkowski's subcommittee Gregson-Obassy contacted Smith and let him view a recent videotape showing 39 American war prisoners still alive and working in gold mining and logging operations in Laos, under guard. In one last effort at making the system work, Smith reported the film's existence to DIA who decided it could be a PLO training film. Smith traveled to the middle east where he reported he made a deal to buy the tape from a foreign government

Smith's and McIntire's attorney, Mark Waple, wrote a letter to President Reagan outlining the proposed deal, which was given to Vice President George Bush by Waple and Congressman William Hendon, who both stated they had seen part of the four-hour long film. According to Texas businessman and long-time POW/MIA advocate Ross Perot, George Bush subsequently contacted him, asking Perot to put up the money for the film, in case the offer was genuine. Perot had sought and received permission from President Reagan to investigate the entire handling of the POW/MIA matter, and had subsequently cooperated with Vice President Bush, who was the administration's pointman for the missing-in-action issue during the 1986-87 period. Gregson/Obassy was reported to have been arrested and jailed in Singapore, where U.S. officials visited him and requested the tape in exchange for his release. Obassy refused on the grounds that he was only acting as an agent for the foreign government that owned the videotape. Reportedly, Ross Perot put up the bail money to gain Obassy's release at George Bush's instigation. Obassy believed that the U.S. government was behind his arrest, but nevertheless agreed to testify at the Murkowski hearings, whereupon his real identity was publicly revealed and he vanished. Author Nigel Cawthorne has reported that at the suggestion of Vice President George Bush, Ross Perot once more broadcast an offer of $4.2 million for the videotape, but Gregson/ Obassy had disappeared, and the CIA was said to have gotten the tape from MOSSAD, the Israeli intelligence service.[24]

Ross Perot had been a supporter of POWs since the war, when in 1969 and 1970 he publicized their inhumane treatment in North Vietnam by attempting to deliver relief supplies to them. Although his efforts were blocked by the Hanoi regime, he succeeded in causing an improvement in the POWs conditions, in line with a simultaneous effort by Secretary of Defense Melvin Laird. In 1971 Perot had told a Chicago Tribune reporter that the American people held the key to the POW/MIA issue because "The Communists have shown they respond to the pressure of world opinion. Perot had served on President Reagan's foreign intelligence advisory board from 1983-1986, and had been disappointed in the slow official response to his concerns about still-living prisoners being reported in Southeast Asia. Later in 1986, a major effort was mounted in the Congress to urge formation of another special Presidential Commission to investigate the POW/ MIA issue, to be headed by Ross Perot. Leaders in this effort were the young Republican members of the House POW/MIA Task Force, Bob Smith and Bill Hendon and Democrats like Frank McCloskey, and although 275 members of Congress eventually supported the plan, the administration opposed the idea and with unlikely allies in the House (including a former antiwar protestor from New York named Stephen Solarz), backed the Pentagon's and NSC's objections and scuttled the initiative. Perot then sought permission from President Reagan personally to analyze the POW/MIA problem, which the President encouraged him to do. Perot subsequently gained access to classified livesightings held by DIA which convinced him beyond a doubt American POWs had been withheld in captivity long after the war, and were probably still alive. The witnesses at the 1986 hearings only substantiated further the information that Perot had already acquired, which he subsequently reported to President Reagan, in April 1987.

Although all three of these courageous Green Beret officers were, in effect, ignored by the Congress after their sworn testimony was debunked by government officials, which took a heavy toll on their personal lives; they were to be vindicated years later, after an exhaustive investigation by a U.S. Army Criminal Investigation Division (CID) agent, Tracy Usry, THE RESULTS OF WHICH WERE KEPT SECRET, however, as the matter was "classified." The U.S. Supreme Court was to refuse to hear the case instituted by Smith and McIntire, after years of being pushed forcibly by the Fayetteville attorney, Mark Waple. The substance of the testimony of these distinguished officers, published by the Baltimore Sun and other conscientious newspapers in 1985 and 1986, spread across the United States and they became a living-legend to hundreds of thousands, if not millions of fellow Vietnam veterans, as Lt. Colonel "Bo" Gritz had done between 1981 and 1983. The Green Berets stand galvanized an already restive population of these now-middle aged men, many of whom had wives and children, but remained bitter about what they perceived to be a betrayal by the nation-at-large that had sent them to war.

This was to take the form of hundreds of individual POW/MIA groups run by veterans in every part of the nation, too numerous to mention herein, but all getting their information from what author Monika Jensen-Stevenson later called a "Samizdat" underground press, which printed otherwise unobtainable POW/MIA information and distributed it to thousands of other veterans. The underground was backed up by what the same author later termed the "telephone tree," comprised of those informed leaders and researchers (including this writer) who were actually uncovering solid information.[25] It must also be noted that the former commander of these men in Vietnam, and as Army Chief of Staff, General William C. Westmoreland, took an active part in many late 1980s "Welcome Home" gatherings of sometimes hundreds of thousands of Vietnam veterans, in places such as Washington, D.C., Chicago, New York and Los Angeles. Westmoreland always acknowledged the importance of the MIA issue, and many times encouraged the author's research, in writing.[26]

Yet, the most important of all the vital testimony at these 1986 Senate hearings, and the most ignored by the American media, was that of retired Air Force Security/NSA analyst Jerry J. Mooney, who had come to Washington because his conscience would not allow him to do otherwise. Mooney had by now decided that he must reveal SOME of what he knew to the American people, through the only legal vehicle available him; the U.S. Congress. The courage exhibited by the three Green Berets, Smith, McIntire and Howard, and the ex-POW Garwood, had also inspired Mooney to carry this risky venture through to a conclusion.

In Mr. Mooney's November 3, 1985 affidavit which was submitted to Congress for the January 1986 Murkowski hearings, and was shortly thereafter made public, he stated under oath:

"During the Vietnam conflict I was assigned to the J.S.P.C. in Sobe Okinawa and later with the 6970th Support Group with duty assignment to the National Security Agency. Based upon my six years of experience in these assignments and based upon my intelligence background, I am presently convinced that there are living Americans who were prisoners of the Vietnam war being held in captivity in Southeast Asia. I base this opinion upon the following: In my job during these assignments I received intercepted North Vietnamese communications and messages directly relating to the command and control military operations of North Vietnamese units operating in North Vietnam, northern South Vietnam and Laos. These communications were both directive and informative in nature and contained orders to 'shoot down the enemy and capture the pilot alive,' and 'shoot down the enemy and execute the pilot.' These were the exact operational orders which I collected.

"Further, other communications were received discussing the handling, disposition and transportation of captured American personnel, both pilots and ground forces people. These messages revealed the delivery of captured Americans from Laos, South Vietnam through the Bankari and Mu Ghia passes to VINH, North Vietnam and probably on to Hanoi. They also revealed the execution of captured Americans and the preplanned execution of impending captured Americans. In my role as a senior analyst and having access to both operational and collateral data from other military organizations and national agencies I was able to associate North Vietnamese references to U.S. official listings of missing U.S. personnel. I COMPILED A LISTING OF OVER THREE HUNDRED U.S. MILITARY PERSONNEL CATEGORIZED AS MIA/POW. AT HOMECOMING ONE LESS THAN FIVE PERCENT OF THOSE ON MY LIST KNOWN TO BE ALIVE WERE RETURNED TO THE UNITED STATES. Further as a a basis for my opinion, North Vietnamese messages revealed a high interest in selecting priority targets to include F-111 aircraft, Airborne Intelligence Collectors, F-4 laser bomb equipped aircraft and electronic support aircraft. IT WAS CLEAR FROM THE INTELLIGENCE COLLECTED THAT THE NORTH VIETNAMESE WERE PARTICULARLY INTERESTED IN CAPTURING THE CREW OR PILOTS OF THESE AIRCRAFT ALIVE. THEY WERE CONSIDERED VERY IMPORTANT PRISONERS.

"In approximately February 1973, while assigned with the 6970th Support Group assigned further to the National Security Agency at Fort Meade, Maryland, my section received, analyzed, evaluated and formally reported the shoot-down of an EC-47Q aircraft in Laos. BASED UPON THE ENEMY MESSAGES WHICH WE COLLECTED THERE WERE AT LEAST FIVE TO SEVEN SURVIVORS WHO WERE IDENTIFIED AS AMERICANS AND TRANSPORTED TO NORTH VIETNAM. THIS IS AN EXAMPLE OF THEIR INTEREST IN AN INTELLIGENCE COLLECTOR AIRCRAFT. Since the aircraft was assigned to the 6994th Security Squadron (an intelligence collection unit) and its members were trained intelligence collectors, this confirms my earlier statement of high North Vietnamese interest in capturing these individuals alive. THIS INFORMATION WAS FORMALLY REPORTED TO INTERESTED CONSUMERS WITH AN ADD-ON OF 'WHITE HOUSE.' I PERSONALLY WROTE THE MESSAGE THAT THESE MEN HAD BEEN CAPTURED ALIVE, THAT THEY WERE AMERICANS AND HAD BEEN TRANSPORTED TO NORTH VIETNAM. IN SECURE PHONE CONVERSATIONS WITH THE DEFENSE INTELLIGENCE AGENCY WE WERE IN TOTAL AGREEMENT THAT THESE WERE THE CREW MEMBERS OF THE

DOWNED EC-47Q. In this regard the attached correspondence from Rear Admiral Jerry O. Tuttle dated 25 February 1981, relates to this specific incident that I reported on."

Four of the ECQ-47 known-POWs in the affidavit were of course, Melton, Cressman, Matejov and Brandenburg, who had been added to the NSA's February 1973 secret "Kissinger list" of 21 names for the then-National Security Advisor, Henry Kissinger, who was still secretly negotiating with the North Vietnamese for non-returned American known-POWs, including 14 who were identified as "Moscow-bound" and seven as "China-bound." (It has been further reported that two other non-crewmembers may have also been on this intelligence aircraft, and may also have been captured alive.) These four names were from the longer 75-name list of Moscow-bound Americans, whom Mooney had isolated ONLY in the panhandle area of North Vietnam and adjacent Laos, which list in turn had been analytically honed from his longer 305-name-list of KNOWN U.S. POWs in 1973 from decryptions of enemy communications, of whom more than 290 Americans were not released from captivity by Hanoi with the 591 who were repatriated at the end of the war. Mooney later clarified this further by stating that his share of the 'analytical pie' on the POW problem in Southeast Asia was between 20% and 25%.

While under oath in secret session at the U.S. Senate in 1986, Mooney revealed that some of the POWs he had tracked were carried as "Moscow-bound" by NSA and analysis indicated that they had been shipped to the Soviet Union. In effect Mooney was revealing in public and in secret warning-sessions that due to his access to the most secret and valuable intelligence of all, decrypted enemy orders and communications intercepted by the NSA, he actually had far more information to reveal. This was truly serious business to the policymakers of the U.S. government, as the security of communications intelligence was paramount in the Cold War face-off between the U.S. and Russia. The codebreaking establishment has always feared establishment of any precedent that could lead to future compromises of their product or by their past, present and future personnel. Republican Senator Murkowski and Democratic Senator Dennis DeConcini of Arizona already knew, but they could be controlled through raising their mandated obligations to maintain silence on "national security" matters.

Of immediate concern to the government was that public POW advocates like Congressmen Billy Hendon and Robert Smith, retired Navy Captain and ex-POW Eugene "Red" McDaniel, and also POW/MIA researchers including the author, John and Margaret Nevin, of the MIA organization Homecoming II and Thomas V. Ashworth had already become aware of the "Moscow-bound" American POWs from Indochina, through listening to Jerry Mooney speak of them by early

1986 in the hall of the Rayburn Building, in the Crystal City Marriott hallways, or in a jammed crowd of veterans in Waple's and Hendon's hotel room. Mooney subsequently told the author that by late-1986, film producer Ted Landreth (a former producer of CBS world news during the war years) and his colleague Ed Tivnin, together with BBC-Lionheart's Chris Ogiatti, were also fully aware of the "Moscow-bound" U.S. POWs. Mooney also related in a later interview with the author: "I went to Dallas and made a tape-recording with Ross Perot for some four and one half hours, of POW information gained from decryptions, including the many known to have disappeared who were confirmed POWs, and I definitely made references to the "Moscow-bound" American POWs." Mooney remembered saying to Perot: "I wouldn't go to Forida to meet with General Tighe, but I'll talk to you and be taped, BECAUSE YOU ARE A PRESIDENTIAL REPRESENTATIVE." Perot continued to communicate with Mooney for years to come.

    In March 1986, in "The Insider," one of the home-edited Veterans POW/MIA newsletters, which was published by Michael Van Atta, a confirming link to Mooney's still secret and little-known testimony on Moscow-bound Americans was revealed. This concerned a former U.S. serviceman named Jon Sweeney who deserted his unit after four days in the field in 1969 but was subsequently captured by the North Vietnamese. He was moved to Hanoi and then on to Moscow and eventually to Sweden, where he had agreed to make antiwar Broadcasts. He escaped to the U.S. Embassy, and was subsequently charged with desertion and court-martialed, but after six days Sweeney was allowed to go free. The U.S. Government had no desire to have the Soviet connection to missing American POWs raised in the media during 1986, with the still-ongoing Cold War.

    Jerry J. Mooney had received only the highest praise in fitness reports from his superiors in Air Force Intelligence and at NSA, which are in the author's possession (and on file at the Senate Select Committee on POW/MIA Affairs). Before going public with his information on "Moscow-bound" U.S. POWs of the Vietnam war in meetings with veterans and family members in Alexandria, Virginia during his January 1986 Senate testimony and less than a year later in a 1987 BBC/ Landreth-Lionheart documentary film (We Can Keep You Forever), Mr. Mooney had tried to work with the U.S. Government through the Senate, as Congress was the only avenue legally open to him as a retired intelligence officer, in revealing what he knew about the U.S. POWs who had not been released in 1973. Senator Frank Murkowski, the freshman Republican from Alaska chairing the 1986 subcommittee on Veterans Affairs, prevented Jerry J. Mooney from revealing all that he knew about the "Moscow-bound" and other POWs of Indochina, although such testimony was heard secretly in the secure area known as the 'bubble.' In a later affidavit of

December 1991 sent to the Congress, Mooney revealed his past attempts to conform to the system:

"With regard to my MIA/POW concerns... In 1972, in an interview with a Department of Defense Inspector General, I expressed my concerns. In 1973, following Homecoming I, I expressed those same concerns to Defense Intelligence Agency. On the latter occasion, DIA expressed concerns as well and stated "they were working on it." Further, since 1985, I have been to the Judicial Branch of Government ('no jurisdiction'), the Executive Branch of Government ('Not Government Business'), and the Legislative Branch of Government (Mostly no response)..."

While courageously facing critics and periodic tormentors over the next six, lonely years, Mooney related to author John M. G. Brown and to other researchers including British researcher Nigel Cawthorne (and later to Senate investigators), many facts about the POW/MIA matter, verbally and in the form of letters and affidavits, some of which have been cited earlier in this book. He also suffered numerous instances of threats against his life and the lives of his family members from 1986-1992, from unknown sources, including several physical assaults and one shooting incident.

Former Director of Defense Intelligence, retired Lt. General Eugene Tighe, had actually set the stage and had upheld the honor of all U.S. intelligence agencies by testifying to the Congress that U.S. analysts had displayed, "A MINDSET TO DEBUNK" livesighting reports and later that there were some indications that U.S. prisoners had been transferred to Soviet control. Tighe had been appointed to head a Senior Review Panel at DIA to investigate that agency's handling of the intelligence on prisoners. Although it remained classified until 1992, the findings of Tighe's report released to the media later in September 1986 were that evidence indicated that U.S. POWs were probably still alive in Southeast Asia. Privately, Tighe was to maintain that Jerry J. Mooney's statements on POWs had substance too, but he was to be debunked and even ridiculed by his own intelligence colleagues for years, for doing so.

Of all the witnesses at the Murkowski hearings, none was to be more controversial than Scott Barnes, the Army veteran discharged in 1974 who had been recruited for Colonel Bo Gritz's 1981 reconnaissance/ rescue mission. Barnes testified under penalty of perjury that after Gritz had returned to the U.S. during the initial stages of the 1981 mission, that he, Barnes, had been ordered to accompany a CIA officer named Michael J. Baldwin, also known as Jerry B. Daniels, on a patrol into communist controlled Laos where they had been guided by free Lao to a prison camp in the Mahaxay region, containing Americans. Barnes stated that from the jungle outside the camp, he had taken many high-speed photographs of two Caucasian prisoners through two telephoto lens cameras, while

Baldwin/Daniels was recording the voices of the prisoners and their guards with a directional microphone and tapes, while listening in.

Barnes claimed that Baldwin/Daniels turned to him saying, "they're American!" and began to cry, saying: "they're here...we really did leave them behind..." When they succeeded in returning to Thailand, Baldwin/Daniels went via Udorn to the U.S. Embassy in Bangkok with his portion of the film and tapes of the American POWs voices, while Barnes went to the team's forward base at Nakom Phanom, where he stated he immediately mailed his portion of the film to a CIA official named Daniel C. Arnold in the Washington, D.C. area, as instructed. The team he was part of then told Barnes they had received a coded order via a U.S. Department of Energy telex from Washington, D.C.: "If merchandise confirmed, then liquidate." Barnes said that his team's codeword for American POWs was "merchandise," and that the CIA and military intelligence personnel involved in the mission confirmed that the orders meant that the team must kill the American prisoners. Barnes claimed that he refused to take part in this, and that his superior whom he named as William Macris, ordered another team member who Barnes named as J.D. Bath, to take him to the United States Embassy in Bangkok, where he would be turned over to officials he had already met there, one of whom he named as Colonel Mike Eiland.

No other witness inspired such a frantic series of denials, abuse and scorn by government officials and their allies in the media as Scott Barnes. A veteran BBC investigative journalist in Washington, D.C., named David Taylor had thoroughly checked his story and had him tested by lie detector and injections of sodium amatyl (known as "truth serum"), and he'd passed several examinations witnessed by Dr. Robert Crummie and Chris Gugas, a polygraph expert formerly with the CIA who had previously tested James Earl Ray, Robert Vesco and others. It is a matter of fact that Jerry B. Daniels, originally from Montana before his recruitment by CIA, and his involvement with the CIAs Laotian army commander, General Vang Pao, was subsequently found dead of gas inhalation in his Bangkok apartment, and his body was flown back to Montana for burial in a sealed casket. Scott Barnes claimed he did not know whether the orders to "liquidate the merchandise" had ever been carried out.

Years later, another witness at a Senate hearing on POWs, Terrell Minarcin, was to state that while serving as an NSA cryptolinguist he had provided support for a 1981 POW rescue mission, during which another NSA analyst received a prearranged signal that a live American POW had been rescued, but that he learned later only POW remains had been reported recovered and sent to the Central Identification Laboratory in Hawaii. The CIL-HI subsequently denied that any American remains had been received by them during the first half of 1981, while in the second half of the year, in July three sets of American remains came in from the Vietnamese.[27] Barnes' 1986 testimony was interrupted by the tragic explosion of the Space Shuttle Challenger, and Senator Frank Murkowski abruptly adjourned the hearings. The mass media pack rushed from the Senate hearing room to cover the newest "big story," and the startling testimony of the witnesses on POWs and MIAs remaining in Asia was largely ignored afterward.

Yet, the powerful testimony of many witnesses at the 1986 Senate POW/MIA hearings, although largely downplayed or ignored by much of the American mass media, had been effective in forcing the U.S. position into the open. In March 1986 Richard Armitage, then Assistant Secretary of Defense for International Security Affairs (ISA), responded to a question on what steps the U.S. Government would take on a report of POWs it believed was credible:

"(The U.S.)...Might get him from an operation...it could be that he would be an individual, held in captivity, in proximity with other individuals who are reported to be in captivity...and we would want to be careful that we did not extract one, to the disbenefit of the others."

This was perceived as a U.S. threat of force in Hanoi, as Vietnam's commentary later blasted Armitage's statement as violating the U.S. pledge to create a favorable atmosphere to discuss the POW/MIA issue. But on May 29, 1986, Henry Kissinger stated on an ABC News 20/20 program: "I DO NOT TO THIS DAY BELIEVE THAT THERE ARE LIVE PRISONERS IN VIETNAM BECAUSE I DO NOT SEE WHAT PURPOSE THEY WOULD SERVE, KEEPING PRISONERS FOR 13 YEARS WITHOUT TRYING TO BARGAIN...THERE WERE 70 NAMES ON OUR LIST WHERE WE HAD REASON TO BELIEVE THEY HAD BEEN ALIVE...THEY WOULD JUST SAY ITS A FABRICATION...I THOUGHT THEY WERE LYING. I THOUGHT PROBABLY THEY HAD KILLED THESE PEOPLE OR THEY HAD DIED IN SOME WAY..."[28]

In response to media exposure of the Murkowski hearings and criticism from veterans, POW families and members of Congress, the DIA initiated yet another investigation of its own management of the prisoner of war matter, which became known as the Tighe Commission for it's chairman, former DIA director Lt. General Eugene Tighe. In a May 1986 letter to the author, Major General John Murray, LeGro's former commander in Saigon and a member of Tighe's commission, wrote:

"As I am sure you understand, I cannot now comment on the MIA study, with General Tighe and his group, but I can promise you that I will look into that Smith, McIntire and Howard case carefully. I will report to you more fully when I am permitted to do so. Rest assured that the people investigating this PW/MIA subject are of the highest quality and integrity. The Senior Review Group is made up of former commanders, General Dougherty of the Strategic Air Command, General Bob Kingston of Central Command...Two other Air Force generals, former POWs (one representing Ross Perot) and a gentleman named Fitzpatrick (Kirkpatrick?) who is known as the 'dean of intelligence professionals..."[29]

Murray served as the counsel and was also involved with actually obtaining the information from DIA for Tighe and the others to review. However, in a meeting later in his suburban Virginia home, near the Capitol, General Murray spoke to the author about "acceptable losses, the higher interests of national security and the enormous casualties of World War II in comparison to the missing of Vietnam." The Tighe commission was not to issue a report until late September, and then it was to remain "classified."

In a July 1986 meeting on the 7th floor of the Department of State, Deputy Secretary of State John Whitehead had admitted to the author, Anthony D. Duke and former Congressman John LeBoutillier that U.S. prisoners were still being held by the Communists, saying: "...WE KNOW THEY ARE ALIVE AND WE KNOW THEY ARE DYING." Whitehead did not elaborate on the source of his information. He went on to explain that POW policy was handled through the National Security Council by Richard Childress, the NSC's Director of Asian Affairs, and in the Pentagon by Assistant Secretary of Defense Richard Armitage, who he said, both actively resisted any change in the current policy toward the MIAs, from a focus on recovering remains of the dead to one of public diplomatic negotiations for living POW survivors, wherever they were held.

At this time Ann Hart, the wife of Captain Thomas Hart III who had been missing since December 1972 when his Air Force AC 130 gunship was shot down in Laos, was confronting the U.S. Government about the return of her husband's alleged remains. Since she had discovered the aerial photograph of the initials "T.H." and "1973" in a friend's casualty file, she had been demanding that his fate be determined. In 1982 she had been permitted to travel with a JCRC team to examine a crash site near Pakse, Laos, and had been shown bone fragments alleged to be from her husband and other crewmembers. When these "remains" were finally returned to the U.S. in July 1985, she was told that two bone fragments she had been shown before were part of her husband, although she had evidence and there were intelligence reports that he had survived the crash. Mrs. Hart and others challenged the identification of their family member's remains by the U.S. Army's Central Identification laboratory in Hawaii (CIL-HI) Independent experts who studied the remains, of Hart, and Master Sergeant James Ray Fuller, such as Dr. John K. Lundy of the Oregon Health Sciences University and Dr. Michael Charney, a recognized forensics expert at the Center of Human Identification at Colorado State University concluded that it was impossible to make identifications from such fragments. When Mrs. Hart refused to accept the "remains" she was told that the Air Force would bury them in Arlington National Cemetery "with full military honors."

Mrs. Hart filed suit in U.S. District court using reports from independent forensic experts, and eventually proved that the Central Identification laboratory, under Lt. Colonel Jonnie Webb and "Dr." Tadao Furue, a Japanese-educated imposter with no University degree, had misidentified many purported "remains" of U.S. servicemen missing in Vietnam and Laos. Dr. Samuel Dunlop, a physical anthropologist under Furue at CIL-HI was told by a special agent of the Army's Criminal Investigation Division (CID) that the Secretary of the Army, John Marsh, had ordered the CID to cease an investigation of the CIL-HI operation, even though he and other independent experts such as Drs. Lundy and Miller and several other forensic scientists believed that Webb, Furue and H. Thorne Helgason were "incompetent at best." Dunlop was willing to testify to the identifications of U.S. MIAs made at CIL-HI were fraudulent. Yet the CID investigation never recommenced and Dr. Dunlop was forced to resign under pressure. Mrs. Hart eventually won her case, and was awarded $632,000. in damages[30] It was a landmark decision that called into question the U.S. Government's identification of MIAs after Korea and World War II, and back to the 1929 VFW-U.S. Graves Registration expedition to north Russia.

In the meantime, on August 19, 1986, the Wall Street Journal published a story saying that the incoming Reagan Administration had declined a 1981 offer by Hanoi to sell the surviving American POWs. The Journal's Staff reporter, Bill Paul, wrote:

"In 1981, just weeks after President Reagan took office, the new administration learned that Vietnam wanted to sell to the U.S. an unspecified number of live POWs still in Southeast Asia for the sum of $4 billion (less than the U.S. had promised Hanoi in the post-war rebuilding aid. It's unclear whether the offer was made to Mr. Reagan or that he had inherited it from President Carter. The proposal was discussed by Mr. Reagan and his advisers at a general meeting on security matters, according to one person who says he was in the room, and whose story is supported by another attendee. During the discussion, it was first decided that the offer was indeed genuine. Then, a number of the President's advisers said they opposed paying for POWs on the ground that it would appear as if the U.S. could be blackmailed. Mr. Reagan concurred...To his credit, however, Mr. Reagan told William Casey, director of the Central Intelligence Agency, and Richard Allen, then the National Security Advisor, to try to find another way to get the men home.

"Mr. Allen subsequently proposed a reconnaissance mission into Laos which, if successful, would lead to a military rescue of prisoners presumed held at a jungle prison camp. The secret mission failed and word of the failure leaked to the press. After that...a new POW policy was concocted by the State Department and the Pentagon. It maintained that the U.S. can neither confirm nor deny that POWs remain in Southeast Asia, but operates on the assumption that they are there."

The author subsequently met with Bill Paul to verify that he stood by his story, for which he had two sources. The author then interviewed President Reagan's 1981 national security advisor, Richard Allen, on September 24, 1986. Several other Vietnam veteran leaders were present, including retired Navy Captain Eugene "Red" McDaniel, Jerry Kiley, William T. Bennett, Bruce Rehmer and J. Thomas Burch of the National Vietnam Veterans Coalition (of which the author was then the POW/MIA Chairman). After receiving the request for a meeting Allen had written to the author on September 8th: "It is my firm belief that the issue of POWs/MIAs is not a closed chapter..."

At the meeting Allen would not deny that the Vietnamese offer had indeed been discussed at the White House after Reagan's inauguration in early 1981. He did say that prior to our meeting he had, "received a call from (former Secretary of State) Al Haig who told me: 'Look Dick, I don't think you should meet with Brown and his group of vets. This POW thing is a can of worms and if it's opened up somebody's going to get hurt.' (The author had previously written to General Haig requesting a meeting with some Vietnam veterans to discuss the reported POW offer, but Haig had declined.)

When asked by the author about whether there were live POWs still being held against there will in Southeast Asia, Allen said:

"I believe they are there...I am sure they are there. I can't prove it today but I believe it, and we must do more to get them out..."

Allen confirmed that U.S. reconnaissance/rescue missions were subsequently dispatched to Southeast Asia, as Bill Paul had reported, and may have hinted that the number of U.S. POWs involved in the offer was 57, by writing this number on a piece of paper many times. U.S. Special Operations officers who took part in these missions, interviewed by this writer, have revealed that rescue missions believed to have a good chance of recovering live U.S. POWs from Southeast Asia in the early 1980s, were cancelled while in progress, for unknown reasons, on orders from high Pentagon officials, including then-Assistant Secretary of Defense Richard Armitage. All U.S. participants were immediately sworn to secrecy. A Vietnam veteran and retired U.S. Special Operations and intelligence officer, who served in the Pentagon under Richard Armitage and his Principal Deputy, is among those who have confirmed details of these matters to the author, which were later revealed to federal investigators. The question raised in the Wall

Street Journal Report about whether President Jimmy Carter had previously received the offer was answered emphatically by that former President, in a note to a private historical researcher and disabled veteran named Charles Bates, in 1991: "Charles—The answer is NO. J. Carter."[31]

Late in September 1986 the Tighe Commission report was finally completed, but it remained "classified," and would not become available to the public for nearly seven more years. This is not surprising in view of some of the findings within, which concerned the livesightings of American prisoners in Southeast Asia long after the war:

"The perceived mission of the PW/MIA center at DIA has changed, officially and unofficially, from analysis of the intelligence flowing into DIA on this issue, to 'resolving the issue,' whereby doubt is cast on the veracity of the intelligence...an example of the effort is one case where four years were spent trying to prove that a re-education camp which was a key part of one livesighting report did not exist (this to disprove the report), only to find that the camp did indeed exist. During the intervening years, the report was not analyzed for its contribution to the overall issue...There is a total absence of rigorous, standard, disciplined, professional, administrative procedures...A...BASIC PROBLEM IS THE BIAS IN EXPECTATIONS THAT REFUGEES ARE NOT RELIABLE REPORTERS UNLESS PROVEN TO BE SO...YET REFUGEE ACCOUNTS ARE THE MAJOR DATABASE.... The refugee community that has provided the bulk of the eyewitness reports strikes us as possibly the finest human intelligence database in the U.S. post World War II experience..."[32]

On October 1st, the New York Times first reported the Tighe Commission's conclusion under the headline: "POWs ALIVE IN VIETNAM," and went on to state that, "A large volume of evidence leads to the conclusion that POWs are alive..." But the initial Times story was based on an earlier interview with General Tighe which had been improperly used by NBC News TV, to reflect conclusions of a report which in fact remained secret. The next day, the New York Times carried a totally opposite report, quoting General Leonard Peroots, Director of DIA, who repeated the government line: "No credible evidence, no strong compelling evidence that prisoners are alive." Peroots was infuriated with Tighe and accused him of leaking classified information. A devoted and honorable man who had distinguished himself above all his colleagues by his honesty and courage, Tighe was already ill and his subsequent treatment by DIA colleagues was yet another shameful episode in the POW/MIA affair.[33]

Director of Central Intelligence William Casey, a veteran of 1945 OSS service in Europe, who was then embroiled in the beginning of the Iran-Contra affair, permitted an interview at CIA Headquarters in Langley, Virginia on October 2, 1986, with the author, Anthony D. Duke, former-Congressman John LeBoutillier, NVVC Chairman J. Thomas Burch and POW/MIA wife Barbara Mullen Keenan. Both the author and Burch raised the issue of the 1981 White House meeting on the offer by Hanoi to sell American POWs, which had been reported by Bill Paul in the Wall Street Journal six weeks before. Several sources had confirmed that this 1981 meeting took place, including the Secret Service agent whom Burch was the legal counsel for, who had been present, but who felt unable to go public with his knowledge. Burch's confirming source who had also been made known to Ross Perot and Congressman William Hendon. To prove the reliability of his sources of information Burch said to Casey:

"Mr. Casey, we understand that this past July (1986) that you gave a briefing at Camp David where you presented three options in dealing with the live POW issue. You stated the U.S. could either buy them out, use a military extraction method or use private sector means to get POWs out..." The CIA Director, who appeared to have been half-asleep, was obviously startled at how confident and precise Burch was in his statement, and mumbled something inaudible to himself. His two CIA aides present (who gave false names at the meeting which were later denied by the Agency to LeBoutillier) were hostile; one of whom, a dark-haired, hostile and scowling man who identified himself as a deputy director specializing in Asia, seemed infuriated that the group had been given access to the Director of Central Intelligence at all.

Casey was asked about actual negotiations with the Communists in Hanoi and Laos for surviving American POWs, and in answer he reiterated the Administration's policy of 'not bargaining with terrorists or paying ransom for prisoners.' The author and others present made it clear that many Americans felt that officials such as Armitage, Childress and Colonel Paul Mather of the JCRC in Bangkok, who were charged with resolving the POW matter, should be replaced by new negotiators who would be trusted by the MIA families and veterans groups and by refugees with information. He said, "I can't do whatever I want on everything. Still, I believe that what you're saying and trying to do is correct," and eventually claimed to all present, "I don't have the kind of authority you think I have...the problem is over at the White House, the Chief of staff (Donald Regan) and the President's other advisers, it's not with me. I would do something more if I could." He invited LeBoutillier and the author in to go over the CIA's POW/MIA files, but CIA never subsequently honored this offer by their director.

Although the two CIA officials present scornfully and totally denied there were still American prisoners in captivity, or that any had been left behind after the war, Casey himself would not deny that there were live POWs, nor, when asked, would he deny that either the 1981 White House meeting on the POW offer from Hanoi or the 1986 Camp David briefing of Reagan on POWs had taken place; but he took the writer's hand privately, at an exit door after the meeting and mumbled words to the effect of: "What do you expect us to do, pay for them? We'd have another hostage-crisis on our hands."[34]

Former Congressman William Hendon later maintained that George Bush, who was not mentioned as present in the published account of the 1981 meeting, but whom the Secret Service agent had said was there, had spent: 'nearly two hours on the phone trying to persuade me that the offer was for the remains of fifty seven prisoners.'[35]

Yet, on that same October 2nd, 1986, Vice President George Bush wrote a letter to Anthony Drexel Duke of New York (also with Casey that day), urging the author of this book (Anthony Duke's son-in-law), to continue his POW/MIA investigation wherever it might lead, and to assist Mr. Ross Perot's POW/MIA investigation, with the encouraging words: "...THE PRESIDENT AND I, ON THE ASSUMPTION THAT OUR MEN **ARE** ALIVE, WANT EVERY AVENUE CHECKED OUT."

At the time, the author considered these to be heartfelt words from a Vice President who was surrounded, as all Presidents and Vice Presidents are, by constraining bonds of U.S. policy, politics and national security, but a man who nonetheless was still personally committed to resolving the problem, in the hope that eventually prisoners could be returned. (The results of the author's initial investigation first appeared in a December 1986 American Foreign Policy Newsletter article on POW/MIAs, published in New York, with information gained through Mooney and others on U.S. POWs taken to the USSR and China from Vietnam, declassified livesightings of POWs and reference to the ransoming of German POWs from Stalin's gulag labor camps by Konrad Adenauer in the 1950s.)

The author had met with investigative journalist Bill Paul of the Wall Street Journal in the latter part of 1986, to confirm various details of his pioneering articles on the subject of POWs, and particularly the reported 1981 White House meeting involving Richard Allen and the North Vietnamese proposal to sell surviving American POWs. Paul stood by all the facts presented in his article and also directed the author's attention to allegations that 10,000 or more American POWs had been held by the Russians after WW II, which information he had obtained from a confidential Soviet source. (Paul was later reported to have been shown a photo by his source, of a reunion of the 1st Infantry Division, "Big Red One," in Siberia.) He strongly encouraged the author to investigate the Soviet connection with the 78,000 American MIAs of WW II.

Former UN Ambassador Jeanne Kirkpatrick, who had also served on President Reagan's foreign intelligence advisory board (as had Ross Perot), stated to author J.M.G. Brown, in a November 7, 1986 interview, that she had "never once heard the POW/MIA issue raised," in the highest government councils she attended, and she said with conviction, "You will never succeed in raising this POW/MIA problem as a national issue unless you can put the whole history of POWs before the public, through the media." She also opined that the major media in America had no desire to face the harsh questions raised by the evidence that U.S. prisoners of war had been left behind in Vietnam, Laos and Cambodia, and that some hundreds had continually been reported alive in captivity ever since.

The author appeared on Tom Brokaw's NBC national news program on Veterans Day, November 11,1986, as a spokesman for Vietnam veterans to President Reagan, urging the President to 'bring in some new people' to manage the POW/MIA matter and appoint a special emissary to Hanoi to negotiate for living U.S. prisoners and MIAs. After the author and New York Vietnam veteran/ Wall Street banker John Molloy met with Cardinal O'Conner of New York (who was a also former military officer), that Catholic leader was contacted by Vice President George Bush. The Cardinal had also conveyed a message through Anthony D. Duke of New York, that he was "fed up and disgusted" with the Administration's policy towards POW/MIAs. Duke delivered the letter at the White House during a "Private Sector Initiatives" conference, to which he had been invited. On December 8, 1986, Donald Gregg, a former CIA official then serving as Mr. Bush's national security advisor (later appointed U.S. Ambassador to Korea), wrote to the author:

"I am sorry it is 'no consolation' hearing from me or hearing the facts on this issue...We continue to monitor the efforts of the intelligence community as they work the field and investigate each live sighting in Southeast Asia...We continue to press the Communist governments for resolution of the unanswered questions...We all remain interested in hard information that will stand up to objective analysis; information that will enable our Government to either bring our men home or to account for them to their families...IF YOU HAVE EVEN A SINGLE PIECE OF REAL EVIDENCE THAT WE CAN USE TO HELP AN AMERICAN SERVICEMAN, GIVE IT TO US. WE WILL ACT ON IT IMMEDIATELY..."

From this point on, the author did what was requested, but never received a response. Gregg's letter of December 8th is evidence of the trap that U.S. security officials were caught in, by complying with the oath taken by all officials with access to classified intelligence, to not divulge security-classified information in any way, shape or form to the public. For some officials this becomes second nature and may not be difficult. For others, this has proven to be a difficult and painful oath to uphold. This is the oath that Jerry J. Mooney, formerly of Air Force Intelligence and NSA, had broken when he revealed parts of his testimony before Senator Murkowski's 1985-86 Committee on POWs carried as Moscow-bound and shipped to the Soviet Union. It is the same oath that had been broken by the distinguished Special Forces officers: Major Mark Smith, Colonel Robert Howard and Sergeant Melvin McIntire, when they had publicly revealed secrets prior to testifying at the same 1985-86 Senate hearings (attended by this writer), to being ordered by superior officers to "SHRED AND FORGET" fresh intelligence of U.S. POWs still held in captivity in Laos during 1984.

As national security adviser to the Vice President and as a long-time CIA official who had served in Indochina, Mr. Gregg was no doubt aware of this testimony and possibly elements of the historical material revealed in this book, when he wrote the letter to the author quoted above. He was also aware by then of the findings of the Tighe Commission, a review panel established to investigate DIA's handling of the POW matter. Although that report remained "classified" from public view (until mid-19921), Gregg also knew that the Tighe Commission had "unanimously" found in June 1986, six months before, that there was a, "STRONG POSSIBILITY THAT LIVE AMERICAN MILITARY POWs REMAIN CAPTIVE IN SOUTHEAST ASIA." And Gregg may have also known that in a last bitter wrangle of the divided Commission, the word "POSSIBILITY" had replaced the previous word: "PROBABILITY." Given his responsibilities to Mr. Bush, Donald Gregg must have also been aware of a classified DIA Memorandum for Brigadier General Shufelt dated September 25, 1985, written by U.S. Navy Commodore Thomas A. Brooks, who was DIA's Assistant Director For Collection Management. This now-declassified (in 1992) confidential memorandum stated that:

"I was not at all pleased with the situation I found when I took over responsibility for the POW/MIA issue. The deeper I looked the less professional the operation appeared. It appeared to be particularly sloppy in the late seventies, but it is by no means a squared-away operation today. As a professional intelligence officer, with a significant portion of my career spent as an analyst, I found the following to be particular problems:

"CASE FILES WERE INCOMPLETE, SLOPPY (ALL MIXED UP, LOOSE PAPERS...MISFILED PAPERS, ETC.) AND GENERALLY UNPROFESSIONAL. THERE WERE NO ACTION LOGS IN THE CASES OR WHERE THERE WERE LOGS, ENTRIES HAD NOT BEEN MADE IN A LONG TIME. FOLLOWUP ACTIONS HAD NOT BEEN PURSUED, OBVIOUS FOLLOWUP ACTIONS WERE CALLED FOR BUT WERE NEVER TAKEN AND YEARS HAD PASSED.

EFFORTS TO RECONTACT SOURCES IN THE US WERE PERFUNCTORY AT BEST...WE HAD NEVER EMPLOYED SOME OF THE MOST BASIC ANALYTIC TOOLS SUCH AS PLOTTING ALL SIGHTINGS ON A MAP TO LOOK FOR PATTERNS, CONCENTRATIONS... WITH REGARD TO THE ALLEGATION OF A 'MINDSET TO DEBUNK', I MUST CONCLUDE THAT THERE IS AN ELEMENT OF TRUTH TO THIS AS WELL...I AM NOT PERSUADED THAT ENOUGH ASSETS ARE BEING DEDICATED TO THIS PROBLEM IF IT IS THE TOP PRIORITY PROBLEM WE CLAIM IT IS.

"I SEE THE MOST IMPORTANT THING WE MUST DO RIGHT NOW IS TO BE CEMENTING RELATIONSHIPS ON THE HILL. WE HAVE NOT DONE AS WELL THERE AS WE SHOULD. IT IS CLEAR THAT CONGRESSMAN HENDON[36] WILL BE USING OUR FILES TO DISCREDIT US (AND HE WILL HAVE LOTS OF AMMUNITION THERE). WE NEED TO ENSURE THAT WE HAVE FORMED THE NECESSARY ALLIANCES WITH HPSCI AND THE ASIAN-PACIFIC AFFAIRS COMMITTEE; THEIR STAFFERS, **AND THEIR CHAIRMEN** (Note, original underlining) THAT WE RECEIVE SUPPORT IN OUR EFFORTS TO DAMAGE-LIMIT HENDON."

A working paper attached to this document, titled "Actions Tasked to DC-2," states: "GET TOGETHER WITH TOM LATIMER AT HPSI AND ULTIMATELY PERHAPS CONGRESSMAN HAMILTON TO LINE UP THEIR SUPPORT VIS-A-VIS CONGRESSMAN HENDON. DO THE SAME THING WITH THE ASIA/PACIFIC AFFAIRS STAFF AND PERSONALLY WITH CONGRESSMEN SOLARZ, SOLOMAN AND GILMAN. GET TOGETHER WITH GARWOOD ASAP TO DEBRIEF HIM...REVIEW THE YEN BAI CASES AFTER TALKING TO HIM...INCREASE THE USE OF POLYGRAPHS. ALL LIVESIGHTINGS SHOULD BE POLYGRAPHED. POLYGRAPH THE SOURCE IN JAIL IN DENMARK RE HIS SIGHTING AT YEN BAI...SEEK SOURCES OF ANALYTIC SUPPORT OUTSIDE OF FBI..."[37]

(Brigadier General Shufelt, the Deputy Director of the Defense Intelligence Agency, in 1987 refused to accept an offer of LIVESIGHTING INTELLIGENCE ON 300 U.S. POWs HELD IN LAOS, COMPLETE WITH CROSS-INDICES, ASSEMBLED OVER THE PREVIOUS FOUR YEARS BY COLONEL ALBERT SHINKLE, THE RETIRED U.S. AIR FORCE INTELLIGENCE OFFICER OPERATING IN THAILAND, made available to DIA by the author and former U.S. Congressman John LeBoutillier through DIA official Sedgewick Tourison in a 10 June 1987 letter, and in person. (Tourison was a Defense Intelligence Agency senior analyst on Laos who in 1992 was on the staff of the Senate Select Committee on POW/MIA Affairs, sharing a crowded office with Chief Counsel William Codinha whom he influenced throughout that Congressional probe.)

When writing to the author on December 8, 1986, Donald Gregg would also have had access to the "GAINES REPORT," which was the "CLASSIFIED" result of an investigation of DIAs handling of the POW/MIA issue by Air Force Colonel Kimball Gaines in 1986. Among Gaines' findings were: "THE CHIEF OF THE ANALYSIS BRANCH FEELS

THAT THERE IS PROBABLY ENOUGH INFORMATION ON HAND ALREADY TO ALLOW A DEFINITIVE JUDGMENT ON THE LIVE-POW ISSUE IN NORTH VIETNAM, BUT THEY JUST CAN'T GET AROUND TO DOING IT...THE GAINES REPORT ALSO FOUND THAT 'A MIND-SET TO DEBUNK,' REPORTS "INTENSE EFFORT IS INITIALLY FOCUSED ON VERACITY OF SOURCES WITH A VIEW TOWARD DISCREDITING THEM. (Senator Robert Smith (formerly a Congressman) later Vice-Chairman of the Senate Select Committee on POW/MIA Affairs tried to obtain a copy of the Gaines Report in 1990, and was told it didn't exist.)

Such classified information and much more to do with U.S. POWs and covert actions, was known to national security adviser (and future Ambassador) Donald Gregg, and to Richard Armitage, the NSC's Childress and others in similar positions), but because of his oath to maintain security Mr. Gregg was required to write to the author, in the way he did, in the same manner that the State Department's Robert Kelley had been required to respond to similar questions from citizens, in 1933. Gregg's letter merely conformed to a historical pattern of official responses to citizens' concerns about the fate of U.S. POWs in Communist control going back over 70 years, as has been previously documented in this book.

Donald Gregg's boss, Vice President George Bush, was cautious about his comments on the August 1986 Wall Street Journal report of athe1981 offer by Hanoi to sell surviving American POWs to the United States. According to information made available to the author in 1986 from the source in the U.S. Secret Service, who had been present during the 1981 White House meeting when the POW offer from Hanoi was discussed, Vice President Bush had been present with Ronald Reagan, William Casey and Richard Allen. Due to the seriousness with which the Secret Service views a breaching of Presidential confidentiality by its employees, this Secret Service agent, who was later identified as John F. Syphrit, was understandably reluctant to go public with his knowledge of the 1981 offer, and it was not to be until 1992, during a Senate select committee investigation, that the Secret Service agent's corroboration of Bill Paul's Wall Street Journal report would become a major issue, as will be seen. George Bush was thus permitted to escape any serious scrutiny on the subject of the 1981 offer during his Vice Presidency. In response to a researcher's questions in 1987, asking the Vice President if he had been present in the 1981 meeting, whether he had advised the President on the offer from Vietnam, and whether he would negotiate with Hanoi for survivors if he was elected President in 1988, Bush avoided specific answers:

"...I am distressed that you do not appear to understand the President's position. You should be aware that when we came to office in 1981, neither the President nor I were satisfied with the effort the U.S. Government had made to that time on behalf of our

POWs and MIAs. Therefore, immediately following his inauguration, the President directed the military and intelligence organizations to coordinate and concentrate their resources on this once-neglected issue. Fundamental to our approach, and contained in the mandate the President issued to his staff and all agencies, was and is the assumption that live American prisoners remain in Southeast Asia. YET, DESPITE SENSATIONAL AND SOMETIMES IRRESPONSIBLE REPORTS TO THE CONTRARY, OUR INTELLIGENCE EFFORT AND FOLLOW-UP ON HUNDREDS OF 'LIVE-SIGHTING' REPORTS HAS PRODUCED NO CREDIBLE, SPECIFIC EVIDENCE THAT A SINGLE AMERICAN IS BEING HELD CAPTIVE IN SOUTHEAST ASIA..."[38]

In October 1986, Vice President George Bush had received a report dated August 29, 1986, from one Arthur Suchesk of California, stating that the Shan States warlord in eastern Burma known variously as Khun Sa and Chan Shee Fu (or Chang Chi-fu) had information and control over some U.S. POWs of the Vietnam war. Khun Sa had been raised by an old Nationalist Chinese general of the Kuomintang (KMT) Army of Chaing Kai Shek. He was Shan-born but had a Chinese name, Chan Si Fu, and spoke Chinese, as did Gritz, who had been trained in Mandarin by the U.S. Army. Khun Sa had survived an opium war in 1967 and now had a 40,000-man army equipped with U.S. weapons. James Mills, a careful researcher who authored a prodigious work entitled "Underground Empire," found evidence of an association between Khun Sa/Chang Chi-fu and the CIA, ostensibly for intelligence-gathering purposes, through the Thai Border Patrol Police, which was believed by U.S. Drug Enforcement Agency (DEA) agents to be 'a wholly owned subsidiary of the CIA.'[39]

Gritz had just finished training a group of Mujahadeen general staff officers from Afghanistan in the southern Nevada desert, under authorization from William Bode, a special assistant to the Undersecretary of State for Security Assistance. The report said that 4 U.S. POWs had drowned in the Mekong River while being transported from Laos to Burma and that Khun Sa had information on 70 Americans being held in Laos. Suchesk said he had met with Khun Sa, who wanted U.S. recognition to save his people. Tom Harvey, an NSC staff officer who had previously been a Military Assistant to the U.S. Senate, contacted retired Lt. Colonel "Bo" Gritz and asked if he would be willing to go to the Shan States to confirm whether the report was true. Harvey was a West Point graduate known to Gritz, who lifted weights in the Pentagon officers athletic club with his friend, Assistant Secretary of Defense Richard Armitage. (Armitage and Gritz shared a common acquaintance with Erich Von Marbod.) Harvey instructed Gritz to use the code name "Tango" when contacting him, and subsequently supplied the veteran POW hunter with letters of verification, on both White House and NSC stationary, authorizing assistance to Gritz in the operation. Money for the mission was supplied by a "foundation" to the United Vietnam Veterans Organization.

Using a false passport because of his earlier difficulties in Thailand, Gritz and team member Lance Trimmer did indeed travel into Burma by truck from Chaing Mai, and then by mule and horseback on a route by which smugglers transported hundreds of tons of opium a year, destined for the U.S. heroin market. They met with Khun Sa at his headquarters at the time of the Shan new year on December 1, 1986, but the warlord denied any knowledge of the reported American prisoners, saying: "I have never heard of this before." This was not surprising to Gritz in view of the remoteness of Khun Sa's domain from the areas of Laos where POWs had been reported held long after the war. Khun Sa acknowledged that Suchesk had come to the border with a Taiwanese to arrange a "trade mission" with the free world, with him as the "exclusive agent." According to Gritz, "Suchesk had used (the reference to POWs in) the letter to Bush as a fuse too get high-level attention on a gambit he hoped to turn into a gold mine." More startling to Gritz, however, Khun Sa then offered to interdict or to sell the next year's entire crop of opium from the Golden Triangle to the U.S. government, for a fraction of its market value, or what was currently spent on drug suppression in the area. Khun Sa also told Gritz that his best customers for the past 20 years had been U.S. Government officials. He wanted in return a U.S. trade mission in the Shan States and Peace Corps volunteers to teach the people crop substitution, but he said that he believed Burmese, Thai and U.S. officials profited too much from the drug trade to shut it down.

After returning to Thailand, Gritz administered a polygraph test to Suchesk in Bangkok, which showed deception. Gritz then returned to the United States just before Christmas 1986, bringing three videotapes proving that Suchesk's claims were fraudulent, that Khun Sa was offering to stop 900 tons of opium, and to expose corrupt U.S. officials whom he said were involved in the drug trade. Gritz maintains that upon receiving the report of Khun Sa's drug interdiction offer, his NSC contact, Tom Harvey, told him: "Bo, there's no one here who supports that...there's no interest here in doing that." He was repeatedly warned to drop the subject, which he stubbornly refused to do. The Vice President's office attempted to distance George Bush from any involvement with the POW hunter.[40] Scott Weekly was subsequently convicted of transporting C-4 explosives used in the government-approved Mujahadeen training project by commercial aircraft and sentenced to five years in prison. Gritz was indicted for U.S. passport violations.

The author had, in early 1986 attended a lecture on the application of military intelligence during the Vietnam War given by Colonel William LeGro, a retired U.S. Army intelligence officer. A friendship grew which resulted in LeGro assisting the author in pursuing an investigation of the POW/MIA matter by introducing several of his former staff officers who had served in the DAO-Saigon, Military Intelligence Branch, from 1973 to 1975, or in a liaison capacity to that branch. Some of these sources, who still remained on active duty in CIA, DIA, and the Pentagon's International Security Affairs, supplied the author with useful information. One of these sources was a military assistant who had served under Armitage in the Pentagon, a field-grade officer who had been wounded in action as an infantry platoon leader in the Tet Offensive of 1968, and subsequently transferred to military intelligence.

Another confidential CIA source who had personally briefed Bush and was identified by the author to the Iran-Contra independent prosecutor and to the FBI special agents assigned to that investigation in 1987 and 1989, provided information on Richard Armitage's role in the intelligence world and in policymaking areas. In two 1986-87 meetings, Mr. Doug Dearth, a DIA official who was then a special assistant to the Director, admitted to the author privately (without revealing classified information), that retired NSA analyst Jerry J. Mooney's 1985-1987 affidavits on the MIAs who were left alive in captivity in Southeast Asia in 1973 and his statements on the "Moscow-bound" POWs had substance, and said: "The Soviet Union could hold the solution to resolving the POW/MIA issue." He advised that the author and other serious researchers would have to expose the very roots of U.S. POW/MIA policy, in order to change that policy towards an active diplomatic negotiation for survivors. The author gained the impression from Dearth that confirmation of Mooney's analysis of the Moscow-bound POWs of Vietnam may have come from Soviet-bloc defectors and other intelligence sources.

At the end of the year, author J.M.G. Brown published the initial results of an investigation that had begun two years before, and had been encouraged by Vice President George Bush, with particular reference to sworn Congressional testimony and affidavits on livesightings of Americans and information on U.S. POWs shipped from Vietnam to the Soviet Union and Communist China. This appeared in a front-page article of the December 1986 American Foreign Policy Newsletter, next to a statement on the fundamentals of American foreign policy by Vernon Walters, then-U.S. envoy to the United Nations. (Walters was the former deputy Director of CIA, who had testified before the Congress during the 1975-76 Montgomery committee hearings, while George Bush was CIA Director, that no evidence of POWs remaining behind in captivity existed).

Ambassador Angier Biddle Duke (a nephew of former Ambassador and SHAEF officer Anthony J. Drexel Biddle), and his brother, Anthony Drexel Duke of the National Committee on American Foreign Policy in New York, encouraged editor George Schwab to publish this article in an edited form entitled: "THE POW/ MIA QUESTION: NATIONAL HONOR AND FOREIGN POLICY." The article reviewed the salient facts known thus far of the POW/MIA matter, urged the appointment of a special Presidential envoy to negotiate for surviving POWs in Indochina—including those offered for sale to the Reagan administration in 1981, mentioned the German Chancellor Adenauer's precedent of trading aid for WW II POWs held in Soviet gulags into the 1950s, and U.S. deals for the last Bay of Pigs prisoners. In addition, using the information obtained during and after the 1986 Senate hearings, through Mooney, Waple and others, the author proposed that the new Presidential negotiator ask Beijing about POWs sent to China from Vietnam, and also:

"...To discuss the tragedy with the Soviet Union. Evidence seems to indicate that the Soviet Union may be holding some of our Vietnam prisoners who were transferred because of their specialized electronic warfare training."

This article, appearing in the journal of New York's National Committee on American Foreign Policy, which had many distinguished citizens on its board ranging from a Nobel Prize-winning humanitarian, Elie Wiesel, to President Reagan's former U.N. Ambassador Jean Kirkpatrick, and often provided a forum for Soviet diplomats in America, must have given the new-style Soviet government of Mikhail Gorbachev an early indication that the American establishment would ultimately deal with the longstanding POW/MIA issue.

Averell Harriman, who had profoundly influenced American-Soviet relations and international negotiations with Communist governments since the early Bolshevik period, had died just as the mid-1980s "perestroika" reforms of Mikhail Gorbachev began to be recognized. He would have probably been astonished by the rapid collapse of the Soviet Union and the Communist system in Russia and eastern Europe between 1986 and 1991. Harriman retained great influence and power until his death, and his funeral service was attended by the high and mighty of the nation and of the world, who shared his views. His memoirs, Special Envoy to Churchill and Stalin, which had been published in 1975, hinted strongly that not all American POWs in Soviet-controlled Poland in 1945 had been released, yet all his life Harriman had retained a conviction that the Communists could be bargained with, and held to their agreements.

Armand Hammer lived on for years to come, still acting as a friendly emissary to Marxist-Leninist regimes, and expending some of his energy and enormous fortune in promoting his own image. After a lifetime of ignoring U.S. war prisoners in the Soviet Union, in 1974 Hammer had assisted Senator Paul Simon of Illinois in raising the case of an American civilian in the USSR named Abe Stolar. Stolar was one of the Americans who had been moved to Russia by his idealistic parents in 1931, and had been unable to leave since. (Stolar was finally to be allowed to leave the USSR in 1989.) In 1979 and 1980 Hammer had also been instrumental at opening Communist China to American business while negotiating contracts for his company, Occidental Petroleum, and its subsidiaries, with Deng Xiaoping, who had been impressed at meeting this American friend of Lenin. (His introduction to the Chinese dictator had been facilitated by President Carter's trade special representative, Robert Strauss, who was himself to be appointed U.S. ambassador to Russia at a critical time, a decade later.) It was Hammer also, who was to obtain from the Soviets the alleged WW II identification card of an American immigrant named John Demjanjuk, who was accused and convicted of war crimes by Israel; but whose conviction was later to be overturned. For the remainder of his life, Hammer continued to ingratiate himself with American and foreign political leaders, while promoting international cooperation and disarmament.[41]

Meanwhile, a January 14, 1987 letter from the National Security Council's Colonel Richard T. Childress, Director of Asian Affairs at the NSC, to the National Commander of the Veterans of Foreign Wars (VFW), Norman G. Staab, illustrates the continual intervention by senior government officials in the resolutions about U.S. POWs passed by the members of the most powerful American veterans organizations:

"Thank you for copies of the VFW's 87th National Convention resolutions (author's note: including a resolution from the membership on live POWs) which we have reviewed carefully...IT IS OUR OPINION THAT THE VFW'S SUPPORT COULD BE EVEN MORE EFFECTIVE THROUGH A GREATER EMPHASIS ON THE NEED FOR NATIONAL UNITY BEHIND THE PRESIDENT'S (ie: the administration's) PROGRAM. SOME DETAILED COMMENTS ARE BEING FORWARDED FROM DEPARTMENT OF DEFENSE THAT MAY BE OF INTEREST WHEN CONSIDERATION IS GIVEN TO NEXT YEAR'S RESOLUTIONS.

"I enclose the latest report on the Central Identification Laboratoy. IN ADDITION WE ARE PREPARED TO CONSULT CLOSELY WITH YOUR NATIONAL STAFF ON WORDING THAT WOULD BE MOST EFFECTIVE, in motivating the Indochinese governments to greater cooperation. This first opportunity would be January 30, 1987, at the League (of POW and MIA Families) sponsored POW/MIA conference to which the VFW has been invited."

Colonel Richard Childress, an Army Foreign Area Officer (FAO) in Southeast Asia in the late 1970s, had joined the National Security Council in August 1981 as an Asian affairs staffer, although he aided the Reagan Administration's task force on the sale of AWACs to Saudi Arabia, in which Air Force Major General Richard Secord was involved. In January 1982 Childress "assumed responsibility for POW/MIA Affairs," in the National Security Council.[42] In a subsequent letter written to POW/MIA researcher Charles Bates, of Albuquerque, New Mexico, Childress had denied that any offer by Hanoi to sell U.S.A. POWs had occurred, as had been reported by Bill Paul in the Wall Street Journal: "There exists a great deal of misinformation circulating throughout the country on this issue. The report you cite is one example. To our chagrin, no such opportunity or offer to negotiate the release of prisoners arose in 1981."[43]

Childress was a senior U.S. government official and spokesman, whose position required that he field questions from reporters or historical researchers. Bill Paul stood by his story, as did the Wall Street Journal, and he subsequently offered to debate Childress at any time or place, an offer the NSC official failed to accept. But for some of the retired military officers staffing the headquarters of the VFW, American Legion, Disabled American Veterans (DAV) and other veterans organizations in Washington DC, it proved difficult to publicly challenge the veracity of senior U.S. officials and thus confront the Reagan Administration on a sensitive national security issue, on the word of a journalist who could not safely reveal the names of his confirming sources in the U.S. government on the story. Thus the major veterans organizations were doing little, while across the country in all 50 states, rank and file members of these organizations took part in a grass roots POW/MIA movement, that led to mass rallies for POW/MIA awareness, involving sometimes hundreds of thousands of veterans marching behind black POW/MIA flags in Chicago, New York, Los Angeles and elsewhere.

From Richard Armitage's Assistant Secretary of Defense for International Security Affairs office, came another January 1987 letter for VFW National Commander Norman Staab, this one also concerned about the VFW membership's resolutions on POWs, and was signed by an Armitage assistant at ISA, Rear Admiral E.B. Baker:

"As mentioned in Mr. Armitage's December 30, 1986 letter to you, WE HAVE TAKEN UNDER ADVISEMENT THE POW/MIA RESOLUTIONS PASSED AT THE VFW'S 87TH NATIONAL CONVENTION...IN REVIEWING THE RESOLUTIONS, WE NOTE THAT SOME APPEAR TO BE BASED UPON A MISUNDERSTANDING OR A MISPERCEPTION OF THE SITUATION AND THAT SOME OF THEM ARE CONTRADICTORY. I AM ENCLOSING APPLICABLE COMMENTS WHICH EXPAND ON THIS ASSESSMENT..."

Thus, for years were the expressed desires and objectives of millions of American veterans who demanded that the VFW, American Legion and DAV take a stronger stand on still-captive U.S. POWs, quietly influenced by powerful officials of the government's national security apparatus. Richard Childress subsequently gave a brief interview to the press during which he claimed that protests by veterans and relatives of men missing since the Vietnam War had harmed the POW/MIA cause. He referred to those protesting the official U.S. policy of negotiating over alleged MIA "remains" rather than surviving POWs as "crazies," citing North Carolina Vietnam veteran and former Green Beret Ted Sampley by name. Sampley had organized a recent demonstration by 25 or more Vietnam War veterans and POW/MIA family members who attempted to deliver more than 1,500 packages of food, medicine and other relief supplies to the Laotian Embassy in Washington, for delivery to U.S. POWs they believed were still held in Laos. (An event witnessed by the author) About 100 of the packages were delivered at that embassy before Washington D.C. police and Secret Service agents halted the demonstration. (The U.S. maintained diplomatic relations with Communist Laos but not Vietnam.) The protestors left the rest of the packages at the homes of Colonel Childress and Reagan aide Frank Carlucci. Childress later told an American Legion Conference that representatives of the Laotian government had complained to the United States Government about the actions of the POW/MIA activists:

"THEY TOLD US THIS KIND OF ACTIVITY CAN INHIBIT COOPERATION...IT'S NOT HELPFUL TO THE PROCESS. AND WE AGREE WITH THEM."

Childress also said that the POW/MIA protests had "stalled somewhat and exacerbated the delays" in obtaining "remains" from Vietnam and Laos. He said: "Hanoi is watching this...Think how it's perceived in Hanoi. They need to know the United States is completely behind what the President is trying to do."[44] By the "President," Childress in reality meant the national security apparatchiks, such as he, Childress, Richard Armitage, and a small coterie of other Pentagon, State Department, CIA and DIA officials who actually controlled American policy towards non-repatriated U.S. prisoners of war.

Meanwhile, former NSA analyst Jerry J. Mooney had come out with another sworn affidavit for the Congress in January 1987, further explaining the special intelligence methods which had enabled him to determine which MIAs had become unacknowledged POWs who had not been repatriated in 1973. In this period, the author had became involved in the POW/MIA investigation of Monika Jensen-Stevenson, a CBS 60 Minutes producer, with her husband William Stevenson, author of "A Man Called Intrepid," "90 Minutes at Entebbe" and other internationally-acclaimed books. The author assisted in the research process for their book, which dealt only with the Vietnam war POWs, "KISS THE BOYS GOODBYE," which the authors basically completed by the fall of 1987, but which, despite their impeccable journalistic credentials, they found great difficulty in publishing, in the United States until September of 1990.

Through Colonel William LeGro, the author also met in early 1987 with Wolf Lehmann of Rockville, Maryland, formerly a Chairman of the 'DCI's HUMINT COMMITTEE,' and thus head of all CIA human intelligence tasking during the Carter transition year and the early years of the Reagan Administration from 1980-1983. He admitted that he had authority in all human intelligence tasking, including directing what the systems and attaches would do. When questioned by the author on the subject of the 1981 White House meeting on POWs, Lehmann went into a visible state of shock. The author conveyed information contained in this book on the subject of POW livesightings and the 1981 White House meeting, which Lehman would not deny had occurred, or that this had resulted in a rapid upgrading of POW intelligence requirements. He would only say it was true that suddenly, "The POW/MIA issue became a priority, and I also remember when Childress from the National Security Council began upgrading the effort." Lehman obstinately refused to discuss the CIA human intelligence reports since then, of live American POWs gathered by CIA agents, or the special intelligence from decoded enemy communications on the Moscow-bound POWs of Vietnam.

While investigating published allegations of official misconduct concerning Assistant Secretary of Defense Richard Armitage (who then controlled U.S. POW policy in the Pentagon), by Benjamin Bradlee Jr. in the Boston Globe in January 1987, and elsewhere, the author interviewed U.S. officials in the National Security Council, Department of State, Pentagon, CIA, DIA and other agencies. The author was in turn interviewed by FBI agents and the Defense Investigative Service on this subject and oh the POW matter several times between 1987 and 1989, and during one such interview, when some of the author's documentary and human sources were revealed in April 1989, he was reminded of the necessity for national security secrecy in certain matters relating to POW/MIAs by the senior FBI agent, several times. The non-blacked-out portions of the FBI report of this interview (which had been witnessed by the author's wife and five children), that was later released through FOIA request, proved almost entirely inaccurate.

In one meeting at the State Department with Assistant Secretary of State Morton Abramowitz on February 27, 1987, the Secretary discussed Armitage's role in a "pharmaceutical importing and bamboo furniture exporting business" in Thailand in the late 1970s, and denied to the author and to Colonel William LeGro (USA Ret'd) that any information on secretly-held U.S. POWs existed at the time he was Ambassador to Thailand, or later when he was Assistant Secretary of State for Intelligence and Research.

Abramowitz was one of the few American officials willing to speak openly about the overriding importance of policy in determining the U.S. response to reports of Americans still held in captivity for what amounted to a ransom demand by Hanoi: "Well, even if there was evidence of live Americans, known by the U.S. Government as you say, our question is, would it be in the best interest of the United States to alter our present policy to gain the return of a few individuals?... We have to look at the broad policy concerns, and it does no good for us to become emotional about things like this." Mr Abramowitz later wrote to the author (on June 12, 1987) concerning Armitage's residence in Thailand, while Abramowitz had been U.S. Ambassador there from 1978-81:

"I don't believe I said that Armitage was living with Heinie Aderholt in Thailand but that he was working with him. My understanding is that Armitage was in business with Aderholt the year he was in Thailand." (Confidential informants within the U.S. Government have stated to the author that Mr. Armitage was a "double-ranker," and was actually a high-level CIA official who was considered "essential to the functioning of the U.S. Government.")

For unknown reasons, Armitage had taken pains to deny to CBS 60-Minutes producer and author Monika Jensen, then investigating the POW/MIA matter, that he'd had any business affiliations with General Hienie Aderholt in Thailand in the late 1970s.[45]

On January 15, 1987 the author, Anthony D. Duke and former Congressman John LeBoutillier met with Senator John Tower of the Presidential Commission investigating the role of the National Security Council in the Iran-Contra affair, that had been authorized by President Ronald Reagan known as the "Tower Commission." The author brought for Senator Tower's attention declassified livesighting reports of American prisoners in Southeast Asia from among the "Uncorrelated Reports" quoted elsewhere in this book, the sworn affidavits of Mooney, Smith, McIntire, Howard, Garwood, Ashworth and other witnesses at the Senate Veterans Affairs Committee hearings, chaired by Senator Murkowski the year before, and other documentary information on the conduct of the National Security Council Affairs Director, Colonel Richard Childress, in implementing U.S. POW/MIA policy. Like Armitage, Childress was a friend and colleague of Colonel Oliver North, who was then under investigation for illegal acts stemming from the covert, White House-run aid program for supporting the anti-Communist Nicaraguan "Contras." Tower appeared weary and depressed about the enormous volume of information on the functioning of the National Security Council that had been brought to his attention. He expressed no doubt that if he could investigate the NSC's conduct in the POW/MIA matter that he would find it to be a "can of worms." But, he said, he was, "as busy as a one-armed paper-hanger with the crabs," and insisted that he was going to have to limit the inquiry to the NSC's role in the Iran-Contra Affair, only. The author also met Edmund Muskie that day, but was unable to gain access to General Brent Scowcroft, the executive branch's man on the Commission, who had gained intimate knowledge of the POW/MIA issue since his service under Kissinger and Nixon in 1973.

Other U.S. officials the author met with in early 1987, and brought the livesightings and sworn affidavits of the witnesses at the Murkowski hearings to, included: National Security Council (NSC) Executive Secretary Grant Green, in the White House, Undersecretary of State Ed Derwinski, Assistant Secretary of State Gaston Sigur and Deputy Assistant Secretary of State David Lambertson, Vietnam, Laos and Cambodia (VLC) Director Shephard Lowman and the State Department's Vietnam Desk officer, Lawrence Kerr, who all stated to the author and to other witnesses present, including retired Army Colonel William LeGro, that no evidence existed that U.S. POWs had been left in captivity after 1973. They also stated that no U.S. POWs had been sent to the USSR, China or North Korea, that no information on still-living POWs existed, and all five vigorously defended Richard Armitage's "integrity" and verified his control of the POW/MIA issue.

In fact he was considered indispensable by some senior US officials, as the brash Texas businessman, MIA/POW advocate, and former U.S. naval officer, Ross Perot, discovered during his own 1986-87 POW investigation. After initiating and financing an inquiry into the Indochina aspect of the POW/MIA matter (which the author had no part then or since), and after being given authority from President Reagan to analyze the matter, Perot personally confronted Richard Armitage with a perceived wrongdoing, involving Armitage's possibly compromising relationship with a female Vietnamese underworld gambling figure named Nguyet Tui O'Rourke, and urged him to resign his position as Assistant Secretary of Defense for International Security Affairs (with policymaking control over the POW/MIA matter). Armitage, who was also chief of all Pentagon covert actions, angrily refused. In a later meeting in the White House, when Perot urged a change in the personnel assigned to the POW matter, the new National Security Advisor, Frank Carlucci, defended Armitage and refused.[46] Informants within the Pentagon's International Security Affairs office have stated to the author that Carlucci, former head of Sears World Trade, was another protege of Erich Von Marbod, and was known within the Pentagon as Armitage's "Godfather."

Anthony D. Duke and his cousin (also a former naval officer) Anthony J. Drexel Biddle III, had already met with long-time Reagan aide and family friend Michael Deaver, a few days before, to impress upon him their belief that President Reagan should receive an alternative view of the current government policy towards the missing prisoners of war. (Biddle was the son of former U.S. Ambassador to Poland and SHAEF staff officer Anthony J. Drexel Biddle Jr., who had been present at the May 1945 Halle Conference.) Another meeting followed in March 1987 during which the author and Colonel William LeGro brought affidavits of Mooney, Garwood, Smith and McIntire and others from the previous year's Senate hearings, and other declassified POW/MIA documents including livesighting reports. Although under fire in the media himself at this time, Deaver was still moved by the information, and he also wrote a letter, on March 18th, to the new White House Chief of Staff, Howard Baker. This was forwarded (with the help of Ken Cribb) to Deputy Chief of Staff Kenneth Duberstein, urging the appointment of a special Presidential envoy to negotiate for living POW survivors in Southeast Asia:

"The continuing issue of American servicemen missing -in-action or prisoners of War still alive in Laos and Vietnam was brought to my attention recently after a visit with several representatives of veterans' groups who sought my advice and help. One was William E. LeGro, retired U.S. Army Colonel who was senior military intelligence officer in Vietnam from early 1973 until our forces left, and also John M.G. Brown, a Vietnam veteran active in MIA/POW activities with Vietnam veteran groups. These gentlemen are concerned that official efforts to obtain the release of Americans...are stymied...with what they believe is a lack of action by your POW/MIA Interagency Group (IAG), and especially with Assistant Secretary of Defense Richard Armitage, who heads the group.

"As an example, they told me that a recent study in the Department of Defense by Lt. General Eugene Tighe concluded that Americans are being held in Laos and Vietnam. According to my visitors, General Tighe's report has been classified and not released because it is counter to U.S. policy. At the heart of the problem... is a standoff between our nation and the Laotians and Vietnamese. We refuse to negotiate with them until all missing Americans are accounted for, a policy that, perversely, prevents negotiations concerning the release of Americans held prisoner. They believe our policy to be especially unfortunate when new leadership in Laos and Vietnam seems anxious to normalize relations between our countries. Because matters of diplomacy are involved, my visitors believe that official action concerning missing Americans should be transferred to the Department of State and handled on the Ambassadorial level, replacing present low-level technical discussions originating in the Department of Defense. I disagree, IF WE ASK ANY OF THE INTERESTED AGENCIES, WE WILL NOT GET ANY MORE ANSWERS THAN WE RECEIVED OVER THE PAST FIFTEEN YEARS."

"What has been suggested to them, I suggest to you, is the appointment of a distinguished American to look into this situation and report back to the President within a specified time. The report's conclusion would then form the basis for future American policy..."

Deaver recommended former-President Richard Nixon for the job, but others must have felt that to be an inappropriate idea, and claimed to have already decided on retired Army General John Vessey, who had once commanded secret U.S. military operations in Laos, and who would presumably have some knowledge of U.S. POWs left behind there in 1973.

The idea of a Presidential emissary to Hanoi, which had been proposed for months by Ross Perot's supporters, by former Congressman John LeBoutillier, and by several veterans and POW/MIA family groups, was diverted to a lower level of diplomacy by the fact that the appointee was not a nationally-known figure, but a former senior U.S. military commander during the CIA's secret war in Laos, who, as the Vietnamese were aware, knew already that not all American POWs and MIAs held in Laos had been released. This doubtless indicated to the Vietnamese that the charade involving the recovery of human skeletal remains, identified as specific U.S. MIAs, would continue. The Armitage and Childress wing of the national security apparatus had thus nipped in the bud any chance that Ross Perot—or any other proven bargainer—would be allowed to discuss the release of surviving POWs with the Vietnamese. Perot had already made several trips to and from Hanoi, while conducting secret talks with the Vietnamese, using as a mandate President Reagan's request of the year before, that he investigate the POW/MIA matter, and the subsequent encouragement from Vice President George Bush. In later testimony before a Senate Select Committee, the National Security Council Asian Affairs director, Richard Childress stated: "The President agreed upon the conceptual recommendation to appoint a Presidential Emissary in October 1986. We felt it should be someone closely identified with the President...General Jack Vessey was asked in January 1987 if he would serve in this capacity and he accepted the job in early February..."[47]

The actions taken by the national security establishment in Washington to head off Perot's investigative-diplomatic effort on behalf of the abandoned POWs in Vietnam and Laos, amounted to a redirecting the President's own desire to resolve the POW/MIA matter into a channel that led to further stagnation of the issue, for years to come. Childress and Armitage had won, although Perot was originally given a personal go-ahead by Reagan, and a subsequent Associated Press story of November 14, 1986 (first reported by the Dallas Morning News), said that both Reagan and Bush had asked

Perot: "to dig into this issue—go all the way to the bottom of it and figure out what the situation was—then come see them and give them my recommendations."

Perot remembered later:

"Reagan talked to me much earlier, and said he'd heard I wanted to get to the bottom of this. Nancy (Reagan) chimed in to ask if anyone was still over there, and I was as forceful as I could be. I said there was no doubt they'd been left behind, and recent good human intelligence indicated they were still there..."[48]

But Richard Childress, Richard Armitage, and later Frank Carlucci apparently feared that Perot was an uncontrollable personality, whose serious investigation of the POW/MIA matter had delved deeply into questionable covert operations that involved Pentagon, NSC and CIA figures connected to both the POW/MIA matter and the ongoing Iran-Contra scandal, which was threatening the Reagan Administration, and which Armitage had been involved with.[49] It may have been feared that Perot would indeed produce dramatic results in negotiations with the Vietnamese, and this might lead to a much wider investigation of National Security Council procedures with, among other things, the POW/MIA matter, and of the Inter Agency Group (IAG) which controlled POW/MIA policy. Thus occurred the appointment of retired Army General John Vessey to the position, supposedly, according to Childress, planned long before by he and other far-seeing bureaucrats.

Before one of his secret trips to Hanoi during the spring of 1987, White House Chief of Staff Howard Baker had warned Perot: "You can go, but go on your own...Try to get them to deal with us, but remember, you have to go on your own." Perot was angry at the fear shown by U.S. officials in negotiations for surviving prisoners of war and answered: "At the end of the war we just abandoned those men. Our government froze like deer then, and it's freezing now..."[50]

Colonel Richard Childress of the NSC later testified to a Senate committee:

"I have no firsthand knowledge of any direct taskings to Mr. Perot by the President or the Vice President. It is my clear impression, however, that Mr. Perot was discouraged from going to Vietnam in the Spring of 1987, and that if he insisted on going it was to be as a private citizen, since the President had already selected General Vessey as the President's emissary to Hanoi. General Vessey had accepted the position and was in the research phase...Upon Mr. Perot's return from Vietnam...he was debriefed in early April concerning his trip. (by the CIA's Jim Wink perhaps under another name among others.-author's note) He also asked to meet with the President, which was scheduled...I was tasked to prepare the meeting documents for the President from the National Security Advisor, WHICH INCLUDED SUGGESTED TALKING POINTS...It is my opinion, which you requested, that Mr. Perot's trip was counterproductive to U.S. efforts...when he returned, derisive press reports about the Department of State and the previous U.S. negotiating team members were published. Confusion reigned for awhile...On the IAG, where policy initiatives were hammered out, we had staunch advocates in Richard Armitage, Paul Wolfowitz, Gaston Sigur, Dave Lambertson and many others...In the middle of us all was Ann Mills Griffiths—prodding, suggesting, criticizing and pitching in to help... Threats by phone and mail were received, harassment calls at home increased...The Lao Embassy was subjected to tremendous harassment..."[51]

The author witnessed several of the demonstrations at the Lao Embassy by Vietnam veterans, led by Ted Sampley, John Molloy, Jerry Kiley and others, who dragged symbolic bamboo cages in front of the doors and demanded the return of their fellow servicemen who had disappeared in Laos. The U.S. Secret Service and the Washington D.C. metropolitan police worked closely with the Lao embassy officials to harass and arrest the protesting veterans, and the only damage done was an occasional disturbance of the otherwise tranquil sleep of the Laotian Communists, who knew only too well that their government had withheld many American war prisoners to gain reparations.

Perot later recounted to authors William Stevenson and Monika Jensen that in his meeting with Reagan the President was manipulated by his advisors, and resorted to "Q cards" (prepared by Childress) containing proposed answers for him to various questions. A man used to getting his own way in business, Ross Perot decided to quit the POW/MIA issue more than two weeks before turning in his April 8th report on POWs to President Reagan, after he learned for certain that General Vessey was to be the Administration's choice of a Presidential negotiator in Hanoi. At 2:00 PM, on March 21st, Perot called George Bush and told the Vice President of his decision, conveying his belief in the importance of the delicate negotiations he had initiated with Hanoi, and reiterating the fact that he didn't believe the POWs were the main concern of U.S. Government officials Richard Armitage and Richard Childress. George Bush recorded the details of the conversation in a personal, self-typed memorandum of March 21st, which recorded that, while announcing his quitting the issue, Perot also complained of being prevented from taking one last trip to Hanoi, to discuss the POWs and set the stage for Vessey:

"Phone call from Ross Perot (he says) 'I'm shutting down my operation.' He is upset because the government's top two people (RR + GB) got me into this. 'I could never get an answer on anything,' he says. "I tried through Carlucci.' Carlucci says, 'Will you get off Armitage's back if we appoint a negotiator?' I reminded Ross that I had told him that **his** suggestion of a special negotiator had been approved. I told him the name of the negotiator. He replied, 'Yes, but I had already been told of both the approval and the name,' (strange twist here). 'The Vietnamese came to me out of the sky,' (Perot said) 'The NSC is trying to move heaven and earth to get Vessey in ahead of me...There is no down side to my meeting, I'll tell Vessey everything. I'd make it clear to them Vessey is our man, and turn it over to him.' Ross then went into his concerns about Childress and Armitage, stating that POWs are not 'their main concern...'"

In a March 23, 1987 letter, Bush wrote to Ross Perot in Texas about his decision:

"I am sorry you feel you have had less than full cooperation; but I do understand your decision relayed to me yesterday, to 'get out of it' and convey whatever information you have to the new negotiator (General Vessey). As I recall you strongly favored this high level negotiator concept. I helped get that concept put into effect and I know we've got a good man..."

"...I have worked with you all along the way hoping that your energy and principled determination would lead to what you and I both want—the return of our POW/MIAs. I will continue to do everything in my power to help gain their return. So, I am sure, will the President. You asked if Craig (Fuller) had told me that you had cancelled your pledge to the Reagan Library. He did tell me this. He has accurately reflected all that you have told him.

"...I ACCEPT YOUR DECISION TO 'GET OUT OF IT'; BUT I HOPE THIS DOES NOT MEAN THAT YOU ARE UNWILLING TO PASS ALONG LEADS ON THIS CRITICAL SUBJECT TO THOSE IN THE UNITED STATES GOVERNMENT WHO ARE WORKING DAY AND NIGHT TO GET THE POW/MIAs RELEASED."

In a hand-written footnote Bush tried to disclaim future responsibility for the issue by adding: "...Given the new White House set-up I will stay out of the 'line,' but will continue to assist the process any way possible."[52]

Deeply disappointed that his efforts at diplomacy had been diverted and subverted and that he was not to be permitted to play a role in the negotiating process, Ross Perot turned in his report on the investigation of the POW/MIA matter to President Ronald Reagan on April 8, 1987. It was a succinct and potentially devastating document given Perot's stature and national prominence:

"My findings on the POW/MIA study are as follows:

1. We left POWs behind at the end of the war in Vietnam.

2. We knew we were leaving men behind.

3. The men left behind were held in Laos.

4. The evidence that men were held in Laos is substantial There are 343 MIAs in Laos..."[53]

Perot went on to outline briefly the original abandonment in

1973 and then bringing the report up to April 1987 with the following:

"...In my recent visits with the Vietnamese in Hanoi, they said, 'Why did your own government declare these men dead immediately after the war? After all these years, how can you expect us to take you seriously about looking for live Americans?'

Perot cautioned Reagan that it was, "...Unrealistic to attempt a military rescue of these men...we must know exactly where they are held to successfully carry out a rescue."

Instead the Texan recommended diplomacy to the President:

"There is only one realistic way to gain the release of these men—through negotiations. Several months ago, I recommended appointing a Presidential negotiator. I urge you to appoint a personal representative to negotiate with the Vietnamese. General Vessey is an excellent choice. He will have my full support...POW/MIA family members and veterans' groups will react positively to this action... At Howard Baker's request, I have strongly endorsed General Vessey to the Vietnamese. They did not seem to know much about him. The fact that he is a soldier, not a diplomat, is a plus to them...

"They were pleased that we had reviewed this matter with them in advance. Apparently we have not cleared such appointments in the past...My recent meetings in Hanoi were with the Foreign Minister, Nguyen Co Thach, who is Vice-Chairman of the Council of Ministers... You should not announce General Vessey as your negotiator until we have received a positive indication from the Vietnamese... During my meetings with the Vietnamese, I carefully postured the conversations so that the MIAs would be found in Laos and returned by the Laotian government. This approach allows Vietnam to avoid criticism for having held the men...ECONOMIC CONDITIONS, AND OTHER CONCERNS EXPRESSED BY VIETNAM'S LEADERS WERE CONVEYED TO ME DURING THE MEETINGS... This may be helpful to General Vessey. Howard Baker has this information. THE PRINCIPAL OBSTACLE IN OBTAINING THE RELEASE OF THESE MEN SINCE THE END OF THE WAR HAS BEEN A LACK OF DILIGENCE AND FOLLOW-THROUGH BY OUR GOVERNMENT."[54]

Such a valuable prelude to substantive talks on the issues important to for so long to both the Vietnamese and the Americans appears to have been ignored in the subsequent efforts of General Vessey, which for years to come would continue to focus on the search for human skeletal remains.

Ross Perot continued to be ridiculed and worse for his stand on the secretly-held U.S. POWs by the mainstream American media, from this point on until the 1992 election campaign. He was baited by the Reagan Administration's Democratic Party apologist in the House, Congressman Stephen Solarz, who tried to force Perot to publicly reveal the names of confidential sources of information on

U.S. POWs, when he, Solarz, had proven to be untrustworthy and biased on the issue. By his very act of never retracting what he knew to be true about the American prisoners left behind, Ross Perot helped keep the issue alive, and he also offered a powerful incentive for the Communists to keep some number of the POWs alive, at least into the foreseeable future. In this manner he had performed an important duty for the country and for all who cherished liberty around the world. Even the United States Government, despite the endless official denials that POWs had been left behind, had used Perot's prodding to obtain concessions from Hanoi.

Lt. Colonel "Bo" Gritz returned to Thailand and reentered Burma with an associate, Vietnam veteran Lance Trimmer, in May 1987. He was determined to interview Khun Sa once again on the allegations the warlord had made about involvement of U.S. officials in the Golden Triangle opium trade. Khun Sa told Gritz that his agents had reported no American POWs were being held in the part of Laos that adjoined Burma, but during the videotaping the warlord, through his interpreter, Dr. Yi Shan Ji, he named former Deputy Director of CIA Theodore Shackley; Santo Trafficante; Richard Armitage, who 'handled financial transactions of the U.S. narcotics trade with the bank in Australia' (Nugan-Hand Bank); Daniel Arnold, a former CIA station chief in Thailand, supposedly recruited by Armitage; and Jerry B. Daniels, who was reported to have assisted Armitage, but had been found dead of gas poisoning in his Bangkok apartment in 1982. During the videotaping a copy of an affidavit of the Christic Institute was visible on part of the film that was viewed by the author. Gritz thereupon returned to the United States and had his colleagues, Chuck Johnson and John Heyer, send copies of the Khun Sa videotape to the Senate Select Committee on Intelligence. Other copies were sent to U.S. Representative Lawrence Smith (from Dade County, Florida), Chairman of the House Task Force on International Narcotics Control. Gritz was called before the Committee to testify on June 30th. Chairman Smith rejected the videos and the entire report of Gritz as nonsense, calling Khun Sa a criminal.[55]

The appearance of the Christic Institute affidavit in the videotape Gritz had taken in Burma on his second trip, raised serious questions as to whether the Shan warlord or his interpreter had been prompted on what to say by the contents of the affidavit. Gritz himself admitted in his memoirs that he suddenly interrupted Khun Sa during the filming because he "didn't want him to get ahead of me and name Erich Von Marbod, a man I have great loyalty to." (Von Marbod had also been named in the Christic affidavit.)

The Christic Institute was a liberal Catholic public-interest law firm in Washington, D.C., headed by Daniel Sheehan, an activist attorney who had achieved prominence in class action suits following the Three Mile Island nuclear accident and the Karen

Silkwood suit against Kerr-McGee Corporation. The Christic affidavit and suit named retired Major General Richard Secord and 30 other Americans and Latin Americans as co-conspirators in a covert CIA-connected arms and drug smuggling operation that had expanded from Southeast Asia to Central America, and in a bombing plot intended to assassinate Nicaraguan Contra leader Eden Pastora (Commandante Zero), who had declined to cooperate with the CIA. Sheehan claimed that POW/MIA activities of several U.S. officials had been used as a cover, after the fall of Vietnam in 1975, to conduct drug and weapons trading. Among those named in Sheehan's suit were Theodore Shackley, Santo Trafficante, Richard Armitage, Daniel Arnold and Jerry Daniels. All of these individuals had been named by Khun Sa on the Gritz videotape.

In his later memoirs, Secord recorded that he hired ex-CIA agent Glen Robinette to uncover derogatory information on the Christic Institute's sources, including the two journalists on whose behalf the suit had been filed: Tony Avirgan and Martha Honey. Secord charged that Avirgan was an anti-American who had written an article in North Vietnam during the war entitled, "From Hanoi With Love," that Honey had no publishing history at all, and that the suit "seemed to be the end product of a concerted effort to undermine U.S. Contra support with roots in the Congressional opposition and possibly assistance from foreign intelligence services—perhaps the KGB, but more likely the DGI, the Cuban intelligence arm. The author and research colleague Thomas V. Ashworth had met with Daniel Sheehan in Washington, D.C., early in 1987 and offered, unsuccessfully, to correct certain historical inaccuracies in the original Christic affidavit. The suit was eventually thrown out of court by a federal judge.

However, the Senate Subcommittee on Terrorism, Narcotics and International Operations had recently revealed classified 1985 memoranda from General Paul F. Gorman, head of the U.S, Southern Command in Central America, to the Department of Defense, in which he warned: "...There is not a single group in unconventional warfare that does not use narcotics to fund itself...We have to think beyond the Communist threat to the more acute problem of international drug rings and their effect upon ourselves..." In his testimony before the subcommittee Gorman said: "There is an artificial membrane between intelligence relating to drugs and intelligence relating to military business."[56]

Gorman's former commander in Vietnam, General William C. Westmoreland, had subsequently written to the author concerning these memos: "Paul Gorman is a very astute officer." In addition, Gorman's assessment had been confirmed to the author in 1986 and 1987 by a U.S. Army Special Operations officer, who had known LeGro while in Saigon and was now serving in the Pentagon's ISA, who revealed that involvement with drug traffickers by the CIA and secret U.S. military intelligence groups had severely compromised their personnel, mission and effectiveness, and had exposed individual officials to possible blackmail by foreign agents. This source maintained that these compromises were affecting U.S. security policy. The author had also obtained information about the shipment of cocaine into the U.S. from Central America during backhauls of the Contra airlift, through personal acquaintances who had been recruited as pilots and aircrew for the operation from Mark Air in Alaska.

The theme of the Christic affidavit was further expanded upon by author Leslie Cockburn, in a book entitled: "Out of Control," which was published by Atlantic Monthly Press. She was the wife of Andrew Cockburn, the author of a prescient book on the flaws within the Soviet military, and was thus the sister-in-law of Alexander Cockburn, who wrote for that venerable left-wing American magazine, The Nation. (A 1985 article by Alex Cockburn in the Nation had questioned the reliability of information from the 1968 Czechoslovak defector, General Jan Sejna, citing unnamed 'CIA' sources.) Although Secord was to launch a defamation suit against her the next year, it was ultimately dismissed because of his failure to demonstrate that Cockburn had "malicious intent" in reporting factual errors.[57] Cockburn subsequently informed the author that the judge in the case had responded to Secord's complaint that U.S. agencies and officials had been charged in the suit with involvement in cocaine trafficking in Central America with the question: "'Is that in dispute?'[58] Still, the appearance of the Christic affidavit in Gritz's videotape of Khun Sa had contributed to the "smoke and mirrors" which surrounded the POW/MIA matter.

Together with Anthony D. Duke, Colonel William LeGro and former U.S. Representative John LeBoutillier, the author had met with Senator Paul Laxalt on April 21st, in Washington. As he was a close friend of Ronald Reagan, there was hope that Laxalt might be induced to give the President an alternative analysis of the POW/MIA matter. Laxalt told the group that the last time he had spoken to the President about the prisoner of war issue, "He was not happy with how it was being handled." Following the briefing on livesightings and special intelligence Laxalt promised to do what he could and instructed the author to contact his aide, Tom Loranger, if any additional information was forthcoming. During this meeting Duke was interrupted by a telephone call informing him that he was not welcome to come to the Pentagon and receive a briefing on the Defense Department's handling of the POW/MIA issue, and that neither Frank Carlucci or General Colin Powell would meet with him on the subject, under any circumstances. The former U.S. Navy commander and trustee of Duke University was further informed that if he came to the Pentagon he would be removed.

Subsequent to this meeting, the author informed Senator Laxalt, through his aide Loranger, that a certain colonel, identified by name, had been placed in the White House Military Office (or WHMO, which among other duties had charge of the nuclear "football") as an "agent," to be always near the President in order to subtly and constantly influence his thinking on certain key issues (one of which was the POW/MIA matter). This was reportedly part of a strategy to surround the President with as many security officials and military officers as possible, who could insulate him from receiving unwanted information. The author's source, a field grade Army officer in the Pentagon's International Security Affairs, expressed his willingness to meet with the President to verify his information that the WHMO officer in question was acting as an "agent" of Richard Armitage, and that the term "Coup d'état" was now in common use in certain circles of the Pentagon.[59]

Loranger contacted the author two more times for additional information on this subject and on May 12th said, "Senator Laxalt passed it all forward; I cannot discuss with you what type of action is being taken." In a subsequent meeting on June 23rd with a confidential source, an active-duty CIA officer who had briefed George Bush in the past, the author was informed that the WHMO officer in question had been removed. The information was corroborated by an Air Force officer then on loan to CIA who was also present. (The author was later questioned by the FBI on this incident on several occasions, and revealed the name of the primary source, who had agreed to speak with federal investigators if required to do so. The FBI subsequently warned the author in April 1989 that public disclosure of this CIA officer's identity would be viewed as a serious "breach of security" and could result in all of the author's declassified U.S. documentary evidence, source material and other research on American POWs used in this book being subpoenaed and seized, "and kept for years.")

Meanwhile, at a press conference in Hanoi on June 18th, Vietnam's Foreign Minister, Nguyen Co Thach had stated, in reference to the MIA question, that Vietnam was, "MANY TIMES MORE CONCERNED ABOUT HEALING ITS OWN WOUNDS OF WAR," than with the MIA issue. Three days later Vietnam Communist Party General Secretary Linh said in an interview with the Italian newspaper L'Unità, that the United States had 'failed to honor' the pledges it made in the Paris accords and is 'not paying the more than $3 billion in war compensation' due Vietnam. In an interview for the Far Eastern Economic Review, Vietnam's Vice Foreign Minister Nguyen Dy Dien stated once more the Vietnamese policy of holding the U.S. POWs as hostages for war reparations by stating that Hanoi could not search for missing American servicemen "FREE OF CHARGE."[60]

The harsh reality of the Vietnam aspect of the POW/MIA matter was first fully exposed in the summer of 1987, when the BBC released across the United States a documentary film entitled "We Can Keep You Forever," which had been produced by the tireless efforts of former CBS news executive Ted Landreth, of Los Angeles, California (who had succeeded at getting NBC News interested in the story also). Retired NSA analyst Jerry J. Mooney was finally permitted to tell at least some of the American people what he knew of the fate of U.S. 'special talent' U.S. POWs of Vietnam, who had been carried on his lists at that agency as "MB" or "Moscow Bound." Alarmed at the breach of "security," the Department of Defense subsequently issued a "Guidance-BBC Documentary, 'MIA': We Can Keep You Forever,' which denied what were called the film's: 'several statements alluding to the possibility that U.S. personnel may have been moved to China or the Soviet Union.'[61]

Another witness recorded in this important news documentary was a Laotian refugee whose identity had long been protected. Somdee (Som dy) Phommachanh had been held in a prison camp in northern Laos long after the war with two American POWs he called Nelson and Smiley. (These names matched with the identities of two Americans missing and presumed killed in Laos, David Nelson on an Army helicopter, and Navy Lieutenant Stanley Smiley) Just prior to the Americans being removed from Laos to Vietnam, David Nelson died, and Somdee later recounted that when he discovered this, the American's face had already been partially eaten by rats. After escaping to a refugee camp in Thailand, Somdee had told his story to a U.S. Embassy official in Bangkok (where the JCRC under Colonel Paul Mather had jurisdiction) but instead of following up, that official took away Somdee's only photograph of himself in a Royal Laotian Army uniform, which caused him to spend two more years in the camp in Thailand.[62]

Documentary film producer Ted Landreth later recounted to the author how, on the eve of the documentary being aired, the Defense Intelligence Agency called the producers claiming that Somdee had recanted his story about the American POWs. The Laotian told about how the DIA had come to the Midwest state where he worked as a janitor in a small-town school, and had demanded that he stay with them in a hotel room, where they kept him a virtual prisoner and abusing their authority as U.S. Government agents, proceeded to threaten him and his family with dire consequences and possible repatriation to Laos if he didn't recant his story. Although he was also contacted by a former Laotian general who had served in the U.S. secret war, Somdee stuck to his story and was taken under the personal protection of the school principal where he worked, who was a Vietnam veteran. But the Department of Defense still stated officially that he had recanted the story in an effort to stop television stations around the country from airing the documentary. The incident indicated what lengths DIA had gone to over the previous years, in terrorizing Indochinese eyewitnesses into silence, if they swore they had seen American POWs held long after the war. How much information had thus been lost and how much damage had been done to potential human sources, can only be conjectured.

A significant report on Soviet-held U.S. POWs of WW II based upon recently-declassified U.S. archival documents, was published in the United States during August 1987 by Wall Street Journal investigative reporter Bill Paul, entitled "POWs: Four Decades of Abandonment." Three months later in the upstart Washington Times, independent researcher James Sanders, published a similar story entitled "Liberated into Legions of the missing?" Sanders, a former policeman who, at the instigation of Paul, had begun an investigation of declassified archival documents on Soviet-held U.S. POWs of WW II. Sanders contacted the author for information on WW II POWs in 1986 and 1987 (as did Thomas V. Ashworth), and the author briefly assisted Sanders in his research and introduced him to Major Carl Heinmiller, of Haines, Alaska, a key, confirming human source of the author's on the Soviet-held U.S. POWs of the WW II period, during the summer of 1987. Also during 1987, at the instigation of a former marine platoon commander in Vietnam and CIA officer in the 1970s named John Biddle Brock, and Assistant Attorney General William Weld (later governor of Massachusetts), the author met with staff investigators of Iran Contra Independent Counsel Lawrence Walsh. These were Victor O'Korne and Jim Beane of the FBI's national headquarters in Washington D.C., who were assigned to the Walsh investigation. The author revealed information from several of his sources noted within this book concerning the conduct of federal officials such as Richard Armitage and Richard Childress, who were involved with the POW/MIA matter in the Pentagon and National Security Council. Based upon responses to later Freedom-of-Information requests to the FBI, this information was classified, and may have been reported incompletely or inaccurately.

While researching Soviet-related U.S. POW/MIA documents during 1988 in the National Archives in Washington, D.C. and Suitland, Maryland, in company with James Sanders and colleague Thomas V. Ashworth, Sanders' maintained that certain documents had been inserted into the archival files which appeared to correct his and Paul's earlier assertions in print that 15,597 U.S. POWs in Soviet-controlled Austria and adjacent SACMED Theater areas, had remained in Russian control after the WW II prisoner repatriations. Earlier SHAEF and British cables in the archives appeared to indicate that the 15,597 POWs "U.S. account" were actually Soviet prisoners in American control awaiting repatriation. Since the discovery of these U.S. and British documents (also cited in this book) involving 15,597 Americans reported in Russian hands in Marshal Tolbukhin's area were prominently reported in both Paul's and Sanders' articles, their thesis was thus being reported as disproven by Admiral E.B. Baker of the Office of the Assistant Secretary of State for International Security Affairs, Richard Armitage. In obtaining and studying all the relevant documents then and since made available in the archives, the author concluded that even if the 15,597 first-reported U.S. POWs in Soviet-controlled Austria were actually Soviet prisoners awaiting repatriation under American control, and even if it was later estimated that only 5,500 Americans were believed to be under Soviet control in Austria on June 5th (as subsequently mentioned in a June 6, 1945 Moscow meeting attended by General John Deane), as noted earlier in this book, the documentary evidence still appears to show that an estimated total of 20,000 or more U.S. POWs and MIAs disappeared inside Soviet-occupied Poland, Germany and Austria in 1945, few of whom were subsequently repatriated. (By January 1946 this officially-estimated number had been reduced, largely through the "Presumptive Finding of Death"(PFOD) process, some identifications of prisoners who died in captivity, and at least several hundred "late returnee" Americans, to a reported 5,414 American known-POWs in the European Theater not returned to military control by January 1946).

General William C. Westmoreland, the former commander of U.S. forces in Vietnam, who had faced the Soviets with the 9th Infantry Division in 1945, and had fought the Communists in Korea, had encouraged the author's research in person and in writing since 1986, and in 1988 wrote:

"Thank you for the file of affidavits which I am reading with interest. The investigative work that you are doing on WW II and Korea sounds interesting. I look forward to learning about that effort and will be alert to what is published."[63]

During 1988 the author made many efforts to interest various influential American journals in publishing an analysis of the recently-declassified U.S. documents on American POWs and MIAs of earlier 20th century wars involving the Communists, and supplied many historically-significant documents to various editors, but none showed any interest. An October 1988 letter from Peter Grose, the managing editor of the Council on Foreign Relation's influential publication Foreign Affairs, in response to the author's query to editor William Hyland (a former assistant of Henry Kissinger), is typical.:

"Thank you for your proposal for an article on POWs and Forced Repatriation. I cannot offer much encouragement that this would be something we could use in Foreign Affairs, since we tend to shy away from basically historical studies in favor of more contemporary political and economic analysis. Yours is a topic of great personal interest to me, however, and I would welcome seeing what your research produces. I am sure you will find other journals in publishing whatever articles emerge from your work, and I would be happy to see copies of them. If I may, I will hold onto the documentation you sent me, for my files."[64]

It was to prove exceedingly difficult however, throughout the coming collapse of the Soviet empire, for the author to find other journals in America willing to publish the documentable facts revealed about U.S. POWs by the declassification process together with substantiating human sources, or willing to even investigate further and study the formerly-secret cables and reports then available.

The author's independent 1988 research into the fate of Soviet-held U.S. POWs of WW I, who had been captured by Lenin's Red Army in north Russia, caused a differing view to arise over the actual historical roots of American policy towards Communist-held U.S. POWs and MIAs; between James Sanders' subsequently well documented WW II origin theory, and Brown's and Ashworth's analysis of the declassified documentary evidence of Bolshevik conduct with U.S. POWs of the WW I American intervention in Russia.

## Notes on Chapter Twelve

1   *October Surprise*, Barbara Honneger

2   Codinha Box 11, Allen Depo.," records of the Select Committee on POW/MIA Affairs, U.S. Senate, courtesy of researcher William T. Bennett

3   These events occurred in 1992. see: p. 284, Report of the Select Committee on POW/MIA Affairs, U.S. Senate, January 13, 1993, U.S. Government Printing Office.

4   *Kiss The Boys Goodbye* and *the Bamboo Cage*; also interviews by the author with Scott Barnes and Lt.Col. James Gritz

5   FBIS analyses group report, February 1982

6   *By Way of Deception, The Making and Unmaking of a Mossad Officer*, by Victor Ostrovsky and Claire Hoy, St. Martin's Press, NY, 1990, p. 251

7   David Dimas, in his later book *Missing in Action-Prisoner of War*.

8   *Bamboo Cage*, p. 165, op. cit.

9   Draft of Senate Select Committee Report, Jan. 1993

10  Interview with Lt. Col. Andrew Genbara

11  Time Magazine, Nov. 9, 1992

12  The Chronology, The National Security Archive, Warner Books, pp. 234 & 231

1987; also: Washington Post 2/17/87 and 2/20/87

13   *The Bamboo Cage*, Nigel Cawthorne, Leo Cooper-Pen & Sword Books, London, 1991

14   *Rice Paddy Grunt*, Regnery-Gateway, Chicago and Lake Bluff, Illinois, 1986, and the Military Book Club, 1986

15   Washington Times, October 9, 1992

16   Interviews with LeBoutillier and Thornton and see: Wall Street Journal, October 15, 1985

17   Syndicated column by Jack Anderson, September 18, 1985.

18   Washington Post, August 31, 1985, "Claim That American POWs Were Sent to Siberia Probed," by George C. Wilson

19   Letter dated 29 January 1986, #-0119/VO-PW, to Department of State, Box 15804, 88D127, declassified through FOIA request

20   DIA Source 2623, sighting 2623A

21   Source 2638, DIA, E42C, obtained by the author through FOIA request

22   Source 2676, DIA, #420, obtained by FOIA request of the author

23   Letter to the author from the Defense Intelligence Agency, dated 24 November 1992, #U-5,179/PL/FOIA, signed by Robert P. Richardson, Chief, Freedom of Information Act Staff-DIA, with enclosures consisting of 43 Firsthand livesightings

24   Conversations between the author and Waple, Smith, Hendon, LeBoutillier and others in 1986-87, also: *The Bamboo Cage*, Nigel Cawthorne, 1991.

25   *Kiss The Boys Goodbye*, op. cit.

26   Letters of 1986-1992 from Westmoreland, in the author's files.

27   Hearings, Senate Select Committee on POW/MIA Affairs, January 21-22, 1992

28   Policy-Related Events, Office of Vice Chairman, Senate Select Committee on POW/MIA Affairs, September 21, 1992

29   Letter of 19 May 1986, from Major General John Murray (AUS-Ret'd) to John M. G. Brown

30   Associated Press, October 21, 1988; for further details on this case see: The Bamboo Cage, op. cit., pp.229-235, also: "An Examination of U.S. Policy Toward POW/MIAs," issued by the Senate Foreign Relations Committee Republican Staff on May 23, 1991.

31   Letter dated June 12, 1991

32   Report of the Select Committee on POW/MIA Affairs, pp. 178-179, January 13, 1993

33   *Kiss The Boys Goodbye*, p. 222, op. cit.

34   Author's and A.D.D. recollections and a letter of J. Thomas Burch, October 19, 1990.

35   *Kiss The Boys Goodbye*, p. 156., op. cit.

36   Congressman William Hendon, R.-NC, a long time advocate for live prisoners of war

37   DIA Memorandum # C-109/DC, September 25, 1985, Brooks to Shufelt

38   Letters from Charles Bates, December 14, 1987 and Vice President George Bush, January 6, 1988.

39   *Underground Empire*, by James Mills, Dell Publishing, NY, 1986, pp. 1064-1073

40  *A Nation Betrayed*, op. cit.

41  *Armand Hammer-The Untold Story*, by Steve Weinberg, op. cit.

42  Testimony of Richard T. Childress, Senate Select Committee, 12 August 1992

43  April 17, 1987 letter from the National Security Council

44  Associated Press report, Thursday February 12, 1987

45  *Kiss The Boys Goodbye*, op. cit.

46  Jack Anderson and Dale Van Atta syndicated column, February 23, 1987

47  Childress testimony, Senate Select Committee. 12 August 1992

48  *Kiss The Boys Goodbye*, op. cit.

49  Tower Commission Report, Washington Post- Walter Pincus, etc.

50  *Kiss The Boys Goodbye*, p. 343, op. cit.

51  12 August 1992 Childress testimony, Senate Select Committee on POW/MIA Affairs

52  Papers of James Cannon, Assistant to Howard Baker, delivered to Chief Counsel J. William Codinha, Senate Select Committee on POW/MIA Affairs, August 7, 1992

53  Copies of the original document obtained from the Select Committee on POW/MIA Affairs, U.S. Senate, in the author's files

54  Ibid.

55  *A Nation Betrayed*, by James Gritz, Lazarus Publishing, Boulder City, Nevada, 1988, 1989

56  The Senate Subcommittee, Terrorism, Narcotics and International Operations, II, p. 44; See also *Kiss the Boys Goodbye*, pp. 256-57, op. cit.

57  *Honored and Betrayed*, p. 349, op. cit.

58  1993 interview by the author, Mattole Valley, CA.

59  Letter of 27 April 1987 to Senator Paul Laxalt from the author

60  Policy Related Events, Office of the Vice Chairman, Robert Smith, Senate Select Committee on POW/MIA Affairs. September 21, 1992

61  *Kiss The Boys Goodbye*, op. cit.

62  *Bamboo Cage*, p. 142, op. cit.

63  Letter from General William C. Westmoreland, dated May 19, 1988, in the author's files.

64  Letter to John M. G. Brown dated October 5, 1988, from Peter Grose, Managing Editor of Foreign Affairs, in the author's files

# CHAPTER THIRTEEN

## THE POW/MIAS AND THE BUSH ADMINISTRATION

George Herbert Walker Bush was inaugurated as President in January 1989. He had been carried into office by a wave of general appreciation for the relative stability of the Reagan years, and despite the unpleasantries revealed about the Iran-Contra affair. The future looked brighter than in decades, and the greatest threat to America, posed by the Soviet Union, seemed to have perceptibly diminished under the leadership of Mikhail Gorbachev. Bush's lifelong friend and biographer, Fitzhugh Green, later wrote of the new President's method of management:

"While Vice President he also was assiduous in tracking the affairs of the world and America, reading everything and absorbing thousands of written and oral briefings. In his other years as diplomat and CIA chief in the executive branch, he was admired as a quick study who could retain what he had read and heard. Furthermore, he displayed a probing curiosity beyond what was merely proffered officially."[1]

If this is a reasonably accurate picture of Bush's method of operation then the events of the coming year, regarding American missing in action and prisoners of war, could not have escaped his attention, and there is some evidence that Bush was indeed planning for a methodical diplomatic process that would lead ultimately to resolving the POW/MIA matter.

Born in Massachusetts in 1924, George Bush was the son of Prescott Bush, a St. George's graduate who served in WW I and became an executive in Averell Harriman's investment bank in the mid-1920s. George Bush graduated from the exclusive Philips Academy at Andover in 1942; the address to the school that year was delivered by Secretary of War Henry Stimson, who advised the graduates to continue their education. But George Bush had determined to enlist, and by the middle of 1943 he had become the U.S. Navy's youngest pilot. He served for 9 months in the Pacific Theater, where he flew 58 missions as a carrier-based Grumman TBM torpedo-bomber pilot, from March to December 1944 and was decorated for valor. In September 1944, he was shot down near Chichi Jima and his two crewmen were lost. He was picked up by a U.S. Navy submarine and spent the next month on it, during a combat cruise. After returning to the U.S. in December 1944, and discharge from the Navy, he entered Yale in the fall of 1945.

At Yale he studied economics, became captain of the baseball team and was invited to join the elite and secretive Skull and Bones Club, which produced many influential American leaders. After graduation in 1948, Bush moved to Texas with his wife Barbara and entered the oil drilling business. His father Prescott was elected to the U.S. Senate from Connecticut in 1952, and in 1964 George Bush ran for the Congress from Texas, but lost. He ran again in 1966 and this time won; he was sworn in during January 1967, as the Vietnam War was becoming a major conflict. A Republican, he supported the war in Vietnam, and developed close ties to several other Congressmen including, Republicans William Steiger of Wisconsin, John Paul Hammerschmidt of Arkansas and several Democrats, including Gillepie V. "Sonny" Montgomery, of Mississippi (who would later head a Committee on the missing American POWs and MIAs of Vietnam.)[2] In 1970 George Bush ran for the Senate but was defeated by Lloyd Bentsen. Richard Nixon appointed Bush as Ambassador to the UN in December 1970 and asked Bush to serve as Chairman of the Republican National Committee in 1972. He was clearly being rapidly groomed for the highest level of public service. From this position, on August 7, 1974, with the Watergate scandal now a national crisis, Bush wrote Nixon and respectfully urged him to resign.

After Gerald Ford was sworn in, he appointed George Bush to represent the U.S. in the People's Republic of China, and late in 1975 Ford asked him to take over as Director of Central Intelligence; he was sworn in during January 1976. As CIA Director, and later as Vice-President under Ronald Reagan from 1981-1989, George Bush must have become aware of many state secrets that concerned U.S. diplomacy and covert actions, as well as classified facts regarding American war prisoners and MIAs. For a time, as Vice-President, Bush had in fact acted as the Administration's pointman on the POW/MIA issue, particularly in dealings with the outspoken Texas native, H. Ross Perot, as noted in the previous chapter.

As the Reagan Administration ended and the Bush era began, a final report of the Inter Agency Group on the POW/MIA situation during Ronald Reagan's term was released, that extolled an increase in the return of human skeletal remains by the Vietnamese which had resulted from the increased diplomatic emphasis resulting from the Vessey mission. This report, entitled the "Final Interagency Report of the Reagan Administration on the POW/MIA Issue in Southeast Asia," stated: "We have yet to find conclusive evidence of the existence of live prisoners, and returnees at Operation Homecoming in 1973 knew of no Americans who were left behind in captivity..." This was a patently untrue statement as was revealed by subsequent Senate testimony.

The surge of Soviet diplomatic activity about their own missing and POWs in Afghanistan, however, from mid-1988 to early 1989, hinted at a new possibility of Russian candor in the POW/MIA matter. A June 1988 report in the Los Angeles Times on Soviet-

Afghan POWs, quoted Alexander Sabov, foreign editor of the Soviet newspaper, LITERARY GAZETTE, as saying, "People want this war to be over and it won't be over they feel until all our boys are home." As Soviet troops were gradually withdrawing from Afghanistan, Moscow had begun a world-wide campaign to recover an estimated 200 or more Soviets actually held as POWs from among the 311 they admitted were missing in action. Mindful of the massacres and imprisonment of past Soviet POWs, Major General Valentin Khrobostov, a top political 'commissar' of the Soviet Armed Forces was quoted as saying in an appeal for the prisoners release: "No one is going to be punished. No one will be prosecuted...everything will be done to help them come home, to settle themselves..." Another official said that the Soviet government had, "A fundamentally new attitude to POWs than we did during WW II, an entirely new approach, and no punishment will be meted out."

In a June 22, 1988 article in "The Nation," Thailand's respected English-language newspaper, Dmitry Benediktov, head of the Soviet Red Cross reported that the Soviet Union believed some Soviet soldiers taken prisoner in Afghanistan were being held in the United States and Canada. At a Moscow news conference Benediktov said that Moscow did not have the exact figures on the number, but that, "There is information that certain POWs can be found in third countries including Iran, the United States and Canada." Moscow had publicly claimed earlier that 311 Soviets had gone missing in Afghanistan and that many were held as POWs by the anti-communist Mujuhedeen guerrillas. A U.S. embassy spokesman in Moscow was quoted as saying: "We are not aware of Soviet prisoners of war in the United States."

The Soviet domestic press was also publishing more of the truth of modern Russian history for their own domestic consumption, and during 1989 the USSR propaganda magazine SOVIET LIFE, published by governmental agreement in the US, admitted some of Stalin's crimes to the American public:

"Stalinism implies mass terror, contempt for human life, the massacre of millions of innocent people on political grounds... Stalinism means fraud on the state level, fabrication of "traitors against the homeland," misrepresented results of collectivization, falsified history of the party, the state and the world." The Soviet Union also publicly announced in April 1989 that Soviet troops sent to North Vietnam as advisers had shot down 24 U.S. aircraft and that some Soviet personnel had been decorated for their actions against American forces in combat in Vietnam.[3]

In Southeast Asia little changed in the U.S. government's glacial efforts to resolve the issue of surviving American prisoners held there. The biggest official news five months into the new administration, in June, was an announcement from the omnipotent

Colonel Paul Mather at JCRC in Bangkok that great progress had been made in the POW/MIA matter by the Vietnamese Communists, during a recent three-day visit to Hanoi by a 5-man U.S. team which he headed. Mather was quoted as stating, "We had a positive and productive meeting focusing on the issue of Americans missing from the Indochina period." Mather also said the Vietnamese had reported: '...THEY ARE NOW INVESTIGATING 32 INSTANCES IN WHICH THEY HAVE RECEIVED INFORMATION OR REMAINS WHICH COULD BE ASSOCIATED WITH LIVING AMERICANS.'[4]

The U.S. Government's problem was whether the families of "identified" MIAs would ever believe whose or what kind of remains they eventually received, or if they consisted of fragments of animal bones, such as had happened previously. If they trusted the Army-run Central Identification Laboratory in Hawaii, under Colonel Jonnie Webb and the forensic anthropologist-impostor, "Dr." Furue, they accepted the proffered remains, but there was always the possibility now, that a family might take the remains to an independent expert like Dr. Charney of Colorado State University, who would make a truly professional analysis and give an honest answer on identifications.

Months before, however, on January 20, 1989, just prior to the inauguration of George Bush as President of the United States, the maliciously clever Vietnamese had decided to send yet another subtle diplomatic signal to America that living POWs still survived in Southeast Asia, by releasing a Japanese monk who had recently been held with American prisoners of war in Vietnam. But, it was not to be until June 1989 that what he had seen during years of semi-starvation and brutality, was finally publicized in the Japanese and American media. His name was Iwanobu Yoshida, he had been held over 13 years in Hanoi-area prisons and he claimed to have been held recently in a cell with three American POWs, and that two other American prisoners were also confined there. The Buddhist monk claimed that in recent times, when he had been too ill to live much longer, the three American POWs brought him bananas they had gathered on labor-details, and thus he was able to survive.

A 65-year-old Zen Buddhist priest, Yoshida had gone to Vietnam as a missionary in 1965. Since being imprisoned after the 1975 Communist victory, he claimed to have seen 'many' Americans in different prisons saying: 'There are still lots of them....He claimed to have heard two officers speaking of 700 Americans still being held in Vietnam. 'I also heard them say that they won't ever release these men.'[5]

Yet, no perceptible alteration in the Bush Administration's POW/MIA policy resulted from Yoshida's information and the official U.S. Government position on his testimony from the living-dead was reflected in an August 17, 1989 letter from Deputy Assistant

Secretary of Defense William E. Hart, to a concerned U.S. Congressman, Gus Yatron of Pennsylvania:

"Mr. Ganshin Yoshida was arrested by Vietnamese authorities in 1975 and released in January 1989. U.S. Government representatives attempted to interview Mr. Yoshida immediately following his arrival in Japan. Because of his poor health and generally fragile condition, interviews were not then possible. In June (1989), Mr. Yoshida's daughter advised the U.S. Embassy in Tokyo that her father had informed her that during his imprisonment he had contact with persons he believed were Americans. Mr. Yoshida's daughter also provided her interpretation of her father's statements to the Japanese press. Upon the publication of these accounts, the Vietnamese embassy in Tokyo issued an immediate denial that Vietnam continues to hold U.S. prisoners of war. On June 12 and 22, U.S. representatives were able to interview Mr. Yoshida in a Sapporo, Japan hospital and reported:

"...HIS ILL HEALTH SERIOUSLY AFFECTED HIS ABILITY TO RECALL EVENTS WITH ANY CLARITY, DETAIL OR CERTAINTY... HAS DIFFICULTY IN SPEAKING AND IS UNABLE TO WRITE...NO CONCRETE DETAILS OF HIS INCARCERATION WERE OBTAINED... HIS REPORTED CONTACT WITH AMERICANS REMAINS UNCLEAR... INVESTIGATIONS BY U.S. AGENCIES HAVE DETERMINED THAT SOME PUBLISHED DETAILS OF MR. YOSHIDA'S IMPRISONMENT ARE ACCURATE AND CORROBORATED BY OTHER FORMER INMATES OF THE VIETNAMESE PRISON SYSTEM. IT IS CLEAR IN SOME INSTANCES HE WAS IMPRISONED WITH AMERICANS, NONE OF WHOM WERE PRISONERS OF WAR. INSTEAD THESE INDIVIDUALS ARE AMERICAN CIVILIANS WHO WERE DETAINED AFTER 1975..."

Thus, were most members of the U.S. Congress lulled into silence by Department of Defense officials and others about the discrediting of human sources of intelligence on live Americans still held in captivity in Vietnam up into 1989. The philosophy of the U.S. Government approach to livesightings of U.S. POWs was essentially unchanged since the 1930s era of Congressman Hamilton Fish and Assistant Secretary of State Robert Kelley: Just because there were prisoners of war alive yesterday does not prove they are alive today. Only in this manner could the U.S. Government avoid one or more diplomatic showdowns over POWs which could lead to a military confrontation with unknown international consequences.

Soviet officials in Afghanistan continued to press for an accounting of their MIAs and the secretly-held POWs in Mujahedeen control in the spring of 1989, during the first few months of George Bush's presidency. In an interview at the Soviet Embassy in Kabul, reported in the Boston Globe in April 1989, Soviet officials said that Soviet leader Mikhail Gorbachev had personally raised the MIA issue with President George Bush, in hopes that the United States would use its influence with the Afghan rebels to find those Soviets held as prisoners of war in Afghanistan and neighboring Pakistan and Iran. A high-ranking Soviet diplomat was quoted in the same report as saying: "Think what an emotional issue the question of MIAs is for Americans. Do you think our missing soldiers mean any less to us? You can say there is no higher priority for us than accounting for these men." An official of the International Red Cross was quoted in the Globe report on how many Soviets were estimated to have survived: "There is reason to believe that at least 100, and perhaps as many as 200, are alive and being held captive...The Russians are extremely keen on finding them." A Soviet embassy official in Kabul also said: "We've applied many times for help from the American government... This is something you Americans could do easily and for which the Soviet people would feel the deepest gratitude."

During this period the author had been obtaining and analyzing thousands more declassified U.S. government documents on POW/MIAs from the National Archives and decided to respond to the publicized Soviet requests for information about their own POWs with a full-length report on American POWs vanishing into Soviet captivity for 70 years. By early May 1989 the author had published, with research-colleague Thomas V. Ashworth, the first fully-documented (25,000-word) report on the history of Soviet-held U.S. prisoners of war to appear in the United States, from then-available declassified U.S. documents and substantiating human sources, which covered U.S. POWs and MIAs of World War I and the American intervention in Russia, World War II, the Cold War, Korea and Vietnam. This report was entitled: "A CHAIN OF PRISONERS—FROM YALTA TO VIETNAM" and subtitled: "A Secret that Shames Humanity," and it appeared as a special Memorial Day 1989 edition. The publisher was U.S. Veteran News and Report of the Homecoming II Project, headed by Vietnam veteran and long-time POW/MIA activist Theodore Sampley, with the assistance of Camp Brandenburg's leader, Vietnam veteran Robert Schmitt. Many other POW groups assisted in distributing it.

This report was noted with interest by several members of Congress, including Dante Fascell of Florida, and was reviewed by author William Stevenson, in the Toronto Sun, on May 27, 1989.[6] Leaders of veterans groups passed copies of it to Senators who kept abreast of the POW/MIA matter, such as Daniel Patrick Moynihan of New York.[7] It was quoted later in 1989 news articles in Canada on the subject of Soviet-held U.S. and British POWs of World War II, by the Toronto Star and in the Canadian Press across that country, but in America the author could not get it reprinted, even in an edited version, by any major publication. Former Ambassador Angier Biddle Duke, a nephew of the 1939-1944 U.S. Ambassador to Poland, Anthony J. Drexel Biddle Jr. (who had later served on Eisenhower's SHAEF staff and was present at the May 17-22, 1945, Halle POW

Conference with Soviet General Golubev), wrote a letter of confirmation to the author at this time: "...I do recall Uncle Tony's concern about the non-return of the POWs taken by the USSR...Your documentation and presentation are first-rate and reveal a chapter or chapters of our history of which we are so ashamed that we seem to prefer to draw a curtain of silence and forgetfulness over all..."

More than 50,000 copies of this May 1989 report were distributed to veterans and POW/MIA family members and other concerned citizens from all 50 states and to many foreign countries, through the efforts of U.S. Veteran News and Report, Homecoming Two, the Release Foundation and many other POW/MIA organizations. The Soviets also requested and received copies of it through their embassies in Ottawa, Canada and in Washington, D.C. Out of respect for the October 2, 1986 note to Anthony Duke, the author's information from the National Archives and some of the sources used within this report had previously been made available to then-Vice-President George Bush, by the author, in the fall of 1988. This occurred during an extended interview in Georgetown, Washington, D.C., with Mr. Bush's old friend and biographer, Fitzhugh Green, which the author attended at the request of Mr. Anthony D. Duke. The author's research colleague, Thomas Ashworth had also supplied two or three of the cited documents on WW II Soviet-held POWs to a reporter from the Arkansas Gazette who showed them to President Bush during a 1988 campaign stop in Arkansas.

Co-author J.M.G. Brown proposed in this May 1989 report, that linkage be established between the Soviet government's expressed interest in recovering their POWs from Afghanistan and the U.S. interest in obtaining information on the fate of U.S. POWs from Vietnam who had disappeared while alive and in captivity. Former Congressman and POW/MIA advocate William Hendon also adopted this theme, and President George Bush accepted the concept of linkage during July 1989, referring to it in a speech to the National League of Families, the substance of which was published in the "Washington Times." Former Congressman Hendon also met with anti-Soviet Mujahedin guerrillas in Pakistan and encouraged them to help the U.S. return or account for the 311 or more Soviet MIAs in Afghanistan.

Investigative reporter Bill Paul challenged Mikhail Gorbachev over the POW/MIA matter by again raising the issue of U.S. WW II and Korean war POWs who had disappeared in Russian control, in a Wall Street Journal article entitled "Will Gorbachev Find WW II's Long-Lost POWs?" In noting Gorbachev's revelations of Soviet complicity in the imprisonment and death of WW II Swedish diplomat Raoul Wallenberg, who had assisted many Jews in escaping the Nazis in Hungary, Paul called attention to the 5,414 U.S. POWs known to have been in German camps at the end of World War II, who were still not accounted for by February 1946. He reiterated the importance for the United States of resolving the continuing livesightings of American POWs in Vietnam and Laos, and emphasized that Hanoi may have stripped American POWs of their citizenship and sentenced them as war criminals.[8]

A great convulsion had meanwhile shaken Beijing, capital of the world's most populous nation, in June 1989, as the Chinese Communist Army, on orders from strongman Deng Xiaoping, attacked and crushed a massive pro-democracy gathering of students and their supporters in Tiananmen Square. The crisis had been building for weeks, and the use of video cameras, fax machines and other high-tech equipment had facilitated communication with the outside world despite all efforts to seal off the nation, vividly recording the massacre and the widespread security police crackdown that followed. On-the-scene videotaped images of idealistic young men and women from Chinese universities grouped around a "goddess of Democracy," and then being machine-gunned or crushed by armored vehicles in the assault were flashed to every corner of the world. Once again the mask of Communist totalitarianism had been torn away and the whole world could see the naked truth. Although some of the major media in America at first downplayed the violence by reporting that only "hundreds" had been killed, it soon became clear from on-the-spot reporting that the dead numbered in the thousands. For months afterwards arrests, interrogations under torture, and public executions by security forces continued in cities across China, and were reported or pictured in the Western media.

Thousands of dissidents were confined in prisons and forced labor camps and thousands more fled China. The Western technology previously supplied to Chinese authorities was shown to have been of great assistance to China's security forces in filming and recording those among the masses who participated in, or showed sympathy for the pro-democracy protests. While the widespread revulsion of the people of America was obvious from the upwelling of public fury against China's rulers, the Bush Administration's response was sadly muted. Once again the dictates of foreign policy affected what should have been a unanimous condemnation by U.S. officials, that would have reflected the basic values that America had long stood for. The overriding security concerns about the Soviet threat which had led to Nixon's surprise 1972 China initiative apparently again influenced the United States Government's relations with Beijing, so that a balance would be maintained in America's relations with the two communist superstates.

However, later in 1989, Secretary of State James Baker raised the issue of linkage between Soviet POW/MIAs of the Afghan war and U.S. POW/MIAs of the Vietnam war, in a Paris meeting on Cambodia with Soviet Foreign Minister Eduard Shevardnadze. Such a message could hardly be ignored by either the beleaguered Soviet government of Mikhail Gorbachev, or by China's hard-line communist rulers.[9] The conservative New American Magazine published an article by the author in September 1989 on the historic origins of the POW issue from and the linkage that had been established between Soviet POWs missing in Afghanistan and American POWs missing in Indochina. By late 1989, with the assistance of a small group of Americans in Pakistan, which included a Vietnam veteran of the 25th Infantry Division and former-policeman from Boston named Thomas Flaherty, two former Soviet POWs who had been held by the Afghan Mujahedeen were freed and returned to the USSR, in a clear diplomatic gesture of cooperation.[10]

The Seattle Times reported on the prisoner of war research of Thomas V. Ashworth and the author on September 25th, in an article entitled: "Dark Chapter in History Revealed?" It was mentioned within that the Defense Intelligence Agency had written to a Texas Republican member of Congress, Richard Armey, that stories of the U.S. Government abandoning American POWs in Soviet control weren't true, and had falsely claimed that U.S. and British contact officers had been assigned to Russian units at the time, to ensure that repatriation of U.S. and Allied POWs went "as smoothly as possible." The newspaper pointed out that recently-declassified U.S. 1945 documents located by Ashworth and the author in the National Archives (quoted earlier in this book) indicated this was not true. In addition the account mentioned the original withholding of American prisoners by the Soviets after World War I.[11]

Not long afterward, in November and December 1989, investigative reporter Mark Sauter of Washington State contributed to the POW/MIA expose by airing a TV documentary about the Korean war POWs who had disappeared in Chinese communist and Soviet control, entitled: "Secret Prisoners," on KIRO-TV, a CBS affiliate, and followed this up with a December 1st article in the upstart capitol newspaper, the "Washington Times" (owned by the South Korean Sun Yung Moon) which had published several significant exposes on POW/MIAs since 1987. This series and earlier research by others encouraged Republican Congressman John Miller of Seattle to become more prominently involved in the POW/MIA issue, as it related to both the USSR and communist China, and substantiated earlier published reports by Paul, Sanders, Brown and Ashworth. U.S. Representative Nancy Pelosi of California also lent strong support to Miller in urging that the People's Republic of China account for American POWs of the Korean War and later.

The Bush Administration unfortunately chose this period to demonstrate U.S. support for the Beijing regime, only six months after the massacre of thousands of pro-democracy demonstrators at Tiananmen Square had shocked the world. Despite widespread feelings of outrage in America among both conservatives and liberals towards the Chinese communist government, certain U.S. leaders were determined to be the first to show support for Deng Xiaoping, among them former President Richard Nixon and former Secretary of State Henry Kissinger, both of whom retained great influence in Washington. Both had wished to be the first to make a high-profile visit to China since the Tiananmen Square massacre. Kissinger had canceled a planned September visit following publication of a Wall Street Journal article about his business interests, which included a $75 million firm called China Ventures.

Nixon began a visit to China three weeks later, during which an aide artfully emphasized that Nixon had no Chinese business interests, thus taking the "moral high ground." This had reportedly infuriated his former Secretary of State, who apparently craved the international attention he had received in the 1970s. Then, at two o'clock in the morning on December 9th, the Bush Administration announced that National Security Advisor Brent Scowcroft and Deputy Secretary of State Lawrence Eagleburger, both former aides to Kissinger at the Paris peace talks, had secretly traveled to China and were already in Beijing. According to Time Magazine, "...Scowcroft addressed the Chinese rulers as 'friends,' referred oh-so-delicately to 'the events at Tiananmen,' and described U.S. critics of the massacre as 'irritants' to Chinese-American relations." Scowcroft and Eagleburger were both former executives in Kissinger's private consulting firm, which in the time-honored Washington D.C. tradition had made millions through the influence of such government officials. (Also, on December 12th, perhaps not coincidentally, forced repatriations to Hanoi of Vietnamese refugees in Hong Kong commenced under British authority.)

The obedient soldier as ever, General Scowcroft embarrassed freedom-loving Americans everywhere by warmly clasping the hand of the bloody dictator Deng Xiaoping in both of his own, while beaming broadly. Almost obscenely, Scowcroft indicated to Beijing's rulers the Bush Administration's view that the major problem in Chinese-American relations was the U.S. Congress! White House spokesman Marlin Fitzwater explained Bush's position:

"The President knew he would be criticized for this, but he feels strongly that it's in our national interest to improve relations with China. He feels he knows China as well as anybody—and better than his critics in Congress."[12] While Scowcroft had already proved his ruthlessness years before when he acquiesced to the secret communist withholding of American POWs in 1973, his 1989 actions in Beijing indicated to many observers just how cut off from American public opinion the Bush White House had become. The Democratic party and prominent individual Democrats with a professed dedication to human rights did not fail to notice the

Orwellian unreality of official White House and presidential statements about Scowcroft's and Eagleburger's mission, which may have actually helped lay the groundwork for the Republican Party's defeat that was to occur in 1992.

As a combat veteran of the Vietnam War, author John M.G. Brown (with fellow researcher and Vietnam veteran Ashworth), concentrated for more than a year on urging the three most powerful national veterans' groups to publish an independent historical analysis of the declassified U.S. documentary evidence and human source confirmation of Soviet-held U.S. POWs, which had been uncovered thus far. These were: the Veterans of Foreign Wars, The American Legion and the Disabled American Veterans, and none had ever dealt with the U.S. POWs secretly withheld by the Soviets after America's 20th Century wars.

The first major crack in the wall of silence on this issue, which had stood unbroken for so many decades in all of these organizations, appeared in the February 1990 Veterans of Foreign Wars-VFW Magazine, with publication of the author's documented research on Soviet-held POWs of WW II, Korea and Vietnam in an article entitled, "HIDDEN POWs OF THE COLD WAR." This was accomplished with the assistance of a fellow Vietnam veteran and nationally-respected VFW officer, Richard Keeton. The article resulted in an objection from the VFW's National Security and Foreign Affairs Director, Kenneth Steadman, a former military officer who had testified at the Murkowski hearings. Steadman was described by the magazine's editor, Richard Kolb, in a note to the author, as being "hopping mad" over its publication. (It is likely that Steadman was simply unfamiliar with the recently-declassified documentary evidence). The VFW's Director of Publications, Wade LaDue, subsequently resigned, effective March 15th, and his duties were assumed by Kolb. The information in this report, drawn from declassified U.S. documents, revealed that many thousands of American, British and French POWs of WW II had been "held as hostages by the Russians" in 1945, followed by the history of similar Soviet actions after Korea and Vietnam, and reached some 2,100,000 American veterans and their families in all 50 states and in many overseas areas. This had the effect of legitimizing the author's and others earlier published research. The Soviet and American governments could not have failed to take note of the influence of the organization that published it. VFW Executive Director, Larry Rivers, subsequently arranged for Vice President Dan Quayle, Secretary of Veterans Affairs Ed Derwinski, Senator Frank Murkowski and others to address the March 1990 VFW National Conference, but the subject of Soviet-held American POWs was avoided.

Although the author had written and documented a longer and

more comprehensive report for the American Legion Magazine on the history of U.S. POWs held by Communist regimes after four wars, beginning in Bolshevik Russia during WW I, the Legion declined to publish it, despite the author's Alaskan years of membership in that organization and the active assistance of the Legion's Oregon POW/MIA Chairman, a dedicated ex-Vietnam war U.S. Marine named Carl Rice. National Commander Miles Epling, also a Marine veteran of Vietnam, wrote to the author on June 5, 1990:

"Upon my return from Europe, I learned that the Editor-in-Chief of the American Legion Magazine had informed you that your article, "An American Sacrifice For Peace," would not be printed in our magazine...MR. WHEELER ADVISES ME THAT HE CONSULTED WITH OUR NATIONAL SECURITY-FOREIGN RELATIONS STAFF BEFORE MAKING HIS DECISION...AS NATIONAL COMMANDER, I DO NOT INTERFERE WITH THE EDITORIAL POLICIES OF THE AMERICAN LEGION MAGAZINE..."

The Magazine's editor, Daniel S. Wheeler, had also written to the author, the day before: ...WE TOOK THE LIBERTY OF FORWARDING IT TO OUR NATIONAL SECURITY DIVISION FOR THEIR COMMENTS... IT WAS WELL WRITTEN AND INFORMATIVE, AND YOU APPEAR TO HAVE DONE A GREAT DEAL OF INVESTIGATION INTO THE U.S. GOVERNMENT'S HANDLING OF THE FATES OF AMERICAN POWs AND MIAs. I'M SORRY TO TELL YOU THAT THE MATERIAL DOES NOT MEET OUR CURRENT EDITORIAL NEEDS...SUCH REPORTS ARE GENERATED AND RESEARCHED BY THE LEGION'S NATIONAL SECURITY DIVISION STAFF, OR OUR OWN MAGAZINE STAFF."[13]

Such an admission by a national magazine that spoke officially for some three million other American veterans, was a clear indication that control of what recently-declassified POW/MIA information would reach the membership of influential veterans' organizations, was still viewed internally by their national headquarters staffs, as a "national security matter," and thus censored, indicating a close cooperation with U.S. intelligence agencies. This longer article, covering the period from the Bolshevik Revolution through Vietnam, was ultimately published by a smaller organization, the Veterans of the Vietnam War, which was headed by a disabled Vietnam Veteran and long-time POW/MIA advocate, Michael Milne of Pennsylvania. Subsequently, Richard Christian, a former Army officer in the American Legion's national leadership who understood the historic significance of the recently declassified documentary evidence on U.S. POWs being published by American researchers, did credit to the organization by urging that the POW/MIA matter be thoroughly investigated, and that the membership be kept fully informed, which was subsequently done.

Other publications which rejected the author's short articles or full-length reports on the history of the POW/MIA matter, even when accompanied by substantiating declassified U.S. documentary evidence, included the Atlantic, the New Republic, the New York Times, National Review, the American Spectator, Commentary, Newsweek, U.S. News and World Report, Time, Reader's Digest, Foreign Affairs, The San Francisco Chronicle and Examiner, The Saint Louis Post-Despatch, The Des Moines Register, The Chicago Tribune, The Los Angeles Times, The Boston Globe, The Anchorage Times and many others across the United States, from the 1988 period, through 1992.

One typical example, from Executive Editor William German, of the San Francisco Chronicle, dated November 21, 1990, concerned an article proposed to that newspaper on November 9th, with substantiating declassified U.S. documentary evidence from the National Archives and human sources, which was subsequently published by The Sunday Oregonian in December 1990: "Thanks for submitting this to us. The editors of several feature sections have read it, but feel it is not the sort of story we can publish in the Chronicle. I think it is more suited to a magazine with a special interest in history."[14] This newspaper, and the others noted previously, initiated no investigation of the author's declassified documentary evidence or human sources, nor did any of them have a reporter contact the author for further clarification of the story.

Yet, in this era of officially-promoted censorship of the POW/ MIA matter, apparently condoned by much of the major media in America, the Soviet government had finally admitted publicly on April 13, 1990, that Stalin's NKVD secret police had shot the thousands of Polish POW officers at Katyn Forest in 1940. The Soviet government blamed the secrecy of the NKVD's records-keeping for hiding the truth so long, in the statement translated by the official news agency Tass:

"The need to elucidate the circumstances of the death of Polish officers interned in 1939 has long been raised at meetings between representatives of the Soviet and Polish leadership in broad public circles. Historians of the two countries have conducted a careful investigation of the Katyn tragedy, including a search for documents.

"Just recently Soviet archive workers and historians discovered some documents concerning Polish servicemen who were kept in the Kozelsk, Starobielsk and Ostashkov camps by the NKVD security police. It follows from these documents that in April-May 1940, 394 of the 15,000 or so Polish officers were transferred to the Gryazovetsk camp. The larger part however, were turned over to the NKVD administrations in the Smolensk, Voroshilograd and Kalinin regions and never mentioned in Statistical accounts since. The discovered archival material puts direct responsibility for the atrocities in the Katyn Forest on (secret police chief) Lavrenti Beria, (V.N.) Merkulov and their henchmen. The Soviet side, expressing profound regret over the Katyn tragedy, declares that this tragedy is one of the gravest crimes of Stalinism. Copies of the discovered documents have been passed over to the Polish side the search for more archival material continues."

The Polish government acknowledged in an official statement that the Soviet admission was long awaited but that "reconciliation can only be built on the truth," and demanded "clarification of all so-called blank spots in our common history." Relatives of thousands of other missing Poles who had disappeared in Beria's camps.[15] Less than two weeks later, on April 27, a senior KGB officer, Major-General Alexander Karbainov, told a Japanese newspaper that 3.5 million people had been executed by the Soviet government since the 1917 Bolshevik Revolution. The Japanese newspaper Sankei Shimbun reported that this was the first time the KGB had publicly revealed how many people had been executed since the revolution. Karbainov also told the Japanese newspaper 29 of 30 spies arrested since the institution of "perestroika" reforms in 1985 had been executed.[16]

In the May and June 1990 issues of the conservative magazine "New American," the author published the full history of Soviet-held POWs from the time of Lenin and Trotsky to the "Moscow-bound" POW/MIAs of Vietnam, entitled: "Mikhail Gorbachev, Let Our People Go," which resulted in further interest on the part of Soviet embassy officials in North America. Between the fall of 1989 and October 1990 the author initially demonstrated and then turned over extensive documentary and human source evidence on POW/MIAs to Senator Charles Grassley of Iowa, and to staff investigators of the Senate Foreign Relations Committee. The author and research colleague Thomas V. Ashworth had met in a Senate office with Horace Coleman, a Secret Service agent assigned to protection of President Bush, in October 1989, along with Army Criminal Investigation Division agent Tracy Usry, Mr. Harvey Andrews and with other officials, to demonstrate hundreds of declassified U.S. POW/MIA documents dated from 1918-1980.

As a result of personal 1987 correspondence on the POW/MIA subject between the author and Washington Post owner and editor-in-chief Donald E. Graham, the author was permitted to respond to an editorial titled, "No MIA Conspiracy," written by the powerful Washington Post's own deputy editorial page editor, Stephen Rosenfeld. The author briefly covered the entire history of Communist-held POWs since 1918 in a short Washington Post column entitled "THE POW CONSPIRACY," on September 15, 1990, emphasizing the irony that the history of U.S. POWs in Soviet control after WW I, WW II, Korea and Vietnam had for long been "classified" and that those uncovering the historical roots of the POW/MIA dilemma were duty-bound to do so, but that those U.S. officers and officials keeping such secrets were also merely doing their duty, by following orders; and that the post-Korean war Soviet nuclear threat against the entire United States had been a consideration in maintaining such secrecy about Soviet-held POWs of Korea and Vietnam.

A few days later, on October 10, 1990, the author turned over some 350 historically-significant, declassified U.S. POW/MIA-related documents dated from 1918 to 1980, to Senator Grassley and his top aide, Kris Kolesnik, and also met with Grassley on the subject of Soviet and Chinese-held U.S. POWs and MIAs. In a letter to the author that evening, Grassley wrote: "I want to thank you for your generosity in contributing your research and many long hours of research to our POW/MIA investigation. The documents you have provided pertaining to POWs of World War II, the Korean war and the Vietnam war will offer us an invaluable historical context as we seek the truth..." A subsequent October 29, 1990, Senate Foreign Relations Committee Republican Staff Interim report on POWs mentioned the historic precedents in the POW/MIA matter and stated that the Defense Intelligence Agency (DIA) knew that at least several hundred U.S. prisoners (of the 1,259 POW/MIAs reported captured but not repatriated, or the 1,100 others listed as killed-body not recovered-KBNR), were alive in enemy control in 1974, after the final prisoner returns. Since the end of the war over 1,400 first-hand eyewitness livesighting reports of American POWs remaining in enemy captivity after 1973 have been recorded by the U.S. Defense Intelligence Agency, but not a single one has been judged to be credible by the U.S. Government, which actually meant, by the POW/MIA Office of the Defense Intelligence Agency.

In the Portland, Oregon, "SUNDAY OREGONIAN," of December 2, 1990, the author published the first article to appear in any major-metro daily newspaper in the United States, that revealed the entire, long-classified history of Soviet-held U.S. prisoners, from WW I in north Russia, through WW II and Korea, to the Moscow-bound Americans of Vietnam, shipped from Vinh to Sam Neua and on to the USSR. Retired NSA analyst Jerry J. Mooney was identified within as a significant source on the longtime existence of an unacknowledged 'wink and nod diplomacy,' between the Soviet and U.S. governments, which had resulted in an official ignoring of most prisoners of war taken by either side, all through the Cold War period.

In the meantime, an Iraqi invasion of Kuwait had led to an international crisis and finally to a vote by the Congress on January 12th, 1991, granting President George Bush the authority to go to war. Bush gave the Iraqi leader until January 15th to comply with U.N. resolutions and withdraw his army, but Saddam Hussein refused. In the brief war that followed, American forces under the command of General H. Norman Schwarzkopf, together with British, French, Saudi and other allies, defeated Saddam Hussein's army and air force and occupied part of southern Iraq to the Euphrates River.

American casualties were extraordinarily light in the 42-day long Persian Gulf War, totaling a reported 137 killed in action and 7 missing in action, including some killed by 'friendly fire.' Tens of thousands of Iraqis were killed, largely by aerial bombardment, during a 38-day air war that was conducted in three phases. The estimates of the total number of Iraqi deaths varied widely, but appear to have been some 80.000 dead. The ground war involving U.S. and allied armor and infantry lasted only four days before President Bush declared a cease-fire which ultimately permitted many of Hussein's elite forces to escape a rapidly approaching encirclement. This permitted him to subsequently conduct offensive operations against the Kurdish minority in northern Iraq, who had been encouraged by the American CIA to rebel against the central authority in Baghdad.

The American public once again viewed films of obviously battered U.S. POWs in enemy control being used for propaganda purposes, female American prisoners of war later admitted to being raped, and questions were raised as to how many U.S. prisoners the Iraqis were actually holding. The author received information from a confidential source that some Americans reported as "killed," including Army, Marine and Navy SEAL reconnaissance personnel and airmen, were actually taken prisoner and failed to reappear. This information could not be verified by the author as much information on the casualties remains classified, although a subsequent Senate Foreign Relations Committee staff report on U.S. POW/MIA policy stated that "the war was so brief and powerful that all prisoners were returned without question." Yet, this report cited specific cases in which the standards set by U.S. officials for "positive identification" of U.S. prisoners in enemy control had been set unrealistically high. Some U.S. personnel who had actually been captured by the enemy were initially listed as "duty status unknown" and then as "missing," but not as "missing in action." ("Missing" is reserved for personnel unaccounted for in non-combat operations.)[17]

According to the Senate report, two drivers of a Heavy Equipment Transport (HET) captured by the enemy, "were never listed as MIA or POW, even though the Army had information that they had been captured under fire." These two missing Americans, Specialist Melissa Rathbun-Nealy, a Caucasian female, and Specialist David Lockett, an African-American male, were subsequently released by the Iraqis. The same Senate report also stated:

"Inaccurate battle casualty reporting resulted in the next-of-kin of Daniel J. Stomaris and Troy A. Dunlop being officially notified by DoD that the soldiers had been killed in Action (KIA); in fact, these men were slightly wounded or taken prisoner by the enemy, but were not listed as POW or MIA or KIA; their subsequent release by the Iraqis came as a surprise to the American public and the national media.[18]

With the increased media attention on POW/MIAs and the unceasing efforts of investigators like Paul, Sanders, Brown, Ashworth and Sauter, together with film producer Ted Landreth's pioneering work, a crack appeared in the armor of the Department of Defense in early 1991. Generations of U.S. officials and military officers involved with resolving the fate of American POWs had faced a painful moral dilemma as to where their duty lay, when remaining silent about such reported U.S. prisoners for security reasons. But the predicted ending of the Cold War and repeated publication of the truth in courageous segments of the American media was beginning to take a toll which was to begin with the February 1991 protest resignation of Defense Intelligence Agency (DIA) POW/MIA Chief, Colonel Millard Peck. This was suppressed at first, but later printed at the end of the final Senate Foreign Relations Committee Republican staff report on U.S. POW/MIA Policy, in May 1991. In his resignation memorandum of February 12th, Peck admitted that as a combat veteran of the war in Vietnam he had felt compelled to accept the position in the hope of ultimately resolving the issue, but instead found:

"...I was not really in charge of my own office, but was merely a figurehead or whipping boy for a larger and totally Machiavellian group of players outside of DIA. What I witnessed during my tenure as the cardboard cut-out "Chief" of POW-MIA could be euphemistically labeled as disillusioning...That National leaders continue to address the prisoner of war and missing in action issue as 'the highest national priority' is a travesty. FROM MY VANTAGE POINT, I OBSERVED THAT THE PRINCIPAL GOVERNMENT PLAYERS WERE INTERESTED PRIMARILY IN CONDUCTING A 'DAMAGE LIMITATION EXERCISE,' AND APPEARED TO KNOWINGLY AND DELIBERATELY GENERATE AN ENDLESS SUCCESSION OF MANUFACTURED CRISES AND 'BUSY WORK'...THE MINDSET TO 'DEBUNK' IS ALIVE AND WELL. IT IS HELD AT ALL LEVELS AND CONTINUES TO PERVADE THE POW-MIA OFFICE, WHICH IS NOT NECESSARILY THE FAULT OF DIA. PRACTICALLY ALL ANALYSIS IS DIRECTED TO FINDING FAULT WITH THE SOURCE. RARELY HAS THERE BEEN ANY EFFECTIVE, ACTIVE FOLLOW THROUGH ON ANY OF THE SIGHTINGS...IT APPEARS THAT THE ENTIRE ISSUE IS BEING MANIPULATED BY UNSCRUPULOUS PEOPLE IN THE GOVERNMENT, OR ASSOCIATED WITH THE GOVERNMENT... THIS ISSUE IS BEING CONTROLLED AND A COVER-UP MAY BE IN PROGRESS..."[19]

After Peck's resignation was made public, Secretary of Defense Richard Cheney (who had been Gerald Ford's Chief of Staff) announced there was "no foundation to support" Colonel Peck's charges. Ronald J. Knecht became the latest in a long line of U.S. intelligence officials assigned to conduct an investigation of DIA's handling of the POW/MIA matter by the Pentagon. Knecht had reported to the Secretary of Defense that Colonel Peck's allegations were groundless. After several members of Congress pressed the Pentagon on Peck's charges, Assistant Secretary of Defense Duane Andrews had ordered this inquiry by his aide Knecht, a former NSA official. Knecht attempted to undermine the credibility of the ex-POW/MIA Chief at DIA, claiming in his report that Peck was a self-serving incompetent, who had no evidence to back up his claims, and that the Colonel had urged a CIA operation to compromise or kidnap a top Hanoi Politburo member, or threaten a U.S. tactical nuclear strike to force the Vietnamese to release the POWs they had been holding.

Since the post-Korean War period, the development of the Soviet nuclear threat against the United States had kept the issue of Soviet and surrogate-held U.S. POWs at the highest level of national security concerns. This insured that the National Security Council remained in control of the POW/MIA issue and coordinated actions regarding POW/MIAs between the Defense Department's DIA and ISA (International Security Affairs) and the State Department, CIA, NSA, FBI and other agencies. This has been confirmed to the writer by letters and documents originating at the NSC.[20]

The Vietnamese Communists had meanwhile been pushing harder for U.S. diplomatic recognition of their regime, and from February 10th to 15th, 1991, Vietnamese Justice Minister Phan Hien and other Vietnamese officials met with U.S. senators, congressmen, and the U.S. State Department's Chairman of the POW/MIA Interagency group (IAG), Kenneth Quinn, in Montego Bay, Jamaica. Vietnamese officials privately expressed the view that they had received the impression from the meetings that the lifting of the U.S. embargo and normalization of relations could be achieved without preconditions on either side. This apparent caving-in of the U.S. position on MIAs leaked to the public, however and caused the U.S. Government to announce on April 8th, with some fanfare, a "Roadmap," which was presented to Hanoi, outlining a step-by-step process to normalize relations with Vietnam, if Vietnam satisfied conditions along the way regarding Cambodia and the POW/MIA issue. The conditions on the POW/MIA issue were reportedly inserted into the "road map" document after strong pressure from Ann Mills Griffiths of the National League of POW/MIA Families, who was a member of the U.S. Government's Interagency Group on POW/MIAs headed by Kenneth Quinn.[21]

In early 1991, at the request of Walter B. O'Reilly of the National Forget-Me-Not Association For POW/MIAs, an organization of POW/MIA family members and veterans (such as Colonel Vincent Donahue, father of MIA Morgan Donahue), the author had assisted Thomas Flaherty, an American Vietnam War veteran representing that organization and living in the USSR, by supplying him with 300 historically-significant declassified U.S. POW/MIA-related documents, dated from 1918-1986.These were intended for delivery by Flaherty to Soviet researchers in Moscow and to members of he Supreme Soviet. (Flaherty had previously assisted in the repatriation of two Soviet POWs from Afghanistan to Russia in November 1989, and with others thereafter.) In a subsequent English edition of the Moscow News Weekly (no. 15), of April 19, 1991, a significant invitation was extended by Magadan authorities (in the Kolyma region of far eastern Siberia) to relatives of American, French and Japanese "former POWs who died in the Soviet concentration camps" to visit their loved ones' protected gravesites. Another article on Soviet-held U.S. POWs appeared later in a September 1991 issue of Moscow's "Nesavisimaya Gazeta."

The author also worked extensively on documenting the existence of Soviet-held U.S. POWs of WW I, WW II, Cold War, Korea, and Vietnam with Senate Foreign Relations Committee investigators Daniel Perrin and Tracy Usry. The original proposed "final report," which consisted largely of undocumented allegations, was received by the author through efforts of Senator Grassley's aide Kris Kolesnik and two retired U.S. Army Intelligence officers, Colonel William LeGro and Lt. Colonel Andrew Gembara. The author fully documented virtually every page of the historical section of the final report with many lineal feet of declassified U.S. documentary evidence attached to each page, sent by express in individual envelopes, and wrote proposed additions, revisions, and corrections for it, which were sent by fax from March through late May 1991. Based upon all the documentary evidence supplied on U.S. POWs of WW I, WW II, Korea and Vietnam, the author urged upon investigators Daniel Perrin and Tracy Usry inclusion of a reference in the report to "Moscow-bound" U.S. POWs from Vietnam, using a declassified 1971 CIA cable on interrogations of Americans by Soviets in North Vietnam and Mr. Jerry J. Mooney as a confirming human source, which was agreed to. The evidence used was essentially the same as had been supplied to the Washington Post Company and Newsweek Magazine, and to the Soviet researchers.

The result of the 1989-1991 Senate Foreign Relations Committee Republican Staff investigation was a May 23, 1991 report entitled: "AN EXAMINATION OF U.S. POLICY TOWARD POW/MIAs." The findings of this report, published in various newspapers across the country, were that the United States Government had administratively written-off thousands of U.S. POWs and MIAs of WW I, WW II, the Cold War, Korea and Vietnam, who had actually fallen under Soviet or Soviet/surrogate control. This was the first time that any official report by any part of the United States Government had drawn such conclusions, and thus the report was historically significant. Since virtually all of the archival documents and human sources used by the Senate investigators in the historical section of the report had been supplied by the author and research-colleague Ashworth, the historical conclusions largely conformed to earlier chapters of this book, and it is therefore not necessary to review them here.[22] The Senate Foreign Relations Committee staff report on POW/MIA policy also recommended the creation of a Senate Select Committee to further investigate the prisoner of war and missing in action issue. This was to be acted upon by the full Senate within ten weeks' time.

In addition to including an important reference to the "Moscow-bound" POWs of Vietnam which had been proposed to Perrin and Usry and documented by the author, the report stated: "IN REVIEWING HUNDREDS OF RAW INTELLIGENCE FILES ON THE 1,400 REPORTS, MINORITY STAFF INVESTIGATORS FOUND A PREDISPOSITION BY DEPARTMENT OF DEFENSE (DIA) EVALUATORS TO IGNORE CORROBORATIVE EVIDENCE, AND LITTLE INTEREST TO FOLLOW UP WHAT NORMAL SEARCHERS WOULD CONSIDER AS GOOD LEADS..."

Postwar U.S. government negotiations with the Vietnamese over the missing POWs had resulted (at least publicly) only in the return of human remains. In specific cases documented for the U.S. Senate Foreign Relations Committee investigation of 1991 by a renowned forensic expert, Dr. Michael Charney, Asiatic and non-human remains have been misidentified as those of specifically-named, missing U.S. POWs. As in the precedent-setting 1919-1929 U.S. MIA cases which the author had fully documented for the Foreign Relations Committee, and which were included in the report, recovered human skeletal remains appear to have once again been assigned to the names of those who remained in Communist control in Vietnam. According to the report:

"Two methods are used by DOD (Department of Defense) to account for missing Americans. One is the statutory presumptive finding of death in individual cases; the other is categorizing casualties as Killed in Action, Body Not Recovered (KIA-BNR). In either case, when human remains are repatriated from Southeast Asia, they are identified against persons in these two categories. When an identification is made, the individual is accounted for as having died while in the Indochina war zone. Individually, members of the military services, or U.S. Government employees who were missing while serving in Indochina and remain unaccounted for, can be declared dead by the secretary of the military service or head of the government agency responsible for that individual."[23]

"...Basically, the U.S. Code permits the secretaries and/or heads of agencies to declare an individual dead after that person has been missing for 12 months under circumstances indicating he or she MAY have died...BOTH PRESUMPTIVE FINDINGS OF DEATH AND KIA-BNR

STATUS STRONGLY PREJUDICE SUBSEQUENT EVALUATIONS OF LIVESIGHTING INFORMATION. FOR EXAMPLE, LIVE-SIGHTING INFORMATIONISMUCHMORELIKELYTOBEDISREGARDEDINTHEFIELD AS A RESULT OF AN INDIVIDUAL HAVING BEEN ALREADY BEEN MOVED TO ONE OF THE LEGAL STATUS OF DEATH CATEGORIES WITHOUT IDENTIFIABLE HUMAN REMAINS TO SUBSTANTIATE THE STATUS."

"Part of DOD's solution to "resolve" POW/MIA or KIA-BNR cases is to identify recovered remains of individuals from Southeast Asia, and match those remains with unaccounted-for or missing Americans on the Vietnam-era casualty lists. However, the Committee has reviewed numerous cases that pieces of bone, or bone fragments were the basis for identification of the remains of POW/MIA or KIA-BNR cases. These cases, if measured against court room body identification and death evidence criteria, would not be acceptable in court proceedings, except to infer, or to provide circumstantial evidence that something happened to a human being(s) at that location. Furthermore, a scientific evaluation of remains identification methodology used by DOD can be most politely described as not being based on any known and accepted forensic procedures...Proof that bone fragments belonging to an individual were recovered is sorely lacking in many instances. In some cases, DOD has made "identifications" of individual servicemen based on less than a handful of bone fragments. FURTHER, IN SOME CASES, THIS FINDING WAS MADE BY DOD, DESPITE LIVE-SIGHTING REPORTS THAT SOME OF THE INDIVIDUALS DECLARED DEAD, AND THEIR REMAINS 'IDENTIFIED' AT A CRASH SITE, WERE SEEN IN CAPTIVITY AFTER THE SUPPOSED DATE OF DEATH."

"For example, on October 5, 1990, at Arlington National Cemetery, DOD buried the 'remains' (bone fragments) of four U.S. servicemen presumed to have died when a helicopter crashed in Laos during the war. These remains were buried with full military honors. Then, their names were taken off the unaccounted-for list, and added to the list of those accounted for from the Second Indochina War. HOWEVER, ACCORDING TO FAMILY MEMBERS, AND ADMITTED BY DOD, TWO OF THE CASKETS OF 'REMAINS' CONTAINED NO BONES AT ALL—NO PHYSICAL MATTER, WHATSOEVER... Scientifically, these remains buried October 5, 1990 were not identifiable by any known or accepted forensic analysis. In the statements released to the press at the time of these 'burials,' DOD referred to 'remains' and new cases 'accounted for'... In 1986, a Laotian eyewitness (Som Dy), a member of the Royal Laotian Army, reported that he had been imprisoned with Captain Nelson, one of the four 'buried' at Arlington National Cemetery. The Laotian stated that he nursed Captain Nelson until he died, and that he was the one who buried Nelson. The Laotian identified a photograph of Captain Nelson and provided DOD specific locations, geographical details as well as a hand-drawn map of the camp."[24]

This was the Laotian refugee Som Dy, or "Somdee," whose story had been featured in Ted Landreth's 1987 documentary, "MIAs, We can Keep You Forever," that had so enraged the Pentagon. The other American POW Som Dy had been held with, Stanley Smiley, who had gone missing while flying an A4 aircraft, has still alive when the Vietnamese took him from the same POW camp, and his real fate remains unknown. The Laotian witness had been viciously harassed and threatened by DIA officials, in an effort to get him to recant his statement about being held with these American POWs long after the war. The DIA had ruled Som Dy's eyewitness account as yet another "fabrication."[25]

In confirming the pattern which had been documented and supplied by the author, of official U.S. actions since 1919 in declaring non-returned POWs and MIAs dead by a presumptive-finding process, and allegedly matching their names to later-recovered human skeletal remains as had been done in north Russia during 1929, the May 1991 Senate report made another important contribution. This involved the management of the Army's Central Identification Laboratory. The report stated:

"The responsibility for forensic identification of remains of U.S. Armed Forces personnel in the Pacific theater rests with the Army Central Identification Laboratory, Hawaii (CIL-HI). According to DOD, by early 1990 CIL-HI had identified 255 sets of repatriated remains from Indochina as the remains of U.S. personnel unaccounted for from the Second Indochina War. For a number of years, CIL-HI has been identifying remains of missing U.S. personnel from the Korean War and World War II's Pacific Theater still being discovered or, in a recent case, returned by foreign governments.[26] A prominent physical anthropologist, Dr. Michael Charney, Professor Emeritus... an internationally recognized expert in the science of forensics, has conducted an extensive review of physical remains 'identified' as missing Americans from Southeast Asia by CIL-HI. He concluded that it was scientifically impossible to have identified the cases he reviewed from bone fragments returned to the next-of-kin. Dr. Charney has levied serious charges against CIL-HI both publicly and to Committee staff. Dr. Charney states, 'This facility, entrusted with the analysis of mostly skeletonized remains of our servicemen and women in the identification process, is guilty of unscientific, unprofessional work.

"The administrative and technical personnel have engaged knowingly in deliberate distortion of details deduced from bones to give credibility to otherwise impossible identification.' Dr. Charney also went on to say that CIL-HI has blatantly and deliberately lied about a large number of the remains CIL-HI has identified. Dr. Charney states that, in his professional opinion, CIL-HI technicians have in some instances made identifications of remains based upon human remains or other material not capable of providing such an identification. He further states that many of the technicians who performed the identifications lacked advanced training in the field of forensic anthropology. Prior to 1986, CIL-HI's technicians referred to themselves as 'doctors,' when, in fact, they had never been awarded doctorates in medicine or any other recognized academic or medical discipline."

The Army's Central Identification Laboratory was a subject of hearings held before the Investigations Subcommittee of the Armed Services Committee, during the second session of the 99th Congress in 1986. The testimony of witnesses who demonstrated that the Army was fraudulently identifying human skeletal fragments as the "remains" of specific missing Americans in Asia, was of historic significance, as such identifications had gone largely unchallenged after World War I, World War II, the Cold War and Korea. The publicity which resulted from exposure of the government's lying to MIA family members about the remains of their loved ones resulted in the Army hiring better qualified personnel, but the senior anthropologist "Dr." Tadao Furue, remained. The May 1991 Senate Foreign Relations Committee report on POW/MIA policy summed up the findings of the independent anthropologists:

"The senior anthropologist, a longtime employee of CIL-HI, did not hold a doctorate in the field of anthropology but had worked in the field of forensic anthropology since the end of World War II. To accomplish his tasks at CIL-HI he insisted on using a theory he developed for the identification (of) human remains, a theory that was rejected by the anthropological scientific community. Between 1985 and 1987, Dr. Charney reviewed CIL-HI's identification of thirty sets of repatriated remains from North Vietnam and he concluded that CIL had wrongly identified these remains as those of individual servicemen from the MIA or KIA-BNR lists. In each of these cases, the material matter available to the CIL forensic examiners (bone parts and fragments) was not sufficient to identify a specific individual by sex, race, height, weight, physical peculiarities, etc.

"CIL technicians...in some instances employed forensic methods and procedures not recognized by the international community of professional forensic anthropologists. ACCORDING TO DR. CHARNEY THE CIL-HI TECHNICIANS DELIBERATELY MIS IDENTIFIED REMAINS AS INDIVIDUAL U.S. SERVICEMEN OFF THE LIST OF UNACCOUNTED-FOR DURING THE U.S. WAR IN SOUTHEAST ASIA. He believes the only conceivable reason for this demonstrable pattern of misidentification was a desire to clear the lists of MIA while deceiving the MIA families through the return of misidentified remains. Dr. (George W.) Gill, former secretary of the physical anthropology section, American Academy of Forensic Sciences, and a member of the board of directors of the American Board of Forensic Anthropology, substantiates Dr. Charney's statements concerning CIL-HI. Dr. Gill has publicly stated: 'IT IS CLEAR FROM THE BONES THAT THE PROBLEM IN THE CIL-HI REPORTS RESULTS EITHER FROM EXTREME CARELESSNESS, INCOMPETENCE, FABRICATION OF DATA, OR SOME COMBINATION OF THESE THINGS.' These charges levied by Dr. Charney and Dr. Gill against CIL-HI have not been refuted by (the) Department of Defense, and this inquiry has found no evidence that contradicts Dr. Charney or Dr. Gill."

This important part of the Senate Foreign Relations Committee Republican staff report on U.S. POW/MIA policy, was not to be refuted by a subsequent Senate Select Committee investigation of the issue, nor was the historical section of the report on American POWs of WW I, WW II and Korea, compiled largely from U.S. documents supplied by the author.

The author disagreed with Senate Foreign Relations Committee investigators Perrin and Usry on their use in the report of the figure "5,000" Americans from the Vietnam war once held as known-POWs, which they had apparently drawn from a newspaper article. The author and research colleague Ashworth had interpreted this 1973 report as actually referring to North Vietnamese prisoners being demanded back from South Vietnam by Hanoi and from all available sources believed that some 1,000 or more Americans had remained in captivity, including an estimated 300 transported to the USSR, and 700 or more who remained captive in Indochina, primarily in North Vietnam or nearby regions of Laos under Hanoi's military control. The subsequent livesightings of captive Americans held long after the war were of unacknowledged POWs from among the total of 3,734 U.S. missing and presumed dead officially reported by the Comptroller, Office of the Assistant Secretary of Defense (Directorate for Information Operations), on December 5, 1973. This number included 3,637 combat missing, which was administratively reduced subsequently to some 2,500 and eventually to the currently reported 2,273, often by matching fragmentary human skeletal remains with the names of American POWs and MIAs as had been done after previous wars.

After the author had fully documented the long-classified history of Soviet-held U.S. POWs, and after many weeks of urging, on July 4th, 1991, the Washington Post published a lead editorial announcing a change in their position on the POW/MIA matter, entitled "Were Some POWs Left Behind?," and introduced the author's historical research on Communist-held American POW/MIAs of every U.S. war since WW I for the Senate Foreign Relations Committee, which appeared in a simultaneous article on the opposite page, entitled: "A-70-YEAR-OLD HOSTAGE CRISIS." The date of publication indicated to all that the Washington Post was indeed still an independent newspaper, which had chosen to send an unmistakable message to the Soviet Union by highlighting the words: "U.S. AID COULD BE MADE TO DEPEND ON HOW HELPFUL THE SOVIET UNION IS ABOUT AMERICAN POW/MIAS WHO DISAPPEARED IN INDOCHINA, AND ANY SURVIVORS OF EARLIER WARS WHO WERE GIVEN SOVIET CITIZENSHIP."

New York Times editors Max Frankel, Michael Levitas and Jack Rosenthal had turned down this article, as did the editors of several other publications, even when supplied with substantiating documentary and human source evidence which had been used by the U.S. Senate. But the influential Washington Post had departed from the pack and now urged the formation of a Senate Select Committee for POW/MIA affairs, to fully investigate the matter. Senate Resolution 82 had been authored by Senator Robert Smith to legislate creation of a Select Committee on POW/MIA Affairs, for conducting an inquiry into "matters pertaining to U.S. military personnel unaccounted for from military conflicts." Within a few short weeks after the Washington Post's Independence Day statement, on August 2, 1991, strong Senate support was mustered to pass legislation creating a select committee to examine the full history of communist-held American prisoners of war.

A few days after the Washington Post editorial and publication of the author's article, on July 11th, Secretary of State James Baker had answered a question by KIRO-TV investigative journalist Mark Sauter, who used the historical evidence published in the May 23rd Senate Foreign Relations POW/MIA report as a lead to ask, "Mr. Secretary, will you also be discussing the American POWs in the USSR?... THE SENATE SAYS THAT OVER 20,000 HAVE BEEN TAKEN SINCE 1945. On 9 April, you asked for them back. Have you gotten an answer from the foreign minister?..."

While other reporters expressed amusement about a question which they themselves had for so long deliberately ignored, Baker acknowledged the validity of the issue by replying: "WE WILL TOUCH ON THAT SUBJECT VERY BRIEFLY..."

At a press conference on July 26, 1991, President Bush's National Security Advisor, General Brent Scowcroft, had falsely stated that there "is still no credible evidence" that MIAs survived after the war and that he was convinced that no MIAs were still alive. Ignoring the supreme embarrassment that would be caused to him and his former masters Nixon and Kissinger if a surviving POW was to appear, Scowcroft said: "There is no incentive on the part of the United States to conceal anything on this tragic issue."[27] Most of the American media ignored the significance of all that had happened in regard to POWs between late May and mid-July 1991, but in early August the full U.S. Senate approved formation of a special Senate

Committee to investigate the POW/MIA matter, not just in regard to Vietnam, but all the way back to the beginning of the problem.

The State Department publicly confirmed that at the end-of-July 1991 Moscow Summit, President Bush handed Mikhail Gorbachev a list containing a request for information on U.S. prisoners of war who came under Soviet control after WW II, Korea and Vietnam. The President acknowledged this on August 2nd. At a press conference in the White House Rose Garden, following the summit, reporter Sarah McClendon asked Bush:

"You were shot down, and you know what it's like, and if you had been captured and they had not come after you, it'd been pretty bad, wouldn't it? I wonder how you feel about the possibility that there is still live people over there who were captured who might be in Cambodia, or Laos, or Vietnam, and didn't you ask the Soviets about any prisoners they might have from past World Wars?"

President Bush replied:

"Yes...yes we raised that with the Soviets. They've maintained before and I...would expect maintain again, that they know of no American prisoners, but look, you're talking to one who was almost taken prisoner, and I think the United States Government should run down every single lead. As General Scowcroft said the other day, and I back him fully, there's no hard evidence of prisoners being alive, and for those who are unscrupulously raising the hopes of families by fraud, that should be really condemned. You talk about something brutal to a family, that's about as cruel as you can do. However if there's any hard evidence, it will be pursued and run to the ground and our policy has always been based on the assumption that until we can account for every... every person missing...that we have to run down these leads to prove that nobody is held... But Sarah, I've got to be careful that I don't do what some have done and maliciously raise the hopes of families, and yet I want to reassure those families, our government, our Defense Department, they're going to go the extra mile to find out if there's anything there, and if anybody has any hard evidence please bring it forward. So you hit me on something that really I feel strongly about in my heart."[28]

In the third week of August, following the Summit meeting, an attempted coup d'état against Mikhail Gorbachev occurred in the USSR, led by elements of his own inner circle including the head of the KGB (Ivan Serov's onetime protégé) Vladimir Kryuchkov, the hardline Communist Interior Minister, Boris Pugo, and the Defense Minister, Dmitry Yazov. The revolt began on Sunday, August 18th, with Gorbachev's telephone communication reportedly being cut off in the Crimea, whereupon he was confronted by some of the rebels and told to cooperate, which he refused to do. Yanayev named himself as acting president, declared a state of emergency and ordered military units into Moscow to enforce his orders; but the operation appeared to be both badly planned and ill-supported.

Thousands of protesters gathered around the Russian parliament building (called the White House), to protect the legally elected government there, and its leader, Boris Yeltsin. Some of the tanks and armored vehicles deployed around this crowd suddenly switched sides and turned their guns around to protect Yeltsin's supporters. Elements of the Soviet Army, including the 106th Guards Airborne Division at the Tula Garrison near Moscow, and others, which had been cultivated by Russian leader Yeltsin, opposed the coup, threatening the actual possibility of civil war for the first time since the 1921 defeat of the White armies and of the Kronstadt revolt. Gorbachev, still under house arrest in the Crimea, reportedly was visited by some of the coup leaders. (This contributed to as yet unproven allegations that the coup had been staged by Mikhail Sergeyavitch in order to win more sympathy and massive aid from the United States and its allies.

In fact there were tense moments of the crisis, particularly when Boris Yeltsin came out of the Parliament accompanied by armed guards and climbed atop a tank on the afternoon of Monday, August 19th, surrounded by a crowd estimated at 150,000. Yeltsin called for a general strike and civil disobedience and appealed to the Soviet Army to support democracy in Russia. Later that same afternoon the United States Government announced it was suspending aid to the Soviet Union and other nations also made statements condemning the rebellion. The crowds around the Russian Parliament building began building tank barriers, and late the next day a dangerous point was reached when three civilians were killed in a confrontation with tanks on one of the approaches to the Russian Parliament building near the old U.S. Embassy. In other locations the masses of people stopped columns of tanks from moving forward, and the soldiers did not fire upon the crowds.

By August 22nd the coup had fallen apart and Gorbachev returned to Moscow just 70 hours after the KGB, military, and Communist Party members had announced his ouster. Interior Minister Pugo committed suicide, the KGB chairman Kryuchkov was arrested, as were Defense Minister Yazov, Vice President Gennaday Yanayev and Prime Minister Valentin Pavlov. In the aftermath of the failed coup, a massive crowd of many thousands of Muscovites advanced into Dzerzhinsky Square where the dreaded KGB headquarters stood as a reminder to the vast number of innocent persons who had been done to death by the secret police since the Bolshevik Revolution. There the crowd turned its rage on the great statue of the Cheka's founder, "Iron Felix" Dzerzhinsky, and pulled it down from its pedestal. (An American who was present when this event occurred was Thomas Flaherty, who had previously helped in repatriating Soviet POWs from Afghanistan and had delivered the author's U.S. POW/MIA documents to Soviet researchers. (Flaherty later married a Ukrainian girl he had met while living in Kiev and at last report had returned to the United States.) On August 23rd, the Communist Party headquarters was sealed, reportedly to protect evidence within.

Although publicly humbled by the triumphant Yeltsin immediately after his return to Moscow, Gorbachev gradually reasserted his role as a leader. He appointed Vadim Bakatin to head the KGB, a liberal who had been forced out of power in December 1990. Time Magazine reported that one of the complaints by the communist old guard around Kryuchkov had been that, "THE SECRET FILES WERE BEING OPENED AND COVERT METHODS EXPOSED," and that the new KGB Chairman, Bakatin, was "expected to move decisively in cleaning up the agency." The KGB was reported at this time to have an estimated 600,000 members, of which some 40,000 were assigned to domestic surveillance, 20,000 to foreign intelligence operations, 265,000 as border guards, and another 230,000 in KGB military units.[29]

Meanwhile, on the night of August 30th, in the Pacific Northwest area of the United States, another retired Air Force Intelligence and NSA analyst, Terrell Minarcin, revealed on a KIRO-TV Seattle television program that he had monitored decoded enemy communications which had revealed that American prisoners had been shipped from Vietnam to the USSR during the Carter Administration. (This was while Robert Garwood was still in Vietnam and seeing substantial numbers of U.S. POWs Minarcin, who had once worked under Jerry J. Mooney at NSA for two years, had remained on active duty a decade after Mooney retired.

Minarcin stated publicly that years after the American involvement in Indochina ended, he had monitored decoded enemy communications which revealed that from December 1977-January 1978, at least 22 more U.S. POWs were flown from Vietnam to the USSR, as a result of further screening by the Soviets of American prisoners left behind in Southeast Asia after the 1975 American withdrawal. Minarcin also stated in a later affidavit for the Senate Select Committee on POW/MIAs, that more such shipments occurred until at least the spring of 1983 when 200-300 American POWs were transported from Vietnam to the Soviet Union, while hundreds more still remained in Southeast Asia. Postwar shipments such as these would explain at least some of the Indochina livesightings of the late 1970s and early 1980s, although it is clear from intelligence sources and continuing eyewitness refugee reports into 1992 that other U.S. POWs remained alive in Indochina.

In a written statement issued on August 30th, 1991, the Pentagon claimed that "No information from U.S. intelligence sources indicates that a movement of POWs from Vietnam to the USSR occurred." This inadequate response to the quality of the evidence so far revealed on the Moscow-bound POWs, countered not only the Minarcin report produced by KIRO-TV investigative journalist Mark A. Sauter, but also the author's July 4th, 1991 Washington Post article, "A 70-Year-Old Hostage Crisis," and a book just-published in England, "The Bamboo Cage," by the British author Nigel Cawthorne, which was as yet unpublished in the United States. What DIA had actually done with the volumes of intelligence it received from NSA on the subject was apparently still a classified matter. Perhaps the intelligence was so highly compartmentalized originally, that once filed, it might have been difficult to locate, and its existence unknown to many Defense Department officials.

In an even more startling development, in early September 1991, just before the formal appointment of the U.S. Senate Select Committee on POW/MIAs, the Soviet Newspaper Nezavisimaya Gazeta, published in English (and several other languages besides Russian), citing some of the declassified documents from the American archives which had been supplied by the author to Soviet researchers, recognized the prisoner of war issue in printing a story of historic significance for the USSR, by Vadim Birshtein, under headlines of "WHAT HAPPENED TO 20,000 AMERICANS? They were released from Nazi Prison Camps by the Red Army. Since then nothing has been heard of them." The article read:

"The problem first arose in late 1944 when Soviet troops liberated many Nazi internment camps in Poland and Germany. Hundreds of thousands of West European, American and Canadian citizens found themselves in the Soviet occupied zone. Under the Yalta agreement signed between Stalin, Roosevelt and Churchill on February 4, 1945, the Western Allies were committed to deport from Europe tens of thousands of Russian emigres and displaced persons, including those released from Nazi camps. In return, Stalin had promised to allow citizens of Allied countries in the Soviet zone to go home.

"But the Soviet authorities were not in a hurry to fulfill the pledge. Six days after VE Day, reports from France said that about one half of the 200,000 British and 76,000 American POWs still in Germany were probably in the Russian occupation zone. One report said that American POWs released by the Red Army were often mistreated and they were beginning to hat Russians. Many were stripped of their possessions that they had kept throughout the long German captivity, watches, rings and so on. American POWs in Odessa (a filtration camp) were guarded by Russian soldiers with loaded rifles and fixed bayonets and the Russian secret service was more strict than the Germans.

"The history of the Cold War is riddled with such cases. Usually the U.S. Administration made one or two queries about the fate of the crews. But after that it was left to the relatives who still believed their relatives were alive. To be sure, the reluctance of the U.S. Administration to admit that it had done little or nothing to save its citizens is a matter for the U.S. electorate and the government. The U.S. Senate is currently engaged in a stormy debate as to whether to set up an inquiry into the POW/MIA issue. It looks as if such a commission is going to be formed.

"There is yet another side to the matter. The information on foreign POWs who found themselves on the territory of the USSR during World War II and before the late 1950s is kept at the Central State Archive of the USSR in the section of the Central Board for the Affairs of POWs and Interns under the USSR Interior Ministry. It is the same fund that contained the lists of Polish officers shot in the woods in Katyn in 1940 and data on Japanese and German POWs which are now being handed over to these countries. So, the names of the 20,000 missing Americans and their fate can easily be found out in Moscow. The trouble is that gaining access to these archives is anything but easy. One needs the permission of two government bodies, the Interior Ministry and the KGB, which are notoriously reluctant to part with their secrets.[30]

The Soviet government recognized the independence of the Baltic republics of Latvia, Lithuania and Estonia on September 6th, thus at once complying with the weight of world opinion and permitting achievement of a half-century-old American policy goal. On that same day, the presidium of the Russian Republic's Supreme Soviet voted to confirm a referendum by the voters of Leningrad to change the city's name back to the pre-Bolshevik, "St. Petersburg," yet another indication of the collapse of public support for the Communist system, and the growing hatred among the people for anything connected to the Bolshevik leaders who had brought dictatorial oppression and economic ruin upon Russia for over seven decades of the 20th century.

Mikhail Gorbachev announced on September 11th that his government was beginning talks with Fidel Castro's government about removing the 11,000 Soviet troops from Cuba. Secretary of State James Baker called this initiative a "substantial gesture," while President George Bush, the inveterate Cold Warrior, was quoted in the press saying, "I wish they'd hurry up."[31] At the time of this exchange, the Associated Press reported from Moscow, on September 12th, that the Soviet government was implementing a plan to remove all obstacles to free emigration for all Soviet citizens, a condition that the United States had pushed for since the post-WW II period. The plan had not been scheduled to be implemented until 1993 but the fast-moving pace of events had changed the Soviet priorities. Adoption of this law was reported to have been a key condition set by the United States and other Western nations before they agreed to allow Moscow to host the human rights conference. Such a policy could allow the free emigration of former prisoners of war who had been given Soviet citizenship, of many nationalities, including Americans, to leave the Soviet Union. In an apparently qualifying warning however, Associated Press reported a statement by Mikhail Gorbachev's new Foreign Minister Boris Pankin: "Pankin said the government will also review cases of Soviets denied permission to emigrate with the exception of special cases where it is necessary to protect state security. Pankin's reference to state security as an exception made it unclear whether any of these cases would be reviewed."

The Soviet Government also announced that charges against the world famous Russian dissident author Alexander Solzhenitsyn had been found to be groundless, and he was free to return to the Soviet Union. At his farm in Cavendish, Vermont, the writer said that he would indeed return to Russia but that he was first going to complete research projects begun in the United States. Between September 11th and the 16th, events moved swiftly, as Secretary of State James Baker repeatedly stated that the Bush Administration was rethinking its policy on aid to the USSR and now favored substantial aid grants in the immediate future.

The Senate Select Committee on POW/MIAs, which the Soviet publication referred to, had been approved in August 1991 by a vote of the full U.S. Senate. Senator John Kerry, a Democrat from Massachusetts who had previously investigated drug trafficking and the BCCI banking scandal, was named as Chairman by the Senate Democratic leadership. Kerry was a Navy veteran of Vietnam who had served in combat operations on gunboats in the Mekong Delta area, and subsequently had become a leading anti-war veteran spokesman in the early 1970s, when he was involved with founding an organization called "Vietnam Veterans Against the War." He gained some prominence by being among a group of veterans who threw their campaign medals back to the government in an anti-war demonstration. He was also however, a Yale graduate who had been asked to join the secretive Skull and Bones Club, as had President George Bush, years before.

Kerry's experiences had paralleled those of many bitter and disaffected veterans of the "no win," but vastly destructive policies of the Indochina war at the time. Kerry had often been accused by conservatives of demonstrating a pro-Hanoi bias, although others argued that he was merely a realist, who believed in putting the wounds of war behind, for both sides, by providing U.S. diplomatic recognition and aid to the Hanoi regime. In the late 1980s and the early 1990s Kerry had gained national prominence through his publicized investigation of narcotics trafficking and money laundering, including facets of what became the BCCI scandal.

During the first informal meeting as the Committee on POWs was forming, Kerry was quoted as saying:

"This is not a partisan issue, it is an American issue. That is why we have six Democrats and six Republicans...The Pentagon, DOD, and the Administration clearly have an interest in resolving this because they have been put on the defensive...We need to go back in time...review past efforts."

Kerry recommended that all the members of the Committee review the still-classified 1986 Tighe Report and the May 23, 1991 Senate Foreign Relations Committee Republican Staff report on U.S. POW/MIA policy since 1918, produced under the guidance of Senators Charles Grassley and Jesse Helms with primary research assistance of the author and colleague Thomas V. Ashworth and some additional information from Mark Sauter. Other nationally prominent Democrats appointed to the Committee included Senator Robert Kerrey of Nebraska, a former Navy SEAL who had won the Medal of Honor in Vietnam and who announced his candidacy for the Democratic Presidential nomination in the same period, and also Senator Charles Robb of Virginia, President Lyndon B. Johnson's son-in-law, who had commanded Marines in Vietnam combat.

The man recommended as Vice Chairman by Republican Senate leader Robert Dole, after some delay, was first-term Senator Robert Smith of New Hampshire, author of the legislation that created the Select Committee. Smith had also served in Vietnam waters with the Navy fleet, and, as a congressman, had been a leading member of the vocal group of POW/MIA advocates in the House, which had included William Hendon, John Miller, Frank McCloskey, Robert Dornan, John LeBoutillier, and others. These honorable men had opposed the pro-DIA/Pentagon faction so strangely exemplified by a former antiwar protester, Democratic Representative Stephen Solarz of New York. (For a decade the Smith group had refused to let the POW/MIA issue die in the House, despite the best efforts of the Pentagon and other agencies to discredit them with assistance from Solarz and others in the Congress.) Other leading Republicans appointed to the Select Committee included Senator Charles Grassley, an Iowa farmer of proven integrity and courage, whose POW/MIA investigation had already been assisted by the author for several years; the conservative North Carolina Senator, Jesse Helms, under whose name the May 23, 1991 Senate Foreign Relations Republican Staff report on POW/MIAs had been released; and John McCain of Arizona, an ex-POW who had been held in Hanoi for over five years, but who had publicly refuted all existing evidence that any other American POWs had been left behind, when he and 590 other prisoners were released in 1973.

November 1991 hearings held by this Select Committee heard testimony from Secretary of Defense Richard Cheney, General John

Vessey, Kenneth Quinn and various assistant secretaries and officials of the Defense Department, including Carl Ford, Duane Andrews, Dennis M. Nagy, Robert Sheetz, Charles Trowbridge and Robert DeStatte, who all maintained that no evidence existed that American POWs had remained captive after the Vietnam War or still remained alive in Indochina, or anywhere else. A single U.S. official of modest standing, Garnett E. Bell, the Chief of the U.S. Office of POW/MIA Affairs in Hanoi, had the courage to admit that some small number of American POWs, perhaps 10, had remained in enemy captivity after the 1973 prisoner exchange. (Bell was later to be relegated to a backwater job in Bangkok for his breach of U.S. policy guidelines.) Senate investigator Tracy Usry, authors Monika Jensen-Stevenson, Nigel Cawthorne and Dr. Jeffery Donahue; activists including Colonel Jack Baily and U.S. Veteran News publisher Ted Sampley; and MIA family members of the Stevens, Robertson and Lundy families, all testified on behalf of the American prisoners of war they were certain had been abandoned. Senator Harry Reid, of Nevada, limited Usry's testimony on the historical precedents of Soviet-held American POWs of Korea, WW II and WW I, while Senator McCain attacked him on Vietnam-era details. Colonel Bui Tinh, once an interrogator of U.S. POWs and the former editor of the Vietnamese Communist Party newspaper Nhan Dan, now styled as a Vietnamese "defector," was warmly embraced by Senator McCain during his appearance and repeated his web of lies and half-truths. Although treated with great deference and respect by the Chairman, Senator Kerry, Bui Tinh sounded more like a controlled agent; a sent-messenger of the Hanoi Politburo:

"I had special authorization from the General Vo Nguyen Giap, then Defense Minister, to go to any camps, to meet with any officers, and to interview any POWs and read their files...the government or any leaders could not hide any information on MIA or POW from me... About this issue, I can say that I know as well as any top leader in Vietnam, and in my opinion, I state categorically that there is not any American prisoner alive in Vietnam. There is the only single case of Robert Garwood who lived freely in South Vietnam (and) in Hanoi and returned to the U.S. ...if there were any American alive and willing to live in Vietnam, I would have known about them... I categorically state that there is not any MIAs or POWs alive in Vietnam today. As for Laos and Cambodia, I do not know the situation as well as in Vietnam. But I trust that the Laotian and Cambodian governments are not interested in keeping any Americans. I have heard that some Cubans and Russians interrogated some American prisoners and treated them badly. But I do not know any information about Russia and Cuba keeping Americans."[32]

*U.S. Veteran News and Report* publisher Ted Sampley later printed a story about then-Navy Commander John McCain's record as a POW in Hanoi, implying that the Arizona Republican had collaborated with his North Vietnamese captors and was subsequently compromised, which explained his conduct on the Select Committee. McCain and his staff, particularly an assistant named Mark Salter, were infuriated and the incident was to lead ultimately to a violent confrontation resulting in Sampley being jailed for allegedly assaulting the staffer. Salter had been one of the Senatorial staff ears for the Defense Intelligence Agency during the Select Committee's investigation. Other POWs who knew McCain in captivity had spoken highly of his conduct while a POW.

Senator Charles Grassley confronted Ann Mills Griffiths, Executive Director of the National League of Families, on November 6th, about her 1990 effort to prevent his and Senator Helms' gaining access to classified POW/MIA intelligence for the Senate Foreign Relations Committee Republican staff POW/MIA investigation:

"Shortly after our visit in my office (in 1990), or about the same time I wrote to the Defense Department to get access to livesighting reports, the very reports that have been discussed before this Committee...It took me two months to get a response. I received a letter from Assistant Secretary of Defense Henry Rowen. He said that I could not see the reports because they are classified...The response basically said that a U.S. Senator cannot review information because it is classified...For the next 2 1/2 months I fought quietly to get a more reasonable response. I kept hearing from inside the Pentagon that there was fierce opposition to my access for some reason, and that for some reason I heard that fierce opposition was coming from you...After 2 1/2 more months of failure to get a reasonable response...I read Mr. Rowen's response to my colleagues in a speech on the Senate floor, and then within 24 hours I suddenly had access to what I wanted... 48 hours later a memo from you to Paul Wolfowitz fell into my possession... I am going to put it in the record..."

Grassley placed in the record an August 3, 1990 letter from Griffiths to senior Pentagon official Paul Wolfowitz (that had also been sent to DIA Director General Soyster and Watson of the National Security Council), which illustrated the official defense mechanism that had been set up, with Griffiths acting as a willing puppet, to keep the elected representatives of the American people from analyzing the eyewitness accounts of Americans held captive in Indochina long after the war ended. The 1990 letter stated (in part):

"On August 1, letters from Assistant Secretary of Defense for Legislative Affairs Dave Gribbin were sent to Senators Jesse Helms (R-NG) and Charles Grassley (R-IA) giving them full access to cases of interest in the POW/MIA issue...Mr. Gribbin suggested that the Senators may wish to invite staff of the Foreign Relations, Armed Services or Intelligence Committees, with the appropriate clearances and with responsibility for the POW/MIA issue, to accompany them to review the classified material. An additional... letter was sent...stating that Congressman Bob Smith (R-NH) would be provided the same opportunity, along with the staff of the Armed Services or intelligence Committees, under the same clearance criteria. Mr. Gribbin's decision to pursue this course (which I understand was strongly opposed by ISA, DIA and the Deputy Chief of Staff for Intelligence, Department of the Army) has far reaching ramifications about which I feel you and the Secretary (Richard Cheney) should be aware... MR. GRIBBIN'S COMMUNICATIONS AUTHORIZE ACCESS TO INDIVIDUAL MEMBERS OF CONGRESS WHO HAVE BEEN LOUDEST IN PROCLAIMING THAT U.S. GOVERNMENT OFFICIALS ARE NOT PURSUING THIS ISSUE WITH INTEGRITY OR PRIORITY. THEIR RECORDS ARE REPLETE WITH IMPLICATIONS, OR OUTRIGHT ASSERTIONS, THAT THE U.S. GOVERNMENT IS INVOLVED IN A CONSPIRACY TO COVER-UP EVIDENCE THAT AMERICANS ARE BEING HELD CAPTIVE IN SOUTHEAST ASIA. I have no concern that Members of Congress or staff will uncover anything which would lend credence to the charges of conspiracy against which we have collectively fought since the 1983-86 time frame. My concern is that if this decision is implemented, the above message—that the POW/MIA issue is one solely of rhetorical priority—will be made clear throughout the U.S. Government. Equally or more important, Hanoi will perceive that stated U.S. policy on this issue is no longer valid...One can anticipate immediate CIA and NSC action to pull their documents and information from the files currently held by DIA. Department of State, the FBI and Drug Enforcement Agency, as well as friendly foreign intelligence services and others with ongoing investigations and cooperative programs will cease all cooperation on POW/MIA related matters..."[33]

Griffiths' somewhat subdued response to Grassley on November 6, 1991, about the contents of this 1990 letter was: "I'm sorry you misinterpreted it. It was never directed at you alone. It was a process that they were talking about, of opening up everything in DIA's files—outside of the committee structure of both houses, of the Senate. It had nothing to do with you..."[34]

Author J.M.G. Brown had published the name Sary Shagan (Kazakhstan) as a suspected/expected location in the USSR for American POWs transferred from Vietnam, based upon information from an ex-NSA source, Jerry J. Mooney, in the American Foreign Policy Newsletter for October 1991, Volume 14, No.5: "A Chain of Prisoners." This report covered the long-classified history of U.S. POWs remaining in Soviet control after four 20th century wars, since the administration of Woodrow Wilson, and also of Chinese-held American POWs of Korea. This particular issue of the National Committee on American Foreign Policy also contained an address by the Ambassador of the People's Republic of China, Li Daoyu, on "China's Foreign Policy on Global and Regional Issues."

Senator John Kerry wrote to the author on November 20, 1991: "Thank you for sending me your article which appears in the October issue of the *American Foreign Policy Newsletter*. We are looking into the questions you raise and welcome your input." The author began to supply documentary and human-source information on POWs to the Committee and urged that retired U.S. Army Colonel William LeGro be engaged as a investigator. In this effort the author enlisted the support of the former Chief of Staff of the U.S. Army, General William Westmoreland. Due to the efforts of Senator Charles Grassley and Senate investigator Tracy B. Usry, 277 pages of the author's declassified U.S. documents on Soviet-held American POWs of WW I, WW II, Korea and Vietnam were accepted as genuine official documents and published in the record of the November 1991 hearings by the Senate Select Committee on POW/MIA Affairs under the title: "POW/MIA Policy And Process, Part II of II"[35] The Chairman expressed no interest in the author's (and researcher James Sanders') offer to work for the Select Committee as an unpaid researcher, indicating that from the beginning, Senator Kerry never intended to allow a comprehensive investigation of the historic precedents in the POW/MIA matter. Despite Kerry's decision, with the support of Senators Grassley and Smith the author continued to provide substantial documentary and analytical information on U.S. POWs during the entire life of the Committee, through staff members.

A significant article by investigative journalist Ed Tivnin, dealing with the Moscow-bound U.S. POWs of Vietnam only, subsequently appeared in the Los Angeles Times magazine, which utilized much information from Mooney, Minarcin and several Soviet sources. Yet, in the aftermath of the abortive Soviet coup, the only information on living U.S. prisoners revealed, was a mid-November report by Russian journalist Yuri Pankov in "Kommersant," of an American POW from Vietnam living at SARYSHAGAN in Kazakhstan; and statements by former KGB Counterintelligence Chief, Oleg Kalugin, to the U.S. Senate Select Committee on POW/MIA affairs, that some U.S. POWs had been interrogated by Soviet intelligence after the Vietnam War and the alleged return of all American POWs by the Vietnamese government. There could be little doubt however, that the Soviet press confirmation of Sary Shagan as a location for U.S. POWs from Vietnam in the USSR, was an important step in resolving the issue.

In a nearly simultaneous development, The Disabled American Veterans (DAV), a million-member organization which ranked third in size and influence among national veterans groups (after the American Legion and VFW), with members in all 50 states, joined the effort to expose the truth about Soviet-held American POWs of past wars. The DAV Magazine published a feature article in the November 1991 issue entitled "THE POW/MIA MYSTERY UNRAVELED," which was substantially a result of the author's two-year effort to supply enough declassified documentary evidence to the three major U.S. veterans organizations to alter their position on the subject of Soviet-held American prisoners and POWs remaining in other communist nations. The DAV's National Commander, a decisive black American combat veteran of Vietnam named Cleveland Jordan, and a scholarly past national commander, the DAV Magazine's editor, David Givans and a disabled Vietnam veteran named Joseph E. Andry, were the force behind this expose of the truth, which utilized over 200 declassified U.S. archival POW/MIA documents of WW II, Korea and Vietnam, supplied to the organization's headquarters by author J.M.G. Brown. These included many of the same declassified documents which the author had supplied to Senator Grassley for the May 1991 Senate Foreign Relations Committee report of Senator Helms, and to Soviet researchers in Moscow; and to the senior editors of the Washington Post and Newsweek Magazine in 1990 and 1991. This powerful report by the DAV helped to galvanize the American Legion and the VFW in their public stances on the POW/MIA issue, in subsequent dealings of all three organizations with the Bush Administration, and indeed, in meetings with that President himself, in the White House.

President Bush subsequently approved 1.4 billion dollars in American aid to Russia, and the U.S. House of Representatives, by now largely aware of the long history of the POW/MIA matter, further demonstrated American compassion for the Russian people by approving most-favored nation status for the Soviet Union. The Mujahedeen rebels of Afghanistan had already promised, on November 15th, to release some Soviet POWs by the end of the year, and they subsequently freed several in December. Yet, U.S. officials from the Moscow Embassy reported on December 8th that they had not been allowed access to the "restricted area" at Sary Shagan where the American prisoner from Vietnam (tentatively identified by the reported date of loss as Captain Eugene McDaniel's co-pilot, Kelly Patterson) was reported to be.[36] This was followed by a request from the U.S. State Department for information from Moscow on 50 U.S. POWs of Korea and WW II once reported in Soviet control. Since the first two men on this list had been returned by the Soviets decades before, while the second two had never been Americans, and the fifth one's remains had been returned, it does not appear to have been a serious diplomatic request.

Shortly thereafter, on December 19, 1991, Russian President Boris Yeltsin took over the Kremlin and almost all control of former Soviet agencies, from Mikhail Gorbachev. Three days later, Afghan|

Mujahedeen leaders returned a Soviet POW who had been captured in Afghanistan to Russian Federation Vice President Alexander Rutskoi (who was himself a former Red Army officer and POW in Afghanistan), on December 22nd. The former Red Army POW was reported to be a native of the Soviet Central Asian Republic of Turkmenia. Two more Soviet POWs were subsequently reported freed, one of whom returned home, while the other elected to remain. On Christmas Day, December 25, 1991, the Soviet Union was officially dissolved and became Russia and the Commonwealth of Independent States. Thus, after seven decades of monolithic police-state rule, the largest empire on earth collapsed of its own inertia and many internal contradictions.

A published report of December 28th on the list of 50 Americans stated: "The action followed U.S. Ambassador Robert Strauss' cable last month from Moscow to the State Department seeking increased diplomatic pressure on Russian authorities regarding the POW issue."[37] The author then sent to Secretary of State James Baker III, on December 31, 1991, the names of the 21 American POWs on the NSA's February 1973 "headliner," or "Kissinger" list, who had been known through intercepted and decoded Vietnamese communications to have been interrogated and processed for shipment to the USSR and China from Vietnam, prior to and just after the Paris Peace Accords of 1973. The Department of State never responded to this letter.

The casefile of just one of the 21 Americans on this list, Lt. Colonel (then Major) Robert M. Brown, who was listed within NSA as "Moscow-bound," would have provided the Department of State with compelling evidence that the Soviet intelligence services would have information on his fate. As with most of the other 20 American POWs on the 21-name "Kissinger list," Brown was not just an ordinary prisoner, he was a prize catch. He had already served an earlier tour of duty in 1966-67, flying close air support for U.S. infantry operations in South Vietnam, and with his impressive qualifications it is astonishing that his government sent him once again into harm's way in 1972, when the war was almost over. Not only was Colonel Brown flying a U.S. Air Force F-111 (with his weapons systems officer Lt. Colonel Robert D. Morrissey, also on the Moscow-bound list), which was then the most advanced fighter-bomber aircraft in the world, he was also an electrical engineer who had worked as a scientist on the U.S. Mercury and Gemini space programs. A top-level graduate of the Naval Academy who earned an Electrical Engineering degree from the University of Michigan, he was the back-up Network Controller for the first Gemini space mission, was also abreast on the Apollo program and knowledgeable of the latest instrumentation improvements in manned space networks and Defense and NASA data-gathering systems, as well as conducting experiments for the U.S. Anti-Ballistic missile (ABM) system program.

Many facts were uncovered by Colonel Brown's son, Bruce Brown, through Freedom of Information requests. The POW's son had received a telephone call from the U.S. Air Force on December 7th, 1990, informing him of a "Dog Tag Report" on his father, which aroused him for the first time to discover whatever he could about his father's real fate.

Flying out of Takhli AFB in Thailand, Brown and Morrissey were shot down in the southern province of North Vietnam, Quang Binh, near the DMZ, on November 7, 1972, on a single-ship strike mission across Laos against a ferry and ford complex near Vinh Linh, North Vietnam. The F-111 aircraft he was flying, with its high-tech navigation and weapons systems, was priceless to the Soviets for their advanced aeronautic program, which made it and its crew a high priority target for the Vietnamese, and it was at the top of the requests listed by their Soviet allies. Capable of flying at Mach 2.5 and carrying a bomb load equal to 20 times the F4 while evading enemy radar, the F-111's mission success rate was 97.5% with an aircraft loss rate of only 0.7%.[38]

U.S. intelligence reports and intercepts available thus far through declassification report the successful downing of the F-111 aircraft on November 7, 1972, one calling it a low-flying F-4 and another reporting the operation to transport the "hulk" of the downed aircraft, which was "to be carried out in secrecy."[39] As retired NSA analyst Jerry J. Mooney had testified: "They trained to spot, track and 'flack-trap' these planes. Shells from a larger gun would force a plane to veer into the sights of smaller guns, increasing the probability that the plane would crash land in pieces big enough for future examination. Their orders were to capture the pilots alive." The area of the shootdown was being monitored by Mooney at the time, who later emphasized to the author that most of the Moscow-bound POWs on his own list, including then-Major Brown, were captured in the panhandle of North Vietnam and adjacent border areas of Laos.

Colonel Robert M. Brown's son, Bruce, a methodical and stubborn young man who was 13 years of age when his father vanished, had inherited some of his father's professionalism and applied it to his investigation. He learned from Jerry J. Mooney that the F-111's and their crews were the highest-priority aircraft target of the Vietnamese and Soviets and that, "the evidence that Colonel Brown was taken alive and the reasons why intelligence listed him as such came from intercepted North Vietnamese coded traffic, Post Fire Reports, and plain text radio communications all reporting the shootdown and capture of an F-111 and its aircrew. The Post Fire reports listed the downed/captured category as 2/2-

two airmen in aircraft, two captured. In other communications Brown's and Morrissey's names were spelled out phonetically by the Vietnamese for their reports. All reports were "signed" by weapons officers not political officers (of the North Vietnamese unit's ADWOC); this was significant since political officers tended to embellish on the facts. The NSA believed that a Khe Sanh "flak trap" unit, either the 284th or 287th AAA Professional Air Defense Regiment (NVA) had moved into position to catch the next F-111."

Mooney said operational and collateral data corroborated their evidence, while another retired NSA cryptolinguist, Terrell Minarcin, later testified before a Senate Select Committee, directly on Colonel Brown's loss incident:

"The original transshipment of the aircraft hulk was initially reported in a cryptsystem used by the Vinh Linh Special Air Defense Zone units. The aircraft was shipped from the crash site to Vinh then up the Muong Sen Valley to Sam Neua (Laos) where the Soviets took control of the aircraft. At that time a Soviet IL-14 flew out to the Soviet Union. Several passengers were manifested on that flight. It was believed that at least Brown and Morrissey were among the passengers."

Years later, a DIA report stated that Colonel Brown's ID card was seen on display in Vinh, North Vietnam, a feeder location for the Tentacle MB interrogation camp at Sam Neua, Laos: "On 20 and 23 July 1991, a joint U.S./Socialist Republic of Vietnam team met with of the Military Region 4 museum in Vinh City, Nghe Tinh Province...items examined (included): "Military identification card of Major Robert M. Brown, good condition." A fellow F-111 pilot (Colonel D.) said: "There is no way that I.D. card would be in good condition if they hit the jungle; that's very significant."

Another DIA report of intercepted enemy communications dated November 9, 1972 quoted enemy communications: "On 07 November our workers shot down a low flying F4, the workers... captured alive the pilot." Another reported that an NVA unit in Quang Binh province "shot down a plane during the night of seven November 1972. The plane was thought to be an F-111."[40] Many other documents concerning intercepted enemy communications remain classified at the time this is written.

Within 18 months of Robert M. Brown's (and his co-pilot Morrissey's) capture, the Soviet space program went through a major upgrade, bringing it to the U.S. Gemini level of complexity. The Soviets also came out with an almost exact replica of the F-111 (the Sukhoi SU-19/24, in full production by the summer of 1975. It is POWs like Colonel Brown who could very well be alive today in the former Soviet republics such as Kazakhstan or Tajikistan, or perhaps in some restricted security area in Siberia. By December 1991, many families of Vietnam MIAs like the Brown's believed that these types of American POWs had been far too valuable and talented for the Soviets to simply kill them, and if rewarded for good work done on Soviet aeronautic or space programs could be expected to remain useful Soviet citizens. It would seem more likely in fact, that some would have married Soviet women, perhaps former gulag inmates, and started Soviet-American families, even if hoping all along, that someday they might be permitted to return to their homeland. Subsequently, an entire U.S. F-111 aircraft which had been downed and captured in Vietnam, was located intact in Moscow, where it had been transported to from Southeast Asia, as was later described in testimony by NSA cryptolinguist Terrell Minarcin.

Meanwhile, a standard bearer of American left once again attacked those who had exposed the crimes of Communist regimes to the world, in the December 1991 issue of the liberal American magazine, *The Atlantic*, in an article entitled: "THE POW/MIA MYTH." This was written by Rutgers University Professor H. Bruce Franklin. (Franklin had previously written a sympathetic introduction to a collection of Josef Stalin's writings, helped found the "Venceremos Brigade," and lost his job at Stanford University in the early 1970s for his radical pro-Communist actions; a difficult thing to have done in the context of the times.) He referred to the United States Government in his 1975 autobiography as the "enemy of mankind, the number-one criminal of the world, wanted dead or alive." In his lengthy Atlantic article, Franklin was permitted to use undocumented sources and unsubstantiated statements which amounted to gossip, to discredit any persons who maintained that American POWs had remained in Vietnamese or other Communist captivity; to blame successive Presidential administrations for falsely promoting that idea, and to defame the character and reputation of several Americans who had refused to let the issue die for decades. The article advanced the current "politically-correct" doctrine that belief in the existence of secretly-held U.S. POWS in Communist prisons was a myth of the white American working-class, perpetuated mostly by disaffected war veterans, of what the professor maintained was an immoral and unjust war.

It was indeed a classic and masterful piece of disinformation, of a type long produced by the KGB and its Cheka ancestor, and for months to come, H. Bruce Franklin was to be quoted extensively in the mainstream American press as an "expert" on the POW/MIA issue, when debunking subsequent startling Russian admissions that American POWs of at least three wars, including Vietnam, had indeed been held in Soviet labor camps.[41] Yet, author J.M.G. Brown had, since 1989, repeatedly proposed to the senior editors of The Atlantic, a major article based upon declassified U.S. archival documents on the historical precedents in the POW/MIA matter, dating back through four major wars to the time of Presidents

Roosevelt and Wilson. Hundreds of these documents were supplied by the author to the editors of the magazine. A March 1, 1989 letter in response to this proposal from Cullen Murphy, Managing Editor of the Atlantic, informed author J.M.G. Brown: "The article you propose IS NOT RIGHT FOR THE ATLANTIC, I FEAR, but thank you for getting in touch." Throughout 1990 and 1991, as the Soviet empire collapsed and it's grisly secrets were revealed by the foreign press and by Russian dissidents, the Atlantic's senior editors, William Whitworth, Cullen Murphy, and C. Michael Curtis, refused to consider a short, or medium-length, or a long, fully documented article on the history of Communist-held POWs, drawn from the U.S. archival records and from still-living human sources, which were accepted by Senate investigators as historically significant.

The Atlantic's Washington editor, James Fallows, widely known as a former anti-war activist and Carter Administration speech writer who subsequently covered Asia for National Public Radio, wrote to the author on August 5, 1991; four months before Franklin's disinformation article appeared:

"This isn't the right kind of article for the Atlantic. I appreciate the political and historical arguments you make for publishing the piece. No one doubts the importance of the subject. But the style of presentation, while systematic and logical, is different from what the magazine usually publishes...The magazine receives many more proposals than it can possibly publish. THEREFORE THE EDITOR USUALLY CONCENTRATES ON TOPICS HE THINKS THE MAGAZINE CAN HANDLE UNIQUELY WELL. I KNOW IT"S DISAPPOINTING FOR YOU THAT THIS SUBJECT IS NOT ONE OF THEM, BUT THAT'S THE CASE..."

In the subsequent December 1991 issue featuring the MIA "Myth," the Atlantic falsely editorialized that Franklin's review of the evidence in the POW/MIA matter, "Shows that it is groundless, a myth created and exploited for political reasons...ONE OF THE MOST STUBBORN MASS IRRATIONALITIES OF OUR TIMES." The editors in fact knew this was false, from information they had received from the author. It was a classic example of the self-censorship widely practised in the American media for decades, on the subject of Communist-held American prisoners. Following the publication of Franklin's article (which must have been in the works for several months), the Atlantic's Senior Editor, C. Michael Curtis, wrote to the author on December 17th: "We are unlikely to want to publish another long piece on the MIA issue...and your thoughts on the subject... have already appeared in the Washington Post, a fact that greatly diminishes their freshness in our eyes." On March 16, 1992, editor Curtis again wrote the author after receiving a submission entitled "Policy, Politics and the POW/MIA Dilemma," drawn from this book, that refuted, through declassified U.S. documents and human sources which were supplied to the magazine, Franklin's distorted version of history: "I don't understand your seeming indifference to our earlier responses to your writing on the POW/MIA issue. And, as you know, we recently published a long piece on the subject..."

Months after the article appeared, and after the author's research colleague, Thomas V. Ashworth, had appeared on a television program with Professor Franklin (to refute some of his disinformation), Ashworth witnessed former CIA official Alan Ptak, who had that year been promoted from the Agency to the Pentagon as a "Deputy Assistant Secretary Of Defense for POW/MIA Affairs," embrace Franklin warmly, and compliment him on how well he had done on the program refuting the POW/MIA "myth." At the time of this incident, Mr. Ptak, who had long before been assigned the 1986 Murkowski hearings and later to the 1991 Senate Select Committee on POW/MIA Affairs staff, was a member of the U.S.-Russian Commission on POWs.

Such powerful propaganda as that produced by H. Bruce Franklin and endorsed by the national security establishment's Alan Ptak, was effective, as subsequent U.S. aid to Russia failed to obtain even a token release of a few American ex-POWs. The indications were that the representatives of the U.S. Government did not want any breakthrough in the POW/MIA matter which might lead to a release of survivors who could greatly embarrass many present and past U.S. officials. (With the assistance of a Marine veteran of Vietnam and former CIA officer, John Biddle Brock, many of the author's historically-significant documents were also given to the Atlantic's owner, Mortimer Zuckerman, a long-time friend of Brock's. who was also Chairman of the Board and editor-in-chief of U.S. News and World Report Magazine. For many months to come, the Washington Whispers column of U.S. News and World Report was to provide comprehensive and timely coverage of the most important new developments in the POW/MIA matter.)

At the beginning of 1992, official U.S. POW/MIA efforts in Indochina remained focused on the recovery of human skeletal remains rather than living U.S. survivors. The New York Times published a news story by Barbara Crossette about the reports of Soviet-held U.S. POWs from Vietnam on January 6th. Rather than introduce the subject as important new evidence on the fate of missing Americans, the Times chose to headline the article with: "NEW INTEREST IN MISSING SERVICEMEN MAY IMPERIL MOVE TOWARD HANOI TIES." Immediately below this, a large photograph of three U.S. prisoners (Robertson, Stevens and Lundy) was displayed with the caption, "Last summer copies of a photograph purporting to show three captive American pilots were circulated, but U.S. authorities, after studying them, decided they were not valid." The New York Times did not inform it's readers that, in fact, according to the

POWs' family members, the fingerprints of these men had accompanied the photograph, and that all records of the three men's fingerprints had subsequently "disappeared" from their military files, from FBI custody, and even from their home states and hospital records.

The Russian Foreign Ministry had announced on January 9th that it was forming a joint parliamentary commission with the United States to examine the topic of missing American prisoners of war which would include U.S. congressmen and members of the Russian parliament. (The author subsequently received information that the Russians had proposed several published American researchers of the POW/MIA matter, including the author, be included on the Commission, but that this had been opposed by the U.S. Government, Department of State.) Meanwhile, in a speech before the World Affairs Council in San Francisco, 'Ex'-KGB Major General Oleg Kalugin, who had once controlled the Walker brothers spy ring for Moscow Center, repeated his statement that American POWs were still alive in Vietnam in 1978, and were interrogated then by Soviet intelligence officers. In a January 11th report on the event, the San Francisco Chronicle quoted the former Soviet counterintelligence chief as saying,

"IT IS NOT AN ALLEGATION. I STATE IT AS A FACT. AMERICAN SERVICEMEN WERE INTERROGATED IN VIETNAM WELL AFTER THE WAR, BUT NOT AFTER 1978."

In another development, the Senate staffers of the Foreign Relations Committee Republican staff, with whom the author had worked closely in producing the May 1991 report on U.S. POW/MIAs policy, Daniel Perrin and Tracy Usry, were dismissed. The reason they were given for this action was: "IT IS FOR THE GOOD OF THE COUNTRY." The Washington Post covered the "purge on Foreign Relations," in its National Weekly edition of January 19th, with barely a comment on the constitutional implications of the firing of the key staff members who had worked on the May 23, 1991 report on POW/MIAs. The new Foreign Relations Republican Staff Director, retired Admiral James W. "Bud" Nance, a former Deputy National Security Adviser under President Reagan and Richard Allen in the early 1980s, was quoted as saying: "I'm just trying to put some order in the chaos."

Admiral Nance was reported to be restructuring the Senate Foreign Relations Committee staff, "TO REFLECT MORE CLOSELY THE WHITE HOUSE NATIONAL SECURITY ORGANIZATION," and to take account of recent upheavals in the world. While mentioning that Tracy E. Usry and James P. Lucier had been removed from the staff (together with Daniel Perrin and others), the Post said nothing about the role of these two men in publishing, through the Senate Foreign Relations Committee, a definitive report on the declassified documentary evidence of Soviet-held U.S. POWs, from 1918-1919 to 1980.

On January 19, 1992, in Washington D.C., Mr. J. William Codinha, Sen. John Kerry's Chief Counsel for the Select Committee on POW/MIA Affairs, personally took 13 hours of sworn testimony from retired Air Force/NSA analyst Jerry J. Mooney. Another Committee staffer was present, who was reported to be a representative of President George Bush. Everything he had brought with him, "was tagged as evidence," Mooney told the author a few days later. He revealed the location of secret raw intelligence files from U.S. intercepts, which would prove his contention on the fate of the Moscow-bound and other abandoned POWs of Vietnam. The location of these classified TOP-SECRET-CRYPTO records of the A Group-NSA, stored at Fort Meade and (earlier at) Fort Holabird, Maryland, and never revealed to the Congress or the American people: "RS"-(Russian Shipping), "RTV"-(Soviet Civil Air), and "RAV"-(Soviet Navigational Air), was made available to the Senate Select Committee on POW/MIA Affairs in early 1992 by Mooney and then by the author through staff investigator William LeGro to Senator Kerry. Kerry, who should have extended every effort to protect these records from politically-motivated destruction by U.S. Security agencies, instead declined to seriously investigate the matter. NSA eventually claimed they couldn't locate many of these records.)

Yet, Codinha shared an office with former DIA analyst Sedgewick "Wick" Tourison, who had been hired as a member of the Committee's investigative staff, but who, in reality, represented the DIA's POW/MIA Office, which had for years committed itself professionally to insisting that no American POWs had been left behind in 1973, and that none of the hundreds of eyewitness livesightings of American POWs in captivity were valid. Tourison was known to the author who had interviewed him in his Annapolis, Maryland home in 1987, and again later. Tourison had been adamant at that time in his belief that all livesightings from human sources in Southeast Asia were "fabrications." Tourison had also published a book expounding on this view and since leaving DIA had been a proponent of increased U.S.-Vietnamese trade. In a book written by an Army warrant officer, Gary L. Smith, who conducted human skeletal remains excavations in Indochina for the Joint Casualty Resolution Center (JCRC), is a colleague's description of Tourison's feelings about the MIA/POW matter:

"I saw the television documentary called 'We Can Keep You Forever," The theme of the program was that American POWs were still being held in Southeast Asia, long after the U.S. withdrawal. I was impressed, After watching the television special, I was convinced that Americans were being held in Indochina, and it was going to be my mission at JCRC to find them. I graduated from the

Vietnamese course at DLI...I took a quick trip to Washington, D.C., to meet some of the folks in the POW/MIA business at the Defense Intelligence Agency (DIA). Two of the guys there, Bob Hyp and Sedgewick "Wick" Tourison, were former prisoner of war interrogators during the Vietnam War. I spent several days getting briefings from them on the MIA issue, as seen through the eyes of DIA. Those two guys were legends in military intelligence circles in the old days and talking to them face to face was a real thrill for me.... Wick was the senior POW/MIA analyst at DIA and probably the most knowledgeable person in the entire federal government about the MIA issue. I talked to him at length about the television documentary I had seen, about MIAs who allegedly had been kept in Laos and Vietnam. Wick told me it just didn't happen. He said there wasn't a shred of proof any Americans had been held against their will after the war. He said the program was full of inaccuracies and was nothing but a con-job getting people sucked into conspiracy theories. I told him it had certainly convinced me..."[42]

It was because of Tourison's predetermined attitude about the reported sightings of Americans in captivity after the war, that many veterans and POW/MIA family members objected to his presence on the staff of the Select Committee on POW/MIA Affairs. Here he occupied a strategic and influential position, and together with other like-minded federal officials on the staff, was able to affect the ultimate outcome of the Committee's supposedly unbiased investigation. In this manner, the input of the author and other independent researchers and investigators was largely nullified in the final report of the Select Committee.

In an extended September 1992 interview with the author, Mooney revealed that he was subsequently 'reindoctrinated' into NSA for life, after making a formal agreement with a Deputy Director of that organization that NSA in turn would promise to:

"1. Investigate the Soviet connection to U.S. POW/MIAs in detail through cryptsystems and machine ciphers.

2. All North Vietnamese, Lao and Viet Cong tactical messages would be recalled for analysis along with all briefing items, reports and memoranda.

3. All data base systems would be checked for any references to POWs, including COIN, disposition of forces, COINS/2, another program designator, and C-Ref, the NSA's Central Reference Files.

4. NSA was to investigate all correlation of Moscow bound information between NSA, CIA and DIA data."

The 1971 CIA cable on the eyewitness account of the Lam Thao Superphosphate plant in Vinh Phu Province being used as a debriefing site for U.S. POWs with Soviet and Chinese officers present, had been supplied to Mooney by the author previously, and he attached this declassified document to the agreement with NSA as an example of correlating intelligence. Mooney was then assigned a young Senate investigator named Thomas Lang, who had done some previous investigative work for Congress, and was planning to go to law school after the Select Committee's job was done. Over the next eight months Mooney proceeded to indoctrinate Lang into a secret world that few Americans even knew existed. On January 22, 1992, Mooney testified publicly before the Senate Select Committee on POW/MIA Affairs, and later in the day Minarcin testified. At the beginning only 7 of the 12 senators on the Committee were present: Kerry, Smith, McCain, Brown, Grassley, Reid and Daschle. Kerry stated at the onset, "the NSA wants to discuss these gentlemen's testimony publicly." Mooney testified (in part):

"Sir, it's been a long journey, 6 years, since I came out, and it took an act of God to bring me out. It's been a journey of 'not Government business,' being told there's no jurisdiction, and mostly no response...There's been a little damage done to me and to my property and to my family...The interest in Moscow Bound is totally surprising to me. I told this to the BBC, I told this to the Australians, I've told it to reporters. Its an intelligence given. The first time I heard the term MB was my last week in 202 Communications School in 1958. MB, Moscow Bound. The Soviets do take our people, also (it) was referenced to a wink and nod diplomacy. So it was nothing unique and new to me...There's two areas right now that are on the public mind. One is the Kelly Patterson incident at Sary Shagansk, and the other is the Kalugin involvement. The Kelly Patterson incident can be resolved in a very short period of time to determine if there's smoke or fire Sary Shagansk is well known. Kelly Patterson must have good classified and unclassified records, all that has to be done is match the two, see if it fits. Kalugin, when he first came out, I was delighted. For the first time somebody above the rank of NCO was providing me with some support...I had two questions. Why did he come out, and what about. The why he came out, I'm still thinking about. The what-about question I can answer. Those 'friends,' those Soviets, that I saw in communications, in the tactical communications of the North Vietnamese, WE DID NOT CONSIDER TO BE KGB. THEY WERE GRU. They were there repairing the equipment, showing the Vietnamese how to operate a knife rest radar, how to shoot a 100 millimeter gun, how to fire an SA7 missile, how to repair a MiG 21 J, how to operate a CCI station. These were not KGB people, these were GRU people, tech reps. They were in the chain of command. They did not have their communications system. We saw them referred to, because of some very good operators, they heard them in the background and put it on paper. In other words, telling the Vietnamese exactly what to do...."

"I'm just the tip of the iceberg. The very, very small tip of the information that exists on this issue. And I have provided your staff with very sensitive places to look, including names and technical details that a Committee of Congress has never had before...Beneath me, there is much information. You have to bring it out. I hope General Tighe is wrong when he says he does not believe people, quote, with regard will step forward, unquote. That was in the L.A. Times article... You have to get up where the intelligence is interpreted and used for policy and politics."

When asked by the Chairman, Senator Kerry, whether he could reconstruct his wartime lists of American POWs left behind in captivity, "if DIA and NSA were cooperating," Mooney testified:

"If I had access to the COINS files...It is a database file and access to the translations again, and had access to certain briefings, I could." Mooney also testified about the American POWs held in a separate, parallel prison system for forced-labor as related previously in this book, and told Senator Kerry the cryptographic titles for the code systems where records of these intercepted communications were located: "Encrypted messages, sir, that were read...The titles are in the VCZD series, they start—and the VCZf series, they run from...VCDZ001 through at least VCZD 300, that represents 300 code systems..."

In response to a question from Senator Brown on whether he believed American POWs were sent to China as well, Mooney answered:

"I believe the following. There is very little intelligence that we saw on the Chinese. What we knew was this. The Chinese were in the Haiphong area in the 1962, 1963, 1964, 1965 area, 1966, with AAA forces. They were shooting down U.S. aircraft. As far as the intelligence goes, and this is as far as I saw, they had the opportunity and they had the motive...They were losing their technological base for aircraft from the Soviet Union and they had to start their own industrial complex. Pilots with experience would represent a quantum leap. So the only intelligence that we had was the opportunity and the motive."

In answer to Senator Charles Grassley's question about these "sighting reports without people," Mooney testified: "I had one brief brush with it...and it has to do with CB, China Bound. What it amounts to is, if you suspect that Americans are being taken for technological reasons, you can watch the growth of the target nation. If there's any unexpected quantum leaps in their development that you could relate to a POW, you then have at least some information that they have a POW in hand." Mooney then went on to cite as an example his 1976 meeting with a CIA National Intelligence Officer (NIO), while accompanying NSA's Whitey Machelski, mentioned previously in this book, in which the coverage of the Olympic Torch airborne resource was altered so as to monitor the activities of the CC-F3 fighter testing program for the People's

Republic of China, in addition to North Korea and Communist China in general. In response to a question from Senator Brown on whether he had "any knowledge of the use of POWs on the Chinese-Vietnamese border," Mooney testified: "I have only hearsay on that, sir. When I was talking to the DIA about the cover name Son Tay, which was friend, Soviet associate. I can recall a vague reference by the DIA analyst saying that we have a similar situation at Long Son, but he never explained what it was. That's near the Chinese-Vietnamese border...When he said, a similar situation, I kind of took it to mean that either there was a Soviet connection there or a China connection there with relationship to POWs...I have no data that I saw that would reflect the movement of American POWs from either Sam Neua...or Hanoi, over the hump, through India, Delhi, and up into the Soviet Union, or by shipping either way they were removed. I have informed your investigators, your Senate committee staff people, that if such evidence existed during or after the war, where it would be..."[43]

Retired NSA crypto-linguist Terrell A. Minarcin testified later the same afternoon of January 22nd, that he had been "an asset of the National Security Agency/Central Security Services from September 14, 1967 until July 9, 1984." When asked by the Chairman, Senator John Kerry, whether he had knowledge of American prisoners left behind in captivity after Operation Homecoming in 1973, and what proof of this he had, Minarcin testified:

"Yes, sir...The proof...I've given to the staff, or how to find the proof...The proof would be in the actual raw traffic that we saw at the time as a matter of routine....In Vietnamese communications that I either heard or saw on my desk...It would vary according to the report. The one report that I mention in my statement listed the 12, or the helicopter flying up to pick 12 American POWs up, to go pick up firewood...The helicopter was an MI-8, which is a Soviet helicopter... The helicopter came out of Quan Long airstrip, went up to Bai Thuong airfield, picked up the 12 Americans, flew out to Muong Sen to pick up the firewood and returned the prisoners and the firewood back to Bai Thuong and then the helicopter returned back to its home base at Quan Long...(on) March, 3, 1978...I was back at NSA proper, up at Fort Meade...

"At the fall of Ban Me Thuot, in 1975, there was an American advisory team that was captured at the command post of the 23rd ARVN Division. We saw many, many references. We saw the initial report...We reported it via SONGBIRD, which is the reporting vehicle for captured Americans...I don't know the distribution list for SONGBIRD...It would be a generic term for the entire distribution list...Beginning in mid-December 1977, we started seeing the special flights coming out of special prisoner of war camps, going into

Hanoi's Gia Lam, where—not necessarily Hanoi's Gia Lam, but other centers throughout North Vietnam, where we knew that there were Soviet military personnel stationed....This came from manual Morse, and it was dealing with the Vietnamese civil aviation network. It was the country-wide flight schedule for that day...They were coming from known prisoner of war camps. These were historically prisoner of war camps, we saw no stand-down of activity...Vietnamese prisoner of war camps where they held prisoners of war...up until 1973, when you would have thought that they would have shut them down, and sent them home, the camps were still up and active. There was no cessation of activity."

Senator Kerry asked, "Now, what proof do you have that the people being held in those camps were either Caucasian or American?" Minarcin responded:

"By the terms that they would use...'Tu binh my,' which means, specifically, American prisoner of war...In 1978, we saw them being turned over to the Soviets at Hanoi's Gia Lam and being flown out of Hanoi's Gia Lam...when I was at Fort Meade, Maryland, and those names have also been turned in to the committee...We were in B15, and we were with the North Vietnamese air shop."

Kerry then expressed astonishment, saying, "And you are saying that in Fort Meade, Maryland, in 1978, right here next door to the Capitol, we had information coming in of American prisoners being turned over to the Soviets?" Minarcin testified:

"They were being flown into Hanoi's Gia Lam, and the only flight activity coming out of Hanoi's Gia Lam would be an IL-62...Based on the analysis of what we saw, sir we drew that conclusion. We saw the aircraft. If they were flying into Gia Lam on an IL-14, when the IL-14 left Gia Lam it was empty. Those people had to go somewhere. The IL-62 when it left later in the day carried the same number of people that were coming in on other flights."

Kerry asked: "You know that they were specifically Americans who were brought into the airport?" Minarcin replied:

"Based on the camps where they came from, yes, sir...Based on the camps and they're saying that they only had American POWs. They did not have French POWs, which would have been 'tu binh phap.' They did not have West German prisoners, which would have been 'tu binh duc.' They'd only have 'tu binh my,' which is American prisoners of war."

A short time later during Minarcin's testimony, Senator Kerry summed up his reaction:

"What I have from you right now is a deduction, that is all this committee has in front of it from you right now is an assumption that Americans were in the camp. And because there is an assumption Americans were in the camp, a flight coming out of the camps means Americans were on the flight. And because there was a

Soviet plane waiting at the airport and it left with a certain number of people in it, that Americans went. That is not good enough for this committee." Minarcin responded:

""Sir, you're asking, did we transcribe this voice over here that said, were there tu binh in this camp, and I said, yes, sir...It was written off the military communications dealing with the logistics report for that specific camp...Again, we have to make the assumption that if this camp is holding American prisoners of war, and we got two sets of passengers coming out of here, and the guards back here in this military network are complaining that they have to go to Hanoi, and all of a sudden we see on the civilian network an aircraft leaving the camp and it's got two sets of passengers on there—I don't know how you would do it sir, but for me, it shows that there's American prisoners leaving that camp."

Kerry then said: "It shows there might be American prisoners. It might be any number of other kinds of prisoners there. There might be Vietnamese." Minarcin countered:

"No, sir. No, sir...The term for Vietnamese nguy, which means puppets, and nguy were never held with the Americans...they do not specifically say on the civilian network that they had American POWs. That came from military communications. The military communications supported the network at the camp. You would have to correlate the two, which is what was done daily....Post-1975, from conversations of the guards off the military networks."[44]

Minarcin's testimony, like Mooney's and others, was largely ignored by the American mass media. The final report of the Senate Select Committee stated that no other NSA employee had come forward to verify Minarcin's statements, and several Senators questioned his veracity, because of his reported failure of a small portion of an extensive polygraph test (according to a January 21, 1992 memorandum to William Codinha, the second page of which criticized the polygrapher), most of which, however, had indicated no deception; and the fact that on December 23, 1983 his security clearance had been downgraded from Sensitive Compartmented Information (SCI or "codeword) when he claimed to have obtained POW information from highly classified communications intercepts. Yet, significant portions of Minarcin's testimony had been backed up by many other sources, over the years. The Committee found no fault with Mooney and praised him for his assistance. In a subsequent interview, Mooney told the author that when Minarcin had worked under him at NSA he had been a very good analyst but occasionally demonstrated a tendency to reach conclusions before others had. Mooney believed that much of what Minarcin said in his testimony matched information he had seen, while other statements were troubling.

Russian Federation President Boris Yeltsin traveled to the

United States at the end of January 1992, to attend a meeting of world leaders at the United Nations. On his way to the United States he visited British Prime Minister John Major, in London, and warned him that without massive Western aid, Russia might succumb to unrest. He was quoted by the New York Times as saying, "General unrest will happen only if our reforms fail. Should the reforms fail, we shall face a new leadership, and Russia will fall into the habits that tortured us for 74 years." Yeltsin proposed huge new arms cuts in nuclear arms, and appealed for massive U.S. and Western aid to prevent chaos in Russia and the other Republics of the Commonwealth. On the day of Yeltsin's arrival, the New York Times reported (on January 30th), that Colonel-General Dmitri Volkogonov, an intelligence adviser to Boris Yeltsin and chairman of the Russian parliamentary commission that oversees the KGB archives, would invite American historians to search for evidence on the fate of American prisoners missing since the Vietnam War.

The massive February 1992 airlift of U.S. aid to Russia, using military aircraft under direction of Richard Armitage, also (at least publicly), failed to procure the release of any American prisoners who had been transported to the Soviet Union from Vietnam, or after earlier wars. A French news report by Michell Honorin of "Antenne 2" on February 13, about 72 American POWs remaining alive in three Vietnamese camps was mentioned in the American press, but the Bush Administration still announced plans for eventual U.S. diplomatic recognition of the Communist regime in Hanoi.[45] American policy toward secretly-held POWs thus appeared to be unchanged since 1933, despite all that had been revealed since the late 1980s. Alan Ptak, a CIA official (who years before had participated in the Murkowski hearings), was temporarily assigned to the Senate Select Committee on POW/MIAs, and not long afterward he was promoted to Deputy Assistant Secretary of Defense for POW/MIA Affairs. He then went to Hanoi to instruct Presidential envoy General John Vessey in current POW policy of the Bush Administration. Shortly thereafter, a Reuters report from Hanoi quoted General Vessey as saying the evidence suggested that nearly all U.S. POWs who could have survived, had died.

The Chief of the U.S. POW/MIA office in Hanoi, Garnett "Bill" Bell was informed shortly thereafter that he was to be replaced at the end of March by a non-Vietnamese-speaking U.S. officer, with no intelligence background (as was the head of the current chief in Vientiane, Laos, Bill Gaddoury, who had formerly served for years under the controversial Colonel Paul Mather, at the Bangkok Embassy-JCRC. Bell wrote his friend, Vietnam author Al Santoli from Hanoi that, "HAMBURGER HILL TYPE OPERATIONS ARE NOW BEING PLANNED TO BLITZKREIG THE COUNTRYSIDE TO DIG UP CRASH SITES WHICH HAVE ALREADY BEEN DUG UP SEVERAL TIMES BY THE VIETS."[46]

CIA official Alan Ptak was then assigned to serve on the newly-formed U.S.-Russian Commission on POWs, chaired by Presidential appointee Ambassador Malcolm Toon of the U.S., and General Volkogonov of Russia, other members included State Department official and former Kissinger aide Kenneth Quinn, who had long denied the existence of U.S. POWs, and others such as Denis Clift of Defense Intelligence Agency (DIA), who as a federal official had, for years, obfuscated and denied release of classified U.S. POW-related documents to bona-fide researchers, including the author. Another member, however, was Trudy Huskamp Peterson of the National Archives, who had, on several occasions assisted the author's research.

Ambassador Toon had been a career State Department Russian expert, who had begun his diplomatic career in 1946 as a Foreign Service officer in Soviet-occupied Poland, where according to declassified U.S. documents turned over to the U.S. Senate by the author in 1991, the search was then still continuing for U.S. POWs who had disappeared behind Soviet lines in 1945. Later Malcolm Toon rose to become Director of Soviet Affairs in the State Department at the height of the Vietnam War (1968-69), and eventually was appointed U.S. Ambassador the Soviet Union. In his various capacities in the Department of State spanning four decades, he would have become aware of continual classified reports of U.S. POWs of WW II, the Cold War, Korea and Vietnam, seen alive in Soviet control by eyewitnesses. The U.S. Government appeared to be 'gridlocked' over the POW issue, and Toon's feigned ignorance of the facts was just one symptom of this. Another example is a response to several Freedom of Information requests by the author to the FBI for release of classified information concerning Soviet/surrogate-held U.S. POWs resulted in a February 28, 1992 letter from J. Kevin O'Brien, Chief of FOIA section for the FBI: "Regarding your Freedom of Information Act request for information on 'Soviet and Soviet/surrogate-held U.S. POW/MIAs,' the FBI is unable to locate a record of having received such a request." The author had already received a previous letter from FBI acknowledging receipt of this FOIA request. Thus did the FBI and other federal agencies stall and delay the research of the author and other researchers, for years.

The Vietnamese had demonstrated for years that they were willing to return human skeletal remains which the U.S. Government had long admitted knowing had been warehoused in Hanoi since the war. (A mortician working there had testified to Congress years before about this, but when he had said also that he had seen live American POWs too, he was not believed.) The real issue was not skeletal remains however, but living prisoners, perhaps judged to be "war criminals" deprived of their native citizenship, and held since 1973-75. Diplomatic recognition by the U.S. would have been premature until living American survivors were released in Indochina, either to the U.S. government or to a private group. The alleged importance of the Vietnamese market to U.S. businessmen eager to exploit the cheap Vietnamese labor was insufficient reason to thus condemn surviving prisoners to the oblivion suffered by their predecessors of WW I, WW II the Cold War and Korea. The Hanoi regime's maintaining that none of the hundreds or more Americans now known to have been left behind in Vietnam and Laos remained alive, by implication admitted that they had been secretly transported elsewhere or killed in one way or another. To many Americans such a regime seemed unworthy of U.S. diplomatic recognition in a post-Communist age and Former President Richard Nixon was very outspoken in his opposition to such recognition.

The author completed a 133-page report, which was condensed from this book in March 1992, entitled: POLICY, POLITICS AND THE POW/MIA DILEMMA, and supplied it as a briefing-book on the history of Communist-held U.S. POWs to Senators John Kerry and Robert Smith. In April the author also sent this report to President Bush, through Mr. Anthony D. Duke, who was not notified of its receipt by White House staffer John R. Nicholson Jr. until June 23rd, after Boris Yeltsin had admitted that American POWs of three wars had been held in Soviet labor camps. Nicholson confirmed that any such POW/MIA information sent to a President through the White House staff system was immediately funnelled into the National Security Council:

"...Thank you for the material on the POW/MIA issue compiled by your son-in-law John Brown. The report is certainly timely. Because the POW/MIA situation is a national security issue, Mr. Petersmeyer has forwarded the information to Douglas Paal at the National Security Council for appropriate action. Mr Paal is Special Assistant to the President and Senior Director Asian Affairs (NSC)..."

Somewhat later, a Senate investigator at the Select Committee on POW/MIA Affairs, Colonel William LeGro, received a request for a copy of this report from the office of Tennessee Senator Albert Gore, a Vietnam veteran who had become the choice of Democratic Presidential contender Bill Clinton to be his vice president. In addition, Mr. Ross Perot also received a copy.

From March through May 1992, the Senate Select Committee on POW/MIA Affairs was torn by internal divisions over interpretations of classified POW/MIA intelligence, and in one case, during May, Senate Security guards invaded the investigators' office and swept up all classified evidence and material, including four safes holding their intelligence files. (This material even included the contents of the desks of the investigators, containing sensitive material supplied by author John M. G. Brown and his research colleague Thomas V. Ashworth, film producer Ted Landreth, and other private and confidential sources, without which, according to POW/MIA Committee Vice-Chairman Robert Smith, in a letter to the author dated June 18, 1992, the committee may not have been formed in the first place:

"...Your seminal work, best expressed in your 'Policy, Politics and the POW/MIA Dilemma,' has provided a solid historical foundation for inquiry and analysis. I am also aware that you contributed significantly to the report of the Republican Staff of the Senate Committee on Foreign Relations. It was this report, 'An Examination of U.S. Policy Toward POW/MIAs,' that provided the compelling rationale for the creation of our Select Committee."[47]

Yet, the only answers the investigators got to their questions about the Senate Security action were allegations about 'leaks' to the media. (Subsequently one of the Committee's investigators, William 'Billy' Hendon, a long time POW-MIA advocate, was dismissed, as was Minority Staff Director Dino Carluccio, for allegedly leaking classified U.S. POW/MIA documents to the media. Retired Army Colonel William LeGro, another staff investigator, was reported in the media to also be under suspicion, but this proved unfounded.[48]

A leaked Defense Department "memorandum for the record" dated April 13, 1992 illustrates the turmoil within the Select Committee on POW/MIAs. It is stamped: "For Official Use Only," under "subject: 9 April Senate Select Committee on POW/MIA Affairs Briefing," and signed by Charles Wells of the Pentagon Central Documentation Office and Mark Bitterman, a Special Assistant to Secretary of Defense Richard Cheney:

"During the week of 30 March, Mark Salter of Senator McCain's office notified the Department that John McCreary and Billy Hendon of the Senate Select Committee would be briefing Senators on 9 April on their so-called 'cluster theory' about POWs/MIAs being alive in Southeast Asia. According to Salter, the conclusion of the 'intelligence' briefing was that Americans were still being held prisoner in Vietnam and Laos as late as 1989. Senator McCain felt it was critical that DIA be present at this briefing to rebut the assertions of McCreary and Hendon. We told Salter, however, that Mr. Ptak would have to make the decision on DoD's participation.

"In the absence of any specific details about the briefing or ability to take DIA staff (in addition to Bob Sheetz), Mr. Ptak decided that Department of Defense should not participate. On the morning of 9 April, however, Salter provided a summary of the McCreary briefing as well as supporting material to DoD. Additionally, with the assistance of Senator Kassebaum, MR. SHEETZ WOULD BE PERMITTED TO HAVE MEMBERS OF HIS STAFF PRESENT. Once in possession of this information, Brigadier General Joersz and Dennis Nagy authorized DIA's participation in the form of being able to comment when asked by the Senators. Approximately 2 hours prior to the scheduled briefing, DoD was informed that Minority Staff Director Dino Carluccio had told Senate Security that the DoD team would not be allowed into the briefing in S-407. Subsequently, the DoD team was told by Staff Director Frances Zwenig to assemble in Senator McCain's office pending a possible vote by the Committee on whether or not DoD could attend. At approximately 1400 we were told to report to S-407 to participate. (The Committee had voted 7-2 in favor of DoD participation, WITH NEGATIVE VOTES FROM SENATORS GRASSLEY AND SMITH.

"The DIA team of Bob Sheetz, Warren Gray and Gary Sydow was accompanied by Mark Bitterman (Office of the Secretary of Defense) and Chuck (Charles) Wells, of ASD/C31, the Central Documentation Office. Senators present included: Kerry, Smith, McCain, Grassley, Kassebaum, Brown, Robb, Daschle, Reid and Kohl (only initially). The briefings were presented by John McCreary, Bob Taylor and Billy Hendon. John McCreary presented a briefing employing a large map of Indochina with numerous color-coded pins inserted throughout... He cited the basis of his findings as 928 live-sighting reports culled from over 6,000 such reports from DIA's automated database. The gist of the McCreary/Hendon theory is that where the pins tend to cluster on the map, THERE HAVE BEEN LIVE AMERICANS AS RECENTLY AS 1989. McCreary explained his alternative theory as emanating from the Gaines Report...Before departing, Senator Kerry asked Bob Sheetz to make a commitment to get together with Committee staff to discuss the livesighting reports. Sheetz explained to Kerry that the dilemma is that committing DIA analysts to work with Committee staff can only be done if DIA sacrifices its support to the Joint Task Force...Hendon continued his briefing (of Committee members only Senator Smith remained). After being challenged by the DIA team for taking part of a livesighting report completely out of context, Hendon and Smith began criticizing DIA. As it became more difficult for DIA to respond, Smith blamed his fellow Senators for DIA having to defend itself. Smith said DoD should never have been allowed in the closed briefing and that his fellow Senators brought this folly about by voting to allow DoD to participate.

"As the discussion grew more contentious, Frances Zwenig approached Mark Bitterman and stated that Senator Kerry had not intended the briefing to be carried on after he left AND THAT DoD SHOULD NOT CONTINUE TO ANSWER QUESTIONS. Bitterman and Sheetz both told Senator Smith that we would defer further comment to a later session. SMITH PROTESTED AS WE LEFT BUT DID ADJOURN THE BRIEFING."[49]

In several written statements between June 14th and 19th, 1992, to Senator John Kerry, Chairman of the Select Committee on POW/MIA Affairs, a former Army Intelligence NCO, Barry A. Toll, of

St. Petersburg, Florida, further substantiated the previous testimony by the NSA's ex-analysts Mooney and Minarcin. After combat infantry and Long Range Reconnaissance patrol duty in Vietnam where he was severely wounded in 1968, Toll had been reassigned and trained for intelligence work with the "World Wide Military Command and control system or ...Variously assigned within the World Wide Airborne Command Post system as an intelligence specialist and Operations Assistant on Battle Staff Teams formulated specifically to implement the nation's highest strategic nuclear policies and plans, and directly assist the President or designated National Command Authority." From June 1973 through July 4, 1975, Mr. Toll was, according to his statement to Senator Kerry:

In his June 14, 1992 statement to Senator Kerry, Mr. Toll revealed that the shipment of U.S. POWs from Vietnam and Laos to the Soviet Union and other Soviet Bloc nations had been made known to President Richard Nixon during the 1973-1974 period:

"I PERSONALLY SAW, DISTRIBUTED AND BRIEFED HIGH RANKING OFFICERS OF THE JOINT STAFF, ON INTELLIGENCE REPORTS, ANALYSES AND OPERATIONS REGARDING THE TRANSFER OF U.S. POWs AND/OR MIAs FROM THE CUSTODY OF NORTH VIETNAMESE OR LAOTIAN AUTHORITIES THROUGH SOVIET BLOC NATIONS, OR DIRECTLY INTO THE USSR. Further, it was the considered opinion of the Joint Chiefs of Staff, and the entire U.S. intelligence community that at the conclusion of Operation Homecoming in 1973, that there were an estimated 290 to 340 U.S. POWs and MIAs alive, and held captive in Laos, and that analyses and reporting was disseminated to the President of the United States, Richard Nixon. I SPECIFICALLY RECALL THAT INFORMATION BEING INCLUDED ON A LIST OF THE PRESIDENT'S DAILY INTELLIGENCE BRIEFING AGENDA MORE THAN ONCE, AND REPORTS ON REAL-TIME TRANSFERS IN PROGRESS (AIRCRAFT BEARING U.S. POWs IN THE AIR ENROUTE TO USSR AND SOVIET BLOC COUNTRIES) BEING PASSED TO THE WHITE HOUSE FOR THE PRESIDENT SEVERAL TIMES IN THIS PERIOD."[50]

If it was based on fact, Toll's information to the Senate Select Committee (when taken together with the substance of Mooney and Minarcin's testimony), indicates that President Gerald Ford, and later President Jimmy Carter would thus have received any similar intelligence on Moscow-bound U.S. POWs.) The Committee later claimed that no other former intelligence person or official had come forward to substantiate Toll's statement.

In subsequently assisting the staff of the Senate Select Committee during June and July 1992, retired NSA analyst Jerry J. Mooney revealed the locations of secret files on U.S. prisoners of war which the NSA had been unable to find for the Committee. When the correct computer discs were finally located, where Mooney had said that they would be, the result was discovery of 15,000 POW-related files of one page or more each, which contained significant data on a large number of known-U.S. POWs who had not been released, and also on POW/MIAs who had subsequently vanished. (The in-house reaction at NSA to this discovery was that the Agency was not even aware that such detailed intelligence information on this subject existed.) Yet it later appeared from the final report of the Select Committee on POW/MIA Affairs, that much of this information was never used.

---

## Notes on Chapter Thirteen

1   *George Bush, An Intimate Portrait*, by Fitzhugh Green, p. 247

2   *George Bush, An Intimate Portrait*, op. cit.

3   Wash. Times, April 1 989

4   Seattle Times June 3, 1988

5   *The Bamboo Cage*, by Nigel Cawthorne, published in England by Cooper- Pen and Sword Books, Copyright 1991.

6   Letter dated July 21, 1989, from U.S. Representative Dante Fascell to John Ordway Duke of Summerland Key, Florida: "I appreciate having the benefit of your views regarding prisoners of war (POWs), and read with interest the articles you forwarded from the Herald and the U.S. Veteran..." also: Toronto Sun, May 27, 1989, by William Stevenson, "Dreadful Military Secrets"

7   Letter of July 12, 1989, to Vietnam veteran John J. Molloy Jr., President of the Release Foundation, from Senator Daniel Patrick Moynihan: "Thank you very much for having taken the time to write and to send the Memorial Day edition of the U.S. Veteran News and Report. I am glad to have it, and I anticipate it will prove to be of use."

8   Wall Street Journal, July 19, 1989

9   Weekend Australian, 12/13 August 1989, see also *Bamboo Cage*

10  1991 interview by author

11  Seattle Times, Monday, September 25, 1989, by Steve Johnston

12  Time, December 25, 1989. (The forced repatriations of Vietnamese refugees were reported by UPI- Hong Kong, Dec. 14,1989, Eureka, CA, Times-Standard)

13  POW/MIA Policy hearings of Select Committee on POW/MIA Affairs, November 1991, U.S. Government Printing Office, 1992.

14  Letters from the editors of the listed publications, in the author's files

15  Text of Tass statement from Reuter's despatch, Warsaw, published in the Toronto Star on April 14, 1990

16  Toronto Star, April 27, 1990

17  Prologue, pp. ii and iII, An Examination of U.S. Policy Toward POW/MIAs, by the U.S. Senate Committee on Foreign Relations Republican Staff, May 23, 1991

18    Ibid., p. ii, Prologue

19    Report of the Senate Foreign Relations Committee Republican Staff on U.S. POW/MIA Policy, May 23, 1991

20    One example is a letter dated June, 23, 1992, from John R. Nicholson of the White House Staff to Anthony Duke: "Mr. Petersmeyer asked me to thank you for your recent letter and the material on the POW/MIA issue compiled by your son-in-law, John Brown. The report is certainly timely. Because the POW/MIA situation is a national security issue, Mr. Petersmeyer has forwarded the information to Douglas Paal at he National Security Council for appropriate action. Mr. Paal is Special Assistant to the President and Senior Director Asian Affairs."

21    Chronology of Policy Related Events and Decisions, Office of the Vice Chairman, Robert Smith, Senate Select Committee on POW/MIA Affairs, September 21, 1992

22    The most accurate press account of the Senate Foreign Relations Committee's findings on Communist-held U.S. POW/MIAs of WW I, WW II, Korea and Vietnam which was seen by the author is an Associated Press report by Jim Abrams, "U.S. has abandoned POWs, MIAs in 20th century, GOP senators say," published in the Arizona Star, May 24, 1991; see also Seattle Post-Intelligencer, report of May 25, 1991 by P-I Military Reporter, Ed Offley

23    Authority for 'presumptive findings of death" are found in Title 5, Section 5565 through 5566 for civilian employees; title 37 USC, Section 555 through 557 for U.S. military personnel. These codified sections of law are implemented through regulations issued by the various departments and agencies responsible.

24    pp. 8-1 to 8-6, "An Examination of U.S. Policy Toward POW/MIAs," May 23, 1991, U.S. Senate Committee Minority Staff

25    Interview of Ted Landreth by the author in 1987, see also: p. 141 The Bamboo Cage

26    In May 1990, North Korea had returned five sets of remains of U.S. servicemen from the Korean War.

27    Washington Post, July 27, 1991

28    Transcribed from C-SPAN video tape by John M. G. Brown

29    Time: Sept. 2

30    September 1991, No.2 Issue, Nezavisimaya Gazeta, Moscow, USSR. The official Soviet government version of the WW II POW repatriations was also published in the same issue of Nezavisimaya Gazeta.

31    Cox News Service, Sept. 12th

32    Hearings of the Select Committee on POW/MIA Affairs, November 1991, op. cit.

33    Ibid.

34    Ibid.

35    U.S. Government Printing Office, 1992- ISBN 0-16-038479-6, Hearings, Nov. 5,6,7, and 15, 1991, Part II of II

36    NYT, USA Today, Arizona Republic and other publications

37    Wash.Times

38    Testimony of Colonel William R. Nelson, etc.

39    Full messages in bibliographical notes by Bruce Brown

40    See others in bibliographical notes, op. cit.

41    Editorials of Edward Epstein in the San Francisco Chronicle

42    The Search For MIAs, p. 17, by Garry L. Smith, op. cit.

43   Hearing Before the Select Committee on POW/MIA Affairs, January 21 and 22, 1992, U.S. Government Printing Office, 1992, ISBN 0-16-038399-4

44   Ibid.

45   Boston Globe, February 14, 1992

46   Letter in the author's files

47   Letter from Senator Robert Smith, dated 18 June 1992, in the author's files

48   1992-93 interviews with Senate investigators John McCreary, William LeGro and John Holstine, Select Committee on POW/MIA Affairs.

49   Defense Department "Memorandum for the Record," dated April 13, 1992, in the author's files.

50   Statements of Barry M. Toll to Chairman John Kerry of the Select Committee on POW/MIA Affairs, U.S. Senate; in the author's files.

# CHAPTER FOURTEEN

## BORIS YELTSIN'S REVELATION

On Monday, June 15, 1992, in an interview aboard his presidential jet while traveling to the United States, Boris Yeltsin said that American POWs from the Vietnam War had been sent to Soviet labor camps and some might still be alive: "OUR ARCHIVES HAVE SHOWN THAT IT IS TRUE—SOME OF THEM WERE TRANSFERRED TO THE TERRITORY OF THE FORMER USSR AND WERE KEPT IN LABOR CAMPS...WE DON'T HAVE COMPLETE DATA AND WE CAN ONLY SURMISE THAT SOME OF THEM MAY STILL BE ALIVE."[1]

According to news reports Yeltsin did not provide any further details but his spokesman, Vyacheslav V. Kostikov later repeated the statement, at the Russian Embassy in Washington, saying:

"After the Vietnam War, a certain number of (American) military prisoners were in Russia...the President and the new democratic government are trying to do their utmost to find those people—to find the memory of those people, because most of them, of course, have already died."

Yeltsin had written a letter to the U.S. Congress the week before about other, earlier American POWs, including several captured after spy plane shootdowns of the Cold War era, but this was the first time he had spoken about American POWs from the Vietnam War being transported to the Soviet Union. Early news reports said that the White House had declined to comment on Yeltsin's statement. A White House spokesman, Doug Davidson, said, "We have been discussing the POW issue with the Russians for quite a while and it will be on the agenda with President Yeltsin on Tuesday."

Yeltsin later said that "...some of them might still be alive, possibly in psychiatric hospitals," in Russia or the Commonwealth of Independent States.

Simultaneous with Yeltsin's announcement about American POWs, a story appeared that indicated the American POWs from Vietnam taken to the USSR may have died in secret Soviet bacteriological warfare tests, and during other experiments, as has been posited by the Czech defector, General Jan Sejna. The subject was introduced in the first news accounts of the day Russian President Yeltsin, "acknowledged that an epidemic of anthrax in the Ural Mountains 12 years ago was caused by military researchers trying to make a germ weapon, not by natural causes, as claimed by senior officials of the former Soviet Union. Yeltsin's unusual public statement came in an interview May 27 with the Russian mass circulation daily, Komsomolskaya Pravda."[2]

Senior officials of the Bush Administration appear to have been surprised by Yeltsin's admission and several U.S. officials interpreted his remarks as not being sufficient evidence that such had indeed been the case for decades. Such statements may have reflected either ignorance of the true facts due to not receiving an independent analysis or fear on the part of some senior U.S. officials whose record on the POW/MIA matter was dismal, but it was this negative information on Yeltsin's statement which was carried in the next few days by much of the American media. President Bush soon announced that he was sending Ambassador Malcolmb Toon to Russia immediately, to investigate Yeltsin's contention that some U.S. POWs from Vietnam might still be alive in Russia.

At an extraordinary meeting in the White House Rose Garden, between Boris Yeltsin and George Bush, the American President's reaction to Yeltsin's admissions was:

"IF ANYONE'S ALIVE THAT PERSON THOSE PEOPLE WILL BE FOUND. And equally as important to the loved ones is the accounting for any MIAs...If any single American is unaccounted for, they will go the extra mile to see that person is accounted for... President Yeltsin informed me for the first time that Russia, may (original emphasis) have information about the fate of some of our servicemen from Vietnam... For him to go back and dig into these records without fear of embarrassment is of enormous consequence to the people of the United States of America...I have every confidence that what he says here is true, that they will get to the bottom of it."

President Yeltsin said, "It's very possible that there are a few of them left alive, even on our own territory perhaps... We are going to carry this all the way to the very ground to find out the fate of every single last American who might be on our territory. Yeltsin also promised that he would open the archives of the KGB and the former Communist Party Central Committee to the joint U.S.-Russian Commission on Soviet held U.S. POWs and MIAs of WW II, the Cold War, Korea, Vietnam and Afghanistan.[3]

This shocking and unexpected story broke across the United States almost immediately, through television and radio news and in newspapers shortly thereafter. In Moscow, Richard L. Armitage, Director of U.S. aid to Russia and the former Soviet Republics (and thereby inheritor of the unfulfilled historic role of Herbert Hoover in 1921) was interviewed after Yeltsin's admission and was quoted by the New York Times of June 15th as saying, "If this is true, it is a fantastic breakthrough and a very positive development." Although he was in charge of POW/MIA policy in the Pentagon for nearly a decade under Presidents Reagan and Bush, Armitage claimed that to his knowledge, the United States had no information like Mr.

Yeltsin's assertions.

Russian President Boris Yeltsin, the former Siberian worker, stunned the world and brought a joint session of the U.S. Congress to its feet on Wednesday, June 17th, 1992, after admitting to the assembled U.S. Senators and Representatives that the Soviet Union had secretly brought American prisoners of war to the Soviet Union after World War II, Korea and Vietnam. A news report of the event recounted:

"Amid chants of 'Boris, Boris, Boris!' reverberating through the House chamber, Yeltsin stirred Congress to 13 standing ovations as he vowed the 'idol of Communism' has 'collapsed, never to rise again...'The most dramatic moment of the day came when Yeltsin strode into the packed chamber of the House to a standing ovation that was more enthusiastic than the reception Yeltsin gets in the Russian Parliament and rivaled Bush's own reception in Congress after the victory in the Persian Gulf War. It was the first speech by a Russian or Soviet president to a joint meeting of Congress...He stood before the House's enormous American flag, with his broad shoulders and swept back silver hair towering over the rostrum. His broad face creasing frequently into his easy grin. Yeltsin alternately flattered, cajoled, pleaded, blustered and sternly lectured the assembled senators and representatives about their responsibility to promote political and economic reform in the former Soviet Union. Drawing a thundering ovation, Yeltsin promised to account for every American POW...saying: 'Even if one American has been detained in my country and can still be found, I will find him...I will get him back to his family.'"[4]

At a news conference following his two-day summit with President Bush, Yeltsin was questioned about whether his predecessor Mikhail Gorbachev or other Soviet officials had known that American war prisoners taken by the Soviets may have survived. Yeltsin answered: "Well, that's just the point. They did know. That's the very point, that they kept it a secret...You have had a chance to ask (him)...why he kept this a secret. I am not responsible for him... The most important thing is that we know the numerical picture. We know how many people there were on the territory, how many were left, what camps the POWs were held in...We know who died, where they are buried...We don't know a certain number of people...We have simply no information about them. And this is why we say that maybe some of them are still alive and still in Russia."

The Socialist Republic of Vietnam meanwhile, rejected Yeltsin's statement that American prisoners were transported to the Soviet Union.

The Pentagon-inspired disinformation apparatus in the United States went into action almost immediately, and reports debunking or casting doubt the Russian President's startling admission appeared in the American media on the 17th and 18th. In one June 18th example headlined "LITTLE EVIDENCE OF POWs IN RUSSIA," and "Doubts Cast On Yeltsin Claim," it was stated that "U.S. and Russian EXPERTS said yesterday there are intriguing leads but little hard evidence to support Russian President Boris Yeltsin's repeated claims about POSSIBLE Soviet complicity in the disappearance of U.S. soldiers from Korea and Vietnam." A key U.S. "expert" quoted in this report was a "Paul Cole," as yet unknown to the handful of American specialists in POW/MIA research, but who would later be used by the U.S. Government in public Senate hearings to counter and debunk the years of investigation by the author and two other researchers, with his own hasty and inaccurate analysis. On June 18th it was reported:

"'IF THERE ARE ANY AMERICANS ALIVE (IN RUSSIA FROM THE KOREAN WAR) THEY DON'T WANT TO BE FOUND," SAID PAUL COLE, A RAND CORPORATION ANALYST WHO IS DIRECTING A STUDY ON THE FATE OF POW-MIAs FROM WW II, THE KOREAN WAR AND COLD WAR PERIODS FOR THE PENTAGON."[5] Thus did the Pentagon establishment and it's paid servants, with the help of the mass media, undercut Yeltsin's risky and courageous stand, and even cast aspersions on the character of any surviving American prisoners in Russia, who were unable to speak for themselves. "Dr." Paul Cole was to play a prominent role in the coming effort by the Pentagon bureaucracy to minimize the number of American prisoners who disappeared into the Soviet gulag after World War II and the Korean War. (Informants to the author who worked on a subsequent Senate Select Committee staff investigating POW/MIAs, maintained that Cole had a close relationship with former DIA analyst Sedgewick Tourison, Jr., who was to be hired through the efforts of Senator John Kerry of Massachusetts.)

Yet, on June 19th, an Associated Press wire story about a U.S. POW reported to have been recently held at an old Stalinist Gulag camp still containing several hundred prisoners under KGB guard was published across America. The report indicated that U.S. investigators had traveled to Pechora Camp # 350, 900 miles north of Moscow in arctic north Russia and were met with, "FRESHLY PAINTED WALLS, SMILING PRISONERS AND OFFICIALS WHO INSISTED THAT THE HUNT WAS POINTLESS." (Hence, no chance of revealing POW graffiti.) The prisoner was believed to be a Korean War or Vietnam War prisoner named Martin, which had been changed to MARKEN, for deniability purposes. The scene was faintly reminiscent of Maxim Gorky's visit to the Solovetski Archipelago 61 years before. Subsequently, Newsweek Magazine became the first of the three major national magazines to publish a serious account of American POWs of several wars held in the USSR, citing the above eyewitness livesighting account, with the source being a secret Pentagon report.[6]

The author had meanwhile forwarded two memorandums by fax (June 18 and June 20, 1992) for Ambassador Robert Strauss and for Ambassador Malcolm Toon in Moscow through Mr. Anthony Drexel Duke of New York, former Ambassador Angier Biddle Duke and General William Westmoreland (who had long encouraged this author's research). The two memorandums consisted of estimates of possible and probable numbers of U.S. POW survivors of several conflicts, in former USSR territory, and possible/probable locations of Americans there, based upon the author's analysis of all sources developed during seven years of investigations. After consultation with research colleague Thomas V. Ashworth the writer estimated that several hundred to 1,000 or more American survivors of WW II, including many native-born Americans, some foreign born and some Jewish Americans; and the Korean War, from the Cold War intelligence shoot-down incidents and U.S. POWs of the Vietnam-Indochina War, should be alive (unless they had been executed). In addition to previously published expected locations for American ex-POWs in former Soviet territory in and around Tyra Tam, Kapostin Yan, Sary Shagan, Novosibirsk, Shuli, Semipalatinsk, and Uralsk; other locations were added to the memorandum; in the vicinity of Krasodar (ICBM Missiles, telemetric work, and a Soviet military/industrial think-tank); BAKU (carrier aircraft landings and other specialties, may be ex-POWs of WW II, Korea, Cold War and Vietnam in vicinity); ALMA ATA and environs, particularly in radio and satellite interception locations; and again in and around NOVOSIBIRSK, which should also have surviving U.S. and other western allied ex-POWs of WW II and Korea.

The June 20th memorandum also noted that the existence of a super-secret military intelligence unit comprised of 50-60 of the best U.S. professionals which was later absorbed by NSA (National Security Agency) and had been unknown to Congress until recent days, had been revealed to the Senate Select Committee, and to the writer by Jerry J. Mooney. This highly secret unit had assembled in-depth studies which were the best-ever total analyses of special and operational intelligence, and from all other sources, of Soviet-related U.S. problems. The location of records and names of (other) knowledgeable, still living personnel involved in this unit had also been revealed to the Senate Select Committee, and to undeclared Presidential candidate Ross Perot, according Mooney, because of Perot's long-time stand on the necessity of recovering all possible U.S. POW survivors, and his personal interest in Mooney's testimony. The author's memoranda were also sent to the Senate Select Committee on POW/MIA Affairs. On July 9th, Ambassador Robert Strauss wrote to author J.M.G. Brown from the U.S. Embassy in Moscow:

"I just received a letter from Mr. Anthony Duke, together with a copy of your recent memorandum on 'Soviet-held POWs,' as well as a copy of your July 4, 1991 Op-Ed piece in the Washington Post, which, by the way, I thought was very powerful. I immediately passed your memo to the appropriate members of my staff, who, you may be interested to know, are working nearly full-time on the POW-MIA issue. I can assure you that we will look at your information very carefully and follow it up as best we can. You should know that we are doing everything humanly possible to assist Ambassador Toon and his people to find answers to the questions that remain and to finally put this matter behind us. I realize you may be skeptical, but I promise you there is no higher priority for us than to face up to this issue and to fully resolve it, if possible. I also believe that President Yeltsin is equally committed to resolving it, openly and honestly however painful it may prove to be for him and his country, although I should add that not all the bureaucrats in his government feel the same way, and so it will be a struggle to get from them all the information we need. We won't give up though."[7]

During this same period, reporter David Hendrix published a story in the Riverside (California) "Press-Enterprise," on June 19th, claiming the existence of a classified witness-protection type program for secretly-returned U.S. POWs from Indochina, in the 1980s, that may have involved from 100-300 Americans brought home clandestinely and reportedly not permitted to notify their families of their survival. The author had been informed of a program of this type during 1986-87, and given lists of 7-10 names of alleged returned prisoners, but had been unable to verify the facts as presented. However, the author had an experience with an individual who claimed to have been one of these American POWs returned clandestinely, long after the war's end, which is recounted later in this chapter.

The New York Times headlined a report on June 23rd: "U.S. Helps Russians Search For Missing in Afghanistan." The story indicated that Afghan guerrillas still held, "30-50 Soviet soldiers, most of them from Russia or Ukraine" of Moscow's estimated 300 POWs and MIAs who had failed to return after the war, and that U.S. officials were assisting in attempts to locate them using photographs and other information. It thus appeared that the diplomatic linkage between the Afghan-held Soviet POWs and the lost U.S. POWs of the Vietnam War, proposed by the author and Bill Hendon and established by President Bush in July 1989, was still functioning.

At public Senate Select Committee hearings on June 24th, Senator John Kerry said that the Pentagon had lied to family members and the public about the fate of at least 244 officially-listed U.S. POWs and the fact that at least 80 Americans were known to be in Vietnamese control, but never returned. The Pentagon claimed to have established that 127 of these 244 POWs had "died in captivity." (According to a source of the author within the Committee, this did not include other American POWs known to have been held in Laos, China, the USSR, and Cambodia, some of whom were captured during covert operations and were unacknowledged as POWs.) During the hearings on the 24th and 25th, the Vice Chairman of the Committee Senator Robert Smith stated: "Speaking for myself, I interpret the evidence as saying that POWs and MIAs may have been alive, or were alive, up through 1989." He said also the evidence he had been permitted to see did not allow him to determine whether any Americans had survived beyond 1989.[8]

At a news conference that same day, as-yet-undeclared Presidential candidate Ross Perot was quoted by the Washington Post as saying: "There has been a large number of sightings, et cetera, over the years. If you look in detail at the records, no significant effort was made to run that down."

Other press reports indicated that Russian intelligence was resisting the POW/MIA inquiry. Quoting "a secret Pentagon report," the Washington Times reported on June 24th that at a meeting of the joint U.S.-Russian Commission on POWs, "The Russians gave the most thorough accounting ever...but still it is only a preliminary accounting...It was clear throughout the course of the meeting that the KGB and GRU (military intelligence), among others, are not at all happy to be part of this process and are cooperating as little as possible...General Volkogonov said the Russians would provide a specific list of POWs in Indochina...Their research found instructions in the Ministry of Defense Archives on the treatment of American servicemen...There were many documents on AAA (Anti-Aircraft Artillery), but no documents on prisoners...Vietnam, according to General Volkogonov, is the most difficult case because much of the information has not reached the archives...But the Russian side is determined to study each case."[9]

The author delivered a letter for President Yeltsin of Russia on June 24th, together with the original edition of this book, and some 300 U.S. POW-related declassified documents, to Aleksandr Suchkov, identified as the "diplomat in charge," at the Russian Consulate-General in San Francisco. (This interview was filmed and recorded by the Russians.) The letter to Mr. Yeltsin also requested information from Russian or C.I.S. officials as to whether they believed any former Soviet POWs or civilian advisers captured by American forces in a past conflict (such as Vietnam) might still remain under American control, and revealed a personal experience of the author's, in capturing a non-Vietnamese prisoner of war in combat along the Song Be River, South Vietnam, during July 1968, who, under interrogation and in broken Vietnamese had identified himself as a Soviet adviser to the North Vietnamese Army, as noted previously in this book). All these communications were also faxed to Colonel Badiychuk of the Russian Military Attaché office in Washington, D.C., as the diplomat-in-charge at the San Francisco Consulate-General had refused to permit a Russian military officer to be present, even after one was pointed out nearby.

An Associated Press report from Moscow published in the San Francisco Chronicle on June 29th, accused the KGB of stonewalling the investigation about possible U.S. POW survivors in Russia, and stated that: "TOON AND VOLKOGONOV SAID PRISONERS MIGHT EVEN BE FOUND ALIVE, A SHIFT FROM THEIR EARLIER POSITION THAT THE PANEL (U.S.-Russian Commission on POWs) WAS UNLIKELY TO FIND SURVIVORS. BUT NOTHING WAS SAID TO INDICATE THAT THEY HAD MORE INFORMATION.

"Ambassador Malcolm Toon returned to the United States for consultation with President Bush on June 30th... The American Co-Chairman of the U.S.-Russian Commission on POW/MIAs stated to the American print media and on the "Today Show" of Wednesday July 1, 1992, that he had "seen no evidence that any U.S. POWs were alive and held in captivity in the former USSR." He said that in "two weeks" the U.S.-Russian Commission would conclude their investigation, and that it might take a long time to get back all human skeletal remains of U.S. POWs in Russia. Ambassador Toon also said however, that prior to his departure for the United States, the Russian President could not talk to him because of other commitments. Three Russian intelligence officials, including Yeltsin's deputy, General Volkogonov, claimed they didn't know what Yeltsin meant about live American POWs still being held prisoner in Russia or former Soviet territory. While the American Ambassador was saying that the investigation must be terminated within two weeks, the Russian officials were requesting they be given another 1-2 months to "thoroughly search their territory."[10]

In speaking to an agricultural group after meeting with the special Ambassador in the Oval Office on June 30, President George Bush said of Toon's investigation: "HIS SEARCH HAS YET TO UNCOVER ANY EVIDENCE THAT AMERICAN MIAs OR POWs ARE CURRENTLY BEING HELD IN RUSSIA."[11]

Ambassador Toon was quoted by reporters after the meeting as saying: "THERE PROBABLY ISN'T ANY LIVE AMERICAN POW BEING DETAINED AGAINST HIS WILL IN RUSSIAN FACILITIES, OR THE FORMER SOVIET UNION," and that he had encountered "some puzzlement" among Russian officials about why President Boris Yeltsin had suggested there were Americans still in captivity. Ambassador Toon said that at his urging, Russian officials had agreed to make a statement within two weeks as to "whether there (are) live American POWs being detained against their will."

The report also revealed that Ambassador Toon had said there could be some former POWs living voluntarily in the former Soviet Union, but that he had found no such information. In addition it was reported from by the Moscow News Agency ITAR-TASS that the grave of a U.S. POW reported missing since WW II, Francesco Luigi Di Bartolomeo, had been located in southern Russia, at a site that was after identified in the media as TAMBOV. This could indicate that Russian intelligence is determined to stick to the old KGB line that U.S. WW II POWs in Soviet camps such as Tambov "had been captured in German uniforms." (It is also noteworthy that the San Francisco Chronicle, commensurate with its' longtime policy of negating the substance of the POW issue, chose to headline this story with the somewhat disingenuous words: "PROBABLY NO U.S. POWS IN RUSSIA, BUSH SAYS." Other television news reports mentioned discovery of much larger numbers of thousands of graves of American, British and French WW II POWs, in former Soviet Territory.

In a news story also published in other newspapers across the country on July 2nd, the Washington Post reported that dismissed Senate Select Committee on POW/MIAs staffer William "Billy" Hendon had said after his firing that the Select Committee had been sharing its information with undeclared Presidential candidate Ross Perot. Hendon stated that he had talked to Perot about the Senate Select Committee on POW/MIA Affairs, but had not given Perot any classified documents.

Late in the afternoon on this same day in Washington, the assembled U.S. Senate unanimously asked in a 96-0 vote that the White House declassify secret eyewitness livesighting reports of American POWs of the Vietnam War held prisoner in Southeast Asia at least until 1989, DIA documents relating to U.S. POWs, the personal papers of Henry Kissinger during and after the Paris Peace negotiations, and the POW/MIA Presidential papers of Presidents Nixon, Ford, Carter, Reagan and Bush.

Senator John Kerry, the Democratic Chairman of the bipartisan Senate POW/MIA Committee was quoted in an Associated Press despatch of July 3rd:

"We believe that the American people have a right to read and evaluate the reports that Americans have been seen alive in Southeast Asia, and a right to read the official evaluations of those reports."

His Republican colleague, Vice Chairman Robert Smith, said that although he hoped the White House would comply with the bipartisan proposal for the July 23rd deadline, set by its authors Senators Grassley and Charles Robb, if the Bush Administration refused to cooperate, he would immediately move that the Senate POW/MIA Select Committee release the classified POW/MIA documents already in its possession. Associated Press reported that an Administration official, who typically spoke only "on condition that he remain anonymous," said that the White House would consider release for any POW/MIA material, through its "normal security review process."

Among the documents declassified by the action of the Committee was a July 1992 copy (#Vientiane 2503) of a November 1991 State Department cable to the Secretary of State from the U.S. Embassy in Vientiane Laos (#2798) concerning an interview of Pathet Lao official Soth Pethrasi by the U.S. Charge d'Affairs in Vientiane Laos, and his deputy chief of mission. In the meeting the Lao Communist repudiated all his wartime and postwar statements about U.S. POWs in Pathet Lao control as propaganda, which he had done to keep the people's spirits up. Moreover, the two senior U.S. representatives in Laos in 1991 seemed willing to take his word for it:

"WE DO NOT KNOW WHETHER OR TO WHAT EXTENT THE MFA MIGHT HAVE PREPPED SOTH FOR THE MEETING. ONLY MFA's SAYAKANE AND BOUNTHON (BOTH RELATIVELY JUNIOR) SAT IN ON THE MEETING AND SOTH SEEMED OBLIVIOUS TO THEIR PRESENCE. IF SOTH WAS REHEARSED, HE PERFORMED CONVINCINGLY. STILL WE HAD THE IMPRESSION WHAT HE WAS SAYING WAS SPONTANEOUS. SOTH MADE A POINT OF EMPHASIZING THAT HE WAS 76 AND VISITED THE HOSPITAL ALMOST DAILY IN CONNECTION WITH BACK AND STOMACH PROBLEMS. THE IMPLICATION SEEMED TO BE WHY SHOULD I LIE WHEN I AM ABOUT TO DIE. ONE MIGHT ARGUE THE PATHET LAO FLACK"S LATEST PERFORMANCE FOR HIS REGIME. BUT THE SINCERITY AND SPONTANEITY OF HIS STATEMENTS DO NOT APPEAR TO SUPPORT THIS ARGUMENT. DEPARTMENT PLEASE CONVEY GIST OF FOREGOING TO CONGRESSMAN FRANK MCCLOSKEY WHO MADE STRONG PITCH TO SEE SOTH DURING HIS VISIT TO VIENTIANE AND WHOSE INTERVENTION WAS IMPORTANT IN PERSUADING THE LAO TO AGREE TO THE MEETING. SIGNED: SALMON (CHARGE D'affairs)[12]

The Los Angeles Times published a report about the use of American POWs as test subjects in germ warfare experiments conducted in Communist China during the Korean War, with Soviet intelligence specialists and scientists present, at a facility believed to have been located in the Manchurian city of Harbin. The tests reportedly attempted to compare how racial differences affected the ability of American prisoners to withstand torture and interrogation. Another source said the testing may have involved germ warfare experiments. The East European source was said to be from Czechoslovakia and probably referred to former Czech Defense Ministry official, Major General Jan Sejna, who had defected to the West in 1968 and was a long-time employee of the U.S. Department of Defense. (Sejna had also alluded to possible biological warfare experiments with U.S. POWs from Vietnam, who he knew had been shipped to the Soviet Union.) According the news story, Secretary of Defense Richard Cheney had received the report from Defense Intelligence Agency investigators in April, and subsequently turned it over to the CIA and NSC for further investigation. The information was reported to come from an eastern European intelligence source and had caused the United States Government to send a high-level delegation to Beijing in May to request further information from the Chinese. The government of the People's Republic of China claimed that it had no evidence such a facility had existed.[13]

On the 216th anniversary of the independence of the United States, Nguyen Thi Binh, known as "Madame Binh" by the American press during the 1968-1973 war period when she was the Viet Cong representative in Paris, a signer of the 1973 Peace Accords, was in California. As the former Minister of Foreign Affairs for the Communist regime after the fall of Saigon in 1975, and later Minister of Education (and re-education) for the Socialist government of Vietnam until her "retirement" in 1987, who was in the United States to meet with receptive and pliable U.S. veterans groups, in a publicized attempt to achieve diplomatic recognition and trade from the United States for the Hanoi regime. A San Francisco Examiner reporter quoted (without comment or dispute) her statement on U.S. POWs, in a prominent and flattering feature article the following day, in a newspaper which, like dozens of other influential U.S. publications approached by the author, had declined to publish the human source reporting or declassified documentary evidence on Communist-held U.S. POWs of four wars, thus far uncovered in the National Archives or through FOIA request. Madame Binh, who as a Vietcong leader during the war was responsible for the deaths of many American servicemen, indicated that all the U.S. Senate's witnesses on American POWs left behind in Vietnam, and President Boris Yeltsin's statement to Congress on American POWs from Vietnam in Soviet labor camps, were wrong, and the San Francisco Examiner printed her words without clarification or rebuke: "Binh dismissed RUMORS that Vietnam is secretly holding American prisoners. 'WE HAVE ON MANY OCCASIONS AFFIRMED THAT THERE ARE NO POWS LIVING IN VIETNAM NOW,' she said. Nor did Vietnam 'SEND ANY POWS TO THE SOVIET UNION DURING THE WAR OR AFTER THE WAR...We have made efforts to cooperate with the U.S. side to search for the remains of American dead...But the public here does not understand that."[14]

The American public couldn't understand because, despite U.S. "national security" secrecy about POW/MIAs, together with the collusion of much of the American mass media in avoiding serious treatment of the subject, the actual fate of thousands of U.S. POWs and MIAs who remained in Communist control after several wars was already known or suspected by many Americans. The apparent collusion between Communist "diplomats," U.S. Government officials involved with POWs, and influential segments of the American media, thus continued in the 216th year of American liberty. It appeared that only a unilateral admission of the truth by the U.S. Government or a release of U.S. POWs could break the deadlock and confirm the full and awful truth about the lost American prisoners of war from 1918-1973.

Despite statements reported in the American media during 1992 by CIA Director Robert Gates and other officials, that all documents of the Cold War era were being declassified by that agency, many thousands of U.S. POW/MIA documents of the OSS, FBI, State and CIA, from WW I, WW II, Korea and later, still remain classified at the time of this writing, despite Freedom of Information requests and appeals by the present writer and other researchers. Appeals to CIA to release information in its possession on other Soviet and Chinese debriefing sites in North Vietnam, Laos and Cambodia, were ignored for many months, as were requests for the full debriefings of NKVD/ KGB defectors Anatoli Granovsky, Ilya Dzhirkvelov, Yuri Rastvorov, Pyotr Deryabin, Anatoli Golitsyn, Yuri Nosenko and others. On a related request, a December 10, 1991 letter from John H. Wright, Information Coordinator for CIA in response to author John M. G. Brown's October 10, 1991, FOIA request:

"...With reference to...paragraph 3 of your letter wherein you asked for certain information regarding 'debriefing sites for American personnel captured...at which Soviet or Chinese personnel were present,' FOIA does not require us to perform research or create records for a requester, nor does it require us to study a body of material...the same applies to that portion of your request for records on 'debriefings of any U.S. or Southeast Asian personnel who escaped or provided information on any such POW debriefing sites with a reported presence of Soviet or Chinese personnel.' Wright invited the author in a letter to review certain POW/MIA material in the CIAs archival reading room, but when the author attempted to go there after testifying under oath to the Senate Select Committee at the end of July 1992, 5 or 6 CIA security guards refused him entrance at Langley and claimed they could not reach Mr. Wright or anyone in his section by telephone, to verify the documents and identification presented.

Yet this same Mr. John H. Wright had traveled all the way across the United States for CIA in 1988 to tell television personality Janice Pennington, the widow of a German mountain climber residing in the United States named Fritz Stammberger, who disappeared on the Pakistani-Afghan-Soviet border in 1975, that she should just accept an off-the-record story of what CIA really knew about her husband, and if she tried to gain CIA documents on her husband's covert operations connections through the courts, that 'no judge in America would rule against the CIA.' The last night Pennington had been with her husband, he had told her that someone might tell her someday that he had been a Russian spy and that it would not be true. A declassified October 28, 1983 Interpol cable cited a source who stated that "Stammberger may have been involved in some sort of covert action along the Pakistan-Afghanistan border for which he was arrested, again he has no proof?" Was Mr. Wright's strange journey some further confirmation that the CIA has indeed never cared much about U.S. military POWs, as Colonel Shinkle and others had testified to, but would send their own top expert out to see the wife of a possible CIA agent.?

Personal appeals by the author to the Director of Central Intelligence, Robert Gates for declassification were ignored. Many priceless historical documents involving Communist-held American POWs of WW I, WW II, Korea and Vietnam may have been destroyed already by federal agencies interested in protecting institutional reputations in a post-Communist era. Other documents may be fabricated by these same agencies to refute the conclusions of the author and other POW/MIA researchers. In this regard it must be pointed out that Gates and other intelligence managers in the CIA, DIA-military intelligence, or at the State Department, have for years been in complicity in keeping this information on American POWs remaining in enemy control secret from POW/MIA family members, from Congressionally-chartered U.S. veterans organizations such as the VFW, American Legion, Disabled American Veterans and American Ex-POWs, and from the American people as a whole.

Undeclared Presidential candidate Ross Perot, who had engendered great disappointment among millions of his supporters that summer by suddenly withdrawing from the campaign, testified to the Senate Select Committee on July 1st in closed session. Perot confirmed under oath that Vice President Bush had asked him in 1986 to obtain for $4.2 million, the videotape showing American POWs as slave laborers in Laos which had been offered to Robin Gregson/John Obassy, then in prison in Singapore, by a foreign government. In his deposition given before Chief Counsel J. William Codinha, Perot said: "The Vice President asked me if I would acquire the tape and, if it were authentic, the U.S. Government was to reimburse me. Perot testified that after he had tried to contact the Vice President to report that the tape had vanished, Bush said he knew nothing about the effort. Perot testified also that Bush had encouraged him to "make one more sweep" through the classified POW/MIA files to re-investigate the matter but Perot refused: "I told him I did not want to because I knew we had left men behind in Laos and I did not need to make a study to prove that, and that I had real concerns whether or not our government would do anything if I did make a study and proved it again."

Perot also stated that as recently as 1991, the CIA had asked him to help the Department of Defense pay for the production of information on POWs demanded by the Senate Select Committee, but when he asked CIA to "have the National Security Council give me a green light and i'll do it, I never heard back." The Texas businessman stated that it was no longer up to the United States to prove that U.S. POWs were still alive in Indochina; that it was the Communist governments of Vietnam and Laos which must reveal what they had done with the prisoners they had held alive for years.[15]

After receiving a subpoena signed by the Senate Select Committee's Chairman, Senator John Kerry, author John M. G. Brown testified under oath on the process of uncovering documentary evidence and human sources concerning Communist-held U.S. POW/MIAs of four wars, for approximately seven hours on July 30, 1992. The author's testimony centered on the years of investigation and research which resulted in the historical evidence obtained and turned over to the Senate from 1989-1992, and recorded in the first edition of this book, which was self-published by the author in July 1992. The deposition was taken by John Erickson, Senator Robert Dole's 'man on the Committee,' who interrupted the proceedings at least ten times to leave the room "to make telephone calls" or to change the subject, and then arbitrarily halted the deposition prior to its proper completion. (Speaking almost entirely from memory, the author was not permitted to make any subsequent corrections, although senior U.S. officials thus deposed were permitted to do so.) The author simultaneously submitted the original, 700-page first edition of this book, *Moscow Bound*, as written testimony to the Select Committee, for its use in understanding the historical precedents in the POW/MIA matter dating back to 1918. A copy of this book was also forwarded by the author to President George Bush through Senator Charles Grassley of Iowa, and to White House Counsel C. Boyden Gray. (In a report from Moscow a few days later, investigative journalist Mark Sauter stated that a Russian Army officer, who asked that his name not be used, revealed Soviet records of WW I and the Russian civil war that confirmed author John M.G. Brown's research on the U.S. POWs and MIAs withheld by Lenin's Bolshevik Army in north Russia after 1919. The Soviet 6th Red Army documents revealed that at least 56 American POWs were in Soviet captivity in 1919 of which 43 U.S. prisoners were not released.[16])

Also in early August 1992, DIA officials who over many years were responsible for resolving livesightings of American prisoners held in captivity in Indochina by Southeast Asia refugees, and even a few American and European eyewitnesses, were required to testify again under oath before the Senate Select Committee. During these two days of public hearings, most of the 12 Senators appointed to the Committee were not present, Senator John Kerry, Senator Robert Smith and Senator Charles Grassley asked most of the questions, while Senators McCain, Robb, Kerrey and a few others put in only minor appearances. The author witnessed the entire two days of testimony, and it was soon apparent that the DIA officials present, Robert DeStatte, Warren Gray, Robert Sheetz, Charles Trowbridge, Gary Sydow and others, were continuing the years of disinformation by DIA which they and others had been responsible for, by debunking all eyewitness livesightings of American POWs seen in captivity over the 19 years since the war ended for the United States.

The DIA officials recapitulated for the Select Committee the status of the only 110 first-hand livesighting reports which were still under question. Little attention was given by the Committee that DIA analysts over the years had "resolved" hundreds of other such reports over the preceding years by highly-questionable analytic methods such as attacking many sources credibility or honesty in every conceivable way or debunking other reports on whimsical claims of geographic error, and through terrorizing informants with threats about their families' safety in Asia, demands that they take repeated lie detector tests, or recant their statements, and debunking all reports by any other method necessary, so that they had judged not a single report of U.S. POWs in captivity after Operation Homecoming to be credible, in the 19 years since the war had ended for the United States.

All the thousands of sightings of U.S. POWs after the war's end were explained away as "fabrications," by Southeast Asian refugees who, they claimed, usually were seeking immigration to America or financial reward, or as actually being sightings of other visiting European or East-bloc Caucasians. Out of thousands of first-hand and second-hand reports of Americans held in Southeast Asia since the war made available to the Committee, its investigators focused on four "clusters" or groups of livesightings, two in Vietnam and two in Laos. This clustering of livesighting reports was an obviously basic and rudimentary analytical intelligence tool which had, in fact, been totally ignored by DIA until attention was called to it by external and internal DIA critics in the late 1980s, the common-sense of which appeared to still be ignored by DIA as the hearings progressed.

Seventy reports by Vietnamese sources of an underground prison holding American POWs in Hanoi, including 22 separate first-hand reports by eyewitnesses, and 48 second-hand or 'hearsay reports, were discredited by DIA officials DeStatte, Sheetz and Gray on the grounds that the 'high water table of Hanoi precluded the existence of such a subterranean prison, reported to be under Ho Chi Minh's mausoleum and adjacent Vietnamese military high command buildings. DIA analyst Robert DeStatte, a cold-eyed, temperamental man who had for many years debunked all reports of living Americans remaining in captivity, told a largely irrelevant and time-consuming story of his recent travails in Hanoi as a member of the U.S. POW/ MIA team there, including details of his solitary bicycle rides past the Mausoleum of Ho Chi Minh and his talks with Vietnamese street urchins in the area, who, he maintained, WOULD HAVE SURELY INFORMED HIM IF INDEED AMERICAN POWs WERE HELD IN A SECRET PRISON BENEATH THE FACILITY. DeStatte was particularly emphatic in his testimony that no underground facility that could hold POWs could have been built in Hanoi because of the water table, yet months later, an October 1992 Reuters despatch from Hanoi stated: "THE RUSSIAN AMBASSADOR, RASHID KHAMIDOULINE, DESCRIBED HOW SOVIET EXPERTS WORKED WITH VIETNAMESE IN THE 1970S TO BUILD AN UNDERGROUND CHAMBER FOR GENERATORS AND OTHER EQUIPMENT TO MAINTAIN THE MAUSOLEUM."[17]

Senate investigators John McCreary, Colonel William Legro, Robert Taylor and other long-time professional intelligence specialists, next presented evidence of a cluster of 90 reports of a prison in the Viengxay area in Laos, a Communist Pathet Lao headquarters where American POWs had been held during the war and reported held since the war. 13 OF THESE REPORTS OF AMERICAN POWS HELD THERE FROM 1970-1980 WERE FIRST-HAND EYEWITNESS ACCOUNTS, WHILE THE REMAINING 77 WERE SECOND-HAND OR HEARSAY REPORTS. The DIA personnel at the hearings CHALLENGED THE TRUTHFULNESS OF ALL THE SEPARATE HUMAN SOURCES OF THESE REPORTS, apparently because that was the only way they could explain away such a nearly impossible set of coincidences. Faced with such a stubborn adherence by the DIA officials present, to the official government 'line' that 'no U.S. POWs were left behind in captivity in Southeast Asia,' Senator Robert Smith eventually became angry and accused the DIA officials of always discrediting the refugee sources rather than making any genuine effort to determine the accuracy of the eyewitness and second-hand reports. He also noted that this was a continuing example of suspicious conduct on the part of DIA analysts for years, that had led to public charges by former DIA Director Lt. General Eugene Tighe, former DIA POW/MIA Chief Colonel Millard Peck and many others, that DIA's analysts 'HAD A MINDSET TO DEBUNK' all livesightings of American prisoners remaining in Indochina since the end of the war.

Following this, the members of the Committee present examined the "Son La Cluster" or group of 19 livesighting reports, including nine by eyewitnesses. The DIA officials present again testified under oath that all these sightings of American POWs held in captivity in this area since the end of the war were "FABRICATIONS... BECAUSE NONE OF THE SOURCES DESCRIPTIONS OF THE AREA MATCHED THE ACTUAL GEOGRAPHY OF THE REGION." Senator

Robert Smith again took the lead in confronting DeStatte, Sheetz, Gray, Sydow and the other DIA officials by accusing them of "ATTACKING THE MESSENGER RATHER THAN DETERMINING THE ACCURACY OF THE MESSAGE..."

THIRTY-TWO SPECIFIC CASES OF LIVESIGHTINGS OF CAUCASIANS UNDER VIETNAMESE ARMED GUARD IN THE OUDOMSAI 'GROUP' OR 'CLUSTER' OF POW SIGHTINGS FROM 1973 TO 1985, FROM SEPARATE, UNRELATED WITNESSES, WERE DISCREDITED BY THESE SAME DIA OFFICIALS ON THE GROUNDS THAT, WHILE UNDISPUTABLY VALID, THEY WERE ACTUALLY ALL SIGHTINGS OF RUSSIAN 'GEOLOGISTS' OR SOVIET MILITARY ADVISERS 'UNDER VIETNAMESE 'ESCORT.' The DIA disinformation specialists present ignored or downplayed the fact that a number of the sightings reported the POWs or Caucasians were wearing prisoner garb or Vietnamese clothing and sometimes "Ho Chi Minh" sandals made from old American rubber tires. When pressed by Senators Kerry, Smith and Grassley for an explanation of such ludicrous analyses, DeStatte and Gray launched into long, rambling tales of their own personal experiences, embellished by self-aggrandizing "war stories" which had little or no bearing on the questions at hand, but succeeded in wasting a cumulative total of hours of the hearings. Their performance was tragically despicable, and clearly intended to confuse the outcome.[18]

The Committee also raised the issue of the EC47-Q shootdown on February 5, 1973, the 'Baron 52' incident, which Mooney and Minarcin had already testified about. DIA official Robert DeStatte (among others) clearly perjured himself before the Committee by denying under oath that any U.S. intelligence existed, either from the NSA's decoded traffic or human sources, that at least four of the crew of this aircraft had been taken prisoner and never subsequently released. He had been in a responsible position on POW/MIAs in 1973 at DIA, and DIA had in fact received this intelligence and more on the POWs, who subsequently failed to return in the prisoner-exchange. These of course, were Matejov, Brandenburg, Melton and Cressman, whom Jerry Mooney had listed as 'Moscow-bound.' In a classic example of DIA's "mindset to debunk," DeStatte testified that this type of intelligence data, drawn from decrypted enemy communications, was only "analyst-to-analyst musings, in which Jerry Mooney took this toothpick and built it into a house."[19]

A book published in 1992 by an Army Warrant Officer involved with the DIA and the JCRC remains-recovery effort in Southeast Asia, Garry L. Smith, gave some indication of the influence of DIA analysts like Robert Destatte and the others who had testified; or in the case of DIA's Sedgewick Tourison, had worked on the Select Committee staff with Chief Counsel Codinha and Chairman Kerry. . Warrant Officer Smith wrote:

"There is one final reason why I don't believe POWs were held against their will in Southeast Asia after 1973. This is because some outstanding senior analysts at the Defense Intelligence Agency, such as Wick Tourison and Bob Destatte don't believe it...They spent years sifting through intelligence reports and have not been able to find one scintilla of convincing proof that any were held..."

Warrant officer Garry Smith also revealed a significant conversation he had with General John Vessey, the President's special envoy to Hanoi for resolving the POW/MIA matter:

"I once had a one-on-one conversation in Bangkok with General John Vessey, Special Envoy for Presidents Reagan and Bush on the POW/MIA issue...General Vessey had been in Hanoi meeting with Foreign Minister Nguyen Co Thach, and stopped over in Bangkok at the American Embassy before returning to Washington to brief senior members of the Bush Administration about his talks. During our meeting, Vessey asked me what I thought of the live-sighting reports that (DIA) interviewers would submit from time to time, Before I could answer, he said: 'Most of these reports are pure baloney, aren't they?'

"I answered: 'Sir, they are all baloney. No American POWs are being held in Southeast Asia.'

"Vessey said nothing, but sat silently staring at me with his arms folded. I had no idea what he was thinking, but there was no absolutely way he could say something like that because of his official position and the potential political fallout that would occur as a result of his comments. Vessey told me that Foreign Minister Thach said in their meeting that he would swear on a bible, or anything else Vessey wanted him to swear on, that no Americans were being held in Vietnam, nor were there any remains being 'warehoused.'"[20] Warehousing was a sensitive issue within the U.S. Government. There had been undeniable evidence—real proof—that remains had been warehoused by the Vietnamese government in past years. However, Foreign Minister Thach may have been telling the truth. I had heard good analysts at JCRC say they believed all warehoused remains had been turned over to the United States in a spectacular series of remains repatriations that had occurred in the few months prior to Vessey's trip to Vietnam..."[21]

Not long after these hearings in Washington, D.C., on August 13th, the Russian co-Chairman of the U.S-Russian joint POW Commission, General Dmitri Volkogonov, released eight more names of American civilians who had been held as prisoners in the USSR, which were added to the list of 39 Americans that had been revealed by the Russians on July 31st. Two elderly men living in the Ukraine were reported to have surfaced from the July list, expressing the desire to visit their relatives in America.[22]

More significantly, on August 7th, the London Evening Standard published the results of a special investigation by Stanislav Kutcher in Moscow and Paul Cheston in London, under the heading: "1,400 Britons Died in Stalin's Gulags." Citing secret records uncovered by KGB officers in the Archive 12 vault of the Soviet Ministry of Defense, the newspaper reported that the 1,400 British POWs now proven to have been secretly withheld by the Russians in 1945, "were mostly pilots, naval officers and technical experts, seized for their vital knowledge of Western military intelligence and technology...a former KGB counterintelligence officer described the Britons who were secretly transported as 'SUPERB MATERIAL FOR TRAINING SOVIET AGENTS FOR FUTURE INTELLIGENCE WORK IN THEIR COUNTRIES... OTHERS WERE USED FOR ENTIRELY TECHNICAL PURPOSES, IN LABORATORIES WHICH STUDIED DETAILS OF ENEMY TANKS, PLANES AND COMMUNICATION SYSTEMS. THOSE OF COURSE WE JUST COULD NOT LET GO.'"

The prison locations in the USSR where some 200 of the British POWs named within the article were held included KARLAG in the Karaganda, Kazakh SSR region, UNZHLAG in eastern Siberia, VORKUTLAG in northern Siberia and YOSSELAG in southern Siberia. The newspaper cited high-level KGB sources to a Russian journalist who had been researching a book, parts of which had been published in the Russian publication Komosmolskaya Pravda: "Retired Soviet army colonel Boris Osipov described how prisoners were told 'You'll never get back home.' He said: 'IN JUNE 1945 I WAS ORDERED TO TRANSPORT ABOUT 1,000 POWs BY TRAIN TO THE SOVIET UNION FROM EASTERN EUROPE. THEY WERE MOSTLY BRITISH, BUT THERE WERE ALSO FRENCH AND DUTCH. THEY WERE CARRIED IN SHEEP TRUCKS NORMALLY USED FOR CRIMINALS...' In an interview with the Evening Standard, a leading British author and professor of international history at the London School of Economics named Cameron Watt said, "fears that British prisoners were being held in Soviet camps played its part in the controversial decision to repatriate Cossacks immediately after the war." (The author's years of research and investigation into the fate of missing American and Allied POWs who disappeared inside Soviet occupied territory in 1945, contained in this book, had resulted in the same conclusion.)

This important revelation was followed by an August 19th report in The Times, a most influential British newspaper, which also reported that Stalin's secret police had secretly withheld British POWs of WW II in Soviet prisons and labor camps, where they subsequently died in captivity: "The KGB has revealed that more than 2,000 foreigners, including hundreds of Britons, were held in Siberian labor camps after the Second World War." The newspaper reported that the British Ministry of Defense was checking a list of 200 British prisoners names supplied by the Russians against those held by the Army Historical Branch of thousands of missing British POWs of WW II and the Korean War, to see if they matched. At a Moscow news conference, a Russian security ministry director named Major-General Anatoli Krayushkin admitted that Stalin's KGB had definitely listed the foreign prisoners in their secret archives, and stated also that, "overall, the number of people who suffered from the purges IS SIGNIFICANTLY HIGHER THAN HAS BEEN REVEALED." The security ministry had also provided the names of two British POWs who died in KGB forced labor camps:

"ONE BRITISH SERVICEMAN, WHOSE NAME WAS SPELLED AS LT. MICHAEL McCAY AFTER BEING RETRANSLATED INTO ENGLISH FROM THE CYRILLIC VERSION, WAS SAID IN MOSCOW TO HAVE ENDED UP IN A KGB PRISON IN THE NORTHERN PORT OF ARCHANGEL BELIEVED TO BE A PILOT ON A CONVOY CARRYING PLANES TO MURMANSK...LATER HE WAS MOVED TO THE UNZHLAG COMPLEX IN MORDOVIA, ...RUSSIA. HE IS BELIEVED TO HAVE DIED IN 1954.

"THE OTHER WAS NAMED AS GERALD PHILLIPS. YURI VIDOVSKY, A RUSSIAN WHO SPENT 20 YEARS IN PRISON CAMPS, IS REPORTED AS SAYING THAT HE HAD MET PHILLIPS IN CAMP NO. 15-63 AT NOVOCHERKASSK IN SOUTHERN RUSSIA IN 1962. IN HIS WORDS, 'HE (PHILLIPS) TOLD ME THAT HE HAD SERVED IN A RECONNAISSANCE SECTION OF A REGIMENT AND WAS TAKEN PRISONER BY THE GERMANS IN LATE 1944 ON THE FRENCH BORDER...HE WAS INTERROGATED BY KGB OFFICERS EVERY DAY FOR FIVE YEARS.' VIDOVSKY SAID HE DID NOT REMEMBER THE NAME OF A SECOND BRITON HE HAD MET, WHO WAS FIGHTER PILOT..."[23]

General Krayushkin was reported by ITAR-TASS news agency as speaking of the British and other foreign POWs after turning over to the Austrian Embassy a list of more than 200 Austrian POWs also withheld by the KGB. Sergei Osipov, an analyst with the Russian parliamentary commission investigating foreign prisoners held in the gulags, was reported to have confirmed the existence of British servicemen held there. Thus, while the U.S. government and the American mass media denied the facts recently revealed by declassified U.S. records and human sources, and confirmed by no less a personage than the President of Russia, Boris Yeltsin, the truth about the disappearance of thousands of Allied POWs into the Soviet Union in 1945 began to emerge on the other side of the Atlantic. A great window of opportunity was thus ignored, and would later be nearly closed.

The author had meanwhile been informed by Committee staffer William LeGro that a special U.S. Army Intelligence team called "Joint Task Force Russia," headed by Major General Bernard Loeffke (formerly the U.S. military attaché in Moscow under Ambassador Malcomb Toon) and by a U.S. Military Intelligence officer, Colonel Stuart Herrington (see Chapter 11), was going to Russia, "to assist

Ambassador Toon and the others of the American delegation of the joint U.S.-Russian Commission on POWs." (This task force had been authorized by the Department of Defense and was implemented by the Department of the Army.) According to LeGro, in line with recent international disclosures in the mass media, Herrington had reportedly already gone to the Pentagon in order to get the official Defense Department policy straight on Soviet-held U.S. POWs of Korea and the two World Wars. At the request of Committee staffer LeGro, the author supplied the results of research contained in this book to the leaders of Task Force Russia, fully cognizant that U.S. officials might subsequently use the book to discredit the evidence and produce countering disinformation on POW/MIA cases specifically cited by the author.

On Capitol Hill, the Senate Select Committee on POW/MIA Affairs was still taking closed-door depositions from witnesses with knowledge about prisoners of war and on September 4th, former CIA Director and Secretary of Defense James R. Schlesinger went under oath in the Senate security area, and somewhat vaguely recounted his memory of a critical time for the Vietnam POW/MIAs, from March through July 1973, when he had been confirmed as Secretary of Defense and the POWs had been written off as all being "dead." For only three months, February, March and April, Schlesinger had effective control as Director of CIA and he maintained that during that time the Pentagon was in control of the POW/MIA matter, although he admitted that Laos was a CIA-run war under his purview, he claimed to remember almost nothing about the hundreds of prisoners reported held there who had never returned, and could not remember being briefed on the missing POWs in Laos at CIA. He thought perhaps the deputy directors had handled it, but after his appointment as Secretary of Defense he remembered:

"I do know when I got to the Pentagon (May 1973) there was a growing exasperation...an anger, dismay about some of the prisoners of war. It was still a period in which information was being absorbed from those who had been released..."

A Senate investigator brought up one of the confirming pieces of evidence that POWs had been left behind: "A witness yesterday... testified that in his recollection...10 to 15 of the returning POWs stated during their debriefings that they had seen other live Americans in captivity who did not return, and he put the number in Vietnam alone as somewhere between 25 and approximately 100... actually mentioned in debriefings...He said....many of the men who did not return...were said to have been seen very recently in captivity, many of them within a couple of days of the release of the returnee..."

Schlesinger responded: "I was not briefed on the detailed read-outs from individual prisoners. I was aware of a good deal of anger and the belief that 'those bastards may have deliberately killed our people at the very close.' ...I think there was a belief that there may have been some that had remained behind alive; as in Korea, a... very small number may have remained behind voluntarily...but others may have been held behind; but we didn't have...we couldn't pin it down... General Haig was frequently consulted in that period by President Nixon, and it was more his desire to consult Haig than Kissinger's desire. These decisions were made on a very, very limited basis, in my judgment, particularly because of the sensitivity of these issues.

"Kissinger would, from time to time, consult with the Russians... with Dobrynin...Kissinger had a complex world to deal with. The President had a complex world to deal with... Kissinger, who would have the lead on this...was not only dealing with the POW situation... He was concerned about not drawing down on his credit with the Russians too frequently, because he had a SALT negotiation under way.... I think that I can recall some documentation...that indicated that Dr. Kissinger had expressed concern about the possibility of the North Vietnamese demanding concessions as a consequence of pressing certain issues too hard. Yes, I think that (POWs) was the context...This is the kind of sensitive thing that Kissinger would frequently not discuss in a broader forum...He would reach out and say Elliott, we'll talk about your memorandum later on, in my office."

"Sometime in early May, I began to spend most of my time at the Pentagon...(after appointment by Nixon)...I was not confirmed by the Senate as Secretary of Defense until sometime in mid-July... there were certain things I did at the Pentagon with the permission of Senator Stennis...I remind you, I didn't get back to the Pentagon until three or four months later, and things were different then. This is reasonably hard evidence you've got here (March-April) I assume that people from Eagleburger, Shields, Richardson, Moorer, other members of the uniformed military, had been frustrated...People had become inured to the fact that we were going to go through a set of procedures through the operation in Thailand...we were going to go down the trail of trying to persuade the North Vietnamese and the Laotians to cooperate and live up to the agreement...When that first set of passions and anger that those bastards aren't going to do it and giving way to frustration...You can't maintain this degree of hope and anger over a period of months." "I got my briefings basically...more or less on the basis of WE THINK THEY'VE GOT, THEY MAY HAVE SOME PEOPLE BACK THERE AND WE ARE MAKING EVERY EFFORT THAT WE KNOW THROUGH INTELLIGENCE...TO GAIN INFORMATION."

When Schlesinger was asked, 'who was Kissinger's man for Vietnam?' He answered: "Well, there was Winston Lord and John

Negroponte...although I think (they) were being moved in other directions in the March, April, May, June time period...My military assistant was...General John Wickham...Haig should know a great deal about this. You know, for this kind of thing there's not a long list of people."

The former CIA Director and Secretary of Defense was asked: 'Were you aware, either in your role as DCI or in your role as Secretary of Defense, of any intelligence information indicating that U.S. prisoners of war may have been moved from Indochina to either the People's Republic of China or to the Soviet Union?'

Schlesinger answered: "Let me be very careful in my answer. I do not recall any such information, but there may have been speculation...It's quite plausible to me...relations between North Vietnam and China were poor even then...so I would not think that there was much likelihood of the North Vietnamese doing anything, including providing prisoners of war to the Chinese. That they might have been moved to the Soviet Union is possible...More likely than that would have been turning them over to the Soviets in-country for interrogation in-country. And to the extent that people may have been tortured or brutalized in the process and subsequently removed or not returned, ANYONE WHO HAD BEEN SUBJECT TO INTERROGATION BY THE SOVIET AUTHORITIES, GRU OR KGB, WOULD MORE LIKELY THAN OTHERS HAVE BEEN SO REMOVED...I myself would speculate that it would have been wise for the Soviets to move American prisoners to the Soviet Union...If the Soviets had desired, I feel reasonably confident that the North Vietnamese would have accommodated that desire."[24]

When he was asked if he'd ever seen Intelligence information on interrogation of U.S. Vietnam POWs by Soviet intelligence, in-country, Schlesinger replied: "No...It seems to me that I either read speculation or that there were hints from this and that quarter, maybe from some of the foreign diplomats assigned to Hanoi...I believe that, given the nature of these regimes and given my clear recollection in the 1973, '74, '75 time period of the treatment that had been meted out...to American prisoners of war, and given the habits of these regimes up to the level of the Soviet Union, that they had been held back, deliberately held back struck me as plausible at the time."[25]

On September 11,1992, a 20/20 television broadcast hosted by Barbara Walters promoted a just-published book by former-policeman Jim Sanders and KIRO-TV journalist Mark Sauter entitled: *Soldiers of Misfortune*, which covered the POW/MIA matter from WW II to Vietnam. In the book's preface (on page 15) the authors state erroneously that the origins of the secret history of U.S. POW retention by the Russians lay in an American failure to act honorably with Soviet POWs in U.S. control at the end of WW II, in 1945: "The

United States and Britain cheated on the hostage exchange, retaining many anti-communist Ukrainians and Byelorussians to help fight the rapidly developing Cold War. Stalin learned of the deceit, and, in return, kept up to 23,500 Americans and 31,000 British and Commonwealth soldiers."[26]

Although it was stated on the 20/20 program, and given as this book's thesis, that the Soviets had withheld American POWs of WW II because the Americans had not repatriated all Soviet POWs in U.S.-Allied control, in fact, as the present writer has demonstrated in this book, the Soviet Communists had actually initiated this reprehensible policy long before, under Lenin and Trotsky in 1919, in order to force U.S. diplomatic recognition of the Bolshevik regime and the repatriation of Russian prisoners, by using illegally-withheld American POWs and MIAs as hostages, and also to subsequently obtain U.S. aid. American (and British and French) POWs/MIAs had actually been held in Moscow prisons and Gulag labor camps since the 1918-1919 period, at the close of WW I and the Allied-American intervention in Russia. This fact was known to senior U.S. and British policymakers of WW II, and had a major impact on all policy decisions relating to American POWs taken under Soviet control in 1945. American and British policymakers of 1945 thus already knew, and fully expected, similar actions with POWs by the Soviet dictator Josef Stalin in 1945, and acted accordingly. Much of the WW II and later research material used in the Sanders-Sauter-Kirkwood book was the same as that which was simultaneously uncovered by the author with research-colleague Thomas V. Ashworth, and was turned over to the U.S. Senate from 1989-1992, and used in writing this volume. By not extending their investigation and research back to the actual historic origins of the problem of Communist-held U.S. POW/MIAs following the Bolshevik revolution, the authors drew an incorrect historical conclusion, and caused 20/20 to present an incorrect version of the history of this tragic matter to the American people. (Author John M. G. Brown subsequently sent the first edition of this book to Barbara Walters, who did not reply.

If the producers and researchers for 20/20 and ABC News had studied the May 23, 1991 Senate Foreign Relations Committee Republican Staff report on U.S. POW/MIA policy, which had been produced largely from the archival research of the present writer and Thomas V. Ashworth, and also Volume II of *POW/MIA Policy and Process*, the November 1991 records of the Senate Select Committee on POW/MIA Affairs which contained nearly 300 pages of declassified U.S. documents supplied by the author to Senator Charles Grassley for the Foreign Relations Committee investigation and reproduced therein, they would have found reference to this important historical evidence on U.S. POWs and MIAs held by the

Soviets from 1918 onward. In addition, 20/20 claimed to have made a significant discovery in locating the graves of former U.S. POWs Timmerman and Yates, buried in Odessa, Russia, giving the impression that they had been secretly-withheld as POWs by the Soviets after WW II, when in fact, they had been accidentally killed by a falling wall that had been previously damaged by German shells while they and hundreds of their comrades had been in the process of being repatriated to Italy on the British freighter Circassia. A source of the author, a still-living ex-POW named Arley Goodenkauf, of Table Rock, Nebraska, was a witness to this event as noted earlier in this book. Their burial in Odessa had actually been known to the U.S. Government since April 1945 and their remains subsequently were repatriated to the U.S. in the late 1940s, for a high price demanded by the Soviet government.

Shortly after this, the combined sworn testimony of former defense secretaries Melvin Laird and James Schlesinger, and of former NSC advisers and secretaries of state Alexander Haig and Henry Kissinger, on September 22nd and 23rd, 1992, confirmed through the lips of senior U.S. officials that American POWs were left behind in captivity in Southeast Asia after the U.S. military withdrawal, and the alleged repatriation of all American POWs in early 1973. The testimony of all these officials taken together also made it clear that this had been a political and policy decision, taken to increase the chances of President Nixon's survival in the office of the Presidency as the Watergate scandal had been in the process of being exposed to the American people during the spring and summer of 1973. James Schlesinger said under oath: "I have a high-probability assessment that people were left behind in Laos ...and a medium-probability assessment with regard to Vietnam, and he also testified that, in 1973, "We were not going to roil the waters."

Former Secretary of Defense Melvin Laird, who left office in January 1973, admitted to being "disappointed" with the list of prisoners presented by Hanoi in 1973: "Had I been secretary of defense at the time, I would have gone public." Laird also testified that the Department of Defense had received under his direction over 5,000 letters from 1969 to 1972 from prisoners of war in Indochina, including some 1,000 made available by American anti-war groups visiting Hanoi, which named within many other U.S. POWs including 474 or 478 named Americans who were never released by the Vietnamese. (Laird also testified that he had been told by some ex-POWs, including later-Senator John McCain, a member of the Committee, that some of these men had died in captivity.) Yet, many others had not.[27]

According to Senate investigators who later spoke with the author, former Nixon aide and Secretary of State Alexander Haig accused the Committee of engaging in a "witchhunt." Described by the Washington Post of September 22, as "pugnacious and defensive" while testifying, Haig made derogatory remarks about Senator Kerry's Vietnam-era experiences, and read a proposal that the Senator had made when a wounded navy lieutenant in his younger anti-war days, instead of dealing with the prisoner issue directly. The Washington Post reported that Haig accused the committee of conducting a "headhunting mission against Nixon, Kissinger and perhaps Al Haig..." Admiral Daniel Murphy was a former military assistant to 1973 defense secretaries Melvin Laird and Elliot Richardson and later a CIA Deputy Director and Chief of Staff to Vice President George Bush. He said: "Mr. Chairman, in my personal view, there were NO confirmed reports of live U.S. military personnel left behind in Vietnam or Laos...I do not recall seeing any such reports..."

Admiral Moorer, Chairman of the Joint Chiefs of Staff in 1973, and responsible for the numbers of POWs to be negotiated for, testified that he had ordered the U.S. withdrawal halted but was overruled, and that it would not have been possible for Nixon to start the war anew over the possible-non return of POWs, given the political situation in America at the time. Senator Robert Smith also raised the issue at the hearings of U.S. POWs known to have been held at or near Sam Neua, Laos, the Pathet Lao headquarters and location of the "Tentacle-MB" interrogation camp for POWs, discovered through the NSA decryptions by Mooney and others:

"In a recent statement from DIA regarding some other things that are going on that I don't wish to disclose here in public, CIA has said in this memorandum, '...historical precedent exists for suggesting the presence of American POWs in the Sam Neua area. Photographs taken by a reconnaissance aircraft in October 1969, show what may be as many as 20 non-Asians accompanied by Pathet Lao guards near caves at Ban Nakai Thua, 20 kilometers east of...Sam Neua."

Former Ambassador G. McMurtie Godley, who had been in Vientiane when estimated 350 American POWs and MIAs captured in Laos were abandoned, self-righteously testified: "I am, to put it frankly, disgusted with the way POW/MIA issues have been handled both in the media and by disreputable Americans who obtain...from the families to finance quote private search and rescue operations..."

Winston Lord, the quintessential State Department bureaucrat who had been Kissinger's aristocratic senior aide in Paris and was subsequently rewarded with appointment as U.S. Ambassador to China, defended his old bosses, saying that Nixon and Kissinger had little choice but to complete the 1973 withdrawal and act like North Vietnam was carrying out the prisoner repatriation agreement, but he admitted some of the truth about the lost American POWs and MIAs: "THE PRESIDENT DECIDED NOT TO SCUTTLE THE (Paris peace)

AGREEMENT OVER THE MIA ISSUE....IT WAS A VERY TOUGH DECISION..."

Lord wasted no words on how much tougher such a decision was for the surviving U.S. POWs and MIAs who had been abandoned while alive in enemy control, and few words on why such information was kept secret from the American people. He said that if Nixon had stopped the American withdrawal or resumed bombing Hanoi because North Vietnam's short POW lists contained... "Very disturbing evidence...of discrepancies," he believed that, "... AMERICAN SOCIETY WOULD HAVE BLOWN APART."

While some other details of this testimony have already been incorporated into the earlier part of this book dealing with the 1973 POW repatriations after the Paris peace agreement, it should be again noted here that in Henry Kissinger's prepared statement of the 22nd of September 1992, to the Senate Select Committee on POW/MIA Affairs, he wrote:

"...WHILE I WAS IN OFFICE, WE DID RECEIVE SOME REPORTS ALLEGING THAT LIVE AMERICANS WERE STILL IN INDOCHINA. I CAN ASSURE THE COMMITTEE THAT ALL THOSE REPORTS WERE TAKEN SERIOUSLY. BUT NO CONFIRMED REPORT OF LIVING AMERICAN PRISONERS EVER CROSSED MY DESK..."

"ON MARCH 29, PRESIDENT NIXON ANNOUNCED THAT 'ALL OF OUR AMERICAN POWS ARE ON THEIR WAY HOME.' THOSE WHO LEAKED THAT PRESIDENT NIXON, OR HIS ADVISERS, IN ANNOUNCING THAT ALL PRISONERS HAD BEEN RELEASED KNEW THAT WE WERE KEEPING PRISONERS BEHIND, DISINGENUOUSLY NEGLECTED TO MENTION PRESIDENT NIXON'S STATEMENT THAT, 'THERE ARE STILL SOME PROBLEM AREAS. THE PROVISION OF THE AGREEMENT REQUIRING AN ACCOUNTING FOR ALL MISSING IN ACTION IN INDOCHINA...(HAS) NOT BEEN COMPLIED WITH...' THAT IS ALL WE KNEW...NO ONE WAS LEFT THERE BY THE DELIBERATE ACT OR NEGLIGENT OMISSION OF ANY U.S. OFFICIAL. LET US STOP TORTURING OURSELVES. THE UNITED STATES GOVERNMENT KEPT FAITH WITH THOSE WHO SERVED THEIR COUNTRY. IF SERVICEMEN WERE KEPT BY OUR ENEMIES, THERE IS ONE VILLAIN AND ONE VILLAIN ONLY: THE COLD HEARTED RULERS IN HANOI. THAT IS THE CONCLUSION YOUR REPORT OWES TO OUR NATION..."

Yet in the hearings which followed it became clear that at least hundreds of American prisoners known to have been alive in captivity in Vietnam and Laos were never released. During the September 22nd hearings, Kissinger later made a last effort to deny any had survived, saying:

"...I think it's improbable that any are alive today. I honestly did not think there were any alive in Vietnam when the war ended, but I have always kept open the possibility in my mind that there were some in Laos...I certainly told them innumerable times that we were not paying ransom, we were not paying reparations...They never said you owe us economic aid, and therefore we are holding prisoners..."

The method of hiding losses of U.S. POWs remaining in enemy control by falsely keeping them in an MIA category or moving them to other categories of loss after the fact, which had been used after 1919, 1945 and 1953, was continued in 1973 with the end of American military involvement in the Vietnam war. A secret U.S. document declassified on September 21, 1992, the same day as the public testimony of former secretaries of defense Schlesinger and Melvin Laird, illustrates the continuity of this seven-decades-old policy. This was a May 22, 1973 memorandum from then-Deputy Secretary of Defense William Clements to all branches of the American military services that ANY RECLASSIFICATION OF A SERVICEMAN FROM MISSING IN ACTION TO POW STATUS MUST BE CLEARED BY SECRETARY CLEMENTS HIMSELF, BEFOREHAND. Former Secretary of Defense Schlesinger, Clements' old boss in the Pentagon, testified that he had just been informed of this memorandum in September 1992, that he considered it highly unusual, and that the "POLITICAL SENSITIVITY" of the POW/MIA issue may have been the motive for Clements actions at the time.

Rather than deal directly with the larger, more accurate number of approximately 1,200 Americans who were believed to have been captives but never returned, the Committee concentrated on some 130 cases of prisoners who were known with certainty to have been captives in Vietnam and Laos but failed to return. In what was perhaps his best performance as Chairman of the Select Committee, Senator Kerry ultimately proved that Kissinger had in fact made an appeal in May 1973 to North Vietnam's Le Duc Tho to help the Nixon administration convince the American people that the 9 Americans released from Laos were the only U.S. POWs alive in that country. It was a classic example of 'wink and nod diplomacy,' in this case involving a Soviet surrogate regime. Kerry read from recently declassified minutes of the exchange:

"This is in respect to Laos. You're talking to Le Duc Tho, and Le Duc Tho says: 'Minister Thach has told Ambassador (William) Sullivan that we will HELP YOU COORDINATE WITH OUR ALLY IN LAOS...' Secretary Kissinger: 'YES, BUT ALSO FOR THE PURPOSE OF REALITY. IF YOU WILL HELP US, IT WILL BE HELPFUL IF YOU WILL GIVE US YOUR ASSISTANCE WITHOUT MAKING A PUBLIC STATEMENT ABOUT IT. You have often told me you could do things that are not written down.'

"Le Duc Tho: 'I agree, but I have to add that we have to cooperate with our Lao friends because it is their sovereignty. You then say: I UNDERSTAND. NOW WE WOULD LIKE A SENTENCE FROM YOU–AND I CAN'T UNDERSTAND WHY YOU CAN'T GIVE IT TO US–WHICH SAYS THAT, QUOTE, THE DEMOCRATIC REPUBLIC OF VIETNAM HAS BEEN INFORMED THAT THERE ARE NO PRISONERS BEING HELD IN LAOS,

THAT ALL THE PRISONERS IN LAOS HAVE BEEN RELEASED. IT WOULD BE VERY IMPORTANT FOR US...'"

Senator Kerry then said to Kissinger:

"So there you are in May (1973), with Le Duc Tho, saying not what happened to John Sparks...we need an accounting, but saying, give us a sentence that says there's nobody alive in Laos, it'll be helpful to us."

At this point in his testimony Kissinger became infuriated and challenged the patriotism of Kerry, a decorated Vietnam veteran who had later opposed the war.

During the hearings on September 22, Republican Senator Nancy Kassebaum of Kansas also questioned Kissinger about the Soviet connection to American POWs who had never returned. She asked:

"In these negotiations regarding the list, was there any consideration taken to whether there might be any in the Soviet Union at that time, any prisoners of war taken to the Soviet Union?

Kissinger answered somewhat evasively: "You know, one of the unsettling things to me seeing all these documents (is) that things appear that I might easily have sworn to never happened. I do not recall any such conversation, and we would have reacted so neuralgically that I don't believe I would have forgotten that. I mean, if we had had any suspicions with all the contacts we had with the Soviet Union they never would have heard the end of it...I DON'T BELIEVE WE EVER RAISED IT WITH THEM...I KNOW WE NEVER RAISED IT WITH THEM, AND THAT HAS TO BE BECAUSE NO ONE EVER ALLEGED IT."[28]

Senator Kerry summed up much of the evidence that POWs had been left alive in Indochina (and mentioned a specific case where POWs had been reported as Moscow-bound by NSA analysts Mooney and Minarcin) in a statement to Admiral Moorer on September 24th:

"Dr. Shields, in his own memorandum written to the Secretary of Defense, sought to correct his earlier statement recognizing that it had become policy. And he specifically said to the Secretary, we've got evidence of three or four people possibly from a Baron 52 flight (the EC47-Q flight of February 5, 1973)...and he said this policy is not correct. So I must say to you that the evidence is overwhelming that there is a gap between the stated public policy and the reality at that point in time."

As the Senate hearings continued in Washington, D.C., the simultaneous half-hearted Russian effort to locate records of American POWs, and the survivors who still remained alive on former USSR territory, digressed into a series of minor revelations on American civilian prisoners held in the gulags, as U.S. Co-Chairman of the Russian-American Joint POW Commission, Ambassador Malcolm Toon again said on September 21st that he saw no evidence of American POWs remaining alive in the former Soviet

Union, but that the Russian statement "wasn't sufficiently specific... Above all, I want a clearer statement from the Russian government that there is no live American POW held in detention in Russian facilities...The American public is getting awfully impatient...The feeling back home is that this exercise is not doing very much."

The Russian intelligence aide to Boris Yeltsin on the POW issue, General Dmitri Volkogonov, announced in Moscow, also on the 22nd, that no American prisoners of war remained alive in the former Soviet Union, and on September 23rd released the names of two more Americans, identified as a "Mr. Oggins and a Mr. Clifford," who had died in Stalin's camps in the 1940s. The American named Oggens (later corrected to Oggins) had reportedly been arrested in 1939 on "espionage" charges, and was believed to have been killed on the orders of Viktor Abakumov, one of the merciless deputies of Stalin's KGB chief, Lavrenti Beria. The reason given in the Soviet documents for the killing of the American, Oggens, was so that his story would not be told to the western world. Vologonov also claimed that some files on American POWs had been falsified, and that, "IN MANY CASES THERE ARE SIMPLY NO TRACES LEFT."

A Washington Post foreign service despatch of September 24th reported that U.S.-Russian POW Commission Co-chairman Volkogonov had revealed that a secret plan discussed between Josef Stalin, Mao Zedong and Kim Il Sung had urged that one-fifth of all U.S. prisoners of war be withheld in captivity after all others had been released. The Russian general claimed that this proposal had originated with the Chinese, and that he didn't know if the plan had been implemented. This last patently untrue statement clearly indicated that Volkogonov was not being truthful in this matter, and probably in all his other dealings with the U.S. representatives on the Commission, when he also said that the chances of finding a live American POW "ARE ALMOST NIL." Volkogonov claimed that "no evidence has been found" that Korean war POWs had been transferred to the USSR, including the 54 admitted by the Russians to have been interrogated by Soviet intelligence, while in Chinese Communist control. This report also indicated that the murdered American reported earlier, (first name, possibly Isaac) Noggins, had been born in Massachusetts in 1897, was arrested in Moscow in 1938 and held in a camp near NORILSK, and executed in January 1947. Five of the reported 49 American civilians held in the former USSR were said to be still alive, and wishing to find their relatives in the U.S.[29]

In the meantime, a leak from the Senate Select Committee on POW/MIAs to syndicated columnists Evans and Novak, which was also published on September 24th, revealed that a highly secret "Eyes Only" memorandum involving U.S. POW/MIAs, written by the current National Security Adviser to President Bush, General Brent

Scowcroft, when he was a senior White House aid to President Nixon in 1973, had been refused to the Senators on the grounds of "EXECUTIVE PRIVILEGE." It appeared that to the bitter end certain senior Bush Administration officials dreaded the publicly acknowledged return of even a few living American POW survivors from Vietnam, Laos, China, Korea or from the Russian republics, as too great of a political embarrassment to endure, after so many official denials that no U.S. POWs were left behind. A former intelligence officer, now a staffer for the Committee, confirmed to the author in October 1992 that Bush Administration stalling tactics on releasing POW/MIA information to the Senate investigation continued after Ross Perot's October 1st reentry into the Presidential race and prior the November election. Many Americans later believed that if Bush had resolved the issue that had plagued virtually every administration since the time of Woodrow Wilson, he would have achieved greatness in history and been reelected.

Near the end of September, with the 1992 election only five weeks ahead, and his popularity sinking, President Bush had vetoed a Congressional bill that would have imposed restrictions on renewing normal trade relations with Communist China. This theme had already been sounded by Democratic candidate Bill Clinton, and expressed a growing liberal outrage in the nation over China's repressive police state policies towards its citizens. The bill would have imposed conditions of granting Most-favored-nation trade status to China related to human-rights concerns such as the freeing of still-imprisoned 1989 pro-democracy demonstrators, barred the import of goods produced by prison or forced labor and according to an Associated Press report of September 29th: "WOULD HAVE REQUIRED CHINA TO COOPERATE IN ATTEMPTS TO LOCATE THOSE MISSING IN THE VIETNAM AND KOREAN WARS, AS AN ADDITIONAL CONDITION OF KEEPING ITS NORMAL TRADE STATUS."

Bush explained that although he deemed China's limited steps on human rights inadequate, he felt the U.S. had a dialogue with China that, "Gives us an avenue to express our views directly to China's leaders...it would be a serious mistake to let our frustration lead us to gamble with policies that would undermine our goals." Thus did the Bush Administration, despite all the increased public awareness on the issue of secretly withheld U.S. POWs in Russia and China, publicly continue to subordinate their stated "highest national priority," of resolving the POW/MIA issue, to the dictates of the administration's foreign and economic policies.

During a continual effort undertaken since the late 1980s to publish the facts on POWs revealed by declassified documents in major American newspapers, the author of this book concentrated unsuccessfully on the two nearest his home, the San Francisco Chronicle and the San Francisco Examiner (owned by the Hearst family). All senior editors of both newspapers had for years rejected, without investigation, the results of the author's research, even when supplied with the declassified archival documents and further human source evidence that the U.S. Senate took seriously. After having seemingly exhausted all avenues, the author wrote a protest letter to William Randolph Hearst Jr., who had once achieved acclaim for his gentle interviews of Soviet leaders. Author J. M. G. Brown received a response on April 17, 1992, from the Washington headquarters of the Hearst newspapers, which went a long way towards explaining his newspaper's and President Bush's position of apparently ignoring the supreme human rights concern about U.S. POWs and MIAs held by the Communist Chinese after Korea and Vietnam:

"Mr. William Randolph Hearst Jr., who suffered a parayzing stroke in February and is not capable of handling correspondence, has requested me to acknowledge receipt of your letter. MR. HEARST'S POSITION IS, AS STATED IN THE EDITOR'S REPORT (by Hearst), MARCH 29TH, THAT IT IS IN AMERICA'S STRATEGIC WORLDWIDE INTERESTS TO MAINTAIN A WORKABLE RELATIONSHIP WITH THE CHINESE COMMUNIST GOVERNMENT. Sincerely, Joseph Kingsbury-Smith, National Editor and Senior Washington Editor."[30] Thus, the self-imposed censorship by much of the American media continued concerning the embarrassing American POWs held by China (and by Russia), until recent times.

Retired NSA analyst Jerry J. Mooney had gotten a new rash of telephone threats before and after former Secretary of State Kissinger's testimony, and he so informed the Senate Select Committee, but nothing, apparently, could be done about such relatively minor annoyances. In early October Mooney received a letter from Tom Lang, the Senate investigator he had indoctrinated in NSA procedures for eight months, who had then suddenly disappeared into Cornell Law School in September 1992. Lang told Mooney that he was going to temporarily come back to the Committee until the mandate ran out at the end of the year and the final report was written. Mooney told the author in early October that his contact at the Committee had revealed details of the Senate investigation which confirmed that, in Mooney's words:

"NSA is playing a game... and they have broken their agreement with me...They are sending to the select Committee only peripheral data from their Central Reference Files...This will only contain the basically 'sanitized' or general information...they are not retrieving and sending to the Committee any of the raw intelligence data from the decrypted North Vietnamese communications and Soviet aircraft flights. Without the signals intelligence they will never uncover the truth..."[31] As a result of the efforts of producer Ted Landreth, Mooney had been filmed by an NBC film crew at his new Montana home in Billings, on the Yellowstone River, and a short take of this interview was aired by the network on October 7th, featured with other information on POW/MIAs.

Senator Charles Grassley of Iowa addressed the Senate floor on October 5th, on the Select Committee's investigation thus far, which he said had proven that American POWs had indeed been left behind in captivity after the war. He also asked the Senate: "Can we trust DIA to competently evaluate evidence of possible survival" of American prisoners or MIAs? Grassley also charged that DIA "hasn't even analyzed the data the data it's been sitting on for years." According to Senator Grassley, the Select Committee on POW/ MIAs "DOES HAVE EVIDENCE OF POSSIBLE SURVIVAL...BECAUSE THE FACT OF THE MATTER IS THAT OUR COMMITTEE DOES HAVE EVIDENCE THAT SUGGESTS SURVIVAL...," from radio transmissions and distress signals, of an unstated number of American prisoners of war. Grassley stated for the Congressional Record:

"The Committee must examine the possibility that a number of symbols and markings may be attempts by possible U.S. POWs to communicate their locations to intelligence collectors. These symbols and markings have been identified through the use of overhead reconnaissance photography...which match pilot distress symbols used during the war, and span a period from 1973 to 1988...The committee must also examine follow-up actions taken by the Government to investigate these symbols. It should be noted that our Government launched a reconnaissance operation to a possible detention site on the basis of just one such symbol" (Nhommoroth-1981).[32]

Other evidence resulting from satellite and aerial photography included a standard "Walking K" marker, made from rice sacks laid on the ground, had been photographed in the Sam Neua area of Laos in January 1988, indicating a downed, live U.S. pilot. This report had resulted in no action by DIA until nearly a year had passed, and Reagan had been succeeded by Bush as President. Investigators at the Committee did not trust DIA to properly follow up on such intelligence, which could still be coming in. Senator Grassley urged an investigation of DIAs handling of the matter and charged: "After the war years, the intelligence community all but stopped looking for distress signals," apparently to conform to the official policy initiated in the 1973-74 period, that all American POWs had been returned at the war's end.

Senator John Kerry, Chairman of the Committee, rose after Grassley to caution the Iowa Republican: "If the Senator means...we left people behind, they were captured and they were alive and we did not get the accounting...that is true...I just want to be very careful... before we come up as a group with our conclusions."[33]

But Grassley's words about taking control of the POW/MIA

issue away from DIA and assigning it to another agency, JSSA, may have been prophetic. JSSA was an acronym Joint Services SERE Agency, and SERE stood for Search, Escape, Rescue and Evasion. This was the agency responsible for creating the distress symbols, but it had been cut out of the loop by DIA and was never asked to evaluate those which had been photographed since the war. Grassley said:

"JSSA has recently been given responsibility for the POW/MIA issue and has been chartered to review the performance of DIA...In my view, it is high time this issue was taken away from DIA to remove once and for all the stigma of the mindset to debunk."

The Select Committee's Vice Chairman, Senator Robert Smith, wrote a memorandum to Senator Kerry charging that two Defense Department officials who had just briefed the senators on the Friday before, assistant secretary Duane Andrews and Deputy Assistant Secretary of Defense Alan Ptak (the ex-CIA official), had tried to cloud the issue once again: "In my opinion, the briefing was the culmination of years of doubletalk, misinformation and obfuscation by officials responsible for the POW/MIA issue."[34]

That same day Senator Kerry joined a long list of American political leaders who, for seven decades, had assisted in suppressing the truth about American POWs who had long remained alive in Communist control. This was recorded in an October 7th "Memorandum for the Record" from the Office of the Assistant Secretary Of Defense for Command Control and Intelligence (Duayne Andrews), signed by Ronald J. Knecht, the same Pentagon official who at the direction of Secretary of Defense Richard Cheney, had debunked Colonel Peck's charges of mismanagement of the POW/MIA issue by the DIA in early 1991, prior to the Senate Foreign Relations Committee Republican staff report on U.S. POW/MIA policy. This document was leaked by a Pentagon staffer and its existence revealed to the author and research colleague Thomas V. Ashworth on November 10, 1992, the day before both testified under oath about their years of investigation and research into the subject of Communist-held American POWs. Under the heading: "Telephone call from Senator Kerry," this memorandum stated:

"At 12:30 PM on October 7, 1992, Senator Kerry. Chairman of the Senate Select Committee on POW/MIA Affairs, called for Assistant Secretary of Defense Andrews. As he was at lunch, Senator Kerry asked for me. This was a conference call and included Ms. Francis Zwenig, Staff Director of the Senate Select Committee on POW/MIA Affairs.

"He cited the POW/MIA intelligence material released on NBC last evening (Dateline) and the leaks in the press (Evans and Novak, etc.) and said that he was VERY upset.

"Senator Kerry was very emphatic that this has gone too far and that the Department has an obligation to take on the issue. We cannot continue to keep our arms tied behind our back. HE SAID WE NEED TO DO AT THE HEARING NEXT WEEK (October 15 and 16) WHAT WE DID AT THE LAST MEETING (Monday, October 5). BRING A BLOW UP OF THE PHOTOGRAPHS OF THE: USA" AND "K" SYMBOLS AND SHOW THE "K" IS NOT A K AND THAT IT DOES NOT HAVE WALKING FEET. HAVE MR. GADOURY EXPLAIN HOW THE USA COULD HAVE BEEN MADE. He stressed that unless we answer this attack directly the leakers will win and they will be able to claim "everyone knows" it was made by a POW. We need to demonstrate reality. 'PUT A LIE TO IT.' The Secretary (Andrews) should make a one-time exception on the photography-show a blow-up of the actual photograph (symbol portion) on TV.

"He said that we have been repeatedly attacked by those who do not want to deal with reality. Each time we answer they find something new to raise. If we are ever going to stop this we need to demonstrate convincingly what the real thing looks like. Openness will take the wind out of their sails.

"He urges the Department (of Defense) to come on VERY STRONG: 'WE ARE APPALLED. These leaks jeopardize any American in captivity who would try to signal. It is dishonest to leak information obtained in closed hearing, knowing the Department cannot discuss intelligence sources and methods in public. We took responsible actions as soon as we found this symbol.

"I raised our concern with discussing intelligence in public as we have global responsibilities that we cannot jeopardize. He agreed that this is difficult but that in this one case we need to find a way to take on the issue, find some way to lay out the issue."

"Concerning the hearing, he agreed that a script would be unwise and recommended Mr. Andrews just COME ON STRONG-APPALLED. HE STRESSED THAT HE WANTS TO WORK WITH US BUT REITERATED THAT IT IS TIME FOR US TO TAKE ON THE ISSUE."[35]

An October 6, 1992 memorandum to Senator John Kerry from Frances A. Zwenig, the Staff Director of the Senate Select Committee, the former antiwar activist who reportedly desired appointment as U.S. Ambassador to Vietnam, revealed how closely Kerry's most trusted staffer was involved with the Defense Intelligence Agency in assuring that information about the abandoned American POWs would not be revealed to the American public in the televised hearings:

"JK (John Kerry) Looks like auto metrics fills the bill for expert on images—I AM WORKING ON THE SCRIPT WITH DIA."[36]

In subsequent hearings on the signals, Assistant Secretary of Defense Duane Andrews did indeed "come on appalled," and he debunked all the reconnaissance photographs except one in 1973 and one in 1988. Of the aerial photos that led to the Nhommoroth mission in 1981, Andrews said, "The exact measures undertaken by the U.S.

Government to inspect the facility and to determine whether American prisoners were being held or not remains classified... Continued analysis since 1980, photographs of this facility and human sources support our best judgment THAT AMERICANS WERE NOT HELD IN THE FACILITY...There is even more reason to believe now that what was interpreted as a possible 52 symbol was nothing more than the irregular furrows of many individual garden plots." Senator Smith and Grassley charged that DIA analysts "made disappear" as many as 19 four-digit authenticator numbers spotted between 1973 and 1988, that could have been the pilots' personal codes, similar to PIN numbers, which had appeared in reconnaissance photographs viewed by Select Committee members and the JSSA.[37]

Among the authenticator codes spotted in satellite imagery photography as recently as June 1992 were two tramped out in the grass near a prison that American investigators had never been allowed to visit, which belonged to MIAs Peter R. Matthes, a U.S. Air Force captain from Toledo, Ohio, missing since November 1969, and Air Force Lt. Colonel Henry M. Serex of New Orleans, missing since April 1972. (This was revealed to the public by former Congressman William Hendon on the "Larry King Live" television program.) This incident was practically unreported as news in the American mass media. "Dateline NBC" subsequently reported more about aerial photos, and showed one on television in April 1993 that clearly showed distinct numbers of a U.S. pilot's authenticator code.

The likelihood that American POWs could have easily been hidden for decades in remote Indochinese jungle prison camps, was buttressed by the startling appearance of a fighting band of 398 anti-communist Vietnamese Montagnard tribesmen in the wilds of eastern Cambodia. Originally supported and trained by U.S. Special Forces and CIA in the 1960s, they had been fighting the Vietnamese communist forces ever since the 1975 American withdrawal, and their number had once been in the thousands when they had first fled Vietnam's Central Highlands. The Montagnards hoped to be permitted to settle in North Carolina where some of their relatives lived, but despite their unequaled loyalty to a cause America had once espoused, U.S. officials announced that these Montagnard would first have to be "Screened" to determine which ones were suitable to immigrate to the United States.[38]

Newsweek Magazine published an expose on the Soviet-held American POWs of WW II and Korea on October 5th, which cited the research in the new book Soldiers of Misfortune. (The author had supplied conclusive documentary evidence on this story to Newsweek a year and a half before, while simultaneously assisting the Senate Foreign Relations Committee investigation.) Newsweek ignored the historic precedent of the original Soviet withholding of

American POWs and MIAs of the World War I Allied intervention in Russia, which had been documented by author John M. G. Brown for the Senate Foreign Relations Committee and the Washington Post Company. Included in the report were the results of the Senate Select Committee's investigation thus far, and quotes from the testimony of Henry Kissinger and other Nixon administration officials.

Washington D.C. reporters Evans and Novak wrote on October 8th that the charge that DIA was suspected of either "not understanding or of routinely sitting on obvious calls for help from prisoners of war," had been kept "under wraps until this week." In their 'Inside Report' the columnists said that this: "...Might partially explain the committee's rare and unpublicized meeting with Defense Secretary Dick Cheney and National Security Adviser Brent Scowcroft in a Capitol hideaway scrubbed for security. Principles told us that Cheney and Scowcroft wanted to 'tie up loose ends' as the panel's sensational probe begins to wind down." Thus did the Bush Administration resist to the bitter end, the inevitable, despite the best efforts of Republican Senators Smith and Grassley and the restrained bipartisan leadership demonstrated by Democrat John Kerry, who had actually attempted to assist the DIA.

The Russians, however, continued to reveal more details of their blackest secrets in the middle of October. Russian President Boris Yeltsin's special envoy, Rudolf Pikhoya, brought copies of Soviet documents bearing the signature of Josef Stalin and ordering the execution of 14,700 Polish POWs at Katyn Forest and elsewhere, to Poland where they were presented to President Lech Walesa. The former shipyard worker and organizer of a once-outlawed union, said, "We are witnessing the handing over of the most important documents concerning the cruel crime against the Polish nation. "Walesa also said he would urge that the originals of the orders be brought to Warsaw. In addition Yeltsin released information on the 1983 shootdown of Korean Airlines Flight (KAL) 007, indicating for the first time officially, that Russia had acknowledged that the 'black boxes' of 007 were in their possession. A recently published Soviet Politburo document revealed that at that time Mikhail Gorbachev had defended the Soviet decision to shoot down the airliner. In the October 15th edition of Izvestia in Moscow, a December 1983 note was published from Soviet Defense Minister Ustinov and KGB Chief Chebrikov to Yuri Andropov in the Kremlin that indicated there was no evidence that KAL 007 was an American spy plane.

The author had previously received substantial information from a valued source in Israel, Avraham Shifrin, indicating that some of the passengers and crew may have survived the 007 flight (on which U.S. Representative Larry McDonald of Georgia was a passenger), which in reality had ended in a forced landing recorded on radar screens, rather than a shootdown. Shifrin's sources among recently-arrived Soviet Jews had revealed that the KGB had covered up their imprisonment of the survivors. This information had been turned over to the Senate Foreign Relations Committee by the author in 1991 and the Select Committee on POW/MIA Affairs in 1992.

While the POW/MIA issue was avoided in the debates between incumbent President George Bush and challengers William Clinton and Ross Perot, CIA Director Robert Gates had flown to Moscow to meet with Russian officials, offering them information on the American burial of the remains of six Soviet sailors from a sunken Russian submarine that had once been recovered by U.S. intelligence in the Pacific Ocean. This appeared however, to be merely an extension of the Bush Administration's policy of confining the "resolution" of the MIA and POW matter to accounting for "remains" of the dead. The ex-CIA man, Alan Ptak, now a senior Defense Department official, traveled to Hanoi at this same time as a member of a sudden diplomatic mission, which included, at the personal request of President Bush, Senator John McCain, and which was headed by General John Vessey. Douglas H. Paal, who held Childress' former position as Director of Asian Affairs on the National Security Council staff, and Ann Mills Griffiths, Executive Director of the National League of Families of American POWs and MIAs in Southeast Asia, who was the U.S. Government's officially-designated liaison to American POW families, also participated in the mission. The Washington Post reported that the trip was a result of a meeting the week before in Washington, involving Secretary of Defense Richard Cheney, Acting Secretary of State Lawrence S. Eagleburger and Vietnamese Foreign Minister Nguyen Manh Canh, who had been attending the UN General Assembly in New York. Administration officials were reported by the Post to be "unusually guarded" about the meeting and about the mission to Hanoi, "saying only that both sessions were arranged to obtain more cooperation from Vietnam in resolving long-standing prisoner of war and missing in action issues."

At this point a major American media campaign was orchestrated around a mass of photographs of dead American servicemen and relics held by the North Vietnamese since the war's end, twenty years before, which had been obtained by a mysterious Defense Department-connected 'researcher,' identified as one Ted Schweitzer. The timing of the "great discovery," just before the presidential election, was clearly contrived to influence the voters, particularly the millions of American veterans who had become discouraged by the Bush Administration's transparent effort to "resolve" the POW matter by artificial means and methods. It turned out that Schweitzer's investigative effort in the North Vietnamese

Archives had been promoted by former Assistant Secretary of Defense Richard Armitage, and others who had long held that no living American POWs were held as captives in Indochina after the war. Schweitzer's findings were widely reported to support this conclusion, but when the photographic evidence he obtained was actually examined by unbiased observers, it was found that only one certain case of an MIA was resolved by a picture obtained by Schweitzer, with a possible handful of two or three other uncertain identifications.

But also in mid-October, other family members of missing Americans, such as the determined young Bruce Brown, son of the Moscow-bound F-111 pilot and Gemini space scientist, Colonel Robert M. Brown, accompanied by Janice Pennington, the former wife of the mysterious missing mountain climber Fritz Stammberger, whose fate was so sensitive to CIA, were on their way to Russia to request assistance in locating their relatives. They would not easily be put off, and it appeared likely that they and others like them would continue to demand answers as the 1992 Presidential election approached, and indeed until some American leader showed the political courage to admit the existence and negotiate the return of all surviving American prisoners.

An important confirmation of the Soviet practise of utilizing American pilots captured in Vietnam to achieve technological breakthroughs beyond the capacity of Russian science, and that some might still be alive in the USSR, came from Alexander Zuyev, a Soviet "top gun" pilot who had defected to the west. In recalling a new Soviet process for arming nuclear weapons carried by Soviet aircraft, he wrote:

"All nuclear-capable fighter regiments practised loading their bombs at least twice a month. This process was always conducted at night, inside a hanger, in order to avoid American spy satellites. During one such training exercise at Vaziani, I explained to a young weapons officer from the RTB that our superiors had never revealed to the pilots just how powerful the RN-40 actually was. I felt we had a right to know. 'The yield is slightly over thirty kilotons,' he said casually...When I commented on the well-conceived arming process, he went on to tell me some fascinating information.

"'Soviet forces adopted this system in the 1970s,' he said, stroking the gray flanks of the practise bomb. 'It's the Americans' own system, but we've added some improvements.'

"He must have noted my confused expression. 'THE AMERICAN METHODS,' he added, 'WE OBTAINED FROM SEVERAL U.S. AIR FORCE 'GUESTS,' NUCLEAR-QUALIFIED PILOTS OUR FRATERNAL SOCIALIST COMRADES IN VIETNAM PROVIDED US DURING THE IMPERIALIST WAR.'

"At the time, I had not wanted to consider the methods the GRU had used to extract such information from professional military pilots. This was a cruel side of war that I hated, but I knew the Americans would do the same to me if I were to parachute into the hands of one of their 'fraternal' Imperialist allies. All that could be said was that the unfortunate American pilots had probably died painlessly soon after they had revealed this vital information. Or maybe not. I had grown up watching television reports of the endless anti-war demonstrations on the streets and university campuses of America. Clever Soviet intelligence officers might have manipulated images of the demonstrations TO PERSUADE AMERICAN PRISONERS TO REMAIN VOLUNTARILY IN THE SOVIET UNION, CONVINCING THESE BATTERED, VULNERABLE PILOTS THAT THEY WOULD BE IMPRISONED AS TRAITORS IF THEY EVER RETURNED TO THEIR COUNTRY.

"It was not surprising that Soviet interrogators had used whatever methods necessary to extract secret nuclear weapon arming procedures from captured American pilots. Prisoners of war were held in contempt by the Soviet military. During my years at the Armavir Academy and the constant training in the regiments, I was never instructed in the Geneva Convention on the Treatment of Prisoners of War. Never; I didn't even know it existed. NO Soviet soldier or officer was ever told that the Imperialist enemy had signed binding international treaties guaranteeing humane treatment of prisoners. I first learned about the Geneva Convention during my debriefings in America. Throughout years of Air Force service, I was constantly taught that my oath of duty to the Motherland bound me from ever surrendering, as long as I was physically capable of fighting. And then, we were told, it was better to use the last bullet or grenade for suicide than to surrender. Many young Soviet soldiers in Afghanistan had chosen death rather than breaking this sacred oath."[39]

Thus, an independent Soviet military source had confirmed the testimony of the American NSA analysts Mooney and Minarcin, and other witnesses appearing before the Senate Select Committee on POW/MIA Affairs, but Alexander Zuyev was not called upon to testify before the American people.

William Jefferson Clinton, Governor of Arkansas, was elected President of the United States on November 4, 1992, by less than half of the popular vote, although George Bush gained few electoral college votes. Clinton had been born in 1946 and as a youth had opposed the Vietnam War while studying in England on a Rhode Scholarship at Oxford University. He and his running mate, Senator Albert Gore, a Vietnam veteran and environmentalist from Tennessee who was also the son of a Senator, represented a younger generation which was, to a significant degree, still divided over the Vietnam War and the 1960s radicalism that characterized the time. (Gore had served as an Army reporter attached to an engineer unit's headquarters in Vietnam, and was the son of Tennessee Senator

Albert Gore, a friend and business partner of Armand Hammer.)

Nearly 20 million Americans voted for the independent candidate, Texas businessman Ross Perot, indicating widespread discontent with the candidates who had been chosen within the two-party system. The alarm felt by some in the American mass media at Perot's popularity was evidenced by a spate of hostile reports questioning his judgment and ability, which appeared in major newsmagazines and on the air just prior to the election. In several cases Perot's past devotion to the POW/MIA cause was used to indicate that he was paranoid and unstable. Newsweek Magazine labeled Perot the "bigmouth billionaire," with a "conspiratorial cast of mind," and "paranoid tendencies," with an "obsession" about American POWs and MIAs of Vietnam. Newsweek reported an interview in which Scott Barnes (who had testified at the Murkowski hearings on POWs in 1986) claimed that a Bush campaign official had hired him to wiretap Perot's office and fax machine, which he agreed to do, but that he also kept Perot informed, which resulted in an FBI undercover sting operation in which the Bush campaign's Texas chairman, James Oberwetter, was videotaped meeting with Barnes in an attempt to peddle illegal tape recordings of Perot's conversations during the election campaign.

A simultaneous Time Magazine article, that was dated November 9th, but which actually appeared across the United States just before the election, was entitled: "Perot-noia," and highlighted Perot's belief that the Bush campaign had planned to smear his daughter Carolyn by publicizing a faked photograph of her involved in a Lesbian act, thus causing his withdrawal from the race on July 16th. The magazine stated that one of Perot's sources for this story was Scott Barnes, the covert operator who had testified before the Senate about seeing U.S. POWs in Laos in 1981. Barnes was identified as a "notorious conspiracy-theory peddler" who had claimed in sworn testimony before the U.S. Senate in 1986 that the CIA had ordered the murder of two U.S. POWs sighted in 1981 and, according to a September 1984 ABC News report, had later been given a job at a Honolulu prison so that he could spy on Michael Rewald, a banker jailed during a fraud investigation, whom Barnes claimed the CIA had subsequently ordered him to kill. ABC had retracted the story three months later because of their inability to substantiate it, and because of CIA's denial.

Rewald's bank, Bishop, Baldwin, Rewald, Dillingham and Wong, was reported to have been a CIA front for laundering illicit funds from many sources, and to be a successor to the CIA-connected Nugan-Hand Bank in Australia, which had collapsed in 1980, ruining many depositors. The author had been questioned about this bank by FBI agents Stan Walker and Dick Miller in April 1989. The Bank's Australian partner Frank Nugan had been found dead of an apparently self-inflicted gunshot wound and the American partner, a former Vietnam Green Beret and CIA officer named Michael Hand, disappeared, never to be heard of again. General LeRoy Manor, who had been decorated by President Nixon for his role in the unsuccessful raid on North Vietnam's Son Tay POW camp, was an officer of the Nugan-Hand Bank, as were Admiral Earl "Buddy" Yates, formerly of Pacific Command during Vietnam, General Erie Cocke, General Edwin Black and others. All claimed to be "true patriots," innocent of any wrongdoing. Former CIA Director William Colby acted as Counsel for the Nugan-Hand Bank, which conducted operations in the Chiang Mai area of Thailand, adjacent to the Golden Triangle opium production region that included Laos and Burma, as well as in the Philippines, the Middle East, the Caribbean and other areas of the world, before it collapsed.

Reports in both the Wall Street Journal and the New York Times had revealed many details of this CIA front company, including involvements of Major General Richard Secord and rogue CIA agent Edwin Wilson with Michael Hand. A 1983 Australian government investigation (the Commonwealth-New South Wales Joint Task Force on Drug Trafficking), parts of which were kept secret, had concluded that other personalities involved with Oliver North in the Iran-Contra Affair, including former CIA officials Theodore Shackley, Thomas Clines and Rafael Quintero, "were involved in military and intelligence-related activities with Hand and other top officers of the Nugan-Hand concern."[40] The Nugan-Hand Bank affair recalled a type of operation that had been promoted long before by the Nazi intelligence chief, Admiral Canaris, and even earlier by the Bolshevik Cheka. Thus had the American intelligence services adopted the methods of Hitler and Stalin in confronting the totalitarian Communist enemy, as had been urged and implemented following World War II by the guiding lights of the Office of Policy Coordination. A December 13, 1984, Los Angeles Times report on the Rewald-Barnes story had stated that CIA had "admitted limited dealings with the firm," and that Rewald had a secrecy agreement with CIA, but the CIA had still filed an FCC complaint against ABC urging that the network's broadcast license be revoked; a serious attack by the Agency. The files of Rewald's firm were reported to have been sealed, for "national security reasons."

Time claimed that Perot was "obsessed with plots" and quoted Bush's White House spokesman, Marlin Fitzwater, as saying that the self-made billionaire from Texas was "a paranoid person who has delusions." Examples given by the magazine included Perot's claim that he had been targeted by a North Vietnamese assassination team twenty years before, and Perot's accusation that Assistant Secretary of Defense Richard Armitage had conducted a "nefarious coverup" of the MIA issue. Time Magazine revealed that the veteran FBI covert operator, Oliver "Buck" Revell, who now headed the FBI's Dallas office (a demotion from his former Washington D.C. posting on the intergovernmental Operations Sub Group (OSG) with Richard Armitage and Oliver North), had initiated the "sting" operation against James Oberwetter, Chairman of the Bush campaign in Texas, allegedly on the information supplied by Perot and Scott Barnes. It was "Buck" Revell who had sent an undercover agent to Oberwetter, offering him wiretap tape recordings of some of Perot's telephone conversations and other documents for a sum of money. Perot had reportedly made the recordings as bait for the FBI's sting operation. The results were so damaging to Perot nationally that questions arose as to who Revell was actually serving, in planning, approving and executing the operation.

Aside from Revell's earlier activities mentioned previously in this book, Time reported the FBI official had been passed over for promotion after admitting at a 1988 Senate hearing, "that the FBI had been misled by an undercover informer whose 'concocted' data led to a two-year surveillance program against Americans opposed to U.S. policies in Central America...Ironically, Revell went public last June on behalf of Richard Armitage, a former Pentagon official whom Perot had accused of complicity in drug smuggling and covering up the existence of Vietnam MIAs. Revell felt strongly that Armitage was a victim of 'wild charges' that the FBI had been unable to substantiate." The Perot-Revell-Barnes story surfaced at a critical time in the presidential campaign, when the FBI's Director, William Sessions, was the subject of an ethics investigation, whom his wife claimed had been illegally wiretapped in their home by other FBI officials. Thus did the same small group of American officials involved in one way or another with the POW/MIA matter, continually reappear in other security-related covert operations.

The author of this book subsequently learned that in statements to the Select Committee on POW/MIA Affairs, Barnes claimed to have been honorably discharged from the U.S. Army, but that his DD-214 had been tampered with, so that he was on record as having received an undesirable discharge. Yet, Senate investigators learned that several months before his deposition Barnes had applied for an upgrade of his undesirable discharge. Barnes subsequently recanted his testimony before the Committee. He remained an elusive and shadowy character; a "mystery box," as described by Ross Perot. (The author interviewed Barnes in 1987 in company with retired U.S. Army Colonel William LeGro; an inconclusive meeting that raised more questions than it answered.)

While working closely with Senate investigators in the fall of 1992, the author had been informed about an interview conducted by two Select Committee staffers, of a high-level Czechoslovakian military defector, General Jan Sejna, who was still employed by the U.S. Department of Defense. In the first meeting with Senate investigators, General Sejna revealed that when he was a member of the Czech Defense Ministry, prior to his defection in 1968, he learned that some 90 American prisoners of war from Vietnam had transited Prague as they were being shipped to the Soviet Union from Indochina. The route through Czechoslovakia was apparently a ploy to obscure the POWs ultimate destination in the USSR. One Senate investigator, a retired U.S. Army intelligence officer, informed the author that during the meeting a Counsel for the Defense Intelligence Agency who was present attempted to halt General Sejna's statement about the American POWs, but was unsuccessful. In a subsequent meeting with the Senate investigators, General Sejna admitted that he himself had seen the American POWs in Prague. Although he was deposed under oath by the Select Committee staff, General Sejna was not called to testify in public as a witness, thus denying the American people the right to hear what had happened to at least some of the missing U.S. POWs from Vietnam. Sejna's testimony was taken by John Erickson, Senator Dole's 'man on the Committee' who had repeatedly interrupted the author's own sworn deposition in July (to leave the room and make 'telephone calls'), and who had later arbitrarily halted it. The decision was up to Senator John Kerry, who was soon to demonstrate publicly that he had little desire to learn more about the fate of Moscow-bound American POWs of several wars.[41]

Author John M. G. Brown testified under oath at a televised hearing of the Select Committee on POW/MIA Affairs on November 11th (in part):

"...My research among thousands of recently declassified U.S. documents revealed that since the 1918-1921 Bolshevik period, the Soviets and their later surrogates always demanded diplomatic concessions or financial aid for the release of U.S. and Allied war prisoners and civilian hostages they publicly denied holding. The repatriation of Russian and Soviet prisoners in Allied control during WW I and WW II was also of great importance, but my research indicates that it was the Soviets who first initiated the reprehensible policy of secretly withholding war prisoners in the 1918-1919 period..."

The author's research colleague, Thomas V. Ashworth eloquently assailed members of the Committee who had been in collusion with DIA during the hearings, as illustrated by the leaked October 7th Pentagon memorandum on Senator Kerry, cited previously in this chapter, which had been shown to Ashworth and to the author prior to the hearings. Senator Kerry became infuriated, and demanded that Ashworth reveal the source of the leak, reminding the researcher that he was under oath. Ashworth refused to divulge the source and Kerry angrily said he did not want to hear anything further from Ashworth. A Vietnam veteran in the audience, later identified as Jerry Kiley, a long-time POW/MIA activist, rose and shouted at Kerry, "You wouldn't be here if it wasn't for these men!" Kerry said nothing in retort.

Journalist Mark Sauter had unfortunately declined to testify under oath before the Committee. However, James D. Sanders testified at the same hearing as the author and Ashworth, on his independent research into the World War II Soviet-American hostage crisis, but he, Ashworth and author John M. G. Brown were repeatedly interrupted, and the testimony was then arbitrarily halted by Senator Kerry (with help from Senator McCain), as was recorded by C-SPAN in the nationally-televised hearings. Although Senators Smith and Grassley attempted to intervene, Kerry was the Chairman, and summarily cut-off the hearings, reportedly to attend a dinner that evening for President George Bush. The author was thus prevented from testifying on much of the historical evidence obtained through the declassification of U.S. POW/MIA records or from human sources and contained in this book, despite having served as an unpaid consultant throughout the life of the Select Committee, and having supplied hundreds of historically-significant U.S. and British POW/MIA documents to the Select Committee's staff investigators. (None of this documentary evidence was used in the final report of the Committee, apparently because of the predetermined attitude of Senators Kerry, McCain and others.)

Seated behind the author, Sanders and Asworth, were Colonel Stuart Herrington and General Loeffke of the Department of the Army's "Task Force Russia," along with various other DIA, CIA and Pentagon officials, in civilian suits. A youthful researcher for the government-contracted RAND Corporation, ceremoniously introduced as Dr. Paul Cole, who testified about his perfunctory and careless examination of U.S. POW/MIA records lasting, by his own account, only a few months, utilized the official versions of POW repatriations in WW II such as the European Command's RAMPs study (cited earlier in this book), and said what was expected of him by the Department of Defense, which had hired RAND:

"Dr. Cole found no evidence to support charges that thousands of American POWs liberated from Nazi German POW camps were never repatriated...the initial RAND review, while incomplete and inconclusive tends to discredit the idea that a substantial number of U.S. POWs were held by the Soviet Union following World War II and not repatriated. In this regard, Dr. Cole took issue with the authors of *Soldiers of Misfortune* and *Moscow Bound* concerning the number of POWs the Red Army 'liberated' from German POW camps and failed to repatriate. His conclusions: 'The number of American POWs who were not repatriated from German POW camps in World War II appears to be less than 200...'"

While the extensive documentary and human source evidence already supplied to the Committee by the author (and later by James Sanders) was almost completely ignored and was not used in the Select Committee's final report, apparently at the direction of the Chairman, Senator Kerry, the RAND Corporation spokesman, Cole, was permitted unlimited time on the witness stand, treated with courtesy by Kerry, and the final report contains much misinformation from Cole and liberal use of quotes from him.[42] Cole's hasty 'research' was unworthy of the national coverage it received, and clouded the POW/MIA issue at a critical time, as the international exposure of the historical roots of the POW/MIA matter was occurring. Dr. Paul Cole had surfaced at a critical juncture earlier in 1992, following Boris Yeltsin's June 15th admission about American POWs of WW II, Korea and Vietnam in Russia, when he had been cited in the mass media as an 'expert,' and used to downplay the possibility that American prisoners were held in the former Soviet Union; saying that if any U.S. POWs were there they didn't want to be found.

Despite contrary testimony of some of the witnesses at these hearings, including the author's and Ashworth's and Sanders', the New York Times published an article on American prisoners in Russia by one of its anointed "Soviet experts," Serge Schmemann, which merely repeated a Soviet explanation that had actually been developed in 1945: That the Soviets had only arrested a few foreign-born or dual-national Americans at the end of World War II, about whom they felt a legitimate argument or suspected of espionage. Thus, the New York Times acted as if Volkogonov's statement, which the newspaper repeated, that, "119 AMERICAN CITIZENS WITH RUSSIAN, UKRAINIAN AND JEWISH NAMES WERE DETAINED AFTER THE WAR," was of great historic significance when it was not. Schmemann said nothing about the testimony by authors, John M. G. Brown, James Sanders and Thomas Ashworth on the declassified U.S. archival documentary evidence and eyewitnesses accounts concerning American POWs of World War II and the World War I U.S. intervention in Russia, secretly imprisoned in Soviet Russia, which had required seven years of research to assemble. Many of the key documents cited in this book had in fact been turned over to Max Frankel and other senior editors of the New York Times by the author long before, who outwardly ignored them. The extraordinary testimony of Lt. Colonel Corso and Lt. Colonel Simpson about American prisoners of war transported from North Korea and Manchuria across the Soviet border into Siberia was also ignored by Schmemann, as was the moving statement by Serban Oprika of his seeing American prisoners still alive in North Korea in 1979, instead he innocently wrote:

"The record may never be complete, and there are certain to be Americans who will not accept Russia's assurances that no prisoners of the Korean or Vietnam conflicts were imprisoned in the Soviet Union."

Senator Robert Smith released a list of 324 unresolved U.S. MIA cases on December 1st, with a report that stated: "Approximately 300 of these personnel were last known alive in captivity in Vietnam and Laos, last known alive out of their aircraft before it crashed, or their names were passed to POWs who later returned." Senator Smith's list gave further credibility to the affidavits and statements of retired NSA analyst Jerry J. Mooney and others who had testified before Smith's committee. Meanwhile, the ponderous Defense Department bureaucracy inexorably continued "resolving" cases of MIAs from Vietnam, in the same manner as had been done for decades. Of the 2,265 Americans still carried officially as missing or presumed dead from the Vietnam War in June 1992, the Pentagon had given the Senate a list of only 135 unresolved MIA cases, an unrealistically low number Joint Task Force Full Accounting had employed more than 50 American investigators to search for these MIAs and the Pentagon claimed that 48 of the MIA cases had already been "resolved" by August. By December 2nd the Defense Department claimed these "cases have been further whittled down to about 35. Also by December, the Pentagon was claiming that 187 U.S. investigators were searching for human skeletal remains, examining the Vietnamese archives or conducting interviews. It was through these interviews with Vietnamese about events of 20 years before that the U.S. Department of Defense was resolving most of these MIA cases. Simply put, the Pentagon was relying on statements of Vietnamese Communists to uncover the fate of American POWs and MIAs.[43]

It was not until December 2nd that the author received confirmation from the White House that a copy of the first edition of this book which had been forwarded to George Bush through the President's Counsel, C. Boyden Gray, over 4 months earlier (and again in October), had been delivered: "On behalf of President Bush, thank you for your letter and the copy of your book, MOSCOW BOUND. You can be assured your book will be given thoughtful consideration... Mark R.A. Paoletta, Assistant Counsel to the President."[44] Whether the President had ever been permitted to see an alternative analysis of the history of the POW/MIA matter was not revealed. Presidential envoy General John Vessey testified before the Select Committee on December 4, but his statements had little significance; largely being devoted to praise about the U.S. Government's efforts and the recovery of human skeletal remains alleged to be those of Americans lost in Vietnam during the war.

During a subsequent trip to North Korea, Senator Robert Smith met with government officials in Pyongyang, who admitted to him that hundreds of American POWs from Korea (and perhaps more) had been imprisoned by their Chinese Communist allies in Manchuria, and not returned after the war. A Chinese government spokesman denied that American POWs were transferred to China. Following his return from Korea, Smith announced on December 22nd, that Deputy Assistant Secretary of State Kenneth Quinn would ask the Chinese leaders about American POWs who had vanished, during a visit to China in January. Smith said that the evidence was clear and that the Chinese owed the United States an answer. President Bush had recently vetoed legislation from Congress which would have tied most-favored-nation status for China to Chinese assistance in resolving the fate of American POWs and MIAs.

President George Bush traveled to Moscow in the last weeks of his administration, to sign a nuclear arms treaty with President Boris Yeltsin of Russia on January 3rd. The treaty promised sweeping reductions in intercontinental ballistic missiles and other nuclear weapons, although the precarious position of Yeltsin's government added a note of uncertainty about the future of such agreements, and indeed, of Russia itself. Back in the United States, on January 11th, Bush continued a series of appointments for his former top aides by naming James Baker III to the board of the Woodrow Wilson International Center for Scholars. Bush also rewarded C. Boyden Gray, his chief legal counsel and close confidante in the White House, by appointing him as one of 10 members of the Council of the Administrative Conference of the United States, for a term of three years. (The Administrative Conference, chartered by Congress in 1964, studies procedural aspects of how government operates, with a special focus on the fairness and efficiency of legal operations.)

To replace Ed Derwinski as Secretary of Veterans Affairs (who had reportedly resigned because of his failure to line up the veterans vote behind George Bush), President-elect Clinton chose Jesse Brown, Executive Director of the million-member Disabled American Veterans (DAV). An African-American Vietnam veteran, Brown had been part of the small group of DAV leaders around National Commander Cleveland Jordan DAV Magazine editor Givans, and past National Commander Joseph Andry, who had demonstrated courage and vision in November 1991, with the publication of an expose on the history of Soviet and Communist-held U.S. POWs of WW II, Korea and Vietnam, in the DAV's national magazine, with a circulation of over 1,000,000 families, citing the author's and James Sanders years of historical research, and hundreds of declassified U.S. POW/MIA documents turned over to the DAV by the author in early 1991.

On January 13, 1993, just a few days prior to William

Clinton's inauguration and George Bush's departure, the Select Committee on POW/MIA Affairs of the U.S. Senate published its final report, a document which was, not surprisingly, heavily influenced by policy and politics. Chairman Senator John Kerry, Staff Director Frances Zwenig and General Counsel William Codinha, who along with Senator John McCain and some of the ex-intelligence operatives on the Committee staff, had curtailed the investigation all along and failed to utilize or follow-up on much historically significant documentary POW/MIA evidence given to the committee by the author, in addition to interrupting and then halting the November 11, 1992 testimony of the author and two other witnesses (James Sanders and Thomas V. Ashworth) on the evidence of Soviet-held American POWs.

The Committee's final report attempted to discredit the testimony and previous affidavits of retired NSA analyst Jerry J. Mooney by stating:

"A number of those who have written books about POW/MIA-related issues, including John M.G. Brown, Thomas Ashworth, Mark Sauter, James Sanders, and Monika Jensen-Stevenson have asserted or speculated that some Americans captured during the Vietnam War were transferred to the Soviet Union. For many the principal source for this allegation has been Mr. Jerry Mooney... Considering the fact that Jerry Mooney was the principal source cited by those who assert that American POWs were 'Moscow Bound,' his testimony was remarkably equivocal on the subject."

The report went on to quote directly from a very small portion of Mooney's testimony, during which Senator John McCain, who had fought frantically against any disclosure that U.S. POWs were left behind in captivity, pressed Mooney about his knowledge of the U.S. prisoners' destination once they had been transported to the Soviet interrogation camp designated 'Tentacle M-B" at Sam Neua, Laos. In his full testimony and in conversations with the author, Mooney had explained that monitoring the Soviet aircraft carrying prisoners from Sam Neua to Hanoi for further transport to the Soviet Union was the responsibility of another group at NSA (A-3) which refused, on a no-need-to-know basis, to share their information with Mooney's section, and thus he had no first hand evidence that had been the case.

The final report quoted from Mooney's testimony about the connections to U.S. POWs in Vietnam by the Soviets—the "friends," as the Vietnamese had referred to them in their secret wartime communications:

"...What I have said is that there was a Tentacle Moscow-bound. The men were collected. There was a connection by the "friends." We knew where they were transported within North Vietnam, I have no knowledge of Laos; and we knew where they went. We knew where the "friends" primary prison camp was and we knew how they were transported from North Vietnam over to Sam Neua, Laos, which we designated as Tentacle MB. I never saw an American prisoner being transported out of Southeast Asia and I have never said that."

In the usual sneering tone which he reserved throughout the hearings for witnesses to the fact that Americans were left behind in captivity in Vietnam McCain, shot back at Mooney: "You either believe that some were taken to the Soviet Union or you do not believe some were taken to the Soviet Union, Mr. Mooney. I think it's a pretty straightforward question."

What the final report of the Senate Select Committee on POW/MIA Affairs deliberately omitted was Mooney's answer to McCain's repeated question, which was:

"Let me explain it to you, sir. Let's lay it out. Unlike World War II and Korea, the Soviets did not need a bunch of people for labor. They were after the minds. This was reflected in the targeting and tasking which we saw in the surface-to-air missile and anti-aircraft artillery messages that we decrypted and read. So we did not anticipate that the Soviets were going to come in there and grab up a bunch of Americans and run them over to Moscow. I mean, that just wasn't going to happen. Further, we did not anticipate that we would see anything. We were dealing with tactical communications. If we saw anything it would have to be by accident or security violation. We did see it. We did see the Soviets, not the Soviets, the 'Friends,' who were identified as Soviets moving American POWs, ME's up the line from capture units in the south, to the southern part of North Vietnam. They were taken to a prison camp northwest of Vinh which had the cover-name Son Tay. Not the Son Tay south of Hanoi, Senator.

"At first we didn't know where this location was, but when you walked out of the fifth floor of the National Security Agency, you run into a Cardex database system and we found the second Son Tay. It described when it was built and what it consisted of. It was mostly underground. We then noted that these POWs or these ME's from cover name Son Tay were moved up to the Bai Thuong air field. Not consistently, this was rare. They were not stealing them by the hundreds. They were few and rare. From there they were put on an IL-14 aircraft. From there they were flown to Sam Neua. The unique thing about that flight, Senator, is that there was no air-to-ground communications. There was mike clicks, to acknowledge their directions, but we did not know if they went on beyond Sam Neua..."

McCain then said: "Now I would like you to answer my question. Do you believe or not believe that American POWs were taken to the Soviet Union?" Mooney responded: "I believe it, sir...Yes sir, absolutely."[45]

Jerry Mooney's honesty in portraying the strict

compartmentalization of intelligence information within NSA, in which a different section, from the A Group, monitored Soviet aircraft flights from Sam Neua and Hanoi, gave the willing American media an excuse to once again ignore his testimony and that of others who substantiated it independently, including Trung Hieu, Terrell A. Minarcin, Jan Sejna, Barry Toll and other sources within the former Soviet Union. The Committee's final report stated: "Trung Hieu, a North Vietnamese who has sought political asylum in the United States, was interviewed by Committee staff in June 1992. In an interview, Hieu said that the entire crew of a downed B-52 was turned over to the Soviet Union in 1972; but he backed away from his assertions during his sworn deposition..."

The final report quoted none of Minarcin's extensive testimony, some of which has been noted previously in this book, but stated only: "Terrell "Terry" A. Minarcin was also in communications intelligence in the Air Force (and had worked under Mooney for two years—author's note). Mr. Minarcin told the Committee that he tracked 'special flights' of Soviet aircraft in 1977 that carried American POWs to the Soviet Union...The Committee found no information to corroborate the reports of Trung Hieu or Mr. Minarcin."

This statement within the final report is incorrect in that former Army intelligence specialist Barry M. Toll had corroborated Mooney's and Minarcin's testimony in his own statements to the Committee, also noted earlier in this book. Although Committee investigators told the author that no high ranking officers of the JCS staff of the time under Admiral Thomas Moorer would confirm Toll's statements about U.S. POWs shipped from Vietnam to the USSR following the end of the war in 1973, the Committee found that Toll had indeed served in the units he specified from 1973-1975. The final report dealt more carefully with General Sejna, who was a long-time employee of the U.S. Department of Defense:

"Jan Sejna, a retired Major General in the Czechoslovakian Army, has testified in a deposition and stated in interviews that American POWs were transported to the Soviet Union, transiting Prague. HE SAID HE HAD PERSONAL KNOWLEDGE OF THE TRANSFER OF UP TO 90 SUCH POWs THROUGH PRAGUE. General Sejna defected from a high-level position in the Ministry of Defense—where he would have had access to such information—in 1968, and is now an employee of the Defense Intelligence Agency...In December 1992, during a visit by the Joint Commission to Prague, Ambassador Toon asked Czech officials whether they had heard of the allegations made by Jan Sejna. None of the officials denounced or discredited Sejna. All promised to research their archives, but referred the U.S. delegation to the Ministry of the Interior for answers. The Federal Minister of the Interior, Mr. Petr Cermak said that the allegations must be taken seriously, that the Communists were capable of anything, and that his Ministry would turn over to the U.S. Government everything it found concerning Czechoslovakia's involvement in the Korean and Vietnam Wars."

The author had learned from a Senate Select Committee staff investigator who requested anonymity that in the initial meeting with General Jan Sejna, a top counsel for the Defense Department was present who repeatedly interrupted Sejna and attempted to stop him from discussing the American prisoners from Vietnam seen transiting Prague on their way to the Soviet Union. The Czech defector at first admitted only that he had been informed about the American POWs, but in his second meeting with Senate investigators in late 1992, to give a deposition, he stated that he personally had also seen the U.S. POWs in Czechoslovakia, and estimated their number as up to 90 Americans.[46]

Other evidence uncovered by the Committee about NSA, which substantiated Mooney's previous testimony, was included in the final report but was virtually ignored by the mass media in America:

"During the Vietnam War, the NSA monitored all available sources of signals intelligence bearing on the loss, capture or condition of American personnel. Such information would sometimes provide a basis for concluding whether or not a missing American had survived his incident and, if so, possibly been taken prisoner. During its investigation, the Committee was disturbed to learn that the NSA and its Vietnam branch were never asked to provide an overall assessment of the status of POW/MIA personnel prior to Operation Homecoming...The Committee also found that neither DIA nor any other agency within the intelligence community placed a formal requirement for collection with NSA concerning POW/MIA related information. In fact, the Committee found that NSA end product reports were not used regularly to evaluate the POW/MIA situation until 1977.

"It was not until 1984 that the collection of information on POW/MIAs was formally established as a matter of the highest priority for SIGINT...In conducting its review of NSA files, the Committee examined more than 3,000 postwar reports and 90 boxes of wartime files...Hundreds of thousands of hard copy documents, memoranda, raw reports, operational messages and possibly tapes from both the wartime and post-war periods remain unreviewed in various archives and storage facilities. MOST TROUBLING, NSA FAILED TO LOCATE FOR INVESTIGATORS ANY WARTIME ANALYST FILES RELATED SPECIFICALLY TO TRACKING POWs, DESPITE THE FACT THAT TRACKING POWs WAS A KNOWN PRIORITY AT THE TIME. THIS FAILURE MADE IT IMPOSSIBLE FOR THE COMMITTEE TO CONFIRM SOME INFORMATION ON DOWNED PILOTS THAT WAS PROVIDED BY NSA EMPLOYEE JERRY MOONEY...The Committee benefitted from the insights of retired NSA SIGINT analyst, Senior Master Sergeant Jerry Mooney. During the war, SMSgt. Mooney maintained detailed personal files concerning losses of aircraft and downed airmen. UNFORTUNATELY, THOSE PERSONAL FILES DID NOT BECOME PART OF THE ARCHIVES FILES MAINTAINED BY NSA AND HAVE BEEN LOST. Although SMSgt. Mooney has sought to reconstruct some of that information from personal memory, the loss of the files makes it impossible to check those recollections against the contemporaneous information. The Committee found no evidence to substantiate claims that signals intelligence gathered during the war constitute evidence that U.S. POWs were transferred to the Soviet Union from Vietnam."[47]

On the subject of historical precedents in the POW/MIA matter, the Senate Select Committee's final report stated (in part):

"John M. G. Brown and James D. Sanders, assisted by Mark A. Sauter, have conducted years of research in the U.S. archives, searching for information relating to U.S. and Allied POWs who fell into the hands of the Soviet Army as it pursued the rapidly retreating Wermacht across eastern Europe in 1945. Thousands of soldiers were moved by rail, truck and foot eastward, not westward, and most ended their cross-country journey at the Port of Odessa, on the Black Sea, there to await transport by sea to their homelands. This much is not in dispute. What is in question is how many of these soldiers were not allowed to board ship, but were destined for the vast Gulag of the Russian-Siberian interior. Mr. Sanders and Mr. Brown estimate that between 20,000 and 23,500 were POWs of the Germans and became prisoners of the Soviets.

"It is Mr. Brown's theory that Communist mistreatment of POWs—that is retaining them as hostages for political purposes—can be traced to the behavior of the Bolsheviks. According to Mr. Brown, the Bolsheviks kept at least 60 American soldiers they captured during the Allied intervention of 1918-1919 at Archangel and a few from the Siberian front. In his view, this was a prelude to the retention by the Soviets of thousands of soldiers taken from the German POW camps after World War II."[48]

In regard to the 1981 White House meeting in which an offer from Hanoi to trade a number of U.S. POWs in return for the aid promised by President Nixon was discussed, Richard Allen changed his previous sworn statements in which he had stated he recalled such a meeting, to saying subsequently that he had first heard of it was during the meeting arranged by the author in September 1986, which was recounted earlier in this book. The Committee declined to subpoena John F. Syphrit, the U.S. Secret Service agent who had overheard the 1981 White House meeting, and one of the original sources about it. According to the final report of the Select Committee:

"The Committee received information that President Reagan had received an 'offer' in early 1981 transmitted through a third country (Canada and/or China)...by the government of Vietnam to sell live POWs to the U.S. for $4.5 billion dollars. The source of this information was a Secret Service agent who allegedly was present and overheard part of a meeting in the White House where this matter was discussed. The agent reportedly overheard President Reagan discussing this offer with Vice-President George Bush, Richard Allen (National Security Adviser and William Casey (CIA Director). The conversation reportedly took place in the Roosevelt Room, as the four were walking from the Oval Office to a meeting in an adjoining conference room.

"The agent reported that James Baker (Chief of Staff), Michael Deaver (Deputy Chief of Staff) and Edwin Meese (Attorney General) were waiting in the area of the conference room for the meeting to begin, but he was unsure whether these individuals would have heard any of the conversation. The Committee treated this report seriously and first attempted to depose the Secret Service agent. Objections were raised by the Department of the Treasury and the Secret Service claiming that such a deposition would forever impair the ability of the Secret Service to guard the President. The attorney for the agent, J. Thomas Burch...explained that the agent would not testify without permission of his agency or a subpoena from the Committee..."

"The Committee regrets that the secret service agent was unwilling, out of concern for his job, to testify concerning his report. Faced with this unwillingness, the Committee was divided about whether to compel the agent's testimony by issuing a subpoena. Some members agreed with the Administration that compelling the testimony of a Secret Service agent concerning a conversation involving the President would set a harmful precedent...Others felt that the agent had waived his claim to special consideration by talking to others about what he reportedly heard, and that his testimony might contribute significantly to the Committee's investigation. After a lengthy debate, the Committee voted 7-4, with one Senator absent, not to subpoena the testimony of the Secret Service agent."

As noted previously in this book, According to Senator Robert Smith, the CIA's Deputy Director for Operations, Max Hugel, admitted that the 1981 White House meeting to discuss the North Vietnamese offer to trade for POWs had indeed taken place, but during a later deposition, he testified he had no knowledge of the meeting, and did not recall telling anyone he had.[49]

The national newspaper USA TODAY reported to the American people on the findings of the 1,223-page report of the Senate Select Committee on POW/MIA Affairs on January 14, 1993, under the headline: "POW/MIA Senate report: GOVERNMENT DIDN'T 'ABANDON' PRISONERS." The wording was similar to that used by most of the major U.S. print and broadcast media at the time, in adopting a pro-government line when the subject was covered at all: "A small number of U.S. prisoners of war may have been left behind after the Vietnam War, but the U.S. government did not 'knowingly abandon' them, a Senate report said Wednesday. If any POWs were left in Indochina after 1973, there's no evidence they're still alive... 'WE ACKNOWLEDGE THAT THERE IS NO PROOF THAT U.S. POWs SURVIVED, BUT NEITHER IS THERE PROOF THAT ALL OF THOSE WHO DID NOT RETURN HAD DIED...' "Nothing was said by the newspaper of the abandonment by the U.S. Government of thousands of American POWs in Communist control after Korea, WW II and WW I, which had been thoroughly documented for Senate Select Committee investigators by the author and two other researchers.

The next day, the Boston Globe headlined their similar story on the Select Committee report with: "NO SIGN MIAs ALIVE, SENATE PANEL REPORTS." The Globe noted: "Only a 'small number' of missing Americans may 'possibly' have survived the rigors of captivity. The newspaper quoted parts of the report for its readers: "While the Committee has some evidence suggesting the possibility a POW may have survived to the present, and while some information remains yet to be investigated, there is, at this time, no compelling evidence that proves that any American remains alive in captivity.'" According to the Boston Globe, "The Committee blasted those who, they said, have exploited the pain of POW families over the years and ridiculed that a government conspiracy took place to conceal the fate of the missing men." The Globe again quoted the report: "'The isolated bits of information out of which some have constructed whole labyrinths of intrigue and deception have not withstood the tests of objective investigation...The vast archives of secret U.S. documents that some felt contained incriminating evidence have been thoroughly examined by the committee only to find that the conspiracy cupboard is bare."

This was a patently untrue statement. As has been noted in this book, many thousands of POW/MIA intelligence documents held by the NSA, CIA and other U.S. agencies had never been released to the committee or declassified. According to the Committee's final report other critical documents had disappeared or been deliberately destroyed:

"The U.S. Army's Deputy Chief of Staff for Intelligence has been unable to locate any of his agency's archival POW/MIA intelligence staff records from the Vietnam war era. This includes internal intelligence reports, memoranda, planning documents and similar records documenting what the Army knew or suspected about personnel captured or missing in Southeast Asia. IT REMAINS

UNKNOWN WHETHER THE RECORDS WERE DESTROYED OR SIMPLY MISPLACED...The U.S. Navy provided a small collection of assorted documents in response to the Committee's request, but advised that nothing further could be located. After repeated prodding from the Committee, THE NAVY REPORTED THAT ALL REMAINING POW/MIA RECORDS HAD BEEN DESTROYED IN 1975. Committee investigators then uncovered extensive Navy records at the Naval Historical Center...including most of the major files of the Chief of Naval Operations Special Assistant for POW/MIA Affairs. THERE ARE INDICATIONS THAT CERTAIN SENSITIVE NAVAL INTELLIGENCE FILES WERE SHIPPED TO DIA IN 1981, WHILE OTHERS APPEAR TO HAVE BEEN DESTROYED... The U.S. Air Force provided no response to the Committee's original request for records. Finally in September 1992, the Committee was provided a printout of a small portion of the archives at the Joint Services SERE (Search, Evasion, Rescue, Escape) Agency (JSSA)...IT APPEARS THAT THE WARTIME AIR INTELLIGENCE FILES WERE TRANSFERRED TO JSSA IN 1974, PUT ON MICROFICHE (WHERE THEY HAVE BECOME LARGELY ILLEGIBLE...) AND THE ORIGINAL DOCUMENTS DESTROYED...Sources indicate that there were some intelligence reports on POW/MIAs collected through MACVSOG during the war, especially in Laos. Unfortunately the Committee was not able to locate these reports..."

"The Joint Task Force Full Accounting (JTFFA) HAS YET TO PROVIDE THE WARTIME PERMANENT RECORDS OF THE PRINCIPAL ORGANIZATION RESPONSIBLE FOR MONITORING THE POW/ MIA PROBLEM ON THE GROUND IN SOUTHEAST ASIA, the special operations related Joint Personnel Recovery Center. JPRC was transformed into the Joint Casualty Resolution Center (JCRC) in January 1973; the Committee has requested, but at publication had yet to receive, an index of its archival files. The Pacific Command has reported it has no documents, even though it was one of the most major command players throughout the Vietnam war. FINALLY, THE COMMITTEE WAS HINDERED IN JUDGING THE ACCURACY OF SERVICEMEN ACCOUNTED FOR AND NOT ACCOUNTED FOR DURING THE WAR BY THE FACT THAT SEARCH AND RESCUE (SAR) REPORTS HAD BEEN DESTROYED FOLLOWING THE WAR.

"We note that General Vessey confirmed to the Committee that these records had been destroyed by 1979...The Committee's investigation disclosed the possible existence of other collections of POW/MIA related files which have been requested for review and declassification, but which at publication time had not been received. These include, but are not limited to, the POW/MIA staff and operational files of the Military Assistance Command Vietnam (MACV), J-2 (Intelligence) staff element responsible for management of POW intelligence in Vietnam, and the Pacific Command's (CINCPAC) POW/MIA staff. The archival POW/MIA intelligence files from the Department of State are also undergoing declassification. However, the Committee has been advised informally by the Department that these files are poorly organized and never have been indexed."[50] But the major media in America ignored such embarrassing facts which had been revealed in the report, allowing the stalling tactics of the national security apparatchiks in Washington to succeed once again.

The Globe's competitor, the Boston Herald, presented a somewhat more objective if sketchy account of the Committee's findings, in a January 14 article by reporter Christopher Cox, who quoted directly from the report: "There is no proof that POWs survived, but neither is there proof that all of these who did not return had died. THERE IS EVIDENCE, MOREOVER, THAT INDICATES THE POSSIBILITY OF SURVIVAL, AT LEAST FOR A SMALL NUMBER, AFTER OPERATION HOMECOMING." The Herald reported: "In a dissenting footnote, Senators Robert Smith, the Committee's Vice Chairman and Charles Grassley, two stalwarts of the live POW lobby, argued that livesighting reports and other sources of intelligence are evidence that POWs may have survived to the present. "The Herald also reported that, "Dan Perrin, principal author of a May 1991 Senate report on the POW issue, said Henry Kissinger's input into the final report 'completely violated' the committee's integrity. After receiving a leaked draft, Kissinger, architect of the 1973 peace treaty with Hanoi, lobbied the Committee for changes. 'It would be like Ollie North getting to write the key paragraphs of the Iran-Contra report,' said Perrin, a former aide to Senator Jesse Helms. Kerry said Kissinger 'had no more input than anybody else.'"[51]

The author believes that the United States Government may have covertly ransomed some small number of the surviving American prisoners of war from the 1980 to 1989 period. As in 1920-21, 1946-48 and 1954-57, a few American POWs from Vietnam were said to have been secretly returned to the U.S. throughout the 1980s, but the author was unable to verify the facts of this matter, as reported by others. The author's investigation uncovered information on reported U.S. ex-POWs, who claimed to have been returned to U.S. control long after the American withdrawal from Indochina, supposedly involved in a classified witness protection-type program, and in several cases personal tragedies, including suicide, ostensibly to facilitate the secret returns of even more U.S. POWs.

In a particular case in Washington State, where the author investigated the POW/MIA matter during 1986-87, an ex-U.S. Navy SEAL named Robert Steelman (whose identity was verified by a retired Navy SEAL Chief Petty Officer named Bud Potter with whom he had served during an earlier tour of duty, and by another retired Navy SEAL Lt. Commander, Clifford Johnson), informed the author that he, Steelman, had later been captured by the Communists in Vietnam and held there as a prisoner until the 1980-1981 period, when he was suddenly released and brought back to the United States, whereupon he was separated in Seattle and encouraged to start a new life. He claimed that his record had been "scrubbed" of everything except reference to "service in Southeast Asia." After revealing this story to the author in 1986-87, Steelman agreed to meet in the future with two journalists who were then researching and writing a book on the Vietnam War POWs and MIAs. (William Stevenson and Monika Jensen-Stevenson) Shortly after this, Steelman was found shot to death near Port Townsend, Washington; the back of his head having been partially blown away. The Port Townsend Police at first said it appeared to be a suicide, but it was officially recorded as an accidental death, although several of his friends believed that he had been murdered, because he had talked about something he shouldn't have. The author returned to Haines, Alaska shortly afterward, and had no way of further verifying the truth of this matter at the time. This information was given to the Senate Select Committee staff investigators with whom the author worked for an entire year, during 1992 and 1993, with a request for further investigation, but for reasons unknown, no investigation occurred, and the author's sources were never contacted.

The author's research-colleague, Thomas V. Ashworth, had testified before a 1985-86 Senate Committee concerning statements made to him by an active duty U.S. military officer, who had been informed by Colonel Mark Richards of DIA that there was 'no reason to worry about the POWs, no expense was being spared and the prisoners would be brought home quietly; they could not be publicly returned so no one would ever hear about it and the POW families would be kept quiet.' Just prior to publication of this book, research colleague Ashworth learned that former Defense Department POW/ MIA expert Bill Bell (now retired in Arkansas) had taken statements from North Vietnamese officers about secret POW exchanges during the Vietnam War, in which American POWs were exchanged for Communist prisoners. Some of the Americans reportedly came out of Laos, while other exchanges even occurred in Saigon. Some of the POWs were reportedly flown down in ICC aircraft. If this type of secret program for POW exchanges was in operation during the war, and has yet to be acknowledged by the U.S. Government, there is reason to suspect that such secret exchanges continued after the war.

A 1992 article by reporter David Hendrix, in the Riverside (California) Press-Enterprise, had stated that up to several hundred American prisoners (some of whom subsequently died) had been secretly returned to the U.S. from Indochina in a classified witness protection-type program, from 1979 on, and particularly around the

1986-87 period, when the Iran-Contra revelations of covert operations temporarily halted the project, although it was later reportedly revived in 1989. Other than the single experience in Washington State noted previously, the author was not able to verify these reports, and one of the primary sources used by Hendrix later denied under oath before the Select Committee on POW/MIA Affairs that he had said certain things to Hendrix that had been reported.

The final report of the Select Committee on POW/MIA Affairs barely touched on this subject. Under the heading: "Systematic Lie Theory," the report states:

"Other stories are more difficult to disprove, but even their defiance of common sense does not stop their spread, which in turn mainstream media, fuels these rumors. For example, one persistent story is that the U.S. Government has been bringing POW/MIAs back secretly and providing them with new identities such as is done in the federal witness protection program or, in the alternative, incarcerating them in mental hospitals. The ostensible reason for this secrecy is presumably to avoid contradicting official policy since 1973 that all live POWs were returned home. Another theory argues that since no amputees or mentally deranged people returned at Operation Homecoming, these men have been smuggled back and are kept hidden. Committee investigators interviewed a newspaper reporter who printed this story as fact, his sources, and others with variations of this story; THEY FOUND NO FACTUAL SUPPORT FOR IT. One supposed source summoned to testify, and subpoenaed, was the victim of his ex-wife's fantasies."

In the author's view however, the Senate Committee did not adequately investigate this matter because they did not want to deal with the subject.

---

## Notes on Chapter Fourteen

1   Santa Rosa Press Democrat, June 16, 1992 and Associated Press

2   Santa Rosa Press-Democrat, Santa Rosa, California, June 16, 1992, Press Democrat News Services

3   Washington Times, June 17, 1992

4   Press Democrat news service, Santa Rosa Press Democrat, Thursday, June 18, 1992

5   Santa Rosa Press Democrat, Thursday, June 18, 1992

6   Associated Press, June 19, 1992

7   Letter dated July 9, 1992 from Ambassador Robert Strauss, U.S. Embassy, Moscow, to John M.G. Brown, in the author's files.

8    Hearings of the Select Committee on POW/MIA Affairs, June 1992

9    Washington Times, June 24, 1992

10   Associated Press report June 28-29, 1992

11   San Francisco Chronicle, Wire Service report, July 1, 1992

12   U.S. Representative Frank McCloskey of Indiana was a long-time Democratic leader in the House on POW/MIA matters, known and respected by the author, who was herein being urged to accept the word of a Communist official that no POW remained in Laos after 1973, when several of that official's own statements to the contrary had long before been broadcast to the world, and a massive amount of operational and special intelligence to the contrary had long existed, and this had long been known to the Congressman. Thus it can be seen that into 1992 U.S. Government policy towards persistantly inquiring Congressmen had also remained unchanged since Hamilton Fish's day, in 1930.

13   Los Angeles Times, July 7, 1992, "MIAs: Korea GIs May Have Been Test Subjects," by LA Times Staff Writer Melissa Healy

14   San Francisco Examiner, July 5, 1992

15   NYT, October 8, 1992

16   Tacoma News Tribune, August 7, 1992.

17   Kathleen Callo, Reuters News Agency, October 1992

18   August 1992 Hearings of the Select Committee on POW/MIA Affairs, U.S. Senate

19   The full record of responses by DIA officials present at the Hearings may be examined in publications of the U.S. Senate Select Committee on POW/MIA Affairs, U.S. Goverment Printing Office, 1992.

20   A meaningless statement from a confirmed atheistic Marxist-Leninist

21   *The Search For MIAs*, p. 166, by Chief Warrant Officer Garry L. Smith, AUS-Ret'd, Honoribus Press, Spartanburg, South Carolina, 1992

22   Associated Press, Aug. 14, 1992

23   The Times (of London)

24   Deposition of James Schlesinger, September 4, 1992, Select Committee on POW/MIA Affairs, U.S. Senate

25   Ibid.

26   Soldiers of Misfortune, by Jim Sanders, Mark Sauter and R. Cort Kirkwood, National Press Books, Bethesda, Maryland, p. 15

27   Testimony of Melvin Laird, September 22 & 23, 1992, Select Committee on POW/MIA Affairs, U.S. Senate

28   Testimony of Henry Kissinger, September 22, 1992, Select Committee on POW/MIA Affairs, U.S. Senate

29   Washington Times 22 Sept., NY Times Sept. 23 & 24, Wash.Post, Sept.24-25th, 1992

30   Letter of April 17, 1992 from Joseph Kingsbury-Smith, in the files

31   October 1992 interview with Jerry J. Mooney, by the author.

32   Congressional Record, October 5, 1992

33   Ibid.

34   Washington Times, Oct. 7

35   Memorandum For The Record, Office of the Assistant Secretary of Defense, subject: Telephone call from Senator Kerry, October 7, 1992, signed by Ronald J.

Knecht; in the author's files

36    Hand-written memorandum dated October 6, 1992, to Senator John Kerry, signed by Frances A. Zwenig, Staff Director, Select Committee on POW/MIA Affairs; in the author's files

37    Statement of Assistant Secretary of Defense Duane P. Andrews Assistant Secretary of Defense Command, Control, Communications and Intelligence, U.S. Senate, October 15, 1992

38    Washington Post, Oct. 11,1992

39    *Fulcrum- A Top Gun Pilot's Escape From The Soviet Empire,* October 1992

40    NYT, March 8, 1987 and Crimes Of Patriots- A True Tale of Dope, Dirty Money and the CIA, by Jonathon Kwitny, W.W. Norton, NY, 1987; also: Kiss The Boys Goodbye

41    Interview by the author; see also: Tacoma News Tribune, November 7th, 1992

42    pp. 417-420, and other references, Report of The Select Committee on POW/MIA Affairs, January 13, 1993, op. cit.

43    Associated Press and The Stars and Stripes National Tribune

44    Letter in the author's files

45    p. 184, Hearing before Select Committee on POW/MIA Affairs, January 21 and 22, 1992

46    Confidential source interview by the author.

47    Select Committee Report, pp. 22-23

48    pp. 419-420, Report of the Select Committee on POW/MIAAffairs, January 13, 1993, op. cit.

49    pp. 282-284, Final Report of the Select Committee on POW/MIA AFFAIRS.

50    Final Report, pp 130-31

51    Boston Herald, January 14, 1993

# CHAPTER FIFTEEN

## THE CLINTON ADMINISTRATION AND THE LOST POW/MIAS

President William Jefferson Clinton was inaugurated on January 20th, 1993, with his Vice President, Albert Gore. Clinton chose Warren Christopher, who had been Cyrus Vance's deputy secretary of State in the Carter Administration, to be his Secretary of State. As Secretary of Defense he appointed Congressman Les Aspin, who had specialized in military matters since the 1960s Vietnam War buildup, when he had been one of Robert McNamara's Pentagon "whiz kids." Former Kissinger aide Anthony Lake became Clinton's national security adviser. (Henry Kissinger himself became the Honorary Chairman of the National Committee on American Foreign Policy, in New York, the same organization that had permitted the author to publish a synopsis of this book in 1991.)

Other important security-related appointments by President Clinton included: former ambassador Frank Wisner Jr. (the son of CIA's deceased OPC chief, Frank Wisner) as Undersecretary of Defense, and another Kissinger protégé, Peter Tarnoff, as Undersecretary of State. The relatively unknown James Woolsey, was to replace Robert Gates as Director of Central Intelligence. Evidence that Clinton would continue to follow the Bush administration's policy towards Iraq's combative leader, Saddam Hussein, came with renewed U.S. airstrikes against Baghdad at the very onset of Clinton's presidency. A significant change in Russian policy occurred soon afterward as the Foreign Ministry in Moscow criticized the continued U.S. air raids on Baghdad as, 'inappropriate" and urged a United Nations review of the American, British and French military actions, to ensure they followed Security Council resolutions. (At this time a few Americans, mostly reconnaissance and commando troops or airmen, still remained missing and presumed killed or unaccounted-for, from the brief 1991 Persian Gulf conflict.)

The neo-Stalinist Russian Vice President, Aleksandr Rutskoi, warned in Moscow that Russia was prepared to use its veto power in the UN Security Council to halt American-led attacks on Iraq.[1] Although Yeltsin's statements were not as confrontational, the change in Russia since General Volkogonov's testimony before the Senate in November, as hard-liners reasserted themselves, boded ill for any release of American POW/MIA survivors from restricted areas in the old Soviet republics.

Following President Clinton's inauguration, the Senate Select Committee report on prisoners of war continued under discussion in the media, but in most cases virtually nothing was said of the Soviet connection to missing American prisoners of Vietnam and earlier wars, which had been largely exposed by witnesses appearing before the Select Committee, such as Eisenhower's former NSC aide, Lt. Colonel Phillip Corso, former U.S. air attaché Lt. Colonel Delk Simpson, and others, but was downplayed in the major American media. The work of the Select Committee was summed up in a Washington Post editorial which, not surprisingly, largely conformed to official U.S. policy statements about American POWs and MIAs that had been repeated for decades:

"The Senate Select Committee, however, has done what duty demanded to wrap up an inquiry that has roiled the national conscience for 20 years. Its conclusion that some Americans may have been left behind but that there is 'no compelling evidence' any are now alive deserves a sober hearing. Some anguished families may be unable to accept it. It is notable, however, that on the committee the unanimous support for this conclusion reached from Chairman John Kerry to Jesse Helms. Much of the public discussion of MIAs has been an intensely partisan inquiry into whether the Nixon administration or the Defense Department abandoned American fighting men and then covered up the abandonment. The Committee found evidence of sloppiness, secrecy and fatigue on the bureaucratic level and of evasion on the political level, but not of a cover-up or conspiracy. Even as they ...The American debate should not impede understanding of where the principal onus for the failure to obtain a full accounting lies: on Vietnam. Hanoi saw in American concern for MIAs a lever with which to bargain successively for: 1) reparations, which the United States flatly refused to pay; 2): economic aid, a tenuous possibility that disappeared when Hanoi broke the peace accords, and 3), more recently, normalization of relations. Its bargaining involved constant lies so that each new slice of disclosure inevitably became a confession of past deception. No one can know what secrets Vietnam may still be hiding..."[2]

Thus, without even hinting at the Committee's initial responsibility to examine the entire U.S. POW/MIA issue, including the historical precedents of Communist-held American POWs of previous U.S. wars, did the Washington Post again put their journalistic seal of approval on a U.S. government POW/MIA study, as they had done nearly 16 years before, after the final report of the Montgomery Committee was issued. The Post chose not to deal with the substantial documentary and human source evidence that had been turned over to Kerry during the life of the committee by author John M. G. Brown and research colleague, Thomas V. Ashworth, and subsequently by James D. Sanders and other private researchers and eyewitnesses, or with the fact that the Select Committee's belated interest in this declassified information came too late, and was finally dealt with in a hurried and slipshod manner. This occurred despite the historic admissions of Boris Yeltsin, Lt. Colonel Philip Corso, General Jan Sejna and many others during 1992, which had penetrated to the very heart of the prisoner of war dilemma. Among the findings of the Committee which were nearly universally ignored by the mainstream American media, were those concerning the non-compliance of various federal departments, agencies and high officials in meeting requests of the Senate Select Committee.:

"The DIA refused to declassify the 'sources' and 'methods' which they had used to build up their files. The Committee understood the grounds for not declassifying these materials to the general public since the sources lives could be endangered, information resources compromised or hard-won crypto-analysis work lost. IT WAS LESS UNDERSTANDABLE WHY THE DIA REFUSED TO DISCLOSE THE NAMES OF SOURCES TO APPROPRIATELY CLEARED STAFF OF THE COMMITTEE SO THE SOURCE'S STORY COULD BE CHECKED WITH THE ORIGINAL SOURCE.

"The CIA initially refused to allow even appropriately cleared members of the Committee staff to review past and current operational files (with the notable exception of a detention camp in Laos). This matter was partially resolved in December WHEN A SINGLE SELECTED STAFF MEMBER WAS ALLOWED TO REVIEW THE FILES. CIA OFFICIALS DID NOT ALLOW THE COMMITTEE TO HAVE ACCESS TO THEIR PRESIDENTIAL DAILY BRIEFS..."

"The NSC refused to allow anyone but Senators to review the current administration files and limited the review of past administration files to the Staff Director (Frances Zwenig), Chief Counsel (William Codinha) and three senior staff members."

"The DoD (Department of Defense) refused to allow anyone but the Chairman and Vice Chairman to review the POW returnee debriefings from 1973. This was in spite of a release that the Committee had obtained from several hundred of the POW returnees involved. The DoD OSD/ISA initially refused to provide certain files to the staff of the Committee, but later allowed access. THE COMMITTEE WAS DISTURBED TO LEARN, THROUGH INTERNAL CDO E-MAIL NOTES, THAT ISA (International Security Affairs) HAD INTENTIONALLY DELAYED PROVIDING FILES IN ORDER TO PRE-SCREEN THEM.

"The Nixon Archives refused to allow access by the Committee to any of the Watergate tapes that had been requested. Former President Nixon's refusal to allow even the most limited access in the face of repeated requests, letters and entreaties at the highest levels CAUSED THE COMMITTEE TO DRAW SOUND UNFAVORABLE INFERENCES ABOUT THE ACTIONS OF THE FORMER PRESIDENT ON THIS ISSUE. It is unfortunate that the former President had the power to limit the access and frustrate the wishes of a constitutionally created Committee of Congress to what was clearly the best evidence available...FORMER PRESIDENT NIXON'S ATTORNEYS WERE ABLE TO FRUSTRATE THE DESIRE OF THE COMMITTEE TO REVIEW THE TAPES FOR POW/MIA DISCUSSIONS."[3]

Sydney H. Schanberg, the nationally-known columnist for Newsday who had been instrumental in exposing the Cambodian Khymer Rouge genocide of the 1970s, and had begun writing about the POW/MIA investigation of the Senate Select Committee during 1992, and commented about the findings of the Committee on the editorial page of the Boston Globe on January 23rd:

"The 2-inch-thick document is a testament to the reliable principle that one should never expect a congressional investigation to get to the bottom of anything. Part of the reason in this case is that the Committee was stampeded by a herd of powerful people and government agencies lobbying to rewrite history—from Henry Kissinger and Richard Nixon to the Pentagon and the intelligence community...In a better world, the senators would have reviewed and endorsed the compelling evidence they had gathered over a period of 15 months that American prisoners were left behind when we signed the peace accords with Hanoi 20 years ago. Yet, when the crunch came, the committee chose to step back from its information. It said only, and weakly, that there was evidence that indicates the possibility of survival, at least for a small number" of prisoners, after Hanoi's repatriation of 591 men in 1973."

In late January President Clinton chose a career diplomat, Timothy Pickering, as U.S. Ambassador to Russia, to replace Robert Strauss, the former U.S. trade negotiator and chairman of the Democratic National Committee who had submitted his resignation to George Bush just before the November election. Pickering had held high-level positions under Presidents Nixon, Ford, Carter, Reagan and Bush, whom he had served as U.N. ambassador during the Persian Gulf campaign, and had recently been the U.S. ambassador to India. Clinton nominated Strobe Talbott as Ambassador at large to the former Soviet republics. Talbott had been an editor at Time Magazine and had written about U.S.-Soviet relations; avoiding the truly serious matters which had divided the two nations for so long, such as the secretly-held American prisoners of World Wars I and II, Korea and Vietnam.[4]

CIA Director Robert Gates had meanwhile left office, amidst a minor furor over the results of CIA's internal investigation that it's officials were responsible for failing to provide information from the Justice Department and a federal judge in Atlanta in the case involving $5 billion in American loans to Iraq prior to the U.S. Persian Gulf War, through an Atlanta branch of an Italian bank, in a brewing scandal that came to be called "Iraqgate." President Clinton had meanwhile appointed James Woolsey to succeed Gates as DCI. Gates had gained substantial media publicity for the CIA during

1992, with his promises to declassify information on American POWs and on U.S.-Soviet relations, but for researchers such as the present author, these promises proved empty, and long-time FOIA requests and appeals by the author for CIA documents concerning Soviet-held U.S. POWs, and debriefings of specific Soviet defectors continued to be ignored by that agency. This was not surprising; in the final report of the Senate Select Committee on POW/MIAs, it was recorded that, "...THE CIA DID NOT ALLOW COMMITTEE INVESTIGATORS TO HAVE ACCESS TO THE EXECUTIVE REGISTRY, WHICH ARE THE DCI's PERSONAL OFFICE FILES."[5]

President Bill Clinton, now the Commander-in-Chief of the U.S. armed forces, had been an anti-war activist in the 1960s, who, like many other young men of the Vietnam generation, had obtained a draft deferment by questionable methods, and had helped organize demonstrations against U.S. military involvement in Vietnam while a student at Oxford University on a Rhodes scholarship. During the election campaign of 1992, Clinton had reluctantly admitted to being a draft evader during the Vietnam War, and to having visited Moscow in 1968. At the start of his Presidency, Clinton avoided the POW/MIA subject, and made clear his number-one priority for the U.S. military soon after the inauguration, was to keep a campaign promise to cancel, by executive order, a Defense Department ban on homosexuals and lesbians serving openly in the U.S. military services. His announcements on the subject in late January caused an immediate negative reaction from the Joint Chiefs of Staff, still under General Colin Powell, from many Senators, Congressmen and veterans' groups, and from the public at large, which resulted in his decision to wait until Congressional hearings on the subject could be held. Eventually he was to pull back from this commitment.

Meanwhile, from late January to mid-February, several reports were published in the American media about sightings of a U.S. POW from the Vietnam War said to have been living in a village in the Ukraine as recently as May or June 1992. The Pentagon had announced: "The investigation is ongoing...we have some leads we're pursuing," for an American prisoner from Vietnam last reported in the Ukrainian town of Upper Khortitsa. The U.S. Embassy in Kiev released a statement to the Ukrainian press seeking information about an American serviceman named "Tom" or "Timhoka." The Washington Times reported that U.S. officials had told the newspaper in late January about an emigre named Sergei Malnikov who claimed to have seen an American pilot from Vietnam in the Ukraine, and who accompanied two U.S. military investigators to the newly independent republic. The first reports indicated the American POW could have two children, and might be held in a mental hospital, which fact would validate the statements made months before by Russian President Boris Yeltsin.[6]

The veteran covert operator Richard Armitage finally fell from grace in late February, after a recording of a speech he'd made before the Vanderbilt Institute for Public Policy Studies was provided to the Associated Press, which published the substance of it on February 21st. According to the New York Times, in his remarks Armitage forecasted that Russian President Boris Yeltsin would be ousted soon: "Not unlike Gorbachev, his days are somewhat numbered...I Think he's about at the end of his usefulness and someone else will step on the scene." Such a statement from a senior U.S. official who was then being called 'the most powerful man in Russia,' could have been interpreted as a kiss of death for the beleaguered Russian president. The New York Times called Armitage's remarks, "HIGHLY UNUSUAL IN THE WORLD OF DIPLOMACY AND COULD BE INTERPRETED AS UNDERCUTTING MR. YELTSIN AT A TIME WHEN HE IS ENGAGED IN A POWER STRUGGLE WITH HIS FRACTIOUS PARLIAMENT." Although Armitage said later that his remarks were "injudicious," the damage was already done.

The New York Times reported that Secretary of State Warren Christopher, when contacted in Cyprus while on a trip to the strife-torn Middle East, responded to questions about Armitage's remarks: "It's an assessment I do not agree with. I think President Yeltsin is the best opportunity the people of Russia have at the present time. We strongly support his leadership." On February 22 the Department of State announced the appointment of a new coordinator of American aid to the former Soviet republics, career diplomat Thomas W. Simons Jr., a Russian and East European specialist who had been Director of the Office of Soviet Union Affairs and was currently serving as U.S. Ambassador to Poland. The Department announced that Simons would report directly to the U.S. Ambassador-at-Large to Russia and the newly independent states, Strobe Talbott, the former Time Magazine editor who had been nominated by Clinton but as yet not confirmed by the U.S. Senate.[7]

During this same period, the journal "France-Amérique" reported that a 74-year old French POW of World War II, originally captured by the Germans in 1940 and overrun by the Soviets in 1945, had been located alive in Ukrainia, and after an absence of more than half-a-century had returned to his native France to visit relatives. Paul Catrain, a native-born Frenchman from the village of Bois-Lès-Pargny (Aisne), had been a farmer when called up to fight the Germans. After his capture he had been moved from one Nazi POW camp to another until he ended up interned under Soviet control in Koenigsberg, East Prussia, now part of the USSR and known as Kaliningrad. According to the French newspaper "La Nouvel Observateur," Catrain explained that when he was liberated from his prison camp at the end of World War II by the Red Army he was shifted back and forth, first to Poland then back to the USSR, and had thus been unable to return to France. The ensuing Cold War then cut off all communication. After he disappeared, Catrain's family and the French government had presumed him dead and his name was inscribed on a war memorial in his native village. While a captive and later as a free citizen in the Soviet Union, Paul Catrain had been called "Paul Catrovitch Flamme," a Russified version of his real name, that incorporated his mother's maiden name: Flammant. This reportedly had led to some confusion when Catrain had first succeeded in informing the French Embassy (in Moscow) of his presence in 1980 at the time of the Olympic Games, when he was reported by his Russified name as being an unknown, from "Bolepargni, France." Catrain had met a Ukrainian woman named Maria while in captivity, and later married her and moved to "Strikhivsti," in the Ukraine, where he was known as "Paul the Frenchman." Although first allowed to meet with French diplomats in 1984 under KGB control, Catrain had not been able to leave Ukraine until after the great Soviet upheaval. He had almost lost the use of his native French language by the time the French cable network "Planète" organized a trip.[8] This extraordinary tale of survival was largely ignored by the American mass-media.

Boris Yeltsin meanwhile faced an increasingly ugly power struggle through the end of March, with the Communists and Russian nationalists controlling the Russian Congress, most of whom had been elected under the old Soviet system in the time of Mikhail Gorbachev, and who resented the collapse of the Soviet empire and institution of free market reforms, not to mention Yeltsin's revelations of Soviet secrets such as the holding of American prisoners of war. Among his chief antagonists were Ruslan Khasbulatov, the hardline parliamentary chairman, and Aleksandr Rutskoi, the militaristic Russian Vice President who had been a POW of the mujahedeen while serving in the Soviet Air Force in Afghanistan. Yeltsin barely survived an impeachment attempt on March 28th, in which the hardline Communists and nationalists failed to reach the necessary two thirds majority by only 72 votes out of 1,033 members.

Yeltsin warned the West of a "slide back to Soviet power," while preparing for an April 3rd and 4th summit meeting with President Clinton in Vancouver, British Columbia. Those concerned with Yeltsin's survival urged that the meeting be held in Moscow to underscore western approval of his reform policies. The forces opposed to Yeltsin in Russia were the same which would continue to conceal surviving American and Allied prisoners of war in the former Soviet republics.

At this time the author learned from several simultaneous sources that between March 25 and March 29, at the U.S. Embassy in Bangkok, Thailand, U.S. Government agents shredded and destroyed thousands of files about American POWs and MIAs, which contained numerous refugee livesighting reports of Americans in captivity long after the war, and years of hand-written notes and comments from investigators. The mass-shredding was conducted by Brigadier General Thomas Needham, commander of "Joint Task Force Full Accounting," a Navy Commander identified as Dale Hayes and an 'unidentified' CIA officer. The destruction of these documents made the task of analyzing which of the POWs and MIAs may have survived in captivity for years, more difficult, and appeared to be an effort to destroy critical evidence that could be used to prove criminal negligence on the part of DIA and JCRC analysts who had for decades been responsible for resolving the fate of over 2,500 missing American servicemen. It is noteworthy that the United States Ambassador to Thailand at this time was David Lambertson, who as a Deputy Assistant Secretary of State, six years before, had denied to the author in a meeting at the Department of State, that any evidence existed concerning American POWs who remained in Communist control after Operation Homecoming in 1973, or that any POW/MIAs had survived in captivity years after the war's end.

On April 9th, the Russian member of the US-Russian Commission on POWs, General Volkogonov, claimed that there was no evidence of any American prisoners of war alive in Russia, but Ambassador Malcolm Toon said that some Russians were hampering efforts to account for missing U.S. soldiers, and reported that he had given the Russians details on the shootdowns of ten U.S. aircraft. He was quoted by the press as saying: "We cannot understand why the Russian government cannot give definitive information about the shootdowns, or whether there were any survivors.[9] The Clinton Administration announced on the same day that General John Vessey would travel to Hanoi again on a fact-finding mission, and that the President would make no decision on relaxing U.S. pressure on Vietnam until he felt convinced that Hanoi was actually cooperating on resolving the fate of American MIAs. It was reported that the International Monetary Fund would decide in late April whether to resume lending to Vietnam, a move supported by France, Germany and Japan.[10] Meanwhile, American business interests, led by major oil companies, kept up pressure for normalization of relations with Vietnam's Communist government, irregardless of the fate of U.S. prisoners. At this point a major story on the missing Americans again brought the issue to national prominence.

The New York Times published a report on April 12th, revealing that a top-secret report on U.S. POWs, written by General Tran Van Quang, Deputy Chief of Staff of the North Vietnamese Army, on September 12, 1972 had been uncovered in the archives of the Soviet Communist party in Moscow, which stated that there were:
"1,205 AMERICAN PRISONERS OF WAR LOCATED IN THE PRISONS

OF NORTH VIETNAM—THIS IS A BIG NUMBER. OFFICIALLY, UNTIL NOW, WE PUBLISHED A LIST OF ONLY 368 PRISONERS OF WAR, THE REST WE HAVE NOT REVEALED. THE GOVERNMENT OF THE U.S.A. KNOWS THIS WELL, BUT IT DOES NOT KNOW THE EXACT NUMBER OF PRISONERS OF WAR, AND CAN ONLY MAKE GUESSES BASED ON ITS LOSSES. THAT IS WHY WE ARE KEEPING THE NUMBER OF PRISONERS OF WAR SECRET, IN ACCORDANCE WITH THE POLITBURO'S INSTRUCTIONS." The U.S. POWs were held in 11 North Vietnamese prisons in the fall of 1972 before the Paris peace agreement was signed. Since more Americans were captured between September 1972 and February 1973 and a total of 591 American POWs were released, the figures indicated that some 700 American POWs were secretly withheld by Hanoi after Operation Homecoming in 1973. (This number closely conforms to the author's estimates for two U.S. Senate investigations of 1990-1993, based on the research for this book. It would not include those Americans already transported to the USSR (or China) from 1965-September 1972, and would not necessarily include all those held in Laos, South Vietnam or Cambodia at that time.)

Consisting of both a Russian translation of Tran Van Quang's report marked "top secret" in Russian, and a summary of it by the Soviet Army Intelligence, the document was found in February 1993 by Stephen J. Morris, an Australian researcher for the Harvard Center for International Affairs and the Russian Research Center at Harvard University, who stated he had first showed it to senior White House officials of the Clinton Administration in February. According to the New York Times, copies of it were subsequently circulated among U.S. Government officials, and on the first page of the summary are handwritten instructions for a 'brief note...on the prisoners of war,' to be sent to the Soviet Politburo. The newspaper reported that General Quang said in the report that the American POWs could only be freed as part of an overall peace settlement, and they could be "used as leverage to obtain compensation for the devastation caused by the war." Members of the U.S.-Russian Commission on POWs said that the document was authentic and some experts were quoted by the newspaper as calling it a "smoking gun." The Russian newspaper Izvestia had reported on April 10th that the document on the U.S. POWs and other recently declassified files had been a topic of a closed meeting of the Commission.

The New York Times said that a report on the Soviet document was provided to President Clinton just before the Vancouver summit with Boris Yeltsin, and quoted an unnamed Clinton Administration official saying of the report of the document, 'We are pursuing it very seriously but are not in a position to evaluate it." The spokesman said that the researcher had informed the Government of the discovery but 'he would not give us the document,' that former Ambassador Malcolm Toon had been requested to follow up on the matter in Moscow, and that he assumed it was 'something Vessey will raise' with the Vietnamese. Apparently caught by surprise, Alan C. Ptak, the CIA and Pentagon official who had for years conducted damage-control for the U.S. Government whenever POW/MIA revelations reached the public, was quoted by the New York Times as saying: 'I always had an inkling that we hadn't been told everything... We had suspicions the Russians knew more than they were telling us.'"

Yet, the debunking and disinformation process which had for decades followed all reports of U.S. POW/MIA information that did not conform with the Pentagon's falsified version of history, began immediately. On the night of April 12th, a retired military officer now working for the U.S. Government's "Task Force Russia," informed the author that within Washington's official circles, "DIA is already tearing holes in the Russian document," and that Acting Deputy Assistant Secretary of Defense for POW/MIAs, Edward Ross (a national security apparatchik who had replaced the ex-CIA man Allan Ptak), had already "classified" the document on behalf of the Clinton Administration. The author's source said that the DIA and Pentagon hierarchy were also unhappy about the way the Army's "Task Force Russia" had become so closely involved with POW/MIA family members and veterans, and that General Loeffke was being retired again, and was likely to be replaced as head of "Task Force Russia" by General Soyster, the former Director of DIA.

This information on DIA's reaction to the Soviet document proved accurate, and one of the first newspapers in America to begin the debunking process was the San Francisco Chronicle, on the morning of the 13th, which attacked the numbers of American prisoners in the document first, by saying that, "U.S. officials (unnamed)...have some questions about the authenticity of the document, which Russian officials turned over to U.S. POW researchers. AS DESCRIBED BY U.S. OFFICIALS WHO HAVE READ IT, THE DOCUMENT APPEARS TO BE RIDDLED WITH ERRONEOUS STATEMENTS...NORTH VIETNAMESE OFFICIALS SAID IN SEPTEMBER 1972 THAT THEY HELD 368 AMERICAN POWs...BUT STATE DEPARTMENT SPOKESMAN RICHARD BOUCHER SAID THE RUSSIAN TRANSLATION OF THE VIETNAMESE DOCUMENT INDICATED THAT HANOI HELD 1205 PRISONERS...IT PURPORTEDLY WAS WRITTEN BY GENERAL TRAN VAN QUANG... SKEPTICS (unnamed) HAVE POINTED OUT SEVERAL DISCREPANCIES IN THE DOCUMENT. THE RUSSIAN VERSION, FOR EXAMPLE, SAYS THAT PRISONERS WERE SEGREGATED BY RANK AND THAT THREE WERE ASTRONAUTS-NEITHER OF WHICH IS TRUE..."[11]

Also on April 13th, not coincidentally, the Vietnamese Communist government announced that the Soviet document was a fraud, intended to sabotage the move towards U.S.-Vietnamese diplomatic relations and an end to the American trade embargo which had been imposed in 1975. The Foreign Ministry in Hanoi said: "Vietnam categorically rejects this ill-intentioned fabrication...in 1973, after the signing of the Paris Agreement, Vietnam handed over to the U.S. all American prisoners captured in Vietnam."[12] In the midst of the media furor caused by the Soviet document the world's seven leading industrial nations, headed by the United States, suddenly announced on April 15th, in Tokyo, that they would extend another $28.4 billion in aid to Boris Yeltsin's Russian government. The aid package was described as "huge" and "massive" in the American media, and touted as being of potential assistance to Yeltsin in winning an upcoming April 25th nationwide referendum.[13] Presidential emissary General John Vessey was then dispatched to Hanoi with an 18-member party, amidst rumors questioning the competency of the 1972 Russian translator of the document.[14] A few days later, the Washington Post reported on General Vessey's appearance with the Vietnamese Foreign Minister at a joint press conference in Hanoi, after two days of talks about American POWs and the newly-discovered Soviet document:

"Retired U.S., General John Vessey, asked at a news conference whether he has more doubts about the document than when he arrived Sunday, replied, 'Yes.' HE SAID HE HAS NO REASON TO DISBELIEVE A RETIRED VIETNAMESE GENERAL WHO TOLD THE U.S. DELEGATION THAT HE NEVER MADE THE REPORT ON AMERICAN POWs ATTRIBUTED TO HIM IN THE RUSSIAN DOCUMENT...Foreign Minister Nguyen Manh Cam, appearing with Vessey in a joint press conference, SAID THAT STEPHEN MORRIS, THE RESEARCHER WHO FOUND THE DOCUMENT, SUPPORTED THE VIETNAM WAR AND 'HAS A LONG HISTORY' POF OPPOSITION TO THE HANOI GOVERNMENT AND CONNECTIONS WITH VIETNAMESE ANTI-COMMUNIST GROUPS IN THE UNITED STATES."[15]

The Orwellian specter of a senior American general (who had once commanded U.S. forces during the CIAs secret war in Laos) saying he could see no reason why a former enemy general would lie about American war prisoners, while standing beside a high Communist official who proceeded to smear the reputation of the researcher on the basis of his opposition to Communism, was perhaps inevitable, given the decades of official lying about the numbers of known-POWs who had been left behind in captivity. Vessey returned to the United States and reported to President Clinton that the document discovered in the Soviet archives was full of errors and that the Vietnamese in Hanoi had presented him with other documents which provided different, much smaller numbers of American POWs in their control in 1972, which (not surprisingly) closely conformed to the Pentagon's fabricated postwar numbers, which were based on the number of POWs released in 1973.

On April 22nd, the nationally-circulated newspaper USA TODAY stated: "Reporting to President Clinton after returning from Hanoi,

Vessey said 'a lot of facts' in the document...are wrong, especially numbers of downed fliers." The newspaper labeled the Harvard researcher an "activist" who hoped to derail the process of U.S. diplomatic recognition of the Hanoi government, who "said he found the document in Communist Party files in Moscow." Stephen Morris was quoted as saying he was "surprised and saddened" by Vessey's statements. Associated Press simultaneously reported "the Vietnamese insisted that the Russian document is a fraud," and quoted Vessey as saying: "'The list of errors is long...We know that some of the facts that are alleged in the Russian document are wrong, a lot of the facts." Vessey claimed that the Vietnamese Communists had given him "important information that deals with our goal of the fullest possible accounting for all missing Americans," including alleged lists of Americans who had died in captivity and locations of alleged grave sites of more American MIAs.[16]

The largest circulation national newsmagazines both assisted the Pentagon's debunking effort on April 26th, with predictably similar conclusions. Time dutifully questioned the accuracy of the Soviet document which it said "purports to be a translation," and stigmatized the Australian discover of the document as "hardly...an enthusiastic supporter of normalization." The magazine also adhered strictly to the U.S. Government line that all U.S. POWs were dead: "... if the Quang report is accurate, more than 600 POWs must have died or been killed between the fall of 1972 and April 1, 1973." No hint was permitted in the article that hundreds of POWs may have survived long after the war, or might still be alive. As a primary source used in discrediting the accuracy of the Soviet document, Time quoted DIA's POW/MIA director Robert Sheetz, whose recent testimony before the Senate Select Committee had been repeatedly discredited. The magazine cited an internal Pentagon memorandum written by Sheetz: "'DIA BELIEVES THE NUMBER 1,205 COULD BE AN ACCURATE ACCOUNTING OF TOTAL PRISONERS HELD,' IF FOREIGNERS WORKING AS U.S. AGENTS ARE INCLUDED. BUT, SHEETZ ADDED IN HIS MEMO, 'THE NUMBERS CANNOT BE ACCURATE IF DISCUSSING ONLY U.S. POWs.'"

Newsweek adopted the same line, reporting: "U.S. government specialists (unnamed) said the letter appears to be authentic, but that the information it contains is wildly inaccurate," and quoting a "Capitol Hill Aide" (unnamed) as saying the figures for POWs were "'totally, totally implausible.'" The magazine also used unidentified sources for the Pentagon's cover-story, that Time had attributed to DIA's Robert Sheetz: "Some suggested that whoever translated the document from Vietnamese to Russian might have confused the term for Asian commandos captured fighting with U.S. forces." Newsweek, however, gave substantial space to a differing view, in an interview with former presidential contender Ross Perot. When asked about the authenticity of the Soviet document Perot responded: "If Saint Peter showed up with it on a scroll, somebody would try to debunk it." The Texas businessman proposed sending a "world-class negotiator" to Hanoi:

"...Treat them with dignity and respect and hold non-stop negotiations until you get it done. Then, if we got the people out, I would be a strong proponent, and say, 'All right, let's normalize... Continuing to punish this country that released the guys is wrong... All I can do is to discuss this publicly across the country, and if Americans care enough they will reflect that to their congressmen or senators and the White House. I should not have to make this a sensitive issue. Could I (block a lifting of the embargo)? Probably. Will I if I have to? Yes. Let's put that in big print: Yes. Because you don't leave a single person behind."

In a parallel effort aimed at discrediting and bankrupting the most active and effective veterans POW/MIA advocacy group in the country, a series of legal actions and financial judgments had been instituted by the National Park Service and the Vietnam Veterans Memorial Fund against Homecoming II and its leader, Vietnam veteran Ted Sampley. The former president's wife, Barbara Bush, served as Honorary Chairwoman of the Vietnam Memorial's advisory board. (Homecoming II's four-member board of Directors was headed by Dr. Bruce Adams of Laramie, Wyoming, whose brother was still listed as an MIA in Southeast Asia. (Other board members included Thomas V. Ashworth, a former Marine pilot in Vietnam, and two Army veterans, Tim Tripp and Larry Bice.)

The origin of this attack was a complaint by Ann Mills Griffiths, the government-sponsored Executive Director of the National League of Families, to the Secretary of the Interior in the Bush Administration, about the POW/MIA awareness campaign carried on by Homecoming II on park property next to the Vietnam memorial. This had included the distribution of more than three-quarters of a million copies of U.S. Veteran News and Report, which had led in the publication of recently declassified or leaked information about missing U.S. prisoners of several wars. Among these were 50.000 copies of "A Secret That Shames Humanity," the first complete and documented history of the disappearance of Communist-held U.S. prisoners of war of WW I, WW II, Korea, the Cold War and Vietnam, written by author John M. G. Brown and research-colleague Thomas V. Ashworth, in May 1989. Griffiths, who had publicly referred to POW/MIA activists as "scum," and various federal officials, had tried for years, unsuccessfully, to shut down this information source, which had reached a circulation of 30,000 a month, largely among veterans and POW/MIA family numbers in all 50 states.

Eventually, an effective tactic was evolved with the help of the Vietnam Veterans Memorial Fund leader, Jan Scruggs, who claimed that the memorial's board and an artist named Frederick Hart owned a copyright on the now-famous statue of the "Three Servicemen," which had been erected facing the Vietnam Memorial Wall in Washington, D.C. Scruggs and Hart claimed that Homecoming II and Sampley had illegally profited from the sale of shirts that pictured the war memorial statue, and obtained a judgment of more than $359,000., entitling them to seize all assets of Homecoming II, including the newspaper, Sampley's construction business and even his home. The artist and the board of the fund were thus attempting to establish a highly questionable precedent that such war memorials or other public monuments no longer became the property of the American people when purchased, but could be copyrighted and thus ensure perpetual royalties for an artist /owner who had already been paid for his work by veterans and other Americans.

In this and other litigation, Sampley's POW/MIA group faced high-powered Washington attorneys such as Terrence O'Donnell, a Director of the Vietnam Veterans Memorial Fund, who also just happened to be a chief counsel for the U.S. Department of Defense. O'Donnell was the General Counsel of the Department of Defense who had investigated Colonel Millard Peck's charges about DIA as part of a "management inquiry team" appointed by Secretary of Defense Richard Cheney. O'Donnell, Assistant Secretary of Defense Duane P. Andrews and others on the team had attempted to publicly discredit Colonel Peck and claimed that the Director of DIA, General Soyster was actually dissatisfied with Peck's performance.

As far back as 1980, O'Donnell had represented the Defense Department in the successful blockage of a House Resolution (H.Res.415) intended to force the Pentagon to furnish certain information relating to Americans in captivity in Southeast Asia to the House Permanent Select Committee on Intelligence and to the Armed Services Committee, on the grounds that it "would serve no useful purpose." Ann Mills Griffiths had also opposed release of this information to the Congress, and her continued opposition to the release of POW/MIA information for over a decade had materially aided the cover-up process, which still continued. A powerful Washington legal firm with old CIA/OSS connections, Donovan, Leisure, Regovin, Huge and Schiller, represented the League of Families in the litigation against Homecoming II and Sampley. Mitchell Regovin had formerly been a counsel for the Central Intelligence Agency, which indicated the high level of official interest in halting Homecoming II's publication of heretofore secret POW/MIA information and it's continual activism in the issue. Two prestigious North Carolina law firms had also been hired to go after Sampley's property and home, where he lived with his wife Robin (the daughter of a U.S. MIA in Laos, S.Sgt. Robert Owen), and their two children.

The courageous North Carolina Vietnam veteran, who had once been asked to infiltrate a Veterans group at the Vietnam Memorial Wall to provide information for the National Security Council's Richard Childress, had enraged federal officials for years. Sampley had taken over "The Last Firebase" near the memorial in 1986, and had sponsored "Truth Litigation," a legal effort seeking the release of U.S. Government information on livesightings of U.S. POWs. In this effort Homecoming II had paid thousands of dollars in attorney fees to POW/MIA family members seeking classified documents about their loved ones. In 1987 Sampley had helped organize and lead a march of Vietnam veterans and POW/MIA family members on the White House (which the author witnessed), and also the delivery of "care packages" to Laos addressed to missing American servicemen who had vanished there. He helped organize a 1987 trip to Thailand by POW/MIA family members and former U.S. Congressman Bill Hendon, during which a reward offer of $2.4 million for the return of a POW was disseminated into Laos.

Sampley subsequently organized and led repeated marches of Vietnam veterans on the Lao Embassy in Washington, year after year, demanding the release of secretly-held Americans who had disappeared in a country that the United States maintained diplomatic relations with, despite U.S. Secret Service and Metropolitan Police efforts to stop them from thus "embarrassing the U.S. Government." In 1988 and 1989 Sampley provided a headquarters in Washington for the author and research colleague Thomas V. Ashworth to conduct a massive research project among hundreds of thousands of recently-declassified POW/MIA documents, dated from 1918-1980, in the U.S. National Archives and the Suitland National Records Center. In addition, Homecoming II had provided the initial start-up money for the National Alliance of Families, a group which fought for the rights of POWs abandoned to the Communists in every major U.S. war of the 20th Century. For these and many other actions, and for the sometimes abrasive manner in which Sampley pursued his POW/MIA mission, he was hated by those who had Assisted in the abandonment of the prisoners, and the legal campaign against him which resulted was perceived to be the only way the government could silence him, and stop the publication of embarrassing facts about American prisoners of war.

By the end of May 1993, the author had finally received responses to various Freedom of Information Act requests for additional declassified documents concerning Soviet-held American POWs in the possession of the CIA and the Department of State, but little substantive or historically significant material was sent. The

Pentagon, NSA-INSCOM, DIA and FBI had long before ceased to cooperate with the author's search for records pertaining to American prisoners of war who failed to return from Communist control, pleading either lack of funds for an adequate FOIA staff, unavoidable delays (of years) due to the backlog of other requests, or inability to locate any documents pertaining to the requests.

*The FBI ludicrously denied having received the author's FOIA requests for access to FBI files containing livesightings of US POWs in the Soviet Union, although they had already responded in writing to one such request.*

When the author attempted to make an appointment with CIA 's FOIA Chief, John Wright, to view certain declassified documents in CIAs research room, CIA failed to respond until the author was forced to cancel reservations for a proposed mid-July trip to Washington D.C. It is interesting to note that approximately 20 minutes after the author had cancelled the trip by a long-distance telephone call to a travel agent, the CIA called the author on the telephone from Langley, Virginia, claiming to be anxious to set up the research visit, after having ignored most of the author's requests for months and even years. A former military officer who had been hired by the Army's Task Force Russia subsequently confirmed to the author that CIA was not going to declassify and release to scholars the debriefings of several important post-WW II Soviet NKVD/KGB defectors. The author had sought these in order to reanalyze these defectors statements and subsequent activities in light of what they had specifically said about U.S. prisoners of war and civilian captives in Russia.

At the end of May, author and journalist Al Santoli, a Vietnam veteran, had written a carefully researched report on the missing POWs of Vietnam for Parade Magazine, which had a wide circulation in Sunday newspapers across the United States.[17] U.S. News and World Report had subsequently published a story on June 7th, stating that two former Vietnamese Army officers had reported to the U.S. Defense Attaché's Office in Phnom Penh on May 18th that, "as many as 150 American prisoners of war remain in two camps near Vietnam's border with China." The Clinton Administration nevertheless went ahead with plans to approve international loans to Vietnam, because of what was termed the cooperative and helpful attitude of the Vietnamese in resolving MIA cases. At the same time, the Administration chose to confront North Korea over the issue of nuclear weapons proliferation. The North Korean regime, which, according to intelligence reports, still held both American POWs from the Korean War and some undetermined number of U.S. prisoners who had been transferred there from Vietnam, responded by releasing 17 sets of human skeletal remains at Panmunjom, on the day following a visit by President Clinton to the tense North Korean border.

Senator Robert Smith became the first member of Congress to urge that criminal charges be brought against U.S. Government officials involved in the POW/MIA matter, in a letter of June 29, 1993, to Attorney General Janet Reno.[18] (The letter was not released by the Senator until early July, in Hanoi, after Smith had discovered prison facilities where DIA had said 'none existed,' in order to debunk reports of Americans in captivity by both American and Vietnamese eyewitnesses.) He requested that the Attorney General investigate "several specific incidents of false testimony and statements made to United States Senators or received by the Senate Select Committee on POW/MIA Affairs during its investigation," also "potential mail fraud and false personation violations," and "potential violations by certain personnel of the Department of State and the Department of Defense of provisions of Executive order 12356 in relation to the classification of an archival document from the former Soviet Union turned over to the United States by Russia on April 8, 1993, and the handling of the disclosure of said document to Congress and the Socialist Republic of Vietnam on April 12, 1993." Senator Smith also wrote:

"I believe the above mentioned potential federal and criminal violations involve the following 10 persons currently and formerly employed by the United States Government:

"Colonel Joseph A. Schlatter, U.S. Army, Deputy to the Acting Deputy Assistant Secretary of Defense for POW/MIA Affairs; Robert Destatte, Senior Analyst, Defense Intelligence Agency, currently detailed to the POW/MIA archival research program being carried out in Vietnam by the Joint Task Force ("Full Accounting"), U.S. Pacific Command, Department of Defense; Robert R. Sheetz, former Chief of the Special Office for Prisoners of War and Missing in Action, Defense Intelligence Agency; Charles F. Trowbridge, Chief, Special Office For POW/MIAs, DIA; Warren Grey, Senior Analyst, Special Office for POW/MIAs, DIA; Duane P. Andrews, former Assistant Secretary of Defense (Command, Control, Communications and Intelligence); Edward W. Ross, Acting Deputy Assistant Secretary of Defense for POW/MIA Affairs; Norman Cass, Assistant to the Acting Deputy Assistant Secretary of Defense for POW/MIA Affairs; Colonel William Jordan, Office of the Assistant Secretary of Defense, International Security Affairs; and Christopher LaFleur, Director of the Office for Vietnam, Laos and Cambodia Affairs, Department of State.

The New York Times later published a report listing all of these names and criticizing Senator Smith, who was characterized as "little known," for upsetting the decorum of the Senate with his charges. Robert DeStatte of the Defense Intelligence Agency was quoted as saying that Senator Smith had been, 'hostile to me, mocking me....I...stand prepared now as always to defend the truth of any statements I have made." Justice Department officials were reported to be "exploring whether the charges, if proved true, would actually constitute a crime." Smith was reported to be regretful that he had released the letter dated June 29th on July 11th in Hanoi, but that he had been "angry to find on his trip to Vietnam, confirmation of the existence of prisons that the DIA said did not exist and because reporters there had pressed him on what he was going to do about it."[19]

In early July, Senator Smith had traveled to Hanoi for yet another effort at trying to gain the cooperation of the Vietnamese government in returning surviving American ex-POWs who had been reported in recent months as still being held in captivity. DIA analyst John McCreary, formerly on the Select Committee staff assisted Smith, and ex-POW Robert Garwood, who had seen many other live American prisoners between 1973 and 1979, also accompanied him. According to a National Public Radio (NPR) report, the Vietnamese were "bemused" by Garwood's statements, and denied any Americans left from the war were alive in Vietnam. Smith had visited the sites of three POW camps and found exact locations described by eyewitnesses who had seen captive Americans, but who had been previously debunked by DIA analysts on the grounds that all three of these prison camps had not existed.[20] In a speech before the U.S. Senate, Senator Smith, described a site where Garwood had said under oath that he had seen 20 other American POWs:

'Thac Ba Lake is a manmade lake made by a Soviet dam in North Vietnam. I was told personally by DIA that, first of all, there was no lake. Then I was told there were no islands in the lake. We now find out there were dozens of islands in the lake. Then I was told there was no prison compound on the lake. Then we find out there was a prison compound on the lake. Then I was told Garwood was never there...Amazingly, the joint task force over there (JTFFA) doing the investigating could not find the island and could not find the prison camp, and yet...ONE OF THE VIETNAMESE OFFICIALS THERE SAID POINT BLANK, YES, THERE WAS A PRISON; YES GARWOOD WAS THERE; YES, HE WAS REPAIRING GENERATORS."[21]

During a July 13th press conference after returning from Vietnam, Senator Smith released an April 22nd DIA "Memorandum for the Record," stating that a reinvestigation of the Lake site a year before by flyovers had revealed nothing on the islands but grass huts; which now was shown to have been yet another lie by DIA to refute an eyewitness livesighting report. At the press conference Smith also reported:

"We went places that no American (investigator) had ever been...We went to places where no American investigator had ever asked to be...The first area is a camp in a town called Phu Ly...the My camp...(The refugee source reporting a live sighting) even gave us grid coordinates of the camp and gave us a sketch of the camp...THIS SOURCE WAS DEBUNKED BY DIA FOR A NUMBER OF REASONS, THE MOST IMPORTANT BEING THERE WAS NO CAMP...THEREFORE, THE MAN WAS A LIAR...WE WENT THERE...THIS CAMP DOES EXIST, AND THE DESCRIPTION THAT WAS GIVEN TO US WAS ACCURATE, RIGHT TO THE LETTER...."

'Thanh Loc Phu Tu...We had a source who said he saw ten American POWs...The source said there was a special detention facility. The DIA said this man was a fabricator because there was no special detention facility...I asked the commander of the camp if there was a special detention facility at the camp. He said, yes, there was... Another thing that the source said was that in this special detention facility was a room, a prison room with a special metal bunk bed...and on the metal bunk bed...was an iron bar which controlled manacles, where a man named Jackson...was manacled to the bed...So we got there and we found: One. YES, THERE WAS A SPECIAL DETENTION FACILITY, WE WENT INTO IT...WAS THERE A METAL BUNK BED?... YES,THERE WAS...WAS THERE AN IRON BAR"...WELL, THE IRON WASN'T ON THE BED, IT WAS IN THE ROOM NEXT DOOR...Everything he said was accurate except for the fact that Jackson wasn't there, and we couldn't find his initials...All the walls in the prisons were whitewashed."[22]

Due to his security clearances, John McCreary had been given access to the best intelligence available to the U.S. Government, from all sources. He informed the author prior to this trip that of approximately 1,300 American POWs and MIAs left behind in captivity at the end of the war, he felt certain, based on the classified U.S. intelligence he had seen, that several hundred U.S. POWs were still alive in 1992-93, in at least three locations in northern Vietnam. Because of his devotion to uncovering the truth about the lost American POWs of Indochina, McCreary had already been threatened while on the Select Committee staff the year before, that he might not have a job any more at DIA after the Committee's work was finished, "because General Clapper (DIA Director) and Mr. Nagy were receiving reports that I was disloyal to DIA."[23] Just prior to the July 1993 trip to Hanoi, McCreary informed the author that his former staff had been taken away from him and he was indeed no longer involved in the highly sensitive national intelligence-level work that he had done prior to his assignment at the Select Committee on POW/MIA Affairs.

While in Vietnam Smith had thus confirmed the existence of prison facilities where Americans had been reported held, but which DIA had claimed, in refuting these reports, did not exist. After Senator Smith's delegation left Vietnam, a trio of high-ranking Vietnamese Communist Army officers held a press conference, in which they once again denied all of Garwood's testimony about other

American POWs, and reiterated that no Americans remained in captivity. Senator John McCain of Arizona contributed to this disinformation campaign in July by publishing an article in the Stars and Stripes-The National Tribune attacking Garwood's character and veracity, with exaggerations and untrue statements about the former Marine POW who had been held in Vietnam for 14 years, while McCain had himself received preferential treatment and given the North Vietnamese military information, and had been allowed to return home in 1973.[24] McCain's use of the term "fairy tale" to describe Garwood's sightings of American prisoners held long after the war by the Communists, was eerily reminiscent of Josef Stalin's use of the same phrase nearly half a century before.

Clinton's new Assistant Secretary of State for East Asian and Pacific Affairs, Winston Lord, who had been one of Kissinger's assistants at the Paris peace talks two decades before and later U.S. Ambassador to the People's Republic of China, was now being used as a pointman for the administration in the POW/MIA matter. In addition to traveling to Hanoi for talks with Vietnamese Communist leaders while the charade of "remains recovery" continued, he had taken part in a meeting between leaders of American veterans' groups and President Clinton and Vice President Gore. The meeting had taken place on August 11th in the little-used Roosevelt Room of the White House, where 12 years before President Reagan, Vice President Bush, William Casey and Richard Allen had discussed the POW offer from Hanoi. Lord later told a Disabled American Veterans (DAV) national convention:

"I am here to express on behalf of President Clinton his wholehearted commitment to the POW/MIA issue and his desire to include you in and your organization in the process...Last Wednesday (August 11) in the Roosevelt Room of the White House, (DAV Magazine Editor) David Givans and I and the rest of our delegation heard the President and the Vice President reaffirm these goals."

But Lord once again made it clear that the real priority in attaining a "full accounting" of U.S. POWs and MIAs was to be the "remains recovery" effort by the U.S. Joint Task Force Full Accounting mission of 100 personnel in Vietnam, not finding out what had happened to hundreds of secretly-withheld American prisoners. He also warned that this so-called accounting would be "a long and painstaking process that will stretch out over many years because the nature of the work inherently defies complete success."[25]

In the meantime, a potential scandal for the new Clinton Administration involving U.S. relations with Vietnam, which had been simmering for months, was revealed to the public. An Associated Press story of August 14th, which was published in newspapers across the nation, reported that Ronald Brown,

Commerce Secretary in President Clinton's cabinet and a former chairman of the Democratic National Committee, had denied allegations printed by U.S. News And World Report Magazine, in their August 23rd issue (which was already being distributed) and in other publications, that he had accepted payoffs or bribes from the Vietnamese in return for using his influence in the Administration to lobby for improved American-Vietnamese relations.[26]

The scheme allegedly involved Brown's business dealings with Vietnamese-American Ly Thanh Binh (also reported as "Binh Ly"), and Nguyen Van Hao, now of Florida, but before 1975 a deputy prime minister in the former South Vietnamese government, who had long been suspected of having been a secret agent for Hanoi. Hao had helped block the shipment of South Vietnam's gold reserves out of the country, and after the fall of Saigon had remained behind and assisted the Communist regime as an economic adviser until he left in the early 1980s, eventually settling in Haiti. According to press reports, he had met Marc Ashton there, a Florida businessman who was a friend and associate of Ronald Brown. (Ashton's sister-in-law was reported to be a close personal friend of Brown's, who lived in a Washington, D.C. townhouse owned by him. At this time Brown was a highly-paid Washington lobbyist for the Duvalier regime which was known for its brutal oppression of the Haitian people.

Ly Tranh Binh said Hao asked him in July 1992 to help him establish the "Vietnam Economic Development Corporation" as a conduit, and that they traveled to Hanoi where they met with senior Vietnamese officials in Hanoi and Ho Chi Minh City. Binh said in interviews with reporter Mike Blair that the scheme he had participated in involved Nguyen Van Hao, Vietnam Communist Party chief Du Muoi, and the Prime minister of Vietnam, Vo Van Kiet. Binh said that Ronald Brown had agreed to accept a payoff of $700,000. to be deposited in an offshore account to lobby within the Clinton Administration for lifting American economic sanctions and embargoes on Vietnam that had been in place since the war's end. Binh claimed that after Ronald Brown was appointed Secretary of Commerce by President Clinton, he Binh, had objected to continuing the dealings because of the conflict of interest inherent in Brown's high official position. He claimed to have had several arguments about this with Nguyen Van Hao, before Binh broke the relationship off.

Reportedly, the FBI had been investigating Ly Thanh Binh's allegations under the supervision of William Sessions from February to mid-July, when Judge Sessions had been removed from his position and replaced with a career FBI man. Brown had told Associated Press that the FBI had never contacted him. (The Wall Street Journal later reported that Brown was to be questioned by Justice Department officials.) Binh alleged that Brown had already found American businesses interested in the categories for the lobbying "contract," including the Rockefeller Group (oil and gas), Apple Computers and AT&T (communications, and Hyatt Hotels (tourism). Binh alleged that Brown was to transfer $50 million from a Swiss bank account maintained by former Haitian dictator Jean-Claude Duvalier.[27]

At a news conference beside Air Force One in California, where the President had been visiting, Brown was quoted as saying (in part): "I have never been involved in any such thing. I have never had any kind of business relationship, any kind of financial relationship... in this matter...It's absolutely and totally absurd. I have never represented Vietnam at any time or interests representing Vietnam, or any individuals or any companies or corporations."[28] A later report by Newsweek Magazine downplayed the impropriety and potential disgrace of a nominated and appointed cabinet member dealing clandestinely for personal gain with a proven high-level agent of a hostile foreign regime, with which the United States Government was attempting to resolve painful diplomatic differences over missing American prisoners of war possibly remaining in captivity. The FBI had reportedly obtained a copy of a fax sent by Hao to Hanoi describing a December 1992 meeting with Brown in the Washington, D.C. townhouse.[29]

Former Senate Select Committee on POW/MIA's staff Director Frances Zwenig was reported to be close to Commerce Secretary Brown and to have played a role in his transition to the office in early 1993. The former chairman of the Select Committee, Democratic Senator John Kerry, along with Republican John McCain, both of whom had long favored normalizing economic and diplomatic relations with Vietnam, helped speed Brown's confirmation through the Commerce Committee. One of the firms set to benefit from relaxed restrictions on Hanoi was Colliers International, based in Boston, which had announced prior to Clinton's partial lifting of the trade embargo, a deal involving almost $1 billion for construction of a new port facility south of Saigon, or Ho Chi Minh City. The American company had gotten around the trade embargo by having its Singapore affiliate, Colliers Jardines, make the deal with Vietnam. According to one published report, Colliers International was headed by Stewart Forbes, a cousin of Senator Kerry of Massachusetts, who was Chairman of the Foreign Commerce and Tourism Committee.[30]

Meanwhile, three POW/MIA family members and a disabled Vietnam infantry veteran began a fast in mid-August, vowing to continue until President Clinton kept his earlier promise made to POW/MIA family members not to lift the trade embargo against Vietnam. Dolores Apodka Alfond, chairman of the National Alliance of POW/MIA Families, sister of Air Force Major Victor Apodoka, Ann

Holland, wife of Air Force Technical Sergeant Melvin A. Holland, Kathy Borah Duez, sister of Navy Lieutenant Daniel Borah, and Jerry Birch, a veteran of the 25th Infantry Division who lost his legs to a Vietcong mine in 1969, began the hunger strike near the Marine Corps base at Camp Lejeune, North Carolina. They protested the U.S. Government's acceptance of minute human skeletal fragments as positive identification of missing American servicemen, and the gradual succombing of the Clinton Administration to pressures from big business to resume normal relations with Vietnam in order to capitalize on another cheap labor force in the third world and access to Vietnam's oil reserves. Former presidential candidate Ross Perot was to visit the four later during their fast, and express support for their demands.

During this period, financier Robert Altman, a partner of one-time Truman naval aide and former Secretary of Defense Clark Clifford, was found not guilty of fraud in the illegal relationship between BCCI and First American Bankshares Inc. The Washington Post editorialized: "Similar charges have been brought against Mr. Altman's partner, Clark Clifford, whose trial was deferred because of his poor health. Since the evidence against him was much the same, the Altman verdict can be applied by inference to Mr. Clifford as well." The BCCI case appeared to be yet another example of the U.S. Government's seeming inability to accept responsibility for its own clandestine actions and to punish those guilty of serious crimes, particularly if there was a "national security" connection. A report by Senators John Kerry and Hank Brown for the Senate Foreign Relations Committee the year before had stated that even after the CIA was aware that BCCI was a criminal enterprise, they continued to use it and First American for CIA operations.[31] The outcome of this episode boded ill for the FBI investigation into the actions of DIAs POW/MIA personnel which had been requested of Attorney General Reno by Senator Smith, or the reported improprieties of Secretary of Commerce Brown's dealings with the Vietnamese.

On August 24th, 800,000 pages of heavily-censored secret papers on the assassination of President John F. Kennedy were made public by the CIA and FBI, as had been mandated by Congress the year before. The declassified documents reportedly contained most of the records of the Warren Commission, the 1975 Rockefeller Commission study of CIA domestic activities, and the 1979 House Select Committee on Assassinations, which had concluded there was evidence of a probable conspiracy involving more than one gunman, and had raised questions about Mafia involvement in the murder. Many other released documents were largely related to the murdered, accused assassin Lee Harvey Oswald's foreign travels and his alleged relations with Soviet and Cuban contacts.[32]

Important evidence on American POWs from Korea being

interrogated by Soviets prior to their disappearance into the USSR was also revealed in late August. Ironically, the information came from the debriefing of an American POW, Marine Corporal Nick Flores, 41 years before. Flores and several other prisoners had escaped from a POW camp in northeastern Korea on July 22, 1952, but was recaptured when he ran into a Soviet anti-aircraft artillery unit.[33] Since Flores was wearing a borrowed Air Force jacket at the time, and the Russian unit had just shot down an F-86 Sabrejet, his captors believed he was the escaped pilot and shipped him to an interrogation center, which was believed to be Antung, China. Flores was interrogated for 48 hours inside a bunker there by men wearing Soviet military uniforms. Apparently convinced that the Marine truck driver was indeed not an F-86 pilot, the Soviets returned him to his former POW camp in North Korea, from which he was fortunate enough to be repatriated in Operation Big Switch on August 20, 1953.

Flores was debriefed at sea on his way back to the U.S., but what he had seen remained classified for more than 40 years to come He had been approached before leaving Korea by high-ranking officers who requested him to volunteer for training to be dropped behind enemy lines but declined The U.S. Marine Corps did not give him credit for the time he spent as a POW and he was denied an honorable discharge in October 1953. He then returned to his home in San Jose, California, and years later wrote about his predicament to his Congressman, but nothing was done. Associated Press reported on August 22nd: "The record was corrected at a quiet Marine ceremony earlier this month, and now Flores has his honorable discharge."

Four days after this report appeared in the press it was announced that North Korea, which had been censured by the Clinton Administration recently about nuclear proliferation, had reached an agreement with U.S. authorities to cooperate for the first time in accounting for thousands of U.S. missing in action of the Korean War. Another Associated Press report on this agreement contained the absurd statement:

"The U.S. Government lists about 8,140 U.S. servicemen as unaccounted for from the 1950-53 Korean conflict. IT SAYS IT HAS NO CREDIBLE EVIDENCE THAT ANY OF THEM WERE ALIVE AT THE END OF THE WAR, although some experts dispute that."[34] It appeared that the U.S. authorities were primarily interested in repatriating human skeletal remains, not in demanding from the North Koreans information on the fate of thousands of Americans who had never been released from captivity. As a subsequent letter to the author from Senator Robert Smith noted:

"President Clinton recently approved the signing of an agreement at Panmunjom which only asks for "remains" of missing servicemen from North Korea, despite the overwhelming evidence that POWs were not returned forty years ago. I have traveled into North Korea twice in the last two years to discuss with Communist officials the fate of our POW/MIAs...I feel my efforts with North Korea, however, have been in vain as President Clinton is apparently not willing to pursue anything other than the return of remains from North Korea....With respect to Communist China, the President has done little if nothing since his inauguration to demand answers from Beijing on American POWs under their control during the Korean and Vietnam conflicts..."[35]

A retired military officer now working at the Washington, D.C. headquarters of the Army's Task Force Russia (recently renamed the Joint Commission Support Branch (JCSB), informed the author during September that he had just returned from a meeting in Moscow with Russian intelligence officials during which his group turned over a new, classified report on Americans POWs of Korea shipped to the Soviet Union, and summaries of interrogations on American prisoners of the Korean War, mostly F-86 pilots, but the Russians had still continued to claim that these interrogations had been carried out by "Koreans." In this they also adhered to their established "line" about American prisoners from Vietnam: That the Vietnamese, not Russians, had carried out interrogations of U.S. POWs, except perhaps on rare occasions. A Task Force Russia staff member also confirmed to the author that Charles Trowbridge of the DIA, who had been involved in debunking eyewitness sightings of American POWs in captivity ever since the war, and Colonel Joseph Schlatter, had been moved to senior positions at Task Force Russia (now JCSB), indicating, almost certainly, a further effort from within the Pentagon hierarchy to suppress or discredit livesighting information on U.S. POWs in the former Soviet Union.

On September 8th, prior to the previously announced date when President Clinton would issue his decision about easing economic sanctions on Hanoi, which had been in place since the end of the war, Senator Robert Smith addressed his colleagues in Congress (in part):

"This past weekend in Moscow, the United States concluded its sixth formal meeting with the Russian government as part of our joint efforts begun 18 months ago to investigate the fate of POWs from past wars...General Dmitri Volkogonov...has turned over to the U.S. side another dramatic and deeply troubling document concerning American POWs from the Vietnam War..."

"Mr. President, the new archival document just acquired last week is a Russian translation of yet another North Vietnamese Politburo presentation, this one from late December 1970, almost three years before the end of the war and the return of POWs at Homecoming. In the presentation, a North Vietnamese is informing his Politburo, in secret session, THAT THEY ARE HOLDING 735 CAPTURED AMERICAN FLIERS IN NORTH VIETNAM., AND THAT THE LIST

368 AMERICAN POWs WHICH THEY HAD JUST PROVISED TO THE STAFF OF THE SENATOR FROM MASSACHUSETTS, SENATOR KENNEDY, IN PARIS IN DECEMBER 1970 WAS FOR DIPLOMATIC PURPOSES ONLY AND DID NOT REPRESENT THE TRUE NUMBER OF AMERICAN POWs AT THE TIME."

"The actual quote from the presentation, given to the Central Committee of the Communist Party of North Vietnam by Hoang Anh, reads:

'Now, I want to stop on one more issue—about the captured American fliers. The total number of captured American fliers in the DRV consists of 735 people. As I have already stated, we published the names of 368 fliers. That's our diplomatic step. If the Americans will agree to withdraw their forces from South Vietnam, we will, for a beginning, return these 368 people to them; and when the Americans finish withdrawing their forces, we will give the rest back to them. The issue of the captured American fliers, by virtue of what has been said above, is of great importance to us.'"

Smith also referred to the difference in the Pentagon's handling of the newest piece of documentary evidence of non-returned prisoners in Vietnam, and their previous act in April 1993 of "classifying" the first significant Soviet document which had been released by the Russians in April, so that it could be given to the Hanoi government before it was released in the United States:

"I congratulate the Pentagon and the State Department for not doing what it did last time around—which was to classify and withhold the earlier document from the American people while at the same time giving it to the Vietnamese. However, I think the Defense Department Press talking points on this document are pitiful..."[36]

This just-declassified report from Moscow confirmed once again that the evidence amassed from human and electronic intelligence during and after the war, some of which has been noted previously in this book, that North Vietnam was holding many hundreds more American prisoners than it admitted, was correct (as it had been correct after earlier wars). Thus, the Vietnamese Communist government had been lying all along, and was continuing to lie, about the fate of over 1,000 missing Americans who had at one time been held in captivity but were never returned. The fact that this document was released by the beleaguered Russian government of Boris Yeltsin, at a time of constitutional crisis in that nation, indicates that Russia desired to deviate attention from the fate of the Moscow-Bound POWs of Vietnam, perhaps because the Russian intelligence organs were refusing to cooperate in uncovering the fate of Americans sent to the Soviet Union from Southeast Asia, or earlier after Korea, World War II and World War I.

The New York Times reported on the new Soviet document the day after Senator Smith's speech and gave the reaction of the

Department of Defense: "The Pentagon, in a statement released Tuesday night, said it could not yet vouch for the statement's authenticity or accuracy. The statement did say it 'came from the files of the G.R.U.,' Soviet military intelligence."[37]

Despite the continuing controversy over the missing in action issue, President Clinton substantially eased the economic embargo against Vietnam on September 13th, for the second time in two months. American companies would now be permitted to bid on construction and development projects in Vietnam being financed by the World Bank and other international financing organizations. On July 2nd the Clinton Administration had announced that it would not block France, Japan and Australia from paying off Vietnam's foreign debts so that it would be eligible for much larger World Bank and Asian Development Bank loans. Following a tradition started by Armand Hammer, Averell Harriman and others in the early 1920s, many American business leaders had been pushing for both of these steps, as had Senators Frank Murkowski, John Kerry, John McCain and others. Although Assistant Secretary of State Winston Lord (accompanied by Presidential veterans representative Herschell Gober), had told Hanoi's leaders earlier in the summer that improved U.S.-Vietnamese relations would "depend on tangible progress on the outstanding POW/MIA cases," the actions of the Clinton Administration indicated otherwise.

Clinton's actions made it clear that his Administration and the U.S. Government's national security apparatus was satisfied with the continual dribble of fragmentary skeletal "remains" alleged to be those of American MIAs, which continued to be doled out by Hanoi, or "found at crashsites." Indeed, mysterious evidence, such as dog tags and personal items "found" at surface level in jungle crash-site investigations by the Pentagon's Joint Task Force Full Accounting in Southeast Asia, attracted press attention.[38] Little or nothing was said about the continuing reports of Americans still being held in the 1989-1992 period, and up to the present. The Vietnamese soon expressed disappointment, however, that all trade restrictions had not been lifted[39] Senator Robert Smith, former Vice-Chairman of the Select Committee on POW/MIA Affairs, wrote to the author (and others) two weeks later:

"I am...outraged by President Clinton's actions and his inaction with respect to the POW/MIA issue. Many of the Committee's most important recommendations have never been seriously acted upon, particularly with respect to Communist China's complicity in not returning POWs from the Korean conflict and Vietnam's complicity in not returning POWs captured in Laos during the Vietnam conflict...With respect to Vietnam, President Clinton is relaxing the U.S. trade embargo at a time when this leverage is critical...The President has referenced the importance he places on the 'continued counsel and advice from the families whose loved ones are missing and the veterans whose fellow soldiers did not come home.' In April 1993, the President further stated that, with respect to Vietnam, he was, 'MUCH MORE HEAVILY INFLUENCED BY THE FAMILIES OF THE PEOPLE WHOSE LIVES WERE LOST OR WHOSE LIVES REMAIN IN QUESTION THAN BY THE COMMERCIAL INTERESTS AND THE OTHER THINGS...'

Smith continued: "It should be noted, however, that the nation's largest veterans' organization, the American Legion, and the nation's largest POW/MIA family organization, the National League of Families, are condemning the President's recent actions concerning Vietnam, in addition to countless other POW/MIA family and veterans' groups across the country." The Clinton Administration thus seemed to be adhering to a long-established American policy, that the steps towards U.S. diplomatic recognition of a formerly-hostile Communist regime (such as Soviet Russia and Communist China, depended largely upon world-wide strategic concerns and economics rather than on the return of, or truthful accounting for American war prisoners known to have been withheld alive in captivity by such regimes.

Yet, during the first week of September at the meeting in Moscow the United States Government had also presented an official report to the Russian government containing detailed information about hundreds of American prisoners (up to 1,350) from the Korean War being sent to the Soviet Union for use by the KGB. This report, entitled: "The Transfer of U.S. Korean War POWs To the Soviet Union," was the product of the Research and Analysis Division of the U.S. Army's Task Force Russia, which had now been renamed the "Joint Commission Support Branch,"(JCSB) and brought under authority of the Department of Defense's POW/MIA office. The well-researched and heavily-documented report was largely the work of a Department of the Army civilian (DAC) employee named Peter Tsouras, who had been ably assisted by Air Force Major Werner Hindrichs, and a Russian scholar and linguist who specialized in tracking F-86 MIAs, Master Sergeant Danz Blasser (USAF), along with four other researchers. Aside from precise intelligence information on the transfer of Americans from Korea to the USSR, the report contained many livesightings of such American prisoners held for decades in Soviet labor camps, together with substantiating cables from U.S. embassies in Moscow, the Baltic republics and Ukraine.

While many important details contained within the report are beyond the scope of this book (aside from those that have been noted in chapter eight), the Executive Summary, created for the perusal of busy policymakers, adequately conveys the contents:

"U.S. Korean War POWs were transferred to the Soviet Union

and never repatriated. This transfer was a highly-secret MGB (KGB) program approved by the inner circle of the Stalinist dictatorship. The rationale for taking SELECTED PRISONERS to the USSR was: To exploit and counter U.S. aircraft technologies; to use them for general intelligence purposes. It is possible that Stalin, given his positive experience with Axis POWs, viewed U.S. POWs as potentially lucrative hostages. THE RANGE OF EYEWITNESS TESTIMONY AS TO THE PRESENCE OF U.S. KOREAN WAR POWs IN THE GULAG IS SO BROAD AND CONVINCING THAT WE CANNOT DISMISS IT.

"The Soviet 64th Fighter Aviation Corps, which supported the North Korean and Chinese forces in the Korean War, had an important intelligence collection mission that included the collection, selection and interrogation of POWs. A (Soviet) General Staff-based analytical group was assigned to the Far East Military District and conducted extensive interrogations of U.S. and other U.N. POWs in Khabarovsk. This was confirmed by a distinguished retired Soviet officer, Colonel Gavriil Korotkov, who participated in this operation. No prisoners were repatriated who related such an experience. PRISONERS WERE MOVED BY VARIOUS MODES OF TRANSPORTATION. LARGE SHIPMENTS MOVED THROUGH MANCHOULI AND POS'YET. KHABAROVSK (Siberia) WAS THE HUB OF A MAJOR INTERROGATION OPERATION DIRECTED AGAINST U.N. POWs FROM KOREA. The MGB controlled these prisoners, but the GRU was allowed to interrogate them. IRKUTSK AND NOVOSIBIRSK WERE TRANSSHIPMENT POINTS, BUT THE KOMI ASSR (Vorkuta area) AND PERM OBLAST WERE THE FINAL DESTINATIONS OF MANY POWs. Other camps where Americans were held were in the Bashkir ASSR, the Kemerovo and Archangelsk (Archangel) Oblasts...Komi Permyastskiy and Taymyskiy..."

"POW transfers also included thousands of South Koreans, a fact confirmed by the Soviet general officer, Kan San Kho, who served as the deputy chief of the Korean MVD. The most highly-sought-after POWs for exploitation were F-86 pilots and others knowledgeable of new technologies...A former Chinese officer stated he turned U.S. pilot POWs directly over to the Soviets as a matter of policy. Missing F-86 pilots, whose captivity was never acknowledged by the Communists in Korea, were identified in recent interviews with former Soviet intelligence officers who served in Korea. Captured F-86 aircraft were taken to at least three Moscow aircraft design bureaus for exploitation. PILOTS ACCOMPANIED THE AIRCRAFT TO ENRICH AND ACCELERATE THE EXPLOITATION PROCESS."[40]

The report had also made an effort to rehabilitate the reputation of the discredited RAND Corporation researcher, Dr. Paul Cole, who had heretofore been used by the U.S. Government to publicly debunk the research of the author and others indicating that large numbers of American war prisoners had been shipped to Russia (or of any remaining alive, after Yeltsin's June 1992 admission), by including some of his interviews and analysis within.

The reorganization of the Army's Task Force Russia to become the JCSB, had resulted in the assignment of Charles Trowbridge and Colonel Joseph Schlatter, two long-time DIA specialists at debunking eyewitness accounts of American POWs and MIAs remaining in captivity, to the staff of what had been known as Task Force Russia. General Loeffke, the former U.S. military attaché under Ambassador Malcolmb Toon (and once a U.S. infantry commander in Vietnam), had been 'retired' again, while Colonel Stuart Herrington had been sent back to Fort Meade, home of the NSA. According to a staff member of the organization who spoke to the author, Trowbridge, Schlatter and others attempted to obstruct release of the Korean War POW report by placing a qualifying cover on it, so that it would not be obtainable by the mass media or the public. This cover-sheet is stamped "Working Papers," and states: "contains subjective evaluations, opinions, and recommendations concerning ongoing analysis that may impact future U.S. foreign policy decisions. This document has not been finalized for public release." In staff meetings, an obviously-worried Trowbridge, who had been among those singled out by Senator Robert Smith for investigation by the Department of Justice, referred to the historically-significant report in obscene terms, categorizing it as animal fecal matter. Yet, when President Clinton's national security adviser, Anthony Lake, heard about the report he had requested that a copy of it be sent to the National Security Council, immediately.

In a meeting in Moscow on or about September 2nd Assistant Secretary of State Richard D. Kauzlarich (European Affairs) presented the Korean War POW report to representatives of the Russian government. According to a Joint Commission Support Branch (JCSB) staff member who was present, but requested anonymity, Secretary Kauzlarich made a forceful presentation of the evidence contained in the 77-page report to the Russian (and former Soviet) diplomatic and intelligence officials who attended the meeting, and backed it up with the authority of the Department of State.

The Department of State had rejected a misguided effort by Trowbridge, Schlatter and Edward Ross to classify the report so that independent researchers and scholars (such as the author) would be unable to obtain it through Freedom of Information (FOIA) request. The State Department's reasoning was that the report was a product of intelligence collection and research by a U. S. Government agency, about an ongoing diplomatic matter of high national priority, and did not need a disclaimer or qualifying cover, that in effect "classified" it. When the State Department sent a copy of the report to the White House, they sent it without the cover sheet. A source stated to the author that the Russians seemed impressed by the report and the supporting material presented at the meeting in Moscow, and they indicated they would "study it."

The State Department's presentation of the report in Moscow amounted to an admission by the U.S. Government that what had been officially denied for 40 years was now accepted as true. The report also implied, by the evidence presented, that some American POWs from the Korean War might be expected to be alive, somewhere in the former USSR. It clearly stated that the American side considered that former Soviet intelligence officials, and other human sources who were still alive, had given sufficient evidence that many of these American soldiers had remained alive in Soviet captivity for years, and accurate information about their fate should be readily available in Soviet records.

The objections of Trowbridge and others resulted in the dispatch of a classified U.S. cable to Moscow requesting information as to whether General Volkogonov or Ambassador Toon of the U.S. Russian Joint Commission on POW/MIAs objected to release of the report to the public. After discussion a reply was cabled saying in effect, that release was approved as long as it was explained as the U.S. position, not the position of the Joint Commission or the Russian government.

After receiving a "leaked" copy (as did the author), Associated Press said that the report was supported by new information from U.S. and Russian sources, and USA Today reported that up to "600" U.S. prisoners, including a group of F-86 Sabrejet pilots and many technical personnel were transported secretly to the Soviet Union for use by the KGB; were subsequently imprisoned and never released.[41]

Contrary to information contained in some news accounts, the report also did mention information contained in State Department documents on the shipping of trainloads of many other U.S. military prisoners to Siberia, and referred to the testimony of Lt. Colonels Delk Simpson and Philip Corso, and other witnesses and documents (referred to in chapter eight of this book).

The American Co-Chairman of the U.S.-Russian Commission on POWs and MIAs, Ambassador Malcolm Toon, had said at a Moscow press conference (also on September 2nd) that Russian intelligence agencies had not cooperated fully in the Commission's work, while praising Boris Yeltsin's candor and assistance in the matter. The Washington Post reported that the Russian member of the Commission, General Dmitry Volkogonov, said that he had asked the U.S. side for information about 11 Soviet aircraft that "perished in mysterious circumstances," which U.S. Co-Chairman Malcolm Toon referred to as "shootdowns," noting that U.S. intelligence agencies had also been reluctant to share information in their possession. Associated Press reported that the Commission "has never held out much hope for finding prisoners of war still alive," and that Volkogonov stated: "We believe that there are no Americans on the territory of the former Soviet Union, except persons who have willingly stayed here...Such people and their addresses are known by the U.S. side." The American side was reported to have provided information on the 11 missing Soviet aircraft and on approximately 20 Soviet soldiers formerly captured in Afghanistan who had settled in the west. Volkogonov stated that 40-50 Soviet soldiers were still listed as missing from the 1980-1989 war.[42]

Yet another livesighting of an American POW from the Korean War held in the Soviet Union was reported in newspapers across the United States on September 16th. This was the case of U.S. Marine Corps Sergeant Philip Vincent Mandra, (cited in the Joint Commission Support Branch report) who was identified from photographs by a former MVD Colonel, Vladimir Malinin (as noted previously in this book). Malinin had reported that he had seen Sergeant Mandra twice and he was still alive in 1966, confined to a prison in the Magadan region of Siberia.[43]

Meanwhile, in the former Soviet Union heavy fighting had been continuing during September in the Armenian-Azerbaijani border area, and in the republic of Georgia, Josef Stalin's tortured homeland. The head of state in Georgia (and former Soviet Foreign Minister), Eduard Shevardnadze, accused the Russian military of having 'masterminded" an assault on the city of Sukhumi by Abkhazian separatist rebels. Shevardnadze accused the rebels of having 'executed scores of officials, policemen and ordinary citizens. A Russian Foreign ministry spokesman said his government was perplexed and concerned about the anti-Russian coloring' of Shevardnadze's statements.[44]

Shevardnadze's reputation among many Georgians, as well as Abkhazians and Russians who were nationalists or wished to continue a socialist system, may have been damaged by media revelations that he had allowed CIA agents into his personal circle of advisers. Earlier in the spring, President Clinton had approved a CIA operation in Georgia to assist in providing security for Shevardnadze, after several assassination plots against him had been foiled. This was the first CIA operation inside the former Soviet Union to be acknowledged by senior U.S. officials. Subsequently, Fred Woodruff, of Herndon, Virginia, who was assigned to Shevardnadze from the U.S. Embassy in Moscow on June 3rd, had been shot to death at Natakhtari, near Tbilisi, in the Caucasus Mountains.[45]

A crisis began in Russia itself after President Boris Yeltsin issued a decree dissolving the Russian Parliament on September 21st, and ordering new elections to be held on December 12th. Within the parliament, a die-hard group of Communists and nationalists united behind Vice President (and Afghan war hero), Aleksandr Rutskoi and Rusian Khasbulatov, the speaker of the now-dissolved parliament, who assembled supporters for a planned revolt; in a what appeared to be a last gasp attempt at restoring the Communist system and the Soviet empire. Rutskoi was declared President of Russia, and much of the parliament, who had been elected under the old one-party system during the Gorbachev regime, barricaded themselves inside the "White House," as the building was called.

After a period of indecision and confrontation, Boris Yeltsin ordered the Russian armed forces to surround the Russian Parliament, where members were refusing to depart in preparation for new elections called for by the president. On CNN News, viewed around the world, Russian troops could be seen marching and in some cases running into position. Within the "White House," parliament members and supporters were said to have "over 1,000 Kalisnikov assault rifles" stored.

On Sunday, October 3, 1993, in the bloodiest popular uprising Moscow had seen since the Bolshevik Revolution of 1917, an estimated 10,000 supporters of the Parliamentary rebels attacked the Ostankino broadcast center (six miles away), the mayor's office and other targets, in an attempt to seize control of the state apparatus. They had been urged on by Rutskoi in a harangue from a window of the parliament building. Heavy fighting broke out between the revolutionists, who were armed with automatic weapons, and Spetsnaz troops, paratroopers and tank forces loyal to Boris Yeltsin. A reported 127 people were killed and hundreds more were wounded as the fighting continued into the morning of October 4th. (Among those killed was an American lawyer from Louisiana named Terry Duncan, who had been caught in the crossfire.) Mistakenly believing the broadcast center had been captured, Khasbulatov told cheering legislators: "Today we must seize the Kremlin."[46] The rebel leaders believed that their own latter-day "October Revolution" was about to take over control of Russia.

Yeltsin ordered an assault on the Russian Parliament by Army troops and riot police on Monday morning, October 4th. The estimated 1,500 people inside the building fought back, and at 7 a.m. the Russian Army began firing tank cannon shells into the White House, in addition to a barrage of automatic weapons fire, which caused severe casualties within, and set the building on fire. On-the-scene film of the fighting, with tracer bullets flying in all directions, was broadcast around the world by CNN News and others.

Among the Parliamentary supporters taking part in the battle were Russian soldiers who had defected with their weapons and joined the revolutionaries. When Rutskoi and Khasbulatov and their supporters finally came out and surrendered, many were wearing

Russian Army camouflage uniforms. After the rebellion was ended a reported 1,400 people had been arrested and taken to prisons and to an open stadium or in some cases released. Rutskoi and Khasbulatov were taken to Moscow's Lefortovo Prison, where generations of political prisoners had suffered and died since the time of Lenin and Stalin. The total number killed was unclear; on Monday, one Russian general was quoted as saying over 500 had already died in the fighting. Later estimates were revised downward to "several hundred" and finally to an official total of 187 killed and 437 wounded; whether this was actually the full number of dead may never be known. (Other reports put the casualties at about 1,000, including some 800 wounded.) After being imposed on the Russian people with bloodshed nearly 76 years before, the last official vestige of Soviet Communism was thus extinguished in another bloodbath, but whether the battle would only lead to another form of dictatorship for Russia was uncertain.

Yeltsin immediately reasserted state control over the media by imposing censorship and closing 10 opposition publications, including Pravda; he banned the country's highest court after it sided with the hardliners and outlawed 11 opposition political parties. Russian newspapers began publishing with blank spaces representing censored reports. For example, a blank spot appeared in the newspaper Sevodyna, where an article by Sergei Parkhomenko had been censored, which reported that confusion had reigned at the Kremlin when Yeltsin's top aides finally assembled hours after the revolt in Moscow had begun. The censored report also claimed that the Russian military, while "entirely loyal to the President," had intentionally delayed intervening in the Moscow revolt, 'for almost 12 hours.' This may have been the final bargaining point of the Soviet military with Yeltsin, in a possible deal to guarantee his survival in return for increased sympathy for Russian military goals.[47]

A few days before the revolt in Moscow, leading generals of the Russian armed forces had prevailed upon Yeltsin to send letters to the governments of Central European nations warning them not to join the NATO alliance. He had also recently increased the military's pay.[48]

With the revolt against Yeltsin in Moscow and a bloody attack on a U.S. Army ranger company and supporting helicopters engaged in the UN humanitarian effort in Somalia, President Bill Clinton was faced with two simultaneous international crises. As the fighting was occurring in Moscow, Clinton expressed strong U.S. support for Boris Yeltsin, saying he believed the Russian president had "bent over backwards" in trying to end the crisis peacefully. In Mogadishu, 17 American servicemen were eventually reported to have been killed and 17 others wounded, in an ambush by Somali gunmen and subsequent combat action.[49] At least one, and reportedly as many as 5 other Americans were missing or had been taken prisoner and were being held as hostages. One dazed U.S. Army helicopter pilot, Warrant Officer Michael Durant, was later filmed in enemy control and appeared to have been severely mistreated. (He was to be released after 11 days in captivity.) Some of the dead Americans shown on newsfilm being dragged through Somali streets in wheelbarrows, or on the ground, had their hands and feet securely tied with heavy knotted ropes, indicating they had been killed after capture. The Pentagon only admitted to "a handful of unaccounted for" (later reduced to one MIA) and "one" American taken prisoner. Whether this affected Somali treatment of other U.S. MIAs remains unknown. Secretary of Defense Les Aspin, who had previously ordered the withdrawal of U.S. tanks and armored vehicles from Somalia over the objections of U.S. commanders who wanted them on hand for just such an emergency, further infuriated the American combat soldiers in Somalia when he referred to the U.S. POW as a "detainee."

In a televised statement during the fighting in Russia and Somalia, that was reminiscent of the beleaguered Lyndon B. Johnson in 1968 or Richard Nixon in 1970, the once-youthful idealist and war protester, William Clinton, who was now the President of the United States, resolutely defended American combat troop presence in the Horn of Africa, and said that American policy would remain the same, except that more U.S. troops were being sent as reinforcements. In a live speech on the 7th he announced that the U.S. combat force in Somalia was to be increased from 4,500 to 10,000 troops. Clinton had already dispatched hundreds of U.S. soldiers to Macedonia, promised 25,000 more American troops for a UN peacekeeping mission in the war-torn former Yugoslav republic of Bosnia-Herzegovina (which Congress had yet to approve), and 600 U.S. troops to intervene in Haiti.

The People's Republic of China had meanwhile taken advantage of the international turmoil over events in Russia and Somalia, by exploding a nuclear bomb on October 5th, thus ignoring a worldwide ban on nuclear weapons testing. Clinton's response was to condemn China and then to direct the Department of Energy to begin preparations for a possible test of U.S. nuclear weapons in the coming year.

On that same day, according to a subsequent White House statement, Clinton had called Yeltsin, "to express the continued, strong support of the United States for President Yeltsin and the Russian government in the wake of the political crisis."[50]

But of the President's concern for the lost American POWs and MIAs of previous wars, Senator Robert Smith had written a sharply critical letter to the author and others just before the calamitous events in Moscow and Somalia:

"With respect to Russia, the President has taken no discernible steps to personally underscore to Russian authorities the importance of receiving full cooperation from their intelligence services, As a member of the Joint Commission with Russia (on POWs), I can attest to the fact that our work will only be successful if the President uses the full weight of his office to convince Moscow to be more forthcoming. We still face many roadblocks in gaining the full cooperation of Soviet military intelligence (GRU) and the KGB... The Russian leadership must be urged directly by the White House to have their intelligence and security organs tell us everything they know about captured Americans..."[51]

Yet, at the time this book went to press all indications were that the Clinton Administration was determined to adhere to the long-established U.S. policy of not allowing the missing American prisoners of war to affect geopolitical decisions.

The effort in subordinating Task Force Russia (JCSB) to the Pentagon-DIA hierarchy (which the Army intelligence agency had, in effect, made an end-run around, in confirming for the first time through an internal U.S. Government agency that many American prisoners of war had been retained alive and transported to Soviet labor camps), was continued. By the end of October the author of this book was requested to supply a copy of the final pre-printing draft, with notes and index, to members of Task Force Russia's staff who were attempting to prevent a misguided, DIA-inspired closure of the World War II U.S. POW/MIA issue, with yet another inaccurate and untrue statement, despite all the contrary revelations of the previous four years. True to form, the DIA group, represented by Schlatter and Trowbridge, also wished to discharge staff members at JCSB who had been instrumental in uncovering facts about American POWs remaining in Communist control, such as retired Army Colonel William LeGro, whom the author had worked with in cooperation for seven years.

A powerful U.S. News and World Report article on U.S. prisoners of war and missing in action subsequently appeared on Veterans Day, 1993, which accurately depicted the demoralization at Task Force Russia stemming from the actions of the Pentagon's Edward Ross. The report also cited sources used in this book which indicated that the North Vietnamese had withheld as many as 700-800 American prisoners after the Operation Homecoming returns of 1973, and that some U.S. POWs had been transported to the Soviet Union. Among the human sources cited was former Army intelligence specialist Barry Toll, who was reported to have attended a meeting with national security adviser Anthony Lake at the White House on November 4th, in company with Carol Hrdlicka, wife of Air Force POW David Hrdlicka missing in Laos since 1965, and George Carver, a former special assistant to 3 CIA directors from 1966-1973. Carver (who was the "man of repute" long anticipated by General Tighe and Jerry Mooney), stated that while he was at CIA he had seen evidence that the Vietnamese and Laotians were holding hundreds more Americans then they admitted to publicly. Toll again revealed that between 1973 and 1975 he had seen top-secret cables about the transfer of 10-20 U.S. POWs from Hanoi to the Soviet Union on diplomatic aircraft., and another one stating that 290-340 U.S. POWs had been reported by the Pentagon as remaining in Laos.

Another sign of a sea-change in the American media's attention to the POW/MIA story was a December 1993 report on the Vietnam prisoners of war in the "American Spectator," which had for years declined the author's submissions and avoided the subject. Vietnam veterans' publications continued to report new discoveries made by investigators. The "Veteran Leader" (Volume 12) reported the uncovering of a top-secret U.S. Navy message dated September 10, 1972, which recorded the movements of U.S. POWs from Vietnam into the People's Republic of China. One group was spread out in small groups in Danam Province and the other in Quang Dong Province. They were to be held in China as long as North Vietnam desired.

Meanwhile, no established U.S. publisher contacted by three literary agents on behalf of the author had expressed interest in publishing this book in any form in the one and a half years following Boris Yeltsin's revelation, although various U.S. Government agencies and other researchers had made use of the first, self-published edition. The last literary agent, Jeff Gerecke of JCA Inc., in New York, approached many of America's major publishing companies with the manuscript over a year's time, to no avail. The author was thus forced to print this revised edition immediately after it had become necessary to voluntarily supply a copy to an agency of the U.S. Government in early November 1993.

A serious U.S. Government investigation into the history of Soviet-held U.S. prisoners of war, and those held by other Communist nations, requires a comparison of Moscow's conduct with secretly-held American POWs and MIAs between 1921 and 1993, and cooperation with the private researchers who first exposed this tragic, hidden history. There are many incidents and cases which the Russian government and the Commonwealth of Independent States should be able to quickly resolve, whether or not the prisoners were subsequently assigned new names and Soviet citizenship, but the full cooperation of former and present Russian intelligence officials must be obtained. However, even if this information is received by the U.S. side, the American people must have some credible guarantee that U.S. intelligence agencies will in turn reveal it to the public.

Intelligence indicates that American POWs almost certainly remained alive in Southeast Asia until the time of this writing, and that the Vietnamese were uncertain about what to do with them. The United States Government appeared to be caught in a trap, of not being able to admit the prisoners were left behind or that many remained alive for two decades, because of the potential embarrassment for many former and present U.S. officials who denied their existence for so long. The granting of full U.S. diplomatic recognition of Vietnam by the United States would likely remove the last incentive for the Hanoi government to keep the surviving prisoners alive, even if they had heretofore been granted citizenship and some measure of protection by the Vietnamese government, after their own government had declared them dead. Therefore, a cautious diplomatic course, as has been followed for so long, would appear to remain the safest one for all possible U.S. survivors.

Yet, American business interests with no concern for the facts of the POW/MIA matter continued to push for diplomatic recognition of the Hanoi regime at the time of this book's publication. Representatives of American corporations desirous of securing profitable contracts, including Pfizer, Occidental Petroleum (the recently-deceased Armand Hammer's company) Mobil and Avon, were reported in the New York Times during October as being feted with the Vietnamese Deputy Prime Minister, Phan Van Khai, by Irwin Jay Robinson, president of the Vietnam-American Chamber of Commerce. Although the reception was temporarily disrupted by a U.S. war veteran who accused the eager American executives of disgracing the United States, the participants were unfazed, knowing they had the support of the Clinton Administration.[52] In keeping with such demonstrations of the continuation of the controlling interests of policy and politics over the unresolved POW/MIA issue, the National Committee on American Foreign Policy in New York, now under the honorary chairmanship of Henry Kissinger, issued a policy recommendation, "for the record," in October, calling for normalization of relations with Vietnam, and stating: "The resolution of remaining differences over POW/MIA issues will be enhanced by normalization."[53]

Similarly, the Clinton Administration had begun to reverse its critical stance towards the People's Republic of China, which had resulted from China's severe human rights violations since the Tiananmen Square massacre of 1989. Despite a widespread disapproval of the Chinese regime by rank and file American Democrats as well as Republicans, geopolitics were now said to be reasserting themselves in the formulation of American policy towards the world's most populous nation. The People's Liberation Army, now the dominant force in China's political life, was said to be viewing America as their primary enemy, and to be aspiring to international military projection capability. The Washington Post reported that a classified U.S. National Intelligence Estimate on China had forecast that the Chinese military leadership "would regard U.S. forces as 'the enemy' unless Washington avoided fostering that impression."[54] The alteration of policy, spelled out in an 'action memorandum' signed by President Clinton in August, was outlined by NSC adviser Anthony Lake to Chinese Ambassador Li Daoyu in late September, and by Secretary of State Warren Christopher in a meeting with Chinese Foreign Minister Qian Qichen in New York at the end of the month. Thus, known human rights violations by the Chinese government, including the decades-long confinement of American prisoners of war, was to be once again overlooked, due to the overriding policy concerns about nuclear proliferation and other national security matters, which were said to require cooperative dialogue and friendly relations.

The overwhelming evidence that many U.S. POWs and MIAs of several wars were held prisoner in the Soviet Union requires that an intensive and exhaustive search, throughout all the republics of the former USSR, should be undertaken for American survivors of World War II, Korea and Vietnam who may be living in remote areas. This can only be successful with the full cooperation of the various Russian intelligence services, which have not been forthcoming thus far, according to Senate investigators and members of Task Force Russia who have spoken with the author. Sadly, as Ambassador Toon pointed out, the U.S. intelligence agencies have also failed to cooperate fully with revealing POW/MIA information in their own files, which sends an unmistakable signal to their die-hard counterparts in the Russian intelligence services.

The individual governments within the Commonwealth of Independent States, led by Russia's, should demonstrate authority and compassion by allowing public contact with any former U.S. prisoners who may be living inside restricted security areas of former Soviet republics, whether in Kazakhstan or Russia and in the vast reaches of Siberia. It is also virtually certain that surviving American prisoners of war remain in North Korea and in the People's Republic of China; thus America should continue to focus on human rights in these nations, requesting truthful information on the fate of all American POWs and MIAs once in their control, and the return of all survivors. If U.S. prisoners of all America's 20th century wars involving the Communists were murdered, the American people need to know where, when and how these deeds were accomplished. Anything less would constitute an abrogation by the Government of the United States of the principles upon which the Nation was founded, and to which the people of America owe their true allegiance.

## Notes on Chapter Fifteen

1    Interfax and SF Chronicle January 28, 1993

2    Washington Post National Weekly Edition, January 25-31, 1993

3    Final Report, pp 243-44

4    Associated Press

5    Final Report p. 243

6    Washington Times, January 28, February 17, 1993

7    NYT, February 23, 1993

8    France-Amerique, 20-26 February 1993

9    The Sunday Oregonian, April 10, 1993

10   NYT, April 10, 1993

11   San Francisco Chronicle, Tuesday, April 13, 1993

12   San Francisco Chronicle Wire Services, April 14, 1993

13   SFChron Apr 16

14   NYT, April 18-19, 1993

15   Washington Post article by William Branigan, reprinted in San Francisco Chronicle, April 20, 1993

16   AP April 22, 1993 S.F. Chronicle

17   "MIA Families Won't Forget," by Al Santoli, Parade Magazine, May 30, 1993

18   Washington Post, July 17, 1993,)

19   NYT, Wednesday, September 8, 1993

20   Manchester, N.H., Union Leader, July 14, 1993, Associated Press report by Jim Abrams, "Senator: Pentagon is Wrong, Smith says he visted sites of 3 POW camps"

21   Congressional Record, June 29,1993, p. S8206

22   C-Span, July 13, 1993

23   Memorandum for the Record, John F. McCreary, April 21, 1992, Codinha Box 3,Ethics Committee Referral.'

24   The Stars and Stripes-The National Tribune, 12-18 July 1993 "ABC's Garwood Story: A Fairy Tale That Shames POWs., by Senator John McCain

25   DAV Magazine, October 1993

26   Associated Press, August 14, 1993, "Brown Denies Getting Viet Cash"

27   U.S. Veteran Dispatch, July and September 1993

28   Associated Press, August 14, 1993, op. cit.

29   Newsweek, October 11,1993, "Dubious Commerce," by Howard Fineman and Bob Cohn

30   U.S. Veteran Dispatch, September 1993

31   Washington Post National Weekly Edition, August 23-29, 1993

32   Los Angeles Times, August 24, 1993

33   Associated Press report by Robert Burns, published in the (Portland) Sunday

Oregonian on August 22, 1993

34 Associated Press reports, San Francisco Chronicle, and Houston Chronicle, August 26, 1993

35 Letter of September 29, 1993, from Senator Robert Smith to the author

36 Statement of Senator Robert Smith of New Hampshire, to the U.S. Senate on September 8, 1993, on a new POW document just released from Moscow to the joint U.S.-Russian Commission on POWs.

37 New York Times, September 9, 1993, "Soviet File Feeds Debate on POW's"

38 Los Angeles Times report by Richard A. Serrano, reprinted in the San Jose Mercury News, September, 17, 1993: "Mysterious evidence arouses new debate on fate of MIAs"

39 NYT & USA Today, September 14, & NYT September 15th 1993

40 "The Transfer of U.S. Korean War POWs To the Soviet Union," by the Joint Commission Support Branch, Research and Analysis Division, Defense Prisoner of War and Missing in Action Office (DPMO) 26 August 1993)

41 USA Today, report of September 27, 1993, by Tom Squitieri, "Russia Kept Korea POWs, U.S. Says," also: Associated Press

42 "U.S. contends Russians impeding hunt for MIAs," by Fred Hiatt of the Washington Post, reprinted in the Houston Chronicle, September 3, 1993; also: Associated Press, "U.S. Russian delegation ends search for POWs," The Oregonian, September 3, 1993

43 Associated Press report published in The Oregonian, September 16, 1993

44 "Claim of Executions In Georgian Enclave, by Michael Hiltzik, Los Angeles Times

45 "CIA Acted To Bolster Shevardnadze," by Knut Royce and Patrick J. Sloyan, New York Newsday

46 Newsweek, October 18, 1993

47 Washington Post, S.F. Chronicle, October 6, 1993

48 San Francisco Chronicle, October 6, 1993, "How Yeltsin Got Military to Attack the Parliament by Alexander MacLeod, reprinted from the Christian Science Monitor"

49 Newsweek, October 18, 1993

50 Washington Post/San Francisco Chronicle, October 6, 1993

51 Letter to the author of September 29, 1993, from Senator Robert Smith, P. 6

52 NYT, October 9, 1993

53 American Foreign Policy Newsletter, October 1993

54 Washington Post, National Weekly Edition, November 8-14, 1993

## POSTSCRIPT

In researching and writing this book the author has completed a mission which he accepted while serving in combat in Vietnam throughout 1968: That if a soldier, sailor, airman or marine has knowledge of a comrade who is in enemy control, he must make every effort to recover that fellow-serviceman (or woman), as he would expect to be done for him. Over 2,000 years before, Athenian youth had taken the Epheboi oath upon reaching manhood, which began with the words, "I will not disgrace the sacred arms. Nor will I abandon the man next to me in battle..." This ancient trust has also been a "covenant" between the United States Government and those who have served it and upheld American foreign policy around the world, through all the severe battles of the 20th century. The author's quest resulted from our government's breaking of that covenant. Over a nine-year period, from 1983-1992, this mission was encouraged by the author's wife of 24 years, Josephine Duke Brown, and by our five children, and by many other persons in different ways, including, perhaps somewhat ironically in light of what the investigation revealed, a future President, George Bush. But the real purpose of those years of research into the long-classified history of this tragic issue, was to force the American democratic system of representative government to work, on behalf of any and all surviving American prisoners of war, from every former enemy nation where they were held; the living-dead who were unable to plead their own cases before the American people.

This has yet to be accomplished at the time of publication of this book, according to the United States Government.

# APPENDIX I

WAR DEPARTMENT
EXECUTIVE DIVISION, GENERAL STAFF
MILITARY INTELLIGENCE BRANCH
WASHINGTON

# TELEGRAM

Archangel
Dated Feb. 4, 1919
Received Feb. 4, 11:58 p.m.

COMBAT.

COIE

Milstaff,

Washington.

No. 159, Feb. 4.

1. Up to and including January 31, the total American casualties were as follows:

|  | Officers | Enlisted Men | Total |
|---|---|---|---|
| Died of sickness | 2 | 64 | 66 |
| Died, killed in action | 3 | 58 | 61 |
| Died of wounds received in action | 1 | 12 | 13 |
| Died accidentally killed | – | 3 | 3 |
| Died, drowned | 1 | 2 | 3 |
| Missing in action | – | 34 | 34 |
| Total | 7 | 174 | 181 |

| | |
|---|---|
| Wounded in action, all ranks | 198 |
| Accidentally wounded | 25 |
| Self inflicted wounds | 6 |
| Total | 229 |

Grand Total, all ranks, 410

2. The total American casualties from January 19 to January 31, inclusive, in the fighting on the Vaga front were as follows:

WAR DEPARTMENT
EXECUTIVE DIVISION, GENERAL STAFF
MILITARY INTELLIGENCE BRANCH
WASHINGTON

# TELEGRAM

-2- Archangel
Dated Feb. 4, 1919
No. 159

|  | Officers | Enlisted men | Total |
|---|---|---|---|
| Died, wounds received in action | 1 | 3 | 4 |
| Killed in action | - | 11 | 11 |
| Missing in action | - | 18 | 18 |
|  | 2 | 28 | 30 |
| Total | 3 | 60 | 63 |

POW's / MIA  (?)

3.  On January 31 the total American forces remaining comprised 161 officers and 4,764 other ranks, making a total of 4,925.

4.  The above has been transmitted to Churchill.

RUGGLES

mep

0437

WAR DEPARTMENT
EXECUTIVE DIVISION, GENERAL STAFF
MILITARY INTELLIGENCE BRANCH
WASHINGTON

# TELEGRAM

Archangel
Dated April 14,
No. 221.

On the morning of the fifth the enemy made an attack
on Sredmekhrenga which resulted in his defeat with the
loss of 100 men and five machine guns. The Allies had
no casualties. The Slavo-British Legion carried out a
successful raid at Bolshayaozerka, taking nine prisoners,
on April 9. During the remainder of the week there were
a number of small attacks which resulted successfully for
the Allied Russians, one at Kodish.

4. Negotiations for the exchange of prisoners
have been terminated by orders from General Pershing,
after having been delayed, although under discussion by
both sides, through failure of the bolshevik commander
to obtain authority from Moscow. As far as the negotiations
went they showed the presence of wounded Allied prisoners,
apparently receiving good treatment (in order to strengthen
the effect of bolshevik propaganda), in hospital at
Vologda and in prison at Moscow.

5. Have informed Warburton.

                              R U G G L E S

rew
CONFIDENTIAL COPIES TO
Chief of Staff   W. P. D.   State Dept.   Cable Office M. I. 2.   Capt. Martin.
P. 1580

DECLASSIFIED BY NND740058 [signature] 7/28/86

In reply refer to
(MID) 2070-1380

OFFICE CHIEF OF STAFF
M. C.    Executive Division
Mil. Int. Branch

May 12 1919

My dear Mr. Secretary:

I have the honor to acknowledge receipt of
your letter ("WE-M"), dated April 28, 1919, regarding
the negotiations with the Bolshevik government in
Russia for the exchange of Allied prisoners, referred
to in cablegram No. 230 from the Military Attache,
Archangel, Russia. In accordance with your suggestion,
a cablegram was sent to the Military Attache on May 1,
reminding him that the United States has not recognized
the Bolshevik regime as a government either de facto or
de jure.

Very sincerely,

Secretary of War

The Honorable Frank L. Polk

Acting Secretary of State.

CHARGE TO Office, Chief of Staff.
GOVERNMENT RATE.                                    Capt. Kenyon, M.I.5.

WAR DEPARTMENT TELEGRAM.

OFFICIAL BUSINESS.                          2070-1400

WASHINGTON.                          WAR DEPARTMENT
——— C O D E ———                          May 1, 1919.

Military Attache,
Archangel, Russia.

2070-1380
Your 230 stop   Under date of April 28th State Department requests
you be informed that United States has not recognized the Bolshevik
regime as a government either de facto or de jure stop

CHURCHILL

V T. C. Cook,
Lieut.Col.,C.A.C. U.S.A.

COPY

U.S. Department of Justice

BUREAU OF INVESTIGATION

Post Office Box 251
Grand Central Sta.
New York City.

EJC-F

October 25, 1930.

PERSONAL AND
CONFIDENTIAL

Mr. J. Edgar Hoover,
Director,
Bureau of Investigation,
Department of Justice,
Washington, D.C.

Dear Mr. Hoover:

Agreeable with instructions received by telephone from you this morning, I contacted with Mr. John B. Trevor who, in turn, placed me in touch with Mr. Gregory G. Bernadsky, president of the Research Publishing Corporation at 11 West 42nd Street, New York City. As a result of arrangements made with Mr. Bernadsky, both he and Alexander Grube, the sailor who desired to furnish certain information concerning Russian prisons, appeared at this office this afternoon and were interviewed by Special Agent F. X. O'Donnell. As a result an affidavit was obtained from Alexander Grube and in triplicate is enclosed herein.

In addition, in an interview had by Agent O'Donnell with Mr. Bernadsky, Mr. Bernadsky furnished certain information concerning one Fritz Karklin which, in memorandum form prepared by Agent O'Donnell, is likewise in triplicate herein forwarded to you.

very truly yours,

E. J. Connelley,
Special Agent in Charge.

DECLASSIFIED

County of New York )
                    ) ss
State of New York  )

October 25, 1930

Alexander Grube, being duly sworn, deposes and says:

That he is a sailor by occupation and is temporarily residing at the
Seamen's Church Institute, 25 South Street, this city, that he is 29
years of age and that he was born in Riga, Latvia; that he first arrived in the United States at the port of New York in 1920; that he
took out first citizenship papers in New York County, New York on November 3, 1923 but that because of long absences from this country he
has thus far been unable to complete his American citizenship.

That on February 2, 1926 at the port of New York, he signed on as an
able seaman aboard the S/S "Winona" of the American Export Lines (25
Broadway, New York City) and sailed aboard said ship for Black Sea
ports.

That having heard that living conditions and employment were very
desirable in Russia, he did, on March 5, 1926, desert the S/S "Winona"
while said ship was harbored in the port of Constantinople, Turkey and
did thereafter, on April 2, 1926, sign on as able seaman aboard the
S/S "dekabrist", a Russian Freighter bound for Russian ports.

That on June 2, 1926 he arrived at the port of Vladivostock, where he
was arrested and thrown into prison charged with being an American spy,
said charge having been based on the suspicion which he had created
by reason of his desertion of an American ship upon which he was being
paid at the rate of $62.50 per month, in order to transfer to a Russian
ship upon which his salary would be only $35.00 per month, and for the
further reason that upon being searched, certain papers were found on
his person indicating prior service on his part on ships of the American
Transport Line, which the Russian officials believed to be part of the
United States forces.

That, thereafter, for nine months he was incarcerated in the G.P.U.
Prison at Vladivostock and that he was beaten every day until he became
unconscious, that on December 22, 1926 he received notice from the
Collegia of the G.P.U. at Moscow that he had been sentenced to serve
ten years at hard labor on Solovetz Island and thereafter five years in
Siberia.

That, thereafter, he was transferred from prison to prison until March
1st, 1927 when he arrived in Lubianka Prison at Moscow where he learned
and occasionally saw four American Army officers and fifteen United
States Army soldiers who had, since 1919, been there in prison.  Of all
these American soldiers he was able to learn the name of only one, that
of Alfred Lindsay whom he met while the latter was employed as a cook
in the prison kitchen.

That, from Lubianka Prison, he was transferred directly to Solovetz
Island where, as fellow prisoners, he met many American soldiers and
civilians who were there in prison, among whom he particularly mentions
the following:  A Mr. Martin or Marten and a Mr. G. Helsinkruk (this
last name being spelled as nearly like the name as it sounded to affiant)
both of whom, he thinks were American Army officers who had been sent to
Solovetz Island from Vladivostock and who were here working as laborers.

COPY

That, in addition, he met one Roy Molner whom, he states, had been a Sergeant in the United States Army, formerly located at Archangel, from which city he had been sent as a prisoner to Solovetz Island. Affiant further states that the terms of imprisonment of all these American soldiers are indefinite. Molner is 35 years of age.

That, any inquiry made at either of the prisons for any of the individuals described by name would result in the immediate death of the individuals asked for.

That, conditions at Solovetz Island are such as to sooner or later guarantee the death of anyone there incarcerated, which accomplishes for the Soviet government the same result as that obtained by a firing squad without the disadvantage which would accrue to publicity concerning the latter method.

That, affiant escaped from Solovetz Island in August 1929 and that thereafter he proceeded to Liverpool, England aboard the S/S "Eric Larson" aboard which he had been smuggled as a stowaway by the German crew of said ship at the port of Kem on the White Sea.

That, if at any time needed in the future, affiant may be reached through Mr. Gregory G. Bernadsky, president of the Research Publishing Corporation at 11 West 42nd Street, New York City, with whom he will maintain contact on the occasions of his future visits to New York.

That, this is a voluntary statement on his part and is made in the presence of Special Agent F. X. O'Donnell of the United States Department of Justice, Bureau of Investigation and of Mr. Gregory G. Bernadsky, of his own free will, without duress and without receipt by him of any promise or any reward from any source whatsoever.

(Signed)_____A. Grube_____

Sworn to before me this 25 day of October 1930

_____Annette E. Pleser_____
Notary Public

Witnessed:-

F. X. O'Donnell

Gregory G. Bernadsky

IN EDGAR HOOVER
    TOR

### U. S. Department of Justice
### Bureau of Investigation
### Washington, D. C.

~10110-2623

1930

October 31, 1930

Director, Military Intelligence Division,
War Department,
Washington, D. C.

Dear Sir: -

      There is tranamitted herewith, for your
information, copy of a letter dated October 25, 1930, which
was addressed to the Bureau by the Special Agent in Charge
of the New York City office of the Bureau, together with
the inclosures which accompanied this communication.

      There is also inclosed a report (copy) submitted
by Special Agent R. W. Finch, July 11, 1918, at New York
City, relative to Ernest Karklin alias Fritz or Fred
Karklin. It appears probable that the Subject of this
report is identical with the Fritz Karklin referred to by
Mr. Gregory G. Bernadsky in his interview with Special
Agent O'Donnell of the New York office.

                Very truly yours,

                         Director.

Incl.136931

10110-2623

10110-2623

WAR DEPARTMENT

CONFIDENTIAL                                         November 6, 1930.

MEMORANDUM FOR THE ADJUTANT GENERAL:

        Subject:   Alleged confinement of American Officers
        and Soldiers in Russian prisons.

    1.   It is requested that a careful search of your records be
made to determine whether or not it can be ascertained if the persons
named in the affidavit of Alexander Grube were ever in the military
service, and if so, when, where, and how separated therefrom.

    2.   This paper was informally referred to the World War Record
Division on November 3d for a quick check among the reported casualties. Nothing was found. However, in view of the seriousness of
the possibilities in this case further search is considered necessary
and desirable.

                      R. C. FOY,
                    Colonel, General Staff,
                    Acting A. C. of S., G-2.

Received A. G. O. NOV 6 1930
    encl. 1
    Affidavit of Alexander Grube.
    eh

CONFIDENTIAL

**WAR DEPARTMENT,**

**THE ADJUTANT GENERAL'S OFFICE.**

———

MEMORANDUM.

November 12, 1930.

To:  The Acting Assistant Chief of Staff, G-2.

Subject:  Alleged confinement of American Officers and Soldiers in
Russian prisons.

The attached affidavit of Alexander Grube, seaman, sets forth
among other statements, that he arrived March 1, 1927, in Lubianka
Prison at Moscow where he saw four (4) American Army Officers and
fifteen (15) American soldiers who had been there since 1919, one
of whom was Alfred Lindsay; that he subsequently was transferred
to Solovetz Island Prison where he met many American soldiers and
civilians, and names two of them as Mr. Martin or Marten and Mr. G.
Neinainkruk, both of whom he thinks were American Army Officers sent
to the Island from Vladivostok. He also mentions one Roy Molner whom
he states had been a sergeant in the U.S. Army at Archangel from which
place he had been sent as a prisoner.

From the information furnished this office is unable to identify
any of the persons named as having served in the Army.

With reference to the statement as to the two men, Martin and
Neinainkruk, being American Army officers sent to the prison from
Vladivostok, the records show that there were one (1) officer and
six (6) enlisted men carried on the records as captured by enemy
from the American Expeditionary Forces in Siberia, all of whom were
subsequently repatriated.

Herewith is a list containing the names, record and personal
description of the officers and enlisted men who were reported as
killed or missing in action in the North Russia Expedition (Archangel) and for whom no grave location has been found. This list
contains the names of two (2) officers and twenty-nine (29) enlisted men.

In the cases of those reported as missing in action an administrative determination had been placed on each of their records
that they were killed in action on the date they were reported as
missing. This action was taken in accordance with the office
precedent based on the opinion of the Judge Advocate General of
the Army under date of April 12,1923,(Dan Mathews,#2,649,999, Co.I,
369th Infantry) that when the latest authentic report indicates
soldier was in action and when sufficient time has elapsed so that
report of him would have been received through various organizations,
and there is no information found as to his status, the only reasonable
presumption is that he is dead.

Form No.57—A.G.O.
Ed. Sept. 20 19—45,000

CONFIDENTIAL

It is noted that among the names given by Grube are those of Mr. Martin or Marten, and one Alfred Lindsay.

The records show that William J. Martin,#2,984,430, and Lindsay Retherford,#2,890,713, both of Company A, 339th Infantry were reported as missing in action January 19,1919, at Ust Padenga,Russia. Determinations were made that these men were killed in action. The Graves Registration Service reports that the bodies of these two men have been recovered.

No record has been found that will connect the other names mentioned by Grube with service in the Army.

Major General,
The Adjutant General.

1 incl.
    List.

Mr. Huckleberry:

Please see inclosed correspondence also slips attached
thereto from Colonel Adams, and a list of names of the supposed
American prisoners in Russian Prisons. The whole number being
apparently fifteen, about eleven being unnamed.

*(only those 19
in the
Lubyanka
Prison, Moscow)*

I have looked into this question and find that at least
one case that has an important bearing on it, namely the case of
William J. Martin, Company A, 339th Infantry, which regiment served
in Archangel or North Russian Expedition. Under date of Feb. 3,
1919 a report from Archangel showed Martin missing in action. This
was the last information received from Europe. Under date of March
14, 1921 we made a determination showing: "Was killed in action
Jan. 19, 1919". This determination was no doubt predicated on the
unexplained absence of the soldier for about two years.

I also find another case which may possibly be involved,
it is that of Lindsay Retherford, up in my mind because of the mention of the Russian sailor of Alfred Lindsay. Lindsay Retherford was
reported missing and a similar determination was made in his case.

I showed to Colonel Parrott the Martin case and he was very
much impressed by the thought that Martin may possibly be alive.

He wished you to obtain from Miss Shehan a list of all
determinations concerning men who served in North Russia or Siberia
in the cases of men missing in action, prisoners of war, deserters, Etc.
With regard to Siberia, I understand that one officer and six enlisted
men were taken prisoners in Siberia, but I understand that they were
repatriated from capture by the forces operating against the Americans
in Siberia. This in order to determine whether there could be any
more cases like the Martin case.

R.J.O
Chief Clerk

11-8-30.

To Mr. Drown
Can you or Mr. Rose see any other line
of approach in this case.—

R.J.

CONFIDENTIAL

G-2/10110-2623

G-2
WDY
NOV 21 10110-2623 1930
5
WAR DEPARTMENT
27

November 14, 1930.

Honorable Hamilton Fish,
    House of Representatives,
        Washington, D. C.

Dear Mr. Fish:

In reply to your telephone enquiry of November 10th concerning the affidavit of Mr. Grube in which he alleged that certain persons named therein are probably American officers and soldiers, you are informed that a careful search of the records in the War Department fails to identify any of the persons named therein as having served in the Army.

The records further show that the one officer and six enlisted men carried on the list of the American Expeditionary Forces in Siberia as "Captured by the enemy" were subsequently repatriated.

There were two officers and twenty-nine enlisted men in the Archangel expedition reported killed or missing in action and for whom no grave location has yet been found. None of their names, however, appear in the affidavit of Mr. Grube.

Sincerely yours,

MAILED G-2 W. D. NOV 15 1930

R. C. FOY,
Colonel, General Staff,
Acting A. C. of S., G-2.

wm

CONFIDENTIAL

DECLASSIFIED

JOHN EDGAR HOOVER
DIRECTOR

**U. S. Department of Justice**

**Bureau of Investigation**

Washington, D. C.

NOV 17   10110-262   1930

WAR DEPARTMENT

November 15, 1930.

Director,
Military Intelligence Division,
War Department,
Washington, D.C.

Dear Sir:

      With further reference to my letter of
October 31, 1930, relative to one Alexander Grube,
who has furnished information concerning Russian
prisons, there are enclosed herewith, for your
information, copies of communications, with their
enclosures, from the New York office of the Bureau
of Investigation, dated October 29, November 1, and
November 8, 1930.

                    Very truly yours,

                    J. Edgar Hoover
                    Director.

Enc. 136872

RECEIVED

24      REPORT OF THE SECRETARY OF WAR.  - 1919

selves may be willing to accept assistance, whether from Vladivostok or from Murmansk and Archangel. The only present object for which American troops will be employed will be to guard military stores which may subsequently be needed by the Russian forces, and to render such aid as may be acceptable to Russians in the organization of their own self-defense. With such objects in view the Government of the United States will cooperate with the Governments of France and Great Britain in the neighborhood of Murmansk and Archangel.

The forces under Col. Stewart remained in North Russia from September, 1918, to June, 1919. During this time they occupied various positions along the Vologda Railroad and the Onega, Dvina, and Vaga Rivers, and engaged in many minor operations against enemy forces. The following is the list of casualties in this force:

Killed in action.................................................109
Died of wounds..................................................  35
Died of disease.................................................  81
Died of accident and other causes..............................  19

   Total deaths................................................  244
Wounded........................................................  305
Taken prisoner (all released)..................................    4

   Total casualties............................................  553

After the armistice was signed, it was impossible to remove this detachment, owing to the climatic conditions which forbade the movement of large bodies of troops or large amounts of supplies. In April, 1919, Brig. Gen. W. P. Richardson arrived at Archangel and took command of all troops of the American Expeditionary Forces in North Russia. The One hundred and sixty-seventh Company, Transportation Corps, came at the same time, but stopped at Murmansk in order to cooperate with the One hundred and sixty-eighth Company, Transportation Corps, which had arrived in March, 1919, and was then engaged in construction on the Murmansk Railroad. During the month of June, 1919, all the forces of the United States were withdrawn, except a small detachment of Engineers. In September the last personnel of the American forces in northern Russia were en route to the United States.

AMERICAN EXPEDITIONARY FORCE—SIBERIA.

During the summer of 1918 the Czecho-Slovak armies operating in Siberia were threatened with destruction by hostile forces apparently organized by, and often largely composed of, enemy prisoners of war. It was of importance to the Government of the United States that these friendly Czecho-Slovak forces be rendered assistance and that the Russian people be aided in every way acceptable to themselves in any and all efforts to regain control of their own affairs and their own territory. As Japan and the United States were the only powers in a position to act in Siberia in sufficient force to accomplish the

--------

Includes 28 presumed killed in action.

MISSING IN ACTION U.S. ARMY PERSONNEL
German and Japanese Theaters
7 December 1941 through 31 December 1945

| | Captured | | | | | Other Missing in Action | | | | Total Missing |
|---|---|---|---|---|---|---|---|---|---|---|
| | Died of Wounds | Died of Other Causes | Returned to Mil.Control | P.O.W. (Curr.Stat) | Total | Declared Dead | Rep't Dead From MIA | Ret'd to Duty | MIA (Current Status) | (Including Captured) |
| German Theatres | 210 | 648 | 90,937 | 5,414 | 97,209 | 11,753 | 98 | 20,932 | 2,997 | 132,929 |
| Japanese Theatres | 48 | 8,436 | 16,961 | 86 | 25,531 | 1,742 | 1,202 | 2,253 | 9,685 | 40,413 |
| Totals(a) | 258 | 9,084 | 107,898 | 5,500(b) | 122,740 | 13,495 | 1,240 | 23,185 | 12,682 | 173,342 |

(a) Grand totals do not include Africa Middle East Theater, En Route and Not Chargeable to any Command, Caribbean been Defense Command and USAFSA.

(b) Since all prisoners of war known to be living have been returned to military control, the status of the above indicated personnel is now being investigated by the War Department.

SOURCE: "Battle Casualties of the Army" 1 January, 1946.

Strength Accounting & Statistics Office - 25 February 1946

~~TOP SECRET~~
~~CONFIDENTIAL~~

HEADQUARTERS
SUB REGION GOEPPINGEN
COUNTER INTELLIGENCE CORPS REGION I

File Nr. I-G- 820

APO 154
FIELD OFFICE ULM/D
4 August 1947

MEMORANDUM TO THE OFFICER IN CHARGE

Subject: American and Allied ~~soldiers held as prisoners~~

RE : Project 154032

### 1. Reason for Investigation.

Reference is made to Orientation and Guidance Report #2 dated 30 April 1947 re: Project 154032.

A casual informant, HOFMANN, Hans Joachim, recently released German Prisoner of War from Russia contacted CIC ULM (I49/x68) thru the local American Red Cross on 2 August 1947 to give information about American and Allied personnel held as prisoners in SIBERIA, USSR.                    (b) (6) & (b) (7) (c)

### 2. Results of Investigative Activity.

a.  Informant HOFMANN was released from the Dachau Prisoner of War Processing Center on or about 17 July 1947. The informant did not contact American authorities there, but German officials in DACHAU told informant to contact UNRRA in ULM. The informant did this and was referred to the American Red Cross Blub in ULM.

b.  The informant gave this agent three (3) pictures of American soldiers atta-ched hereto as exhibit A,B, and C.

c.  The informant claims to have talked with the soldier shown on the picture marked exhibit "A" whose name is BOEHM, Viktor ▮▮▮▮▮▮▮▮▮▮▮▮ emigrated to the United States of America in 1928, lived in ▮▮▮▮▮▮▮▮▮▮▮▮ According to the informant BOEHM served in the Army of the United States, was taken prisoner by the Germans and then taken over by the Russians on or about 5 May 1945 in OELMUETZ, Silisia, Germany and transported to SIBERIA on 17 July 1945 to the NOROSIBIRSK Camp 311 near KRASNOE, Siberia where BOEHM is presently working in a tank factory. The informant worked with BOEHM who speaks German. BOEHM gave HOFMANN the attached pictures.

d.  The informant claims there are two hundred (200) American soldiers working in this plant and about nine hundred (900) Allied soldiers, mostly English and French soldiers.

COL D. G. ▮▮▮▮▮▮▮
▮▮▮ HQ 65 ▮ CIC Det, 14 JULY ▮▮▮▮

~~CONFIDENTIAL~~

CON█████ENTIAL

HEADQUARTERS
SUB-REGION GOEPPINGEN
COUNTER INTELLIGENCE CORPS REGION I
EUROPEAN COMMAND
APO 154

File Nr. I-G-820
MOIC Subject: Americans and Allied Soldiers held by Soviet

e. On the back of the picture attached as exhibit
"B" is the address of Mr. JAMES E. GREEN ████████████
██████████████ who is also working in said tank factory
in Siberia.
    f. The picture attached as exhibit "C" could not
be identified by the informant.

    g. The informant knows the names of the following
Allied personnel:
    1. Sergeant RIEDEL,
    2. Schanno AMBROSINI,

also working in said tank factory.

    3. Agent's Notes.

    a. Evaluation of above information: F-6.

    b. The informant, ████████████████████████████
The informant claims to be the sons of a German colonel
Hofmann killed in TUNISIA by Gestapo agents.

    c. It is recommended to have CIC WIESBADEN, Region
III recheck informants statements thru a discreet investigat

    d. It is further recommended above information be
made available to the Adjutant General, Washington, D.C., the
F.B.I. and the American Red Cross. If the F.B.I. confirms
informants statements this matter should be of interest to the
State Department.

                                    John J Menken
                                    John J Menken
                                    Special Agent
                                    CIC
3. Incl.
    Exhibit A (Picture of Viktor Boehm)
       "    B (   "    of James E. Green)
       "    C (   "    of unidentified American soldier)

APPROVED:

                            REGRADED UNCLASSIFIED        48
                            ON   2 0 SEP 1991
            HERRING/        BY CDR USAINSCOM F01/PO
                            AUTH Para 1-603 DOD 5200.1R

ipo:s Of Harry S. Truman

ychological Strategy

ard

DECLASSIFIED    **TOP SECRET**    NLT (PSB) 306

85-62 (ASS    SECURITY INFORMATION
1-21-86)

DCD    10 2-10-86    UNC    D    28 December 1951

D - Mr. Gordon Gray

P - Wallace Carroll

Repatriation of Prisoners of War

1.  You will recall that in October we developed and sent to the Secretary
of State, Secretary of Defense, and Director of the Joint Staff a staff study
on the question of repatriation of prisoners of war.

2.  There can be no doubt that the forcible repatriation of prisoners of
war in Korea would have serious effects upon U.S. psychological warfare operations for many years to come.

3.  Our treatment of Soviet and satellite expatriates has an unfortunate
history, as you will recall. As a result of an agreement at Yalta, the United
States in the years immediately after World War II assisted the Soviet Union
in the repatriation of various categories of Soviet bloc persons—chiefly
prisoners of war, escapees, and displaced persons. The result of our cooperation was that more than four million Soviet citizens were returned to the
Soviet Union and that thousands were executed or punished in other ways without
regard to the conditions which caused their displacement from Soviet-controlled
territory.

4.  In addition, persons escaping from the Soviet area after World War II
were forcibly returned to Soviet control as a matter of U.S. policy up until
well into 1946. This treatment of Soviet expatriates became well known to
the populations within the Soviet area and, as has been well documented,
became the cause of widespread despair. It practically stopped the flow
of defectors, and it would make it very difficult to wage effective psychological
warfare against the Red Army in event of war.

5.  Our policy was subsequently changed so that persons entering the
Allied Zones of Germany and Austria are no longer forcibly returned. The
treatment we have given defectors has, however, not been sufficiently good
or well publicized to erase the former picture.

6.  This is the background within which the question of forcible repatriation of Chinese and North Korean prisoners of war must be examined.
Repetition of our previous mistake would discourage defection by Chinese
communist forces in any future conflict. It would therefore in the long run
cost us more American lives than are involved in the exchange of prisoners
problem.

7.  The latest JCS directive to General Ridgway on the subject indicates
that we are making an effort to exchange prisoners on a voluntary basis. (We
are not giving publicity to this fact, probably because we are not yet ready to

**TOP SECRET**

FEB 27 1959

UNCLASSIFIED
55

pers Of Harry S. Truman

ychological Strategy

ard

_J.H.C.B._

TOP SECRET

SECURITY INFORMATION

UN...

-2-

make the final decision).  The instructions to General Ridgway require that
should he be compelled to examine the desirability of forcible repatriation,
he must refer the matter for decision to Washington.

10.  I have learned by hearsay that the President is informed on this
question and is inclined to oppose forcible repatriation.

11.  Recognizing that there are factors to consider in making a decision
on this question other than the psychological, I believe nevertheless that
the great importance of the psychological factor should be emphasized at the
time when a decision is about to be reached.  I recommend that you discuss
this problem with the Board members and consider also discussing it with the
President.

R.F. Copy of this memo sent to J. Phillips,
State Dept. on 1/16/52 per Col. Davis.
cc

UNCLASSIFIED

TOP SECRET

SECURITY INFORMATION

P:PCDavis
JWCarroll:mas

# AMERICAN
# FOREIGN POLICY
## NEWSLETTER

NATIONAL COMMITTEE
ON AMERICAN
FOREIGN POLICY, INC.
211 EAST 43rd STREET
SUITE 2302
NEW YORK, N.Y. 10017
Telephone: (212) 687-9332

| OL. 9 — NO. 6 | $2.00 A COPY | DECEMBER 1986 |
|---|---|---|

Professor Hans J. Morgenthau
*Founder*

Professor George Schwab
*Editor*

## PROFESSOR ELIE WIESEL
(a member of the Executive Committee of the National Committee on American Foreign Policy)

*For your exemplary contributions to the cause of*

### HUMANITARIANISM AND PEACE
the National Committee on American Foreign Policy congratulates you
for having been awarded the

### NOBEL PEACE PRIZE

---

## Guiding Principles of American Foreign Policy
*Remarks by Ambassador Vernon A. Walters Before the National Committee on American Foreign Policy New York City October 15, 1986*

Thank you very much, Mr. Ambassador [Kellogg]. I am delighted to be here. This is an extraordinarily prestigious group, and I feel very honored that you have asked me to speak to you. I see in the audience many old friends and a number of colleagues from the United Nations here, and I will try to be as brief as possible because my experience has been that the speaker speaks about what interests him and the questions are what forces him to speak about what the audience wants to hear. I will, therefore, briefly run across the field of American foreign policy. .

First, my instructions from the President as far as the United Nations is concerned are to reform it and not to kill it. I think that's very important. If the United Nations did not exist, it would have to be invented. It has a number of shortcomings, but it also is unique and offers an opportunity to so many nations to keep contact with others. The smaller nations have relatively few diplomatic posts, and yet at the United Nations they can be in

(continued on page 2)

## United States Policy Toward South Africa: Why Did It Fail?
*by Rhoda Plotkin*

American foreign policy in southern Africa has failed to achieve its objectives. Originally hailed by its proponents as a realistic approach to the problems of political reform and peaceful change in South Africa and the region, the Reagan policy of constructive engagement now shares the unhappy fate of its failed predecessors. In terms of

(continued on page 5)

## The POW/MIA Question: National Honor and Foreign Policy
*A Letter to the National Committee on American Foreign Policy by John M. G. Brown*

National honor and national interests are not mutually exclusive. To the extent that national honor is undermined by actions that threaten the morale of citizen-soldiers, their families, friends, and others it becomes a matter of security and therefore also a foreign policy concern. What I am

(continued on page 12)

---

The POW/MIA Question: (cont'd from pg. 1)

alluding to is, of course, the POW/MIA tragedy in Southeast Asia.

As is well known, Washington's decision to quit Vietnam was accompanied by an endeavor to forget the immediate past that led to the U.S. involvement in Southeast Asia. In this attempt Washington violated the tradition of the citizen-soldier who, in return for performing honorably the profession of soldiering, with all that that implies, especially the willingness to die and the expectation of being honored by the nation instead of being abandoned. By continuing to transgress the tradition by not resolving the POW/MIA tragedy honorably, the United States is continuing to turn its back on the people who defend the country

and is thus continuing to court disaster. To prevent this from happening it is vital to recognize the tragedy, to recognize once and for all that for millions of Americans the war is not yet over, and to act accordingly.

Before suggesting a course of action, I shall briefly review several salient facts: (1) Of more than 500 Americans held prisoners by the Pathet Lao in Laos, not one has returned; (2) of hundreds captured or missing in South Vietnam, many have never been freed; (3) beyond those returned from North Vietnam in Operation Homecoming in 1973, many others are known to have been captured. (Of those returned, none was missing an arm or a leg or was disfigured by burns, ejections from flaming aircraft notwithstanding.) The presence of POWs

*[Handwritten note on Office of the Vice President stationery]*

From: Vice President [stamp] George Bush

Anthony (m law) to my Sep 8    Oct 2, 1986

(Save Bay —
A.D. Dek
& J.M.C.
Brown
sent with
William
Casey
at CIA HQ
Langley —

Dear Tony,

I was glad to get your letter. I am an admirer of Ross Perot and I'm glad your son-in-law is helping Ross.

The President & I, on the assumption that our men are alive, want every avenue checked out.

Thanks so much for that good letter. Sincerely, Geo Bush

---

In Vietnam has been corroborated by, among others, hundreds of live sightings on the part of U.S. prisoners, refugees, camp inmates, former South Vietnamese soldiers and officers, by Robert Garwood, the last U.S. prisoner to leave Vietnam in 1979, by highly suggestive satellite evidence and communications interceptions, and finally, the evidence submitted by the highly decorated U.S. officers Mark Smith, Robert Howard, and Melvin McIntire, orders to destroy incriminating evidence or risk destruction of career and character notwithstanding.

Because the present administration and past administrations have done next to nothing to resolve the problem I, a wounded veteran of the Vietnam War, a member of the National Advisory Board of Skyhook II Project—Account for POW/MIA's, and the chairman of the POW/MIA Committee of the National Vietnam Veterans Coalition, urge the President to appoint immediately a capable, prominent citizen to negotiate a successful resolution of the tragedy. That citizen must be helped by a competent office staff and by the vast intelligence capabilities at the disposal of the country.

The negotiator must be instructed to (1) reopen the bona fide offer made by Vietnam to "sell" an unspecified number of U.S. POWs, something President Reagan rejected in 1981 (see the editorial in *The Wall Street Journal*, August 19, 1986); (2) discuss the issue with Communist China; (3) discuss the tragedy with the Soviet Union. Evidence seems to indicate that the Soviet Union may be holding some of our Vietnam prisoners who were transferred because of their specialized electronic warfare training.

In short, what is being suggested here is nothing that is out of line. There are ample precedents. For example, from the depth of a far more severe defeat Chancellor Konrad Adenauer obtained the release of German army prisoners from Soviet labor camps. Now the last of the Bay of Pigs prisoners are being released.

What is unacceptable is a do-nothing policy or one that aims at a token release of- only two or three Americans who may well be branded as former deserters or collaborators.

---

## TORONTO SUN. 27 MAY, 1989

The Saturday Sun, May 27, 1989 11

# Dreadful military secrets

**WILLIAM STEVENSON**

"It seems unbelievable," begins a feature in the *U.S. Veteran News and Report.* "Since the end of the Vietnam war, the Vietnam veterans have contended that Americans were left behind. Scoffed at and harrassed, the veterans have been accused by government officials of using the myth of PoWs to hold onto the war. But if the United States was willing to forsake more than 20,000 men while possessing the most powerful military force in history ... "

Let me cut in here to say the authors of *A Secret that Shames Humanity* are no screwballs. One is a valued friend: John Brown, a combat infantryman in Vietnam; and Tom Ashworth, a former Marine helicopter pilot. They have been digging through bureaucratic secrecy and now come up with troubling questions:

"Where are the secret records concerning thousands of live Canadian, American and British fighting men who disappeared into the Soviet prison system at the end of World War II?

"Were those U.S. officers who handled the coverup on abandoned Allied prisoners then promoted to high rank in policy-making and intelligence positions because of their dangerous knowledge and the need to ensure silence?

"Did this become a pattern, ensuring that similarly involved officials in later wars would also rise rapidly and even become assigned to questionable covert operations of recent years?

"Have the Soviets and other communist governments used the complicity inherent in this coverup to compromise or blackmail key U.S. officials, thus endangering U.S. security ?"

To get such questions before the public, the authors have been obliged to publish some 12,000 words in *The U.S. Veteran*, having been opposed in their efforts to find someone bold enough to print their book. Their theme is said to be too potentially damaging to national interests. Their answer is that nothing can damage a nation more than the practise of abandoning its own men on the battlefield.

"The Soviet Union kidnapped and held in perpetual captivity tens of thousands of Canadian, American, British and Commonwealth PoWs," says Brown. "It was done with the knowledge of the highest level decision-makers in the United States and Great Britain ...

"Allied prisoners were the principle component in forcing mass-repatriations of Russians, Ukrainians, Cossacks and other Soviet citizens and in helping ensure British and American recognition of puppet Soviet governments in East Europe."

Brown and his co-author spent years burrowing through files labelled TOP SECRET long after there was any apparent need for secrecy. They began by looking for hard evidence of U.S. secret-intelligence on Americans left behind in Vietnam and found themselves on a paper-chase back through history.

They retrieved confidential messages between Moscow, London and Washington following the Yalta Conference in February, 1945 when Stalin, Roosevelt and Churchill met for the last time. Brown believes "it was then that the the West agreed to the forced repatriation of two million 'Soviet citizens' who were in Allied hands. The price was the return by Stalin of some 20,000 Americans and some 30,000 British in Stalin's hands."

Stalin never paid the agreed price.

Now the authors disclose cables between British Field Marshal Alexander and the supreme headquarters of Allied command in London (SHAEF) that show Alexander resisting orders to forcibly deliver up some 38,000 Cossacks to the Russian NKVD (now the KGB).

In the end, Alexander complied in order to get the Russians to release British and American soldiers. The bloody operations resulted in some British troops throwing down their rifles and refusing to carry out orders to force the Cossack families — wives and children as well as the soldiers — into the waiting Russian arms.

Alexander's cables in late spring 1945 reflect his growing realization that the terrible deed — Operation Keelhaul — served little purpose. For instance: "In Russian zone difficulty experienced in tracing British PoWs ... unconfirmed reports British/U.S. PoWs being evacuated Odessa by rail in boxcars ... " There were many top-secret messages like that.

Ten years later, British security files still bore traces of sensitivity in the form of a red tag: "This document is to be handled for official use only ... The term Forcible Repatriation cannot be used for dissemination to the public ... "

U.S. Gen. George C. Marshall was less willing to gloss things over. A month after victory in Europe, he cabled Gen. Dwight Eisenhower, the Allied supreme commander: "Concerned your report S-88613 that 25,000 U.S. prisoners still in Russian hands." But Marshall was thrown a distracting bone, and told to forget what had become "a national security matter."

Armed with paper clues that soon filled their basements, Brown and Ashworth concluded that the communist countries have always used prisoners as pawns, to be discarded when no longer playable. They depict world statesmen sucked into a game that, in Vietnam, gave the Hanoi government a strong bargaining hand before letting U.S. President Richard Nixon withdraw from that war.

They claim to have accumulated documentary evidence of the continued existence of thousands of American prisoners in enemy hands following the Korean and Vietnam wars.

Brown says, "Federal agents came to see me last month (April) and reminded me of the need for national security secrecy in these matters."

From this he concluded if U.S. publishers were being told the same yarn, he and Ashworth might as well print the bare details in the veterans' newsletter. "Then," he tells me, "maybe someone will let us publish the whole truth."

THE TORONTO SUN

Dear Mr. Brown,

Bill Stevenson wanted you to have a copy of his column.

Regards,

U.S. VETERAN NEWS AND REPORT

# A SECRET THAT SHAMES HUMANITY

**By John M.G. Brown**
**and Thomas W. Ashworth**

*ople want this war to be over and it won't
er, they feel, until all our boys are home."*

*otional, heartfelt, patriotic, the above
could have been uttered by any American,
rned over the continued plight of more
2,000 servicemen who never came home
Vietnam. It could have been any American,
wasn't. Instead, a Russian, an editor of a
newspaper, simply echoing the concerns
people in the aftermath of their war in
nistan.*

*Russians are in a quandary, unable to
, much less bring home, their fighting men.
of the missing may be in the United States
Canada, the Soviets claim, but these men
not fear punishment. "Think what an
onal issue the question of MIA's is for
icans...Do you think our missing soldiers
any less to us?" a Soviet diplomat asked in
nt Boston Globe article.*

*rching for its lost men as with the eyes of
, the Soviet Union may be sincere in its
ence. Still the irony is almost obscene. For
as Soviet Premier Mikhail Gorbachev
aches President Bush for assistance in his
MIA dilemma, U.S. Veteran brings you this
l documented report of a well-hidden
agenda which abducted and enslaved
than 20,000 U.S. servicemen and
onal Allied troops at the end of World War*

*evelation of Soviet malevolence? To some,
ps. To American officials, it has been a
ept secret, a labyrinth of deception and
al, a child born of compliance, locked
and grown evil in its whitewashed
nment. They all knew, all those heroes of
and freedom: Eisenhower, Truman,
hill, Marshall.*

*eems unbelievable. Since the end of the
am War, veterans have contended that
icans were left behind. Scoffed at and
ssed, the Vietnam veteran has been
ed, by government officials and World War
rans, of using the myth of the POWs to hold
he war. But if the United States was willing
rsake more than 20,000 men while
sing the most powerful military force in
y, it hardly seems inconceivable that it
leave men behind when running away
a war it didn't want to be in anymore.*

*accept the following report is to forego
n instinct, for it invites belief in a history
ut illusion. But this is the way the world
ted then, and this is the way the world
tes today.*

*re, then, are the facts to undo the lie
ne wanted to believe.*

the end of the Second World War, the Soviet Union,
ith the knowledge of the highest level decision-
akers in the United States and Great Britain, kid-
and held in perpetual captivity tens of thousands of
an, British and Commonwealth POW's, along with
ds of thousands of Western Europeans.

unknown number (with smaller augmentations from
nerican wars in Korea and Indochina) could still be

The calculated Soviet action of withholding U.S. and British POWs for blackmail purposes had its roots in the Russian use of the slave labor empire known as the GULAG in which millions of anti- communist Russians, Ukrainian peasants and nationalists, Baltic peoples and other nationalities had already been shot, starved or labored to death before the Second World War.

In relation to this era, official U.S. documents uncovered indicate that American prisoners, including soldiers of the 339th U.S. Infantry captured while serving in the 1919 "Intervention" forces near Archangel, Russia, were also retained in the GULAG for the rest of their lives-with the knowledge of the U.S. government. Livesighting intelligence of American POWs in Soviet prisons such as the Lubianka in Moscow and Solovetz Island in the White Sea and near Vladivostock in Siberia continued at least into the 1930's. (Note: Orig. War Dept. Doc.)

It is also now clear that Lenin's intelligence commissar, Felix Dzerzhinsky, had specifically targeted these men for capture and had assigned Aleksandr Eiduk as the chief Chekist agent for operations against Americans in the Archangel area during 1918- 19. (Note: see for example, Secret Intelligence, pp 3-15, Volkman and Baggett, N.Y. Doubleday, 1989).

The American National Archives have now been searched by researchers, including the authors, originally involved in investigating the fate of American prisoners left behind in Indochina.

Many records are still security-classified after 44 years, but a large volume of evidence, aggregating several thousand formerly- secret documents, some possibly misfiled, have been obtained, nevertheless.

This research, coupled with information obtained from interviews with participants, chronicles a long-hidden Soviet blackmail operation at the end of the war, utilizing tens of thousands of American, British and other Western Allied POWs as pawns in a postwar power struggle. Those involved include Joseph Stalin, Franklin Delano Roosevelt, Winston Churchill, Harry Truman, Dwight D. Eisenhower, Clement Atlee, Anthony Eden, Edward Stettinius, Henry Stimson, Earnest Bevin, George C. Marshall, Sir Archibald Clark Kerr, Charles Bohlen, George Kennan, Vyacheslav Molotov, Andrei Vyshinsky, Laurenti Beria, Andrei Gromyko, "Kim" Philby, Guy Burgess and many other Allied functionarys.

Allied prisoners served as a principle component in forcing the mass-repatriations of Russians, Ukrainians, Cossacks and other Soviet citizens and in helping to ensure American and British recognition of Soviet puppet governments in Poland and elsewhere in eastern Europe.

The Western Allied concern over the future liberation and return of POWs by the Russians inspired an American and British decision on the methods of eventual repatriation of Russians under their control. By the fall of 1944, British Foreign Secretary Eden was already prominent among those advocating future forcible repatriation of Soviet citizens to obtain Stalin's reciprocation with U.S. and British POWs.

After the agreement signing at the Yalta Conference in 1945 by an exhausted and ill Pres. Roosevelt-whose advisors included the aforementioned Stettinius and Bohlen, as well as Averell Harriman and Alger Hiss-British and American officials initiated the repatriations-forced and otherwise-of over two million Soviets despite fears that Stalin might renege his promise to return all Allied prisoners-particularly after learning that the Allied POW's repatriation would be under control of Beria's NKVD (KGB).

In a short time, the fears became reality and by June, 1945 American and British citizens were informed that all prisoners were repatriated-notwithstanding the estimated 20,000 Americans and 30,000 British still held under Soviet Control.

The Soviet betrayal of its promise poisoned its future relationships with western countries and helped to inspire the sudden onset of the Cold War. Although there is some evidence that a small number of Americans and British were secretly returned in later years, the majority of the prisoners-including hundreds of thousands of French, Belgian, Dutch and other western Europeans-were never returned.

## KIDNAPPINGS IN POLAND

In January and February, 1945, the Soviet Army overran Poland, relinquishing the Nazi grip on the small country whose invasion had invited Britain and France to war.

The Russians had spent part of the war preparing for this moment, gradually forcing defacto Allied recognition of their communist puppet regime. The Soviet's groundwork paid off, even causing Anthony Biddle, the U.S. ambassador to Poland since 1939, to resign in 1944 over his disenchantment with the Roosevelt administration's gradual submission to Stalin's desires on Poland following revelation of the Katyn Forest massacre by the Russians and diplomatic disagreements

The United States had been fortunate in successfully
cuing by air-evacuation (They received local assistance f
King Michael of Romania) about 1,000 of the Ameri
prisoners in Romania and Bulgaria before the takeover by
Red Army.

The Allies left behind stay-in-place agents of the Offic
Strategic Studies and a British Special Operations Execu
to monitor the Soviet takeover. Angered, the Russians
dered them to leave, wanting no witnesses to their merci
methods of establishing total communist control in areas t
occupied. This may have influenced later Soviet conduc
Poland, while also giving the Soviets a better chance to s
U.S. and British POWs as hostages.

In Poland and East Prussia the Germans forced hundr
of thousands of Allied POWs and civilians to retr
westward with them but many were left behind and taker
the Red Army. In January and early February, 1945, prio
the Yalta Agreement on repatriation of prisoners,
American Embassy in Moscow began receiving reports of
prisoners in Soviet hands through U.S. intelligence, Sen. C
nally and the Polish ambassador to the United States,
Ciechanowski. Ambassador Harriman, his No. 2 George K
nan and Gen. John Deane, Chief of the U.S. Military Miss
had already begun responding with demands for informa
from the Russians.

Upon receiving evidence that many Americans had b
held incommunicado by the Soviets, who hadn't revealed
fact, Kennan, as charge d'affairs, demanded contact with
prisoners and subsequently suggested a stiff protest
Foreign Commissar Molotov.

Another foretaste of trouble occurred when the Emb
received a note dated February 14, 1945, signed by
Brooks, which read, "Here in "Nowy Sacz" are eight offi
and twelve NCO's (non- commissioned officers) all of wh
are mostly Americans and British airmen. At the moment
are with the Russian Army in Nowy Sacz for over three we
and no responsible parties have been notified of our p
cence and whereabouts. We have no freedom and have b
told we are internees."

Contrary to the Yalta Agreements, the Soviets attemp
to hide their possession of these and thousands of ot
prisoners until a lucky three Americans escaped eastwa
eventually reaching the Military Mission at the U.S. Emb
sy in Moscow where they were debriefed by Gen. John Dea
More handwritten notes from other groups of prisoners c
tinued reaching the American and British embassies i
March, 1945.

There followed a state of continual crisis in the Emba
for over three months as the Soviets under Stalin and Be
particularly Foreign Minister Molotov, his deputy (Stal
1930s purge prosecutor) Andrei Vishinsky and repatriati
generals Golubev and Golikov (a high level intelligence a
to Stalin), lied, stalled and accused the Americans them
ves of perfidy in POW handling.

(The Russians also simultaneously demanded recognit
of their Polish regime and indicated their expectations c
postwar $6 billion war recontruction "credit".)

U.S. and British POWs from German camps such as Sta
II-B, III- C, Stalags VIII-A, B and C, Stalag XX-A and OFL
64, among others had been left behind or escaped during
German retreat and overtaken by the Red Army. Thousa
were being withheld by the Soviets instead of be
repatriated through Odessa, on the Black Sea, as had b
promised following the Yalta Agreements.

The withholding of American prisoners had become
blatant that Harriman prevailed upon Pres. Roosevelt
make a personal appeal to Stalin on March 4 to all
American planes into Poland to rescue U.S. POWs. St
refused, denying that any groups of Americans remained
Poland.

On March 8, 1945 Harriman cabled the president
23119): "...information received from our liberated prison
indicates there have been four or five thousand officers
enlisted men freed. The Russians claim today there are o
2100..." In a March 14 cable to Stettinius, also read by P
Roosevelt and Gen. Marshall, Ambassador Harriman clea
saw the political implications, (#738):

"I assume the Department has been informed by the W
Department of the great difficulties General Deane and I h
been having in regard to the care and repatriation of
liberated prisoners of war....I FEEL THAT THE SOVI
GOVERNMENT IS TRYING TO USE OUR LIBERAT
PRISONERS OF WAR AS A CLUB TO INDUCE US TO GI
INCREASED PRESTIGE TO THE (communist) PROVISION
POLISH GOVERNMENT BY DEALING WITH IT IN THIS CO
NECTION..."

At this time Stalin and Molotov were, in fact, sim
taneously demanding American and British diplomatic rec
nition of the Russian-sponsored Polish Lublin-Wars

Veterans of Foreign Wars of the United States Magazine • February 1990 • Vol. 77 No. 6

## VETERAN VIEWS

# Hidden POWs of the Cold War: One Man's View

By John M.G. Brown

A series of signals from the Soviet Union asking for U.S. assistance in repatriating some 300 Soviets missing in Afghanistan and Pakistan offers a unique opportunity.

Reports from Soviet diplomatic sources appearing in the American press in 1989, indicating that Soviet leader Mikhail Gorbachev had discussed this matter with President George Bush, demonstrates possible linkage in accounting for Russian MIAs in Afghanistan and American POWs and MIAs from the Vietnam War as well as previous wars.

A number of U.S. prisoners may have been transferred to the Soviet Union from Indochina, and documentary evidence exists of the transfer of large numbers of U.S. POWs of the Korean War to the Soviet Union. The U.S. demanded their return in May 1954, but the Soviet Union denied holding American POWs "under guard."

And the existence and disappearance of American and British POWs under Soviet control in 1945 has been documented by this writer through declassification of thousands of formerly secret U.S. cables and reports in the National Archives.

Although many Freedom of Information Act (FOIA) requests have been denied me for "national security" reasons, some records have been obtained through FOIA. Interviews with 1945 participants have clarified other aspects of this hidden aspect of the Cold War.

In 1945 the Soviet Army liberated over a million Western Allied POWs and displaced persons in Nazi camps in Poland, eastern Germany and Austria. According to the Yalta Agreements they were to be repatriated to their native lands.

While some were relatively well-treated by front-line Russian troops, documentary evidence details how over 500,000 of these prisoners were forcibly retained by Soviet authorities. Included were hundreds of thousands of French, Belgians, Dutch and ethnic Jews who had survived the Nazi concentration camps.

Among them were tens of thousands of American, British and Commonwealth POWs whose kidnapping became a high-level national security secret. American and British prisoners were also among the more than 4,500,000 German and Japanese POWs retained as forced labor by the Russians.

On May 23, 1945, two weeks after the end of the war in Europe, Gen. Dwight D. Eisenhower's chief POW negotiator with the Soviets, Maj. Gen. Ray Barker, stated in a secret report: "The SHAEF (Supreme Headquarters Allied Expeditionary Force) representatives came to the firm conviction that British and American prisoners of war were, in effect, being held as hostages by the Russians ... "

Recently declassified U.S. and British documents record details of a Soviet "blackmail" operation executed by Soviet leader Joseph Stalin and his cohorts Molotov, Beria, Golikov and others.

Their object was to obtain Allied compliance in the coerced and forcible repatriations of more than 5,000,000 Russians, Ukrainians, Cossacks, Poles, Balts and other eastern Europeans. They also demanded U.S. diplomatic recognition of Communist regimes in Poland and elsewhere in Eastern Europe.

An additional benefit to the Russians was to be the labor of thousands of technologically advanced (by Soviet standards) Allied prisoners. The actions of highly placed Soviet spies in the British and American governments severely compromised the Allies and greatly assisted the Soviets in this operation.

Repatriation of POWs was a secret topic of the 1945 founding United Nations Conference in San Francisco, the 1945 Hopkins-Stalin meetings and the Potsdam Conference. It appears that Field Marshal Alexander, Gen. George Patton and to a lesser extent others, including Gen. Eisenhower, resisted this chain of events prior to and following the unsuccessful September 1945 foreign ministers meetings.

Well after the war, Alexander cabled London: "Difficulty in tracing BR/US PWs in Russian zone." He reported that U.S. and British

Though most American prisoners held by the Germans were liberated by U.S. troops (as shown here), possibly as many as 20,000 U.S. POWs were not repatriated from Soviet-controlled territories. U.S. Army

POW's were being shipped east into Russia "*in boxcars with German PWs.*"

Meanwhile, Patton's Third Army had located thousands more American and British prisoners being withheld after the war by the Soviets in Austria. According to once-top secret U.S. documents, decisions were first made by President Franklin D. Roosevelt and then by President Harry Truman that the potential losses of a 1945-46 war with Russia were not worth the possible recovery of some 20,000 American and 20,000 or more British and Commonwealth POWs.

Implementation of this policy was passed on to the responsible departments and agencies. Evidence has surfaced of the secret return of Allied prisoners for the forcibly repatriated Soviet citizens of whom Soviet dissident Alexander Solzhenitsyn, Lord Bethell and Nikolai Tolstoy have written so eloquently.

Repatriations to Stalin's GULAG were marred by mass suicides. Stalin reneged on full reciprocation and most of the Allied POWs disappeared into secret, special camps. Stalin's failure to reciprocate in the repatriations of 1945-47 resulted in an official tendency to minimize the extent of Allied losses.

In classified U.S. and British documents, the total number of prisoners known to be in Soviet hands was scaled down to some 5,500 Americans and 8,500 British and Commonwealth troops. Continuing research indicates that these figures reflect a shifting of known prisoners into other categories. All told, some 78,000 Americans remain unaccounted for from WWII.

The acute need for intelligence inside Soviet territory to locate the large number of missing Allied prisoners for possible secret rescues or negotiations led to U.S.-Soviet air battles in Europe and U.S. naval threats to the Russians in Manchuria.

This crisis led to the first major American covert actions after WWII inside Soviet territory executed by the Office of Strategic Services (OSS), Army G-2/SSU, the State Department's Office of Policy Coordination, and later the CIA.

Among other assets, they used existing German intelligence networks. The perceived need for extreme security in this matter helped create a chain of events leading to the present debate on the limits of secrecy in a free society.

The latest Soviet statement of concern for the return of their missing Afghan War prisoners should be viewed as a public diplomatic signal of their willingness to consider an exchange. This should not, however, be permitted to become a forced-repatriation of those 40-50 former Soviet prisoners known to be living in freedom in the U.S. and Canada.

What would be in order, however, is the use of American influence to gain control of those Soviet POWs in Afghanistan and Pakistan who wish to return home.

Given the present *glasnost* mood and Gorbachev's statements on Stalin's crimes, it might finally be to the Soviet Union's advantage to reveal the truth about American POWs from WWII.

---

*John M.G. Brown, a former combat infantryman with the 1st and 9th Infantry Divisions in Vietnam, is a Life Member of Post 2886 in Seattle, Wash. and author of Rice Paddy Grunt (1986).*

---

THE NEW AMERICAN / MAY 21, 1990

AMERICAN OPINION

## Mikhail Gorbachev, Let Our People Go

### First of a two-part series

Soldiers in the American intervention forces were targeted for capture by Lenin

J O H N   M. G.   B R O W N
with research colleague Thomas V. Ashworth

*John M.G. Brown, a combat infantryman with the 1st and 9th Infantry Divisions in Vietnam, has been investigating the POW/MIA issue for seven years. He is the author of Moscow Bound: Policy, Politics, and the POW/MIA Dilemma of the Korean/Vietnam War and the anti-war movement in the 1960s.* *Mr. Brown's book, published in 1989 by Veteran's Press and recently updated, dramatically expands the information in this subject, for inclusion in a forthcoming book, at Veteran's Press. (Additional information contact Rice Paddy Grunt. A complete list of additional persons and documents of historical significance should be sent certified, return receipt.)*

Holding American, Allied, and other prisoners as hostages — for intelligence purposes — has been a Soviet-communist policy since 1919. Hundreds of American civilians and U.S. military hostages of the victorious Bolsheviks, along with others arrested or detained later, were used by the Soviets in 1921 to obtain $100 million worth of desperately needed food supplies under the authority of war relief director (later President) Herbert Hoover. In his memoirs Hoover listed as the first minimum condition for U.S. aid: "freedom of all American prisoners in Russia."

According to an October 21, 1920 Associated Press report, the total number of American citizens in Russia was estimated at 3000 (including some captured American soldiers). Certainly, at this time, hundreds of native-born American civilians and U.S. military prisoners were being held hostage by the Soviets. One of them, a Red Cross official [and U.S. Army officer in WWI] named Em-

mett Kilpatrick, captured with men of Wrangel's White Army,* smuggled a letter out of a Cheka (secret police) prison: "... I am now held in prison as a hostage for one Jim Larkin [a communist agitator imprisoned in the U.S.] now serving a sentence of twenty years, and the same has been awarded me."

Western European war prisoners and civilians were also being held hostage by the Soviets at this time, in an effort to gain concessions from their home governments. The U.S. State Department sought to put the matter to rest with a public announcement on October 22, 1920, a day after the above quoted AP story: "Bolshevist figures on the number of Americans in Russia are 'off' by about 3000, according to

*Baron Pyotr Nikolaevich Wrangel led a White Russian force of counterrevolutionaries fighting against the regime established in Russia by the Bolsheviks in 1917. Following his appointment as commander-in-chief of the monarchist forces in April 1920, Wrangel established a provisional government in the Crimea and scored several victories against the Soviet troops. His defeat late in 1920 marked the end of the White Russian opposition to the Soviet revolutionary government.*

State Department records, which put the correct number at about 35."

### Death Warrant

The State Department claimed that the others were Russian-Americans, and that "all the Americans in Russia remained there of their own free will when Ambassador Francis left that country [July 1918]." Recently uncovered official documents clearly indicate that this 1920 statement served as a death warrant for those American military POWs eventually left behind in Soviet prisons.

Although many reports and live sightings remain classified or hidden, official War Department and FBI-related documents, recently uncovered in the National Archives, indicate that at least 40-50 American prisoners of war — including soldiers of the 339th U.S. Infantry, captured while serving in the 1918-1919 Intervention forces near Archangel, Russia — remained imprisoned in Soviet gulags for the rest of their lives. Live sightings of American officers and enlist- 4 military POWs in Soviet prisons — such as the Lublianka in Moscow, and the Solovetski Islands slave-labor camps in the White Sea area — continued at least into the 1930s.

Some of the soldiers were identified by name, by an eyewitness who eventually reached the U.S. 1919 U.S. casualty reports. It is now clear that Lenin's intelligence commissar, Felix Derzhinsky, had specifically targeted these men for capture, and had assigned Aleksandr Eiduk as the chief Chekist (later GPU-NKVD) agent for operations against: Americans in the Archangel area during 1918-1919. The kidnapping of hostages was a Cheka-Bolshevik trademark.

During the December 1918 founding meeting of the Comintern, Soviet sources record that American, English, and Scottish POWs captured on the Archangel front were used for propaganda purposes. At least partially in retaliation, in 1919-1920 the Allies withheld thousands of Russian POWs on the Western Front.

On August 17, 1921 Lewis S. Gannett wrote that (contrary to State Department claims) at least 15,000 Russian-Americans were being held as hostages in Russia. The actual numbers of native-born Americans in Russia appears to have been kept secret by the State Department and U.S. Army Intelligence.

Approximately 100 of the

Americans — including Emmett Kilpatrick, and a former U.S. Army intelligence officer, Captain W. H. Estes — were eventually released from Russia following the promise of ransom shipments from the U.S. In September 1921, whereupon they revealed horrifying details of mass executions in Soviet prisons. Others, including secretly withheld U.S. Army prisoners of the 1919 Intervention, never reappeared. By retaining military and civilian prisoners, the Soviets were able to thoroughly analyze their adversaries' armed forces and intelligence services. The Soviets also kidnapped Bavarian negotiators and demanded the repatriations of Russian prisoners then in Germany.

### "You No Longer Exist"

British prisoners were also held hostage by the Soviets, among them a British Embassy official, R.H. Bruce Lockhart. Some, including Lockhart, were eventually exchanged for Soviet officials then held in England, including the later Soviet Foreign Minister Maxim Litvinov. Some were declared dead, though they were in fact being used by Soviet intelligence. KGB defector Ilya Dzhirkvelov wrote from Britain in 1987 of his work in the secret KGB Archives in the early 1950s. He came across clear evidence that the famous British secret agent of this period, Sidney Reilly, long remained alive in Soviet secret police control after news of his death had been published and accepted in Britain. These reports were shown to him to gain his ultimate capitulation. Artuzov, his secret police interrogator, had then told Reilly: "You no longer exist in this world. Your hopes of being freed will come to nothing...."

This tactic for gaining the cooperation of prisoners was repeatedly used by the NKVD-KGB to succeed in getting, all the time, every reason imaginable, that it was also used on the abandoned U.S. prisoners of the 1918-19 Intervention and on those abandoned in each succeeding war involving Americans.

Throughout the 1920s and 30s, Lenin's successor, dictator Josef

Stalin, imprisoned ever-increasing numbers of Russian "counterrevolutionaries." Western Europeans, and other foreigners. The great majority of the estimated five million pre-WWII prisoner-slaves who died in the forced labor camps were Russian and Ukrainian peasants and other ethnic minority peoples of the Soviet Union who had resisted collectivization and harsh communist rule.

The growing use of costless state slaves provided the Communist Party elite with a vast source of export and consumer goods requiring no initial capital investment. U.S. media apologists for the communist regime in Moscow, and powerful and influential businessmen who desired to trade with Russia, discounted reports of slave labor. During this same period, Stalin also greatly expanded the Soviet foreign intelligence network inside the western democratic nations. This was to become highly significant during and after WWI.

Following the secret 1939 Hitler-Stalin Pact, which divided up Eastern Europe, Germany and Russia both attacked eastern Poland. In the ensuing occupation of eastern Poland, Stalin's NKVD secret police deported two million "anti-communist" Poles to forced labor camps in Siberia. According to documents in our National Archives, the American OSS and Military Intelligence and their British counterparts carefully studied top-secret intelligence reports of this action, and of the Soviet massacre of 15,000 Polish Army officers at Katyn Forest, to assess Stalin's possible future conduct with military POWs. The Soviet Union had refused to sign the pre-war Geneva convention on treatment of POWs.

By 1944 British Foreign Secretary Anthony Eden, Winston Churchill, and others (including, to a lesser extent, American government officials) were already worried about the repatriation of hundreds of thousands of Allied POWs held by the Germans in the east, who would ultimately be liberated by the Red Army. These concerns increased following the merciless German

THE NEW AMERICAN / MAY 21, 1990

19

20

# The Washington Post

SATURDAY, SEPTEMBER 15, 1990     A21

## The POW Conspiracy

At the close of his article "No MIA Conspiracy" [op-ed, Aug. 24], Stephen Rosenfeld wrote "my own sense of things is that the Americans who were taken prisoner in Vietnam have been honorably served by those charged with pursuing their fate. Does that make me part of the conspiracy?"

The answer is that nearly all of the "hundreds and perhaps thousands of officials" Rosenfeld referred to did play a part, usually minor, in ensuring that the full truth on American and Allied POWs held by the Soviets and their surrogates after World War II, Korea and Vietnam would not be revealed to the American public. The matter has been a national security concern since the development of the Soviet nuclear threat to the very existence of the United States. All the officials and officers who have taken part in suppressing the truth of Soviet-surrogate-held American POWs have faced a painful moral choice while doing their lawfully ordered duty.

Based on thousands of declassified official documents and supporting eyewitness accounts obtained during seven years of investigation and research, it is clear that thousands of American POWs have been secretly held by the Soviets and their surrogates since 1945. The numbers of Americans thus illegally withheld appear to have been approximately 20,000 in 1945, 4,000 to 5,000 (or more) after 1953 and perhaps as many as 1,000 after 1973. These Americans, and many thousands more Allied POWs, were withheld for ransom and blackmail, used as forced labor and for espionage purposes. In addition I can document the secret withholding of American POWs of the 1918-1920 intervention in Russia in Lubyanka prison in Moscow and the Solovetsky Islands GPU forced-labor camps until at least the 1929-1930 period.

In answer to Rosenfeld's final question: All major-circulation U.S. publications I have contacted, with the exception of the VFW Magazine, have declined to publish documentable facts of the history of U.S. prisoners of war held by Communist nations. These publications have apparently declined to even investigate the matter seriously. It is therefore difficult to escape the conclusion that the editors of such publications have avoided publishing such historically significant material in the interest of some higher political purpose.

Rosenfeld accused people like me of "turning on our own government." It might be in order to state that my family helped found and protect the revolutionary government of the United States in 1776 that I in my turn served that government in front-line combat and that it was the intention of the Founding Fathers that our government be constantly held accountable to the people. That is the principle for which the tens of thousands of missing American POWs fought, and that is why people such as myself consider that we are merely doing our duty toward our fellow American servicemen by uncovering the truth about their fate.

—*John M. G. Brown*

REPLY TO:

☐ 135 HART SENATE OFFICE BUILDING
   WASHINGTON, DC 20510-1501
   (202) 224-3744
   TTY: (202) 224-4479

☐ 721 FEDERAL BUILDING
   210 WALNUT STREET
   DES MOINES, IA 50309-2140
   (515) 284-4890

☐ 206 FEDERAL BUILDING
   101 1ST STREET S.E.
   CEDAR RAPIDS, IA 52401-1227
   (319) 363-6832

**United States Senate**

CHARLES E. GRASSLEY

WASHINGTON, DC 20510-1501

REPLY TO

☐ 103 FEDERAL COURTHOUSE BUILDING
   320 6TH STREET
   SIOUX CITY, IA 51101-1244
   (712) 233-1860

☐ 210 WATERLOO BUILDING
   531 COMMERCIAL STREET
   WATERLOO, IA 50701-5497
   (319) 232-6657

☐ 116 FEDERAL BUILDING
   131 E. 4TH STREET
   DAVENPORT, IA 52801-1513
   (319) 322-4331

October 10, 1990

Mr. John M.G. Brown
Box 30
Petrolia, CA  95558

Dear Mr. Brown:

I wish to thank you for your generosity in contributing your research and many long hours of work to our POW/MIA investigation. The documents you provided pertaining to POWs of World War II, the Korean War and the Vietnam War will offer us an invaluable historical context for understanding this issue as we seek the truth and a resolution of the matter.

As you request, let me give you my assurance that I and my staff will not permit any copies to be made of these documents, nor will we permit access to them, without your express approval. We are grateful that you have donated this effort to our cause and I will do whatever I can to ensure that your request is honored.

Once again, thank you for your outstanding work and generosity. Please let me know if I can be of help in your future endeavors.

Sincerely,

*Chuck*

Charles E. Grassley
United States Senator

Committee Assignments:

APPROPRIATIONS          JUDICIARY          SPECIAL COMMITTEE ON AGING
BUDGET                  SMALL BUSINESS     OFFICE OF TECHNOLOGY ASSESSMENT

# The Sunday Oregonian

FORUM

PORTLAND, OREGON, DECEMBER 2, 1990

# U.S. has historically neglected POW/MIAs

Documents describe
incidents back as far
as formation of U.S.S.R.

By JOHN M.G. BROWN

The revelations in both the new book, "Kiss the Boys Goodbye: How the United States Betrayed Its Own POWs in Vietnam," and a recently released Senate report charging that the United States abandoned POW/MIAs after the Vietnam War, while important, lack the broader historical perspective.

Such actions would be consistent with U.S. policy in dealing with communist regimes dating to the formation of the Soviet Union and continuing through the aftermath of World War II, Korea and Vietnam.

● 1918-1919 — A study of coded cables and other now-declassified documents in the National Archives indicates that many of the roughly 125 Americans missing in action from the 1918-1919 U.S. intervention forces in Archangel, Russia, were secretly confined as prisoners of war.

About 100 U.S. civilian hostages seized during the Bolshevik Revolution and Soviet civil war were released in a 1921 grain-for-hostages exchange negotiated by future President Herbert Hoover, but the U.S. military POWs in Russia disappeared.

A released prisoner of the time, Emmitt Kilpatrick, reported, "Prisoners captured on the battlefields of the late war are rotting in prisons, all records of them being destroyed and their existence forgotten."

In 1930, about 40 to 50 American POWs of the intervention force were reported by an eyewitness escapee as being alive in the Lubyanka Prison in Moscow and the Solovetsky Islands forced-labor camps in the White Sea.

*John M.G. Brown of Petrolia, Calif., is a Vietnam veteran and the author of "Rice Paddy Grunt," (Regnery-Gateway, 1986). He has been investigating POW/MIA matters for seven years.*

FBI Director J. Edgar Hoover pursued the matter with the War Department and Army G-2 intelligence, but the Army had already assigned human skeletal remains found by U.S. Graves Registration to the names of American POWs now reported alive in Stalin's gulag prisons. In addition, the eyewitness reported that requesting the return of the prisoners by name would result in their immediate execution by the Soviets.

In 1933 the United States extended diplomatic recognition to the Soviet regime, without recovering the prisoners.

● World War II — Thousands of recently declassified World War II-era documents uncovered by this writer (and independent researcher Thomas V. Ashworth) in the National Archives indicate that at the end of that war as many as 60,000 American POWs held by the Germans were overrun by the Soviet Army in Poland, eastern Germany and Austria. About 20,000 American POWs and hundreds of thousands of Allied POWs in Soviet control subsequently disappeared behind the Iron Curtain.

In March and April 1945, President Roosevelt and, subsequently, President Truman made policy decisions against the use of military force to recover U.S. POWs in Soviet control. The decisions were based on American hope for Soviet participation in finishing the Japanese war and in the United Nations.

Formerly secret U.S. and British documents indicate that weeks after VE Day, thousands of American and British POWs were being, in the words of Gen. Dwight Eisenhower's chief POW negotiator, Maj. Gen. Ray W. Barker, "held as hostages by the Russians." Barker noted in a secret report of the time that "it would be necessary for visits of inspection to 'uncover' our men."

Other documents state that this constituted a Soviet effort to blackmail the Allies on Stalin's demands for coerced and forcible repatriations of all Soviet citizens, including millions of Ukrainians, Poles, Balts and other East Europeans, and for Allied diplomatic recognition of Soviet puppet regimes in Poland and elsewhere in Eastern Europe. The Soviets simultaneously demanded

$6 billion in war reconstruction aid from the United States.

The Soviets later refused U.S. contact officers access to all but one of the many camps. Simultaneous intelligence reports indicated the shipment of American, British, French, Belgian and Dutch prisoners of war, sometimes with German POWs, east into the Soviet Union, where many hundreds of thousands disappeared.

Stalin had reneged on reciprocal POW repatriations.

This documentary evidence on a secret cause of the Cold War has now been substantiated by information from still-living American eyewitnesses who escaped from Soviet-controlled prison camps, but were ordered to remain silent about many U.S. POWs left behind. They were told that their reports of American POWs in Soviet control would be investigated.

Some American POWs were reported alive in the late 1940s and early 1950s in Stalin's gulag camps, in now-declassified live-sighting reports from returning German and Japanese POWs. Other American POWs were reportedly held in secret, "unregistered lagers ... from which confinees will never be repatriated." A number of live-sightings of American prisoners in Soviet gulags have remained classified.

● Korea — More than 8,000 Americans still remain missing from the Korean War. Gen. James A. Van Fleet, retired commander of the U.S. 8th Army in Korea, said on Aug. 7, 1953: "A large percentage of the 8,000 American soldiers listed as missing in action in Korea are still alive."

The Pentagon at one time acknowledged that just under 1,000 prisoners were known to have been alive in captivity but never released. U.S. documents reveal that this already minimal figure was then gradually reduced over the years to 389, by matching names with human skeletal remains later found on South Korean battlefields by U.S. Graves Registration teams (and recently a handful of remains from North Korea).

Please turn to
POWS, Page D4

PAUL KOLSTI

HE SUNDAY OREGONIAN, DECEMBER 2, 1990

Founded Dec. 4, 1850. Established as a daily Feb. 4, 1861. The Sunday Oregonian established Dec. 4, 1881. Published daily and Sunday by the Oregonian Publishing Co., 1320 S.W. Broadway, Portland, Oregon 97201

# POWs: Manchurian 'peace' and 'reform' camps staffed by Soviets

■ Continued from Page D1

Documents from the CIA and other sources indicate that from 1951 to 1954, about 4,000 to 5,000 (American prisoners were in the process of being transferred from North Korea to Manchuria and into the Soviet Union.

The Chinese and North Koreans demanded forcible repatriations by the United States of many thousands of their nationals who refused to return to communist control. Recently declassified top-secret U.S. Psychological Strategy Board reports indicate that Stalin's previous reneging on reciprocal returns made U.S. officials reluctant.

One such document, dated Dec. 28, 1951, states: "Repetition of our previous mistake would discourage defection by Chinese communist forces in any future conflict. It would therefore in the long run cost us more American lives than are involved in the exchange of prisoners problem."

Despite initial Pentagon resistance, a new U.S. policy of "voluntary repatriation" was agreed on by early 1952. But the communists, in any case, already had decided to keep many American prisoners captured in Korea.

U.S. intelligence was aware of secret, special interrogation "peace" and "reform" camps inside Chinese Manchuria for Americans judged "pro-communist" and "anti-communist" staffed by Soviet intelligence officers. According to a July 1952 declassified CIA report, "Prisoners in peace and reform camps will not be repatriated."

Now-declassified CIA reports indicate that large numbers of U.S. POWs were known to have been shipped from Korea to areas of Communist China.

An eyewitness reported American POWs being shipped into Siberia from Manchuria, as of March 23, 1954, Foreign Service report (generated by Hong Kong U.S. air attaché [Col. Delk Simpson]) indicates:

"A recently arrived ... refugee from Manchuria has reported seeing several hundred American prisoners of war being transferred from Chinese trains to Russian trains at Manchouli near the border of Manchuria and Siberia ... large numbers of negroes among POWs."

The Eisenhower administration demanded the return of Korean War prisoners in the Soviet Union on May 5, 1954. On May 13, the Soviets denied holding American POWs "under Soviet guard."

A few American civilian prisoners returned by the Soviet Union after the mid-1950s, including John Noble, have reported learning from other gulag inmates of 3,000 surviving American military POWs of Korea and World War II held in special camps.

In 1975 a Romanian POW since 1945 named George Risiou escaped from the Soviet Union with five others and reported 900 American POWs still held in a secret KGB prison camp. Soviet authorities had assigned Russian names to American, English, French and other POWs there, for deniability.

● Vietnam — Following the Vietnam War, the long-established policy of high-level national security secrecy remained in effect concerning non-returned U.S. POWs in Vietnam, Laos, Cambodia, China and the Soviet Union. Precise and detailed intelligence on U.S. POWs disappearing in enemy hands remains classified.

Only 591 American prisoners were released after the Paris peace accords. The Vietnamese demanded both a ransom of $3 billion to $4 billion and U.S. diplomatic recognition for the release of prisoners whom they publicly denied holding.

Robert Garwood, the last acknowledged American POW to leave Vietnam (in 1979) and many Indochinese eyewitnesses in the 1970s and 1980s have reported seeing numbers of U.S. prisoners remaining in captivity following the 1973-75 U.S. withdrawal. Hundreds of Americans known to be POWs in Laos were not repatriated, but were written off by the U.S. government.

According to the recently released Senate report, there is some documentation that 1,259 POW/MIAs of the Vietnam War were captured but never repatriated. Additional prisoners lost during covert operations could well equal this number.

At least several hundred U.S. POWs were known to be alive in enemy control in 1974, after the final prisoner returns. Questionably identified skeletal remains appear to have been assigned once again to names of those who remained missing after being reported alive in communist control.

Lists of captured Americans with special training were carried in a category known as "MB" or "Moscow-bound." A now-declassified 1971 CIA cable of the Vietnam War reports interrogations of American POWs by Soviet and Communist Chinese personnel and that the POWs changed into civilian clothes for transport to unknown destinations.

The Vietnamese were known by U.S. intelligence analysts of the time such as Jerry Mooney (then of Air Force Intelligence and the National Security Agency) to assemble pre-selected U.S. prisoners from South and North Vietnam at a special POW camp near Vinh, where they would be interrogat-

ed by the "friends" (Soviet intelligence).

From there, many U.S. POWs were transferred to another special prison camp near Sam Neua, Laos, where they joined other American POWs. Here the final interrogations by Soviet intelligence officers decided which Americans would be the most useful.

From this point (and from other special collection points near Hanoi), Moscow-bound U.S. POWs were shipped by air to the high-tech development areas of the Soviet Union where their skills and training would be most valuable, and their only hope of survival would lie in cooperation.

This was merely a continuation of the decadeslong history of "wink-and-nod diplomacy."

□

Because of a mandate to preserve world peace, even at great cost, the security agencies of the U.S. government classified from public view information on thousands of American POWs once held by the Soviets and their Asian surrogates.

With the announced end of the Cold War, perhaps Soviet President Mikhail S. Gorbachev (or Russian leader Boris Yeltsin) could now ask some hard questions of surviving KGB hands from the times of Lavrenti Beria and Yuri Andropov, leading to a release of survivors.

*The Sunday Oregonian*

FRED A. STICKEL, President and Publisher
PATRICK F. STICKEL
WILLIAM A. HILLIARD, Editor

FRONT PAGE:

**WASHINGTON**

**POWs-MIAs.** The U.S. government in the 20th century has ignored the plight of its prisoners of war and soldiers missing in action to appease other countries, charges a Republican report.

Tucson, Friday, May 24, 1991

*The Arizona Daily*

WASHINGTON

# U.S. has abandoned POWs, MIAs in 20th century, GOP senators say

## Report states GIs are political pawns

**By Jim Abrams**
The Associated Press

WASHINGTON — The U.S. government throughout the 20th century has ignored thousands of American prisoners of war left to languish in Soviet labor camps and North Korean and Vietnamese prisons, a Senate staff report said yesterday.

The 112-page report compiled for Republicans on the Senate Foreign Relations Committee concluded that government pledges to give POW-MIA issues "the highest national priority" were a sham.

The committee's minority staff said its findings were "remarkably similar" to those of Army Col. Millard A. Peck, who recently resigned as head of the Defense Intelligence Agency's Special Office for Prisoners of War and Missing in Action in protest over what he called the government "charade" in trying to resolve the MIA issue.

The report, released by North Carolina Sen. Jesse Helms, said Washington, for political and diplomatic reasons, had acquiesced in a policy of Communist regimes throughout the 20th century to use POWs as leverage for political bargaining and for forced labor.

"Any evidence that suggested an MIA might be alive was uniformly and arbitrarily rejected, and all efforts were directed towards finding and identifying remains of dead personnel, even though the U.S. government's techniques of identification were inadequate and deeply flawed," it said.

The report said the Pentagon,

**1987 AP photo**

**Sen. Jesse Helms**

which denies any evidence of live POWs in Vietnam, has acted "contrary to common sense" in rejecting as spurious all 1,400 first-hand sightings of what were believed to be Americans in Vietnam.

Pentagon spokesman Pete Williams, questioned at a news conference yesterday on the Peck resignation, said there had been seven formal investigations over the past decade into the government's handling of the POW-MIA issue.

"They have all concluded that there is no conspiracy, that there is no cover-up, that the government is undertaking a good-faith effort to account for prisoners-of-war and those Americans listed as missing in action," Williams said.

The Pentagon lists 2,273 as miss-

ing as a result of the war in Southeast Asia. The MIA issue has again come to the forefront as the Bush administration considers establishing diplomatic relations with the Hanoi government.

The Senate report, citing classified documents and personal accounts, said the Soviet Red Army captured and never returned dozens of soldiers from the American Expeditionary Force that intervened in the Russian Revolution in 1918. Nineteen were spotted a decade later in Moscow's Lubyanka Prison, and many others were seen in a Siberian labor camp.

The report said Gen. Dwight Eisenhower was told at the end of World War II that 20,000 U.S. POWs were being held in territory overrun by the Red Army but acknowledged publicly that "only small numbers" were still in Soviet hands right after the German surrender in May 1945.

It said the Soviets used the Americans to gain economic concessions as blackmail on political issues, and as a source of slave labor.

During the 1950-1953 Korean War, it said, North Korean and Chinese forces shipped American POWs to Siberia in railroad cars. Many of the 8,000 American soldiers listed as MIAs in Korea were still alive and never repatriated, it said, citing secret government documents.

The State Department last month made a formal request to the Soviet government for any information it had about unreturned Americans.

The Republican staff report quoted a National Security Agency employee as stating that the 591 POWs returned by North Vietnam in 1973 represented only 15 percent of American servicemen held in captivity.

# The Washington Post

THURSDAY, JULY 4, 1991

*John M. G. Brown.*

# A 70-Year-Old Hostage Crisis

It is heartening that the United States has now formally asked Moscow for information on American prisoners who came under Soviet control during World War II, the Cold War, Korea and Vietnam. The request should help bring to broad attention a problem that has festered since World War I. It is fairly introduced in a new report by the Senate Foreign Relations Committee Republican staff, which examines American prisoners of war and Americans missing in action in Soviet and Soviet-surrogate control going back more than 70 years.

According to declassified U.S. cables and other documents in the National Archives and to eyewitness accounts, in 1919, Lenin's new Bolshevik regime used hundreds of secretly held American, British and French POW-MIAs (members of a force sent to suppress the regime) as hostages in a demand for U.S. diplomatic recognition. Many of the 70 U.S MIAs and 57 "presumed" dead continued to be withheld after an exchange of American aid for prisoners in 1921. In 1930, approximately 40 to 50 U.S. military POWs were reported by an eyewitness, Latvian-American escapee Alexander Grube, to be alive and in the Lubyanka prison and Solovetsky Islands forced-labor camps. After the United States recognized the Soviet regime in 1933, some skeletal remains were returned and assigned to the names of missing men. The hundreds of missing Allied prisoners of 1919, and many others, disappeared.

In 1945, declassified records of the Supreme Headquaters, Allied Expeditionary Forces, indicate, the Red Army overran an estimated 1.5 million Allied POWs and displaced persons held by the Germans. Contrary to the Yalta Agreements, American officers were refused access to nearly all Soviet-controlled POW camps in Europe. According to recently declassified U.S. documents, approximately 500,000 of these prisoners subsequently disappeared. The documents indicate that about 20,000 U.S. POWs and MIAs were held as "hostages" after VE Day by Stalin to ensure Allied repatriation of millions of Soviet citizens and to obtain U.S. and Allied aid and diplomatic recognition of Soviet puppet regimes in East Europe. Declassified U.S documents reveal that in March and April 1945, the Roosevelt and Truman administrations decided against the use of force to recover American POWs in order to ensure Soviet participation in the Japanese war and the postwar United Nations.

Still-living U.S. ex-POWs such as Martin Siegel, who escaped from Soviet-controled Stalag IV-B Muhlberg and reported many Americans left behind, were instructed to remain silent and told the matter was being "investigated." Of 90,000 U.S. POWs officially reported returned in Europe by June 1945, 25,000 had been MIAs not known to be prisoners. Documents reveal that administrative "presumptive findings of death" reduced the official number of "currently held" known U.S. POWs from 12,500 or more in June 1945 to 5,414 by the following January. Tens of thousands more U.S. MIAs were also presumed dead.

*U.S. aid could be made to depend on how helpful the Soviet Union is about living American POW-MIAs who disappeared in Indochina and any survivors of earlier wars who were given Soviet citizenship.*

THURSDAY, JULY 4, 1991                                    THE WASHINGTON POST

# A 70-Year-Old Hostage Crisis

Some American POWs were reported in declassified live sightings as alive in forced-labor camps and "unregistered lagers." Soviet citizenship and Russian names were reportedly assigned to American, British and French survivors in secret Soviet camps. Some 78,000 Americans remain missing from World War II. Additional U.S. prisoners captured during Cold War intelligence missions or kidnapped in Europe were also retained and reported alive in gulag camps. Information on several of these cases was specifically sought in the U.S. diplomatic note of last April.

During the Korean War, the North Koreans and Chinese demanded repatriation of tens of thousands of their nationals who refused to return to Communist control. Unwilling to repeat the errors of 1945-48, the Truman administration adopted a new policy of voluntary repatriation. The Communists retaliated by secretly withholding thousands of U.S. and U.N. POW-MIAs.

In August 1953, Gen. James Van Fleet, retired commander of the U.S. 8th Army in Korea, said: "A large percentage of the 8,000 American soldiers listed as missing in action in Korea are still alive." According to declassified CIA and other sources, an estimated 2,000 or more U.S. POW-MIAs were reported transferred to the Soviet Union from secret Soviet-run camps in China. In May 1954 the United States demanded their return of the new, post-Stalin government, which denied holding "American" prisoners "under Soviet guard." Sightings of Korean War POWs in the Soviet Union, China and North Korea have persisted into recent years. Presumptive findings of death and the recovery of remains (which were assigned to the names of the missing) gradually reduced the already minimal number of U.S. prisoners acknowledged by U.S. authorities to be in Communist control from just under 1,000 in 1954 to 389.

Persistent reports from Indochinese refugees and from Marine Pvt. Robert Garwood, the last acknowledged U.S. POW to leave Vietnam (in 1979), indicated that perhaps 1,000 or more American prisoners from among the more than 2,500 U.S. missing and presumed dead in Vietnam and Laos remained in captivity following the 1973 return of 591 POWs. The Vietnamese demanded both U.S. diplomatic recognition and $3 billion to $4 billion in postwar aid—as secretly promised by President Nixon but rejected by Congress—for the release of prisoners they publicly denied holding. U.S. diplomatic efforts have thus far produced largely fragmentary human skeletal remains from Southeast Asia, later assigned to the names of missing Americans.

Through their intercepted, decoded orders, the North Vietnamese were known by a U.S. Air Force and National Security Agency analyst, whose name must be withheld, to have assembled pre-selected U.S. prisoners from North and South Vietnam at a camp near Vinh, where they were interrogated by Soviet intelligence officers. These Americans were categorized within NSA as "MB" or Moscow Bound. A declassified CIA cable confirms that U.S. POWs in North Vietnam were interrogated by Soviet and Chinese personnel and dressed in civilian clothing for transport to unknown destinations. From Vinh, the prisoners were shipped to another camp near the Pathet Lao headquarters of Sam Neua, Laos, where Soviets determined which Americans had technological or intelligence value, and these POWs, according to the NSA analyst, were transported to the Soviet Union. Those deemed uncooperative but also unreturnable, such as Navy Cdr. Larry Van Renselaar, were sent to almost certain death at forced labor on the Ho Chi Minh trail.

The NSA analyst has offered testimony and sworn affidavits to Congress on the Moscow-bound POWs and on the fact that only 5 percent of the hundreds of U.S. POWs he tracked through intercepts were returned by the Communists. Approximately 300 or more POWs were reported transferred by air and sea to high-tech development areas in the Soviet Union.

On the Soviet side, there are a few recent positive glimmers. An American who assisted in repatriating two Soviet POWs from Afghanistan in 1989 recently delivered hundreds of declassified U.S. POW-related documents (dated 1919-80) from the National Archives to researchers in Moscow and to three members of the Supreme Soviet. In the Moscow News of April 20, authorities in the Siberian town of Magadan invited relatives of American, French and Japanese "former POWs who died in Soviet concentration camps" to visit their kin's protected graves.

Mikhail Gorbachev's hopes of receiving massive U.S. aid could well be made to depend on how helpful the Soviet Union is about living American POW-MIAs who disappeared in Indochina and about allowing emigration of any survivors of earlier wars who were given Soviet citizenship.

The easing of Cold War tensions should encourage President Bush to address the American people on the history of the POW-MIA issue. A presidential decision to disclose long-classified documentary information could begin to heal a deep national wound.

*The writer, a combat infantryman in Vietnam, is author of the forthcoming book on American prisoners, "Moscow Bound."*

A14 THURSDAY, JULY 4, 1991

# The Washington Post

### AN INDEPENDENT NEWSPAPER

## Were Some POWs Left Behind?

THE FOURTH of July is the right day to consider whether all those who contributed in their time to maintaining the independence and freedom that the holiday marks have received their due. We refer specifically to the American fighting men, captured or missing in the country's wars, who either may have been left behind alive and in enemy hands or may not have been fully accounted for. For decades the official government position has been that there is no credible evidence that American prisoners of war from the world wars, Korea and Vietnam are still detained. Yet the expanding public record of inquiry into the situation of these thousands of men suggests that it is too soon to put their fate out of mind. While it seems true that there is no credible positive evidence that any are still alive, there still remains a jarring measure of arbitrariness about the basis on which the Pentagon has officially pronounced them dead or has otherwise accounted for them. The matter cannot yet be put finally to rest.

It is a deadly serious thing to suggest that the country may not have used every resource at its command to pursue American prisoners and to ensure American prisoners and re-ported prisoners differently. But two sorts of evidence have inclined us to edge from skepticism to agnosticism on this issue. The first includes historical documentation such as that presented in an article on the opposite page today. The gist of it is that for various political and bureaucratic reasons, the minding of prisoners has not always been the top American priority.

The second sort of evidence includes the discrepancies in the accounting of POWs, MIAs and KIA-BNRs (killed in action-body not recovered) among combatants in the Vietnam War. In a typical instance, a Vietnamese refugee's sighting of a 1967 shootdown was matched up with the

shootdown of a surviving Navy pilot who was then repatriated in 1973. But that pilot subsequently contradicted the specific details of the refugee's report. Where, then, is the other pilot whose shootdown correlates much more closely with the refugee's account? This example comes from a sober new report by the Senate Foreign Relations Committee's Republican staff—a Jesse Helms operation that presses questions that will not go away.

Partly in response to such goading, the Bush administration has gotten Vietnam to let it open an office in Hanoi to investigate reports of live sightings. The administration has also asked the Soviet Union for any help it can give in accounting for missing servicemen from the world wars, Korea, Vietnam and Cold War shooting incidents. Some 16 senators are now calling for formation of a select committee to assess the government's efforts to track down POWs and MIAs. There have been previous probes, but not of the kind to bolster public confidence. The recent charges of an official coverup by the colonel formerly in charge of the Pentagon's MIA office, for example, were brushed off by an "internal management inquiry." The Helms report makes the further suggestion that Vietnam's offer to open its territory to American search should prompt the Pentagon to reciprocate and open its "territory"—its POW-MIA files.

One does not have to believe that the U.S. government has engaged in a devious and misguided conspiracy in order to admit the possibility that things might have been done differently at many points along the way. International relaxation now provides new opportunities abroad for search and review. The Pentagon has its own profound obligation to respond.

# The Washington Post

DONALD E. GRAHAM *Publisher*
KATHARINE GRAHAM *Chairman of the Board*
PHILIP L. GRAHAM, 1915-1963

**BENJAMIN C. BRADLEE** *Executive Editor*
**LEONARD DOWNIE JR.** *Managing Editor*

**MEG GREENFIELD** *Editorial Page Editor*
**STEPHEN S. ROSENFELD** *Deputy Editorial Page Editor*

---

A12  SUNDAY, JULY 21, 1991  ·  THE WASHINGTON POST

## U.S. to Ask Soviets About POWs

### GOP Congressman Says Issue Will Be Raised at Moscow Summit

The United States will seek an accounting from the Soviet Union at the summit in Moscow later this month of U.S. POWs believed to have been held in the Soviet Union following World War II and the Korean and Vietnam wars, according to Rep. John Miller (R-Wash.).

"It looks like we are going to pop the question and push the issue," said Miller, who sent President Bush a letter last week requesting the matter be placed on the summit agenda.

Miller said he received confirmation in a telephone conversation Thursday with Deputy Secretary of State Lawrence S. Eagleburger that

the issue would be raised at the Moscow summit. A State Department spokesman declined to comment.

In March, the United States sent a note to the Soviet Ministry of Foreign Affairs asking the Soviets to review their records for information about U.S. air crews shot down in or near the Soviet Union in the early 1950s and about U.S. soldiers and citizens who were not repatriated following World War II, Korea and Vietnam.

Tens of thousands of Americans were listed as missing in action from World War II, 8,172 from Korea and 2,273 from Vietnam.

The Soviet Interior Ministry has records of tens of thousands of American, Japanese, German, French and other POWs they've held over the years, Miller said. "In April, they gave the Japanese a list of 60,000 Japanese POWs who had died in the Soviet Union. In this era of *glasnost*, the United States should get an accounting too."

He said the Soviets are believed to have rounded up U.S. soldiers when they overran German POW camps in the late stages of World War II. Difficulties with repatriation of those soldiers was one of the early signs that the former allies had entered the Cold War era.

---

## Soviet help sought
## for locating POWs

July 20
1991

WASHINGTON (AP)—President Bush will ask Soviet President Mikhail Gorbachev for help in accounting for American soldiers held in the Soviet Union after World War II, Rep. John Miller said Friday.

Miller, R-Wash., said he was assured by Deputy Secretary of State Lawrence Eagleburger that Bush has placed the question on his agenda for his meeting with Gorbachev in Moscow.

The State Department asked the Soviet Foreign Ministry for help in the matter earlier this year.

State Department spokesman Denny Denny said Friday night the department maintains that the U.S. Embassy in Moscow to repeat a request for access to the Interior Ministry Central State Archives for POWs and civilians.

Miller wrote to Bush on Wednesday asking that the president take advantage of Soviet requests for economic assistance by asking in return for information on the status of American POWs who may have come home from World War II. Some 20,000 were classified as

missing in action from that conflict.

Miller said the Soviets after World War II held thousands of American soldiers, many of them former prisoners of war in German camps overrun by the Red Army. There have been several reports of American POWs being shipped to the Soviet Union after the Korean and Vietnam wars.

"I've been very careful to never say that I believe people are still alive today. I have no evidence to the contrary," Miller said.

"What I do believe, because there is substantial corroborating evidence, is that literally thousands of Americans were alive after World War II in Soviet hands and were not returned, and that a lesser number were probably in Soviet hands after the Korean War and were not returned.

"I think the relatives of these people want to find out what happened. Of course, they hope somebody is alive, but even if they are not alive they would still like to know how they died, where they died."

ISSN: 0738-3169

# AMERICAN FOREIGN POLICY
## NEWSLETTER

NATIONAL COMMITTEE
ON AMERICAN
FOREIGN POLICY, INC.
SUITE 615
220 MADISON AVENUE
NEW YORK, N.Y. 10016
Telephone (212) 685-3411

VOL. 14 — NO. 5

OCTOBER 1991

$5.00 A COPY

Professor Hans J. Morgenthau
Founder

Professor George Schwab
Cofounder and Editor

Ambassador Li Daoyu
Permanent Representative of China
to the United Nations

## China's Foreign Policy on Global and Regional Issues

### Address by
### Li Daoyu
### Permanent Representative of China to the United Nations

Before the National Committee on American Foreign Policy

New York, September 4, 1991

Mr. President, Ladies and Gentlemen: It gives me great honor and pleasure to meet and speak to you today. I shall take this opportunity to brief you on China's policies on some global and regional issues.

In recent years the world situation has undergone profound changes, the global structure formed after World War II has crumbled, and the tendency of multipolarization in the world has been fostered. On the one hand, people have seen improvements in U.S.–

(Continued on page 2)

## Population Movements: An Emerging Concern of American Foreign Policymakers

### by
### Gallya Lahav

The issue of immigration is becoming salient and merits attention, for it is becoming increasingly apparent that population flows affect the internal stability of states as well as international security. A product of disparate prosperity, contemporary population movements represent a complex global challenge because they pose a series of questions about economic, social, and political stability. The permeability of borders that has become evident in the twentieth century demands a reevaluation of security.

This article reviews demographic trends in Europe and the Middle East and examines their implications for U.S. national interests and foreign policy. In particular, they show that a fundamental shift in emphasis about immigration is occurring based on the realization that immigration is a key issue for the West. Consequently, U.S. policymakers have been forced to shift their focus from the consequences of direct immigration to the United States to global flows of population that

(Continued on page 3)

--- HOLD DATE ---

## FOREIGN POLICY BRIEFING

"Turkey and Cyprus"

by
Ambassador Mustafa Aksin
Permanent Representative of Turkey
to the United Nations

November 4, 1991  •  5:00 p.m.

The LOTOS CLUB  •  5 E. 66th St., N.Y.C.

OCTOBER 1991

## A Chain of Prisoners

### by
### John M. G. Brown

The Soviet Union's response to the U.S. diplomatic note of April 9, 1991 and President Bush's request in July at the Moscow summit for information about American prisoners who came under Soviet control during World War II and the cold war, including Korea, and Vietnam, will be of great significance to future Soviet-American relations. On April 4 President George Bush pledged to "renew our commitment to secure the release of any American still held in captivity," but history indicates that the heartfelt wishes of an American President can often be thwarted by considerations of foreign policy. Recovering American MIAs and POWs in hostile captivity has been a goal of almost every President since Woodrow Wilson, but foreign policy considerations have usually prevented its achievement.

A report recently prepared and issued by the Republican staff of the Senate Foreign Relations Committee on U.S. POWs/MIAs from the Vietnam War reveals many details about the seventy-two-year history of the disappearance of American prisoners of war and those missing in action while being held under Soviet and Soviet-surrogate control. Present U.S. policy in dealing with communist regimes that held American POWs/ MIAs after World War II, Korea, and Vietnam was shaped by the events of the 1917 Bolshevik Revolution and the Russian Civil War of 1918–1921. According to declassified U.S. cables and other documents in the national archives, many of the 127 U.S. MIAs and those presumed killed as a result of the 1918–1920 American intervention in the Soviet Union were secretly held as prisoners of war by Lenin's regime. The Soviet demand conveyed to presidential envoy William C. Bullitt for U.S. diplomatic recognition of the regime in early 1919 was subsequently linked with Bolshevik promises to provide information about several hundred American, British, and French MIAs who had disappeared on the Archangel front and to demands for the repatriation of thousands of Soviet POWs in western Europe, some of whom did not wish to return to communist rule. That heavy-handed Soviet approach was rejected by U.S. policymakers despite their desire to recover American POWs. At the behest of Secretary of State Robert Lansing, Secretary of War Baker and General Pershing ordered the termination of negotiations that had been undertaken to secure the release of POWs/MIAs. Weeks of local negotiations resulted in the return of a mere handful of prisoners. Consequently, U.S. diplomatic recognition of the Soviet communist regime was not granted.

Although the 1919 annual report of the Secretary of War listed no U.S. soldiers as missing in action in north

(Continued on page 11)

JOHN KERRY
MASSACHUSETTS

# United States Senate
### WASHINGTON, DC 20510

20 November, 1991

Mr. John Brown
Box 30
Petrolia, CA
95558

Dear Mr. Brown:

Thank you for sending me your article which appears in the
October issue of the American Foreign Policy Newsletter.

We are looking into the questions you raise and welcome your
input.

Sincerely,

John F. Kerry
United States Senator

JFK/jl

William Childs Westmoreland
General, United States Army, Retired
Box 1059
Charleston, South Carolina 29402

February 18, 1992

Dear John:

Congratulations!  Your persistence has paid off.
You have been the major factor in bringing "front and
center" the POW/MIA dilemma.  Bill LeGro will also
help.

Your letter of 6 February and enclosures have
been read with interest.

Best wishes.

Sincerely,

W.C. WESTMORELAND

John M. G. Brown
Box 30
Petrolia, California 95558

**United States Senate**

SELECT COMMITTEE ON POW/MIA AFFAIRS

WASHINGTON, DC 20510-6500

18 June 1992

John M.G. Brown
Box 30
Petrolia, CA 95558

Dear Mr. Brown: *John —*

On behalf of the Select Committee on POW/MIA Affairs, I wish to
express my gratitude to you for your vital contributions to our
investigation of this most difficult, complex matter.  Your
seminal work, best expressed in your "Policy, Politics and the
POW/MIA Dilemna," has provided a solid historical foundation for
inquiry and analysis.

I am also aware of the fact that you contributed significantly to
the report of the Republican Staff of the Senate Committee on
Foreign Relations.  It was this report, "An Examination of U.S.
Policy Toward POW/MIAs," that provided the compelling rationale
for the creation of our Select Committee.

Sincerely,

*Bob Smith*

BOB SMITH
Vice Chairman

BS/kb

*Thanks John!*

---

A24 WEDNESDAY, JUNE 17, 1992                                      THE

# The Washington Post

## AN INDEPENDENT NEWSPAPER

---

## Mr. Yeltsin's Revelations

BORIS YELTSIN'S trip to Washington comes decked in a series of stunning revelations that are bound to have a lasting effect on the tone of Russian-American relations even as the heavy-duty political exchanges, including yesterday's arms control announcements, are absorbed into regular state practice.

Russia's first elected democratic president in a millennium of history chose as a kind of special gift to the American democracy a selection of documents culled from the Kremlin's awesomely capacious secret files. On display at the Library of Congress, these papers tell of tales of misconduct and treachery not previously heard in the 70-odd communist and Soviet years. Their presentation now suggests a commitment to openness and honest inquiry on the international as well as the national level. This is exactly the basis on which a deepening Russian-American connection must be built.

With Mr. Yeltsin arrived word of a recent Russian account of a truly appalling incident—the Soviet Union's concealment of an accident at

a biological warfare works in Smolensk in 1979 when hundreds were killed in an anthrax epidemic that Moscow falsely attributed to naturally tainted meat. The suggestion—proven only now—that the Soviets were prepared to chea and lie to hide a secret program, a terribl tragedy and a violation of a formal arms contro treaty, had cast a heavy pall over Soviet-Ameri can deliberations as a whole into the '80s. Th latest disclosure is encouraging as a concret sign of a new day.

Even more gratifying is Mr. Yeltsin's acknowl edgment that American POWs from World Wa II, the Korean War, the Cold War period an even from Vietnam were secretly stashed awa in the Soviet Union and that some from Vietnam and perhaps others, may still be alive. Thes disclosures reach into a matter of great sadness bitterness and fellow feeling on the part of man Americans. The electrifying possibility that som of those who were left behind in America's war may yet be restored to their country and kin ca now be pursued with full vigor.

EMBASSY OF THE
UNITED STATES OF AMERICA
MOSCOW

OFFICE OF THE AMBASSADOR

July 9, 1992

Mr. John Brown
P.O. Box 30
Petrolia, CA  95558

Dear Mr. Brown:

      I just received a letter from Mr. Anthony Duke,
together with a copy of your recent memorandum on
"Soviet-held POWs," as well as a copy of your July 4, 1991
Op-Ed piece in the Washington Post, which, by the way, I
thought was very powerful.  I immediately passed your memo
to the appropriate members of my staff, who, you may be
interested to know, are working nearly full-time on the
POW-MIA issue.  I can assure you that we will look at your
information very carefully and follow it up as best we
can.  You should know that we are doing everything humanly
possible to assist Ambassador Toon and his people to find
answers to the questions that remain and to finally put
this matter behind us.

      I realize you may be skeptical, but I promise you,
there is no higher priority for us than to face up to this
issue and to fully resolve it, if possible.  I also
believe that President Yel'tsin is equally committed to
resolving it, openly and honestly however painful it may
prove to be for him and his country, although I should add
that not all the bureaucrats in his government feel the
same way, and so it will be a struggle to get from them
all the information we need.  We won't give up though.

      With best wishes.

Sincerely

Robert S. Strauss

THE WHITE HOUSE
WASHINGTON

June 23, 1992

Dear Mr. Duke:

Mr. Petersmeyer asked me to thank you for
your recent letter and the material on the
POW/MIA issue compiled by your son-in-law,
John Brown.  The report is certainly timely.

Because the POW/MIA situation is a national
security issue, Mr. Petersmeyer has forwarded
the information to Douglas Paal at the
National Security Council for appropriate
action.  Mr. Paal is Special Assistant to the
President and Senior Director, Asian Affairs.

Mr. Petersmeyer sends you his best wishes.

Sincerely,

John R. Nicholson, Jr.
Office of National Service
(The White House Points of Light Office)

Mr. Anthony D. Duke
3 Pound Hollow Road
Glen Head, L.I., New York 11545

S. Hrg. 102-351 Pt. 2

POW/MIA POLICY AND PROCESS

'POW/MIA POLICY & PROCESS)   896

## HEARINGS
BEFORE THE

## SELECT COMMITTEE ON
## POW/MIA AFFAIRS
## UNITED STATES SENATE

ONE HUNDRED SECOND CONGRESS

FIRST SESSION

ON

THE U.S. GOVERNMENT'S EFFORTS TO LEARN THE FATE OF AMERICA'S
MISSING SERVICEMEN

### PART II OF II

NOVEMBER 5, 6, 7, AND 15, 1991

Printed for the use of the Select Committee on POW/MIA Affairs

U.S. GOVERNMENT PRINTING OFFICE
WASHINGTON : 1992

For sale by the U.S. Government Printing Office
Superintendent of Documents, Congressional Sales Office, Washington, DC 20402
ISBN 0-16-058679-6

897

---

Intelligence Information Report

2-1877

PAGE 1 OF 3 PAGES

REPORT NO. CS-311/04439-71

DATE DISTR. 10 JUNE 1971

CIA Report

COUNTRY  NORTH VIETNAM
DOI  1965-JUNE 1967
SUBJECT  PRELIMINARY DEBRIEFING SITE FOR CAPTURED U.S.
PILOTS IN VINH PHU PROVINCE AND PRESENCE OF SOVIET
AND COMMUNIST CHINESE PERSONNEL AT THE SITE

ACQ  VIETNAM, SAIGON (18 APRIL 1971)
SOURCE   898

PAGE 3 OF 3 PAGES
CS-311/04439-71   234
J.E.

2. AFTER SHAKING HANDS WITH THE SOVIET AND CHINESE, THE
PRISONERS WERE LED TO A DIFFERENT VEHICLE FROM THE ONE WHICH
BROUGHT THEM TO THE SITE. THEY WERE ESCORTED FROM THE PLANT BY
A DIFFERENT SET OF GUARDS WHO WORE YELLOW AND WHITE UNIFORMS AND
WERE ARMED WITH RIFLES AND PISTOLS.   DID NOT KNOW THE DESTI-
NATION OF THE PRISONERS.

3. ABOUT 30 STUDENTS FROM THE LAM THAO SECOND-LEVEL
SCHOOL WERE SELECTED BY THE SCHOOL SUPERINTENDENT, DAO KHAC
T R U N G, FOR GUARD DUTY AT THE SUPERPHOSPHATE PLANT. EACH
STUDENT WAS CAREFULLY SCREENED AND THEN TRAINED IN THE
PROCEDURES FOR HANDLING U.S. PRISONERS. A PAMPHLET ENTITLED
POLICY ON TREATMENT OF AMERICAN PRISONERS /CHINH SACH DOI
XU TU BINH MY/ WAS USED AS A TRAINING DOCUMENT, AND STUDENTS
WERE GIVEN SPECIFIC INSTRUCTIONS ON THE RECEPTION OF THE
PRISONERS AT THE DEBRIEFING SITE. ONE OF THE FUNCTIONS OF
THE STUDENT GUARDS WAS TO DISARM ANY VILLAGERS WITH WEAPONS
BEFORE ALLOWING THEM CLOSE TO THE AMERICANS. AS LONG AS NO
PHYSICAL CONTACT WAS MADE WITH THE AMERICANS, THE VILLAGERS
WERE PERMITTED TO APPROACH THE PILOTS AND MAKE ANY REMARK
SUITABLE TO THE OCCASION.

4. FIELD DISSEM: STATE USMACV 7TH AIR FORCE NAVFORV
CINCPAC PACFLT PACAF ARPAC   899

---

1. A PRELIMINARY DEBRIEFING POINT FOR U.S. PILOTS SHOT
DOWN OVER VINH PHU PROVINCE, NORTH VIETNAM /NVN/, WAS LOCATED
AT THE LAM THAO SUPERPHOSPHATE PLANT /VJ581589/ NEAR THACH SON
VILLAGE, LAM THAO DISTRICT, VINH PHU PROVINCE. TWO U.S. PILOTS
WERE TAKEN TO THE DEBRIEFING POINT ON ONE OCCASION IN 1965;
EIGHT, IN 1966; AND AN UNKNOWN NUMBER, IN 1967. THE PRISONERS
WERE ESCORTED TO THE SITE BY PERSONNEL OF THE ARMED PUBLIC
SECURITY FORCES /APSF/, AND STUDENTS FROM A NEARBY SCHOOL
SERVED AS PERIMETER GUARDS. EACH TIME PRISONERS WERE BROUGHT TO
THE SITE THEY RODE IN AN OPEN CAR OF CHINESE ORIGIN RESEMBLING
AN AMERICAN JEEP. SOME OF THE ESCORT GUARDS RODE IN A LEAD CAR
AND OTHERS RODE IN TWO CARS FOLLOWING THE PRISONERS. UPON THEIR
ARRIVAL AT THE PLANT, THE GUARDS LINED UP, FORMING A CORRIDOR
THROUGH WHICH THE PILOTS ENTERED THE BUILDING. AT THIS POINT
A SOVIET, A CHINESE AND A VIETNAMESE GREETED THE PILOTS AND
LED THEM INTO THE BUILDING. THE PILOTS USUALLY REMAINED IN
THE BUILDING FOR SEVERAL HOURS. WHEN THEY EMERGED THEY HAD CHANGED
FROM UNIFORMS INTO CIVILIAN CLOTHING.
SAID
HAD TOLD HIM THE FOREIGNERS WERE SOVIET AND COMMUNIST
CHINESE. SOVIET PERSONNEL HAD BEEN STATIONED AT THE PLANT SINCE
ITS CONSTRUCTION IN 1963, BUT IN 1965 THE NUMBER OF SOVIETS WAS
REDUCED TO THREE OR FOUR, AND IT REMAINED AT THAT LEVEL AS OF
JUNE 1967. ABOUT 20 COMMUNIST CHINESE PERSONNEL ARRIVED AT THE
PLANT IN 1966 AND THERE WERE STILL ABOUT 20 THERE AS OF JUNE 1967
AS FAR AS   KNEW, THE SOVIET AND COMMUNIST CHINESE PERSONNEL
GOT ALONG WELL.

---

MEMORANDUM   286

NATIONAL SECURITY COUNCIL

Carter Library
386

ACTION

January 21, 1980

MEMORANDUM FOR:  ZBIGNIEW BRZEZINSKI

FROM:  MICHEL OKSENBERG

SUBJECT:  Renewed League of MIA Families Request
for Appointment

Once again, the National League of Families of American
Prisoners and Missing in Southeast Asia seeks to meet
you (Tab B).

They have nothing new to say, and I am capable of summarizing
any developments for you. So I recommend turning down the
request, and I will call Ann Griffiths separately to say
you have instructed me to see her.

However, a Seeger from you is important to indicate that
you take Soviet refugee reports of sighting of live Americans
seriously. This is simply good politics: DIA and State
are playing this game, and you should not be the whistle
blower. The idea is to say that the President is determined
to pursue any lead concerning possible live MIAs.

Do not offer an opinion as to whether these leads are realistic.
Apparently you revealed skepticism to Congressman Gilman, and
my recommended letter to the League walks you back from that.

RECOMMENDATION:

That you sign the letter at Tab A to Ann Griffiths.

JOHN M.G. BROWN
P.O. BOX 30
PETROLIA, CA. 95558
(707) 629-3547

June 26, 1992

Attn: personal and confidential
Mr. Boyden Gray
Counsel to the President
The White House
Washington, D.C.

Dear Mr. Gray,

    I enclose a confidential 3-page memorandum for the President dated June 25, 1992 regarding the current U.S./Russian POW/MIA affair, which was also forwarded by Mr. Anthony Drexel Duke of New York, through an assigned Presidential aide. Also enclosed are a memorandum concerning declassification and historic precedents in the POW/MIA matter, requested of me on May 7th by the Senate Select Committee on POW/MIA Affairs; an 18 June 1992 letter from the Committee's Vice Chairman, Senator Bob Smith, regarding my research for the U.S. Senate; an 18 June memorandum for Ambassador Robert Strauss in Moscow on possible/probable numbers of U.S. ex-POW survivors in C.I.S. territory forwarded by my wife's uncle, Ambassador Angier Biddle Duke; a June 20, 1992 memorandum on U.S. POW locations in the former USSR for Ambassadors Strauss and Toon forwarded by Mr. Anthony Duke and General William Westmoreland to the Ambassadors; and a June 21, 1992 letter from this writer to President Boris Yeltsin of Russia. This was delivered by me and my wife, Josephine Duke Brown, at the Russian Consulate General in San Francisco, to an official who identified himself as Aleksandr Suchkov ('diplomat-in- charge') on June 24, 1992 together with some 300 declassified U.S. POW/MIA- related documents, and the 133-page red-bound report POLICY, POLITICS AND THE POW/MIA DILEMMA, which I had sent previously to President Bush in care of Mr. Skinner and Mr. Gates. I have received no acknowledgement of delivery to date.

    I must say to you that I hope the President is being permitted to see such an alternative analysis of this decades-old problem, which he himself encouraged me in 1986 to compile, and which problem Mr. Yeltsin recently brought to the fore. Failure of the staff system in this regard could have historic implications that could be detrimental to the President and to the United States.

Sincerely,

| 103d Congress | | SENATE | | Report |
| 1st Session | | | | 103–1 |

# POW/MIA'S

---

# R E P O R T

OF THE

## SELECT COMMITTEE ON POW/MIA AFFAIRS
## UNITED STATES SENATE

JANUARY 13, 1993.—Ordered to be printed
Filed pursuant to Senate Resolution 10

---

U.S. GOVERNMENT PRINTING OFFICE
WASHINGTON : 1993

62–704

419

and buried at his family's request in the United States in 1957.
There was no record with respect to the other individuals identified
by the U.S.
   On July 30, 1992, Gen. Dimitri Volkogonov, chairman of the Rus-
sian Delegation to the U.S.-Russian Commission on MIA-POWs,
published an article in *Izvestia* listing the names of 39 American
citizens who had been illegally detained by the Soviet government.
According to Dr. Cole, however, none of the 39 was an American
POW.
   In summary, the initial phase of the Rand review, while incomplete and inconclusive, tends to discredit the idea that a substantial number of U.S. POWs were held by the Soviet Union following
World War II and not repatriated.
   In this regard, Dr. Cole took issue with the authors of Soldiers of
Misfortune and Moscow Bound concerning the number of POWs
the Red Army "liberated" from German POW camps and failed to
repatriate. His conclusions:

> The number of American POW's who were not repatriat-
> ed from German POW camps in World War II appears to
> be less than 200. Assertions that tens of thousands of
> American POW's were abandoned are "inconsistent with
> the historical record."
> U.S. and Soviet Archives suggest that fewer than 100
> American POW's, perhaps 50 or fewer, were held on the
> territory of the U.S.S.R. after World War II.
> An undetermined number of American air crews—not
> POWs—were detained by the U.S.S.R. after making forced
> landings on territory it controlled. Most, if not all, of these
> crews were repatriated from the U.S.S.R. Some others may
> not have been repatriated from Soviet-occupied territory,
> but answering this question requires further research.
> The U.S. government located the graves of hundreds of
> American servicemen on Soviet-controlled territory. These
> were not POWs; most were on the territory of Soviet-occu-
> pied Germany. Records show few of these remains were re-
> covered from the territory of the U.S.S.R.[599]

*Sanders, Sauter, and Brown*

   John M.G. Brown and James D. Sanders, assisted by Mark A.
Sauter, have conducted years of research in U.S. archives, searching for information relating to U.S. and allied POWs who fell into
the hands of the Soviet Army as it pursued the rapidly retreating
Wehrmacht across Eastern Europe in 1945. Thousands of soldiers
were moved by rail, truck and foot eastward, not westward, and
most ended their cross-country journey at the port of Odessa, on
the Black sea, there to await transport by sea to their homelands.
This much is not in dispute. What is in question is how many of
these soldiers were not allowed to board ship, but were destined for
the vast Gulag of the Russian-Siberian interior. Mr. Sanders and
Mr. Brown estimate that between 20,000 and 23,500 were POWs of
the Germans and became prisoners of the Soviets.

---

[599] Select Committee hearing, 11/10/92.

420

It is Mr. Brown's theory that Communist mistreatment of POWs—that is, retaining them as hostages for political purposes—can be traced to the behavior of the Bolsheviks. According to Mr. Brown, the Bolsheviks kept at least 60 American soldiers they captured during the Allied intervention of 1918-1919 at Archangel, and a few from the Siberian front. In his view, this was a prelude to the retention by the Soviets of thousands of soldiers taken from the German POW camps after World War II.

Mr. Sanders furnished the Committee with a critique of Dr. Cole's research in a letter on November 15.[600] Pertinent excerpts follow:

> Let me start by stating that the World War II portion of Dr. Cole's report is hopelessly incompetent. Any investigator/analyst/historian researching a possible Government cover-up of historic proportions, would begin by testing the official Government history against the available data. Dr. Cole, however, failed to do this.
>
> Instead, he relied exclusively on the RAMPs Report (Recovered Allied Military Personnel) to formulate his working hypothesis. Since the RAMPs report, completed in 1946, is the official Government version of the recovery of POWs, a competent historian would first demonstrate that the official history is correct. It is incorrect in virtually all critical areas.
>
> Cole quotes the RAMPs disinformation line that only "76,854 were estimated to be in German POW camps." Here are the correct confirmed American POWs held by the Germans:

| | |
|---|---:|
| European Theater | 76,474 |
| Mediterranean Theater | 20,171 |
| North African Theater | 1,667 |
| Total | 98,312 |

Mr. Sanders went on to say that his archival research turned up "Battle Casualties of the Army," which support his figures. He also asserts that his research shows that the U.S. actually expected 106-107,000 POWs to be returned, which included between 8,000 and 9,000 men carried as MIA but not definitely known to be in captivity. On May 19, 1945, a document found by Mr. Sanders—signed by Gen. Eisenhower—shows that 105,000 returnees were expected.

How many returned? Dr. Cole, using the RAMPs report, says 91,252. Mr. Sanders says that his research shows that the number did not exceed 85,000.

Mr. Sanders letter continued with its summary of his findings:

> Between February and April 1945, 5,159 Americans should have been evacuated through Odessa. . . . Only 2,858 were recovered, however. At least 2,301 Americans disappeared. A June 1945, State Department study in the MIS–X files confirms this, stating that 5,200 Americans should have come out through Odessa.

---

[600] Letter to Select Committee from James D. Sanders. 11/15/92.

S. HRG. 102-1130

# HEARINGS ON COLD WAR, KOREA, WWII POWS

# HEARINGS

BEFORE THE

## SELECT COMMITTEE ON POW/MIA AFFAIRS

## UNITED STATES SENATE

ONE HUNDRED SECOND CONGRESS

SECOND SESSION

ON

COLD WAR, KOREA, WWII POWS

NOVEMBER 10 and 11, 1992

Printed for the use of the Select Committee on POW/MIA Affairs

U.S. GOVERNMENT PRINTING OFFICE

61-958 ≈            WASHINGTON : 1993

For sale by the U.S. Government Printing Office
Superintendent of Documents, Congressional Sales Office, Washington, DC 20402
ISBN 0-16-041612-4

61-958 O - 93 - 1

# CONTENTS

### TUESDAY, NOVEMBER 10, 1992

### WEDNESDAY, NOVEMBER 11, 1992

(III)

277

Chairman KERRY. Let me ask you gentlemen if you would stand, if all of you would stand, so that I could swear you since you were unsworn.
Mr. ASHWORTH. We were sworn.
Chairman KERRY. You were sworn, you were sworn yesterday?
Mr. ASHWORTH. Yes.
Mr. SANDERS. Yes.
Chairman KERRY. All right, that is fine. Let me just say that I appreciate it. I was just given a note saying that you were unsworn and I forgot about yesterday.
Why don't we run through each of the testimonies and then we will come back and ask you our questions. Who is next, Mr. Brown.

## TESTIMONY OF JOHN M.G. BROWN, AUTHOR, MOSCOW BOUND

Mr. BROWN. Is this loud and clear?
Chairman KERRY. Yes.
Mr. BROWN. To the chairman and members of the select committee, my name is John M.G. Brown. Like several members of the committee, I served in Vietnam, in my case as a rifleman with the 1st and 9th infantry divisions, from 1967 to late 1968, when I was wounded in action.

Prior to my Vietnam service I had been made aware of Korean war veterans—by Korean war veterans that many American POWs had been withheld by the North Koreans and by the Chinese communists in Manchuria. Between 1979 and 1983, while a commercial fisherman and guide in Alaska, I learned that the communists had withheld U.S. POWs in Vietnam and Laos.

I began to investigate the POW/MIA subject while researching my first non-fiction book on Vietnam, *Rice Paddy Grunt*, and helping to support a private intelligence-gathering network in Southeast Asia which was run by Colonel Al Shinkle at the time. I also acquired much information from witnesses at previous Senate hearings, from declassified U.S. documents, and from a series of 1986–1987 meetings with senior U.S. officials.

On October 2, 1986, then Vice President George Bush encouraged me, in a letter to my father-in-law, Mr. Anthony Drexel Duke, a trustee of Duke University, to assist Mr. Ross Perot and to pursue my POW/MIA investigation wherever it might lead, and I quote President Bush, on the assumption that our men are alive.

At the time I hoped that these were Mr. Bush's heartfelt words. I subsequently published a series of reports and articles on Soviet-held U.S. POWs, beginning with the December 1986 American Foreign Policy Newsletter, under the chairmanship of former ambassador Angier Biddle Duke. From 1986 to 1988, my archival and human source research extended back through the cold war, Korea, and World War II to the World War I American intervention in Russia when U.S. documents and human sources reveal that American military POWs and civilians were first secretly withheld by the Soviets under Lenin and Trotsky.

My research among thousands of recently-declassified U.S. documents revealed that since the 1918–1921 Bolshevik period, the Soviets and their later surrogates always demanded diplomatic conces-

278

sions or financial aid for the release of U.S. and Allied war prisoners and civilian hostages they publicly denied holding.

The repatriation of Russian and Soviet prisoners in Allied control during World War I and World War II was also of great importance, but my research indicates that it was the Soviets who first initiated the reprehensible policy of secretly withholding war prisoners in the 1918-1919 period. Although many of these American POWs were seen alive by an eyewitness a decade later in Moscow's Lubianka Prison and the Solovetski Archipelago forced labor camps, Congress was misled by U.S. intelligence about their existence and they were never freed, even after the United States recognized Stalin's Soviet regime in 1933 under President Franklin D. Roosevelt.

Virtually every American President since Woodrow Wilson has had to deal with classified information and decrypted communications on Soviet and surrogate-held U.S. prisoners of war. Thus, no one President could or should be blamed for this 7-decades-old problem.

In May 1989, with my research colleague Thomas V. Ashworth, I wrote the first documented report on the entire history of Soviet and surrogate-held U.S. POWs of World War I, World War II, the cold war, Korea, and Vietnam, entitled A Chain of Prisoners, a Secret that Shames Humanity. This was followed by an article on Soviet-held U.S. POWs in the February 1990 Veterans of Foreign Wars Magazine.

In the Sunday Oregonian on December 2, 1990, I wrote the first article to appear in a major metro daily American newspaper on the full history of Soviet-held U.S. POWs from World War I, World War II, Korea, to the Moscow-bound American POWs of Vietnam, assembled from a secret parallel prison system at the Soviet interrogation camp, Tentacle MB at Sam Neua, Laos.

From 1989 to 1991, at the instigation of Senator Charles Grassley and others, I assisted the Senate Foreign Relations Committee Republican staff in their POW/MIA investigation with my research and human sources. And I thoroughly documented the historical section of their May 23, 1991 final report on U.S. POW/MIA policy with declassified U.S. records dated from 1918 to 1980.

Approximately 300 of my declassified U.S. documents were turned over to investigative journalists in the old Soviet Union at the beginning of 1991, and this resulted in several articles published in the Soviet press before Mikhail Gorbachev left office. And later in the year I turned over these same documents to the Disabled American Veterans.

Many of these documents supplied by me were subsequently published by your select committee in November 1991. On July 4, 1991, American Independence Day, I had published an article in the Washington Post, a 70-year-old hostage crisis, which the Post introduced with their own editorial the same day calling for the formation of this select committee.

Of some 25,000 or more American POWs of three major wars since 1945 who came under Soviet and Soviet surrogate control but were never released, many no doubt died in captivity. But to claim all of them are dead defies logic unless they were murdered by Stalin's or Mao Tse-tung's or Ho Chi Minh's successors. The same

279

must be said for many thousands more British, Commonwealth, French, Dutch, Belgian, and other Allied POWs and displaced persons who vanished in Soviet control in 1945.

The outgoing Bush-Quayle administration should focus on the recovering of live ex-prisoners in its remaining months and the new Clinton-Gore administration should end this tragedy by openly negotiating for all surviving U.S. ex-POWs. It would be premature to reward the Hanoi regime with U.S. diplomatic recognition until the prisoner and missing issue is truthfully resolved.

If, as Ambassadors Strauss and Toon have indicated, the Russian intelligence organs are not cooperating with President Yeltsin in locating American ex-prisoners of war on former Soviet territory, he should exert his authority. Former Soviet officials should not fear retribution if all survivors are allowed to return to America. The United States should continue to supply information on Soviet MIAs of Afghanistan and of earlier cold war conflicts.

I have written a fully documented book on communist-held U.S. POWs since 1918 entitled *Moscow Bound*, an edited version of which I published myself and turned over to this select committee when I testified in July of this year. I have been informed that it has been used by members of the U.S.-Russian Commission on POWs and others, so I hope that my statement here today will call attention to vital unpublished information about U.S. POWs and MIAs, and that my work will get the attention it deserves, even if it might be perceived by some as embarrassing to current or past U.S. officials.

Thank you for giving me the opportunity to appear before you.

[The prepared statement of Mr. Brown follows:]

THE WHITE HOUSE
WASHINGTON

November 12, 1992

Dear Mr. Brown:

Thank you for your letter and book, <u>Policy, Politics and the POW/MIA Dilemma</u>, which has been referred to me for reply and review. Thank you for sharing your views on this sensitive issue.

Thank you again for writing.

Very truly yours,

Mark R. A. Paoletta
Assistant Counsel to the President

Mr. John M. G. Brown
P.O. Box 30
Petrolia, CA. 95558

THE WHITE HOUSE
WASHINGTON

December 2, 1992

Dear Mr. Brown:

On behalf of President Bush, thank you for your letter and the copy of your book, <u>Moscow Bound</u>. You can be assured your book will be given thoughtful consideration.

Thank you again for writing.

Very truly yours,

Mark R. A. Paoletta
Assistant Counsel to the President

Mr. John M. G. Brown
P.O. Box 30
Petrolia, CA. 95558

# APPENDIX II

8 MAR, 1945

ACWAR, Washington

Personal Message for the President

Harriman                    White

U R G E N T

TOP SECRET

TOP SECRET

Personal and Top Secret for the PRESIDENT from HARRIMAN.  TOPSEC

In light of Marshal Stalin's reply to your message regarding
our liberated prisoners of war I feel you will be interested to have
from me a brief review of the situation.

Our information received from our liberated prisoners indicates
that there have been four or five thousand officers and enlisted men
freed.  The Russians today claim that there are only 2,100 of whom 1,350
have arrived at Odessa and the balance being en route by train.

*Actually more than
5,000,
of whom 2,858
were eventually
repatriated through
Odessa by late-
April – early May '45*

Russian information is based on reports from concentration
points within Poland where our prisoners have been collected.  Meantime
there appear to be hundreds of our prisoners wandering about Poland trying to locate American contact officers for protection.  I am told that
our men don't like the idea of getting into a Russian camp.  The Polish
people and Polish Red Cross are being extremely hospitable, whereas food
and living conditions in Russian camps are poor.  In addition we have
reports that there are a number of sick and wounded who are too ill to
move.  These Stalin does not mention in his cable.  Only a small percentage of those reported sick or wounded have arrived at Odessa.

For the past ten days the Soviets have made the same statement
to me that Stalin has made to you, namely, that all prisoners are in

TOP SECRET

Page #1

## TOP SECRET

Odessa or entrained thereto, whereas I have now positive proof that this was not repeat not true on February 26, the date on which the statement was first made. This supports my belief that Stalin's statement to you is inaccurate.

I am glad to say that the reports from our contact officers in Odessa indicate that the Russians have done a first rate job in providing quickly a reasonably adequate camp in Odessa and our prisoners are reasonably well provided with food, etc. Our officers there also are allowed to communicate with us daily. I have no present reason to complain about the situation in Odessa or about the speed with which our prisoners have been moved from Poland by train, considering the shortage of transportation.

I am outraged, however, that the Soviet Government has declined to carry out the agreement signed at Yalta in its other aspects, namely, that our contact officers be permitted to go immediately to points where our prisoners are first collected, to evacuate our prisoners, particularly the sick, in our own airplanes, or to send our supplies to points other than Odessa, which is 1,000 miles from point of liberation, where they are urgently needed.

Since the Yalta Conference General Deane and I have been making constant efforts to get the Soviets to carry out this agreement in full. We have been baffled by promises which have not been fulfilled or have

## TOP SECRET

been subsequently withdrawn. We succeeded after considerable delay in getting one contact team of an officer and a doctor to Lublin but they have not been permitted to move to other points and our infrequent communications with them have been largely through the friendly intervention of the Polish Embassy here.

Ten days ago the Soviet Foreign Office finally authorized General Deane to go to Poland to review the situation but no action has been taken so far. I pressed it again last night and hope to hear today. I have proposed that he go with a Russian officer and report jointly to the Soviet authorities and myself as to whether their information or ours is correct.

I am not so worried about our prisoners who are well. These, I believe, will gradually be assembled and shipped to Odessa. I am extremely concerned, however, over the sick and wounded. I hope to get an answer today about Deane's trip. If it is not satisfactory I will recommend that you cable Stalin again.

## TOP SECRET

San Francisco U.N. Conf

APRIL 19, 1945.

Mr. Secretary (of State Stettinius)

The following is a list of questions which Mr. Molotov may raise or which you might wish to raise depending upon the course which the discussions take:

1. Implementation of the Crimea Agreement on liberated areas in Russia.

2. Implementation of the Crimea Agreement on liberated areas in Bulgaria.

3. Need for improvement of the American position on Allied Control Commissions in satellite states.

4. The failure of the Soviet Government to provide facilities for the entry of UNRRA and Red Cross personnel into Poland and the Balkans.

5. Failure of the Soviet Government to satisfy our requirements for full advance information concerning transfers to third countries of supplies furnished to the Soviet Union under Lend-Lease or supplies of Soviet origin similar to Lend-Lease materials.

6. The implementation of the Crimea agreement regarding the exchange of liberated prisoners of war and civilians.

7. The Kravchenko case.

8. The Soviet request for a six billion dollar loan.

9. Forthcoming meeting of the Reparations Commission in Moscow.

---

BRITISH EMBASSY W.D.C.

CYPHER (O.T.P.)

Telegram No.3936
of April 20th, 1945

From: Acting Secretary of State
To: Lord Halifax

Desp: 6.36 pm April 20th, 1945
Recd: 2.56 pm April 20th, 1945

CHANCERY (P.O.W's)
DISTRIBUTION
IMMEDIATE

Chancery - Action
Secretary of State
T. File
Spares (3)
Red Box A
Red Box B
Red Box C
H.E.
J.H.Magowan
H.B.Butler
R.Makins
Sir G.Sansom
Sir G.Campbell
A.D.Marris
R.Opie
B.Cockram
G.W.McKenzie
F.Healey
A.McD.Gordon
P.H.Gore-Booth
J.M.Russell
J.2.M.(8)
Comm.Coleridge
Roger Stevens
Col.G.D.Loup (2)
G.R.Rankan
Major Berkeley
J.Underwood

Addressed to Washington telegram No.3936 of April 20th repeated to Moscow,

Your telegram No.2659.

Following for Secretary of State.

We are repeating to you Moscow telegrams Nos.1173,1205 and Foreign Office telegram to Moscow No.1922.

It is clear that Soviet Government will not allow our contact teams into Poland. "The Russians deny the existence of any British prisoners of war in Poland but we have evidence of their being concentrated at Cracow(?) and in hospital. This is a clear breach of Yalta agreement and we ought to maintain our position on this. The important thing at the moment however, is to get medical supplies and comforts to our men in Poland. We have therefore turned to the Red Cross channel (see Foreign Office telegram No.1922 to Moscow) meanwhile reports from our contact officers, who have been at Lwow-Volkovsk which are contact points designated by the Russians themselves and are East of Curzon line disclosed serious breaching of the agreement which merit strong protest. A reply Foreign Office telegram No.1958 has been sent.

CYPHER (OTP)

Telegram No. 3924.
of April 20th 1945.

Desp: 8.30 pm April 20th 1945.
Recd: 4.46 pm April 20th 1945.

Following received from Moscow
telegram No. 1205 of April 10th.

CHANCERY (P.O.W.'s)
DISTRIBUTION

Chancery - Action
Secretary of State
T. File
Spares (5)
Red Box A
Red Box B
Red Box C
H.B.
H.B. Makins
H.B. Butler
J.H. Magowan
Sir G. Sansom
Sir G. Campbell
A.D. Marris
R. Ople
B. Cockram
G.G. McKenzie
F.W. McCombe
F.W. Healey
A. McD. Gordon
P.H. Gore-Booth
J.W. Russell
J.S.M. (8)
Comm. Coleridge
R. Stevens
Col. G.D. Loup (2)
G.R. Ranken
Major Berkeley
J.W. Underwood.

Begins.

(c).

Your telegram No. 1325, paragraph 2

1. The Russians have already undertaken
to arrange for distribution of supplies to
collection camps which they have established
on Soviet territory.

2. Although Russians deny there are any
of our men in hospitals in Poland General
Younger intends to confront Soviet Red Cross
with statements from our liberated prisoners
in Odessa and to press them to arrange for
despatch of supplies to Polish Red Cross for
distribution amongst our men.

3. I shall of course support his
approach but I expect he will either be told
that the matter is outside competence of the
Soviet Red Cross or he will be referred to
the Polish Red Cross.

4. In the latter case he would have
to make arrangements through the Polish
Embassy in Moscow and I should be grateful
if you would telegraph at once to confirm
that he may so act.

5. Since we do not know and cannot
find out how many men we have in Poland it
is difficult to decide how much it is
necessary to send to Poland in the way of
stores. General Younger would therefore
suggest that a tour of hospitals in Poland
where our men are thought to be and
quantities of stores despatched would be
based on report.

9340:OFW

---

CYPHER (O.T.P.)

Telegram No. 3923
of April 20th, 1945

From: Acting Secretary of State
To: Lord Halifax

Desp: 8:25 p.m., April 20th, 1945
Recd: 5:48 p.m., April 20th, 1945

BRITISH EMBASSY - D.C.

Following received from Moscow
telegram No. 1663 of April 8th.

CHANCERY (P.O.W.'s)
DISTRIBUTION

Chancery - Action
Secretary of State
T. File
Spares (3)
Red Box A
Red Box B
Red Box C
H.B.
H.B. Makins
H.B. Butler
J.H. Magowan
Sir G. Sansom
Sir G. Campbell
A.D. Marris
R. Ople
B. Cockram
G.G. McKenzie
F.W. McCombe
F.W. Healey
A. McD. Gordon
P.H. Gore-Booth
J.W. Russell
J.S.M. (8)
Comm. Coleridge
Roger Stevens
Col. G.D. Loup (2)
G.R. Ranken
Major Berkeley
J. Underwood.

Begins.

Your telegram No. 1663.

1. Soviet authorities at Lwow
have refused to extend permits of our
officers beyond April 5th and have ordered
them to return to Moscow. Admiral Archer
has instructed our officers to stay where
they are but I understand that they have
been confined to their hotel. The reason
given by the Soviet authorities was that
the camp at Lwow had been cleared. They
promised new passes if more prisoners arrived.

2. Owing to lack of communications
with Volkovysk we do not know whether the
same has happened there but it seems likely.

3. Before the receipt of your
telegram under reference I had written to
M. Molotov pressing strongly for our officers
to be allowed to remain until they have completed the duties which they were sent from
England to perform and pointing out to him
that the British public would never understand withdrawal of our officers until all
our missing prisoners had been accounted for,
particularly because a great number must
have been liberated by recent operations of
the Red Army or would shortly be liberated
which were about to be undertaken.

4. While I agree that we should
now await reports from our officers at Lwow
and Volkovysk I doubt if these will in fact
provide much additional information.

TELEPHONE CONVERSATION ... AT ODESSA

1 - There are 27 officers, 42 enlisted men and two civilians now at Odessa waiting for shipment. Of these, 34 officers, and 37 enlisted men are Air Corps non-prisoners of war from the 15th Air Force.

2.- There are three liberated prisoners of war in the hospital at Odessa.

3.- It is rumored that 3,000 more liberated prisoners of war are in Poland awaiting shipment to Odessa. (Information from ex-prisoner of war now in Odessa.)

4.- Colonel Fennell stated it was difficult to visit Odessa camps and carry out his duties because of restrictions and necessity for movement passes every time he desires to visit the camps.

5.- Colonel Fennell was informed again that partial payments would be made by him to all prisoners of war prior to embarkation and that no more letters to ship captains requesting payment on the ship would be made in the future. Colonel Fennell requested that pay data cards and payroll forms for enlisted men be supplied him.

6.- Colonel Fennell was informed that a French officer was on his way to Odessa and that he (Colonel Fennell) should supply approximately 25 percent of our supplies to him for ex-prisoner of war use.

7.- Colonel Fennell was informed that broadcasts from the Moscow station have been heard in which there were transmitted interviews with American ex-prisoners of war. He was informed that no such interviews should be allowed in the future, as they are contrary to War Department security regulations.

8.- Colonel Fennell was informed that the Soviets have approved the shipment of ex-prisoners of war on liberty ships.

9.- Colonel Fennell stated that blankets were not being supplied to enlisted men by the Soviets, and asked that we take action here, if possible.

10.- Colonel Fennell asked me to pass on two questions to the Navy Division:
(1) Where is personal mail for the Navy.
(2) When will Lt. Oram arrive.

---

MEMORANDUM FOR GENERAL DEANE:

12 April 1945

Status of Prisoner of War evacuation as of this date:

| | PW and Air Crews | American Civilians | Foreign Civilians |
|---|---|---|---|
| From Odessa by ship | 2,666 | 21 | 8 |
| From Poltava by air | 20(PW) | 0 | 0 |
| From Moscow by air | 9 | 0 | 0 |
| At Odessa | 79 | 3 | 0 |
| At Poltava | 2(PW) | 0 | 0 |
| | 2,776 | 24 | 8 |

GERALD C. RICE,
Major, S.C.

RS-710
This telegram must be
closely paraphrased be-
fore being communicated
to anyone. (CONFIDENTIAL)

Rec'd 2:48 p.m.

SPECIAL WAR PROBLEMS DIVISION
MAY 1 - 1945
DEPARTMENT OF STATE

Secretary of State
Washington

1425, April 30, 9 p.m.

I am wiring immediately upon completion full summary of interview with Col. General Golikov, representative of Soviet of Peoples Commissars for Repatriation Affairs, which was featured in today's PRAVDA (To Paris for Murphy and Rober as 86 and to AMPOLAD Caserta as 75) Golikov makes starting allegations regarding mistreatment of Soviet citizens in British and American prisoner of war camps and contrasts unfavorably treatment accorded them with generous treatment allegedly accorded American and British prisoners of war liberated by Red Army. He alleges that all liberated British and American prisoners have been repatriated except for small groups and complains of delays in repatriation of liberated Russian prisoners of war.

Suggest this telegram and summary of interview which will follow be brought to attention of Ambassador Harriman and General Deane.

KENNAN

THE INSTITUTE FOR ADVANCED STUDY

PRINCETON, NEW JERSEY 08540

Telephone (609) 734-8000    Telex: 2297744AS UR    Fax: 609-924-8399

SCHOOL OF HISTORICAL STUDIES

December 13, 1988

Dear Mr. Brown:

I have your further letter, of December 5, concerning your study of questions connected with the American prisoners of war who fell into Russian hands in the final months of the Second World War. I have also received the many copies of documents that accompanied your letter, as well as the volume of reminiscences of your Vietnam service, the gift of which I appreciate.

I thought I had made it plain in my earlier letter (the text of which I do not have at hand at this moment) that I was never officially involved or even regularly informed about this matter. In the 170-odd document copies of which you sent to me, I can find only three that went out over my own signature. Two of these concerned a query Senator Connally had addressed to the Department State and which the latter had passed on to the Moscow Embassy, in response to which I sent such information as I could obtain from General Deane. The other was a message, sent to the Department of State for the information of Ambassador Harriman and General Deane (both at the time apparently in the United States), summarizing the content of an article that had appeared in the Soviet press which I thought would be of interest to them. Beyond that, there was one despatch signed by Ambassador Harriman, reporting further items from the Soviet press, which despatch had evidently been prepared in the Embassy chancery and submitted by me to the Ambassador for his approval and signature. None of these, as you will readily see, show me to have been in any way involved in this problem except in relaying information to others where this seemed indicated.

To the first three of your questions the answer is: I was simply insufficiently informed to permit me to arrive at useful judgements on these matters. I was aware that the repatriation of these prisoners was under discussion between General Deane and his official contacts in the Soviet military establishment (General Golikov was the one he frequently mentioned.) I assumed from the duration of these exchanges that these exchanges were not easy ones. This did not surprise me. Very few of the questions we had under discussion with the Russians at that time were easy ones.

To the remaining questions the answer, I am afraid, is simply: no. I may from time to time have heard gossip about these matters, but I cannot recall doing so.

It would seem to me that the sort of documentation you have sent to me should permit you to form a pretty fair picture of the course of these

---

Mr. John M. G. Brown                    December 13, 1988

-2-

exchanges with the Russians on the part of both the British and ourselves, and of the results obtained.

I can only give you one memory that may be of interest to you. I thought it, at the time, highly probable that General Golikov, while wearing a regular military uniform, was actually a high official of the then N.K.V.D. and not of the regular Russian military authorities. I believe I made my view plain to General Deane, but that he felt that he had no choice but to deal with the man assigned for this purpose, and had no formal grounds for questioning his status. I could understand his position, but I resented this sort of obfuscation and saw it as an attempt to prevent and contact between our Military Mission and the regular Russian military establishment.

Very sincerely,

George Kennan

John M. G. Brown
477 Mattole Road
trolia, California 95558

REPRODUCED AT THE NATIONAL ARCHIVES

1 MAY 1945

MEMORANDUM: To General Roberts

SUBJECT : Prisoner of War Status

1. Number of PW and Air Crew shipped from Poltava or Lwow
   Number of PW and Air Crew shipped from Odessa
   Number of American civilians shipped from Odessa
   Number of Air Crew now-PW at Odessa
   Number of PW now at Odessa
   Number of civilians now at Odessa

| | |
|---|---|
| | 2,882 |
| | 34 |
| | 25 |
| | 25 |
| | 2 |
| | 11 |
| Total | 2,984 |

2. 25% of supplies previously received at Odessa have been turned over to the French. This leaves us with 4,500 units. In addition 12,000 units have just been received by ship at Odessa and 25% of these will be turned over to the French, leaving us with a grand total of 13,500 units.

3. One car load of supplies has been received at Lwow and has been stored. The car planned for Volkovisk has not yet been sent and the Soviets suggested to us on 3 May that these supplies not be sent inasmuch as they plan to abandon the Lwow and Volkovisk collection points.

4. Last ship, Bergensfjord, sailed May 3 - carrying 92 POWs.

JAMES C. CROCKETT,
Colonel, G. S. C.

---

REPRODUCED AT THE NATIONAL ARCHIVES

## SHAEF
### STAFF MESSAGE CONTROL
## OUTGOING MESSAGE

S E C R E T

P R I O R I T Y

| | | |
|---|---|---|
| TO | : | MILITARY MISSION MOSCOW FOR DEANE |
| FOR INFO | : | SHAEF FWD |
| FROM | : | SHAEF MAIN, FROM BARKER, SIGNED EISENHOWER |
| REF NO | : | S-87705 ........ 1118158 |

This reference to our number FWD-21010.

Information received from prisoner of war camps in rear of Russian lines indicates thousands of United States and British prisoners of war held in close confinement under unsatisfactory conditions. We could have evacuated them by air a week or ten days ago if the Russians would agree to cooperate by permitting us to land planes at nearby airfield, as specifically authorized by Article 4 of YALTA agreement.

Unless this evacuation can be effected promptly, there may well ensue most undesirable consequences.

Please express to Russians the urgency of this matter and ask that they direct their Commanders on western front to permit the landing of United States and British planes at fields near camp for evacuation of our prisoners of war.

FWD-21010 is not identified in SHAEF SMC files

| | | |
|---|---|---|
| ORIGINATOR | : G-1 | AUTHENTICATION: ROBERT L MAY, MAJOR |
| INFORMATION | : SGS | |
| | G-4 | COORDINATED: DEPUTY CHIEF |
| | G-5 | OF STAFF |
| SUMMARY | | |
| AG RECORDS | | |

SMC OUT 1076    11 May 45 . 1835B    JOB/lp    REF NO: S-87705

15 0031    TOO: 1118158    COPY NO 3

---

SHAEF FWD 117/10
FOR 10135B MAY
eel 10141 5B MAY

JEFF
TOO 10131B MAY

C O N F I D E N T I A L

O P E R A T I O N A L   P R I O R I T Y

FROM : CG NINTH US ARMY

TO FOR ACTION : CG SHAEF FWD ATTN G-1 FWX INFO ROOM

FOR INFO : CG TWELFTH ARMY GP ATTN G-1 FWX

REF NO : KX-21203, 10 MAY 1945      CITE: GNMDA

Stalag II A located at NEUBRANDENBURG now under Russian control, contains 1100 Americans in the camp with 2600 Americans attached to Stalag II A located within radius of 50 km or camp. 650 British attached to camp. 200 Americans hospitalized, 50 being seriously ill.

Present conditions of camp: No camp control, Russian officer in charge usually drunk, drunken Russian soldiers molesting American sick, Americans robbed by some Russian soldiers making it difficult to avoid trouble. Food supply not critical on 7 May, but running low.

ACTION : G-1

INFORMATION : AG RECORDS
SGS

(400 (100 (100 (10S)

PS IN 2699    10 May 45    1517B    10W/lf    REF NO:    KX-21203

36    576

C O N F I D E N T I A L

THE MAKING OF AN EXACT COPY OF THIS MESSAGE IS FORBIDDEN

COPY NO.    5

---

SHAEF FWD 278/17
FOR 17125B MAY
fv, 17200B MAY

JEFF
TOO 17174OB MAY

C O N F I D E N T I A L

O P E R A T I O N A L   P R I O R I T Y

FROM : CG NINTH UNITED STATES ARMY

TO FOR ACTION: CG SHAEF FORWARD ATTENTION G-1 FWX

FOR INFO : CG VII CORPS ATTENTION G-1 FWX, CG TWELFTH ARMY GROUP ATTENTION G-1 FWX

REF NO : KX-21617, 17 MAY 1945      Skley iv-8

GNMDA.

Following VII Corps message quoted:

"Reports received that 7,000 United States and British ex-PWs formerly in MUHLBURG and NOR RIESA 8715-E need medical supplies, additional medical attention and food. Many have left because of conditions. Reports indicate camp leader doing all in his power to enforce stay-put order. Russians alleged to have threatened to use force to prevent escape. Suggest immediate evacuation of both camp and hospital."

Liaison being carried on with Russians with view to allowing United States medical supplies and ambulances to enter camp.

Request that your Headquarters establish liaison with Russian SHAEF Representative with view to negotiating return of these United States and British ex-PWs to Ninth Army area.

ACTION : G-1

INFORMATION : SGS    G-2    AG RECORDS

36    550

PS IN 4738    17 May 45    2335B    AGD/lr    REF NO: KX-21617

C O N F I D E N T I A L

THE MAKING OF AN EXACT COPY OF THIS MESSAGE IS FORBIDDEN

JMSS
TOO 1307452 MAY

SHAEF FWD 120/13
TOR 131124B MAY
eel 13225B MAY

CONFIDENTIAL
PRIORITY

FROM : MILITARY MISSION TO MOSCOW FROM DEANE
TO : SHAEF FWD TO EISENHOWER
REF NO : M-24293, 13 MAY 1945

prisoner of war at LUECKENWALD, South of BERLIN, has arrived in MOSCOW.

He states that conditions in LUECKENWALD under Russian control are extremely bad. Camp COMMANDANT is a major who is usually intoxicated and treatment of British and American prisoners is not good. He states that recently a great many British and American prisoners were flown out in American transports against the Russian's wishes and that the Russians have vented their resentment on the remainder.

The Norwegian does not want his name used for fear of the effect it might have on Russian treatment of other Norwegian prisoners; however, if you have any supporting data that you can furnish, I would appreciate your sending it to me so that I can put pressure on the Russian Repatriation Committee.

In any event, when GOUSEV meets your representatives in LEIPZIG, we should be prepared to confront him with all factual data possible concerning mistreatment of our prisoners by the Russians.

FS IN 3492
1295

CONFIDENTIAL
THE MAKING OF AN EXTRACT FROM THIS MESSAGE IS FORBIDDEN
COPY No.7

36

---

JETE
TOO 191045B MAY

SHAEF FWD 142/1
TOR 191506B MAY
fw 191613B MAY

SECRET
PRIORITY

FROM : SHAEF MAIN SIGNED EISENHOWER
TO FOR ACTION : AGWAR
FOR INFO : EXFOR, MFOUSA, SHAEF FWD EISENHOWER
REF NO : S-88613;   CITE: SHGAP

Reference WX-82478.

Estimate number of German personnel now held by United States is subject.

Best information available at moment follows:

A. In prisoner of war status outside GERMANY 880,000. This does not include 142,000 evacuated to US and 39,000 evacuated to MFOUSA November; but does include 34,000 in UK. Inside GERMANY or AUSTRIA 1,000,000.

B. As disarmed German troops, either within or outside GERMANY or AUSTRIA 2,000,000. Above figures are subject to considerable uncertainty.

Reports have been requested of Army Groups and data will be transmitted to you as soon as available.

New Subject: Number of US prisoners of war recovered from German custody NOW in US or British hands 80,000. This figure includes 27,000 already evacuated to US.

Numbers of US prisoners estimated in Russian control 25,000.

FS IN 5235

530

3

# INCOMING MESSAGE

SHAEF FWD 76/23
FOR 230630B MAY
1kJ 230754B MAY

TOO 222022B MAY

SECRET

PRIORITY

FROM: • AFHQ SIGNED ALEXANDER (first two words)

TO FOR ACTION: • TROOPERS FWD

FOR INFO: • SHAEF FWD, SHAEF MAIN

REF NO: • FX-80335, 22 MAY 45, CITE, FHQAB

1. WOLFSBERG and SPITTAL Camps now clear of British Commonwealth US PW. AFHQ occupied AUSTRIA completely covered by detachment of Repatriation Unit. Evident that no British Commonwealth US PW remaining except for few stragglers who may be unwilling to declare themselves. Total of 265 evacuated by air to ITALY as of 21 May. All in good condition. 65 hr Commonwealth sick have been evacuated to ITALY by air. 45 of these due to fly UK ETD 23 May.

2. Agreement reached with Major SKVORTZOFF [DEVICHSOFF?] Repatriation Staff TOLBUKHIN's Hq and General GRZEKIN 57 Army GRAZ for handover to them of all Soviet(?) ex PW(?) in British Zone. 2847 transferred 17/18 May.

3. Permission obtained from local Soviet Commander for Repatriation Detachment to enter Russian Zone but difficulty experienced in tracing Br/US PW. Approx 300 have exfiltrated into KLAGENFURT. Unconfirmed reports suggest Br/US PW still being evacuated ODESSA by rail in box cars with German PW. Many instances of theft of clothing and personal effects.

4. Jugoslav occupied AUSTRIA believed clear of PW. Local representatives in the main friendly.

5. Approx 12000 French PW now uncovered presenting accommodation problem. 3900(3970?) being shipped to MARSEILLES and May.

37
FS IN 60239

SECRET
THE MAKING OF AN EXTRACT OF THIS MESSAGE IS FORBIDDEN

COPY NO.

May 17, 1990

Mr. John M.G. Brown
Box 30
Petrolia, CA 95558

Dear Mr. Brown:

I received your letter and the articles you enclosed, and have read them with great interest in light of my own experience when I was liberated by Russian forces in late April, 1945. For many years after the war I thought about the circumstances of my own exfiltration from Stalag IV B in eastern Germany. I tried, unsuccessfully, to communicate my concerns to American officers when I reached Frankfurt's repatriation Center and to again when I was flown to LeHavre, France. My concerns for other prisoners left behind at IV B were treated with initial scepticism, then annoyance at my persistance, and finally with reassurances that the matter "would be investigated".

When I finally came back to the States, recovery from wounds that I had received at the time of my capture, the malnutrition and amoebic dysentary I developed, and the sheer exhilaration of surviving and being home took precedence and life soon resumed some semblance of normalcy.

Your articles, and the enormity of the possible number of prisoners that suffered a quite different fate have served to bring that traumatic period back in memory, tho' possibly dimmed by the passage of the years.

At any rate, the following are my recollections to the best of my ability.

When we were liberated by a Russian tank battalion, a group of us enthusiastically sought out a Russian officer to try to find out what was happening, and when we could expect to be repatriated back to American control. This officer spoke no english, but did have a smattering of German and since my mother had spoken German/Yiddish and my father had come from Russia and had taught me some of that language, I became sort of an unofficial translator for some of the men in my section.

This is exactly what I was told by a Major Vasilii Vershenko.

The Russians were first concerned about the repatriation of the Russian prisoners held in a separate compound at IVB, and the Major indicated that they had to be interviewed individually since they felt that there were many "cowards, traitors and deserters among them and they had to be dealt with expeditiously". Then he told me that the Russians and

the Americans had agreed to a pact wherein the Russians would receive "credits" for each American POW returned. This, he explained was a complex logistical matter, best handled by sending us to Odessa for treatment and repatriation. The callousness of his response and the officious tone in which this information was given, gave me real pause and I tried to explain to others that I was suspicious of their methods and motivations. Most of the men, however, felt that the Russians were our allies and that we were going to be well treated and returned home shortly. That night, my bunkmate, Cpl. William Smith of the 9th Division shared our mutual concerns and decided to take off on our own. The next evening, we "liberated" two Russian bicycles, got thru a gap in the wire where a Russian tank was parked and took off toward the west where we believed the American army would be. Our subsequent journey back thru Russian held territory, our capture by a band of fanatical "Hitler Youth" and subsequent escape, is a two week adventure that is an unbelievable story that must wait for another time to relate. As I said earlier, I tried to tell American Intelligence officers about the Majors comments. But was told that we were probably told that version in order to preserve order in the camp.

I'm happy that you have and are pursuing the matter of missing POW'S, and would appreciate hearing more about your efforts.

I'm truly sorry that I can't shed more light on the subject, but as you know, generally a G.I.'s view of war is limited to what he directly sees, feels and experiences.

Sincerely,

(Martin Siegel)
MHB

From a Survivor of Stalag IV-B who escaped fro Russians

FROM: Major General Ray W. Barker
to: Gen. Bedell Smith

6 SEP 1945

25 May 1945

SUBJECT: Report on Conference with Russian Officials Relative to Repatriation of Prisoners of War and Displaced Persons.

TO: The Chief of Staff
Supreme Headquarters, AEF

1. In accordance with your oral instructions of 13 May 1945, a party of SHAEF officers and enlisted men proceeded by air transport to Halle, Saalkreis, Germany, on 16 May, 1946, for the purpose of conferring with representatives of the Russian High Command on the matter of repatriation of prisoners of war and displaced persons. This party was accompanied from Versailles to the Russian Major Generals DRAGUN and RATOV and their secretaries.

A Russian party arrived from Moscow via Berlin about 1530 hours, same date. This party was headed by Lieutenant General GOLUBEV. When the Russian Mission was finally assembled it numbered some forty officers and forty to fifty enlisted men. Among the Russian officers were one Lieutenant General and six Major Generals.

The Russian party arrived in requisitioned German vehicles of all makes, an American-type armored car, fully equipped, and a radio truck, which latter was in operation during most of the time. All Russian male personnel were heavily armed with pistols, submachine guns and rifles. The first conference was held in a former Luftwaffe officers club at the Luftwaffe school, at about 2100 hours, 16 May, the date of arrival. The respective Missions were headed by General GOLUBEV and the undersigned. After opening statements by various officers, I proposed the immediate initiation of steps looking toward prompt release and return to Allied control of all British and American prisoners of war then in Russian custody, using air and motor transport. This proposal was firmly resisted by General GOLUBEV, who cited all manner of local administrative difficulties which precluded the operation. He stated that serviceable air fields did not exist, which was known by myself to be not the case and I so informed him. The Russian position was very clear that neither now, nor at any time in the future, would they permit airplanes to be used for the movement into or out of their territory of prisoners

-1-

37 230

---

of war or displaced persons, except "Distinguished persons, sick and wounded." From the discussions on this first point, which lasted some two hours, the SHAEF representatives came to the firm conviction that British and American prisoners of war were, in effect, being held as hostages by the Russians, until deemed expedient to them to permit their release. This latter point was further borne out by subsequent events.

As soon as it became apparent that the above proposition could not be effected, the conference proceeded to a discussion of the broad problem of repatriation procedure. This first meeting closed about 0115 hours by which time the general principles upon which repatriation could be effected had been agreed. It was apparent to the SHAEF officers, however, that extreme difficulty lay ahead, because the Russian Mission had brought with them a very detailed plan which contained many items wholly unacceptable to us. Both parties agreed to appoint a drafting sub-committee, which would prepare a joint draft plan, for submission to the Chiefs of the Missions for final approval.

The drafting committee met on 17 May and worked through that day and into the night and on the morning of the 18th. This drafting committee was, however, wholly ineffective, since the Russian members were not authorized to depart in any way from the text given them by Moscow. The work of the drafting committee having proven futile, it became necessary thenceforth for all discussions to be carried on directly between the heads of the Missions. To certain members of the drafting committee, those parties in attendance. On the Russian side, those representatives normally from twenty to twenty-five, including several general officers. The SHAEF representative staff ranged normally representatives of technical services.

2. The text of the proposal brought from Moscow by General GOLUBEV was drawn in the form of a quasi-legal document, and called an "Agreement". Its tenor was clearly that of a party who felt obliged to protect themselves in every way from evasion. The tendency was to extract pledges of compliance to the last degree with the terms of this "Agreement" and the YALTA Agreement. The SHAEF Mission informed General GOLUBEV that since an agreement had already been signed between the three Governments on this subject, it was superfluous to enact any further ones, and that it was the purpose of this conference to evolve a working plan under which the YALTA Agreement could be made effective on the Western German Front. This view was finally accepted by General GOLUBEV.

37 231

The Moscow proposal contained a number of provisions which were either unacceptable to SHAEF or required modification before acceptance, for example:

a. The number of reception/delivery points proposed by the Russians was not sufficient, and did not conform to our road/railway net. It made no provision for points in American territory.

b. It proposed that no limit be placed on the amount of personal effects which each Russian repatriates would be allowed to carry away.

c. It proposed that Russian repatriates should have in their possession at time of ultimate delivery three (3) days rations.

d. An unreasonable and wholly unworkable system of medical documentation, examination and disinfestation was proposed.

e. A complicated and cumbersome method of last-minute documentation and nominal rolls was desired.

f. The wording of their sub-paragraph on transportation was such as to preclude any movement of Russians by marching in Allied territory. Previous study had shown that repatriation in reasonable time made a certain amount of marching mandatory.

5. Conferences tended to be long drawn out, due, first to the language difficulty; second, the apparent fear of the Russian officers to depart from the letter of their Moscow instructions, thus necessitating constant reference to higher headquarters on small matters; third, their very apparent lack of faith in our sincerity or good intentions, thus, causing them to insist on much greater detail in the plan than we felt necessary.

The issue on sub-paragraph f, above, was so strongly debated that the conferees came to a complete impasse on the evening of 19 May. However, we were able to resume discussions at about 2150 hours on the 20th, although this question was not finally determined until the last session, during the night of 21-22 May.

4. A strong effort was made by the SHAEF Mission to establish on the spot, at the conclusion of this conference, what would ultimately become a permanent liaison committee which could thenceforth resolve routine questions of detail with regard to this and other Allied/Russian matters. The Russian Mission could not agree to this, apparently in response to instructions from Moscow.

5. Final agreement on a plan was reached at 0415 hours, 22 May. Signatures were affixed thereto at 1500 hours, same date. Copy of the Agreed Plan is attached hereto (Appendix "A").

6. Although the Plan indicates that its provisions are to become effective twenty-four hours after signature, it actually began to operate on the afternoon of the 20th, with the transfer of some 2,200 US/British prisoners from the camp at Luckenwalde.

a. There is every indication that the Russians intend to make a big show of rapid repatriation of our men, although I am of the opinion that we may find a reluctance to return them all, for an appreciable time to come, since those men constitute a valuable bargaining point. It will be necessary for us, therefore, to arrange for constant liaison and visits of inspection to "uncover" our men.

7. On the afternoon of 22 May, two groups of SHAEF officers, by arrangement with General COLUMBY, went on a tour of camps in the Russian area where US/British prisoners of war are thought to be located. They will investigate conditions and gather data thereon. This is the first instance of either American or British officers being permitted to visit such camps in the forward Russian area.

R. R. BARKER
Major General, GSC
A. C. of S., G-1

## CIPHER TELEGRAM

This message will not be distributed outside British Government Departments
or Headquarters or re-transmitted, even in cipher, without being paraphrased.
(Messages Marked O.T.P. need not be paraphrased.)

To : Troopers Info : 30 Mission - AC? British Delegation Hungary
AC? Roumania.

From : A.F.H.Q. Signed ALEXANDER CINC FWEAR.

FX 82606    26 May.    1859 B.hrs.

**SECRET**

DECLASSIFIED
NND750115
Authority
By _____ NARA, Date _____

1. Agreement with Russians at GRAZ only applies to handing over of Soviet citizens in British zone Austria. NO request NO reciprocal guarantee in respect of British P.Ws obtained apart from hair hearted promise which so far has NOT been honoured. Evacuation to USSRA still continuing from this area.

2. Pressure to plan on overland exchange on a local contact basis till Moscow issue directive to GRAZ commander. - Marshal Tolbukhin's ...

3. Considered essential Moscow be asked to announce their agreement to local overland exchange, as there are 15,597 U.S.A. account, 8,462 British account awaiting repatriation in this theatre.

FW:
U.S.Mission
Embassy
Canadian Dept. Embassy.
Australian Legation
New Zealand Legation.
(India)
(Poland)

---

**SHAEF**
STAFF MESSAGE CONTROL

## OUTGOING MESSAGE

CONFIDENTIAL

ROUTINE

| | | |
|---|---|---|
| TO | : | US MILITARY MISSION MOSCOW FOR DEANE |
| SHOW | : | SHAPE MAIIN SIGNED SCARF |
| REF NO | : | S-99942    CITE SHGS    TOO:3114508 |

Latest available displaced persons and prisoners of war figures show about 1,500,000 Western European (French, Belgian, Dutch and Luxembourgeois) either repatriated from or at present held in SHAEF area. Soviet delegates at MOSCOW conference ...

| ORIGINATOR | : | G-1 |
| | | SGS |
| | | AG RECORDS |

37   181   31 May 45   1555Z    EB/em    REF SGS-99942
TOO:3114508

AUTHENTICATION: E.A. MURPHY
Lt Col

29 OCT 1949

STAFF MESSAGE CONTROL

# INCOMING MESSAGE

JEBR

TOO 302000B MAY

SHAEF FWD  9/31
TOR 310010B MAY
1th 310156B MAY

SGS-SHAEF File No:

S E C R E T

R O U T I N E

FROM          : SHAEF MISSION FRANCE SIGNED LEWIS

TO FOR ACTION : SHAEF FORWARD G-5 DP BRANCH

FOR INFO      : SHAEF MAIN

REF NO        : MF-14427,   30 MAY 1945

      Accordance your telephone request, cable from
Fifteenth Army French Detachment to General CHERRIERE MMFA
Hotel CONTINENTAL PARIS of 25 May is paraphrased for your
information.

      Report of Lt D HAVERNAS according to confirmed
reports, Russians still do not release thousands of French
ex-PW's and civilians, forcing them to work. Many transferred
eastwards to unknown destination. Please inform high authority.
700 ex-PW's are evacuated daily from this area to UDINE. Civilians held under difficult food and accomodation conditions.
Mission CHERRIERE local staff still badly needed.

      Please dispatch instructions concerning VICHY
Consul NOEL HENRY and staff who claim to have been approved
by French Government. Signal Captain SAVARY (SACARY?).

ACTION        : G-5

INFORMATION   : SGS
                G-1
                SUSPENSE
                AG RECORDS

FS IN 7747   31 May 1945   0225B  JOB/feh  Ref No: MF-14427

36  523

S E C R E T

8

## SECRET

COPY NO.

30 may

30C

THE MAKING OF AN EXACT COPY OF THIS MESSAGE IS FORB***

SHAEF 383.6-2 Md

30 May 1945

MEMORANDUM FOR GENERAL _____

SUBJECT: Displaced Persons, Allied ex PW and German PW.

I. Object of Memorandum.

1. To obtain the general numerical picture of displaced persons and Allied ex PW in Germany and German PW on Continent.

2. To obtain information on the progress of repatriation of displaced persons and Allied ex PW and the disbandment of the German Army.

3. To note any fact of medical importance discovered.

II. Tactical Boundary and Occupation Zones.

The date of withdrawal from present tactical boundary to final occupation zones is under discussion at the highest levels. The final decision may be left to the Quadripartite Control Council.

In the meantime transfers from present SHAEF Area to Russian Area will occur across the present tactical line.

III. Disbandment of German Armed Forces.

The present approximate strength of German armed forces in SHAEF Area on the Continent is 5,300,000.

... the following classes has been completed and is proceeding ...

Agricultural workers
Miners
Transportation workers
Men over 50
Women
Other essential workers

The following classes are being considered for discharge:

Enemy nationals other than Germans
Nationals of Allied Nations incorporated in German forces,

except if belonging to ... formations or War Criminals. The resultant decrease in strength is not known nor is an estimate obtainable.

IV. Allied ex PW remaining in Shaef Sphere.

Accurate figures are not obtainable in many cases as the transfer to and from the Russian Sphere has commenced.

Figures obtainable:

| | |
|---|---|
| US/Br | 8,000 |
| Poles | 50,000 |
| Russian | 400,000 * |
| Yugo Slav | 50,000 |
| Italians | 50,000 * |

V. Displaced Persons.

1. SHAEF Area.

| Nationality | On hand in Germany | Repatriated |
|---|---|---|
| French | 200,000 | 882,000 |
| Belgian | 70,000 | 147,000 |
| Dutch | 56,000 | 120,000 |
| Russians | 1,284,000 | 160,000 |
| Poles | 651,000 | |
| Yugo Slavs | 56,000 | |
| Czechs | 31,000 | |
| Greeks | 7,000 | |
| Italians | 242,000 | - under discussion |
| Other Allied | 106,000 | |
| Other ex enemy | 85,000 | |

Approximate Total 3,055,000

Repatriation of Western Nationals is proceeding rapidly these figures are the latest obtainable but are already out of date.

No figures of Displaced persons hospitalized in Germany are available.

2. Russian Sphere.

| | PW | | DP |
|---|---|---|---|
| Belgian | 50,000 | | 115,000 |
| Dutch | 4,000 | | 140,000 |
| British | 20,000 | | |
| U.S. | 20,000 | | |
| French | 250,000 | | 850,000 |
| Approx. Total | 1,145,000 | | |

No information as to repatriation obtainable, subject under discussion.

- 1 -

- 2 -

STAFF MESSAGE CONTROL

# OUTGOING MESSAGE

CONFIDENTIAL

ROUTINE

U. S. MILITARY MISSION
MOSCOW.

TO        : US MILITARY MISSION MOSCOW FOR DEANE

FROM      : SHAEF MAIN SIGNED SCAEF (Eisenhower)

REF NO    : S-89942    CITE SCGS    TO: 3114303

Date      : 12 June 1945

Distribution: Ambr;
Hq;
G2S;
Air;    Number:

Subject   : Evacuation of POW's

Originator: Deane

Precedence: PRIORITY

Security  : TOP SECRET

Typist: McCoy...

Approved:

Cryptographer:

Cryptographed At:

M 24524

TOP SECRET

To: ADMIN for GENERAL MARSHALL from DEANE: TOPSEC....

: In further reply to MARK-90429 of 30 May, General Golikov' head
of the Soviet Repatriation Commission, assured me last night that
they would make every effort to locate individual Americans who might
be stranded in Soviet-occupied territory and when located, they would
be evacuated by any route that I requested. I informed him that I
would like to have them evacuated westward, in accordance with the
agreements made at Halle. I do not believe there are very many, and
in any event, although I do know of a few and have already furnished
the information to the Soviet Repatriation Commission.

: Concerning those liberated prisoners of war in Marshal Tolbukhin's
area, estimated in excess of 15,000, Golikov assured me that they
would be evacuated westward in accordance with the Halle agreement.
He confirmed my previous belief that Odessa was to be abandoned as
a transfer camp for the repatriation of American prisoners, of war.

Latest available displaced persons and
prisoners of war figures show almost 1,600,000 Western
Europeans (French, Belgian, Dutch and Luxembourgeois)
either repatriated from or at present held in SHAEF
area. Soviet delegates at YALTA conference
listed only 300,000 Western Europeans in their area.
Combined Working party on European food supplies,
composed of representatives from UNRRA, SHAEF, USSR, UK,
and USA, including Soviet delegate LINSCHENKO
estimated approximately 3,000,000 displaced Western Europeans
in enemy-held territory at beginning 1944. This
discrepancy of over 1,000,000 Western Europeans is
causing the Dutch and French Governments considerable
anxiety. In further breakdown of all Western European
displaced persons and prisoners of war now held by
USSR, Soviets help to clarify situation by
giving nationality breakdown of all Western European
displaced persons and prisoners of war now held by
them and gathered in GERMANY, POLAND, USSR, BULGARIA,
HUNGARY, ROUMANIA, AUSTRIA and CZECHOSLOVAKIA.

This copy only MAY BE SHOWN to only
accredited unofficial researchers under
See/Army by RAG/_____ date _____
2 9 DEC 1949

AUTHENTICATION: L.A.TROSKY
Lt Col

REF HQ:S-89994
TO:3114303

ALL (36 S.H.)

CONFIDENTIAL

ORIGINATOR   : G-1
INFORMATION  : G-5
                BGS
                AG RECORDS

37   181
SMG 104 33MA   31 May 45   1555E   SL/mm

NND750115

TOP SECRET

FOR GERMANY

SECRET

June 1, 1945.

No. 449.

Col. Biddle:

Subject: Overland Exchange of Ex-Prisoners of War
and Displaced Persons, Liberated by the Allied
Expeditionary Force and the Red Army.

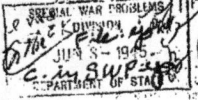

JUN 13 1945

JUL 26 1945

The Secretary of State,
Washington.

Sir:

With reference to my telegrams Nos. 2625 and 2936
of May 15 and 24, 1945, respectively, I have the honor
to transmit a copy of the plan agreed by representatives
of Supreme Headquarters, Allied Expeditionary Force,
and Supreme Command, Red Army, at Halle, Germany, May 22,
1945, for the most expeditious overland delivery of Allied
and Soviet ex-prisoners of war and displaced persons
liberated by the Allied Expeditionary Force and the Red
Army. The two delegations were headed by Lieutenant
General K. D. Golubev, Red Army, Soviet Assistant Administrator for Repatriation, and Major General R. W. Barker,
U.S.A., Assistant Chief of Staff, G-1, SHAEF.

As reported in my telegram No. 2936 of May 24, 4 p.m.,
the plan, in its final form, was only agreed after six
days of conversations. Allied desiderata which were
unacceptable to the Red Army delegation included permission to fly transport aircraft into Soviet-occupied
territory and the establishment of a standing working
committee or commission to be set up at some convenient
location near the demarcation line, such as Leipzig or
Halle, charged with handling day-to-day details of the
exchange. Although General Golubev would not agree to
the incorporation of a paragraph providing first priority
delivery of U.S. and U.K. ex-prisoners of war, he gave
his most solemn personal assurances that all U.S. and
U.K. ex-prisoners of war would, in fact, be given preferential treatment. A request for second priority for
Western European ex-political deportees, in accordance
with the desires of the Western European governments
that such persons be repatriated before their respective
ex-prisoners of war and other displaced persons, was
countered by the flat assertion that all political prisoners held in German concentration camps overrun by the
Red Army had been released and that there were, accordingly, no more political prisoners in Soviet-occupied
territory. With respect to this category of displaced
persons, not even verbal assurances were to be had.

The Red Army delegation, for its part, had arrived
in Halle with a set document which contained numerals
provisions unacceptable to SHAEF. In form, the document

DECLASSIFIED
E.O. 12356, Sec. 3.3
NND 760080
By ___ NARA, Date 2/17/87

SECRET

VERY
IMP.

331/6/29

By .......... WOJE
MEMORANDUM: 5 NOV 1945
Date

11 June 1945

Repatriation of British, US, and Other United Nations
Prisoners of War as of 7 June 1945

1. Figures of British and US recovered Prisoners of War shown in
this memorandum represent personnel handled by PWX Organization. The same is
true to a large extent of the numbers of Dutch and Belgians. Russians reported
delivered to the Russians have been handled through both DP and PW channels.

It is believed that there are more British and US PWs to be recovered and the War Office/War Department are endeavoring to ascertain the
number and if possible the location of such personnel. As regards the number
of US PWs the figure is somewhat complicated by the large number of missing in
action personnel recovered as PW but never reported as such due to lack of
time between their capture and recovery.

From the SHAEF area, practically 90% of BR/US recovered personnel
were evacuated by air, the majority of the British being flown to the UK.
Of the remaining 10% evacuated by land/sea a large number were medical
evacuees. A large proportion of French, Dutch, and Belgians and some Polish
recovered PWs have been evacuated by air as well as a number of Russians flown
to the vicinity of the Russian lines for handing over to the Russians. Air
evacuations is still continuing for allied personnel.

2. Under the provisions of the Crimean Agreement, negotiations were
completed 22 May 1945, between the Soviet High Command and SHAEF for delivery
of liberated former Prisoners of War and citizens to the respective forces
for repatriation. Certain points along the Soviet-SHAEF line in Germany and
Austria were agreed as assembly points to which Soviet and Allied subjects
would be collected for delivery to Soviet or Allied forces as the case might
be.

3. BRITISH: The total number of British and Dominion Prisoners of War
repatriated is less than the total expected which is (175,000.) Information has

18 0005  *Already reduced secretly from 199,500 Brit/Commonwealth known - PW*

CONFIDENTIAL

*The Cover-up :*

## NEW YORK HERALD TRIBUNE.

### WEDNESDAY, JUNE 6, 1945

## 25,000 Missing U. S. Soldiers Turn Up Alive

### Men, Found in Nazi Prison Camps, on Way Home; 89,776 Prisoners Found

By Carl Levin

By Wireless to the Herald Tribune
Copyright, 1945, New York Tribune Int.
SUPREME HEADQUARTERS,
Allied Expeditionary Force, Paris,
June 5. — Twenty-five thousand
American soldiers who had been
listed as missing in action have
turned up as prisoners of war and
now are on their way back to the
United States, it was disclosed
here today by the Recovered
Allied Military Personnel Division
of the Office of the European
Theater of Operations Provost
Marshal.

Lieutenant Colonel W. P.
Schweitzer, of Elizabeth, N. J.,
chief of the R. A. M. P. Division,
said there was little likelihood that
any others would be found, except
stragglers who may yet come in
from Russian-occupied Germany.

It also is believed, he said, that
the 89,776 American prisoners of
war recovered from camps in Germany constitute the entire American prisoner group which was in
Nazi hands during the final stages
*(Continued on page 8, column 4)*

## 25,000 Missing

*(Continued from page one)*

of the war. Of this number, which
includes those previously reported
missing in action, all but about
20,000 are either already in the
United States or at sea en route.
The 20,000 have been processed
and are waiting at Lucky Strike,
the R. A. M. P. camp near Le
Havre, for ships home.

Schweitzer, a New York paper
manufacturer in civilian life, said
most of the recovered prisoners
will suffer no permanent health
defects as a result of their experiences. Surprisingly, he said, those
who had been captives for the
shortest periods were found to be
in the worst condition.

This was particularly true of the
men captured in the Ardennes
battle last December. Many of
these, Schweitzer said, had no rest
and almost no food for weeks as
the Nazis tried to keep them from
falling into Allied hands as the
end neared. A slice of bread a day
and a little soup, if it was available, was a common ration for
Ardennes captives during their
forced marches to the Nazis' rear.

The weakest of these, as well as
a considerable number of other
wounded or ill prisoners who
needed hospitalization and could
not go through the established repatriation procedure, are not included in the R. A. M. P. figure.

THE REAL STORY :

By December 19, 1945:

(1) 11,800 POWs were declared dead

(2) 5,414 still listed as POW

(3) 3,000 MIAs, known or suspected
    of being POWs had been
    declared dead.

    20,000 + POWs declared dead or
          still listed as POW
    7 months after end of war.

THIS IS FOR EUROPE ONLY!!!

ALSO, THIS 29,000 FIGURE IS IN
ADDITION TO THE 800 AMERICAN POWs
WHOSE DEATHS WERE REPORTED BY THE
GERMAN GOVERNMENT, AND THE 38 POWs
WHOSE DEATHS WERE REPORTED BY OTHER
AMERICANS WHO WERE REPATRIATED.

On my first leave I was told, "Don't talk to anyone."
was confused: "Why?" "If you say anything at all about wh
happened overseas, Red will get court-martialed." Red Le
son, because he was blind and had beriberi so bad that even
fly landing on his feet made him cry, was left behind by t
Japs in Cabanatuan. When he was released by the Rangers ear
in 1945, he got home to our little town before everyone els
There was a big City Hall meeting. Everyone asked question
about what happened to their sons or husbands. Red new
lied. The War Department had never told the mothers anythi
about what Red was now telling them. This was the first tim
Some of the mothers threw a fit. Washington, D.C., got a floo
of indignant letters, "Why haven't we been told? What are yo
hiding?" The chair-bound brass in the Pentagon were there fi
glory, not lip. Red got an airline ticket to Washington. Messag
said, "Urgent!" For one week they took turns screaming at Re
Court-martial papers were drawn up, "Revealing secrets in tim
of war." A Jap's favorite pastime was to stand you at attentio
and then slap you silly. In Washington, Red found the sam
game, but with a different opponent. The generals said Re
could go home and they would hold his court-martial paper
in his file. Just one more letter from Janesville and they woul
activate them. He still will never talk about what happened
They really made a believer out of him. He remained a prisone
in his own home. He never went out because peop
wanted the truth of what happened.

Col. Biddle:    Date Subject: Overland Exchange of Ex-Prisoners of War
and Displaced Persons Liberated by the Allied
Expeditionary Force and the Red Army.

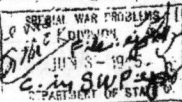

JUN 13 1945     SPECIAL WAR PROBLEMS     JUN 8 1945     WAR REFUGEE BOARD     JUL 26 1945

The Secretary of State,
Washington.

Sir:

With reference to my telegrams Nos. 2625 and 2936
of May 15 and 24, 1945, respectively, I have the honor
to transmit a copy of the plan agreed by representatives
of Supreme Headquarters, Allied Expeditionary Force,
and Supreme Command, Red Army, at Halle, Germany, May 22,
1945, for the most expeditious overland delivery of Allied
and Soviet ex-prisoners of war and displaced persons
liberated by the Allied Expeditionary Force and the Red
Army. The two delegations were headed by Lieutenant
General K. D. Golubev, Red Army, Soviet Assistant Administrator for Repatriation, and Major General R. W. Barker,
U.S.A., Assistant Chief of Staff, G-1, SHAEF.

As reported in my telegram No. 2936 of May 24, 4 p.m.
the plan, in its final form, was only agreed after six
days of conversations. Allied desiderata which were
unacceptable to the Red Army delegation included permission to fly transport aircraft into Soviet-occupied
territory and the establishment of a standing working
committee or commission to be set up at some convenient
location near the demarcation line, such as Leipzig or
Halle, charged with handling day-to-day details of the
exchange. Although General Golubev would not agree to
the incorporation of a paragraph providing first priority
delivery of U.S. and U.K. ex-prisoners of war, he gave
his most solemn personal assurances that all U.S. and
U.K. ex-prisoners of war would, in fact, be given preferential treatment. A request for second priority for
Western European ex-political deportees, in accordance
with the desires of the Western European governments
that such persons be repatriated before their respective
ex-prisoners of war and other displaced persons, was
countered by the flat assertion that all political prisoners held in German concentration camps overrun by the
Red Army had been released and that there were, accordingly, no more political prisoners in Soviet-occupied
territory. With respect to this category of displaced
persons, not even verbal assurances were to be had.

The Red Army delegation, for its part, had arrived
in Halle with a set document which contained numerous
provisions unacceptable to SHAEF. In form the document

SECRET

SECRET
-3-

Following the conclusion of the meeting at Halle, certain members of the Red Army delegation were taken on a fairly extensive tour of prisoner of war and displaced person facilities on SHAEF territory in the vicinity of Halle, Leipzig, and Magdeburg. Contrary to the report transmitted in my telegram No. 2936 that three small parties of Allied officers would be given facilities to inspect camps in the proximity of the demarcation line on Red Army territory, information has now been received that, in fact, only one party was received in the Soviet zone and its movements were limited to a visit to the camp at Riesa, just east of Leipzig.

The distribution list attached to the enclosed copy of the "PLAN" represents, with the exceptions indicated, the personnel which composed the SHAEF mission to the meeting. The Red Army delegation, as already reported, included, in addition to General Golubev, four major generals and some 40 other officers, not to mention the security guard of 51 officers and men equipped with an armored car and a mobile radio transmitter.

The unannounced and unauthorized arrival of several Red Army communications aircraft during the course of the meeting at Halle is understood to be under examination by Air Staff of SHAEF. What action is contemplated in this connection is not yet known to this office.

According to all available reports the overland exchange is progressing satisfactorily.

Respectfully yours,

Donald R. Heath
Deputy U.S. Political Adviser

Enclosure:
   Copy of plan.

Copies:
   Amembassy, Moscow; U.S. Political Adviser, AFHQ.

9 June 1946

331/6/28

Major General John R. Deane,
Military Mission, Moscow,
Moscow, Russia

My dear Deane:

Enclosed herewith is a copy of my reply to General Golubev's letter of 18 May, relative to the dispatch of General Vershinin and some 162 Russian officers to the SHAEF area.

For your information, following is some of the background of this matter. At the Halle Conference in May, General Golubev read this proposal to me and asked for my acquiescence. I informed him that the matter was of such far-reaching consequence that I should have to take it under study with the SHAEF Staff and with 21 Army Group. At the time, I realized that, of course, the Russians would never permit any such a scheme to be put into effect within their own area, and I realized that so far as the British and Americans were concerned, we could not reasonably ask for it, because all of our people would be withdrawn within a very short time. However, in order to put him on the spot, I asked him if he was willing to grant similar privileges to the French who would have great numbers of their people in the Russian Zone. As expected, his reply was non-committal, in that he said he would have to refer that to his Government.

As you can imagine, a scheme such as this amounts to the creation of a Soviet "Empire" within the SHAEF area, which would produce an impossible situation, administratively, as well as giving them boundless opportunity for criticism. My personal feeling is that the primary purpose of this is in the field of Intelligence.

It is our view that the task of repatriation is ours and entails a responsibility which we could not share with the Russians. We take the firm position that any Russian officers on repatriation duty with the AEF must work, in effect, as semi-staff officers of the Headquarters to which they are accredited; that their work is in extension of that of our staffs and not along independent lines. We attempt to follow this principle, although we realize

383 6/2

SECRET
(TOP)

17 1052

that they always do a great deal of "free wheeling", on their own. We have allowed the Russian officers an unprecedented amount of freedom of movement and action, far beyond anything ever permitted our people in the Russian area.

You may be interested in our efforts to make some sort of an inspection of camps in the Russian zone where our men were kept. We had repeatedly tried to send officers in to visit these camps and had met with firm refusal from Russian officials, including General Golubev, in person. At the end of our conferences, General Golubev asked me for permission to visit some of our camps where Russians were being kept. I told him we would be glad to have him visit them, but stated that we also wanted to visit some of the Russian camps. He said he was in full accord with this and promised to give one of my officers an order authorizing him to visit five different camps. The next morning when our officer set upon his journey he was accompanied by a Russian Major, who stated he had the necessary orders in his pocket. After visiting the first and nearest camp, our officer then stated he wished to go on to the next camp, whereupon the Major produced an order from General Golubev which restricted our officer's visit to this one camp only. No amount of persuasion would prevail upon them to extend the visit, in spite of the promise of General Golubev. This is the only instance of the Soviet authorities permitting British or American officers to visit camps in their area, which certainly is in sharp contrast with the liberal policy pursued by us, although I must admit that our own policy is dictated in very large measure by our own convenience, since we need these Russian officers to help us in handling the huge numbers of their people whom we have on hand.

General Golubev's latest letter to you on the situation in our area contains the usual unfounded allegations and, in addition, the ridiculous charge that Soviet citizens have been poisoned in a way which cannot be considered accidental. We are preparing a reply and propose to make it a very sharp one.

Again, with reference to the proposal about General Vershinin and his 182 officers, be assured that we have no intention, whatever, of ever granting it.

With kindest regards,

Sincerely,

(TOP)

17 1053

R. W. BARKER
G-1

SHAEF MAIN                                                  S-91662          274

Repatriation                                               19 June 1945

Hqs ✓ GAS   Rmb

WIM                                                        191925Z

Carter                   **SECRET**                        191600B

6487                                                       SECRET
                                                           ROUTINE

U S MILITARY MISSION MOSCOW for DEANE from SHAEF MAIN signed

(Eisenhower) → SCAEF cite SHX

    1.  Reference our S-89942 of 31 May and your M-24569 of
3 June.

    2.  A further approach to the Soviets regarding numbers of
western Europeans in Soviet-occupied area of Eastern Europe is
urgently necessary.  About 1,200,000 French have been repatriated.
Less than 100,000 remain in SHAEF-occupied area.  French insist
total POW and displaced persons is 2,300,000.  Even allowing for
several hundred thousand unaccounted trekkers, discrepancy is still
very great.  About 170,000 Dutch have been repatriated, with less
than 25,000 in SHAEF area.  Total Dutch estimate of deportees is
540,000.  Belgian discrepancy is comparatively smaller.

    3.  Have the Soviets any further evidence of Western Europeans
who are accessible to Allied lines?  Is it possible to obtain
figures for those not so accessible - that is, Western Europeans
in the USSR, Poland, Austria, Czechoslovakia, and Balkan countries?

Action: Hy

**SECRET**

JAJA
TOO  081500B JUNE

SHAEF FWD 144/9
TOR  091113B JUNE
fw,  091256B JUNE

## SECRET
## PRIORITY

FROM  :  PRISTERN AUDLEY LONDON
TO    :  SHAEF FORWARD PWX
REF NO :  T/108/PW5, 8 JUNE 1945

Your FWD-22864, 31 May.

1. Your 1. Following is summary of information received or officially notified PW not yet evacuated from Continent.

A. UK Army      6050
B. RAF           900 approx
C. CANADA        101
D. AUSTRALIA     454
E. S AFRICA      701
F. N. Z.         174
G. Merchant      171
   Navy

Total to date: 8551

159

Other figures not yet available, will forward soonest together with amendments to above.

FS IN 2563

**SECRET**
- 1 -
**SECRET**

THE MAKING OF AN EXACT COPY OF THIS MESSAGE IS FORBIDDEN

COPY NO.

---

29 JUNE 45 ... (continued)

"Before The HALLE Conference we had made numerous attempts to visit PW Camps in the Russian Zone and always met a fl... refusal. After the HALLE Conference General GOLUBEV asked to visit Camps where Russians were being kept. We agreed and asked him for permission to visit Camps in the Russian Zone. He agreed to allow 1 of our Officers to visit 5 Camps. One of my representatives started on the trip accompanied by a Russian Major who stated he h... the necessary orders. After visiting the first and nearest Camp t... Russian Officer produced orders signed by General GOLUBEV restrict... our Officers visit to the one Camp. This is the only instance of Soviet authorities permitting U S or British Officers to visit Camps in their area, which is in sharp contrast to the liberal policy pursued us."

"With reference to the proposal of sending General VERSHININ and his 162 Officers, we have no intention what- ever of granting it."

WX-23/28 is SMC IN 8425, 20/6/45, G-1.

ORIGINATOR:  G-1            AUTHENTICATION:  R.H.S. VENABLES, BRIGADIER

INFORMATION:  SGS
              G-4
              G-3
              MR MURPHY
              MR KIRKPATRICK
              G-5
              **
              AG RECORDS

egos to the French who would have many citizens in the Russian Zone. His reply was non-committal stating he would have to refer t to his Government.

39 1400        1709B     KBH/ra     REF NO: S-94080
                                     - TOO, 291500B

- 2 -
**SECRET**

POW's      ~~GNNP750118~~     25 June 1995

Authority_____

BY_____RARS, Date     2̲5̲ ̲J̲U̲N̲E̲ ̲,̲ ̲1̲9̲4̲5̲

Hqs / CAS

JM                              251130Z

Feldstein                      251030B

6590     **SECRET**         SECRET PRIORITY

To USABM MOSCOW, HQ 21st ARMY GROUP, CG 12th ARMY GROUP, CG 6th ARMY GROUP cite SHAEF signed SCAEF from SHAEF MAIN

The following communication has been received from Gen. Conrad Actg Deputy A.C.ofS, G-2, ETOUSA.

"Possibility that several hundred American prisoners of war liberated from Stalag Luft 1, Barth, are now confined by the Russian Army in the Rostock area pending identification as Americans is reported by an American who recently returned from such confinement.

S/Sgt Anthony Sherg was one of 1000 air force officers and non-commissioned officers who left Stalag Luft 1 immediately prior to assumption of control in Barth by the Red Army in order to obtain rumored air transport from Wismar. The group of ten in which Sgt. Sherg travelled was arrested by Russian soldiers and held in jails in Bad Doberan, then Rostock. Ten other Americans were seen under similar circumstances in Rostock.

Russian authorities demanded identification papers, which no prisoner possessed, and refused to consider dog tags proof of the Americans' status. The Americans were well fed and well treated but Sherg complains there was no disposition to speed identification and evacuation. After 25 days he escaped from jail and made his way to British Forces.

From his own observations and conversations with other former prisoners he believes several hundred Americans may be held in like circumstances in the Wismar-Bad Doberan-Rostock Area."

2. Will you please expedite any action that can be taken by ___ regarding the above and also possibility that same situation
SECRET

No. 04997

Major General John R. Deane,
Chief, American Military Mission
in Moscow

21 July 1945

Dear General Deane:

In answer to your letter of 13 July, I can state the following:

1. During the entire time of repatriation, the following numbers of foreigners were handed over to the Allied Command:

| | |
|---|---|
| Americans | — 22,010 |
| English | — 20,483 |
| French | — 225,111 |
| Belgian | — 31,530 |
| Dutch | — 29,558 |
| Luxembourg | — 1,246 |
| Norwegian | — 1,011 |

2. Inasmuch as a check has not yet been finished, I am unable to tell you the precise number of Western Europeans who have remained in the Soviet zone of occupation in Germany after Allied Anglo-American troops withdrew.

However, we know exactly that in this zone after the withdrawal of troops there were more than 100,000 foreigners of which more than 70,000 were Poles, 80 Americans and 15 English.

3. We do not have precise information concerning the number of Soviet citizens who have been returned to us by Anglo-American troops and left by them in the zone of occupation. Therefore, we would be very appreciative if you could tell us.

Respectfully,

/s/ Golubev
Lt. General

Translated by Major Hall
vlm

SECRET

---

6 June 1945, 1300 hours.

Present:

| Americans | Russians |
|---|---|
| General Deane | General Golubev and Staff, including |
| Captain Ware | General Vershinin |
| | Lieutenant Tarasov |

GENERAL GOLUBEV reported that the meeting at Halle had been satisfactory where conversations lasted for about a week.

GENERAL DEANE asked, if General Golubev could give him the breakdown by countries of the Allied liberated prisoners of war still in German-occupied areas.

GENERAL GOLUBEV read off from his notes a few figures for various nationalities. He said there were over 50,000 Americans in all, 3,000 of whom had been evacuated through Odessa, and 17,000 westward through the Allied lines. He mentioned some 300 who had been sent through Budapest leaving. He claimed only 10 or 15 Americans who had not been evacuated. He said there had been over 100,000 French, 28,000 of whom had been repatriated through Odessa. He said there were 50,000 Italians who were not being returned yet.

GENERAL GOLUBEV said the original estimate of 300,000 Allied prisoners for registration was low. He insisted that 500,000 already have been evacuated and that there are probably 100,000 more.

GENERAL DEANE indicated that the representatives of Holland and Belgium believed that they had more war prisoners to be repatriated than the combined numbers indicated by both the Allies and the Russians. General Golubev said Belgium should not worry about repatriation of their citizens, but that it would be well for them to concern themselves with speeding up repatriation of Soviet citizens from their countries. He said the Russians had turned over the whole Belgian General Staff, but that no Soviet citizens had been returned from either of those countries. General Golubev said that he was supplied to bear the claim that their right be 5,300 U.S. prisoners still in Marshal Robokin's army, and that their right be supposedly being evacuated through Odessa. General Golubev said this was not so.

GENERAL DEANE said he had had an opportunity to speak with General Golubev the other night at the reception which was given for Mr. Hopkins, and that General Golubev said the remaining U.S. stragglers and prisoners of war being released from hospitals may be evacuated westward by any means which General Deane might desire.

DEPARTMENT
The Adjutant General's Office
Washington

_Biello_
_8 August 1945_

AGPC-R 383.6
(212047)

1 August 1945

SUBJECT: Return to United States of Military Personnel Liberated in Europe.

TO:      All Officers on Distribution List.

1. The following twenty-six (26) military personnel, having been returned to military control after being released or having escaped from enemy territory are being evacuated to the United States. This movement constitutes a permanent change of station for each individual with temporary duty enroute to U. S. Port, Reception Station, and Redistribution Station. The latter will designate the new permanent station for each individual.

2. ETA is 8 August 1945 and Port of Debarkation is New York, New York.

| NO | RANK OR GRADE | NAME | A OR S | ASN |
|----|---------------|------|--------|-----|
| E 1098-2A | | | | |
| 1. | Pfc | WIERZECHOWSKI, Alexander P. | Cav | 32896914 |
| E 1098-6A | | | | |
| 2. | Pfc | CRAYMER, Dennis G. | Inf | 35142455 |
| 3. | Pvt | WEISSMAN, Val E. | Inf | 35809731 |
| E 1098-21A | | | | |
| 4. | 2d Lt | BURNS, William U. | Inf | 01312404 |
| E 1099-3A | | | | |
| 5. | Pfc | CERRA, Robert J. | 6 Armd | 35375212 |
| E 1099-14A | | | | |
| 6. | Pfc | RUEFLI, Albert F. | 13 Inf | 39046521 |
| E 1100-1A | | | | |
| 7. | 2d Lt | LIMON, Gerald R. | Bm Gp | 0809658 |
| E 1100-2A | | | | |
| 8. | Pvt | PETRELLA, Anthony G. | CE | 32194451 |
| E 1100-3A | | | | |
| 9. | Tec 5 | LIVELY, Denver C. | Inf | 35200763 |
| E 1100-21A | | | | |
| 10. | T Sgt | MC ELYEA, Ralph W. | Bm Gp | 33479223 |
| E 1101-2A | | | | |
| 11. | Pfc | NEGLI, Rosario A. | Inf | 32823822 |
| E 1101-2B | | | | |
| 12. | Maj | TAKACS, William J. | FA | 01172689 |
| 13. | Cpl | CASO, Daniel R. | FA | 32108885 |
| 14. | Pfc | BACAS, Lee L. | FA | 32091301 |
| 15. | Pfc | PETRELL, Joseph J. | FA | 33553607 |
| 16. | Pfc | PRATE, Ralph | FA | 32954177 |
| 17. | Pvt | BARRY, Harold J. | FA | 32194721 |
| E 1101-3A | | | | |
| 18. | Pfc | JONES, Van L. | Armd | 35212725 |

SECRET

THE TOTAL NUMBER OF P.O.W. DEATH AND THE NUMBER OFFICIALLY REPORTED
(Information from Japanese sources)
(The plain figures indicate the number of deaths)
(The figures in parentheses indicate the number officially reported)
As of Sept. 20, 1945

| CAMP | AMER. | BRIT. | AUST. | CANA. | DUTCH | OTHERS | TOTAL | NOT REPORTED |
|---|---|---|---|---|---|---|---|---|
| Hakodate | 2 (2) | 94 (94) | | | 82 (66) | | 178 (162) | 16 |
| Sendai | 20 (1) | 9 | 4 | | 26 (4) | 1 | 60 (5) | 55 |
| Tokyo | 173 (69) | 198 (106) | 79 (64) | 100 (46) | 110 (44) | 7 | 667 (329) | 338 |
| Nagoya | 21 | 5 | 1 | | 15 | 1 | 43 | 43 |
| Osaka | 493 (397) | 584 (211) | 30 (12) | 23 (15) | 163 (72) | 9 | 1102 (787) | 395 |
| Hiroshima | 11 (6) | 49 (7) | 14 )5) | | 11 (2) | | 45 (20) | 25 |
| Fukuoka | 363 (185) | 300 (271) | 52 (15) | 10 (2) | 414 (359) | 30 | 1172 (830) | 342 |
| Chosen | 2 | 23 (12) | 1 | | | | 26 (12) | 14 |
| Formosa | 25 (20) | 313 (216) | 3 | 2 | 9 (4) | | 350 (242) | 108 |
| Mukden | 254 (222) | 2 (2) | | | | | 256 (224) | 32 |
| Shanghai | 20 (8) | 16 | | | | 2 | 38 (8) | 30 |
| Hongkong | 5 | 236 (129) | | 171 (107) | | 5 | 417 (236) | 181 |
| P. I. | 5020 (4961) | 49 (24) | 1 (1) | 2 (2) | 6 (4) | 16 (15) | 5085 (5007) | 78 |
| Siam | 134 (94) | 5807 (5369) | 2167 (2084) | | 3110 (2335) | 11 (2) | 11230 (9885) | 1345 |
| Malay | 13 | 3255 (802) | 1223 (490) | 3 (1) | 558 (112) | 859 (1) | 5911 (1406) | 4505 |
| Java | 40 (39) | 1652 (927) | 206 (201) | | 2154 (1734) | 49 (1) | 4101 (2902) | 1199 |
| Borneo | | 411 (210) | 100 (37) | | 4 | 2 | 517 (247) | 270 |
| TOTAL | 6599 (6002) | 12754 (8380) | 3881 (2912) | 310 (173) | 6662 (4736) | 992 (19) | 31198 (22222) | 8976 |

POW INFORMATION BUREAU

OSS Lt.Col. James F. Donovan 033835 FA / 22 Sept. 1945

Mukden, Manchuria

SECRET

JAMES DONOVAN, Lt. Col. 7 OSS

It is an understandable reaction, of course. They wonder why their American Allies are spying on them in territory occupied by the Red Army in fulfillment of an agreement to help the two Allies regain some of these ideas were ever openly stated, but oblique remarks carried the above impressions.

b.　Two incidents at Dairen did not help to relieve the slight tension. On one day between the 2nd and 8th of September an American Naval carrier aircraft made a "show of force" over the harbor of Dairen according to Commodore C.C. Wood, who witnessed it. This excited the Russian authorities in Dairen a great deal and they asked Commodore Wood what it was all about. As he had not been given any advance information from the Navy he was just as puzzled as the Russians and could offer only a weak explanation. On the afternoon of the 13th of Sept. (after the ex-POWs had sailed) Admiral Settle was informed by radio that another "show of force" would be made that afternoon, this time by cruisers and destroyers of the Seventh Fleet. This worried Admiral Settle considerably and he sent a strong protest to his headquarters. He was very puzzled by this show of force which was duly made outside the harbor about 1500 hours 13 Sept.

Although some Russian officers came to see him after the officers and destroyers had departed Admiral Settle was ashore and did not talk with them so I do not know if they were lodging a protest against the second Naval demonstration or not. The Admiral expected a protest, however, and was at a loss for a reasonable explanation.

His information from Fleet headquarters stated that the show of force would be repeated off Port Arthur though I do not know whether it was or not. I returned to Mukden on the 14th and Admiral Settle radioed me that he had sailed from Dairen that afternoon. As the Admiral had told me that a "show of force" is technically rather a belligerent action, I was interested to see if the Russians would increase the light pressure on us to leave Manchuria now that the POWs had sailed and we had no excuse (or actually any intention) of remaining in Manchuria. I did not notice any increase in Russian curiosity as to the date of our final departure. In any case we wasted no time in completing arrangements to leave.

1.　It is perhaps worth mentioning that the Russians were either notably lacking in friendly feelings toward the British or openly contemptuous of them. Some made very injurious remarks, such as describing England

SECRET

FOREIGN RELATIONS, 1945, VOLUME V

THE SOVIET UNION

1105

Please clarify my understanding of our policy: Did we at Yalta assume the specific obligation to return these Russians by force if necessary? In the protocol which I have seen informally I find no reference to this subject but understand the Chiefs of Staff made an agreement on this subject.

Gray [20] has suggested that where force is necessary we might wish to permit Russian troops to enter our zone for the purpose of removing these individuals. G–5 [21] estimates there are from twenty to thirty thousand of them.

MURPHY

---

740.62114/8–2745: Telegram

*The Secretary of State to the United States Political Adviser for Germany (Murphy), at Berlin*

WASHINGTON, August 29, 1945—3 p. m.

363. From Matthews. Reurtel 383, August 27, 7 p. m. Full text agreement signed Yalta being forwarded by airmail. Briefly agreement provides all Soviet citizens liberated by US forces be separated from enemy POW's and maintained separately until landed over to Soviet authorities. Military authorities each contracting party required without delay to inform competent authorities other party regarding citizens found by them and at same time to take necessary measures to implement all provisions of agreement.

While agreement makes no mention of Soviet citizens captured in German uniforms nor of the use of force, Soviets have consistently claimed it covers all their citizens and Department has interpreted it as covering POW's of Soviet nationality prior to 1939 and in concurrence with War Department ordered return to Europe and turned over to Soviet authorities a number of POW's brought to this country and later ascertained after thorough screening to be Soviet citizens. Incidents involving resistance requiring use of force by our military authorities occurred in connection with this group.

For your confidential information, Department has been anxious in handling these cases to avoid giving Soviet authorities any pretext for delaying return of American POW's of Japanese now in Soviet occupied zone, particularly in Manchuria. [Matthews.]

BYRNES

[20] Cecil W. Gray, Counselor of Mission, Office of United States Political Adviser for Austria Affairs.
[21] Civil Affairs Division.

---

CAPTURED OR INTERNED UNITED STATES ARMY PERSONNEL

7 December 1941 – 31 October 1945

| | Total Prisoners Taken | Prisoners Died While In Captivity | Prisoners Returned to Military Control | Prisoners Not Yet Returned to Military Control | Prisoners Return to Military Control |
|---|---|---|---|---|---|
| **A. GERMAN THEATERS** | | | | | |
| European Theater | 77,528 | 624 | | | 70,779 |
| Medit. Theater | 20,351 | 128 | | | 19,753 |
| Sub Total: | 97,879 | 752 | | | 90,532 |
| **B. JAPANESE THEATERS** | | | | | |
| China, Burma, India | 679 | 269 | | 33 | 377 |
| USAF Pacific | 22,518 | 6,665 | | 1,179 | 44,674 |
| U.S. Army Strategic Air Forces | 251 | 7 | | 9 | 235 |
| Alaskan Dept. | 12 | | | 2 | 10 |
| Sub Total: | 23,460 | 6,941 | 6,595 | 1,223 | 15,296 |
| **C. OTHER:** | 1,725 | 6 | | 100 | 4,619 |
| **GRAND TOTAL:** | 123,064 | 7,699 | 6,125 470 | 7,918 (a) | 107,447 |

(a) Since practically all living known prisoners of war have been returned to military control, the status of the above personnel is now being audited by the War Department.

Statistics-Branch G.S.
7 January 1946

WAR DEPARTMENT
OFFICE OF THE CHIEF OF STAFF
Washington, 25, D. C.

25 February, 1946

Maurice Pate, Esq., Director
Relief to Prisoners of War
National Headquarters
American Red Cross
Washington 13, D. C.

Dear Sir:

In accordance with your oral request, supplemented by recent conversation
with Major Grenshaw of this office, inclosed herewith is chart showing Missing
in Action (including captured) U.S. Army personnel for the period 7 December 1941
through 31 December 1945. This will provide you with the best recent figures
available in published form, for purposes of comparing German and Japanese theaters.

It will be noted that the items "Prisoners of War (Current Status)" and
"Missing in Action (Current Status)" are still large. The reason of course is
that as of 31 December 1945 these categories reflected latest definite reports
available for statistical compilation, and the situation to date has not
materially changed. You will appreciate that for statistical purposes these
casualties cannot be moved to other categories until detailed disposition records
have been processed. In many cases, final disposition must await a legal
determination of death under PL 490 which may take up to next September, even
though investigation to date leaves little logical doubt that a given man is
permanently lost. There is the further factor that the processing of individual
records for final statistical purposes is necessarily much slower than a numerical
count of unprocessed cases in theaters, creating a time lag. As further data is
available in final form, this office will be glad to make it available to you.

The foregoing data was classified "Restricted", but has been approved
for release to you.

Very truly yours,

(Signed)
J. L. Ballard Jr.
Lt. Col., GSC
Chief, Strength Accounting
and Statistics Office, OCS

1 Incl
Chart, "Missing in Action
U.S. Army Personnel"

National Archives
Red Cross Files

J.M.C. Bean:

Jan 1, 1946

## MISSING IN ACTION U.S. ARMY PERSONNEL
### German and Japanese Theaters
### 7 December 1941 through 31 December 1945

| | Died of Wounds | Died of Other Causes | Captured | | | Other Missing in Action | | | | Total Missing (Including Captured) |
|---|---|---|---|---|---|---|---|---|---|---|
| | | | Returned to Mil.Control | P.O.W. (Curr.Stat) | Total | Declared Dead | Dep't Dead From MIA | Ret'd Dead to Duty | MIA (Current Status) | |
| German Theatre | 210 | 648 | 90,937 *(includes 25,000 MIAs)* | 5,414 | 97,209 | 11,753 | 38 | 20,932 | 2,997 | 132,929 |
| Japanese Theatre | 48 | 8,436 | 16,961 | 86 | 25,531 | 1,742 | 1,202 | 2,253 | 9,685 | 40,443 |
| Totals(a) | 258 | 9,084 | 107,898 | 5,500 | 122,740 | 13,495 | 1,240 | 23,185 | 12,682 | 173,342 |

(a) Grand totals do not include Africa Middle East Theater, En Route and Not Chargeable to any Command, Caribbean been Defense Command and USAFPA.

(b) Since all prisoners of war known to be living have been returned to military control, the status of the above indicated personnel is now being investigated by the War Department.

SOURCE: "Battle Casualties of the Army" 1 January, 1946.
Strength Accounting & Statistics Office - 25 February 1946

(FROM: OSS - CIG FILES) U.S.S.R.

IMBOV is:
E. of Moscow

P.O.W. and Internee Camp near

TAMBOV.

Copy. 1 of 1
Recvd. DEC 20 1945

Date: April - May 1945.

1.   Informant, a Pole forced to serve in the German Army, was taken prisoner by the Russians in 1944. He was kept for a time in the Transit Camp in KAUNAS, then in MINSK until he was deported across SIBERIA to the SEVINSKAYA camp near VLADIVOSTOK. At the end of 1945 - April, he escaped and tried to get to Europe. He was, however, arrested by the NKVD after he had got beyond MOSCOW, and placed in the P.O.W. and Internee Camp in TAMBOV, which was occupied by Germans, French, Americans, British, Dutch, Belgians, and even a few citizens of LUXEMBOURG; also Estonians, Letts, Roumanians and Yugoslavs. The prisoners, numbered, in informant's estimation, well over 20,000; they were both military and civilian, most likely over-run by the Russians during the offensive.

2.   All prisoners were forced to work, and the food they were given was very bad and monotonous. They were housed not in huts but in dug-outs.

Hungeroedema or dropsy

3.   The monotonous food caused some strange disease which made the legs and arms swell; the swellings spread and the swollen places developed sores and wounds. After a time men afflicted with this disease died. Informant was told that more than 23,000 Italians, more than 2,500 French and approximately 10,000 Roumanian and Hungarian prisoners had died in this manner. There were also many casualties among Poles and the other nationalities.

4.   Prisoners in this camp included men of very high culture and learning and great experts in many fields of science. Informant observed that German engineers were employed on a special task - the drawing up of blue-prints for a four-engined aircraft, which would carry about 500 men and achieve a speed - it was alleged - of 1,000 kilometres per hour. The Russians were extremely interested in these blue-prints and men working on the invention were granted all possible facilities both in work and the conditions of life in the camp. Informant was also told that a group of technicians and engineers of all nationalities was busy working out plans for new types of tanks and other technical machines and installations, but he was unable to learn any details about them.

5.   When informant left the camp there were still numerous Frenchmen due to be repatriated in the next batch; there were also some Belgians and Dutch, and others, including some Englishmen and several score Americans, the presence of whom in this camp is probably unknown to the British and the U.S.A. authorities. When he was leaving, these Englishmen and Americans asked him urgently (as did the French officers and men) to notify the Allied authorities of their plight. Informant succeeded in reaching France with a convoy of Allied nationals.

NEW YORK TIMES/Oct. 31, 1945

**SAYS SOVIET BARS SEARCH**

**Briton Asserts Hunt for War Prisoners in Reich Is Held Up**

By Wireless to The New York Times.

LONDON, Oct. 30—War Secretary James Lawson told the House of Commons today that the Soviet Government had not yet given permission for British search teams to enter Russian-occupied Germany to look for missing British and Dominions prisoners of war.

Asked whether the Soviet Government had given any reason "for this extraordinary refusal," the Secretary said that it was a matter for the Foreign Office. He refused to answer additional questions.

TRANSLATION

NATIONAL CONSTITUENT ASSEMBLY                    FRENCH REPUBLIC

Paris, August 17, 1946

Excellency,

I have brought to the attention of the Minister
for ex-Prisoners of War the testimony of Mr. Joseph Bogenschutz, 55 Grand'Rue, at Mulhouse, (Haut Rhin), who was
repatriated on last July 7 from Russia, from Camp 199-6
at Inskaya, which is 75 kilometers from Nowosibirsk, and
who states that about 20 Alsatians and Lorrains were still
in this Camp at the time of his departure.

Bogenschutz states that he wrote at least three cards
a month through the Red Cross (Red Crescent) since September 1944 and that none of these cards ever arrived. He
states that the Camp Commander is a German and that instead
of repatriating Alsatians, he selects Germans for repatriation.

Bogenschutz, in addition thereto, alleges that there
still remain American, British, Belgian, Polish, Rumanian
Luxemburg, etc. nationals in the Camp.

I am bringing this matter to your attention more particularly because of the latter information.

Please accept, Excellency, the assurance of my
highest consideration.

(signed) HENRI MECK
Deputy of
the Bas-Rhin

# The Wisconsin State Journal

MADISON, SUNDAY, DECEMBER 1, 1946

## Iron Curtain Shrouds Lee's Fate

### Parents Believe Russians Hold Him With 20,000 Other Yanks

## Iron Curtain Shrouds Fate of Lee

(Continued From Page One)

UNITED STATES FORCES, ~~~~~~~~~
**REGION VI (BAMBERG)**
APO 139 U. S. ARMY

RE: Case No: 0507

~~~~~~RG FIELD OFFICE~~~~~~
20~~ August    1946

SUBJECT:    Information Received from Illegal Border Crosser.

TO      :   Commanding Officer, Bayreuth Sub-Region CIC Office.

RE      :   Three American Soldiers supposedly held by Russians.

1. On 16 August 1946, GRZYBECK, Friedrich, ~~~~~~~~~~~~~~~ ~~~~~~~~~~ crossed the border into American Occupied Territory from the Russian Occupied Zone and was arrested for illegal border crossing. During routine questioning GRZYBECK told the following story.

2. During the war, GRZYBECK was a member of the LUFTWAFFE Ground Force Division Number 21 and was taken prisoner by the Russians on or about 10 April 1945. From 10 April 1945 to 11 August 1946 GRZYBECK was a Russian Prisoner of War and was held at the Former German Prisoner of War Camp, DRESDEN/ Ost Number 11. This same had had been used for Russian Civilian workers.

3. There were about 800 men in that camp and they were transported daily to an airfield near DRESDEN to work.

4. GRZYBECK states that in May 1945 three United State Soldiers were brought to the Camp as prisoners. He gave their names as follows:

a. Pfc TAYLOR, Olen of
b. Cpl OKRANA, Ducki of
c. Pvt KAPLAN, Billy of

These names were written by GRZYBECK and may not be spelled correctly, except for TAYLOR, Olen as GRZYBECK has in his possession a picture of TAYLOR in a United States Army uniform, signed by TAYLOR.(See attached photographs).

5. The three United States soldiers and GRZYBECK were confined in the camp near DRESDEN from May 1945 to August 1946. Before being confined to this camp near DRESDEN, the three United States soldiers were German Prisoners of War, having been captured by the Germans shortly after the Normandy Invasion near CHERBOURG, France. According to GRZYBECK, the three American Soldiers were in an Airborne Division which called itself "King of the Air".

6. During the morning of the 13 August 1946, three airplanes crashed at the airfield near DRESDEN and the Russians, suspecting sabotage loaded all 800 prisoners in railroad cars in preparation for shipping them East.

REGRADED UNCLASSIFIED
ON 2 USEP 1991
BY OUR USA~~~~~~~~ F91/PO
AUTH Para 1-603 DOD 5200.1R

120

GRZYBECK and seven others escaped from the train and GRZYBECK made his way into the United States Zone of Occupation.

~~~~~~~~~~~~~~~~~~~~~~~~~~~~~~~~~~~~~~~~~~~~~~~~~~~~ to go on with him because his two buddies were sick and would not have been able to walk.

DISTRIBUTION:
4-Forward
1-File

/s/ Manfred Maier
/t/ MANFRED MAIER
Special Agent CIC

1st. Ind                                                     RLK/ts
Headquarters, Sub-Region Bayreuth, Counter Intelligence Corps, Region VI, APO 139,
U S Army, 26 August 1946.

TO:  Commanding Officer, Counter Intelligence Corps, Region VI, 970th CIC Detachment, A PO 139, U S Army.

1.  Forwarded.

Phone: BAYREUTH 3476
Incls:
ltr.dtd 20 Aug 46

HAROLD L. ~~~~~~~~~~
Special Agent CIC
Commanding

SECRET

OFFICE OF MILITARY GOVERNMENT FOR BAVARIA

AG 000.12 MGBI                                        26 August 1946

SUBJECT:  Special Report  (Friedrich Grzybeck)

TO        :  Office of Military Government for Germany (U.S.)
             Office of the Director of Intelligence     5 USC 552 (b) (6) & (b) (7) (c)
             APO 742

1.  Detachment B-225 Liaison and Security Office, Coburg, submitted
the following report obtained from Friedrich Grzybeck now under arrest
in Zinkenwehr jail in Coburg, charged with illegal crossing of the border.

2.  Subject was taken prisoner by the Russians on or about 10 April
1945 and escaped on 12 August 1945.  From date of capture to date of
escape he was at the Deutsche Kriegsgefangenlager Dresden/East #11.  A
total of about 1600 prisoners were in this camp.  During the last days of
May 1945, 3 United States soldiers (a) Pvt. Glen Taylor,
_____, (b) Cpl. Bucki Chkane, _____ and (c)
Pvt. Billy Hafers, Texas (German-American) came to the camp.  Grzybeck
claims that Americans were treated the same way as the Germans, most
likely because they attempted to escape once but the Russians succeeded
in recapturing them.  Although subject never saw the uniforms of the
three soldiers, he claims they belonged to the 101st Airborne Division.

3.  On an airport near the camp on which the prisoners worked,
there was a total of about 200 planes of which 125 were four-motored.
There were some U.S. fortresses too and Russian B-34s.  On 12 August 1946
three Russian planes crashed.  The same day 800 prisoners were supposed to
be shipped to Varsovia as the Russians suspected sabotage.  At 2300, 12 August,
subject and seven other prisoners (3 Austrians and 4 from the British zone)
escaped.  The train was then still at the railway station at Dresden and subject
told the three U.S. soldiers about his plans.  Chkane and Hafers were sick but
Glen Taylor gave subject his picture and asked subject to try to help all
three of them as soon as he could.  Taylor did not want to escape because
he did not want to desert his two comrades.

4.  Picture of Taylor is attached hereto.

5.  Subject crossed the border illegally on 15 August 1946 at Hildburg.

6.  Comment - The above information is transmitted to you as it was
received from Detachment B-225.  No check has been made as to whether the
three U.S. soldiers mentioned above are missing.

FOR THE DIRECTOR                                              115

Tel:  Munich Military 334    056                    French Chief of Int

CONFIDENTIAL

HEADQUARTERS
SUB REGION GOEPPINGEN
COUNTER INTELLIGENCE CORPS REGION I

File Nr. I-G- 820

MEMORANDUM TO THE OFFICER IN CHARGE

APO 154
FIELD OFFICE ULM/D
4 August 1947

Subject:  American and Allied soldiers held by Soviet

RE    :  Project 154032

1.  Reason for Investigation.

Reference is made to Orientation and Guidance Report #2
dated 30 April 1947 re: Project 154032.

A casual informant, HOFMANN, Hans Joachim, recently
released German Prisoner of War from Russia contacted CIC
ULM (I49/x68) thru the local American Red Cross on 2 August
1947 to give information about American and Allied personnel
held as prisoners in SIBERIA, USSR.

(b) (6) & (b) (7) (c)

2.  Results of Investigative Activity.

a.  Informant HOFMANN was released from the Dachau
Prisoner of War Processing Center on or about 17 July 1947.
The informant did not contact American authorities there, but
German officials in DACHAU told informant to contact UNRRA in
ULM. The informant did this and was referred to the American
Red Cross Club in ULM.

b.  The informant gave this agent three (3) pictures
of American soldiers attached hereto as exhibit A, B, and C.

c.  The informant claims to have talked with the
soldier shown on the picture marked exhibit "A" whose name
is BOEHM, Viktor                                          emigrated to the United States of America in 1928, lived in
                                      According to the informant BOEHM
served in the Army of the United States, was taken prisoner
by the Germans and then taken over by the Russians on or
about 5 May 1945 in OELMUETZ, Silisia, Germany and transported
to SIBERIA on 17 July 1945 to the NOROSIBIRSK Camp 311 near
KRASNOE, Siberia where BOEHM is presently working in a tank
factory. The informant worked with BOEHM who speaks German.
BOEHM gave HOFMANN the attached pictures.

d.  The informant claims there are two hundred
(200) American soldiers working in this plant and about
nine hundred (900) Allied soldiers, mostly English and French
soldiers.

CONFIDENTIAL

CONFIDENTIAL

HEADQUARTERS
SUB-REGION GOEPPINGEN
COUNTER INTELLIGENCE CORPS REGION I
EUROPEAN COMMAND
APO 154

File Nr. I-G-820
MOIC Subject: Americans and Allied Soldiers held by Soviet

e. On the back of the picture attached as exhib
"B" is the address of Mr. JAMES E. GREEN ████████ who is also working in said tank fact
in Siberia.

f. The picture attached as exhibit "C" could no
be identified by the informant.

g. The informant knows the names of the followi
Allied personnel:
    1. Sergeant RIEDEL, ████████
    2. Schanno AMBROSINI, ████████

also working in said tank factory.

3. Agent's Notes.

a. Evaluation of above information: F-6.

b. The informant. ████████████████
The informant claims to be the song of a German colonel
Hofmann killed in TUNISIA by Gestapo agents.

c. It is recommended to have CIC WIESBADEN, Reg
III recheck informants statements thru a discreet investig

d. It is further recommended above information
made available to the Adjutant General, Washington, D.C.,
F.B.I. and the American Red Cross. If the FBI. confirms
informants statements this matter should be of interest to
State Department.

John J Menken
Special Agent
CIC

3. Incl.
Exhibit A (Picture of Viktor Boehm)
    "    B (   "   of James E. Green)
    "    C (   "   of unidentified American soldier)

APPROVED:

HERRING

48

CONFIDENTIAL

**HEADQUARTERS**
**COUNTER INTELLIGENCE CORPS**
(430th CIC Detachment)
**UNITED STATES FORCES IN AUSTRIA**
APO 777                              U.S. Army

29 June 1947
Vienna City Section

SUBJECT:  Possible Detention of American and Allied Personnel at Concentra-

SUMMARY OF INFORMATION

The following information was indirectly received from an Informant who escaped from Subject Camp in May 1947. (Evaluation: F-6)

Subject Camp, near BUDAPEST Hungary and under the direct administration of the Russian NVD, contains approximately 35,000 persons, the
majority of which are German and Austrian. There are, however, a considerable
number of American, British, French, Italian, Dutch and Belgian nationals in
confinement at the same.

The American and British personnel remain in the camp for approximately eight days and are then dispatched to other, more permanent places
of detention.

Postal addresses of Subject Camp are as follows:

C.R. MOSKAU P.F. 3578/049
P.C. MOSKAU P.F. 3578/049
C.R. LENINGRAD P.F. 6114/3
P.C. LENINGRAD P.F. 6114/3

Informant further stated that among American internees at the camp
was MCC-MORRAN, Fred, 1st Lieutenant; approximately 35 years old; native of
FLORIDA U.S.A.  MCC-MORRAN arrived at Subject Camp near the end of January
1947 and was supposed to have been formerly stationed with United States
Forces in Austria.

AGENTS NOTE:

Available personnel records in VIENNA were checked with no
evidence disclosed that MCC-MORRAN was ever a member of USFA.

CORNELIUS S MUIST
Special Agent, CIC

JOHN W SCHILTER
Special Agent, CIC

SECRET

(CIC/US/CIC)
Possible Detention of American and Allied Personnel
at Concentration Camp 4108/S STARNOVII                    19 July 1947

1   OFFI Director  19 July    Attached correspondence forwarded as a matter
        of         47         pertaining to your Office.
    Personnel
        &                              For the Deputy Director of Intelligence
    Admini-
    stration

1 incl:  Summary of Information, 450 CIC Det., USFA,
25 June 1947, Subject                                    A. C. MOE
Possible Detention of Amer-                              1st Lt      WAC
ican and allied Personnel                               Asst Exec...
at concentration camp
4108/S STARNOVII

Telephone:
6436 (Maj HUTSCH)

M/R

        S/I from CIC USFA dated 25 June 1947, reports that in Camp 4108/S STARNOVII,
near BUDAPEST, Hungary, which is under the direct administration of the Russian Army
there are approximately 35,000 prisoners, the majority of which are German and
Austrian. There are however, a considerable number of American, British, French,
Italian, Dutch and Belgian nationals in confinement at the camp. American and
British personnel remain in the camp for approximately eight (8) days, and then
dispatched to other, more permanent places of detention. It is alleged that among
American internees at the camp was MEC-MORRAN, Fred, 1st Lt., approximately 35 years
old, native of FLORIDA, USA. MEC-MORRAN arrived at subject camp near the end of
January 1947 and was supposed to have formerly been stationed with USFA. Records in
VIENNA disclosed that MEC-MORRAN was never a member of USFA. Subject S/I sent to G-1
for information and any necessary action.
Coordination:  PFC SILVESTRI for Maj HUTSCH/ Lt Col KINZIG (G-1, Mil Pers Services Br

SECRET

FRANCE STILL SEEKS 100,000 EX-CAPTIVES

FRENCH LOSING HOPE FOR MANY DEPORTEES

France and Russia Sign Accord To Return 500,000 Frenchmen

## The Military and UNRRA

losis surveys of all centre residents. Penicillin was used to reduce the allegedly high incidence of gonorrhoea and syphilis. Supplies of standard medicines and surgical equipment were made available in ample quantities, backed by large military reserves, and the American Red Cross released large stocks of unused 'standard medical kits' originally earmarked for prisoners of war. Forty-seven thousand of these kits were distributed in Western Germany, and 7,400 in Austria.

Because of the concern felt by UNRRA medical personnel over the general reduction in rations during 1946, UNRRA made substantial distributions of cod-liver oil and allotted 900,000 tablets, containing standard dosages of vitamins C and D, to all children and adolescents under twenty years of age, to all pregnant and nursing women, and to all persons suffering from vitamin deficiency. Simultaneously, UNRRA 'weighing teams' working in the British and United States Zones ascertained that displaced persons averaged from 10 to 20 pounds more than the German civilians, so that fears on this score were to some extent allayed.[1]

Dental care, too, was provided for displaced persons by UNRRA, although a continuing shortage was reported of toothbrushes and acrylic resin needed to make dentures. By July 1946 there were 76 dental centres in the British Zone, with 130 dentists, 54 dental assistants, and 22 technicians, practically all recruited from the displaced persons; 16 German dentists, and 2 UNRRA dentists. During June comparable personnel in the United States Zone attended to about 40,000 patients. Dental services in the French Zone were considered inadequate, being described as 'rudimentary in personnel and equipment'.[2]

### TRACING OF MISSING PERSONS

During 1946 European tracing bureaux prepared what were then considered careful estimates of the number of persons who had lost their lives as a result of the war, and in addition com-

[1] Hospitalized displaced person received letter and, more abundant rations, varying from 2,550 to 3,350 calories per day, and there undoubtedly were many improperly hospitalized persons seeking more food. The British authorities estimated in April 1946 that four times too many people were hospitalized in their zone and they instituted a careful check with a view to weeding out impostors.

[2] A school for dental technicians was established at a hospital in Hohne near Belsen in the British Zone, and during June 1946 pupils were enrolled.

270

## The Military and UNRRA

piled records pertaining to individuals using personal files, assessments, although probably no more than guesses for Poland and Yugoslavia, were given as follows for September 1946:[1]

| | |
|---|---|
| Belgium* | 69,000 |
| France | 590,000 |
| Greece | 100,000 |
| Italy | 490,000 |
| Luxembourg | 3,876 |
| Netherlands | 190,000 |
| Poland | 3,000,000 |
| Yugoslavia | 1,100,000 |

At this time missing Danes and Norwegians probably only numbered a few hundred persons. Missing Czechoslovaks, including Jews, numbered tens of thousands, as did the Rumanians and the Hungarians, Bulgarians and Greeks, apart from the Jews, probably numbered no more than a few thousand. Soviet nationals, for whom no official estimates were given, probably equalled several million persons.

'The work of tracing missing persons in Western Germany, which had been an UNRRA responsibility from its inception, was transferred to the UNRRA Central Headquarters by the end of September 1945. Here it became the responsibility of the newly created Central Tracing Bureau.[2] By June 1946 this bureau had been expanded and employed 263 persons.[3]

The Central Tracing Bureau relayed requests for information regarding missing persons between the four zonal tracing bureaux, Austria, Italy, and more than twenty national tracing bureaux.[4] In addition to searching for individuals, lists of centre residents were obtained, duplicated, and distributed for general

[1] Quoted from M. Painalevone, the Belgian Director-General of an International Congress of Tracing Bureaux (*Le Soir*, 20 September 1946).

[2] For approximately one year in Berlin the Soviet authorities operated a tracing bureau under the direction of a Major Gregorov. This bureau cooperated with the UNRRA Central Tracing Bureau, but with the passage of time it became more and more one-sided, with the Soviet office receiving information but giving none.

[3] Other tracing bureaux were located as follows: for the British Zone, at Goettingen which was operated by a combination of military, British voluntary relief agency and UNRRA personnel; for the United States Zone at Wiesbaden, operated by UNRRA personnel under military supervision; for the French Zone at Haslach where it was operated by the French Ministry of Prisoners, Deportees, and Refugees under nominal military supervision.

[4] In Austria, national tracing bureaux were consolidated during 1946 into a national tracing bureau. In Italy, the tracing work, initiated by British voluntary relief agencies, was taken over during 1946 by UNRRA, and subsequently operated as a national tracing bureau by the Italian Red Cross.

271

CENTRAL INTELLIGENCE AGENCY

# INFORMATION REPORT

COUNTRY  USSR

REPORT NO.

SUBJECT  Location of Certain Soviet Transit Camps for
Prisoners of War from Korea

PLACE ACQUIRED
(BY SOURCE)

DATE ACQUIRED
(BY SOURCE)

DATE (OF INFO.)  Jul 51 - Apr 52

DATE DISTR.  25 Sep 1952

NO. OF PAGES  3

NO. OF ENCLS.

SUPP. TO
REPORT NO.

THIS IS UNEVALUATED INFORMATION

SOURCE

1. In December it was known that transit camps for prisoners of war captured by the Communists in Korea had been established in: Komsomolsk on the river Amur, Magadan on Nagaevo Bay in the Sea of Okhotsk, Chita and Irkutsk. Through these transit camps were passing not only Korean P.O.W.s but also American P.O.W.s.

2. Since July 1951, according to new information, several transports of Korean P.O.W.s have passed through the ports of Nakhta (near Vladivostok), Okhotsk and Magadan. Each ship has contained 1,000 or more prisoners. Between the end of November 1951 and April 1952 transports of P.O.W.s were sent by rail from the Pogoi railway junction on the Chinese-Soviet frontier. Some were directed to Chita in Eastern Siberia and some to Molotov, European Soviet Russia, East of Ural Mountains.

3. Those P.O.W.s who arrived by ship in the ports of Nakhta, Okhotsk and Magadan were then transported by train, or by trucks or by motor driven barges, to Vaikarem on the Chukotsk Sea, to Ust Maist on the river Aldan and to Takutsk on the river Lena.

4. P.O.W.s shipped to Vaikarem were sent to a network of camps in the Nizhni Kolymsk region on the East Siberian Sea, to be employed building roads, electric power plants and airfields. Their number varies considerably due to high mortality and to transfer to other camps on the Chukotski Peninsula. All these camps are under supervision of NVD and are entirely isolated. There were about 12,000 Korean P.O.W.s in April 1952 in the Nizhno Kolymsk camp network. The camps were under the charge of (fnu) Skretchuk, a major of NVD and (fnu) Chiabo, a civilian Party functionary, probably an employee of MGB. Chiabo was in charge of education and political indoctrination.

DISTRIBUTION

This report is for the use within the USA of the intelligence components of the Departments or Agencies indicated above. It is not to be transmitted overseas without the concurrence of the originating office through the Assistant Director of the Office of Collection and Dissemination, CIA.

CLASSIFICATION

CENTRAL INTELLIGENCE AGENCY

**INFORMATION REPORT**

THIS IS UNEVALUATED INFORMATION

| COUNTRY | Korea/China | | CD NO. | |
| SUBJECT | Prisoner-of-War Camps in North Korea and China | | REPORT NO. SO 91634 | |
| | | | DATE DISTR. 17 July 1952 | |
| DATE OF INFO. | January - May 1952 | | NO. OF PAGES 3 | |
| | | | NO. OF ENCLS. | |
| PLACE ACQUIRED | | | SUPPLEMENT TO REPORT NO. | |

* Except as noted

SOURCE

War Prisoner Administrative Office and Camp Classification

1. In May 1952 the War Prisoner Administrative Office (Chan Fu Kuan Li Ch'u), under Colonel Su-man-ch'i-fu (2069/0192/4619/9310/9710) in Pyongyang, was an intelligence officer attached to the general headquarters of the Soviet Far Eastern Military District, controlled prisoner of war camps in Manchuria and North Korea. The office, formerly in Mukden, held two thousand prisoners, several of whom were English-speaking Soviets.

2. The office had developed three types of prisoner-of-war camp. Camps termed "peace camps", detaining persons who exhibited pro-Communist leanings, were characterized by considerable treatment of the prisoners and the staging within the camp of Communist rallies and meetings. The largest peace camp, which held two thousand prisoners, was at Chungchon. Peace camps were also held at P'yongan Namdo (126-05, 43-36) and Pzon'i (120-45, 41-40).

3. Before camps, all of which were in Manchuria, detained anti-Communist prisoners possessing certain technical skills. Emphasis at these camps was on reindoctrination of the prisoners.

4. North Korean prisoner-of-war camps, all of which were in North Korea, detained prisoners whom the Communists will exchange. Prisoners in the peace and reform camps will not be exchanged.

5. Officials of North Korean prisoner of war camps sent reports on individual prisoners to the War Prisoner Administrative Office. Cooperative prisoners were being transferred to peace camps. ROK army officers were being asked by ROK army soldiers who were being reindoctrinated and assimilated into the North Korean Army.

Kangdong Camp

6. In May the largest North Korean prisoner of war camp, detaining twelve hundred prisoners, was near Tzai Ling (1132/2548) mountain, five miles southwest of the Kangdong (126-05, 39-09) railroad station. The compound, divided with barbed wire and sand embankments into four partitions for American, English, and Turkish prisoners and prisoners of other nationalities, held 840 American, 100 English, 60 Turkish, and 200 Kempsel French, Dutch, and Canadian troops. Most of the United States prisoners were members of the 1 Cavalry Division and taken by the 24 Infantry Division. General William Dean was held from Korea and taken to this camp in 1951.

7. The Kangdong camp, organized into study management sessions, and these sections, compelled the prisoners to study for three hours, to labor for four hours, and to discuss political problems for two hours.

8. On 1 May nine thousand (sic) ROK army prisoners and fifty United Nations prisoners were in caves (sic) at the Kangdong camp, extending from approximately Ni-149-939 in a valley at Adu-di, Kangdong-gun (126-05, 39-09) (BU-637). Of the ROK army prisoners 10 percent were officers, 50 percent enlisted men, and 40 percent were enlisted men. Of the United Nations prisoners, 10 percent were officers. The prisoners bailed 600 gross at each day out of the cave. An average of two prisoners were dying daily from malnutrition and erupting typhus. The majority of prisoners at this camp were extremely anti-Communist in thinking. Three North Korean army guards, armed with PPSh's and rifles, were at the entrance of each cave.

Camp Number 106, Maria

9. On 1 May approximately sixteen hundred ROK army prisoners of war, including one hundred officers and five hundred non-commissioned officers, were at the Maria prisoner of war camp, approximately 106 kilometers northeast of the Maria railroad station (125-51, 39-01) (TD-622). Prisoners held here, having been processed through five ideological screenings, were believed to be potential converts to Communism. Members of Political security detachments maintained strict surveillance of the prisoners. The surveillance often was carried out by members of these bureau who entered the camps disguised as prisoners.

10. Each prisoner receives 650 mu monthly, 1 kilogram of grain and 45 grams of soy bean oil, vegetable salt, and soy bean paste daily. The prisoners wore North Korean army uniforms. The prisoners slept in trenches near the Maria airfield ten hours each day. Two hours of indoctrination lectures were also held daily. The prisoners had been organized into squads each of ten men. Each of the camp's four battalions had three platoons and each platoon, four squads. A guard platoon, armed with M-1's, carbines, and PPSh's, was at the camp.

SO-91634

-2-

4. Normal prisoner-of-war camps, all of which were in North Korea, detained prisoners whom the Communists will exchange. Prisoners in the peace and reform camps will not be exchanged.

5. Officials of North Korean prisoner of war camps sent reports on individual prisoners to the War Prisoner Administrative Office. Cooperative prisoners were being transferred to peace camps. ROK army officers were being shot; ROK army soldiers were being reindoctrinated and assimilated into the North Korean army.

### Kangdong Camp

6. In May the largest North Korean prisoner of war camp, detaining twelve hundred prisoners, was near T'ai Ling (1132/1545) mountain, six miles southeast of the Kangdong (126-05, 39-09) (BU-4837) railroad station.[1] The compound, divided with barbed wire and mud embankments into four partitions for American, English, and Turkish prisoners and prisoners of other nationalities, held 840 American, 100 English, 60 Turkish, and 200 hundred French, Dutch, and Canadian troops. Most of the United States prisoners were members of the 1 Cavalry Division and the 24 Infantry Division. General William Dean was moved from Karbin and Mukden to this camp in 1951.

7. The Kangdong camp, organized into study, management, sanitation, and finance sections, compelled the prisoners to study for three hours, to labor for four hours, and to discuss political problems for two hours.

F-3 8. On 1 May nine thousand (sic) ROK army prisoners and fifty United Nations prisoners were in caves at the Kangdong camp, extending from approximately BU-492363 to BU-494368 in a valley at Ada'-ni, Kangdong-myon (126-05, 39-09) (BU-4837).[2] Of the ROK army prisoners 10 percent were officers, 50 percent non-commissioned officers, and 40 percent privates. Of the United Nations prisoners 10 percent were Negroes. The prisoners, who received 600 grams of cereal and salt each day, were not required to work and spent only two hours each day out of the caves. An average of two prisoners were dying daily from malnutrition and eruptive typhus. The majority of prisoners at this camp were extremely anti-Communist in thinking. Three North Korean army guards, armed with PPSh's and rifles, were at the entrance of each cave.

### Camp Number 106, Mirim

F-3 9. On 1 May approximately sixteen hundred ROK army prisoners of war, including one hundred officers and five hundred non-commissioned officers, were at the North Korean prisoner of war camp Number 106 at approximately YD-47221A, 1.6 kilometers southwest of the Mirim railroad station (125-51, 39-01) (YD-4722). Prisoners held here, having been processed through five ideological screenings, were believed to be potential converts to Communism. The prisoners believed that they were to be assimilated into the North Korean army. Members of political and security detachments maintained strict surveillance of the prisoners. The surveillance often was carried out be members of these bureaus who entered the camps disguised as prisoners.

F-3 10. Each prisoner receive d 50 won monthly, 1 kilogram of grain and 45 grams of soy bean oil, vegetables, salt, and soy bean paste daily. The prisoners were wearing North Korean army uniforms. The prisoners were constructing air raid shelters near the Mirim-ni airfield ten hours each day. Two hours of indoctrination lectures were also held daily. The prisoners had been organized into squads of ten men. Each of the camp's four battalions had three platoons and each platoon, four squads. A guard platoon, armed with M-1's, carbines, and PPSh's, was at the camp.

technical jobs. These camps were completely isolated from any civilian camps located in neighborhood. Political control was carried out by the local Party organization, headed by (fnu) Edorin, a delegate from the Obkom of the Komi-Pezmyak National District. All these camps were under the charge of (fnu) Kalypin, a Soviet officer of unknown rank who was sent from Molotov in February 1952.

12.   In some camps situated near the Gubakha railway, which are called 'Zapretchdelanki', (Russian term difficult to translate—means 'isolated plots') about 150 Americans were kept, probably soldiers and N.C.Os. An interesting thing was that from these camps one to three P.O.Ws were taken every few days by officers of the MVD for transportation to Gubakha or Molotov. They never returned to their camps and their camps and their fate remained unknown. According to the supposition of persons acquainted with MVD methods these P.O.Ws had been observed in their camps by specially assigned agents of MVD, who knew the English language and thus were able to find out those who were very hostile to the Communist regime and ideology and those who could be considered sympathetic. Those belonging to the first group were most probably sent either to prison or to especially hard labor camps for extermination; the others were probably sent to special political courses in Molotov.

( CCRAK M-101 )
X 9 M- 101 dtd 24 February 1953 .(Cont'd)

4. .About 7,200 artillery men and engineers attached to the 2nd field Army now stationed in the SACHON area have recently returned from the front in Korea. The North Korean puppet government has secretly left PYONGYANG since 22 December. The command of PYONGYANG was turned over to KANG DOK JI , Chief of Staff of CCF. The Russian naval vessel Stalin (an old DD) is reported to have appeared in the gulf of PO-HAI recently. It is also reported that Russia has turned over to China many submarines. Naval vessels turned over to China by Russia are the "Hubeh" 1509 tons, the "Changchubelts" 1478 tons, 3 landing craft and 2 small gun boats. It is reported that if the Korean war expands, Iahitopu (?) Chief of Staff of the Russian Far East Forces will take over the position of Supreme Chief of Staff now held by Mao Tse Tung and he (Iahitopu) will command the CCF (F).

CCRAK Comment: Command (Reference Paragraph 2). This office has received sporadic reports of PoW being moved to the USSR since the very inception of the hostilities in Korea. These reports came in great volume through the earlier months of the war, then tapered off to a standstill in early 1951; being revived by a report from January of this year. It is definitely possible that such action is being taken as evidenced by past experience with Soviet authorities. All previous reports state PoW who are moved to the U.S.R are technical specialists who are employed in mines, factories, etc. This is the first report that they are being used as espionage agents that is carried by this office.

NOTE: Any portion of this report reproduced will be identified by CCRAK # M-101 (ROK CNI).

DISTRIBUTION:
A

8,000 Missing, Van Fleet Says
BELMONT, Mass., Aug. 7 —
Gen. James A. Van Fleet, retired commander of the United States Eighth Army in Korea, estimated tonight that a large percentage of the 8,000 American soldiers listed as missing in Korea were still alive. He said in an interview that Gen. Clark's estimate of at least 3,000 G.I.'s still held captive by the Communists was "conservative."

CONTINUATION of PAGE 1033

CENTRAL INTELLIGENCE AGENCY

INFORMATION REPORT

COUNTRY    USSR

SUBJECT    Location of Certain Soviet Transit Camps for
           Prisoners of War from Korea

PLACE ACQUIRED
(BY SOURCE)

DATE ACQUIRED
(BY SOURCE)

DATE (of info.)  Jul 51 - Apr 52

REPORT NO.

NO. OF PAGES

DATE DISTR.  2 Sept 1952

# Bound

- - - Air Pouch - - - ┌─────────────────┐   ┌ DO NO  IT  E IN THIS SPACE ┐
                        (Security Classification)   │ 6.11.45ₐ 241/ 3 ᴄ │
**FOREIGN SERVICE DESPATCH**

FROM  :  AMCONGEN, Hong Kong  - - _1716_
                                    DESP. NO.
TO    :  THE DEPARTMENT OF STATE, WASHINGTON.  - _March 23, 1954_ .

REF

SUBJECT:   American POWs Reported en route to Siberia

A recently arrived Greek refugee from Manchuria has reported
seeing several hundred American prisoners of war being transferred
from Chinese trains to Russian trains at Manchouli near the border
of Manchuria and Siberia. The POWs were seen late in 1951 and in
the spring of 1952 by the informant and a Russian friend of his.
The informant was interrogated on two occasions by the Assistant
Air Liaison Officer and the Consulate General agrees with his
evaluation of the information as probably true and the source as
of unknown reliability. The full text of the initial Air Liaison
Office report follows:

First report dated March 16, 1954, from Air Liaison Office,
Hong Kong, to USAF, Washington, C2.

"This office has interviewed refugee source who states that
he observed hundreds of prisoners of war in American uniforms being
sent into Siberia in late 1951 and 1952. Observations were made
at Manchouli (Lupin), 49°50'-117°30' Manchuria Road Map, AMSL 201
First Edition, on USSR-Manchurian border. Source observed POWs
on railway station platform loading into trains for movement into
Siberia. In railway restaurant source closely observed three POWs
who were under guard and were conversing in English. POWs wore
sleeve insignia which indicated POWs were Air Force noncommissioned
officers. Source states that there were a great number of Negroes
among POWs shipments and also states that at no time later were any
POWs observed returning from Siberia. Source does not wish to be
identified for fear of reprisals against friends in Manchuria,
however is willing to cooperate in answering further questions and
will be available Hong Kong for questioning for the next four days."

Upon receipt of this information, USAF, Washington, requested
elaboration of the following points:

1. Description of uniforms or clothing worn by POWs including
   ornaments.
2. Physical condition of POWs.

DMForman/RDYoder/jmo        SECRET

By ~~BOYLAN~~ NARA, Date 1/29/86

Page 3 of 1716
Encl. No.
Desp. No.
From

(Classification)

(7) Three POWs observed in station restaurant appeared
to be 30 to 35. Source identified Air Force non-
commissioned officer sleeve insignia of Staff
Sergeant rank, stated that several inches above
insignia there was a propeller but says that all
three did not have propeller. Three POWs accompanied
by Chinese guard. POWs appeared thin but in good
health and spirits, were being given what source
described as good food. POWs were talking in English
but did not converse with guard. Further information
as to number of POWs observed source states that
first observation filled a seven passenger car train
and second observation about the same. Source
continues to emphasize the number of Negro troops,
which evidently impressed him because he had seen so
few Negroes before. Source further states that his
Russian railroad worker friend was attempting to
obtain a visa to Canada and that he could furnish
more information, the railroad workers name is Leon
Strelnikov whose mothers sister lives in Canada and
is applying for a visa for Strelnikov (phonetic).
Comment Reporting Officer: Source is very careful
not to exaggerate information and is positive of
identification of American POWs. In view of information contained in Charity Interrogation Report No. 619
dated 5 February 54, Reporting Officer gives above
information rating of F-2. Source departing Hong Kong
today by ship. Future address on file this office."

In this connection the Department's attention is called to
Charity Interrogation Report No. 619, forwarded to the Department under cover of a letter dated March 1, 1954, to Mr. A. Sabin
Chase, DRF. Section 6 of this report states, "On another
occasion source saw several coaches full of Europeans who were
also taken to USSR. They were not Russians. Source passed the
coaches several times and heard them talk in a language unknown
to him."

(handwritten left margin) Contact (British?) Embassy - to run down

cc: Taipei
    Moscow
    London
    Paris

SECRET

Page 2 of
Disp. No. 1716
From Hong Kong

Page of
Encl. No.
Disp. No.
From

3. Nationality of guards.
4. Specific dates of observations.
5. Destination in Siberia.
6. Presence of Russians in uniform or civilian clothing accompanying movement of POWs.
7. Complete description of three POWs specifically mentioned.

The Air Liaison Office complied by submitting the telegram quoted below.

"FROM USAIHLO SGN LACKEY. CITE C 4. .EUR 53737 following answers submitted to seven questions.

(1) POWs wore OD outer clothing described as not heavy inasmuch as weather considered early spring. Source identified from pictures service jacket, field, M1943. No belongings except canteen. No ornaments observed.

(2) Condition appeared good, no wounded all ambulatory.

(3) Station divided into two sections with tracks on each side of loading platform. On Chinese side POWs accompanied by Chinese guards. POWs passed through gate bisecting platform to Russian train manned and operated by Russians. Russian trainmen wore dark blue or black tunic with silver colored shoulder boards. Source says this regular train uniform but he knows the trainmen are military and wearing regular train uniforms.

(4) Interrogation with aid of more fluent interpreter reveals source first observed POWs in railroad station in spring 1951. Second observation was outside city of Manchouli about three months later with POW train headed towards station where he observed POW transfer. Source was impressed with second observation because of large number of Negroes among POWs. Source states that he was told by a very close Russian friend whose job was numbering railroad cars at Manchouli every time subsequent POW shipments passed through Manchouli. Source says these shipments were reported often and occurred when United Nation forces in Korea were on the offensive.

(5) Unknown.

(6) Only Russian accompanying POWs were those who manned train.

#1
1954

## Document 1 (telegram)

Secret

ACTION COPY

666  FROM DEPT

SENT APRIL 19 1 pm

SECRET

ACTION: DRM

RECD:  APRIL NO 1 am

DULLES

According Despatch 1716 from Hong Kong airpouched you, a recently arrived Greek refugee from Manchuria reported seeing several hundred American POW's being transferred Chinese trains to Russian trains Manchouli late 1951 and early 1952. Some POW's were aiborne insignia indicating they were Air Force non-coms. Great number Negro troops also observed. This report corroborates previous indications UNC POW's might have been shipped to Siberia during Korean hostilities.

United States has been freely concerned general subject UNC personnel who may still be Communist custody. Department has just accepted British offer maxi representations behalf UNC personnel who may be Chinese Communist custody. Question raising this matter informally Geneva under careful consideration.

Unless you perceive objection, request you approach highest available level Foreign Ministry and leave Aide Memoire indicating reports have now come attention United States Government which support earlier indications that American Prisoners of war Korea had been transported into Soviet Union and are now Soviet custody. Request fullest possible information these POW's and their reparation earliest possible time.

In your discussion with Foreign Office, you may desire inform Soviets without revealing source that we have reliable accounts transfers POW's Manchouli.

RL:gr

AMB CHAR: CHRON

SECRET

## Document 2 (memorandum)

MEMORANDUM    UNITED STATES GOVERNMENT

DATE: January 4, 1953

FE/NA - Mr. Drew

EUR/EE - Mr. Walter J. Stoessel, Jr.

Article in December 18, 1953 issue of U.S. News and World Report re Americans held by Communists.

The attached letter dated December 21, 1953 from Senator Lyndon B. Johnson is referred to NA for handling since the enclosure apparently refers to an article concerning missing Korean War GI's and not to American soldiers in Soviet custody, as the writer alleges. The writer is apparently referring to the article on page 27 of the December 18th issue of U.S. News and World Report entitled "where are 944 Missing GI's?". NE has been unable to find an article referring to six or eight hundred American soldiers in Soviet custody.

It is suggested that the following be included in the letter which NA drafts in reply to Senator Johnson:

[The Department of State has no information to the effect that there are approximately six or eight hundred American soldiers in the custody of the Soviet Government.] A few of the prisoners-of-war of other nationalities recently released by the Soviet Government have made reports alleging that American citizens are imprisoned in the Soviet Union. All of these reports are being investigated by this Department with the cooperation of other agencies of the Government.

You are probably aware that representations which the United States Government recently made to the Soviet Government resulted in the release in Berlin on December 29 of Homer H. Cox and Leland Towers, two Americans reported by returning prisoners-of-war as being in Soviet custody. The Department will investigate, as it has done in the past, every report indicating that American citizens are held in the custody of foreign governments.

Senator Johnson's letter has been acknowledged by telephone.

Please return to the library the attached copy of U.S. News and World Report.

EUR:EE:WJStoessel:jo 1/4/54

With regard to question
to which these are of

THE NEW YORK TIMES
JAN. 5, 1954

## THE OTHER PRISONERS

While world attention has been focused on the Korean war prisoners comparatively little heed has been paid to many other prisoners from many nations who languish in Communist jails or are being worked to death in Communist slave labor camps. Their plight has been brought into sharp relief by the reported death of the two Americans, Homer H. Cox and Leland Towers, who have just been freed by the Soviets, and of Arnold M. Klein, the American business man released from a Shanghai prison in northeast, from a partial paralysis he suffered before.

All three confirm that the Soviet bloc and the Chinese Communists are holding in their jails and slave camps many foreigners, including soldiers and civilians, women and children. Cox and Towers, who were freed because they had been identified by returning Austrian prisoners, have identified at least four more Americans and reported on six others. But, according to State Department figures, the total number of Americans held by the Soviets and their European satellites exceeds 5,000, in addition to several times that number who are possible claimants to American citizenship, and the Chinese Communists are known to hold about 100 Americans. The latter figure may be increased by those unidentified prisoners of the Korean war reportedly deported to China.

Many of these Americans, like many West Europeans, were residents in the Iron Curtain countries caught by the Communist tide; others were deported from German war prison camps; some, like Cox, were simply kidnapped. How many have since died is uncertain, but the known death toll among German and Japanese war prisoners held in their own grim work of Beria, and that the latter's liquidation might bring them freedom in keeping with Premier Malenkov's New Year's wishes to the American and all Western people. But with this hope, it would be some comfort to think that their imprisonment was the Communist system. Nothing can be taken for granted, and it is the duty of the American Government to use all available means, through diplomatic channels and the United Nations, to press for their liberation.

*Tom — How does this relate to the hostages gathered near Dresden in late April? — [initials]*

---

OPERATIONS MEMORANDUM

CONFIDENTIAL

TO: Department of State.

FROM: Amconsul Hamburg

SUBJECT: WELFARE — WHEREABOUTS: William George H[...]

REF:

Mr. Fritz BAUER, a German national, recently released[...]

In March 1955 at International Camp No. 668/VIII[...]

Roberts[...]

Roberts[...]

Paisy in reporting the matter has been due to confusion with regard to Mr. Bauer's own status. Because he was born in the Tyrol he was erroneously released to the Italians and much delay was before his German citizenship was established.

cc Amembassy Moscow, Bonn

255-RICHARDSON, William George
HQrahl4l.ser

OFFICE OF
SPECIAL CONSULAR SERVICES
DEC 28 1955
DEPARTMENT OF STATE

CONFIDENTIAL FIL

UNCLASSIFIED

*From a source at Bolling AFB to Tom Ashworth* Table 51

Number of Casualties Incurred by U. S. Military Personnel
In Connection With the Conflict in Vietnam

Cumulative from January 1, 1961 through December 1, 1973

| | Army | Navy a/ | Marine Corps | Air Force | Total |
|---|---|---|---|---|---|
| **A. CASUALTIES FROM ACTIONS BY HOSTILE FORCES** | | | | | |
| 1. Killed | 25,341 | 1,095 | 11,477 | 504 | 38,417 |
| 2. Wounded or Injured | | | | | |
| a. Died of wound | 3,520 | 146 | 1,452 | 48 | 5,166 |
| b. Nonfatal wounds | | | | | |
| Hospital care required | 96,809 | 4,178 | 51,391 | 933 | 153,311 |
| Hospital care not required | 104,723 | 5,898 | 37,202 | 2,518 | 150,341 |
| 3. Missing | | | | | |
| a. Died while missing | 1,731 | 221 | 10 | 559 | 2,521 (1 |
| b. Returned to control | 54 | 5 | 2 | 35 | 96 |
| c. Current missing c/ | 229 | 104 | 86 | 588 | 1,007 (2 |
| 4. Captured or Interned | | | | | |
| a. Died while captured or interned | 32 | 6 | 7 | 5 | 50 (3 |
| b. Returned to control | 134 | 145 | 38 | 333 | 650 |
| c. Current captured or interned c/ | 13 | 30 | 3 | 13 | 59 (4 |
| 5. Deaths | | | | | |
| a. From aircraft accidents/incidents | | | | | 3734 |
| Fixed Wing | 91 | 208 | 148 | 883 | 1,330 |
| Helicopter | 2,396 | 66 | 434 | 85 | 2,981 |
| b. From ground action | 28,137 | 1,194 | 12,364 | 148 | 41,843 |
| TOTAL DEATHS b/ | 30,624 | 1,468 | 12,946 | 1,116 | 46,154 |
| **B. CASUALTIES NOT THE RESULT OF ACTIONS BY HOSTILE FORCES** | | | | | |
| 6. Current Missing c/ | 83 | -- | 14 | -- | 97 (5 |
| 7. Deaths | | | | | |
| a. From aircraft accidents/incidents | | | | | |
| Fixed Wing | 278 | 186 | 46 | 283 | 793 |
| Helicopter | 1,890 | 56 | 242 | 19 | 2,207 |
| b. From other causes | 4,999 | 638 | 1,393 | 290 | 7,320 |
| TOTAL DEATHS | 7,167 | 880 | 1,681 | 592 | 10,320 |

a/ Navy figures include a small number of
Coast Guard casualties.
b/ Sum of lines 1, 2a, 3a and 4a.
c/ Technical status at time of ceasefire.
Have not been released to U.S. control
and are currently unaccounted for.

Department of Defense
OASD (Comptroller)
Directorate for Information Operations
December 5, 1973

UNCLASSIFIED

Intelligence Information Report

2-1877

██████████████████████████████████

DIRECTORATE FOR PLANS ██████████████████████ PAGE 1 OF 3 PAGES

THIS IS AN INFORMATION REPORT, NOT FINALLY EVALUATED INTELLIGENCE

████████████████████ REPORT NO. CS-311/04439-71

DATE DISTR. 10 JUNE 1971

C.I.A. Report #

COUNTRY NORTH VIETNAM

DOI 1965-JUNE 1967

SUBJECT PRELIMINARY DEBRIEFING SITE FOR CAPTURED U.S. PILOTS IN VINH PHU PROVINCE AND PRESENCE OF SOVIET AND COMMUNIST CHINESE PERSONNEL AT THE SITE

ACQ VIETNAM - SAIGON - 28 APRIL 1971

SOURCE ████████████████████████████████████████

████████████████████████████████████████

████████████████████████████████████████

████████████████████████████████████████

████████████████████████████████████████

████████████████████████████████████████

████████████████████████████████████████

████████████████████████████████████████

████████████████████████████████████████

████████████████████████████████████████

████████████████████████████████████████

APPROVED FOR RELEASE

1. A PRELIMINARY DEBRIEFING POINT FOR U.S. PILOTS SHOT
DOWN OVER VINH PHU PROVINCE, NORTH VIETNAM /NVN/, WAS LOCATED
AT THE LAM THAO SUPERPHOSPHATE PLANT /WJ301589/ NEAR THACH SON
VILLAGE, LAM THAO DISTRICT, VINH PHU PROVINCE. TWO U.S. PILOTS
WERE TAKEN TO THE DEBRIEFING POINT ON ONE OCCASION IN 1965;
EIGHT, IN 1966; AND AN UNKNOWN NUMBER, IN 1967. THE PRISONERS
WERE ESCORTED TO THE SITE BY PERSONNEL OF THE ARMED PUBLIC
SECURITY FORCES /APSF/, AND STUDENTS FROM A NEARBY SCHOOL
SERVED AS PERIMETER GUARDS. EACH TIME PRISONERS WERE BROUGHT TO
THE SITE THEY RODE IN AN OPEN CAR OF CHINESE ORIGIN RESEMBLING
AN AMERICAN JEEP. SOME OF THE ESCORT GUARDS RODE IN A LEAD CAR
AND OTHERS RODE IN TWO CARS FOLLOWING THE PRISONERS. UPON THEIR
ARRIVAL AT THE PLANT, THE GUARDS LINED UP, FORMING A CORRIDOR
THROUGH WHICH THE PILOTS ENTERED THE BUILDING. AT THIS POINT
A SOVIET, A CHINESE AND A VIETNAMESE GREETED THE PILOTS AND
LED THEM INTO THE BUILDING. THE PILOTS USUALLY REMAINED IN
THE BUILDING FOR SEVERAL HOURS. WHEN THEY EMERGED THEY HAD CHANGED
FROM UNIFORMS INTO CIVILIAN CLOTHING. ███████████████████

███████ SAID ████████████████████████████████████

███████ HAD TOLD HIM THE FOREIGNERS WERE SOVIET AND COMMUNIST
CHINESE. SOVIET PERSONNEL HAD BEEN STATIONED AT THE PLANT SINCE
ITS CONSTRUCTION IN 1963, BUT IN 1965 THE NUMBER OF SOVIETS WAS
REDUCED TO THREE OR FOUR, AND IT REMAINED AT THAT LEVEL AS OF
JUNE 1967. ABOUT 20 COMMUNIST CHINESE PERSONNEL ARRIVED AT THE
PLANT IN 1966 AND THERE WERE STILL ABOUT 20 THERE AS OF JUNE 1967
AS FAR AS ███ KNEW, THE SOVIET AND COMMUNIST CHINESE PERSONNEL
GOT ALONG WELL.

2. AFTER SHAKING HANDS WITH THE SOVIET AND CHINESE, THE PRISONERS WERE LED TO A DIFFERENT VEHICLE FROM THE ONE WHICH BROUGHT THEM TO THE SITE. THEY WERE ESCORTED FROM THE PLANT BY A DIFFERENT SET OF GUARDS WHO WORE YELLOW AND WHITE UNIFORMS AND WERE ARMED WITH RIFLES AND PISTOLS. ▓▓▓ DID NOT KNOW THE DESTI- NATION OF THE PRISONERS.

3. ABOUT 30 STUDENTS FROM THE LAM THAO SECOND-LEVEL SCHOOL WERE SELECTED BY THE SCHOOL SUPERINTENDENT, DAO KHAC T R U N G, FOR GUARD DUTY AT THE SUPERPHOSPHATE PLANT. EACH STUDENT WAS CAREFULLY SCREENED AND THEN TRAINED IN THE PROCEDURES FOR HANDLING U.S. PRISONERS. A PAMPHLET ENTITLED "POLICY ON TREATMENT OF AMERICAN PRISONERS" /CHINH SACH DOI XU TU BINH MY/ WAS USED AS A TRAINING DOCUMENT, AND STUDENTS WERE GIVEN SPECIFIC INSTRUCTIONS ON THE RECEPTION OF THE PRISONERS AT THE DEBRIEFING SITE. ONE OF THE FUNCTIONS OF THE STUDENT GUARDS WAS TO DISARM ANY VILLAGERS WITH WEAPONS BEFORE ALLOWING THEM CLOSE TO THE AMERICANS. AS LONG AS NO PHYSICAL CONTACT WAS MADE WITH THE AMERICANS, THE VILLAGERS WERE PERMITTED TO APPROACH THE PILOTS AND MAKE ANY REMARK SUITABLE TO THE OCCASION.

4. FIELD DISSEM: STATE USMACV 7TH AIR FORCE NAVFORV CINCPAC PACFLT PACAF ARPAC / ▓▓▓▓▓▓

· Foreign Intelligence Information Report  /7

WARNING NOTICE— INTELLIGENCE SOURCES AND METHODS INVOLVED

DIRECTORATE OF OPERATIONS

FURTHER DISSEMINATION AND USE OF THIS INFORMATION SUBJECT TO CONTR
STATED AT BEGINNING AND END OF REPORT            PAGE 1 OF  4  PA
THIS IS AN INFORMATION REPORT, NOT FINALLY EVALUATED INTELLIGENCE

REPORT CLASS. C O N F I D E N T I A L —WNINTEL—
NOFORN

COUNTRY    USSR                      DATE DISTR.    12 March 1982

SUBJECT    Alleged Soviet Incarceration of    REFERENCES.
U.S. Vietnam Prisoners of War
(DOI: 1970)

SOURCE

SUMMARY: According to
Grigoriyev, specially selected U.S. prisoners of war were being
received into the Soviet Union circa 1970 for long term or
lifetime incarceration and "ideological retraining." He implied
the number involved to be about 2,000. The goal of the program
was indefinite, but involved intensive psychological investigation of the prisoners and retraining to make them available
as required to serve the needs of the Soviet Union.

1. (Headquarters Comment: This report should be read with
caution. CIA records contain no confirmation of the alleged intelligence
affiliation of the subsource cited below, despite the source's assertion
that Grigoriyev held a leading position in the KGB. Several other
persons named in the text likewise cannot be identified. We have
never before encountered even vague rumors among Soviet dissidents or
other informants that any U.S. POW's from Vietnam are incarcerated in

C O N F I D E N

FORM 2290

C O N F I D E N T I A L

the USSR, much less that 2,000 such individuals are leading "reasonably normal lives" in the same region where numerous Soviet political prisoners have resided in exile. In short, while the source may be reporting his recollection of an actual conversation, we strongly believe that this report merits little if any credence from analysts. However, in light of continuing high interest in the question of U.S. personnel still listed as missing in action in Southeast Asia, this report is being disseminated with appropriate caveats to concerned members of the U.S. Intelligence Community.)

2. In a private conversation which was held circa 1970, KGB Lieutenant General Petr Ivanovich ((Grigoriyev)) stated that many specially selected U.S. prisoners of war were being received from North Vietnam for long term or lifetime custody and "ideological re-training" in the Soviet Union. (Source Comment: Grigoriyev did not state specifically the number of prisoners involved. The term he used was "v poryadke neskol'kikh tysyach v nas toshe yest'" which translates as "on the order of several thousand," implying the number to be about 2,000). The prisoners were destined for confinement at a facility near Perm. Grigoriyev, who learned of the program from an unnamed high level KGB colleague, understood that Soviets rather than North Vietnamese were involved in the initial selection process and that participants were to be continually assessed for suitability. He implied that individuals determined to be unsuitable would be eliminated and replaced with other candidates. (Source Comment: Grigoriyev made his comment while serving as a political ideologist and personnel officer at the All-Union Scientific-Technical Information Center of the State Committee for Science and Technology in Moscow. He had previously served as Chief of the KGB's Personnel Directorate and in that capacity would have very likely made contacts among KGB officials subsequently responsible for organizing any such prisoner program.)

3. According to Grigoriyev, the goals of the U.S. prisoner program were indefinite but involved intensive psychological studies of the individuals and utilization of them as required to serve the needs of the Soviet Union. Grigoriyev understood that the detention facility was not a standard prison, but rather one in which inmates could lead reasonably normal lives. During the conversation Grigoriyev recalled that precedents existed for such a program in the Soviet Union and cited similar previous efforts with Spanish, Japanese, and Chinese nationals. He stated that in past programs, participants were encouraged to marry Soviet women.

C O N F I D E N T I A L

C O N F I   E N T I A L

4. ▮▮▮▮ Comment: Source described Grigoriyev as a very professional and security-conscious person who confided in source because of their unique personal relationship. Grigoriyev, in his capacity as an institute personnel officer, was the first individual to interview source upon his transfer to the information center Grigoriyev requested basic biographic data and acknowledged being acquainted with several individuals listed by source, particularly ▮▮▮▮ General Feodor Petrovich ((Skrynnik)), who had a GRU officer serve as Deputy Chief of Intelligence for the Far Eastern Military District in the early 1950's. Skrynnik and Grigoriyev owned dachas near each other and while not close friends, held each other in high respect. In addition to Skrynnik, Grigoriyev was acquainted with (FNU) ((Prudnikov)) who was active in Western Europe and Germany for the KGB, (FNU) ((Gradoselskiy)) who had served as a KGB official in Poland and Germany, and (FNU) ((Gridniyev)) who served as KGB Deputy Chief for Administration and Supply. A sense of trust had therefore been developed over many years of mutual association with top level KGB and GRU officers.) (Headquarters Comment: Prudnikov may be identical with Mikhail Sidorovich Prudnikov, dob circa 1912, a senior Soviet intelligence official whose memoirs of operations during and after World War Two have been published in the USSR. CIA records do not identify any individual named Grodoselskiy or Gridniyev as having served in Soviet intelligence.)

5. Grigoriyev volunteered the information regarding the Vietnam prisoners during one of many private conversations during the late 1960's and early 1970's. His duties were not particularly demanding after his years as an administrator in the KGB. He was often finished with his work in the early afternoon and, rather than go home or engage in outside interests, held informal discussions in his office. ▮▮▮▮ Comment: Source stated that he was the person most frequently chosen by Grigoriyev for private conversations.) Topics primarily involved Grigoriyev's personal affairs and health, but also included political topics. During one of these sessions the subject of prison camps arose, in particular those which furnished labor for Siberian economic development. The conversation then shifted to Vietnam and the apparent increase in strength of South Vietnam at the time and the apparent instability in the North. Grigoriyev agreed, citing the massive U.S. commitment to the South, but added that the Soviets were also making gains. He then described the program involving U.S. prisoners.

C O N F I   E N T I A L

6. Grigoriyev was trained as a professional military officer and
served in the tank troops during World War II. After the war he was
assigned to the Party Central Committee as an army representative.
In the period 1953-1954 he became KGB Deputy Chief for Personnel. He
subsequently became critical of the recruitment policies of KGB head
Vladimir Yefimovich ((Semichastnyy)) and was transferred from his
position to that of KGB Security Chief for Soviet Bloc nations. Soon
thereafter he developed a heart ailment and retired. In the late 1960's
he accepted the position at the Information Center.

7. General Skrynnik joined the Russian cavalry in 1917 and
subsequently entered the Odessa artillery school. Upon graduation he
was assigned to the Zhitomir military district. In 1931 he entered the
Frunze Military Academy. He advanced rapidly and in the 1933-1934
period was sent to China as Deputy Military Attache. He joined Mao's
long march and began to establish intelligence agent networks for the
Soviet Union. He remained in China until 1942 except for a brief
return in 1939 to establish an intelligence school in Moscow for
China operations. In the spring of 1942 he was recalled from China
to become chief of intelligence on the northwestern front, where he
remained for the duration of the war. After the war he was assigned
as Soviet representative to the Berlin Joint Commission for Repatriation.
After serving in Berlin from 1945 to 1949 he returned to Moscow as
either chief or deputy chief for intelligence at the Frunze Academy.
He then served as Deputy Intelligence Chief of the Far Eastern Military
District. He retired from the military in 1953. Skrynnik was
subsequently recalled to duty to re-establish agent networks in China
after the China-USSR split but refused to leave retirement. (Headquarters
Comment: CIA records contain no independent confirmation of the details
of Skrynnik's career provided here.)

Statement of Jan Sejna Prepared for the

Senate Select Committee on POW/MIA Affairs

November 5, 1992


Gentlemen, it is a pleasure to be with you today. I want to
thank you for the opportunity to help this committee and,
especially, the POW and MIA families across America. It is a
terrible thing to lose a loved one and never know what happened.

My name is Jan Sejna. I left my home country, Czechoslovakia
in late February, 1968, and requested political asylum in the
United States, which was subsequently granted. I am now a citizen
of the United States.

Before I left Czechoslovakia, I was an Army officer with the
rank of General Major, which is equivalent to a U.S. Brigadier
General. From 1954 until I left in 1968, I held a variety of
official positions that brought me into direct contact with various
Soviet operations involving U.S. POWs in the Korean and Vietnam
Wars. The most important such positions were chief of staff to the
Minister of Defense, member of the Minister's Kollegium, and
secretary of the Defense Council, which was where key decisions on
issues of national security and intelligence were made. In these
official capacities I attended many meetings in the Soviet Union
and in Czechoslovakia at which Soviet instructions affecting U.S.
POWs were received, implementing decisions were made, and where
briefings on various projects that used POWs were presented.
Additionally I took part in discussions respecting these programs
and discussed them with Czechoslovak, Soviet, North Korean, and
North Vietnamese officials. As secretary of the Defense Council,
it was my responsibility to prepare directives based on such
discussions and decisions.

The events I am about to describe all happened roughly twenty-
five to thirty-eight years ago. As you can appreciate, it is an
arduous and time-consuming task to reconstruct the events that

1

decision to continue such experiments in Vietnam was made by Soviet
General Secretary Khrushchev before he was ousted from his
leadership position. Specifically, he directed Czechoslovakia to
negotiate with the North Vietnamese arrangements for conducting
medical experiments on POWs. As secretary of the Defense Council,
I was aware of this direction and of the negotiations, which again
involved General Babka, who was then chief of Foreign Assistance
Administration of the General Staff, Prime Minister Joseph Lenart,
and General Vaclav Prchlik, who was chief of the Main Political
Administration.

While the operations in Korea and North Vietnam were similar,
there were several important differences. First, many more
experiments were conducted during the Vietnam War. Second, in the
Vietnam War the North Vietnamese maintained tight control over the
conduct of the war and over Soviet operations in the war, including
those in the medical experimentation project. North Vietnamese
hospitals were used and North Vietnamese doctors participated in
the experiments. Third, and perhaps to circumvent the North
Vietnamese penchant for control, POWs also were shipped to
Czechoslovakia and thence to the Soviet Union during the course of
the Vietnam War for experimental purposes. Because I left
Czechoslovakia before the end of the war, I can not tell you how
many POWs were shipped through Czechoslovakia to the Soviet Union.

As I said before, I am now working to recall with greater
precision the various details about these operations. Because this
effort has just begun, it will be several months before we are able
to reconstruct the next level of detail. At that time, I hope to
be able to provide better answers to those questions that I am not
prepared to answer today. I certainly encourage your questions
because they are certain to help us in our effort to develop the
most complete picture we can.

DEFENSE INTELLIGENCE AGENCY

WASHINGTON, D.C. 20340

27 APR 1992

S/NF-0466/POW-MIA

MEMORANDUM FOR THE UNDER SECRETARY OF DEFENSE FOR POLICY
    THE ASSISTANT SECRETARY OF DEFENSE (COMMAND,
    CONTROL, COMMUNICATIONS AND INTELLIGENCE)

SUBJECT: Defense Intelligence Agency Report S/NF-0418 (U)
Information Memorandum

1.     The enclosed intelligence report summarizes the results of a DIA investigation into possible drug experimentation on U.S. Prisoners of War during the Korean War carried out by Soviet and Czechoslovakian personnel. The purpose of this program was to develop comprehensive interrogation techniques involving medical, psychological and drug-induced behavior modification. Information uncovered by DIA indicates that up to "several dozen" unwilling participants in this program may have been executed upon its conclusion in North Korea.

2.     The source was well placed in that he personally saw progress reports on the work in North Korea that were forwarded to top leadership in the Czech Central Committee and Ministry of Defense. He remains a very sensitive source who has provided reliable information to the U.S. intelligence community for many years. The source is most reluctant to have his identity become known or to be tied to the information he provided. It should be noted that the source did submit to polygraph examination during which no deception was indicated. This report is classified both to protect the source's identity and to ensure proper security is maintained during possible demarche and follow-up investigative activity.

3.     I have furnished the attached report to the Secretary and Deputy Secretary of Defense for their information. Normally, intelligence reports concerning American prisoners of war are distributed within the Government to the Military departments, the intelligence agencies, the Department of State, the temporary Senate Select Committee on POW/MIA Affairs, the House POW/MIA

CLASSIFIED BY: DIRECTOR, DIA
DECLASSIFY: OADR

DELIVER BY HAND TO
ADDRESSES

October 30, 1992

MEMORANDUM FOR THE RECORD *(Chief of four Sept Intelligence - Pentagon for Joint Chiefs of Staff on loan to U.S. Senate now retired)*

FROM: John F. McCreary

SUBJECT: *Obstruction of the Investigation*

1. *I am concerned that recent lines of investigation have been seriously compromised by leaks of sensitive information by the Committee Staff Director to the Department of Defense. Leaks to the Department of Defense or other agencies of the Executive Branch of my Memoranda for the Record are interfering with follow-up discussions with useful witnesses. Moreover, they are endangering the lives and livelihood of two witnesses.*

*Leak of Information on Jan Sejna*   *aide to Sen Smith*   *Chief counsel*

2. *My MFR concerning discussions with former Czech Gen Maj Sejna have ended up in the hands of private citizen and Sejna's co-author Joseph Douglas and the LA Times. I provided copies of that memorandum to Carluccie, Codinha, and Kolesnick* — *aide to Sen Grass*

3. *Irrespective of leaks outside the government, Bill LeGro attended a meeting of the US-Russia Joint Commission group in Washington on 28 October 1992 at the Department of State. The discussion featured information provided by Sejna. LeGro stated that Ambassador Malcolm Toon called for his dismissal. DIA personnel defended Sejna as to his expertise on Central Europe, but not as to his information on other areas, particularly POW-related.*   *(Col. U.S. Mil Intell friend of mine)*

4. *On 30 October 1992, I learned from Bill LeGro that he was directed to read a letter from the Central Intelligence Agency to the Select Committee that discredits Sejna's information. The letter reportedly indicates that Sejna's information has been checked and not been confirmed by his former government. At the time this letter was received, the Staff had decided to take Sejna's deposition but had not yet scheduled a deposition of Sejna. In addition, my MFR was written from memory, and did not do justice to all that Sejna stated, either in detail or in context. As of this writing, we do not know what Sejna knows or will say under oath, yet his testimony has already been written off. This anticipatory discrediting of a Select Committee potential witness is tantamount to tampering with the evidence.*

*Suspected Leak of Information on Le Quang Khai*

5. *The second issue of suspected misconduct concerns witness Le Quang Khai. Although Le made a public statement concerning POWs on 12 September 1992, no agency of the US government contacted him concerning his POW information. He told me on 26 October that some men who represented themselves as FBI agents contacted him to attempt to recruit him to return to Vietnam as a US intelligence agent for six months. After which*

1 of 2

his request for asylum would be favorably considered.

6. On 30 October, Mr. Robert Egan of Hackensack, New Jersey, who is a close friend of Mr. Le and the intermediary whereby the Committee Staff met Mr. Le, informed McCreary and LeGro that the FBI had again contacted Mr. Le. A person representing himself as an FBI person called on 30 October to set up a meeting with Le to discuss Le's working as an intelligence agent for the FBI's POW/MIA office.

7. So far informal checks indicate there is no such office. Secondly, this contact occurred three days after my return from taking Le's deposition in Hackensack on 26 October after which I wrote another MFR. This MFR was sent only to JW Codinha on 28 October. I observed a copy of the MFR with apparent routing designators written in the top margin on the desk of Frances Zwenig on 28 October.

8. The contact with Le two days after preparation of my MFR, despite the passage of a month since his public declarations, is highly suspicious and more than coincidental. The circumstances of both contacts in which persons identifying themselves as FBI without showing credentials or other evidence of authenticity or authority and also making a pitch to recruit Le are also highly suspicious.

9. An internal Department of Defense Memorandum identifies Frances Zwenig as the conduit to the Department of Defense for the acquisition of sensitive and restricted information from this Committee. Based on the above sequences of events, I must conclude that Frances Zwenig continues to leak all of my papers to the Defense Department. Her flagrant disregard of the rules of the Senate and her oath of office are now jeopardizing the livelihood, if not the safety , of Senate witnesses. In addition, the Department of Defense's continuing access to sensitive Committee Staff papers is resulting in obstructions of the investigations by the Senate Select Committee by various agencies of the Executive Branch.

*/ʷ ᵒᴰ 6*

Statement of Jan Sejna
Before the Subcommittee on Military Personnel
of the House National Security Committee

September 17, 1996

Chairman Dornan, ladies and gentlemen, it is a privilege to be here this
afternoon.

It is heartwarming for me after so many years to find people who are sincerely
interested in events that actually happened in various communist countries that
were under the rule of the Soviet Union.

In 1968 I was forced to choose between following instructions I received from
Moscow and doing what I believed to be best for my country, Czechoslovakia.
At the time, I was first secretary of the Party at the Ministry of Defense and
chief of staff to the Minister of Defense, in addition to numerous other
positions. The Soviet Union was preparing to invade Czechoslovakia, and I
choose to alert the Czech leadership and refused to follow the Soviet plan as
directed.

A week later, I learned that my immunity from arrest as a member of the
Parliament had been lifted and I was about to be arrested. I believe my arrest
had been directed personally by Soviet General Yepishev. After thirteen years
in high-level positions, I knew precisely what that meant, and along with my
son and his girl friend, who later became his wife, I fled through Yugoslavia to
Trieste, where I went to the U.S. consulate and requested political asylum. In
two days I was in the United States.

While what I have just said describes what happened in Korea, I want to point
out that the same things happened in Vietnam and Laos during the Vietnam
War. The only difference is the operation in Vietnam was better planned and
more Americans POWs were used, both in Vietnam and Laos and in the Soviet
Union.

On several occasions my office was responsible for organizing the shipments of
POWs and their housing in Prague before they were shipped to the Soviet
Union. I personally was present when American POWs were unloaded from
planes, put on buses whose windows had been painted black, and then driven to
Prague where they were placed in various military intelligence barracks and
other secure buildings until they were shipped to the Soviet Union.

Between 1961 and 1968 when I left Czechoslovakia, I would estimate at least
200 American POWs were shipped to the Soviet Union through
Czechoslovakia.

I believe there were others who were shipped to the Soviet Union through
North Korea and East Germany, although I have no first hand knowledge of
those transfers. I know that many were given to the Chinese for experiments
during the Korean War, and Czech intelligence reported that the North
Vietnamese also provided American POWs to the Chinese.

In closing I want to emphasize that this operation was conducted at the highest
level of secrecy. Information on this operation was labeled State Secret, which
was higher than Top Secret, and no one who did not have a real need to know
was aware of the operation. When I was there, my estimate is that fewer than
15 people in all of Czechoslovakia were aware of the transfer of American
POWs to the Soviet Union. I will never forget the written directions on the
original Soviet order that started the operation in 1951. It said that the operation
was to be conducted in such a way that "no one would ever know about it."

SEATTLE TIMES

SATURDAY, AUGUST 31, 1991

## POWs sent from Hanoi to Soviets, man says

**Associated Press**

At least 22 U.S. prisoners of war were sent from North Vietnam to the Soviet Union in 1977 and 1978, a retired intelligence official told a Seattle television station last night.

KIRO-TV also quoted a Soviet Embassy spokesman, George Oganov, as saying his country's prisons were being checked for any American POWs. So far none has been found, he said.

POW movements were revealed in intercepted radio transmissions, said retired Air Force Tech. Sgt. Terry Minarcin, a Vietnamese language specialist and intelligence expert. He served with Air Force intelligence and National Security Agency units, KIRO said.

"Beginning in late '77, around mid-December of '77, we started noticing flights from various parts of the country carrying specific types of POWs into Hanoi to be handed over to the Russians to be flown out of Hanoi into Moscow," Minarcin said.

He said the prisoners were escorted by KGB agents on Soviet planes and included all those then in Vietnam with communications and signals-intelligence expertise.

At least 22 prisoners were flown to Moscow in December 1977 and January 1978, and other intelligence officials told him the shipments continued at least through the next fall and perhaps into 1983, Minarcin said.

Minarcin's account was supported by Jerry Mooney, another former Air Force and NSA employee, but was denied in a written statement issued yesterday by the Pentagon.

"No information from U.S. intelligence sources indicates that a movement of POWs from Vietnam to the U.S.S.R. occurred," the statement said.

SEPTEMBER 1, 1991
THE SUNDAY OREGONIAN

## Ex-official says Vietnam sent POWs to Soviet Union

SEATTLE — At least 22 U.S. prisoners of war were sent from North Vietnam to the Soviet Union in 1977 and 1978, a retired intelligence official told a television station.

In a copyright report Friday evening, KIRO-TV also quoted a Soviet Embassy spokesman, George Oganov, as saying his country's prisons were being checked for any American POWs. So far, none has been found, he said.

POW movements were revealed in intercepted radio transmissions, said retired Air Force Tech. Sgt. Terry Minarcin, a Vietnamese language specialist and intelligence expert. He served with Air Force intelligence and National Security Agency units, KIRO said.

# New tale of Americans in Vietnam

## Ex-Viet officer says POWs were sent to USSR in 1986

BY KEN McLAUGHLIN
AND DE TRAN
Mercury News Staff Writers

A former officer in Vietnam's internal security police claims that colleagues in his unit transported dozens of American prisoners of war in northern Vietnam until 1986, when the prisoners reportedly were sent to the Soviet Union.

Trinh Do, who escaped from Vietnam in 1988 and now lives in Hong Kong, told the Mercury News in telephone interviews that several colleagues often escorted groups of four or five Americans to old military installations in Da Nang and Saigon. The colleagues told him the prisoners were used for their technical expertise in helping to repair and operate equipment left behind by Americans, Do said.

"Sometimes the American pilots and my friends traveled by auto, sometimes by plane," said Do, now a fairly well-known Hong Kong artist.

Do's claims come as the spotlight is focused on the Senate Select Committee on POW-MIA Affairs, which begins hearings today. For months, the panel has been sifting through eyebrow-raising reports from U.S. intelligence sources and Vietnamese refugees, among others. The committee's goal: to separate the fraudulent and flaky from the factual.

### Reinforcing Yeltsin's claim

However extraordinary, Do's claims are consistent with last week's startling statement by Russian President Boris Yeltsin, who said just before his Washington summit with President Bush that American prisoners of war may have been sent from Vietnam to the old Soviet Union.

According to Do, a second lieutenant, his job was to guard the Foreign Ministry and the private homes of Communist Party officials in and around Hanoi. He said he never escorted American prisoners himself, but learned of their presence from two good friends in the security force and five or six other colleagues.

Do, 36, said his colleagues told him the Americans were kept in groups of three to 15 in "very nice houses" outside the Vietnamese capital and in the nearby provinces of Ha Son Binh and Vinh Phu.

In 1986, the colleagues were suddenly "promoted" and sent to the Soviet Union and East Germany "to study" in a transparent attempt to keep them quiet, Do said. The colleagues were told that the Americans had been sent to the Soviet Union, he said.

The colleagues have since returned to Vietnam and "now hold commanding posts in the security police," said Do, who left the force in 1979.

### No names

Do said he would not reveal their names because he feared for their safety. He promised, however, to take U.S. MIA-POW investigators to their homes in Vietnam if requested.

"We're absolutely interested" in hearing Do's story," said Deborah DeYoung, spokeswoman for the Senate panel. "The sooner the better."

Sen. Bob Smith, R-N.H., the committee's vice chairman, told a news conference Tuesday that his reading of classified Pentagon documents indicates that Vietnam-era prisoners of war "may have been alive through 1989" in North Vietnam and Laos. But he said some senators who have reviewed the same information disagree.

Senators have good reasons to be skeptical. Some "live sightings" have been obvious fabrications by Vietnamese boat people seeking to gain favor with American officials in the hope that the

Americans will ease their entry into the United States. And in recent years, a number of faked photographs purportedly showing Americans held in captivity have surfaced in Southeast Asia.

Do, asked why he had never revealed the information to U.S. investigators, said: "I just didn't think it was that important."

### Unaware of controversy

He said many Vietnamese, particularly in the north, saw Americans after the war but thought little of it because they were unaware of the controversy swirling around the issue of missing Americans.

Authorities routinely told Vietnamese that the Americans "are here voluntarily," Do said.

He was told exactly that by officials one day in late March 1979, he said, after he saw a truck carrying about 70 white people in the city of Bac Giang in Ha Bac Province. Do said several of the men were shouting in Vietnamese, "Toi la nguoi my," which means: "I am American."

Doan Van Ban, a diplomat at Vietnam's mission at the United Nations in New York, flatly denied Do's claims.

"Vietnam released all of its prisoners of war after 1973," he said. "We've received thousands of reports about MIA sightings, and up until now nothing has been true or can be confirmed. ... There's no one alive in Vietnam. The question of whether their bodies will be found remains to be seen."

Do apparently first made his claims in July 1991 in a brief conversation with Thang Nguyen, executive director of Boat People SOS, a Virginia-based advocacy group for refugees.

Do "told me this story last summer, long before there were any real clues about Americans prisoners having been sent to the Soviet Union," Nguyen said.

### Americans in Hanoi

The conversation took place several months before Maj. Gen. Oleg Kalugin, who headed Soviet counterintelligence in the late '70s, claimed that KGB agents had interrogated American prisoners in Vietnam in 1978. Kalugin said the agents questioned "at least three Americans" for about a month in or near Hanoi.

Former Air Force Master Jerry Mooney, who spent decades working for the National Security Agency, said in a telephone interview from his Montana home that Do's story didn't surprise him.

"I think the American people are going to be quite shocked at what they hear at the Senate hearings," said Mooney, who culled information from Vietnamese radio transmissions.

In January, Mooney testified before the Senate panel that he decided to break his 1987 vow of secrecy because he felt his government had abandoned servicemen in the "hells of Southeast Asia."

Mooney, who estimates 300 POWs were left behind in 1973, testified that the Soviets had a clear motive in keeping pilots. The men included veterans of the space program and others specializing in electronic warfare, Mooney said. He called them "MEs."

The initials stood for Mission Essential or Mission Bound.

CONFIDENTIAL INTELLIGENCE REPORT

NACHOG

1. COUNTRY: COMMUNIST CHINA
2. SUBJECT: (C) PW Camp in K'UN-MING (250AN/1021ET, NIS), Yun-nan Province
5. REC NUMBER: 723.600
7. SOURCE: 15310005/Friendly foreign government
8. EVALUATION: SOURCE B/F

DATE OF INFORMATION: February 1971
PLACE AND DATE OF ACQ: Taiwan / March 1971
PREPARED BY: Hq 1021st USAF Fld Acty Sq

RUDOLPH W. KELLER JR.
Colonel, USAF

CHINA — BOUND
from Roger Arra
8113 Fort Arra, Apt. 821
Silver Spring, MD 20910
301-585-8361

RECD CO-3 21 MAY 1971

EXCLUDED FROM AUTOMATIC REGRADING: DOD DIR 5200.10

(U) Information concerns a PW camp in K'UN-MING, Yun-nan Province, allegedly used for the detention of U.S. PW's. The caveat NO FOREIGN DISSEM is applied to this report because the disclosure of the information to foreign governments could adversely affect the productivity of a sensitive U.S. intelligence project.

CIRCUMSTANCES OF ACQUISITION:

1. (C) The numbered Source acquired the information in the form of three documents produced by a friendly foreign government. The foreign government attributed the information to both controlled and uncontrolled Sources and evaluates it as C-3. Due to the sensitivity of this collection effort, the collector cannot obtain further clarification or amplification of this information. This report is a translation of significant information in the original documents.

2. (C) The first document states that a U.S. PW camp is located at HSI-PA.

(STC 6607/1109), which is located in the southern suburbs of K'UN-MING, on the northern bank of the TIEN CHIH (Lake) (250AN/3329/3069) 3 kilometers from the K'UN-MING Military Region Headquarters at HSI-PA. (STC 3944/1100). The camp was used to intern Indian prisoners after the U.S.-Indian border clashes. When the U.S. began bombing North Vietnam in 1964, the ChiCom Defense Ministry ordered the camp reactivated for the internment of U.S. PW's and directed the K'UN-MING Military Region Headquarters to assume responsibility for the camp under the direct supervision of the ChiCom Central Committee. Representatives of the political, operational, and organizational departments of the K'UN-MING Military Region Headquarters were sent to work with the PW's. One platoon of guards is responsible for camp security. There are more than 20 interpreters are assigned to the camp. The interpreters are fluent in English and are graduates of the ChiCom Foreign Language School. There are more than 120 PW's at the camp.

3. (C) The prisoners arise at 0530 hours daily and begin physical exercises at 0730 hours. All prisoners are dressed in ordinary Chinese clothing. Breakfast consists of steamed buns, rice soup, or milk, and bread. Lunch is served at 1200 hours and dinner at 1800 hours. Meals consist of food and sometimes depending upon the extent of their lack of cooperation. Classes are held daily between 1000 and 1200 hours. Except for these classes, the prisoners are permitted to move about the camp freely between 0800 and 1200 hours daily. Study materials for the classes consist of the English translation of MAO Tse-tung's quotations; statements made to the Anna Louise (ED: sic; refer to COMMENTS), an American female reporter, by MAO Tse-tung; the history of the U.S. invasion of China (ED: sic); Victory in the Vietnam War; The Failure of U.S. Imperialism; "Long Live the Victory of the People's War"; "People's Pictorial"; and the "PEI-CHING Weekly". Prisoners are interrogated between 1400 and 1500 hours.

4. (C) The second document states that a secret U.S. PW camp is located at TIEN-PA-CHING of HSI-PA in the southern suburbs of K'UN-MING (approximately 3 kilometers from the K'UN-MING Military Region Headquarters compound on HU KUO (STC 3337/0948)] Road). The PW camp is surrounded by ponds and a barbed wire fence. The site is used to imprison U.S. military personnel.

5. (C) The third document, which is being published as IR I 773 0025 71, Subject: Town Plan of K'UN-MING, China, identifies Point 39 of Figure 1 (attached town plan sketch) as: "K'UN-MING Military Region PW Detention Center (Secret installation). This center detained USAF higher-ranking personnel, who were captured in the Vietnam War by the Communist Chinese."

COMMENTS OF THE APPROVING OFFICIAL:

6. (U) A review of collateral information failed to confirm the existence of the facility as described. Based on available information, it appears improbable that the facility described would be used as a PW camp.

7. (U) The mention of "Anna Louise", an American female reporter, is probably a reference to Anna Louise STRONG, Miss STRONG, born 24 November 1885 in FRIEND, Nebraska, died of a heart attack in PEI-CHING on 30 March 1970. She first arrived in China in 1927 to report on the revolution, briefly revisited China in the late 1930's, and in 1960 finally returned to China where she resided until her death. Although she was not a member of the Communist Party, she was sympathetic to the Party and was publicly protected by MAO. She lived in YEN-AN (3638N/10937E, NIS) prior to WWII and in PEI-CHING from 1960 to 1970.

8. (U) The identity of the Source of the first document is unknown. Source 15310005 has identified the Source of the second document as a Chinese Nationalist agent in the Yun-nan area. The Source of the third document is identified as a 29-year-old refugee from CHEN-HSIUNG, Yun-nan Province, who was employed for several years in K'UN-MING and vicinity. In February 1970, he and his nephew escaped to Burma.

POLICY, POLITICS AND THE POW/MIA DILEMMA

by

John M. G. Brown
Unpublished Manuscript, 1992

A Brief Review by William E. LeGro
13 May 1992

Mr. Brown has conducted a massive and exhaustive research
effort to document his thesis which is that the United States has
abandoned its fighting men in Communist prisons at the conclusion
of our four major wars in this century.

Under British command, the 339th Infantry went into battle
against the Bolsheviks in northern Russia in 1918. Mr. Brown
asserts that up to 60 or so were actually captured, imprisoned,
and written off as killed-in-action, body-not-recovered.
Negotiations to obtain their release foundered when the
Bolsheviks demanded diplomatic recognition in return.

In 1945, the Red Army swept through Eastern Europe and overran numerous German POW camps, "liberating" hundreds of thousands
of Allied prisoners, among them up to 60,000 Americans. Mr.
Brown's research reveals that up to 20,000 of these were never
repatriated. This, despite the announcement on May 31, 1945, by
the Undersecretary of War, Robert Patterson, that all U.S.
prisoners of war had been returned. (Sounds familiar.) Forced
repatriation of millions of Russians and East Europeans by the
Allies to unquestionably dismal fates in the Soviet Union was
recompensed by the release of a few thousand Americans by the
Soviets. They kept the rest. Mr. Brown's documentation in this
chapter is impressive, but he admits to its limitations because
much of the material remains (incredibly) classified and beyond
his reach. (In this regard, Mr. Brown's research, which is only
touched on in this manuscript, should certainly be seen by the
U.S.-Russian Commission or, even better, Mr. Brown should be
engaged as a consultant to the Commission.)

Live-sighting reports", the grist of the current
investigation into the Indochina MIA mystery, were significant
intelligence sources in Korea as well. Thousands of Americans
were taken into Manchuria and other parts of China where they
were seen in camps and laboring in mines and factories. CIA
sources also reported that some Americans captured in Korea were
shipped to the Soviet Union. Mr. Brown's research into these
matters has also been obstructed by agencies' refusals to
declassify documents. The Korean War experience, as it pertains
to missing-in-action, accounting for and identifying remains,
findings of death, and official relations with next-of-kin,
appears as a most remarkable overture to the Indochina
experience. Something like a Verdi overture in which all of the
most important arias appear here first.

And so Mr. Brown brings us to Indochina. His understanding
of the POW/MIA issue in this war is remarkably complete, despite
the fact that he has seen only unclassified materials. And his
insights are buttressed by his grasp of the historical
precedents.

Mr. Brown has made, in this short work, an invaluable
contribution to the store of knowledge about missing-in-action
matters in four wars. This should inspire determined efforts to
expedite the declassification of the thousands of documents he
and other dedicated researchers and authors have been trying to
see--but have been refused--for years.

# POCKET BOOKS

Simon & Schuster Consumer Group
1230 Avenue of the Americas
New York, NY 10020
212-698-7580

Paul D. McCarthy
Senior Editor

January 19, 1993

Mr. Jeff Gerecke
JCA
27 West 20th Street
New York, NY 10011

Dear Jeff:

Thanks for the extra time on MOSCOW BOUND by John Brown. It was what I needed to give this truly extraordinary book the attention it deserved. The sheer depth, scope and successful execution of an immensely ambitious project is most impressive, and this book is likely to be the authoritative text on the POW/MIA issue for years to come.

It's the kind of book that I wish we could do more of because of its importance and significant contribution to the issue but after rather exhaustive discussions here it was finally decided that despite the book's quite apparent qualities and likely critical success, it didn't have the vast commercial appeal that would make it right for the increasingly selective Pocket Hardcover list and make it a lead mass market paperback, which is what we look for with hard/soft deals.

I regret that after all this time, I wasn't able to bring matters to a more positive conclusion but I'm glad I had the opportunity to read the book and make the effort.

I hope to hear from you again soon.

Best,

Enc.

**FP**

THE FREE PRESS
A DIVISION OF MACMILLAN, INC.
866 Third Avenue
New York, N.Y. 10022

---

December 16, 1992

Jeff Gerecke
JCA
27 West 20th Street
New York NY 10011

Dear Jeff:

I have given the Brown manuscript a quick read and was very impressed by the level of detail, the mastery of historical and political context, and the authoritative tone. This is a very good book. However, I don't really think its going to work for us. My guess is that sales would be limited, despite press interest in the topic.

I return the manuscript herewith. Very best of luck with it.

Sincerely,

Adam Bellow

THE SECRETARY OF STATE

WASHINGTON

March 18, 1994

Dear Mr. Brown:

Thank you ever so much for Moscow Bound:
Policy, Politics and the POW/MIA Dilemma. I
appreciate your sending it to me, and for the very
thoughtful inscription.

I already have requests from Strobe Talbott
and Malcolm Toon to borrow the book.

Again, with thanks and best regards.

Sincerely,

Warren Christopher

*Embassy of the United States of America*

Moscow, Russia

June 10, 1994

Mr. John M.G. Brown
P.O. Box 30
Petrolia, CA  95558

Dear Mr. Brown:

I want to thank you very much for the book Moscow Bound,
which you sent to me through the kindness of George Schwab of the
National Committee on American Foreign Policy. I have received
the book now here in Moscow and will look forward to reading it
with a great deal of interest.

I congratulate you for having prepared it, and thank you
most sincerely for sending it on to me. The issue which you
addressed directly -- POW/MIA questions -- is of special interest
to us here, as you know better than anyone. I follow it closely,
work with Mac Toon and the Russia Task Force people, and consider
it a major item of responsibility for this Embassy.

Again, thank you very much for your kindness in sending me a
copy. With warm good wishes,

Sincerely

Thomas R. Pickering
Ambassador

GEORGE BUSH

May 23, 1994

Dear Mr. Brown,

I am delighted to have the personally inscribed copy
of *Moscow Bound: Policy, Politics and the POW/MIA
Dilemma.* Thank you for thinking of me.

Best wishes.

Sincerely,

*[signature: George Bush]*

Mr. John M. G. Brown

---

THE WHITE HOUSE
WASHINGTON

May 2, 1994

Dear Mr. Brown:

Thanks so much for the copy of your book,
"Moscow Bound." It was very thoughtful
of you to send it.

I hear very good things about it and look
forward to reading it.

Best,

*[signature: Tony Lake]*

Anthony Lake

# Indochina Chronology

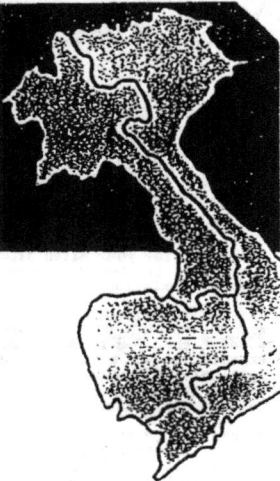

"TEAR SHEET"

Volume XIII, Number 2, April-June 1994

*Moscow Bound: Policy, Politics and the POW/MIA Dilemma*, by John M.G. Brown (Vietnam veteran/-POW · researcher). The vanished POW is not a phenomenon unique to the Vietnam War as this lengthy (1,055 page) work testifies. Beginning with

the Allied intervention in the Bolshevik revolution (in the early 1920's) on through the Korea and Vietnam Wars, American POW's were known to have been held in the USSR but not their ultimate fate. A highly literate work and basically objective. Available from author: John M.G. Brown, P.O. Box 30, Petrolia, CA  95558. $38.25 ($40 for first class mail) (hc). (1994).

Indochina Chronology is a quarterly publication of the Institute of East Asian Studies of the University of California, Berkeley. Editor is Douglas Pike. Address Indochina Chronology, Institute of East Asian Studies, 2223 Fulton St., 6th Floor, University of California, Berkeley, CA  94720  USA.

# National Archives

Washington, DC 20408

MAY † † 1995

Mr. John H.G. Brown
P. O. Box 30
Petrolia, CA  95558

Dear Mr. Brown:

Thank you for your March 31, 1995, letter and your interest in the World War II Working Group, which I co-chair, of the Joint US Russia Commission on POW/MIAs.

The additional copy of your book, Moscow Bound: Policy, Politics, and the POW/MIA Dilemma, has proved most useful in our work. The account that you forwarded by Mr. F. Eugene Liggett describing his experiences as a POW from January to May 1945 was interesting, particularly regarding the movement of prisoners among several German POW camps and the chaos that ensued when the Soviet Army reached those camps.

By the end of this year, our World War II Working Group hopes to produce a research report on prisoner-of war issues of mutual interest to the US and Russian sides. This will be a good faith effort by both parties to explain the facts of what happened and to provide documentation in support of those facts. Disagreements between US and Russian staff analysts will not be glossed over, but described and explained.

Again, thank you for your interest and the information you have provided.

Sincerely,

TRUDY HUSKAMP PETERSON
Acting Archivist
of the United States

National Archives and Records Administration
Washington, DC 20408

Archivist of the United States

---

July 3, 1995

Mr. John H.G. Brown
Box 30
Petrolia, CA  95558

Dear Mr. Brown:

Thank you for sending me a copy of your book, Moscow Bound: Policy, Politics, and the POW/MIA Dilemma. I have it proudly displayed in the bookcase in my office.

Best wishes.

Sincerely,

JOHN W. CARLIN
Archivist of the United States

THE AMERICAN

LEGION

FOR GOD AN

September 1995

VICTORY

1945

August 17, 1995

John M.G. Brown:

Here is our special World
War II issue. Hope you like
it. Thanks for sharing in
making this happen.

Best,

STEVE SALERNO
Editor

Volume 139, Number 3

September 1995

# VICTORY 1945

CIRCULATION : 3 Million

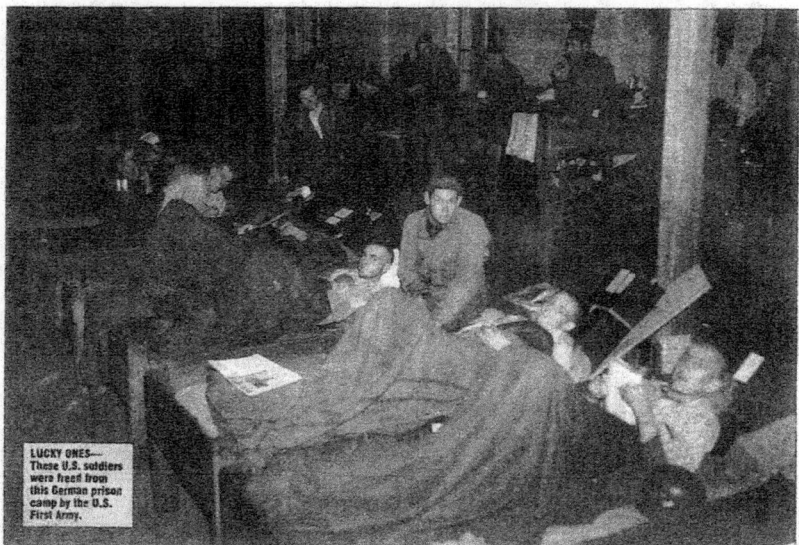

# THE TRUTH IS OUT: THE SOVIET UNION KEPT U.S. POWS AS BARGAINING CHIPS AFTER WORLD WAR II.

# OUR UNKNOWN POWs

LUCKY ONES— These U.S. soldiers were freed from this German prison camp by the U.S. First Army.

## By John M.G. Brown

ONE OF THE most striking revelations to emerge from the collapse of the Soviet Union was Boris Yeltsin's confirmation in 1992 that U.S. POWs from three major wars had been held inside Soviet borders.

While Yeltsin's frank admission gave new impetus to the efforts of the many tireless crusaders seeking a resolution to the Vietnam POW/MIA issue, it also stirred memories of a group of earlier POWs on whom the books have never been closed—the thousands of GIs in German camps overrun by the Red Army in the final stages of World War II.

All told, after World War II, half a million Allied POWs and displaced Europeans were swallowed up into Soviet-occupied territory, including an estimated 20,000 American POWs and MIAs. About 4.5 million Axis and Japanese POWs were also held by the Soviets in a nationwide gulag slave empire containing 20 million of the USSR's own citizens.

While 28,662 American POWs and MIAs in German captivity were repatriated, many other MIAs and about 12,000 U.S. POWs in German camps "liberated" by the Red Army simply vanished. No U.S. POW of World War II—or later conflicts—has ever returned from the former USSR.

Half a century later, the question remains: *What became of these men?*

*Legionnaire John M.G. Brown, an Army veteran of the Vietnam War, has served as a POW/MIA expert for the Senate Select Committee on POW/MIA Affairs. He is also author of Moscow Bound: Policy, Politics and the POW/MIA Dilemma.*

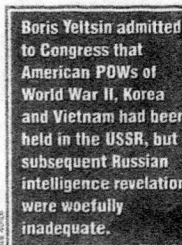

According to declassified documents in the records of the American Red Cross, our own government made them "disappear" in two ways:

• through an accounting process that reclassified them as MIAs instead of POWs, making it appear less likely that there were live people missing.

• through a Presumptive Finding of Death (PFOD). This means that they were classified as dead merely because of the passage of time or the lack of any new evidence suggesting they remained alive.

By Oct. 31, 1945, these methods had allowed the number of U.S. Army "prisoners not yet returned from German camps" to be reduced to 6,595. Two months later, the number was cut to 5,414 U.S. Army POWs not repatriated from Soviet-occupied territory. Meanwhile, Russian leader Josef Stalin refused access to Soviet territory for U.S. observers trying to ascertain the fate of the 78,750 Americans missing and presumed dead.

Stalin had his reasons for stonewalling—specifically, according to British and U.S. documents, he wanted several demands met. The demands included U.S. diplomatic recognition

of Soviet puppet regimes in Poland and Eastern Europe, the repatriation of millions of reluctant Red Army POWs, and $6 billion in postwar reconstruction aid.

The Russian dictator wasted little time putting his blackmail scheme into effect. Formerly secret documents of Gen. Dwight D. Eisenhower's SHAEF Headquarters, declassified in the late 1980s, reveal that two weeks after V-E Day in May 1945, 32,000 U.S. and British POWs in Eastern Germany were being "held as hostages by the Russians" in order to obtain the return of all Soviet citizens without exception. This, of course, was while the Pacific fighting continued.

By war's end, more than 50,000 U.S. POWs and 1.5 million Allied troops and civilians were held in German stalags under control of the victorious Red Army in Poland, Eastern Germany, Austria and the Balkans. In violation of the February 1945 Yalta Agreements, the Soviet government refused to grant Allied officers access to most of the prisoners.

ONLY 5,241 Americans were reported released by the Russians after the 1945 Halle agreement in Germany outlining prisoner exchange. Repatriation of POWs became a secret topic of the 1945 founding conference of the United Nations in San Francisco, the 1945 Stalin-Hopkins meetings in Moscow, and the Potsdam Conference. Despite the resulting 1945-46 repatriations of millions of Soviet citizens, Stalin did not reciprocate by freeing U.S. POWs under his control.

Ironically, it was the United States' own interests that handed Stalin the leverage he needed to get away with his brazen behavior. He knew that the Roosevelt and Truman administrations placed overriding importance on

avoiding confrontations with the Soviets. As the Nazis collapsed between the Anglo-American and Soviet fronts during April 1945, the United States began seeking Soviet participation in the war against Japan. In addition, Stalin knew Truman courted the Soviet Union's approval for the postwar United Nations. As a result, Stalin was able to use the POW issue as a bargaining chip to stave off retaliation for holding onto U.S. POWs.

The stage for such concessions actually had been set long before. In fact, if the War Department had listened to American Legion founder Hamilton Fish in 1930, the United States may have brought home more of its POWs after both world wars and avoided POW debacles in later conflicts. Instead, a lack of honesty by the U.S. government at this early date inevitably left our nation with more than 78,000 missing from World War II, over 8,000 from the Korean War and more than 2,000 from Vietnam.

After serving in World War I, Legionnaire Fish was elected to the House of Representatives, where he discovered an eyewitness report of 40 to 50 live U.S. POWs still in Stalin's gulag prisons after the 1918-20 American intervention in Russia. This information was confirmed by a Bureau of Investigation (later the FBI) report on specific American MIAs reported

*Please turn to page 106*

# ALL FOR AMERICA

*Continued from page 70*

Out for Victory," captured the Legion's overwhelming commitment: "There is nothing half-way about the Legion's participation in the victory effort of our common country. The American Legion embodies the most victory-conscious group in the United States of America."

And when the war was over, The American Legion was again called into service to help veterans who were jobless, disabled or without homes. To pick up the pieces and rebuild the lives of veterans and the nation, The American Legion authored and introduced one of the greatest pieces of legislation ever: the GI Bill.

Officially called The Serviceman's Readjustment Act of 1944, it provided education and housing benefits for veterans and virtually created a new society of citizens—middle-class America.

Eight million WWII veterans went to school on the GI Bill. As a result, America gained 450,000 engineers; 180,000 medical doctors, dentists and nurses; 107,000 attorneys; 200,000 accountants; 36,000 clergymen; 17,000 writers and journalists; 700,000 mechanics; 383,000 construction workers; 83,000 policemen and firemen; 200,000 metal workers; 500,000 electrical workers; and 61,000 printers and typesetters. Another 700,000 GIs took business courses.

Thanks to the GI Bill, millions of veterans were prepared to compete for good-paying jobs and also were able to buy homes under the bill's home-loan provisions.

Still, there were other battles to be waged. The Legion fought for and won more hospitals for veterans and more job-placement assistance for disabled veterans. Legion spokesmen were constantly before Congress seeking help for those who had borne the burden of war.

"When historians look at the history of World War II, they see the role of The American Legion as an important one," says Library Director Hovish. "Every Legionnaire should be proud of what we did during that war—for our families, for our nation and for the world."

As the revered Army Gen. Douglas MacArthur once said with pride: "The American Legion is particulary well-fitted to stand guard over our heritage of American liberty."    □

# POWs

*Continued from page 76*

alive and under GPU (later the KGB) control in 1929.

When called on to investigate Fish's discovery, the War Department told Fish the Soviets had repatriated all U.S. POWs and that any prisoners in Russia had not "served in the Army." Fish had little choice but to accept the War Department's tale, and the facts remained classified.

In 1933, the United States, concerned about the resurgent power of Germany and Imperial Japan, extended diplomatic recognition to the Soviet government. Although the decision to recognize the Soviet Union made political sense, the ease with which the U.S. government erased the missing WWI GIs from its list of concerns established an unfortunate pattern for World War II, Korea and Vietnam.

Recently declassified U.S. intelligence documents reveal the following:

• After World War II, the U.S. Counter-Intelligence Corps (CIC) debriefed German POWs released from the USSR, and assembled top-secret reports on American and Allied POWs in Siberia—some of which remain classified to this day. U.S. POWs captured by the Germans at Normandy and the Bulge were still being sent to the Soviet Union from Eastern Europe during 1946-47 and were treated worse than German POWs, according to CIC records declassified in 1991. Japanese POWs repatriated from Siberia in 1947 and 1948 also reported U.S. WWII POWs in Soviet labor camps in Kazakhstan.

• During the Korean War, the surviving American prisoners of World War II in Siberia were joined by 1,200 to 2,000 more U.S. POWs. U.S. POWs in KGB control reported by eyewitnesses throughout the 1950s and '60s included American POWs from World War II, Korean War and Cold War reconnaissance missions. Much POW information held by the U.S. government remained classified until the 1990s. All the while, the Soviets publicly denied holding U.S. POWs whenever asked.

Sadly, on the 50th anniversary of the end of World War II, the United States has still not established a national priority of accounting for thousands of American POWs, or locating and returning survivors.    □

THE VVA VETERAN

SEPTEMBER 1996

VIETNAM VETERANS OF AMERICA
national magazine

The VVA Veteran,
Vietnam Veterans of America

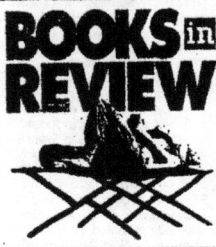

**BOOKS in REVIEW**

## Shipping American POWs to the Soviets

REVIEWED BY MARC LEEPSON

John M. G. Brown's massive *Moscow Bound* (Veteran Press, 1,055 pp., $40) contains a mountain of convincing evidence suggesting that the former Soviet Union made use of American prisoners of war during World War II, Korea, and Vietnam. Brown is a U.S. Army Vietnam veteran who has done extensive research on American POWs in recent years with the help of Thomas V. Ashworth, a former Marine helicopter pilot who served in Vietnam.

Brown is one of the best-known researchers on the POW/MIA issue. His research was a large part of the 1991 Senate Foreign Relations Committee staff report on American POW/MIA policy. His work has also been cited in the 1993 report issued by the U.S. Senate Select Committee on POW/MIA Affairs.

In *Moscow Bound,* Brown sketches the unsavory history of the Soviet Union and western POWs, dating from 1918 when Lenin's Bolshevik government took hundreds of American, British, and French prisoners hostage but did not release the men as promised in 1921 and 1933. Brown cites evidence showing that after World War II, the Soviet Union under Stalin held thousands of American prisoners, keeping the POWs as ransom to try to force the United States to come up with billions of dollars in promised postwar aid and to force this country to recognize the newly created Soviet-dominated regimes in Central and Eastern Europe.

The same story was replayed following the Korean War. This time, the culprits were the North Koreans and the Chinese as well as the Soviets. As many as 2,000 American Korean War prisoners, Brown believes, were shipped from China to the Soviet Union following the war in retaliation for the tens of thousands of North Korean and Chinese POWs who refused to be repatriated to their countries following the end of the Korean War. Brown also cites cases of American intelligence agents who were kidnaped or taken prisoner during various cold war intelligence-gathering missions around the world and shipped to concentration-like gulags in the former Soviet Union.

As for Vietnam, Brown believes that North Vietnam did not return all American prisoners in 1973 as it claimed. During the war, Brown says, about 1,000 American POWs were held in a secret prison-camp system staffed by Soviet intelligence officers, and none of those prisoners were returned. "In several of these secret Vietnamese prison camps," Brown says, "selections were made among POWs for shipment to the Soviet Union or China, in much the same manner as had previously been done in communist China during the Korean War."

The reason these prisoners were either killed in Vietnam or shipped to the Soviet Union, Brown says, was "the North Vietnamese could not afford to release any of these specially interrogated POWs because of their dangerous knowledge of the Soviet intelligence connection, and because their observations would impact North Vietnamese national security interests, which remained in effect after the [1973] American withdrawal and until the final collapse of South Vietnam in 1975." ∎

*To order this thought-provoking book, write: Veteran Press, Box 30, Petrolia CA 95558.*

S. Hrg. 102-351, Pt. 2

# POW/MIA POLICY AND PROCESS

## HEARINGS

BEFORE THE

### SELECT COMMITTEE ON
### POW/MIA AFFAIRS
### UNITED STATES SENATE

ONE HUNDRED SECOND CONGRESS

FIRST SESSION

ON

THE U.S. GOVERNMENT'S EFFORTS TO LEARN THE FATE OF AMERICA'S
MISSING SERVICEMEN

#### PART II OF II

NOVEMBER 5, 6, 7, AND 15, 1991

Printed for the use of the Select Committee on POW/MIA Affairs

*[handwritten notes]*

— John M. G. Brown — documents.

See pp : 654 to 931, inclusive

U.S. - Russia Conference

— Halle, p. 744.
929-930 — Conference.

Washington Post — July 4, 1991 — John M.G. Brown

Mike Siegel / Seattle Times

Tom Ashworth, a former captain in the Marine Corps, says his research into records from World War II, the Korean War and Vietnam War, leads him to believe that the U.S. government has a history of abandoning both military and civilian personnel in hostile countries.

The Seattle Times   Monday, September 25, 1989

## NAZI PRISON CAMPS BY THE RED ARMY. SINCE THEN NOTHING HAS BEEN HEARD OF THEM

### By *Vadim BIRSHTEIN*

Inside the Capitol in Washington, paintings on the walls feature scenes from American history. Between them stand statues of US statesmen. High above is the statue of George Washington. Only one detail is a reminder of the present day: a Black flag near the Lincoln statue. It has a bowed man's head with a prison watchtower in the background. The letters read POW/MIA. Prisoners of War and Missing in Action. American families lost 30,000 of their fathers and sons who did not return from World War II and the Vietnam War. The members of the POW/MIA associations spend all their free time and a lot of money to try to tease information on the fate of their relatives from the US government. One such group, the National Union of Families, is based in Seattle and is headed by Dolores Alfond whose air force pilot brother disappeared in Indochina 24 years ago.

The problem first arose in late 1944 when Soviet troops liberated many nazi internment camps in Poland and Germany. Hundreds of thousands of West European, American and Canadian citizens found themselves in the Soviet occupation zone. Under the Yalta agreement signed between Stalin, Roosevelt and Churchill on February 4, 1945, the Western allies were committed to deport from Europe tens of thousands of Russian emigres and displaced persons, including those released from nazi camps, in return, Stalin had promised to allow citizens of Allied countries in the Soviet zone to go home.

But the Soviet authorities were not in a hurry to fulfil the pledge. Six days after VE day reports from France said that about one half of the 200,000 British and 76,000 Americans POWs still in Germany were probably in the Russian occupation zone. One report said that American POWs released by the Red Army were often maltreated and they were beginning to hate Russians. Many were stripped of their possessions that they had kept throughout the long German captivity, watches, rings and so on. American POWs in Odessa (a filtration camp — VB) were guarded by Russian soldiers with loaded rifles and fixed bayonets and the Russian secret service was more strict than the German.

On May 19, 1945 General Eisenhower, in a cable to the Allied headquarters, put the number of American POWs under Russian control at 25,000. Reports that reached Eisenhower's headquarters at the end of May spoke of 20,000. Contrary to the facts, Eisenhower made an official statement on July 1, 1945 that "only a small number" of American POWs were still in Russian hands. As a result, thousands of American families are still waiting for their relatives who never returned from World War II. All in all, there are 20,000 missing servicemen.

Twenty thousand is a large number. Where were they kept? This from a former internee about a camp near Tambow:

"At the end of 1945 there were Germans, French, Americans, Belgians — more than 20,000 — servicemen and civilians. All the inmates were made to work, the food was poor and monotonous. They were kept in dugouts, not in barracks.

"The consequence of the monotonous food was a strange disease. Those who contracted it got swollen hands and feet. Eventually they died. More than 23,000 Italian, about 2,500 French and about 10,000 Rumanian and Hungarian prisoners died. Several Poles and members of other nationalities also died from the disease.

"The inmates were highly educated and knowledgeable people, brilliant scientists. The German engineers were used for specialized work designing an aircraft that could carry 500 passengers at the speed of about 1,000 km an hour. The Russians were keen on these projects and the people involved in them enjoyed better working and living conditions."

A book has been published in France about the French inmates of the Tambov camp. But what about Americans?

Reminiscences of people who have survived the GULAG in the post-war years occasionally mention Americans in various camps: in Vorkuta, Inta, Dzhezkazgan, Magadan and other places. Not only POWs, but some members of the US Embassy staff in Moscow are known to have been inmates of these camps. But not a single account mentions thousands of American POWs. The lists of American MIAs are still secret in the US. Where are they?

Americans feel at least as strongly about the problem of the servicemen who did not return from the Korean and Vietnam wars. The official figure is that 1,000 POWs were not repatriated from Korea, with another 8,000 reported missing. There is evidence that some of these people had been moved to camps in China, Manchuria and the USSR. This is how it happened in the winter of 1951 to spring 1952.

"The railway station was divided into two sections with tracks on either side. On the Chinese side, the American POWs were escorted by Chinese soldiers. The prisoners went through the gates separating the platform toward the train served by a Russian team. The Russians wore railwaymen's uniforms, but they were disguised soldiers."

On May 5, 1954 the US Embassy sent a note to the USSR Foreign Affairs Ministry asking about the fate of American POWs. The Foreign Ministry replied on May 12, 1954 that "There are no such people in the Soviet Union and never have been."

POW/MIA families believe that at many as 2,000 Americans were sent from Korea to the USSR.

The representatives of the governments of the USA, North and South Vietnam and the Provisional Revolutionary Government of South Vietnam signed a peace treaty in Paris on January 27, 1973 which envisaged, among other things, bilateral exchange of POWs. In spite of the accord, the US never received the complete list of American POWs kept in North Vietnam and neighbouring Laos. On March 26 of that year Vietnam declared that the last American soldier would be repatriated on March 27. And the US Defense Department too announced that there were no more POWs in Southeast Asia, that they were all dead.

According to POW/MIA families, 591 POWs were repatriated from Vietnam, which accounts for a mere 12 per cent of the total number of 5,000 American POWs who were alive in 1973.

"As long as a situation is tolerated in which American citizens are imprisoned abroad, the freedom of all Americans is in danger." This is a sentence from a leaflet published by the relatives of the crew of a RB-89 plane which was shot down over neutral waters by the Japanese on June 13, 1952 while on a routine reconnaissance mission. The leaflet goes on to claim that the crew of 12 was apparently picked up by the Soviet naval ships which were there during the incident. Perhaps the members of the crew are still alive in Soviet captivity. Here are their names: S.R.Bush, a Major; G.A.Scully, S.D.Service and R.G.McDonnell, First Lieutenants; W.B.Homer, D.L.Moore, Sergeants First Class; M.V.Monserrat, E.R.Berg, L.E.Bonnita and V.G.Becker, Sergeants Second Class; D.N.Pilbourne, Private.

The history of the Cold War is riddled with such cases. Usually the US Administration made one or two queries about the fate of the crews. But after that, it was left to the relatives who still believed their relatives were alive.

To be sure, the reluctance of the US Administration to admit that it had done little or nothing to save its citizens is a matter for the US electorate and the government. The US Senate is currently engaged in a stormy debate as to whether or to set up an inquiry into the POW/MIA issue. It looks as if such a commission is going to be formed.

There is yet another side to the matter. The information on foreign POWs who found themselves on the territory of the USSR during World War II and before the late 1950s is kept at the Central State Archive of the USSR in the section of the Central Board for the affairs of POWs and internees under the USSR Interior Ministry. It is the same fund that contained the lists of Polish officers shot in the woods in Katyn in 1940 and data on Japanese and German POWs which are now being handed over to these countries. So, the memoirs of the 20,000 missing Americans and their fate can easily be found out in Moscow. The trouble is that gaining access to these archives is anything but easy. One needs the permission of two government bodies, the Interior Ministry and the KGB, which are notoriously reluctant to part with their secrets.

It is far easier to establish mutual understanding between those who share the same misfortune: the relatives of Soviet and American MIAs and the relatives of Soviet POWs who were taken prisoner in Afghanistan. The American POW/MIA families are willing and able to help the Soviet relatives of missing "Afghans."

The American National Alliance of Families would appreciate any news of relatives. Human contacts can and must be established.

INDEPENDENT NEWSPAPER FROM RUSSIA

*History*

# WHAT HAPPENED TO 20,000 AMERICANS?

## THEY WERE RELEASED FROM NAZI PRISON CAMPS BY THE RED ARMY. SINCE THEN NOTHING HAS BEEN HEARD OF THEM

*By Vadim BIRSHTEIN*

Inside the Capitol in Washington, paintings on the walls feature scenes from American history. Between them stand statues of US statesmen. High above is the statue of George Washington. Only one detail is a reminder of the present day: a black flag near the Lincoln statue. It has a bowed man's head with a prison watchtower in the background. The letters read POW/MIA. Prisoners of War and Missing in Action. American families lost 30,000 of their fathers and sons who did not return from World War II and the Vietnam War. The members of the POW/MIA associations spend all their free time and a lot of money to try to tease information on the fate of their relatives from the US government. One such group, the National Union of Families, is based in Seattle and is headed by Dolores Alfond whose air force pilot brother disappeared in Indochina 24 years ago.

The problem first arose in late 1944 when Soviet troops liberated many nazi internment camps in Poland and Germany. Hundreds of thousands of West European, American and Canadian citizens found themselves in the Soviet occupation zone. Under the Yalta agreement signed between Stalin, Roosevelt and Churchill on February 4, 1945, the Western allies were committed to deport from Europe tens of thousands of Russian emigres and displaced persons, including those released from nazi camps. In return, Stalin had promised to allow those citizens of Allied countries in the Soviet zone to go home.

But the Soviet authorities were not in a hurry to fulfil the pledge. Six days after VE day reports from France said that about one half of the 200,000 British and 76,000 Americans POWs still in Germany were probably in the Russian occupation zone. One report said that American POWs released by the Red Army were often maltreated and they were beginning to hate Russians. Many were stripped of their possessions that they had kept throughout the long German captivity, watches, rings and so on. American POWs in Odessa (a filtration camp — VB) were guarded by Russian soldiers with loaded rifles and fixed bayonets than the German.

On May 19, 1945 General Eisenhower, in a cable to the

'The consequence of the monotonous food was a strange disease. Those who contracted it got swollen hands and feet. Eventually they died. More than 23,000 Italian, about 2,500 French and about 10,000 Rumanian and Hungarian prisoners died. Several Poles and members of other nationalities also died from the disease.

The inmates were highly educated and knowledgeable people, brilliant scientists. The German engineers were used for specialized work designing an aircraft that could carry 500 passengers at the speed of about 1,000 km an hour. The Russians were keen on these projects and the people involved in them enjoyed better working and living conditions.'

A book has been published in France about the French inmates of the Tambov camp. But what about Americans?

Reminiscences of people who have survived the GULAG in the post-war years occasionally mention Americans in various camps: in Vorkuta, Inta, Dzhezkazgan, Magadan and other places. Not only POWs, but some members of the US Embassy staff in Moscow are known to have been inmates of these camps. But not a single account mentions thousands of American POWs. The lists of American MIAs are still secret in the US. Where are they?

...the servicemen who did not return from the Korean and Vietnam wars. The official figure is that 1,000 POWs were notre- ... There is evidence that some of these people had been moved to camps in China, Manchuria and the USSR. This is how it happened in the winter of 1951 to spring 1952.

'The railway station was divided into two sections with tracks on either side. On the China side, the American POWs were escorted by Chinese soldiers. The prisoners

26 of that year Vietnam declared that the last American soldier would be repatriated on March 27. And the US Defense Department too announced that there were no more POWs in Southeast Asia, that they were all dead.

According to POW/MIA families, 591 POWs were repatriated from Vietnam, which accounts for a mere 12 per cent of the total number of 5,000 American POWs who were alive in 1973.

'As long as a situation is tolerated in which American citizens are imprisoned abroad, the freedom of all American is in danger. This is a sentence from a leaflet published by the relatives of the crew of a RB-89 plane which was shot down over neutral waters by the Japanese on June 13, 1952 while on a routine reconnaissance mission. The leaflet goes on to claim that the crew of 12 was apparently picked up by the Soviet naval ships which were there during the incident. Perhaps the members of the crew are still alive in Soviet captivity. Here are their names: S.R.Bush, a Major; G.A.Sculley, S.D.Service and R.G.McDonnel, First Lieutenants; W.B.Horner, D.L.Moore, Sergeants First Class; W.A.Blizzard, M.V.Monserrat, E.R.Berg, L.E.Bonine and V.G.Becker, Sergeants Second Class; D.N.Pilbourne, Private.

The history of the Cold War is riddled with such sas as. Usually, the US Administration made one or two queries about the time of the crash. But after that, it was left to the relatives who still believed their relatives were alive.

To be sure, the reluctance of the US Administration to admit that it had done little or nothing to save its citizens is a matter for the US electorate and the government. The US Senate is currently engaged in a stormy debate as to whether to set up an inquiry into the POW/MIA issue. It looks as if such a commission is going to be formed.

There is yet another side to the matter. The information on foreign POWs who found themselves on the territory of the USSR during World War II and before the late 1950s is kept at the Central State Archive of the USSR in the section on the POW/MIA Interior Ministry. It is the same fund that contained the lists of Polish officers shot in the woods in Ka...

# ALL AMERICAN POWs WERE SENT HOME, BUT....

*(The official Soviet version)*

### By Vladimir GALITSKY

The disappearance of 20,000 American prisoners of war freed from Nazi camps by the Red Army in 1945 continues to puzzle the American public. Congress has formed a subcommittee with the power to subpoena former high-ranking officials and ask for any document from CIA and Pentagon files. According to one White House source, President Bush raised the issue at the Moscow Summit.

*Nezavisimaya Gazeta* spoke to Captain Vladimir Galitsky, who spent more than fifteen years in Soviet archives researching the problem of POWs.

The Red Army set free a total of 1,021,455 foreign nationals from Nazi prison camps. Of that number, according to official data, 1,016,566 persons of 33 nationalities were repatriated, including 22,481 Americans (22,429 POWs and 52 internees). Between 1945 and 1947, all the American POWs and internees were sent back to the U.S. (22,449 in 1945, 22 in 1946, and 10 in 1947).

The repatriation process was monitored on Soviet territory by representatives of the United States: first by General Dean, of the U.S. military mission in the USSR; from March 27, 1945, he was aided by a group of ten American officers headed by Colonel Crocket. This group left the USSR on June 18. All in all, twenty American officers supervised the repatriation of U.S. citizens.

There was plenty of good will in U.S.-Soviet relations, and those directly in charge of this important mission developed close personal contacts. All information about the U.S. citizens freed at the time was passed on to the American side. All organizational discrepancies were settled on the spot without any bureaucratic rigmarole.

Archive documents show that the Soviet government spent 253,099,590 rubles to repatriate foreigners freed from Nazi camps, including 5,580,482 rubles on U.S. citizens (cost of food, clothing, transportation, accommodation, and pocket money). In return, the U.S. administration provided similar services in repatriating Soviet citizens freed by U.S. forces from Nazi

captivity following the mutual repatriation agreement of February 11, 1945.

Large-scale repatriation of U.S. citizens started on March 7, 1945, via Odessa and Murmansk. In the transit camps they were treated well. Later, considering that the journey home through Odessa and Murmansk would be long and arduous, an agreement was reached in May 1945 between the Commander-in-Chief of the Allied expeditionary force and Supreme Commander-in-Chief of the Red Army to repatriate foreigners directly across the front-lines. Transfer posts could process up to 5,000 persons a day. In addition, after the defeat of Japan, the Soviets handed over to the U.S. Britain and Holland 2,100 nationals freed by the Red Army from Japanese captivity (the exact number of U.S. POWs and internees is unknown).

A large number of foreign nationals freed from German captivity were handed over to the Allied powers without being registered in any document. For example, the command of the 2nd Belorussian Front, in the course of one day, turned over to the American command 12,000 former POWs and internees from different countries, who were then evacuated by U.S. aircraft without any registration.

Conditions in Soviet transit camps for American POWs and internees were good, by the standards of those days. The food they received was just as good as, and sometimes better than, what Red Army servicemen had. The Soviet side received no complaints on this score from U.S. officials.

P.S. However, the problem cannot in any way be considered resolved. It follows from Capt. Galitsky's account that no lists of names of those repatriated exist. But without such lists, there can be no serious research.

And finally, the fact remains that these Americans never did return to their families.

•

# INDEX

Yalu dams and Hydroelectric plants
448
Yalu River 419, 423
Yanayev, Gennaday 797, 798
Yarborough, Gen. William 648
Yardley, Maj. Herbert 77, 88, 520
Yates, Ted (POW) 188, 384, 856
Yates, Adm. Earl 873
Yatron, U.S. Rep. Gus 776
Yazov, Dimitry 797, 798
Yekaterinburg 29
Yellow Sea 437
Yellow Sea missile testing modes
668
Yeltsin, Boris 149, 798, 799, 808,
822, 832, 834, 835, 838, 851,
868, 877, 879, 895, 897, 898,
899, 918, 923, 924, 926, 927
Yeltsin's contention that some U.S.
POWs from Vietnam might still
be alive in Russia 833
Yeltsin's June 1992 admission 921,
929
YEN BAI, North Vietnam 593, 744
Yen Bai Camps (YEN BAY) 593, 689,
723, 744
Yenan 120
Yerevan (Erevan) area of Armenia
521
Yeserskiy, Gen. Yuriy Filipovich 487
Yi Shan Ji 762
Yong-do Island 436
Yongsandong 431
Yorkshire Regiment of British
Infantry 47
Yoshida, Iwanobu 775
YOSSELAG in southern Siberia 850
Yost, Charles 676

Young, Evan E. 68
Young, Stanley 313, 404
Young, James R. (POW) 434
Young, Herbert L. 643
Young, David 647
Young, Myron 685
Yudenich's Northwestern White
Army 53
Yugoslavia 359, 361, 381
Yugoslavs 253, 359, 366
Yunan Province 667
Yunnan Province 512
Yuri, Grand Prince 14
Zabaikal Province 35
Zablocki, U.S. Rep. Clement J. 489
Zablocki Committee hearings 491
Zablocki congressional committee
493
Zemke, Col. Hubert 231
Zen Buddhist 775
Zhukov, Marshal Georgi
Konstantinovich 295
Zhukovsky Central
Aerohydrodynamics Institute
448
Zigerman, "Gunner" of Philadelphia
526
Zimmerman, Stanley G. 421,422
Zinoviev, Gregori E. 39, 70, 120, 122
"Zionist conspiracy" 451
Znoj, Alex 642
Zoo 543, 583, 684
Zuberbuhler, Rudolph 685
Zuckerman, Mortimer 814
Zuyev, Alexander 870
Zwenig, Frances 827, 865, 866,
880, 895, 914

# MOSCOW BOUND
*Policy, Politics, and the POW/MIA Dilemma*

★

# John M.G. Brown

**Moscow Bound** is the most thoroughly documented and comprehensive book ever written on the subject of American and Allied prisoners of war who disappeared in Soviet captivity from the time of the 1918 Allied intervention in Russia to the Vietnam War. It is the result of a ten-year investigation that extended from the CIA, White House, State Department, and Pentagon offices, to military and intelligence files in the National Archives, and to hearing rooms of the U.S. Congress. During this period author John M.G. Brown and his research colleague, Thomas V. Ashworth, assisted the U.S. Government while simultaneously confronting it in a reluctant American print media with documentable facts about POWs that had long remained classified and highly compartmentalized. The revelation of these facts was to alter history at a time when the Soviet Union was collapsing and Communism was being overthrown in eastern Europe.

**Moscow Bound** documents how, following World War I and World War II, the secret withholding of U.S. prisoners of war by Lenin and Stalin was not revealed to the American people, to avoid prolonged warfare that would have resulted in many more casualties. The classified information about the missing American and Allied prisoners of two world wars constituted a hidden history of U.S.-Soviet relations in which the fundamental cause of hostility remained a state secret. In the post-Korean War era, after thousands more American prisoners were withheld in captivity by Soviet Russia and Communist China, the POW/MIA matter was viewed by the U.S. Government as a national security issue that could upset the hair-trigger relations between the nuclear-armed superpowers.

By the time of the Vietnam War a long-established U.S. policy dictated that intelligence information on American POWs remaining in enemy control in North Vietnam, Laos, Cambodia, China, Cuba and the Soviet Union after the 1973 Paris Peace Agreement and prison exchange, would remain "classified." This book resulted in part from a declassification process for obtaining formerly secret U.S. Government documents, and from interviews with human sources, who in some cases 'broke security' to record their part in the events involving the lost American POW/MIAs.

Of necessity **Moscow Bound** delves into various covert operations and political decisions related in one way or another to the prisoner of war/missing in action issue, and certain revelations within could be perceived as embarrassing to some past and present US. Officials and to government agencies which have been proven to be unable to objectively investigate themselves. This no doubt contributed to the difficulties experienced by the author in finding an American publisher for this book. The author recorded this history in the hope that by understanding the past the people of the United States can better affect the future course of the nation.

**John M.G. Brown** served as an infantry rifleman in Vietnam and authored a memoir of that war (Rice Paddy Grunt). He subsequently conducted extensive documentary and human-source research on U.S. POWs with colleague Thomas V. Ashworth, a former Marine captain who served as a helicopter pilot in Vietnam. The author contributed significantly to the 1991 Senate Foreign Relations Committee staff report on U.S. POW/MIA policy and was cited as an authority on POWs and MIAs in the 1993 report of the U.S. Senate's Select Committee on POW/MIA Affairs. Brown has written about American war prisoners for the *Washington Post*, the *Oregonian*, *VFW Magazine, American Foreign Policy Newsletter* and others.